ATLANTIS STUDIES IN MATHEMATICS
VOLUME 1

SERIES EDITOR: J. VAN MILL

# Atlantis Studies in Mathematics



(ISSN: 1875-7634)

**Aims and scope of the series**

The series 'Atlantis Studies in Mathematics' (ASM) publishes monographs of high quality in all areas of mathematics. Both research monographs and books of an expository nature are welcome.

All books in this series are co-published with World Scientific.

For more information on this series and our other book series, please visit our website at:

*www.atlantis-press.com/publications/books*

ATLANTIS PRESS

AMSTERDAM – PARIS

World Scientific

# Topological Groups and Related Structures

Alexander Arhangel'skii,
Ohio University, Athens, Ohio, U.S.A.

Mikhail Tkachenko,
Universidad Autónoma Metropolitana, Mexico City, Mexico

ATLANTIS PRESS
AMSTERDAM – PARIS

World Scientific

**Atlantis Press**
29 avenue Laumière
75019 Paris, France

For information on all Atlantis Press publications, visit our website at: www.atlantis-press.com.



ISBN: 978-90-78677-06-2
ISSN: 1875-7634

# Foreword

Algebra and topology, the two fundamental domains of mathematics, play complementary roles. Topology studies continuity and convergence and provides a general framework to study the concept of a limit. Much of topology is devoted to handling infinite sets and infinity itself; the methods developed are qualitative and, in a certain sense, irrational. Algebra studies all kinds of operations and provides a basis for algorithms and calculations. Very often, the methods here are finitistic in nature.

Because of this difference in nature, algebra and topology have a strong tendency to develop independently, not in direct contact with each other. However, in applications, in higher level domains of mathematics, such as functional analysis, dynamical systems, representation theory, and others, topology and algebra come in contact most naturally. Many of the most important objects of mathematics represent a blend of algebraic and of topological structures. Topological function spaces and linear topological spaces in general, topological groups and topological fields, transformation groups, topological lattices are objects of this kind. Very often an algebraic structure and a topology come naturally together; this is the case when they are both determined by the nature of the elements of the set considered (a group of transformations is a typical example). The rules that describe the relationship between a topology and an algebraic operation are almost always transparent and natural — the operation has to be continuous, jointly or separately. However, the methods of study developed in algebra and in topology do not blend so easily, and that is why at present there are very few systematic books on topological algebra, probably, none which can be qualified as a reasonably complete textbook for graduate students and a source of references for experts. The need for such a book is all the greater since the last half of the twentieth century has witnessed vigorous research on many topics in topological algebra. Especially strong progress has been made in the theory of topological groups, going well beyond the limits of the class of locally compact groups. The excellent book [**236**] by E. Hewitt and K. A. Ross just sketches some lines of investigation in this direction in a short introductory chapter dedicated to topological groups.

In the 20th century and during the last seven years many topologists and algebraists have contributed to Topological Algebra. Some outstanding mathematicians were involved, among them J. Dieudonné, L. S. Pontryagin, A. Weil, and H. Weyl. The ideas, concepts, and constructions that arise when topology and algebra come into contact are so rich, so versatile, that it has been impossible to include all of them in a single book; we have made our choice. What we have covered here well may be called "topological aspects of topological algebra". This domain can be characterized as the study of connections between topological properties in the presence of an algebraic structure properly related to the topology.

A. D. Alexandroff, N. Bourbaki, M. I. Graev, S. Kakutani, E. van Kampen, A. N. Kolmogorov, A. A. Markov, and L. S. Pontryagin were among the first contributors to the theory of topological groups. Among those who contributed greatly to this field are W. W. Comfort, M. M. Choban, E. van Douwen, V. I. Malykhin, J. van Mill, and B. A. Pasynkov. These mathematicians have also contributed greatly to other aspects of topological algebra and of general topology.

Though the theory of topological groups is a core subject of topological algebra, a considerable attention has been given to the development of the theory of universal topological algebras, where topology and most general algebraic operations are blended together. This subject started to gain momentum with the works of A. I. Mal'tsev and in later years, some aspects of Mal'tsev's work especially close to general topology were developed by M. M. Choban and V. V. Uspenskij.

The fundamental topic of various types of continuity of algebraic operations was developed in the works of A. Bouziad, R. Ellis, D. Montgomery, I. Namioka, J. Troallic, and L. Zippin. The recent excellent book [**241**] of N. Hindman and D. Strauss contains a wealth of material on algebraic operations on compacta satisfying weak continuity requirements.

One of the leading topics in the general theory of topological groups was that of free topological groups of Tychonoff spaces. It is well represented in our book. A. A. Markov, S. Kakutani, T. Nakayama, and M. I. Graev were at the origins of this chapter. In later years S. A. Morris, P. Nickolas, O. G. Okunev, V. G. Pestov, O. Sipacheva, K. Yamada, and some other mathematicians have worked very successfully in this field.

Our book also contains a very brief introduction to topological dynamics. Among the first who worked in this field were W. Gottshalk, V. V. Niemytzki, and R. Ellis. J. de Vries, one of the later contributors to the subject, wrote the basic monograph [**530**] which gave a strong impuls to its further development. Some recent successes in the field are connected with the names of S. A. Antonyan, V. G. Pestov, S. Glasner, M. Megrelishvili, and V. V. Uspenskij. Well-written, very informative surveys [**379, 378**] by Pestov will orient the reader on this topic.

In this book we refer also to the works of many other excellent mathematicians, among them O. Alas, T. Banakh, L. G. Brown, R. Z. Buzyakova, D. Dikranjan, S. García-Ferreira, P. Gartside, I. I. Guran, K. P. Hart, S. Hernández, G. Itzkowitz, P. Kenderov, K. Kunen, W. B. Moors, P. Nyikos, E. Martín-Peinador, I. Prodanov, I. V. Protasov, D. A. Raĭkov, E. A. Reznichenko, D. Robbie, M. Sanchis, D. B. Shakhmatov, A. Shibakov, L. Stoyanov, A. Tomita, F. J. Trigos-Arrieta, N. Ya. Vilenkin, S. Watson, and E. Zelenyuk. We should also mention that the development of topological algebra was strongly influenced by survey papers [**109, 110, 113**] of W. W. Comfort (and coauthors), and by the books [**410**], [**249**] of W. Roelke and S. Dierolf and of T. Hussain, respectively.

We do not mention here the names of those who have worked recently in the theory of locally compact topological groups. This vast subject is mostly beyond the scope of this book, we have provided only a brief introduction to it.

This book is devoted to that area of topological algebra which studies the influence of algebraic structures on topologies that properly fit the structures. This domain could be called *"Topological invariants under algebraic boundary conditions"*. The book is by no means complete, since this area of mathematics is now rapidly developing in many directions. The central theme in the book is that of general (not necessarily locally compact) topological groups. However, we do not restrict ourselves to this main topic; on the contrary, we try to use it as a starting point in the investigation of more general objects, such as semitopological groups or paratopological groups, for example.

While not striving for completeness, we have made an attempt to provide a representative sample of some old and of some relatively recent results on general topological groups, not restricting ourselves just to two or three topics. The areas covered to a lesser or greater extent are cardinal invariants in topological algebra, separate and joint continuity of group operations, extremally disconnected and related topologies on groups, free topological groups, the Raĭkov completion of topological groups, Bohr topologies, and duality theory for compact Abelian groups.

One of the generic questions in topological algebra is how the relationship between topological properties depends on the underlying algebraic structure. Clearly, the answer to this should strongly depend on the way the algebraic structure is related to the topology. The weaker the restrictions on the connection between topology and algebraic structure are, the larger is the class of objects entering the theory. Because of that, even when our main interest is in topological groups, it is natural to consider more general objects with a less rigid connection between topology and algebra. Examples we encounter in such a theory help us to better understand and appreciate the fruits of the theory of topological groups.

Chapter 1 is of course, of an introductory nature. We define, apart from topological groups, the main objects of topological algebra such as semitopological groups, quasitopological groups, paratopological groups, and present the most elementary and natural examples and the most general facts. Some of these facts are non-trivial, even though they are easy to prove. For example, we establish that every open subgroup of a topological group is closed, and that every discrete subgroup of a pseudocompact group is finite. It is proved in this chapter that every infinite Abelian group admits a non-discrete Hausdorff topological group topology. Quotients, products, and $\Sigma$-products are also discussed in Chapter 1, as well as the natural uniformities on topological groups and their quotients.

In the course of the book, we introduce and study several important classes of topological groups. In particular, in Chapter 3 we study systematically $\omega$-narrow topological groups which can be characterized as topological subgroups of arbitrary topological products of second-countable topological groups. An elementary introduction to the theory of locally compact groups is also given in Chapter 3. Then this topic is developed in Chapter 9, where an introduction into the theory of characters of compact and locally compact Abelian groups is to be found. Since there are several good sources covering this subject (such as [**236**], [**243**], and [**327**] just to mention a few), we do not pursue this topic very far. However, elements of the Pontryagin–van Kampen duality theory are presented, and the exposition is elementary and practically self-contained.

The celebrated theorem of Ivanovskij and Kuz'minov on the dyadicity of compact groups is proved in Chapter 4. Again, the proof is elementary (though not simple), polished, and self-contained. We apply Pontryagin–van Kampen duality theory to continue the study of the algebraic and topological structures of compact Abelian groups in Chapter 9. The book [**243**] by K.-H. Hofmann and S. A. Morris provides those readers who are interested in the duality theory with considerably more advanced material in this direction.

In Chapter 4 we consider the class of extremally disconnected groups, the class of Čech-complete groups, as well as the classes of feathered groups and $P$-groups. For each of these classes of groups we prove original and delicate theorems and then establish non-trivial relations between them. Feathered topological groups (called also $p$-groups) present a natural generalization of locally compact groups and of metrizable groups, that makes them especially interesting.

One of the unifying themes of this book is that of completions and completeness. One can look at completeness in topological algebra either from a purely topological point of view or from the point of view of the theory of uniform spaces; this latter takes into account the algebraic structure much better than the purely topological one. The basic construction of the Raĭkov completion of an arbitrary topological group is presented in Chapter 3; later on, it has many applications. Čech-completeness of topological groups is studied in Chapter 4, and the relationship of Dieudonné completion of a topological group with the group structure is a subject of a rather deep investigation in Chapter 6. In particular, we learn in Chapter 6 that under very general assumptions it is possible to extend continuously the group operations from a topological group to its Dieudonné completion. We also establish that this is not always possible. The class of Moscow groups is instrumental in the theory developed in Chapter 6. The class of $\mathbb{R}$-factorizable groups is studied in Chapter 8. This class serves as a bridge from general topological groups to second-countable groups via continuous real-valued functions. It also turns out to be important in the study of completions of topological groups.

Chapter 5 is devoted to cardinal invariants of topological groups. Invariants of this kind (which associate with topological spaces cardinal numbers "measuring" the space under consideration in one sense or another), play an especially important role in general topology; probably, this happens because the techniques they provide fit best the set-theoretic nature of general topology. So one may expect that in the study of non-compact topological groups cardinal invariants should occupy a prominent place. The following phenomenon makes the situation even more interesting: The structure of topological groups turns out to be much more sensitive to restrictions in terms of cardinal invariants than the structure of general topological spaces. For example, metrizability of a topological group depends only on whether the group is first-countable or not (Birkhoff–Kakutani's theorem). For paratopological groups the statement is no longer true (the Sorgenfrey line witnesses this); however, a weaker theorem holds: every first-countable paratopological group has a $G_\delta$-diagonal. How delicate problems involving cardinal invariants of compact groups can be, is shown by the following simple result: It is not possible either to prove, or to disprove in $ZFC$ that every compact group of cardinality not greater than $\mathfrak{c} = 2^\omega$ is metrizable.

In Chapter 7, a very powerful and delicate construction is presented — that of a free topological group over a Tychonoff space. Under this construction, any Tychonoff space $X$ can be represented as a closed subspace of a topological group $F(X)$ in such a way that every continuous mapping of a space $X$ into a space $Y$ can be uniquely extended to a continuous homomorphism of $F(X)$ to $F(Y)$. The set $X$, of course, serves as an algebraic basis of $F(X)$. However, the relationship between the topology of $X$ and that of $F(X)$ is the most intriguing; there are many unexpected and subtle results on free topological groups and quite a few unsolved problems. One of the most important theorems here states that the cellularity of the free topological group $F(X)$ of an arbitrary compact Hausdorff space $X$ is countable. Curiously, one can demonstrate that this happens not because of the existence of a regular measure on $F(X)$ (in contrast with the case of compact groups). In fact, such a measure on $F(X)$ exists if and only if $X$ is discrete.

Yet another major topic in the book is that of transformation groups, and the closely associated concepts of homogeneous spaces and of groups of homeomorphisms. A section is devoted to these matters in Chapter 3, where it is established that the group of isometries of a metric space is a topological group, when endowed with the topology of pointwise convergence. It is proved in this connection that every topological group is topologically isomorphic to a subgroup of the group of isometries of some metric space. This provides an important technical tool for some arguments.

Frequently, results on topological groups are followed by a discussion of other structures of topological algebra, such as semitopological and paratopological groups. This is done in almost every chapter. However, we have also devoted the whole of Chapter 2 to basic facts regarding such objects. A *paratopological group* is a group $G$ with a topology such that the multiplication mapping of $G \times G$ to $G$ is jointly continuous. A *semitopological group* $G$ is a group $G$ with a topology such that the multiplication mapping of $G \times G$ to $G$ is separately continuous. A *quasitopological group* $G$ is a group $G$ with a topology such that the multiplication mapping of $G \times G$ to $G$ is separately continuous and the inverse mapping of $G$ onto itself is continuous. A natural example of a paratopological group is obtained by taking the group of homeomorphisms of a dense-in-itself locally compact zero-dimensional non-compact space, with the compact-open topology. The Sorgenfrey line under the usual addition is a paratopological group which is hereditarily separable, hereditarily Lindelöf and has the Baire property.

In 1936, D. Montgomery proved that every semitopological group metrizable by a complete metric is, in fact, a paratopological group. In 1957, R. Ellis showed that every locally compact semitopological group is a topological group. In 1960, W. Zelazko established that each completely metrizable semitopological group is a topological group. Later, in 1982, N. Brand proved that every Čech-complete paratopological group is a topological group. Recently A. Bouziad made a decisive contribution to this topic. He proved that every Čech-complete semitopological group is a topological group. This theorem naturally covers and unifies both principal cases, those of locally compact semitopological groups and of completely metrizable semitopological groups.

Since each Čech-complete topological group is paracompact, Bouziad's theorem implies that every Čech-complete semitopological group is paracompact. These and related results, with applications, are presented in Chapter 2. In this same chapter we construct an operation of a rather general nature on the Čech–Stone compactification $\beta G$ of an arbitrary discrete group $G$. With this operation, the compact space $\beta G$ becomes a right topological group. This structure has interesting applications; we mention some of them in problem sections. The reader who wants to learn more on this subject is advised to study the recent book [**241**] by N. Hindman and D. Strauss.

We formulate and discuss quite a few open problems, many of them are new. Each section is followed by a list of exercises and problems, including open problems. However, we should warn the reader that some of the new open problems might turn out not to be difficult after all. That does not necessarily mean that they should have been discarded. The main interest of many of the new questions we have posed lies in the fact that they delineate a new direction of research. On many occasions exercises and problems are provided with hints, references, and comments. In this way, many additional directions and topics are introduced. Here are two outstanding examples of old unsolved questions. Is it possible to construct in $ZFC$ a non-discrete extremally disconnected topological group? Is it possible to construct in $ZFC$ a countable non-metrizable topological group $G$ such that $G$ is a Fréchet–Urysohn space?

We would be grateful for the information on the progress of open problems posed in this book.

We hope that this book will achieve several goals. First, we believe that it can be used as a reasonably complete introduction to the theory of general topological groups beyond the limits of the class of locally compact groups. Second, we expect that it may lead advanced students to the very boundaries of modern topological algebra, providing them with goals and with powerful techniques (and maybe, with inspiration!). The exercise and problem sections can be especially useful in that respect. One can use this book in a research seminar on topological algebra (with an eye to unsolved problems) and also in advanced courses — at least four special courses can be arranged on the basis of this book. Fourth, we expect that the book will serve quite effectively as a reference, and will be helpful to mathematicians working in other domains of mathematics.

The standard reference book for general topology is R. Engelking's book *General Topology* [**165**]. We expect that the reader either knows the basic facts from general topology that we need, or that he/she will not find it too difficult to extract the corresponding information from [**165**].

We wish to express our deep gratitude to the second author's former students Constancio Hernández García and Yolanda Torres Falcón for their continued help in our work on this book over many years. We are also indebted to Richard G. Wilson whose comments enabled us to improve the text.

*A. V. ARHANGEL'SKII,* *M. G. TKACHENKO*
arhangel@math.ohiou.edu mich@xanum.uam.mx

February 21, 2008

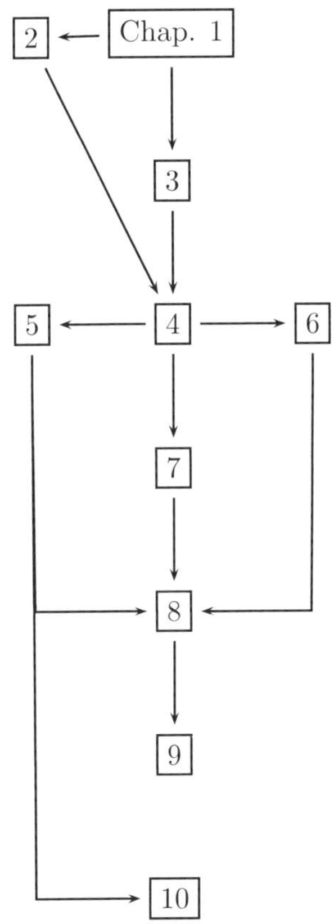


FIGURE 1. Logical dependence of chapters.

# Contents

Chapter 1

# Introduction to Topological Groups and Semigroups

**Notation.** We write $\mathbb{N}$ for the set of positive integers, $\omega$ for the set of non-negative integers, and $\mathbb{P}$ for the set of prime numbers. The set of all integers is denoted by $\mathbb{Z}$, the set of all real numbers is $\mathbb{R}$, and $\mathbb{Q}$ stands for the set of all rational numbers.

The symbols $\tau$, $\lambda$, $\kappa$ are used to denote infinite cardinal numbers. A cardinal number $\tau$ is also interpreted as the smallest ordinal number of cardinality $\tau$. Each ordinal is the set of all smaller ordinals. Thus, $\omega$ is both the smallest infinite ordinal number and the smallest infinite cardinal number.

All topologies considered below are assumed to satisfy $T_1$-separation axiom, that is, we declare all one-point sets to be closed. The closed unit interval $[0, 1]$ of the real line $\mathbb{R}$, with its usual topology, is denoted by $I$, and $Sq$ stands for the convergent sequence $\{1/n : n \in \mathbb{N}\} \cup \{0\}$ with its limit point 0, also taken with the usual topology. We use the symbol $\mathbb{C}$ to denote the complex plane with the usual sum and product operations, while $\mathbb{T} = \{z \in \mathbb{C} : |z| = 1\}$ is the unit circle with center at the origin of $\mathbb{C}$.

## 1.1. Some algebraic concepts

In this section we establish the terminology and notation that will be used throughout the book.

In dealing with groups, we will adhere to the multiplicative notation for the binary group operation. In discussions involving a multiplicative group $G$, the symbol $e$ will be reserved for the identity element of $G$.

We are very much concerned with groups in this course. For many purposes, however, it is natural and convenient considering semigroups. A *semigroup* is a non-void set $S$ together with a mapping $(x, y) \to xy$ of $S \times S$ to $S$ such that $x(yz) = (xy)z$ for all $x, y, z$ in $S$. That is, a semigroup is a non-void set with an associative multiplication. Given an element $x$ of a semigroup $S$, one inductively defines

$$x^2 = xx,\ x^3 = xx^2,\ \ldots,\ x^{n+1} = xx^n,$$

for every $n \in \mathbb{N}$. The associativity of multiplication in $S$ implies the equality $x^n x^m = x^{n+m}$ for all $x \in S$ and $n, m \in \mathbb{N}$.

An element $e$ of a semigroup $S$ is called an *identity* for $S$ if $ex = x = xe$ for every $x \in S$. Not every semigroup has an identity (see items 4) and 6) of Example 1.1.1). However, if a semigroup $S$ has an identity, then it is easy to see that this identity is unique. Whenever we use the symbol $e$ without explanation, it always stands for the identity of the semigroup under consideration.

A semigroup with identity is called *monoid*. An element $a$ of a monoid $M$ is said to be *invertible* if there exists an element $b$ of $M$ such that $ab = e = ba$. Note that if $a$ is an invertible element of a monoid $M$, then the element $b \in M$ such that $ab = e = ba$ is unique. Indeed, suppose that $ab = e$, $ba = e$, $ac = e$, and $ca = e$. Then we have

$$c = ec = (ba)c = b(ac) = be = b.$$

This fact enables us to use notation $a^{-1}$ for such an element $b$ of $M$. We also say that $b$ is the *inverse* of $a$. It is clear that $(a^{-1})^{-1} = a$ for each invertible element $a \in M$. Further, one can define negative powers of an invertible element $a \in M$ by the rule $a^{-n} = (a^{-1})^n$, for each $n \in \mathbb{N}$. It is a common convention to put $a^0 = e$. We leave to the reader a simple verification of the equality $a^n a^m = a^{n+m}$ which holds for all $n, m \in \mathbb{Z}$.

If every element $a$ of a monoid $M$ is invertible, then $M$ is called a *group*.

Let $S$ be a semigroup. For a fixed element $a \in S$, the mappings $x \mapsto ax$ and $x \mapsto xa$ of $S$ to itself are called the *left* and *right actions* of $a$ on $S$, and are denoted by $\lambda_a$ and $\varrho_a$, respectively.

If $G$ is a group, the mapping $x \mapsto x^{-1}$ of $G$ onto itself is called *inversion*. Left and right actions of every element $a \in G$ on $G$ are, in this case, bijections. They are called *left* and *right translations* of $G$ by $a$.

EXAMPLE 1.1.1. Each of the following is a semigroup but not a group.

1) The set $\mathbb{Z}$ of all integers with the usual multiplication.
2) The set $\mathbb{Q}$ of all rational numbers with the usual multiplication.
3) The set $\mathbb{R}$ of all real numbers with the usual multiplication.
4) The set of all positive real numbers with the usual addition in the role of the product operation.
5) The set $\mathbb{N}$, in which the product of $x$ and $y$ is defined as $\max\{x, y\}$.
6) The set $\mathbb{N}$, in which the product of $x$ and $y$ is defined as $\min\{x, y\}$.
7) Any set $S$ with $|S| > 1$, where the product $xy$ is defined as $y$.
8) Any set $S$ with $|S| > 1$, where the product $xy$ is defined as $x$.
9) The set $S(X, X)$ of all mappings of a set $X$ to itself with the composition of mappings in the role of multiplication, where $|X| > 1$. □

In items 4) and 6) of the above example, the corresponding semigroups have no identity. The semigroups in 1)–3), 5), and 9) are monoids.

Now we present a few standard examples of groups.

EXAMPLE 1.1.2. Each of the following is a group:

1) The set $\mathbb{Z}$ of all integers with the usual addition in the role of multiplication.
2) The set $\mathbb{Q} \setminus \{0\}$ of all non-zero rational numbers with the usual multiplication.
3) The set $\mathbb{R} \setminus \{0\}$ of all non-zero real numbers with the usual multiplication.
4) The set of all positive real numbers with the usual multiplication.
5) The set $\{0, 1\}$ with the binary operation defined as follows:

$$0 + 0 = 0, \quad 0 + 1 = 1 + 0 = 1, \quad 1 + 1 = 0.$$

This group is denoted by $\mathbb{Z}(2)$ or by $D$; it is called the *cyclic group* of two elements, or the *two-element group*. More generally, for an integer $n > 1$, let $\mathbb{Z}(n) = \{0, 1, \ldots, n-1\}$

be the set of all non-negative residues modulo $n$ with addition modulo $n$. For example, $2 + (n - 1) = 1$ in $\mathbb{Z}(n)$; of course, $n$ must be greater than or equal to 3 in order for 2 to be an element of $\mathbb{Z}(n)$. An easy verification shows that $\mathbb{Z}(n)$ with the addition just defined is a commutative group. It is called the *cyclic group of order n*.

6) The set $\mathbb{T}$ of all complex numbers $z$ such that $|z| = 1$ with respect to the usual multiplication of complex numbers ($|z|$ denotes here the modulus of $z$).
7) The set of all $n$ by $n$ matrices, where the coefficients are real numbers, with non-zero determinant, and with the usual matrix multiplication. This group is called the *general linear group* of degree $n$ over $\mathbb{R}$.
8) If $X$ is any non-void set, then the set of all one-to-one mappings of $X$ onto $X$ forms a group $\mathcal{S}_X$ under the operation of composition. This group is called the *permutation group* on $X$. If $X$ is finite and has $n$ elements, then $\mathcal{S}_X$ is denoted by $\mathcal{S}_n$ and is called the *symmetric group* of degree $n$. □

A semigroup (monoid, group) $S$ is called *Abelian* or *commutative* if $xy = yx$, for all $x, y \in S$. Clearly, the semigroups in items 7), 8), 9) of Example 1.1.1 and the groups in 7) of Example 1.1.2 (with $n \geq 2$ and $|X| \geq 3$, respectively) are not Abelian.

Let $A$ and $B$ be subsets of a semigroup $G$. Then $AB$ denotes the set $\{ab : a \in A, b \in B\}$, and, if $G$ is a group, $A^{-1}$ denotes the set $\{a^{-1} : a \in A\}$. A subset $A$ of a group $G$ is called *symmetric* if $A^{-1} = A$.

We write $aB$ for $\{a\}B$ and $Ba$ for $B\{a\}$. We abbreviate $AA$ as $A^2$, $AAA$ as $A^3$, *etc.* Similarly, $A^{-2}$ is a substitute for $A^{-1}A^{-1}$, *etc.*

A non-empty subset $H$ of a semigroup $S$ is called a *subsemigroup* of $S$ if $xy \in H$, for all $x, y$ in $H$. A non-empty subset $H$ of a group $G$ is called a *subgroup* of $G$ if $xy^{-1} \in H$, for all $x, y$ in $H$.

Clearly, a subset $H$ of a group $G$ is a subgroup of $G$ if and only if for each $x, y \in H$, $xy \in H$ and $x^{-1} \in H$. Every group contains at least two subgroups — the whole group $G$ and the subgroup consisting of the identity only. These subgroups are called *trivial*. Note that the empty set is not a group and, therefore, is not a subgroup of any group. Obviously, for any subgroup $H$ of a group $G$ we have $H^2 = HH = H$. However, in general, the same is not true for subsemigroups of semigroups.

In any group $G$, if $a, b \in G$ then $(ab)^{-1} = b^{-1}a^{-1}$. If $H$ is any subgroup of $G$ then, for any $a \in G$, $a^{-1}Ha$ is also a subgroup of $G$. If $H$ is a subgroup of a group $G$ such that $a^{-1}Ha = H$ for each $a \in G$, then $H$ is said to be an *invariant* or *normal* subgroup of $G$. Since in Topology "normal" refers to a separation property of spaces, we will use the term "invariant" to denote this property of subgroups. Of course, in any Abelian group, every subgroup is invariant.

If $H$ is a subgroup of a group $G$ and $a \in G$, then the sets $aH$ and $Ha$ are called *left* and *right cosets* of $H$ in $G$, respectively. The element $a$ is a *representative* of both cosets.

For any two right cosets $Ha$ and $Hb$, either they are disjoint or coincide. Furthermore, $Ha = Hb$ if and only if $ab^{-1} \in H$. Indeed, $ab^{-1} \in H$ implies that $Hab^{-1} \subset H^2 = H$. Hence, $Ha \subset Hb$. Similarly, since $(ab^{-1})^{-1} = ba^{-1} \in H$, it follows that $Hb \subset Ha$. Therefore, $Ha = Hb$. Conversely, if $Ha = Hb$, then $h_1a = h_2b$ for some $h_1, h_2 \in H$, whence $ab^{-1} = h_1^{-1}h_2 \in H$.

Let $H$ be an invariant subgroup of a group $G$. Then $aH = Ha$ for each $a \in G$. In other words, the left cosets of $H$ are the same as the right cosets of $H$. On the set of all cosets of

$H$ we define multiplication by the rule $aHbH = abH$. It is easy to see that our definition of multiplication of cosets is correct. Indeed, suppose that $aH = a_1H$ and $bH = b_1H$ for some $a, a_1 \in G$ and $b, b_1 \in G$. Then $aa_1^{-1} \in H$ and $bb_1^{-1} \in H$ whence it follows, by the invariance of $H$ in $G$, that

$$ab(a_1b_1)^{-1} = abb_1^{-1}a_1^{-1} \in aHa_1^{-1} = (aHa^{-1})aa_1^{-1} = Haa_1^{-1} = H.$$

Therefore, $aHbH = abH = a_1b_1H = a_1Hb_1H$, thus showing that the result of multiplication of two cosets does not depend on the choice of representatives in these cosets.

For each $aH$, we have that $(aH)H = (aH)(eH) = aH$ and $(a^{-1}H)(aH) = (a^{-1}a)H = eH = H$. This shows that $H$ plays the role of the identity in the set of all cosets, and $a^{-1}H$ is the inverse of $aH$. Hence, the set of all cosets of $H$ is a group with respect to the multiplication defined above. This group is called the *quotient group* of $G$ and denoted by $G/H$. Note that if $G$ is an Abelian group, then the quotient group $G/H$ is defined for each subgroup $H$ of $G$.

If $G$ is a semigroup and $H$ is a subsemigroup of $S$, then it may happen that, for some $a$ and $b$ in $S$, the sets $aS$ and $bS$ do not coincide and, nonetheless, are not disjoint.

We can also think of subgroups of semigroups. Let $S$ be a semigroup. We will call $G$ a *subgroup* of $S$ if $G \subset S$ and $G$ is a group under the restriction of the product operation in $S$ to $G$.

Let $S$ be a semigroup. An element $x$ of $S$ is called an *idempotent* if $xx = x$. The set of all idempotents of $S$ is denoted by $E(S)$. Every idempotent of a group $G$ coincides with the identity $e$ of $G$. Indeed, if $x^2 = x$ for some $x \in G$, then $x^2x^{-1} = xx^{-1} = e$, that is, $x = e$.

EXAMPLE 1.1.3. Let $X$ be a non-empty set. Then the idempotents of the semigroup $S(X, X)$ of all mappings of $X$ to itself are precisely the mappings $f\colon X \to X$ satisfying the condition $f(x) = x$ for every $x \in f(X)$. □

Now we give several definitions and present some special results on groups which will be used in the sequel.

A *homomorphism* of a semigroup (monoid, group) $G$ to a semigroup (monoid, group) $F$ is a mapping $f\colon G \to F$ such that $f(ab) = f(a)f(b)$ for all $a, b \in G$. Given a homomorphism of monoids $f\colon G \to H$, the set $\{x \in G : f(x) = e_H\}$ is called the *kernel* of $f$ and denoted by $\ker f$. It follows immediately from the definition that $\ker f$ is a subsemigroup of $G$.

A homomorphism of a semigroup (monoid, group) $G$ onto a semigroup (monoid, group) $F$ which is a one-to-one mapping is called an *isomorphism*.

If $G$ and $H$ are monoids with respective identities $e_G$ and $e_H$ and $f\colon G \to H$ is a homomorphism onto $H$, then $f(e_G) = e_H$. Indeed, put $h = f(e_G)$ and for an arbitrary element $b \in H$, take $a \in G$ with $f(a) = b$. Then $bh = f(ae_G) = f(a) = b$ and $hb = f(e_Ga) = f(a) = b$. Since the identity of a monoid is unique, we infer that $h = e_H$.

If $G$ and $H$ are groups, then the equality $f(e_G) = e_H$ holds for every homomorphism $f$ of $G$ to $H$. Indeed, the above argument gives the equality $bh = b$ for an arbitrary element $b \in f(G)$, where $h = f(e_G)$. Since $H$ is a group, we have that $b^{-1}bh = b^{-1}b = e_H$ and, hence, $h = e_Hh = e_H$.

Furthermore, the kernel of $f$ is a subgroup of the group $G$. For every $a \in \ker f$, we have that $e_H = f(e_G) = f(aa^{-1}) = f(a)f(a^{-1}) = f(a^{-1})$, whence it follows that $a^{-1} \in \ker f$. Since $\ker f$ is a subsemigroup of $G$, it must be a subgroup of $G$.

An isomorphism of a group $G$ onto itself is called an *automorphism* of $G$.

EXAMPLE 1.1.4. Let $Aut(G)$ be the set of all automorphisms of a group $G$ with the operation of composition, $(g \circ f)(x) = g(f(x))$ for all $f, g \in Aut(G)$ and $x \in G$. Evidently, $Aut(G)$ is a group. The group $Aut(G)$ need not be commutative, even if $G$ is commutative.

There exists a natural homomorphism $\varphi$ of $G$ onto a subgroup of $Aut(G)$ defined as follows. For arbitrary $a, x \in G$, put $\varphi_a(x) = axa^{-1}$. It is clear that

$$\varphi_a(xy) = axya^{-1} = axa^{-1}aya^{-1} = \varphi_a(x)\varphi_a(y)$$

for all $x, y \in G$, so $\varphi_a$ is a homomorphism of $G$ to $G$. For an element $y \in G$, put $x = a^{-1}ya$. Then $\varphi_a(x) = y$ and, therefore, $\varphi_a(G) = G$. One easily verifies that the mapping $\varphi_a$ is one-to-one, so it is an automorphism of $G$. The correspondence $a \mapsto \varphi_a$ is the homomorphism $\varphi$ of $G$ to $Aut(G)$ we are looking for. Indeed,

$$(\varphi_a \circ \varphi_b)(x) = \varphi_a(\varphi_b(x)) = \varphi_a(bxb^{-1}) = abxb^{-1}a^{-1} = \varphi_{ab}(x)$$

for all $a, b, x \in G$. Therefore, $\varphi_a \circ \varphi_b = \varphi_{ab}$ or, equivalently, $\varphi(ab) = \varphi(a) \circ \varphi(b)$.

The automorphisms $\varphi_a$ of $G$, with $a \in G$, are called *inner automorphisms*. If the group $G$ is Abelian, then $\varphi_a$ is the identity mapping of $G$ for each $a \in G$. In other words, $\ker \varphi$ coincides with the group $G$ in this case. In general, the kernel of $\varphi$ coincides with the *center* $Z(G)$ of $G$ defined by $Z(G) = \{a \in G : ax = xa \text{ for all } x \in G\}$. □

Given a subgroup $H$ of a group $G$ with identity $e$, we say that $H$ is a *central subgroup* of $G$ if $H \subset Z(G)$, that is, $hx = xh$ for all $h \in H$ and $x \in G$. An element $a$ of $G$ is said to be an element *of finite order* or, equivalently, a *torsion element* if $a^n = e$, for some $n \in \mathbb{N}$. If this is the case, then the smallest $n \in \mathbb{N}$ for which $a^n = e$ is called the *order* of $a$ and is denoted by $o(a)$. If all elements of $G$ have finite orders, we say that $G$ is a *torsion group*. If the group $G$ has no elements of finite order, except for $e$, then it is called *torsion-free*.

The *cyclic subgroup* of $G$ generated by an element $a \in G$ is the set $\{a^k : k \in \mathbb{Z}\}$. This subgroup is also denoted by $\langle a \rangle$. Every cyclic group is evidently commutative. If $a \in G$ is a torsion element and $o(a) = n$, then the cyclic subgroup $\langle a \rangle$ has exactly $n$ elements or, more precisely, $\langle a \rangle = \{a^k : 1 \le k \le n\}$. The set of all elements $a \in G$ of finite order is called the *torsion part* of $G$ and is denoted by $tor(G)$. If the group $G$ is commutative, the torsion part $tor(G)$ is a subgroup of $G$. Indeed, if $x, y \in tor(G)$, there exist integers $m, n \in \mathbb{N}$ such that $x^m = e$ and $y^n = e$. Set $N = mn$. Since $G$ is commutative, it follows that $(xy)^N = x^N y^N = (x^m)^n (y^n)^m = e^n e^m = e$, whence $xy \in tor(G)$. Similarly, $(x^{-1})^m = (x^m)^{-1} = e^{-1} = e$. If $G$ is commutative, we will call $tor(G)$ the *torsion subgroup* of $G$.

Suppose that $A$ is a non-empty subset of a semigroup $S$. Then $\langle A \rangle$ denotes the smallest subsemigroup of $S$ which contains the set $A$. It is clear that every element $b \in \langle A \rangle$ has the form $b = a_1 \dots a_n$, where $a_1, \dots, a_n$ are arbitrary elements of $A$. If $A$ is a non-empty subset of a group $G$, we use the same symbol $\langle A \rangle$ to denote the smallest subgroup of $G$ that contains $A$. Similarly, every element $g \in \langle A \rangle$ has the form $g = a_1^{\varepsilon_1} \dots a_n^{\varepsilon_n}$ for some $a_1, \dots, a_n \in A$ and $\varepsilon_1, \dots, \varepsilon_n = \pm 1$, where $n$ is an arbitrary positive integer. Therefore, the cyclic subgroup $\langle a \rangle$ of $G$ is generated by the one-point set $\{a\}$.

In the next lemma we give a necessary and sufficient condition for a homomorphism defined on a subgroup of an Abelian group to admit an extension over a bigger subgroup.

LEMMA 1.1.5. *Let $H$ be a subgroup of an Abelian group $G$, $f$ a homomorphism of $H$ to an Abelian group $F$, and let $x \in G$ and $y \in F$. Define a number $m \in \mathbb{N}^* = \mathbb{N} \cup \{\infty\}$ by $m = \min\{n \in \mathbb{N}^* : nx \in H\}$ (thus, $m = \infty$ if $nx \notin H$ for each $n \in \mathbb{N}$). Then $f$ admits an extension to a homomorphism $h\colon \langle H \cup \{x\}\rangle \to F$ satisfying $h(x) = y$ if and only if either $m = \infty$ or $m \in \mathbb{N}$ and $my = f(mx)$.*

PROOF. Necessity is evident, so we only prove sufficiency. Let $H_0 = \langle H \cup \{x\}\rangle$. It suffices to consider the following two cases.
*Case 1.* $m = \infty$. Then $kx$ is not in $H$ for each $k \in \mathbb{N}$, so every element $z \in H_0$ has a unique representation in the form $z = kx + a$ with $k \in \mathbb{Z}$ and $a \in H$. Put $h(kx+a) = ky + f(a)$ for all $k \in \mathbb{Z}$ and $a \in H$. This defines a mapping $h\colon H_0 \to F$. If $z_1 = kx + a$ and $z_2 = lx + b$ are two elements of $H_0$, then $h(z_1 + z_2) = (k + l)y + f(a + b) = h(z_1) + h(z_2)$, so $h$ is a homomorphism.
*Case 2.* $m \in \mathbb{N}$. Then $mx \in H$, by the definition of $m$. Put $h(x) = y$ and, more generally, $h(kx + a) = ky + f(a)$ for all $k \in \mathbb{Z}$ and $a \in H$. Let us verify that the mapping $h\colon H_0 \to F$ is correctly defined. Indeed, suppose that $kx + a = lx + b$ for some $k, l \in \mathbb{Z}$ and $a, b \in H$. Then $(k - l)x = b - a \in H$, so $m$ divides $k - l$ by our choice of $m$. Hence $k - l = mp$, for some $p \in \mathbb{Z}$. It follows that

$$(ky + f(a)) - (ly + f(b)) = (k - l)y - f(b - a) = mpy - f(mpx) = 0,$$

or, in other words, $h(kx + a) = h(ly + b)$. Thus, the value $h(kx + a)$ does not depend on the choice of $k \in \mathbb{Z}$ and $a \in H$.

One easily verifies that, in either case, $h$ is a homomorphism of $H_0$ to $F$ that extends $f$. □

A group $G$ is said to be *divisible* if $G^n = G$ for each $n \in \mathbb{N}$. In other words, given $x \in G$ and $n \in \mathbb{N}$, there is an element $y \in G$ such that $y^n = x$. The group $\mathbb{T}$ of complex numbers $z \in \mathbb{C}$ with $|z| = 1$ is, obviously, divisible. The additive group of real numbers is also a divisible group. On the other hand, the group $\mathbb{Z}(2) = \{0, 1\}$ is not divisible. A fundamental property of divisible groups is the following one.

THEOREM 1.1.6. *Let $H$ be a subgroup of an Abelian group $G$. Then every homomorphism $f$ of $H$ to any divisible group $F$ can be extended to a homomorphism of $G$ to $F$.*

PROOF. We argue with the aim to use Zorn's lemma. Denote by $\mathcal{P}$ the family of the pairs $(K, g)$ such that $K$ is a subgroup of $G$ containing $H$ and $g\colon K \to F$ a homomorphism such that the restriction of $g$ to $H$ coincides with $f$. Given two elements $(K, g)$ and $(K_1, g_1)$ in $\mathcal{P}$, we put $(K, g) \leq (K_1, g_1)$ if $K \subset K_1$ and $g_1$ extends $g$. This gives us a partially ordered set $(\mathcal{P}, \leq)$. Suppose that $\mathcal{C} \subset \mathcal{P}$ is a *chain* in $\mathcal{P}$, that is, a subset of $\mathcal{P}$ linearly ordered by the order $\leq$ of $\mathcal{P}$. Put

$$P^* = \bigcup\{P : (P, g) \in \mathcal{C} \text{ for some homomorphism } g\colon P \to F\}$$

and define a mapping $g^*\colon P^* \to F$ by the rule $g^*(x) = g(x)$, where $x \in P$ and $(P, g) \in \mathcal{C}$. Since $\mathcal{C}$ is a chain in $(\mathcal{P}, \leq)$, it follows that $P^*$ is a subgroup of $G$, $H \subset P^*$, and $g^*$ is a homomorphism of $P^*$ to $F$. If follows from the definition of $(P^*, g^*)$ that $(P, g) \leq (P^*, g^*)$ for each $(P, g) \in \mathcal{C}$. We have proved that every chain in $(\mathcal{P}, \leq)$ has an upper bound in $(\mathcal{P}, \leq)$.

Therefore, by Zorn's lemma, the partially ordered set $(\mathcal{P}, \leq)$ has a maximal element $(K, h)$. It remains to verify that $K = G$. Suppose to the contrary that $K \neq G$ and choose an element $a \in G \setminus K$. Since $F$ is divisible, Lemma 1.1.5 guarantees the existence of a homomorphism $h_0 \colon K_0 \to F$ extending $h$, where $K_0 = \langle K \cup \{a\} \rangle$. It is clear that $(K, h) < (K_0, h_0)$ and that $(K_0, h_0) \in \mathcal{P}$. This contradicts the maximality of $(K, h)$ and finishes the proof. □

PROPOSITION 1.1.7. *Let $G$ be any group, and $b$ any element of $G$ distinct from the identity $e$ of $G$. Then there exists a cyclic subgroup $H$ of $G$ containing $b$ and isomorphic to a subgroup of the circle group $\mathbb{T}$.*

PROOF. Let us consider two cases.

*Case 1.* $b$ is of finite order. Let $p$ be the order of $b$. Put $\varphi = 2\pi/p$, and $a = \cos\varphi + i \sin\varphi$. Then $a \in \mathbb{T}$, and the order of $a$ in the group $\mathbb{T}$ is $p$. Put $H = \langle b \rangle$, $K = \langle a \rangle$, and $g(b^n) = a^n$, for each $n = 1, \ldots, p$. Then $H$ is a subgroup of $G$ containing $b$, $K$ is a subgroup of $\mathbb{T}$, and $g$ is an isomorphism of $H$ onto $K$.

*Case 2.* $b$ is not of finite order. There exists $\varphi$ such that $0 < \varphi < \pi$ and for any pair $(n, k) \in \mathbb{N} \times \mathbb{N}$, $n\varphi$ is not equal to $2k\pi$. Put $a = \cos\varphi + i \sin\varphi$ and $g(b^n) = a^n$, for each $n \in \mathbb{Z}$. Then $g$ is an isomorphism of the subgroup $H = \langle b \rangle$ of $G$ onto the subgroup $K = \langle a \rangle$ of $\mathbb{T}$, and $b \in H$. □

From Theorem 1.1.6 and Proposition 1.1.7 we obtain:

COROLLARY 1.1.8. *For any Abelian group $G$, and any element $a$ of $G$ distinct from the identity $e$ of $G$, there exists a homomorphism $g$ of $G$ to the circle group $\mathbb{T}$ such that $g(a) \neq 1$.*

PROOF. By Proposition 1.1.7, there exists a subgroup $H$ of $G$ which contains $a$ and is isomorphic to a subgroup $K$ of $\mathbb{T}$. Fix an isomorphism $f$ of $H$ onto $K$. Clearly, $f(a) \neq 1$. It remains to apply Theorem 1.1.6. □

To finish this section we present two interesting and important examples of groups the first of which is called the *group of quaternions* and the second one is the group of *r-adic numbers*. In fact, the group of quaternions has a richer structure. Here are the necessary definitions.

A non-empty set $S$ with two binary operations $+$ and $\cdot$ called *addition* and *multiplication*, respectively, is said to be a *ring* if the following conditions are satisfied:

(R1) $(S, +)$ is a commutative group;

(R2) $(S, \cdot)$ is a monoid;

(R3) $x \cdot (y + z) = x \cdot y + x \cdot z$ and $(y + z) \cdot x = y \cdot x + z \cdot x$ for all $x, y, z \in S$.

Condition (R3) expresses the distributive law that relates addition and multiplication in $S$. It is common practice to abbreviate $(S, +, \cdot)$ simply as $S$ when there is no confusion with the operations in $S$. Let 0 and 1 be neutral elements of the group $(S, +)$ and monoid $(S, \cdot)$, respectively. We leave to the reader the simple verification of the equality $0 \cdot x = x \cdot 0 = 0$, which holds for each $x \in S$. Notice that if $0 = 1$, then $S$ contains only the element 0.

If the multiplication in a ring $S$ is commutative, then the ring is called *commutative.* A ring $S$ in which $0 \neq 1$ and every non-zero element $x \in S$ is invertible with respect to multiplication is called a *skew field.* A commutative skew field is a *field.*

Clearly, $\mathbb{R}$ and $\mathbb{C}$ are fields when considered with their usual addition and multiplication. The set $M(n, \mathbb{R})$ of all $n$ by $n$ matrices with real entries and the usual matrix addition and multiplication is a non-commutative ring, for each $n > 1$. The set $P[x_1, \ldots, x_n]$ of all polynomials of mutually commuting variables $x_1, \ldots, x_n$ with real coefficients and the usual sum and multiplication is an example of a commutative ring. For every integer $r > 1$, the set $\mathbb{Z}(r)$ of non-negative residues modulo $r$ with addition and multiplication modulo $r$ is another example of a commutative ring (we noted in item 5) of Example 1.1.2 that $\mathbb{Z}(r)$ is a commutative group with respect to addition). It is well known (and easy to verify) that $\mathbb{Z}(r)$ is a field if and only if the number $r$ is prime. Yet another example of a commutative ring is the set $\mathbb{R}^X$ of all real-valued functions on a non-empty set $X$, with the pointwise operations of addition and multiplication of functions.

A non-trivial example of a skew field is presented below.

EXAMPLE 1.1.9. Denote by $\mathbf{Q}$ the set of all linear combinations $q = a + \mathbf{i}b + \mathbf{j}c + \mathbf{k}d$, where $a, b, c, d$ are real numbers and $\mathbf{i}, \mathbf{j}, \mathbf{k}$ are special symbols satisfying the equalities $\mathbf{i}^2 = \mathbf{j}^2 = \mathbf{k}^2 = -1$ and $\mathbf{ij} = -\mathbf{ji} = \mathbf{k}$, $\mathbf{jk} = -\mathbf{kj} = \mathbf{i}$, $\mathbf{ki} = -\mathbf{ik} = \mathbf{j}$. We introduce the usual coordinatewise addition in $\mathbf{Q}$ by the rule

$$[a + \mathbf{i}b + \mathbf{j}c + \mathbf{k}d] + [a' + \mathbf{i}b' + \mathbf{j}c' + \mathbf{k}d'] = (a + a') + \mathbf{i}(b + b') + \mathbf{j}(c + c') + \mathbf{k}(d + d').$$

With this addition, $\mathbf{Q}$ is a commutative group called the *additive group of quaternions*.

The product of two quaternions $q = a + \mathbf{i}b + \mathbf{j}c + \mathbf{k}d$ and $q' = a' + \mathbf{i}b' + \mathbf{j}c' + \mathbf{k}d'$ is formed by multiplying out the formal linear polynomials and applying the above equalities and the commutativity rules $x\mathbf{i} = \mathbf{i}x$, $x\mathbf{j} = \mathbf{j}x$, $x\mathbf{k} = \mathbf{k}x$ for each $x \in \mathbb{R}$:

$$\begin{aligned}(a + \mathbf{i}b + \mathbf{j}c + \mathbf{k}d)(a' + \mathbf{i}b' + \mathbf{j}c' + \mathbf{k}d') = (&aa' - bb' - cc' - dd') \\ &+ \mathbf{i}(ab' + ba' + cd' - dc') \\ &+ \mathbf{j}(ac' - bd' + ca' + db') \\ &+ \mathbf{k}(ad' + bc' - cb' + da').\end{aligned}$$

We leave to the reader the routine verification of the fact that the multiplication in $\mathbf{Q}$ is associative. It is clear that $\mathbf{Q}$ has the neutral element $\mathbf{1} = 1 + \mathbf{i}0 + \mathbf{j}0 + \mathbf{k}0$ with respect to multiplication, so $(\mathbf{Q}, \cdot)$ is a non-commutative monoid. In addition, the sum and multiplication in $\mathbf{Q}$ satisfy the distributive law, whence it follows that $\mathbf{Q}$ is a non-commutative ring.

Let $\mathbf{Q}^* = \mathbf{Q} \setminus \{\mathbf{0}\}$, where $\mathbf{0} = 0 + \mathbf{i}0 + \mathbf{j}0 + \mathbf{k}0$ is the *zero element* of $\mathbf{Q}$. It turns out that every non-zero element $q \in \mathbf{Q}$ is invertible. Indeed, for $q = a + \mathbf{i}b + \mathbf{j}c + \mathbf{k}d$, put $\bar{q} = a - \mathbf{i}b - \mathbf{j}c - \mathbf{k}d$. An easy calculation shows that $q\bar{q} = \bar{q}q = a^2 + b^2 + c^2 + d^2$. Therefore, if $q \neq \mathbf{0}$, then $r = \bar{q}\alpha$ is an inverse of $q$, where $\alpha = 1/(a^2 + b^2 + c^2 + d^2)$. Since $\mathbf{Q}$ is a monoid, the inverse of $q$ is unique, and $\mathbf{Q}^*$ is a multiplicative group. It follows that $\mathbf{Q}$ is a skew field. Sometimes $\mathbf{Q}^*$ is called the *multiplicative group of quaternions*. Notice that $U = \{\pm 1, \pm\mathbf{i}, \pm\mathbf{j}, \pm\mathbf{k}\}$ is a subgroup of $\mathbf{Q}^*$ called the *group of quaternion units*. □

In the following example we construct, for every integer $r > 1$, the additive group $\Omega_r$ of $r$-adic numbers. Later on, we shall define multiplication in $\Omega_r$, thus making the group $\Omega_r$ into a commutative ring (see Example 3.1.31).

EXAMPLE 1.1.10. Let $r$ be an integer with $r > 1$. Denote by $A$ the set $\{0, 1, \ldots, r - 1\}$ and consider the Cartesian product $P = A^{\mathbb{Z}}$ of infinitely many copies of the set $A$ enumerated by the integers. In other words, $P$ consists of the sequences $\mathbf{x} = (\ldots, x_{n-1}, x_n, x_{n+1}, \ldots)$ infinite in both sides, where $x_n \in A$ for each $n \in \mathbb{Z}$. Let $\Omega_r$ be the subset of $P$ formed by all sequences $\mathbf{x}$ such that $x_n = 0$ for all $n < k$, where $k$ is an integer depending upon $\mathbf{x}$.

Our aim is to define addition in $\Omega_r$ which, in a sense, mimics the decomposition of a natural number $M$ into powers of $r$. More precisely, every $M \in \mathbb{N}$ can be represented in the polynomial form

$$M = x_0 + x_1 r + x_2 r^2 + \cdots + x_m r^m, \tag{1.1}$$

where $x_0, x_1, x_2, \ldots, x_m \in A$, $x_m \neq 0$ and $m \in \omega$. It is easy to verify that such a representation of $M$ is unique. This gives us a bijection between $\mathbb{N}$ and all finite sequences $\mathbf{x} = (x_0, x_1, x_2, \ldots, x_m)$ such that $x_i \in A$ for each $i \le m$ and $x_m \neq 0$.

Suppose that for a positive integer $N$, we have a decomposition

$$N = y_0 + y_1 r + y_2 r^2 + \cdots + y_n r^n \tag{1.2}$$

similar to (1.1). Then the number $M + N$ can also be decomposed into powers of $r$, and the coefficients of the corresponding decomposition admit an explicit expression in terms of $x_i$ and $y_i$. Indeed, let

$$M + N = z_0 + z_1 r + z_2 r^2 + \cdots + z_k r^k, \tag{1.3}$$

where $z_0, z_1, \ldots, z_k \in A$, $z_k \neq 0$ and $k \in \omega$. It follows from (1.1), (1.2) and (1.3) that $x_0 + y_0 = z_0 + t_0 r$, where $t_0$ is either 0 or 1. Clearly, $t_0 = 1$ iff $x_0 + y_0 \ge r$, so that $z_0$ is the least non-negative residue of $x_0 + y_0$ modulo $r$. Again, we apply (1.1)–(1.3) and the above equality $x_0 + y_0 = z_0 + t_0 r$ to deduce that $z_1 r + r^2 P = (x_1 + y_1 + t_0) r + r^2 Q$ for some non-negative integers $P$ and $Q$. Therefore, $x_1 + y_1 + t_0 = z_1 + t_1 r$, where $t_1 = P - Q$. Since $0 \le x_1 + y_1 + t_0 \le 2(r-1) + 1 < 2r$ and $z_1 \ge 0$, it follows that $t_1$ is either 0 or 1, and that $z_1$ is the least non-negative residue of $x_1 + y_1 + t_0$ modulo $r$.

Suppose that we have defined $z_0, z_1, \ldots, z_s \in A$ and $t_0, t_1, \ldots, t_s \in \{0, 1\}$ for some integer $s \ge 1$ such that the equality $z_i + t_i r = x_i + y_i + t_{i-1}$ holds for each $i = 0, 1, \ldots, s$ (where $t_{-1} = 0$). Taking the sum of (1.1) and (1.2) compared with (1.3) and using the above equalities with $i = 0, 1, \ldots, s$, we obtain that

$$x_{s+1} + y_{s+1} + t_s = z_{s+1} + t_{s+1} r, \tag{1.4}$$

where $z_{s+1} \in A$ and $t_{s+1}$ is either 0 or 1. It is clear that the integer $k$ in the equality (1.3) that determines the expansion of $M + N$ satisfies $k \le \max\{m, n\} + 1$. Summing up, the equalities (1.4), together with $x_0 + y_0 = z_0 + t_0 r$, enable us to define inductively the numbers $z_0, z_1, \ldots, z_k \in A$ satisfying (1.3) in terms of $x_i$ and $y_i$.

We are now in position to give a formal definition of the addition in $\Omega_r$ which is based on the equalities (1.4). Denote by $\mathbf{0}$ the zero sequence in $\Omega_r$ each element of which is zero. First, we set $\mathbf{x} + \mathbf{0} = \mathbf{x}$ and $\mathbf{0} + \mathbf{x} = \mathbf{x}$ for each $\mathbf{x} \in \Omega_r$. Let $\mathbf{x} = (x_n)$ and $\mathbf{y} = (y_n)$ be arbitrary elements of $\Omega_r$, both distinct from $\mathbf{0}$. Choose integers $m_0$ and $n_0$ such that $x_{m_0} \neq 0$ and $x_n = 0$ if $n < m_0$ and, similarly, $y_{n_0} \neq 0$ and $y_n = 0$ if $n < n_0$. Set $k_0 = \min\{m_0, n_0\}$. We define a sequence $\mathbf{z} = (z_n) \in \Omega_r$ as follows. First, we choose $z_{k_0} \in A$ and $t_{k_0} \in \{0, 1\}$ satisfying $x_{k_0} + y_{k_0} = z_{k_0} + t_{k_0} r$. Clearly, the numbers $z_{k_0}$ and $t_{k_0}$ are uniquely defined by this equality and the restrictions on them. Suppose that $z_{k_0}, z_{k_0+1}, \ldots, z_k$ and $t_{k_0}, t_{k_0+1}, \ldots, t_k$ have been defined for some integer $k \ge k_0$. Then we choose $z_{k+1} \in A$ and $t_{k+1} \in \{0, 1\}$ satisfying $x_{k+1} + y_{k+1} + t_k = z_{k+1} + t_{k+1} r$. Again, such a choice is always possible and is unique. This inductive procedure defines a sequence $\mathbf{z} = (z_n) \in \Omega_r$, where $z_n = 0$ if $n < k_0$. It remains to put $\mathbf{x} + \mathbf{y} = \mathbf{z}$. This finishes our definition of addition in $\Omega_r$.

It turns out that the set $\Omega_r$ with the addition just defined is a commutative group called the *group of r-adic numbers*. Indeed, it is clear from the above definition and the commutativity of the addition in $\mathbb{Z}$ that $\mathbf{x}+\mathbf{y} = \mathbf{y}+\mathbf{x}$ and that $\mathbf{x}+\mathbf{0} = \mathbf{x}$ for all $\mathbf{x}, \mathbf{y} \in \Omega_r$. Given an arbitrary non-zero element $\mathbf{x} = (x_n)$ in $\Omega_r$, we define an element $\mathbf{y} \in \Omega_r$ as follows. Suppose that $x_m \neq 0$ and $x_n = 0$ for each $n < m$. Set $y_n = 0$ if $n < m$, $y_m = r - x_m$, and $y_n = r - x_n - 1$ if $n > m$. Then the element $\mathbf{y} = (y_n) \in \Omega_r$ satisfies $\mathbf{x} + \mathbf{y} = \mathbf{y} + \mathbf{x} = \mathbf{0}$.

It remains to verify that the addition in $\Omega_r$ is an associative operation. Let $\mathbf{x}, \mathbf{y}, \mathbf{z}$ be arbitrary elements of $\Omega_r$ and suppose that at least one of them is different from $\mathbf{0}$. Take the least integer $m$ such that one of the values $x_m, y_m, z_m$ is distinct from zero. Then $x_n = y_n = z_n = 0$ for each $n < m$ and it is clear that the values of $(\mathbf{x}+\mathbf{y})+\mathbf{z}$ and $\mathbf{x}+(\mathbf{y}+\mathbf{z})$ at the index $n$ are equal to zero, for each $n < m$. To show that the values of the two elements coincide at every index $n \geq m$, we argue as follows. Let $\Sigma_m$ be the set of all elements $\mathbf{t} \in \Omega_r$ such that $t_n = 0$ if $n < m$ and only finitely many values $t_n$ with $n \geq m$ are distinct from zero. Consider the mapping $\varphi$ of $\Sigma_m$ to the set of non-negative integers defined by the rule

$$\varphi(\mathbf{t}) = t_m + t_{m+1}r + t_{m+2}r^2 + \cdots .$$

Since $\mathbf{t}$ is in $\Sigma_m$, the above sum is finite. It follows from our definition of the addition in $\Omega_r$ that $\varphi$ has the property $\varphi(\mathbf{x} + \mathbf{y}) = \varphi(\mathbf{x}) + \varphi(\mathbf{y})$ for all $\mathbf{x}, \mathbf{y} \in \Sigma_m$. Informally speaking, $\varphi$ preserves addition or, equivalently, $\varphi$ is a homomorphism of $\Sigma_m$ onto the additive semigroup of non-negative integers. It is also clear that $\varphi(\mathbf{t}) = 0$ iff $\mathbf{t} = \mathbf{0}$. It follows that $\varphi$ is a bijection preserving operation. Since addition of integers is associative, the same is true for addition in $\Sigma_m$. Finally, we return back to the elements $\mathbf{x}, \mathbf{y}, \mathbf{z} \in \Omega_r$ considered above. For any index $k \geq m$, consider the "truncated" elements $\mathbf{x}', \mathbf{y}', \mathbf{z}'$ coinciding with $\mathbf{x}, \mathbf{y}, \mathbf{z}$, respectively, at every index $n \leq k$, and whose values are equal to zero for each index $n > k$. Then $\mathbf{x}', \mathbf{y}', \mathbf{z}'$ are elements of $\Sigma_m$, whence it follows that $(\mathbf{x}' + \mathbf{y}') + \mathbf{z}' = \mathbf{x}' + (\mathbf{y}' + \mathbf{z}')$. Evidently, the values of the elements $(\mathbf{x}' + \mathbf{y}') + \mathbf{z}'$ and $(\mathbf{x} + \mathbf{y}) + \mathbf{z}$ at the index $k$ coincide, and the same is valid for the elements $\mathbf{x}' + (\mathbf{y}' + \mathbf{z}')$ and $\mathbf{x} + (\mathbf{y} + \mathbf{z})$. Since $k \geq m$ is arbitrary, we conclude that $(\mathbf{x} + \mathbf{y}) + \mathbf{z} = \mathbf{x} + (\mathbf{y} + \mathbf{z})$. Therefore, the addition in $\Omega_r$ is associative and $\Omega_r$ is a commutative group with neutral element $\mathbf{0}$.

Denote by $\mathbb{Z}_r$ the set of all elements $\mathbf{x} \in \Omega_r$ such that $x_n = 0$ for each $n < 0$. Omitting the values of $\mathbf{x}$ at negative indices, we may rewrite every element $\mathbf{x} \in \mathbb{Z}_r$ as $\mathbf{x} = (x_0, x_1, x_2, \ldots)$, thus identifying $\mathbb{Z}_r$ with the corresponding subset of $A^{\omega}$. We leave to the reader a simple verification of the fact that $\mathbb{Z}_r$ is a subgroup of the additive group $\Omega_r$. We will call $\mathbb{Z}_r$ the *group of r-adic integers*. □

## Exercises

1.1.a. Prove that if $H$ is a subgroup of a group $G$, then, for any $a \in G$, $a^{-1}Ha$ is also a subgroup of $G$.

1.1.b. Verify that for every Abelian group $G$ and every $n \in \mathbb{N}$, the set $G[n] = \{x \in G : nx = 0_G\}$ is a subgroup of $G$.

1.1.c. Let $H$ be a subgroup of a group $G$ such that $G = H \cup aH$, for some $a \in G$. Show that $H$ is an invariant subgroup of $G$.

1.1.d. Let $r > 1$ be an integer and $G$ a torsion group such that for every element $x \in G$, the order $o(x)$ of $x$ and the number $r$ are mutually prime. Prove that the mapping $\varphi_r \colon G \to G$ defined by $\varphi_r(x) = x^r$ is a bijection of $G$ onto $G$. Show that $\varphi_r$ is a homomorphism if the group $G$ is commutative.

1.1.e. Let $G$ be an Abelian torsion group. Verify that every finite subset $A$ of $G$ generates a finite subgroup $\langle A \rangle$ of $G$. Show that the conclusion is no longer valid for non-commutative groups.

1.1.f. Give an example of an infinite Abelian group all proper subgroups of which are finite.

1.1.g. Give an example of a semigroup $G$ and a subsemigroup $S$ of $G$ such that, for some $a$ and $b$ in $G$, the sets $aS$ and $bS$ do not coincide and are not disjoint.

1.1.h. Does Theorem 1.1.6 remain valid if we drop the assumption that the group $G$ is Abelian?

1.1.i. Give an example of a semigroup $G$ such that $\varnothing = \bigcap_{n=1}^{\infty} G^n$.

1.1.j. For every quaternion $q = a+\mathbf{i}b+\mathbf{j}c+\mathbf{k}d$ (see Example 1.1.9), put $|q| = \sqrt{a^2+b^2+c^2+d^2}$. Show that the set $\{q \in \mathbf{Q} : |q| = 1\}$ is a subgroup of $\mathbf{Q}^*$.

1.1.k. Let $\Omega_r$ be the group of $r$-adic numbers defined in Example 1.1.10, where $r > 1$.
   (a) Verify that the "multiplication" of a non-zero element $\mathbf{x} \in \Omega_r$ by $r$ moves every value $x_k$ of $\mathbf{x}$ to the next position on the right. In other words, if $m$ is the least integer such that $x_m \neq 0$, then the values $y_n$ of the element $\mathbf{y} = r\mathbf{x}$ satisfy $y_n = 0$ for each $n \leq m$ and $y_{n+1} = x_n$ for $n \geq m$. Deduce that the mapping $\varphi_r$ of $\Omega_r$ to $\Omega_r$ defined by $\varphi_r(\mathbf{x}) = r\mathbf{x}$ is an isomorphism of $\Omega_r$ onto itself.
   (b) Use a) to solve the equation $2x = \mathbf{a}$ in the group $\Omega_6$, where $\mathbf{a} = (\ldots, 0, \ldots, 0, 1, 1, 1, \ldots)$ is an element of the subgroup $\mathbb{Z}_6$ of $\Omega_6$ with $a_0 = 1$.

## Problems

1.1.A. Show that the solutions of the equation $X^2 = E_2$ in the multiplicative group $GL(2,\mathbb{R})$ of all invertible 2 by 2 matrices with real entries do not form a subgroup of the group $GL(2,\mathbb{R})$, where $E_2$ is the identity matrix (i.e., the neutral element of $GL(2,\mathbb{R})$). Deduce that the conclusion in Exercise 1.1.b is no longer valid in the non-Abelian case.

1.1.B. Let $\mathbf{Q}^*$ be the multiplicative group of non-zero quaternions.
   (a) Verify that the solutions of the equation $q^2 = \mathbf{1}$ in $\mathbf{Q}$ form a two-element subgroup of $\mathbf{Q}^*$.
   (b) Show that the solutions of the equation $q^2 = -\mathbf{1}$ in $\mathbf{Q}$ can be naturally identified with the points of the unit sphere in $\mathbb{R}^3$.
   (c) How many solutions does the equation $q^3 = \mathbf{1}$ have in $\mathbf{Q}$? Does the set of solutions form a submonoid of $\mathbf{Q}$?

1.1.C. If $G$ is a group and $a, b \in G$, then the element $[a,b] = aba^{-1}b^{-1}$ of $G$ is called the *commutator* of $a$ and $b$. Is the set $\{[x,y] : x, y \in G\}$ a subgroup (submonoid) of $G$?

1.1.D. How many subgroups does the symmetric group $S_4$ contain (see item 8) of Example 1.1.2)? How many of them are invariant in $S_4$?

1.1.E. Let $H$ be an invariant subgroup of a group $G$ and $K$ an invariant subgroup of $H$. Is $K$ then invariant in $G$?

1.1.F. Let $r$ and $k$ be mutually prime natural numbers, where $r > 1$. Prove that for every element $\mathbf{a}$ of the group $\mathbb{Z}_r$ of $r$-adic integers defined in Example 1.1.10, the equation $k\mathbf{x} = \mathbf{a}$ has a solution in $\mathbb{Z}_r$. Deduce that the group $\Omega_r$ is divisible, for each $r$.
   *Hint.* To prove the first assertion, take any $\mathbf{a} = (a_n)_{n\in\omega}$ in $\mathbb{Z}_r$, and define by induction on $n \in \omega$ an element $\mathbf{x} = (x_n)_{n\in\omega}$ in $\mathbb{Z}_r$ and a function $t \colon \omega \to \{0,1\}$ such that $kx_0 \equiv a_0 \pmod{r}$ and $kx_{n+1} + t(n) \equiv a_{n+1} \pmod{r}$ for each $n \geq 0$. To guarantee the existence of solutions $x_0, x_1, \ldots$ of the congruences, apply the conclusion of Example 1.1.d to the cyclic group $\mathbb{Z}(r)$. The second assertion of the problem follows from the first one if one applies Exercise 1.1.k.

1.1.G. The construction of the group $\Omega_r$ of $r$-adic numbers described in Example 1.1.10 admits a natural generalization as follows. First, we fix an element $\mathbf{a} = (\ldots, a_{-1}, a_0, a_1, \ldots)$ of $\mathbb{N}^{\mathbb{Z}}$ such that $a_n \geq 2$, for each $n \in \mathbb{Z}$. For every $n \in \mathbb{Z}$, put $A_n = \{0, 1, \ldots, a_n - 1\}$, and consider the product $\Pi = \prod_{n\in\mathbb{Z}} A_n$. Denote by $\Omega_{\mathbf{a}}$ the set of all elements $\mathbf{x} = (\ldots, x_{-1}, x_0, x_1, \ldots)$ of $\Pi$ such that $x_n = 0$ for each $n < n_0$, where $n_0 \in \mathbb{Z}$ depends on $\mathbf{x}$. Given two elements $\mathbf{x} = (x_n)$ and $\mathbf{y} = (y_n)$ of $\Omega_{\mathbf{a}}$, we define by induction an element $\mathbf{x} + \mathbf{y} = \mathbf{z} = (z_n) \in \Omega_{\mathbf{a}}$ as follows. If $\mathbf{x}$ (or $\mathbf{y}$) contains only zero entries, put $\mathbf{z} = \mathbf{y}$ ($\mathbf{z} = \mathbf{x}$, respectively). Otherwise put $z_n = 0$ and $t_n = 0$ for each $n < m_0$, where $m_0$ is the maximal integer with the property that $x_n = 0 = y_n$ for all $n < m_0$. Choose $z_{m_0} \in A_{m_0}$ and an integer $t_{m_0} \geq 0$ such that $x_{m_0} + y_{m_0} = z_{m_0} + t_{m_0}a_{m_0}$. If we have defined $z_k$ and $t_k$ for all $k < n$, where $m_0 < n$, then there exist $z_n \in A_n$ and a non-negative integer $t_n$ such that $x_n + y_n + t_{n-1} = z_n + t_n a_n$ (note that the numbers $z_n$ and $t_n$ are uniquely determined by these conditions). Each element of $\Omega_{\mathbf{a}}$ is called an **a**-*adic number*.

a) Verify that $\Omega_{\mathbf{a}}$ is an Abelian group (called the *group of* **a**-*adic numbers*).
b) Show that, for certain **a**, the group $\Omega_{\mathbf{a}}$ can have elements of finite order distinct from the neutral element of the group.
c) Characterize the sequences **a** such that the corresponding group $\Omega_{\mathbf{a}}$ is torsion-free.
d) Verify that the set of $\mathbf{x} \in \Omega_{\mathbf{a}}$ with $x_n = 0$, for each $n < 0$, is a subgroup of $\Omega_{\mathbf{a}}$; this group is called the *group of* **a**-*adic integers* and is denoted by $\mathbb{Z}_{\mathbf{a}}$.
e) Prove that the group $\mathbb{Z}_{\mathbf{a}}$, for $\mathbf{a} = (2, 3, 4, \ldots)$, is divisible and torsion-free (notice that in the case of $\mathbb{Z}_{\mathbf{a}}$, we do not have to specify the entries $a_n$ of **a** with $n < 0$).

## 1.2. Groups and semigroups with topologies

A *right topological semigroup* consists of a semigroup $S$ and a topology $\mathcal{T}$ on $S$ such that for all $a \in S$, the right action $\varrho_a$ of $a$ on $S$ is a continuous mapping of the space $S$ to itself.

A *left topological semigroup* consists of a semigroup $S$ and a topology $\mathcal{T}$ on the set $S$ such that for all $a \in S$, the left action $\lambda_a$ of $a$ on $S$ is a continuous mapping of the space $S$ to itself.

A *semitopological semigroup* is a right topological semigroup which is also a left topological semigroup.

A *topological semigroup* consists of a semigroup $S$ and a topology $\mathcal{T}$ on $S$ such that the multiplication in $S$, as a mapping of $S \times S$ to $S$, is continuous when $S \times S$ is endowed with the product topology.

A *right topological monoid* is a right topological semigroup with identity. Similarly, a *topological monoid* is a topological semigroup with identity, and a *semitopological monoid* is a semitopological semigroup with identity.

A *left (right) topological group* is a left (right) topological semigroup whose underlying semigroup is a group, and a *semitopological group* is a left topological group which is also a right topological group.

A *paratopological group* $G$ is a group $G$ with a topology on the set $G$ that makes the multiplication mapping $G \times G \to G$ continuous, when $G \times G$ is given the product topology.

For a group $G$, the inverse mapping $In\colon G \to G$ is defined by the rule $In(x) = x^{-1}$, for each $x \in G$. A semitopological group with continuous inverse is called a *quasitopological group*.

A *topological group* $G$ is a paratopological group $G$ such that the inverse mapping $In\colon G \to G$ is continuous. An easy verification shows that $G$ is a topological group if and only if the mapping $(x, y) \mapsto xy^{-1}$ of $G \times G$ to $G$ is continuous.

It is evident that every topological group is a topological semigroup, every topological semigroup is a semitopological semigroup, and every semitopological semigroup is both a left and right topological semigroup.

EXAMPLE 1.2.1. Let $\mathcal{T}$ be the topology on $\mathbb{R}$ with the base $\mathcal{B}$ consisting of the sets $[a, b) = \{x \in \mathbb{R} : a \leq x < b\}$, where $a, b \in \mathbb{R}$ and $a < b$. With this topology, and the natural addition in the role of multiplication, $\mathbb{R}$ is a paratopological group and, therefore, a topological semigroup. However, $(\mathbb{R}, \mathcal{T})$ is not a topological group since the inverse operation $x \mapsto -x$ is discontinuous. This paratopological group is called the *Sorgenfrey line*. $\square$

EXAMPLE 1.2.2. Let $S = \mathbb{R} \cup \{\alpha\}$ be the one-point compactification of the usual space $\mathbb{R}$ of real numbers. Define multiplication on $S$ by the rule $xy = x + y$ if $x$ and $y$ are in $\mathbb{R}$, and $xy = \alpha$, otherwise. With this operation, $S$ is a semitopological semigroup. However, $S$ is not a topological semigroup, since the multiplication mapping of $S \times S \to S$ is not (jointly) continuous at the point $(\alpha, \alpha)$. $\square$

Note that every group (semigroup) can be turned into a topological group (semigroup) by providing it with the discrete topology. However, the problem of existence of non-discrete Hausdorff topologies on infinite groups which would make them into topological groups is a delicate one. We will discuss it for Abelian groups in Section 1.4.

A useful series of right topological semigroups comes when considering semigroups of the form $S(X, X)$ with weak topologies.

For a topological space $X$, let $S_p(X, X)$ be the semigroup $S(X, X)$ of all mappings of the set $X$ to $X$, taken with the topology of pointwise convergence. This topology has the standard base $\mathcal{B}$ which consists of the sets

$$O(x_1, \ldots, x_n, U_1, \ldots, U_n) = \{f \in S(X, X) : f(x_i) \in U_i \text{ for } i = 1, \ldots, n\},$$

where $x_1, \ldots, x_n$ are pairwise distinct points of $X$ and $U_1, \ldots, U_n$ are non-empty open sets in $X$, for some $n \in \mathbb{N}$.

THEOREM 1.2.3. *For any topological space $X$, $S_p(X, X)$ is a right topological semigroup. Further, for any $f \in S(X, X)$, the left action $\lambda_f$ of $f$ on $S_p(X, X)$ is continuous if and only if the mapping $f\colon X \to X$ is continuous.*

PROOF. Take any $f, g \in S(X, X)$ and a finite subset $K$ of $X$. Put $L = f(K)$. For each $x \in K$ take an open neighbourhood $O_x$ of $gf(x)$ in $X$. Let $V$ be the set of all $h \in S(X, X)$ such that $h(x) \in O_x$ for each $x \in K$, and let $U$ be the set of all $g' \in S(X, X)$ such that $g'f(x) \in O_x$ for each $f(x) \in L = f(K)$. Then, clearly, $V$ is a standard open neighbourhood of $gf$ in $S_p(X, X)$, and $U$ is an open neighbourhood of $g$ in $S_p(X, X)$ such that $\varrho_f(U) \subset V$ (that is, $Uf \subset V$). Therefore, $S_p(X, X)$ is a right topological semigroup.

To deduce the last statement of the theorem, we consider the left action $\lambda_f$ for some $f \in S(X, X)$. Take an arbitrary point $a \in X$ and a non-empty open set $V$ in $X$. It is easy to see that

$$\lambda_f^{-1}[O(a, V)] = \{g \in S(X, X) : f(g(a)) \in V\} = \{g \in S(X, X) : g(a) \in f^{-1}(V)\}.$$

Therefore, if the mapping $f: X \to X$ is continuous, the set $\lambda_f^{-1}[O(a, V)]$ is open in $S_p(X, X)$. Since the sets of the form $O(a, V)$ constitute a subbase for the topology of $S_p(X, X)$, we conclude that the left action $\lambda_f$ is continuous.

Conversely, suppose that the mapping $f$ is discontinuous. Then we can find a point $a \in X$ and an open neighbourhood $V_0$ of $b = f(a)$ in $X$ such that $f(U) \setminus V_0 \neq \varnothing$ for each neighbourhood $U$ of $a$. Let $1_X$ be the identity mapping of $X$ onto itself. Evidently, $f = \lambda_f(1_X)$ is in $O(a, V_0)$. Take an arbitrary basic neighbourhood $O = O(x_1, \ldots, x_n, U_1, \ldots, U_n)$ of $1_X$ in $S_p(X, X)$, where the points $x_1, \ldots, x_n$ are pairwise distinct. Then $x_i \in U_i$ for each $i \leq n$. We claim that the image $\lambda_f(O)$ is not a subset of $O(a, V_0)$, so that $\lambda_f$ is discontinuous at $1_X$.

Indeed, if $a \in \{x_1, \ldots, x_n\}$, we can assume that $a = x_1$. Then $W = U_1 \setminus \{x_2, \ldots, x_n\}$ is an open neighbourhood of $a$ in $X$ and, hence, $f(W) \setminus V_0 \neq \varnothing$. Choose a point $y_0 \in W$ such that $f(y_0) \notin V_0$ and take an arbitrary $g \in S(X, X)$ such that $g(x_1) = y_0$ and $g(x_i) = x_i$ for each $i$ with $1 < i \leq n$. Then $g \in O$, and $\lambda_f(g) = f \circ g \in \lambda_f(O) \setminus O(a, V_0)$ since $f(g(a)) = f(y_0) \notin V_0$. Similarly, if $a \notin \{x_1, \ldots, x_n\}$, we can choose a point $y_0 \in W = X \setminus \{x_1, \ldots, x_n\}$ such that $f(y_0) \notin V_0$ and take a mapping $g: X \to X$ with $g(a) = y_0$ and $g(x_i) = x_i$ for each $i \leq n$. Then again $g \in O$ and $f \circ g = \lambda_f(g) \in \lambda_f(O) \setminus O(a, V_0)$.

Thus, the left action $\lambda_f$ of $f$ on $S_p(X, X)$ is discontinuous for every discontinuous mapping $f$. □

COROLLARY 1.2.4. *Let $X$ be a topological space. The following statements are equivalent:*

1) *$S_p(X, X)$ is a topological semigroup.*
2) *$S_p(X, X)$ is a semitopological semigroup.*
3) *The space $X$ is discrete.*

PROOF. If $X$ is discrete, then every mapping $f$ of $X$ to $X$ is continuous, so the left action $\lambda_f$ is continuous by Theorem 1.2.3, and $S_p(X, X)$ is a semitopological semigroup by the same theorem. Hence, 3) implies 2). Conversely, if $S_p(X, X)$ is a semitopological semigroup then all left actions $\lambda_f$ are continuous, which implies, by Theorem 1.2.3, that all mappings $f: X \to X$ are continuous. Since the one-point sets in $X$ are closed, the space $X$ must be discrete. Therefore, 2) and 3) are equivalent.

Clearly, 1) implies 2). We leave it to the reader to verify that 3) implies 1). □

Some more examples are in order.

EXAMPLE 1.2.5. We present here several types of topologies on groups (semigroups).

a) Let $G$ be an arbitrary group (semigroup), and let $\mathcal{T}$ be the family of all subsets of $G$, i.e., the discrete topology of $G$. With this topology, $G$ is a topological group (semigroup). We shall often refer to such a $G$ as a *discrete group.*
b) Let $G$ be an arbitrary infinite group and let $\mathcal{T}$ consist of $G$ and the subsets of $G$ having finite complements. Then $G$ is not a paratopological group. However, $G$ is a semitopological group with continuous inverse, that is, a quasitopological group. Note that $G$ with this topology is a $T_1$-space but not Hausdorff.
c) The additive group $\mathbb{R}$ of all real numbers with its usual topology is a locally compact, non-compact Abelian group.

d) The multiplicative circle group $\mathbb{T}$ with the topology inherited from the complex number field $\mathbb{C}$ is a compact Abelian group.

e) Let $G$ be the group $GL(n, \mathbb{R})$ of all invertible $n$ by $n$ matrices with real entries (see also item 7) of Example 1.1.2). We endow $G$ with the topology of a subspace of the $n^2$-dimensional Euclidean space. Then $G$ is a topological group. Indeed, the formula for multiplying two matrices and the formula for inverting a matrix employ only continuous functions of the entries of the matrices. This group is called the *general linear group* of degree $n$ over $\mathbb{R}$. Similarly, the general linear group $GL(n, \mathbb{C})$ over the field $\mathbb{C}$ of complex numbers, with the matrix multiplication and the topology induced from $\mathbb{C}^{n^2}$ is again a topological group.

f) Let $\mathbf{Q}$ be the additive group of quaternions (see Example 1.1.9). Consider the natural mapping $f\colon \mathbf{Q} \to \mathbb{R}^4$ defined by the rule $f(q) = (x, y, z, t)$, for each $q = x+\mathbf{i}y+\mathbf{j}z+\mathbf{k}t \in \mathbf{Q}$. Clearly, $f$ is a bijection of $\mathbf{Q}$ onto $\mathbb{R}^4$. We topologize $\mathbf{Q}$ by declaring the mapping $f$ to be a homeomorphism. In other words, a subset $U$ of $\mathbf{Q}$ is open if and only if the image $f(U)$ is open in $\mathbb{R}^4$. This agreement makes $\mathbf{Q}$ into a locally compact, second-countable Hausdorff topological group. Clearly, the restriction of $f$ to $\mathbf{Q}^* = \mathbf{Q} \setminus \{\mathbf{0}\}$ fails to be a homomorphism of the multiplicative group $\mathbf{Q}^*$ to the additive group $\mathbb{R}^4$. However, $\mathbf{Q}^*$ with this topology (called Euclidean) turns out to be a topological group. This fact follows easily from the definition of the multiplication and the procedure of inversion in $\mathbf{Q}^*$ given in Example 1.1.9. Therefore, the multiplicative group of quaternions $\mathbf{Q}^*$ with the Euclidean topology is a locally compact Hausdorff topological group. □

EXAMPLE 1.2.6. Let $G$ and $H$ be right topological semigroups, and $G \times H$ their product. For each $x \in G$, the set $\{x\} \times H$ is called a *vertical fiber* of $G \times H$. For each $y \in H$, the set $G \times \{y\}$ is called a *horizontal fiber* of $G \times H$. Every vertical fiber can be considered as a copy of the right topological semigroup $H$. Similarly, every horizontal fiber can be interpreted as a copy of the right topological semigroup $G$. Therefore, we can treat every fiber as a topological space.

Now we will define two new topologies on $G \times H$, each of which contains the product topology on $G \times H$.

The first one is the *cross* topology. A subset $W$ of $G \times H$ belongs to it if and only if the intersection of $W$ with every fiber of $G \times H$ (horizontal and vertical) is open in the fiber. It is easy to verify that the semigroup $G \times H$ with the usual (coordinatewise) multiplication and with the cross topology is again a right topological semigroup.

The second new topology on $G \times H$ is defined as follows. Take the family $SC(G \times H)$ of all real-valued functions $f$ on $G \times H$ such that the restriction of $f$ to each fiber is a continuous function on this fiber. It is well known that the functions in $SC(G \times H)$ need not be continuous on the space $G \times H$. Now let $\sigma$ be the smallest topology on $G \times H$ which makes all functions $f \in SC(G \times H)$ continuous.

It is easy to see that if $G$ and $H$ are Tychonoff spaces, then the topology $\sigma$ on $G \times H$ contains the usual topology of $G \times H$ and also is Tychonoff. However, the cross topology on $G \times H$ need not be regular, even if $G$ and $H$ are second-countable spaces. □

The usual product topology on products of finitely many factors can be used to obtain many interesting examples of topological groups and semigroups. Especially, if we combine the product operation with the operation of taking a topological subgroup (subsemigroup) of a topological group (semigroup). Obviously, every subgroup of a topological group, endowed with the subspace topology, is again a topological group.

Let us describe briefly the much more general operation of *topological product* of an arbitrary family of topological groups (semigroups, and so on).

In the next theorem we restrict ourselves to the case of topological groups. A similar result holds for topological semigroups, semitopological semigroups, right topological groups, *etc.*

THEOREM 1.2.7. *Suppose that $\{G_\alpha : \alpha \in A\}$ is a family of topological groups, $e_\alpha$ is the neutral element of $G_\alpha$ for each $\alpha \in A$, $G = \Pi_{\alpha\in A} G_\alpha$ is the Cartesian product of the sets $G_\alpha$, with the Tychonoff product topology and the product operation defined coordinatewise. Then $G$ is also a topological group, with neutral element $e = (e_\alpha)_{\alpha\in A}$; this topological group $G$ is called the topological or direct product of the family $\{G_\alpha : \alpha \in A\}$.*

PROOF. Let $x = (x_\alpha)_{\alpha\in A}$ and $y = (y_\alpha)_{\alpha\in A}$ be arbitrary elements of $G$ and $O$ a be neighbourhood of $z = xy^{-1}$ in $G$. If $z = (z_\alpha)_{\alpha\in A}$ then, clearly, $z_\alpha = x_\alpha y_\alpha^{-1}$ for each $\alpha \in A$. Since $G$ carries the Tychonoff product topology, we can find pairwise distinct elements $\alpha_1, \ldots, \alpha_n$ of the index set $A$ and an open neighbourhood $W_{\alpha_k}$ of the point $z_{\alpha_k}$ in the group $G_{\alpha_k}$, for each $k = 1, \ldots, n$ such that $W = \prod_{\alpha\in A} W_\alpha \subset O$, where $W_\alpha = W_{\alpha_k}$ if $\alpha = \alpha_k$ for some $k \leq n$, and $W_\alpha = G_\alpha$ otherwise.

Since each $G_{\alpha_k}$ is a topological group, there exist open neighbourhoods $U_{\alpha_k}$ and $V_{\alpha_k}$ of $x_{\alpha_k}$ and $y_{\alpha_k}$, respectively, in $G_{\alpha_k}$ such that $U_{\alpha_k} V_{\alpha_k}^{-1} \subset W_{\alpha_k}$. Also, put $U_\alpha = V_\alpha = G_\alpha$ for each $\alpha \in A \setminus \{\alpha_1, \ldots, \alpha_n\}$. Then the sets $U = \prod_{\alpha\in A} U_\alpha$ and $V = \prod_{\alpha\in A} V_\alpha$ are open neighbourhoods of $x$ and $y$, respectively, in the product group $G$. It follows immediately from our definition of the sets $U$ and $V$ that $UV^{-1} \subset W \subset O$. Therefore, $G$ is a topological group. □

A shorter alternative proof of Theorem 1.2.7 goes like this. For every $\alpha \in A$, let $p_\alpha : G_\alpha \times G_\alpha \to G_\alpha$ be the product operation in the group $G_\alpha$, and $p: G \times G \to G$ the product operation in $G$. Clearly, the mapping $p$ can be represented as the Cartesian product of the mappings $p_\alpha$. It follows that $p$ is continuous. Similarly, the inversion in $G$ is the Cartesian product of inverse operations in the groups $G_\alpha$. Therefore, the inversion in $G$ is also continuous, and $G$ is a topological group with neutral element $e$.

What is particularly interesting and instructive in the examples below is their common feature — we start with some discrete topological group, and using products and subgroups, we obtain some topological groups of a highly non-trivial nature.

EXAMPLE 1.2.8. Let $\mathbb{T}^\tau$ be the topological product of $\tau$ copies of the circle group $\mathbb{T}$. Then $\mathbb{T}^\tau$ is a compact Hausdorff topological group. This group resembles the Tychonoff cube $I^\tau$. However, in the realm of topological groups, $\mathbb{T}^\tau$ does not play such a fundamental role as the space $I^\tau$ does in the theory of Tychonoff spaces. Indeed, it is no longer true that every compact group is topologically isomorphic to a topological subgroup of $\mathbb{T}^\tau$ (the non-commutativity of many classic compact groups is responsible for this phenomenon). However, every compact commutative group is topologically isomorphic to a subgroup of $\mathbb{T}^\tau$, for some cardinal $\tau$ (see Exercise 9.4.b). □

EXAMPLE 1.2.9. Denote by $D$ the discrete two-element group $\{0, 1\}$.

a) Let $D^\tau$ be the topological product of $\tau$ copies of the group $D$. Then $D^\tau$ is a compact zero-dimensional topological group. Note that $a^2 = e$, for every element $a$ of $D^\tau$. Thus, each element of $D^\tau$ is its own inverse. Such groups are called *Boolean*.
b) Take only those elements of $D^\tau$ which are non-zero at countably many coordinates at most. They form a topological subgroup of $D^\tau$. This subgroup is called the *Σ-product* of $\tau$ copies of the group $D$ and is denoted by $\Sigma D^\tau$. Clearly, the topological group $\Sigma D^\tau$ is dense in $D^\tau$. Since $\Sigma D^\tau$ is a proper subgroup of $D^\tau$ if $\tau > \omega$, it follows that $\Sigma D^\tau$ is not compact in this case.
c) Similarly, we could define the *Σ-product* of $\tau$ copies of the group $\mathbb{T}$ denoted by $\Sigma\mathbb{T}^\tau$. This topological group is not compact if $\tau > \omega$; however, it is countably compact (see Corollary 1.6.34).
d) Another interesting topological subgroup of $D^\tau$ is the $\sigma$-product $\sigma D^\tau$ of $\tau$ copies of the discrete group $D$; it consists of points with at most finitely many non-zero coordinates. A curious property of this subgroup is that it is $\sigma$-compact, that is, the union of countably many compact subspaces (see Proposition 1.6.41 below). □

We will continue the study of Σ-products and $\sigma$-products in Section 1.6.

EXAMPLE 1.2.10. For a topological space $X$, let $C_p(X, X)$ be the semigroup $C(X, X)$ of all continuous mappings of $X$ to $X$, with the topology of pointwise convergence. In other words, $C_p(X, X)$ is a subsemigroup of the right topological semigroup $S_p(X, X)$ considered in Theorem 1.2.3. Then $C_p(X, X)$ is a semitopological semigroup. Again, this follows from Theorem 1.2.3. □

EXAMPLE 1.2.11. Let $G$ be a topological group (semigroup) and $X$ a topological space. Consider the set of all continuous mappings of $X$ to $G$, with the product operation defined coordinatewise and with the topology of pointwise convergence. This object, denoted by $C_p(X, G)$, is a topological group (semigroup). Note that $C_p(X, G)$ is a topological subgroup (subsemigroup) of the topological product $G^X$ of $|X|$ copies of the group (semigroup) $G$. □

EXAMPLE 1.2.12. Let $X$ be a topological space and $Homeo_p(X)$ the group of all homeomorphisms of $X$ onto itself, with the pointwise convergence topology.

a) $Homeo_p(X)$ is a semitopological group, for every space $X$ (see Example 1.2.10).
b) $Homeo_p(X)$ need not be a topological group (though, obviously, $Homeo_p(X)$ is a group).
c) If the space $X$ is discrete, then $Homeo_p(X)$ is a topological group. □

Item c) of Example 1.2.12 is generalized as follows:

EXAMPLE 1.2.13. Let $X$ be a metric space, with metric $\varrho$. Denote by $Is_p(X)$ the set of all isometries of $X$ onto itself (that is, the set of all one-to-one mappings of $X$ onto itself preserving distance), endowed with the topology of pointwise convergence. Then $Is_p(X)$ is a topological group. The groups of the form $Is_p(X)$ will be studied in Section 3.5. □

## Exercises

1.2.a. Prove the statements in Examples 1.2.1 and 1.2.2.
1.2.b. Prove the part of Corollary 1.2.4 left unproved.
1.2.c. Prove all statements in Examples 1.2.12 and 1.2.13.
1.2.d. Verify that every infinite subgroup of the topological group $\mathbb{T}$ is dense in $\mathbb{T}$.

1.2.e. Verify that Theorem 1.2.7 remains valid for left (right) topological semigroups, topological semigroups, paratopological and quasitopological groups.

1.2.f. Let $S = \{q \in \mathbf{Q}^* : |q| = 1\}$ (see Exercise 1.1.j). Show that $S$ is a compact subgroup of the group $\mathbf{Q}^*$ when the latter carries the Euclidean topology (see item f) of Example 1.2.5).

1.2.g. Suppose that $H$ is a dense subgroup of a paratopological group $G$ with identity $e$. Show that if $H$ is commutative, then so is $G$. Verify that if $n \in \mathbb{N}$ and every element $x \in H$ satisfies $x^n = e$, then all elements of $G$ satisfy the same equation.

1.2.h. Prove that for every element $g$ of a Hausdorff paratopological group $G$, the set $G_g = \{x \in G : xg = gx\}$ is a closed subgroup of $G$. Show that $G_g$ need not be invariant in $G$, even if $G$ is a topological group.

1.2.i. Let $D$ be a dense subset of a topological group $G$. Verify that the equalities $UD = G = DU$ hold for every neighbourhood $U$ of the neutral element in $G$. Is the conclusion valid for paratopological (or quasitopological) groups?

1.2.j. Let $G$ be an arbitrary topological group. Is the set of commutators $\{[x, y] : x, y \in G\}$ (see Problem 1.1.C) closed in $G$?

1.2.k. Let $I$ be the closed unit interval with the usual topology. Prove that $Homeo_p(I)$ (see Example 1.2.12) is a topological group.

## Problems

1.2.A. Let $S$ be a subgroup of $\mathbf{Q}^*$ considered in Exercise 1.2.f. Prove that there exist elements $q_1, \dots, q_n \in S$ such that the subgroup $\langle q_1, \dots, q_n \rangle$ generated by these elements is dense in $S$. What is the minimum value of $n$?

1.2.B. Prove that the topological group $\mathbf{Q}^*$ defined in item f) of Example 1.2.5 is topologically isomorphic to a subgroup of the topological group $GL(4, \mathbb{R})$.

1.2.C. Let $G$ be an abstract group and $n$ a positive integer. Prove that if $\mathcal{T}$ is a Hausdorff paratopological group topology on $G$, then the set $G[n] = \{x \in G : x^n = e_G\}$ is closed in $(G, \mathcal{T})$. Does the conclusion remain valid for semitopological (quasitopological) group topologies on $G$?

1.2.D. For a given integer $n \geq 1$, we consider the following subsets of the general linear group $GL(n, \mathbb{R})$: The set $SL(n, \mathbb{R})$ of all matrices with determinant equal to 1, the set $TS(n, \mathbb{R})$ of all triangular superior matrices with the elements on the main diagonal equal to 1, and the set $O(n, \mathbb{R})$ of all orthogonal matrices. Prove that each of the three sets is a closed subgroup of $GL(n, \mathbb{R})$. Verify that $SL(n, \mathbb{R})$ is invariant in $GL(n, \mathbb{R})$, while $TS(n, \mathbb{R})$ and $O(n, \mathbb{R})$ are not if $n \geq 2$.

1.2.E. Let $G$ and $H$ be topological groups and $\varphi$ be a homomorphism of $H$ to the group $Aut(G)$ of automorphisms of $G$ (see Example 1.1.4). We define the multiplication $\circ$ in $G \times H$ by the rule

$$(g_1, h_1) \circ (g_2, h_2) = (g_1\varphi(h_1)(g_2), h_1h_2)$$

for all $g_1, g_2 \in G$ and $h_1, h_2 \in H$. Prove that $G \times H$ with the product topology and multiplication just defined is a topological group. Verify that the group $G$ is topologically isomorphic to the closed invariant subgroup $G \times \{e_H\}$ of $G \times H$.

## 1.3. Neighbourhoods of the identity in topological groups and semigroups

Given a semigroup $S$ or a group $G$, what are the ways to topologize $S$, respectively, $G$? Of course, this informal question can be given different formal interpretations; for example, we could look for topologies which would make $S$ into a topological semigroup, or a semitopological semigroup, or a right topological semigroup. It is also clear that we can always take the discrete topology on $S$ or $G$, which makes $S$ into a topological semigroup and $G$ into a topological group. To exclude this trivial solution, we should look for non-discrete topologizations of $S$ and $G$, also called *non-trivial*.

In the matters of topologizations, the following simple fact is instrumental.

PROPOSITION 1.3.1. *Let $G$ be a right topological group and $g$ be any element of $G$. Then:*

a) *the right translation $\varrho_g$ of $G$ by $g$ is a homeomorphism of the space $G$ onto itself;*
b) *for any base $\mathcal{B}_e$ of the space $G$ at $e$, the family $\mathcal{B}_g = \{Ug : U \in \mathcal{B}_e\}$ is a base of $G$ at $g$.*

PROOF. Clearly, b) follows from a). To prove a), we just observe that, in a right topological group, every right translation $\varrho_g$ is a continuous bijection. Since $\varrho_g \circ \varrho_{g^{-1}}$ is the identity mapping, it follows that the inverse of $\varrho_g$ is also continuous, that is, $\varrho_g$ is a homeomorphism of $G$ onto itself. □

COROLLARY 1.3.2. *In every semitopological group $G$ all, right and left, translations are homeomorphisms.*

COROLLARY 1.3.3. *Suppose that a subgroup $H$ of a right (or left) topological group $G$ contains a non-empty open subset of $G$. Then $H$ is open in $G$.*

PROOF. Let $U$ be an open non-empty subset of $G$ with $U \subset H$. For every $a \in H$, the set $\varrho_a(U) = Ua$ is open in $G$ by a) of Proposition 1.3.1. Therefore, the set $H = \bigcup_{a\in H} Ua$ is open in $G$. □

The following simple fact is useful in many occasions.

PROPOSITION 1.3.4. *Let $f\colon G \to H$ be a homomorphism of left (right) topological groups. If $f$ is continuous at the neutral element $e_G$ of $G$, then $f$ is continuous.*

PROOF. Let $x \in G$ be arbitrary, and suppose that $O$ is an open neighbourhood of $y = f(x)$ in $H$. Since the left translation $\lambda_y$ is a homeomorphism of $H$, there exists an open neighbourhood $V$ of the neutral element $e_H$ in $H$ such that $yV \subset O$. It follows from the continuity of $f$ at $e_G$ that $f(U) \subset V$, for some open neighbourhood $U$ of $e_G$ in $G$. Again, since $\lambda_x$ is a homeomorphism of $G$ onto itself, the set $xU$ is an open neighbourhood of $x$ in $G$, and we have that $f(xU) = yf(U) \subset yV \subset O$. Hence, $f$ is continuous at the point $x \in G$. □

Let $\gamma$ be a family of subsets of a set $X$. We say that $\gamma$ is a *covering* of $X$ or, equivalently, that $\gamma$ *covers* $X$ if $X = \bigcup \gamma$. In addition, if $X$ is a topological space and all elements of $\gamma$ are open (closed) in $X$, then $\gamma$ is called an *open covering* (respectively, a *closed covering*).

A simple theorem below complements Corollary 1.3.3.

THEOREM 1.3.5. *Every open subgroup $H$ of a right (or left) topological group $G$ is closed in $G$.*

PROOF. The family $\gamma = \{Ha : a \in G\}$ of all right cosets of $H$ in $G$ is a disjoint open covering of $G$ (it is here that it is important that $G$ be a group and $H$ be its subgroup). Therefore, every element of $\gamma$ is closed in $G$. In particular, $H = He$ is closed in $G$. By the symmetry argument, the same holds true in the case when $G$ is a left topological group. □

Recall that a topological space $X$ is said to be *homogeneous* if for each $x \in X$ and each $y \in X$, there exists a homeomorphism $f$ of the space $X$ onto itself such that $f(x) = y$. From Proposition 1.3.1 and Corollary 1.3.2 we easily obtain the next result:

COROLLARY 1.3.6. *Every right topological group $G$, in particular, every semitopological group is a homogeneous space.*

PROOF. Take any elements $x$ and $y$ in $G$, and put $z = x^{-1}y$. Then $\varrho_z(x) = xz = xx^{-1}y = y$. Since, by Proposition 1.3.1, $\varrho_z$ is a homeomorphism, the space $G$ is homogeneous. □

On the contrary, a topological semigroup, even if it has identity and is Abelian, need not be homogeneous.

EXAMPLE 1.3.7. Take the closed unit interval $I = [0, 1]$, and put $xy = \max\{x, y\}$, for all $x, y \in I$. Clearly, $I$ with the usual topology and this product operation, is a topological Abelian semigroup (with 0 in the role of identity). However, $I$ is not a homogeneous space since no homeomorphism of $I$ takes 0 to $1/2$. □

Given a group $G$, it follows from Corollary 1.3.6 that to make $G$ into a right topological group, we can only use homogeneous topologies. Now, one of the main features of homogeneous spaces is that they behave in the same way at any point. It follows that if we know how the topology of a right topological group behaves at the identity, we know this topology everywhere (see Proposition 1.3.1). This observation suggests a certain approach to topologizing a group $G$ — first we have to define a family of basic neighbourhoods at the identity, then move this family around the group by means of translations, and generate a topology on $G$ by declaring the family thus obtained to be a base for the topology.

Let us see how this approach can be used to turn an arbitrary Abelian group $G$ into a semitopological group (not necessarily Hausdorff). Recall that a family $\xi$ of non-empty subsets of a set $X$ is called a *prefilter* on $X$ if $X \in \xi$ and, for each finite collection $A_1, \ldots, A_k$ of elements of $\xi$, there exists $B \in \xi$ such that $B \subset \bigcap_{i=1}^k A_i$. If, in addition, from $A \in \xi$ and $A \subset B \subset X$ it follows that $B \in \xi$, then $\xi$ is called a *filter* on $X$.

CONSTRUCTION 1.3.8. (Right topology on a monoid.) Let $S$ be a monoid, $e$ the identity of $S$, and $\mathcal{F}_e$ any prefilter on $S$ such that $e \in \bigcap \mathcal{F}_e$. For each $a \in S$, put $\mathcal{F}_a = \{Pa : P \in \mathcal{F}_e\}$. Then $\mathcal{F}_a$ is a prefilter on $S$ such that $a \in \bigcap \mathcal{F}_a$. Call a set $U \subset S$ open if for each $a \in U$ there exists $F \in \mathcal{F}_a$ such that $F \subset U$. Then the family of all open subsets forms a topology on $S$. We claim that, with this topology, $S$ becomes a right topological monoid (not necessarily Hausdorff).

Indeed, take $a \in S$ and an open set $V \subset S$. We have to show that the set $U = \varrho_a^{-1}(V)$ is open in $S$. Take any $b \in U$. Then $c = ba = \varrho_a(b) \in V$. Since $V$ is open and $c \in V$, there exists $P_c \in \mathcal{F}_c$ such that $P_c \subset V$. Then $P_c = Pc$, for some $P \in \mathcal{F}_e$, and $b \in P_b = Pb \in \mathcal{F}_b$. We have $\varrho_a(Pb) = Pba = Pc \subset V$. It follows that $Pb \subset U$, which implies that $U$ is open. Hence, the mapping $\varrho_a$ is continuous, and $S$ is a right topological monoid. □

Note that the elements of the family $\mathcal{F}_e$ need not belong to the topology defined in Construction 1.3.8. To see this, it is enough to consider the cross topology defined in Example 1.2.6; obviously, this definition is a particular case of Construction 1.3.8. Now we formulate a restriction on the family $\mathcal{F}_e$ that guarantees that all elements of $\mathcal{F}_e$ belong to the topology defined above and that the family serves as a base for this topology.

PROPOSITION 1.3.9. *Suppose that $S$ is a monoid and $\mathcal{F}_e$ is a prefilter on $S$ such that $e \in \bigcap \mathcal{F}_e$ and the next condition (t) is satisfied:*

(t) *for each $U \in \mathcal{F}_e$ and each $x \in U$, there exists $V \in \mathcal{F}_e$ such that $Vx \subset U$.*

*Then, for each $U \in \mathcal{F}_e$ and every $a \in S$, the set $Ua$ belongs to the topology on $S$ defined in Construction 1.3.8, and the family $\mathcal{F}_a = \{Ua : U \in \mathcal{F}_e\}$ is a base of this topology at $a$. In particular, it follows that every right action $\varrho_a$ is an open continuous mapping of $S$ to itself.*

PROOF. Let $a \in S$ and $U \in \mathcal{F}_e$ be arbitrary. Take any $y \in Ua$. Then $y = xa$, for some $x \in U$. According to (t), there is $V \in \mathcal{F}_e$ such that $Vx \subset U$. It follows that $y = xa \in Vxa = Vy \subset Ua$. Since $Vy \in \mathcal{F}_y$, we conclude that the set $Ua$ is open.

It follows from the definition of the topology on $S$ that the family $\mathcal{F}_a = \{Ua : U \in \mathcal{F}_e\}$ is a base of the space $S$ at $a$, for every $a \in S$. This statement, together with the obvious equality $\varrho_a(Ub) = Uba$, implies that each $\varrho_a$ is an open continuous mapping. □

Notice that if $S$ is a right topological monoid with topology $\mathcal{T}$, and $\mathcal{F}_e$ is a base of the space $S$ at $e$, then the prefilter $\mathcal{F}_e$ satisfies condition (t). Therefore, we can introduce a new topology $\mathcal{T}_0$ on $S$, following Construction 1.3.8. This new topology does not have to coincide with the original topology, but it is equivalent to it at the identity $e$. Of course, $\mathcal{T} \subset \mathcal{T}_0$. With the new topology $S$ may be a better right topological monoid than with the old one, since all right actions $\varrho_a$ are now open.

Let us now assume that $S = G$ is a group. Which restrictions on the prefilter $\mathcal{F}_e$ guarantee that the inverse is continuous? Which restrictions are necessary to make sure that the multiplication operation is jointly continuous?

The answers to these questions are not completely straightforward. For example, *it is not enough to require that all elements of $\mathcal{F}_e$ be symmetric sets* (that is, satisfy the condition $A = A^{-1}$) *to ensure that the inverse be continuous.* It is also not enough to assume that for each $U \in \mathcal{F}_e$ there be $V \in \mathcal{F}_e$ such that $V^2 \subset U$ to make the product operation jointly continuous, even in the presence of condition (t).

However, for Abelian groups the following results are available. The first of them is almost obvious.

THEOREM 1.3.10. *Let $G$ be an Abelian group, $e$ the neutral element of $G$, and $\mathcal{F}_e$ a prefilter on $G$ such that $e \in \bigcap \mathcal{F}_e$ and each $A \in \mathcal{F}_e$ is symmetric. Then the topology on $G$ generated by $\mathcal{F}_e$ as in Construction 1.3.8, turns $G$ into a quasitopological group.*

PROOF. Since in the Abelian group $G$ each left translation $\lambda_a$ coincides with the right translation $\varrho_a$, the right semitopological group $G$ is automatically a semitopological group. Suppose that $U$ is an open subset of $G$. To show that the set $U^{-1}$ is open in $G$, take an arbitrary point $x \in U$ and choose $P \in \mathcal{F}_e$ such that $Px \subset U$. Since $G$ is Abelian and $P$ is symmetric, we have that $Px^{-1} = P^{-1}x^{-1} = x^{-1}P^{-1} = (Px)^{-1} \subset U^{-1}$. Thus, the set $U^{-1}$ is open in $G$, so the inverse in $G$ is continuous. □

THEOREM 1.3.11. *Let $G$ be an Abelian group, $e$ the identity of $G$, and $\mathcal{F}_e$ a prefilter on $G$ such that $e \in \bigcap \mathcal{F}_e$ and the condition (t) is satisfied. Assume also that for each $U \in \mathcal{F}_e$ there exists $V \in \mathcal{F}_e$ such that $V^2 \subset U$. Then the topology on $G$ generated by $\mathcal{F}_e$ as in Construction 1.3.8, turns $G$ into a paratopological group.*

PROOF. It follows from Proposition 1.3.9 and the commutativity of $G$ that $G$ is a semitopological group. Let us verify that the multiplication in $G$ is jointly continuous. Take arbitrary elements $x, y \in G$ and an open neighbourhood $O$ of $z = xy$ in $G$. We use property (t) of the prefilter $\mathcal{F}_e$ to choose $U \in \mathcal{F}_e$ such that $Uz \subset O$. By assumptions of the theorem, there exists $V \in \mathcal{F}_e$ satisfying $V^2 \subset U$. Then $VxVy = VVxy = VVz \subset Uz \subset O$. Since, by Proposition 1.3.9, $Vx$ and $Vy$ are open neighbourhoods in $G$ of $x$ and $y$, respectively, this proves that the multiplication in $G$ is continuous. Hence, $G$ is a paratopological group. □

Using Construction 1.3.8 together with Theorems 1.3.10 and 1.3.11, one can effectively 'semitopologize' or 'paratopologize' many Abelian groups. However, to make a group into a topological group is a much more complicated matter. This is so because to ensure the joint continuity of multiplication turns out to be much more difficult than to guarantee the separate continuity only. Indeed, there exists an infinite abstract group $G$ which can be made into a Hausdorff topological group only by taking the discrete topology on $G$ (see Problem 1.3.F).

Let $G$ be an infinite abstract group. To understand what conditions should be imposed on $\mathcal{F}_e$ to guarantee that the topology on $G$ defined in Construction 1.3.8 will turn $G$ into a Hausdorff topological group, we first consider a simpler question: Given a topological group $G$, what are the properties of an open base at the identity $e$ of $G$? The answer is contained in the next statement.

THEOREM 1.3.12. *Let $G$ be a topological group and $\mathcal{U}$ an open base at the identity $e$ of $G$. Then:*

i) *for every $U \in \mathcal{U}$, there is an element $V \in \mathcal{U}$ such that $V^2 \subset U$;*
ii) *for every $U \in \mathcal{U}$, there is an element $V \in \mathcal{U}$ such that $V^{-1} \subset U$;*
iii) *for every $U \in \mathcal{U}$ and every $x \in U$, there is $V \in \mathcal{U}$ such that $Vx \subset U$;*
iv) *for every $U \in \mathcal{U}$ and $x \in G$, there is $V \in \mathcal{U}$ such that $xVx^{-1} \subset U$;*
v) *for $U, V \in \mathcal{U}$, there is $W \in \mathcal{U}$ such that $W \subset U \cap V$;*
vi) $\{e\} = \bigcap \mathcal{U}$.

*Conversely, let $G$ be a group, and let $\mathcal{U}$ be a family of subsets of $G$ satisfying conditions i)–vi). Then the family $\mathcal{B}_{\mathcal{U}} = \{Ua : a \in G,\ U \in \mathcal{U}\}$ is a base for a $T_1$-topology $\mathcal{T}_{\mathcal{U}}$ on $G$. With this topology, $G$ is a topological group, and the family $\{aU : a \in G,\ U \in \mathcal{U}\}$ is a base for the same topology on $G$.*

PROOF. If $G$ is a topological group, then i) and ii) follow from the continuity of the mappings $(x, y) \mapsto xy$ and $x \mapsto x^{-1}$ at the identity $e$. Property iii) follows from the continuity of left translations in $G$. Similarly, iv) follows from the fact that $x \mapsto ax$ and $ax \mapsto axa^{-1}$ are homeomorphisms of $G$. Property v) is clear since $\mathcal{U}$ is an open base at $e$. Property vi) is also clear since $G$ is a $T_1$-space and $\mathcal{U}$ is an open base at $e$.

To prove the converse, let $\mathcal{U}$ be a family of subsets of $G$ such that conditions i)–vi) hold. Let $\mathcal{T}$ be the family of all subsets $W$ of $G$ satisfying the following condition:

(n) for each $x \in W$, there is $U \in \mathcal{U}$ such that $Ux \subset W$.

**Claim 1.** *$\mathcal{T}$ is a topology on $G$.*

Indeed, it is clear that $\bigcup \gamma \in \mathcal{T}$ for any subfamily $\gamma$ of $\mathcal{T}$. Assume now that $W_1 \in \mathcal{T}$ and $W_2 \in \mathcal{T}$, and put $W = W_1 \cap W_2$. We have to prove that $W \in \mathcal{T}$. Take any $x \in W$. There exist $U_1 \in \mathcal{U}$ and $U_2 \in \mathcal{U}$ such that $U_1x \subset W_1$ and $U_2x \subset W_2$. From v) it follows that there is $U \in \mathcal{U}$ such that $U \subset U_1 \cap U_2$. Then, clearly, $Ux \subset W_1 \cap W_2 = W$. Hence, $W \in \mathcal{T}$, and $\mathcal{T}$ is a topology on $G$.

**Claim 2.** *$Ux \in \mathcal{T}$, for each $x \in G$ and each $U \in \mathcal{U}$.*

Take any $y \in Ux$. Then $yx^{-1} \in U$. By property iii), there is an element $V \in \mathcal{U}$ such that $Vyx^{-1} \subset U$. It follows that $Vy \subset Ux$. Hence, $Ux \in \mathcal{T}$.

Note that Claim 2 and iii) together imply that the following is true:

**Claim 3.** *The family $\mathcal{B}_{\mathcal{U}} = \{Ua : a \in G, \ U \in \mathcal{U}\}$ is a base for the topology $\mathcal{T}$. Hence, $\mathcal{T} = \mathcal{T}_{\mathcal{U}}$.*

**Claim 4.** *The multiplication in $G$ is jointly continuous with respect to the topology $\mathcal{T}$.*

Let $a$ and $b$ be arbitrary elements of $G$, and $O$ be any element of $\mathcal{T}$ such that $ab \in O$. Then there exists $W \in \mathcal{U}$ such that $Wab \subset O$. To prove Claim 4, it suffices to find $U \in \mathcal{U}$ and $V \in \mathcal{U}$ such that $UaVb \subset Wab$ or, equivalently, $UaV \subset Wa$ which is, in its turn, equivalent to $U(aVa^{-1}) \subset W$. Now we can see how to choose $U$ and $V$ in $\mathcal{U}$.

First, apply i) to choose $U \in \mathcal{U}$ such that $U^2 \subset W$. After that, use iv) to choose $V \in \mathcal{U}$ such that $aVa^{-1} \subset U$. Then, by the choice of $U$ and $V$, we have $U(aVa^{-1}) \subset U^2 \subset W$ which implies that $UaVb \subset Wab$. Thus, the multiplication in $G$ is continuous with respect to the topology $\mathcal{T}$. In particular, all right translations of $G$ are continuous, and the space $(G, \mathcal{T})$ is homogeneous.

**Claim 5.** *$bV \in \mathcal{T}$, for all $b \in G$ and $V \in \mathcal{U}$.*

Take any $y \in bV$. We have to find $U \in \mathcal{U}$ such that $Uy \subset bV$. Clearly, we have $b^{-1}y \in V$. By iii), there is an element $W \in \mathcal{U}$ such that $Wb^{-1}y \subset V$. It follows from iv) that there is $U \in \mathcal{U}$ such that $b^{-1}Ub \subset W$. Therefore, $b^{-1}Ubb^{-1}y \subset V$, that is, $b^{-1}Uy \subset V$. Hence, $Uy \subset bV$. It follows that $bV \in \mathcal{T}$.

**Claim 6.** *The mapping $In$ of $G$ onto $G$ given by $In(x) = x^{-1}$ is continuous with respect to the topology $\mathcal{T}$.*

Indeed, it follows from Claim 3 that we have only to show that the set $a^{-1}U^{-1}$ is in $\mathcal{T}$ for all $a \in G$ and $U \in \mathcal{U}$, since $In^{-1}(Ua) = a^{-1}U^{-1}$. By Claim 5, it suffices to verify that $U^{-1} \in \mathcal{T}$. Take an arbitrary point $x \in U^{-1}$. Then $x^{-1} \in U$, so iii) implies that $Vx^{-1} \subset U$ for some $V \in \mathcal{U}$. Apply ii) to choose $W \in \mathcal{U}$ such that $W^{-1} \subset V$. Then $W^{-1}x^{-1} \subset Vx^{-1} \subset U$, whence it follows that $xW = (W^{-1}x^{-1})^{-1} \subset U^{-1}$. Again, by Claim 5, $xW$ is an open neighbourhood of $x$ in $(G, \mathcal{T})$, so we conclude that $U^{-1}$ is an element of $\mathcal{T}$. This proves Claim 6.

Finally, vi) and the homogeneity of $G$ imply that the topology $\mathcal{T}$ satisfies the $T_1$-separation axiom. This finishes the proof of the theorem. □

THEOREM 1.3.13. *Every topological group $G$ has an open base at the identity consisting of symmetric neighbourhoods.*

PROOF. For an arbitrary open neighbourhood $U$ of the identity $e$ in $G$, let $V = U \cap U^{-1}$. Then $V = V^{-1}$, the set $V$ is an open neighbourhood of $e$, and $V \subset U$. □

THEOREM 1.3.14. *Every topological group $G$ is a regular space.*

PROOF. Let $U$ be an open neighbourhood of the identity $e$ in $G$. By i) of Theorem 1.3.12 and Theorem 1.3.13, there is an open neighbourhood $V$ of $e$ such that $V^{-1} = V$ and $V^2 \subset U$. Then if $x \in \overline{V}$, we have $Vx \cap V \neq \varnothing$. Hence $a_1x = a_2$ for some $a_1, a_2$ in $V$, and thus

$x = a_1^{-1}a_2 \in V^{-1}V = V^2 \subset U$. This implies that $\overline{V} \subset U$. Since $G$ is a homogeneous space by Corollary 1.3.6, the regularity of $G$ is now immediate. □

EXAMPLE 1.3.15. Let $G$ be an abstract group and let $\mathcal{F}$ be a family of invariant subgroups of $G$ closed under the formation of finite intersections and such that $\{e\} = \bigcap \mathcal{F}$. Then the family of all sets of the form $Hx$, where $H \in \mathcal{F}$ and $x \in G$, is a topology $\mathcal{T}$ on $G$. This follows immediately from Theorem 1.3.12. With this topology, $G$ is zero-dimensional. Indeed, each $H \in \mathcal{F}$ is both open and closed since the complement of $H$ is the union of cosets disjoint from $H$. □

Unfortunately, Theorem 1.3.12 does not provide us with a clear recipe for defining non-discrete Hausdorff group topologies on abstract groups. But it does provide us with a general strategy for searching for such topologies. We present below an important example of such a topologization of the group of $r$-adic numbers.

EXAMPLE 1.3.16. Let us show that for every integer $r > 1$, the group $\Omega_r$ of $r$-adic numbers defined in Example 1.1.10 admits a non-discrete locally compact Hausdorff group topology. For every $m \in \mathbb{Z}$, denote by $\Lambda_m$ the set of all $\mathbf{x} \in \Omega_r$ such that $x_n = 0$ for each $n < m$. Clearly, $\Lambda_m$ is a subgroup of $\Omega_r$ and $\Lambda_{m+1} \subset \Lambda_m$ for each $m \in \mathbb{Z}$.

We claim that the family $\mathcal{U} = \{\Lambda_m : m \in \mathbb{Z}\}$ satisfies conditions i)–vi) of Theorem 1.3.12 and, hence, constitutes a local base at the neutral element $\mathbf{0}$ for a Hausdorff topological group topology on $\Omega_r$. Since each $\Lambda_m$ is a subgroup of $\Omega_r$, conditions i)–iii) are evident. Condition iv) holds trivially since the group $\Omega_r$ is commutative, while (v) follows from the inclusions $\ldots \subset \Lambda_{m+1} \subset \Lambda_m \subset \Lambda_{m-1} \subset \cdots$. Finally, it is easy to see that the intersection of the sets $\Lambda_m$ with $m \geq 0$ contains only the element $\mathbf{0}$, whence vi) follows. Thus, $\mathcal{U}$ is a local base at $\mathbf{0}$ for a Hausdorff group topology $\mathcal{T}$ on $\Omega_r$. Since $\mathcal{U}$ is a decreasing chain of non-trivial subgroups of $\Omega_r$, the topology $\mathcal{T}$ is non-discrete. Clearly, each $\Lambda_m$ is open in the group $(\Omega_r, \mathcal{T})$ which will be denoted simply by $\Omega_r$. Since every open subgroup is closed, we conclude that the neutral element $\mathbf{0}$ of $\Omega_r$ has a local base of open and closed neighbourhoods. Therefore, the homogeneity of the space $\Omega_r$ implies that it is zero-dimensional.

Let us verify that every group $\Lambda_m$ is compact as a subspace of $\Omega_r$. Identify $\Lambda_m$ with a subset of $A^{J_m}$ by means of truncating the values of $\mathbf{x}$ at all positions less than $m$, where $A = \{0, 1, \ldots, r-1\}$ and $J_m = \{n \in \mathbb{Z} : m \leq n\}$ (see Example 1.1.10). Then the group $\Lambda_m$ inherits a topology $\mathcal{T}_m$ from the space $A^{J_m}$ when the latter carries the usual Tychonoff product topology (and $A$ is discrete). We claim that $\mathcal{T}_m$ coincides with the restriction of the topology $\mathcal{T}$ to $\Lambda_m$. Indeed, for every $\mathbf{x} \in \Lambda_m$ and every $k > m$, the sum $\mathbf{x} + \Lambda_k \subset \Lambda_m$ consists of all sequences $(\ldots, 0, \ldots, 0, x_m, x_{m+1}, \ldots, x_{k-1}, y_k, y_{k+1}, \ldots)$, where the value $x_m$ stands at the $m$th position in the sequence and $y_k, y_{k+1}, \ldots$ are arbitrary elements of $A$. It follows that $\mathbf{x} + \Lambda_k$ is a canonical open set in $A^{J_m}$ under our identification. This proves that $\mathcal{T}_m = \mathcal{T} \restriction \Lambda_m$. Since the product space $A^{J_m}$ is compact and every canonical open set in $A^{J_m}$ is closed, it follows that $\Lambda_m$ is a compact subgroup of $\Omega_r$, for each $m \in \mathbb{Z}$. In particular, the group of $r$-adic integers $\mathbb{Z}_r = \Lambda_0$ is an open compact subgroup of $\Omega_r$.

By our definition of the topology in $\Omega_r$, each $\Lambda_m$ is an open subgroup of $\Omega_r$. We conclude, therefore, that the group $\Omega_r$ is locally compact. Notice that the quotient group $\Omega_r/\Lambda_m$ is countable for each $m \in \mathbb{Z}$, so the group $\Omega_r$ is a countable union of cosets of the compact subgroup $\Lambda_m$. It follows that the group $\Omega_r$ is $\sigma$-compact.

Summing up, the group of $r$-adic numbers $\Omega_r$ is locally compact, $\sigma$-compact, zero-dimensional, and has a local base at zero consisting of compact open subgroups. □

## Exercises

1.3.a. Is the usual convergent sequence $Sq$ homeomorphic to a topological semigroup with an identity, that is, to a topological monoid? Is it homeomorphic to a semitopological monoid?

1.3.b. Show that the neutral element $e$ is not isolated in the space $S$ defined in Construction 1.3.8 if and only if each $P \in \mathcal{F}_e$ contains more than one element.

1.3.c. Verify that if in a paratopological group $G$ all elements have orders less than or equal to a given integer $n \in \mathbb{N}$, then $G$ is a topological group. Does the same assertion remain valid for any commutative paratopological group $G$ such that all elements of $G$ have finite orders?

1.3.d. Give an example of a Hausdorff paratopological group that fails to be regular.

1.3.e. Show that the Niemytzki plane (see [**165**, Example 1.2.4]) admits a binary operation that makes $X$ into a commutative monoid with jointly continuous multiplication.

1.3.f. (A. V. Arhangel'skii and M. Hušek [**54**]) Let $f\colon G \to H$ be a continuous homomorphism of right topological groups, and let $X$ be a dense subspace of $G$. Suppose also that the space $G$ is regular and that the restriction of $f$ to $X$ is a topological embedding. Show that $f$ is a topological embedding of $G$ into $H$.

1.3.g. Show that for every integer $m$, the subgroup $\Lambda_m$ of the group of $r$-adic numbers $\Omega_r$ (see Example 1.3.16) contains a dense cyclic subgroup.

1.3.h. For any distinct $\mathbf{x}, \mathbf{y} \in \Omega_r$, set $\sigma(\mathbf{x}, \mathbf{y}) = 2^{-m}$, where $m$ is the least integer such that $x_m \neq y_m$. Also set $\sigma(\mathbf{x}, \mathbf{x}) = 0$ for each $\mathbf{x} \in \Omega_r$. Verify that $\sigma$ is an invariant metric on the group $\Omega_r$ which generates the topology of this group defined in Example 1.3.16. Show that $\sigma(\mathbf{0}, \mathbf{x}+\mathbf{y}) \leq \max\{\sigma(\mathbf{0}, \mathbf{x}), \sigma(\mathbf{0}, \mathbf{y})\}$ for all $\mathbf{x}, \mathbf{y} \in \Omega_r$. Metrics with this property are called *non-archimedian*.

1.3.i. Define a topology on the group $\Omega_{\mathbf{a}}$ of $\mathbf{a}$-adic numbers (see Problem 1.1.G) similarly to that in Example 1.3.16, by declaring the corresponding subgroups $\Lambda_n$ of $\Omega_{\mathbf{a}}$, with $n \in \mathbb{Z}$, to be basic open neighbourhoods of the neutral element in $\Omega_{\mathbf{a}}$. Show that the group $\Omega_{\mathbf{a}}$ with this topology is locally compact, and that the group $\mathbb{Z}_{\mathbf{a}}$ of $\mathbf{a}$-adic integers is a compact and open subgroup of $\Omega_{\mathbf{a}}$.

## Problems

1.3.A. Apply Theorem 1.3.12 to define a Hausdorff topological group topology on the real line $\mathbb{R}$ strictly coarser than its usual topology.

1.3.B. Prove that there exists a topological group topology on the real line $\mathbb{R}$ strictly finer than its usual topology that makes $\mathbb{R}$ into a group topologically isomorphic to a dense subgroup of the Euclidean plane $\mathbb{R}^2$.

1.3.C. Does there exist a continuous homomorphism of $\mathbb{R}$ onto $\mathbb{R}^2$ (both groups carry the usual Euclidean topologies)?

1.3.D. Let us say that a space $X$ *admits the structure of a topological group* (*of a paratopological group*) if there exists a continuous binary operation on $X \times X$ that makes $X$ into a topological group (paratopological group).

(a) Prove that the Cantor set and the set of irrational numbers, both considered as subspaces of the real line $\mathbb{R}$, admit the structure of a topological group.

(b) Show that the Sorgenfrey line does not admit the structure of a topological group. Does it admit (in a natural sense) the structure of a quasitopological group?

1.3.E. (E. Hewitt and K. A. Ross [**236**]) Let $p$ be a prime number. Prove that every closed subgroup of the group $\Omega_p$ of $p$-adic numbers is either trivial or coincides with one of the groups $\Lambda_m$ defined in Example 1.3.16. Show that for every composite integer $r > 1$, the group $\Omega_r$ contains a compact open subgroup distinct from $\Lambda_m$ for each $m \in \mathbb{Z}$.
*Hint.* To deduce the second claim of the problem, suppose that $r = ab$ for some integers $a, b > 1$. Let $\mathbf{x}$ be an $r$-adic integer defined by $x_0 = a$ and $x_l = 0$ if $l \neq 0$. Denote by $H$ the minimal closed subgroup of $\Omega_r$ that contains $\mathbf{x}$. Verify that $\Lambda_1 \subset H$ and, hence, $H = \bigcup_{k=0}^{r-1}(k\mathbf{x} + \Lambda_1)$. Therefore, $\Lambda_0 \subsetneqq H \subsetneqq \Lambda_1$.

1.3.F. Prove that there exists a countable infinite group $G$ such that the only Hausdorff topological group topology on $G$ is discrete.
*Hint.* An example of such a group is given by A. I. Ol'shanskiĭ in [**361**].

## Open Problems

1.3.1. Is every regular paratopological group completely regular?

1.3.2. Are there Abelian topological groups $G$ and $H$ such that the product groups $G \times \mathbb{Z}$ and $H \times \mathbb{Z}$ are topologically isomorphic while $G$ and $H$ are not? The group $\mathbb{Z}$ carries the discrete topology.

## 1.4. Open sets, closures, connected sets and compact sets

We describe here several simple properties of the families of open, closed, and compact sets in topological groups and semigroups. Though many of these properties are almost evident and easy to formulate, they form a solid basis for constructing the edifice of topological algebra.

PROPOSITION 1.4.1. *Let $G$ be a left (right) topological group, $U$ an open subset of $G$, and $A$ any subset of $G$. Then the set $AU$ (respectively, $UA$) is open in $G$.*

PROOF. Every left translation of $G$ is a homeomorphism, by Proposition 1.3.1. Since $AU = \bigcup_{a \in A} \lambda_a(U)$, the conclusion follows. A similar argument applies in the case when $G$ is a right topological group. □

COROLLARY 1.4.2. *If $G$ is a semitopological group then, for any open subset $U$ of $G$ and any subset $A$ of $G$, the sets $UA$ and $AU$ are open.*

The next example shows that the situation with topological semigroups is different from what we have just seen in Proposition 1.4.1 and Corollary 1.4.2. First, we introduce the *Vietoris topology* on the family $Exp(X)$ of all non-empty closed subsets of a space $X$. Let $U$ and $V_1, \ldots, V_n$ be non-empty open sets in $X$. Then we put

$$\langle U, V_1, \ldots, V_n\rangle = \{F \in Exp(X) : F \subset U,\ F \cap V_i \neq \varnothing \text{ for each } i = 1, \ldots, n\}.$$

The family $\mathcal{B}$ of all sets of the form $\langle U, V_1, \ldots, V_n\rangle$ constitutes a base of a topology on the set $Exp(X)$ which is called the *Vietoris topology* (see [**165**, 2.7.20]). It is easy to verify that $Exp(X)$ with the Vietoris topology is a $T_1$-space provided that $X$ is a $T_1$-space.

EXAMPLE 1.4.3. Consider the space $Exp(\mathbb{R})$ of all non-empty closed subsets of the usual real line $\mathbb{R}$, in the Vietoris topology, and with the product operation defined as the union of sets. Then, clearly, $Exp(\mathbb{R})$ is a topological semigroup. Put $a = \{0\}$, $U = \{x \in \mathbb{R} : x > 1\}$, and $W = \{P \in Exp(\mathbb{R}) : P \subset U\}$. Then $a \in Exp(\mathbb{R})$ and $W$ is an open subset of $Exp(\mathbb{R})$. However, $aW$ is not open in $Exp(\mathbb{R})$. Indeed, the point $b = \{0, 2\}$ of $Exp(\mathbb{R})$ is in $aW$; however, no neighbourhood of $b$ is contained in $aW$. Of course, this example also shows that a left action in a topological semigroup need not be a one-to-one mapping. □

In left topological groups with continuous inverse there is an intimate relationship between the sets of the form $AU$, where $U$ is open, and the closure operation.

PROPOSITION 1.4.4. *Let $G$ be a left topological group with continuous inverse. Then, for every subset $A$ of $G$ and every open neighbourhood $U$ of the neutral element $e$, $\overline{A} \subset AU$.*

PROOF. Since the inverse is continuous, we can find an open neighbourhood $V$ of $e$ such that $V^{-1} \subset U$. Take any $x \in \overline{A}$. Then $xV$ is an open neighbourhood of $x$; therefore, there is $a \in A \cap xV$, that is, $a = xb$, for some $b \in V$. Then $x = ab^{-1} \in AV^{-1} \subset AU$; hence, $\overline{A} \subset AU$. □

A similar statement holds for right topological groups with continuous inverse. Let us show that the conclusion in Proposition 1.4.4 can be considerably strengthened.

THEOREM 1.4.5. *Let $G$ be a left topological group with continuous inverse, and $\mathcal{B}_e$ a base of the space $G$ at the neutral element $e$. Then, for every subset $A$ of $G$,*

$$\overline{A} = \bigcap\{AU : U \in \mathcal{B}_e\}.$$

PROOF. In view of Proposition 1.4.4, we only have to verify that if $x$ is not in $\overline{A}$, then there exists $U \in \mathcal{B}_e$ such that $x \notin AU$. Since $x \notin \overline{A}$, there exists an open neighbourhood $W$ of $e$ such that $(xW) \cap A = \varnothing$. Take $U$ in $\mathcal{B}_e$ satisfying the condition $U^{-1} \subset W$. Then $(xU^{-1}) \cap A = \varnothing$, which obviously implies that $AU$ does not contain $x$. □

Similarly, the equality $\overline{A} = \bigcap\{UA : U \in \mathcal{B}_e\}$ holds for right topological groups with continuous inverse.

In what follows we often use the next obvious statement:

PROPOSITION 1.4.6. *For any subsets $A$, $B$, and $C$ of a group $G$, $AB \cap C = \varnothing$ if and only if $A \cap CB^{-1} = \varnothing$.*

Curiously enough, Theorem 1.4.5 can be partially reversed. First, we establish the following:

PROPOSITION 1.4.7. *Let $G$ be a semitopological group such that for each open neighbourhood $U$ of the neutral element $e$, there exists an open neighbourhood $V$ of $e$ satisfying the condition $V^{-1} \subset U$. Then the inverse operation in $G$ is continuous and, therefore, $G$ is a quasitopological group.*

PROOF. Take an element $x \in G$ and any open neighbourhood $W$ of $x^{-1}$. Since $G$ is a left topological group, there exists an open neighbourhood $U$ of $e$ such that $x^{-1}U \subset W$. By the assumption, there exists an open neighbourhood $V$ of $e$ such that $V^{-1} \subset U$. Then $Vx$ is an open neighbourhood of $x$, since $G$ is a right topological group. Now we have $(Vx)^{-1} = x^{-1}V^{-1} \subset x^{-1}U \subset W$. It follows that the inverse mapping on $G$ is continuous, so the group $G$ is quasitopological. □

PROPOSITION 1.4.8. *Let $G$ be a semitopological group such that for each closed subset $A$ of $G$ and every point $x \in G \setminus A$, $x \notin AU$ for some open neighbourhood $U$ of the neutral element $e$. Then the inverse mapping of $G$ onto itself is continuous and $G$ is a quasitopological group.*

PROOF. By Proposition 1.4.7, it suffices to check the continuity of the inverse mapping at the neutral element $e$. Take any open neighbourhood $U$ of $e$ and put $A = G \setminus U$. Then $e \notin A$ and, by the assumption, there is an open neighbourhood $V$ of $e$ such that $AV$ does not contain $e$. Then, by Proposition 1.4.6, $A \cap V^{-1} = \varnothing$, that is, $V^{-1} \subset U$. Hence, the inverse mapping is continuous. □

In connection with the above statements we should also mention the following simple result:

PROPOSITION 1.4.9. *Every left topological group $G$ with continuous inverse is a semitopological group and, hence, a quasitopological group.*

PROOF. We can get from $x$ to $xa$ in three steps — from $x$ to $x^{-1}$, then from $x^{-1}$ to $a^{-1}x^{-1}$, and, finally, from $a^{-1}x^{-1}$ to $xa$. This means simply that $\varrho_a = In \circ \lambda_{a^{-1}} \circ In$. Since all the mappings on the right side of the above equality are continuous, it follows that the mapping $\varrho_a$ is also continuous. □

Some delicate situations arise when we take closures of subsemigroups or subgroups in semitopological semigroups and groups.

PROPOSITION 1.4.10. *Let $G$ be a semitopological semigroup, and $H$ a subsemigroup of $G$. Then the closure $\overline{H}$ of $H$ in $G$ is a (semitopological) subsemigroup of $G$.*

PROOF. Take any $y \in \overline{H}$ and any $x \in H$. Then $xy \in \overline{xH}$, since the left translation $\lambda_x$ is continuous. We have $xH \subset H$, since $x \in H$ and $H$ is a subsemigroup of $G$. It follows that $xy \in \overline{H}$, for each $x \in H$. Now take any $z \in \overline{H}$. By the continuity of $\varrho_y$, it follows that $zy \in \overline{H}$. Hence, $\overline{H}$ is a subsemigroup of $G$. □

The next example shows that Proposition 1.4.10 cannot be extended to right topological semigroups.

EXAMPLE 1.4.11. Let $Z = \alpha\mathbb{N} = \mathbb{N} \cup \{\alpha\}$ be the one-point compactification of the discrete space $\mathbb{N}$ of positive natural numbers. Recall that $S_p(Z, Z) = Z^Z$ is the right topological semigroup of all mappings of $Z$ to itself, endowed with the topology of pointwise convergence (see Theorem 1.2.3). The product operation in $Z^Z$ is the composition of mappings. Since $Z$ is compact, the right topological semigroup $Z^Z$ is also compact. Consider the set $B$ consisting of all one-to-one mappings of $Z$ to itself. Clearly, $B$ is closed under compositions; therefore, $B$ is a subsemigroup of $Z^Z$. Denote by $C$ the closure of $B$ in $Z^Z$. Let us show that $C$ is not closed under the product operation and, therefore, is not a subsemigroup of $Z^Z$.

For each $k \in \mathbb{N}$, define $f_k\colon Z \to Z$ as follows: $f_k(x) = x + k$ if $x \in \mathbb{N}$, and $f(\alpha) = \alpha$. Let $g$ be the constant mapping of $Z$ to itself that brings each point of $Z$ to the point $\alpha$. Clearly, $f_k \in B$, for each $k \in \mathbb{N}$, and the sequence $\{f_k(x) : k \in \mathbb{N}\}$ converges to $g(x)$ in $Z$, for each $x \in Z$. Therefore, the sequence $\{f_k : k \in \mathbb{N}\}$ converges to $g$ in the space $Z^Z$, which implies that $g$ is in the closure of $B$, that is, $g \in C$ but, clearly, $g$ is not in $B$.

Now define $h \in Z^Z$ by the rule $h(x) = x + 1$, for each $x \in \mathbb{N}$, and $h(\alpha) = 1$. Since $h$ is one-to-one, we have $h \in B \subset C$. Thus, both $g$ and $h$ are in $C$.

Let us show that $hg$ is not in $C$ and, therefore, $C$ is not a subsemigroup of $Z^Z$. Indeed, $hg(x) = h(g(x)) = h(\alpha) = 1$, for each $x \in Z$. Since the element 1 is isolated in $Z$, the set $V = \{f \in Z^Z : f(1) = 1 \text{ and } f(3) = 1\}$ is open in the space $Z^Z$. Obviously, $V$ contains $hg$ and none of elements of $V$ is a one-to-one mapping, that is, $V \cap B = \varnothing$ and, hence, $V \cap C = \varnothing$. It follows that $hg$ is not in $C$. □

The next statement is almost obvious.

PROPOSITION 1.4.12. *Let $G$ be an abstract group with a topology $\mathcal{T}$ such that the inverse mapping $In$ is continuous. Then, for any symmetric subset $A$ of $G$, the closure of $A$ in $G$ is also symmetric.*

PROOF. Since the inverse mapping $In$ is continuous and the composition $In \circ In$ is the identity mapping of $G$ onto itself, $In$ is a homeomorphism. Hence, the closure of $A^{-1}$ coincides with the image of the closure of $A$ under $In$. □

From Propositions 1.4.10 and 1.4.12 we obtain immediately the next result:

PROPOSITION 1.4.13. *Let $G$ be a quasitopological group, and $H$ an algebraic subgroup of $G$. Then the closure of $H$ in $G$ is also a subgroup of $G$.*

COROLLARY 1.4.14. *Let $G$ be a topological group and $H$ a subgroup of $G$. Then $\overline{H}$ is also a subgroup of $G$.*

Several properties of a subgroup $H$ of a topological group $G$ pass on to the closure of $H$ in $G$. In Proposition 1.4.16 below we present one of such properties (see also Theorem 3.4.9 and Exercise 3.4.d in this respect). First, we establish a useful fact about the *character* of regular spaces.

Suppose that $X$ is a space and $x \in X$. A family $\mathcal{B}$ of subsets of $X$ is called a *base for $X$ at the point $x$* if all elements of $\mathcal{B}$ contain $x$ and, for every neighbourhood $O$ of $x$ in $X$, there exists $U \in \mathcal{B}$ such that $U \subset O$. The minimal cardinality of a base of $X$ at $x$ is called the *character of $X$ at $x$* and is denoted by $\chi(x, X)$. If $X$ has a countable base at each point $x \in X$, we say that the space $X$ is *first-countable* (see [**165**, Section 1.1]).

LEMMA 1.4.15. *If $Y$ is a dense subspace of a regular space $X$, then $\chi(y, Y) = \chi(y, X)$ for each $y \in Y$.*

PROOF. If $\mathcal{B}$ is a base for $X$ at a point $y \in Y$, then the family $\mathcal{B}_Y = \{U \cap Y : U \in \mathcal{B}\}$ is evidently a base for $Y$ at $y$. Therefore, $\chi(y, Y) \le \chi(y, X)$. Conversely, let $\mathcal{B}_Y$ be a base for $Y$ at a point $y \in Y$ such that $|\mathcal{B}_Y| = \chi(y, Y)$. For every $U \in \mathcal{B}_Y$, choose an open set $V_U$ in $X$ such that $V_U \cap Y = U$. We claim that the family $\mathcal{B} = \{V_U : U \in \mathcal{B}_Y\}$ is a base for $X$ at the point $y$. Indeed, take an arbitrary neighbourhood $O$ of $y$ in $X$. There exists an open neighbourhood $W$ of $y$ in $X$ such that $\overline{W} \subset O$. Since $\mathcal{B}_Y$ is a base for $Y$ at $y$, we can find $U \in \mathcal{B}_Y$ such that $U \subset W \cap Y$. The set $Y$ being dense in $X$, the intersection $V_U \cap Y = U$ is dense in $V_U$. Hence $\overline{V_U} = \overline{U} \subset \overline{W} \subset O$; it follows that $y \in V_U \subset O$. This proves that the family $\mathcal{B}$ is a base for $X$ at $y$. It is immediate from the definition of $\mathcal{B}$ that $|\mathcal{B}| \le |\mathcal{B}_Y| = \chi(y, Y)$, so that $\chi(y, X) \le \chi(y, Y)$. □

PROPOSITION 1.4.16. *If $H$ is a first-countable subgroup of a topological group $G$, then the closure of $H$ is also first-countable.*

PROOF. By Corollary 1.4.14, $K = \overline{H}$ is a subgroup of $G$ and, hence, is a homogeneous space. It suffices, therefore, to verify that the neutral element $e$ of $K$ has a countable local base in $K$. Since $\chi(e, H) \leq \aleph_0$ and the space $K$ is regular, by Theorem 1.3.14, it follows from Lemma 1.4.15 that $\chi(e, K) \leq \aleph_0$. Thus the group $K$ is first-countable. □

In contrast to the case of topological groups, in a paratopological group the closure of a subgroup need not be a subgroup.

EXAMPLE 1.4.17. Let $G$ be the group $\mathbb{Z}^\omega$, and $a_i$ be the element of $G$ such that the $i$-th coordinate of $a_i$ is 1 and all other coordinates of $a_i$ are 0. For $n \in \mathbb{N}$, let $V_n$ be the set of all $z = (z_k)_{k\in\omega} \in G$ such that $z_k = 0$ for every $k \leq n$, and $z_k \geq 0$ for every $k > n$. The prefilter $\xi = \{V_n : n \in \mathbb{N}\}$ satisfies condition (t) of Proposition 1.3.9, so Theorem 1.3.11 implies that $\xi$ generates a topology $\mathcal{T}_\xi$ on $G$ turning $G$ into a paratopological group. Notice that $\xi$ is a local base at the neutral element of $G$. Consider the subgroup $H$ of $G$ generated by the set $A = \{a_0 + a_n : n \in \mathbb{N}\}$ (recall that $\mathbb{N}$ does not include 0), and let $U = V_1 \cap H$. Then $U$ is an open subset of $H$ containing the neutral element $e$ of the group $H$.

We claim that $U = \{e\}$. Indeed, this follows from the next obvious property of elements of $H$: If $z \in H$, $z \neq e$ and the 0-th coordinate $z_0$ of $z$ is 0, then at least one coordinate of $z$ is strictly negative. Thus, $e$ is isolated in $H$. Since $H$ is a paratopological group, it follows that $H$ is a discrete subspace of $G$.

However, $H$ is not closed in $G$. Indeed, let $b$ the element of $\mathbb{Z}^\omega$ all coordinates of which are equal to $-1$. For every integer $n \geq 1$, we have that

$$\begin{aligned} &-(a_0 + a_1) - (a_0 + a_2) - \cdots - (a_0 + a_{n+1}) + n(a_0 + a_{n+2}) = \\ &-(a_0 + a_1 + a_2 + \cdots + a_{n+1}) + na_{n+2} \in H \cap (b + V_{n+1}). \end{aligned}$$

It follows that the element $b$ belongs to the closure of $H$, while, obviously, $b$ is not in $H$.

Furthermore, the closure $S = \overline{H}$ of $H$ in $G$ is not a subgroup of $G$. Suppose to the contrary that $S$ is a subgroup of $G$. Then, with the topology inherited from the paratopological group $G$, $S$ is a paratopological group, and $H$ is its open subgroup, since the discrete subspace $H$ is dense in $S$. Applying Theorem 1.3.5, we conclude that $H$ is closed in $S$ and in $G$, a contradiction. □

Example 1.4.17 is also complemented by the next corollary to Theorem 1.3.5.

COROLLARY 1.4.18. *If $H$ is a discrete subgroup of a quasitopological group $G$, then $H$ is closed in $G$.*

PROOF. By Proposition 1.4.13, the closure $\overline{H}$ of $H$ in $G$ is a subgroup of $G$. With the topology inherited from $G$, $\overline{H}$ is a quasitopological group. Since $H$ is discrete in itself, it is an open subgroup of $\overline{H}$. From Theorem 1.3.5 it follows that $H$ is closed in $\overline{H}$. Hence, $H$ is closed in $G$. □

The above result admits a considerable generalization in the case of topological groups.

PROPOSITION 1.4.19. *If $H$ is a locally compact subgroup of a topological group $G$, then $H$ is closed in $G$.*

PROOF. Let $K$ be the closure of $H$ in $G$. Then $K$ is a subgroup of $G$ by Corollary 1.4.14. Since $H$ is a dense locally compact subspace of $K$, it follows from [**165**, Theorem 3.3.9] that $H$ is open in $K$. However, an open subgroup of a topological group is closed by Theorem 1.3.5. Therefore, $H = K$, that is, $H$ is closed in $G$. □

COROLLARY 1.4.20. *Every discrete subgroup $H$ of a countably compact quasitopological group $G$ is finite.*

PROOF. By Corollary 1.4.18, $H$ is closed in $G$. Therefore, $H$ is countably compact. Since $H$ is discrete, it follows that $H$ is finite. □

A similar argument shows that the next statement is true:

COROLLARY 1.4.21. *Every discrete subgroup $H$ of a Lindelöf quasitopological group $G$ is countable.*

Corollary 1.4.20 can be complemented in a non-trivial way. This requires a new concept. Suppose that $U$ is a neighbourhood of the neutral element of a topological group $G$. A subset $A$ of $G$ is called *$U$-disjoint* if $b \notin aU$, for any distinct $a, b \in A$.

LEMMA 1.4.22. *Let $U$ and $V$ be open neighbourhoods of the neutral element in a topological group $G$ such that $V^4 \subset U$ and $V^{-1} = V$. If a subset $A$ of $G$ is $U$-disjoint, then the family of open sets $\{aV : a \in A\}$ is discrete in $G$.*

PROOF. It suffices to verify that, for every $x \in G$, the open neighbourhood $xV$ of $x$ intersects at most one element of the family $\{aV : a \in A\}$. Suppose to the contrary that, for some $x \in G$, there exist distinct elements $a, b \in A$ such that $xV \cap aV \neq \varnothing$ and $xV \cap bV \neq \varnothing$. Then $x^{-1}a \in V^2$ and $b^{-1}x \in V^2$, whence $b^{-1}a = (b^{-1}x)(x^{-1}a) \in V^4 \subset U$. This implies that $a \in bU$, thus contradicting the assumption that the set $A$ is $U$-disjoint. □

THEOREM 1.4.23. *Every discrete subgroup $H$ of a pseudocompact topological group $G$ is finite.*

PROOF. By Corollary 1.4.18, $H$ is closed in $G$. However, pseudocompactness is not inherited by closed subspaces, so we cannot conclude at this point that $H$ is pseudocompact and, therefore, finite. Thus, our argument has to be more delicate than that in the proof of Corollary 1.4.20.

Fix an open neighbourhood $U$ of the identity $e$ in $G$ such that $U \cap H = \{e\}$. Since $G$ is a topological group, there is a symmetric open neighbourhood $V$ of $e$ in $G$ such that $V^4 \subset U$. It follows from Lemma 1.4.22 that the family $\eta = \{hV : h \in H\}$ is discrete in $G$. Since $\eta$ is a family of non-empty open subsets of the pseudocompact space $G$, it follows that the index set $H$ in the definition of the family $\eta$ is finite. □

It is worth noting an interesting result which follows easily from Theorem 1.4.23:

COROLLARY 1.4.24. *Every infinite pseudocompact topological group $G$ contains a non-closed countable subset.*

PROOF. Take any infinite countable subset $A$ of $G$, and let $H$ be the subgroup of $G$ algebraically generated by $A$. Then $H$ is countable and infinite. Therefore, by Theorem 1.4.23, $H$ cannot be discrete. It follows that the subset $B = H \setminus \{e\}$ of $H$ is not closed in $H$ and, therefore, not closed in $G$. Clearly, $B$ is countable. □

Corollary 1.4.24 cannot be extended to infinite pseudocompact spaces (see Problem 1.4.F), while it is obviously true for all infinite countably compact spaces. A non-trivial generalization of Corollary 1.4.24 will be given in Theorem 3.7.27.

We now present a theorem on non-discrete topologizations of infinite Abelian groups. This important result is easily obtained on the basis of techniques developed in this section.

THEOREM 1.4.25. [**A. Kertész and T. Szele**] *Every infinite Abelian group $G$ admits a non-discrete Tychonoff topology under which it is a topological group.*

PROOF. By Corollary 1.1.8, for each $a \in G$ distinct from the identity $e$ of $G$, we can fix a homomorphism $f_a$ of $G$ to the topological group $\mathbb{T}$ such that $f_a(a) \neq 1$ (see item d) of Example 1.2.5). Let $f$ be the diagonal product of the family $\{f_a : a \in G,\ a \neq e\}$. Then $f$ is a one-to-one homomorphism of the group $G$ to the topological group $\mathbb{T}^G$, which is the product of $|G|$ copies of the group $\mathbb{T}$. Therefore, $G$ is algebraically isomorphic to the subgroup $H = f(G)$ of the group $\mathbb{T}^G$.

Now, with the product topology, $\mathbb{T}^G$ is a compact topological group. Since $H$ is an infinite subgroup of $\mathbb{T}^G$, the topological subgroup $H$ of $\mathbb{T}^G$ cannot be discrete, by Corollary 1.4.20. Denote by $\mathcal{P}$ the topology on $H$ inherited from $\mathbb{T}^G$. Then, since $f$ is an isomorphism of $G$ onto $H$, the family $\mathcal{T} = \{f^{-1}(V) : V \in \mathcal{P}\}$ is a topology on $G$ turning $G$ into a Tychonoff non-discrete topological group, topologically isomorphic to $H$. □

Let $G$ be a topological group with neutral element $e$. The *connected component* or, simply, the *component* of $G$ is the union of all connected subsets of $G$ containing $e$. Since the union of any family of connected subspaces containing a given point is connected, the connected component of $G$ can be described as the biggest connected subspace of $G$ containing $e$ (see [**165**, Section 6.1]). Even more is true:

PROPOSITION 1.4.26. *The connected component $H$ of any topological group $G$ is a closed invariant subgroup of $G$, that is, $aHa^{-1} = H$, for each $a \in G$.*

PROOF. Indeed, since the left multiplication by $a \in G$ is a homeomorphism of $G$ onto itself, $aH$ is homeomorphic to $H$. Hence, $aH$ is connected. Similarly, $aHa^{-1}$ is connected. Since $e \in aHa^{-1}$, and $H$ is the biggest connected subset of $G$ containing $e$, it follows that $aHa^{-1} \subset H$. Replacing $a$ by $a^{-1}$ in this inclusion, we obtain that $a^{-1}Ha \subset H$ or, equivalently, $H \subset aHa^{-1}$. Thus, $aHa^{-1} = H$. Since the closure of a connected subset is connected [**165**, Corollary 6.1.11] and $H$ is the biggest connected subset of $G$ containing $e$, it follows that $H$ is closed in $G$. □

Theorem 1.4.28 below can be used to find discrete invariant subgroups of a connected topological group — all such subgroups lie in the center of the group. Let us first prove a simple auxiliary result on generating connected groups.

LEMMA 1.4.27. *Let $U$ be an arbitrary open neighbourhood of the neutral element $e$ of a connected topological group $G$. Then $G = \bigcup_{n=1}^{\infty} U^n$.*

PROOF. Choose an open symmetric neighbourhood $V$ of $e$ in $G$ such that $V \subset U$. Then $H = \bigcup_{n=1}^{\infty} V^n$ is an open subgroup of $G$, and Theorem 1.3.5 implies that $H$ is closed in $G$. Since $G$ is connected, we must have $H = G$. As $V \subset U$, it follows that $G = \bigcup_{n=1}^{\infty} U^n$. □

THEOREM 1.4.28. *Let $K$ be a discrete invariant subgroup of a connected topological group $G$. Then every element of $K$ commutes with every element of $G$, that is, $K$ is contained in the center of the group $G$.*

PROOF. If $K = \{e\}$, there is nothing to prove. Suppose, therefore, that the subgroup $K$ is not trivial. Take an arbitrary element $x \in K$ distinct from the identity $e$ of $G$. Since the group $K$ is discrete, we can find an open neighbourhood $U$ of $x$ in $G$ such that $U \cap K = \{x\}$. It follows from the continuity of the multiplication in $G$ and the obvious equality $exe = x$ that there exists an open symmetric neighbourhood $V$ of $e$ in $G$ such that $VxV \subset U$. Let $y \in V$ be arbitrary. Since $K$ is an invariant subgroup of $G$, we have that $yxy^{-1} \in K$. It is also clear that $yxy^{-1} \in VxV^{-1} = VxV \subset U$. Therefore, $yxy^{-1} \in U \cap K = \{x\}$, that is, $yxy^{-1} = x$. This implies that $yx = xy$ for each $y \in V$.

Since the group $G$ is connected, Lemma 1.4.27 implies that the sets $V^n$, with $n \in \mathbb{N}$, cover the group $G$. Therefore, every element $g \in G$ can be written in the form $g = y_1 \cdots y_n$, where $y_1 \ldots, y_n \in V$ and $n \in \mathbb{N}$. Since $x$ commutes with every element of $V$, we have:

$$gx = y_1 \cdots y_n x = y_1 \ldots x y_n = \cdots = y_1 x \cdots y_n = x y_1 \cdots y_n = xg.$$

We have thus proved that the element $x \in K$ is in the center of the group $G$. Since $x$ is an arbitrary element of $K$, we conclude that the center of $G$ contains $K$. □

In conclusion of this section, we present a few important general facts concerning separation of compact subsets from closed sets in topological and paratopological groups.

THEOREM 1.4.29. *Let $G$ be a paratopological group, $F$ be a compact subset of $G$, and $P$ be a closed subset of $G$ such that $F \cap P = \varnothing$. Then there exists an open neighbourhood $V$ of the neutral element $e$ such that $FV \cap P = \varnothing$ and $VF \cap P = \varnothing$.*

PROOF. Since the left translations in $G$ are continuous, we can choose, for every $x \in F$, an open neighbourhood $V_x$ of the neutral element $e$ in $G$ such that $xV_x \cap P = \varnothing$. Using the joint continuity of the multiplication in $G$, we can also take an open neighbourhood $W_x$ of $e$ such that $W_x^2 \subset V_x$. The open sets $xW_x$, with $x \in F$, cover the compact set $F$, so there exists a finite set $C \subset F$ such that $F \subset \bigcup_{x\in C} xW_x$. Put $V_1 = \bigcap_{x\in C} W_x$. We claim that $FV_1 \cap P = \varnothing$. Indeed, it suffices to verify that $yV_1 \cap P = \varnothing$, for each $y \in F$. Given an element $y \in F$, we can find $x \in C$ such that $y \in xW_x$. Then

$$yV_1 \subset xW_xV_1 \subset xW_xW_x \subset xV_x \subset G \setminus P,$$

by our choice of the sets $V_x$ and $W_x$. This proves that the sets $FV_1$ and $P$ are disjoint.

Similarly, one can find an open neighbourhood $V_2$ of $e$ in $G$ satisfying $V_2F \cap P = \varnothing$. Then the set $V = V_1 \cap V_2$ is as required. □

The next result is closely related to Theorem 1.4.29, though this is not obvious immediately.

THEOREM 1.4.30. *Let $G$ be a topological group, $F$ a compact subset of $G$, and $P$ a closed subset of $G$. Then the sets $FP$ and $PF$ are closed in $G$.*

PROOF. We will show that $FP$ is closed in $G$. The case of $PF$ differs only in trivial details. Take any point $a \notin FP$. It follows that the sets $F^{-1}a$ and $P$ are disjoint. Clearly, the set $F^{-1}a$ is compact. Therefore, by Theorem 1.4.29, there is an open neighbourhood $U$ of $e$ such that $F^{-1}aU$ and $P$ are disjoint. It follows from Proposition 1.4.6 that the sets $aU$

and $FP$ are disjoint. Since $aU$ is an open neighbourhood of $a$, we conclude that $a$ is not in the closure of $FP$. Hence, $FP$ is closed in $G$. □

Easy examples show that the similar statement about the product of arbitrary two closed subsets of a topological group is not true — the product of two such sets may even be a proper dense subset of the group (see Exercise 1.4.h).

The above theorem is complemented by the following simple result:

PROPOSITION 1.4.31. *For any two compact subsets $E$ and $F$ of a paratopological group $G$, their product $EF$ in $G$ is a compact subspace of $G$.*

PROOF. Since multiplication in a paratopological group is jointly continuous, the subspace $EF$ of $G$ is a continuous image of the Cartesian product $E \times F$ of the spaces $E$ and $F$. Since $E \times F$ is compact, by Tychonoff's theorem, the space $EF$ is also compact. □

Another interesting property of compact sets in topological groups is presented in the next proposition.

PROPOSITION 1.4.32. *Let $B$ be a compact subset of a topological group $G$. Then, for every neighbourhood $U$ of the identity $e$ in $G$, there exists a neighbourhood $V$ of $e$ in $G$ such that $bVb^{-1} \subset U$, for each $b \in B$.*

PROOF. Let $U$ be a neighbourhood of $e$ in $G$. Choose an open symmetric neighbourhood $W$ of $e$ in $G$ such that $W^3 \subset U$. Since $B$ is compact, we can find a finite set $F \subset B$ such that $B \subset WF$. Put $V = \bigcap_{x \in F} x^{-1}Wx$. Then $V$ is an open neighbourhood of $e$ in $G$. If $b \in B$, then $b = wx$ for some $w \in W$ and $x \in F$. Therefore,

$$bVb^{-1} = wxVx^{-1}w^{-1} \subset wWw^{-1} \subset W^3 \subset U,$$

as required. □

A space $X$ is called *pathwise connected* if for any points $x, y \in X$, there exists a continuous mapping $f$ of the closed unit interval $I$ to $X$ such that $f(0) = x$ and $f(1) = y$ (see [**165**, 6.3.9]). Here is an important example of a compact pathwise connected topological group.

EXAMPLE 1.4.33. Given an $n \times n$ matrix $A = (a_{i,j})_{i,j=1}^n$ with complex entries $a_{i,j}$, we denote by $A^t = (a_{j,i})_{j,i=1}^n$ the *transpose* of $A$, and by $\overline{A} = (\overline{a}_{i,j})_{i,j=1}^n$ the *conjugate* of $A$. We define $A^*$ to be $\overline{A}^t$ which is evidently coincides with $\overline{A^t}$.

Let $U(n)$ be the set of all $n \times n$ matrices $A$ with complex entries satisfying $AA^* = E_n$, where $E_n$ is the identity $n \times n$ matrix. Each matrix $A \in U(n)$ is called *unitary*. Since the determinant Det of matrices is a multiplicative function, $\mathrm{Det}(BC) = \mathrm{Det}\, B \cdot \mathrm{Det}\, C$, we have that

$$1 = \mathrm{Det}\, E_n = \mathrm{Det}(AA^*) = \mathrm{Det}\, A \cdot \mathrm{Det}\, A^*. \tag{1.5}$$

Hence, every unitary matrix $A$ is invertible, i.e., $\mathrm{Det}\, A \neq 0$. In fact, from the equalities $\mathrm{Det}\, A^t = \mathrm{Det}\, A$, $|\mathrm{Det}\, \overline{A}| = |\mathrm{Det}\, A|$, and (1.5) it follows that $|\mathrm{Det}\, A| = 1$, for every $A \in U(n)$. Furthermore, if $A$ is unitary, then $A^{-1} = A^*$ and $A^*A = E_n$. This implies that the matrix $A^{-1}$ is unitary as well. If, in addition, $A, B$ are unitary, then $(AB)(AB)^* = ABB^*A^* = AA^* = E_n$, and we conclude that $U(n)$ is a subgroup of the general linear group $GL(n, \mathbb{C})$ (see item e) of Example 1.2.5) which is called the *unitary group* of degree $n$ over $\mathbb{C}$.

We consider $U(n)$ with the topology inherited from the group $GL(n, \mathbb{C})$. Then $U(n)$ is a topological group. Let us verify that $U(n)$ is compact. If $A \in U(n)$, then $AA^* = E_n$, so that the rows of $A$ form an orthonormal basis of the complex space $\mathbb{C}^n$. In particular, the entries $a_{i,j}$ of $A$ satisfy $|a_{i,j}| \leq 1$ for all $i, j \leq n$. It follows that $U(n)$, considered as a space, can be identified with a subspace of $E^{n^2} \subset \mathbb{C}^{n^2}$, where $E = \{z \in \mathbb{C} : |z| \leq 1\}$. Evidently, the equality $AA^* = E_n$ is equivalent to a system of $n^2$ scalar equations of order 2, while the latter defines a closed subspace of the spaces $\mathbb{C}^{n^2}$ and $E^{n^2}$. Therefore, $U(n)$ is homeomorphic to a closed subspace of the compact space $E^{n^2}$, whence the required conclusion follows.

It remains to show that the topological group $U(n)$ is pathwise connected. It follows from [**288**, Theorem 2.10.2] that every unitary matrix $A \in U(n)$ is *unitary equivalent* to a diagonal matrix, that is, there exist $n \times n$ matrices $U$ and $D$ such that $UAU^* = D$, where $U$ is unitary and $D$ is diagonal. Without loss of generality we can assume that $A \neq E_n$. Since $A, U \in U(n)$, it follows that $D \in U(n)$ and $\mathrm{Det}\, D = 1$. Therefore, the diagonal elements $d_{k,k}$ of $D$ satisfy $|d_{k,k}| = 1$, and we can write $d_{k,k} = e^{2\pi i \lambda_k}$, where $0 \leq \lambda_k < 1$ for each $k = 1, \ldots, n$. For every real number $t$ and $k \leq n$, put $d_{k,k}(t) = 2^{2\pi i t \lambda_k}$ and consider the diagonal matrix $D(t)$ with diagonal elements $d_{k,k}(t)$, $1 \leq k \leq n$. Clearly, each matrix $D(t)$ is unitary, $D(0) = E_n$, and $D(1) = D$. It is also clear that the mapping $t \mapsto D(t)$ of $I = [0, 1]$ to $U(n)$ is continuous. Finally, put $\varphi(t) = UD(t)U^*$, for each $t \in I$. Then $\varphi(t) \in U(n)$ for each $t \in I$, the mapping $\varphi$ is a continuous bijection (hence, a topological embedding), $\varphi(0) = UE_nU^* = E_n$, and $\varphi(1) = UDU^* = A$. We have thus proved that the neutral element $E_n$ of $U(n)$ can be connected by a copy of the unit interval $I$ with an arbitrary element $A \neq E_n$ of $U(n)$. This implies immediately that the space $U(n)$ is pathwise connected.

Obviously, the group $U(n)$ is connected. In fact, it is locally connected and even locally pathwise connected (see Problem 1.4.K). □

## Exercises

1.4.a. Prove the claim in Example 1.4.3 left unproved.

1.4.b. Answer the following questions:

  (a) Is it true that for any compact subsets $E$ and $F$ of a semitopological group $G$, their product $EF$ in $G$ is a compact subspace of $G$?

  (b) Is the analogous statement true for quasitopological groups?

1.4.c. Can Corollaries 1.4.20, 1.4.21, and Theorem 1.4.23 be generalized to semitopological groups (with continuous inverse)? To paratopological groups?

1.4.d. Show that Theorem 1.4.29 is no longer valid for regular quasitopological groups.

*Hint.* For every $\varepsilon > 0$, let $U_\varepsilon = \{(0,0)\} \cup (0,\varepsilon)^2 \cup (-\varepsilon, 0)^2$. Then the family $\mathcal{B} = \{B_\varepsilon : \varepsilon > 0\}$ of symmetric sets in the plane forms a base for a regular quasitopological group topology $\mathcal{T}$ at the neutral element $(0,0)$ of the additive group $\mathbb{R}^2$. Verify that $F = \{(x,x) : |x| \leq 1\}$ is a compact subset of the regular quasitopological group $G = (\mathbb{R}^2, \mathcal{T})$, the set $P = (\mathbb{R} \times \{0\}) \setminus \{(0,0)\}$ is closed in $G$ and disjoint from $P$, but $F + U_\varepsilon$ intersects $P$, for each $\varepsilon > 0$.

1.4.e. Show that Theorem 1.4.30 cannot be extended to paratopological groups.

1.4.f. Show that every regular paratopological group algebraically generated by a countable family of separable metrizable symmetric subspaces has a countable network.

1.4.g. Generalize Theorem 1.4.29 as follows. If $F$ and $P$ are disjoint closed subsets of a topological group $G$ and $F$ is compact, then there exists an open neighbourhood $V$ of the identity in $G$ such that $VFV \cap P = \emptyset$.

1.4.h. Give an example of a topological Abelian group $G$ and closed subsets $A$ and $B$ of $G$ such that the product $AB$ is a proper dense subset of $G$. Show that one can choose $A$ and $B$ to be closed subgroups.

1.4.i. Let $G$ be a compact topological Abelian group. Show that if $G$ is a torsion group, then there exists a positive integer $m$ such that $x^m = e$, for each $x \in G$. Extend the conclusion to pseudocompact topological Abelian groups.

1.4.j. If $H$ is a dense subgroup of a connected topological group, then every neighbourhood $U$ of the identity element in $H$ algebraically generates the group $H$ (this generalizes Lemma 1.4.27.)

1.4.k. Suppose that $m$ and $n$ are integers with $2 \leq m < n$. Show that no dense subgroup of a connected topological group is isomorphic to the group $\mathbb{Z}(m)^{(\omega)} \oplus \mathbb{Z}(n)$, the direct sum of $\omega$ copies of the cyclic group $\mathbb{Z}(m)$ and of the cyclic group $\mathbb{Z}(n)$.

1.4.l. A space $X$ is called *resolvable* if there exist dense disjoint subsets $A$ and $B$ of $X$. Verify that the following statements are valid:
   (a) The groups $\mathbb{R}$, $\mathbb{Q}$, $\mathbb{T}$, and $\mathbb{C}$, with their usual topologies, are resolvable.
   (b) If a subgroup $H$ of a topological group $G$ is resolvable, then so is $G$.
   (c) If a topological group $G$ contains a proper dense subgroup, then $G$ is resolvable.
   (d) If a topological group $G$ contains a non-closed subgroup, then $G$ is resolvable.
   (e) Suppose that $\mathcal{T}$ is a non-discrete quasitopological group topology on the group of integers $\mathbb{Z}$. Show that the space $(\mathbb{Z}, \mathcal{T})$ is resolvable.

## Problems

1.4.A. Let $G$ be a topological group and $A$, $B$ (closed) subgroups of $G$. Let, further, $H$ be the closure of the product $AB$ in $G$.
   (a) Is the space $H$ necessarily homogeneous?
   (b) Show that if $A$ or $B$ is compact, then $H$ is homogeneous.
   (c) Show that if $G$ is Abelian, then $H$ is homogeneous.

1.4.B. Let $G$ be a topological group with neutral element $e$. Show that the space $X = G \setminus \{e\}$ can fail to be homogeneous.

*Hint.* Consider the product of the group of reals with the group $\mathbb{Z}^\omega$, both endowed with their usual topologies.

1.4.C. Let $G$ be a zero-dimensional topological group with neutral element $e$. Show that the space $X = G \setminus \{e\}$ is homogeneous.

*Hint.* Use Ford's lemma in [**173**] on strong local homogeneity.

1.4.D. Give an example of a homogeneous Tychonoff space $X$ that does not have a closed separable homogeneous subspace. Show that $X$ cannot be a paratopological group. Can such a space $X$ be a semitopological group?

1.4.E. Let $G$ be a pseudocompact topological group with the property that all proper subgroups of $G$ are finite. Prove that $G$ is finite. (See also Exercise 1.1.f.)

1.4.F. Give an example of an infinite pseudocompact Tychonoff space $X$ such that all countable subsets of $X$ are closed (compare this with Corollary 1.4.24).

*Hint.* Such a space is constructed by E. A. Reznichenko in [**403**].

1.4.G. Does the assertion of Exercise 1.4.i remain valid for compact semitopological semigroups? For compact semitopological monoids?

1.4.H. Prove that every closed non-discrete subgroup of the $n$-dimensional Euclidean vector space $\mathbb{R}^n$ (considered as an Abelian topological group) contains a straight line that passes through the origin, where $n \in \mathbb{N}$.

1.4.I. Does the Euclidean plane $\mathbb{R}^2$ contain a proper dense connected and locally connected subgroup?

1.4.J. Let $G$ be an Abelian group with $|G| \leq 2^\omega$. Prove that $G$ is (algebraically) isomorphic to a dense subgroup of the group $\mathbb{T}^\omega$ if either $G$ has an element of infinite order or $G$ is a torsion group of unbounded period. The circle group $\mathbb{T}$ carries the usual compact group topology.

1.4.K. Prove the following:
(a) Each of the topological groups $GL(n, \mathbb{R})$, $SL(n, \mathbb{R})$, $TS(n, \mathbb{R})$, $O(n, \mathbb{R})$, and $U(n)$ (see Problem 1.2.D and Example 1.4.33) has a base of pathwise connected open sets.
(b) The groups $SL(n, \mathbb{R})$ and $TS(n, \mathbb{R})$ are pathwise connected, while the groups $GL(n, \mathbb{R})$ and $O(n, \mathbb{R})$ are disconnected.

### Open Problems

1.4.1. Let $G$ be a zero-dimensional topological group with neutral element $e$. Must the space $X = G \setminus \{e\}$ be homeomorphic to a topological group? (See Problems 1.4.B and 1.4.C.)

1.4.2. Is it true that every infinite "punctured" topological group, that is, a topological group without the neutral element, has an infinite closed separable homogeneous subspace? (See Problem 1.4.D.)

1.4.3. Does there exist an infinite (Abelian, Boolean) topological group $G$ such that all dense subgroups of $G$ are connected?

1.4.4. Does there exist an infinite (Abelian) group $G$ such that all closed subgroups of $G$ are connected?

## 1.5. Quotients of topological groups

One of the main operations on topological groups is that of taking quotient groups. Many non-trivial examples and counterexamples arise as quotients of relatively simple and well-known topological groups. This operation has been the subject of an intensive and thorough study; but there exists still a wealth of interesting open problems related to the behaviour of different topological and algebraic properties under taking quotients.

We start with a general statement about quotients of left topological groups.

THEOREM 1.5.1. *Suppose that $G$ is a left topological group with identity $e$ and a topology $\mathcal{T}$, and $H$ is a closed subgroup of $G$. Denote by $G/H$ the set of all left cosets $aH$ of $H$ in $G$, and endow it with the quotient topology with respect to the canonical mapping $\pi\colon G \to G/H$ defined by $\pi(a) = aH$, for each $a \in G$. Then the family $\{\pi(xU) : U \in \mathcal{T},\ e \in U\}$ is a local base of the space $G/H$ at the point $xH \in G/H$, the mapping $\pi$ is open, and $G/H$ is a homogeneous $T_1$-space.*

PROOF. Clearly, the set $xUH$ is the union of a family of left cosets $yH$, with $y \in xH$. Therefore, $\pi^{-1}\pi(xUH) = xUH$. Since the set $xUH$ is open in $G$ and the mapping $\pi$ is quotient, it follows that $\pi(xUH)$ is open in $G/H$.

Now take any open neighbourhood $W$ of $xH$ in $G/H$ and put $O = \pi^{-1}(W)$. The set $O$ is open, since $\pi$ is continuous. Clearly, $x \in O$. There is an open neighbourhood $U$ of $e$ in $G$ such that $xU \in O$. Then $\pi(xU) \subset W$ and, hence, $\pi^{-1}\pi(xU) \subset O$. Since $xUH = \pi^{-1}\pi(xU)$, it follows that $\pi(xUH) \subset W$. Thus, the first two statements of the theorem are proved.

Let us now prove the homogeneity of $G/H$. For any $a \in G$, define a mapping $h_a$ of $G/H$ to itself by the rule $h_a(xH) = axH$. Since $axH \in G/H$, this definition is correct. Since $G$ is a group, the mapping $h_a$ is evidently a bijection of $G/H$ onto $G/H$. In fact, $h_a$ is a homeomorphism. This can be seen from the following argument.

Take any $xH \in G/H$ and any open neighbourhood $U$ of $e$. Then $\pi(xUH)$ is a basic neighbourhood of $xH$ in $G/H$. Similarly, the set $\pi(axUH)$ is a basic neighbourhood of $axH$ in $G/H$. Since, obviously, $h_a(\pi(xUH)) = \pi(axUH)$, it easily follows that $h_a$ is a homeomorphism. It is also clear that $h_a(xH) = axH$. Now, for any given $xH$ and $yH$ in $G/H$, we can take $a = yx^{-1}$. Then $h_a(xH) = yH$. Hence, the quotient space $G/H$ is homogeneous. It is a $T_1$-space, since all right cosets $xH$ are closed in $G$ and the mapping $\pi$ is quotient. □

Let $G$ be a left topological group and $H$ a closed subgroup of $G$. Then the space $G/H$ defined in Theorem 1.5.1 is called the *left coset space* of $G$ with respect to $H$. Similarly, if $G$ is a right topological group, one can define the *right coset space* $H\backslash G = \{Hx : x \in G\}$ whose topology is determined by the requirement that the canonical mapping $p\colon G \to H\backslash G$, where $p(x) = Hx$ for each $x \in G$, be quotient. As in Theorem 1.5.1, the mapping $p$ turns out to be open and the sets $p(Ux)$ constitute a local base for $H\backslash G$ at the point $Hx$, where $U$ runs over open neighbourhoods of the neutral element in $G$.

In the proof of Theorem 1.5.1 we have defined, for each $a \in G$, a mapping $h_a$ of $G/H$ onto itself by the rule $h_a(xH) = axH$. It was shown that $h_a$ is a homeomorphism of the space $G/H$ onto itself. This homeomorphism is called the *left translation of $G/H$ by $a$*. In fact, we can say a little more. Here is an obvious statement which is sometimes quite useful.

PROPOSITION 1.5.2. *Suppose that $G$ is a topological group, $H$ is a closed subgroup of $G$, $\pi$ is the natural quotient mapping of $G$ onto the left quotient space $G/H$, $a \in G$, $\lambda_a$ is the left translation of $G$ by $a$ (that is, $\lambda_a(x) = ax$, for each $x \in G$), and $h_a$ is the left translation of $G/H$ by $a$ (that is, $h_a(xH) = axH$, for each $xH \in G/H$). Then $\lambda_a$ and $h_a$ are homeomorphisms of $G$ and $G/H$, respectively, and $\pi \circ \lambda_a = h_a \circ \pi$.*

If $G$ is a left topological group and $H$ is a closed invariant subgroup of $G$, then each left coset of $H$ in $G$ is also a right coset of $H$ in $G$, and a natural multiplication of cosets in $G/H$ is defined by the rule $xHyH = xyH$, for all $x, y \in G$. It is well known that this operation turns $G/H$ into a group called the *quotient group* of $G$ with respect to $H$. After these definitions and the proof of Theorem 1.5.1, the next statement is obvious.

THEOREM 1.5.3. *Suppose that $G$ is a left topological group, and that $H$ is a closed invariant subgroup of $G$. Then $G/H$ with the quotient topology and multiplication is a left topological group, and the canonical mapping $\pi\colon G \to G/H$ is an open continuous homomorphism. If $G$ is a topological (semitopological) group, then $G/H$ is a topological (semitopological) group.*

The next statement is also obvious.

PROPOSITION 1.5.4. *Suppose that $G$ is a left topological group and $H$ is a closed subgroup of $G$. Then $G/H$ with the quotient topology is discrete if and only if $H$ is open in $G$.*

Quotient spaces of topological groups have nice separation properties. To see this, we establish first a useful property of the closure operator in quotient spaces.

LEMMA 1.5.5. *Suppose that $G$ is a topological group, $H$ is a closed subgroup of $G$, $\pi$ is the natural quotient mapping of $G$ onto the quotient space $G/H$, and let $U$ and $V$ be open neighbourhoods of the neutral element $e$ in $G$ such that $V^{-1}V \subset U$. Then $\overline{\pi(V)} \subset \pi(U)$.*

PROOF. Take any $x \in G$ such that $\pi(x) \in \overline{\pi(V)}$. Since $Vx$ is an open neighbourhood of $x$ and the mapping $\pi$ is open, $\pi(Vx)$ is an open neighbourhood of $\pi(x)$. Therefore, $\pi(Vx) \cap \pi(V) \neq \varnothing$. It follows that, for some $a \in V$ and $b \in V$, we have $\pi(ax) = \pi(b)$, that is, $ax = bh$, for some $h \in H$. Hence, $x = (a^{-1}b)h \in UH$, since $a^{-1}b \in V^{-1}V \subset U$. Therefore, $\pi(x) \in \pi(UH) = \pi(U)$. □

THEOREM 1.5.6. *For any topological group $G$ and any closed subgroup $H$ of $G$, the quotient space $G/H$ is regular.*

PROOF. Let $\pi$ be the natural quotient mapping of $G$ onto the quotient space $G/H$, and let $W$ be an arbitrary open neighbourhood of $\pi(e)$ in $G/H$, where $e$ is the neutral element of $G$. By the continuity of $\pi$, we can find an open neighbourhood $U$ of $e$ in $G$ such that $\pi(U) \subset W$. Since $G$ is a topological group, we can choose an open neighbourhood $V$ of $e$ such that $V^{-1}V \subset U$. Then, by Lemma 1.5.5, $\overline{\pi(V)} \subset \pi(U) \subset W$. Since $\pi(V)$ is an open neighbourhood of $\pi(e)$, the regularity of $G/H$ at the point $\pi(e)$ is verified. Now it follows from the homogeneity of $G/H$ that the space $G/H$ is regular. □

In the case of topological spaces, quotient mappings with compact fibers need not be closed. It is remarkable that for topological groups the situation is quite different. Let us recall that closed continuous mappings with compact preimages of points are called *perfect* (see [**165**, Section 3.7]).

THEOREM 1.5.7. *Suppose that $G$ is a topological group and that $H$ is a compact subgroup of $G$. Then the quotient mapping $\pi$ of $G$ onto the quotient space $G/H$ is perfect.*

PROOF. Take any closed subset $P$ of $G$. Then, by Theorem 1.4.30, $PH$ is closed in $G$. However, $PH$ is, obviously, the union of a certain family of left cosets, that is, $PH = \pi^{-1}\pi(P)$. It follows, by the definition of a quotient mapping, that the set $\pi(P)$ is closed in the quotient space $G/H$. Thus $\pi$ is a closed mapping. In addition, if $y \in G/H$ and $\pi(x) = y$ for some $x \in G$, then $\pi^{-1}(y) = xH$ is a compact subset of $G$. Hence the fibers of $\pi$ are compact and $\pi$ is perfect. □

If $f: X \to Y$ is a perfect mapping then, for any compact subset $F$ of $Y$, the preimage $f^{-1}(F)$ is a compact subset of $X$ [**165**, Theorem 3.7.2]. Therefore, the next statement follows from Theorem 1.5.7.

COROLLARY 1.5.8. *Suppose that $G$ is a topological group and $H$ is a compact subgroup of $G$ such that the quotient space $G/H$ is compact. Then $G$ is also compact.*

The next example shows, in particular, that Corollary 1.5.8 cannot be extended to quasitopological groups.

EXAMPLE 1.5.9. Suppose that $G$ is an infinite compact topological group with neutral element $e$. Take the square $G \times G$ with the usual product operation and the cross topology $\mathcal{T}_{cr}$, that is, the strongest topology on $G \times G$ which generates the usual topology on each "vertical" subspace $\{a\} \times G$ and each "horizontal" subspace $G \times \{a\}$, for every $a \in G$ (see Example 1.2.6). From this definition it follows directly that the product group $G \times G$ with the cross topology is a quasitopological group. The natural projection $p_1$ of $G \times G$ onto the first factor $G$ is, obviously, an open continuous homomorphism with compact fibers (homeomorphic to the original compact group $G$). It follows (see Theorem 1.5.13 below) that $p_1$ can be identified with the quotient mapping $\pi\colon G \times G \to (G \times G)/N$, where $N = \{e\} \times G$ is a closed invariant subgroup of $(G \times G, \mathcal{T}_{cr})$. However, the mapping $p_1$ (and, therefore, $\pi$) is not closed. Indeed, the set $A = \{(a, a) : a \in G,\ a \neq e\}$ is evidently closed in the space $(G \times G, \mathcal{T}_{cr})$ while $p_1(A) = G \setminus \{e\}$ is not closed in the space $G$, since $G$ is not discrete. Thus, Theorem 1.5.7 does not extend to arbitrary left topological groups. The space $(G \times G, \mathcal{T}_{cr})$ is not compact — otherwise the continuous mapping $p_1$ would be closed, since $G$ is Hausdorff. Hence, Corollary 1.5.8 is no longer valid for quasitopological groups. □

Given a homomorphism $f\colon G \to H$ of abstract groups $G$ and $H$, we denote by $\ker f$ the *kernel* of $f$, that is, the preimage of the identity $e_H$ of the group $H$ under the mapping $f$. It is easy to see that the kernel of $f$ is an invariant subgroup of $G$. Indeed, if $a, x \in G$ and $f(x) = e_H$, then $f(axa^{-1}) = f(a)f(x)f(a^{-1}) = f(a)e_H f(a)^{-1} = e_H$. In other words, if $N = \ker f$, then $aNa^{-1} \subset N$ for each $a \in G$. Hence, the subgroup $N$ of $G$ is invariant.

In the following proposition, we compare two homomorphisms of a given group according to the size of their kernels.

PROPOSITION 1.5.10. *Suppose that $G$, $H$, and $K$ are abstract groups and that $\varphi\colon G \to H$ and $\psi\colon G \to K$ are homomorphisms such that $\psi(G) = K$ and $\ker\psi \subset \ker\varphi$. Then there exists a homomorphism $f\colon K \to H$ such that $\varphi = f \circ \psi$. If in addition, $G$, $H$, $K$ are left topological groups, $\varphi$ and $\psi$ are continuous, and for each neighbourhood $U$ of the identity $e_H$ in $H$ there exists a neighbourhood $V$ of the identity $e_K$ in $K$ such that $\psi^{-1}(V) \subset \varphi^{-1}(U)$, then $f$ is continuous.*

PROOF. The algebraic part of the proposition is well known. Let us verify the continuity of $f$ in the second part of the proposition. Suppose that $U$ is a neighbourhood of $e_H$ in $H$. By our assumption, there exists a neighbourhood $V$ of $e_K$ in $K$ such that $W = \psi^{-1}(V) \subset \varphi^{-1}(U)$. Then $f(V) = \varphi(W) \subset U$, that is, $f$ is continuous at the identity of $K$. By Proposition 1.3.4, $f$ is continuous. □

COROLLARY 1.5.11. *Let $\varphi\colon G \to H$ and $\psi\colon G \to K$ be continuous homomorphisms of left topological groups $G$, $H$, and $K$ such that $\psi(G) = K$ and $\ker\psi \subset \ker\varphi$. If the homomorphism $\psi$ is open, then there exists a continuous homomorphism $f\colon K \to H$ such that $\varphi = f \circ \psi$.*

PROOF. The existence of a homomorphism $f\colon K \to H$ satisfying $\varphi = f \circ \psi$ follows from Proposition 1.5.10. To show that $f$ is continuous, take an arbitrary open set $V$ in $H$. Then $f^{-1}(V) = \psi(\varphi^{-1}(V))$. Since $\varphi$ is continuous and $\psi$ is open, we conclude that the set $f^{-1}(V)$ is open in $K$. Therefore, $f$ is continuous. □

PROPOSITION 1.5.12. *Let $G$ and $H$ be left topological groups and $p$ be a topological isomorphism of $G$ onto $H$. If $G_0$ is a closed invariant subgroup of $G$ and $H_0 = p(G_0)$, then the quotient groups $G/G_0$ and $H/H_0$ are topologically isomorphic. The corresponding isomorphism $\Phi\colon G/G_0 \to H/H_0$ is given by the formula $\Phi(xG_0) = yH_0$, where $x \in G$ and $y = p(x)$.*

Proof. Let $\varphi\colon G \to G/G_0$ and $\psi\colon H \to H/H_0$ be the quotient homomorphisms. An easy verification shows that $\Phi$ is a homomorphism of $G/G_0$ onto $H/H_0$. From the definition of $\Phi$ it follows that $\psi \circ p = \Phi \circ \varphi$.

$$\begin{array}{ccc} G & \xrightarrow{\ p\ } & H \\ {\scriptstyle\varphi}\downarrow & & \downarrow{\scriptstyle\psi} \\ G/G_0 & \xrightarrow{\ \Phi\ } & H/H_0 \end{array}$$

Since $p$, $\varphi$, and $\psi$ are open continuous homomorphisms, so is $\Phi$. It remains to verify that $\Phi$ is an isomorphism. Let $xG_0$ be an arbitrary element of $G/G_0$. Set $y = p(x)$. If $\Phi(xG_0) = H_0$ then $\psi(y) = H_0$, so that $y \in H_0$ and $x \in G_0$. This shows that the kernel of $\Phi$ is trivial, that is, $\Phi$ is an isomorphism. We conclude, therefore, that $\Phi$ is a topological isomorphism. □

The next result is known as the *first isomorphism theorem.*

Theorem 1.5.13. *Let $G$ and $H$ be left topological groups with neutral elements $e_G$ and $e_H$, respectively, and let $p$ be an open continuous homomorphism of $G$ onto $H$. Then the kernel $N = p^{-1}(e_H)$ of $p$ is a closed invariant subgroup of $G$, and the fibers $p^{-1}(y)$ with $y \in H$ coincide with the cosets of $N$ in $G$. The mapping $\Phi\colon G/N \to H$ which assigns to a coset $xN$ the element $p(x) \in H$ is a topological isomorphism.*

Proof. The assertions of the theorem about the kernel $N$ and the fibers $p^{-1}(y)$ are well known, so that $\Phi$ is correctly defined, and all we have to verify is that $\Phi$ is a topological isomorphism.

From the definition of the multiplication in the quotient group $G/N$ given before Theorem 1.5.3 it follows that $\Phi$ is a homomorphism. Denote by $\pi$ the quotient homomorphism of $G$ onto $G/N$. Again, the definition of $\Phi$ implies that $p = \Phi \circ \pi$ and, since $\pi$ is open, the homomorphism $\Phi$ is continuous. In addition, if $x \in G$ and $\Phi(xN) = e_H$, then $p(x) = \Phi(\pi(x)) = e_H$. Hence $x \in N$ and $xN = N$, that is, $\Phi$ is an isomorphism. Finally, if $U$ is an open set in $G/N$, then the image $\Phi(U) = p(\pi^{-1}(U))$ is open in $H$. Thus, $\Phi$ is an open continuous isomorphism and, hence, a homeomorphism. □

We present, in connection with Theorem 1.5.13, the following two sufficient conditions for a continuous homomorphism of (left) topological groups to be open:

Proposition 1.5.14. *Let $G$ and $H$ be left topological groups with neutral elements $e_G$ and $e_H$, respectively, and let $p$ be a continuous homomorphism of $G$ onto $H$ such that, for some non-empty subset $U$ of $G$, the set $p(U)$ is open in $H$ and the restriction of $p$ to $U$ is an open mapping of $U$ onto $p(U)$. Then the homomorphism $p$ is open.*

Proof. It suffices to show that if $x \in W$, where $W$ is an open neighbourhood of $x$ in $G$, then $p(W)$ is a neighbourhood of $p(x)$ in $H$. Fix a point $y$ in $U$, and let $l$ be the left translation of $G$ by $yx^{-1}$. Then $l$ is a homeomorphism of $G$ onto itself such that $l(x) = y$, so $V = U \cap l(W)$ is an open neighbourhood of $y$ in $U$. Then $p(V)$ is an open subset of $H$. Consider now the left translation $h$ of $H$ by the inverse to $p(yx^{-1})$, that is, by $p(xy^{-1})$. Clearly, $h \circ p \circ l = p$. Hence, $h(p(l(W))) = p(W)$. However, $h$ is a homeomorphism of $H$ onto itself. Since we already know that $p(V)$ is open in $H$ it follows that $h(p(V))$ is also open in $H$, that is, $p(W)$ contains the open neighbourhood $h(p(V))$ of $p(x)$ in $H$. Hence, $p(W)$ is a neighbourhood of $p(x)$ in $H$. □

The set $U$ in Proposition 1.5.14 need not be open in $G$, which makes the result fairly general.

PROPOSITION 1.5.15. *Let $p\colon G \to H$ be a continuous homomorphism of topological groups. Suppose that the image $p(U)$ contains a non-empty open set in $H$, for each open neighbourhood $U$ of the neutral element $e_G$ in $G$. Then the homomorphism $p$ is open.*

PROOF. First, we claim that the neutral element $e_H$ of $H$ is in the interior of $p(U)$, for each open neighbourhood $U$ of $e_G$ in $G$. Indeed, choose an open neighbourhood $V$ of $e_G$ such that $V^{-1}V \subset U$. By our assumption, $p(V)$ contains a non-empty open set $W$ in $H$. Then $W^{-1}W$ is an open neighbourhood of $e_H$, and we have that $W^{-1}W \subset p(V)^{-1}p(V) = p(V^{-1}V) \subset p(U)$.

Choose an arbitrary element $y \in p(U)$, where $U$ is an arbitrary non-empty open set in $G$. We can find $x \in U$ with $p(x) = y$ and an open neighbourhood $V$ of $e_G$ in $G$ such that $xV \subset U$. Let $W$ be an open neighbourhood of $e_H$ with $W \subset p(V)$. Then the set $yW$ contains $y$, it is open in $H$, and $yW \subset p(xV) \subset p(U)$. This implies that $p(U)$ is open in $H$. □

In the next theorem we consider conditions under which the restriction to a dense subgroup of the canonical mapping of a topological group $G$ onto a quotient space $G/H$ remains open.

THEOREM 1.5.16. *Let $G$ be a topological group, $H$ a closed subgroup of $G$, and $\pi\colon G \to G/H$ be the canonical mapping. If $K$ is a dense subgroup of $G$, then the restriction $r = \pi\restriction K$ is an open mapping of $K$ onto $\pi(K)$ if and only if the intersection $K \cap H$ is dense in $H$.*

PROOF. Suppose that $K \cap H$ is dense in $H$. Take an arbitrary non-empty open set $U$ in $K$. Then $U = V \cap K$, for some open set $V$ in $G$. Since the mapping $\pi$ is open, the set $O = \pi(V) \cap \pi(K)$ is open in $\pi(K)$, and we claim that $r(U) = O$. Indeed, if $y \in O$, take a point $x \in K$ with $\pi(x) = y$. Then $xH \cap V = \pi^{-1}(y) \cap V \neq \varnothing$. Since $K \cap H$ is dense in $H$, the set $x(K \cap H) = K \cap xH$ is dense in $xH$. Hence $(K \cap xH) \cap V \neq \varnothing$, so we can pick a point $x' \in K \cap xH \cap V$. Then $x' \in U$ and $r(x') = \pi(x') = \pi(x) = y$, which implies that $r(U) = O$. Therefore, the mapping $r\colon K \to r(K)$ is open.

Conversely, suppose that $r$ is open as a mapping of $K$ onto $r(K) = \pi(K)$. Let $U$ be a neighbourhood of the neutral element $e$ in $G$. We claim that $H \subset U(K \cap H)$. Indeed, take an open symmetric neighbourhood $V$ of $e$ in $G$ such that $V^2 \subset U$. Then $r(V \cap K) = \pi(V \cap K)$ is an open neighbourhood of the point $\bar{e} = \pi(e)$ in $\pi(K)$, so there exists an open set $W$ in $G/H$ such that $\bar{e} \in W \cap \pi(K) \subset \pi(V \cap K)$. This implies that

$$\pi^{-1}(W) \cap KH \subset (V \cap K)H. \tag{1.6}$$

Take an arbitrary element $h \in H$. By the density of $K$ in $G$, there exists $g \in K$ such that $gh^{-1} \in V \cap \pi^{-1}(W)$. Then $gh^{-1} \in \pi^{-1}(W) \cap KH$ and, by (1.6), we can choose $x \in V \cap K$ and $y \in H$ such that $gh^{-1} = xy$. It follows that $x^{-1}g = yh \in K \cap H$ and $y^{-1} = (hg^{-1})x \in V^{-1}V \subset U$. So $h = y^{-1}(x^{-1}g) \in U(K \cap H)$ and, therefore, $H \subset U(K \cap H)$. This proves our claim.

Since the latter inclusion holds for every neighbourhood $U$ of $e$ in $G$, the set $K \cap H$ must be dense in $H$. Indeed, otherwise we can find an element $h \in H$ and a symmetric neighbourhood $U$ of $e$ in $G$ such that $Uh \cap (K \cap H) = \varnothing$, whence $h \in H \setminus U(K \cap H) \neq \varnothing$, a contradiction. □

For certain pairs of left topological groups $G$ and $H$, every continuous homomorphism of $G$ onto $H$ is open. Let us show that this is true for compact groups $G$ and $H$.

PROPOSITION 1.5.17. *Let $f\colon G \to H$ be a continuous onto homomorphism of left topological groups. If $G$ is compact and $H$ is Hausdorff, then $f$ is open.*

PROOF. By [**165**, Theorem 3.1.12], the mapping $f$ is closed and, hence, it is quotient. Let $K$ be the kernel of $f$. If $U$ is open in $G$, then $f^{-1}(f(U)) = KU$ is open in $G$, by Proposition 1.4.1. Since $f$ is quotient, it follows that the image $f(U)$ is open in $H$. Therefore, $f$ is an open mapping. □

Evidently, Proposition 1.5.17 remains valid for right topological groups.

The next result is known as the *second isomorphism theorem.* It is useful in situations when iterated quotient groups are involved.

THEOREM 1.5.18. *Let $G$ and $H$ be left topological groups with neutral elements $e_G$ and $e_H$, respectively, and let $p\colon G \to H$ be an open continuous homomorphism of $G$ onto $H$. Let $H_0$ be a closed invariant subgroup of $H$, $G_0 = p^{-1}(H_0)$, and $N = p^{-1}(e_H)$. Then the left topological groups $G/G_0$, $H/H_0$, and $(G/N)/(G_0/N)$ are topologically isomorphic.*

PROOF. Denote by $\varphi$ the quotient homomorphism of $H$ onto $H/H_0$. By Theorem 1.5.1, $\varphi$ is open, so the composition $\varphi \circ p$ is an open continuous homomorphism of $G$ onto $H/H_0$ with kernel $G_0 = p^{-1}(H_0)$. Hence the quotient group $G/G_0$ is topologically isomorphic to $H/H_0$, by Theorem 1.5.13. It is clear that $G_0$ is a closed invariant subgroup of $G$. Observe that the mapping $\Phi$ assigning to every coset $xN$ of $N$ in $G$ the element $p(x) \in H$ is a topological isomorphism of $G/N$ onto $H$, by Theorem 1.5.13, and $\Phi(G_0/N) = H_0$. Therefore, applying Proposition 1.5.12, we conclude that the group $(G/N)/(G_0/N)$ is topologically isomorphic to $H/H_0$. □

Let us prove one more theorem on quotients of topological groups and topological isomorphisms known as the *third isomorphism theorem.*

THEOREM 1.5.19. *Suppose that $G$ is a topological group, $H$ is a closed invariant subgroup of $G$, and $M$ is any topological subgroup of $G$. Then the quotient group $MH/H$ is topologically isomorphic to the subgroup $\pi(M)$ of the topological group $G/H$, where $\pi\colon G \to G/H$ is the natural quotient homomorphism.*

PROOF. Clearly, $MH = \pi^{-1}(\pi(M))$. Since the mapping $\pi$ is open and continuous, the restriction $\psi$ of $\pi$ to $MH$ is an open continuous mapping of $MH$ onto $\pi(M)$. Since $M$ is a subgroup of $G$ and $\pi$ is a homomorphism of $G$ onto $G/H$, it follows that $\pi(M)$ and $MH$ are subgroups of the groups $G/H$ and $G$, respectively, and $\psi$ is a homomorphism of $MH$ onto $\pi(M)$. Let $e$ be the neutral element of $G$. Clearly, $\psi^{-1}(\psi(e)) = \pi^{-1}(\pi(e)) = H$, that is, the kernel of the homomorphism $\psi$ is $H$. Now it follows from Theorem 1.5.13 that the topological groups $MH/H$ and $\pi(M)$ are topologically isomorphic. □

To conclude the section, we prove two statements on connections of cardinal invariants of a topological group $G$ with cardinal invariants of a closed subgroup $H$ of $G$ and the quotient space $G/H$.

THEOREM 1.5.20. *Suppose that $G$ is a topological group, $H$ is a closed subgroup of $G$, $X$ is a subspace of $G$, $\pi$ is the natural homomorphism of $G$ onto the quotient space $G/H$, and $Y = \pi(X)$. Suppose also that the space $H$ and the subspace $Y$ of $G/H$ are first-countable. Then $X$ is also first-countable.*

PROOF. By Proposition 1.5.2, we can assume that the neutral element $e$ of $G$ is in $X$ and, for the same reason, it suffices to verify that $X$ is first-countable at $e$. Let us fix a sequence of symmetric open neighbourhoods $W_n$ of $e$ in $G$ such that $W_{n+1}^2 \subset W_n$, for each $n \in \omega$, and $\{W_n \cap H : n \in \omega\}$ is a base for the space $H$ at $e$. We also fix a sequence of open neighbourhoods $U_n$ of $e$ in $G$ such that $\{\pi(U_n) \cap Y : n \in \omega\}$ is a base for $Y$ at $\pi(e)$. Now put $B_{i,j} = W_i \cap U_j \cap X$, for $i, j \in \omega$. To finish the proof, it suffices to establish the following:

**Claim.** *The family $\eta = \{B_{i,j} : i, j \in \omega\}$ is a base for $X$ at $e$.*

Clearly, each $B_{i,j}$ is open in $X$ and contains $e$. Now take any open neighbourhood $O$ of $e$ in $G$. Let us show that some element of $\eta$ is contained in $O$. There exists an open neighbourhood $V$ of $e$ in $G$ such that $V^2 \subset O$. Choose $m \in \omega$ such that $W_m \cap H \subset V$. Further, there exists $k \in \omega$ such that

$$\pi(U_k) \cap Y \subset \pi(V \cap W_{m+1}).$$

Let us verify that

$$B_{m+1,k} \subset O.$$

Take any $z \in B_{m+1,k} = W_{m+1} \cap U_k \cap X$. Then $z \in U_k \cap X \subset (V \cap W_{m+1})H$, since $\pi(z) \in \pi(U_k) \cap Y \subset \pi(V \cap W_{m+1})$. However, $z$ does not belong to $W_{m+1}(G \setminus W_m)$, since $W_{m+1}^2 \subset W_m$ and $z \in W_{m+1} = W_{m+1}^{-1}$. Hence, $z \in (V \cap W_{m+1})(H \cap W_m)$. Since $W_m \cap H \subset V$, we conclude that $z \in V^2 \subset O$. Thus, $B_{m+1,k} \subset O$, and $\eta$ is a base for $X$ at $e$. Since $\eta$ is countable, it follows that $X$ is first-countable at $e$. □

COROLLARY 1.5.21. [**N. Ya. Vilenkin**] *Suppose that $G$ is a topological group and $H$ is a closed subgroup of $G$. If the spaces $H$ and $G/H$ are first-countable, then the space $G$ is also first-countable.*

It turns out that Corollary 1.5.21 remains valid if one replaces first countability by separability. To prove this, we need an auxiliary topological result.

LEMMA 1.5.22. *Suppose that $f\colon X \to Y$ is an open continuous mapping of a space $X$ onto a space $Y$, $x \in X$, $B \subset Y$, and $f(x) \in \overline{B}$. Then $x \in \overline{f^{-1}(B)}$. In particular, $\overline{f^{-1}(B)} = f^{-1}(\overline{B})$.*

PROOF. Put $y = f(x)$, and let $O$ be any open neighbourhood of $x$. Then $f(O)$ is an open neighbourhood of $y$. Therefore, $f(O) \cap B \neq \varnothing$ and, hence, $O \cap f^{-1}(B) \neq \varnothing$. It follows that $x \in \overline{f^{-1}(B)}$. The equality $\overline{f^{-1}(B)} = f^{-1}(\overline{B})$ is now evident. □

THEOREM 1.5.23. *Suppose that $G$ is a topological group and $H$ is a closed subgroup of $G$. If the spaces $H$ and $G/H$ are separable, then the space $G$ is also separable.*

PROOF. Let $\pi$ be the natural homomorphism of $G$ onto the quotient space $G/H$. Since $G/H$ is separable, we can fix a dense countable subset $B$ of $G/H$. Since $H$ is separable and every coset $xH$ is homeomorphic to $H$, we can fix a dense countable subset $M_y$ of $\pi^{-1}(y)$, for each $y \in B$. Put $M = \bigcup\{M_y : y \in B\}$. Then $M$ is a countable subset of $G$ and $M$ is dense in $\pi^{-1}(B)$. Since $\pi$ is an open mapping of $G$ onto $G/H$, it follows from Lemma 1.5.22 that $\overline{\pi^{-1}(B)} = G$. Hence, $M$ is dense in $G$ and $G$ is separable. □

## Exercises

1.5.a. Can Theorem 1.5.7, Corollary 1.5.8, or Proposition 1.5.15 be generalized to paratopological groups?

1.5.b. Let $G$ be a paratopological group and $N$ a closed invariant subgroup of $G$ such that both $N$ and the quotient paratopological group $G/N$ are topological groups. Prove that $G$ is also a topological group.

1.5.c. Let $G = \mathbb{R} \setminus \{0\}$ be the multiplicative group of non-zero real numbers with the topology inherited from $\mathbb{R}$, and let $H = \{-1, 1\}$ and $K = \{x \in \mathbb{R} : x > 0\}$. Verify the following:
   (a) $H$ and $K$ are closed subgroups of $G$;
   (b) the quotient group $G/H$ is topologically isomorphic to $K$;
   (c) the groups $G$ and $H \times K$ are topologically isomorphic.

1.5.d. Suppose that $H$ is a proper non-trivial closed subgroup of the group $\mathbb{R}$. Show that the quotient group $\mathbb{R}/H$ is topologically isomorphic to the circle group $\mathbb{T}$.

1.5.e. Let $H$ be a closed invariant subgroup of a topological group $G$. Show that if both groups $H$ and $G/H$ are connected, then so is $G$.

1.5.f. Let $G$ be a topological group, not necessarily Hausdorff, and let $H$ be the closure of the set $\{e\}$ in $G$, where $e$ is the neutral element of $G$. Prove the following:
   (a) $H$ is an invariant subgroup of $G$;
   (b) for every continuous homomorphism $f\colon G \to K$ to a Hausdorff topological group $K$, there exists a continuous homomorphism $h\colon G/H \to K$ satisfying $f = h \circ \pi$, where $\pi$ is the canonical homomorphism of $G$ onto $G/H$.

1.5.g. Which homogeneous spaces can be represented as quotients of right topological groups with respect to a closed subgroup?

1.5.h. Can every compact Hausdorff homogeneous space be represented as a quotient of a right topological group with respect to a closed subgroup?

1.5.i. Let $H$ be a closed nowhere dense subgroup of a topological group $G$, and suppose that the quotient space $G/H$ is resolvable (see Exercise 1.4.l). Show that $G$ is resolvable.

## Problems

1.5.A. Let $GL(n, \mathbb{R})$ be the general linear group with the topology inherited from $\mathbb{R}^{n^2}$ (see e) of Example 1.2.5), and $E_n$ the neutral element of $GL(n, \mathbb{R})$, where $n \in \mathbb{N}$.
   (a) Verify that $K_n = \{E_n, -E_n\}$ is an invariant subgroup of $GL(n, \mathbb{R})$ and prove that the quotient group $GL(n, \mathbb{R})/K_n$ is pathwise connected and locally pathwise connected.
   (b) Prove that the quotient group $GL(n, \mathbb{R})/K_n$ is topologically isomorphic to an open subgroup of $GL(n, \mathbb{R})$.
   (c) Denote by $N$ the subgroup of $GL(n, \mathbb{R})$ consisting of all diagonal matrices. Show that the groups $GL(n, \mathbb{R})/N$ and $SL(n, \mathbb{R})$ are topologically isomorphic.

1.5.B. Let $H$ be an open divisible subgroup of an Abelian topological group $G$. Prove that the groups $G$ and $H \times G/H$ are topologically isomorphic.

1.5.C. Suppose that $A$ and $B$ are closed subgroups of a topological group $G$, and let $p$ be the canonical mapping of $G$ onto the quotient space $G/B$. Show that the closure of $p(A)$ in $G/B$ can fail to be homogeneous. (See also Problem 1.4.A.)

1.5.D. (S. Dierolf and U. Schwanengel [**133**]) Let $H$ be a closed subgroup of a topological group $G$ with topology $\tau$ and $\pi\colon G \to G/H$ the canonical mapping. Suppose that $\sigma$ is a group topology on $G$ weaker than $\tau$ satisfying $\sigma{\restriction}H = \tau{\restriction}H$ and $\pi(\sigma) = \pi(\tau)$, where $\pi(\gamma) = \{\pi(U) : U \in \gamma\}$ for any family $\gamma$ of subsets of $G$. Prove that the topologies $\sigma$ and $\tau$ coincide.

1.5.E. (W. W. Comfort and J. van Mill [**115**]) An abstract group $G$ is called *strongly resolvable* if for every non-discrete Hausdorff group topology $\mathcal{T}$ on $G$, the space $(G, \mathcal{T})$ is resolvable. Prove the following:
  (a) Let $H$ be a subgroup of an Abelian group $G$. If $H$ is not strongly resolvable, then neither is $G$.
  (b) If $H$ is an invariant subgroup of an abstract group $G$ and both groups $H$ and $G/H$ are strongly resolvable, then so is $G$.

1.5.F. Suppose that $H$ is a closed $\sigma$-compact subgroup of a topological Abelian group $G$ and that the quotient group $G/H$ is compact. Is the group $G$ then $\sigma$-compact?
*Hint.* To give the answer, modify the construction in [**511**].

1.5.G. Let $H$ be a closed invariant subgroup of a topological group $G$. Suppose that $H$ is countably compact and the quotient group $G/H$ is compact and metrizable. Is $G$ countably compact?
*Hint.* The answer to the question can be found in [**89**].

1.5.H. Can Theorem 1.5.23 be generalized to paratopological groups?

### Open Problems

1.5.1. Characterize (or find the typical properties) of compact spaces that can be represented as quotients of topological groups with respect to closed metrizable subgroups.

## 1.6. Products, $\Sigma$-products, and $\sigma$-products

Cartesian products of algebraic objects with topology, in particular, of topological groups play a fundamental role in topological algebra. Starting with a certain class $\mathcal{P}$ of topological groups (second-countable groups or compact groups, for example), it is natural to form topological products of arbitrary collections of groups in this class. Then, passing to arbitrary topological subgroups of the topological groups so obtained, we introduce some of the most important classes of topological groups. This approach is implemented in Sections 3.4 and 3.7. Besides, many interesting examples of topological groups with non-trivial combination of properties are defined in this way. In this section we present some basic facts on topological products of topological groups, semigroups and other objects of topological algebra, and specify certain standard subobjects of these products, called $\Sigma$-products and $\sigma$-products. Taking $\Sigma$-products and $\sigma$-products expand further our opportunities to construct non-trivial examples of topological groups.

It is well known that topological products can be used to represent and study the operation of union of a family of topologies given on the same set. Indeed, we have the following simple statement:

PROPOSITION 1.6.1. *Suppose that $\{\mathcal{T}_\alpha : \alpha \in A\}$ is a family of topologies on a set $X$, and let $\mathcal{T}$ be the smallest topology on $X$ containing $\mathcal{T}_\alpha$, for each $\alpha \in A$. Let, further, $\Pi$*

*be the topological product of the family* $\mathcal{F} = \{(X, \mathcal{T}_\alpha) : \alpha \in A\}$. *For each* $x \in X$, *put* $d(x) = z \in \Pi$, *where* $z_\alpha = x$ *for each* $\alpha \in A$. *Then* $d$ *is a homeomorphism of the space* $(X, \mathcal{T})$ *onto the subspace* $\Delta_{\mathcal{F}} = \{d(x) : x \in X\}$ *of* $\Pi$ *called the diagonal of* $\mathcal{F}$. *Suppose also that there exists* $\alpha_0 \in A$ *such that* $\mathcal{T}_{\alpha_0} \subset \mathcal{T}_\alpha$, *for each* $\alpha \in A$, *and that the topology* $\mathcal{T}_{\alpha_0}$ *is Hausdorff. Then the subspace* $\Delta_{\mathcal{F}}$ *is closed in the product space* $\Pi$.

PROOF. The mapping $d\colon (X, \mathcal{T}) \to \Pi$ is continuous as the diagonal product of the family of continuous mappings $\{i_\alpha : \alpha \in A\}$, where $i_\alpha\colon (X, \mathcal{T}) \to (X, \mathcal{T}_\alpha)$ is the identity mapping for each $\alpha \in A$. It is also clear that $d(X) = \Delta_{\mathcal{F}}$. For every $\alpha \in A$, let $p_\alpha\colon \Pi \to (X, \mathcal{T}_\alpha)$ be the natural projection. It follows from the definition of $\mathcal{T}$ that this topology is the smallest one that makes each mapping $p_\alpha \circ d$ continuous. Therefore, the inverse mapping $d^{-1}\colon \Delta_{\mathcal{F}} \to (X, \mathcal{T})$ is continuous and, hence, $d$ is a homeomorphism of $(X, \mathcal{T})$ onto $\Delta_{\mathcal{F}}$.

To prove the rest of the theorem, take an arbitrary point $x \in \Pi \setminus \Delta_{\mathcal{F}}$. Then there exist two distinct indices $\alpha, \beta \in A$ such that $p_\alpha(x) \neq p_\beta(x)$. Since $y = p_\alpha(x)$ and $z = p_\beta(x)$ are distinct points of $X$ and the topology $\mathcal{T}_{\alpha_0}$ is Hausdorff, we can choose disjoint open neighbourhoods $U$ and $V$ of $y$ and $z$, respectively, in $(X, \mathcal{T}_{\alpha_0})$. It follows from $\mathcal{T}_{\alpha_0} \subset \mathcal{T}_\alpha$ and $\mathcal{T}_{\alpha_0} \subset \mathcal{T}_\beta$ that $W = p_\alpha^{-1}(U) \cap p_\beta^{-1}(V)$ is an open neighbourhood of $x$ in $\Pi$ disjoint from $\Delta_{\mathcal{F}}$. This proves that the complement $\Pi \setminus \Delta_{\mathcal{F}}$ is open in $\Pi$ and, therefore, $\Delta_{\mathcal{F}}$ is closed. □

The topology $\mathcal{T}$ on $X$ considered in Proposition 1.6.1 is usually denoted by $\bigvee_{\alpha\in A} \mathcal{T}_\alpha$ and is called the *join* of the topologies $\mathcal{T}_\alpha$, with $\alpha \in A$.

Formulating definitions and results below, we restrict ourselves to the case of topological groups. Similar definitions and statements obviously hold for topological semigroups, right topological groups, semitopological semigroups, and so on.

We now present an analog of Proposition 1.6.1 for topological groups. It follows directly from Proposition 1.6.1 and Theorem 1.2.7.

THEOREM 1.6.2. *Suppose that* $\{\mathcal{T}_\alpha : \alpha \in A\}$ *is a family of topologies on a group* $G$ *such that* $(G, \mathcal{T}_\alpha)$ *is a topological group, for each* $\alpha \in A$, *and let* $\mathcal{T}$ *be the smallest topology on* $G$ *containing each* $\mathcal{T}_\alpha$. *Let, further,* $\Pi$ *be the topological product of the family* $\mathcal{F} = \{(G, \mathcal{T}_\alpha) : \alpha \in A\}$. *For every* $g \in G$, *put* $d(g) = z \in \Pi$, *where* $z_\alpha = g$, *for each* $\alpha \in A$. *Then* $d$ *is a topological isomorphism of* $(G, \mathcal{T})$ *onto the subgroup* $\Delta_{\mathcal{F}} = \{d(g) : g \in G\}$ *of the topological group* $\Pi$. *In particular, it follows that* $(G, \mathcal{T})$ *is a topological group. If there exists* $\alpha_0 \in A$ *such that* $\mathcal{T}_{\alpha_0} \subset \mathcal{T}_\alpha$ *for each* $\alpha \in A$ *and the topology* $\mathcal{T}_{\alpha_0}$ *is Hausdorff, then* $\Delta_{\mathcal{F}}$ *(also called the diagonal of* $\mathcal{F}$*) is a closed subgroup of the Hausdorff topological group* $\Pi$.

Similarly to the case of topologies on a set, the topology $\mathcal{T}$ on $G$ in Theorem 1.6.2 is called the *join* of the topologies $\mathcal{T}_\alpha$, $\alpha \in A$. The above theorem can be reformulated by saying that the join of a family of topological group topologies on a given group is again a topological group topology.

Since a countable product of metrizable spaces is metrizable, Theorem 1.6.2 implies the following:

COROLLARY 1.6.3. *Let* $\{\mathcal{T}_n : n \in \omega\}$ *be a family of metrizable topological group topologies on a group* $G$. *Then the topological group topology* $\mathcal{T} = \bigvee_{n\in\omega} \mathcal{T}_n$ *is also metrizable.*

Note that Theorem 1.6.2 is applicable to families of non-Hausdorff group topologies as well. In particular, we have:

COROLLARY 1.6.4. *Suppose that G is a group and* $\{h_\alpha : \alpha \in A\}$ *is a family of homomorphisms* $h_\alpha$ *of G to topological groups* $H_\alpha$. *Then there exists a smallest topology* $\mathcal{T}$ *on G such that each* $h_\alpha\colon (G, \mathcal{T}) \to H_\alpha$ *is continuous. Moreover, this topology* $\mathcal{T}$ *turns G into a topological group topologically isomorphic to the diagonal of* $\mathcal{F}$, *where* $\mathcal{F} = \{\mathcal{T}_\alpha : \alpha \in A\}$ *and* $\mathcal{T}_\alpha = \{{h_\alpha}^{-1}(V) : V \text{ is open in } H_\alpha\}$, *for each* $\alpha \in A$.

We will now present some results on products of topological spaces which are not always covered by standard courses of general topology, though they play an important role in the study of products of topological groups and their topological subgroups. Our aim is to consider several topological properties defined in terms of the closure operator in a space (such as the tightness, sequentiality, Fréchet–Urysohn property, *etc.*) and study the behaviour of these properties under the product operation.

We recall that the *tightness* of a space $X$ is the minimal cardinal $\tau \geq \omega$ with the property that for every point $x \in X$ and every set $P \subset X$ with $x \in \overline{P}$, there exists a subset $Q$ of $P$ such that $|Q| \leq \tau$ and $x \in \overline{Q}$. The tightness of $X$ is denoted by $t(X)$. All first-countable spaces have countable tightness; furthermore, every Fréchet–Urysohn space and every sequential space has countable tightness (see [**165**, 1.7.13 (c)]).

Here are two natural modifications of the concept of tightness. Let $\tau$ be an infinite cardinal. We say that the *$G_\delta$-tightness* of a space $X$ is not greater than $\tau$, and write $get(X) \leq \tau$, if whenever a point $x \in X$ belongs to the closure of $\bigcup \gamma$, where $\gamma$ is a family of $G_\delta$-sets in $X$, then there exists a subfamily $\eta$ of $\gamma$ such that $x$ is in the closure of $\bigcup \eta$ and $|\eta| \leq \tau$. If under the same assumptions about $x$ and $\gamma$ there exists a subset $M$ of $\bigcup \gamma$ such that $x \in \overline{M}$ and $|M| \leq \tau$, we say that the *$\delta$-tightness* of $X$ is not greater than $\tau$ and write $t_\delta(X) \leq \tau$. As usual, the $G_\delta$-tightness of $X$, $get(X)$, is the minimal cardinal $\tau \geq \omega$ satisfying $get(X) \leq \tau$, and the $\delta$-tightness of $X$ is defined similarly.

Note that if the tightness of $X$ is countable then the $\delta$-tightness of $X$ is countable, and if the $\delta$-tightness of $X$ is countable then the $G_\delta$-tightness of $X$ is countable. The converse to each of the two implications is false, even for topological groups (see Example 1.6.13 and Problem 6.6.D). If, however, the space $X$ has countable pseudocharacter, then the three cardinal functions of $X$ coincide:

PROPOSITION 1.6.5. *Suppose that all points of a space X are $G_\delta$-sets. Then* $t(X) = t_\delta(X) = get(X)$.

PROOF. Since, by our assumption, each point of $X$ is a $G_\delta$-set, the family $\gamma$ in the above definition of the $G_\delta$-tightness and $\delta$-tightness may consist of singletons. In this case, the invariants $get(X)$ and $t_\delta(X)$ become equal to $t(X)$. □

It turns out that both the $G_\delta$-tightness and $\delta$-tightness of arbitrarily big products with "good" factors are countable. To deduce this important result, we introduce several technical concepts related to product spaces.

Let $\{X_a : a \in A\}$ be a family of topological spaces and $X = \Pi_{a\in A} X_a$ be their topological product. Then an *$\omega$-cube* in $X$ is any subset $B$ of $X$ that can be represented as the product $B = \Pi_{a\in A} B_a$, where $B_a$ is a non-empty subset of $X_a$, for each $a \in A$, and the set $A_B = \{a \in A : B_a \neq X_a\}$ is countable. The set $A_B$ in this case will be called the *core* of the $\omega$-cube $B$. We also put $X_K = \Pi_{a\in K} X_a$, for every non-empty subset $K$ of $A$, and denote by $p_K$ the natural projection mapping of $X$ onto $X_K$. If $B$ is an $\omega$-cube with core $K$ such that the image $p_K(B)$ under the projection $p_K$ consists only of one point, we say that $B$ is an *elementary $\omega$-cube*.

Suppose that $B = \prod_{a\in A} B_a$ is an $\omega$-cube in $X$ with core $K$ such that $B_a$ is a $G_\delta$-set in $X_a$, for each $a \in K$. Then $B$ is called a *canonical $G_\delta$-set* in $X$. Note that, for every $J \subset A$ and every canonical $G_\delta$-set $F$ in $X$, the projection $p_J(F)$ is a $G_\delta$-set in $X_J$.

PROPOSITION 1.6.6. *Let $\{X_a : a \in A\}$ be a family of topological spaces such that the tightness of $X_K = \prod_{a\in K} X_a$ is countable, for every countable subset $K$ of $A$, and let $X = \prod_{a\in A} X_a$ be their topological product. Then, for any family $\gamma$ of $\omega$-cubes in $X$ and for any point $x \in \overline{\bigcup \gamma}$, there exists a countable subfamily $\eta$ of $\gamma$ such that $x \in \overline{\bigcup \eta}$.*

PROOF. We are going to define, by induction, an increasing sequence of countable subsets $A_n$ of $A$ and a sequence of countable subfamilies $\gamma_n$ of $\gamma$ in the following way.

Let $A_0$ be a non-empty countable subset of $A$. Assume that for some $n \in \omega$, a countable set $A_n \subset A$ is already defined, and put $K = A_n$. By the assumption, $t(X_K) \le \omega$. Therefore, there exists a countable subfamily $\gamma_n$ of $\gamma$ such that $p_K(x)$ is in the closure of the set $\bigcup\{p_K(V) : V \in \gamma_n\}$ in the space $X_K$. Now put $A_{n+1} = A_n \cup \bigcup\{A_B : B \in \gamma_n\}$. The inductive step is complete.

Put $M = \bigcup\{A_n : n \in \omega\}$ and $\eta = \bigcup\{\gamma_n : n \in \omega\}$. Clearly, $\eta$ is a countable subfamily of $\gamma$. Let $H$ be the closure of $\bigcup \eta$. We have to show that $x \in H$, that is, we have to verify that every canonical open neighbourhood $O_1$ of $x$ in $X$ has a common point with $H$. Clearly, $O_1 = p_S^{-1} p_S(O_1)$, for some finite set $S \subset A$. Put $F = S \cap M$ and $O = p_F^{-1} p_F(O_1)$. Then $x \in O_1 \subset O = p_F^{-1} p_F(O)$. It follows from the definition of $M$ and $\eta$ that $\bigcup \eta = p_M^{-1} p_M(\bigcup \eta)$ and, since the projection $p_M$ of $X$ onto $X_M$ is open and $H = \overline{\bigcup \eta}$, Lemma 1.5.22 implies that $p_M^{-1} p_M(H) = H$. Therefore, the conditions $O \cap H \neq \varnothing$ and $O_1 \cap H \neq \varnothing$ are equivalent.

Since the sequence $\{A_n : n \in \omega\}$ is increasing, there exists $n \in \omega$ such that $F \subset A_n$. By the choice of $\gamma_n$, the point $p_F(x)$ is in the closure of the set $\bigcup\{p_F(V) : V \in \gamma_n\}$ in the space $X_F$. Therefore, there exists a point $z \in \bigcup \gamma_n \subset \bigcup \eta$ such that $p_F(z) \in p_F(O)$. It follows that $z \in O \cap \bigcup \eta$. Hence, $x \in H$. □

The next assertion is close to Proposition 1.6.6, but its proof requires some extra argument.

PROPOSITION 1.6.7. *Suppose that $\{X_a : a \in A\}$ is a family of topological spaces such that the $G_\delta$-tightness of $X_K = \prod_{a\in K} X_a$ is countable, for every countable subset $K$ of $A$, and let $X = \Pi_{a\in A} X_a$ be their topological product. Then, for any family $\gamma$ of $G_\delta$-sets in $X$ and for any point $x$ in the closure of the set $\bigcup \gamma$, there exists a countable subfamily $\eta$ of $\gamma$ such that $x \in \overline{\bigcup \eta}$, that is, $get(X) \le \omega$.*

PROOF. Clearly, for each point $y$ in $\bigcup \gamma$ there exists a canonical $G_\delta$-set $B$ in $X$ such that $y \in B \subset \bigcup \gamma$. Therefore, we may assume that $\gamma$ consists of canonical $G_\delta$-sets in $X$.

Now we can repeat the proof of Proposition 1.6.6 with a minor modification: to conclude, at the inductive step, that there exists a countable subfamily $\gamma_n$ of $\gamma$ such that $p_K(x)$ is in the closure of the set $\bigcup\{p_K(V) : V \in \gamma_n\}$ in the space $X_K$, we have to refer to the obvious fact that $p_K(V)$ is an $\omega$-cube of type $G_\delta$ in the space $X_K$ for each $V \in \gamma_n$, and apply the assumption that the $G_\delta$-tightness of $X_K$ is countable. □

Since the class of spaces with a countable network is closed under countable products, and every space with a countable network has countable tightness, Proposition 1.6.7 implies the following corollaries.

COROLLARY 1.6.8. *Let $X$ be the product of some family of spaces each of which has a countable network. Then the $G_\delta$-tightness of $X$ is countable.*

COROLLARY 1.6.9. *The product of any family of first-countable spaces has countable $G_\delta$-tightness.*

The last two statements can be strengthened with the help of the next simple lemma.

LEMMA 1.6.10. *Suppose that $X = \Pi_{a \in A} X_a$ is the topological product of a family of topological spaces, $\gamma$ is a countable family of elementary $\omega$-cubes in $X$, and $x \in \overline{\bigcup \gamma}$. For every $B \in \gamma$, let $y(B)$ be any point of $B$ such that $y(B)_\alpha = x_\alpha$, for each $\alpha \in A \setminus A_B$, where $A_B$ is the core of $B$. Put $M = \{y(B) : B \in \gamma\}$. Then $M$ is countable, $M \subset \bigcup \gamma$, and $x \in \overline{M}$.*

PROOF. Suppose that $O$ is a canonical open neighbourhood of the point $x$ in $X$. Since $x \in \overline{\bigcup \gamma}$, there exists $B \in \gamma$ such that $O \cap B \neq \varnothing$. For every $a \in A$, let $p_a$ be the projection of $X$ onto the factor $X_a$. Since $B$ is an elementary $\omega$-cube, we must have $y(B)_a \in p_a(O)$, for each $a \in A_B$. Taking into account that $y(B)_a = x_a$ for each $a \in A \setminus A_B$, we conclude that $y(B) \in O \cap M \neq \varnothing$. Therefore, $x \in \overline{M}$. □

THEOREM 1.6.11. *Suppose that $\{X_a : a \in A\}$ is a family of first-countable spaces, and let $X = \Pi_{a \in A} X_a$ be the topological product of this family. Then the $\delta$-tightness of $X$ is countable.*

PROOF. Let $\gamma$ be an arbitrary family of $G_\delta$-sets in $X$ and $x$ a point in the closure of $\bigcup \gamma$. Since the product of any countable family of first-countable spaces is first-countable, every element of $\gamma$ is the union of a family of elementary $\omega$-cubes in $X$. Therefore, we may assume that $\gamma$ consists of elementary $\omega$-cubes. By Corollary 1.6.9, the $G_\delta$-tightness of $X$ is countable. Therefore, there exists a countable subfamily $\eta$ of $\gamma$ such that $x \in \overline{\bigcup \eta}$. Since $\eta$ consists of elementary $\omega$-cubes, Lemma 1.6.10 implies that there exists a countable subset $M$ of $\bigcup \eta$ such that $x \in \overline{M}$. Clearly, $M \subset \bigcup \gamma$. Hence, $t_\delta(X) \leq \omega$. □

As every Hausdorff space with a countable network has countable pseudocharacter, a similar argument proves the next statement:

THEOREM 1.6.12. *Let $X$ be the product of some family of Hausdorff spaces each of which has a countable network. Then the $\delta$-tightness of $X$ is countable.*

In the next example we show that topological groups of countable $G_\delta$-tightness or even countable $\delta$-tightness may have arbitrary big tightness. Furthermore, our groups will be compact.

EXAMPLE 1.6.13. For every infinite cardinal $\tau$, there exists a compact Abelian topological group $G$ such that the $\delta$-tightness of $G$ is countable, but $t(G) \geq \tau$.

Indeed, taking a bigger cardinal, if necessary, we can assume that $\tau$ is regular, i.e., the cofinality of $\tau$ is equal to $\tau$. Let $D = \{0, 1\}$ be the two-element discrete group. Then the compact Abelian group $G = D^\tau$ has countable $\delta$-tightness by Theorem 1.6.11 or Theorem 1.6.12. To show that $t(G) \geq \tau$, we argue as follows. Denote by $\bar{0}$ the point of $D^\tau$ with zero coordinates, i.e., the neutral element of the group $D^\tau$. For every ordinal $\beta < \tau$, let an element $x_\beta \in D^\tau$ be defined by $x_\beta(\alpha) = 0$ if $\alpha \leq \beta$ and $x_\beta(\alpha) = 1$ otherwise. It is clear that $\bar{0}$ is in the closure of the set $X = \{x_\beta : \beta < \tau\}$. To deduce the inequality $t(G) \geq \tau$, it suffices to verify that $\bar{0}$ is not in the closure of each $Y \subset X$ with $|Y| < \tau$. Since $\tau$ is a regular cardinal, for every such a subset $Y \subset X$ there exists $\alpha < \tau$ such that the inequality $\beta < \alpha$ holds for each $x_\beta \in Y$. Then $x_\beta(\alpha) = 1$ for all $x_\beta \in Y$ and, therefore, $\bar{0}$ does not belong to the closure of $Y$ in $G$. □

A space $X$ is called *$G_\delta$-preserving* if for each family $\gamma$ of $G_\delta$-sets in $X$, the closure of the set $\bigcup\gamma$ is again the union of some family of $G_\delta$-sets in $X$. Evidently, every space of countable pseudocharacter is $G_\delta$-preserving. In particular, every Hausdorff space with a countable network is $G_\delta$-preserving. We shall see in Proposition 5.5.5 and Corollary 5.5.6 that, under mild restrictions, topological groups become $G_\delta$-preserving when considered as topological spaces.

Let us show that, under a natural additional assumption, the property of being $G_\delta$-preserving becomes productive.

THEOREM 1.6.14. *Suppose that $\{X_a : a \in A\}$ is a family of spaces such that the space $X_K = \prod_{a\in K} X_a$ is $G_\delta$-preserving and has countable $G_\delta$-tightness, for every countable subset $K$ of $A$. Then the product space $X = \prod_{a\in A} X_a$ is $G_\delta$-preserving.*

PROOF. As in the proof of Proposition 1.6.7, we can assume that the family $\gamma$ consists of canonical $G_\delta$-sets in $X$. Put $R = \bigcup\gamma$ and take an arbitrary point $x \in \overline{R}$. Again, $get(X) \leq \omega$ by Proposition 1.6.7, so $x \in \overline{\bigcup\mu}$ for some countable subfamily $\mu$ of $\gamma$. For every $B \in \mu$, let $A_B$ be the core of $B$. Clearly, $|A_B| \leq \omega$ and $p_{A_B}(B)$ is a $G_\delta$-set in $X_{A_B}$. Put $K = \bigcup\{A_B : B \in \mu\}$. Then $K$ is a countable subset of $A$ and $p_K(B)$ is a $G_\delta$-set in $X_K$, for each $B \in \mu$. By our assumption, the space $X_K$ is $G_\delta$-preserving, so the set $\overline{p_K(\bigcup\mu)}$ is the union of a family of $G_\delta$-sets in $X_K$. Since the projection $p_K\colon X \to X_K$ is open, we conclude, by Lemma 1.5.22, that

$$p_K^{-1}\left(\overline{p_K(\bigcup\mu)}\right) = \overline{p^{-1}(p_K(\bigcup\mu))} = \overline{\bigcup\mu}.$$

Clearly, the preimage of a $G_\delta$-set in $X_K$ under the mapping $p_K$ is a $G_\delta$-set in $X$, so the set on the left hand side of the above equalities is the union of a family of $G_\delta$-sets in $X$. Hence, there exists a $G_\delta$-set $P$ in $X$ such that $x \in P \subset \overline{\bigcup\mu} \subset \overline{\bigcup\gamma}$. This proves that $X$ is $G_\delta$-preserving. □

The next two results are now immediate from Corollaries 1.6.8, 1.6.9, and Theorem 1.6.14.

COROLLARY 1.6.15. *The product of an arbitrary family of Hausdorff spaces each of which has a countable network is $G_\delta$-preserving.*

COROLLARY 1.6.16. *The product of any family of first-countable spaces is $G_\delta$-preserving.*

We recall (see [**165**, 1.7.12]) that the *Souslin number* or, equivalently, *cellularity* of a space $X$ is the minimal infinite cardinal $\kappa$ such that every family of pairwise disjoint open sets in $X$ has cardinality less than or equal to $\kappa$. The cellularity of $X$ is denoted by $c(X)$. A space $X$ has the *Souslin property* if $c(X) = \aleph_0$. Clearly, every separable space has the Souslin property.

For every infinite cardinal $\tau$, we introduce a cardinal function $cel_\tau$ analogous to the cellularity. As usual, a $G_\tau$-*set* in a topological space is the intersection of a family of at most $\tau$ open subsets of the space. Given a space $X$, we define $cel_\tau(X)$ to be the minimal cardinal $\lambda \geq \aleph_0$ such that every family $\mathcal{G}$ consisting of $G_\tau$-sets in $X$ contains a subfamily $\mathcal{H}$ satisfying $\overline{\bigcup \mathcal{H}} = \overline{\bigcup \mathcal{G}}$ and $|\mathcal{H}| \leq \lambda$. A space $X$ with $cel_\tau(X) \leq \tau$ is called $\tau$-*cellular*. Note that every $\omega$-cellular space has countable $G_\delta$-tightness, but the converse is false (the Niemytzki plane is a counterexample).

It follows immediately from the above definition that $c(X) \leq cel_\tau(X) \leq cel_\kappa(X) \leq hd(X)$ if $\aleph_0 \leq \tau \leq \kappa$, and $cel_\kappa(X) = hd(X)$ for every $\kappa \geq \psi(X)$, where $\psi(X)$ and $hd(X)$ are the pseudocharacter and the hereditary density of $X$, respectively (see [**165**, 3.1.F, 2.7.9] or [**262**] for the definition of the functions $\psi$ and $hd$).

The following lemma shows that a product space is $\omega$-cellular if and only if every countable subproduct is.

LEMMA 1.6.17. *Let $X = \prod_{i\in I} X_i$ be a product space and suppose that the subproduct $X_K = \prod_{i\in K} X_i$ satisfies $cel_\omega(X_K) \leq \omega$, for each countable set $K \subset I$. Then $X$ also satisfies $cel_\omega(X) \leq \omega$.*

PROOF. Let $\gamma$ be a family of $G_\delta$-sets in $X$. Since every element of $\gamma$ is the union of a family of canonical $G_\delta$-sets in $X$, we can assume without loss of generality that all elements of $\gamma$ are canonical $G_\delta$-sets in $X$. Take an arbitrary countable subfamily $\gamma_0$ of $\gamma$ and denote by $K_0$ the union of the cores of the elements of $\gamma_0$. Then $K_0$ is a countable subset of $I$. Suppose that we have defined, for some $n \in \omega$, a countable subfamily $\gamma_n$ of $\gamma$ and a countable set $K_n \subset I$. Since $cel_\omega(X_{K_n}) \leq \omega$, there exists a countable family $\gamma_{n+1} \subset \gamma$ such that $p_{K_n}(\bigcup \gamma_{n+1})$ is dense in $p_{K_n}(\bigcup \gamma)$. Denote by $K_{n+1}$ the union of $K_n$ with the cores of the elements of $\gamma_{n+1}$. Clearly, $K_{n+1}$ is countable and $K_n \subset K_{n+1}$. In addition, our definition of the set $K_{n+1}$ implies that $F = p_{K_{n+1}}^{-1} p_{K_{n+1}}(F)$, for each $F \in \gamma_n$.

Put $\gamma^* = \bigcup_{n\in\omega} \gamma_n$ and $K = \bigcup_{n\in\omega} K_n$. Then $\gamma^*$ is a countable subfamily of $\gamma$ and it follows from our construction that $p_K(\bigcup \gamma^*)$ is dense in $p_K(\bigcup \gamma)$. Since $K_{n+1} \subset K$ for all $n \in \omega$, the equality $F = p_K^{-1} p_K(F)$ holds for each $F \in \gamma^*$. Finally, since the projection $p_K\colon X \to X_K$ is open, we conclude that $\bigcup \gamma^* = p_K^{-1} p_K(\bigcup \gamma^*)$ is dense in $p_K^{-1}(\overline{p_K(\bigcup \gamma)}) \supseteq \bigcup \gamma$. In other words, $\bigcup \gamma^*$ is dense in $\bigcup \gamma$ and, therefore, $cel_\omega(X) \leq \omega$. □

Let us say that $X$ is an *Efimov space* if, for every family $\gamma$ of $G_\delta$-sets in $X$, the closure of the set $\bigcup \gamma$ is again a $G_\delta$-set in $X$. Evidently, if every closed subset of $X$ is a $G_\delta$-set, i.e., if $X$ is *perfect*, then $X$ is an Efimov space. We will show below that the class of Efimov spaces is much wider than the class of perfect spaces; for example, the former includes arbitrary products of spaces with a countable network (see Corollary 1.6.19). Note that every Efimov space is $G_\delta$-preserving, but not vice versa (every first-countable compact space which fails to be perfectly normal is a counterexample).

THEOREM 1.6.18. *Let $\{X_i : i \in I\}$ be a family of spaces such that for every countable set $K \subset I$, the product $X_K = \prod_{i\in K} X_i$ is an Efimov space and satisfies $cel_\omega(X_K) \leq \omega$. Then $X = \prod_{i\in I} X_i$ is an Efimov space. Furthermore, every closed $G_\delta$-set $P$ in $X$ has the form $P = p_J^{-1} p_J(P)$ for some countable set $J \subset I$.*

PROOF. Let $\mathcal{F}$ be a family of $G_\delta$-sets in $X$. We can assume that all elements of $\mathcal{F}$ are canonical $G_\delta$-sets. In particular, every $F \in \mathcal{F}$ satisfies $F = p_C^{-1} p_C(F)$, where $C$ is the core of $F$. Since $cel_\omega(X) \leq \omega$ by Lemma 1.6.17, one can find a countable subfamily $\mathcal{G}$ of $\mathcal{F}$ such that $\bigcup \mathcal{G}$ is dense in $\bigcup \mathcal{F}$. All elements of $\mathcal{G}$ are canonical $G_\delta$-sets, so the union of the cores of the elements of $\mathcal{G}$, say, $J$ is countable. Clearly, $F = p_J^{-1} p_J(F)$ for each $F \in \mathcal{G}$. Therefore, the set $E = \bigcup \mathcal{G}$ satisfies $E = p_J^{-1} p_J(E)$. Since the mapping $p_J$ is open, we have that

$$\overline{\bigcup \mathcal{F}} = \overline{E} = \overline{p_J^{-1} p_J(E)} = p_J^{-1}\left(\overline{p_J(E)}\right). \tag{1.7}$$

It follows from the choice of $\mathcal{G}$ that the family $\mathcal{G}_J = \{p_J(F) : F \in \mathcal{G}\}$ consists of $G_\delta$-sets in the Efimov space $X_J$. Thus, the closure of the set $p_J(E) = \bigcup \mathcal{G}_J$ is of type $G_\delta$ in $X_J$. Then (1.7) implies that $\overline{\bigcup \mathcal{F}} = p_J^{-1} \overline{p_J(E)}$ and, hence, $\overline{\bigcup \mathcal{F}}$ is a $G_\delta$-set in $X$.

Let us prove the second claim of the theorem. A closed $G_\delta$-set $P$ in $X$ is the union of a family of canonical $G_\delta$-sets in $X$, say, $P = \bigcup \gamma$. Since $cel_\omega(X) \leq \omega$, $\gamma$ contains a countable subfamily $\mu$ such that $\bigcup \mu$ is dense in $P$. Since every $F \in \mu$ has a countable core, there exists a countable set $J \subset I$ such that $F = p_J^{-1} p_J(F)$ for each $F \in \mu$. Then $P = p_J^{-1} p_J(P)$, due to the fact that $p_J$ is an open mapping. □

COROLLARY 1.6.19. *The product of an arbitrary family of Hausdorff spaces each of which has a countable network is an Efimov space and, hence, is $G_\delta$-preserving.*

PROOF. A countable product $P$ of Hausdorff spaces with a countable network is Hausdorff and has a countable network, so $P$ is a perfect, hereditarily separable space. In particular, $P$ is an Efimov space and $cel_\omega(P) \leq \omega$. The required conclusion now follows from Theorem 1.6.18. □

Let $\gamma$ be a family of sets. Then $\gamma$ is called a Δ-*system* if there exists a set $R$ (called the *root* of $\gamma$) such that $A \cap B = R$ for all distinct $A, B \in \gamma$. It is clear that if $R$ is the root of $\gamma$, then the family $\{A \setminus R : A \in \gamma\}$ is disjoint.

The next assertion is known as the Δ-*lemma* (see [**285**, Chap. 2, Theorem 1.5] or [**413**, Section 2]).

THEOREM 1.6.20. *Let $\gamma$ be a family of finite sets, and suppose that the cardinality of $\gamma$ is an uncountable regular cardinal. Then $\gamma$ contains a subfamily $\mu$ of the same cardinality which forms a Δ-system.*

Theorem 1.6.21 below states that the cellularity of a product space $X = \prod_{i\in I} X_i$ is defined by the values of cellularity on finite subproducts of $X$.

THEOREM 1.6.21. *Let $X = \prod_{i\in I} X_i$ be a product space and suppose that the subproduct $X_K = \prod_{i\in K} X_i$ satisfies $c(X_K) \leq \tau$ for every finite $K \subset I$, where $\tau$ is an infinite cardinal. Then $c(X) \leq \tau$.*

PROOF. Suppose to the contrary that $c(X) > \tau$. Then $X$ contains a disjoint family $\{U_\alpha : \alpha < \tau^+\}$ of non-empty canonical open sets. For every $\alpha < \tau^+$, choose a finite set $A_\alpha \subset I$ such that $U_\alpha = \pi_{A_\alpha}^{-1}\pi_{A_\alpha}(U_\alpha)$. Let $\gamma = \{A_\alpha : \alpha < \tau^+\}$. If $|\gamma| \leq \tau$, then there exists a set $C \subset \tau^+$ with $|C| = \tau^+$ such that $A_\alpha = A_\beta = J$ for all $\alpha, \beta \in C$. Denote by $\pi_J$ the projection of $X$ onto $X_J$. Then $\{\pi_J(U_\alpha) : \alpha \in C\}$ is a disjoint family of open sets in $X_J$ which has cardinality $\tau^+$, thus contradicting $c(X_J) \leq \tau$. We can assume, therefore, that $\gamma$ has cardinality $\tau^+$. Choosing a subfamily of $\gamma$, we may assume additionally that $A_\alpha \neq A_\beta$ for all distinct $\alpha, \beta < \tau^+$.

By Theorem 1.6.20, one can find $B \subset \tau^+$ with $|B| = \tau^+$ and a finite set $R \subset I$ such that $A_\alpha \cap A_\beta = R$ whenever $\alpha, \beta \in B$ and $\alpha \neq \beta$. By hypothesis, $c(X_J) \leq \tau$, so there exist distinct $\alpha, \beta \in B$ such that $V = \pi_R(U_\alpha) \cap \pi_R(U_\beta) \neq \varnothing$. Since the sets $A_\alpha \setminus R$ and $A_\beta \setminus R$ are disjoint, we can take a point $x \in X$ such that $\pi_R(x) \in V$, $\pi_i(x) \in \pi_i(U_\alpha)$ for each $i \in A_\alpha \setminus R$, and $\pi_i(x) \in \pi_i(U_\beta)$ for each $i \in A_\beta \setminus R$. Clearly, $x \in U_\alpha \cap U_\beta \neq \varnothing$, which is a contradiction. So, we have proved that $c(X) \leq \tau$. □

Let $\tau$ be an infinite cardinal. A space $X$ is said to be *pseudo-$\tau$-compact* if every discrete in $X$ family of open sets has cardinality strictly less than $\tau$. It is easy to see that $X$ is pseudo-$\tau$-compact iff every family $\gamma$ of open sets in $X$ (not necessarily pairwise disjoint) with $|\gamma| \geq \tau$ has an accumulation point in $X$. It follows from the definition that pseudo-$\tau$-compactness implies pseudo-$\lambda$-compactness whenever $\tau < \lambda$. Evidently, pseudo-$\omega$-compactness and pseudocompactness coincide in the class of Tychonoff spaces. Hence, pseudocompact spaces are pseudo-$\aleph_1$-compact. All Lindelöf spaces and all separable spaces are also pseudo-$\aleph_1$-compact. More generally, every space of countable cellularity is pseudo-$\aleph_1$-compact.

PROPOSITION 1.6.22. *Let $X = \prod_{i \in I} X_i$ be a Tychonoff product such that every finite subproduct $X_J = \prod_{i \in J} X_i$ is pseudo-$\tau$-compact, where $\tau$ is a regular uncountable cardinal. Then $X$ is also pseudo-$\tau$-compact.*

PROOF. For every finite $J \subset I$, denote by $\pi_J$ the projection of $X$ onto $X_J$. To show that $X$ is pseudo-$\tau$-compact, consider a family $\gamma = \{U_\alpha : \alpha < \tau\}$ of canonical open sets in $X$. Every $U_\alpha$ has the form $U_\alpha = \pi_{J_\alpha}^{-1}(V_\alpha)$, where $J_\alpha$ is a finite subset of $I$ and $V_\alpha$ is a rectangular open set in $X_{J_\alpha}$. Since $X_J$ is pseudo-$\tau$-compact for each finite $J \subset I$, we can assume (choosing a set $A \subset \tau$ with $|A| = \tau$ and considering the family $\{U_\alpha : \alpha \in A\}$ in place of $\gamma$) that $J_\alpha \neq J_\beta$ if $\alpha < \beta < \tau$.

Apply the $\Delta$-lemma to find a finite set $J \subset I$ and a set $B \subset \tau$ such that $|B| = \tau$ and $J_\alpha \cap J_\beta = J$, for all distinct $\alpha, \beta \in B$. For every $\alpha \in B$, put $O_\alpha = \pi_J(U_\alpha)$. Since the space $X_J$ is pseudo-$\tau$-compact, the family $\{O_\alpha : \alpha \in B\}$ has an accumulation point $x_J \in X_J$. Take an arbitrary point $x \in X$ with $\pi_J(x) = x_J$. We claim that $x$ is an accumulation point for the family $\{U_\alpha : \alpha \in B\}$.

Indeed, let $W$ be a basic neighbourhood of $x$ in $X$, say, $W = \pi_K^{-1}(W_0)$, where $W_0$ is a rectangular open in $X_K$ and $K \subset I$ is finite. Then $\pi_J(W)$ is a neighbourhood of $x_J$ in $X_J$, so $\pi_J(W)$ intersects $O_\alpha$ for each $\alpha \in C$, where $C \subset B$ is infinite. Since the family $\{J_\alpha \setminus J : \alpha \in B\}$ is disjoint, there exists an infinite set $D \subset C$ such that $(J_\alpha \setminus J) \cap K = \varnothing$ for each $\alpha \in D$. Let us verify that $W \cap U_\alpha \neq \varnothing$ for each $\alpha \in D$. For $\alpha \in D$, take $p \in \pi_J(W) \cap \pi_J(U_\alpha)$ and choose a point $q \in X$ such that

$$\pi_J(q) = p, \quad \pi_{J_\alpha \setminus J}(q) \in \pi_{J_\alpha \setminus J}(U_\alpha) \text{ and } \pi_{K \setminus J}(q) \in \pi_{K \setminus J}(W).$$

It is easy to see that $q \in W \cap U_\alpha \neq \varnothing$. Therefore, $x$ is an accumulation point for $\{U_\alpha : \alpha \in B\}$ and for $\gamma$. This proves that the product space $X$ is pseudo-$\tau$-compact. □

Many important examples of topological groups are defined as topological subgroups of products of very simple topological groups. Because of that, we have to introduce and study some standard subspaces of topological products of topological spaces.

Suppose that $\eta = \{X_\alpha : \alpha \in A\}$ is a family of topological spaces, $X = \Pi\{X_\alpha : \alpha \in A\}$ is the topological product of the family $\eta$, and $b$ is a point in $X$. Then the *$\Sigma$-product of $\eta$ with basic point $b$* is the subspace of $X$ consisting of all points $x \in X$ such that only countably many coordinates $x_\alpha$ of $x$ are distinct from the corresponding coordinates $b_\alpha$ of $b$. This subspace is denoted by $\Sigma\Pi\{X_\alpha : \alpha \in A\}$ or by $\Sigma\Pi\eta$. For $x \in X$, the set $\{\alpha \in A : x_\alpha \neq b_\alpha\}$ is called the *support of $x$* (with respect to $b$) and is denoted by $supp_b(x)$. Clearly, the $\Sigma$-product is the set of all $x \in X$ such that $supp_b(x)$ is countable. Sometimes $\Sigma\Pi\eta$ will be called the *$\Sigma$-product of $\eta$ at $b$.*

Similarly, the *$\sigma$-product of $\eta$ with basic point $b$* is the subspace of $X$ consisting of all points $x \in X$ such that only finitely many coordinates $x_\alpha$ of $x$ are distinct from the corresponding coordinates $b_\alpha$ of $b$. This subspace is denoted by $\sigma\Pi\{X_\alpha : \alpha \in A\}$ or simply by $\sigma\Pi\eta$; note that the basic point is usually not shown in the notation.

Clearly, the $\sigma$-product is a subspace of the $\Sigma$-product, and both subspaces are dense in the product. For those interested primarily in topological algebra, the importance of the operations of taking $\Sigma$-product and $\sigma$-product is based, in particular, on the following simple fact:

PROPOSITION 1.6.23. *Suppose that $\eta = \{G_\alpha : \alpha \in A\}$ is a family of topological groups, $e_\alpha$ is the neutral element of $G_\alpha$, and $G = \Pi\{G_\alpha : \alpha \in A\}$ is the product of the family $\eta$. Then the $\Sigma$-product $\Sigma\Pi\eta$ and the $\sigma$-product $\sigma\Pi\eta$ with basic point $e = (e_\alpha)_{\alpha \in A}$ (the neutral element of $G$) are dense topological subgroups of the topological group $G$.*

In what follows when we consider the $\Sigma$-product or $\sigma$-product of a family of topological groups, we always assume that the basic point is the neutral element $e$ of the product.

The above statement as well as Example 1.2.9 provide us with motivation to learn more about topological properties of $\Sigma$-products and $\sigma$-products of families of topological spaces and topological groups. We start with a few results on $\Sigma$-products.

THEOREM 1.6.24. *Suppose that $\eta = \{X_\alpha : \alpha \in A\}$ is a family of topological spaces, $X = \Pi\{X_\alpha : \alpha \in A\}$ is the topological product of the family $\eta$, and $b$ is a point of $X$. Suppose also that, for each countable subset $C$ of $A$, the tightness of the product space $\prod\{X_\alpha : \alpha \in C\}$ is countable. Then the tightness of the $\Sigma$-product $\Sigma\Pi\eta$ of $\eta$ with basic point $b$ is also countable.*

PROOF. We put $Y = \Sigma\Pi\eta$. Take a point $y \in Y$ and consider a set $P \subset Y$ such that $y \in \overline{P}$. We can assume that $b \notin P$. Put $S = supp_b(y)$ and $K_p = supp_b(p) \cup S$, for each $p \in P$. Then $K_p$ is a countable non-empty subset of $A$. Let $B_p$ be an elementary $\omega$-cube in $X$ with the core set $K_p$ such that $p \in B_p$. Put $\gamma = \{B_p : p \in P\}$. Clearly, $P \subset \bigcup\gamma$. Therefore, $y \in \overline{\bigcup\gamma}$. It follows from Proposition 1.6.6 that there exists a countable subset $M$ of $P$ such that $y \in \overline{\bigcup\{B_p : p \in M\}}$. As in Lemma 1.6.10, for every $p \in M$, choose a point $z(p) \in B_p$ such that $z(p)_\alpha = y_\alpha$ for each $\alpha \in A \setminus K_p$. Since $S \subset K_p$ for each $p \in M$, we have that $z(p) = p$. Therefore, Lemma 1.6.10 implies that $y \in \overline{M}$. □

The above theorem has several important corollaries.

COROLLARY 1.6.25. *The $\Sigma$-product of any family of first-countable spaces is a space of countable tightness.*

Clearly, Corollary 1.6.25 covers the case of metrizable spaces.

A space $X$ is called *cosmic* if it has a countable network. All second-countable spaces as well as all continuous images of second-countable spaces are cosmic.

COROLLARY 1.6.26. *The $\Sigma$-product of any family of cosmic spaces is a space of countable tightness.*

We will now establish another series of very useful facts on $\Sigma$-products. The most general of them is the next lemma.

LEMMA 1.6.27. *Suppose that $\eta = \{X_\alpha : \alpha \in A\}$ is a family of topological spaces, $X = \Pi\{X_\alpha : \alpha \in A\}$ is the topological product of $\eta$, and $b \in X$. Then, for each countable subset $M$ of the $\Sigma$-product $Y = \Sigma\Pi\eta$ of $\eta$ with basic point $b$, the closure of $M$ in $Y$ coincides with the closure of $M$ in $X$ and is naturally homeomorphic to a closed subspace of the topological product of some countable subfamily of $\eta$.*

PROOF. Put $K = \bigcup\{supp_b(x) : x \in M\}$. Then $K$ is a countable subset of $A$ and, clearly, for any $y$ in the closure of $M$ in $X$, we have that $y_\alpha = b_\alpha$, for each $\alpha \in A \setminus K$. Therefore, the closure of $M$ in $X$ is contained in $Y$ and is homeomorphic to the closure of $p_K(M)$ in $X_K = \Pi\{X_\alpha : \alpha \in K\}$, where $p_K : X \to X_K$ is the projection; the restriction of $p_K$ to the closure of $M$ in $X$ is the natural homeomorphism we are looking for. □

The following four more concrete statements are immediate corollaries of Lemma 1.6.27. The first of them will be strengthened in Theorem 1.6.32.

PROPOSITION 1.6.28. *Suppose that $\eta = \{X_\alpha : \alpha \in A\}$ is a family of topological spaces, $X = \Pi\{X_\alpha : \alpha \in A\}$, and $b \in X$. Suppose also that, for each countable subset $C$ of $A$, the product space $\prod\{X_\alpha : \alpha \in C\}$ is Fréchet–Urysohn. Let $Y$ be the $\Sigma$-product of $\eta$ at $b$. Then, for each countable subset $M$ of $Y$, the closure of $M$ in $Y$ is also Fréchet–Urysohn.*

PROPOSITION 1.6.29. *Let $\eta = \{X_\alpha : \alpha \in A\}$ be a family of metrizable spaces, $X = \Pi\{X_\alpha : \alpha \in A\}$, and $b \in X$. If $Y$ is the $\Sigma$-product of $\eta$ at $b$ then, for each countable subset $M$ of $Y$, the closure of $M$ in $Y$ is metrizable.*

PROPOSITION 1.6.30. *Suppose that $\eta = \{X_\alpha : \alpha \in A\}$ is a family of compact spaces, $X = \Pi\{X_\alpha : \alpha \in A\}$, and $b \in X$. Let $Y$ be the $\Sigma$-product of $\eta$ at $b$. Then, for each countable subset $M$ of $Y$, the closure of $M$ in $Y$ is compact.*

PROPOSITION 1.6.31. *Suppose that $\eta = \{X_\alpha : \alpha \in A\}$ is a family of first-countable spaces, $X = \Pi\{X_\alpha : \alpha \in A\}$, and $b \in X$. Let $Y$ be the $\Sigma$-product of $\eta$ at $b$. Then, for each countable subset $M$ of $Y$, the closure of $M$ in $Y$ is first-countable.*

Here is the promised generalization of Proposition 1.6.28.

THEOREM 1.6.32. *Let $\eta = \{X_\alpha : \alpha \in A\}$ be a family of topological spaces, $X = \Pi\{X_\alpha : \alpha \in A\}$ be the product of $\eta$, and $b \in X$. Suppose also that, for each countable subfamily $\xi$ of $\eta$, the product $\prod \xi$ is a Fréchet–Urysohn space. Then the $\Sigma$-product $Y$ of $\eta$ with basic point $b$ is also Fréchet–Urysohn.*

PROOF. Suppose that $Z \subset Y$, $y \in Y$, and $y \in \overline{Z}$. Note that every Fréchet–Urysohn space has countable tightness. Hence the tightness of $Y$ is countable, by Theorem 1.6.24. Therefore, there exists a countable subset $M$ of $Z$ such that $y \in \overline{M}$. However, the closure $\overline{M}$ of $M$ in $Y$ is Fréchet–Urysohn, by Proposition 1.6.28. It follows that some sequence of points of $M$ converges to $y$. Hence, the space $Y$ is Fréchet–Urysohn. □

COROLLARY 1.6.33. *The $\Sigma$-product of any family of first-countable spaces is a Fréchet–Urysohn space.*

COROLLARY 1.6.34. *The $\Sigma$-product $Y$ of any family of compact spaces is countably compact. Moreover, the closure of any countable subset $M$ of $Y$ in $Y$ is compact.*

PROOF. The first part of the statement obviously follows from the second part. And the second part is a direct corollary from Proposition 1.6.30, since the product of any family of compact spaces is compact. □

Combining Corollaries 1.6.33 and 1.6.34, we obtain:

COROLLARY 1.6.35. *The $\Sigma$-product of any family of compact first-countable spaces is a countably compact Fréchet–Urysohn space.*

Here is another useful general fact. It concerns a relationship between the product of a family of spaces and the $\Sigma$-product of the same family.

PROPOSITION 1.6.36. *For any family $\eta = \{X_\alpha : \alpha \in A\}$ of spaces and any $b \in X = \prod \eta$, the $\Sigma$-product $Y$ of $\eta$ at $b$ is $G_\delta$-dense in $X$, that is, every non-empty $G_\delta$-set in $X$ intersects $Y$.*

PROOF. The statement is almost obvious. Indeed, $P$ is the union of a family of non-empty $\omega$-cubes, since $P$ is a non-empty $G_\delta$-set in $X$. However, every $\omega$-cube intersects $Y$. Hence, $P \cap Y \neq \varnothing$. (Observe that, by a standard argument, the intersection $Y \cap P$ is dense in $P$.) □

THEOREM 1.6.37. *Suppose that $\eta = \{X_\alpha : \alpha \in A\}$ is a family of first-countable spaces, and let $X = \Pi_{\alpha \in A} X_\alpha$ be their topological product. Then, for any family $\gamma$ of $G_\delta$-sets in $X$ and for any point $b$ in the closure of the set $U = \bigcup \gamma$, there exists a sequence $\{x_n : n \in \omega\}$ of points of $U$ converging to $b$.*

PROOF. Let $Y$ be the $\Sigma$-product of $\eta$ at $b$, and put $B = U \cap Y$. It follows from Proposition 1.6.36 that $B$ is dense in $U$. Therefore, $b \in \overline{B}$. Clearly, $b \in Y$, $B \subset Y$, and the space $Y$ is Fréchet–Urysohn, according to Corollary 1.6.33. It follows that some sequence of points of $B$ converges to $b$. Since $B \subset U$, we are done. □

With the help of the above constructions and results, we can define some non-trivial topological groups and identify certain non-trivial properties of these groups.

THEOREM 1.6.38. *Suppose that $A$ is an uncountable set and that, for each $\alpha \in A$, $G_\alpha$ is a compact metrizable topological group containing at least two points. Let $G = \Pi_{\alpha \in A} G_\alpha$ be the product of topological groups $G_\alpha$. Then the $\Sigma$-product $H = \Sigma\Pi_{\alpha \in A} G_\alpha$ is a countably compact, Fréchet–Urysohn, non-compact, non-metrizable topological subgroup of $G$.*

PROOF. Since $A$ is uncountable, $H$ is a proper subset of $G$. However, $H$ is dense in $G$. Therefore, $H$ is not closed in $G$. Since the space $G$ is Hausdorff, it follows that the subspace $H$ is not compact. By Corollary 1.6.35, $H$ is countably compact and Fréchet–Urysohn. Since $H$ is not compact, and every countably compact metrizable space is compact, we conclude that $H$ is not metrizable. Finally, $H$ is a topological subgroup of $G$, by Proposition 1.6.23. □

It will be shown in Corollary 4.2.2 that, in contrast with the above statement, every compact topological group of countable tightness is metrizable and, hence, the cardinality of such a group does not exceed the power of the continuum. However, the cardinalities of countably compact groups of countable tightness (or even countably compact Fréchet–Urysohn groups) are unbounded, as Theorem 1.6.38 shows.

EXAMPLE 1.6.39. Suppose that $A$ is an uncountable set.

a) For every $\alpha \in A$, let $G_\alpha = \mathbb{T}$ be the circle group. Then the $\Sigma$-product $\Sigma\Pi_{\alpha\in A} G_\alpha$ is denoted by $\Sigma\mathbb{T}^A$. By Theorem 1.6.38, $\Sigma\mathbb{T}^A$ is a countably compact, non-metrizable, Fréchet–Urysohn topological group which is not compact.
b) Similarly, if $G_\alpha = D$ for each $\alpha \in A$, where $D = \{0, 1\}$ is the discrete Boolean group, then the $\Sigma$-product $\Sigma\Pi_{\alpha\in A} G_\alpha$ is denoted by $\Sigma D^A$. By Theorem 1.6.38, $\Sigma D^A$ is a countably compact, non-metrizable, Fréchet–Urysohn Boolean topological group which is not compact. Clearly, the groups $D^A$ and $\Sigma D^A$ are zero-dimensional. □

The above example is instrumental in many situations when compact and countably compact topological groups are involved.

Let us now make a few general observations on the structure of $\sigma$-products of spaces and topological groups.

PROPOSITION 1.6.40. *Let $X = \prod_{\alpha\in A} X_\alpha$ be a product space and $Y \subset X$ be the corresponding $\sigma$-product with center at $b \in X$. Then $Y$ is the union of a countable family $\{Y_n : n \in \omega\}$ of closed subspaces $Y_n$ of $X$ such that $Y_0 = \{b\}$ and, for each $n \in \omega$, $Y_n \subset Y_{n+1}$, and $Y_{n+1} \setminus Y_n$ admits a disjoint open covering $\lambda_n$ such that each $U \in \lambda_n$ is homeomorphic to an open subspace of the product of a finite subfamily of $\{X_\alpha : \alpha \in A\}$.*

PROOF. For every $y \in Y$, only finitely many coordinates $y_\alpha$ of $y$ are distinct from the corresponding coordinates $b_\alpha$ of $b$. We denote by $r(y)$ the number of coordinates of a point $y \in Y$ distinct from those of $b$. Let $Y_n = \{y \in Y : r(y) \le n\}$, for $n \in \omega$. Then, clearly, $Y = \bigcup_{n\in\omega} Y_n$, where each $Y_n$ is closed in the product space $X$.

For every $n \in \mathbb{N}$, put $U_n = Y_n \setminus Y_{n-1}$, and let $B_n$ be the set of all finite subsets of $A$ of cardinality $n$. Given $K \in B_n$, let $W_K$ be the set of all $y \in Y$ such that $\{\alpha \in A : y_\alpha \neq b_\alpha\} = K$. Clearly, $W_K$ is open in $Y_n$, $W_K \cap W_L = \varnothing$, for any two distinct $K, L \in B_n$, and $U_n = \bigcup \lambda_n$, where $\lambda_n = \{W_K : K \in B_n\}$. It is also clear that each $W_K$ is homeomorphic to an open subspace of $\prod_{\alpha\in K} X_\alpha$. □

PROPOSITION 1.6.41. *The $\sigma$-product of any family of $\sigma$-compact spaces is $\sigma$-compact.*

PROOF. Note first that the $\sigma$-product of any family of compact spaces is $\sigma$-compact. This follows immediately from Proposition 1.6.40, since the product of any family of compact spaces is compact.

Let $\eta = \{X_\alpha : \alpha \in A\}$ be any family of $\sigma$-compact spaces. Then, for each $\alpha \in A$, $X_\alpha = \bigcup_{n\in\omega} X_{\alpha,n}$, where each $X_{\alpha,n}$ is compact. We can also assume that $X_{\alpha,i} \subset X_{\alpha,j}$ whenever $i < j$. Let $b$ be any point of the product of the family $\eta$, and $Y$ be the $\sigma$-product of $\eta$ at $b$. We can assume that $b_\alpha \in X_{\alpha,0}$, for each $\alpha \in A$. Indeed, $b_\alpha \in X_{\alpha,k}$ for some $k \in \omega$, and we can place $X_{\alpha,k}$ in the role of $X_{\alpha,0}$ changing the enumeration accordingly.

Put $\eta_n = \{X_{\alpha,n} : \alpha \in A\}$, and let $Y_n$ be the $\sigma$-product of $\eta_n$ at $b$, for each $n \in \omega$. Clearly, $Y = \bigcup_{n\in\omega} Y_n$. However, each $Y_n$ is $\sigma$-compact since all elements of $\eta_n$ are compact. □

The next result follows directly from Propositions 1.6.41 and 1.6.23.

PROPOSITION 1.6.42. *The product and the Σ-product of an arbitrary family of $\sigma$-compact topological groups contain a dense $\sigma$-compact subgroup.*

EXAMPLE 1.6.43. Let $M = \sigma D^\tau$ be the $\sigma$-product of $\tau$ copies of the Boolean group $D = \{0, 1\}$, where $\tau$ is an uncountable cardinal. Then $M$ is a $\sigma$-compact, Fréchet–Urysohn, non-compact, non-metrizable Boolean topological group. Indeed, since $\tau$ is uncountable and the character of the space $\chi(D^\tau)$ is uncountable, the product space $D^\tau$ cannot contain a dense first-countable subspace, by Lemma 1.4.15. However, $M$ is dense in $D^\tau$. Therefore, the space $M$ is not metrizable. Since a subspace of a Fréchet–Urysohn space is also Fréchet–Urysohn, the rest follows from Propositions 1.6.41 and 1.6.23. □

Sometimes it is useful to consider the so-called *$\omega$-box topology* on product spaces. Let $X = \prod_{i\in I} X_i$ be the product of spaces $X_i$, with $i \in I$. A standard base of the $\omega$-box topology on $X$ consists of the $\omega$-cubes $B = \prod_{i\in I} B_i$, where each $B_i$ is open in $X_i$ (and, clearly, the number of indices $i \in I$ with $B_i \neq X_i$ is countable). It follows from the definition that the $\omega$-box topology on $X$ is finer than the usual Tychonoff product topology. It is easy to see that for every set $J \subset I$, the projection $p_J : X \to X_J$ of $X$ onto the subproduct $X_J$ is open provided that both spaces carry the $\omega$-box topology.

The following facts complement our knowledge of the properties of $\sigma$-products. As usual, $l(Y)$ denotes the Lindelöf number of a space $Y$ (see [**165**, Section 3.8]).

PROPOSITION 1.6.44. *Let $X = \prod_{i\in I} X_i$ be a product space such that for every finite $J \subset I$, the subproduct $X_J = \prod_{i\in J} X_i$ satisfies $l(X_J) \le \tau$. Then the $\sigma$-product $\sigma(p) \subset X$, with the $\omega$-box topology inherited from $X$ and a basic point $p$, satisfies $l(\sigma(p)) \le \tau$, for any $p \in X$.*

PROOF. First, we introduce some notation. For every non-empty set $J \subset I$, put

$$\sigma_J^* = \{x \in \sigma(p) : \pi_i(x) = \pi_i(p) \text{ for each } i \in I \setminus J\},$$

where $\pi_i : X \to X_i$ is the projection, $i \in I$. If $J \subset I$ is finite, then $\sigma_J^* \cong X_J$, so that $l(\sigma_J^*) \le \tau$. Observe that if $J \subset I$ and $|J| \le \tau$, then

$$\sigma_J^* = \bigcup\{\sigma_K^* : K \subset J,\ |K| < \omega\}.$$

Since the family of finite subsets of $J$ has cardinality $\le \tau$, we have $l(\sigma_J^*) \le \tau$ for such a set $J$.

Suppose that $\gamma$ is an open covering of $\sigma(p)$. Without loss of generality we can assume that every element $V \in \gamma$ has the form $V = U_V \cap \sigma(p)$, for some canonical open $\omega$-cube $U_V$ in $X$. Hence there exists a countable set $B(V) \subset I$ such that $U_V = \pi_{B(V)}^{-1}\pi_{B(V)}(U_V)$.

Choose an arbitrary element $W_0$ of $\gamma$ and put $B_0 = B(W_0)$ and $\mu_0 = \{W_0\}$. Suppose that for some $n \in \omega$, we have defined a set $B_n \subset A$ and a subfamily $\mu_n$ of $\gamma$ such that $|B_n| \leq \tau$ and $|\mu_n| \leq \tau$. Since $l(\sigma^*_{B_n}) \leq \tau$, we can find a subfamily $\mu_{n+1}$ of $\gamma$ such that $\sigma^*_{B_n} \subset \bigcup \mu_{n+1}$ and $|\mu_{n+1}| \leq \tau$. It remains to put $B_{n+1} = B_n \cup \bigcup\{B(V) : V \in \mu_{n+1}\}$. This finishes our inductive construction of the sequences $\{B_n : n \in \omega\}$ and $\{\mu_n : n \in \omega\}$.

Consider the set $B = \bigcup_{n=0}^{\infty} B_n$ and the subfamily $\mu = \bigcup_{n=0}^{\infty} \mu_n$ of $\gamma$. It is clear that $|B| \leq \tau$ and $|\mu| \leq \tau$. We claim that $\mu$ covers $\sigma(p)$. Indeed, let $x \in \sigma(p)$ be arbitrary. Denote by $y$ the element of $\sigma(p)$ such that $\pi_i(y) = \pi_i(x)$ for each $i \in B$ and $\pi_i(y) = \pi_i(p)$ for all $i \in I \setminus B$. Then $K = \{i \in I : \pi_i(v) \neq \pi_i(p)\}$ is a subset of $B$. Since $|K| < \omega$ and $B_0 \subset B_1 \subset B_2 \subset \cdots$, there exists $n \in \omega$ such that $K \subset B_n$. Therefore, $y \in \sigma^*_{B_n} \subset \bigcup \mu_{n+1}$ and, hence, $y \in V$ for some $V \in \mu_{n+1}$. It follows from our inductive construction that $B(V) \subset B_{n+1} \subset B$ and, since $\pi_i(x) = \pi_i(y)$ for each $i \in B$, we conclude that $x \in U_V \cap \sigma(p) = V$. Thus, $\sigma(p) = \bigcup \mu$. This proves that the space $\sigma(p)$ satisfies $l(\sigma(p)) \leq \tau$. □

Since the Tychonoff product topology is weaker than the $\omega$-box topology, we obtain the following corollary.

COROLLARY 1.6.45. *Let $X = \prod_{i \in I} X_i$ be a product space such that for every finite $J \subset I$, the subproduct $X_J = \prod_{i \in J} X_i$ satisfies $l(X_J) \leq \tau$. Then the $\sigma$-product $\sigma(p) \subset X$, with the subspace topology and center at $p$, satisfies $l(\sigma(p)) \leq \tau$ for any $p \in X$.*

## Exercises

1.6.a. Verify that Theorem 1.6.2 remains valid for monoidal topologies on monoids, quasitopological topologies and paratopological topologies on groups, ring topologies on rings and field topologies on fields.

1.6.b. Give an example of an infinite group $G$ and two Hausdorff topological group topologies $\mathcal{T}_1$ and $\mathcal{T}_2$ on $G$ such that the groups $(G, \mathcal{T}_1)$ and $(G, \mathcal{T}_2)$ are compact and the diagonal $\Delta_G = \{(x, x) : x \in G\}$ is dense in the product group $(G, \mathcal{T}_1) \times (G, \mathcal{T}_2)$. Explain why this does not contradict Theorem 1.6.2.

1.6.c. Show that the join of two compact metrizable topological group topologies on a group need not be either locally compact or $\sigma$-compact.

1.6.d. Construct a topological group $K$ and a subgroup $L$ of $K$ satisfying the following conditions:

(a) $L$ is closed in $K$ and $\omega = get(K) < get(L)$;

(b) $L$ is dense in $K$ and $\omega = get(K) < get(L)$.

1.6.e. Show that if a subgroup $L$ of a topological group $K$ intersects every non-empty $G_\delta$-set in the group $K$, then $get(L) \leq get(K)$.

1.6.f. Let $N$ be a closed subgroup of a topological group $G$ and $\pi\colon G \to G/N$ the corresponding quotient mapping. Show that $get(G/N) \leq get(G)$ and $t_\delta(G/N) \leq t_\delta(G)$.

1.6.g. Is the square of an Efimov space $X$ necessarily an Efimov space? What if $X$ is a topological group?

1.6.h. Let $p$ be a prime number and $G = \mathbb{Z}(p)^\omega$ be the product group with the usual Tychonoff product topology, where $\mathbb{Z}(p)$ is the discrete cyclic group with $p$ elements. Show that for every element $a \in G$ distinct from the neutral element of $G$, there exists a dense subgroup $H$ of $G$ such that $G = H \oplus \langle a \rangle$.

1.6.i. Let $\bar{0}$ be the neutral element of the product group $\mathbb{Z}(2)^\omega$, and $\sigma(\bar{0})$ be the corresponding $\sigma$-product in $\mathbb{Z}(2)^\omega$ with center at $\bar{0}$. Construct a discontinuous isomorphism of $\sigma(\bar{0})$ onto itself. Does there exist a continuous isomorphism $f: \sigma(\bar{0}) \to \sigma(\bar{0})$ which fails to be a homeomorphism?

1.6.j. Give an example of a $\sigma$-compact topological group $G$ which is Fréchet–Urysohn but is not metrizable.

*Hint.* Apply Corollary 1.6.33 and Proposition 1.6.41.

## Problems

1.6.A. Give an example of a dense subgroup $G$ of the product group $\mathbb{T}^{\omega_1}$ such that $G$ has countable pseudocharacter and satisfies $t(G) > \omega$.

1.6.B. (M. G. Tkachenko and Y. Torres [**491**]) Let $G$ be a non-trivial second-countable Abelian topological group and $\kappa$ be a cardinal satisfying $\omega < \kappa \leq 2^{\omega}$. Prove that the following are equivalent in the case when $G$ is torsion-free:

a) The $\sigma$-product $\sigma G^{\kappa}$ of $\kappa$ copies of the group $G$ contains a dense subgroup of countable pseudocharacter.

b) $|G| \geq \kappa$.

Give an example of a torsion Abelian second-countable topological group $G$ with $|G| = 2^{\omega}$ such that $\sigma G^{\kappa}$ does not contain a dense subgroup of countable pseudocharacter, for any cardinal $\kappa$ satisfying $\omega < \kappa \leq 2^{\omega}$.

1.6.C. Let $\Sigma = \Sigma\Pi_{i \in I} G_i$ be the $\Sigma$-product of compact metrizable topological groups, and suppose that $f: \Sigma \to K$ is a continuous homomorphism onto a compact topological group $K$. Prove that $K$ is metrizable.

*Remark.* The conclusion remains valid for an arbitrary continuous mapping of $\Sigma$ onto $K$; this is Efimov's theorem in [**158**].

1.6.D. (W. W. Comfort and J. van Mill [**115**]) Prove the following statements (see also Problem 1.5.E):

(a) If $G$ and $H$ are strongly resolvable groups, then so is the product group $G \times H$.

(b) Let $\{G_i : i \in I\}$ be a family of strongly resolvable groups and suppose that the direct sum $G = \bigoplus_{i \in I} G_i$ does not contain a copy of the group $\mathbb{Z}(2)^{(\omega)}$, the direct sum of $\omega$ copies of the group $\mathbb{Z}(2) = \{0, 1\}$. Then $G$ is strongly resolvable.

(c) For every prime $p$, the subgroup $\mathbb{Z}(p^{\infty}) = \{x \in \mathbb{T} : x^{p^n} = 1 \text{ for some } n \in \mathbb{N}\}$ of the circle group $\mathbb{T}$ is strongly resolvable.

(d) The additive group of rationals $\mathbb{Q}$ is strongly resolvable.

(e) The abstract groups $\mathbb{R}$, $\mathbb{C}$, and $\mathbb{T}$ are strongly resolvable (compare with (a) of Exercise 1.4.l).

## Open Problems

1.6.1. Find out whether the join of two Fréchet–Urysohn (sequential, of countable tightness) Hausdorff topological group topologies on a group is Fréchet–Urysohn (sequential, of countable tightness).

1.6.2. Does the product of two topological groups of countable $G_\delta$-tightness ($\delta$-tightness) have countable $G_\delta$-tightness ($\delta$-tightness)? What is the answer if, additionally, both factors have countable pseudocharacter?

1.6.3. Is the product of two (or countably many) $G_\delta$-preserving topological groups $G_\delta$-preserving?

1.6.4. Suppose that $\pi: G \to H$ is a continuous open homomorphism of a $G_\delta$-preserving topological group $G$ onto a topological group $H$. Is $H$ then $G_\delta$-preserving?

## 1.7. Factorization theorems

In this short section we study the question of when a given continuous function defined on a product space $\prod_{i\in I} X_i$ depends only on a "small" subset of the index set $I$. The results of this section will be applied many times in the following chapters, and this material will play an especially important role in Chapter 8.

Let $f: X \to Z$ and $p: X \to Y$ be two continuous mappings. We say that *$f$ factorizes through $p$* if there exists a mapping $g: Y \to Z$, not necessarily continuous, such that $f = g \circ p$. Clearly, $f$ factorizes through $p$ if and only if $p(x_1) = p(x_2)$ implies that $f(x_1) = f(x_2)$, for all $x_1, x_2 \in X$. We also say that *$f$ admits a continuous factorization through $p$* if there exists a continuous mapping $g: Y \to Z$ such that $f = g \circ p$. If this is the case, we write $p \prec f$. The next simple fact will be frequently used in the sequel.

LEMMA 1.7.1. *Let $f: X \to Z$ and $p: X \to Y$ be continuous mappings, where $p(X) = Y$. If $f$ factorizes through $p$ and the mapping $p$ is quotient, then there exists a unique mapping $g: Y \to Z$ satisfying $f = g \circ p$, and $g$ is continuous. Hence, $p \prec f$.*

PROOF. The uniqueness of $g$ is evident. To verify that $g$ is continuous, take an arbitrary open set $U \subset Z$ and note that $p^{-1}(g^{-1}(U)) = f^{-1}(U)$ is open in $X$, by the continuity of $f$. Since $p$ is quotient, it follows that the inverse image $g^{-1}(U)$ is open in $Y$. Therefore, $g$ is continuous. □

Let $X = \prod_{i\in I} X_i$ be a product space and $f: X \to Y$ a continuous mapping. Given a set $J \subset I$, we say that *$f$ depends only on the set $J$* (equivalently, *$f$ does not depend on $I \setminus J$*) if $f$ factorizes through the natural projection $\pi_J: X \to X_J = \prod_{i\in J} X_i$. In other words, $f$ depends only on $J$ if for all $x, y \in X$, $\pi_J(x) = \pi_J(y)$ implies $f(x) = f(y)$. If such a set $J$ is countable, we say that *$f$ depends on countably many coordinates*. For a cardinal $\tau$, the expression "$f$ depends on less than $\tau$ coordinates" is self-explanatory.

We also say that *$f$ depends on an index $j \in I$* if there exist points $x, y \in X$ such that $\pi_i(x) = \pi_i(y)$ for each $i \in I \setminus \{j\}$ and $f(x) \neq f(y)$, where $\pi_i: X \to X_i$ is the projection.

The next general result on factorization of continuous real-valued functions on product spaces is known as Glicksberg's theorem.

THEOREM 1.7.2. **[I. Glicksberg]** *Let $X = \prod_{i\in I} X_i$ be a product space. If $X$ is pseudo-$\tau$-compact, for some cardinal $\tau$ with $cf(\tau) > \omega$, then every continuous real-valued function on $X$ depends on less than $\tau$ coordinates. Furthermore, there exist a set $J \subset I$ with $|J| < \tau$ and a continuous function $h: X_J \to \mathbb{R}$ such that $f = h \circ \pi_J$, where $X_J = \prod_{i\in J} X_i$ and $\pi_J: X \to X_J$ is the projection.*

PROOF. Let $f$ be a continuous real-valued function on $X$. Suppose to the contrary that $f$ depends on at least $\tau$ pairwise distinct coordinates. Denote by $J$ the subset of $I$ consisting of all indices $i \in I$ the function $f$ depends on. We claim that $f$ depends only on $J$.

Indeed, denote by $\pi_J$ the projection of $X$ onto $X_J = \prod_{i\in J} X_i$ and take arbitrary points $x, y \in X$ such that $\pi_J(x) = \pi_J(y)$. If $f(x) \neq f(y)$, we can choose canonical open sets $U$ and $V$ in $X$ such that $x \in U$, $y \in V$ and $f(U) \cap f(V) = \varnothing$. There exists a non-empty finite set $C \subset I$ such that $U = \pi_C^{-1}\pi_C(U)$ and $V = \pi_C^{-1}\pi_C(V)$. Since $\pi_i(x) = \pi_i(y)$ for each $i \in J$, we can assume that $\pi_i(U) = \pi_i(V)$ for all $i \in J$. It is clear that the sets $U$ and $V$ are disjoint, so $D = C \setminus J \neq \varnothing$. Let $D = \{i_1, \ldots, i_n\}$ for some integer $n \geq 1$. It is easy to define points

$z_0, z_1, \ldots, z_n \in X$ such that $z_0 = x$, $z_n = y$ and $\pi_i(z_{k-1}) = \pi_i(z_k)$ for all $i \in I \setminus \{i_k\}$, where $k = 1, \ldots, n$. Since $D \cap J = \varnothing$, we have $f(x) = f(z_0) = f(z_1) = \cdots = f(z_n) = f(y)$. Hence $f(U) \cap f(V) \neq \varnothing$, which is a contradiction. This proves that $f$ depends only on $J$.

Since the projection $\pi_J$ of $X$ onto $X_J$ is open (hence, quotient), it follows from Lemma 1.7.1 that there exists a continuous real-valued function $h$ on $X_J$ such that $f = h \circ \pi_J$.

It remains to show that $|J| < \tau$. Suppose to the contrary that $|J| \geq \tau$. For every $i \in J$, choose points $x_i, y_i \in X$ such that $\pi_j(x_i) = \pi_j(y_i)$, for each $j \neq i$, and $f(x_i) \neq f(y_i)$. Since $cf(\tau) > \omega$, we can find $\varepsilon > 0$ and a set $J' \subset J$ with $|J'| = \tau$ such that $|f(x_i) - f(y_i)| \geq \varepsilon$ for each $i \in J'$. Then, for every $i \in J'$, we choose open neighbourhoods $U_i$ and $V_i$ of $x_i$ and $y_i$, respectively, such that $|f(x) - f(x_i)| < \varepsilon/3$ for all $x \in U_i$ and $|f(y) - f(y_i)| < \varepsilon/3$ for all $y \in V_i$. Put $J_i = J \setminus \{i\}$ and $W_i = V_i \cap \pi_{J_i}^{-1}(\pi_{J_i}(U_i) \cap \pi_{J_i}(V_i))$, where $i \in J'$. Then the sets $W_i$ are open, non-empty, and satisfy $\pi_{J_i}(W_i) = \pi_{J_i}(U_i) \cap \pi_{J_i}(V_i)$. We claim that the family $\{W_i : i \in J'\}$ is locally finite in $X$. Let $x \in X$ be arbitrary. Choose a canonical neighbourhood $O$ of $x$ in $X$ such that $|f(x') - f(x)| < \varepsilon/6$ for each $x' \in O$. Then there exists a finite set $C \subset I$ such that $O = \pi_C^{-1}\pi_C(O)$. If $O \cap W_i \neq \varnothing$ for some $i \in J'$, choose points $y \in O \cap W_i$ and $z \in U_i$ such that $\pi_{J_i}(z) = \pi_{J_i}(y)$. It follows from the choice of the sets $O$, $U_i$, and $V_i$ that $O \cap U_i = \varnothing$, so that $\pi_C(O) \cap \pi_C(U_i) = \varnothing$ and $\pi_C(y) \neq \pi_C(z)$. This implies immediately that $i \in C$, that is, $O$ can intersects only the sets $W_i$ with $i \in C$. Thus, the family $\{W_i : i \in J'\}$ has cardinality $\tau$ and is locally finite in $X$, which contradicts the pseudo-$\tau$-compactness of $X$. The theorem is proved. □

Theorem 1.7.2 implies the following result which is useful on many occasions. In fact, the major part of applications of Glicksberg's theorem requires Theorem 1.7.3 in the special case $\tau = \aleph_0$:

THEOREM 1.7.3. *Let $X = \prod_{i \in I} X_i$ be a product space and $f\colon X \to Z$ a continuous mapping to a Tychonoff space $Z$ satisfying $w(Z) \leq \tau$. If $X$ is pseudo-$\tau^+$-compact, then there exist a set $J \subset I$ with $|J| \leq \tau$ and a continuous mapping $h\colon X_J \to Z$ such that $f = h \circ \pi_J$, where $X_J = \prod_{i \in J} X_i$ and $\pi_J\colon X \to X_J$ is the projection.*

PROOF. Let us identify $Z$ with a subspace of the Tychonoff cube $I^\tau$, where $I = [0, 1]$. For every $\alpha \in \tau$, denote by $p_\alpha$ the projection of $I^\tau$ to the $\alpha$'s factor. Then $f_\alpha = p_\alpha \circ f$ is a continuous real-valued function on $X$. For every $\alpha < \tau$, apply Theorem 1.7.2 to find a set $J_\alpha \subset I$ with $|J_\alpha| \leq \tau$ and a continuous function $g_\alpha\colon X_{J_\alpha} \to I$ satisfying $f_\alpha = g_\alpha \circ \pi_{J_\alpha}$, where $\pi_{J_\alpha}\colon X \to X_{J_\alpha}$ is the projection. Clearly, the cardinality of the set $J = \bigcup_{\alpha<\tau} J_\alpha$ is not greater than $\tau$. Let $\pi^J_{J_\alpha}\colon X_J \to X_{J_\alpha}$ be the projection, $\alpha < \tau$. Then $h_\alpha = g_\alpha \circ \pi^J_{J_\alpha}$ is a continuous real-valued function on $X_J$. Denote by $h$ the diagonal product of the family $\{h_\alpha : \alpha < \tau\}$. The mapping $h\colon X_J \to I^\tau$ is continuous and satisfies the equality $h_\alpha = p_\alpha \circ h$ for each $\alpha < \tau$.

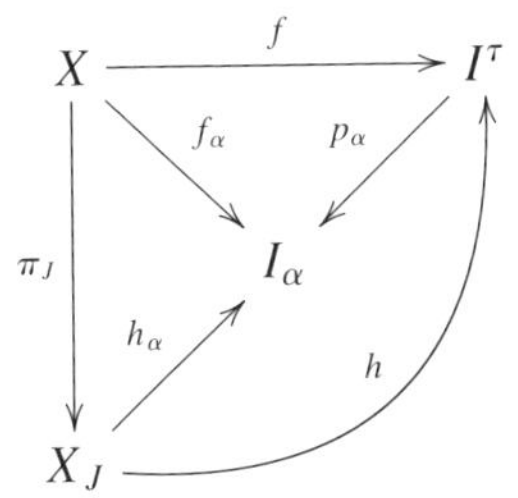

Since $f$ is the diagonal product of the family $\{f_\alpha : \alpha < \tau\}$ and $f_\alpha = h_\alpha \circ \pi_J$ holds for each $\alpha < \tau$, we conclude that $f = h \circ \pi_J$. □

We recall that, for a space $Y$ and an infinite cardinal $\tau$, the fact that $Y$ contains a dense subset $S$ with $|S| \leq \tau$ abbreviates to $d(Y) \leq \tau$ (see [**165**, Section 1.3]). Clearly, the space $Y$ is separable iff $d(Y) \leq \omega$.

COROLLARY 1.7.4. *Let $X = \prod_{i\in I} X_i$ be the product of spaces satisfying $d(X_i) \leq \tau$ for each $i \in I$ and $f\colon X \to Z$ be a continuous mapping to a Tychonoff space $Z$, where $w(Z) \leq \tau$. Then there exist a set $J \subset I$ with $|J| \leq \tau$ and a continuous mapping $h\colon X_J \to Z$ such that $f = h \circ \pi_J$, where $X_J = \prod_{i\in J} X_i$ and $\pi_J\colon X \to X_J$ is the projection.*

PROOF. It is clear that every space $Y$ with $d(Y) \leq \tau$ is pseudo-$\tau^+$-compact. Since the class of spaces that have a dense subset of cardinality $\leq \tau$ is finitely productive, Proposition 1.6.22 implies that the product space $X$ is pseudo-$\tau^+$-compact. The required conclusion now follows from Theorem 1.7.3. □

COROLLARY 1.7.5. [**S. Mazur**] *Every continuous real-valued function on a product of separable spaces depends on countably many coordinates.*

In Theorems 1.7.2, 1.7.3, and Corollary 1.7.4, continuous functions were defined on product spaces. It turns out that, under some additional assumptions, every continuous real-valued function on a dense subspace of a product space admits a continuous factorization via the projection to a countable subproduct. In Theorem 1.7.7 below we present an important result going in this direction. First, we need an auxiliary result about partial factorizations of continuous functions.

LEMMA 1.7.6. *Let $S$ be a dense subspace of a space $X$, $f\colon S \to Y$ and $g\colon X \to Z$ continuous onto mappings, and $T \subset Z$. Suppose that the space $Y$ is regular and that for every $t \in T$, there exists $y_t \in Y$ such that $g^{-1}(t) \subset cl_X f^{-1}(y_t)$. Suppose also that the mapping $g$ is open and put $S_0 = S \cap cl_X g^{-1}(T)$. Then $g{\restriction}S_0 \prec f{\restriction}S_0$.*

PROOF. For every open set $V$ in $Y$, let $e(V) = X \setminus cl_X(S \setminus f^{-1}(V))$. Then $e(V)$ is open in $X$, $S \cap e(V) = f^{-1}(V)$, and $e(V_1) \cap e(V_2) = \varnothing$ if $V_1$, $V_2$ are disjoint open subsets of $Y$. Put $T_0 = g(S_0)$. It is clear that $T_0 \subset cl_Z T$.

**Claim 1.** *If open subsets $V_1$ and $V_2$ of $Y$ satisfy $cl_Y V_1 \cap cl_Y V_2 = \varnothing$, then the intersection $g(e(V_1)) \cap g(e(V_2)) \cap T_0$ is empty.*

Suppose the contrary and consider the open set $O = g(e(V_1)) \cap g(e(V_2))$ in $Z$. Since $O \cap T_0 \neq \varnothing$ and $T_0 \subset cl_Z T$, we can choose a point $t \in O \cap T$. It follows from $g^{-1}(t) \subset cl_X f^{-1}(y_t)$ that $e(V_i) \cap f^{-1}(y_t) \neq \varnothing$ for $i = 1, 2$. If $y_t \notin cl_Y V_i$ for some $i \in \{1, 2\}$, then there exists an open neighbourhood $W$ of $y_t$ in $Y$ such that $W \cap V_i = \varnothing$. Therefore, $e(W) \cap e(V_i) = \varnothing$. It is clear that $f^{-1}(y_t) \subset e(W) \subset X \setminus e(V_i)$, so $f^{-1}(y_t) \cap e(V_i) = \varnothing$, which is impossible. We conclude that $y_t \in cl_Y V_1 \cap cl_Y V_2 \neq \varnothing$, thus contradicting the choice of the sets $V_1$ and $V_2$. This proves Claim 1.

**Claim 2.** *If $x_1, x_2 \in S_0$ and $g(x_1) = g(x_2)$, then $f(x_1) = f(x_2)$.*

Suppose the contrary and choose open neighbourhoods $V_1$ and $V_2$ of the points $f(x_1)$ and $f(x_2)$ in $Y$ such that $cl_Y V_1 \cap cl_Y V_2 = \varnothing$. Then $x_i \in e(V_i) = O_i$ for $i = 1, 2$. Therefore, $g(x_1) \in g(O_1) \cap g(O_2) \cap T_0$, which contradicts Claim 1. So, Claim 2 is proved.

By Claim 2, there exists a mapping $h\colon T_0 \to Y$ such that $f \restriction S_0 = h \circ g \restriction S_0$. Let us prove that $h$ is continuous. Take a point $t \in T_0$ and an open subset $V$ of $Y$ such that $y = h(t) \in V$. Then choose an open neighbourhood $W$ of $y$ with $cl_Y W \subset V$ and consider the sets $O = e(W)$ and $U = g(O) \cap T_0$. Clearly, $t \in U$, and we claim that $h(U) \subset cl_Y W \subset V$. Suppose that $h(U) \setminus cl_Y W \neq \varnothing$. Then there exist points $y_1 \in Y \setminus cl_Y W$ and $t_1 \in U$ such that $h(t_1) = y_1$. We choose an open set $W_1$ containing $y_1$ such that $cl_Y W_1 \cap cl_Y W = \varnothing$, and put $O_1 = e(W_1)$. Pick a point $x \in S_0$ such that $g(x) = t_1$. Then $f(x) = hg(x) = h(t_1) = y_1 \in W_1$, whence $x \in O_1$. Thus, $t_1 \in g(O) \cap g(O_1) \cap T_0$, which contradicts Claim 1. This proves that $h(U) \subset V$ and, hence, $h$ is continuous. Therefore, $g \restriction S_0 \prec f \restriction S_0$. □

THEOREM 1.7.7. *Let $X = \prod_{i \in I} X_i$ be a product space with $cel_\omega(X) \leq \aleph_0$ and suppose that $S$ is a dense subset of $X$. Then, for every continuous mapping $f\colon S \to Y$ to a regular first-countable space $Y$, there exist a countable set $K \subset I$ and a continuous mapping $h\colon p_K(S) \to Y$ such that $f = h \circ p_K \restriction S$, where $p_K\colon X \to \prod_{i \in K} X_i$ is the projection.*

PROOF. By the continuity of $f$, for every point $x \in S$ we can find a canonical $G_\delta$-set $F$ in $X$ such that $x \in F$, $f$ admits a continuous extension over $S \cup F$, and this extension is constant on $F$. Denote by $\mathcal{F}$ the family of these sets $F$. Since $S$ is dense in $X$, we can apply [**165**, 3.2.A(b)] to deduce that there exists a continuous extension of $f$ over $P = \bigcup \mathcal{F}$, and this extension (denoted by the same letter $f$) is constant on each $F \in \mathcal{F}$. It follows from $cel_\omega(X) \leq \omega$ that we can also find a countable subfamily $\mathcal{F}_0$ of $\mathcal{F}$ such that $Q = \bigcup \mathcal{F}_0$ is dense in $P$. Denote by $K$ the union of the cores of the elements of $\mathcal{F}_0$. Then $K$ is a countable subset of $I$ and $Q = p_K^{-1} p_K(Q)$.

It remains to apply Lemma 1.7.6 (with $X = P$, $g = p_K$ and $T = p_K(Q)$) to define a continuous mapping $h\colon p_K(P) \to Y$ such that $f = h \circ p_K \restriction P$. Since $S \subset P$, this finishes the proof. □

COROLLARY 1.7.8. [**A. V. Arhangel'skii**] *Let $S$ be a dense subspace of the product $X = \prod_{i \in I} X_i$, where each space $X_i$ is cosmic. Then, for every continuous real-valued function $f$ on $S$, there exist a countable set $K \subset I$ and a continuous real-valued function $h$ on $p_K(S)$ such that $f = h \circ p_K \restriction S$.*

PROOF. If $J \subset I$ is countable, then the space $X_J = \prod_{i \in J} X_i$ has a countable network and, hence, $cel_\omega(X_J) \leq \omega$. It follows from Lemma 1.6.17 that $cel_\omega(X) \leq \omega$, so Theorem 1.7.7 implies the required conclusion. □

## Exercises

1.7.a. Let $\tau$ be an infinite cardinal. A space $Z$ is said to have a *strong $G_\tau$-diagonal* if there exists a family $\gamma$ of open neighbourhoods of the diagonal $\Delta_Z = \{(x,x) : x \in Z\}$ in $Z^2$ such that $\Delta_Z = \bigcap\{\overline{U} : U \in \gamma\}$ and $|\gamma| \leq \tau$. Note that every regular space $Z$ with $w(Z) \leq \tau$ has a strong $G_\tau$-diagonal. Modify the argument in the proof of Theorem 1.7.2 to show that Theorem 1.7.3 remains valid for any continuous mapping $f\colon X \to Z$ to a space $Z$ with a strong $G_\tau$-diagonal.

1.7.b. Verify that the Sorgenfrey line cannot be represented as a continuous image of a dense subspace of a product of second-countable spaces.

1.7.c. Show that Corollary 1.7.5 can be generalized as follows: Every continuous real-valued function defined on an open subset of a product of separable spaces depends on at most countably many coordinates.

1.7.d. Will Theorem 1.7.7 remain valid if one weakens the first countability of $Y$ to countable pseudocharacter?

## Problems

1.7.A. (A. H. Stone [**461**]) Let $\mathbb{Z}$ be the discrete additive group of integers. Prove that the product group $\mathbb{Z}^{\omega_1}$ is not a normal space.
*Hint.* For every $i = 1, 2$, consider the subset $F_i$ of $G = \mathbb{Z}^{\omega_1}$ which consists of all elements $x = (x_\alpha)$ in $G$ such that for every $j \in \mathbb{Z} \setminus \{i\}$, the equality $x_\alpha = j$ holds for at most one $\alpha < \omega_1$. Verify that $F_1$ and $F_2$ are closed disjoint subsets of $G$. Then apply Corollary 1.7.5 to show that the images $f(F_1)$ and $f(F_2)$ are not disjoint for each continuous real-valued function on $G$. Deduce that the space $G$ is not normal.

1.7.B. (M. G. Tkachenko [**466**]) Let $X = \prod_{i \in I} X_i$ be a product space with countable Souslin number, and $S$ a dense subspace of $X$. Prove that every continuous real-valued function on $S$ admits a continuous factorization through the projection of $S$ to a countable subproduct. In particular, the conclusion holds if $X$ is a product of separable spaces. (This generalizes Corollaries 1.7.5 and 1.7.8.)

## 1.8. Uniformities on topological groups

The continuity of multiplication and inversion in a topological group allows us to supply the group with an additional structure of a *uniform space* which, in its turn, permits us to make use of uniformly continuous functions and apply a well developed theory of uniform spaces (see [**165**, Chap. 8]) to the study of topological groups. We shall give several examples of such applications in Sections 3.3, 3.5, 6.5, 6.6, 6.10, 7.1, 7.7, 7.9, and 7.10.

Let $G$ be a topological group with identity $e$ and $\mathcal{N}_s(e)$ the family of open symmetric neighbourhoods of $e$ in $G$. For an element $V \in \mathcal{N}_s(e)$, we define three subsets $O^l_V$, $O^r_V$ and $O_V$ of $G \times G$ as follows:

$$O^l_V = \{(g, h) \in G \times G : g^{-1}h \in V\}, \tag{1.8}$$

$$O^r_V = \{(g, h) \in G \times G : gh^{-1} \in V\}, \tag{1.9}$$

$$O_V = O^l_V \cap O^r_V. \tag{1.10}$$

Denote by $\Delta_G$ the diagonal of $G \times G$, that is, the set $\Delta_G = \{(x, x) : x \in G\}$. A subset $B$ of $G \times G$ is called *symmetric* if $(y, x) \in B$ for each $(x, y) \in B$. The next result is immediate.

LEMMA 1.8.1. *The sets $O^l_V$, $O^r_V$, and $O_V$ are open symmetric entourages of the diagonal $\Delta_G$ in $G \times G$, for each $V \in \mathcal{N}_s(e)$.*

To introduce three natural uniform structures on $G$, we need one more auxiliary fact. Given two subsets $A$ and $B$ of $G \times G$, the *composition* $A + B$ of $A$ and $B$ is defined to be the set

$$A + B = \{(x, z) \in G \times G : (x, y) \in A \text{ and } (y, z) \in B \text{ for some } y \in G\}.$$

Observe that, in general, $A + B \neq B + A$, even if the sets $A$ and $B$ are symmetric. For a set $A \subset G \times G$ and an integer $n \geq 1$, we define inductively a set $nA \subset G \times G$ by letting $1A = A$, $2A = A + A$ and, in general, $(n + 1)A = nA + A$ for each $n \geq 1$.

LEMMA 1.8.2. *Suppose that $G$ is a topological group, $U$ and $V$ are elements of $\mathcal{N}_s(e)$ in $G$, $n \in \mathbb{N}$, and $V^n \subset U$. Then $nO^l_V \subset O^l_U$, $nO^r_V \subset O^r_U$ and $nO_V \subset O_U$.*

PROOF. We will only verify the inclusion $nO^l_V \subset O^l_U$, the rest is analogous. The case $n = 1$ is trivial, so we may assume that $n \geq 2$.

If $(x, y) \in nO^l_V$, then there exist elements $z_1, \ldots, z_{n-1}$ in $G$ such that $(z_i, z_{i+1}) \in O^l_V$ for each $i = 0, \ldots, n-1$, where $z_0 = x$ and $z_n = y$. Hence, $z_i^{-1} z_{i+1} \in V$ if $0 \leq i \leq n-1$, whence it follows that

$$x^{-1}y = \prod_{i=0}^{n-1} z_i^{-1} z_{i+1} \in V^n \subset U.$$

We thus have $x^{-1}y \in O^l_U$, which implies the inclusion $nO^l_V \subset O^l_U$. □

Let $X$ be a space and $\mathcal{U}$ a uniformity on the underlying set $X$. We say that $\mathcal{U}$ is a *compatible uniformity* on $X$ if the topology induced by $\mathcal{U}$ on $X$ coincides with the original topology of $X$. One can reformulate the definition of compatibility as follows. For every $U \in \mathcal{U}$ and $x \in X$, put

$$U[x] = \{y \in X : (x, y) \in U\}.$$

The set $U[x]$ is called the *$U$-ball* with center at $x$. The uniformity $\mathcal{U}$ on $X$ is compatible with $X$ if $U[x]$ is a neighbourhood of $x$ in $X$ for all $x \in X$ and $U \in \mathcal{U}$, and the family of all $U$-balls forms a neighbourhood base for the original topology of $X$.

We are now in position to define three natural uniformities on a given topological group $G$. Consider the following families:

$$\mathcal{B}^l_G = \{O^l_V : V \in \mathcal{N}_s(e)\}, \tag{1.11}$$

$$\mathcal{B}^r_G = \{O^r_V : V \in \mathcal{N}_s(e)\}, \tag{1.12}$$

$$\mathcal{B}_G = \{O_V : V \in \mathcal{N}_s(e)\}, \tag{1.13}$$

where $\mathcal{N}_s(e)$ denotes, as above, the family of open symmetric neighbourhoods of the identity $e$ in $G$. By Lemma 1.8.1, each of the families $\mathcal{B}^l_G$, $\mathcal{B}^r_G$, and $\mathcal{B}_G$ consists of open symmetric entourages of $\Delta_G$ in $G \times G$. Denote by $\mathcal{D}_G$ the family of symmetric subsets of $G \times G$. Finally, we put:

$$\mathcal{V}^l_G = \{D \in \mathcal{D}_G : O^l_V \subset D \text{ for some } V \in \mathcal{N}_s(e)\}, \tag{1.14}$$

$$\mathcal{V}^r_G = \{D \in \mathcal{D}_G : O^r_V \subset D \text{ for some } V \in \mathcal{N}_s(e)\}, \tag{1.15}$$

$$\mathcal{V}_G = \{D \in \mathcal{D}_G : O_V \subset D \text{ for some } V \in \mathcal{N}_s(e)\}. \tag{1.16}$$

It is clear from the definition that $\mathcal{V}^l_G \subset \mathcal{V}_G$ and $\mathcal{V}^r_G \subset \mathcal{V}_G$. The next theorem explains the role of the six families introduced above.

THEOREM 1.8.3. *Let $G$ be an arbitrary topological group. The families $\mathcal{V}^l_G$, $\mathcal{V}^r_G$, and $\mathcal{V}_G$ are uniformities on the space $G$ with the respective bases $\mathcal{B}^l_G$, $\mathcal{B}^r_G$, and $\mathcal{B}_G$. Each of the three uniformities is compatible with $G$.*

PROOF. We verify the first claim of the theorem only for $\mathcal{V}^l_G$, leaving similar verifications for $\mathcal{V}^r_G$ and $\mathcal{V}_G$ to the reader. According to [**165**, Section 8.1], we have to show that $\mathcal{V}^l_G$ satisfies the following four conditions:

(U1) If $O \in \mathcal{V}^l_G$ and $O \subset W \in \mathcal{D}_G$, then $W \in \mathcal{V}^l_G$.

(U2) If $O_1, O_2 \in \mathcal{V}^l_G$, then $O_1 \cap O_2 \in \mathcal{V}^l_G$.
(U3) For every $O \in \mathcal{V}^l_G$, there is $W \in \mathcal{V}^l_G$ such that $2W \subset O$.
(U4) $\bigcap \mathcal{V}^l_G = \Delta_G$.

Clearly, (U1) follows directly from (1.14). To verify (U2), take arbitrary elements $O_1, O_2 \in \mathcal{V}^l_G$. Again, it follows from (1.14) that there exist $V_1, V_2 \in \mathcal{N}_s(e)$ such that $O^l_{V_i} \subset O_i$ for $i = 1, 2$. Put $V = V_1 \cap V_2$. Then $V \in \mathcal{N}_s(e)$ and $O^l_V \in \mathcal{V}^l_G$. It is clear that $O^l_V \subset O^l_{V_1} \cap O^l_{V_2} \subset O_1 \cap O_2$ and $O_1 \cap O_2 \in \mathcal{D}_G$, so (U1) implies that $O_1 \cap O_2 \in \mathcal{V}^l_G$.

Let us show that (U3) holds. Given an element $O \in \mathcal{V}^l_G$, we can find $U \in \mathcal{N}_s(e)$ such that $O^l_U \subset O$. Choose $V \in \mathcal{N}_s(e)$ satisfying $V^2 \subset U$. Then $W = O^l_V \in \mathcal{V}^l_G$ and Lemma 1.8.2 implies that $2W \subset O^l_U \subset O$.

Finally, let $x$ and $y$ be distinct elements of $G$. Take an element $V \in \mathcal{N}_s(e)$ such that $y \notin xV$. Then $(x, y) \notin O^l_V$, thus proving that $\bigcap \mathcal{V}^l_G = \Delta_G$. This gives (U4).

We conclude that $\mathcal{V}^l_G$ is a uniformity on $G$. From (1.11) and (1.14) it follows that $\mathcal{B}^l_G$ is a base for the uniformity $\mathcal{V}^l_G$. This proves the first part of the theorem for the uniformity $\mathcal{V}^l_G$.

Let us show that the three group uniformities are compatible with $G$. We start with $\mathcal{V}^l_G$. Let $O \in \mathcal{V}^l_G$ and $x \in G$ be arbitrary. Choose $V \in \mathcal{N}_s(e)$ such that $O^l_V \subset O$. Clearly, $O^l_V[x] \subset O[x]$. From (1.8) it follows that $O^l_V[x] = xV$, so the latter set is open in $G$ and we have that $x \in xV \subset O[x]$. Hence $O[x]$ is a neighbourhood of $x$ in $G$ and the family $\{O[x] : O \in \mathcal{V}^l_G\}$ is a neighbourhood base for $G$ at $x$. This implies that the uniformity $\mathcal{V}^l_G$ is compatible with $G$. A similar argument shows that the same remains valid for the uniformity $\mathcal{V}^r_G$.

Finally, an easy verification shows that $O_V[x] = xV \cap Vx$, for all $V \in \mathcal{N}_s(e)$ and $x \in G$. Since the sets $xV \cap Vx$ are open in $G$, this gives the compatibility of the uniformity $\mathcal{V}_G$ with the group $G$. □

In what follows we call $\mathcal{V}^l_G$, $\mathcal{V}^r_G$, and $\mathcal{V}_G$ the *left group uniformity*, *right group uniformity*, and *two-sided group uniformity* on $G$, respectively.

Given a subgroup $H$ of a topological group $G$, one can consider the *induced left uniformity* $\mathcal{V}^l_{G,H}$ on the group $H$ which consists of the intersections $V \cap H^2$, with $V \in \mathcal{V}^l_G$. Similarly, $H$ inherits from $G$ the *right*and *two-sided induced uniformities* denoted respectively by $\mathcal{V}^r_{G,H}$ and $\mathcal{V}_{G,H}$. This, together with the three group uniformities of $H$, increases the number of the natural uniformities on $H$ up to six. Fortunately, the corresponding pairs of these uniformities coincide.

Proposition 1.8.4. *The equalities $\mathcal{V}^l_{G,H} = \mathcal{V}^l_H$, $\mathcal{V}^r_{G,H} = \mathcal{V}^r_H$, and $\mathcal{V}_{G,H} = \mathcal{V}_H$ are valid for each subgroup $H$ of a topological group $G$.*

Proof. It suffices to verify the equality $\mathcal{V}^l_{G,H} = \mathcal{V}^l_H$, the rest is evident. Take an arbitrary open symmetric neighbourhood $V$ of the neutral element $e$ in $G$ and put $U = V \cap H$. Then

$$\begin{aligned} O^l_V \cap (H \times H) &= \{(x, y) \in H \times H : x^{-1}y \in V\} \\ &= \{(x, y) \in H \times H : x^{-1}y \in U\} = O^l_U. \end{aligned}$$

Since $U$ is an open symmetric neighbourhood of $e$ in $H$, the set $O_U^l$ is a basic element of the left group uniformity $\mathcal{V}_H^l$ on $H$. Finally, since the sets $O_V^l$ form a base of the left group uniformity $\mathcal{V}_G^l$ on $G$, we conclude that the uniformities $\mathcal{V}_H^l$ and $\mathcal{V}_{G,H}^l$ coincide. □

The above result shows, in particular, that the induced uniformities on subgroups behave very much like the induced topologies on subsets of a topological space do — given a subset $Y$ of a space $X$ and two subspaces $S, T$ of $X$ with $Y \subset S \cap T$, the topology $Y$ inherits from $S$ is the same one that $Y$ inherits from $T$.

The definition of the two-sided uniformity $\mathcal{V}_G$ on a topological group $G$ given in (1.16) suggests a certain relation between $\mathcal{V}_G^l$, $\mathcal{V}_G^r$ and $\mathcal{V}_G$. This relation is explicitly given in the next theorem.

THEOREM 1.8.5. *For every topological group $G$, the two-sided group uniformity $\mathcal{V}_G$ is the coarsest uniformity on $G$ finer than each of the uniformities $\mathcal{V}_G^l$ and $\mathcal{V}_G^r$.*

PROOF. Since $O_V = O_V^l \cap O_V^r$ for each $V \in \mathcal{N}_s(e)$, from (1.14), (1.15), and (1.16) it follows that $\mathcal{V}_G$ is finer than $\mathcal{V}_G^l$ and $\mathcal{V}_G^r$. Conversely, suppose that $\mathcal{U}$ is a uniformity on $G$ finer than both $\mathcal{V}_G^l$ and $\mathcal{V}_G^r$. Take an arbitrary element $O \in \mathcal{V}_G$. There exists $V \in \mathcal{N}_s(e)$ such that $O_V \subset O$. Since $\mathcal{U}$ is finer than both $\mathcal{V}_G^l$ and $\mathcal{V}_G^r$, we can find $U_1, U_2 \in \mathcal{U}$ such that $U_1 \subset O_V^l$ and $U_2 \subset O_V^r$. Then $U = U_1 \cap U_2 \in \mathcal{U}$ and $U \subset O_V^l \cap O_V^r = O_V \subset O$. Therefore, $\mathcal{U}$ is finer than $\mathcal{V}_G$. □

It is clear that, for an Abelian topological group $G$, the three uniformities $\mathcal{V}_G^l$, $\mathcal{V}_G^r$, and $\mathcal{V}_G$ coincide. It is also well known that every compact Hausdorff space $X$ admits a unique uniformity compatible with it [**165**, Theorem 8.3.13]. Therefore, by Theorem 1.8.3, the three group uniformities coincide for every compact topological group $G$. In fact, the topological groups with this coincidence property admit a complete characterization given in Theorem 1.8.8 below. First, we need two definitions.

Let $G$ be a topological group. A subset $A$ of $G$ is said to be *invariant* if $xAx^{-1} = A$, for each $x \in G$. It is clear that all subsets of Abelian groups are invariant. The group $G$ is called *balanced* if it has a local base at the neutral element consisting of invariant sets. A balanced group is also called a *group with invariant basis.*

It is clear from the above definitions that every Abelian topological group is balanced. The next result gives a useful alternative characterization of balanced groups.

LEMMA 1.8.6. *A topological group $G$ is balanced if and only if, for every neighbourhood $U$ of the identity $e$ in $G$, there exists a neighbourhood $V$ of $e$ such that $xVx^{-1} \subset U$, for each $x \in G$.*

PROOF. Only the sufficiency of the condition requires a proof. Let $U$ be an arbitrary neighbourhood of $e$ in $G$. Choose an open neighbourhood $O$ of $e$ such that $xOx^{-1} \subset U$ for each $x \in G$. Then the set $V = \bigcup_{x \in G} xOx^{-1}$ is open in $G$, contains the identity of $G$ and, clearly, $V \subset U$. It remains to verify that $V$ is invariant. Indeed, take an arbitrary element $y \in G$. Then

$$yVy^{-1} = \bigcup_{x \in G} yxOx^{-1}y^{-1} = \bigcup_{z \in G} zOz^{-1} = V.$$

The above equality shows that the set $V$ is invariant, so the group $G$ has a base of open invariant sets. □

Combining Lemma 1.8.6 and Theorem 1.4.32, we deduce the following fact:

COROLLARY 1.8.7. *Every compact topological group is balanced.*

The next theorem shows that the topological groups with coinciding group uniformities are exactly the balanced groups.

THEOREM 1.8.8. *For a topological group $G$, the uniformities $\mathcal{V}^l_G$, $\mathcal{V}^r_G$, and $\mathcal{V}_G$ coincide if and only if the group $G$ is balanced. Therefore, the three uniformities coincide for every compact topological group $G$.*

PROOF. Suppose that $G$ is balanced. Denote by $\mathcal{N}$ the family of open, symmetric, invariant neighbourhoods of the identity $e$ in $G$. By the assumptions of the theorem, $\mathcal{N}$ is a local base for $G$ at $e$. Hence, the families $\gamma^l = \{O^l_V : V \in \mathcal{N}\}$ and $\gamma^r = \{O^r_V : V \in \mathcal{N}\}$ are bases for the uniformities $\mathcal{V}^l_G$ and $\mathcal{V}^r_G$, respectively. Since each $V \in \mathcal{N}$ is an invariant symmetric subset of $G$, we have:

$$\begin{aligned}(x, y) \in O^l_V &\Leftrightarrow x^{-1}y \in V \Leftrightarrow yx^{-1} \in xVx^{-1} = V \\ &\Leftrightarrow xy^{-1} \in V^{-1} = V \Leftrightarrow (x, y) \in O^r_V.\end{aligned}$$

Therefore, $O^l_V = O^r_V$ for each $V \in \mathcal{N}$, that is, the uniformities $\mathcal{V}^l_G$ and $\mathcal{V}^r_G$ have the same base $\gamma^l = \gamma^r$. This gives the equality $\mathcal{V}^l_G = \mathcal{V}^r_G$. Hence, Theorem 1.8.5 implies that the two-sided uniformity $\mathcal{V}_G$ on $G$ coincides with each of the uniformities $\mathcal{V}^l_G$ and $\mathcal{V}^r_G$.

Conversely, suppose that $\mathcal{V}^l_G = \mathcal{V}^r_G$. Since $\mathcal{B}^l_G$ and $\mathcal{B}^r_G$ are bases for $\mathcal{V}^l_G$ and $\mathcal{V}^r_G$, respectively, for every $U \in \mathcal{N}_s(e)$ there exists $V \in \mathcal{N}_s(e)$ such that $O^l_V \subset O^r_U$. Hence, $x^{-1}y \in V$ implies $xy^{-1} \in U$ for arbitrary $x, y \in G$. In other words, the inclusion $xV \subset Ux$ holds for each $x \in G$. Therefore, $G$ is balanced by Lemma 1.8.6.

The last assertion of the theorem now follows from Corollary 1.8.7. □

The following simple example is a base for many applications.

EXAMPLE 1.8.9. Since the additive topological group of reals $\mathbb{R}$ is commutative, the three group uniformities of $\mathbb{R}$ coincide. Therefore, we denote each of them by $\mathcal{U}$. According to Theorem 1.8.5, the standard base of the uniformity $\mathcal{U}$ is formed by the sets

$$U(\varepsilon) = \{(x, y) \in \mathbb{R}^2 : |x - y| < \varepsilon\},$$

where $\varepsilon > 0$ (see also (1.11)). It is clear that the family $\{U(1/n) : n \in \mathbb{N}\}$ is also a base for $\mathcal{U}$, so the uniformity $\mathcal{U}$ has a countable base. □

In the sequel the space $\mathbb{R}$ will always carry the uniformity $\mathcal{U}$ defined in Example 1.8.9, unless it is explicitly specified otherwise.

Let $G$ be a topological group. A real-valued function $f$ on $G$ is called *left uniformly continuous* if $f$ is a uniformly continuous mapping of $(G, \mathcal{V}^l_G)$ to $(\mathbb{R}, \mathcal{U})$. This means that for every $\varepsilon > 0$, there exists $O \in \mathcal{V}^l_G$ such that $|f(x) - f(y)| < \varepsilon$ whenever $(x, y) \in O$ (see Example 1.8.9). Similarly, $f$ is called *right uniformly continuous* if $f$ is a uniformly continuous mapping of $(G, \mathcal{V}^r_G)$ to $(\mathbb{R}, \mathcal{U})$. Notice that $f$ need not be a homomorphism of $G$ to $\mathbb{R}$, the group structure of $\mathbb{R}$ has been used only to define the uniformity $\mathcal{U}$.

LEMMA 1.8.10. *A real-valued function $f$ on a topological group $G$ is left uniformly continuous if and only if, for every $\varepsilon > 0$, there exists a neighbourhood $V$ of the identity in $G$ such that $|f(xv) - f(x)| < \varepsilon$ for all $x \in G$ and $v \in V$. Similarly, $f$ is right uniformly* *continuous iff for every $\varepsilon > 0$, there exists a neighbourhood $W$ of the identity in $G$ such that $|f(wx) - f(x)| < \varepsilon$ for all $x \in G$ and $w \in W$.*

Proof. By the symmetry argument, it suffices to prove the first claim of the lemma. Suppose that $f$ is left uniformly continuous. Then, for every $\varepsilon > 0$, there exists a neighbourhood $V$ of the identity in $G$ such that $|f(y) - f(x)| < \varepsilon$ for each pair $(x, y) \in O^l_V$. Observe that $(x, y) \in O^l_V$ if and only if $y \in xV$. Therefore, the inequality $|f(xv) - f(x)| < \varepsilon$ holds for all $x \in G$ and $v \in V$. The converse is now immediate. □

We say that a real-valued function $f$ defined on a topological group $G$ is *uniformly continuous* if $f$ is both left uniformly continuous and right uniformly continuous. In Abelian topological groups, every left (right) uniformly continuous function is uniformly continuous. It turns out that an even stronger assertion is valid for compact groups.

Proposition 1.8.11. *Every continuous real-valued function on a compact topological group is uniformly continuous.*

Proof. Let $f\colon G \to \mathbb{R}$ be a continuous function defined on a compact topological group $G$. Since, by Theorem 1.8.8, the left and right group uniformities of $G$ coincide, it suffices to verify that $f$ is left uniformly continuous. Let $\varepsilon > 0$ be a real number. For every $x \in G$, choose a neighbourhood $U_x$ of the neutral element $e$ in $G$ such that $|f(xy) - f(x)| < \varepsilon/2$ for each $y \in U_x$. Then take an open neighbourhood $V_x$ of $e$ such that $V_x^2 \subset U_x$. Since $G$ is compact, there exist elements $x_1, \ldots, x_n$ in $G$ such that $G = \bigcup_{i=1}^n x_i V_{x_i}$. Put $V = \bigcap_{i=1}^n V_{x_i}$. We claim that $|f(xv) - f(x)| < \varepsilon$ for all $x \in G$ and $v \in V$ which, by Lemma 1.8.10, implies the left uniform continuity of the function $f$.

Indeed, given $x \in G$, there exists $i \le n$ such that $x \in x_i V_{x_i}$. Clearly, $x \in x_i V_{x_i} \subset x_i U_{x_i}$, so the choice of $U_{x_i}$ implies that $|f(x) - f(x_i)| < \varepsilon/2$. In addition, if $v \in V$, then $xv \in x_i V_{x_i} V \subset x_i V_{x_i}^2 \subset x_i U_{x_i}$. This gives the inequality $|f(xv) - f(x_i)| < \varepsilon/2$. We thus have

$$|f(xv) - f(x)| \le |f(xv) - f(x_i)| + |f(x_i) - f(x)| < \varepsilon/2 + \varepsilon/2 = \varepsilon,$$

which finishes the proof. □

The above result will be extended to pseudocompact topological groups in Corollary 6.6.9.

Let $f\colon G \to H$ be a mapping of topological groups. We say that $f$ is *left uniformly continuous* if $f$ is uniformly continuous as a mapping of the uniform space $(G, \mathcal{V}^l_G)$ to $(H, \mathcal{V}^l_H)$. Similarly, $f$ is called *right uniformly continuous* if $f$ is uniformly continuous as a mapping of the uniform space $(G, \mathcal{V}^r_G)$ to $(H, \mathcal{V}^r_H)$. If $f$ is both left and right uniformly continuous, we say that it is a *uniformly continuous mapping.*

The following result shows a big difference between continuous mappings and continuous homomorphisms of topological groups (see Exercise 1.8.a).

Proposition 1.8.12. *Every continuous homomorphism $f\colon G \to H$ of topological groups is uniformly continuous.*

Proof. It suffices to verify that $f$ is left uniformly continuous — the right uniform continuity of $f$ will follow by a similar argument. Take an arbitrary basic entourage $O^l_U \in \mathcal{V}^l_H$ of the diagonal in $H \times H$, where $U$ is an open symmetric neighbourhood of the neutral element in $H$. Since $f$ is continuous, there exists an open neighbourhood $V$ of the neutral element in $G$ such that $f(V) \subset U$. Suppose that $(x, y) \in O^l_V$, that is, $x^{-1}y \in V$. Then $f(x)^{-1}f(y) = f(x^{-1}y) \in U$ and, hence, $(f(x), f(y)) \in O^l_U$. This implies that $f$ is left uniformly continuous. □

By Theorem 1.2.7, the product of any family of topological groups has a natural structure of a topological group. In particular, given a topological group $G$ and a positive integer $n$, $G^n$ is a topological group when considered with coordinatewise multiplication and the usual product topology. This enables us considering uniformly continuous mappings of $G^n$ to $G$.

COROLLARY 1.8.13. *Suppose that $G$ is an Abelian topological group. For every $n \in \mathbb{N}$, let $f_n : G^n \to G$ be the multiplication mapping defined by $f_n(x_1, \ldots, x_n) = x_1 \cdots x_n$. Then $f_n$ is uniformly continuous.*

PROOF. Since the topological group $G$ is Abelian, the mapping $f_n$ is a continuous homomorphism. Therefore, the conclusion follows from Proposition 1.8.12. □

Suppose that $(X, \mathcal{U})$ and $(Y, \mathcal{V})$ are uniform spaces. Then the *product* of $(X, \mathcal{U})$ and $(Y, \mathcal{V})$ is a uniform space $(Z, \mathcal{W})$ with the underlying set $Z = X \times Y$ and the uniformity $\mathcal{W}$ on $Z$ whose base consists of the sets

$$W_{U,V} = \{((x, y), (x', y')) \in Z \times Z : (x, x') \in U,\ (y, y') \in V\}, \tag{1.17}$$

where $U \in \mathcal{U}$ and $V \in \mathcal{V}$ (see [**165**, Section 8.2]). The uniformity $\mathcal{W}$ is called the *product* of $\mathcal{U}$ and $\mathcal{V}$ and is written as $\mathcal{W} = \mathcal{U} \times \mathcal{V}$.

In the case of topological groups, there exists an intimate relation between the products of groups and products of their left (right, two-sided) uniformities.

PROPOSITION 1.8.14. *Let $G$ and $H$ be topological groups. Then the left (right, two-sided) group uniformity of the product group $G \times H$ coincides with the product of the left (right, two-sided) group uniformities of $G$ and $H$.*

PROOF. Again, we prove the proposition only for left group uniformities, leaving the rest to the reader. Clearly, in the product group $Z = G \times H$, the sets of the form $U \times V$ constitute a base of open neighborhoods at the neutral element $(e_G, e_H)$, where $U \in \mathcal{N}_s(e_G)$ and $V \in \mathcal{N}_s(e_H)$. It is immediate from the definition of the left group uniformity $\mathcal{V}^l_Z$ on $Z$ that a basic entourage of the diagonal in $Z^2$ has the form

$$O^l_{U,V} = \{((x, y), (x_1, y_1)) \in Z \times Z : x^{-1}x_1 \in U,\ y^{-1}y_1 \in V\},$$

where $U \in \mathcal{N}_s(e_G)$ and $V \in \mathcal{N}_s(e_H)$. An easy calculation also shows that the set $O^l_{U,V}$ coincides with the set $W_{U^*,V^*}$ defined in (1.17), where $U^* = O^l_U \in \mathcal{V}^l_G$ and $V^* = O^l_V \in \mathcal{V}^l_H$. Since the sets $U^*$ and $V^*$, with $U \in \mathcal{N}_s(e_G)$ and $V \in \mathcal{N}_s(e_H)$, form a base for the uniformities $\mathcal{V}^l_G$ and $\mathcal{V}^l_H$, respectively, we infer that the corresponding sets $W_{U^*,V^*}$ form a base for the product uniformity $\mathcal{W} = \mathcal{V}^l_G \times \mathcal{V}^l_H$ on $Z$. Therefore, the equality $O^l_{U,V} = W_{U^*,V^*}$ implies that the uniformities $\mathcal{V}^l_Z$ and $\mathcal{W}$ coincide. □

Apart from the three group uniformities introduced above, every topological group admits a fourth natural uniformity which is called the *Roelcke uniformity*. Again, denote by $\mathcal{N}_s(e)$ the family of open symmetric neighbourhoods of the identity $e$ in a topological group $G$. For an element $V \in \mathcal{N}_s(e)$, let

$$O^t_V = \{(x, y) \in G \times G : y \in VxV\}. \tag{1.18}$$

Obviously, $O^t_V$ is an open symmetric entourage of the diagonal in $G \times G$. Similarly to the case of left and right uniformities, we define two families $\mathcal{B}^t_G$ and $\mathcal{V}^t_G$ as follows:

$$\mathcal{B}^t_G = \{O^t_V : V \in \mathcal{N}_s(e)\} \tag{1.19}$$

and

$$\mathcal{V}^t_G = \{D \in \mathcal{D}_G : O^t_V \subset D \text{ for some } V \in \mathcal{N}_s(e)\}, \tag{1.20}$$

where $\mathcal{D}_G$ is the family of symmetric sets in $G \times G$.

The next result is analogous to Theorem 1.8.3.

THEOREM 1.8.15. *The family $\mathcal{V}^t_G$ is a uniformity compatible with $G$ and $\mathcal{B}^t_G$ is a base for $\mathcal{V}^t_G$. Furthermore, $\mathcal{V}^t_G$ is the finest uniformity on $G$ which is coarser than each of the uniformities $\mathcal{V}^l_G$ and $\mathcal{V}^r_G$.*

PROOF. The verification of the fact that $\mathcal{V}^t_G$ is a uniformity on $G$ goes almost exactly like we argued in the proof of Theorem 1.8.3 in the case of the left group uniformity $\mathcal{V}^l_G$. Therefore, we will only verify the respective conditions (U3) and (U4) for $\mathcal{V}^t_G$:

(U3) For every $O \in \mathcal{V}^t_G$, there is $W \in \mathcal{V}^t_G$ such that $2W \subset O$.

(U4) $\bigcap \mathcal{V}^t_G = \Delta_G$.

Let $O$ be an arbitrary element of $\mathcal{V}^t_G$. By (1.20), there exists $V \in \mathcal{N}_s(e)$ such that $O^t_V \subset O$. Choose $U \in \mathcal{N}_s(e)$ such that $U^2 \subset V$ and put $W = O^t_U$. To see that $2W \subset O$, take two elements $(x, y) \in W$ and $(y, z) \in W$. Then $y \in UxU$ and $z \in UyU$. Hence $z \in U^2xU^2 \subset VxV$, whence it follows that $(x, z) \in O^t_V \subset O$. This gives the inclusion $2W \subset O$, thus implying (U3).

Let $x_1$ and $x_2$ be distinct elements of $G$. Choose disjoint neighbourhoods $U_1$ and $U_2$ of $x_1$ and $x_2$, respectively. Since the right and left translations in $G$ are homeomorphisms, there exist $V_1, V_2 \in \mathcal{N}_s(e)$ such that $V_1x_1 \subset U_1$ and $x_2V_2 \subset U_2$. Let $V = V_1 \cap V_2$. Then $V \in \mathcal{N}_s(e)$ and $Vx_1 \cap x_2V = \varnothing$, so that $x_2 \notin Vx_1V$ and, consequently, $(x_1, x_2) \notin O^t_V$. This implies (U4). We conclude, therefore, that $\mathcal{V}^t_G$ is a uniformity on $G$.

Our next step is to verify that the uniformity $\mathcal{V}^t_G$ is compatible with $G$. Let $O \in \mathcal{V}^t_G$ and $x \in G$ be arbitrary. Choose $V \in \mathcal{N}_s(e)$ such that $O^t_V \subset O$. Clearly, $O^t_V[x] \subset O[x]$. From (1.18) it follows that $O^t_V[x] = VxV$, so the set $O^t_V[x]$ is open in $G$, and we have $x \in VxV \subset O[x]$. Hence $O[x]$ is a neighbourhood of $x$ in $G$. Suppose now that $U$ is an open set in $G$ containing $x$. By the continuity of the mapping $f : G \times G \times G \to G$ defined by $f(x, y, z) = xyz$, there exists an open symmetric neighbourhood $V$ of the identity in $G$ such that $f(V \times \{x\} \times V) \subset U$ or, equivalently, $VxV \subset U$. Therefore, $O^t_V[x] \subset U$ and, hence, the family $\{O[x] : O \in \mathcal{V}^t_G\}$ is a neighbourhood base for $G$ at the point $x$. This implies that the group uniformity $\mathcal{V}^t_G$ is compatible with $G$.

Since $O^l_V \subset O^t_V$ and $O^r_V \subset O^t_V$ for each $V \in \mathcal{N}_s(e)$, it follows that $\mathcal{V}^t_G$ is coarser than $\mathcal{V}^l_G$ and $\mathcal{V}^r_G$. Suppose that $\mathcal{U}$ is a uniformity on $G$ such that $\mathcal{U} \subset \mathcal{V}^l_G$ and $\mathcal{U} \subset \mathcal{V}^r_G$. Let $O$ be an arbitrary element of $\mathcal{U}$. Choose $O_1 \in \mathcal{U}$ such that $O_1 + O_1 \subset O$. Since $\mathcal{U}$ is coarser than both $\mathcal{V}^l_G$ and $\mathcal{V}^r_G$, there exists $V \in \mathcal{N}_s(e)$ in $G$ such that $O^l_V \subset O_1$ and $O^r_V \subset O_1$. Let $x \in G$ and $v, w \in V$ be arbitrary. Then $(vx, x) \in O^r_V \subset O_1$ and $(x, xw) \in O^l_V \subset O_1$, whence it follows that $(vx, xw) \in O_1 + O_1 \subset O$, for all $x \in G$ and $v, w \in V$. Put $y = vx$. Then $(y, v^{-1}yw) \in O$ for all $y \in G$ and $v, w \in V$, so that $O^t_V \subset O$. We have thus proved that $\mathcal{V}^t_G$ is finer than $\mathcal{U}$, whence the second assertion of the theorem follows. □

The next simple fact complements Theorem 1.8.8.

COROLLARY 1.8.16. *The following conditions are equivalent for a topological group $G$:*

a) $\mathcal{V}^l_G = \mathcal{V}^r_G$;

b) $\mathcal{V}^t_G = \mathcal{V}_G$;

c) $\mathcal{V}^t_G = \mathcal{V}^l_G = \mathcal{V}^r_G = \mathcal{V}_G$;

d) *the group $G$ is balanced.*

PROOF. It is clear that c) implies b). The Roelcke uniformity $\mathcal{V}^t_G$ of the group $G$ is coarser than each of the uniformities $\mathcal{V}^l_G$ and $\mathcal{V}^r_G$, by Theorem 1.8.15, while Theorem 1.8.5 implies that the two-sided uniformity $\mathcal{V}_G$ is finer than each of the uniformities $\mathcal{V}^l_G$ and $\mathcal{V}^r_G$. Therefore, b) implies a). In addition, Theorem 1.8.8 gives the equivalence a) $\Leftrightarrow$ d), so it remains to show that d) $\Rightarrow$ c).

Suppose that the group $G$ is balanced. Then $\mathcal{V}_G = \mathcal{V}^l_G = \mathcal{V}^r_G$ by Theorem 1.8.8, so all we need to verify is the equality $\mathcal{V}^t_G = \mathcal{V}^l_G$. Since $\mathcal{V}^t_G$ is coarser than $\mathcal{V}^l_G$, it suffices to show that $\mathcal{V}^l_G \subset \mathcal{V}^t_G$. Let $O \in \mathcal{V}^l_G$ be arbitrary. We can assume without loss of generality that $O = O^l_V$ for some $V \in \mathcal{N}_s(e)$. The group $G$ being balanced, the open symmetric invariant neighbourhoods of the neutral element $e$ constitute a local base for $G$ at $e$. Hence, there exists an invariant set $U \in \mathcal{N}_s(e)$ such that $U^2 \subset V$. If $(x, y) \in O^t_U$, then $y \in UxU = xU^2 \subset xV$, that is, $(x, y) \in O^l_V$. This implies that $O^t_U \subset O^l_V$, whence the inclusion $\mathcal{V}^l_G \subset \mathcal{V}^t_G$ follows. The proof is complete. □

The next construction has several interesting applications. Let $H$ and $K$ be subgroups of an abstract group $G$. The *double coset space* is the family

$$K\backslash G/H = \{KxH : x \in G\}$$

of subsets of $G$. It is clear that for any $x, y \in G$, either $KxH = KyH$ or $KxH \cap KyH = \varnothing$. Therefore, $K\backslash G/H$ is a partition of the group $G$. Denote by $\pi$ the canonical mapping of $G$ onto $K\backslash G/H$ defined by $\pi(x) = KxH$, for every $x \in G$. Then the double coset space $K\backslash G/H$ coincides with the set of fibers of $\pi$.

Suppose now that $H$ and $K$ are closed subgroups of a topological group $G$, and that $K$ is compact. Then the sets $KxH$ are closed in $G$, by Theorem 1.4.30. We topologize $K\backslash G/H$ by declaring the mapping $\pi\colon G \to K\backslash G/H$ quotient.

PROPOSITION 1.8.17. *Let $H$ and $K$ be subgroups of a topological group $G$ such that $K$ is compact and $H$ is closed. Then the double coset space $K\backslash G/H$ is regular and the mapping $\pi\colon G \to K\backslash G/H$ is open.*

PROOF. If $U$ is open in $G$, then the set $\pi^{-1}\pi(U) = KUH$ is open in $G$. Hence $\pi(U)$ is open in $Z = K\backslash G/H$ since the mapping $\pi$ is quotient. This implies that $\pi$ is open.

Let $z$ be an arbitrary point of $Z$. Take $x \in G$ such that $\pi(x) = z$. Then $\pi^{-1}(z) = KxH$ is a closed subset of $G$, by Theorem 1.4.30. Hence the singleton $\{z\}$ is closed in $Z$, for every $z \in Z$, and $Z$ is a $T_1$-space.

Suppose that $O$ is a neighbourhood of a point $z = \pi(x)$ in $Z$. There exist open neighbourhoods $U$ and $V$ of the identity $e$ in $G$ such that $\pi(Ux) \subset O$ and $V^2 \subset U$. Apply Proposition 1.4.32 to choose an open symmetric neighbourhood $W$ of $e$ in $G$ such that $W \subset V$ and $x^{-1}Wx \subset V$, for each $x \in K$; then $WK \subset KV$. Clearly, $\pi(Wx)$ is an open neighbourhood of $\pi(x)$ in $Z$ satisfying $\pi(Wx) \subset O$, and we claim that the closure of $\pi(Wx)$ is contained in $O$.

Indeed, suppose that $\pi(Wy) \cap \pi(Wx) \neq \varnothing$ for some $y \in G$. Then $Wy \cap KWxH \neq \varnothing$ and, consequently, we have:

$$y \in WKWxH \subset KVWxH \subset KV^2xH \subset KUxH = \pi^{-1}\pi(Ux).$$

It follows that $y \in \pi(Ux)$. Since $\pi(Wy)$ is an open neighbourhood of $\pi(y)$ in $Z$, we conclude that all accumulation points of $\pi(Wx)$ lie in $O$, as claimed. Thus the space $Z$ is regular. □

In what follows we denote by $\pi_K$ and $\pi_H$ the natural quotient mappings of $G$ onto the right coset space $K\backslash G$ and the left coset space $G/H$, respectively. The next result generalizes Theorem 1.5.7.

PROPOSITION 1.8.18. *Suppose that $G$, $K$ and $H$ are as in Proposition 1.8.17. Then the natural mapping $q_K\colon G/H \to K\backslash G/H$ defined by $q_K(xH) = KxH$, for each $x \in G$, is open and perfect.*

PROOF. Put $Z = K\backslash G/H$ and let $\pi\colon G \to Z$ be the canonical mapping, $\pi(x) = KxH$ for each $x \in G$. Then $\pi = q_K \circ \pi_H$ and, since the mappings $\pi_H$ and $\pi$ are open and continuous, so is $q_K$.

Let $z = \pi(x)$ be an arbitrary point of $Z$, where $x \in G$. Then $\pi^{-1}(z) = KxH$ and the fiber $q_K^{-1}(z) = \pi_H(KxH) = \pi_H(Kx)$ is compact as a continuous image of the compact set $Kx \subset G$.

It remains to verify that $q_K$ is a closed mapping. Let $z = \pi(x)$ be a point of $Z$. Then $q_K^{-1}(z) = \pi_H(Kx)$ is a compact subset of $G/H$. Take an arbitrary open neighbourhood $O$ of $\pi_H(Kx)$ in $G/H$. Then $Kx \subset \pi_H^{-1}(O)$, so Theorem 1.4.29 implies that there exists an open neighbourhood $V$ of the identity in $G$ such that $KxV \subset \pi_H^{-1}(O)$. In particular, we have that $\pi_H(KxV) \subset O$. Evidently, $W = \pi(xV)$ is an open neighbourhood of $z$ in $Z$ which satisfies

$$\begin{aligned} q_K^{-1}(W) &= q_K^{-1}(\pi(xV)) = \pi_H(\pi^{-1}(\pi(xV))) \\ &= \pi_H(KxVH) = \pi_H(KxV) \subset O. \end{aligned}$$

We have thus proved that, for every neighbourhood $O$ of the fiber $q_K^{-1}(z)$ in $G/H$, there exists an open neighbourhood $W$ of $z$ in $Z$ satisfying $q_K^{-1}(W) \subset O$. Therefore, the mapping $q_K$ is closed. □

Our next goal is to introduce natural uniform structures for double coset spaces. This requires the concept of a *neutral subgroup* of a topological group.

Let us say that a subgroup $H$ of a topological group $G$ is *neutral* in $G$ if for every neighbourhood $U$ of the identity $e$ in $G$, there exists a neighbourhood $V$ of $e$ such that $VH \subset HU$. Clearly, $H$ is neutral in $G$ if and only if for every neighbourhood $U$ of $e$, there exists a neighbourhood $V$ of $e$ satisfying $HV \subset UH$. It follows immediately from the definition that every open subgroup as well as every invariant subgroup of $G$ is neutral in $G$. Therefore, all subgroups of an Abelian topological group are neutral. More generally, every subgroup of a balanced topological group is neutral. More facts on neutral subgroups are given in Exercises 1.8.h–1.8.k.

The next result provides us with a sufficient condition for a subgroup $H$ to be neutral in $G$.

LEMMA 1.8.19. *Let $H$ be a subgroup of a topological group $G$. Suppose that, for every open neighbourhood $U$ of the identity $e$ in $G$, there exists an open neighbourhood $V$ of $e$ in $G$ such that $xVx^{-1} \subset U$ whenever $x \in H$. Then $H$ is neutral in $G$.*

PROOF. Given a neighbourhood $U$ of $e$ in $G$, take an open neighbourhood $V$ of $e$ satisfying $xVx^{-1} \subset U$, for all $x \in H$. Then $xV \subset Ux$ for each $x \in H$, whence it follows that $HV \subset UH$. Therefore, $H$ is neutral in $G$. □

Applying Lemma 1.8.19 and Proposition 1.4.32, we deduce the following fact:

COROLLARY 1.8.20. *Every compact subgroup of a topological group $G$ is neutral in $G$.*

As in the case of topological groups, a double coset space $K\backslash G/H$ can admit natural uniform structures induced, in a sense, by the corresponding uniform structures of the topological group $G$. However, the existence of such uniformities on $K\backslash G/H$ depends on the subgroups $K$ and $H$. We will show how to define two natural uniformities on $K\backslash G/H$ in the case when the subgroup $K$ is compact and $H$ is neutral in $G$. We will also see that the neutrality of $H$ in $G$ is necessary only for the existence of one of these uniformities.

Again, let $\mathcal{N}_s(e)$ be the family of all open symmetric neighbourhoods of the identity $e$ in $G$. Denote by $\pi$ the natural mapping of $G$ onto the quotient space $Z = K\backslash G/H$, $\pi(x) = KxH$ for all $x \in G$. For every $V \in \mathcal{N}_s(e)$, put

$$E^r_V = \{(\pi(x), \pi(y)) : y \in Vx\}$$

and

$$E^l_V = \{(\pi(x), \pi(y)) : y \in xV\}.$$

Since the mapping $\pi\colon G \to Z$ is open, $E^l_V$ and $E^r_V$ are open symmetric entourages of the diagonal in the square of the space $Z$. In the theorem that follows we give some conditions under which the families $\mathcal{B}^r_Z = \{E^r_V : V \in \mathcal{N}_s(e)\}$ and $\mathcal{B}^l_Z = \{E^l_V : V \in \mathcal{N}_s(e)\}$ are bases for the *right uniformity* $\mathcal{E}^r_Z$ and the *left uniformity* $\mathcal{E}^l_Z$, respectively, on the space $Z$.

THEOREM 1.8.21. *Let $K$ and $H$ be subgroups of a topological group $G$, where $H$ is closed and $K$ is compact. Then the family $\mathcal{B}^r_Z$ is a base for the right uniformity $\mathcal{E}^r_Z$ on the double coset space $Z = K\backslash G/H$, and the uniformity $\mathcal{E}^r_Z$ is compatible with $Z$. Furthermore, if the subgroup $H$ is neutral in $G$, then the same assertions are valid for $\mathcal{B}^l_Z$ and $\mathcal{E}^l_Z$.*

PROOF. We will prove the two assertions of the theorem for the family $\mathcal{E}^l_Z$, leaving the rest to the reader.

First, one has to verify that conditions (U1)–(U4) listed in the proof of Theorem 1.8.3 hold valid for the family $\mathcal{E}^l_Z$. Since $\mathcal{B}^l_Z$ is a base for $\mathcal{E}^l_Z$, it suffices to check (U2)–(U4) for the family $\mathcal{B}^l_Z$. Condition (U2) is evident, so we start with (U3). Take an arbitrary open symmetric neighbourhood $V$ of the neutral element $e$ in $G$. There exists $O \in \mathcal{N}_s(e)$ such that $O^2 \subset V$ and, since $H$ is neutral in $G$, we can find $W \in \mathcal{N}_s(e)$ satisfying $W \subset O$ and $HW \subset OH$. Let us show that $2E^l_W \subset E^l_V$.

Suppose that $x, y, y_1, z$ are elements of $G$ such that $y \in xW$, $z \in y_1W$, and $\pi(y) = \pi(y_1)$. In other words, we assume that $(\pi(x), \pi(y)) \in E^l_W$ and $(\pi(y_1), \pi(z)) \in E^l_W$. Then, clearly, $(\pi(x), \pi(z)) \in 2E^l_W$, and we claim that $(\pi(x), \pi(z)) \in E^l_V$. Indeed, it follows from $\pi(y) = \pi(y_1)$ that $y_1 \in KyH$ and, therefore, $y_1 \in KxWH$ and $z \in KxWHW$. Further, it follows from our choice of the sets $O$ and $W$ that

$$WHW \subset WOH \subset O^2H \subset VH.$$

We conclude that $z \in KxVH$ and, hence, $\pi(z) = \pi(xv)$ for some $v \in V$. In its turn, this implies that

$$2E^l_W \subset \{(\pi(x), \pi(xv)) : x \in G,\ v \in V\} = \{(\pi(x), \pi(y)) : y \in xV\} = E^l_V.$$

This gives (U3).

To verify (U4), take distinct elements $\pi(a)$ and $\pi(b)$ in $K\backslash G/H$. Then, clearly, $b \notin KaH$. The set $KaH$ is closed in $G$, by Theorem 1.4.30. Hence, we can find an open symmetric neighbourhood $U$ of $e$ in $G$ such that $bU \cap KaH = \varnothing$. Choose an open neighbourhood $V$ of $e$ in $G$ such that $VH \subset HU$. Then $b \notin KaHU \supseteq KaVH = \pi^{-1}\pi(aV)$. In other words, we have that $b \notin \pi^{-1}\pi(aV)$ or, equivalently, $\pi(b) \notin \pi(aV)$. Now, choose $W \in \mathcal{N}_s(e)$ such that $HW \subset VH$. We claim that $(\pi(a), \pi(b)) \notin E^l_W$.

Indeed, suppose to the contrary that $(\pi(a), \pi(b)) \in E^l_W$. Then there exist $x, y \in G$ such that $\pi(x) = \pi(a)$, $\pi(y) = \pi(b)$ and $y \in xW$. It follows that $x \in KaH$ and $b \in KyH$, whence

$$\begin{aligned} b \in KyH \subset KxWH \subset KKaHWH &= KaHWH \subset KaVHH \\ &= KaVH = \pi^{-1}\pi(aV). \end{aligned}$$

Therefore, $\pi(b) \in \pi(aV)$, which is a contradiction. This proves that $(\pi(a), \pi(b)) \notin E^l_W$, as claimed. So, the intersection of the sets $E^l_W$, where $W \in \mathcal{N}_s(e)$, coincides with the diagonal in the square of the space $Z = K\backslash G/H$, thus implying (U4).

Finally, we show that the uniformity $\mathscr{E}^l_Z$ is compatible with $Z$. It is easy to see that $z \in \pi(xV) \subset E^l_V[z]$ for all $z \in Z$ and $V \in \mathcal{N}_s(e)$, where $x \in G$ satisfies $\pi(x) = z$. Since, by Proposition 1.8.17, the mapping $\pi$ of $G$ to $Z$ is open, $E^r_V[z]$ is a neighbourhood of $z$ in $Z$. Conversely, let $O$ be a neighbourhood of an arbitrary point $z \in Z$. Take $x \in G$ with $\pi(x) = z$ and a symmetric open neighbourhood $V$ of $e$ in $G$ such that $\pi(xV) \subset O$. Then choose $W \in \mathcal{N}_s(e)$ such that $HW \subset VH$. An easy verification shows that $z \in E^l_W[z] \subset \pi(xV) \subset O$. Therefore, the quotient topology of $Z$ is coarser than the topology on $Z$ induced by the uniformity $\mathscr{E}^l_Z$. Thus the two topologies on $Z$ coincide, and the proof is complete. □

Corollary 1.8.22. *Let $G$, $K$, $H$, and $Z$ be as in the first part of Theorem 1.8.21. Then the natural quotient mapping $\pi$ of $(G, \mathscr{V}^r_G)$ to $(Z, \mathscr{E}^r_Z)$ is uniformly continuous. If, in addition, the subgroup $H$ is neutral in $G$, then $\pi\colon (G, \mathscr{V}^l_G) \to (Z, \mathscr{E}^l_Z)$ is also uniformly continuous.*

Proof. Let $U$ be an arbitrary open symmetric neighbourhood of the neutral element in $G$. It follows from our definition of the sets $O^r_G$ and $E^r_U$ that $(\pi \times \pi)(O^r_G) = E^r_U$ and, similarly, if $H$ is neutral in $G$ then $(\pi \times \pi)(O^l_G) = E^l_U$. The conclusion now is immediate. □

In the special case when the subgroup $K$ of the group $G$ is trivial, that is, $K = \{e\}$, Theorem 1.8.21 acquires the following form.

Corollary 1.8.23. *Let $H$ be a closed subgroup of a topological group $G$ and $\pi\colon G \to G/H$ the natural quotient mapping. Then the family $\{(\pi \times \pi)(O^r_V) : V \in \mathcal{N}_s(e)\}$ is a base for a right uniformity $\mathscr{E}^r_M$ of the quotient space $M = G/H$, where $\mathcal{N}_s(e)$ is the collection of all open symmetric neighbourhoods of the identity $e$ in $G$, and $\mathscr{E}^r_M$ is compatible* *with $M$. Furthermore, if $H$ is neutral in $G$, then the family $\{(\pi \times \pi)(O^l_V) : V \in \mathcal{N}_s(e)\}$ is a base for a left uniformity $\mathscr{E}^l_M$ on $M$ compatible with $M$.*

The next example shows that we cannot do without the assumption about the neutrality of $H$ in $G$ in Theorem 1.8.21 and Corollary 1.8.23.

EXAMPLE 1.8.24. Consider the subgroup $G$ of $GL(2, \mathbb{R})$ defined by

$$G = \left\{ \begin{pmatrix} x & y \\ 0 & 1 \end{pmatrix} : x, y \in \mathbb{R}, \ x \neq 0 \right\}.$$

Then

$$H = \left\{ \begin{pmatrix} x & 0 \\ 0 & 1 \end{pmatrix} : x \in \mathbb{R}, \ x \neq 0 \right\}$$

is a closed subgroup of $G$. Denote by $\pi$ the natural mapping of $G$ onto the quotient space $M = G/H$ of right cosets of $H$ in $G$. It follows from Proposition 1.8.17 that the mapping $\pi$ is open and $M$ is a regular space. The topological group $G$ is second-countable (as a subgroup of the second-countable group $GL(2, \mathbb{R})$), and so is the space $M = \pi(G)$ as a continuous open image of $G$.

Suppose that $\mathcal{U}$ is a uniformity on the space $M$ which makes the mapping $\pi\colon (G, \mathcal{V}^l_G) \to (M, \mathcal{U})$ uniformly continuous. We claim that $\mathcal{U}$ is the coarsest uniformity of $M$, that is, $\mathcal{U}$ contains the unique element $M \times M$. To this end, it suffices to show that $(\pi \times \pi)(O^l_V) = M \times M$, for each open symmetric neighbourhood $V$ of the identity $e$ in $G$. It is clear that

$$\begin{aligned}(\pi \times \pi)(O^l_V) &= \{(\pi(x), \pi(y)) : y \in xV\} \\ &= \{(\pi(x'), \pi(y')) : (\exists h_1, h_2 \in H)(\exists x, y \in G) \\ &\qquad\qquad (x' = xh_1, \ y' = yh_2, \ y \in xV)\}.\end{aligned}$$

It follows that $(\pi \times \pi)(O^l_V) = (\pi \times \pi)(O^l_{HVH})$, for each $V \in \mathcal{N}_s(e)$. Hence, all we need to verify is that $HVH = G$, for each $V \in \mathcal{N}_s(e)$. Clearly, for every $V \in \mathcal{N}_s(e)$, there exists $\varepsilon > 0$ such that every matrix

$$\begin{pmatrix} x & y \\ 0 & 1 \end{pmatrix}$$

with $|x - 1| < \varepsilon$ and $|y| < \varepsilon$ is in $V$. In particular, $V$ contains a matrix as above with $xy \neq 0$. A simple calculation shows that

$$\begin{pmatrix} a & 0 \\ 0 & 1 \end{pmatrix} \begin{pmatrix} x & y \\ 0 & 1 \end{pmatrix} \begin{pmatrix} b & 0 \\ 0 & 1 \end{pmatrix} = \begin{pmatrix} axb & ay \\ 0 & 1 \end{pmatrix},$$

so that $HVH = G$ for each $V \in \mathcal{N}_s(e)$. We have thus proved that the unique element of the uniformity $\mathcal{U}$ is $M \times M$. The explanation of this phenomenon is quite clear — the subgroup $H$ of $G$ fails to be neutral in $G$ (see also Problem 1.8.A). □

One more result about the coincidence of natural uniformities on topological groups is in order.

THEOREM 1.8.25. *Let $H$ be a closed invariant subgroup of a topological group $G$ and $M = G/H$ be the quotient group. Then the right uniformity $\mathcal{V}^r_M$ of the topological group $M$ coincides with the right uniformity $\mathscr{E}^r_M$ on $M$ when the latter is considered as the quotient space of $G$ (see Corollary 1.8.23). The corresponding uniformities $\mathcal{V}^l_M$ and $\mathscr{E}^l_M$ on $M$ coincide as well.*

PROOF. Let $\pi: G \to G/H$ be the quotient homomorphism. Take an arbitrary element $U \in \mathcal{N}_s(e)$, where $\mathcal{N}_s(e)$ is the family of all open symmetric neighbourhoods of the neutral element $e$ in $G$. Since the homomorphism $\pi$ is open, $V = \pi(U)$ is an open symmetric neighbourhood of the neutral element in $M = G/H$. Therefore, according to our definitions of the uniformities $\mathcal{V}^r_M$ and $\mathcal{E}^r_M$, it suffices to verify that $(\pi \times \pi)(O^r_U) = O^r_V$.

If $x, y \in G$ and $xy^{-1} \in U$, then $\pi(x)\pi(y)^{-1} = \pi(xy^{-1}) \in \pi(U) = V$. Hence, $(\pi \times \pi)(O^r_U) \subset O^r_V$. Conversely, if $z, t \in M$ and $(z, t) \in O^r_V$, then $zt^{-1} \in V$. Choose $y \in G$ and $u \in U$ such that $\pi(y) = t$ and $\pi(u) = zt^{-1}$. Put $x = uy \in G$. Then $xy^{-1} = u \in U$, that is, $(x, y) \in O^r_U$. We also have that $\pi(x) = \pi(uy) = \pi(u)\pi(y) = zt^{-1}t = z$, whence $(\pi \times \pi)(x, y) = (z, t)$. Therefore, $O^r_V \subset (\pi \times \pi)(O^r_U)$, and the equality $(\pi \times \pi)(O^r_U) = O^r_V$ follows.

A similar argument shows that $(\pi \times \pi)(O^l_U) = O^l_V$ for the same sets $U \in \mathcal{N}_s(e)$ and $V = \pi(U)$, which implies the equality $\mathcal{V}^l_M = \mathcal{E}^l_M$. □

## Exercises

1.8.a. Extend the conclusions of Propositions 1.8.4 and 1.8.14 to the Roelcke uniformity.

1.8.b. Give an example of a continuous function $f: \mathbb{R} \to \mathbb{R}$ which is not uniformly continuous.

1.8.c. Verify that the following conditions are equivalent for any topological group $G$:
   (a) the multiplication mapping $f_2: G^2 \to G$ (see Corollary 1.8.13) is uniformly continuous;
   (b) the multiplication mapping $f_n: G^n \to G$ is uniformly continuous for each $n \geq 2$;
   (c) the group $G$ is balanced.

1.8.d. Show that the general linear groups $GL(n, \mathbb{R})$ and $GL(n, \mathbb{C})$ defined in e) of Example 1.2.5 are balanced iff $n = 1$.

1.8.e. Give an example of topological groups $G$ and $H$ such that the spaces $G$ and $H$ are homeomorphic, but no homeomorphism $f: G \to H$ is uniform.

1.8.f. Let $H$ and $K$ be subgroups of a topological group $G$, where $K$ is compact and $H$ is closed and neutral in $G$. Define the corresponding Roelcke uniformity $\mathcal{E}^t_Z$ on the double coset space $Z = K\backslash G/H$ and show that the uniformity $\mathcal{E}^t_Z$ is compatible with $K\backslash G/H$ (see also Theorem 1.8.21).

1.8.g. Extend Theorem 1.8.25 to the Roelcke uniformity: If $H$ is a closed invariant subgroup of a topological group $G$, then the Roelcke uniformity $\mathcal{V}^t_M$ of the quotient group $M = G/H$ coincides with the Roelcke uniformity $\mathcal{E}^t_M$ on $M$ mentioned in 1.8.f.

1.8.h. (W. Roelcke and S. Dierolf [**410**]) Let $H$ be a closed subgroup of a topological group $G$. Verify that if the left and right uniformities $\mathcal{E}^l_K$ and $\mathcal{E}^r_K$ of the quotient space $K = G/H$ coincide, then $H$ is neutral in $G$.

1.8.i. Find an example of a topological group $G$ and a discrete subgroup $H$ of $G$ such that $H$ is not neutral in $G$.

1.8.j. Verify that the closure $\overline{H}$ of a neutral subgroup $H$ of a topological $G$ is again a neutral subgroup of $G$.

1.8.k. (W. Roelcke and S. Dierolf [**410**]) Let $K$ and $L$ be subgroups of a topological group $G$ such that $L \subset K \subset G$. Then the following assertions are valid:
   (a) If $L$ is a neutral subgroup of $G$, then $L$ is also a neutral subgroup of $K$.
   (b) If $L$ is a neutral subgroup of $K$ and $K$ is dense in $G$, then $L$ is a neutral subgroup of $G$.
   (c) If $L$ is closed and neutral in $G$ and the image $\pi_L(K)$ is a compact subset of the quotient space $G/L$, where $\pi_L: G \to G/L$ is the quotient mapping, then $K$ is neutral in $G$.
   (d) If $L$ is a closed invariant subgroup of $G$ and $K$ is neutral in $G$, then $K/L$ is a neutral subgroup of $G/L$ (see Theorem 1.5.19).

1.8.l. A subset $X$ of a topological group $G$ is called *neutral* if for every neighbourhood $U$ of the identity $e$ in $G$, there exists a neighbourhood $V$ of $e$ in $G$ such that $VX \subset XU$. Show that all compact and all pseudocompact subsets of a topological group $G$ are neutral in $G$.

1.8.m. Suppose that $G$ is a paratopological group. For every open neighbourhood $V$ of the neutral element $e$ in $G$, let

$$O_V = \{(x, y)) \in G \times G : x^{-1}y \in V \text{ and } y^{-1}x \in V\}.$$

Verify that the family $\{O_V : V \in \mathcal{N}(e)\}$ forms a base of a uniformity on $G$, where $\mathcal{N}(e)$ is the family of all open neighbourhoods of $e$ in $G$. Show that this uniformity need not be compatible with $G$.

## Problems

1.8.A. Describe in topological terms the quotient space $G/H$ in Example 1.8.24 and give a direct proof of the fact that $H$ is not neutral in $G$.

1.8.B. (I. V. Protasov [**389**]) A topological group $G$ is called *functionally balanced* if every left uniformly continuous real-valued function on $G$ is right uniformly continuous. Prove the following assertions:

(a) Every balanced topological group is functionally balanced.

(b) A topological group $G$ is functionally balanced iff every subset of $G$ is neutral in $G$ (see Exercise 1.8.l).

(c) Every subgroup of a functionally balanced group is functionally balanced.

(d) A quotient group of a functionally balanced group is functionally balanced.

(e) If a topological group $G$ contains a dense functionally balanced subgroup, then $G$ is functionally balanced itself.

(f) If every neighbourhood of the identity in a group $G$ contains a closed invariant subgroup $N$ such that the quotient group $G/N$ is functionally balanced, then $G$ is functionally balanced as well.

(g) If a functionally balanced group $G$ has a local base at the identity consisting of open balanced subgroups, then $G$ is balanced.

1.8.C. (W. W. Comfort and K. A. Ross [**122**]) A topological group $G$ is called *fine* if every continuous real-valued function on $G$ is left uniformly continuous. Similarly, $G$ is called *b-fine* if every bounded continuous real-valued function on $G$ is left uniformly continuous. Prove the following assertions about a topological group $G$:

(a) If $G$ is metrizable and $b$-fine, then $G$ is compact or discrete.

(b) If $G$ is fine ($b$-fine) and $H$ is a closed invariant subgroup of $G$, then the quotient group $G/H$ is fine ($b$-fine).

1.8.D. (W. W. Comfort and K. A. Ross [**122**]) Show that there exists an Abelian topological group $G$ metrizable by a complete metric such that $G$ is $b$-fine but is not compact.
*Hint.* Consider the abstract group $G = \mathbb{T}^\omega$, where $\mathbb{T}$ is the circle group with the usual *invariant* metric $d$ (that is, $d(ax, ay) = d(x, y)$ for all $a, x, y \in \mathbb{T}$). For two points $x = (x_n)_{n\in\omega}$ and $y = (y_n)_{n\in\omega}$ in $\mathbb{T}^\omega$, let $\varrho(x, y) = \sup_{n\in\omega} d(x_n, y_n)$. Then $\varrho$ is a complete invariant metric on $G$ that generates a topology under which $G$ is a topological group with the required properties. (It is worth noting that there exists an open neighbourhood $U$ of the neutral element in $(G, \varrho)$ such that the group $G$ cannot be covered by less than $2^\omega$ translates of $U$.)

1.8.E. (M. Rice [**406**]) Let $\mathcal{U}$ be a uniformity on a set $X$. A filter $\mathcal{F}$ in $(X, \mathcal{U})$ is said to be *weakly Cauchy* if for every $U \in \mathcal{U}$, there exists $x_U \in X$ such that $U[x_U] \cap F \neq \varnothing$, for each $F \in \mathcal{F}$. A uniformity $\mathcal{U}$ on a set $X$ is called *cofinally complete* if every weakly complete Cauchy filter in $(X, \mathcal{U})$ has a cluster point in $X$. Prove that every first-countable topological group that admits a compatible cofinally complete metric is locally compact.

## Open Problems

1.8.1. (Implicitly, W. W. Comfort and K. A. Ross [**122**]; for locally compact groups, G. Itzkowitz [**257**]) Is every functionally balanced topological group balanced?

## 1.9. Markov's theorem

In this section we will prove that every Tychonoff space $X$ is homeomorphic to a closed subspace of a topological group. This theorem was first proved by A. A. Markov in 1945 with the help of the theory of free topological groups which he created for this purpose. The proof presented below is much simpler and more elementary than the original proof given by Markov. Markov's theorem has fundamental implications. It shows that, in many respects, the topology of a topological group can be as complex as the topology of any Tychonoff space. Later we will show that every topological group is a Tychonoff space (see Theorem 3.3.11). This lead to a conjecture by P. S. Alexandroff that the space of every topological group is normal. However, from Markov's theorem it follows that a topological group need not be normal, since there are non-normal Tychonoff spaces and normality is inherited by closed subspaces.

Let us fix a Tychonoff space $X$ and a topological group $G$. We preserve this notation throughout this section. The space $C_p(X, G)$ of all continuous mappings of $X$ to $G$, with the topology of pointwise convergence and pointwise defined natural operation, is a topological subgroup of the topological group $G^X$ when the latter carries the usual Tychonoff product topology. Therefore, $C_p(X, G)$ itself is a topological group. If $G$ is the additive topological group $\mathbb{R}$ of real numbers, then we write $C_p(X)$ instead of $C_p(X, G)$. In this case, $C_p(X)$ is called the *space of continuous real-valued functions* on $X$.

Let $Y$ be an arbitrary subspace of $C_p(X, G)$. To every $x \in X$ there corresponds the evaluation mapping $\hat{x}\colon C_p(X, G) \to G$ defined by the rule $\hat{x}(f) = f(x)$, for each $f \in C_p(X, G)$. In particular, the value $\hat{x}(f)$ is defined for each $f \in Y$. Obviously, the evaluation mappings $\hat{x}$ are continuous. Thus, we have a mapping $\Psi_Y$ of $X$ to $C_p(Y, G)$ that assigns to each $x \in X$ the restriction of $\hat{x}$ to $Y$. We will call $\Psi_Y$ the *reflection mapping*. It is worth noting that $C_p(Y, G)$ is a topological subgroup of the product group $G^Y$.

The following statement is immediate from the definitions of the reflection mapping and the diagonal product of a family of mappings.

PROPOSITION 1.9.1. *Let $Y$ be an arbitrary subset of $C_p(X, G)$. Then the reflection mapping $\Psi_Y$ coincides with the diagonal product of $Y$.*

COROLLARY 1.9.2. *The reflection mapping $\Psi_Y$ is continuous, for every subset $Y$ of $C_p(X, G)$.*

We say that a set $Y \subset C_p(X, G)$ *separates points of* $X$ if for any distinct points $x_1, x_2 \in X$, there exists $f \in Y$ such that $f(x_1) \neq f(x_2)$.

COROLLARY 1.9.3. *The reflection mapping $\Psi_Y$ is one-to-one if and only if $Y$ separates points of $X$.*

The following simple fact will be used on many occasions.

PROPOSITION 1.9.4. *Let $B$ be a compact subset of a topological group $G$. Then the smallest subgroup $H$ of $G$ containing $B$ is $\sigma$-compact.*

PROOF. Evidently, the subgroup $H$ of $G$ is algebraically generated by $B$. Put $C = B \cup \{e\} \cup B^{-1}$, where $e$ is the neutral element of $G$. Then $C$ is a compact subset of $G$. For every integer $n \in \mathbb{N}$, consider the multiplication mapping $f_n : G^n \to G$ defined by $f_n(x_1, \dots, x_n) = x_1 \cdots x_n$, for all $x_1, \dots, x_n \in G$. Since $G$ is a topological group, the mappings $f_n$ are continuous. Therefore, for every $n \in \mathbb{N}$, $K_n = f_n(C^n)$ is a compact subset of $G$. It is easy to see that $H = \bigcup_{n=1}^{\infty} K_n$, so the group $H$ is $\sigma$-compact. □

We are now ready to prove Markov's theorem in a somewhat stronger form.

THEOREM 1.9.5. *Every Tychonoff space $X$ is homeomorphic to a closed subspace of a topological group which, in its turn, is a dense subgroup of a $\sigma$-compact topological group.*

PROOF. Let $F$ be a Hausdorff compactification of $X$, and set $\Psi = \Psi_{C_p(F)}$. It follows from Corollaries 1.9.3 and 1.9.2 that $\Psi$ is a one-to-one continuous mapping of $F$ to $C_p(C_p(F))$. Since $C_p(C_p(F))$ is a topological subgroup of the group $\mathbb{R}^{C_p(F)}$, it is a Hausdorff space. Therefore, since $F$ is compact, $\Psi$ is a homeomorphism of $F$ onto the subspace $B = \Psi(F)$ of $C_p(C_p(F))$. Hence, $B$ is compact and closed in $C_p(C_p(F))$, and the subspace $M = \Psi(X)$ of $C_p(C_p(F))$ is homeomorphic to $X$.

Let $H$ be the subgroup of the group $C_p(C_p(F))$ algebraically generated by $M$. Clearly, $H$ is dense in the subgroup $P$ of the group $C_p(C_p(F))$ generated by $B$. The group $P$ is $\sigma$-compact, by Proposition 1.9.4.

**Claim.** *The equality $H \cap B = M$ holds.*

Observe that $M \subset H \cap B$. Take an arbitrary point $b \in B \setminus M$. Assume to the contrary that $b \notin H$. Then $b = n_1 y_1 + \cdots + n_k y_k$, where $y_1 = \Psi(x_1), \dots, y_k = \Psi(x_k)$, for some finite collection of distinct points $x_1, \dots, x_k$ in $X$ and some integers $n_1, \dots, n_k$. From the choice of $b$ it follows that $b = \Psi(a)$, for some $a \in F \setminus X$. Clearly, $a \neq x_j$ for each $j \leq k$.

By the definition of operations in the group $C_p(C_p(F))$, we have that $b(f) = n_1 y_1(f) + \cdots + n_k y_k(f)$, for every $f \in C_p(F)$. Hence, $f(a) = n_1 f(x_1) + \cdots + n_k f(x_k)$, for all $f \in C_p(F)$. We can choose $f \in C_p(F)$ such that $f(a) \neq 0$. Therefore, there exists $i \leq k$ such that $n_i \neq 0$. Using the fact that $F$ is compact and, hence, Tychonoff, we select $g \in C_p(F)$ such that $g(a) = 0$ and $g(x_j) = n_j$, for every $j \leq k$. We now have that $g(a) = n_1 g(x_1) + \cdots + n_k g(x_k) = n_1^2 + \cdots + n_k^2 \geq n_i^2 > 0 = g(a)$, which is a contradiction. Thus, our Claim is proved.

It follows that $M$ is closed in $H$, since $B$ is closed in $C_p(C_p(F))$. Since $M$ is homeomorphic to $X$, and $H$ is a dense subgroup of the $\sigma$-compact topological group $P$, the argument is complete. □

THEOREM 1.9.6. [**A. A. Markov**] *There exists an Abelian topological group $H$ such that the space $H$ is not normal.*

PROOF. Take a non-normal Tychonoff space $X$ (for example, we can take as $X$ the Niemytzki plane or the square of the Sorgenfrey line, see [**165**, Examples 1.5.9, 2.3.12]). By Theorem 1.9.5, there exists a topological group $G$ containing $X$ as a closed subspace. Then the space $G$ cannot be normal, since normality is inherited by arbitrary closed subspaces. □

COROLLARY 1.9.7. *There exists an Abelian topological group $H$ such that the space $H$ is not Hewitt–Nachbin complete (that is, the space $H$ is not realcompact).*

PROOF. Take a Tychonoff space $X$ which is not realcompact (for example, we can use the space $\omega_1$ of countable ordinals with the order topology in the role of $X$). There exists a topological group $G$ containing $X$ as a closed subspace, by Theorem 1.9.5. Then the space $G$ cannot be realcompact, since realcompactness is inherited by arbitrary closed subspaces. □

A very interesting general question arises in connection with Theorem 1.9.5: Given a topological property $\mathcal{P}$, when can every Tychonoff space with the property $\mathcal{P}$ be embedded as a closed subspace into a topological group with the property $\mathcal{P}$? Many natural concrete questions of this pattern remain unsolved (see Problems 1.9.1, 1.9.2, 1.9.3, and 9.12.3). However, we do know that every Hewitt–Nachbin complete (i.e., realcompact) space $X$ is homeomorphic to a closed subspace of a Hewitt–Nachbin complete topological group (see Problem 8.8.C).

We now present a few results specifying when the reflection mapping $\Psi_Y$ is a topological embedding (compactness of $X$ is not assumed). Though we did not need these results to prove Markov's theorem, they clarify the situation, and we will refer to some of them on a few occasions later.

It is said that a subset $Y$ of $C_p(X, G)$ *generates the topology of* $X$ if for every $x \in X$ and every closed subset $F$ of $X$ which does not contain $x$, there exists $f \in Y$ such that $f(x)$ is not in the closure of $f(F)$ in $G$.

COROLLARY 1.9.8. *The reflection mapping $\Psi_Y$ embeds $X$ in $C_p(Y, G)$ topologically if and only if the family $Y$ generates the topology of $X$.*

The next statement easily follows from Corollary 1.9.8.

PROPOSITION 1.9.9. *Suppose that $G$ contains a topological copy of the space $\mathbb{R}$, and let $Y = C_p(X, G)$. Then the reflection mapping $\Psi_Y$ is a topological embedding of $X$ into $C_p(Y, G)$.*

In what follows we denote by $C_p^*(X)$ the subspace of $C_p(X)$ consisting of all continuous bounded functions on $X$. Corollary 1.9.8 implies the following statement:

COROLLARY 1.9.10. *Suppose that $G = \mathbb{R}$ and $Y$ is either $C_p(X)$, or $C_p^*(X)$. Then the reflection mapping $\Psi_Y$ embeds $X$ in $C_p(Y)$ topologically.*

Similarly, from Proposition 1.9.9 and Corollary 1.9.8 we obtain:

COROLLARY 1.9.11. *The reflection mapping $\Psi_{C_p(X,\mathbb{T})}$ embeds $X$ in the topological group $C_p(C_p(X, \mathbb{T}), \mathbb{T})$ as a subspace.*

An important complement to Corollary 1.9.10 is the next statement:

PROPOSITION 1.9.12. *For any Tychonoff space $X$, the image of $X$ under the reflection mapping $\Psi = \Psi_{C_p(X)}$ is closed in $C_p(C_p(X))$.*

PROOF. Assume the contrary, and fix $\phi \in C_p(C_p(X))$ such that $\phi \in \overline{\Psi(X)} \setminus \Psi(X)$. A function $f \in C_p(X)$ will be called *special* if there exists an open neighbourhood $U_f$ of $\phi$ in $C_p(C_p(X))$ such that $f(x) = 1$ whenever $\Psi(x) \in U_f$. Let $A$ be the set of all special functions $f \in C_p(X)$. Fix $f \in A$, and let $V$ be a neighbourhood of $\phi$. Since $f \in A$ and $\phi \in \overline{\Psi(X)}$, there exists $b \in X$ such that $\Psi(b) \in V \cap U_f$. Then $f(b) = 1$ and $\Psi(b)(f) = f(b) = 1$.

Since $\Psi(b) \in \Psi(X) \cap V$, it follows, by the continuity of $\phi$, that $\phi(f) = 1$. Therefore, $\phi(f) = 1$ for each $f \in A$.

However, since $\phi$ does not belong to $\Psi(X)$, for every $g \in C_p(X)$ and every finite subset $K$ of $X$, we can find a special function $f \in A$ such that $f$ and $g$ coincide on $K$. Indeed, it follows from $\phi \notin \Psi(X)$ that for every $x \in K$, there exists a function $q_x \in C_p(X)$ such that $\phi(q_x) \neq \Psi(x)(q_x)$ or, equivalently, $\phi(q_x) \neq q_x(x)$. Choose $\varepsilon > 0$ such that $|\phi(q_x) - q_x(x)| \geq 2\varepsilon$ for all $x \in K$. Let $V$ be the family of all functions $h$ in $C_p(C_p(X))$ such that $|h(q_x) - \phi(q_x)| < \varepsilon$ for each $x \in K$. Then $V$ is an open neighbourhood of $\phi$ in $C_p(C_p(X))$ and, clearly, $|\Psi(x)(q_x) - h(q_x)| \geq \varepsilon$ for all $h \in V$ and $x \in K$. In particular, $\Psi(x) \notin \overline{V}$ for each $x \in K$, where the closure of $V$ is taken in $C_p(C_p(X))$. Since the space $C_p(C_p(X))$ is Tychonoff, we can find a continuous real-valued function $f_0$ on $C_p(C_p(X))$ such that $f_0(h) = 1$ for each $h \in \overline{V}$, and $f_0(\Psi(x)) = g(x)$ for each $x \in K$. Then the function $f = f_0 \circ \Psi \in C_p(X)$ is special (it suffices to take $U_f = V$), and $f(x) = g(x)$ for each $x \in K$.

Therefore, $A$ is dense in $C_p(X)$, and, by the continuity of $\phi$, we have $\phi(f) = 1$, for each $f \in C_p(X)$.

On the other hand, the equality $\Psi(x)(\theta) = \theta(x) = 0$ is obviously valid for each $x \in X$, where $\theta$ is the constant zero-function on $X$. Since $\phi \in \overline{\Psi(X)}$, it follows, by the continuity of $\phi$, that $\phi(\theta) = 0$. This contradiction completes the proof. □

Clearly, Corollary 1.9.10 and Proposition 1.9.12 together provide another proof of Markov's theorem.

If a topological space $X$ is endowed with some algebraic structure, then there is a natural way to specify certain subspaces $Y$ of $C_p(X, G)$, and we can consider reflection mappings with respect to them. This turns out to be the first step towards a duality theory, an important chapter in the theory of topological groups. We present in this section only a few elementary results in this direction that concern topological groups (postponing a detailed development of the Pontryagin–van Kampen duality theory for locally compact Abelian groups till Chapter 9). Similar statements hold for topological semigroups and other objects of topological algebra. First, we introduce some notation. For topological groups $H$ and $G$ we denote by $Hom_p(H, G)$ the subspace of the topological group $C_p(H, G)$ consisting of all continuous homomorphisms of $H$ to $G$. Here are two simple statements, the first of which is almost obvious, and the second one is easy to prove.

PROPOSITION 1.9.13. *For any topological groups $H$ and $G$, the subspace $Hom_p(H, G)$ is closed in $C_p(H, G)$.*

PROOF. We will use the additive notation for the group operation in $G$. Take arbitrary elements $x, y \in H$, and let

$$F(x, y) = \{f \in C_p(H, G) : f(xy^{-1}) = f(x) - f(y)\}.$$

Then $F(x, y)$ is a closed subspace of $C_p(H, G)$ for all $x, y \in H$. Evidently, $Hom_p(H, G)$ is the intersection of the sets $F(x, y)$, with $x, y \in H$. Therefore, $Hom_p(H, G)$ is closed in $C_p(H, G)$. □

PROPOSITION 1.9.14. *Suppose that $H$ is a topological group and $G$ is an Abelian topological group. Then $Hom_p(H, G)$ is a topological subgroup of the topological group $C_p(H, G)$.*

PROOF. Again, we use the additive notation for the group operations in $G$ and $C_p(H, G)$. Let $a, b \in H$ and $f, g \in Hom_p(H, G)$. Then, for the element $f + g \in C_p(H, G)$, we have:

$$\begin{aligned}(f+g)(ab) &= f(ab) + g(ab) = f(a) + f(b) + g(a) + g(b)\\ &= (f(a) + g(a)) + (f(b) + g(b)) = (f+g)(a) + (f+g)(b).\end{aligned}$$

Similarly,

$$\begin{aligned}(f+g)(-a) &= f(-a) + g(-a) = -f(a) - g(a) = -(f(a) + g(a))\\ &= -((f+g)(a)).\end{aligned}$$

Hence, $f + g \in Hom_p(H, G)$. We also have:

$$\begin{aligned}(-f)(ab) &= -(f(ab)) = -(f(a) + f(b)) = -f(a) - f(b)\\ &= (-f)(a) + (-f)(b)\end{aligned}$$

and

$$(-f)(-a) = -(f(-a)) = -(-(f(a))) = -((-f)(a)).$$

Therefore, $-f \in Hom_p(H, G)$. It follows that $Hom_p(H, G)$ is closed in $C_p(H, G)$ under the operations of the sum and the inverse. Hence, $Hom_p(H, G)$ is a topological subgroup of the topological group $C_p(H, G)$. □

Evidently, the assumption that the group $G$ is Abelian plays an essential role in the proof of Proposition 1.9.14.

THEOREM 1.9.15. *Let $H$ be a topological group, $G$ an Abelian topological group, and let $\Psi$ be the reflection mapping $\Psi_Y$, where $Y = Hom_p(H, G)$. Then $\Psi$ is a continuous homomorphism of the topological group $H$ to the Abelian topological group $Hom_p(Hom_p(H, G), G)$.*

PROOF. By Corollary 1.9.2, the mapping $\Psi$ is continuous. From Proposition 1.9.14 it follows that $Hom_p(Hom_p(H, G), G)$ is a topological group. This group is Abelian, since the topological group $C_p(X, G)$ is Abelian, for any Tychonoff space $X$. It remains to show that $\Psi$ is a homomorphism.

Take any $a, b \in H$. First, we check that $\Psi(ab) = \Psi(a)\Psi(b)$. Let $f \in Y = Hom_p(H, G)$. Then $\Psi(ab)(f) = f(ab) = f(a)f(b) = (\Psi(a)(f))(\Psi(b)(f))$. Since $f$ is an arbitrary element of $Hom_p(H, G)$, it follows that $\Psi(ab) = \Psi(a)\Psi(b)$. Similarly, one can verify that $\Psi(-a) = -\Psi(a)$. Hence, $\Psi$ is a homomorphism. □

An important direction of research in any kind of a duality theory for topological groups is finding the conditions under which the continuous homomorphism $\Psi$ considered in the last statement happens to be a topological isomorphism of $H$ onto the topological group $Hom_p(Hom_p(H, G), G)$. The topological groups $\mathbb{T}$ and $\mathbb{R}$ in the role of $G$ play a prominent role in this line of investigation. Of course, the group $H$ has to be Abelian if we want $\Psi$ to be a monomorphism. Note also that, in a duality theory, the group $Hom(H, G)$ of all continuous homomorphisms of $H$ to $G$ is often taken with the stronger compact-open topology.

## Exercises

1.9.a. Let $X$ be a Tychonoff space and $Y = C_p(X)$. Verify that the set $\Psi_Y(X)$ is *linearly independent* in $C_p(C_p(X))$, that is, the equality $k_1 y_1 + \cdots + k_n y_n = \bar{0}$ holds for some $k_1, \ldots, k_n \in \mathbb{R}$ and pairwise distinct $y_1, \ldots, y_n \in \Psi_Y(X)$ iff $k_1 = \cdots = k_n = 0$.

1.9.b. Show that every Tychonoff $\sigma$-compact space is homeomorphic to a closed subspace of a $\sigma$-compact topological group.

1.9.c. Show that for every Tychonoff space $X$, $C_p(X)$ is a dense subgroup of the product group $\mathbb{R}^X$.

1.9.d. Let $X$ and $Y$ be topological spaces and $C(X, Y)$ the family of all continuous mappings of $X$ to $Y$. If $C \subset X$ and $W \subset Y$, we put $\langle C, W\rangle = \{f \in C(X, Y) : f(C) \subset W\}$. Then the family

$$\{\langle C, W\rangle : C \subset X \text{ is compact and } W \subset Y \text{ is open}\}$$

forms a subbase for the *compact-open* topology on $C(X, Y)$ (see [**165**, Section 3.4]). The space $C(X, Y)$ endowed with the compact-open topology is denoted by $C_c(X, Y)$. Verify the following assertions:

(i) The compact-open topology is finer than the topology of pointwise convergence on $C(X, Y)$.

(ii) If $X$ and $Y$ are topological groups, then $Hom(X, Y)$ is a closed subset of the space $C_c(X, Y)$.

(iii) If $X$ is a space and $Y$ a topological group, then $C_c(X, Y)$ is also a topological group.

(iv) If $X$ and $Y$ are topological groups, $X$ is locally compact (or, more generally, a $k$-space) and $Y$ is Abelian, then the reflection mapping $\Psi_Y$, with $Y = C_c(X, Y)$, is a continuous homomorphism of $H$ to the Abelian topological group $Hom_c(Hom_c(X, Y), Y)$. (Compare this with Theorem 1.9.15.)

## Problems

1.9.A. Let $X$ be a space and $G$ a topological group. Given a compact set $C \subset X$ and a symmetric open neighbourhood $W$ of the neutral element $e$ in $G$, we put

$$V^l_{C,W} = \{(f, g) \in C_c(X, G) : f(x)^{-1} g(x) \in W \text{ for each } x \in C\}.$$

Prove that the sets $V^l_{C,W}$ constitute a base for a uniformity on $C_c(X, G)$ and this uniformity is compatible with $C_c(X, G)$.

1.9.B. Give an example of a locally compact space that cannot be embedded as a closed subspace into a locally compact topological group.

1.9.C. Can every locally compact $\sigma$-compact space be embedded as a closed subspace into a locally compact $\sigma$-compact topological group?

1.9.D. Give an example of a commutative topological group $G$ algebraically generated by a metrizable subspace such that the space $G$ is not normal.

*Hint.* Let $A$ and $B$ be discrete spaces satisfying $|A| = \omega$, $|B| > \omega$, and $A \cap B = \varnothing$. Denote by $\alpha A$ and $\alpha B$ one-point compactifications of $A$ and $B$, respectively, and let $\alpha A = A \cup \{x_0\}$ and $\alpha B = B \cup \{y_0\}$. It is easy to see that the subspace $Z = \alpha A \times \alpha B \setminus \{(x_0, y_0)\}$ of the compact space $X = \alpha A \times \alpha B$ is not normal. Denote by $G$ the subgroup of $C_p(C_p(X))$ generated by the set $\Psi_Y(Z)$, where $Y = C_p(X)$. By Corollary 1.9.10, $\Psi_Y$ is a topological embedding of $X$ into $C_p(C_p(X))$, so $\Psi_Y(Z)$ is a topological copy of $Z$ in $G$. Apply Exercise 1.9.a to show that $\Psi_Y(X) \cap G = \Psi_Y(Z)$, whence it follows that $\Psi_Y(Z)$ is closed in $G$. In particular, the group $G$ is not normal. Finally, put $P = \alpha A \times B$, $Q = A \times \{y_0\}$, and $c = (x_0, y_0)$. Verify that $M = \Psi_Y(P) \cup (\Psi(Q) + \Psi_Y(c))$ is a metrizable subspace of $G$ (homeomorphic to the topological sum of $P$ and $Q$) and that $G = \langle M\rangle$.

## Open Problems

1.9.1. Suppose that $X$ is a normal topological space. Is it possible to represent $X$ as a closed subspace of a topological group $G$ such that the space $G$ is normal?

1.9.2. Can the Sorgenfrey line be topologically embedded as a closed subspace into a normal (or even Lindelöf) topological group?

1.9.3. Is it true that every paracompact space is homeomorphic to a closed subspace of a paracompact topological group?

## 1.10. Historical comments to Chapter 1

For the definitions and the history of purely algebraic notions, the reader can consult [**409**]. In particular, one can find there a discussion of quaternions. The construction of $r$-adic numbers goes back to K. Hensel [**224**].

Ideas behind the general concept of a topological group have their roots in the works of S. Lie (see [**294**]), who considered groups defined by analytic relations. Another source of topological groups is transformation groups that usually bear a natural topology. In this regard, see also [**293**]. At the beginning of the 20th century D. Hilbert and L. Brouwer showed interest in more general topological groups than just Lie groups. In particular, L. Brouwer demonstrated that the Cantor set can be made, in a natural way, into an Abelian topological group [**87**].

The general notion of topological group was introduced by F. Leja in [**291**] and by O. Schreier in [**423**]. Locally Euclidean topological groups were considered by É. Cartan in [**96**].

An elegant axiomatization of the topology of a general topological group was given by A. Weil [**532**]. Soon after that A. N. Kolmogorov observed that every topological group is a regular space (see a comment by L. S. Pontryagin in [**387**]). Pontryagin proved later that every topological group is a Tychonoff space. It is still an open question whether every regular paratopological group is a Tychonoff space (see Problem 1.3.1).

Associative separately continuous multiplication in the general form was considered, probably for the first time, by D. Montgomery in [**324**]. Semitopological semigroups had already been present in R. Ellis's paper [**159**]. Afterwards they were systematically treated in T. Husain's book [**249**]. Among the first to consider semigroups with multiplication that is only one-sided continuous were R. Ellis [**159**] and W. Gottshalk [**162**]. One of the reasons why objects such as these should not be considered as something too exotic, too special, is that natural topologies of transformation groups very often turn them into semitopological groups only.

Connectedness properties of topological groups were already considered in some of the first articles on topological groups. In particular, Proposition 1.4.26 can be found in [**423**], while Lemma 1.4.27 appeared in [**291**]. In connection with Example 1.4.33, see [**99**].

H. Freudenthal [**177**] and A. A. Markov [**304**] generously contributed to the theory of subgroups and quotient groups of topological groups. Theorem 1.3.14 was established by Markov in [**304**]. Propositions 1.4.1, 1.4.4 and Theorem 1.4.30 are also largely due to Markov [**304**]. The quotient topology on $G/H$, where $G$ is a topological group and $H$ is a subgroup of $G$, was introduced in 1931 in the dissertation of van Dantzig. Markov [**304**] and H. Freudenthal [**177**] established Theorem 1.5.1 in a less general setting. H. Freudenthal proved Corollary 1.5.8 in [**177**]. One finds Theorem 1.5.6 in earlier editions of [**387**]. Theorem 1.5.7 is presented in [**236**]. An improvement of Theorem 1.4.25 appeared in [**277**], where it was shown that every infinite Abelian group admits a non-discrete metrizable group topology. For generalizations of this result see [**121**]. Theorem 1.4.29 was obtained by E. van Kampen [**266**]. In [**325**], a special case of Proposition 1.4.32 can be found.

Theorems 1.5.13, 1.5.18, and 1.5.19 appear, for example, in [**236**], where their origins are discussed as well. Corollary 1.5.21 is a classical result of M. Ya. Vilenkin [**525**]. Exercise 1.5.e is taken from [**99**].

M. Fréchet was the first to consider finite products of abstract spaces in [**175**]. Finite and countable products of metric spaces were already a part of the folklore of the 1920's. However, the addiction to convergent sequences in the study of the general idea of convergence (similar to the still quite common belief that the only topological spaces worthy of some attention are those that are regular and second-countable) prevented for a long time the birth of the general concept of the topological product of spaces. Only in 1930, A. N. Tychonoff gave in [**504**] the general definition of the topological product of an arbitrary family of topological spaces. This was a real breakthrough, a major contribution of General Topology to Mathematics, that widely opened the doors to the future innumerable successes in Functional Analysis, Topological Algebra, Mathematical Logic, and Category Theory. In particular, the general definition of the Tychonoff product of spaces made it immediately clear how to define the topological product of any family of topological groups (paratopological groups, *etc.*). L. S. Pontryagin used countable products of topological groups as a principal tool in his fundamental paper [**385**]. In [**387**], he already used the general topological product of topological groups. The construction of the $\Sigma$-product of spaces described in b) of Example 1.2.9 was also introduced by Pontryagin [**387**] in the context of the theory of topological groups. Years later this construction found numerous applications in General Topology. Proposition 1.6.1, Theorem 1.6.2, and Corollaries 1.6.3 and 1.6.4 have their roots in the Tychonoff theory of products and constitute a standard part of the present day technique. For Proposition 1.6.6, see [**537**]. The concepts of $\delta$-tightness and $G_\delta$-tightness were introduced by A. V. Arhangel'skii in [**41**] and [**39**], respectively, where Proposition 1.6.7, Corollaries 1.6.8, 1.6.9, and Theorems 1.6.11 and 1.6.12 were proved. However, these results were made possible by the work of many authors; see, in particular, [**350**], [**537**], and the article [**468**], where the notion of $o$-tightness was introduced and studied. Historical comments on this subject are given in [**41**]. The class of $G_\delta$-preserving spaces was introduced in [**41**] under a different name, as weakly Klebanov spaces. Theorem 1.6.14 and Corollary 1.6.15 appeared in [**41**]. For Lemma 1.6.17, see [**484**]. The notion of a $\Delta$-system and the techniques related to it (in particular, Theorems 1.6.20 and 1.6.21) are classical; see in this connection [**285**], [**263**], or [**30**].

Pseudo-$\aleph_1$-compact spaces are also known as *DCCC-spaces*. They are especially important in connection with $\mathbb{R}$-factorizable topological groups and free topological groups, which we consider in the forthcoming chapters. Regarding Proposition 1.6.22, see [**484**]. A prototype of Proposition 1.6.23 is found in [**387**]. For Theorems 1.6.24, 1.6.32 and Corollaries 1.6.25 and 1.6.34, see [**350**] and [**351**]. The first part of Corollary 1.6.34 was already known to Pontryagin (see [**387**]). One can find a version of Theorem 1.6.37 in [**165**]. Theorem 1.6.38 is, as we have seen, a work of many mathematicians, beginning with Pontryagin.

Factorization theorems presented and applied in this book are of two kinds — those which deal with "large" products of "small" spaces, and those which refer to topological groups satisfying certain conditions. Among such results are Theorem 1.7.2, due to I. Glicksberg's [**197**], and following from it Theorem 1.7.3 and Corollary 1.7.4. Corollary 1.7.5 is an old classical result of S. Mazur [**313**].

In connection with Theorem 1.7.7 and Corollary 1.7.8, see [**27**], where some of the first factorization theorems for dense subsets of products were established, intended for applications in $C_p$-theory (observe that the space $C_p(X)$ is dense in the product space $\mathbb{R}^X$, for every Tychonoff space $X$). Several important theorems on factorization of mappings defined on subspaces of product spaces were obtained in [**466**] and [**465**] (see also Historical Comments to Chapter 8).

The "germs" of the theory of uniform spaces are found in the theory of metric spaces. The concepts of a uniformly continuous mapping, of a Cauchy sequence, of a natural completion of a metric space gradually lead to understanding that there are situations where metrics are absent or are not adequate but similar ideas are present and lunge to be expressed by means of more general structures. The situation is analogous to the situation in topology — metrics, in general, are not adequate to express convergence properly. A general framework for the ideas of uniform continuity, uniform convergence, completion, *etc.*, was provided by the theory of uniform spaces created by A. Weil in [**532**]. Weil's article also treats the theory of topological groups from the position of the theory of uniform spaces. A basic fact is that the algebraic structure of a topological group, interacting with the topology, imposes two natural uniform structures on the group, generating its topology, in very much the same way as the metric imposes a natural uniform structure generating the topology. For more on that see [**109**] and [**80**]. In connection with Lemma 1.8.6, Corollary 1.8.7, and Theorem 1.8.8 see [**202**] and [**532**]. Proposition 1.8.11 appears in [**387**]. The majority of results in Section 1.8 can be found, with historical comments, in [**410**].

Theorem 1.9.6 is a result of A. A. Markov [**305, 308**]. His proof is quite different; it is based on the much more complicated theory of *multinorms*.

The reflection mapping, also called the *evaluation mapping* used here to prove Markov's results, was available in this general form ever since Tychonoff proved his embedding theorem [**504**]. Pontryagin used reflection mappings in his duality theory. The proof of Markov's theorem, based on $C_p$-theory and reflection mappings, appears now in print for the first time. Proposition 1.9.1 and Corollaries 1.9.2 and 1.9.3 were, for sure, already known to Tychonoff. Corollary 1.9.8, Proposition 1.9.9, Corollaries 1.9.10 and 1.9.11, Proposition 1.9.12, or some very similar statements, can be found in various books and articles on topological function spaces published after 1940. In particular, a systematic treatment of $C_p$-theory is given in [**32**]. Theorem 1.9.5 is a new result.

Further references and comments can be found in [**325**], [**236**], [**410**], [**109**], [**122**], [**110**], and [**80**].

# Chapter 2

# Right Topological and Semitopological Groups

It frequently happens in Mathematics that the study of certain rich structures (like Hilbert or Banach spaces, or bounded linear operators acting on these spaces) requires a detailed knowledge of some weaker structures (locally convex linear topological spaces or, respectively, topological semigroups). The theory of topological groups is not an exception, some subtle properties of certain topological groups can only be established with the use of a well-developed machinery of right topological semigroups.

In the section that follows we show that the Čech–Stone compactification of a discrete semigroup admits a natural structure of a compact right topological semigroup. In particular, each of the spaces $\beta\mathbb{N}$ and $\beta\mathbb{Z}$ admits an associative binary operation which extends the usual sum (or product) operation in $\mathbb{N}$ and $\mathbb{Z}$, respectively. Then we establish some simple algebraic properties of the Čech–Stone compactification of a discrete semigroup $S$ and express in an explicit form "multiplication" of ultrafilters in the compact semigroup $\beta S$.

One of the most important results of contemporary mathematics, Ellis' theorem, is proved in Section 2.2. It simply says that every compact right topological semigroup $S$ has an idempotent, that is, there exists an element $p \in S$ satisfying $pp = p$. This fact has an enormous amount of applications in the theory of numbers, infinite combinatorics, topological algebra, *etc.* One of such applications is Example 2.2.25, where we present a non-discrete extremally disconnected quasitopological group. Another application will be given in Section 4.5 where we construct, following V. I. Malykhin, a maximal topology on a countable infinite Boolean group $G$ under the additional assumption of (something weaker than) Martin's Axiom, that makes $G$ into an extremally disconnected topological group.

In the last three sections of this chapter we present several conditions which imply the continuity of the multiplication and the inverse in semitopological and paratopological groups. For example, we show in Section 2.3 that every locally compact Hausdorff semitopological group is a topological group. The influence of pseudocompactness and Čech-completeness on the continuity of operations in semitopological and paratopological groups is studied in Section 2.4, where we show that every paratopological group is a topological group provided it is pseudocompact or Čech-complete. In fact, every Čech-complete semitopological group is again a topological group (see Theorem 2.4.12). In Section 2.5 we consider the so called *cancellative* semigroups and prove that every compact cancellative topological semigroup is a topological group.

Throughout this chapter, all spaces are assumed to be Hausdorff.

## 2.1. From discrete semigroups to compact semigroups

In this section we show that there is a very natural way to turn the Čech–Stone compactification of a discrete semigroup into a right topological semigroup. Though the construction is standard, it provides us with a large supply of compact right topological semigroups of a highly sophisticated structure from both the topological and the algebraic points of view. In fact, our construction is even more general, as it is clear from the definitions below.

An *operoid* is a non-empty set $O$ with a binary operation called *product* or *multiplication.* Formally, the operation is a mapping $\phi$ of $O \times O$ to $O$; for $(x, y) \in O \times O$ we write $xy$ instead of $\phi(x, y)$ and call $xy$ the product of $x$ and $y$. In general, the associativity of multiplication or the existence of the identity in $O$ are not assumed. Given an operoid $O$ with a topology $\mathcal{T}$, we will call it a *topological operoid* if the multiplication is jointly continuous. As in the case of semigroups, we define the mappings $\varrho_a$ and $\lambda_a$ of $O$ to $O$ by $\varrho_a(x) = xa$ and $\lambda_a(x) = ax$, for any $a$ and $x$ in $O$. The mappings $\varrho_a$ and $\lambda_a$ are called the *right action* and the *left action* by $a$ on the operoid $O$.

If the topology $\mathcal{T}$ on an operoid $O$ is such that $\varrho_a$ is continuous for each $a \in O$, we say that $O$ is a *right topological operoid.* Similarly, if all left actions $\lambda_a$ are continuous, $O$ is called a *left topological operoid*. If $O$ with a topology $\mathcal{T}$ is both a left topological operoid and a right topological operoid, we say that $O$ is a *semitopological operoid*. An element $e$ of an operoid $O$ is called a *right identity* (*left identity*) if $xe = x$ ($ex = x$), for each $x \in O$. If $e \in O$ is both a right identity and a left identity of an operoid $O$, we say that $e$ is an *identity* of $O$. Clearly, an operoid can have at most one identity.

The next result is a starting point for a rich and profound theory of compact right topological operoids and semigroups.

Theorem 2.1.1. **[E. van Douwen]** *Let $O$ be an operoid with the discrete topology and $\beta O$ the Čech–Stone compactification of the discrete space $O$. Then the product operation in $O$ can be extended to a product operation in $\beta O$ in such a way that $\beta O$ becomes a right topological operoid. This can be done in such a way that the left action on $\beta O$ by any element of $O$ be continuous. Furthermore, under the last condition this extension is unique, and if $O$ has a left (right) identity $e$, then $e$ is also a left (right) identity of the operoid $\beta O$.*

Proof. For an element $a$ of $O$, let $\lambda_a$ be the left action by $a$ on $O$. Since $\lambda_a$ is a continuous mapping of $O$ to $O$, we can extend it to a continuous mapping of $\beta O$ to $\beta O$. We also denote the latter mapping by $\lambda_a$, and put $aq = \lambda_a(q)$, for each $q \in \beta O$. Thus, the product $aq$ in $\beta O$ is defined for each $a \in O$ and each $q \in \beta O$.

Now fix $q \in \beta O$ and put $\varrho_q(x) = xq$, for each $x \in O$. In this way a mapping $\varrho_q$ is defined on $O$, with values in $\beta O$. Since $O$ is discrete, $\varrho_q$ is continuous. Therefore, $\varrho_q$ can be extended to $\beta O$; we denote the extension also by $\varrho_q$. Now for any $p$ in $\beta O$, put $pq = \varrho_q(p)$. The definition of the product operation is complete and, since the mapping $\varrho_q$ is continuous for every $q \in \beta O$, $\beta O$ with this product operation is a right topological operoid. Almost all other statements in Theorem 2.1.1 are clearly true. In particular, the statement about identities follows from the continuity of $\lambda_a$ and $\varrho_a$, for each $a \in O$.

The above construction of the extension of the multiplication in $O$ over $\beta O$ shows that this extension is unique. Indeed, suppose that $\psi\colon \beta O \times \beta O \to \beta O$ is a mapping whose restriction to $O \times O$ coincides with the multiplication in $O$ and which makes continuous all right actions $\varrho_a^*$ with $a \in \beta O$, and all left actions $\lambda_a^*$ with $a \in O$, where $\varrho_a^*(x) = \psi(x, a)$ and $\lambda_a^*(x) = \psi(a, x)$ for each $x \in \beta O$. By the assumption, the left actions $\lambda_a^*$ and $\lambda_a$ coincide on the dense subset $X$ of the space $\beta O$ for each $a \in O$, whence it follows that $aq = \psi(a, q)$ for all $a \in O$ and $q \in \beta O$. Equivalently, we have that $\varrho_q^*(a) = \varrho_q(a)$ for each $a \in O$, and the density argument together with the continuity of the right actions $\varrho_q^*$ and $\varrho_q$ for $q \in \beta O$ imply that these actions coincide on $\beta O$. Therefore, $pq = \varrho_q^*(p) = \psi(p, q)$ for all $p, q \in \beta O$. Thus, $\psi$ coincides with the multiplication in $\beta O$ defined at the beginning of the proof. □

Whenever $O$ is a discrete operoid, we consider $\beta O$ as the right topological operoid with the product operation defined in the proof of Theorem 2.1.1. The next result serves to prove Theorem 2.1.3 showing that the compact operoid $\beta O$ plays a main role among all compact right topological operoids containing $O$ as a discrete *suboperoid*. Naturally, a *suboperoid* $S$ of an operoid $O$ is a non-empty subset $S$ of $O$ closed under the multiplication in $O$. In other words, $xy \in S$ for all $x, y \in S$. Then the multiplication in $S$ is, of course, the restriction to $S$ of the multiplication in $O$.

A mapping $h: S \to T$ of operoids $S$ and $T$ is called a *homomorphism* if it satisfies $h(xy) = h(x)h(y)$ for all $x, y \in S$. It is easy to see that if $e$ is the identity of $S$, then $h(e)$ is the identity of $T$ for every homomorphism $h$ of $S$ to $T$.

PROPOSITION 2.1.2. *Let $S$ and $T$ be compact right topological operoids, $D$ a dense suboperoid of $S$, and $h$ a continuous mapping of the space $S$ to the space $T$ satisfying the following conditions:*

a) *the left action $\lambda_a$ is continuous on $S$, for every $a \in D$;*
b) *the restriction of $h$ to $D$ is a homomorphism of $D$ to $T$;*
c) *the left action $\lambda_{h(a)}$ is continuous on $T$, for every $a \in D$.*

*Then $h$ is a homomorphism of $S$ to $T$.*

PROOF. It follows from a)–c) of the proposition that, for each $a \in D$, $h \circ \lambda_a$ and $\lambda_{h(a)} \circ h$ are continuous mappings of $S$ to $T$ coinciding on the dense subset $D$ of $S$. Therefore, they coincide on the whole of $S$, that is, $h(ay) = h(a)h(y)$, for all $a \in D$ and $y \in S$. However, $h(ay) = h(\varrho_y(a))$, and $h(a)h(y) = \varrho_{h(y)}(h(a))$. It follows that the mappings $h \circ \varrho_y$ and $\varrho_{h(y)} \circ h$ coincide on the dense subset $D$ of $S$. Since the mappings $h \circ \varrho_y$ and $\varrho_{h(y)} \circ h$ are continuous, we conclude that they coincide on $S$, that is, $h(x)h(y) = h(xy)$, for all $x, y \in S$. Thus, $h$ is a homomorphism. □

A characteristic property of the Čech–Stone compactification $\beta X$ of a Tychonoff space $X$ is that every continuous mapping $f: X \to K$ to a compact space $K$ can be extended to a continuous mapping $\tilde{f}: \beta X \to K$. The next result has a similar nature and takes into a account the algebraic structure of the Čech–Stone compactification of a discrete operoid $O$.

THEOREM 2.1.3. *Assume that $O$ is a discrete operoid, and $g$ is a homomorphism of $O$ to a compact right topological operoid $T$ such that for each $a \in O$, the left action $\lambda_{g(a)}$ on $T$ is continuous. Then there exists a continuous homomorphism $h$ of the compact right topological operoid $\beta O$ to $T$ such that $h(x) = g(x)$ for each $x \in O$, that is, the restriction of $h$ to $O$ coincides with $g$.*

PROOF. The mapping $g$ is continuous, since the space $O$ is discrete. It follows, by the main property of Čech–Stone compactifications, that $g$ can be extended to a continuous mapping $h$ of $\beta O$ to $T$ (it is here the compactness of $T$ is essential). Now it follows from Proposition 2.1.2 and the properties of the operoid $\beta O$ that $h$ is a homomorphism. □

THEOREM 2.1.4. *If $O$ is a discrete semigroup, that is, a discrete operoid with associative multiplication, then $\beta O$ is a right topological semigroup.*

PROOF. We have to show that the multiplication in $\beta O$ defined in Theorem 2.1.1 is associative, that is, $(pq)r = p(qr)$, for any $p$, $q$, and $r$ in $\beta O$.

We obviously have $p(qr) = \varrho_{qr}(p)$ and $(pq)r = \varrho_r(\varrho_q(p))$. Since all the right actions on $\beta O$ are continuous, it suffices to show that the mappings $\varrho_{qr}$ and $\varrho_r \circ \varrho_q$ coincide on $O$. Let $a \in O$. Then $\varrho_r(\varrho_q(a)) = (aq)r = \varrho_r(\lambda_a(q))$ while $\varrho_{qr}(a) = a(qr) = \lambda_a(\varrho_r(q))$. Since $\lambda_a$ and $\varrho_r$ are continuous on $\beta O$, it suffices to show that $\lambda_a \circ \varrho_r$ and $\varrho_r \circ \lambda_a$ coincide on $O$. Take any $b \in O$. Then $\lambda_a(\varrho_r(b)) = a(br) = \lambda_a(\lambda_b(r))$ while $\varrho_r(\lambda_a(b)) = (ab)r = \lambda_{ab}(r)$. Since the mappings $\lambda_a$, $\lambda_b$, and $\lambda_{ab}$ are continuous on $\beta O$, it remains to check that $\lambda_a(\lambda_b(c)) = \lambda_{ab}(c)$, for each $c \in O$. Since the product operation in $O$ is associative, we have

$$\lambda_a(\lambda_b(c)) = a(bc) = (ab)c = \lambda_{ab}(c),$$

for each $c \in O$. Hence, the product operation in $\beta O$ is associative, and $\beta O$ is a right topological semigroup. □

Theorem 2.1.4 and the method of its proof may suggest the conjecture that if $O$ is commutative then $\beta O$ is also commutative. In general, this is not the case, even if we assume that $O$ is a group. It can be shown that, even for the discrete group $\mathbb{Z}$ of integers, the compact right topological semigroup $\beta\mathbb{Z}$ is not commutative. Even more, $\beta\mathbb{Z}$ is not a semitopological semigroup. We will discuss reasons for that later. However, we have the following partial facts.

PROPOSITION 2.1.5. *If $O$ is a commutative discrete operoid, then $O$ is contained in the center of $\beta O$, that is, $ap = pa$ for all $a \in O$ and $p \in \beta O$.*

PROOF. For each $a$ in $O$, the mappings $\lambda_a$ and $\varrho_a$ are continuous, by their definitions in the proof of Theorem 2.1.1. Since the operoid $O$ is commutative, the restrictions of the mappings $\lambda_a$ and $\varrho_a$ to $O$ coincide. Since $\beta O$ is Hausdorff and $O$ is dense in $\beta O$, it follows that the mappings $\lambda_a$ and $\varrho_a$ coincide everywhere on $\beta O$. That means precisely that $ap = pa$, for all $a \in O$ and $p \in \beta O$. □

PROPOSITION 2.1.6. *If $O$ is a commutative discrete operoid, then the right topological operoid $\beta O$ is commutative if and only if $\beta O$ is semitopological, that is, all left actions $\lambda_a$ are continuous.*

PROOF. Assume that $\beta O$ is commutative. Then $\lambda_p$ coincides with $\varrho_p$, for each $p \in \beta O$. Therefore the operoid $\beta O$ is semitopological, since it is right topological.

Conversely, assume that $\lambda_p$ is continuous, for each $p \in \beta O$. Every right action $\varrho_p$ is also continuous, by Theorem 2.1.1. From Proposition 2.1.5 it follows that $\varrho_p$ restricted to $O$ coincides with the restriction of $\lambda_p$ to $O$. Since $O$ is dense in $\beta O$, it follows that $\varrho_p = \lambda_p$, for each $p \in \beta O$. □

It is helpful to describe the product operation in $\beta O$ in more direct terms involving the actual structure of the ultrafilters multiplied. This is done in the following two statements, in which $O$ is a discrete operoid, and $\beta O$ is the compact operoid generated by it. For any subset $A$ of $O$ we denote by $U_A$ the set of all $p \in \beta O$ such that $A \in p$. It is well known that each $U_A$ is open and the family $\mathcal{T}_p = \{U_A : A \in p\}$ is a base for the space $\beta O$ at the point $p$ (see [**165**, Section 3.6]). We use these simple facts below.

Given an operoid $S$, a set $B \subset S$, and an element $a \in S$, we put $aB = \lambda_a(B)$ and $Ba = \varrho_a(B)$.

PROPOSITION 2.1.7. *Let $a \in O$ and $q \in \beta O$. Then $A \in aq$ if and only if there is $B \in q$ such that $aB \subset A$.*

PROOF. Assume that $A \in aq$. Then $U_A = \{p \in \beta O : A \in p\}$ is an open neighbourhood of $aq$. Since $\lambda_a$ is continuous on $\beta O$, and $\lambda_a(q) = aq$, there exists $B \in q$ such that $\lambda_a(U_B) \subset U_A$. Since $\lambda_a(O) \subset O$, it follows that $aB = \lambda_a(B) \subset U_A \cap O = A$.

Now we prove the converse statement. Assume that $aB \subset A$, for some $B \in q$. Then $q \in \overline{B}$ which implies, by the continuity of $\lambda_a$ on $\beta O$, that $aq = \lambda_a(q) \in \overline{\lambda_a(B)} = \overline{aB}$. Since, by the assumption, $aB \subset A$, we conclude that $aq \in \overline{A}$. However, this implies that $A \in aq$, by the definition of the topology on $\beta O$. □

Note that we can reformulate Proposition 2.1.7 as follows:

$$aq = \{A \subset O : \text{ there exists } B \in q \text{ such that } aB \subset A\},$$

for all $a \in O$ and $q \in \beta O$.

PROPOSITION 2.1.8. *Let $p \in \beta O$ and $q \in \beta O$. Then the following three statements are equivalent:*

a) *$A \in pq$;*
b) *there exists $B \in p$ such that $A \in bq$, for each $b \in B$;*
c) *there exists $B \in p$ such that for each $b \in B$ there exists $C \in q$ with $bC \subset A$.*

PROOF. It follows from Proposition 2.1.7 that b) and c) are equivalent.

Now assume that $A \in pq$. Then $\varrho_q(p) \in U_A$ and, since $U_A$ is open and $\varrho_q$ is continuous, it follows that there exists $B \in p$ such that $\varrho_q(U_B) \subset U_A$. In particular, since $B \subset U_B$, we have $bq = \varrho_q(b) \in U_A$, for each $b \in B$. By the definition of $U_A$, this means that $A \in bq$, for each $b \in B$. Thus, we have shown that a) implies b).

It remains to check that c) implies a). Assume that a subset $A$ of $O$ satisfies c) but not a). Thus, $A$ is not in $pq$. Since $pq$ is an ultrafilter on $O$, it follows that $O \setminus A \in pq$. Then $O \setminus A$ satisfies a) and, therefore, b). Hence, there is $B_1 \in p$ such that, for each $b \in B_1$, $O \setminus A \in bq$. On the other hand, since $A$ satisfies b), there exists $B_2 \in p$ such that $A \in bq$, for every $b \in B_2$. Then $B = B_1 \cap B_2 \in p$, $B$ is not empty, and for every $b$ in $B$ we have $A \in bq$ and $O \setminus A \in bq$, a contradiction. □

Proposition 2.1.7 can be improved slightly and simplified in a special case. Indeed, we have

PROPOSITION 2.1.9. *Suppose that $a \in O$ and $q \in \beta O$. Then:*

a) *if $A \subset O$ and $A \in q$, then $aA \in aq$;*
b) *if $aO = O$ (i.e., the mapping $\lambda_a$ is surjective), then $aq = \{aA : A \in q\}$.*

PROOF. Item a) follows directly from Proposition 2.1.7. Let us prove b). Because of a), it suffices to show that every $C \in aq$ is of the form $aA$, for some $A \in q$. Put $A = \{x \in O : ax \in C\}$. By Proposition 2.1.7, there is $B \in q$ such that $aB \subset C$. Then

$B \subset A$, and since $q$ is an ultrafilter on $O$, it follows that $A \in q$. Obviously, $aA = C$, since $C \subset O = aO$. □

Let us show that the commutativity of a discrete semigroup $S$ does not imply that the operoid $\beta S$ is commutative. Actually, the examples when this happens are very easy to come about; however, to establish the non-commutativity of $\beta S$ is very often a non-trivial task. Because of this, we present the simplest example, where this phenomenon happens and can be easily observed.

As usual, an ultrafilter $p$ on an infinite set $A$ is called *free* if the intersection of the elements of $p$ is empty.

EXAMPLE 2.1.10. Let $\mathbb{N}$ be the discrete semigroup of positive natural numbers where the product $mn$ is defined as $\max\{m,n\}$. Take any $m \in \mathbb{N}$ and any $q \in \beta\mathbb{N} \setminus \mathbb{N}$. Fix an element $A$ of $q$. Then $A$ is infinite, and $mn = n$ for all but finitely many elements of $A$. It follows that the set $K = A \setminus mA$ is finite. Then $A \setminus K \in q$ and $A \setminus K \subset mA$, which implies that $mA \in q$, since $q$ is a free ultrafilter on $\mathbb{N}$.

Now it follows from Proposition 2.1.7 that $mq = q$, that is, $\varrho_q(m) = q$. Since this is true for each $m \in \mathbb{N}$, and $\varrho_q$ is continuous on $\beta\mathbb{N}$, it follows that $\varrho_q(p) = q$ for every $p \in \beta\mathbb{N}$. Thus, we have established that $pq = q$, for every $p \in \beta\mathbb{N}$ and every $q \in \beta\mathbb{N} \setminus \mathbb{N}$.

Now, for any two different elements $p$ and $q$ of $\beta\mathbb{N} \setminus \mathbb{N}$, from the formula proved above, we have that $pq = q \neq p = qp$. Hence, the operoid $\beta\mathbb{N}$ is not commutative. □

A natural question to consider is when, for a discrete operoid $O$, $\beta O \setminus O$ is a suboperoid of $\beta O$. This is not always the case.

EXAMPLE 2.1.11. Let $O$ be any infinite set, $a \in O$ a fixed element of $O$, and the multiplication on $O$ is defined as follows: $xy = a$, for any $x$, $y$ in $O$. Then, for the operoid $\beta O$, we obviously have that $pq = a$, for any $p$ and $q$ in $\beta O$. Since $O$ is infinite, the set $\beta O \setminus O$ is non-empty; clearly, it is not closed under the multiplication, since $a$ is not in it. Hence, $\beta O \setminus O$ is not a suboperoid of $\beta O$. □

Now we give a sufficient condition for $\beta O \setminus O$ to be a suboperoid of $\beta O$.

PROPOSITION 2.1.12. *Suppose that $O$ is a discrete operoid such that, for any infinite subset $C$ of $O$ and any $b \in O$, the set $bC$ is also infinite. Then $\beta O \setminus O$ is a suboperoid of $\beta O$.*

PROOF. We only have to show that the set $\beta O \setminus O$ is closed under multiplication. Assume the contrary. Then we can fix $p$ and $q$ in $\beta O \setminus O$ such that $pq \in O$. Then $\{a\} \in pq$, for some $a \in O$. By c) of Proposition 2.1.8 applied to the set $A = \{a\}$, we can fix $B \in p$ such that for each $b \in B$ there exists $C$ in $q$ such that $bC \subset \{a\}$. Now fix $b \in B$ and $C \in q$ such that $bC \subset \{a\}$. Therefore, the set $bC$ is finite, and it follows from our assumption that $C$ is also finite. Since $C \in q$, we conclude that $q$ is in $O$ and not in $\beta O \setminus O$, a contradiction. □

In the above argument it is not important whether $p$ is in $\beta O \setminus O$ or just in $\beta O$. In fact, under the restrictions imposed on the product operation in Proposition 2.1.12, $pq \in \beta O \setminus O$ for all $p \in \beta O$ and $q \in \beta O \setminus O$. Thus, $(\beta O)(\beta O \setminus O) \subset \beta O \setminus O$. A subset $S$ of an operoid $O$ is called a *left ideal* of $O$ if $OS \subset S$. Similarly, a subset $S$ of an operoid $O$ is called a *right ideal* of $O$ if $SO \subset S$. Thus, Proposition 2.1.12 can be strengthened as follows:

PROPOSITION 2.1.13. *Suppose that $O$ is a discrete operoid such that the set $bC$ is infinite for every infinite subset $C$ of $O$ and for every $b \in O$. Then $\beta O \setminus O$ is a left ideal of the operoid $\beta O$.*

If $S$ is both a left ideal of an operoid $O$ and a right ideal of it, then we say that $S$ is a *two-sided ideal* of $O$ or simply an *ideal* of $O$.

We now present a sufficient condition for $\beta O \setminus O$ to be a two-sided ideal of $\beta O$.

THEOREM 2.1.14. *Suppose that $O$ is a discrete operoid such that the sets $bC$ and $Cb$ are infinite, for every infinite subset $C$ of $O$ and for every $b \in O$. Then $\beta O \setminus O$ is a two-sided ideal of the operoid $\beta O$.*

PROOF. By Proposition 2.1.13, $\beta O \setminus O$ is a left ideal of $\beta O$. Let us show that it is a right ideal of $\beta O$.

Take any $p \in \beta O \setminus O$ and any $q$ in $\beta O$. We have to show that $pq$ is in $\beta O \setminus O$. Since we already know that $\beta O \setminus O$ is a left ideal of $\beta O$, it remains to consider the case when $q \in O$. Then $q = \{C \subset O : c \in C\}$, for some $c \in O$.

Assume now that $pq$ is not in $\beta O \setminus O$. Then $pq \in O$, and therefore, there is $a \in O$ such that $\{a\} \in pq$.

By c) of Proposition 2.1.8 applied to the set $A = \{a\} \in pq$, we can fix $B \in p$ such that for each $b \in B$, there exists $C_b \in q$ satisfying the condition $bC_b \subset \{a\}$. Since $c \in C_b$ for each $b \in B$, we have $bc = a$, for every $b \in B$. Hence, $Bc = \{a\}$, that is, the set $Bc$ is finite. By the assumption, this implies that $B$ is finite. Since $B \in p$, it follows that $p$ is in $O$, a contradiction. □

COROLLARY 2.1.15. *If $O$ is a discrete operoid such that left and right actions $\lambda_a$ and $\varrho_a$ are injective, for each $a \in O$, then $\beta O \setminus O$ is a two-sided ideal of the operoid $\beta O$.*

COROLLARY 2.1.16. *If $G$ is an infinite discrete group, then $\beta G \setminus G$ is a two-sided ideal of the right topological semigroup $\beta G$. In particular, $\beta G \setminus G$ is a compact right topological semigroup.*

## Exercises

2.1.a. Suppose that $S$ is a right topological semigroup and $I$ is a right ideal of $S$. Verify that the closure of $I$ in $S$ is also a right ideal. Give an example of a left ideal $J$ of a compact right topological semigroup $T$ such that the closure of $J$ in $T$ fails to be a left ideal.

2.1.b. (See [**241**, Coro. 2.6]) Let $S$ be a compact right semitopological semigroup. Show that every left ideal of a $S$ contains a minimal (by inclusion) left ideal, and that all minimal left ideals of $S$ are closed.

2.1.c. (See [**241**, Theorem 2.17]) Let $S$ be a compact right topological semigroup such that the set $D = \{x \in S : \lambda_x \text{ is continuous}\}$ is dense in $S$. Show that if $I$ is an arbitrary left ideal of $S$, then the closure of $I$ in $S$ is again a left ideal.

## Problems

2.1.A. Prove that none of the compact semigroups $\beta\mathbb{N}$, $\beta\mathbb{Z}$ is commutative, no matter which of the two natural operations on $\mathbb{N}$ or $\mathbb{Z}$ is extended over the Čech–Stone compactification of the corresponding semigroup.

*Hint.* Let $\circ$ denote the sum or multiplication operation in $\mathbb{N}$. Define by induction two sequences $A = \{a_n : n \in \omega\}$ and $B = \{b_n : n \in \omega\}$ of pairwise distinct elements of $\mathbb{N}$ such that the sets $\{a_k \circ b_n : k < n\}$ and $\{b_k \circ a_n : k < n\}$ are disjoint. Take elements $p, q \in \beta\mathbb{N}\setminus\mathbb{N}$ such that $A \in p$ and $B \in q$ and apply Proposition 2.1.8 to show that $p \circ q \neq q \circ p$. The same argument applies to $\mathbb{Z}$ in place of $\mathbb{N}$.

2.1.B. Let $e$ be an element of a compact right topological semigroup $S$ satisfying $ee = e$. Prove that the following are equivalent:
(i) $Se$ is a minimal left ideal;
(ii) $eS$ is a minimal right ideal;
(iii) $eSe$ is a subgroup of $S$.

## 2.2. Idempotents in compact semigroups

An element $p$ of an operoid $O$ is called an *idempotent* if $pp = p$. The set of all idempotents of an operoid $O$ will be denoted by $E(O)$. The next result is a celebrated theorem of R. Ellis from [**159**]. It is one of mathematical principles in which compactness is blended into an algebraic structure in such an ingenious way that they serve as pillars for the whole edifice of the contemporary mathematics and provide for its unity.

THEOREM 2.2.1. [**R. Ellis**] *Let $S$ be a compact right topological semigroup. Then there exists an idempotent in it, that is, $E(S) \neq \varnothing$.*

PROOF. Let $\mathcal{P}$ be the family of all closed subsemigroups of $S$, and $\mathcal{C}$ any chain in $\mathcal{P}$. Then, since $S$ is compact, and semigroups are always non-empty sets, the intersection of $\mathcal{C}$ is non-empty and is a closed subsemigroup of $S$. Now, by Zorn's Lemma, we are entitled to conclude that there exists a minimal element in $\mathcal{P}$, that is, there exists $A \in \mathcal{P}$ such that any proper closed subset $B$ of $A$ is not a subsemigroup of $S$. Take any element $a \in A$.

**Claim.** *The element $a$ of $A$ is an idempotent.*

Consider the set $B = Aa$. Since $A$ is compact and $B$ is a continuous image of $A$, it follows that $B$ is compact and non-empty. Therefore, since $S$ is Hausdorff, $B$ is closed in $S$. It is also clear that $B$ is contained in $A$. If $x$ and $y$ are in $B$, then $y = za$, for some $z \in A$. Since $x$ and $z$ are in $A$, and $A$ is a subsemigroup of $S$, we have $xz \in A$ and, therefore, $xy = xza \in Aa = B$. Hence, $B$ is a closed subsemigroup of $S$ contained in $A$. By the minimality of $A$, it follows that $B = A$, that is, $Aa = A$. Therefore, since $a \in A$, there exists $c \in A$ such that $ca = a$. It follows that the set $C$ of all $x \in A$ such that $xa = a$ is non-empty. Clearly, $C = A \cap \varrho_a^{-1}(a)$, where $\varrho_a(x) = xa$ for each $x \in S$. Hence, $C$ is closed and compact. Now, $C$ is closed under multiplication. Indeed, if $y$ and $z$ are any elements of $C$, then $yz \in A$ (since $C \subset A$ and $A$ is a subsemigroup of $S$). We also have $yza = ya = a$, since $y$ and $z$ are in $C$. Therefore, $yz \in C$, and $C$ is a closed subsemigroup of $S$ contained in $A$. By the minimality of $A$, it follows that $C = A$. Hence $a \in C$, that is, $aa = a$. Our Claim and the theorem are proved. □

Given an operoid $O$, we denote by $\beta O$ the Čech–Stone compactification of the discrete space $O$ considered as a right topological operoid (see Theorem 2.1.1). Recall that if $S = O$ is a semigroup, then $\beta S$ is a right topological semigroup, according to Theorem 2.1.4.

THEOREM 2.2.2. *If $G$ is an infinite discrete group, and $\beta G$ is the compact right topological semigroup generated by $G$, then there exists an idempotent $p$ in $\beta G \setminus G$.*

PROOF. Indeed, by Corollary 2.1.16, $\beta G \setminus G$ is a compact subsemigroup of $\beta G$. Therefore, by Theorem 2.2.1, there exists $p \in \beta G \setminus G$ such that $pp = p$. □

Points in $\beta G \setminus G$ are free ultrafilters on $G$. Let us clarify what it means for an ultrafilter $p$ to be an idempotent in the compact semigroup $\beta G$.

PROPOSITION 2.2.3. *Suppose that $p$ is a free ultrafilter on a discrete group $G$. Then $p$ is an idempotent in $\beta G$ if and only if for each $A \in p$ there exists $B \in p$ such that $A \in bp$, for each $b \in B$.*

PROOF. This is a direct corollary of the equivalence of a) and b) of Proposition 2.1.8. □

The following statement is also a corollary of Proposition 2.1.8. It can be considered as a modification of Proposition 2.2.3.

PROPOSITION 2.2.4. *Suppose that $O$ is a discrete operoid and $p$ an idempotent in $\beta O$. Then, for each $A \in p$, there exists $B \in p$ such that $B \subset A$ and for each $b \in B$ there exists $C_b \in p$ satisfying $bC_b \subset A$.*

PROOF. Since $pp = p$, we can apply c) of Proposition 2.1.8 with $q = p$. To ensure that $B \subset A$, we just have to replace $B$ with $A \cap B \in p$. □

It follows from Theorem 2.1.1 that if the operoid $O$ has an identity $e$, then $e$ will also play the role of an identity in the operoid $\beta O$. However, it is natural to ask if the subsemigroup $\beta G \setminus G$ will have its own identity, for example, when $G$ is a group. The answer is "no".

EXAMPLE 2.2.5. Let $\mathbb{Z}$ be the discrete group of integers, with the usual addition as multiplication. Then $S = \beta\mathbb{Z} \setminus \mathbb{Z}$ is a compact subsemigroup of the right topological semigroup $\beta\mathbb{Z}$ which has no right identity.

Indeed, take any element $p \in S$. We have to show that there is $q \in S$ such that $qp \neq q$. Assume the contrary. Then $\varrho_p(S) = S$ and, since $\mathbb{Z}$ is a group, Corollary 2.1.16 implies that $\varrho_p(\mathbb{Z}) \subset S$. Hence $\varrho_p(\beta\mathbb{Z}) = S$. Since the mapping $\varrho_p$ is continuous on $\beta\mathbb{Z}$, and $\mathbb{Z}$ is dense in $\beta\mathbb{Z}$, it follows that $\varrho_p(\mathbb{Z})$ is dense in $\varrho_p(\beta\mathbb{Z}) = S$. Therefore, the space $S = \beta\mathbb{Z} \setminus \mathbb{Z}$ is separable. However, $S$ contains an uncountable pairwise disjoint family of non-empty open sets by [**165**, Example 3.6.18]. This contradiction completes the argument. □

If $S$ is a semigroup and $p \in S$, the set $T_p = \{q \in S : qp = p\}$ will be called the *left tail* of $p$ (in $S$). It is natural to ask if Theorem 2.2.2 can be strengthened in the following way: Under the assumptions in Theorem 2.2.2, there exists $p \in \beta G \setminus G$ such that $qp = p$ for each $q \in \beta G \setminus G$, that is, $\beta G \setminus G \subset T_p$. The answer is "no", as the next result shows. First we need a set-theoretic lemma and a corollary to it.

LEMMA 2.2.6. *Suppose that $h$ is a one-to-one mapping of a set $O$ to itself without fixed points. Then $O$ can be decomposed into three sets $A_1$, $A_2$, and $A_3$ in such a way that $h(A_i) \cap A_i = \varnothing$, for every $i \in \{1, 2, 3\}$.*

PROOF. A subset $A$ of $O$ will be called $h$-*simple* if $h(A) \cap A = \varnothing$. By Zorn's Lemma, there exists a maximal chain $\mathcal{C}$ of $h$-simple subsets of $O$. Put $U = \bigcup \mathcal{C}$. Clearly, $U$ is also $h$-simple, that is, the sets $U$ and $h(U)$ are disjoint. It follows that the sets $U$ and $h^{-1}(U)$ are also disjoint (that is, the set $h^{-1}(U)$ is $h$-simple). By the construction, $U$ is a maximal $h$-simple subset of $O$.

**Claim 1.** *The equality $O = U \cup h^{-1}(U) \cup h(U)$ holds.*

We can assume without loss of generality that the set $X = O \setminus (U \cup h^{-1}(U))$ is not empty. Take any $x \in X$. Then the set $U \cup \{x\}$ is not $h$-simple, since it is strictly larger than $U$. Hence, the set $P = (U \cup \{x\}) \cap (h(U) \cup \{h(x)\})$ is not empty. Since $U$ and $h(U)$ are disjoint and $x \neq h(x)$, the set $P$ can be non-empty only if either $x \in h(U)$ or $h(x) \in U$. However, the last case is impossible, since $x$ is not in $h^{-1}(U)$. It follows that $x \in h(U)$, and Claim 1 is proved.

**Claim 2.** *If $x \in O \setminus (U \cup h^{-1}(U))$, then $h(x) \notin (U)$.*

Indeed, if $h(x) \in h(U)$ then $x \in U$, since $h$ is one-to-one. This contradicts our choice of the element $x$.

Put $A_1 = h^{-1}(U)$, $A_2 = U$, and $A_3 = h(U) \setminus (U \cup h^{-1}(U)) = O \setminus (U \cup h^{-1}(U))$. Now, in view of Claims 1 and 2 it is clear that $O = A_1 \cup A_2 \cup A_3$, the sets $A_1$, $A_2$, and $A_3$ are disjoint, and $h(A_i) \cap A_i = \varnothing$, for every $i \in \{1, 2, 3\}$. □

Later we will improve Lemma 2.2.6.

PROPOSITION 2.2.7. *Suppose that $O$ is a discrete operoid and $a \in O$ an element such that $ax \neq x$, for each $x \in O$, and the left action $\lambda_a$ is one-to-one on $O$. Then $aq \neq q$, for each $q \in \beta O$.*

PROOF. By Lemma 2.2.6, there are disjoint subsets $A_1$, $A_2$, and $A_3$ of $O$ such that $O = A_1 \cup A_2 \cup A_3$, and $A_i \cap aA_i = \varnothing$, for each $i \in \{1, 2, 3\}$. Take an arbitrary element $q \in \beta O$. Since $O = A_1 \cup A_2 \cup A_3$, $A_i \in q$, for some $i \in \{1, 2, 3\}$. We can assume that $A_1$ is in $q$.

Now assume that $aq = q$. Then, by a) of Proposition 2.1.9, $aA_1 \in aq = q$. Thus, $q$ contains disjoint sets $A_1$ and $aA_1$, a contradiction. □

We apply Proposition 2.2.7 in the proof of the next result.

THEOREM 2.2.8. *Let $G$ be a countable infinite discrete group and $\beta G$ the compact right topological semigroup generated by it. Then, for each $q \in \beta G \setminus G$, there exists $p \in \beta G \setminus G$ such that $pq \neq q$.*

PROOF. Assume the contrary, and fix $q \in \beta G \setminus G$ such that $pq = q$ for every $p \in \beta G \setminus G$. Since $G$ is a group, Proposition 2.2.7 is applicable. Therefore, for each $a \in G \setminus \{e\}$, $aq \neq q$.

Consider the right action $\varrho_q$ on $\beta G$. By the assumption, $\varrho_q(\beta G \setminus G) = \{q\}$. On the other hand, by Proposition 2.2.7, the set $\varrho_q(G \setminus \{e\})$ does not contain the point $q$. We put $F = \beta G \setminus G$, $A = G \setminus \{e\}$, and $B = \varrho_q(A)$.

Clearly, every open neighbourhood of $F$ contains all but finitely many points of the set $A$. It follows, by the continuity of $\varrho_q$, that every open neighbourhood of $q$ contains all but finitely many points of the set $B$. Since $q$ is not in $B$, we conclude that $\Phi = B \cup \{q\}$ is an infinite compact subspace of $\beta G$. It is also clear that $\Phi$ is countable. It follows that $\Phi$ is metrizable and not discrete. Hence, $\Phi$ contains a non-trivial convergent sequence (in fact, it is almost obvious that $\Phi$ itself, properly enumerated, is such a sequence). This, however, contradicts [**165**, Coro. 3.6.15]. □

For any discrete group $G$, idempotents in $\beta G$ can be used to produce natural topologies on the group $G$ itself. The next statement describes this connection in a somewhat more general situation.

THEOREM 2.2.9. *Suppose that $S$ is a discrete monoid with identity $e$, and $p \in \beta S \setminus S$ is an idempotent of the compact semigroup $\beta S$. Put $\mathcal{F}_p = \{\{e\} \cup A : A \in p\}$. Then there exists a topology $\mathcal{T}_p$ on $S$ such that $S$ endowed with $\mathcal{T}_p$ is a left topological semigroup, the identity $e$ is not isolated in the space $(S, \mathcal{T}_p)$, and $\{\mathrm{Int}(P) : P \in \mathcal{F}_p\}$ is a base of $(S, \mathcal{T}_p)$ at the identity $e$.*

PROOF. Clearly, $\mathcal{F}_p$ is a prefilter on $S$ such that $\{e\} = \bigcap \mathcal{F}_p$. Call a set $U \subset S$ open if for every $a \in U$, there exists $P \in \mathcal{F}_p$ such that $aP \subset U$. By the "left" version of Construction 1.3.8, the set of all open subsets so defined forms a topology $\mathcal{T}_p$ on $S$ such that $S$, with this topology, is a left topological semigroup.

Notice that every open set $U$ containing $e$ must also contain some $A \in p$. Since every element of the free ultrafilter $p$ is an infinite set, it follows that every open neighbourhood of $e$ contains infinitely many points; therefore, $e$ is not isolated in the space $(S, \mathcal{T}_p)$.

Since $p$ is an idempotent, from Proposition 2.2.4 it follows that the prefilter $\mathcal{F}_p$ satisfies the next condition:

(int) For each $U \in \mathcal{F}_p$, there exists $V \in \mathcal{F}_p$ such that for each $x \in V$, there exists $W \in \mathcal{F}_p$ with $xW \subset U$.

Indeed, we just apply Proposition 2.2.4 with $A = U \setminus \{e\}$, choose the corresponding elements $B \in p$ and, for every $C_b \in p$, $C_b \in p$, and put $V = \{e\} \cup B$ and $W_b = \{e\} \cup bC_b$. Since $eU = U$, it is clear that condition (int) is satisfied.

Let us check that $\{\mathrm{Int}(P) : P \in \mathcal{F}_p\}$ is a base of the space $(S, \mathcal{T}_p)$ at the identity $e$. By the definition of $\mathcal{T}_p$, every open neighbourhood $O$ of $e$ contains an element $P$ of the family $\mathcal{F}_p$. Then $\mathrm{Int}(P) \subset O$. Since the set $\mathrm{Int}(P)$ is obviously open, it remains to show that $e \in \mathrm{Int}(P)$, for each $P \in \mathcal{F}_p$.

Take any $P \in \mathcal{F}_p$ and put

$$I_P = \{x \in P : xW \subset P \text{ for some } W \in \mathcal{F}_p\}.$$

**Claim.** *The set $I_P$ is open in $(S, \mathcal{T}_p)$ and $e \in I_P$.*

Clearly, $e \in I_P$ since $eP = P$ and $P \in \mathcal{F}_p$. To show that $I_P$ is open, take any $a \in I_P$. There exists $U \in \mathcal{F}_p$ such that $aU \subset P$ (by the definition of $I_P$). Since $\mathcal{F}_p$ satisfies condition (int), we can choose $V \in \mathcal{F}_p$ such that for each $x \in V$ there exists $W_x \in \mathcal{F}_p$ with $xW_x \subset U$. Let us show that $aV \subset I_P$. Take any $y \in aV$. Then $y = ax$, for some $x \in V$. By the choice of $V$, there exists $W_x \in \mathcal{F}_p$ such that $xW_x \subset U$. Then we have $yW_x = axW_x \subset aU \subset P$. It follows that $y \in I_P$, that is, in view of the choice of $y$, $aV \subset I_P$, and $I_P$ is open. □

In the case when $S = G$ is a group we can add more information on the topology $\mathcal{T}_p$ defined above.

We recall that a space $X$ is *extremally disconnected* if the closure of any open subset of $X$ is open. Every discrete space is clearly extremally disconnected. There are lots of compact extremally disconnected spaces — one can take, for example, the Čech–Stone compactification $\beta D$ of any discrete space $D$. More generally, the Čech–Stone compactification $\beta X$ of every extremally disconnected Tychonoff space $X$ is extremally disconnected [**165**, Theorem 6.2.27].

It is still an open question, formulated for the first time in **[17]**, in 1967, whether there exists in $ZFC$ a non-discrete extremally disconnected topological group. The next theorem supplies us with a series of extremally disconnected left topological groups.

THEOREM 2.2.10. *Suppose that $G$ is an infinite discrete group, $p$ an idempotent in $\beta G \setminus G$, and $G$ is endowed with the topology $\mathcal{T}_p$ defined in the proof of Theorem 2.2.9. Then:*

1) *for each $a \in G$, the family $\eta_a = \{\mathrm{Int}(aP) : P \in \mathcal{F}_p\} = \{\{a\} \cup \mathrm{Int}(aU) : U \in p\}$ is a base of the space $G$ at $a$;*
2) *the space $G$ is Hausdorff;*
3) *$G$ is extremally disconnected;*
4) *$G$ is a left topological group and, hence, is homogeneous;*
5) *there are no isolated points in $G$;*
6) *the identity $e$ of $G$ belongs to the closure of a set $A \subset G \setminus \{e\}$ if and only if $A \in p$;*
7) *for every topology $\mathcal{T}$ on $G$ that is strictly larger than the topology $\mathcal{T}_p$, there exists an isolated point in $(G, \mathcal{T})$ (this means that the topology $\mathcal{T}_p$ is maximal).*

PROOF. With the topology $\mathcal{T}_p$, $G$ is a left topological group and, therefore, every left action $\lambda_a$ on $G$ is a homeomorphism of the space $G$ onto itself. Together with Theorem 2.2.9, this implies 1) and 4), since every left topological group is homogeneous.

Let us prove 2). Take any $a \in G$ different from $e$. By Proposition 2.2.7, $ap \neq p$ (since $\lambda_a$ is an injection on $G$). Since $\lambda_a(G) = G$, we have $ap = \{aV : V \in p\}$, according to Proposition 2.1.9. Since $p$ and $ap$ are distinct ultrafilters on $G$, it follows that we can find $U \in p$ and $V \in p$ such that $U \cap aV = \varnothing$. Clearly, we may also assume that $e$ is not in $aV$ and $a$ is not in $U$, since $p$ and $ap$ are both free ultrafilters. Now it follows from 1) that $\{a\} \cup \mathrm{Int}(aV)$ and $\{e\} \cup \mathrm{Int}(U)$ are disjoint open neighbourhoods of $a$ and $e$, respectively, in the space $(G, \mathcal{T}_p)$. Hence, $G$ is Hausdorff.

By Theorem 2.2.9, the point $e$ is not isolated in $G$. Now from 4) it follows that there are no isolated points in the space $G$.

Let us prove 6). Take any $A \subset G$ such that $e \in \overline{A}$ and $e \notin A$, and let $P$ be any element of $p$. By 1), $\{e\} \cup P$ contains an open neighbourhood of $e$; therefore, $P \cap A \neq \varnothing$. Since this is true for each $P \in p$, and $p$ is an ultrafilter, it follows that $A \in p$.

To prove the converse, take any $A \in p$, and let $V$ be any open neighbourhood of $e$. From 1) it follows that there exists $P \in p$ such that $P \subset V$. Since $A$ and $P$ are both in $p$, we conclude that $P \cap A \neq \varnothing$. Therefore, $V \cap A \neq \varnothing$. Hence, $e \in \overline{A}$.

To prove 7), assume that $\mathcal{T}$ is a topology on $G$ which properly contains the topology $\mathcal{T}_p$. Then there exist a set $A \subset G$ and a point $a \in G$ such that $a$ is in the closure of $A$ in the space $(G, \mathcal{T}_p)$, while $a$ is not in the closure of $A$ in the space $(G, \mathcal{T})$. Since the space $(G, \mathcal{T}_p)$ is homogeneous, we can assume that $a = e$. Since $e$ is not in the closure of $A$ in the space $(G, \mathcal{T})$, there exists $W \in \mathcal{T}$ such that $e \in W$ and $W \cap A = \varnothing$. On the other hand, since $e$ is in the closure of $A$ in $(G, \mathcal{T}_p)$, it follows from 6) that $A \in p$. Therefore, by 1), $A \cup \{e\}$ contains a set $U \in \mathcal{T}_p$ such that $e \in U$. Then $U$ is an open neighbourhood of $e$ in the space $(G, \mathcal{T})$ as well, since $\mathcal{T}_p \subset \mathcal{T}$. Then, clearly, $U \cap W = \{e\}$, that is, the point $e$ is isolated in $(G, \mathcal{T})$.

Finally, let us prove that $(G, \mathcal{T}_p)$ is extremally disconnected. Take any open subset $V$ of $G$ and assume that $\overline{V}$ is not open. Then there exists $a \in \overline{V} \cap \overline{G \setminus \overline{V}}$. Since $(G, \mathcal{T}_p)$ is homogeneous, we can again assume that $a = e$. Then, by 6), both $V$ and $G \setminus \overline{V}$ are in $p$. However, these two sets are, obviously, disjoint. This contradiction completes the proof of 3) and of the theorem. □

In connection with Theorems 2.2.9 and 2.2.10 it is natural to ask what happens if we assume $p$ to be any free ultrafilter on the monoid $S$ (not necessarily an idempotent) What remains of Theorems 2.2.9 and 2.2.10 in this case? Here is a partial answer to this question.

THEOREM 2.2.11. *Suppose that $G$ is a discrete group, $e$ is the identity of $G$, and $p \in \beta G \setminus G$. Put $\mathscr{F}_p = \{\{e\} \cup U : U \in p\}$ and $\mathscr{F}_a p = \{\{a\} \cup aU : U \in p\}$, for each $a \in G$. Call a set $U \subset G$ open if, for every $a \in U$, there exists $P \in \mathscr{F}_a p$ such that $P \subset U$. Then:*

1) *the set of all open subsets so defined forms a topology $\mathscr{T}_p$ on $G$ such that $G$, with this topology, is a left topological group;*
2) *the space $(G, \mathscr{T}_p)$ is extremally disconnected, homogeneous, and satisfies the $T_1$-separation axiom;*
3) *the space $(G, \mathscr{T}_p)$ is dense in itself, that is, there are no isolated points in it.*

PROOF. By the "left" version of Construction 1.3.8, the statement 1) is true. The homogeneity of $(G, \mathscr{T}_p)$ follows from 1). There are no isolated points in $(G, \mathscr{T}_p)$, since the family $\mathscr{F}_a p$ does not contain singletons. Since $\bigcap \mathscr{F}_a p = \{a\}$ for each $a \in G$, it follows from the definition of the topology $\mathscr{T}_p$ that the set $G \setminus \{b\}$ is open, for each $b \in G$. Hence, $(G, \mathscr{T}_p)$ is a $T_1$-space. It remains to establish that $(G, \mathscr{T}_p)$ is extremally disconnected.

**Claim 1.** *Suppose $H$ is a closed subset of $(G, \mathscr{T}_p)$ and $a \in H$ a non-isolated point of $H$. Then $H \in ap$.*

Assume the contrary. Then $G \setminus H \in ap$, since $ap$ is an ultrafilter on $G$. The set $U = G \setminus H$ is open, therefore, for each $b \in U$, there exists $B \in \mathscr{F}_b p$ such that $B \subset U$. Since $U \in ap$, we have $U \cup \{a\} \in \mathscr{F}_a p$. It follows from our definition of the topology $\mathscr{T}_p$ that the set $W = \{a\} \cup U$ is open in $G$. Then $W \cap H$ is open in $H$. Since, obviously, $W \cap H = \{a\}$, the point $a$ is isolated in $H$, a contradiction.

**Claim 2.** *Every infinite dense in itself closed subset $H$ of $G$ is open.*

Assume that $H$ is not open. Then there exists a point $a \in H$ such that each element of the family $\mathscr{F}_a p$ meets $G \setminus H$. Since $ap$ is an ultrafilter, it follows that $G \setminus H \in ap$. On the other hand, $H \in ap$ by Claim 1, since $H$ is closed and $a$ is not isolated in $H$. Thus, the ultrafilter $ap$ contains two disjoint sets, $H$ and $G \setminus H$, a contradiction.

Now, to show that $(G, \mathscr{T}_p)$ is extremally disconnected, take any non-empty open subset $V$ of $G$ and put $H = \overline{V}$. Then $H$ is an infinite dense in itself closed subset of $(G, \mathscr{T}_p)$, since there are no isolated points in $(G, \mathscr{T}_p)$. Hence, by Claim 2, $\overline{V} = H$ is an open set, that is, $(G, \mathscr{T}_p)$ is extremally disconnected. □

There is yet another natural way to use the topology of $\beta G$ to produce some natural topologies on $G$ itself. To describe it, we need a couple of elementary facts about right actions on $\beta G$.

PROPOSITION 2.2.12. *Suppose that $G$ is a group, $p \in \beta G$, and $b, c$ are two distinct elements of $G$. Then $bp \neq cp$.*

PROOF. Assume the contrary, and put $a = b^{-1}c$. From $bp = cp$ it follows that $b^{-1}bp = b^{-1}cp$. Thus, $p = ep = ap$. However, since $G$ is a group, it follows from Proposition 2.2.7 that $a = e$, a contradiction, since $a = b^{-1}c \neq e$. □

COROLLARY 2.2.13. *Let $G$ be a discrete group. Then, for any $q$ in $\beta G \setminus G$, the restriction of the mapping $\varrho_q$ to $G$ is one-to-one.*

PROOF. Indeed, in view of the definition of $\varrho_q$, this is simply a reformulation of Proposition 2.2.12. □

If $G$ is a discrete group, then we put $G_q = \varrho_q(G)$ and $F_q = \varrho_q(\beta G)$. We use this convention from now on. Of course, $G_q$ and $F_q$ are taken with the topology inherited from $\beta G$. Clearly, $G_q$ is dense in $F_q$, since $\varrho_q$ is a continuous mapping. The subspace $G_q$ is called the *orbit* of $q$ in $\beta G$ under the left action of $G$ (or the $G$-*orbit* of $q$). Clearly, the orbit of $q$ always contains $q$.

PROPOSITION 2.2.14. *Suppose that $G$ is a discrete group, $a \in G$, and $q \in \beta G$. Then:*

1) *The mapping $\lambda_a$ restricted to $G_q$ is a homeomorphism of $G_q$ onto itself;*
2) $\varrho_q \lambda_a \restriction G = \lambda_a \varrho_q \restriction G$;
3) $\lambda_a(\varrho_q(b)) = \varrho_q(ab) = abq$, *for each $b \in G$;*
4) *The space $G_q$ is homogeneous.*

PROOF. We first prove 2). Take any $b \in G$. Then $\lambda_a(b) = ab$. Therefore, $\varrho_q \lambda_a(b) = abq$. On the other hand, $\lambda_a(\varrho_q(b)) = \lambda_a(bq) = abq$. It follows that $\varrho_q(\lambda_a(b)) = abq = \lambda_a(\varrho_q(b))$, for each $b \in G$.

Since $abq = \varrho_q(ab)$, 3) follows from 2).

From 3) we see that $\lambda_a(G_q) \subset G_q$. It also follows from 3) that $\lambda_a(\varrho_q(a^{-1}b) = \varrho_q(b)$. Since $G_q = \{bq : b \in G\}$, we conclude that $\lambda_a(G_q) = G_q$. However, $\lambda_a$ is a homeomorphism of $\beta G$ onto itself, since the translations $\lambda_a$ and $\lambda_{a^{-1}}$ are continuous and $\lambda_a \circ \lambda_{a^{-1}} = \lambda_{a^{-1}} \circ \lambda_a = id_{\beta G}$. Therefore, the mapping $\lambda_a$ restricted to $G_q$ is a homeomorphism of the subspace $G_q$ onto itself, which is statement 1).

Now take any $bq \in G_q$ and $cq \in G_q$, and put $a = cb^{-1}$. Then, by 3), $\lambda_a(bq) = abq = cb^{-1}bq = cq$. Therefore, $G_q$ is homogeneous. □

After Proposition 2.2.14 it is natural to ask if $\varrho_q$ restricted to $G$ is actually a homomorphism of the group $G$ to the semigroup $\beta G$. However, $\varrho_q(a)\varrho_q(b) = aqbq$ and $\varrho_q(ab) = abq$. Since there is no reason to believe that $aqbq = abq$, we should not also expect $\varrho_q \restriction G$ to be a homomorphism.

The reasoning above also shows that the subspace $G_q = \{bq : b \in G\}$ is not, in general, a subgroup of the semigroup $\beta G$. Now we are going to show that we can introduce a new product operation on the set $G_q \subset \beta G$ in such a way that, with the subspace topology, $G_q$ will become a left topological group, and $\varrho_q$ will become an isomorphism of the group $G$ onto the group $G_q$. In fact, if the last condition is to be satisfied, there is only one way to define the new operation $\times$ on $G_q$ — we have to put $aq \times bq = abq$. Since $\varrho_q$ is one-to-one on $G$ (see Corollary 2.2.13) and $aq = \varrho_q(a)$, $G_q$ becomes a group, and $\varrho_q$ becomes an isomorphism of $G$ onto $G_q$. Obviously, the left action by the element $aq$ on the group $G_q$ so defined coincides with the restriction of $\lambda_a$ to $G_q$ and is therefore, by Proposition 2.2.14, a homeomorphism of the space $G_q$ onto itself. Hence, $G_q$ is a left topological group. We sum up the information obtained so far in the next statement.

THEOREM 2.2.15. *Suppose that $G$ is a discrete group, $q \in \beta G$, and the product operation $\times$ on $G_q$ is defined by the formula $aq \times bq = abq$. Then $G_q$, with this operation and with the subspace topology, is a left topological group, and the mapping $\varrho_q \restriction G$ is an isomorphism of the group $G$ onto the group $G_q$. Furthermore, there exists a unique topology $\mathcal{T}'$ on $G$ such that $G$ with this topology is a left topological group and $\varrho_q$ is a topological isomorphism of $(G, \mathcal{T}')$ onto the left topological group $G_q$.*

PROOF. Indeed, in view of the argument above, we only have to prove the last statement. There is only one way to define the topology $\mathcal{T}'$, if we wish the mapping $\varrho_q \restriction G$ to be a homeomorphism — $\mathcal{T}'$ must be the family of all sets $\varrho_q^{-1}(V)$, where $V$ is any open subset of the space $G_q$. This definition automatically turns $\varrho_q$ into a homeomorphism. Since each $\lambda_a \restriction G_q$ is a homeomorphism, from 2) and 3) of Proposition 2.2.14 it follows that $G$ with the topology $\mathcal{T}'$ is a left topological group, a copy of $G_q$. □

Let us consider some particular cases of the construction described in Proposition 2.2.14 and in Theorem 2.2.15.

THEOREM 2.2.16. **[V. I. Protasov]** *Suppose that $G$ is a discrete group and $q \in \beta G \setminus G$, $q$ is an idempotent. Then:*

1) *the subspace $G_q = \varrho_q(G)$ of $\beta G$ is extremally disconnected and has no isolated points;*
2) *the closure $F_q$ of $G_q$ in $\beta G$ is also extremally disconnected;*
3) *$F_q = \overline{G_q}$ is the Čech–Stone compactification of the space $G_q$.*

To prove Theorem 2.2.16, we need the following three general results.

LEMMA 2.2.17. *If $S$ is a right topological semigroup and $q$ is an idempotent in $S$, then the subspace $Sq$ of $S$ is a retract of $S$.*

PROOF. The mapping $\varrho_q$ is continuous and is a retraction of $S$ onto $Sq$. Indeed, take any $y \in Sq$. Then $y = xq$, for some $x \in S$, and we have $\varrho_q(y) = yq = xqq = xq = y$. □

PROPOSITION 2.2.18. *A dense subspace of an extremally disconnected space is extremally disconnected.*

PROOF. Let $S$ be a dense subspace of an extremally disconnected space $X$. If $V$ is an open subset of $S$, choose an open set $U$ in $X$ such that $V = U \cap S$. The closure $\overline{U}$ of $U$ in $X$ is open, so the closure of $V$ in $S$, which coincides with $\overline{U} \cap S$, is open in $S$. Hence $S$ is extremally disconnected. □

THEOREM 2.2.19. *Suppose that $X$ is an extremally disconnected compact space, $Y$ is a retract of $X$, $r\colon X \to Y$ is a retraction, and that $D$ is a dense subspace of $X$ such that $X$ is the Čech–Stone compactification of $D$. Then:*

1) *$Y$ is extremally disconnected;*
2) *$Y$ is the Čech–Stone compactification of the subspace $r(D)$.*

PROOF. Take any open subset $V$ of the space $Y$, and put $U = r^{-1}(V)$. Since $r$ is continuous and a retraction of $X$ onto $Y$, the set $U$ is open in $X$ and $V \subset U$. Let $P$ be the closure of $U$ in $X$. Since $X$ is extremally disconnected, the set $P$ is open in $X$. Put $H = P \cap Y$.

**Claim.** *$H$ is the closure of $V$ in $Y$.*

We denote by $T$ the closure of $V$ in $Y$. Clearly, $H$ is open and closed in $Y$, and $T \subset H$. It remains to show that $H \subset T$. Take any $y \in H$. Then $y \in Y$ and, therefore, $r(y) = y$. On the other hand, $y \in P = \overline{U}$, which implies, by the continuity of $r$, that $y = r(y) \in \overline{V}$, since $r(U) \subset V$. Since $y \in Y$, it follows that $y \in T$. Hence, $H = T$, and $T$ is open in $Y$.

It remains to prove the second statement of Theorem 2.2.19, which will obviously imply item 3) of Theorem 2.2.16.

Let $g$ be any bounded real-valued function continuous on the space $r(D)$. Then $f = gr$ is a bounded continuous real-valued function on the subspace $Z = r^{-1}(r(D))$ of the space $X$. Clearly, $D \subset Z$. We also have $r(D) \subset Z$ and $f \restriction r(D) = g$, since $r$ is the identity mapping on $r(D)$. From $X = \beta D$ and $D \subset Z \subset X$ it follows that $X = \beta Z$. Therefore, $f$ can be extended to a continuous real-valued function $f_X$ on $X$. However, $Y$ is a subspace of $X$. Therefore, the restriction of $f_X$ to $Y$ is a continuous real-valued function extending the function $g$. □

**Proof of Theorem 2.2.16.** Note that $q$ is in the closure of $G \setminus \{e\}$. The mapping $\varrho_q$ is continuous and one-to-one on $G$, by Corollary 2.2.13. It follows that the point $\varrho_q(q)$ is in the closure of the set $G_q \setminus \{q\} = \varrho_q(G \setminus \{e\})$. However, $\varrho_q(q) = qq = q$, since $q$ is an idempotent. Hence, $q \in \overline{G_q \setminus \{q\}}$, and the point $q$ is not isolated in $G_q$. Since the space $G_q$ is homogeneous by Proposition 2.2.14, $G_q$ is dense in itself.

Since $G$ is dense in $\beta G$ and $\varrho_q$ is continuous, we have $G_q \subset \varrho_q(\beta G) = F_q$. Moreover, since $\beta G$ is compact and Hausdorff, $F_q$ is compact and closed in $\beta G$.

By Lemma 2.2.17, it follows that $F_q$ is a retract of $\beta G$. Now, as $G$ is discrete, the space $\beta G$ is extremally disconnected (see [**165**, Coro. 6.2.28]). Since every dense subspace of an extremally disconnected space is extremally disconnected by Proposition 2.2.18, items 1)–3) of the theorem follow from Theorem 2.2.19. □

Combining Theorems 2.2.15, 2.2.16, and 2.2.2, we obtain immediately the following corollary:

COROLLARY 2.2.20. *Every infinite discrete group $G$ admits an extremally disconnected Tychonoff topology $\mathcal{T}$ such that $(G, \mathcal{T})$ is a left topological group without isolated points.*

The advantage of Corollary 2.2.20 compared to Theorem 2.2.10 is that the topology $\mathcal{T}$ on $G$ in Corollary 2.2.20 is automatically completely regular, while the topology $\mathcal{T}_p$ in Theorem 2.2.10 need not be even regular.

Here is an important special case of Theorem 2.2.15:

THEOREM 2.2.21. *Suppose $G$ is a discrete Abelian group, and $q \in \beta G \setminus G$ is an idempotent. Then $h_q = \varrho_q \restriction G$ is a monomorphism of the group $G$ to the semigroup $\beta G$, and the image $G_q = h_q(G)$ is an extremally disconnected semitopological Abelian subgroup of the right topological semigroup $\beta G$.*

PROOF. The new element in this statement, compared to Theorem 2.2.15, is that to make the mapping $\varrho_q \restriction G$ into a homomorphism, we do not have to change the product operation on $G_q$ — the multiplication which is already there fits well.

Indeed, $\varrho_q(xy) = xyq = xyqq = xqyq = \varrho_q(x)\varrho_q(y)$ (since $yq = qy$, for each $y \in G$, according to Proposition 2.1.5). Thus, $h_q$ is a homomorphism of $G$ to $\beta G$. Therefore, $G_q = h_q(G)$ is a subgroup of the semigroup $\beta G$. Clearly, $h_q = \varrho_q \restriction G$ is a monomorphism (see Corollary 2.2.13). By Theorem 2.2.16, $G_q$ is extremally disconnected. Finally,

$G_q$ is a left topological group, since it is a subgroup of the right topological semigroup $\beta G$. However, $G_q$ is commutative, since it is isomorphic to $G$. It follows that $G_q$ is a semitopological Abelian subgroup of the right topological semigroup $\beta G$. □

We will now show how Theorem 2.2.10 can be considerably strengthened. First, we introduce some terminology and notation.

Suppose $S$ is a discrete space and $\xi$ a filter base on $S$. Then $F_\xi$ stands for the subspace of $\beta S$ consisting of all $p \in \beta S$ such that $\xi \subset p$. Clearly, $F_\xi$ is a closed subspace of $\beta S$. Therefore, $F_\xi$ is compact. We use this notation below.

PROPOSITION 2.2.22. *Suppose that $G$ is an infinite discrete group and $\xi$ is a filter base on $G$ such that for each $U \in \xi$ there exists $V \in \xi$ satisfying $VV \subset U$. Then $F_\xi$ is a compact subsemigroup of the semigroup $\beta G$.*

PROOF. We already mentioned that $F_\xi$ is compact. It is also clear that $F_\xi$ is not empty. It remains to show that $F_\xi$ is closed under multiplication.

Take any $p$ and $q$ in $F_\xi$. We have to show that $\xi \subset pq$. Assume that $U$ is any element of $\xi$. By the restriction on $\xi$, there exists $V \in \xi$ such that $VV \subset U$. Then $V \in p$ and $V \in q$. Now from the equivalence of a) and c) of Proposition 2.1.8 it follows that $VV \in pq$. Hence, $U \in pq$, that is, $\xi \subset pq$ and $pq \in F_\xi$. □

Let $G$ be a paratopological group and $\xi$ be the family of all open neighbourhoods of the identity $e$ in $G$. Take an ultrafilter $p$ on $G$. We will say that $p$ *converges* to $e$ if $\xi \subset p$.

COROLLARY 2.2.23. *Let $(G, \mathcal{T})$ be a non-discrete paratopological group, where $\mathcal{T}$ is the topology on it. Then the set $F$ of all free ultrafilters on $G$ converging to the identity $e$ of $G$ is a compact subsemigroup of the semigroup $\beta G \setminus G$, where $\beta G$ is the Čech–Stone compactification of the discrete group $G$.*

PROOF. First, we note that $\beta G \setminus G$ is indeed a subsemigroup of $\beta G$, since $G$ is a group (see Corollary 2.1.16). The family $\xi = \{U \in \mathcal{T} : e \in U\}$ satisfies the restrictions imposed on $\xi$ in Proposition 2.2.22, since $(G, \mathcal{T})$ is a paratopological group. Therefore, by Proposition 2.2.22, $F_\xi$ is a compact subsemigroup of $\beta G$. It is easy to see that $F = F_\xi \cap (\beta G \setminus G)$. It is also clear that $F \neq \varnothing$, since the paratopological group topology $\mathcal{T}$ is not discrete. Since the intersection of two compact subsemigroups is a compact subsemigroup whenever this intersection is not empty, it follows that $F$ is a compact subsemigroup of the semigroup $\beta G \setminus G$. □

Here is a strengthening of Theorem 2.2.10.

THEOREM 2.2.24. **[V. I. Protasov]** *Let $(G, \mathcal{T})$ be a non-discrete paratopological group. Then there exists a maximal non-discrete Hausdorff topology $\mathcal{T}'$ on $G$ such that $(G, \mathcal{T}')$ is a left topological group and $\mathcal{T} \subset \mathcal{T}'$.*

PROOF. Let $\beta G$ be the Čech–Stone compactification of the discrete group $G$ and $F$ the set of all free ultrafilters on $G$ converging to $e$ in the topology $\mathcal{T}$. According to Corollary 2.2.23, $F$ is a compact subsemigroup of $\beta G$ and $F \subset \beta G \setminus G$. Therefore, there exists an idempotent $p$ in $F$. By Theorem 2.2.10, $\mathcal{T}_p$ is a maximal Hausdorff topology on $G$ such that $(G, \mathcal{T}_p)$ is a non-discrete left topological group. Let us show that $\mathcal{T} \subset \mathcal{T}_p$.

Take any $U \in \mathcal{T}$ such that $e \in U$. Since both $(G, \mathcal{T}_p)$ and $(G, \mathcal{T})$ are homogeneous, it is enough to find $V \in \mathcal{T}_p$ such that $e \in V \subset U$. Since $p$ converges to $e$ in the space $(G, \mathcal{T})$, there exists $P \in p$ such that $P \subset U$. By item 1) of Theorem 2.2.10, $e \in V \subset \{e\} \cup P$, for some $V \in \mathcal{T}_p$. Hence, $e \in V \subset U$, and the argument is complete. □

EXAMPLE 2.2.25. Using Theorem 2.2.21, we can easily construct a non-discrete extremally disconnected quasitopological group. Indeed, let $G$ be the $\sigma$-product of $\omega$ copies of the Boolean group $\mathbb{Z}(2) = \{0, 1\}$. We endow $G$ with the discrete topology, and apply Theorem 2.2.21 to it. Clearly, $G$ and $G_q$ are Boolean groups, that is, the inverse operation in both groups is the identity mapping. Hence, the inverse operation in $G_q$ is continuous, and $G_q$ is the quasitopological group we are looking for. □

It is still an open problem whether there exists in $ZFC$, without additional set-theoretic assumptions, a non-discrete extremally disconnected topological group (see Problem 4.5.4). In Theorem 4.5.22 we shall prove the existence of such a group under the assumption of Martin's Axiom.

To conclude this section, we give a quite unexpected application of Theorem 2.2.1 to finite partitions of semigroups. A semigroup $S$ is called *cancellative* if $zx = zy$ implies that $x = y$, and $xz = yz$ implies that $x = y$. The additive semigroup of positive integers $\mathbb{N}$ is an example of a cancellative commutative semigroup. The same set $\mathbb{N}$ with multiplication operation is again a cancellative commutative semigroup.

THEOREM 2.2.26. [**N. Hindman**] *Let $S$ be an infinite cancellative semigroup, and $S = P_1 \cup P_2 \cup \ldots \cup P_m$ be a finite partition of $S$. Then there exist a sequence $D = \{x_n : n \in \omega\}$ of pairwise distinct elements of $S$ and an integer $k \leq m$ such that all finite products $x_{i_1} x_{i_2} \ldots x_{i_n}$ with $i_1 < i_2 < \cdots < i_n$ lie in $P_k$.*

PROOF. Consider the Čech–Stone compactification $\beta S$ of the discrete semigroup $S$. By Theorem 2.1.4, $\beta S$ has the natural structure of a compact right topological semigroup. Since the semigroup $S$ is cancellative, for any infinite subset $C$ of $S$ and any $a \in S$, the set $aC$ is also infinite. Therefore, by Proposition 2.1.12, $\beta S \setminus S$ is a suboperoid of $\beta S$. Since $\beta S$ is a right topological semigroup, so is $\beta S \setminus S$. Since $S$ is discrete, we conclude that $\beta S \setminus S$ is compact. Therefore, it follows from Theorem 2.2.1 that there exists an idempotent $p \in \beta S \setminus S$.

Since $p$ is an ultrafilter on $S$, there exists an integer $k \leq m$ such that $P_k \in p$. Put $A_0 = P_k$. By Proposition 2.2.4, there exists $B_0 \in p$ such that $B_0 \subset A_0$ and for every $b \in B_0$, there exists $C \in p$ with $bC \subset A_0$. Pick a point $x_0 \in B_0$ and take $C_0 \in p$ such that $x_0 C_0 \subset A_0$.

Suppose we have defined sequences $A_0, \ldots, A_n$ and $C_0, \ldots, C_n$ and elements $x_0, \ldots, x_n$ of $S$ satisfying the following conditions for each $i \leq n$:

(i) $A_i \in p$ and $C_i \in p$;
(ii) $x_i \in A_i$;
(iii) $A_i = A_{i-1} \cap C_{i-1}$ if $i \geq 1$;
(iv) $x_i C_i \subset A_i$;
(v) $x_i \neq x_j$ if $i \neq j$.

Put $A_{n+1} = A_n \cap C_n$. Then $A_{n+1} \in p$, by (i). Again, we apply Proposition 2.2.4 to choose an element $B_{n+1} \in p$ such that $B_{n+1} \subset A_{n+1}$ and for every $b \in B_{n+1}$, there exists $C \in p$ with $bC \subset A_{n+1}$. It remains to pick a point $x_{n+1} \in B_{n+1}$ distinct from each $x_i$ with $i \leq n$ and choose an element $C_{n+1} \in p$ such that $x_{n+1} C_{n+1} \subset A_{n+1}$. Clearly, the sequences

$\{A_i : i \leq n+1\}$, $\{C_i : i \leq n+1\}$ and $\{x_i : i \leq n+1\}$ satisfy conditions (i)–(v) at the stage $n+1$. This finishes our construction of the elements $A_n$, $C_n$ in $p$ and points $x_n \in S$.

We claim that the set $D = \{x_n : n \in \omega\}$ is as required. First, all points of $D$ are pairwise distinct, by (v). Therefore, to complete the proof, it suffices to show that $x_{i_1}x_{i_2}\ldots x_{i_n} \in A_0 = P_k$ for every increasing sequence $0 \leq i_1 < i_2 < \cdots < i_n$ of integers. We will show by induction on $n = |F|$, where $F = \{i_1, i_2, \ldots, i_n\}$, that $x_{i_1}x_{i_2}\ldots x_{i_n} \in A_{i_1}$. It is clear that if $|F| = 1$, that is, $F = \{i_1\}$, then $x_{i_1} \in A_{i_1}$, by (ii). Suppose that $|F| > 1$, and consider the set $G = F \setminus \{i_1\}$. Then $i_2 = \min G$, so by the inductive hypothesis, $x_{i_2}\ldots x_{i_n} \in A_{i_2}$. Since $i_1 < i_2$, it follows from (iii) that $A_{i_2} \subset A_{i_1+1} \subset C_{i_1}$, and (iv) implies that $x_{i_1}C_{i_1} \subset A_{i_1}$. Therefore, $x_{i_1}x_{i_2}\ldots x_{i_n} \in x_{i_1}A_{i_2} \subset x_{i_1}C_{i_1} \subset A_{i_1}$. The theorem is proved. □

COROLLARY 2.2.27. *Suppose that an infinite Boolean group $G$ with zero element $0$ is the union of finitely many subsets $P_1, \ldots, P_m$. Then there exists a positive integer $k \leq m$ such that the set $P_k \cup \{0\}$ contains an infinite subgroup of $G$.*

PROOF. Taking smaller sets, if necessary, we can assume that the sets $P_i$ are disjoint. It follows from Theorem 2.2.26 that there exist an integer $k \leq m$ and an infinite subset $D = \{x_n : n \in \omega\}$ of $G$ such that all finite sums $x_{i_1} + \ldots + x_{i_n}$ with $i_1 < \cdots < i_n$ are in $P_k \cup \{0\}$. Since the group $G$ is Boolean (hence, Abelian), these sums form an infinite subset of $G$ which coincides with the subgroup of $G$ generated by $D$. □

## Exercises

2.2.a. Is Theorem 2.2.1 valid for locally compact topological semigroups?

2.2.b. Refine the conclusion of Theorem 2.2.8 and prove that for every countable infinite group $G$ and every $q \in \beta G \setminus G$, the set $\{pq : p \in \beta G \setminus G\} = \varrho_q(\beta G \setminus G)$ has cardinality $2^{\mathfrak{c}}$, where $\mathfrak{c} = 2^{\omega}$.

2.2.c. Suppose that $p$ is a free ultrafilter on a group $G$. Is it true that $p$ is an idempotent in $\beta G$ if and only if for each $A \in p$ there exists $a \in A$ such that $A \in ap$ (or, equivalently, $a^{-1}A \in p$)?

2.2.d. Give an example of a group $G$ and $q \in \beta G$ such that $\varrho_q \restriction G$ is not a homomorphism when considered as a mapping onto its image.

## Problems

2.2.A. Prove that the set $E(\beta\mathbb{Z})$ of idempotents of the compact additive semigroup $\beta\mathbb{Z}$ is contained in $\bigcap_{n=1}^{\infty} cl_{\beta\mathbb{Z}} n\mathbb{Z}$. Deduce that $E(\beta\mathbb{Z})$ is not dense in $\beta\mathbb{Z} \setminus \mathbb{Z}$.
*Hint.* Use Proposition 2.1.2 to show that, for every integer $n \geq 1$, the natural homomorphism $\varphi_n : \mathbb{Z} \to \mathbb{Z}_n \cong \mathbb{Z}/n\mathbb{Z}$ admits an extension to a continuous homomorphism of $\beta\mathbb{Z}$ to $\mathbb{Z}_n$.

2.2.B. Show that each of the additive semigroups $\beta\mathbb{N}$, $\beta\mathbb{Z}$ has $2^{\mathfrak{c}}$ idempotents.

2.2.C. (See [**241**, Theorem 2.7 (d)]) Suppose that $S$ is a compact right topological semigroup, $L$ is a minimal left ideal of $S$, and $R$ is a minimal right ideal of $S$. Prove that there is an idempotent $e \in R \cap L$ such that $R \cap L = eSe$ and $eSe$ is a group.

2.2.D. Is there an example in $ZFC$ of a non-discrete extremally disconnected regular (Hausdorff) paratopological group?

2.2.E. Let $S = P_1 \cup \ldots \cup P_m$ be a partition of a commutative semigroup $S$ and $\{x_n : n \in \omega\}$ be a sequence in $S$. Prove that there exist an integer $k \leq m$ and a sequence $D = \{z_n : n \in \omega\}$ of elements of $S$ satisfying the following conditions:

(a) all finite sums of pairwise distinct elements of $D$ lie in $P_k$;

(b) each $z_n \in D$ has the form $z_n = \sum_{i \in F_n} x_i$, where every $F_n$ is a finite subset of $\omega$ and $F_p \cap F_q = \varnothing$ for all distinct $p, q \in \omega$.

*Hint.* Modify the argument in the proof of Theorem 2.2.26.

2.2.F. (Y. Zelenyuk [**548**]) Show that every countable infinite group $G$ admits a non-discrete zero-dimensional Tychonoff topology such that $G$ with this topology is a quasitopological group.

2.2.G. Show that every countable infinite metrizable space $X$ without isolated points admits a group structure turning $X$ into a topological group.

*Hint.* Find a "canonical" topological group homeomorphic to $X$.

2.2.H. Give an example of a metrizable compact homogeneous space which is not homeomorphic to any left topological group.

*Hint.* The Hilbert cube $H$ is homogeneous [**275**]. On the other hand, every continuous mapping of $H$ into itself has a fixed point [**505**]. Therefore, there can be no non-trivial "translations" on $H$.

2.2.I. (E. G. Zelenyuk [**546**]) Show that every countable infinite regular space $X$ admits a group structure turning $X$ into a left topological group.

2.2.J. Let $G$ be a topological group. Then there exists a Hausdorff compactification $S(G)$ of the space $G$ such that the multiplication in $G$ admits a natural extension to a multiplication in $S(G)$ turning $S(G)$ into a right topological group.

*Hint.* Take the natural right uniformity on $G$ and the strongest precompact uniformity $\mathcal{U}$ contained in it. The completion of $G$ with respect to $\mathcal{U}$ is the space $S(G)$ we are looking for.

### Open Problems

2.2.1. When is the space $(G, \mathcal{T}_p)$ constructed in Theorem 2.2.11 Hausdorff? Can it be Hausdorff when $p$ is not an idempotent?

2.2.2. Is every extremally disconnected Hausdorff (regular) paratopological group a topological group?

2.2.3. Let $G$ be an infinite discrete group. Can an idempotent $p \in \beta G \setminus G$ be chosen in such a way that $(G, \mathcal{T}_p)$ or $(G, \mathcal{T}'_p)$ will become a topological group?

## 2.3. Joint continuity and continuity of the inverse in semitopological groups

We have already described several constructions leading to natural topologies on groups. However, in many cases these topologies turn the group only into a right topological group or a semitopological group. In this section we present certain topological conditions under which a semitopological group becomes a paratopological group or even a topological group. First, we consider the case of regular second-countable semitopological groups where the arguments are less abstract and simpler, and their essence is more transparent. We start with notation which is used throughout the section.

Let $X$ be a semitopological group. For $A \subset X$ and $B \subset X$ we put $\langle A, B \rangle = \{x \in X : Ax \subset B\}$. Let $I(A, B)$ be the interior of the set $\langle A, B \rangle$ in the space $X$, and $\Phi(A, B) = \langle A, B \rangle \setminus I(A, B)$. Note that if $B$ is closed, then $\langle A, B \rangle$ is closed and $\Phi(A, B)$ is closed and nowhere dense in $X$.

Proposition 2.3.1. *Let $X$ be a semitopological group such that the space $X$ is regular and has a countable base $\mathcal{B}$. Put $\mathcal{F} = \{\overline{V} : V \in \mathcal{B}\}$, $\mathcal{E} = \{\langle V, F \rangle : V \in \mathcal{B},\ F \in \mathcal{F}\}$, $\mathcal{M} = \{\Phi(V, F) : V \in \mathcal{B},\ F \in \mathcal{F}\}$, $Z = \bigcup \mathcal{M}$, and $Y = X \setminus Z$. Then, for each $b \in Y$, the multiplication is jointly continuous at $b$, that is, given any $a \in X$ and any neighbourhood*

*$O(c)$ of the point $c = ab$, there exist open sets $O(a)$ and $O(b)$ in $X$ such that $a \in O(a)$, $b \in O(b)$, and $O(a)O(b) \subset O(c)$.*

Proof. Clearly, the family $\mathcal{E} = \{\langle V, F\rangle : V \in \mathcal{B},\ F \in \mathcal{F}\}$ is countable and consists of closed sets. Hence $\mathcal{M} = \{\Phi(V, F) : V \in \mathcal{B},\ F \in \mathcal{F}\}$ is a countable family of closed nowhere dense sets in $X$.

Suppose that $a \in X$, $b \in Y$, and $c = ab$. Since $X$ is regular and $\mathcal{B}$ is a base of the space $X$, we can fix $W \in \mathcal{B}$ such that $c \in W \subset \overline{W} \subset O(c)$. Since the multiplication in $X$ is separately continuous, there exists $V \in \mathcal{B}$ such that $a \in V$ and $Vb \subset W$. Then $b \in \langle V, \overline{W}\rangle \in \mathcal{E}$. Since $b \notin Z = \bigcup \mathcal{M}$, $b$ is not in $\Phi(V, \overline{W})$. It follows that $b \in I(V, \overline{W})$. The set $I(V, \overline{W})$ is open and, by the definition of it, $Vx \subset \overline{W} \subset O(c)$, for each $x \in I(V, \overline{W})$. Hence, $O(a) = V$ and $O(b) = I(V, \overline{W})$ are the neighbourhoods of $a$ and $b$ we were looking for. □

Proposition 2.3.1 plays the key role in the proof of the following result:

Theorem 2.3.2. *If a regular semitopological group $X$ has a countable base, and is not of the first category (in itself), then $X$ is a paratopological group, that is, the multiplication in $X$ is jointly continuous at every point.*

Proof. In the notation of Proposition 2.3.1, the set $Y = X \setminus Z$ is not empty, since $Z$ is of the first category in $X$. Therefore, there exists a point $b \in X$ such that the multiplication is jointly continuous at $b$. It remains to derive from this that the multiplication is jointly continuous at every point $y \in X$.

Take any $x \in X$, and let $z = xy$. Let $W$ be any open neighbourhood of $z$. Since $G$ is a group, there exists $h \in G$ such that $b = hy$. Then $y = h^{-1}b$. Put $a = xh^{-1}$. Then $ab = xh^{-1}hy = xy = z$. Since the multiplication is jointly continuous at $b$, there are open sets $U$ and $V$ such that $a \in U$, $b \in V$, and $UV \subset W$. Put $U_1 = Uh$ and $V_1 = h^{-1}V$. Then $U_1$ and $V_1$ are open, since the multiplication is separately continuous, $x = ah \in Uh = U_1$, $y = h^{-1}b \in h^{-1}V = V_1$, and $U_1V_1 = Uhh^{-1}V = UV \subset W$. This completes the proof. □

Now we present our first non-trivial statement on the automatic continuity of the inverse mapping in paratopological groups.

Proposition 2.3.3. *Let $X$ be a compact Hausdorff paratopological group. Then the inverse operation in $X$ is continuous and, therefore, $X$ is a topological group.*

Proof. Let $e$ be the neutral element of $X$. Since $X$ is Hausdorff, the set $M = \{(x, y) \in X \times X : xy = e\}$ is closed in $X \times X$.

Now, let $F$ be any closed subset of $X$, and $P = (X \times F) \cap M$. Then $F$ and $X \times F$ are compact, $P$ closed in $X \times F$, since $M$ is closed, and, therefore, $P$ is compact. Now, $(x, y) \in P$ if and only if $y \in F$ and $xy = e$, that is, $x = y^{-1}$. It follows that the image of $P$ under the natural projection of $X \times X$ onto the first factor $X$ is precisely $F^{-1}$. Since $F$ is compact and the projection mappings is continuous, we conclude that $F^{-1}$ is compact, and therefore, closed in $X$. Thus, the inverse operation in $X$ is continuous. □

Since a compact Hausdorff space is never of the first category in itself [**165**, Th. 3.9.3], the next result follows directly from Theorem 2.3.2 and Proposition 2.3.3.

THEOREM 2.3.4. *Every compact metrizable semitopological group is a topological group.*

By a somewhat more involved argument we can improve both Proposition 2.3.3 and Theorem 2.3.4. First, we need a lemma:

LEMMA 2.3.5. *Suppose that $X$ is a Hausdorff paratopological group. Then, for each compact subset $F$ of $X$, the set $F^{-1}$ is closed in $X$.*

PROOF. Let $x$ be in the closure of $F^{-1}$. There exists an ultrafilter $\xi$ on $F^{-1}$ converging to $x$. Then $\eta = \{P^{-1} : P \in \xi\}$ is an ultrafilter on $F$. Since $F$ is compact, there exists a point $y \in F$ such that $\eta$ converges to $y$. From the continuity of the multiplication in $X$ it follows that the family $\gamma = \{PP^{-1} : P \in \xi\}$ converges to the point $z = xy$. On the other hand, the neutral element $e$ of $X$ clearly belongs to all elements of $\gamma$. Since $X$ is Hausdorff, we conclude that $z = e$, which implies that $x = y^{-1} \in F^{-1}$. Thus, $F^{-1}$ is closed in $X$. □

PROPOSITION 2.3.6. *Suppose that $X$ is a paratopological group, which is a Hausdorff locally compact space with a countable base. Then the inverse operation on $X$ is continuous.*

PROOF. Let $\mathscr{B}$ be a countable base in $X$ such that for each $V \in \mathscr{B}$, the closure of $V$ is compact. For $V$ in $\mathscr{B}$, let $I(V)$ be the interior of the set $(\overline{V})^{-1}$ and $\Phi(V) = (\overline{V})^{-1} \setminus I(V)$.

By Lemma 2.3.5, $(\overline{V})^{-1}$ is closed; therefore, $\Phi(V)$ is closed and nowhere dense in $X$, for each $V$ in $\mathscr{B}$. It follows that the set $Z = \bigcup\{\Phi(V) : V \in \mathscr{B}\}$ is of the first category in $X$. Since the space $X$ is locally compact and Hausdorff, it has the Baire property. It follows that the set $Y = X \setminus Z$ is not empty.

Fix $y \in Y$ and put $x = y^{-1}$. Let us check that the inverse operation is continuous at $y$. Take any open set $W$ containing $x$. By the regularity of $X$, there exists $V \in \mathscr{B}$ such that $x \in V$ and $\overline{V} \subset W$. Then $y = x^{-1} \in (\overline{V})^{-1}$. Since $y$ is not in $Z$, it follows that $y \in I(V)$. Now, $I(V) \subset W^{-1}$, since $\overline{V} \subset W$. This completes the proof of the continuity of inverse at $y$. Obviously, it follows that the inverse operation is continuous at all points of $X$. □

Let us prove that the multiplication in every locally compact Hausdorff semitopological group is jointly continuous. For this, we need some more sophisticated techniques.

A mapping $f$ of a product space $X \times Y$ to a space $Z$ is called *strongly quasicontinuous* at a point $(a, b) \in X \times Y$ if for each open neighbourhood $W$ of $f(a, b)$ in $Z$ and every open neighbourhood $U$ of $a$ in $X$, there exist a non-empty open set $U_1$ in $X$ and an open set $V$ in $Y$ such that $U_1 \subset U$, $b \in V$, and $f(U_1 \times V) \subset W$.

LEMMA 2.3.7. *Suppose that $X$ and $Y$ are Čech-complete spaces and $f$ is a separately continuous mapping of $X \times Y$ to a regular space $Z$. Then $f$ is strongly quasicontinuous at every point of $X \times Y$.*

PROOF. Fix $a \in X$, $b \in Y$, put $c = f(a, b)$, and let $W_1$ be a neighbourhood of $c$ in $Z$. Since $Z$ is regular, we can find open neighbourhoods $W_0$ and $W$ of $c$ such that $\overline{W_0} \subset W \subset \overline{W} \subset W_1$.

Take the Čech–Stone compactifications $\beta X$ and $\beta Y$ of $X$ and $Y$, and fix a decreasing sequence $\{P_n : n \in \mathbb{N}\}$ of open sets in $\beta X$ and a decreasing sequence $\{H_n : n \in \mathbb{N}\}$ of open sets in $\beta Y$ such that $X = \bigcap_{n=1}^{\infty} P_n$ and $Y = \bigcap_{n=1}^{\infty} H_n$. A subset $O$ of $X$ (of $Y$) will be called *$n$-small* for some $n \in \mathbb{N}$ if the closure of $O$ in $\beta X$ (in $\beta Y$) is contained in $P_n$ (in $H_n$, respectively).

We will construct by induction certain decreasing sequences of open sets $\{U_n : n \in \mathbb{N}\}$ and $\{V_n : n \in \mathbb{N}\}$ in $X$ and $Y$, respectively, and sequences $\{x_n : n \in \mathbb{N}\} \subset X$ and $\{y_n : n \in \mathbb{N}\} \subset Y$. We proceed as follows.

*Step 1.* Put $M_1 = \{x \in X : f(x, b) \in W_0\}$. Then $M_1$ is open, by the separate continuity of $f$, and $a \in M_1$, since $f(a, b) \in W_0$. Let $U_1$ be a 1-small open subset of $X$ such that $a \in U_1 \subset M_1$, and let $V_1$ be a 1-small open subset of $Y$ such that $b \in V_1$. If $f(U_1 \times V_1) \subset W_1$, we are done. Assume now that $f(U_1 \times V_1) \setminus W_1$ is not empty. Then we can choose $x_1 \in U_1$ and $y_1 \in V_1$ such that $f(x_1, y_1) \in Z \setminus \overline{W}$.

*Step 2.* Put $M_2 = \{x \in U_1 : f(x, y_1) \in Z \setminus \overline{W}\}$. Then $x_1 \in M_2$, and $M_2$ is open in $X$, by the separate continuity of $f$. Let $U_2$ be a 2-small open neighbourhood of $x_1$ in $X$ such that $\overline{U_2} \subset U_1 \cap M_2$. Put $L_2 = \{y \in V_1 : f(x_1, y) \in W_0\}$. It follows from $x \in U_1 \subset M_1$ that $b \in L_2$, and $L_2$ is open in $Y$, by the separate continuity of $f$. Let $V_2$ be a 2-small open neighbourhood of $b$ in $Y$ such that $V_2 \subset L_2$ and $\overline{V_2} \subset V_1$. If $f(U_2 \times V_2) \subset W_1$, we are done. Assume now that $f(U_2 \times V_2) \setminus W_1$ is not empty. Then we can choose $x_2 \in U_2$ and $y_2 \in V_2$ such that $f(x_2, y_2) \in Z \setminus \overline{W}$.

*Step $n + 1$.* Assume that we have already defined $n$-small open sets $U_n$ and $V_n$ in $X$ and $Y$, respectively, and points $x_n \in U_n$ and $y_n \in V_n$ such that $U_n \subset M_1$, $b \in V_n$, and $f(x_n, y_n) \in Z \setminus \overline{W}$. Then we put $M_{n+1} = \{x \in U_n : f(x, y_n) \in Z \setminus \overline{W}\}$. Clearly, the set $M_{n+1}$ is open and $x_n \in M_{n+1}$. Now we let $U_{n+1}$ be any $(n + 1)$-small open neighbourhood of $x_n$ in $X$ such that $U_{n+1} \subset M_{n+1}$ and $\overline{U_{n+1}} \subset U_n$. Put $L_{n+1} = \{y \in V_n : f(x_n, y) \in W_0\}$. Clearly, this set is open and contains the point $b$, since $x_n \in U_n \subset M_1$. Now let $V_{n+1}$ be any $(n + 1)$-small open neighbourhood of $y_n$ in $Y$ such that $V_{n+1} \subset L_{n+1}$ and $\overline{V_{n+1}} \subset V_n$.

Again, if $f(U_{n+1} \times V_{n+1}) \subset W_1$, we are done. Assume that this is not the case, and choose points $x_{n+1} \in U_{n+1}$ and $y_{n+1} \in V_{n+1}$ such that $f(x_{n+1}, y_{n+1}) \in Z \setminus \overline{W}$. Step $n + 1$ is complete.

The sequences just constructed have the following properties: $\overline{U_{n+1}} \subset U_n$, $\overline{V_{n+1}} \subset V_n$, $x_n \in U_n$, $y_n \in V_n$, and $U_n$, $V_n$ are $n$-small, for each $n \in \omega$. It follows from the last property that the sets $P = \bigcap_{n=1}^{\infty} \overline{U_n}$ and $H = \bigcap_{n=1}^{\infty} \overline{V_n}$ are compact subsets of $X$, that the family $\{U_n : n \in \mathbb{N}\}$ is a base of neighbourhoods of the set $P$ in $X$, and the family $\{V_n : n \in \mathbb{N}\}$ is a base of neighbourhoods of the set $H$ in $Y$. Therefore, there exists an accumulation point $x^*$ for the sequence $\{x_n : n \in \mathbb{N}\}$ in $X$. Since $\overline{U_{n+1}} \subset U_n$, the point $x^*$ belongs to each $U_n$. By a similar reason, some point $y^*$ in $Y$ is an accumulation point of the sequence $\{y_n : n \in \mathbb{N}\}$. Now, from $x^* \in U_{n+1} \subset M_{n+1}$ it follows that $f(x^*, y_n) \in Z \setminus \overline{W}$, for each $n \in \mathbb{N}$. Since $f$ is separately continuous, we conclude from this that $f(x^*, y^*) \in Z \setminus W$.

Fix $k \in \mathbb{N}$, and take any $n \in \mathbb{N}$ such that $n > k$. Then $y_n \in V_n \subset V_{k+1} \subset \{y \in Y : f(x_k, y) \in W_0\}$. Therefore, $f(x_k, y_n) \in W_0$. Since this holds for each $n > k$, it follows, by the separate continuity of $f$, that $f(x_k, y^*) \in \overline{W_0}$. Then again, by the separate continuity of $f$, $f(x^*, y^*) \in \overline{W_0} \subset W$, a contradiction. The proof of Lemma 2.3.7 is complete. □

LEMMA 2.3.8. *Suppose that $X$ is a locally compact Hausdorff semitopological group. Then, for any points $a \in X$ and $b \in X$, there are open sets $U$ and $V$ in $X$ such that $a \in U$, $b \in V$, and the closure of $UV$ is compact.*

PROOF. Fix an open set $W$ in $X$ such that $ab \in W$ and $\overline{W}$ is compact. By Lemma 2.3.7, the multiplication in $X$ is a strongly quasicontinuous mapping of $X \times X$ to $X$. Therefore, one can find a non-empty open subset $U_1$ of $X$ and an open subset $V$ of $X$ such that $b \in V$ and $U_1 V \subset W$. There is $h \in X$ such that $a \in hU_1$. Then $U = hU_1$ is an open neighbourhood of $a$ such that $UV = hU_1 V \subset hW$. Since the multiplication in $X$ is separately continuous, $h\overline{W}$ is compact and $\overline{hW} = h\overline{W}$. Therefore, $\overline{hW}$ is compact, and the closure of $UV$ is compact. □

Here is the promised result on the joint continuity of the multiplication in locally compact semitopological groups which, in its turn, is an important step towards the proof of Theorem 2.3.12.

PROPOSITION 2.3.9. *Suppose that $X$ is a locally compact Hausdorff semitopological group. Then the multiplication in $X$ is continuous, that is, $X$ is a paratopological group.*

PROOF. Fix $a$ and $b$ in $X$, and put $c = ab$. Let $W$ be an open set in $X$ such that $c \in W$, and fix $z \in X \setminus W$. We claim that there are open sets $U$ and $V$ in $X$ such that $a \in U$, $b \in V$, and $z$ is not in $\overline{UV}$.

We can assume that $z$ is not in the closure of $W$ — otherwise, by the regularity of $X$, we can replace $W$ by a smaller open neighbourhood of $c$ the closure of which is contained in $W$. Fix an open neighbourhood $H$ of the neutral element $e$ in $X$ such that $Hz \cap \overline{W} = \varnothing$. The set $Ha$ is also open, and $a \in Ha$.

The multiplication mapping of $X \times X$ to $X$ is strongly quasicontinuous at each point of $X \times X$, by Lemma 2.3.7. Therefore, we can find a non-empty open set $U_1$ in $X$ and an open set $V$ in $X$ such that $U_1 \subset Ha$, $b \in V$, and $U_1 V \subset W$. There is $h \in H$ such that $ha \in U_1$, and there is an open set $U$ in $X$ such that $a \in U$ and $hU \subset U_1$. Then we have:

$$(hU)V \subset U_1 V \subset W.$$

Therefore, $h\overline{UV} \subset \overline{W}$, by the separate continuity of multiplication.

Assume that $z \in \overline{UV}$. Then $hz \in h\overline{UV} \subset \overline{W}$, and $hz \in Hz$. It follows that the set $\overline{W} \cap Hz$ is not empty, a contradiction. This proves our claim.

By virtue of Lemma 2.3.8, we can fix open sets $U_0$ and $V_0$ in $X$ such that $a \in U_0$, $b \in V_0$, and the closure of $U_0 V_0$ is compact. Let $\mathcal{B}_a$ be the family of all open neighbourhoods of $a$ contained in $U_0$, and $\mathcal{B}_b$ the family of all open neighbourhoods of $b$ contained in $V_0$. For subsets $U$ and $V$ of $X$, we put $F_{U,V} = (X \setminus W) \cap \overline{UV}$. Clearly, $F_{U,V}$ is always closed.

Now let $\eta = \{F_{U,V} : U \in \mathcal{B}_a,\ V \in \mathcal{B}_b\}$. Obviously, each $P \in \eta$ is compact. If $F_{U,V}$ is empty, for some $F_{U,V} \in \eta$, then $UV \subset W$, and the continuity of $f$ at $(a, b)$ is verified. Assume that all elements of $\eta$ are non-empty. Then the family $\eta$ is centered, and since the elements of $\eta$ are closed compact sets, there is $z \in \bigcap \eta$. On the other hand, it was shown above that there are $U \in \mathcal{B}_a$ and $V \in \mathcal{B}_b$ such that $z$ is not in the closure of $UV$. Then $z \notin F_{U,V} \in \eta$, a contradiction. The proof is complete. □

Let us now give an especially short and elegant proof of the continuity of the inverse operation in every locally compact Hausdorff paratopological group. Our argument makes use of the following lemma:

LEMMA 2.3.10. *Suppose that $X$ is a semitopological group, $\{U_n : n \in \omega\}$ is a sequence of open neighbourhoods of the neutral element $e$ of $X$, and $\{x_n : n \in \omega\}$ is a sequence of points in $X$ such that $x_n \in U_n$, for each $n \in \omega$, and the next conditions are satisfied:*

a) $\overline{U_{n+1}^2} \subset U_n$ *for each* $n \in \omega$;

b) *the sequence* $\{y_k : k \in \mathbb{N}\}$, *where* $y_k = x_1 \cdots x_k$, *has an accumulation point* $y$ *in* $X$.
*Then there exists* $k \in \omega$ *such that* $x_{k+1}^{-1} \in U_0$.

PROOF. Since $yU_1$ is a neighbourhood of $y$, there exists $k \in \mathbb{N}$ such that $y_k \in yU_1$. Put $z = y_{k+1}^{-1}y$. Then

$$x_{k+1}^{-1} = y_{k+1}^{-1}y_k \in y_{k+1}^{-1}yU_1 = zU_1,$$

and $z$ is an accumulation point of the sequence $\{y_{k+1}^{-1}y_m : m \in \mathbb{N}\}$, by the separate continuity of multiplication in $X$.

It follows from condition a) of the lemma that, for each $m > k+2$,

$$y_{k+1}^{-1}y_m = x_{k+2} \cdots x_m \in U_{k+2} \cdots U_m \subset U_{k+1}.$$

Therefore, $z \in \overline{U_{k+1}} \subset U_k$, which implies that

$$x_{k+1}^{-1} \in zU_1 \subset U_kU_1 \subset U_0.$$

This finishes the proof. □

PROPOSITION 2.3.11. *If* $X$ *is a locally compact Hausdorff paratopological group, then the inverse operation in it is continuous, that is,* $X$ *is a topological group.*

PROOF. It is sufficient to check the continuity of the inverse at the neutral element $e$ of $X$. Assume the contrary. Then we can find an open neighbourhood $U$ of $e$ such that for each open set $V$ containing $e$, $V^{-1}$ is not a subset of $U$. Using the regularity of $X$ and the continuity of multiplication, we can define a sequence of open sets $\{U_n : n \in \omega\}$ in $X$ satisfying condition a) of Lemma 2.3.10. Since $X$ is locally compact, we can also assume that the closure of $U_0$ is compact and contained in $U$. Now, by the choice of $U$, we can find a point $x_n \in U_n$ such that $x_n^{-1}$ is not in $U$, for each $n \in \omega$. Put $y_k = x_1 \cdots x_k$, for each $k \in \mathbb{N}$. Then it easily follows from condition a) that all elements $y_k$ are in $U_0$. Since the closure of $U_0$ is compact, there exists an accumulation point $y$ for the sequence $\{y_k : k \in \mathbb{N}\}$ in $X$. Thus all conditions of Lemma 2.3.10 are satisfied; applying it, we obtain $k \in \omega$ such that $x_{k+1}^{-1} \in U_0$, contradicting $U_0 \subset U$ and $x_{k+1}^{-1} \in X \setminus U$. This finishes the proof. □

We can sum up the results obtained in Propositions 2.3.9 and 2.3.11 in the following theorem:

THEOREM 2.3.12. [**R. Ellis**] *Every locally compact Hausdorff semitopological group is a topological group.*

Let us give now an alternative proof of the continuity of the inverse in any locally compact Hausdorff paratopological group. We do this since the techniques involved are of independent interest and can be helpful in other situations.

One of the ideas in the proof is contained in the next lemma:

LEMMA 2.3.13. *If* $X$ *is a Hausdorff paratopological group, and* $B$ *is a compact subset of* $X$ *such that the inverse* $B^{-1}$ *is compact, then the inverse mapping restricted to* $B$ *is a homeomorphism of* $B$ *onto* $B^{-1}$.

PROOF. Let $h$ be the inverse mapping in $X$ restricted to $B$. By Lemma 2.3.5, the image of any closed subset of $B$ under $h$ is closed, and the same is true for the mapping $h^{-1}$ of $B^{-1}$ onto $B$. Therefore, both $h$ and $h^{-1}$ are closed bijections, which means that $h$ is a homeomorphism. □

In fact, what we will need is not Lemma 2.3.13 itself, but a statement of a similar nature (see Proposition 2.3.15 below). Recall that a space $X$ is said to be a *k-space* [**165**, Section 3.3] if $X$ is Hausdorff and a subset $A$ of $X$ is closed in $X$ if and only if $A \cap F$ is compact, for each compact subset $F$ of $X$.

LEMMA 2.3.14. *Let $f$ be a mapping of a Hausdorff k-space $X$ onto a Hausdorff k-space $Y$ such that $f^{-1}(F)$ is compact for each compact subset $F$ of $Y$. Then $f$ is continuous.*

PROOF. It follows from the assumptions of the lemma that for every compact subset $K$ of $Y$, the restriction of $f_K$ of $f$ to the preimage $f^{-1}(K)$ is continuous. Indeed, if $F$ is closed in $K$, then $F$ is compact and the set $f^{-1}(F)$ is compact. Since $X$ is Hausdorff, $f^{-1}(F)$ is closed in $X$ and in $f^{-1}(K)$, which implies the continuity of $f_K$.

Now we claim that the image $f(C)$ is closed in $Y$, for every compact subset $C$ of $X$. Indeed, otherwise we can use the $k$-property of $Y$ to choose a compact subset $K$ of $Y$ such that $f(C) \cap K$ is not compact. However, since the restriction of $f$ to $f^{-1}(K)$ is continuous, the image $f(C) \cap K$ of the compact set $f^{-1}(K) \cap C$ is compact, a contradiction.

The above claim implies immediately that the restriction of $f$ to every compact subset $C$ of $X$ is continuous. In particular, the image $f(C)$ is compact.

To deduce the continuity of $f$, it suffices to verify that the set $f^{-1}(K)$ is closed in $X$, for every closed subset $K$ of $Y$. If $C$ is a compact subset of $X$, then $f(C)$ is a closed, compact subset of $Y$. Then $F = f(C) \cap K$ is a closed, compact subset of $K$. It follows that $f^{-1}(K) \cap C = f^{-1}(F) \cap C$ is a compact subset of $X$. Since $X$ is a $k$-space, we conclude that $f^{-1}(K)$ is closed in $X$. This completes the proof. □

The result that follows is a corollary to Lemma 2.3.14.

PROPOSITION 2.3.15. *If $X$ is a semitopological group such that $X$ is a Hausdorff k-space, and, for each compact subset $B$ of $X$, the set $B^{-1}$ compact, then the inverse mapping is continuous on $X$. Hence, $X$ is a quasitopological group.*

The general assertion on the continuity of the inverse in locally compact paratopological groups will be reduced to the following special case of it.

PROPOSITION 2.3.16. *Suppose that $X$ is a separable locally compact Hausdorff paratopological group. Then the inverse operation is continuous.*

PROOF. Fix a countable subset $A$ of $X$ such that $\overline{A} = X$, and let $V$ be an open neighbourhood of the neutral element $e$ such that $\Phi = \overline{V}$ is compact. We are going to show that the interior of the set $\Phi^{-1}$ is not empty.

First, let us check that $AV^{-1} = X$. Take any $x \in X$. The set $xV$ is open. Therefore, $xV \cap A$ is not empty, that is, there exists $a \in A$ such that $a = xb$ for some $b \in V$. Then $x = ab^{-1} \in aV^{-1}$. It follows that $X = \bigcup\{aV^{-1} : a \in A\} = AV^{-1}$. Put $F_a = a\Phi^{-1}$. Since $aV^{-1} \subset a\Phi^{-1}$, we have that

$$X = \bigcup\{F_a : a \in A\}.$$

Observe that since $\Phi$ is compact, the set $\Phi^{-1}$ is closed, by Lemma 2.3.5. Therefore, $F_a$ is closed, for each $a \in A$.

Since $A$ is countable, and $X$ has the Baire property, it follows that there is $a \in A$ such that the interior of $F_a$ is not empty. Since all translations in $X$ are homeomorphisms, we conclude that the interior of $\Phi^{-1}$ is not empty. Thus, there exists a non-empty open set $W$ in $X$ such that $\overline{W}$ is compact and $\overline{W^{-1}} \subset \overline{V} = \Phi$. Clearly, both $\overline{W}$ and $\overline{W^{-1}}$ are compact subsets of $X$.

Take now any point $y \in X$. Then $Wy$ is an open neighbourhood of $y$, $\overline{Wy}$ is compact, and $(Wy)^{-1} = y^{-1}W^{-1}$, which implies that the closure of $(Wy)^{-1}$ is the compact set $y^{-1}\overline{W^{-1}}$. This argument shows that there is a base $\mathcal{B}$ of $X$ such that, for each $V \in \mathcal{B}$, the sets $\overline{V}$ and $\overline{V^{-1}}$ are compact. From this it obviously follows that for every compact subset $B$ of $X$, the set $B^{-1}$ is compact.

Since every locally compact Hausdorff space $X$ is, obviously, a $k$-space [**165**, p. 201], it follows from Proposition 2.3.15 that the inverse mapping in $X$ is continuous. □

THEOREM 2.3.17. *Suppose that $X$ is a locally compact Hausdorff paratopological group. Then the inverse operation on $X$ is continuous, so $X$ is a topological group.*

PROOF. As we have already observed, the locally compact space $X$ is a $k$-space. Therefore, in view of Proposition 2.3.15, it suffices to show that if $B$ is a compact subset of $X$, then $B^{-1}$ is compact.

Assume the contrary, and fix an open neighbourhood $V$ of the neutral element $e$ in $X$ such that $\overline{V}$ is compact. Since, for each $x \in X$, $Vx$ is open, and the closure of $Vx$ is compact, we can define by induction a sequence $\{x_n : n \in \omega\}$ of points in $B^{-1}$ such that $x_{n+1}$ is not in $\bigcup_{i=0}^{n} Vx_i$.

Denote by $A$ the subgroup of $X$ algebraically generated by the elements of the sequence $\{x_n : n \in \omega\}$, that is, the smallest subgroup of $X$ containing the sequence. Obviously, $A$ is countable, and $Y = \overline{A}$ is a separable, locally compact, Hausdorff paratopological group.

Now, from Proposition 2.3.16 it follows that $Y$ is a topological group. The sequence $\{x_n^{-1} : n \in \omega\}$ is contained in $B \cap Y$. Therefore, since $B$ is compact, there exists an accumulation point $b$ for $\{x_n^{-1} : n \in \omega\}$ in $X$. Clearly, the sequence $\{x_n^{-1} : n \in \omega\}$ accumulates at $b^{-1}$. Put $W = V \cap Y$. Then $W^{-1}$ is an open neighbourhood of $e$ in the space $Y$. Therefore, $W^{-1}b^{-1}$ is an open neighbourhood of $b^{-1}$ in $Y$, and there exists $k \in \omega$ such that $x_k \in W^{-1}b^{-1}$. Then $b^{-1} \in Wx_k$. Since $Wx_k$ is open in $Y$, there is $m \in \omega$ such that $m > k$ and $x_m \in Wx_k$. Then $x_m \in Vx_k$, a contradiction. □

In the next section, Theorem 2.3.17 will be extended to pseudocompact (and regular countably compact) paratopological groups.

Some further sufficient conditions for a paratopological group to be a topological group can be obtained with the help of the following lemmas.

LEMMA 2.3.18. *Suppose that $G$ is a paratopological group, and $U$ any open neighbourhood of the neutral element $e$ in $G$. Then $\overline{M} \subset MU^{-1}$, for each subset $M$ of $G$.*

PROOF. Put $A = \{g \in G : gU \cap M = \varnothing\}$ and $F = G \setminus AU$. Then, clearly, $F$ is a closed subset of $G$ and $M \subset F$. Therefore, $\overline{M} \subset F$. Take any $y \in F$. Then $yU \cap M \neq \varnothing$, that is, $yh = m$, for some $h \in U$ and $m \in M$. Hence, $y = mh^{-1} \in MU^{-1}$. Thus, $F \subset MU^{-1}$. Since $\overline{M} \subset F$, it follows that $\overline{M} \subset MU^{-1}$. □

LEMMA 2.3.19. *Suppose that $G$ is a paratopological group and not a topological group. Then there exists an open neighbourhood $U$ of the neutral element $e$ of $G$ such that $U \cap U^{-1}$ is nowhere dense in $G$, that is, the interior of the closure of $U \cap U^{-1}$ is empty.*

PROOF. The inverse operation in $G$ is discontinuous. Therefore, it is discontinuous at $e$, and we can choose an open neighbourhood $W$ of $e$ such that $e \notin \mathrm{Int}(W^{-1})$. Since multiplication is continuous in $G$, we can find an open neighbourhood $U$ of $e$ such that $U^3 \subset W$. We claim that the set $U \cap U^{-1}$ is nowhere dense in $G$.

Assume the contrary. Then there exists a non-empty open set $V$ in $G$ such that $V \subset \overline{U \cap U^{-1}}$. From Lemma 2.3.18 it follows that $V \subset \overline{U \cap U^{-1}} \subset (U \cap U^{-1})U^{-1} \subset U^{-2}$. Then $VU^{-1} \subset U^{-3} \subset W^{-1}$. Clearly, $V \cap U \neq \varnothing$, and the set $VU^{-1}$ is open in $G$. Therefore, $e \in VU^{-1} \subset \mathrm{Int}(W^{-1})$, a contradiction. □

The next corollary follows immediately from Lemma 2.3.19.

COROLLARY 2.3.20. *Suppose that $G$ is a paratopological group such that $e \in \mathrm{Int}\,\overline{U^{-1}}$, for each open neighbourhood $U$ of the neutral element $e$ of $G$. Then $G$ is a topological group.*

For the next result, we need a generalization of Lemma 2.3.5.

LEMMA 2.3.21. *Suppose that $G$ is a paratopological group. Then, for each compact subset $F$ of $G$ such that $e \notin F$, there exist an open neighbourhood $O(F)$ of $F$ and an open neighbourhood $O(e)$ of $e$ such that $O(F) \cap O(e)^{-1} = \varnothing$.*

PROOF. For each $x \in F$, we select an open neighbourhood $V_x$ of $e$ such that $x^{-1} \notin V_x{}^2$. Then $V_x x \cap V_x{}^{-1} = \varnothing$. Since $\gamma = \{V_x x : x \in F\}$ is a family of open sets in $G$ covering the compact set $F$, there exists a finite subset $K$ of $F$ such that $F \subset \bigcup_{x\in K} V_x x$. Put $O(e) = \bigcap_{x\in K} V_x$ and $O(F) = \bigcup_{x\in K} V_x x$. Then $O(e)$ is an open neighbourhood of $e$, $O(F)$ is an open neighbourhood of $F$, and $O(F) \cap O(e)^{-1} = \varnothing$. □

THEOREM 2.3.22. *Suppose that $f$ is a perfect homomorphism of a paratopological group $G$ onto a topological group $H$. Then $G$ is also a topological group.*

PROOF. Put $F = f^{-1}f(e)$, where $e$ is the neutral element of $G$. Assume that $G$ is not a topological group. Then, according to Corollary 2.3.20, there exists an open neighbourhood $U$ of $e$ in $G$ such that $e$ is not in $\mathrm{Int}(U^{-1})$. Put $F_1 = F \setminus U$. Since $F_1$ is compact and $e$ is not in $F_1$, Lemma 2.3.21 implies that there exist an open neighbourhood $O(F_1)$ of $F_1$ and an open neighbourhood $O(e)$ of $e$ in $G$ such that $O(F_1) \cap O(e)^{-1} = \varnothing$.

Since $O = O(F_1) \cup U$ is an open neighbourhood of $F$ and the mapping $f$ is closed, there exists an open neighbourhood $V$ of $f(e)$ in $H$ such that $f^{-1}(V) \subset O$. We can also assume that $V^{-1} = V$, since $H$ is a topological group. Then $(f^{-1}(V))^{-1} = f^{-1}(V) \subset O$. Finally, put $W = f^{-1}(V) \cap O(e) \cap U$. Clearly, $W$ is an open neighbourhood of $e$ contained in $U$. We also have $W^{-1} \subset (f^{-1}(V))^{-1} \subset O$ and $W^{-1} \subset O(e)^{-1}$. Since $O(F_1) \cap O(e)^{-1} = \varnothing$, it follows that $W^{-1} \subset U$. Therefore, $e \in W \subset \mathrm{Int}(U^{-1})$, a contradiction. □

Of course, the statement that every compact Hausdorff paratopological group is a topological group is a direct corollary of Theorem 2.3.22.

A left semitopological group $G$ is called *$\omega$-narrow* if for every open neighbourhood $V$ of the neutral element in $G$, there exists a countable subset $A$ of $G$ such that $AV = G$. Similarly, a right semitopological group $G$ is called *$\omega$-narrow* if for every open neighbourhood $V$ of the neutral element in $G$, there exists a countable subset $A$ of $G$ such that $VA = G$. A semitopological group $G$ is said to be *$\omega$-narrow* if for every open neighbourhood $V$ of the neutral element in $G$, there exists a countable set $A \subset G$ such that $VA = G = AV$.

In the case of topological groups, this concept was introduced by I. I. Guran who called such groups *ω-bounded.* To avoid ambiguity (this term has several different meanings in topology), we change the terminology.

If the set $A$ in the above definition can be taken to be finite, the (left, right) semitopological group $G$ is said to be *precompact* or *totally bounded.* In what follows, the concept of precompactness will be applied almost exclusively to topological groups.

THEOREM 2.3.23. *Suppose that $G$ is a topological group, $H$ is a paratopological group and that $\varphi\colon G \to H$ is a continuous onto homomorphism. Suppose also that at least one of the following conditions holds:*

1) *$G$ is a precompact group;*
2) *$G$ is $\omega$-narrow group and $H$ has the Baire property.*

*Then $H$ is a topological group.*

PROOF. Assume that $H$ is not a topological group. By Lemma 2.3.19, there is an open neighbourhood $U$ of the neutral element $e$ of $H$ such that $U \cap U^{-1}$ is nowhere dense. Let $W$ be a symmetric neighbourhood of the identity in $G$ such that $\varphi(W) \subset U$.

Case 1. Since $G$ is a precompact group, there is a finite set $A \subset G$ such that $G = AW$. Then $M = \varphi(W) \subset U \cap U^{-1}$ is nowhere dense, and $BM = H$, where $B = \varphi(A)$. Hence, $H$ is the union of a finite family of nowhere dense sets, which is a contradiction.

Case 2. Since $G$ is a $\omega$-narrow group, there is a countable set $A \subset G$ such that $G = AW$. Then $M = \varphi(W) \subset U \cap U^{-1}$ is nowhere dense and $BM = H$, where $B = \varphi(A)$. Again, $H$ is the union of a countable family of nowhere dense sets, thus contradicting the Baire property of $H$. □

Some interesting applications of Theorem 2.3.23 will be given in Section 5.7. They are based on the following key result:

PROPOSITION 2.3.24. *Suppose that $G$ is a Hausdorff paratopological group with neutral element $e$, and let $H$ be a paratopological group which as a topological space coincides with the topological space $G$ and whose multiplication is given by the rule $g \times h = hg$, for all $g, h \in H$. Put $T = \{(g, g^{-1}) \in G \times H : g \in G\}$. Then:*

1) *$T$ is closed in the space $G \times H$ and is a subgroup of the group $G \times H$;*
2) *$T$ is a topological group;*
3) *the natural projection $(g, g^{-1}) \to g$ is a continuous isomorphism of the topological group $T$ onto the paratopological group $G$.*

PROOF. It is trivially verified that $T$ is algebraically a subgroup of $G \times G$. Let us show that $T$ is closed in the space $G \times H$. Assume that $(x, y) \in (\overline{T} \setminus T)$. Then $y \neq x^{-1}$ and $xy \neq e$. Since $G$ is a paratopological group, we can find open neighbourhoods $U$ and $V$ of $x$ and $y$, respectively, in $G$ such that $e \notin UV$. Then $(U \times V) \cap T = \varnothing$ and $(x, y) \in U \times V$. Hence, $T$ is closed in $G \times H$, and 1) follows.

Clearly, $T$ is a paratopological group. The statement 3) is also obvious. It remains to verify the continuity of the inverse in $T$. It suffices to do this at the neutral element $(e, e)$ of $T$. A basic neighbourhood $W$ of $(e, e)$ in $T$ is of the form $T \cap (U \times U)$, where $U$ is an open neighbourhood of $e$ in $G$. Take any element $(g, g^{-1})$ of $W$. Then $g \in U$ and $g^{-1} \in U$. It follows that $(g, g^{-1})^{-1} = (g^{-1}, g) \in (U \times U) \cap T = W$, that is, $W^{-1} = W$, and the continuity of the inverse mapping in $T$ at $(e, e)$ is verified. □

We conclude this section with a corollary to Proposition 2.3.24.

COROLLARY 2.3.25. *If a Hausdorff paratopological group $G$ is $\sigma$-compact, then there exists a $\sigma$-compact topological group $T$ and a continuous isomorphism of $T$ onto $G$.*

PROOF. Let $G = \bigcup_{n=0}^{\infty} F_n$, where each $F_n$ is a compact subset of a Hausdorff paratopological group $G$. As in Proposition 2.3.24, denote by $H$ a paratopological group which as a topological space coincides with the topological space $G$ and whose multiplication is given by the rule $g \times h = hg$, for all $g, h \in H$. Clearly, the space $H$ is $\sigma$-compact. Then the product space $G \times H$ and its closed subspace $T = \{(g, g^{-1}) : g \in G\}$ are also $\sigma$-compact. According to Proposition 2.3.24, $T$ is a topological group and the mapping $\pi\colon T \to G$, $\pi(g, g^{-1}) = g$, is a continuous isomorphism of $T$ onto $G$. □

## Exercises

2.3.a. Can one weaken 'semitopological group' to 'right topological group' in Theorem 2.3.2?

2.3.b. Let $G$ be a paratopological group and $H$ a dense subgroup of $G$. Show that if $H$ with the topology inherited from $G$ is a topological group, then so is $G$.

2.3.c. Suppose that $f\colon G \to H$ is a continuous isomorphism of a paratopological group $G$ onto a compact topological group $H$. Is $G$ then a topological group? What if $G$ is first-countable?

2.3.d. Is every precompact paratopological group a topological group?

2.3.e. Give an example of a first-countable, countably compact, locally compact, normal topological semigroup with identity which is commutative, but is neither compact nor (algebraically) a group.

2.3.f. Suppose that $G$ is a topological group, $H$ is a Tychonoff paratopological group and that $\varphi\colon G \to H$ is a continuous onto homomorphism. Suppose also that $G$ is separable and that $H$ has a dense subspace homeomorphic to the space of irrational numbers. Prove that $H$ is a topological group.

2.3.g. Let $G$ be a paratopological group with topology $\tau$, and let $\tau^{-1} = \{U^{-1} : U \in \tau\}$ be the *conjugate* topology of $G$. A real-valued function $f$ on $G$ is called *bicontinuous* if $f^{-1}(-\infty, r) \in \tau$ and $f^{-1}(r, \infty) \in \tau^{-1}$, for each $r \in \mathbb{R}$. The group $G$ is said to be 2-*pseudocompact* if every bicontinuous real-valued function on $G$ is bounded.

a) Give an example of a pseudocompact paratopological group which fails to be 2-pseudocompact.

b) Show that a countably compact paratopological group is 2-pseudocompact.

*Hint.* Let $G = \mathbb{Z}$ be the additive group of integers with paratopological group topology $\tau$ whose base at zero consists of the sets $U_n = \{0\} \cup \{k \in \mathbb{Z} : k \geq n\}$, $n \in \mathbb{N}$. Verify that $(G, \tau)$ is as required.

## Problems

2.3.A. (S. Romaguera and M. Sanchis [**412**]) Prove that every compact paratopological group satisfying the $T_1$ separation axiom is a Hausdorff topological group.

2.3.B. Suppose that $G$ is a regular paratopological group such that some neighbourhood of the neutral element in $G$ is countably compact. Is $G$ a topological group? (See also Problem 2.4.5.)

2.3.C. Suppose that $f$ is an open continuous homomorphism of a regular paratopological group $G$ onto a topological group $H$, and that the kernel of $f$ is compact. Is $G$ a topological group?

2.3.D. Suppose that $f$ is a continuous isomorphism of a metrizable paratopological group $G$ onto a metrizable topological group $H$. Is $G$ a topological group?

2.3.E. Suppose that $G$ is a Tychonoff paratopological group such that the inverse to every compact subset of $G$ is $\sigma$-compact. Is $G$ a topological group? Must $G$ be Lindelöf?
*Hint.* Consider the Sorgenfrey line or its square.

2.3.F. (O. V. Ravsky [**398**]) Prove that every commutative Hausdorff paratopological group admits a continuous isomorphism onto a Hausdorff topological group.
*Hint.* Let $\mathcal{B}$ be a local base at zero of a commutative paratopological group $G$. Show that the family $\{U - U : U \in \mathcal{B}\}$ is a local base at zero for a Hausdorff topological group topology on $G$.

2.3.G. (S. García-Ferreira, S. Romaguera, and M. Sanchis [**186**]) Let $G$ be a paratopological group with topology $\tau$, and $\tau^{-1}$ be the conjugate topology of $G$ (see Exercise 2.3.g). Prove that $G$ is 2-pseudocompact iff for every decreasing sequence $\{U_n : n \in \omega\}$ of non-empty $\tau$-open sets in $G$ and every decreasing sequence $\{V_n : n \in \omega\}$ of non-empty $\tau^{-1}$-open sets in $G$, the intersections $\bigcap_{n\in\omega} cl_{(G,\tau^{-1})} U_n$ and $\bigcap_{n\in\omega} cl_{(G,\tau)} V_n$ are non-empty.

2.3.H. (T. G. Raghavan and I. L. Reilly [**395**]) Prove that every countably compact paratopological group of countable pseudocharacter is a topological group.

## Open Problems

2.3.1. Let $G$ be a $\sigma$-compact paratopological group. Does there exist a continuous isomorphism of $G$ onto a topological group? (See Corollary 2.3.25 and Problems 2.3.F and 2.4.2.)

2.3.2. (O. Alas and M. Sanchis [**4**]) Is every 2-pseudocompact paratopological group $G$ of countable pseudocharacter a topological group provided that $G$ satisfies the $T_1$ separation axiom? (See also Exercise 2.3.g and Problems 2.3.G and 2.3.H.)

## 2.4. Pseudocompact semitopological groups

Here we show that every pseudocompact paratopological group is a topological group, but this assertion cannot be extended either to semitopological or quasitopological groups. It is well known that a Tychonoff space $X$ is pseudocompact iff every locally finite family of open sets in $X$ is finite. In fact, pseudocompactness is defined only in the realm of Tychonoff spaces since its definition makes use of continuous real-valued functions (see [**165**, Section 3.10]). To present results in a general form, we recall that a topological space $X$ is called *feebly compact* if every locally finite family of open sets in $X$ is finite. Therefore, 'feebly compact' is equivalent to 'pseudocompact' for Tychonoff spaces.

THEOREM 2.4.1. [**A. V. Arhangel'skii and E. A. Reznichenko**] *Suppose that $G$ is a paratopological group such that $G$ is a dense $G_\delta$-set in a regular feebly compact space $X$. Then $G$ is a topological group.*

PROOF. Assume the contrary. Then, by Lemma 2.3.19, there exists an open neighbourhood $U$ of the neutral element $e$ of $G$ such that $U \cap U^{-1}$ is nowhere dense. Let $W$ be an open neighbourhood of $e$ such that $\overline{WW} \subset U$. Put $O = W \setminus \overline{U \cap U^{-1}}$. Then, clearly, $O \subset W \subset \overline{O}$ and $O^{-1} \cap U = \varnothing$.

First, we fix a sequence $\{M_n : n \in \omega\}$ of open sets in $X$ such that $G = \bigcap_{n=0}^{\infty} M_n$. We are going to define a sequence $\{U_n : n \in \omega\}$ of open subsets of $X$ and a sequence $\{x_n : n \in \omega\}$ of elements of $G$ such that $x_n \in U_n$, for each $n \in \omega$. Put $U_0 = O$, and pick a point $x_0 \in U_0 \cap G$.

Assume now that, for some $n \in \omega$, an open subset $U_n$ of $X$ and a point $x_n \in U_n \cap G$ are already defined. Since $e \in W \subset \overline{O}$, we have $x_n \in x_n\overline{O} = \overline{x_n O}$. Since $U_n$ is an open neighbourhood of $x_n$, it follows that $U_n \cap x_n O \neq \varnothing$. We take $x_{n+1}$ to be any point of $U_n \cap x_n O$. Note that $x_{n+1} \in G$, since $x_n O \subset G$.

Using the regularity of $X$, we can find an open neighbourhood $U_{n+1}$ of $x_{n+1}$ in $X$ such that the closure of $U_{n+1}$ is contained in $U_n \cap M_n$, and $U_{n+1} \cap G \subset x_n O$. The definition of the sets $U_n$ and points $x_n$, for each $n \in \omega$, is complete. Note that $\overline{U_i} \subset U_j$ whenever $j < i$. We also have $x_{n+1} \in x_n O$, for each $n \in \omega$.

Put $F = \bigcap_{n\in\omega} \overline{U_n}$. Clearly, $F \subset G$, and $F \neq \varnothing$ since $X$ is feebly compact. The set $FW$ is an open neighbourhood of $F$ in $G$. Consider the closure $P$ of $FW$ in $X$, and let $H$ be the closure of $X \setminus P$ in $X$. Then $H$ is a regular closed subset of $X$, so that $H$ is feebly compact.

We claim that $H \cap F = \varnothing$. Indeed, assume the contrary, and fix $x \in F \cap H$. Since $FW$ is an open neighbourhood of $F$ in $G$, from $x \in F$ it follows that there exists an open neighbourhood $V$ of $x$ in $X$ such that $V \cap G \subset FW$. Then the density of $G$ in $X$ implies that $V \subset P$, while $x \in V \cap H$ implies that $V \setminus P \neq \varnothing$, which is a contradiction. Thus, $H \cap F = \varnothing$.

Since $H$ is feebly compact, our definition of $F$ implies that $U_k \cap H = \varnothing$, for some $k \in \omega$ (we use that $\overline{U_i} \subset U_j$ whenever $j < i$). Then $U_k \subset P$. Since $x_k \in U_k \cap G$, it follows that $x_k \in \overline{FW}$. However, $F \subset U_{k+2} \cap G \subset x_{k+1}O \subset x_{k+1}W$. Hence,

$$x_k \in \overline{FW} \subset x_{k+1}\overline{WW} \subset x_{k+1}U.$$

Taking into account that $x_{k+1} \in x_k O$, we obtain that $x_k \in x_k OU$. Hence, $e \in OU$ and $O^{-1} \cap U \neq \varnothing$, which is again a contradiction. □

Naturally, the following two corollaries of Theorem 2.4.1 refer to Tychonoff spaces.

COROLLARY 2.4.2. *Every pseudocompact paratopological group is a topological group.*

COROLLARY 2.4.3. *Every Čech-complete paratopological group is a topological group.*

Since countably compact space are feebly compact, Theorem 2.4.1 also implies the next fact.

COROLLARY 2.4.4. *Every regular countably compact paratopological group is a topological group.*

Let us now consider the case of pseudocompact semitopological groups. Below we will see that not all of them are paratopological groups. However, under some additional restrictions on the topology of a semitopological group, some strong positive results in this direction can be obtained. To present one of them, we need a few elementary facts.

LEMMA 2.4.5. *Let $G$ be a semitopological group with quasicontinuous multiplication. Then, for each non-empty open $U \subset G$, there exist $g \in U$ and an open neighbourhood $V$ of the neutral element $e$ in $G$ such that $gV^3 \subset U$.*

PROOF. Since the multiplication in $G$ is quasicontinuous, there exist non-empty open sets $V_1, W \subset G$ such that $V_1 W \subset U$, and there exist non-empty open sets $V_2, V_3 \subset G$ such that $V_2 V_3 \subset W$. Clearly, we have $V_1 V_2 V_3 \subset U$. Fix $g_i \in V_i$ for $i = 1, 2, 3$. Put $g = g_1 g_2 g_3$ and

$$V = g_3^{-1} g_2^{-1} g_1^{-1} V_1 g_2 g_3 \cap g_3^{-1} g_2^{-1} V_2 g_3 \cap g_3^{-1} V_3.$$

Then $g \in U$, $V$ is an open neighbourhood of $e$ in $G$, and an easy calculation shows that $gV^3 \subset U$, as required. □

We also need the following simple lemma which is a part of the folklore.

LEMMA 2.4.6. *Let $X$ be a Hausdorff space, $\mathcal{F}$ a family of compact subsets of $X$, $K = \bigcap \mathcal{F}$, and $W$ an open neighbourhood of $K$ in $X$. Then for every $F \in \mathcal{F}$, there exists an open neighbourhood $O(F)$ of $F$ in $X$ such that $\bigcap\{O(F) : F \in \mathcal{F}\} \subset W$.*

PROOF. Since all elements of $\mathcal{F}$ are compact, we can assume that the family $\mathcal{F}$ is finite, by the Shura-Bura lemma (see [**165**, Corollary 3.1.5]). We apply induction on $n = |\mathcal{F}|$. First, if $\mathcal{F} = \{F_1, F_2\}$, we put $K_1 = F_1 \setminus W$ and $K_2 = F_2 \setminus W$. Then $K_1$ and $K_2$ are disjoint compact sets in $X$, so there are disjoint open neighbourhoods $U_1$ and $U_2$ of $K_1$ and $K_2$, respectively [**165**, Theorem 1.3.6]. Then $O_1 = U_1 \cup W$ and $O_2 = U_2 \cup W$ are open neighbourhoods of $F_1$ and $F_2$, respectively, and $O_1 \cap O_2 = W$.

Suppose that the conclusion of the lemma is valid for every family of compact sets in $X$ with at most $n$ elements ($n \geq 2$), and that $\mathcal{F} = \{F_1, \ldots, F_n, F_{n+1}\}$. Put $P = F_1 \cap \cdots \cap F_n$. Then $P \cap F_{n+1} \subset W$, so there exist open neighbourhoods $W_1$ and $W_2$ of $P$ and $F_{n+1}$, respectively, in $X$ such that $W_1 \cap W_2 \subset W$. By the inductive assumption, one can find open sets $O_1, \ldots, O_n$ in $X$ such that $F_i \subset O_i$ for each $i \leq n$ and $O_1 \cap \ldots \cap O_n \subset W_1$. Then $\bigcap_{i=1}^{n+1} O_i \subset W$, where $O_{n+1} = W_2$. This obviously completes the proof. □

In Lemmas 2.4.7–2.4.11 below we assume that $G$ is a semitopological Čech-complete group with topology $\mathcal{T}$. In particular, the space $G$ is Tychonoff. Let $\mathcal{T}_e$ be the family of open neighbourhoods of the neutral element $e$ in $G$, and $\mathcal{K}$ the family of all non-empty compact subsets of $G$ such that the family $\{KU^2 : U \in \mathcal{T}_e\}$ constitutes a base for the space $G$ at the set $K$.

Let us define a partial order $\leq$ on the topology $\mathcal{T}$ of $G$ as follows: For $U, V \in \mathcal{T}$, $U \leq V$ iff there exists $g \in U$ such that $\overline{g(g^{-1}U)^3} \subset V$. It is clear that $U \leq V$ implies $\overline{U} \subset V$.

LEMMA 2.4.7. *For every non-empty $V \in \mathcal{T}$, there exists a non-empty $U \in \mathcal{T}$ such that $U \leq V$.*

PROOF. Fix a non-empty open set $W$ such that $\overline{W} \subset V$. Since $G$ is Čech-complete, it follows from Lemma 2.3.7 that the multiplication in $G$ is quasicontinuous. Lemma 2.4.5 implies that there exist $g \in W$ and $W_1 \in \mathcal{T}_e$ such that $g{W_1}^3 \subset W$. Hence, for $U = gW_1$ we have $U \leq V$. □

The next statement plays the key role in the proof of the main result of this section, Theorem 2.4.12.

LEMMA 2.4.8. *The family $\mathcal{K}$ is a $\pi$-network in $G$, that is, every non-empty open subset of $G$ contains an element of $\mathcal{K}$.*

PROOF. Let $W$ be a non-empty open subset of $G$. We have to show that there exists $K \in \mathcal{K}$ contained in $W$. Since the space $G$ is Čech-complete, there exists a sequence $\{B_n : n \in \omega\}$ of open subsets of the Čech–Stone compactification $\beta G$ of the space $G$ such that $G = \bigcap_{n\in\omega} B_n$.

Lemma 2.4.7 implies the existence of a sequence $\{U_n : n \in \omega\}$ of non-empty open subsets of $G$ such that $U_0 \subset W$, $U_{n+1} \leq U_n$ and $U_n \subset F_n \subset B_n$ for each $n \in \omega$, where $F_n$

is the closure of $U_n$ in $\beta G$. Put $K = \bigcap_{n\in\omega} U_n$. Clearly, $K \subset W$ and $K = \bigcap_{n\in\omega} F_n$, since $U_n$ contains the closure of $U_{n+1}$ in $G$. Therefore, $K$ is compact and non-empty, and every open neighbourhood of $K$ in $\beta G$ contains some $F_n$ (notice that $F_{n+1} \subset F_n$ for each $n \in \omega$). It follows that the family $\{U_n : n \in \omega\}$ is a base for $G$ at the set $K$.

Finally, let us show that $K \in \mathcal{K}$. Since $U_{n+1} \leq U_n$ for $n \in \omega$, there exists $g_{n+1} \in U_{n+1}$ such that $g_{n+1}(g_{n+1}^{-1}U_{n+1})^3 \subset U_n$. Let also $g_0$ be a point of $U_0$. For $n \in \omega$, put $W_n = g_n^{-1}U_n$. Then $W_n \in \mathcal{T}_e$ and

$$KW_{n+1}^2 \subset U_{n+1}W_{n+1}^2 = g_{n+1}W_{n+1}^3 \subset U_n.$$

Since $\{U_n : n \in \omega\}$ is a base for $G$ at the compact set $K$, it follows that $\{KW_n^2 : n \in \omega\}$ is also a base for $G$ at $K$. Hence, $K \in \mathcal{K}$. □

We also need the following very simple lemma.

LEMMA 2.4.9. *For every open neighbourhood $V$ of the neutral element $e$ of a semitopological group $G$ and every $g \in G$, there exists an open neighbourhood $U$ of $e$ such that $gU^2g^{-1} \subset V^2$.*

PROOF. Put $U = g^{-1}Vg$. Clearly, $U$ is open and $e \in U$. We have: $gU^2g^{-1} = gg^{-1}Vgg^{-1}Vgg^{-1} = V^2$. □

LEMMA 2.4.10. *For any $g \in G$ and any $K \in \mathcal{K}$, $gK \in \mathcal{K}$ and $Kg \in \mathcal{K}$.*

PROOF. Since the family $\mathcal{O} = \{KU^2 : U \in \mathcal{T}_e\}$ is a base for $G$ at the set $K \in \mathcal{K}$, the family $\{gKU^2 : U \in \mathcal{T}_e\} = \{gW : W \in \mathcal{O}\}$ is a base for $G$ at the set $gK$. Hence, $gK \in \mathcal{K}$. Also, it is easy to see that the family $\{KgU^2 : U \in \mathcal{T}_e\}$ is a base for $G$ at $Kg$. Indeed, take an arbitrary neighbourhood $O$ of the set $Kg$ in $G$. Then $K \subset Og^{-1}$ and, since $K \in \mathcal{K}$, there exists $V \in \mathcal{T}_e$ such that $KV^2 \subset Og^{-1}$. Hence, $KV^2g \subset O$. By Lemma 2.4.9, we can choose $U \in \mathcal{T}_e$ with $gU^2g^{-1} \subset V^2$, which gives the inclusion $KgU^2 \subset O$. □

LEMMA 2.4.11. *For any subfamily $\mathcal{F}$ of the family $\mathcal{K}$, if the set $K = \bigcap \mathcal{F}$ is non-empty, then $K \in \mathcal{K}$.*

PROOF. Let $W$ be a neighbourhood of the set $K$ in $G$. There exists a finite family $\{K_1, \ldots, K_n\} \subset \mathcal{F}$ such that $\bigcap_{i=1}^n K_i \subset W$. Now, by Lemma 2.4.6, we can find an open neighbourhood $V_i$ of $K_i$, for each $i \leq n$, such that $\bigcap_{i=1}^n V_i \subset W$. From the definition of $\mathcal{K}$ it follows that there exist sets $U_1, \ldots, U_n \in \mathcal{T}_e$ such that $\bigcap_{i=1}^n K_iU_i^2 \subset W$. For $U = \bigcap_{i=1}^n U_i$, we have that $KU^2 \subset W$. □

THEOREM 2.4.12. [**A. Bouziad**] *Let $G$ be a Čech-complete semitopological group. Then $G$ is a topological group.*

PROOF. First we will prove that $G$ is a paratopological group. Lemmas 2.4.8 and 2.4.10 imply that the family $\{K \in \mathcal{K} : e \in K\}$ is non-empty. Put $H = \bigcap\{K \in \mathcal{K} : e \in K\}$. By Lemma 2.4.11, $H \in \mathcal{K}$. We claim that $H = \{e\}$.

Suppose that $H \neq \{e\}$, i.e., there exists $g \in H \setminus \{e\}$. Then we can find $U_1, U_2 \in \mathcal{T}_e$ such that $U_1 \cap gU_2 = \varnothing$. Put $U = U_1 \cap U_2$. Lemma 2.4.8 implies that there exists $K \in \mathcal{K}$ such that $K \subset U$. Take any $h \in K$. By Lemma 2.4.10, $e \in Kh^{-1} \in \mathcal{K}$, so $H \subset Kh^{-1}$ and $Hh \subset K \subset U$. Therefore, $gh \in U \cap gU \subset U_1 \cap gU_2$, which is a contradiction.

We have established that $G$ is paratopological group. However, every Čech-complete paratopological group is a topological group, by Corollary 2.4.3. □

A natural question which arises at this point is whether the last statement can be extended to pseudocompact semitopological groups. To answer this question, we present a general method allowing to construct semitopological groups with some interesting combinations of properties.

Suppose that $X$ is a topological space, and $G$ is an Abelian group. Let $X^G$ be the space of all mappings of $G$ to $X$, with the pointwise convergence topology (which in this case coincides with the Tychonoff product topology of $X^G$). For $a \in G$, $f \in X^G$ and each $b \in G$, put $s(a, f)(a+b) = s_a(f)(a+b) = f(b)$. Then $s_a(f) \in X^G$, and $s$ is a mapping of $G \times X^G$ to $X^G$ called the *$G$-shift* on $X^G$. The mapping $s_a\colon X^G \to X^G$ is called the *$a$-shift* of $X^G$, or the shift of $X^G$ by $a$. For each $f \in X^G$, the subspace $s(G \times \{f\})$ of $X^G$ is called the *orbit of $f$ under the shift $s$*, or simply the *orbit* of $f$. A mapping $f\colon G \to X$ is called a *Korovin mapping* and the orbit of $f$ is said to be a *Korovin orbit* if for every countable subset $M$ of $G$ and every mapping $h\colon M \to X$, there exists $a \in G$ such that $s(a, f)(m) = h(m)$ for each $m \in M$, that is, the restriction of the mapping $s(a, f)$ (which is an element of the orbit of $f$) to $M$ coincides with $h$.

THEOREM 2.4.13. [**A. V. Korovin**] *Let $X$ be a topological space satisfying $1 < |X| \leq \mathfrak{c} = 2^\omega$, and $G$ an Abelian group such that $|G| = \mathfrak{c}$. Then there exists a Korovin mapping $f\colon G \to X$, that is, there exists a Korovin orbit in $X^G$.*

PROOF. To construct a Korovin orbit, it is convenient to consider partial shifts. For an arbitrary countable subset $B$ of $G$, let $\mathscr{E}_B$ be the set of all mappings of $B$ to $X$. Let $a$ be any element of $G$, $B$ a countable subset of $G$, and $h \in \mathscr{E}_B$. Then $s_{a,B}(h)$ is an element of $\mathscr{E}_{a+B}$ defined by the rule $(s_{a,B}(h))(a+b) = h(b)$, for each $b \in B$. We will say that $h' = s_{a,B}(h)$ is obtained by a shift (or by an $a$-shift) from $h$. Note that if $h'$ is obtained by an $a$-shift from $h$, then $h$ is obtained by an $-a$-shift from $h'$. Clearly, $f\colon G \to X$ is a Korovin mapping if and only if for each countable subset $M$ of $G$ and every mapping $h\colon M \to X$, there exists a countable subset $B \subset G$ and an element $a \in G$ such that $a + B = M$ and $s_{a,B}(f{\restriction}B) = h$. This tells us how to construct the required $f$. We will enumerate all possible $h$ and will select corresponding $B$ as disjoint subsets of $G$, which will allow us to define $f$ properly. Here are the details of this approach.

Let $\mathscr{E}$ be the union of all sets $\mathscr{E}_B$, where $B$ is a countable subset of $G$. Clearly, $|\mathscr{E}_B| = \mathfrak{c}$, for each countable $B \subset G$. Since $|G^\omega| = \mathfrak{c}$, we have $|\mathscr{E}| = \mathfrak{c}$. We well order $\mathscr{E}$ in type $\mathfrak{c}$, so that $\mathscr{E} = \{f_\alpha : \alpha < \mathfrak{c}\}$, and let $B_\alpha$ be the domain of the mapping $f_\alpha$, that is, $Dom(f_\alpha) = B_\alpha$.

We are going to define, by recursion, a transfinite sequence $\xi = \{h_\alpha : \alpha < \mathfrak{c}\}$ of elements of $\mathscr{E}$ such that:

1) $Dom(h_\alpha) \cap Dom(h_\beta) = \varnothing$, whenever $\alpha < \beta < \mathfrak{c}$;
2) for each $\alpha < \mathfrak{c}$, $h_\alpha$ is obtained by an $a_\alpha$-shift from $f_\alpha$, for some $a_\alpha \in G$.

We can start with $h_0 = f_0$. Then conditions 1) and 2) are trivially satisfied at the level 0. Assume that for some $\beta < \mathfrak{c}$, $h_\alpha \in \mathscr{E}$ are already defined, for all $\alpha < \beta$. Let us define $h_\beta \in \mathscr{E}$.

Since $\beta < \mathfrak{c}$ and $Dom(h)$ is countable, for each $h \in \mathscr{E}$, we have $|\bigcup\{Dom(h_\alpha) : \alpha < \beta\}| < \mathfrak{c}$. Put $H_\beta = Dom(f_\beta) \cup (\bigcup\{Dom(h_\alpha) : \alpha < \beta\})$, and let $G_\beta$ be the smallest subgroup of $G$ containing $H_\beta$. Then, obviously, $|G_\beta| < \mathfrak{c} = |G|$. Take any element $a_\beta \in G \setminus G_\beta$, and define $h_\beta$ as the $a_\beta$-shift of $f_\beta$. Then $h_\beta(a_\beta + b) = f_\beta(b)$, for each $b \in B_\beta = Dom(f_\beta)$, and $Dom(h_\beta) = a_\beta + Dom(f_\beta) \subset a_\beta + G_\beta$. Since $G_\beta$ is a subgroup of $G$ and $a_\beta \notin G_\beta$, the sets $G_\beta$ and $a_\beta + G_\beta$ are disjoint. However, $Dom(h_\alpha) \subset G_\beta$, for each $\alpha < \beta$. It follows that $Dom(h_\alpha) \cap Dom(h_\beta) = \varnothing$, whenever $\alpha < \beta$. Condition 2) at the level $\beta$ is also satisfied, since $h_\beta$ was defined as a shift of $f_\beta$. The transfinite recursion is complete — the sequence $\xi$ is defined, and it satisfies conditions 1) and 2).

We define a mapping $f: G \to X$ as follows. Take any $a \in G$. If there exists $\alpha < \mathfrak{c}$ such that $a \in Dom(h_\alpha)$, then such an $\alpha$ is unique, by condition 1), and we put $f(a) = h_\alpha(a)$. Otherwise, let $f(a)$ be the neutral element of $G$. Then it is clear from 2) that $f$ is a Korovin mapping, and the orbit of $f$, that is, the subspace $s(G \times \{f\})$ of $X^G$, is a Korovin orbit. □

We call a subset $Y$ of a space $X$ *$G_\delta$-dense* in $X$ if every non-empty $G_\delta$-set in $X$ intersects $Y$ (see also Proposition 1.6.36). This concept is helpful in many respects; it also appears in the context of Korovin orbits.

PROPOSITION 2.4.14. *Suppose that $X$ is a Hausdorff topological space, $G$ is an Abelian group, $f \in X^G$ is a Korovin mapping, and that $K_f = s(G \times \{f\})$ is the corresponding Korovin orbit, with the topology of the subspace of $X^G$. Put $k(a) = s_a(f)$, for each $a \in G$. Then:*

1) *$k$ is a one-to-one mapping of $G$ onto $K_f$;*
2) *for every $a \in G$, the restriction of the $a$-shift $s_a$ to $K_f$ is a homeomorphism of the space $K_f$ onto itself;*
3) *the subspace $K_f \subset X^G$ with the group operation, under which the mapping $k$ is an isomorphism of the group $G$ onto the group $K_f$, is a semitopological group;*
4) *if, in addition, the group $G$ is Boolean, then $K_f$ with the operation and topology described in 3) is a quasitopological group;*
5) *the space $K_f$ is $G_\delta$-dense in $X^G$; even more, the image of $K_f$ under the natural projection of $X^G$ onto $X^B$, for any countable subset $B \subset G$, is the whole $X^B$;*
6) *if $G$ is Boolean, then the quasitopological group $K_f$ is not a topological group.*

PROOF. First, we prove 1). Assume that $k(a) = k(b)$, for some distinct $a$ and $b$ from $G$. Then $s_a(f) = s_b(f)$, which implies that $s_c(f) = f$, where $c = a - b$. Hence, $g(a + c) = g(a)$, for each $a \in G$ and each $g \in K_f$. It follows that for any $g \in K_f$ and any mapping $h: \{e, c\} \to X$ such that $h(e) \neq h(c)$, the restriction of $g$ to the set $\{e, c\}$ does not coincide with $h$. Thus, $f$ is not a Korovin mapping, a contradiction.

To prove 2), we observe that, obviously, $s_a(K_f) = K_f$, and that $s_a$ is a homeomorphism of $X^G$ onto $X^G$, by the definition of the topology of pointwise convergence, which coincides with the Tychonoff product topology of $X^G$.

To prove 3), consider the action $l$ on the group $K_f$ by an arbitrary element $s_a(f)$ of $K_f$. Take any element $s_b(f)$ of $K_f$. Then $l(s_b(f)) = s_a(f) + s_b(f) = s_{a+b}(f) = s_a(s_b(f))$. Hence, $l$ is the restriction of the shift $s_a$ to $K_f$. It follows from 2) that $l$ is a homeomorphism.

The statement 4) is clear, since the inverse mapping in a Boolean group is the identity mapping.

The statement 5) is immediate, since $f$ is a Korovin mapping.

Let us show that the addition is not continuous at the neutral element of $K_f$. First, we note that the space $K_f$ is not discrete — this follows from 5), since $X^G$ is not discrete. Clearly, the neutral element of $K_f$ is $k(0) = s_0(f) = f$, where 0 is the neutral element of $G$. Put $x_0 = f(0)$, and let $z$ be any element of $X \setminus \{x_0\}$. Put $U = X \setminus \{z\}$, and let $W$ be the set of all $g \in K_f$ such that $g(0) \in U$. Clearly, $W$ is an open neighbourhood of $f$ in $K_f$.

Let $O(f)$ be an arbitrary open neighbourhood of $f$ in $K_f$. By the definition of the product topology in $X^G$, there exists a finite subset $A \subset G$ such that the set

$$P_A = \{g \in K_f : g\restriction A = f\restriction A\}$$

is contained in $O(f)$. Note that the sets $O(f)$ and $k^{-1}(O(f))$ are infinite, since $K_f$ is Hausdorff and not discrete. Fix any $c \in k^{-1}(O(f)) \setminus A$, put $C = A \cup \{c\}$, and define $h_C \in E_C$ by the rule $h_C\restriction A = f\restriction A$ and $h_C(c) = z$. Since $f$ is a Korovin mapping, there exists $a \in G$ such that $s_a(f)\restriction C = h_C$. Then $s_a(f)\restriction A = h_C\restriction A = f\restriction A$; therefore, $s_a(f) \in P_A \subset O(f)$. We also have $s_{-c}(f) = k(-c) = k(c) \in O(f)$. Let us show that $s_a(f) + s_{-c}(f) \notin W$. Indeed, $s_a(f) + s_{-c}(f) = s_{-c}(s_a(f))$. Hence,

$$(s_a(f) + s_{-c}(f))(0) = (s_{-c}(s_a(f)))(c - c) = s_a(f)(c) = h_C(c) = z \notin U.$$

Thus, $(O(f))^2 \setminus W \neq \varnothing$, for each open neighbourhood $O(f)$ of $f$, that is, $X_f$ is not a paratopological group. □

To draw some interesting conclusions from Proposition 2.4.14, we need the following theorem about subspaces of Tychonoff products of compact metrizable spaces.

THEOREM 2.4.15. *Suppose that $S$ is a subspace of the topological product $X = \prod_{\alpha\in A} X_\alpha$ of compact metrizable spaces such that $p_B(S) = X_B$ for every countable subset $B$ of $A$, where $p_B\colon X \to X_B$ is the natural projection of $X$ onto the subproduct $X_B = \prod_{\alpha\in B} X_\alpha$. Then $X$ is the Čech–Stone compactification of $S$ and the space $S$ is pseudocompact. Furthermore, if each $X_\alpha$ is (locally) connected, then $S$ is (locally) connected as well.*

PROOF. Since, by assumptions of the theorem, the projections of $S$ fill in all countable subproducts $X_B$, it follows that $S$ is dense in the product space $X$. Take an arbitrary continuous real-valued function $f$ on $S$. Since each $X_\alpha$ is compact, the product space $X$ is compact, and every continuous real-valued function on $X$ is bounded. Therefore, both assertions of the theorem about the pseudocompactness of $S$ and the equality $X = \beta S$ will follow if we establish that $f$ admits a continuous extension over $X$, that is, there exists a continuous real-valued function $f^*$ on $X$ such that $f^*(s) = f(s)$, for each $s \in S$.

Each space $X_\alpha$ is second-countable, since every compact metrizable space has a countable base. Therefore, by Corollary 1.7.8, we can find a countable subset $B$ of $A$ and a continuous function $g\colon p_B(S) \to \mathbb{R}$ such that $f(s) = g(p_B(s))$, for each $s \in S$, that is, $f = g \circ p_B$. However, by the assumption, $p_B(S) = X_B$. The space $X_B$ is compact, by the Tychonoff theorem. It follows that $g$ is bounded on $X_B$, which implies that $f$ is bounded on $S$, that is, $S$ is pseudocompact.

Define a function $f^*\colon X \to \mathbb{R}$ by $f^*(x) = g(p_B(x))$, for each $x \in X$. Then $f^* = g \circ p_B$ and, clearly, $f^*$ is continuous on $X$ as the composition of two continuous mappings, and the restriction of $f^*$ to $S$ is $f$.

Now suppose that the spaces $X_\alpha$ are connected. If $S$ were disconnected, there would exist a continuous function $f\colon S \to \mathbb{R}$ satisfying $f(S) = \{0, 1\}$. Therefore, to show that $S$ is connected, it suffices to verify that the image $f(S)$ is connected, for every continuous real-valued function $f$ on $S$. However, for every such a function $f$, we can apply Corollary 1.7.8 to find a countable set $B \subset A$ and a continuous real-valued function $g$ on $p_B(S)$ such that $f(x) = g(p_B(x))$ for each $x \in S$. Since $p_B(S) = X_B$ and the space $X_B$ is connected by [**165**,

Theorem 6.1.15], we conclude that the image $f(S) = g(X_B)$ is connected. This proves that $S$ is connected as well.

Finally, suppose that the spaces $X_\alpha$ are locally connected. Given a point $x \in S$ and a neighbourhood $O$ of $x$ in $X$, it suffices to find a closed connected neighbourhood $Q$ of $x$ in $S$ such that $Q \subset O$. We can assume without loss of generality that $O$ is a canonical open set in $X$. Therefore, $O$ has the form $O = \prod_{\alpha\in A} O_\alpha$, where each $O_\alpha$ is open in $X_\alpha$ and there are only finitely many indices $\alpha \in A$ with $O_\alpha \neq X_\alpha$. For every such an $\alpha \in A$, choose a closed connected neighbourhood $P_\alpha$ of $x_\alpha$ in $X_\alpha$ such that $P_\alpha \subset O_\alpha$, and let $P_\alpha = X_\alpha$ otherwise. Then $P = \prod_{\alpha\in A} P_\alpha$ is a closed connected neighbourhood of $x$ in $X$ and, clearly, $P \subset O$. Put $Q = S \cap P$. Then $Q$ fills in all countable subproducts of the product space $P$, so we can apply the assertion proved in the previous paragraph to conclude that $Q$ is connected. □

Now we have all the instruments we need to demonstrate that neither Corollary 2.4.2 nor Theorem 2.4.12 can be extended to pseudocompact quasitopological groups.

Example 2.4.16. There exists a pseudocompact quasitopological group that fails to be a topological group. Indeed, put $\mathfrak{c} = 2^\omega$, and let $D^\mathfrak{c}$ be the topological product of $\mathfrak{c}$ copies of the discrete Abelian group $D = \{0, 1\}$. Below, $D^\mathfrak{c}$ taken as a group is denoted by $G$, and $D^\mathfrak{c}$ taken as a topological space is denoted by $X$. Clearly, $G$ is a Boolean group, and $|G| = |X| = \mathfrak{c}$. Therefore, there exists a Korovin orbit $K_f$ in the space $X^G$. It follows from 5) of Proposition 2.4.14 and Theorem 2.4.15 that the subspace $K_f$ of $X^G$ is pseudocompact, while 4) and 6) of Proposition 2.4.14 ensure that $K_f$ is a quasitopological group and is not a topological group. □

## Exercises

2.4.a. Let $U$ and $V$ be open neighbourhoods of the neutral element in a paratopological group $G$. Show that if $U^2 \subset V$, then $(\overline{U^{-1}})^{-1} \subset V$.

2.4.b. Let us call a semitopological group $S$ *topologically periodic* if the sequence $\{x^n : n \in \mathbb{N}\}$ accumulates at the neutral element of $S$, for each $x \in S$. Verify that every sequentially compact paratopological group is topologically periodic.

2.4.c. Give an example of a closed subgroup $H$ of a commutative Hausdorff paratopological group $G$ such that the quotient space $G/H$ is not Hausdorff.

## Problems

2.4.A. (B.M. Bokalo and I.I. Guran [**78**]) Prove that every sequentially compact Hausdorff paratopological group $G$ is a topological group.
*Hint.* It suffices to verify the continuity of the inverse in $G$. Take an open neighbourhood $U$ of the neutral element $e$ in $G$ and define a sequence $\{V_i : i \in \omega\}$ of open neighbourhoods of $e$ in $G$ such that $V_0 = U$ and $V_{i+1}^2 \subset V_i$, for each $i \in \omega$. Then the set $F = \bigcap_{i=0}^\infty \overline{V_i^{-1}}$ is non-empty and $F \subset U$. Indeed, if $x \in F$, then $x \in \overline{V_i^{-1}}$ and, by Exercise 2.4.a, $x^{-1} \in (\overline{V_{i+1}^{-1}})^{-1} \subset V_i$, for each $i \in \omega$. According to Exercise 2.4.b, $G$ is topologically periodic, so there exists an integer $n \geq 1$ such that $x^n \in V_1$. Choose $k \in \mathbb{N}$ such that $V_k^{n-1} \subset V_1$. It then follows that $x = x^n(x^{-1})^{n-1} \in V_1 V_k^{n-1} \subset V_1 V_1 \subset U$. Thus, $F \subset U$. Since the space $G$ is countably compact and $\overline{V_{i+1}^{-1}} \subset \overline{V_i^{-1}}$ for each $i \in \omega$, there exists an integer $i_0 \geq 1$ such that $V_{i_0}^{-1} \subset \overline{V_{i_0}^{-1}} \subset U$. Therefore, $G$ is a topological group.

2.4.B. Is every separable pseudocompact semitopological group a topological group? (See also Theorem 2.3.23, Example 2.4.16, and Problems 2.4.C and 2.4.D).

2.4.C. Is every first-countable pseudocompact semitopological group a topological group?

2.4.D. Let us call a space $X$ *weakly countably compact* if $X$ contains a dense subset $D$ such that every infinite subset of $D$ has an accumulation point in $X$. Find out whether every regular (or Tychonoff) weakly countably compact semitopological (quasitopological) group $X$ is a topological group. What if $X$ is countably compact?

2.4.E. Consider the space obtained when we join every two consecutive countable ordinals by a copy of the closed interval $I = [0, 1]$. The linearly ordered set so defined is taken with the topology generated by this linear order. Removing the initial point 0 from the space so obtained, we arrive at the space $\mathbb{R}_{\omega_1}$ called the *Alexandroff line*. The Alexandroff line is easily seen to be homogeneous.

a) Is the Alexandroff line $\mathbb{R}_{\omega_1}$ homeomorphic to some semitopological group?

b) Is the Alexandroff line $\mathbb{R}_{\omega_1}$ homeomorphic to some left topological group?

2.4.F. (A. V. Arhangel'skii and M. Hušek [**54**]) Present an example of a pseudocompact quasitopological group which is neither precompact nor $\omega$-narrow.

2.4.G. (C. Hernández and M. G. Tkachenko [**229**]) Show that the product of two pseudocompact quasitopological groups need not be pseudocompact. (See Problems 2.4.6 and 6.6.8.)
*Hint.* Take two subspaces $X$ and $Y$ of $\beta\omega$ such that $X \cap Y = \omega$ and both spaces $X^\omega$ and $Y^\omega$ are countably compact. In addition, we can choose $X$ and $Y$ to satisfy $|X| = |Y| = \mathfrak{c}$. Denote by $G$ any Boolean group of cardinality $\mathfrak{c}$. By Theorem 2.4.13, there exist Korovin orbits $S$ and $T$ in $X^G$ and $Y^G$, respectively. It follows from Proposition 2.4.14 that $S$ and $T$ are quasitopological groups when endowed with the corresponding subspace topologies and group multiplications. Then apply Problem 1.7.B (and the choice of $X$ and $Y$) to deduce that $S$ and $T$ are pseudocompact spaces. Since $X$ is a continuous image of $S$ and $Y$ a continuous image of $T$, the product $X \times Y$ is a continuous image of $S \times T$. However, since $X \cap Y = \omega$, the space $X \times Y$ contains an infinite discrete family of non-empty open sets (a copy of $\omega$). Hence, neither $X \times Y$ nor $S \times T$ is pseudocompact.

2.4.H. (O. V. Ravsky [**401**]) Apply Martin's Axiom to construct a Hausdorff countably compact paratopological group which is not a topological group.

## Open Problems

2.4.1. Does there exist in $ZFC$ a Hausdorff countably compact paratopological group that fails to be a topological group? Is it topologically periodic? (See Corollary 2.4.4, Exercise 2.4.b, Problem 2.4.A, and Problem 2.4.H.)

2.4.2. (I. I. Guran [**210**]) Let $G$ be a Hausdorff (regular, Tychonoff) paratopological group. Does there exist a continuous isomorphism $f\colon G \to H$ onto a Hausdorff topological group $H$? (See also Problem 2.3.1.)

2.4.3. (I. I. Guran [**210**]) Is every Hausdorff countably compact paratopological group precompact?

2.4.4. Does there exist a Hausdorff countably compact paratopological group $G$ such that $G \times G$ is not countably compact?

2.4.5. A semitopological group $G$ is called *locally pseudocompact* if the space $G$ is Tychonoff and there exists a pseudocompact neighbourhood of the neutral element in $G$. Is every locally pseudocompact paratopological group a topological group?

2.4.6. Does there exist a pseudocompact semitopological group $G$ such that $G \times G$ is not locally pseudocompact? (See Problem 2.4.G.)

## 2.5. Cancellative topological semigroups

In this section we consider topological semigroups, and present an important case when a topological semigroup automatically turns out to be a topological group.

A semigroup $G$ is *cancellative* if $zx = zy$ implies that $x = y$, and $xz = yz$ implies that $x = y$, whenever $x, y, z \in G$.

As in a group, if $S$ is a semigroup and $a \in S$, we can define the right translation $\varrho_a$ on $S$ by the rule $\varrho_a(x) = xa$, for each $x \in S$. Similarly one defines the left translation $\lambda_a$. But unlike the case of groups, these translations need not be bijections; they do not have to be either one-to-one, or onto. This explains why topological semigroups do not have to be homogeneous spaces, which is witnessed by the next example (see also Example 1.3.7).

EXAMPLE 2.5.1. Let $S$ be the closed unit interval $I = [0, 1]$, with multiplication defined by the rule $xy = \min\{x, y\}$. Clearly, $S$ is a topological semigroup, which is not a homogeneous space, and the translations in $S$ are obviously not one-to-one. □

Clearly, cancellative semigroups are precisely the semigroups for which all the translations are one-to-one (but they still need not be onto). Thus, the topological semigroup in Example 2.5.1 is not cancellative.

Now, we have the following interesting fact:

THEOREM 2.5.2. [**K. Numakura**] *Every compact Hausdorff cancellative topological semigroup $S$ is a topological group.*

PROOF. As the first step, let us show that $Sa = S$ and $aS = S$, for each $a \in S$, i.e., that every translation in $S$ is a mapping onto $S$. It suffices to establish that $aS = S$.

Put $S_0 = S$, $S_1 = aS$, and, in general, $S_{n+1} = a^{n+1}S$ for each $n \in \omega$. From $aS \subset S$ it follows by induction that $S_{n+1} \subset S_n$, for each $n \in \omega$. Hence, the sequence $\xi = \{S_n : n \in \omega\}$ is decreasing. Since $S$ is compact, Hausdorff, and translations are continuous, all $S_n$ are closed compact subsets of $S$.

Consider the sequence $\eta = \{a^n : n \in \omega\}$. Since $S$ is compact, $\eta$ accumulates at a point $y \in S$. Take any open set $V$ such that $yaS \subset V$. By the continuity of multiplication and the compactness of $aS$, there exists an open neighbourhood $W$ of $y$ such that $WaS \subset V$. Now take any $k \in \omega$ such that $a^k \in W$. Then $a^k aS \subset V$, that is, $S_{k+1} \subset V$. Since the sets $S_n$ are decreasing, we have that $S_n \subset V$, for each $n \geq k + 1$.

Now take any $b \in S$. By the continuity of the translations, $yb$ is an accumulation point of the sequence $\eta_b = \{a^n b : n \in \omega\}$. Clearly, $a^n b \in S_n \subset V$, for each $n \geq k+1$. Therefore, $yb \in \overline{V}$. Thus, we have shown that $yb$ belongs to the closure of any neighbourhood of the set $yaS$. Since $yaS$ is compact and $S$ is Hausdorff, it follows that $yb \in yaS$. The semigroup $S$ being cancellative, we conclude that $b \in aS$. Thus $aS = S$ and, similarly, $Sa = S$.

Let us show that $S$ is a group. Take any $a \in S$. Since $Sa = S$, there exists $e \in S$ such that $ea = a$. Now let $x \in S$. Since $aS = S$, there exists $y \in S$ such that $ay = x$. Then $ex = eay = ay = x$. Therefore, $e$ is a left-side unit for $S$. Since $S$ is cancellative, this left-side unit is unique. Similarly, there exists a right-side unit $e'$ in $S$. A standard argument shows that $e = e'$, i.e., $S$ is a monoid. We can also find $b \in S$ such that $ba = e$. This element $b$ is a left inverse of $a$. Similarly, there exists a right inverse $b'$ of $a$ and, again, it is easy to see that $b = b'$. Now it is clear that $S$ is a group. Since the multiplication in $S$

is continuous, $S$ is a paratopological group. Finally, since $S$ is compact, Proposition 2.3.3 implies that the inverse is also continuous, that is, $S$ is a topological group. □

It is natural to ask whether every regular countably compact cancellative topological semigroup is a topological group. The reader will find some information on this question, known as *Wallace problem*, in Exercise 2.5.b and in Problems 2.5.A–2.5.C and 2.5.1 (as well as in Section 2.6).

The example below shows that the compactness of $S$ in Theorem 2.5.2 cannot be weakened to local compactness, even if the semigroup $S$ is first-countable. The semigroup we have in mind is neither metrizable nor paracompact.

EXAMPLE 2.5.3. There exists a non-metrizable locally compact, first-countable, cancellative Abelian topological semigroup $S$. In particular, $S$ fails to be a topological group.

Indeed, let $H$ be a Hamel basis for the additive group $\mathbb{R}$ of real numbers over the field $\mathbb{Q}$ such that $H \subset \mathbb{R}^+$, where $\mathbb{R}^+$ is the set of positive real numbers, $1 \in H$, and every open interval in $\mathbb{R}^+$ contains $\mathfrak{c} = 2^\omega$ points of $H$. It follows from $1 \in H$ that $H \cap \mathbb{Q} = \{1\}$. Let $H_0$ be the union of $H$ and the set $\mathbb{Q}^+$ of positive rational numbers and $S$ be the smallest subsemigroup of the additive group $\mathbb{R}$ such that $H_0 \subset S$.

Let us now define a topology $\mathcal{T}$ on $S$. All points of $\mathbb{Q}^+$ will be isolated in $S$. For each $x \in H \setminus \{1\}$ we pick an increasing sequence $\{q_n(x) : n \in \omega\}$ in $\mathbb{Q}^+$ converging to $x$, and for each $n \in \omega$, let $V_n(x) = \{x\} \cup \{q_i(x) : i > n\}$. Suppose now that $y$ is an arbitrary point of $S \setminus \mathbb{Q}^+$. Then $y = z_1 + \cdots + z_n$, for some $z_i \in H$, and, for any choice $k_1, \ldots, k_n$ of natural numbers, we put $W_{k_1,\ldots,k_n} = V_{k_1} + \cdots + V_{k_n}$, and declare the set $W_{k_1,\ldots,k_n}$ to be a basic open neighbourhood of $y$. These requirements define a topology $\mathcal{T}$ on $S$ which is finer than the restriction to $S$ of the usual Euclidean topology of the reals. Note that if $a$ and $b$ belong to $S \setminus \{0\}$, then $(a + b) \notin H$ unless $a + b = 1$. Note also that no point of $\mathbb{Q}^+$ is the sum of points from $S$ none of which belongs to $\mathbb{Q}^+$. It follows that the operation $+$ on $S$ is continuous, and $S$ is a topological semigroup. Clearly, $S$ is cancellative, since $S$ is a subsemigroup of the additive group $\mathbb{R}$. Obviously, $(S, \mathcal{T})$ is locally compact and locally metrizable. Hence, $(S, \mathcal{T})$ is first-countable. Since the topology $\mathcal{T}$ on $S$ is finer than the restriction to $S$ of the usual Euclidean topology, it follows that the space $S$ is Hausdorff. Hence, the space $S$ is Tychonoff. The countable set $\mathbb{Q}^+$ is clearly dense in $S$. Thus, $S$ is separable. However, $H$ is an uncountable discrete subspace of $S$. Therefore, $S$ does not have a countable base. Since $S$ is separable, it follows that $S$ is not metrizable. Note that $S$ is not paracompact, since every locally metrizable paracompact space is metrizable [**165**, 5.4.A]. Thus, neither local compactness nor first countability in cancellative topological semigroups implies paracompactness or metrizability. □

## Exercises

2.5.a. Let $S$ be a cancellative semigroup with the property that $aS = S = Sa$, for each $a \in S$. Show that $S$ is a group.

2.5.b. (A. Mukherjea and N. Tserpes [**333**]) Let $S$ be a cancellative Hausdorff semitopological semigroup. Prove that if $S$ is sequentially compact, then $xS = S = Sx$, for each $x \in S$.
*Hint.* To deduce the equality $xS = S$, it suffices to verify that, for every $a \in S$, there exists $b \in S$ such that $ab = a$. Consider the set $S(a) = \{a^n : n \in \mathbb{N}\}$ and let $K(a)$ be the sequential closure of $S(a)$ in $S$, that is, the set of all limit points of convergent sequences lying in $S(a)$. Clearly, $a \in S(a) \subset K(a)$ and $aK(a) \subset K(a)$. Let us show that $K(a) \subset aK(a)$. Take an

arbitrary element $y \in K(a)$ and choose a subsequence $\{a^n : n \in A\}$ of $S(a)$ converging to $y$, where $A \subset \omega$ is infinite. Since $S$ is sequentially compact, there exists an infinite set $B \subset A \setminus \{0, 1\}$ such that the sequence $\{a^{n-1} : n \in B\}$ converges to some element $z \in K(a)$. Then $aa^{n-1} \to az$ when $n \to \infty$ and $n \in B$. By the continuity of multiplication by $a$ and the Hausdorff property of $S$, $az = y$. Hence, $aK(a) = K(a)$ and $ab = a$, for some $b \in K(a)$. A similar argument shows that $K(a)a = K(a)$, whence the equality $Sx = S$ follows.

2.5.c. We say that $X$ is a *Moore space* if there exists a sequence $\mathcal{W}_0, \mathcal{W}_1, \ldots$ of open coverings of $X$ such that for every point $x \in X$ and every neighbourhood $U$ of $x$ in $X$, one can find $n \in \omega$ such that $\bigcup\{W \in \mathcal{W}_n : x \in W\} \subset U$. Verify that the topological semigroup $S$ defined in Example 2.5.3 is a Moore space. Show that the Hamel basis $H$ is a closed discrete subspace of $S$ and deduce that the space $S$ is not normal.

## Problems

2.5.A. (B. M. Bokalo and I. I. Guran [**78**]; for countably compact sequential semigroups, A. Yur'eva [**544**]) Prove that every sequentially compact Hausdorff cancellative topological semigroup $S$ is a topological group.
*Hint.* Apply Exercises 2.5.a and 2.5.b to deduce that $S$ is algebraically a group. It follows that $S$ is a paratopological group, so the conclusion follows from Problem 2.4.A.

2.5.B. (A. H. Tomita [**496**]; under the Continuum Hypothesis, D. Robbie and S. Svetlichny [**408**]) Prove that under Martin's Axiom, there exists a countably compact subsemigroup $S$ of the compact group $\mathbb{T}^{\mathfrak{c}}$ that fails to be a group. Notice that $S$, being a subsemigroup of a group, is cancellative. Deduce that Theorem 2.5.2 cannot be extended in $ZFC$ to countably compact cancellative topological semigroups.

2.5.C. (A. H. Tomita [**496**]) Let $p$ be a free ultrafilter on $\omega$. A space $X$ is called *p-compact* if every sequence $\{x_n : n \in \omega\} \subset X$ has a *p-limit point* $y$ in $X$, that is, a point $y \in X$ with the property that for every neighbourhood $U$ of $y$, the set $\{n \in \omega : x_n \in U\}$ belongs to $p$. Notice that $p$-compactness implies countable compactness and prove that every regular, $p$-compact, cancellative topological semigroup is a topological group.

## Open Problems

2.5.1. (A. D. Wallace [**531**]) Does there exist in $ZFC$ a regular countably compact cancellative topological semigroup that fails to be a topological group?

2.5.2. Let $G$ be a regular countably compact cancellative topological semigroup. Is the cellularity of $G$ countable?

2.5.3. Does there exist in $ZFC$ a regular countably compact cancellative topological semigroup $G$ such that the product $G \times G$ is not countably compact?

## 2.6. Historical comments to Chapter 2

The recent monograph [**241**] treats the subject of the first two sections of Chapter 2 in much greater detail, contains many references and, most intriguing, shows how these abstract techniques can be applied to obtain some important theorems on combinatorics involving natural numbers, theorems which seem to have nothing to do with General Topology. The method of ultrafilters developed in [**241**] lead to the discovery of a fruitful link between combinatorics of numbers and topological algebra. The reader is warmly advised to consult the book [**241**] for comments on the history of the results. The less exhaustive but concentrated and interesting short books [**392**] by I. V. Protasov, and [**394**] by I. V. Protasov and E. G. Zelenyuk are also highly recommended to the reader.

The first prototypes of Theorem 2.1.1 were given by M. Day in [**129**] (implicitly, using methods of R. Arens from [**13**]), and in [**104**]. In [**161**], the technique based on ultrafilters was used to prove a version of Theorem 2.1.1 for the case when the operoid is a discrete group. In connection with Theorem 2.1.4 and for further references, see [**73**]. The general approach, when associativity is not assumed and arbitrary operoids are considered, was developed by E. van Douwen [**151**]. Theorems 2.1.1, 2.1.3 and Propositions 2.1.2, 2.1.5–2.1.9 all originated in his paper [**151**]. As for Example 2.1.10, see [**65**].

Theorem 2.2.1, in the case of the joint continuity of multiplication, is due to K. Numakura [**352**] and A. D. Wallace [**531**], and due to R. Ellis [**161**] in the case of separate continuity. In connection with Propositions 2.2.3, 2.2.4 and Example 2.2.5, see [**241**]. I. V. Protasov used idempotents in remainders of Čech–Stone compactifications to produce non-discrete Tychonoff homogeneous maximal topologies on the set of integers [**393**]. In particular, Theorems 2.2.10, 2.2.16, and 2.2.24 are due to I. V. Protasov. Theorem 2.2.26 is a result of N. Hindman (see [**239**], [**241**], and [**392**] for the history of its proof). These results are also discussed in [**42**] and [**240**]. For some further results in this direction, see [**545**] and [**546**].

D. Montgomery established in [**324**] that if a semitopological group $G$ is metrizable by a complete metric, then it is a paratopological group, and that if $G$ is, in addition, separable, then it is a topological group. Another early result of this kind was obtained by R. Ellis [**159**], who proved that every locally compact Hausdorff semitopological group is a topological group (our Theorem 2.3.12). In 1960 W. Zelazko established that each completely metrizable paratopological (semitopological) group is a topological (quasitopological) group. In 1982, N. Brand proved that every Čech-complete paratopological group is a topological group. The final result in this series was obtained by A. Bouziad in [**85**], who proved that every Čech-complete semitopological group is a topological group (Theorem 2.4.12). Some further results in this direction can be found in [**84**] and [**269**]. Lemmas 2.3.18, 2.3.19, 2.3.21, Corollary 2.3.20, and Theorem 2.3.22 were all proved in [**62**]. Theorem 2.3.23, Proposition 2.3.24, and Corollary 2.3.25 are also taken from [**62**]. Further important results and references on separate continuity versus joint continuity can be found in [**339**] and [**289**].

Theorem 2.4.1 was proved in [**62**]. Corollary 2.4.2 is due to E. A. Reznichenko [**404**]. Corollary 2.4.3 is a weaker form of Bouziad's principal result that appears here as Theorem 2.4.12. Theorem 2.4.13, Proposition 2.4.14, and the construction involved in them can be found in A. V. Korovin's article [**282**]. Korovin also applied his construction to obtain a pseudocompact semitopological group such that its Čech–Stone compactification is not a semitopological group. E. A. Reznichenko observed in [**404**] that Korovin's construction, when a Boolean group is used, yields an example of a pseudocompact quasitopological group which is not a topological group (Example 2.4.16). Further applications of this construction were given in [**54**] and [**229**]. In particular, it was shown in [**54**] that there is a pseudocompact quasitopological group $G$ such that the Dieudonné completion of $G$ is not homogeneous (and even has points of different characters), while [**229**] contains a construction of two pseudocompact quasitopological groups whose product fails to be pseudocompact.

Theorem 2.5.2 on compact cancellative topological semigroups was established in [**352**]. In [**333**], this result was extended to countably compact first-countable cancellative topological semigroups. Further interesting results on cancellative topological semigroups appeared in [**222**]. Wallace's problem [**531**] whether every Hausdorff countably compact cancellative topological semigroup is a topological group was given a negative answer (under the Continuum Hypothesis) in [**408**]. Some interesting results in this direction were obtained by A. Tomita in [**496**].

Further progress in studying paratopological groups, also from the bitopological point of view, can be found in [**4**], [**54**], [**78**], [**210**], [**398**], [**401**], and [**412**].

# Chapter 3

# Topological groups: Basic constructions

In this chapter we introduce several most important notions and constructions concerning topological groups and operations with them. In particular, we develop the technique of prenorms on groups, describe in detail the construction of the Raĭkov completion of a topological group, prove a theorem on embeddings of topological groups into groups of isometries in the topology of pointwise convergence. We also introduce the class of locally compact topological groups and provide the reader with the first basic facts about groups in this class (deeper results in this direction will be presented in Chapter 9). Then we define and study the important classes of $\omega$-narrow topological groups and of precompact (equivalently, totally bounded) topological groups. We finish this chapter with the Hartman–Mycielski construction of an embedding of an arbitrary topological group into a connected, locally pathwise connected topological group.

## 3.1. Locally compact topological groups

This section contains theorems on topological properties of locally compact groups, which are not necessarily the properties of all locally compact spaces. Though the proofs of these theorems are much easier than those of the main results about compact topological groups in the forthcoming Sections 4.1 and 4.2, the theorems themselves are important, they are applied very often and help to discover many unusual facts concerning locally compact groups.

There are many natural examples of locally compact groups. They include the groups $\mathbb{T}$, $\mathbb{R}$, and $\mathbb{Z}$, each one carrying the usual topology. Every compact group is, of course, locally compact. Clearly, any discrete group is locally compact, while a discrete group is compact if and only if it is finite. Unlike the case of compact groups, the product of a family of locally compact groups need not be locally compact. However, the product of any finite collection of locally compact groups is a locally compact group. Using this fact, and also the fact that every closed subgroup of a locally compact group is locally compact, one can construct many interesting examples of locally compact topological groups. We study in this section the most elementary general properties of locally compact groups.

Recall that a topological space $X$ is *paracompact* if every open covering of $X$ can be refined by a locally finite open covering [**165**, Section 5.1]. A somewhat stronger property of the same type is that of *strong paracompactness*. A family $\gamma$ of sets is *star-finite* if every element of $\gamma$ intersects only finitely many elements of $\gamma$. A space $X$ is said to be *strongly* *paracompact* if every open covering of $X$ can be refined by a star-finite open covering [**165**, Section 5.3]. It is obvious that every strongly paracompact space is paracompact. A useful fact which we will use in the sequel is that every regular Lindelöf space is strongly paracompact [**165**, Corollary 5.3.11]. Since $\sigma$-compact spaces are Lindelöf, all regular $\sigma$-compact spaces are strongly paracompact.

Finally, we say that a space $X$ is *locally $\sigma$-compact* if for every point $x$ of $X$, there exists an open neighbourhood $V$ such that the closure of $V$ is $\sigma$-compact.

The following two theorems are typical for locally $\sigma$-compact groups.

THEOREM 3.1.1. *Every locally $\sigma$-compact topological group $G$ is strongly paracompact.*

PROOF. Take a symmetric open neighbourhood $V$ of the identity $e$ in $G$ such that $F = \overline{V}$ is $\sigma$-compact, and put $H = \bigcup_{n\in\omega} F^n$. Clearly, $H$ is a subgroup of $G$, and the interior of $H$ contains $V$. Therefore, $H$ is an open and closed subgroup of $G$ by Corollary 1.3.3 and Theorem 1.3.5. It is also clear that each $F^n$ is compact; therefore, the space $H$ is $\sigma$-compact and, hence, Lindelöf. The space $G$ is the free topological sum of the subspaces homeomorphic to $H$ (of the right cosets of $H$, see Theorem 1.5.1). Since every regular Lindelöf space is strongly paracompact by [**165**, Corollary 5.3.11], it follows that the space $G$ is strongly paracompact. □

THEOREM 3.1.2. *Every connected locally $\sigma$-compact topological group $G$ is $\sigma$-compact (and, hence, Lindelöf).*

PROOF. We repeat the argument in the proof of Theorem 3.1.1. The open and closed subgroup $H$ of $G$ constructed there has to coincide with $G$, since $H$ is non-empty and the space $G$ is connected. Since $H$ is $\sigma$-compact, it follows that $G$ is $\sigma$-compact. □

Note that the proof of Theorem 3.1.1 actually shows that the following statement is true:

PROPOSITION 3.1.3. *Every locally $\sigma$-compact topological group $G$ contains an open and closed subgroup which is $\sigma$-compact.*

COROLLARY 3.1.4. *Every locally compact topological group $G$ is strongly paracompact.*

COROLLARY 3.1.5. *Every connected locally compact topological group $G$ is $\sigma$-compact.*

Notice that neither Corollary 3.1.4 nor Corollary 3.1.5 remains valid for locally compact spaces. Indeed, consider the subspace $X = I^\kappa \setminus \{p\}$ of the Tychonoff cube $I^\kappa$ of weight $\kappa > \omega$, where $p \in I^\kappa$ is an arbitrary point. Clearly, $X$ with the topology inherited from $I^\kappa$ is a locally compact, connected, locally connected space. However, $X$ is not normal (hence, not paracompact) and, therefore, is not $\sigma$-compact.

THEOREM 3.1.6. *Let $H$ be the connected component of a topological group $G$. Then every non-empty connected subset of the quotient group $G/H$ consists of one point, that is, the space $G/H$ is totally disconnected.*

PROOF. Since $H$ is a closed invariant subgroup of $G$, the quotient $G/H$ is a topological group, by Theorem 1.5.3. Let $\hat{e}$ be the neutral element of $G/H$ and $K$ be a connected subset of $G/H$ such that $\hat{e} \in K$. Clearly, it suffices to show that $K = \{\hat{e}\}$. Assume the contrary, and take the preimage $B = p^{-1}(K)$ under the natural quotient mapping $p\colon G \to G/H$. Then $H \subset B$ and $B \neq H$, so $B$ is not connected and we can fix $b \in B$ and a non-empty subset $U$ of $B$ such that $b$ is not in $U$, and $U$ is open and closed in the subspace $B$. We also have $p^{-1}p(U) = U$, since the fibers of $p$ are homeomorphic to $H$ and are, therefore, connected. Since the restriction of $p$ to $B$ is an open mapping of $B$ onto $K$, it follows that $p(U)$ is open and closed in $K$. However, $p(b) \notin p(U)$, since $b \notin U = p^{-1}p(U)$. On the other hand, $p(b) \in p(B) = K$, and $p(U)$ is non-empty. It follows that the subspace $K$ is not connected, a contradiction. □

The next result reveals a fundamental property of totally disconnected locally compact Hausdorff spaces. Recall that a space $X$ is said to be *zero-dimensional* (notation: ind $X = 0$) if it has a base $\mathcal{B}$ consisting of sets which are both open and closed in $X$.

PROPOSITION 3.1.7. *Every totally disconnected locally compact Hausdorff space $X$ is zero-dimensional.*

PROOF. Since $X$ is regular, we can assume that $X$ is compact. Fix a point $a \in X$, and let $\mathcal{P}$ be the family of all open and closed subsets of $X$ which contain $a$. Put $P = \bigcap \mathcal{P}$. Clearly, $P$ is closed in $X$ and $a \in P$. Notice also that the family $\mathcal{P}$ is closed under finite intersections.

**Claim.** *For every closed subset $F$ of $X$ disjoint from $P$, there exists $W \in \mathcal{P}$ such that $W \cap F = \varnothing$.*

Indeed, otherwise $\eta = \{U \cap F : U \in \mathcal{P}\}$ is a centered family of non-empty closed subsets of $F$. Since $F$ is compact, we have $\bigcap \eta \neq \varnothing$, which implies that $P \cap F \neq \varnothing$, a contradiction.

Let us show that $P = \{a\}$. Assume the contrary. Then $P$ is disconnected, since $X$ is totally disconnected. Therefore, there exist disjoint non-empty closed subsets $A$ and $B$ of $P$ such that $P = A \cup B$ and $a \in A$. Since $X$ is normal, we can find disjoint open sets $U$ and $V$ in $X$ such that $A \subset U$ and $B \subset V$. Then $F = X \setminus (U \cup V)$ is closed in $X$, and $P \cap F = \varnothing$. By our Claim, there exists $W \in \mathcal{P}$ such that $W \cap F = \varnothing$. The open set $G = U \cap W$ is also closed in $X$. Indeed, $\overline{G} \subset \overline{U} \cap W \subset X \setminus (F \cup V) \subset U$; therefore, $\overline{G} \subset U \cap W = G$, which implies that $\overline{G} = G$. Since $a \in G$, we have $G \in \mathcal{P}$. However, $G \cap B = \varnothing$. Hence, $G$ does not contain $P$, a contradiction. Therefore, $P = \{a\}$.

Now it follows from the above Claim that every open neighbourhood $O$ of $a$ contains some $V \in \mathcal{P}$, since the set $X \setminus O$ is compact and disjoint from $P = \{a\}$. □

PROPOSITION 3.1.8. *Suppose that $G$ is a topological group and $F$ is an open compact neighbourhood of the neutral element $e$ in $G$. Then there exists an open compact subgroup $H$ of $G$ such that $H \subset F$.*

PROOF. Since $F$ is an open neighbourhood of itself, Theorem 1.4.29 implies that there exists an open neighbourhood $V$ of $e$ such that $FV \subset F$. Then $V \subset F$ and we can assume that $V = V^{-1}$. Clearly, $VV \subset FV = F$. Arguing by induction, we conclude that $V^n \subset F$, for each positive $n \in \omega$. Then $H = \bigcup\{V^n : n \in \omega\}$ is an open subgroup of $G$ and $H \subset F$. The set $H$ is also closed in $G$, since every open subgroup of a topological group is closed. Hence, $H$ is compact. □

THEOREM 3.1.9. *In every locally compact totally disconnected topological group $G$ there exists a local base $\mathcal{B}$ of $G$ at the neutral element $e$ such that each element of $\mathcal{B}$ is an open compact subgroup of $G$.*

PROOF. By Proposition 3.1.7, there exists a base $\mathcal{P}$ of $G$ at $e$ consisting of open compact subsets of $G$. From Proposition 3.1.8 it follows that, for each $V \in \mathcal{P}$, there exists an open compact subgroup $H_V$ of $G$ such that $H_V \subset V$. Clearly, the family $\mathcal{B} = \{H_V : V \in \mathcal{P}\}$ is a base of $G$ at $e$ we are looking for. □

For compact groups, the conclusions in Proposition 3.1.8 and Theorem 3.1.9 can be strengthened as follows.

PROPOSITION 3.1.10. *Suppose that $G$ is a compact topological group and $F$ is an open and closed neighbourhood of the neutral element $e$ in $G$. Then there exists an open invariant subgroup $K$ of $G$ such that $K \subset F$.*

PROOF. Let $U$ be an open and closed neighbourhood of the neutral element $e$ in $G$. Then $U$ is compact, so Theorem 3.1.9 implies that there exists an open subgroup $H$ of $G$ such that $H \subset U$. Clearly, $K = \bigcap_{x \in G} xHx^{-1}$ is an invariant subgroup of $G$ and $K \subset H \subset U$, so it remains to verify that $K$ is open in $G$. To this end, apply Proposition 1.4.32 to choose an open neighbourhood $V$ of $e$ in $G$ such that $xVx^{-1} \subset H$ for each $x \in G$. It follows from the definition of $K$ that $V \subset K$, so that $K$ is open in $G$. □

The next theorem is immediate from Proposition 3.1.10.

THEOREM 3.1.11. *In a compact totally disconnected group $G$, the family of open invariant subgroups forms a local base at the neutral element of $G$.*

Theorem 3.1.9 above has several important applications. The first of them complements the theorem, while the second one is, in a sense, a counterpart of Lemma 1.4.27 for locally compact groups.

COROLLARY 3.1.12. *In a locally compact group topological $G$, the connected component of $G$ is the intersection of all open subgroups of $G$.*

PROOF. Let $C$ be the connected component of $G$. By Proposition 1.4.26, $C$ is a closed invariant subgroup of $G$. Let $G/C$ be the quotient group and $\pi\colon G \to G/C$ the quotient homomorphism. Since the mapping $\pi$ is open, the group $G/C$ is locally compact. It follows from Theorem 3.1.6 that the group $G/C$ is totally disconnected, so Theorem 3.1.9 implies that $G/C$ has a base $\mathcal{B}$ at the neutral element which consists of open subgroups. Clearly, the intersection of the inverse images of elements of $\mathcal{B}$ under $\pi$ is the component $C$ of $G$. □

COROLLARY 3.1.13. *For a locally compact group topological $G$, the following conditions are equivalent:*

a) *$G$ is connected;*
b) *$G$ has no proper open subgroups;*
c) *every neighbourhood of the identity in $G$ algebraically generates the group $G$.*

PROOF. Clearly, a) implies b) and c) follows from b). By Corollary 3.1.12, c) implies a). Hence, the three conditions a), b), and c) are equivalent for the group $G$. □

Here is another interesting application of Theorem 3.1.9:

THEOREM 3.1.14. *If $G$ is a locally compact totally disconnected topological group $G$ and $H$ is a closed subgroup of $G$, then the quotient space $G/H$ is zero-dimensional.*

PROOF. Let $\phi$ be the natural projection of $G$ onto $G/H$, and let $W$ be any open neighbourhood of $\phi(e)$ in $G/H$, where $e$ is the neutral element of $G$. Then the set $U = \phi^{-1}(W)$ is an open neighbourhood of $e$ and, by Theorem 3.1.9, there exists a compact open subgroup $V$ of $G$ such that $V \subset U$. Since the mapping $\phi$ is open and continuous, $\phi(V)$ is an open compact subset of $G/H$. Since the space $G/H$ is Hausdorff, $\phi(V)$ is closed in $G/H$. Clearly, $\phi(V) \subset W$. Since the space $G/H$ is homogeneous, it follows that $G/H$ is zero-dimensional. □

In fact, the above argument shows more than it was claimed. Indeed, $\phi(V)$ is an image of a zero-dimensional compact group $V$ under an open and continuous mapping (the restriction of $\phi$ to the space $V$). Let us call Hausdorff continuous images of compact topological groups *dyadic compacta.* [The reader accustomed to a more traditional definition of dyadic compacta as, for example, in [**165**, 3.12.12], will find a complete justification of our approach to this concept in Section 4.1.] In particular, for every cardinal $\kappa$, continuous images of the space $\{0, 1\}^\kappa$ are dyadic compacta [**165**, 3.12.12]. Thus, in Theorem 3.1.14 we have established the following fact:

THEOREM 3.1.15. *Let $G$ be a locally compact totally disconnected topological group, and let $H$ be a closed subgroup of $G$. Then the quotient space $G/H$ has a base consisting of dyadic compacta. In particular, the group $G$ itself has a base of dyadic compacta.*

The local compactness assumption in the above statements is of crucial importance.

EXAMPLE 3.1.16. Let $G$ be the additive group of all convergent sequences of rational numbers, with coordinatewise defined addition and the natural topology generated by the norm on $G$ defined as the supremum of absolute values of the elements of the sequence. Clearly, the space $\mathbb{Q}^\omega$ is zero-dimensional. Therefore, $G$ is also zero-dimensional [**165**, 6.2.11]. Let $H$ be the set of all sequences of rational numbers converging to zero. Then $H$ is a closed subgroup of $G$, and the topological group $G/H$ is topologically isomorphic to the additive topological group $\mathbb{R}$ of real numbers (endowed with the usual topology). However, $\mathbb{R}$ is not zero-dimensional. □

Let $G$ be a topological group. We say that $G$ is a group with *no small subgroups* or, for brevity, an *$NSS$-group* if there exists a neighbourhood $V$ of the neutral element $e$ such that every subgroup $H$ of $G$ contained in $V$ is trivial, that is, $H = \{e\}$. For example, the group $\mathbb{R}$ of real numbers with the usual topology is a locally compact $NSS$-group. Example 3.1.16 underlines non-triviality of the next statement:

COROLLARY 3.1.17. *Every totally disconnected locally compact $NSS$-group $G$ is discrete.*

PROOF. Take an open neighbourhood $U$ of the neutral element $e$ which does not contain non-trivial subgroups. By Theorem 3.1.9, there exists an open subgroup $H$ of $G$ such that $H \subset U$. Clearly, $H = \{e\}$. Thus, the point $e$ is isolated in $G$, and the space $G$ is discrete. □

The following corollary to Theorem 3.1.9 is especially interesting. It can, however, be considerably strengthened by means of certain algebraic methods blended with the topological considerations.

THEOREM 3.1.18. *Every locally compact totally disconnected Abelian topological group $G$ is topologically isomorphic to a closed subgroup of a topological product of discrete Abelian groups.*

PROOF. Let $\mathcal{F}$ be a base of $G$ at the neutral element $e$. It follows from Theorem 3.1.9 that, for each $U \in \mathcal{F}$, there exists a continuous homomorphism $f_U$ of $G$ onto a discrete Abelian group $H_U$ such that $\ker(f_U) \subset U$. Then the diagonal product $\phi$ of the family $\{f_U : U \in \mathcal{F}\}$ is a topological isomorphism of $G$ onto the subgroup $\phi(G)$ of the topological product $H = \Pi_{U \in \mathcal{F}} H_U$. It follows that $\phi(G)$ is also locally compact and, according to Proposition 1.4.19, is closed in $H$. □

The next statement is a far reaching extension of Proposition 3.1.8 to $G_\delta$-sets in topological groups.

PROPOSITION 3.1.19. *Let $G$ be a topological group, and let $F$ be a non-empty compact $G_\delta$-set in $G$. Then there exists a $G_\delta$-set $P$ in $G$ such that $e \in P$ and $FP \subset F$.*

PROOF. We have $F = \bigcap \gamma$, where $\gamma$ is a countable family of open subsets of $G$. Take any $U \in \gamma$. By Theorem 1.4.29, there exists an open neighbourhood $V_U$ of $e$ such that $FV_U \subset U$. Put $P = \bigcap\{V_U : U \in \gamma\}$. Then $P$ is a $G_\delta$-subset of $G$ such that $e \in P$ and $FP \subset \bigcap \gamma = F$. □

PROPOSITION 3.1.20. *Suppose that $G$ is a topological group and that $F$ is a non-empty compact $G_\delta$-set in $G$ containing the neutral element $e$ of $G$. Then there exist a $G_\delta$-set $P$ in $G$ and a closed subgroup $H$ of $G$ such that $e \in P \subset H \subset F$.*

PROOF. By Proposition 3.1.19, there exists a $G_\delta$-set $P$ in $G$ such that $e \in P$ and $FP \subset F$. Clearly, we may assume that $P = P^{-1}$. From $e \in F$ and $FP \subset F$ it follows, by induction, that $P^n \subset F$, for each $n \in \omega$. Then $S = \bigcup_{n=0}^{\infty} P^n$ is a subgroup of $G$ and $e \in P \subset S \subset F$. The closure of $S$ is the closed subgroup $H$ of $G$ we are looking for. □

THEOREM 3.1.21. *If a topological group $G$ is an NSS-group, then the following two conditions are equivalent:*

1) *there exists a non-empty compact $G_\delta$-set $F$ in $G$;*
2) *the neutral element $e$ of $G$ is a $G_\delta$-point in $G$.*

PROOF. Clearly, the second condition implies the first. Let us show that the first condition implies the second one. Since the space $G$ is homogeneous, we can assume that $e \in F$. By the regularity of $G$, we can also assume that $F \subset U$, where $U$ is an open neighbourhood of $e$ such that all subgroups of $G$ contained in $U$ are trivial. By Proposition 3.1.20, there exist a $G_\delta$-set $P$ in $G$ and a subgroup $H$ of $G$ such that $e \in P \subset H \subset F$. Since $F \subset U$, it follows that $H \subset U$. Therefore, $H = \{e\}$, by the choice of $U$. Then $P = \{e\}$. Since $P$ is a $G_\delta$-set in the group $G$, we are done. □

An interesting application of Theorem 3.1.21 is the next result:

THEOREM 3.1.22. *Every locally compact NSS-group $G$ is first-countable.*

PROOF. Clearly, every locally compact regular space contains a non-empty compact $G_\delta$-set. Therefore, by Theorem 3.1.21, the neutral element $e$ of $G$ is a $G_\delta$-point in $G$. Since the space $G$ is locally compact and Hausdorff, it follows that $G$ is first-countable at $e$. Hence, by homogeneity, the space $G$ is first-countable. □

Now we are going to present several basic results on continuous homomorphisms of locally compact topological groups. Again, these results do not admit an extension to locally compact spaces, except for very special cases.

PROPOSITION 3.1.23. *Suppose that $G$ is a locally compact topological group and $H$ is a closed subgroup of $G$. Then the quotient space $G/H$ is locally compact.*

PROOF. Open continuous mappings preserve local compactness (in the class of Hausdorff spaces). It remains to refer to the fact that the natural quotient mapping of $G$ onto $G/H$ is open and continuous. □

LEMMA 3.1.24. *Suppose that $G$ is a topological group, $H$ is a compact subgroup of $G$, and $P$ is a non-empty $G_\delta$-set in $G$ contained in $H$. Then $H$ also is a $G_\delta$-set in $G$, and every point in the quotient space $G/H$ is a $G_\delta$-set.*

PROOF. We can assume that the neutral element $e$ of $G$ belongs to $P$. Since the space $G$ is regular and $P$ is a $G_\delta$-set in $G$, we can fix a sequence $\{V_n : n \in \omega\}$ of open neighbourhoods of $e$ in $G$ such that $Q = \bigcap\{V_n : n \in \omega\} \subset P$ and $\overline{V_{n+1}} \subset V_n$, for each $n \in \omega$. Take any $x \in G \setminus H$. Then $xH$ is compact and $H \cap xH = \varnothing$. Therefore, $xH \cap Q = \varnothing$. Since $Q = \bigcap\{V_n : n \in \omega\} = \bigcap\{\overline{V_n} : n \in \omega\}$ and the sequence $\{V_n : n \in \omega\}$ is decreasing, we have $xH \cap V_k = \varnothing$, for some $k \in \omega$. It follows that $\{\phi(e)\} = \bigcap_{n=0}^{\infty} \phi(V_n)$, where $\phi$ is the natural quotient mapping of $G$ onto the quotient space $G/H$. Since $\phi$ is open, each $\phi(V_n)$ is an open set in $G/H$. Hence, $\phi(e)$ is a $G_\delta$-point in $G/H$, and $H = \phi^{-1}(\phi(e))$ is a $G_\delta$-set in $G$. By the homogeneity argument, all points of $G/H$ are $G_\delta$'s. □

Lemma 3.1.24 permits us to improve Proposition 3.1.20 as follows.

THEOREM 3.1.25. *Suppose that $G$ is a topological group and $F$ is a compact $G_\delta$-set in $G$ containing the neutral element $e$ of $G$. Then there exists a compact subgroup $H$ of $G$ such that $H$ is a $G_\delta$-set in $G$ contained in $F$ and every point in the quotient space $G/H$ is a $G_\delta$-set.*

In the case when the group $G$ is locally compact, one can go even further:

THEOREM 3.1.26. *Suppose that $G$ is a locally compact topological group and $U$ is an open neighbourhood of the neutral element $e$ of $G$. Then there exists a compact subgroup $H$ of $G$ contained in $U$ such that the quotient space $G/H$ is first-countable.*

PROOF. Clearly, we can assume that the closure of $U$ is compact. It is also obvious that $G$ contains a closed $G_\delta$-set $F$ such that $e \in F \subset U$. Then $F$ is compact, and Theorem 3.1.25 implies that there exists a compact subgroup $H$ of $G$ such that $H \subset F$ and $H$ is a $G_\delta$-set in $G$. Since the space $G$ is locally compact and regular, it follows that there exists a countable base $\gamma$ of open neighbourhoods of $H$ in $G$ (that is, $F = \bigcap \gamma$, and every open neighbourhood of $H$ contains some $U \in \gamma$). Let $\phi$ be the natural mapping of $G$ onto the quotient space $G/H$. Then the countable family $\{\phi(U) : U \in \gamma\}$ is a base of $G/H$ at $\phi(e)$, since $\phi$ is continuous and open. □

Every continuous homomorphism of a compact group onto another compact group is open, by Proposition 1.5.17. A similar statement for locally compact topological groups does not hold. Indeed, every non-discrete topological group is an image under a continuous isomorphism of a discrete group, and this isomorphism cannot be an open mapping.

However, under certain circumstances, a generalization from the compact case to the $\sigma$-compact case is possible.

Recall that a space $X$ *has the Baire property* if $X$ cannot be represented as the union of a countable family of closed nowhere dense subspaces [**165**, Section 3.9].

THEOREM 3.1.27. *Suppose that $G$ is a $\sigma$-compact topological group and that $f$ is a continuous homomorphism of $G$ onto a locally compact topological group $M$. Then $f$ is open.*

PROOF. Put $H = f^{-1}f(e)$, where $e$ is the neutral element of $G$. Then $H$ is an invariant closed topological subgroup of $G$ and, by Corollary 1.5.11, there exists a continuous isomorphism $g$ of the quotient group $G/H$ onto $M$ such that $f = g \circ \pi$, where $\pi$ is the natural quotient homomorphism of $G$ onto $G/H$. Since $G$ is $\sigma$-compact, the quotient group $G/H$ is also $\sigma$-compact. Hence, $G/H = \bigcup_{n\in\omega} F_n$, where each $F_n$ is compact. Since $g$ is one-to-one and continuous, the restriction of $g$ to $F_n$ is a homeomorphism of $F_n$ onto $g(F_n)$, for each $n \in \omega$. We have $M = \bigcup_{n\in\omega} g(F_n)$, where $g(F_n)$ is compact and, hence, closed in $M$. However, the space $M$ is locally compact and Hausdorff; therefore, $M$ has the Baire property [**165**, Theorem 3.9.3]. It follows that there exist $k \in \omega$ and a non-empty open subset $V$ of $M$ such that $V \subset g(F_k)$. Put $U = f^{-1}(V) = \pi^{-1}(g^{-1}(V))$. Then $U$ is a non-empty open set in $G$, and the restriction of $\pi$ to $U$ is an open continuous mapping of $U$ onto $g^{-1}(V)$, since $\pi$ is open and continuous, and $\pi^{-1}(\pi(U)) = U$. Since $f = g \circ \pi$ and the restriction of $g$ to $\pi(U) = g^{-1}(V) \subset F_k$ is a homeomorphism, it follows that the restriction of $f$ to $U$ is an open continuous mapping of $U$ onto $V$. However, $U$ is open in $G$ and $V$ is open in $M$. Now Proposition 1.5.14 implies that the homomorphism $f$ is open. □

The above theorem combined with Corollary 3.1.5 implies the following:

COROLLARY 3.1.28. *Every continuous homomorphism of a connected locally compact topological group onto a locally compact topological group is open.*

The following example shows that none of the assumptions in Theorem 3.1.27 can be omitted, even if $G$ is taken to be the groups of reals.

EXAMPLE 3.1.29. There exists a continuous isomorphism $f$ of the additive group of reals $\mathbb{R}$ with the usual interval topology onto a second-countable topological group $M$ that fails to be open.

To construct such an isomorphism, consider the product group $\mathbb{R} \times \mathbb{R}$ and its closed subgroup $H = \{(x, ax) : x \in \mathbb{R}\}$, where $a \in \mathbb{R}$ is a fixed irrational number. It is clear that the group $H$ with the topology induced from $\mathbb{R} \times \mathbb{R}$ is topologically isomorphic with $\mathbb{R}$. Denote by $P$ the discrete subgroup $\mathbb{Z} \times \mathbb{Z}$ of the group $\mathbb{R} \times \mathbb{R}$, and let $\pi$ be the quotient homomorphism of $\mathbb{R}\times\mathbb{R}$ onto the group $\mathbb{R}\times\mathbb{R}/P$ which is topologically isomorphic to $\mathbb{T}\times\mathbb{T}$. Put $M = \pi(H)$ and denote by $f$ the restriction of $\pi$ to the group $H$. Clearly, $f\colon H \to M$ is a continuous isomorphism and the group $M$ is second-countable as a subgroup of $\mathbb{T} \times \mathbb{T}$. Finally, $f$ is not open, since no continuous isomorphism of a locally compact, non-compact group onto a subgroup of a compact group is open. The group $M$ is known as the *irrational winding* of the circle group $\mathbb{T} \times \mathbb{T}$. □

To finish this section, we consider more complicated objects of topological algebra known as *topological rings and (skew) fields* and show that several topological groups considered earlier have a natural structure of a topological ring or a topological (skew) field.

Let $R$ be a ring and $\mathcal{T}$ a topology on the set $R$. We say that $\mathcal{T}$ is a *ring topology* provided that $R$ is a topological group with respect to addition, and multiplication is a continuous mapping of $R \times R$ to $R$ when $R$ carries the topology $\mathcal{T}$. The pair $(R, \mathcal{T})$ is called a *topological ring.*

Similarly, if $R$ is a field with a given topology $\mathcal{T}$, we say that $\mathcal{T}$ is a *field topology* provided that $(R, \mathcal{T})$ is a topological ring and inversion in the multiplicative group $R \setminus \{0_R\}$ is continuous, where $0_R$ is the zero element of $R$. Again, we call $(R, \mathcal{T})$ a *topological field.* Clearly, $(R, \mathcal{T})$ is a topological field if and only if both the additive group $R$ and multiplicative group $R \setminus \{0_R\}$ are topological groups with respect to the corresponding topologies $\mathcal{T}$ and its restriction to $R \setminus \{0_R\}$. A *topological skew field* is, naturally, a skew field with a topology satisfying the same restrictions as in the case of a topological field.

Here are some simple examples of topological rings and (skew) fields.

EXAMPLE 3.1.30. It turns out that some topological groups that appeared in the previous chapters admit richer algebraic and topological structures.

a) It follows from the above definitions that the complex plane $\mathbb{C}$ with the usual addition, multiplication, and Euclidean topology is a topological field since all algebraic operations, the addition and multiplication in $\mathbb{C}$, and the inverse in $\mathbb{C} \setminus \{0\}$, are continuous.
b) The skew field $\mathbf{Q}$ of quaternions with the Euclidean topology defined in item f) of Example 1.2.5 is a topological skew field. Again, this follows from the fact that the additive group $\mathbf{Q}$ and the multiplicative group $\mathbf{Q}^*$ are topological groups.
c) Suppose that $X$ is a non-empty set and $\mathbb{R}^X$ the corresponding product space. Clearly, $\mathbb{R}^X$ is a commutative topological group with respect to the pointwise addition. It is easy to see that the pointwise multiplication in $\mathbb{R}^X$ is also continuous. This follows immediately from the continuity of multiplication in $\mathbb{R}$ and our definition of the topology and multiplication in $\mathbb{R}^X$. Therefore, $\mathbb{R}^X$ is a commutative topological ring.
d) Let $C_p(X)$ be the space of continuous real-valued functions on a Tychonoff space $X$, endowed with the pointwise convergence topology. We know that $C_p(X)$, with the usual pointwise addition of functions, is a subgroup of the topological group $\mathbb{R}^X$ (see page 81) and, therefore, is itself a topological group. Consider the product operation in $C_p(X)$, also defined pointwise. Clearly, $C_p(X)$ is a *subring* of the ring $\mathbb{R}^X$, that is, $C_p(X)$ is a subgroup of the additive group $\mathbb{R}^X$ and $fg \in C_p(X)$ for all $f, g \in C_p(X)$, provided that $\mathbb{R}^X$ is considered with the pointwise multiplication. Since $\mathbb{R}^X$ is a topological ring, we conclude that $C_p(X)$ is a *subring* of $\mathbb{R}^X$. Hence, $C_p(X)$ itself is a commutative topological ring.

   Finally, let $C_p^*(X)$ be the subspace of $C_p(X)$ which consists of all functions $f$ such that $f(x) \neq 0$, for each $x \in X$. Evidently, $fg \in C_p^*(X)$ and $f^{-1} \in C_p^*(X)$ for all $f, g \in C_p^*(X)$, so $C_p^*(X)$ is a multiplicative subgroup of the ring $C_p(X)$. Notice that $C_p^*(X)$ coincides with the subgroup of all invertible (with respect to multiplication) elements of $C_p(X)$. Since $\mathbb{R}$ is a topological field, it is easy to see that inversion in $C_p^*(X)$ is continuous and, hence, $C_p^*(X)$ is a topological group. □

It was shown in Example 1.3.16 that for every integer $r > 1$, the group of $r$-adic numbers $\Omega_r$ admits a Hausdorff topology making it into a locally compact, $\sigma$-compact topological group. Our next step is to show that the group $\Omega_r$ admits a natural multiplication that makes it into a locally compact topological ring and, if $r$ is a prime power then, in fact, $\Omega_r$ is a topological field.

EXAMPLE 3.1.31. To define multiplication in $\Omega_r$, take any elements $\mathbf{x}, \mathbf{y} \in \Omega_r$. If either $\mathbf{x} = \mathbf{0}$ or $\mathbf{y} = \mathbf{0}$, set $\mathbf{xy} = \mathbf{0}$. If both $\mathbf{x}$ and $\mathbf{y}$ are distinct from zero, let $k$ and $l$ be the least integers such that $x_k \neq 0$ and $y_l \neq 0$, respectively. We define an element $\mathbf{z} = (z_n)$ in $\Omega_r$ as follows. First, we set $z_n = 0$ for each $n < k + l$. Then we write the product $x_k y_l$ in the form $x_k y_l = z_{k+l} + t_{k+l} r$, where $z_{k+l}$ and $t_{k+l}$ are integers and $z_{k+l} \in \{0, 1, \ldots, r-1\} = A$. In other words, $z_{k+l}$ is the least non-negative residue of $x_k y_l$ modulo $r$. Suppose we have defined the integers $z_{k+l}, z_{k+l+1}, \ldots, z_{k+l+s-1} \in A$ and $t_{k+l}, t_{k+l+1}, \ldots, t_{k+l+s-1}$ for some $s \geq 1$. Then we can write

$$x_{k+s} y_l + x_{k+s-1} y_{l+1} + x_{k+s-2} y_{l+2} + \cdots + x_k y_{l+s} + t_{k+l+s-1} = z_{k+l+s} + t_{k+l+s} r, \qquad (3.1)$$

where $z_{k+l+s} \in A$ and $t_{k+l+s}$ is an integer. Clearly, such a representation is unique.

In fact, the above formula for calculation of the entries $z_n$ comes from the usual multiplication of numbers written in the decimal system and differs from it in three unessential details. First, we write $p$-adic numbers from the left to right, starting with smaller entries. Second, the base of $r$-adic numbers is $r$ in place of 10. And the third difference is that we perform all necessary arithmetic operations with entries of $r$-adic numbers from the left to right.

Here is a numerical example which shows in detail the routine of multiplication in the case when $r = 3$. We multiply two 3-adic integers written in the first and the second lines, so all zero entries of both factors and of the product (corresponding to the positions $-1, -2, \ldots$) are suppressed.

| | | | | | | | |
|---|---|---|---|---|---|---|---|
| | 1 | 2 | 0 | 1 | 2 | 0 | ...... |
| × | 1 | 2 | 1 | 2 | 1 | 2 | ...... |
| | 1 | 2 | 0 | 1 | 2 | 0 | ...... |
| | 0 | 2 | 4 | 0 | 2 | 4 | ...... |
| | 0 | 0 | 1 | 2 | 0 | 1 | ...... |
| | 0 | 0 | 0 | 2 | 4 | 0 | ...... |
| | 0 | 0 | 0 | 0 | 1 | 2 | ...... |
| | 0 | 0 | 0 | 0 | 0 | 2 | ...... |
| | ... | ... | ... | ... | ... | ... | ...... |
| = | 1 | 1 | 0 | 1 | 2 | 0 | ...... |

We claim that $\Omega_r$ with addition described in Example 1.1.10 and multiplication given by formula (3.1) is a commutative ring. Notice that the element $\mathbf{u}$ of $\Omega_r$ defined by the rule $u_n = 1$ if $n = 0$ and $u_n = 0$ otherwise, satisfies $\mathbf{xuux} = \mathbf{x}$, for each $\mathbf{x} \in \Omega_r$. Hence, $\mathbf{u}$ is the neutral element of $\Omega_r$ with respect to multiplication. Since $\Omega_r$ is a commutative group with respect to addition, it suffices to verify that the identities $\mathbf{xy} = \mathbf{yx}$, $(\mathbf{xy})\mathbf{z} = \mathbf{x}(\mathbf{yz})$, and $\mathbf{x}(\mathbf{y} + \mathbf{z}) = \mathbf{xy} + \mathbf{xz}$ hold for all $\mathbf{x}, \mathbf{y}, \mathbf{z} \in \Omega_r$. One can easily verify these in the special case when each of the elements $\mathbf{x}, \mathbf{y}, \mathbf{z}$ has only finitely many non-zero entries. In the general case, it suffices to "truncate" these elements at some entry $n$ similarly to that in Example 1.1.10, and then use the above identities for truncated elements to show that the elements on both sides of every identity coincide at each entry less than or equal to $n$. This implies that $\Omega_r$ is a commutative ring, as claimed.

It remains to verify that $\Omega_r$ is a topological ring or, equivalently, multiplication in $\Omega_r$ is continuous when $\Omega_r$ carries the topology defined in Example 1.3.16. Let $\mathbf{a}$ and $\mathbf{b}$ be arbitrary elements of $\Omega_r$ and $\Lambda_m$ a basic open neighbourhood of zero element in $\Omega_r$, where $m \in \mathbb{Z}$. Then $\mathbf{ab}+\Lambda_m$ is a basic open neighbourhood of the element $\mathbf{ab}$ in $\Omega_r$. If $\mathbf{a}$ is distinct from zero element, let $k$ be the least integer with $a_k \neq 0$; if $\mathbf{a} = \mathbf{0}$, set $k = 0$. Similarly, we define $l$ to be the least integer such that $b_l \neq 0$ if $\mathbf{b} \neq \mathbf{0}$ and $l = 0$ otherwise. Notice that, in any case, $\mathbf{a} \in \Lambda_k$ and $\mathbf{b} \in \Lambda_l$. Let $n$ be a non-negative integer greater than or equal to each of the three numbers $m - k,\ m - l,\ m$. The continuity of multiplication in $\Omega_r$ will follow if we show that

$$(\mathbf{a} + \Lambda_n)(\mathbf{b} + \Lambda_n) \subset \mathbf{ab} + \Lambda_m. \tag{3.2}$$

Take any $\mathbf{x}, \mathbf{y} \in \Lambda_n$. It follows from our choice of the number $n$ that $n+k \geq m$ and $n+l \geq m$. Recall that the groups $\Lambda_i$ satisfy $\Lambda_i \subset \Lambda_j$ whenever $i < j$ and the definition of these groups given in Example 1.3.16 implies that $\Lambda_i\Lambda_j \subset \Lambda_{i+j}$ for all $i, j \in \mathbb{Z}$. Therefore, we have that $\mathbf{ay} \in \Lambda_k\Lambda_n \subset \Lambda_{n+k} \subset \Lambda_m$, $\mathbf{xb} \in \Lambda_n\Lambda_l \subset \Lambda_{n+l} \subset \Lambda_m$, and $\mathbf{xy} \in \Lambda_n\Lambda_n = \Lambda_{2n} \subset \Lambda_n \subset \Lambda_m$. This implies immediately that

$$(\mathbf{a} + \mathbf{x})(\mathbf{b} + \mathbf{y}) = \mathbf{ab} + (\mathbf{ay} + \mathbf{xb} + \mathbf{xy}) \in \mathbf{ab} + \Lambda_m,$$

and the continuity of multiplication in $\Omega_r$ is proved. Hence, $\Omega_r$ is a topological ring.

Now suppose that $r = p^m$, where $p$ is a prime number and $m \geq 1$ is an integer. To show that $\Omega_r$ is a topological field in this case, it suffices to verify that every non-zero element $x \in \Omega_r$ is invertible and the mapping $\mathbf{x} \mapsto \mathbf{x}^{-1}$ of $\Omega_r \setminus \{\mathbf{0}\}$ onto itself is continuous. Take an arbitrary non-zero element $\mathbf{x}$ of $\Omega_r$ and let $k$ be the least integer with $x_k \neq 0$. We consider two possible cases.

*Case 1.* If $(x_k, p) = 1$, that is, $p$ does not divide $x_k$, then the inverse $\mathbf{y} = \mathbf{x}^{-1}$ exists and has the form $(\ldots, 0\ldots, 0, y_{-k}, y_{-k+1}, \ldots)$, where $(y_{-k}, p) = 1$. Indeed, there exists $y_{-k} \in \{1, 2, \ldots, r-1\}$ such that $x_k y_{-k} = 1 + v_0 p^m$ for some integer $v_0$. In fact, these conditions define $y_{-k}$ uniquely and imply that $(y_{-k}, p) = 1$. Suppose that for some $n \geq 1$, we have defined numbers $y_{-k}, y_{-k+1}, \ldots, y_{-k+n-1} \in \{0, 1, \ldots, r-1\}$ and non-negative integers $v_0, v_1, \ldots, v_{n-1}$. Then there exists a unique integer $y_{-k+n} \in \{0, 1, \ldots, r-1\}$ such that $x_k y_{-k+n} + x_{k+1} y_{-k+n-1} + \cdots + x_{k+n} y_{-k} + v_{n-1} = v_n p^m$ for some integer $v_n$. This inductive definition gives us the element $\mathbf{y}$ with entries $y_l$, where $y_l = 0$ for each $l < -k$. It follows from our definition of multiplication in $\Omega_r$ that $\mathbf{xy} = \mathbf{u}$.

*Case 2.* Suppose that $p^v$ is the maximal common divisor of $x_k$ and $r$, where $1 \leq v < m$. Then $x_k = p^v x_k'$, where $(x_k', p) = 1$. Let us define an auxiliary element $\mathbf{f} = (f_n) \in \Omega_r$ by the rule $f_0 = p^{m-v}$ and $f_n = 0$ otherwise. In other words, $\mathbf{f} = p^{m-v}\mathbf{u} \in \Lambda_0$. Consider the element $\mathbf{z} = \mathbf{fx}$. It is clear that $z_n = 0$ for each $n < k$. From the equalities $p^{m-v}x_k = p^m x_k' = t_k p^m + z_k$ it follows that $z_k = 0$ and $t_k = x_k'$. In addition, calculating the entry $z_{k+1}$ of $\mathbf{z}$, we have that $p^{m-v}x_{k+1} + x_k' = t_{k+1}p^m + z_{k+1}$. Since $m - v > 0$ and $(x_k', p) = 1$, this implies that $z_{k+1} \neq 0$ and $(z_{k+1}, p) = 1$. So the element $\mathbf{z}$ satisfies the conditions of Case 1 and, hence, it has the multiplicative inverse $\mathbf{z}^{-1} \in \Omega_r$. It follows that $\mathbf{u} = \mathbf{zz}^{-1} = \mathbf{fxz}^{-1} = \mathbf{x}(\mathbf{fz}^{-1})$ or, equivalently, $\mathbf{fz}^{-1}$ is the inverse of $\mathbf{x}$ in $\Omega_r$. We conclude, therefore, that $\Omega_r$ is a field.

It remains to verify the continuity of inverse in $\Omega_r \setminus \{\mathbf{0}\}$. Again, take an arbitrary element $\mathbf{a} \in \Omega_r$ distinct from $\mathbf{0}$ and let $k$ be least integer with $a_k \neq 0$. First, we mention the following property of inversion in $\Omega_r \setminus \{\mathbf{0}\}$:

**Claim.** *If* $\mathbf{b} \in \Omega_r \setminus \{\mathbf{0}\}$, $s \in \mathbb{Z}$, $s \geq k+3$, *and* $b_l = a_l$ *for each* $l < s$, *then the entries of the multiplicative inverses* $\mathbf{c} = \mathbf{a}^{-1}$ *and* $\mathbf{d} = \mathbf{b}^{-1}$ *satisfy* $d_l = c_l$, *for each* $l < s - 2k - 2$.

Indeed, we have $d_l = c_l = 0$ for each $l < -k$. In Case 1, when $(a_k, p) = 1$, the $s - k$ equalities

$$b_k = a_k,\ b_{k+1} = a_{k+1},\ \ldots,\ b_{s-1} = a_{s-1} \tag{3.3}$$

together with (3.1) imply inductively the corresponding $s - k$ equalities for the entries of $\mathbf{c}$ and $\mathbf{d}$:

$$d_{-k} = c_{-k},\ d_{-k+1} = c_{-k+1},\ \ldots,\ d_{s-2k-1} = c_{s-2k-1}. \tag{3.4}$$

In Case 2, when $(a_k, r) = p^v$ and $1 \leq v < m$, we put $\mathbf{f} = p^{m-v}\mathbf{u}$, $\mathbf{z} = \mathbf{fa}$, and $\mathbf{w} = \mathbf{fb}$. Then, by (3.1) and (3.3), the entries of the elements $\mathbf{z}$ and $\mathbf{w}$ satisfy $w_l = z_l = 0$ for each $l \leq k$, and the $s - k - 1$ equalities

$$w_{k+1} = z_{k+1},\ w_{k+2} = z_{k+2},\ \ldots,\ w_{s-1} = z_{s-1}. \tag{3.5}$$

Put $\mathbf{g} = \mathbf{z}^{-1}$ and $\mathbf{h} = \mathbf{w}^{-1}$. Since the elements $\mathbf{z}$ and $\mathbf{w}$ satisfy the conditions of Case 1 (with $k+1$ in place of $k$), we apply (3.5) to conclude, as above, that the entries of the elements $\mathbf{g}$ and $\mathbf{h}$ have to satisfy $g_l = h_l = 0$, for each $l < -k - 1$, and the $s - k - 1$ equalities

$$h_{-k-1} = g_{-k-1},\ h_{-k} = g_{-k},\ \ldots,\ h_{s-2k-3} = g_{s-2k-3}. \tag{3.6}$$

According to Case 2, we have that $\mathbf{c} = \mathbf{fg}$ and $\mathbf{d} = \mathbf{fh}$. It remains to repeat the argument applied earlier to deduce (3.5) from (3.3) and conclude that (3.6) implies the corresponding equalities $d_l = c_l = 0$, for each $l \leq -k - 1$, and the $s - k - 2$ equalities

$$d_{-k} = c_{-k},\ d_{-k+1} = c_{-k+1},\ \ldots,\ d_{s-2k-3} = c_{s-2k-3}. \tag{3.7}$$

This finishes the proof of Claim.

We turn back to the proof of the continuity of the inverse in $\Omega_r \setminus \{\mathbf{0}\}$. Consider a basic open neighbourhood $\mathbf{a} + \Lambda_n$ of the element $\mathbf{a} \in \Omega_r \setminus \{\mathbf{0}\}$, where $n \geq |k|$. Set $s = 2k + n + 2$. Notice that $s \geq k + 3$, and if $\mathbf{x} \in \Lambda_s$, then the entries $b_l$ of the element $\mathbf{b} = \mathbf{a} + \mathbf{x}$ satisfy $b_l = a_l$ for each $l < s$. Hence, the above Claim implies that the entries of the inverses $\mathbf{c} = \mathbf{a}^{-1}$ and $\mathbf{d} = \mathbf{b}^{-1}$ satisfy $d_l = c_l$, for each $l < n$. In its turn, this gives $\mathbf{d} - \mathbf{c} \in \Lambda_n$ or, equivalently, $\mathbf{d} \in \mathbf{c} + \Lambda_n$. Therefore, the inverse in $\Omega_r \setminus \{\mathbf{0}\}$ is continuous.

Summing up, for each prime power number $r$, $\Omega_r$ is a locally compact, $\sigma$-compact, zero-dimensional topological field. □

## Exercises

3.1.a. Show that the product of a countable family of locally compact groups need not be locally compact.

3.1.b. Verify that every locally compact group $G$ is Dieudonné complete (that is, complete with respect to the maximal uniformity on the space $G$ compatible with the topology of $G$).

3.1.c. Prove that a locally compact group is Lindelöf if and only if it is $\sigma$-compact.

3.1.d. Apply Exercise 1.2.i to show that every separable locally compact group is Lindelöf and, therefore, is $\sigma$-compact.

3.1.e. Give an example of a non-separable Lindelöf locally compact group.

3.1.f. Prove that every countable locally compact group is discrete.

3.1.g. Let $GL(n, \mathbb{R})$ be the general linear group with real entries. The topology on $GL(n, \mathbb{R})$ is generated by the metric defined as the maximum of distances between corresponding entries (see item e) of Example 1.2.5). Then, with this topology and the usual multiplication of matrices, $GL(n, \mathbb{R})$ is a locally compact, locally connected topological group. Calculate the connected component of $GL(n, \mathbb{R})$.

3.1.h. Verify that the subgroup $H + P$ of the group $\mathbb{R} \times \mathbb{R}$ in Example 3.1.29 is dense in $\mathbb{R} \times \mathbb{R}$. Therefore, $M$ is a proper dense subgroup of $\mathbb{T} \times \mathbb{T}$ and, in particular, $M$ is not locally compact.

3.1.i. Show that for every Tychonoff space $X$, the multiplicative subgroup $C_p^*(X)$ of the topological ring $C_p(X)$ considered in Example 3.1.30 is dense in $C_p(X)$.

3.1.j. Find the inverses of the "periodic" 3-adic integers $(1, 2, 0, 1, 2, 0, \ldots)$ and $(1, 2, 1, 2, 1, 2, \ldots)$ in the field $\Omega_3$. Are the inverses "periodic"?

3.1.k. Let $r > 1$ be an integer. Verify the following:

(a) If $r$ has two distinct prime divisors, then there exist two non-zero $r$-adic integers $\mathbf{a}$ and $\mathbf{b}$ such that $\mathbf{ab} = \mathbf{0}$. In other words, the subring $\mathbb{Z}_r$ of $\Omega_r$ contains non-trivial divisors of zero. Deduce that $\Omega_r$ is a field if and only if $r$ is a prime power (cf. Example 3.1.31).

(b) Let us call an $r$-adic number $\mathbf{a}$ *infinite* if $\mathbf{a}$ contains infinitely many non-zero entries. Show that if the equality $\mathbf{ab} = \mathbf{0}$ holds for non-zero $r$-adic numbers $\mathbf{a}$ and $\mathbf{b}$, then both numbers are infinite.

## Problems

3.1.A. Prove that Corollaries 3.1.12, 3.1.13, and Theorem 3.1.14 are valid for the class of countably compact topological groups.

*Remark.* Connectedness and total disconnectedness in countably compact and pseudocompact topological groups were studied by D. Dikranjan in [**134, 135, 136**].

3.1.B. A topological group $G$ is called *locally pseudocompact* if $G$ contains a pseudocompact neighbourhood of the neutral element. Which of Corollaries 3.1.12, 3.1.13, and Theorem 3.1.14 remain valid for locally pseudocompact topological groups?

*Remark.* The articles [**136**] and [**477**] contain a helpful information on the subject.

3.1.C. Let $G$ be a connected topological group such that the complement of the neutral element of $G$ is not connected. Show that there exists a continuous isomorphism of $G$ onto the group of reals $\mathbb{R}$ endowed with the usual topology.

3.1.D. Let $G$ be a connected locally compact topological group such that the complement of the neutral element of $G$ is not connected. Show that $G$ is topologically isomorphic to the group $\mathbb{R}$ with the usual topology.

3.1.E. Show that there is a Hausdorff topology on $\mathbb{R}$ turning $\mathbb{R}$ into a connected, locally compact, locally connected topological group such that the complement of the neutral element is connected.

3.1.F. Let $G$ be a locally connected topological group such that, for some connected open neighbourhood $U$ of the neutral element $e$ of $G$, the subspace $U \setminus \{e\}$ is disconnected. Prove that $G$ is locally compact.

3.1.G. (A. A. Markov [**306**]) Give an example of a non-trivial connected topological group $G$ all elements of which have order 2. Can such a group be locally connected?

3.1.H. Prove that the topological ring $\mathbb{R}^\tau$ contains a dense subfield if and only if $\tau \leq 2^{2^\omega}$.

3.1.I. Let $\mathbb{Z}_r$ be the additive group of $r$-adic integers (see Example 1.1.10). Prove that for every continuous homomorphism $h$ of the group $\mathbb{Z}_r$ to itself, there exists $\mathbf{a} \in \mathbb{Z}_r$ such that $h(\mathbf{x}) = \mathbf{ax}$, for each $\mathbf{x} \in \mathbb{Z}_r$.

3.1.J. (a) For which integers $r, s > 1$, the product group $\mathbb{Z}_r \times \mathbb{Z}_s$ is monothetic, that is, contains a dense cyclic subgroup?

(b) Is the compact group $\mathbb{Z}_r \times \mathbb{T}$ monothetic for any integer $r > 1$? For all integers $r > 1$?

3.1.K. A subgroup $I$ of the additive group of a commutative ring $R$ is called *ideal* if $ax \in I$ for all $a \in I$ and $x \in R$. Prove that every ideal $I$ of the ring $\mathbb{Z}_p$ with prime $p$ is either trivial, that is, $I = \{\mathbf{0}\}$, or has the form $I = p^k\mathbb{Z}_p$ for some integer $k \geq 0$. Describe all ideals of the ring $\mathbb{Z}_r$ in the case of a composite integer $r$.

## Open Problems

3.1.1. Characterize Tychonoff spaces $X$ such that the topological ring $C_p(X)$ contains a dense subfield (see also Problem 3.1.H).

3.1.2. Find an internal characterization of subgroups of products of locally compact topological groups.

3.1.3. When a locally compact space can be represented as a retract of a locally compact topological group?

## 3.2. Quotients with respect to locally compact subgroups

In this section we are going to establish that for a locally compact subgroup $H$ of a topological group $G$, the natural quotient mapping $\pi$ of $G$ onto the quotient space $G/H$ has some rather nice properties locally. This will lead us to some interesting results on how properties of $G$ depend on the properties of $G/H$ when $H$ is locally compact. We recall that, by Proposition 1.4.19, every locally compact subgroup $H$ of a topological group $G$ is closed in $G$.

PROPOSITION 3.2.1. *Suppose that $G$ is a topological group, $H$ is a locally compact subgroup of $G$, $P$ is a closed symmetric subset of $G$ such that $P$ contains an open neighbourhood of the neutral element $e$ in $G$, and that $\overline{P^3} \cap H$ is compact. Let $\pi \colon G \to G/H$ be the natural quotient mapping of $G$ onto the quotient space $G/H$. Then the restriction $f$ of $\pi$ to $P$ is a perfect mapping of $P$ onto the subspace $\pi(P)$ of $G/H$.*

PROOF. Clearly, $f$ is continuous. First, we show that $f^{-1}f(a)$ is compact, for any $a \in P$. Indeed, from the definition of $f$ we have $f^{-1}f(a) = aH \cap P$. The subspaces $aH \cap P$ and $H \cap a^{-1}P$ are obviously homeomorphic and closed in $G$. Since $a^{-1} \in P^{-1} = P$, we have $H \cap a^{-1}P \subset H \cap P^2 \subset \overline{P^3} \cap H$. Hence, $H \cap a^{-1}P$ is compact and so is the set $f^{-1}f(a)$.

It remains to prove that the mapping $f$ is closed. Let us fix any closed subset $M$ of $P$, and let $a$ be a point of $P$ such that $f(a) \in \overline{f(M)}$. We have to show that $f(a) \in f(M)$ or, equivalently, that $aH \cap M \neq \varnothing$. Assume the contrary. Then $(aH \cap \overline{P^2}) \cap M = \varnothing$. Note that $aH \cap \overline{P^2}$ is compact, since $aH \cap \overline{P^2}$ is homeomorphic to $H \cap a^{-1}\overline{P^2}$, which is a closed subset of the compact space $H \cap \overline{P^3}$.

Since $aH \cap \overline{P^2}$ is compact and $M$ is closed and disjoint from $aH \cap \overline{P^2}$, there exists an open neighbourhood $W$ of $e$ in $G$ such that $(W(aH \cap \overline{P^2})) \cap M = \varnothing$. Clearly, we can assume that $W$ is symmetric and $W \subset P$.

Since the quotient mapping $\pi$ is open and $Wa$ is an open neighbourhood of $a$, the set $\pi(Wa)$ is an open neighbourhood of $\pi(a)$ in $G/H$. Therefore, the set $\pi(Wa) \cap \pi(M)$ is not empty, and we can fix $m \in M$ and $y \in W$ such that $\pi(m) = \pi(ya)$, that is, $mH = yaH$. Then $m = yah$, for some $h \in H$. However, $ah = y^{-1}m \in \overline{P^2}$, since $y^{-1} \in W^{-1} = W \subset P$ and $M \subset P$. Besides, $ah \in aH$. Hence, $ah \in (aH \cap \overline{P^2})$ and $m = yah \in W(aH \cap \overline{P^2})$, since $y \in W$. Thus, $m \in M \cap W(aH \cap \overline{P^2})$, which contradicts the choice of $W$. Hence, $f(a) \in f(M)$, and $f(M)$ is closed in $f(P)$. □

The next statement is just a smoother way to formulate what was achieved in Proposition 3.2.1.

THEOREM 3.2.2. [**A. V. Arhangel'skii**] *Suppose that $G$ is a topological group, $H$ a locally compact subgroup of $G$, and $\pi\colon G \to G/H$ is the natural quotient mapping of $G$ onto the quotient space $G/H$. Then there exists an open neighbourhood $U$ of the neutral element $e$ such that $\pi(\overline{U})$ is closed in $G/H$ and the restriction of $\pi$ to $\overline{U}$ is a perfect mapping of $\overline{U}$ onto the subspace $\pi(\overline{U})$ (thus, $\pi$ is an open locally perfect mapping of $G$ onto $G/H$).*

PROOF. First, $H$ is closed in $G$ by Proposition 1.4.19. Since $H$ is locally compact, there exists an open neighbourhood $V$ of $e$ in $G$ such that $\overline{V \cap H}$ is compact. Since the space $G$ is regular, we can select an open neighbourhood $W$ of $e$ such that $\overline{W} \subset V$. Then $\overline{W} \cap H$ is compact, since $\overline{W} \cap H$ is a closed subset of the compact subspace $\overline{V \cap H}$. Let $U_0$ be any symmetric open neighbourhood of $e$ such that $U_0^3 \subset W$. Since $\overline{U_0}^3 \subset \overline{U_0^3}$, the set $P = \overline{U_0}$ satisfies all restrictions on $P$ in Proposition 3.2.1. Therefore, by Proposition 3.2.1, the restriction of $\pi$ to $\overline{U_0}$ is a perfect mapping of $\overline{U_0}$ onto the subspace $\pi(\overline{U_0})$. We will now modify $U_0$ to make sure that the other condition is satisfied.

Since $\pi$ is an open mapping, the set $\pi(U_0)$ is open in $G/H$. Since the space $G/H$ is regular, we can take an open neighbourhood $V_0$ of $\pi(e)$ in $G/H$ such that $\overline{V_0} \subset \pi(U_0)$. Then $U = \pi^{-1}(V_0) \cap U_0$ is an open neighbourhood of $e$ contained in $P$, and the restriction $f$ of $\pi$ to the closure of $U$ is a perfect mapping of $\overline{U}$ onto the subspace $\pi(\overline{U})$. However, $\pi(\overline{U})$ is closed in $\pi(P)$, and $\pi(\overline{U}) \subset \overline{V_0} \subset \pi(U_0) \subset \pi(P)$. Therefore, $\pi(\overline{U})$ is closed in the closed set $\overline{V_0}$, which implies that $\pi(\overline{U})$ is closed in $G/H$. □

Notice that if $H$ is compact, then we can put $U = G$ in Theorem 3.2.2. That means that the quotient mapping $\pi$ is perfect in this case (see Theorem 1.5.7). However, in the general case, the mapping $\pi$ in Theorem 3.2.2 need not be closed.

EXAMPLE 3.2.3. Let $\mathbb{R}$ be the topological group of real numbers, and let $\mathbb{Z}$ be its subgroup consisting of integers. Then the quotient mapping $\pi\colon \mathbb{R} \to \mathbb{R}/\mathbb{Z} = \mathbb{T}$ is open, locally perfect, but not closed, since otherwise the kernel $\mathbb{Z}$ of $\pi$ would be compact by [**165**, Theorem 4.4.16]. Another way to see that $\pi$ is not closed is to consider the image under $\pi$ of the closed subset $\{(n^2+1)/n : n \in \mathbb{N}\}$ of $\mathbb{R}$.

THEOREM 3.2.4. *Suppose that $G$ is a zero-dimensional topological group and that $H$ is a locally compact subgroup of $G$. Then the quotient space $G/H$ is also zero-dimensional.*

PROOF. Let $\pi\colon G \to G/H$ be the natural quotient mapping of $G$ onto the quotient space $G/H$. According to Theorem 3.2.2, we can fix an open neighbourhood $U$ of the neutral element $e$ of $G$ such that $\pi(\overline{U})$ is closed in $G$ and the restriction of $\pi$ to $\overline{U}$ is a perfect mapping of $\overline{U}$ onto the subspace $\pi(\overline{U})$. Take any open neighbourhood $W$ of $\pi(e)$ in $G/H$. Since the space $G$ is zero-dimensional, we can fix an open and closed neighbourhood $V$ of $e$ such that $V \subset U \cap \pi^{-1}(W)$. Then $\pi(V)$ is an open subset of $G/H$, since the mapping $\pi$ is open. On the other hand, $\pi(V)$ is closed in $G/H$, since the restriction of $\pi$ to $\overline{U}$ is a closed mapping and $\pi(\overline{U})$ is closed in $G/H$. Clearly, $\pi(V) \subset W$. Hence, $G/H$ is zero-dimensional. □

One can derive a wealth of other corollaries from Theorem 3.2.2. First, we formulate a general statement unifying many of them.

Recall that a *regular closed set* in a space is the closure of an open subset of this space.

COROLLARY 3.2.5. *Suppose that $\mathcal{P}$ is a topological property preserved by preimages of spaces under perfect mappings (in the class of Tychonoff spaces) and also inherited by regular closed sets. Suppose further that $G$ is a topological group, $H$ is a locally compact subgroup of $G$, and that the quotient space $G/H$ has the property $\mathcal{P}$. Then there exists an open neighbourhood $U$ of the neutral element $e$ such that $\overline{U}$ has the property $\mathcal{P}$.*

PROOF. This is an immediate corollary from Theorem 3.2.2. □

Given a space $X$ and a property $\mathcal{P}$, if every point $x$ of $X$ has an open neighbourhood $O(x)$ such that the closure of $O(x)$ in $X$ has $\mathcal{P}$, then we say that $X$ has the property $\mathcal{P}$ *locally*. Local compactness, countable compactness, pseudocompactness, paracompactness, the Lindelöf property, $\sigma$-compactness, Čech-completeness, the Hewitt–Nachbin completeness, and the property of being a $k$-space are all inherited by regular closed sets and preserved by perfect preimages [**165**, Sections 3.7, 3.10, 3.11]. Thus, Corollary 3.2.5 is applicable. This observation proves the following statement:

COROLLARY 3.2.6. *Suppose that $G$ is a topological group, and that $H$ is a locally compact subgroup of $G$ such that the quotient space $G/H$ has some of the following properties:*

1) *$G/H$ is locally compact;*
2) *$G/H$ is locally countably compact;*
3) *$G/H$ is locally pseudocompact;*
4) *$G/H$ is locally paracompact;*
5) *$G/H$ is locally Lindelöf;*
6) *$G/H$ is locally $\sigma$-compact;*
7) *$G/H$ is locally Čech-complete;*
8) *$G/H$ is locally realcompact.*

*Then the space $G$ also has the same property, that is, in case 1) $G$ is locally compact, in case 2) $G$ is locally countably compact, etc.*

COROLLARY 3.2.7. *Suppose that $G$ is a topological group, and that $H$ is a locally compact subgroup of $G$ such that the quotient space $G/H$ is a $k$-space. Then $G$ is also a $k$-space.*

PROOF. This statement also follows from Corollary 3.2.5, since the property of being a $k$-space is invariant under taking perfect preimages and a locally $k$-space is, obviously, a $k$-space [**165**, Section 3.3]. □

## Exercises

3.2.a. A space $X$ is said to be *subparacompact* if every open covering of $X$ can be refined by a $\sigma$-discrete covering. Let $G$ be a topological group, and $H$ be a locally compact subgroup of $G$ such that the quotient space $G/H$ is subparacompact. Show that $G$ is locally subparacompact.

3.2.b. Let $G$ be a topological group, and $H$ be a locally compact subgroup of $G$ such that the quotient space $G/H$ is strongly paracompact. Show that $G$ is locally strongly paracompact.

3.2.c. Let $H$ be a locally compact subgroup of a topological group $G$, and suppose that the quotient space $G/H$ is metacompact. Show that $G$ is locally metacompact. (Recall that a space $X$ is *metacompact* if every open covering can be refined by a point-finite open covering).

3.2.d. Let $G$ be a topological group, and $H$ be a locally compact subgroup of $G$ such that the quotient space $G/H$ is countably paracompact. Show that $G$ is locally countably paracompact.

3.2.e. Let $H$ be a locally compact subgroup of a topological group $G$, and suppose that the quotient space $G/H$ is normal. Show that $G$ need not be locally normal.
*Hint.* Consider a topological group $G$ such that the space $G$ is normal but the space $G \times [0, 1]$ is not normal (see [**217**]).

## Problems

3.2.A. Is every locally Lindelöf topological group paracompact?
*Remark.* The answer is "yes", see [**47**].

3.2.B. (V. V. Uspenskij [**63**]) Show that every locally paracompact topological group is paracompact.
*Hint.* This easily follows from the next two general statements.

3.2.C. Let $G$ be a topological group, and $U$ be a non-empty open subset of $G$. Then there exists a locally finite open covering $\gamma$ of $G$ such that $\overline{V}$ is homeomorphic to a closed subspace of $\overline{U}$, for every $V \in \gamma$.
*Hint.* Consider $\xi = \{xU : x \in G\}$. Clearly, $\xi$ is an open covering of $G$ belonging to the left uniformity of $G$. Therefore, there exists a locally finite open covering $\gamma$ of $G$ refining $\xi$ (see [**165**, Section 8.1]). Since the closure of $xU$ is homeomorphic to the closure of $U$, $\gamma$ is the covering we are looking for.

3.2.D. Let $X$ be a topological space, and $\gamma$ be a locally finite open covering of $X$ such that $\overline{V}$ is paracompact, for each $V \in \gamma$. Then $X$ is paracompact.

3.2.E. Let $G$ be a topological group, and $H$ be a locally compact subgroup of $G$ such that the quotient space $G/H$ is first-countable. Let, further, $F$ be a closed subgroup of $G$ such that $H \cap F = \{e\}$, where $e$ is the neutral element of $G$. Show that $F$ is first-countable.

3.2.F. Let $H$ be a locally compact subgroup of a topological group $G$, and suppose that the tightness of the quotient space $G/H$ is countable. Let, further, $F$ be a closed subgroup of $G$ such that $H \cap F = \{e\}$, where $e$ is the neutral element of $G$. Prove that the tightness of $F$ is countable.

3.2.G. Let $G$ be a topological group, and $H$ be a locally compact subgroup of $G$ such that the quotient space $G/H$ is zero-dimensional. Let, further, $F$ be a closed subgroup of $G$ such that $H \cap F = \{e\}$, where $e$ is the neutral element of $G$. Show that $F$ is zero-dimensional.

3.2.H. Suppose that a locally compact subgroup $H$ of a topological group $G$ is such that the quotient space $G/H$ is normal. Let, further, $F$ be a closed subgroup of $G$ such that $H \cap F = \{e\}$, where $e$ is the neutral element of $G$. Prove that $F$ is locally normal.

3.2.I. Show that every locally normal topological group $G$ is normal.
*Hint.* From Problem 3.2.C it follows that every locally normal topological group $G$ can be covered by a locally finite family of closed normal subspaces. It follows that $G$ is an image of a normal space under a closed continuous mapping, which implies that $G$ is normal.

3.2.J. Prove that every locally metacompact topological group is metacompact.
*Hint.* Apply Problem 3.2.C in the same way as in the solution of Problem 3.2.B.

3.2.K. Show that every locally subparacompact topological group $G$ is subparacompact.

3.2.L. Verify that every locally countably paracompact topological group is countably paracompact.

3.2.M. Let $G$ be a topological group, and $H$ a locally compact subgroup of $G$ such that the quotient space $G/H$ is paracompact (metacompact, subparacompact, countably paracompact, or normal). Show that $G$ is also paracompact (respectively, metacompact, subparacompact, countably paracompact, or normal).

## Open Problems

3.2.1. Is every locally strongly paracompact topological group strongly paracompact?
3.2.2. Is every locally Dieudonné complete topological group Dieudonné complete?
3.2.3. Is every locally paracompact paratopological (quasitopological) group paracompact?
3.2.4. Is every locally normal paratopological (quasitopological) group normal?
3.2.5. Is every locally metacompact (subparacompact) paratopological (quasitopological) group metacompact (subparacompact)?
3.2.6. Let $f$ be an open continuous homomorphism of a paratopological group $G$ onto a paratopological group $F$ with locally compact kernel. Must $f$ be locally perfect?
3.2.7. Suppose that $f$ is an open continuous homomorphism of a paratopological group $G$ onto a paracompact paratopological group $F$ with locally compact fibers. Must $G$ be locally paracompact?

## 3.3. Prenorms on topological groups, metrization

In this section we consider continuous prenorms on topological groups, and present some of their applications. We follow A. A. Markov's approach in this, though we call a *prenorm* what he called a *norm*.

At first, we do not assume that there is any topology on $G$. Let $G$ be a group, with neutral element $e$, and let $N$ be a real-valued function on $G$. We shall call $N$ a *prenorm* on $G$ if the following conditions are satisfied for all $x, y \in G$:

(PN1) $N(e) = 0$;
(PN2) $N(xy) \leq N(x) + N(y)$;
(PN3) $N(x^{-1}) = N(x)$.

PROPOSITION 3.3.1. *If $N$ is a prenorm on $G$, then $N(x) \geq 0$ for each $x \in G$, that is, $N$ is non-negative.*

PROOF. Indeed, since $e = xx^{-1}$, we have from conditions (PN1), (PN2) and (PN3): $0 = N(e) \leq N(x) + N(x^{-1}) = 2N(x)$. Therefore, $0 \leq N(x)$. □

PROPOSITION 3.3.2. *If $N$ is a prenorm on a group $G$, then $|N(x) - N(y)| \leq N(xy^{-1})$, for any $x$ and $y$ in $G$.*

PROOF. From (PN2) we have: $N(y) \leq N(x) + N(x^{-1}y)$. Besides, from (PN2) and (PN3) it follows that $N(x) = N(x^{-1}) \leq N(y^{-1}) + N(x^{-1}y) = N(y) + N(x^{-1}y)$. The two inequalities clearly imply the inequality in our proposition. □

The next two assertions are obvious.

PROPOSITION 3.3.3. *If $N$ is a prenorm on a group $G$ and $\alpha$ is a non-negative real number, then the function $\alpha N$ on $G$ defined by the formula $(\alpha N)(x) = \alpha(N(x))$, for each $x \in G$, is a prenorm on $G$.*

The following simple fact will be used in the next section.

PROPOSITION 3.3.4. *For any prenorm $N$ on a group $G$, the set $Z_N = \{x \in G : N(x) = 0\}$ is a subgroup of $G$.*

PROOF. If $x \in Z_N$, then $N(x) = 0$. Therefore, $N(x^{-1}) = 0$, which implies that $x^{-1} \in G$. Let $x, y \in Z_N$. Then $0 \leq N(xy) \leq N(x) + N(y) = 0$, which implies that $N(xy) = 0$ and $xy \in Z_N$. It follows that $Z_N$ is a subgroup of $G$. □

PROPOSITION 3.3.5. *The sum of two prenorms on a group $G$ is a prenorm on $G$.*

There is a very simple method for constructing prenorms on topological groups. It is described in the following lemma:

LEMMA 3.3.6. *Let $f$ be a bounded real-valued function on a group $G$. Then the function $N_f$ on $G$, defined by the formula:*

$$N_f(x) = \sup\{|f(yx) - f(y)| : y \in G\},$$

*for each $x \in G$, is a prenorm on $G$.*

PROOF. Clearly, condition (PN1) is satisfied. One can also check, though this requires some computation, that conditions (PN2) and (PN3) are satisfied as well. □

In general, a prenorm on a topological group need not be continuous. The next assertion, though simple, is useful.

PROPOSITION 3.3.7. *A prenorm $N$ on a topological group $G$ is continuous if and only if for each positive number $\varepsilon$ there exists a neighbourhood $U$ of the neutral element $e$ such that $N(x) < \varepsilon$, for each $x \in U$.*

PROOF. The necessity is clear. Let us prove the sufficiency. Suppose $z$ is any point of $G$, and $\varepsilon$ is a positive number. Take a neighbourhood $U$ of $e$ such as in Proposition 3.3.7. The set $V = zU$ is an open neighbourhood of $z$. Take any $y \in zU$. Then $z^{-1}y \in U$, and therefore, $N(z^{-1}y) < \varepsilon$. From Proposition 3.3.2 it now follows that $|N(z) - N(y)| < \varepsilon$. Thus, the function $N$ is continuous at $z$. □

Notice that the proof of Proposition 3.3.7 also brings us to the following conclusion:

PROPOSITION 3.3.8. *Every continuous prenorm on a topological group $G$ is a uniformly continuous function with respect to both left and right group uniformities on $G$.*

Now we are going to show how to construct continuous prenorms on a topological group $G$. The importance of the construction lies in the fact that it provides us with a very rich family of continuous prenorms on any topological group.

If $N$ is a prenorm on a group $G$, let us define the *unit ball* of $N$ as the set $B_N = \{x \in G : N(x) < 1\}$. Clearly, if $N$ is a continuous prenorm, then the unit ball $B_N$ is an open subset of $G$. We also put $B_N(\varepsilon) = \{x \in G : N(x) < \varepsilon\}$, where $\varepsilon$ is a positive number; we call $B_N(\varepsilon)$ the *$N$-ball* of radius $\varepsilon$. The sets $B_N(\varepsilon)$ are also open when $N$ is a continuous prenorm.

THEOREM 3.3.9. [**A. A. Markov**] *For each open neighbourhood $U$ of the neutral element $e$ of a topological group $G$, there exists a continuous prenorm $N$ on $G$ such that the unit ball $B_N$ is contained in $U$.*

This theorem will be derived from a more elaborate technical assertion, fixing more relevant details of the situation.

LEMMA 3.3.10. *Let $\{U_n : n \in \omega\}$ be a sequence of open symmetric neighbourhoods of the neutral element e in a topological group G such that $U_{n+1}^2 \subset U_n$, for each $n \in \omega$. Then there exists a prenorm N on G such that the next condition is satisfied:*

(PN4) $\{x \in G : N(x) < 1/2^n\} \subset U_n \subset \{x \in G : N(x) \le 2/2^n\}$.

*Therefore, this prenorm N is continuous. If, in addition, the sets $U_n$ are invariant, then the prenorm N on G can be chosen to satisfy $N(xyx^{-1}) = N(y)$ for all $x, y \in G$.*

PROOF. Put $V(1) = U_0$, fix $n \in \omega$, and assume that open neighbourhoods $V(m/2^n)$ of $e$ are defined for each $m = 1, 2, \dots, 2^n$. Put then $V(1/2^{n+1}) = U_{n+1}$, $V(2m/2^{n+1}) = V(m/2^n)$ for $m = 1, \dots, 2^n$, and

$$V((2m+1)/2^{n+1}) = V(m/2^n) \cdot U_{n+1} = V(m/2^n) \cdot V(1/2^{n+1}),$$

for each $m = 1, 2, \dots, 2^n - 1$. This defines open neighbourhoods $V(r)$ of $e$ for every positive dyadic rational number $r \le 1$. We also put $V(m/2^n) = G$ when $m > 2^n$. It is easy to derive from this definition that the following condition is satisfied:

(p) $V(m/2^n) \cdot V(1/2^n) \subset V((m+1)/2^n)$, for all integers $m > 0$ and $n \ge 0$.

Notice that (p) is obviously true if $m+1 > 2^n$. It remains to consider the case when $m < 2^n$. Let us prove (p) for this case by induction on $n$.

If $n = 1$, then the only possible value for $m$ is also 1, and we have:

$$V(1/2)V(1/2) = U_1^2 \subset U_0 = V(1).$$

Assume that (p) holds for some $n$. Let us verify it for $n+1$. If $m$ is even, then (p) turns into the formula by means of which $V((2m+1)/2^{n+1})$ was defined.

Assume now that $0 < m = 2k + 1 < 2^{n+1}$, for some integer $k$. Then

$$\begin{aligned} V(m/2^{n+1}) \cdot V(1/2^{n+1}) &= V((2k+1)/2^{n+1}) \cdot U_{n+1} \\ &= V(k/2^n) \cdot U_{n+1} \cdot U_{n+1} \subset V(k/2^n) \cdot U_n \\ &= V(k/2^n) \cdot V(1/2^n). \end{aligned}$$

But by the inductive assumption, we have

$$V(k/2^n) \cdot V(1/2^n) \subset V((k+1)/2^n = V((2k+2)/2^{n+1}) = V((m+1)/2^{n+1}),$$

which completes the proof of (p).

Now we define a real-valued function $f$ on $G$ as follows:

$$f(x) = \inf \{r > 0 : x \in V(r)\},$$

for each $x$ in $G$. The function $f$ is well-defined, since $x \in V(2) = G$, for each $x \in G$. From condition (p) it follows that if $0 < r < s$ for positive dyadic rational numbers $r$ and $s$, then $V(r) \subset V(s)$. Let us agree that $r$ and $s$, in the argument below, stand only for positive dyadic rational numbers. Thus, we have:

(1) If $f(x) < r$, then $x \in V(r)$.

Then $f$ is a non-negative function, bounded from above by 2. Therefore, by Lemma 3.3.6, the function $N$ defined by the formula

$$N(x) = \sup_{y \in G} |f(yx) - f(y)|$$

for each $x \in G$, is a prenorm on $G$.

Let us show that $N$ satisfies condition (PN4). Notice that $f(e) = 0$. Assume that $N(x) < 1/2^n$, for some $x \in G$. Then $f(x) = |f(ex) - f(e)| \leq N(x) < 1/2^n$, which implies, by (1), that $x \in V(1/2^n) = U_n$. This proves the first part of (p), namely, that $\{x \in G : N(x) < 1/2^n\} \subset U_n$.

Let us prove the remaining part of (p), which obviously implies the continuity of $N$. Let $x$ be any point of $V(1/2^n)$. Clearly, for any point $y \in G$ there exists a positive integer $k$ such that $(k-1)/2^n \leq f(y) < k/2^n$. Then $y \in V(k/2^n)$, by (1). Since $x \in V(1/2^n)$ and $x^{-1} \in V(1/2^n)$, it follows that $yx$ and $yx^{-1}$ are in $V(k/2^n)V(1/2^n) \subset V((k+1)/2^n)$. Therefore, $f(yx) \leq (k+1)/2^n$ and $f(yx^{-1}) \leq (k+1)/2^n$. From this and the inequality $(k-1)/2^n \leq f(y)$ we obtain: $f(yx) - f(y) \leq 2/2^n$ and $f(yx^{-1}) - f(y) \leq 2/2^n$. Substituting $yx$ for $y$ in the last inequality, we get: $f(y) - f(yx) \leq 2/2^n$. Together with the previous inequality, this implies that $|f(yx) - f(y)| \leq 2/2^n$, for each $y \in G$. Therefore, $N(x) \leq 2/2^n$.

Finally, suppose that the sets $U_n$ are invariant, that is, $xU_nx^{-1} = U_n$ for all $x \in G$ and $n \in \omega$. Since the product of finitely many invariant sets is invariant, it follows that the set $V_r$ is also invariant for each dyadic rational number $r > 0$. In its turn, this implies that $f(xyx^{-1}) = f(y)$ for all $x, y \in G$. Therefore, given elements $x, y \in G$, we obtain the equalities

$$\begin{aligned} N(xyx^{-1}) = \sup_{z\in G} |f(zxyx^{-1}) - f(z)| &= \sup_{z\in G} |f(x^{-1}zxy) - f(z)| \\ &= \sup_{t\in G} |f(ty) - f(xtx^{-1})| \\ &= \sup_{t\in G} |f(ty) - f(t)| = N(y), \end{aligned}$$

where $t = x^{-1}zx$. To write the second line of the above equalities we use the fact that the conjugation mapping $\varphi\colon G \to G$ defined by $\varphi(z) = x^{-1}zx$ for each $z \in G$ is a bijection of $G$ onto itself. This finishes the proof. □

Proof of Theorem 3.3.9. Using the axioms of a topological group, one easily constructs a sequence of open neighbourhoods $\{U_n : n \in \omega\}$ of the identity $e$ in $G$ satisfying all the conditions in Lemma 3.3.10 and such that $U_0 = U$. Take a prenorm $N$ on $G$ satisfying (PN4) of Lemma 3.3.10. Then $N$ is continuous and the unit ball $B_N$ of $N$ is contained in $U_0 = U$. □

A semitopological group $G$ will be called *left uniformly Tychonoff* if for every open neighbourhood $V$ of the neutral element $e$, there exists a left uniformly continuous function $f$ on $G$ such that $f(e) = 0$ and $f(x) \geq 1$, for each $x \in G \setminus V$. Similarly, one defines the concept of a *right uniformly Tychonoff* semitopological group. Clearly, if $G$ is left (or right) uniformly Tychonoff, then the space $G$ is Tychonoff. Finally, if for every open neighbourhood $V$ of the neutral element $e$ in a semitopological group $G$, there exists a real-valued function $f$ on $G$ satisfying $f(e) = 0$ and $f(x) \geq 1$ for each $x \in G \setminus V$ which is simultaneously left and right uniformly continuous, then $G$ will be called *uniformly Tychonoff*.

Theorem 3.3.11. *Every topological group $G$ is uniformly Tychonoff.*

Proof. Let $U$ be any open neighbourhood of the identity $e$ in $G$. By Theorem 3.3.9, there exists a continuous prenorm $N$ on $G$ such that $B_N \subset U$. Then we have that $N(x) = 0$

and $N(x) \geq 1$, for each $x \in G \setminus U$. Since every continuous prenorm is left and right uniformly continuous by Proposition 3.3.8, $G$ is uniformly Tychonoff. □

Here is another important application of Lemma 3.3.10 — a very quick proof of the Birkhoff–Kakutani metrization theorem for topological groups. It is worth noting that Theorem 3.3.9 does not work so well in this situation compared to Lemma 3.3.10.

THEOREM 3.3.12. [**G. Birkhoff**, **S. Kakutani**] *A topological group $G$ is metrizable if and only if it is first-countable.*

PROOF. The necessity is obvious. Let us prove the sufficiency. Fix a countable base $\{W_n : n \in \omega\}$ of the space $G$ at the point $e$. By induction, we obtain a sequence $\{U_n : n \in \omega\}$ of symmetric open neighbourhoods of $e$ such that $U_n \subset W_n$ and $U_{n+1}^2 \subset U_n$, for each $n \in \omega$. This sequence is also a base of $G$ at $e$. By Lemma 3.3.10, there exists a continuous prenorm $N$ on $G$ such that $B_N(1/2^n) \subset U_n$ for each $n \in \omega$. It follows that the open sets $B_N(1/2^n)$ also form a base of $G$ at $e$.

Now, for arbitrary $x$ and $y$ in $G$, put $\varrho_N(x, y) = N(xy^{-1})$. Let us show that $\varrho_N$ is a metric on $G$ generating the original topology on $G$. Clearly, $\varrho_N(x, y) = N(xy^{-1}) \geq 0$, for all $x, y \in G$. It is also clear that $\varrho_N(x, x) = 0$, for each $x \in G$. Assume now that $\varrho_N(x, y) = 0$. Then $xy^{-1} \in B_N(1/2^n) \subset U_n$, for each $n \in \omega$. Since $\{e\} = \bigcap_{n\in\omega} U_n$, it follows that $xy^{-1} = e$, that is, $x = y$.

Let us verify the triangle inequality. Take any three points $x$, $y$, and $z$ in $G$. Then we have:

$$\begin{aligned} \varrho_N(x, z) &= N(xz^{-1}) \\ &= N(xy^{-1}yz^{-1}) \leq N(xy^{-1}) + N(yz^{-1}) \\ &= \varrho_N(x, y) + \varrho_N(y, z). \end{aligned}$$

Thus, $\varrho_N$ is a metric on $G$.

Notice that the metric $\varrho_N$ is *right-invariant*, that is, $\varrho_N(x, y) = \varrho_N(xz, yz)$, for all $x, y, z \in G$. Indeed,

$$\varrho_N(xz, yz) = N(xzz^{-1}y^{-1}) = N(xy^{-1}) = \varrho_N(x, y).$$

Since $B_N(\varepsilon)$ is obviously the spherical $\varrho_N$-neighbourhood of $e$ of radius $\varepsilon$, it follows that the spherical $\varrho_N$-neighbourhood of any point $x$ of $G$ of radius $\varepsilon$ is precisely the set $B_N(\varepsilon)x$. Take any point $x \in G$. Since the sets $B_N(1/2^n)$ form a base of $G$ at $e$, and $G$ is a topological group, the sets $B_N(1/2^n)x$ constitute a base of $G$ at $x$, that is, the spherical $\varrho_N$-neighbourhoods of $e$ of radius $1/2^n$ form a base of the space $G$ at the point $x$. Thus, the metric $\varrho_N$ generates the original topology of the space $G$, that is, $G$ is metrizable. □

One can complement Theorem 3.3.12 as follows:

COROLLARY 3.3.13. *Every first-countable topological group $G$ admits a right-invariant metric $\varrho$ and a left-invariant metric $\lambda$, both generating the original topology of $G$.*

PROOF. Take a continuous prenorm $N$ on $G$ as in the proof of Theorem 3.3.12 and put $\varrho(x, y) = N(xy^{-1})$ and $\lambda(x, y) = N(x^{-1}y)$ for all $x, y \in G$. As it was shown above, $\varrho$ is a right-invariant metric on $G$ that generates the topology of $G$. Since the inverse on $G$ is a homeomorphism of $G$ onto itself, a similar assertion for $\lambda$ is clear. □

Not every metrizable topological group admits an invariant metric that generates the topology of the group. The groups with this property are necessarily balanced:

COROLLARY 3.3.14. *A metrizable topological group $G$ admits an invariant metric generating its topology if and only if $G$ is balanced.*

PROOF. Suppose that $\varrho$ is an invariant metric on $G$ that generates the topology of $G$. For every $n \in \mathbb{N}$, denote by $U_n$ the $1/n$-ball with center at the neutral element $e$ of $G$ with respect to $\varrho$. If $x \in U_n$ and $y \in G$ are arbitrary elements, then the two sides invariance of $\varrho$ implies that

$$\varrho(e, yxy^{-1}) = \varrho(y, yx) = \varrho(e, x) < 1/n,$$

whence $yxy^{-1} \in U_n$. It follows that $yU_ny^{-1} = U_n$ for all $y \in G$ and $n \in \mathbb{N}$, so the family $\{U_n : n \in \mathbb{N}\}$ is an invariant base for $G$ at the neutral element $e$. Hence, the group $G$ is balanced.

Conversely, suppose that the group $G$ is balanced. Since $G$ is first-countable, there exists a sequence $\xi = \{U_n : n \in \omega\}$ of open, symmetric, invariant neighbourhoods of $e$ in $G$ satisfying $U_{n+1}^2 \subset U_n$ for each $n \in \omega$ and such that $\xi$ forms a local base for $G$ at $e$. Therefore, by Lemma 3.3.10, we can find a prenorm $N$ on $G$ satisfying (PN4) of the lemma and the equality $N(xyx^{-1}) = N(y)$ for all $x, y \in G$. Then $N$ is continuous and the open balls $B_N(1/n) = \{x \in G : N(x) < 1/n\}$, with $n \in \mathbb{N}$, form a base for $G$ at the neutral element $e$. Hence the metric $d$ on $G$ defined by $d(x, y) = N(x^{-1}y) = N(xy^{-1})$ is invariant and generates the topology of $G$. □

Applying Theorem 3.3.9 in almost the same way as in the proof of Theorem 3.3.12, we obtain the following important result:

THEOREM 3.3.15. *Every Abelian topological group $G$ is topologically isomorphic to a subgroup of the product of some family of metrizable Abelian topological groups.*

PROOF. Let $U$ be an open neighbourhood of the neutral element $e$ of $G$. According to Theorem 3.3.9, we can fix a continuous prenorm $N_U$ on $G$ such that the unit ball with respect to $N_U$ is contained in $U$.

Put $H_U = \{x \in G : N_U(x) = 0\}$. Since $N_U$ is continuous, $H_U$ is closed in $G$. From conditions (PN1), (PN2), and (PN3) it follows that $H_U$ is a subgroup of $G$. Since $G$ is Abelian, it follows that the quotient set $G_U = G/H_U$ is an Abelian group. We denote by $f_U$ the natural quotient homomorphism of $G$ onto $G_U$, and define a function $P_U$ on the group $G_U$ as follows: $P_U(y) = N_U(x)$, where $x$ is any element of $f_U^{-1}(y)$. Obviously, $P_U(y)$ does not depend on the choice of $x$ in $f_U^{-1}(y)$. Then $P_U$ is a prenorm on the group $G_U$, and $P_U(y) = 0$ if and only if $y$ is the neutral element $e_U$ of $G_U$.

It follows from conditions (PN1), (PN2), and (PN3) that the $1/n$-balls with respect to the prenorm $P_U$ form a base of a topology $\mathcal{T}_U$ on $G_U$ with respect to which $G_U$ is a topological group and the mapping $f_U$ is continuous. It is also clear that the preimage of the 1-ball with respect to $P_U$ is contained in $U$. Note that $G_U$ is first-countable and, therefore, metrizable.

By the standard Tychonoff type argument, it follows that the diagonal product of the mappings $f_U$, where $U$ runs over a basic family $\mathcal{B}$ of open neighbourhoods of $e$ in the group $G$, is a topological isomorphism of $G$ onto a subgroup of the product $\Pi_{U\in\mathcal{B}} G_U$. Since every $G_U$ is an Abelian metrizable group, this completes the argument. □

We shall see later that Theorem 3.3.15 is no longer valid in the non-Abelian case. Notice that the proof of Theorem 3.3.12, practically without any change, can be applied to the situation when the neutral element of a topological group $G$ is a $G_\delta$-set in $G$. Then the metric $\varrho$ on $G$ constructed in the same way as above no longer generates the original topology of $G$ but a weaker topology. This weaker topology is still left-invariant and, therefore, the left translations are homeomorphisms; but in general it need not be a group topology. Thus, we have:

THEOREM 3.3.16. *If $G$ is a topological group such that its neutral element is a $G_\delta$-set in $G$, then there exists a weaker metrizable topology on $G$ with respect to which $G$ is topologically homogeneous (by means of left translations).*

COROLLARY 3.3.17. *If $G$ is a topological group such that the singleton $e$ is a $G_\delta$-set, then every compact subspace $F$ of the space $G$ is metrizable.*

PROOF. Apply Theorem 3.3.16 together with the fact that every one-to-one continuous mapping of a compact space onto a metrizable space is a homeomorphism. □

Theorem 3.3.12 has important applications to quotients groups.

COROLLARY 3.3.18. *Suppose that $f$ is an open continuous homomorphism of a metrizable topological group $G$ onto a topological group $H$. Then $H$ is also metrizable.*

PROOF. Since $f$ is open and continuous, and the space $G$ is first-countable, the space $H$ is also first countable. It remains to apply Theorem 3.3.12. □

We note in connection with Corollary 3.3.18 that metrizability of topological spaces, in general, is not preserved by open continuous mappings. Indeed, every first-countable space can be represented as an image of a metrizable space under an open and continuous mapping [**165**, 4.2.D].

Here is a generalization of Corollary 3.3.18 to quotient spaces of metrizable topological groups:

PROPOSITION 3.3.19. *Let $H$ be a closed subgroup of a metrizable topological group $G$. Then the quotient space $G/H$ is also metrizable.*

PROOF. By Corollary 3.3.13, there exists a right-invariant metric $d$ on $G$ which generates the topology of $G$. For arbitrary points $x, y \in G$, define a number $\varrho(xH, yH)$ by the rule:

$$\varrho(xH, yH) = \inf\{d(xh_1, yh_2) : h_1, h_2 \in H\}.$$

Since $d$ is right-invariant, we have that $\varrho(xH, yH) = d(x, yH) \geq 0$ for all $x, y \in G$. The function $\varrho$ is symmetric:

$$\begin{aligned}\varrho(yH, xH) = d(y, xH) &= \inf_{h\in H} d(y, xh) = \inf_{h\in H} d(yh^{-1}, x) \\ &= \inf_{h\in H} d(x, yh^{-1}) = d(x, yH) = \varrho(xH, yH).\end{aligned}$$

Since $H$ is closed in $G$, we also have that $\varrho(xH, yH) = d(x, yH) = 0$ if and only if $x \in yH$, that is, $xH = yH$. Let us verify that the function $\varrho$ on $G/H$ satisfies the triangle inequality and, hence, $\varrho$ is a metric.

Suppose that $x, y, z$ are arbitrary points of $G$ and $\varepsilon > 0$ is a real number. By the definition of $\varrho$, we can find $h_1, h_2 \in H$ such that $\varrho(xH, yH) < d(x, yh_1) + \varepsilon/2$ and $\varrho(yH, zH) < d(y, zh_2) + \varepsilon/2$. We then have:

$$\begin{aligned}\varrho(xH, zH) &\leq d(x, zh_2h_1)\\ &\leq d(x, yh_1) + d(yh_1, zh_2h_1) = d(x, yh_1) + d(y, zh_2)\\ &< \varrho(xH, yH) + \varrho(yH, zH) + \varepsilon.\end{aligned}$$

Since $\varepsilon$ is an arbitrary positive number, the above inequality implies that $\varrho(xH, zH) \leq \varrho(xH, yH) + \varrho(yH, zH)$. Thus, $\varrho$ is a metric.

To finish the proof, we have to show that $\varrho$ generates the topology of the quotient space $G/H$. For $x \in G$ and $\varepsilon > 0$, let

$$O_\varepsilon(x) = \{y \in G : d(x, y) < \varepsilon\} \text{ and}$$
$$B_\varepsilon(xH) = \{yH : y \in G,\ \varrho(xH, yH) < \varepsilon\}.$$

Denote by $\pi$ the quotient mapping of $G$ onto $G/H$, $\pi(x) = xH$ for each $x \in G$. It follows from the definition of the metric $\varrho$ that $\pi(O_\varepsilon(x)) = B_\varepsilon(xH)$ for all $x \in G$ and $\varepsilon > 0$. Since the sets $O_\varepsilon(x)$ form a base for $G$ and the mapping $\pi\colon G \to G/H$ is continuous and open, we conclude that the sets $B_\varepsilon(xH)$ constitute a base for the original topology of the space $G/H$. The proof is complete. □

Suppose that $H$ is a closed invariant subgroup of a topological group $G$, and $G/H$ is the corresponding quotient group. Then $G$ is called an *extension of the group $H$ by $G/H$*. It turns out that metrizability is stable with respect to extensions of topological groups:

COROLLARY 3.3.20. [**N. Ya. Vilenkin**] *Suppose that $G$ is a topological group, and $H$ a closed metrizable subgroup of $G$ such that the quotient space $G/H$ is first-countable. Then $G$ is also metrizable.*

PROOF. By Corollary 1.5.21, the space $G$ is first-countable. It remains to refer to Theorem 3.3.12. □

COROLLARY 3.3.21. *Suppose that $G$ is a topological group, $H$ is a second-countable subgroup of $G$, and the quotient space $G/H$ is second-countable. Then $G$ is also second-countable.*

PROOF. The space $G$ is separable by Theorem 1.5.23, and Corollary 3.3.20 tells us that $G$ is metrizable. □

Reformulating Corollary 3.3.20, we can say that extensions of topological groups preserve metrizability. Similarly, Corollary 1.5.8 implies that extensions of topological groups preserve compactness. In particular, an extension of a compact metrizable group by another such a group is also compact and metrizable. This fact admits a natural generalization via considering compact metrizable subspaces of topological groups (see Theorem 3.3.24 below).

We start with two preliminary results; the first of them is interesting by itself.

LEMMA 3.3.22. *The following conditions are equivalent for a topological group $G$:*

a) *every compact subspace of $G$ is first-countable;*
b) *every compact subspace of $G$ is metrizable.*

Proof. It suffices to show that a) implies b). Suppose that $X$ is a non-empty compact subset of $G$. Consider the mapping $j\colon G \times G \to G$ defined by $j(x, y) = x^{-1}y$ for all $x, y \in G$. Clearly, $j$ is continuous, so the image $F = j(X \times X)$ is a compact subset of $G$ which contains the identity $e$ of $G$. By our assumption, the space $F$ is first-countable, so $\chi(e, F) \leq \omega$. Denote by $f$ the restriction of $j$ to $X \times X$. Then $f^{-1}(e) = \Delta_X$ is the diagonal in $X \times X$. Since $f$ is a closed mapping, we have $\chi(\Delta, X \times X) = \chi(e, F) \leq \omega$. Therefore, the compact space $X$ is metrizable by [**165**, 4.2.B]. □

Now we need the following property of closed continuous mappings.

Lemma 3.3.23. *Let $f\colon X \to Y$ be a closed continuous mapping of regular spaces. Suppose that a point $x \in X$ satisfies $\chi(f(x), Y) \leq \omega$ and $\chi(x, C) \leq \omega$, where $C = f^{-1}f(x)$. Then $\chi(x, X) \leq \omega$.*

Proof. Let $\{U_i : i \in \omega\}$ be a countable base for $C = f^{-1}f(x)$ at the point $x$, and let $\{V_j : j \in \omega\}$ be a countable base for $Y$ at $y = f(x)$. For every $i \in \omega$, choose an open neighbourhood $O_i$ of $x$ in $X$ such that $\overline{O_i} \cap (C \setminus U_i) = \varnothing$. Then $\overline{O_i} \cap C \subset U_i$. Let us verify that the family $\{O_i \cap f^{-1}(V_j) : i, j \in \omega\}$ is a base for $X$ at $x$. Suppose that $U$ is an arbitrary open neighbourhood of $x$ in $X$. Choose $i \in \omega$ such that $U_i \subset C \cap U$. Then $F = \overline{O_i} \setminus U$ is a closed subset of $X$ disjoint from $C$, so $K = f(F)$ is closed in $Y$ and $y \notin K$. Choose $j \in \omega$ with $V_j \cap K = \varnothing$. Then $O_i \cap f^{-1}(V_j) \subset O_i \setminus F \subset U$, as required. □

Theorem 3.3.24. *Let $H$ be a closed subgroup of a topological group $G$, and suppose that all compact subspaces of $H$ and $G/H$ are metrizable. Then all compact subspaces of $G$ are metrizable as well.*

Proof. Denote by $\pi$ the canonical mapping of $G$ onto the quotient space $G/H$ of left cosets. We claim that all compact subsets of the fibers of $\pi$ are metrizable. Indeed, let $y \in G/H$ be arbitrary. Choose a point $x \in G$ such that $\pi(x) = y$. Then the fiber $\pi^{-1}(y) = xH$ is homeomorphic to the group $H$, whence our claim follows.

Given a compact subset $X$ of $G$, let $f$ be the restriction of $\pi$ to $X$. The compact subspace $Y = f(X)$ of the space $G/H$ is metrizable by our hypothesis, and all compact subsets of the fibers of $f$ are metrizable since they lie in the fibers of $\pi$. Therefore, Lemma 3.3.23 implies that all compact subsets of $G$ are first-countable, and the metrizability of $X$ follows from Lemma 3.3.22. □

Let $\mathcal{P}$ be a topological or algebraic property. It is said that $\mathcal{P}$ is a *three space property* if, given an arbitrary topological group $G$ and a closed invariant subgroup $H$ of $G$ such that both $H$ and $G/H$ have $\mathcal{P}$, it follows that $G$ also has $\mathcal{P}$. Now we can reformulate Corollaries 3.3.20, 3.3.21, and Theorem 3.3.24 by saying that the following are three space properties: a) metrizability; b) to be second-countable; c) "all compact subsets are metrizable".

## Exercises

3.3.a. Show that a first-countable paratopological group need not be metrizable.

3.3.b. Prove that every compact subspace of the Sorgenfrey line is countable (hence, metrizable).

3.3.c. Show that the Sorgenfrey line is not homeomorphic to any topological group.

3.3.d. Verify that the group $GL(2, \mathbb{R})$ does not admit a continuous invariant metric that generates the topology of this group.

3.3.e. Prove that the topology of any compact first-countable group is generated by a continuous invariant metric.

3.3.f. Give an explicit formula for a left-invariant continuous metric that generates the topology of the subgroup

$$H = \left\{ \begin{pmatrix} a & b \\ 0 & d \end{pmatrix} : a, b, c \in \mathbb{R},\ ad = 1 \right\}$$

of the group $GL(2, \mathbb{R})$.

3.3.g. Let $H$ be a closed subgroup of a topological group $G$ and suppose that $d$ is a right-invariant continuous pseudometric on $G$. Define $\sigma$ on the quotient space $G/H$ by

$$\sigma(xH, yH) = \inf\{d(a, b) : a \in xH,\ b \in yH\}.$$

Prove that $\sigma$ is a continuous pseudometric on $G/H$. Use this fact to give an alternative proof of Corollary 3.3.18.

3.3.h. Let $H$ be a closed subgroup of a topological group $G$, and suppose that all compact subsets of the group $H$ and of the quotient space $G/H$ are finite. Prove that all compact subsets of $G$ are finite as well. Extend this result to countably compact sets.

3.3.i. Give an example of a topological group $G$ algebraically generated by a discrete subspace such that $G$ contains a compact set $K$ with $\chi(K) > \omega$. (See Problem 3.3.6.)

## Problems

3.3.A. Give an example of a countable non-discrete topological group $G$ without non-trivial convergent sequences (notice that such a group $G$ cannot be metrizable).

3.3.B. Does there exist an infinite topological group $G$ such that all metrizable subgroups of $G$ are finite? Can such a group be connected or locally connected?

3.3.C. Show that the Sorgenfrey line does not admit an open continuous homomorphism onto a metrizable paratopological group with metrizable fibers.

3.3.D. Let $H$ be a closed invariant subgroup of a topological group $G$ and suppose that both groups $H$ and $G/H$ admit continuous invariant metrics generating their topologies. Does $G$ then have the same property?

3.3.E. Prove that the group $H$ in Exercise 3.3.f does not admit an invariant metric that generates the topology of $H$.
*Hint.* Define two sequences $\{x_n : n \in \omega\}$ and $\{y_n : n \in \omega\}$ of pairwise distinct elements of the group $H$ such that the sequence $\{x_n y_n : n \in \omega\}$ converges to the neutral element of $H$, while $\{y_n x_n : n \in \omega\}$ is a closed discrete subset of $H$.

3.3.F. Let $H$ be a closed subgroup of a topological group $G$. Prove that the quotient space $G/H$ is Tychonoff.
*Hint.* Let $\pi\colon G \to G/H$ be the natural projection. Since the space $G/H$ is homogeneous, it suffices to find, for a given neighbourhood $U$ of the neutral element $e$ in $G$, a continuous function $\sigma$ on $G/H$ such that $\sigma(\pi(e)) = 0$ and $\sigma(y) = 1$, for each $y \in G/H \setminus \pi(U)$. To construct such a function $\sigma$, apply Exercise 3.3.g.

3.3.G. Let $G$ be a topological group. Prove that if the product group $G \times G$ is functionally balanced (see Problem 1.8.B), then $G$ is balanced.
*Hint.* Let $U$ be a symmetric open neighbourhood of the identity $e$ in $G$. Apply Theorem 3.3.9 to choose a continuous prenorm $N$ on $G$ such that $N(x) < 1$, for each $x \in U$ and $N(x) = 1$ for each $x \in G \setminus U$. Then the function $\varrho$ defined by $\varrho(x, y) = N(x^{-1}y)$ for $x, y \in G$ is a continuous left-invariant pseudometric on $G$ bounded by 1. In particular, $\varrho$ is left uniformly continuous. Since $G \times G$ is functionally balanced, $\varrho$ is right uniformly continuous and, therefore, there exists an open symmetric neighbourhood $V$ of $e$ in $G$ such

that $|\varrho(x, vx) - \varrho(x, x)| < 1$, for all $x \in G$ and $v \in V$. Deduce that $yVy^{-1} \subset U$ for each $y \in G$ and, therefore, $G$ is balanced.

3.3.H. Let $H_1$ and $H_2$ be compact topological groups. We can assume without loss of generality that the only common element of the groups $H_1$ and $H_2$ is the identity of both groups. Prove that there exists a Hausdorff topological group $G$ satisfying the following conditions:

i) $G$ contains $H_1$ and $H_2$ as topological subgroups;
ii) the set $H_1 \cup H_2$ algebraically generates $G$;
iii) given a topological group $K$ and continuous homomorphisms $\varphi_1 \colon H_1 \to K$ and $\varphi_2 \colon H_2 \to K$, there exists a continuous homomorphism $\varphi \colon G \to K$ such that $\varphi \restriction H_i = \varphi_i$, for $i = 1, 2$.

The group $G$ is called the *free topological product* of $H_1$ and $H_2$. Prove that the group $G$ is metrizable iff one of the groups $H_1$, $H_2$ is trivial and the other is metrizable.

3.3.I. Show that a topological group algebraically generated by two separable metrizable subgroups is not necessarily metrizable, even if the generating subgroups are compact.

3.3.J. Show that a topological group $G$ algebraically generated by two metrizable subgroups may fail to be paracompact. Is $G$ necessarily a normal space?

## Open Problems

3.3.1. Suppose that $G$ is a Hausdorff (regular) paratopological group in which every point is a $G_\delta$-set. Is $G$ submetrizable? (A space is said to be submetrizable if the topology of the space contains a metrizable topology.)

3.3.2. Let $f \colon G \to F$ be an open continuous homomorphism of a metrizable paratopological group $G$ onto a Hausdorff paratopological group $F$. Must $F$ then be metrizable?

3.3.3. Can the Sorgenfrey line, considered as a paratopological group, be represented as an image of a metrizable paratopological group under an open continuous homomorphism?

3.3.4. Let $G$ be a paratopological group such that every compact subspace of $G$ is first-countable. Is every compact subspace of $G$ metrizable?

3.3.5. Let $G$ be a quasitopological group such that every compact subspace of $G$ is first-countable. Is every compact subspace of $G$ metrizable?

3.3.6. Let $G$ be a topological group algebraically generated by two metrizable subgroups. Is every compact (countably compact) subspace of $G$ metrizable?

3.3.7. Let $G$ be a topological group algebraically generated by two metrizable subgroups. Is $G$ subparacompact? What if $G$ is Abelian? (See Problems 3.3.I and 3.3.J.)

3.3.8. Let $f \colon G \to F$ be an open continuous homomorphism of a paratopological group $G$ onto a metrizable paratopological group $F$, and suppose that the kernel of $f$ is metrizable. Must $G$ be metrizable?

3.3.9. Characterize the paratopological groups that admit an open continuous homomorphism with metrizable kernel onto a metrizable paratopological group.

3.3.10. Let $S$ be the Sorgenfrey line. Is there a cardinal number $\tau > \omega$ such that $S^\tau$ is homeomorphic to a topological group? Notice that $S^\omega$ is not homeomorphic to any topological group.

3.3.11. Is the Sorgenfrey line homeomorphic to any quasitopological group?

## 3.4. $\omega$-narrow and $\omega$-balanced topological groups

In this section we study the class of $\omega$-narrow topological groups. These groups are characterized as subgroups of topological products of (possibly, uncountable) families of second-countable topological groups.

We recall that a left semitopological group $G$ is called *$\omega$-narrow* if, for every open neighbourhood $V$ of the neutral element $e$ in $G$, there exists a countable subset $A$ of $G$ such that $AV = G$ (see Section 2.3).

The following proposition shows that the algebraic asymmetry of the above definition disappears in the case of quasitopological groups:

PROPOSITION 3.4.1. *The following conditions are equivalent for a quasitopological group $G$:*

1) *$G$ is $\omega$-narrow;*
2) *For every open neighbourhood $V$ of the neutral element $e$ in $G$, there exists a countable set $B \subset G$ such that $G = VB$;*
3) *For every open neighbourhood $V$ of the neutral element $e$ in $G$, there exists a countable set $C \subset G$ such that $CV = G = VC$.*

PROOF. It is clear that both 1) and 2) follow from 3). Also, 3) follows from the conjunction of 1) and 2) since the equalities $G = AV$ and $G = VB$ imply that $CV = G = VC$, where $C = A \cup B$. Clearly, the set $C$ is countable if so are $A$ and $B$.

It remains to show that 1) and 2) are equivalent. Let $G$ be an $\omega$-narrow quasitopological group. Given an open neighbourhood $V$ of $e$, one can find an open neighbourhood $U$ of $e$ such that $U^{-1} \subset V$. Choose a countable set $A \subset G$ such that $AU = G$. Then the countable set $B = A^{-1}$ satisfies $G = G^{-1} = (AU)^{-1} = U^{-1}A^{-1} \subset VB$, that is, $G = VB$. This proves the implication 1) $\Rightarrow$ 2). Inverting the above argument, one obtains the implication 2) $\Rightarrow$ 1). Therefore, the three conditions on the group $G$ are equivalent. □

The following two statements are almost obvious, so we leave their proofs to the reader.

PROPOSITION 3.4.2. *If a topological group $H$ is a continuous homomorphic image of an $\omega$-narrow topological group $G$, then $H$ is also $\omega$-narrow.*

PROPOSITION 3.4.3. *The topological product of an arbitrary family of $\omega$-narrow topological groups is an $\omega$-narrow topological group.*

It is clear that Propositions 3.4.2 and 3.4.3 remain valid for left semitopological groups. The next result is just a trifle less trivial, but we shall prove it for the sake of completeness.

THEOREM 3.4.4. *Every subgroup $H$ of an $\omega$-narrow topological group $G$ is $\omega$-narrow.*

PROOF. Let $W$ be an open neighbourhood of the identity $e$ in $H$. Choose an open symmetric neighbourhood $V$ of $e$ in $G$ such that $V^2 \cap H \subset W$. Since $G$ is $\omega$-narrow, there exists a countable subset $B$ of $G$ such that $BV = G$. Let $C$ be the set of all $c \in B$ such that $cV \cap H$ is not empty. Then $|C| \leq |B| \leq \omega$ and, obviously, $H \subset CV$. For each $c \in C$ fix $a_c \in cV \cap H$, and put $A = \{a_c : c \in C\}$. Since $C$ is countable, $A$ is a countable subset of $H$. We claim that $AW = H$.

Indeed, since $H$ is a subgroup of $G$ and $V^2 \cap H \subset W \subset H$, we have: $(AV^2) \cap H \subset AW$. It remains to show that $H \subset AV^2$. Clearly, $A \subset H \subset CV$. Since $V$ is symmetric, it follows that $C \subset AV$, which implies that $H \subset CV \subset AV^2$. The proof is complete. □

The above results show that the class of $\omega$-narrow topological groups is extremely stable under operations, which is, of course, a great advantage. This stability also implies that the class must be rather wide. But how wide is it, really? Which other naturally defined classes of topological groups are contained in it? Here are some results in this direction.

First, it is easy to give an example of a topological group which is not $\omega$-narrow — just take any uncountable discrete group. This simple observation can be given a more general form:

PROPOSITION 3.4.5. *Every first-countable $\omega$-narrow topological group has a countable base.*

PROOF. Let $\{U_n : n \in \omega\}$ be a countable base at the identity $e$ of an $\omega$-narrow topological group $G$. For every $n \in \omega$, choose a countable set $C_n \subset G$ such that $C_n \cdot U_n = G$. Then the family $\mathcal{B} = \{xU_n : x \in C_n,\ n \in \omega\}$ is countable, and we claim that $\mathcal{B}$ is a base for the group $G$.

Indeed, let $O$ be a neighbourhood of a point $a \in G$. One can find $k, l \in \omega$ such that $aU_k \subset O$ and $U_l^{-1} U_l \subset U_k$. There exists $x \in C_l$ such that $a \in xU_l$, whence $x \in aU_l^{-1}$. We have

$$xU_l \subset (aU_l^{-1})U_l = a(U_l^{-1}U_l) \subset aU_k \subset O,$$

that is, $xU_l$ is an open neighbourhood of $a$ and $xU_l \subset O$. It remains to note that $xU_l \in \mathcal{B}$. □

Note that the Sorgenfrey line is a first-countable $\omega$-narrow paratopological group of uncountable weight, so Proposition 3.4.5 cannot be extended to paratopological groups.

The following statement is obvious:

PROPOSITION 3.4.6. *Every Lindelöf topological group is $\omega$-narrow.*

THEOREM 3.4.7. *If the cellularity of a topological group $G$ is countable, then $G$ is $\omega$-narrow.*

PROOF. Let $U$ be an open neighbourhood of the neutral element $e$. Take a symmetric open neighbourhood $V$ of $e$ such that $V^2 \subset U$. Recall that a subset $P$ of $G$ is called *$V$-disjoint* if $xV \cap yV = \varnothing$, for every distinct points $x, y \in P$ (see page 31).

The family $\mathcal{E}$ of all $V$-disjoint subsets of $G$ is (partially) ordered by inclusion, and the union of any chain of $V$-disjoint sets is again a $V$-disjoint set. Therefore, according to Zorn's Lemma, there exists a maximal element $A$ of the ordered set $\mathcal{E}$. Clearly, $\{aV : a \in A\}$ is a disjoint family of non-empty open sets in $G$; since the Souslin number of $G$ is countable, it follows that this family is countable, that is, the set $A$ is countable.

It follows from the maximality of $A$ that for every $x \in G$, there exists $a \in A$ such that the set $xV \cap aV$ is not empty. Then $x \in aVV^{-1} = aV^2 \subset aU$. Therefore, $AU = G$. □

COROLLARY 3.4.8. *Every separable topological group is $\omega$-narrow.*

THEOREM 3.4.9. *If a topological group $G$ contains a dense subgroup $H$ such that $H$ is $\omega$-narrow, then $G$ is also $\omega$-narrow.*

PROOF. Let $U$ be any open neighbourhood of $e$ in $G$. There exists a symmetric open neighbourhood $V$ of $e$ in $G$ such that $V^2 \subset U$. Since $H$ is $\omega$-narrow, we can find a countable subset $A$ of $H$ such that $H \subset AV$. Then, by Proposition 1.4.4, $G = \overline{H} \subset AVV \subset AU$. Therefore, $AU = G$, and $G$ is $\omega$-narrow. □

Let us say that the *invariance number* $inv(G)$ of a semitopological group $G$ is countable (notation: $inv(G) \leq \omega$) if for each open neighbourhood $U$ of the neutral element $e$ in $G$, there exists a countable family $\gamma$ of open neighbourhoods of $e$ such that for each $x \in G$, there exists $V \in \gamma$ satisfying $xVx^{-1} \subset U$. Any such family $\gamma$ will be called *subordinated to* $U$. Topological groups $G$ such that $inv(G) \leq \omega$ are also called *$\omega$-balanced.* Clearly, that every subgroup of an $\omega$-balanced group is also $\omega$-balanced.

The next assertion turns out to be most helpful.

PROPOSITION 3.4.10. *If $G$ is an $\omega$-narrow topological group, then the invariance number of $G$ is countable, that is, $G$ is $\omega$-balanced.*

PROOF. Let $U$ be an open neighbourhood of the neutral element $e$ in $G$. There exists a symmetric open neighbourhood $V$ of $e$ such that $V^3 \subset U$. Since $G$ is $\omega$-narrow, we can find a countable subset $A$ of $G$ such that $VA = G$. Then for each $a \in A$, there exists an open neighbourhood $W_a$ of the neutral element $e$ such that $aW_aa^{-1} \subset V$. We claim that $\gamma = \{W_a : a \in A\}$ is the family we are looking for.

Indeed, $\gamma$ is a countable family of open neighbourhoods of $e$. Now, let $x$ be any element of $G$. Then $x \in Va$, for some $a \in A$, and therefore, $xW_ax^{-1} \subset VaW_aa^{-1}V^{-1} \subset VVV^{-1} \subset V^3 \subset U$, that is, $\gamma$ is subordinated to $U$. □

The converse to the previous statement is not true. Indeed, every discrete group is obviously $\omega$-balanced, while a discrete group is $\omega$-narrow if and only if it is countable.

THEOREM 3.4.11. *The invariance number of an arbitrary first-countable semitopological group $G$ is countable.*

PROOF. Let $\{V_n : n \in \omega\}$ be a countable base of the space $G$ at the neutral element $e$ of the group $G$. Take any open neighbourhood $U$ of $e$. Then $Ux$ is an open neighbourhood of $x$. Since the left translation by $x$ is continuous and $xe \in Ux$, there exists $n \in \omega$ such that $xV_n \subset Ux$. It follows that $xV_nx^{-1} \subset Uxx^{-1} = U$. Hence, $inv(G) \leq \omega$. □

COROLLARY 3.4.12. *Every metrizable topological group is $\omega$-balanced.*

Here is one more technical result we need to prove before a basic fact on continuous prenorms on $\omega$-narrow topological groups will be established.

LEMMA 3.4.13. *Let $G$ be an $\omega$-balanced topological group, and let $\gamma$ be a countable family of open neighbourhoods of the neutral element $e$ in $G$. Then there exists a countable family $\gamma^*$ of open neighbourhoods of $e$ with the following properties:*

1) $\gamma \subset \gamma^*$;
2) *the intersection of any finite subfamily of $\gamma^*$ belongs to $\gamma^*$;*
3) *for each $U \in \gamma^*$, there exists a symmetric $V \in \gamma^*$ such that $V^2 \subset U$;*
4) *for every $U \in \gamma^*$ and every $a \in G$, there exists $V \in \gamma^*$ such that $aVa^{-1} \subset U$.*

PROOF. For each $U \in \gamma$, fix a symmetric open neighbourhood $V_U$ of $e$ such that $V_U^2 \subset U$. Fix also a countable family $\nu_U$ of open neighbourhoods of $e$ subordinated to $U$. Now let

$$\phi(\gamma) = \{\bigcap \lambda : \lambda \subset \gamma,\ |\lambda| < \omega\} \cup \bigcup \{\nu_U : U \in \gamma\} \cup \{V_U : U \in \gamma\}.$$

We put $\gamma_0 = \gamma$, $\gamma_1 = \phi(\gamma_0)$, and repeat this operation, defining by induction countable families $\gamma_2$, $\gamma_3$, and so on, by the rule $\gamma_{n+1} = \phi(\gamma_n)$, for each $n \in \omega$.

Let $\gamma^* = \bigcup_{n\in\omega} \gamma_n$. Since $\gamma_n \subset \gamma_{n+1}$ for each $n \in \omega$, and in view of the above definition, $\gamma^*$ satisfies conditions 1)–4). □

Now we easily obtain the next lemma designed for a direct application.

LEMMA 3.4.14. *Let $G$ be an $\omega$-balanced topological group, and $U$ an open neighbourhood of the neutral element $e$ in $G$. Then there exists a sequence $\{U_n : n \in \omega\}$ of open neighbourhoods of $e$ such that, for each $n \in \omega$, the following conditions are satisfied:*

a) $U_0 \subset U$;
b) $U_n^{-1} = U_n$;
c) $U_{n+1}^2 \subset U_n$, *and*
d) *for each $x \in G$ and each $n \in \omega$, there is $k \in \omega$ such that $xU_kx^{-1} \subset U_n$.*

PROOF. Put $\gamma = \{U\}$, and take a countable family $\gamma^*$ of open neighbourhoods of $e$ satisfying conditions 1)–4) of Lemma 3.4.13. Then $U \in \gamma^*$. We are going to define, by induction, a sequence of elements of $\gamma^*$.

Let us first enumerate the elements of $\gamma^*$, say, $\gamma^* = \{W_n : n \in \omega\}$. Choose $U_0$ to be any symmetric element of $\gamma^*$ such that $U_0 \subset U \cap W_0$. Since $\gamma^*$ satisfies conditions 2) and 3) of Lemma 3.4.13, this is obviously possible. Now assume that for some $n \in \omega$, an element $U_n \in \gamma^*$ has already been defined. Then, since $\gamma^*$ satisfies conditions 2) and 3), we can choose a symmetric element $V$ of $\gamma^*$ such that $V^2 \subset U_n \cap \bigcap_{i=0}^n W_i$. Put $U_{n+1} = V$. The definition is complete.

It is immediate from the construction that the sequence $\{U_n : n \in \omega\}$ satisfies conditions a)–c). Let us show that condition d) is also satisfied. Fix $n \in \omega$ and $x \in G$. Since the family $\gamma^*$ satisfies condition 4) of Lemma 3.4.13, there exists $j \in \omega$ such that $xW_jx^{-1} \subset U_n$. Put $k = \max\{n, j\}$. Then $U_{k+1} \subset U_{k+1}^2 \subset W_j$, by the inductive definition of $U_{k+1}$. Therefore, $xU_{k+1}x^{-1} \subset x^{-1}W_jx \subset U_n$. Condition d) is verified, so the lemma is proved. □

THEOREM 3.4.15. *Let $G$ be an $\omega$-balanced topological group. Then, for every open neighbourhood $U$ of the neutral element $e$ in $G$, there exists a continuous left-invariant pseudometric $\varrho$ on $G$ such that the following conditions are satisfied:*

(p1) $\{x \in G : \varrho(e, x) < 1\} \subset U$;
(p2) $\{x \in G : \varrho(e, x) = 0\}$ *is a closed invariant subgroup of $G$;*
(p3) *for any $x$ and $y$ in $G$, $\varrho(e, xy) \le \varrho(e, x) + \varrho(e, y)$.*

PROOF. By Lemma 3.4.14, we can find a sequence $\{U_n : n \in \omega\}$ of open neighbourhoods of $e$ in $G$ satisfying conditions a)–d) of that lemma. According to Lemmas 3.4.13 and 3.3.10, there exists a continuous prenorm $N$ on $G$ such that the next condition is satisfied:

(PN4) $\{x \in G : N(x) < 1/2^n\} \subset U_n \subset \{x \in G : N(x) \le 2/2^n\}$.

Now, for arbitrary $x$ and $y$ in $G$, put $\varrho(x, y) = N(x^{-1}y)$. Then the continuity of $N$ implies that $\varrho$ is also continuous. It is also clear from (PN4) and condition a) of Lemma 3.4.14 that (p1) is satisfied.

**Claim 1.** *$\varrho$ is a pseudometric on the set $G$.*

Indeed, for any $x$ and $y$ in $G$ we have: $\varrho(x, y) = N(x^{-1}y) \geq 0$, and $\varrho(y, x) = N(y^{-1}x) = N((y^{-1}x)^{-1}) = N(x^{-1}y) = \varrho(x, y)$. Also $\varrho(x, x) = N(x^{-1}x) = N(e) = 0$. Further, for any $x, y, z$ in $G$, we have:

$$\begin{aligned}\varrho(x, z) &= N(x^{-1}z) = N(x^{-1}yy^{-1}z)\\ &\leq N(x^{-1}y) + N(y^{-1}z) = \varrho(x, y) + \varrho(y, z).\end{aligned}$$

Hence $\varrho$ satisfies the triangle inequality.

**Claim 2.** *The pseudometric $\varrho$ is left-invariant.*

Indeed, $\varrho(zx, zy) = N(x^{-1}z^{-1}zy) = N(x^{-1}y) = \varrho(x, y)$, for arbitrary $x$, $y$, and $z$ in $G$. This evidently implies Claim 2.

Put $Z = \{x \in G : N(x) = 0\}$. Notice that $\varrho(e, x) = N(x)$, for each $x \in G$, since $\varrho(e, x) = N(e^{-1}x) = N(x)$. Therefore, we have that $Z = \{x \in G : \varrho(e, x) = 0\}$.

**Claim 3.** $Z = \bigcap_{n\in\omega} U_n$.

This clearly follows from condition (PN4).

**Claim 4.** *$Z$ is a closed invariant subgroup of $G$.*

Since the prenorm $N$ is continuous, the set $Z$ is closed in the space $G$. The fact that $Z$ is a subgroup of $G$ follows from Proposition 3.3.4.

It remains to show that the subgroup $Z$ of $G$ is invariant. Take any $x \in G$. We have to check that $xZx^{-1} \subset Z$. In view of Claim 3, it suffices to show that $xZx^{-1} \subset U_n$, for each $n \in \omega$. Fix $n \in \omega$. From condition d) of Lemma 3.4.14 it follows that there exists $k \in \omega$ such that $xU_kx^{-1} \subset U_n$. Since $Z \subset U_k$, we conclude that $xZx^{-1} \subset U_n$, that is, $Z$ is invariant.

It remains to notice that condition (p3) is obviously satisfied, since $N$ is a prenorm and $\varrho(e, x) = N(x)$. □

We need the following two simple lemmas:

LEMMA 3.4.16. *If $N$ is a prenorm on a group $G$, and $z$ is an element of $G$ such that $N(z) = 0$, then $N(zx) = N(x) = N(xz)$, for each $x \in G$.*

PROOF. We have: $N(zx) \leq N(z) + N(x) = N(x)$. Similarly, $N(x) = N(z^{-1}zx) \leq N(zx)$, since $N(z^{-1}) = N(z) = 0$. Therefore, $N(zx) = N(x)$. The equality $N(x) = N(xz)$ follows in a similar way. □

LEMMA 3.4.17. *In the notation of the proof of Theorem 3.4.15, let $a$ and $b$ be any two elements of $G$, and let $a_1 \in aZ$, $b_1 \in bZ$. Then $\varrho(a_1, b_1) = \varrho(a, b)$.*

PROOF. We may assume that $b = b_1$ (otherwise we simply repeat the argument twice). Clearly, $a_1 = az$, for some $z \in Z$. Then $N(z^{-1}) = N(z) = 0$ and, using Lemma 3.4.16, we obtain: $\varrho(a_1, b) = N(z^{-1}a^{-1}b) = N(a^{-1}b) = \varrho(a, b)$. □

We continue to use the objects constructed in the proof of Theorem 3.4.15, and in particular, we fix a pseudometric $\varrho$ on $G$ constructed above.

Let $H = G/Z$ be the quotient group, and let $\pi$ be the canonical homomorphism of $G$ onto $H$. Let $A, B$ be any elements of $H$ (that is, $A$ and $B$ are cosets of $H$ in $G$). Choose any $a \in A$ and $b \in B$ and put $d(A, B) = \varrho(a, b)$. By Lemma 3.4.17, the definition of $d(A, B)$ does not depend on the choice of $a$ in $A$ and $b$ in $B$.

We also define a function $N_H$ on $H$ by the rule $N_H(A) = N(a)$, for all $A \in H$ and $a \in A$. From Lemma 3.4.16 it follows that this definition does not depend on the choice of $a$ in $A$.

According to these definitions, we have: $d(\pi(a), \pi(b)) = \varrho(a, b)$, for any $a, b$ in $G$, and $N_H(\pi(a)) = N(a)$, for any $a \in G$. Since $\pi$ is a homomorphism of $G$ onto $H$, and $\pi(Z)$ is the neutral element $E$ of the group $H$, it follows, by the definitions of $Z$ and $\varrho$, that $d$ is a metric on $H$ and $N_H$ is a prenorm on $H$ satisfying the additional condition:

(PN5) If $N_H(A) = 0$, then $A$ is the neutral element $E$ of $H$.

For $\varepsilon > 0$, we put $B(\varepsilon) = \{x \in G : N(x) < \varepsilon\}$ and $O(\varepsilon) = \{X \in H : N_H(X) < \varepsilon\}$. Clearly, $\pi(B(\varepsilon)) = O(\varepsilon)$, for each $\varepsilon > 0$. Notice that the prenorm $N$ also satisfies the next condition:

(PN6) For each $\varepsilon > 0$ and each $x \in G$, there exists $\delta > 0$ such that $xB(\delta)x^{-1} \subset B(\varepsilon)$.

Since, obviously, $\pi(B(\varepsilon)) = O(\varepsilon)$, it follows that for the metric $d$ on $H$, we have:

(d1) For every $\varepsilon > 0$ and for every $X \in H$, there exists $\delta > 0$ such that $XO(\delta)X^{-1} \subset O(\varepsilon)$.

Since $\varrho(e, x) = N(x) = N(x^{-1}) = \varrho(e, x^{-1})$, from the definition of $d$ it follows that:

(d2) $d(E, X) = d(E, X^{-1})$, for each $X \in H$.

It follows from (d2) that $O(\varepsilon) = (O(\varepsilon))^{-1}$, for each $\varepsilon > 0$. Also, from (p3) and the definition of $d$ we obtain:

(d3) $(O(1/2^{n+1}))^2 \subset O(1/2^n)$.

Using (PN5), we conclude that

(d4) $\{E\} = \bigcap_{n\in\omega} O(1/2^n)$.

Now let $\mathcal{T}_H$ be the topology generated by the metric $d$ on $H$. Let us show that $H$ with this topology is a topological group.

Indeed, since $d$ is left-invariant, it is enough to observe that the family $\{O(1/2^n) : n \in \omega\}$, which is a base of the space $H$ at the neutral element $E$, satisfies the axioms for a base of a group topology at the neutral element (see Theorem 1.3.12). And this is exactly what conditions (d1)–(d4) guarantee, as is routinely verified. Thus, $H$ with the topology $\mathcal{T}_H$ is a topological group.

Finally, the equality $\pi(B(\varepsilon)) = O(\varepsilon)$, where $\varepsilon > 0$, implies that the homomorphism $\pi$ of $G$ onto $H = G/Z$ is continuous at the neutral element. Since $G$ and $H$ are topological groups, it follows that $\pi$ is continuous. Notice also that if $x \in G$, $X = \pi(x)$, and $\varepsilon > 0$, then $N(x) < \varepsilon$ is equivalent to $N_H(X) < \varepsilon$. Therefore, $\pi^{-1}(O(\varepsilon)) = B(\varepsilon)$ for each $\varepsilon > 0$. In particular, $\pi^{-1}(O(1)) = B(1) \subset U$.

Thus, the next theorem is established:

THEOREM 3.4.18. *If the invariance number of a topological group $G$ is countable, then for each open neighbourhood $U$ of the neutral element $e$ in $G$, there exists a continuous* *homomorphism* $\pi$ *of* $G$ *onto a metrizable group* $H$ *such that* $\pi^{-1}(V) \subset U$, *for some open neighbourhood* $V$ *of the neutral element* $e_H$ *of* $H$.

The above theorem has an important corollary:

COROLLARY 3.4.19. *Let* $G$ *be an* $\omega$*-narrow group. Then for every neighbourhood* $U$ *of the identity in* $G$, *there exists a continuous homomorphism* $\pi$ *of* $G$ *onto a second-countable topological group* $H$ *such that* $\pi^{-1}(V) \subset U$, *for some open neighbourhood* $V$ *of the identity in* $H$.

PROOF. By Theorem 3.4.18, one can find a continuous homomorphism $\pi$ of $G$ onto a metrizable topological group $H$ and an open neighbourhood $V$ of the identity in $H$ such that $\pi^{-1}(V) \subset U$. From Proposition 3.4.2 it follows that the group $H$ is $\omega$-narrow, so Proposition 3.4.5 implies that $H$ is second-countable. □

Now it is convenient to introduce the following concept. A topological group $G$ is called *range-metrizable* if it satisfies the conclusion of Theorem 3.4.18, that is, for every open neighbourhood $U$ of the neutral element $e$ of $G$, there exists a continuous homomorphism $p$ of $G$ onto a metrizable group $H$ such that $\pi^{-1}(V) \subset U$, for some open neighbourhood $V$ of the neutral element of $H$.

Clearly, Theorem 3.4.18 can be reformulated as follows: *If the invariance number of a topological group* $G$ *is countable, then* $G$ *is range-metrizable.*

Now let $\mathcal{P}$ be any class of topological groups (of semitopological groups, of paratopological groups), and let $G$ be any topological group (paratopological group, semitopological group). Let us say that $G$ is *range-*$\mathcal{P}$ if for every open neighbourhood $U$ of the neutral element $e$ of $G$, there exists a continuous homomorphism $p$ of $G$ to a group $H \in \mathcal{P}$ such that $\pi^{-1}(V) \subset U$, for some open neighbourhood $V$ of the neutral element $e_H$ of $H$. It follows immediately from the definition that every subgroup of a range-$\mathcal{P}$ group is also range-$\mathcal{P}$.

Similarly, Corollary 3.4.19 is equivalent to saying that every $\omega$-narrow group is range-$\Omega$, where $\Omega$ is the class of second-countable topological groups. The next fact follows from the definition of the product topology.

PROPOSITION 3.4.20. *Let* $\mathcal{P}$ *be any class of topological groups (of paratopological groups, of semitopological groups) closed under finite products, and let* $H$ *be the topological product of a family* $\{H_a : a \in A\}$ *of groups in the class* $\mathcal{P}$. *Then every subgroup of* $H$ *is range-*$\mathcal{P}$.

THEOREM 3.4.21. *Let* $\mathcal{P}$ *be a class of topological groups (paratopological groups, or semitopological groups),* $\tau$ *an infinite cardinal number, and* $G$ *a topological group (paratopological group, semitopological group), which is range-*$\mathcal{P}$ *and has a base* $\mathcal{B}$ *of open neighbourhoods of the neutral element such that* $|\mathcal{B}| \leq \tau$. *Then* $G$ *is topologically isomorphic to a subgroup of the product of a family* $\{H_a : a \in A\}$ *of groups such that* $H_a \in \mathcal{P}$, *for each* $a \in A$, *and* $|A| \leq \tau$.

PROOF. The argument is practically identical with the proof of the Tychonoff Embedding Theorem [**165**, Theorem 2.3.20]. We fix a base $\mathcal{B}$ of open neighbourhoods of the neutral element in $G$ such that $|\mathcal{B}| \leq \tau$. Then we choose, for every $U \in \mathcal{B}$, a continuous homomorphism $f_U$ of $G$ to a group $H_U \in \mathcal{P}$ such that $(f_U)^{-1}(V) \subset U$, for some open neighbourhood $V$ of the neutral element in $H$. The diagonal product $h$ of the family

$\{f_U : U \in \mathcal{B}\}$ is a topological isomorphism of $G$ onto a topological subgroup of the topological product of the family $\{H_U : U \in \mathcal{B}\}$ (recall that, for arbitrary $x \in G$ and $U \in \mathcal{B}$, the $U$th coordinate of $h(x)$ is the element $f_U(x)$ of $H_U$). □

THEOREM 3.4.22. [**G. I. Katz**] *For every topological group $G$, the following three conditions are equivalent:*

1) *$inv(G) \leq \omega$;*
2) *$G$ is range-metrizable;*
3) *$G$ is topologically isomorphic to a subgroup of a topological product of metrizable groups.*

PROOF. It follows from Proposition 3.4.20 and Theorem 3.4.21 that 2) and 3) are equivalent. Theorem 3.4.18 gives the implication 1) $\Rightarrow$ 2). To show that 2) $\Rightarrow$ 1), take an open neighbourhood $U$ of the neutral element $e_G$ in a range-metrizable group $G$ and consider a continuous homomorphism $p\colon G \to H$ of $G$ onto a metrizable group $H$ such that $p^{-1}(V) \subset U$ for some open neighbourhood $V$ of $e_H$ in $H$. Let also $\mathcal{B}$ be a countable base at $e_H$ in $H$. Then it is easy to verify that the countable family $\{p^{-1}(O) : O \in \mathcal{B}\}$ of open neighbourhoods of $e_G$ is subordinated to $U$, so that $inv(G) \leq \omega$. □

The following special case of Theorem 3.4.21 is particularly important.

THEOREM 3.4.23. [**I. I. Guran**] *A topological group $G$ is topologically isomorphic to a subgroup of the topological product of some family of second-countable groups if and only if $G$ is $\omega$-narrow.*

PROOF. Let $\Omega$ be the class of second-countable topological groups. Since, by Corollary 3.4.19, every $\omega$-narrow group $G$ is range-$\Omega$, it follows from Theorem 3.4.21 that $G$ is topologically isomorphic to a subgroup of the product of a family of groups from $\Omega$. Conversely, every subgroup of a topological product of second-countable topological groups is $\omega$-narrow, by Proposition 3.4.3 and Theorem 3.4.4. □

COROLLARY 3.4.24. *If the invariance number of a topological group $G$ is countable, and the neutral element of $G$ is a $G_\delta$-set in $G$, then there exists a continuous isomorphism of $G$ onto a metrizable topological group.*

PROOF. Let $\{U_n : n \in \omega\}$ be a family of open neighbourhoods of the identity $e$ in $G$ such that $\{e\} = \bigcap_{n\in\omega} U_n$. For every $n \in \omega$, take a continuous homomorphism $p_n : G \to H_n$ onto a metrizable topological group $H_n$ such that $p_n^{-1}(V_n) \subset U_n$ for some open neighbourhood $V_n$ of the identity in $H_n$. Since the product of a countable family of metrizable groups is a metrizable group and a subgroup of a metrizable group is also metrizable, it follows that the homomorphism $p = \Delta_{n\in\omega} p_n$ of $G$ to $\prod_{n\in\omega} H_n$ and the group $H = p(G)$ are as required. □

Similarly, from Corollary 3.4.19 (or applying Corollary 3.4.24 and Proposition 3.4.5) we obtain:

COROLLARY 3.4.25. *If the neutral element of an $\omega$-narrow topological group $G$ is a $G_\delta$-set in $G$, then there exists a continuous isomorphism of $G$ onto a second-countable topological group.*

Since the invariance number of Abelian topological groups is countable, Corollary 3.4.19 can be reformulated for Abelian groups as follows:

COROLLARY 3.4.26. *Suppose that $G$ is an Abelian topological group such that the neutral element of $G$ is a $G_\delta$-set. Then there exists a continuous isomorphism of $G$ onto a metrizable topological Abelian group.*

COROLLARY 3.4.27. *If a topological group $G$ has a countable network, then there exists a continuous isomorphism of $G$ onto a second-countable topological group.*

PROOF. This follows from Proposition 3.4.6 and Corollary 3.4.25, since every space with a countable network is Lindelöf and every point in such a space is a $G_\delta$-set. □

COROLLARY 3.4.28. *If a paratopological group $G$ is topologically isomorphic to a subgroup of the product of some family of metrizable paratopological groups, and $G$ is first-countable, then $G$ is metrizable.*

PROOF. This follows from Theorem 3.4.21, since the product of a countable family of metrizable paratopological groups is metrizable. □

Since the Sorgenfrey line is a non-metrizable paratopological group, we obtain from Corollary 3.4.28:

COROLLARY 3.4.29. *The Sorgenfrey line is not topologically isomorphic to any subgroup of a topological product of metrizable paratopological groups (and, therefore, is not range-metrizable).*

The last assertion is especially interesting since the Sorgenfrey line, being separable, is $\omega$-narrow. This shows that paratopological groups behave very differently with respect to embeddings compared to topological groups.

Applying Theorem 3.4.21, we can establish that certain semitopological groups are, in fact, topological groups.

THEOREM 3.4.30. *Let $G$ be a pseudocompact semitopological group. If $G$ is range-metrizable, then $G$ is a topological group.*

PROOF. By Theorem 3.4.21, $G$ is topologically isomorphic to a subgroup of the product of a family $\gamma$ of metrizable semitopological groups. Since metrizability is inherited by subgroups, we can assume that each factor $H_\alpha \in \gamma$ coincides with the projection of $G$ to $H_\alpha$. Since $G$ is pseudocompact and every metrizable pseudocompact space is compact, we conclude that every $H_\alpha \in \gamma$ is a compact semitopological group. However, every compact semitopological group is a topological group, by Theorem 2.3.12. Therefore, the product of $\gamma$ is a topological group, and every subgroup of this product is also a topological group. □

In Proposition 3.4.31 below we unify some previous statements and then deduce an important property of pseudocompact topological groups. Let $\tau$ be an infinite cardinal. We recall that a space $X$ is said to be *pseudo-$\tau$-compact* if every discrete in $X$ family of open sets has cardinality strictly less than $\tau$ (see page 54).

PROPOSITION 3.4.31. *Every pseudo-$\tau^+$-compact topological group $G$ is $\tau$-narrow.*

PROOF. Let $U$ be an arbitrary open symmetric neighbourhood of the neutral element in $G$. By Zorn's Lemma, there exists a maximal $U$-disjoint subset $A$ of $G$. Then $AU = G$, by the maximality of $A$. Take an open symmetric neighbourhood $V$ of $e$ such that $V^4 \subset U$. By Lemma 1.4.22, the family $\{aV : a \in A\}$ is discrete in $G$. Therefore, $|A| \leq \tau$. Since $AU = G$, we are done. □

Here is an interesting result about pseudocompact topological groups that will be generalized and strengthened in Section 3.7.

THEOREM 3.4.32. *Every pseudocompact topological group $G$ is topologically isomorphic to a dense subgroup of a compact topological group.*

PROOF. By Theorem 3.4.23 and Proposition 3.4.31, $G$ is topologically isomorphic to a subgroup of the product $P$ of a family $\gamma = \{H_\alpha : \alpha \in A\}$ of metrizable topological groups. Since $G$ is pseudocompact, and every metrizable pseudocompact space is compact, we can assume that each $H_\alpha \in \gamma$ is a compact topological group. Then the product group $P$ is compact and the closure of $G$ in $P$ is also a compact group which contains $G$ as a dense subgroup. □

It is worth noting that the proof of Theorem 3.4.32, which we just presented, does not depend on the existence of the Raĭkov completion of an arbitrary topological group (see Theorem 3.6.10).

One special class of $\omega$-narrow topological groups is described in the next theorem. Later, in Theorem 5.1.19, we will prove a stronger result by a more sophisticated argument. In particular, we will see that the theorem below remains valid for non-Abelian groups.

THEOREM 3.4.33. *Suppose that $G$ is an Abelian topological group and $M$ is a pseudo-$\aleph_1$-compact subspace of $G$ such that $M$ algebraically generates $G$. Then $G$ is $\omega$-narrow.*

PROOF. Clearly, we may assume that $M$ is symmetric (otherwise replace $M$ by the union $M \cup M^{-1}$ which is again pseudo-$\aleph_1$-compact).

**Claim 1.** *For each open neighbourhood $V$ of the neutral element $e$ in $G$, there exists a $V$-disjoint subset $A_V$ of $M$ such that $M \subset A_V V$.*

To establish the claim, it suffices to assume that $V$ is symmetric and apply Zorn's Lemma as in the proof of Proposition 3.4.31.

Fix $n \in \mathbb{N}$ and an open neighbourhood $U$ of $e$, and let $M_n$ be the set of all $g \in G$ such that $g = b_1 b_2 \ldots b_n$, for some $b_1, b_2, \ldots, b_n \in M$. Since $M$ is symmetric, we obviously have $G = \bigcup_{n=1}^{\infty} M_n$. Therefore, to prove the theorem, it is enough to verify the following:

**Claim 2.** *For each open neighbourhood $U$ of the neutral element $e$ in $G$ and each $n \in \mathbb{N}$, there exists a countable subset $B_n$ of $G$ such that $M_n \subset B_n U$.*

We may assume that the set $U$ in Claim 2 is symmetric. Since $G$ is Abelian, the natural mapping of $G^n$ to $G$ given by the product operation is uniformly continuous. Hence there exists an open symmetric neighbourhood $V_n$ of $e$ such that for any $b_1, b_2, \ldots, b_n \in G$,

$$b_1 V_n b_2 V_n \ldots b_n V_n \subset b_1 b_2 \ldots b_n U.$$

According to Claim 1, there exists a $V_n$-disjoint subset $A_n$ of $M$ such that $M \subset A_n V_n$. Since $M$ is pseudo-$\aleph_1$-compact, Lemma 1.4.22 implies that the set $A_n$ is countable. We denote by $B_n$ the subgroup of $G$ algebraically generated by the set $A_n$. Clearly, $B_n$ is countable.

Now take any $h \in M_n$. By the definition of $M_n$, there exist elements $b_1, b_2, \dots, b_n$ in $M$ such that $h = b_1 b_2 \dots b_n$. Since $V_n$ is symmetric, we can find $a_1, a_2, \dots, a_n \in A_n$ such that $a_i \in b_i V_n$, for each $i = 1, \dots, n$. Then $a_1 a_2 \dots a_n \in b_1 V_n b_2 V_n \dots b_n V_n \subset b_1 b_2 \dots b_n U = hU$. Since $U$ is symmetric, we conclude that $h \in a_1 a_2 \dots a_n U \subset B_n U$. Hence, $M_n \subset B_n U$. Claim 2 and the theorem are proved. □

Since every Lindelöf space is pseudo-$\aleph_1$-compact, the next fact follows immediately from Theorems 3.4.33 and 3.4.9:

COROLLARY 3.4.34. *Suppose that an Abelian topological group $G$ contains a Lindelöf subspace that algebraically generates a dense subgroup of $G$. Then the group $G$ is $\omega$-narrow.*

Again, one can drop "Abelian" in the above result (see Corollary 5.1.20).

## Exercises

3.4.a. Provide a complete proof of Proposition 3.4.2: If $f$ is a continuous homomorphism of an $\omega$-narrow topological group $G$ onto a topological group $H$, then $H$ is also $\omega$-narrow.

3.4.b. Provide a detailed proof of Proposition 3.4.3: The topological product of an arbitrary family of $\omega$-narrow topological groups is an $\omega$-narrow topological group.

3.4.c. Prove that every subgroup of an $\omega$-balanced group is $\omega$-balanced.

3.4.d. Show that the closure of an $\omega$-balanced subgroup $H$ of a topological group $G$ is again an $\omega$-balanced group.

3.4.e. Verify that $H$ with the topology $\mathcal{T}_H$ defined on page 167, before Theorem 3.4.18, is a topological group.

3.4.f. Give a detailed proof of Proposition 3.4.20.

3.4.g. Verify that the mapping constructed in the proof of Theorem 3.4.21 is indeed a topological isomorphism onto a subgroup of the product.

3.4.h. Give an example of a non-discrete topological group which is not homeomorphic to any $\omega$-narrow topological group.

3.4.i. Let $\{\mathcal{T}_i : i \in I\}$ be a family of $\omega$-narrow ($\omega$-balanced) group topologies on a group $G$. Show that the join of this family is again an $\omega$-narrow ($\omega$-balanced) group topology on $G$.

3.4.j. Show that a first-countable $\omega$-narrow paratopological group need not be Lindelöf.

3.4.k. Let $X$ be an arbitrary Tychonoff space. Prove that every discrete subgroup of the additive group $C_p(X)$ is countable.

## Problems

3.4.A. Give an example of a topological group $P$ in which every point is a $G_\delta$-set and such that $P$ cannot be mapped by a continuous isomorphism onto a metrizable group.
*Hint.* We follow the argument in [**370**]. Let $K$ be the general linear group $GL(2, \mathbb{R})$ (see item e) of Example 1.2.5). Consider the *box topology* $\mathcal{T}$ on the product group $K^{\omega_1}$ whose base consists of the sets $\prod_{\alpha<\omega_1} U_\alpha$, where each $U_\alpha$ is open in $K$. Verify that $\mathcal{T}$ is a Hausdorff group topology on $K^{\omega_1}$ and the neutral element of the group $P = (K^{\omega_1}, \mathcal{T})$ is a $G_\delta$-set. Use the sequences $\{x_n : n \in \omega\}$ and $\{y_n : n \in \omega\}$ of elements of $K$ mentioned in the hint to Problem 3.3.E to show that the group $P$ does not admit a continuous isomorphism onto a metrizable topological group.

3.4.B. Let $H$ be a closed subgroup of a topological group $G$, and suppose that the groups $H$ and $G/H$ are $\omega$-narrow ($\omega$-balanced). Prove that $G$ is also $\omega$-narrow ($\omega$-balanced).

3.4.C. Let $G$ be a cosmic completely regular (regular, Hausdorff) paratopological group. Prove that there exists a continuous isomorphism of $G$ onto a completely regular (regular, Hausdorff) paratopological group with a countable base.
*Hint.* See [**427**] for the Hausdorff case; for regular and Tychonoff paratopological groups, see [**226**].

3.4.D. Prove that every uncountable $\omega$-narrow topological group is resolvable (see Exercise 1.4.l).

3.4.E. (A. V. Arhangel'skii and D. K. Burke [**51**]) Show that a regular first-countable separable $\omega$-narrow paratopological group need not be Lindelöf.

3.4.F. (A. V. Arhangel'skii and A. Bella [**50**]) Show that the cardinality of every $\omega$-narrow first-countable Hausdorff paratopological group is not greater than $2^\omega$.

## Open Problems

3.4.1. Is every first-countable $\omega$-narrow (Tychonoff, regular, Hausdorff) paratopological group separable?

3.4.2. Let $G$ be a first-countable $\omega$-narrow paratopological group. Is the cellularity of $G$ countable?

3.4.3. Let $G$ be a regular $\omega$-narrow first-countable paratopological group. Does there exist a continuous isomorphism of $G$ onto a regular (Hausdorff) second-countable paratopological group?

3.4.4. Let $G$ be a first-countable paratopological group. Is $G$ submetrizable?

3.4.5. Is every topological group homeomorphic to a range-metrizable (or even to a balanced) topological group?

3.4.6. Is every connected topological group homeomorphic to an $\omega$-narrow topological group?

## 3.5. Groups of isometries and groups of homeomorphisms

In this section we consider groups of isometries of a metric space onto itself in the topology of pointwise convergence and groups of homeomorphisms of a topological space onto itself in various topologies.

Let $M$ be a metric space, with a metric $\varrho$ and a topology $\mathcal{T}$ generated by this metric. An *isometry* of $M$ is a mapping $f$ of $M$ onto itself preserving distances, that is, such that $\varrho(x, y) = \varrho(f(x), f(y))$, for every $x, y \in M$. Clearly, the inverse of an isometry of $M$ is defined and is also an isometry of $M$. The composition of two isometries is also an isometry. Therefore, the set of all isometries of $M$, with multiplication defined as composition of isometries, is a group. We denote this group by $Is(M)$ and endow it with the topology of pointwise convergence. A subbasic open neighbourhood of any $f \in Is(M)$ can be described as follows.

Fix any point $x$ in $M$ and a positive real number $\varepsilon$, and put

$$B(x, f, \varepsilon) = \{g \in Is(M) : \varrho((g(x), f(x)) < \varepsilon\}.$$

Intersections of finite families of sets of this type form a standard base of the topology of pointwise convergence on $Is(M)$. We use the terminology and notation described above throughout this section.

THEOREM 3.5.1. *The group $Is(M)$, with the topology of pointwise convergence, is a topological group.*

PROOF. First, we check that multiplication is continuous. Let $h = gf$, for some $f, g \in Is(M)$, and let $U$ be any open neighbourhood of $h$. We have to find open neighbourhoods $V$ and $W$ of $f$ and $g$, respectively, such that $WV \subset U$. Obviously, we may assume that $U = B(x, h, 2\varepsilon)$, for some $\varepsilon > 0$ and for some $x \in M$. Put $y = f(x)$, $V = B(x, f, \varepsilon)$, and $W = B(y, g, \varepsilon)$. Clearly, $V$ and $W$ are open neighbourhoods of $f$ and $g$, respectively.

Let us show that $WV \subset U$. Take any $f_1 \in V$ and $g_1 \in W$, and consider $h_1 = g_1 f_1$. We have to show that $\varrho(h_1(x), h(x)) < 2\varepsilon$. Indeed, we have: $\varrho(f_1(x), f(x)) < \varepsilon$ and $\varrho(g_1(y), g(y)) < \varepsilon$. Since $y = f(x)$ and $g_1$ is an isometry, we obtain: $\varrho(g_1 f_1(x), g_1(y)) < \varepsilon$. On the other hand, $g_1 \circ f_1(x) = h_1(x)$ and $g(y) = h(x)$. Therefore, by the triangle inequality,

$$\begin{aligned}\varrho(h_1(x), h(x)) &\le \varrho(h_1(x), g_1(y)) + \varrho(g_1(y), h(x)) \\ &= \varrho(g_1 f_1(x), g_1(y)) + \varrho(g_1(y), g(y)) < 2\varepsilon.\end{aligned}$$

The continuity of multiplication is established.

To check the continuity of the inverse operation, it is clearly enough to show that the set $(B(x, f, \varepsilon))^{-1}$ is open, for each $B(x, f, \varepsilon)$. Put $y = f(x)$, and take any $g \in Is(M)$.

**Claim.** *$g \in B(x, f, \varepsilon)$ if and only if $g^{-1} \in B(y, f^{-1}, \varepsilon)$.*

Indeed, if $\varrho(g(x), f(x)) < \varepsilon$, then $\varrho(g^{-1}(g(x)), g^{-1}(y)) < \varepsilon$, since $g^{-1}$ is an isometry. Since $g^{-1}(g(x)) = x = f^{-1}(y)$, it follows that $g^{-1} \in B(y, f^{-1}, \varepsilon)$. Similarly (the argument is symmetric) the "if" part is proved. Therefore, $(B(x, f, \varepsilon))^{-1} = B(f(x), f^{-1}, \varepsilon)$, where the set on the right side is open, by the definition of the topology on $Is(M)$. □

The (algebraic) group $Homeo(X)$ of all homeomorphisms of a topological space $X$ onto itself is similarly defined, but the topology of pointwise convergence on $Homeo(X)$ does not, in general, turn $Homeo(X)$ into a topological group (see Exercise 3.5.a). On the other hand, for any metric space $M$, $Is(M)$ is an algebraic subgroup of the group $Homeo(M)$.

Now, let $X$ be a topological space and $G$ a topological group which, algebraically, is a subgroup of $Homeo(X)$. Then we say that $G$ is a topological group of homeomorphisms of the space $X$. For each $x \in X$, the subspace $G_x = \{f(x) : f \in G\}$ of $X$ is called the *orbit of the point $x$ in $X$ under the action of $G$*. We also say that the group $G$ *acts on $X$ transitively* (or is *transitive on* $X$) if $G_x = X$ for some (and, therefore, for each) $x \in X$. Obviously, if this is the case, then the space $X$ is topologically homogeneous.

Every topological subgroup $G$ of the topological group $Is(M)$ of isometries of a metric space $M$ onto itself can serve as an example of a topological group of homeomorphisms of $M$, but not necessarily transitive.

Now let $X$ be a space and let $G = Homeo(X)$ be the group of homeomorphisms of $X$ onto itself. For any subsets $K$, $W$ of $X$ we put $\langle K, W\rangle = \{h \in G : h(K) \subset W\}$. Let $\mathcal{S}$ be the family of all sets $\langle K, W\rangle$ such that $K$ is closed in $X$ and $W$ is open in $X$. Then $\mathcal{S}$ is a subbase of some topology $\mathcal{T}$ on $G$, and $\mathcal{B} = \{\bigcap \lambda : \lambda \subset \mathcal{S},\ |\lambda| < \omega\}$ is a base of $\mathcal{T}$. We will call this topology the *closed-based topology* of $Homeo(X)$.

THEOREM 3.5.2. *If $X$ is a normal space, then the group $G = Homeo(X)$ with the closed-based topology is a topological group.*

PROOF. Clearly, $\langle K, W\rangle^{-1} = \langle X \setminus W, X \setminus K\rangle$, for any two subsets $K$ and $W$ of $X$. It easily follows from this that if $\langle K, W\rangle \in \mathcal{S}$, then $\langle K, W\rangle^{-1} \in \mathcal{S}$, and that if $U \in \mathcal{B}$, then $U^{-1} \in \mathcal{B}$. Therefore, the inverse operation is continuous.

To show that the multiplication in $G$ is continuous, take arbitrary elements $f, g \in G$, and let $h = fg \in U$, where $U$ is an open subset of $G$. Obviously, we can assume that $U = \langle K, W \rangle \in \mathcal{S}$. This means that $f(g(K)) \subset W$, that is, $g(K) \subset f^{-1}(W)$. Since $g$ and $f$ are homeomorphisms of $X$ onto $X$, $g(K)$ is closed in $X$, and $f^{-1}(W)$ is an open neighbourhood of $g(K)$. Since $X$ is normal, there exists an open set $V$ such that $g(K) \subset V \subset \overline{V} \subset f^{-1}(W)$. Then $\langle K, V \rangle$ and $\langle \overline{V}, W \rangle$ are open neighbourhoods of $g$ and $f$, respectively, and it is easy to see that $\langle \overline{V}, W \rangle \langle K, V \rangle \subset \langle K, W \rangle$. Thus, the multiplication is (jointly) continuous. □

Similarly to the closed-based topology, the *compact-based* or *compact-open* topology on $Homeo(X)$ is defined as follows. The sets $\langle K, W \rangle$, where $K$ is compact and $W$ is open in $X$, constitute a standard subbase for this topology on $Homeo(X)$. It is clear that if $X$ is compact, then the compact-based topology and closed-based topology on $Homeo(X)$ coincide, but in general these two topologies are different, and one cannot prove in this case an analog of Theorem 3.5.2, even for second-countable spaces. So that we have to be content with the following corollary of Theorem 3.5.2:

COROLLARY 3.5.3. *If $X$ is a compact Hausdorff space, then $Homeo(X)$ with the compact-open topology is a topological group.*

THEOREM 3.5.4. *If $X$ is a locally compact Hausdorff space, then the group $Homeo(X)$ with the compact-open topology is a paratopological group.*

PROOF. The proof of the joint continuity of the multiplication in $Homeo(X)$ is essentially the same as the proof of Theorem 3.5.2, but instead of the normality of $X$ we have to use the following simple fact: every compact subset of a locally compact Hausdorff space is contained in an open set $U$ such that the closure of $U$ is compact. □

Here is an example which shows that the conclusion in Theorem 3.5.4 cannot be strengthened to the conclusion that $Homeo(X)$ is a topological group. First we present the next general statement:

THEOREM 3.5.5. *If $X$ is a locally compact Hausdorff space with a countable base, then the space $Homeo(X)$, with the compact-open topology, also has a countable base.*

PROOF. Choose a countable base $\mathcal{B}$ for $X$; we can assume that $\mathcal{B}$ is closed under finite unions. Denote by $\mathcal{K}$ the family of the closures of elements of $\mathcal{B}$. It is clear that $\mathcal{K}$ is countable. In addition, if $C$ is a compact subset of $X$ and $C \subset O$, for some open set $O$ in $X$, then there exists $K \in \mathcal{K}$ such that $C \subset K \subset O$. It follows that $\{\langle K, W \rangle : K \in \mathcal{K},\ W \in \mathcal{B}\}$ is a countable base for the compact-open topology on $Homeo(X)$. □

EXAMPLE 3.5.6. Let $X = \mathbb{N} \cup \{0\} \cup \{1/n : n \in \mathbb{N}\}$, with the topology inherited from the real line. The only non-isolated point of $X$ is 0. For an arbitrary $k \in \mathbb{N}$, define a homeomorphism $h_k$ of the space $X$ onto itself in the following way: $h_k(n) = n$ if $n = 0, 1, \ldots, k-1$; $h_k(k) = 1/k$; $h_k(n) = n - 1$ if $n \geq k+1$; $h_k(1/n) = 1/n$ if $n = 2, 3, \ldots, k-1$, and $h_k(1/n) = 1/(n+1)$ if $n \geq k$.

It is easy to see that, endowed with the compact-open topology, the sequence $\{h_k : k \in \mathbb{N}\}$ converges to the identity mapping of $X$ onto itself, while the sequence of the inverses of these homeomorphisms does not converge to the identity mapping, since the open set $\langle P, P \rangle$ in $Homeo(X)$ does not contain any $h_k$ with $k > 1$, where $P = [0, 1] \cap X$ is a compact and open subset of $X$. Therefore, $Homeo(X)$ is a paratopological group which is not a topological group. Notice that the space $Homeo(X)$ is second-countable. Indeed, $X$ is a locally compact Hausdorff space with a countable base, so it remains to apply Theorem 3.5.5. □

The group $Homeo(X)$ in Example 3.5.6 is a metrizable paratopological group which is not metrizable by a left-invariant or right-invariant metric (we leave the verification of this to the reader; see Problem 3.5.A).

EXAMPLE 3.5.7. Let $A$ be a set, and let $G$ be the set of all bijections of $A$ onto itself. We endow $G$ with the topology of pointwise convergence and define multiplication in $G$ as composition of bijections. Then, obviously, $G$ is a group and a topological space. We claim that $G$ is, in fact, a topological group. Indeed, we can treat $A$ as a discrete metric space, with the standard metric $\varrho$ defined by $\varrho(x, y) = 1$ iff $x \neq y$. Every bijection of $A$ onto itself becomes an isometry after this agreement, so that this interpretation turns $G$ into the group of isometries of the metric space $A$. Therefore, by Theorem 3.5.1, $G = Is(A)$ is a topological group. □

Let $X$ be a homogeneous space. A subgroup $G$ of $Homeo(X)$ will be called *rich* if the orbit $G(x) = \{g(x) : g \in G\}$ of $x$ under $G$ coincides with $X$ for some (equivalently, for each) $x \in X$. We will call a space $X$ *p-homogeneous* if there exists a rich subgroup $G$ of the group $Homeo(X)$ such that $G$, with the pointwise convergence topology, is a topological group.

PROPOSITION 3.5.8. *Let $G$ be a topological group. Then:*

a) *the group $G$ is topologically isomorphic to a subgroup $H$ of the group $Homeo(G)$, where $H$ is taken with the topology of pointwise convergence;*
b) *the group $G$ is a p-homogeneous space.*

PROOF. a) For every $a \in G$, let $\lambda_a$ be the left translation of $G$ by $a$, that is, $\lambda_a(x) = ax$, for every $x \in G$. Then $\lambda_a$ is a homeomorphism of the space $G$ onto itself, by Corollary 1.3.2. Put $\phi(a) = \lambda_a$, for every $a \in G$. Clearly, $\phi$ is a one-to-one mapping of $G$ to the group $Homeo(G)$. We also have $\phi(ab) = \lambda_{ab} = \lambda_a\lambda_b = \phi(a)\phi(b)$, for all $a, b \in G$. Indeed, $\lambda_{ab}(x) = (ab)x = a(bx) = \lambda_a(bx) = \lambda_a(\lambda_b(x))$, for each $x \in G$. Hence, $\phi$ is a homomorphism of the group $G$ to $Homeo(G)$.

We endow the set $Homeo(G)$ with the topology of pointwise convergence and consider the subspace $H = \phi(G)$ of the space $Homeo(G)$. Let us show that $\phi$ is a homeomorphism of the space $G$ onto $H$. From this it will clearly follow that $H$ is a topological group, and that $\phi$ is a topological isomorphism between the topological groups $G$ and $H$.

Take any $x \in G$ and any open set $U$ in $G$. If $b \in G$, then $\phi(b)(x) = \lambda_b(x) = bx \in U$ if and only if $b \in Ux^{-1}$. It follows that the preimage under $\phi$ of the subbasic open set $O(U, x) = \{h \in Homeo(G) : h(x) \in U\}$ in $Homeo(G)$ is the set $Ux^{-1}$ which is clearly open in $G$. Therefore, $\phi$ is a homeomorphism of $G$ onto $H = \phi(G)$. This proves a).

b) Clearly, $H$ is a rich subgroup of $Homeo(G)$. Therefore, the space $G$ is $p$-homogeneous, as claimed. □

A space $X$ will be called *homogeneously metrizable* or *m-homogeneous* if there exists a metric $\varrho$ on $X$ generating the topology of $X$ such that for every two points $x$ and $y$ of $X$, there exists a bijection $f$ of $X$ onto itself which is an isometry with respect to $\varrho$ and satisfies the condition $f(x) = y$. The next assertion follows from Theorem 3.5.1 and the definition above:

COROLLARY 3.5.9. *Every homogeneously metrizable space is p-homogeneous.*

Let $G$ be a topological group, and let $M_G$ be the metric space of all bounded left uniformly continuous real-valued functions on $G$, with the uniform convergence metric $\varrho$ given by the formula:

$$\varrho(f, g) = \sup\{|f(x) - g(x)| : x \in G\}.$$

By Theorem 3.5.1, the group $Is(M_G)$ of all isometries of the space $M_G$ onto itself, with the topology of pointwise convergence, is a topological group.

For $f \in M_G$ and $a \in G$, we put $f_a(x) = f(ax)$ for each $x \in G$, and $h_a(f) = f_a$. Clearly, $f_a \in M_G$. In this way we defined a mapping $h_a\colon M_G \to M_G$, which is evidently an isometry of the metric space $M_G$ onto itself. Now we introduce the canonical mapping $\psi$ of $G$ into $Is(M_G)$ defined by the rule $\psi(a) = h_a$, for each $a \in G$.

**Claim 1.** *$\psi$ is a homomorphism.*

**Claim 2.** *$\psi$ is one-to-one.*

Claims 1 and 2 are easily verified. Notice that, in view of Claim 1, to prove Claim 2 it suffices to show that if $a$ is not the neutral element $e$ of $G$, then the mapping $h_a$ of $M_G$ onto $M_G$ is not the identity mapping.

**Claim 3.** *$\psi$ is continuous.*

Indeed, since both $G$ and $Is(M_G)$ are topological groups, it suffices to verify the continuity of $\psi$ at the neutral element $e$ of $G$. Fix $f \in M_G$ and $\varepsilon > 0$, and put $\langle f, \varepsilon\rangle = \{h \in Is(M_G) : \varrho(h(f), f) < \varepsilon\}$. Then $\langle f, \varepsilon\rangle$ is a subbasic open neighbourhood of the neutral element of $Is(M_G)$, and to check the continuity of $\psi$ it is enough to show that there exists an open neighbourhood $V$ of $e$ in $G$ such that $h_a \in \langle f, \varepsilon\rangle$, for each $a \in V$. But this obviously follows from the left uniform continuity of the function $f$.

**Claim 4.** *$\psi$ is a homeomorphism of $G$ onto the subspace $\psi(G)$ of $Is(M_G)$.*

Again, it suffices to verify that if $V$ is any open neighbourhood of $e$ in $G$, then the identity isometry $id$ of $M_G$ is not in the closure of the set $\psi(G \setminus V)$. To do this, we have to use the fact that every topological group $G$ is left uniformly Tychonoff, that is, for each open neighbourhood $V$ of the neutral element $e$ of $G$, there exists a left uniformly continuous function $f$ on $G$ such that $f(e) = 0$ and $f(x) = 1$, for each $x \in G \setminus V$ (see Theorem 3.3.11).

So, let us fix such a function $f$ on $G$. Then $\varrho(f_a, f) \geq 1$, that is, $\varrho(h_a(f), f) \geq 1$ for each $a \in G \setminus V$. This implies that the set $\psi(G \setminus V)$ does not meet the open neighbourhood $\langle f, 1\rangle$ of $id$. Therefore, $\psi$ is a homeomorphism of $G$ onto its image in $Is(M_G)$. This proves Claim 4.

From the four above claims, we obtain the following interesting result immediately:

THEOREM 3.5.10. [**V. V. Uspenskij**] *Every topological group $G$ is topologically isomorphic to a subgroup of the group of isometries $Is(M)$ of some metric space $M$, where $Is(M)$ is taken with the topology of pointwise convergence.*

One can effectively use this theorem, and its proof, to obtain some other general results on topological groups. For example, we have:

THEOREM 3.5.11. *If $G$ is a left uniformly Tychonoff left semitopological group, then $G$ is a topological group.*

PROOF. Let $M_G$ be the metric space of all bounded left uniformly continuous real-valued functions on $G$. Now we can repeat the argument in the proof of Theorem 3.5.10 which provides us with a topological isomorphism $\psi$ between $G$ and a topological subgroup of the topological group $Is(M_G)$. Therefore, $G$ itself is a topological group.

Let us now present an alternative proof. First, we show that the inverse in $G$ is continuous. Assume the contrary. Then there exist an open neighbourhood $V$ of the neutral element $e$ in $G$ and a subset $A$ of $V$ such that $A^{-1} \cap V = \varnothing$ and $e \in \overline{A}$. Since $G$ is left uniformly Tychonoff, we can fix a left uniformly continuous function $f$ on $G$ such that $f(e) = 0$ and $f(x) \geq 1$, for each $x \in G \setminus V$. Let $W$ be any open neighbourhood of $e$. Then $A \cap W$ is not empty, so there exists $a \in A \cap W$. We have: $(a^{-1})^{-1}e = a \in W$, and $|f(a^{-1}) - f(e)| \geq 1$, which implies, since $W$ is an arbitrary open neighbourhood of $e$, that $f$ is not left uniformly continuous, a contradiction.

To show that the multiplication in $G$ is jointly continuous, we fix an open neighbourhood $V$ of $e$, take a bounded left uniformly continuous real-valued function $f$ on $G$ such as above, and define a function $N_f$ on $G$ by the standard rule:

$$N_f(x) = \sup\{|f(ax) - f(a)| : a \in G\},$$

for each $x \in G$. Then $N_f$ is a prenorm on $G$, by Lemma 3.3.6, and we claim that $N_f$ is continuous on $G$. Indeed, since $f$ is left uniformly continuous, for an arbitrary $\varepsilon > 0$ there exists an open neighbourhood $U$ of $e$ in $G$ such that $|f(x) - f(y)| < \varepsilon$ whenever $x^{-1}y \in U$. Hence, the definition of $N_f$ implies that $N_f(x) \leq \varepsilon$ for each $x \in U$, so the continuity of $N_f$ follows from Proposition 3.3.7.

Now put $B(r) = \{x \in G : N_f(x) < r\}$ for each $r > 0$ and let $U = B(1/2)$. Then $U$ is an open neighbourhood of $e$ in $G$ and $U^2 \subset B(1) \subset V$, which proves the joint continuity of multiplication in $G$. Notice also that $U^{-1} = U$, so we again proved the continuity of the inverse in $G$ as well. □

Important natural examples of homogeneous spaces are provided by quotients of topological groups with respect to closed subgroups. Though these quotients need not be topological groups themselves, they are always homogeneous topological spaces. A natural question for consideration is the following one: which homogeneous spaces can be represented as quotients of topological groups with respect to closed subgroups? A partial answer to this question is given below.

A space $X$ is said to be *strongly locally homogeneous* if for each $x \in X$ and for every open neighbourhood $U$ of $x$, there exists an open neighbourhood $V$ of $x$ such that $x \in V \subset U$ and, for every $z \in V$, there exists a homeomorphism $h$ of $X$ onto $X$ such that $f(x) = z$ and $h(y) = y$, for each $y \in X \setminus V$.

PROPOSITION 3.5.12. **[L. R. Ford]** *If a zero-dimensional $T_1$-space $X$ is homogeneous, then it is strongly locally homogeneous.*

PROOF. Note first that $X$ is Hausdorff, since it is $T_1$ and zero-dimensional. Fix a point $x \in X$, and take any open neighbourhood $U$ of $x$. Take arbitrary $z \in U$ distinct from $x$. Since $X$ is homogeneous, there exists a homeomorphism $f$ of $X$ onto itself such that $f(x) = z$. Since $X$ is Hausdorff, we can find disjoint open sets $U_1$ and $U_2$ such that $x \in U_1 \subset U$

and $z \in U_2 \subset U$. By the continuity of $f$, there exists an open neighbourhood $W$ of $x$ such that $W \subset U_1$ and $f(W) \subset U_2$. Then $W$ and $f(W)$ are disjoint. Since the space $X$ is zero-dimensional, we can also assume that $W$ is closed in $X$. Then $f(W)$ is also open and closed in $X$, since $f$ is a homeomorphism of $X$ onto itself. Now we define a mapping $h\colon X \to X$ by the following requirements:

1) $h$ coincides with $f$ on $W$;
2) $h$ coincides with $f^{-1}$ on $f(W)$;
3) $h(y) = y$, for each $y \in X \setminus (W \cup f(W))$.

Clearly, $h$ is a homeomorphism, $h(x) = f(x) = z$, and $h$ coincides with the identity mapping on $X \setminus U$. Hence, $X$ is strongly locally homogeneous. □

Suppose that $X$ is a space. A topology $\mathcal{T}$ on the group $G = Homeo(X)$ of all homeomorphisms of $X$ onto itself will be called *acceptable* if it turns $G$ into a topological group such that the action correspondence $(g, x) \to g(x)$, considered as a mapping of $G \times X$ to $X$, is continuous with respect to the first variable, and for every open neighbourhood $U$ of the neutral element $e$ of $G$ and for each $b \in X$, there exists an open neighbourhood $V$ of $b$ such that $Id_V \subset U$, where

$$Id_V = \{h \in Homeo(X) : h(x) = x \text{ for each } x \in X \setminus V\}.$$

The next statement is obvious, in view of Corollary 3.5.3 and of the definition of acceptable topologies.

PROPOSITION 3.5.13. *For any compact Hausdorff space $X$, the compact-open topology on the group $Homeo(X)$ of all homeomorphisms of $X$ onto itself is acceptable.*

PROPOSITION 3.5.14. *Suppose that $X$ is a homogeneous strongly locally homogeneous space, and $G = Homeo(X)$ is endowed with an acceptable topology. Then $X$ is canonically homeomorphic to the quotient space $G/G_a$, where $a$ is a point of $X$ and $G_a$ is the stabilizer of $a$ in $G$, that is, $G_a = \{g \in G : g(a) = a\}$.*

PROOF. Obviously, $G_a$ is a subgroup of $G$. Notice that $G_a$ is closed in $G$, since the action correspondence is continuous with respect to the first variable.

Denote by $\pi$ the quotient mapping of $G$ onto the quotient space $G/G_a$ of the left cosets of $G_a$ in $G$. Then $\pi$ is open and continuous. Now we define a mapping $p$ of $G$ to $X$ by the obvious rule: $p(g) = g(a)$, for each $g \in G$. The mapping $p$ is also continuous, since the action correspondence is continuous with respect to the first variable. Since $G$ acts transitively on $X$, we have $p(G) = X$.

Let us show that $p$ is open. Take any $g \in G$ and any open neighbourhood $O(g)$ of $g$ in $G$. Put $b = g(a)$. Since the topology on $G$ is acceptable, there exists an open neighbourhood $V$ of $b$ in $X$ such that $(Id_V)g \subset O(g)$. Since $X$ is strongly locally homogeneous, there exists an open neighbourhood $W$ of $b$ such that $W \subset V$ and, for each $y \in W$, there exists $h \in Id_V$ with $h(b) = y$. Then $hg \in O(g)$ and $p(hg) = hg(a) = h(b) = y$. Hence, $b = p(g) \in W \subset p(O(g))$, and the mapping $p$ is open.

Clearly, $p^{-1}p(g) = p^{-1}(g(a)) = gG_a$ for each $g \in G$, that is, the family of fibers of the mapping $p$ coincides with the family of left cosets of $G_a$ in $G$ or, equivalently, with the family of fibers of the mapping $\pi$. Therefore, there exists a natural bijection $i\colon G/G_a \to X$

satisfying $p = i \circ \pi$.

$$\begin{array}{ccc} G & \xrightarrow{\pi} & G/G_a \\ {\scriptstyle p}\downarrow & \swarrow{\scriptstyle i} & \\ X & & \end{array}$$

Since both mappings $p$ and $\pi$ are continuous, open, and onto, it follows that the mapping $i$ of $G/G_a$ onto $X$ is a homeomorphism. □

THEOREM 3.5.15. [**N. Bourbaki**] *Every homogeneous zero-dimensional compact Hausdorff space $X$ can be represented as the quotient space of a topological group with respect to a closed subgroup.*

PROOF. Indeed, by Proposition 3.5.12, the space $X$ is strongly locally homogeneous. By Proposition 3.5.13, the compact-open topology on the group $Homeo(X)$ of all homeomorphisms of $X$ onto itself is acceptable. Now it follows from Proposition 3.5.14 that $X$ is homeomorphic to a quotient space of the group $Homeo(X)$ endowed with this topology, with respect to a closed subgroup. □

EXAMPLE 3.5.16. Let $A_2$ be the two arrows space [**165**, 3.10.C]. Then $A_2$ is a zero-dimensional, homogeneous, first-countable, non-metrizable compact space. By Theorem 3.5.15, $A_2$ is homeomorphic to a quotient space of a topological group with respect to a closed subgroup. Therefore, Theorem 3.3.12 on the metrizability of first-countable topological groups does not admit a generalization to quotients of topological groups with respect to closed subgroups, even if quotient spaces are compact.

## Exercises

3.5.a. Prove that the group $Homeo(\mathbb{R}^2)$ of all homeomorphisms of the Euclidean plane $\mathbb{R}^2$ onto itself, with the topology of pointwise convergence, is neither a paratopological group nor a quasitopological group.

3.5.b. Prove that if $X$ is a locally compact countable Hausdorff space, then the space $Homeo(X)$, with the compact-open topology, has a countable base.

3.5.c. Show that the conclusion in Theorem 3.5.2 cannot be extended to the class of all Tychonoff spaces.

3.5.d. Prove Claim 1 and Claim 2 on page 177, preceding Theorem 3.5.10.

## Problems

3.5.A. Prove that the paratopological group $G = Homeo(X)$ in Example 3.5.6 is not metrizable, neither by a left-invariant metric nor by a right-invariant metric.

3.5.B. Suppose that $M$ is a metric space such that the group of isometries $Is(M)$, with the topology of pointwise convergence, is $\omega$-narrow. Must $M$ be separable? What if $M$ is, in addition, *metrically homogeneous*, that is, any point $x \in M$ can be brought to any other point $y \in M$ by an isometry of $M$ onto itself?

3.5.C. Let $X$ be the space of countable ordinals with the usual order topology. Describe in topological terms the group $Homeo(X)$ when this group is endowed with:

a) the compact-open topology;

b) the topology of pointwise convergence.

Is $Homeo(X)$ a topological group in case a)?

3.5.D. Let $D = \{0, 1\}$ be a discrete two-point space, and $A$ be an infinite set. Suppose that the group $G = Homeo(D^A)$ is endowed with the compact-open topology or with the topology of pointwise convergence, and consider the subgroup $H$ of $G$ which consists of the homeomorphisms of $D^A$ generated by permutations of the index set $A$. Thus, every element $h \in H$ has the form $h(x)(\alpha) = x(f(\alpha))$, for all $x \in D^A$ and $\alpha \in A$, where $f : A \to A$ is a bijection. For each of the two standard topologies on $G$, answer the following:
   (a) Is $H$ closed in $G$?
   (b) Is $H$ connected? Is $H$ zero-dimensional?
   (c) Is $H$ invariant in $G$?

3.5.E. Let $(X, d)$ be a metric, connected, locally compact space. Prove that the group $Is(X, d)$ with the compact-open topology is a metrizable, locally compact topological group.
*Hint.* Apply Theorem 3.5.5 to deduce that the space $Is(X, d)$ with the compact-open topology has a countable base. Apply this fact to verify the continuity of the inverse in $Is(X, d)$ and then use Theorem 3.5.4.

3.5.F. Prove that the group of homeomorphisms of the closed unit interval $[0, 1]$ endowed with the topology of uniform convergence is homeomorphic to the space $\{0, 1\} \times [0, 1]^\omega$.

3.5.G. Prove that the group $Homeo(\mathbb{R}^2)$ of all homeomorphisms of the Euclidean plane $\mathbb{R}^2$ onto itself, with the compact-open topology, is a topological group. Is a similar statement true for $\mathbb{R}^n$, where $n$ is an arbitrary positive integer?

## Open Problems

3.5.1. Characterize in internal terms $p$-homogeneous Tychonoff spaces.

3.5.2. Characterize in internal terms compact Hausdorff $p$-homogeneous spaces.

3.5.3. Let $F$ be a compact Hausdorff $p$-homogeneous space. Is the Souslin number of $F$ not greater than $2^\omega$?

3.5.4. When is the group of isometries $Is(M)$ of a metric space $M$, in the topology of pointwise convergence, Dieudonné complete?

## 3.6. Raĭkov completion of a topological group

In this section we embed an arbitrary topological group into a nicer group, in which all Cauchy filters converge. Topological groups with this property are called *Raĭkov complete.*

Throughout the section, $G$ is a topological group, and $e$ is its neutral element. For each $x \in G$, let $B_x$ be the family of all open sets in $G$ containing $x$.

A *filter* on $G$ is a family $\eta$ of non-empty subsets of $G$ satisfying the next two conditions:

(F1) If $U$ and $V$ are in $\eta$, then $U \cap V$ is also in $\eta$;
(F2) If $U \in \eta$ and $U \subset W \subset G$, then $W \in \eta$.

A family $\xi$ is called an *open filter* on $G$ if there exists a filter $\eta$ on $G$ such that $\xi$ is the intersection of $\eta$ with the family of all open subsets of $G$. Of course, this definition is equivalent to the following one: $\xi$ is an open filter on $G$ if $\xi$ is a family of non-empty open subsets of $G$ such that the intersection of any finite number of elements of $\xi$ is also in $\xi$, and for each $U \in \xi$ and for every open subset $W$ of $G$ such that $U \subset W$, $W$ also belongs to $\xi$.

A family $\eta$ of non-empty sets is called a *filter base* if for any elements $U, V \in \eta$, there exists $W \in \eta$ such that $W \subset U \cap V$. An *open filter base* on $G$ is a filter base all elements of which are open subsets of $G$.

For a family $\eta$ of subsets of $G$, we denote by $o(\eta)$ the family of all open subsets of $G$ containing at least one element of $\eta$. Clearly, if all elements of $\eta$ are open, then $\eta \subset o(\eta)$.

A family $\eta$ of subsets of $G$ is said to be a *Cauchy family* if for every open neighbourhood $V$ of $e$ in $G$, there exist $a, b \in G$ and $A, B \in \eta$ such that $A \subset aV$ and $B \subset Vb$. Cauchy filters and Cauchy open filters play an important role in the construction we are going to describe. Of course, an open filter $\eta$ is a Cauchy family if and only if for each open neighbourhood $V$ of $e$, one can find $a, b \in G$ such that the sets $aV$ and $Vb$ are in $\eta$.

A family $\eta$ of sets will be called *shrinking* if for every $B \in \eta$, there exist $A \in \eta$ and open neighbourhoods $U$ and $V$ of $e$ such that $UAV \subset B$. A *canonical filter* is an open filter which is both shrinking and Cauchy.

Let us list some simple facts concerning the concepts just introduced.

**Fact 1.** If $\eta$ is an open filter base, then $o(\eta)$ is an open filter containing $\eta$.

**Fact 2.** If $\eta$ is a shrinking family of open sets, then $o(\eta)$ is also a shrinking family of open sets.

**Fact 3.** If $\eta$ is a Cauchy family of sets, then $o(\eta)$ is also a Cauchy family of sets.

From Facts 1, 2, and 3 we get:

**Fact 4.** If $\eta$ is an open filter base which is both Cauchy and shrinking, then $o(\eta)$ is a canonical filter containing $\eta$.

For any two families $\xi$ and $\eta$ of subsets of $X$, we put $[\eta\xi] = \{AB : A \in \eta,\ B \in \xi\}$.

**Fact 5.** If $\eta$ and $\xi$ are open filter bases, then $[\eta\xi]$ is also an open filter base.

Though the next assertion is also almost obvious, we provide a short proof of it.

**Fact 6.** If $\eta$ and $\xi$ are shrinking families of sets, then $[\eta\xi]$ is also a shrinking family of sets.

PROOF. Let $W \in [\eta\xi]$. Then $W = AB$, for some $A \in \eta$, $B \in \xi$. Since $\eta$ and $\xi$ are shrinking, there are $A_1 \in \eta$, $B_1 \in \xi$, and open neighbourhoods $U_1, U_2, V_1, V_2$ of $e$ in $G$ such that $U_1A_1V_1 \subset A$ and $U_2B_1V_2 \subset B$. Then $U_1A_1B_1V_2 \subset U_1A_1V_1U_2B_1V_2$, since $e \in V_1$ and $e \in U_2$. Taking into account that $A_1B_1 \in [\eta\xi]$, we conclude that $A_1B_1$ is the element of $[\eta\xi]$ we were looking for. Thus, $[\eta\xi]$ is shrinking. □

**Fact 7.** If $\eta$ and $\xi$ are Cauchy families, then $[\eta\xi]$ is also a Cauchy family.

PROOF. Let $U$ be an open neighbourhood of $e$ in $G$. Take an open set $V$ containing $e$ such that $V^2 \subset U$. Since $\xi$ is a Cauchy family, one can find $B \in \xi$ and $b \in G$ such that $B \subset bV$. Now, $bVb^{-1}$ is an open neighbourhood of $e$. Since $\eta$ is Cauchy, there exist $A \in \eta$ and $a \in G$ such that $A \subset abVb^{-1}$. Then we have:

$$AB \subset abVb^{-1}bV \subset abVV \subset abU.$$

Since $AB \in [\eta\xi]$, this completes the first part of the proof. Similarly, one can find $A_1 \in \eta$, $B_1 \in \xi$ and $a_1, b_1 \in G$ such that $A_1B_1 \subset Ub_1a_1$. Hence $[\eta\xi]$ is a Cauchy family. □

The next statement follows directly from Facts 4–7:

PROPOSITION 3.6.1. *If $\eta$ and $\xi$ are canonical filters, then $[\eta\xi]$ is an open filter base which is both shrinking and Cauchy, and $o([\eta\xi])$ is a canonical filter.*

Now we have all tools to define the Raĭkov completion of a topological group $G$. Let $G^*$ be the family of all canonical filters on $G$. Note that $B_x$ is a canonical filter on $G$, for each $x \in G$. Put $i(x) = B_x$, for each $x \in G$. Thus, we have defined a one-to-one mapping $i$ of $G$ into $G^*$.

Our program is as follows. First, we will introduce operations on $G^*$ which will turn $G^*$ into a group. Then we will introduce a topology on $G^*$ which will turn this group into a topological group. And after all that is done, we will check that $i$ is a topological isomorphism of $G$ onto the topological subgroup $i(G)$ of $G^*$.

We define a multiplication $\circ$ on $G^*$ as follows: for any $\eta, \xi \in G^*$, let $\eta \circ \xi = o([\eta\xi])$. Note that the associativity of $\circ$ follows from the next obvious assertion:

**Fact 8.** If $\eta$ and $\xi$ are open filter bases, then $o([o(\eta)o(\xi)]) = o([\eta\xi])$.

The inverse of $\eta \in G^*$ is defined by the rule: $\eta^{-1} = \{U^{-1} : U \in \eta\}$. Clearly, $\eta^{-1}$ is in $G^*$. One can also easily check the next two facts:

**Fact 9.** $(B_x)^{-1} = B_{x^{-1}}$, for each $x$ in $G$.

**Fact 10.** $B_x \circ B_y = B_{xy}$, for any $x$ and $y$ in $G$.

The property of being a Cauchy family can be given the following equivalent form:

LEMMA 3.6.2. *A family $\xi$ of subsets of a topological group $G$ is a Cauchy family if and only if for each open neighbourhood $U$ of the neutral element $e$, there exists $P \in \xi$ such that $PP^{-1} \subset U$ and $P^{-1}P \subset U$.*

PROOF. Assume that $\xi$ is Cauchy. Take an open neighbourhood $V$ of $e$ such that $VV^{-1} \subset U$. There exist $P \in \xi$ and $a \in G$ such that $P \subset Va$. Then we have:

$$PP^{-1} \subset Vaa^{-1}V^{-1} \subset U.$$

Conversely, assume that $PP^{-1} \subset U$, for some $P \in \xi$. Since $P$ is not empty, we can fix $a \in P$. Then $Pa^{-1} \subset U$ and, therefore, $P \subset Ua$. The rest is obvious. □

The lemma above implies:

**Fact 11.** $\eta \circ \eta^{-1} = B_e = \eta^{-1} \circ \eta$, for any $\eta \in G^*$.

Let us call two filter bases $\eta$ and $\xi$ *synchronous*, or *meshing*, if for every $A \in \eta$ and for every $B \in \xi$, the intersection $A \cap B$ is not empty. The next simple result reveals one of basic properties of canonical filters — their minimality in a certain rather strong sense.

PROPOSITION 3.6.3. *If $\eta$ is a canonical filter and $\xi$ is a Cauchy open filter synchronous with $\eta$, then $\eta \subset \xi$.*

PROOF. Take any $U \in \eta$. Since $\eta$ is shrinking, there exist $P \in \eta$ and an open neighbourhood $V$ of $e$ in $G$ such that $PV \subset U$. We can choose an open neighbourhood $W$ of $e$ such that $W^{-1}W \subset V$.

Since $\xi$ is Cauchy, there exists $b \in G$ such that $bW \in \xi$. The open filters $\eta$ and $\xi$ are synchronous; therefore, $P \cap bW$ is not empty. Then $b \in PW^{-1}$, which implies that $bW \subset PW^{-1}W \subset PV \subset U$. Thus, $bW \subset U$, whence it follows that $U \in \xi$, since $\xi$ is an open filter. Hence, $\eta \subset \xi$. □

COROLLARY 3.6.4. *If two canonical filters are synchronous, they coincide.*

COROLLARY 3.6.5. *If $\eta$ and $\xi$ are any two different canonical filters, then there exist $U \in \eta$ and $V \in \xi$ such that $U \cap V = \varnothing$.*

Now we are going to apply the last statements to study $G^*$.

**Fact 12.** $\eta \circ B_e = \eta$, for each $\eta \in G^*$.

PROOF. Since $e \in V$ for each $V \in B_e$, the canonical filter $\eta$ is contained in the canonical filter $\xi = \eta \circ B_e$. Therefore, $\eta$ and $\xi$ are synchronous, and $\xi = \eta$ by Corollary 3.6.4. □

It is clear from the above facts that $\circ$ is a group operation on the set $G^*$, and that the canonical filter $B_e$ plays the role of the neutral element in $G^*$. It is also clear (see Facts 9 and 10) that the mapping $i$ is an isomorphism of the group $G$ onto the subgroup $i(G) = \{B_x : x \in G\}$ of $G^*$.

It remains to define a topology on $G^*$ which would make $G$ and $i(G)$ homeomorphic and multiplication continuous. We denote by $\mathcal{T}$ the topology of $G$. Take any open set $U$ in $G$, and put $U^* = \{\eta \in G^* : U \in \eta\}$. Clearly, we have:

**Fact 13.** For any open sets $U$ and $V$ in $G$, we have: $U^* \cap i(G) = i(U)$ and $U^* \cap V^* = (U \cap V)^*$.

It follows from Fact 13 that the family $\mathcal{B} = \{U^* : U \in \mathcal{T}\}$ is a base of a topology $\mathcal{T}^*$ on $G^*$. Note that $\mathcal{T}^*$ generates on $i(G)$ the topology consisting precisely of the sets $i(W)$, where $W \in \mathcal{T}$, since the trace of the base $\mathcal{B}$ on $i(G)$ is this family of sets. Thus, the mapping $i$ is a homeomorphism of $G$ onto the subspace $i(G)$ of the space $G^*$.

Obviously, the inverse operation on $G^*$ is continuous, since $(U^*)^{-1} = (U^{-1})^*$, for each open set $U$ in $G$.

Finally, let us show that the multiplication on $G^*$ is continuous. It suffices to consider basic open sets. Let $\eta$, $\xi$ be any elements of $G^*$, and let $W^*$ be any basic open neighbourhood of $\eta \circ \xi$ in $G^*$, where $W$ is an open set in $G$. Then $W \in \eta \circ \xi$. By the definition of multiplication, $\eta \circ \xi = o([\eta\xi])$. Therefore, there exist open sets $U \in \eta$ and $V \in \xi$ such that $UV \subset W$. Clearly, $\eta \in U^*$ and $\xi \in V^*$. We claim that $U^* \circ V^* \subset W^*$. Indeed, let $\eta_1 \in U^*$ and $\xi_1 \in V^*$. Then $UV \in [\eta_1\xi_1]$, which implies that $W \in o([\eta_1\xi_1]) = \eta_1 \circ \xi_1$, that is, $\eta_1 \circ \xi_1 \in W^*$. The continuity of the multiplication is checked.

Since the trace of each non-empty element of the base $\mathcal{B}$ on $i(G)$ is not empty, $i(G)$ is dense in $G^*$.

Let us now show how to construct canonical filters, starting from any Cauchy filter on $G$. This is, obviously, an important step — so far the only examples of canonical filters we explicitly mentioned were the trivial, fixed, canonical filters $B_x$, where $x \in G$.

Below we use notation: If $\xi$ is a family of subsets of $G$, then $s(\xi)$ is the family of all subsets of $G$ of the form $UPV$, where $U$ and $V$ are any open neighbourhoods of $e$, $P \in \xi$, and $c(\xi) = o(s(\xi))$.

PROPOSITION 3.6.6. *Let $\xi$ be a Cauchy filter on $G$. Then $c(\xi)$ is a canonical filter on $G$ contained in $\xi$.*

PROOF. It is clear that $s(\xi)$ is an open filter base and $s(\xi) \subset \xi$. Let us check that the family $s(\xi)$ is shrinking. Indeed, take any $A \in s(\xi)$. Then there are open neighbourhoods $U$ and $V$ of $e$ and $P \in \xi$ such that $A = UPV$. We can find open neighbourhoods $U_1$

and $V_1$ of $e$ such that $U_1^2 \subset U$ and $V_1^2 \subset V$. Then we have: $B = U_1PV_1 \in s(\xi)$, and $U_1BV_1 = U_1^2PV_1^2 \subset UPV = A$. It follows that $s(\xi)$ is shrinking.

Let us check that $s(\xi)$ is Cauchy. Take any open neighbourhood $V$ of $e$, and choose an open neighbourhood $W$ of $e$ such that $W^3 \subset V$. There exists $b$ in $G$ such that $Wb \in \xi$, since $\xi$ is a Cauchy open filter. Then $b^{-1}Wb$ is an open neighbourhood of $e$, and $W(Wb)b^{-1}Wb = W^3b$ is in $s(\xi)$, by the definition of $s(\xi)$. Since $W^3b \subset Vb$, it follows that $s(\xi)$ is Cauchy.

Now, by Fact 4, we conclude that $o(s(\xi))$, that is, $c(\xi)$ is a canonical filter. Since $\xi$ is an open filter, and $s(\xi) \subset \xi$, we have that $c(\xi) = o(s(\xi)) \subset \xi$. This finishes the proof. □

PROPOSITION 3.6.7. *Let $G$ be a dense subgroup of a topological group $H$, and let $\eta$ be an open filter on $H$. Put $\eta_G = \{W \cap G : W \in \eta\}$. Then:*

1) *$\eta_G$ is an open filter on $G$ synchronous with $\eta$;*
2) *if $\eta$ is Cauchy in $H$, then $\eta_G$ is Cauchy in $G$;*
3) *if $\eta_G$ converges in the space $G$, then $\eta$ converges in the space $H$ to the same point;*
4) *if $\eta$ is canonical in $H$, then $\eta_G$ is canonical in $G$.*

PROOF. Both parts of 1) easily follow from the assumption that $G$ is dense in $H$. To prove 3), it is sufficient to note that the space $H$ is regular, and that $W \subset \overline{W \cap G}$, for each $W \in \eta$ (the closure is taken in $H$).

Now we prove 2). Let $W$ be an open neighbourhood of the neutral element $e$ in $G$. There exists an open neighbourhood $U$ of $e$ in $H$ such that $U \cap G = W$. By Lemma 3.6.2, there exists $F \in \eta$ such that $FF^{-1} \subset U$ and $F^{-1}F \subset U$. Put $P = F \cap G$. Then $P \in \eta_G$ and $PP^{-1} \subset G \cap U = W$. Similarly, $P^{-1}P \subset W$. Therefore, $\eta_G$ is a Cauchy filter in $G$.

The proof of 4) is similar to the proof of 2) and is left to the reader. □

Notice that the proof of 2) is necessary since being a Cauchy family is formally, according to the definition, a relative property — we have to specify with respect to which topological group the family is Cauchy.

The lemma below is obvious; it complements Proposition 3.6.7.

LEMMA 3.6.8. *If $\eta$ is a Cauchy filter on $G$ and the filter $o(\eta)$ converges in $G$, then $\eta$ converges in $G$ as well.*

The next result specifies one of the basic properties of $G^*$:

PROPOSITION 3.6.9. *Every Cauchy filter $\eta$ on $G^*$ converges.*

PROOF. By Lemma 3.6.8 and by Facts 1 and 3, it suffices to consider the case when $\eta$ is an open Cauchy filter. It is convenient to identify any $x \in G$ with $i(x)$. So now $i(G)$ becomes $G$, so that $G$ is a dense subgroup of $G^*$. Put $\eta_G = \{W \cap G : W \in \eta\}$. Then $\eta_G$ is an open filter on $G$ synchronous with $\eta$. By Proposition 3.6.7, since $\eta$ is Cauchy in $G^*$, the open filter $\eta_G$ is Cauchy in $G$. Put $\xi = c(\eta_G)$. Then, by Proposition 3.6.6, $\xi$ is a canonical filter on $G$ contained in $\eta_G$.

An arbitrary basic open neighbourhood of the point $\xi$ in $G^*$ is of the form $U^*$, where $U$ is an element of $\xi$. Then $U \in \eta_G$ and, since $U \subset U^*$, this implies that $\eta_G$ converges to the point $\xi$ in $G^*$. By 3) of Proposition 3.6.7, this means that the filter $\eta$ also converges to $\xi$. □

A topological group $G$ such that every Cauchy filter on $G$ converges is called *Raĭkov complete*. Thus, we can sum up the results of our construction and of arguments above as follows:

THEOREM 3.6.10. **[D. A. Raĭkov]** *For every topological group $G$, there exists a Raĭkov complete topological group $G^*$ and a canonical topological isomorphism $i$ of $G$ onto a dense subgroup $i(G)$ of $G^*$.*

We are now going to establish a general result which implies that a topological group $G^*$ such as in Theorem 3.6.10 is, in a natural sense, unique. This requires two preliminary facts the first of which is obvious.

LEMMA 3.6.11. *Under a continuous homomorphism of a topological group $G$ into a topological group $H$, the image of each Cauchy filter on $G$ is a Cauchy filter on $H$.*

In the next proposition we present one of the most important properties of Raĭkov complete topological groups.

PROPOSITION 3.6.12. *Let $G$ be a dense subgroup of a topological group $H$ and $f: G \to K$ a continuous homomorphism of $G$ to a Raĭkov complete group $K$. Then $f$ admits an extension to a continuous homomorphism $f^*: H \to K$.*

PROOF. For every $z \in H$, let $\eta_z$ be the family of all open neighbourhoods of $z$ in $H$. Put $\xi_z = \{U \cap G : U \in \eta_z\}$. Then, obviously, $\eta_z$ is a Cauchy filter on $H$ and, therefore, $\xi_z$ is a Cauchy filter on $G$.

By Lemma 3.6.11, the family $f(\xi_z) = \{f(P) : P \in \xi_z\}$ is a Cauchy filter on the Raĭkov complete group $K$. Therefore, it converges to some $y \in K$. Put $f^*(z) = y$. Clearly $f^*$ is correctly defined, and to prove that the mapping $f^*$ of $H$ to $K$ is continuous, it suffices to establish that if $z$ is in the closure of a subset $A$ of $G$, then $y = f^*(z)$ is in the closure of $f(A)$ (see [**165**, 3.2.A]). Let us do this.

Assume the contrary, and put $\delta_A = \{U \cap A : U \in \eta_z\}$. Then $\delta_A$ is a Cauchy filter on $G$. Therefore, by Lemma 3.6.11, the family $f(\delta_A)$ is a Cauchy filter on $K$. It follows that the filter $f(\delta_A)$ converges to some $y_1 \in K$. Then $y_1$ is in the closure of $f(A)$ and, therefore, $y_1 \neq y$. However, every element of the filter $\delta_A$ is contained in some element of the filter $\eta_z$ and, hence, in some element of $\xi_z$. Therefore, the filters $\delta_A$ and $\xi_z$ are meshing. It follows that the filters $f(\delta_A)$ and $f(\xi_z)$ are also meshing, which implies that the points to which they converge must coincide. This contradiction shows that the mapping $f^*$ is continuous.

It remains to verify that $f^*$ is a homomorphism. If not, there exist elements $a, b \in H$ such that $f^*(ab) \neq f^*(a)f^*(b)$. Let $O$ and $W$ be disjoint open in $K$ neighbourhoods of $f^*(ab)$ and $f^*(a)f^*(b)$, respectively. Choose open neighbourhoods $U$ and $V$ of $f^*(a)$ and of $f^*(b)$, respectively, such that $UV \subset W$. By the continuity of $f^*$ and of the multiplication in $H$, one can find open neighbourhoods $U_1$ and $V_1$ of $a$ and $b$, respectively, in $H$ such that $f^*(U_1) \subset U$, $f^*(V_1) \subset V$ and $f^*(U_1V_1) \subset O$. Since $G$ is dense in $H$, there exist $a_1 \in U_1 \cap G$ and $b_1 \in V_1 \cap G$. Since $f$ is a homomorphism and $f^*$ coincides with $f$ on $G$, we have, on one hand, that

$$f^*(a_1b_1) = f(a_1b_1) = f(a_1)f(b_1) = f^*(a_1)f^*(b_1) \in UV \subset W.$$

On the other hand, $f^*(a_1b_1) \in f^*(U_1V_1) \subset O$. It now follows that $f^*(a_1b_1) \in O \cap W \neq \varnothing$, thus contradicting the choice of the sets $O$ and $W$. The proof is complete. ☐

Proposition 3.6.13. *Let $f: G \to H$ be a topological isomorphism of topological groups. Suppose that $G$ and $H$ are dense subgroups of Raĭkov complete groups $G^*$ and $H^*$, respectively. Then $f$ admits a continuous extension to a topological isomorphism $f^*: G^* \to H^*$.*

Proof. By Proposition 3.6.12, one can extend $f$ to a continuous homomorphism $\varphi: G^* \to H^*$. Similarly, $f^{-1}$ admits an extension to a continuous homomorphism $\psi: H^* \to G^*$. Then the restriction of $\psi \circ \varphi$ to $G$ coincides with the identity mapping $i_G$ of $G$ onto itself and, similarly, the restriction of $\varphi \circ \psi$ to $H$ is the identity mapping $i_H$ of $H$ onto itself. Hence, $\psi \circ \varphi$ is the identity mapping of $G^*$ onto itself and $\varphi \circ \psi$ is the identity mapping of $H^*$. This implies that $\varphi$ is a topological isomorphism of $G^*$ onto $H^*$. □

The following theorem on the uniqueness of the Raĭkov completion of a given a topological group is now immediate.

Theorem 3.6.14. *Let $G$ be a topological group, and let $H_1$ and $H_2$ be Raĭkov complete topological groups such that $G$ is a dense topological subgroup of both of them. Then there exists a topological isomorphism $\varphi$ of $H_1$ onto $H_2$ such that $\varphi(g) = g$, for each $g \in G$.*

Theorems 3.6.10 and 3.6.14 enable us to call the topological group $G^*$ constructed above the *Raĭkov completion* of the group $G$. In the sequel, the group $G^*$ is denoted by $\varrho G$, and $G$ will be always identified with the dense subgroup $i(G)$ of $\varrho G$. Clearly, we have:

Proposition 3.6.15. *For any topological group $G$, $\varrho\varrho G = \varrho G$. In particular, $G$ is Raĭkov complete if and only if $\varrho G = G$.*

Corollary 3.6.16. *Let $G$ be a dense subgroup of a topological group $H$. Then there exists a topological isomorphism $i_H$ of $H$ onto a subgroup of the Raĭkov completion $\varrho G$ of $G$ such that $i_H(g) = g$, for each $g \in G$.*

Proof. Let $\varrho H$ be the Raĭkov completion of $H$. Then $G$ is a dense subgroup of $\varrho H$, and both topological groups $\varrho G$ and $\varrho H$ are Raĭkov complete. Therefore, Theorem 3.6.14 applies, and there exists a topological isomorphism $\varphi$ of $\varrho H$ onto $\varrho G$ such that $\varphi(g) = g$, for each $g \in G$. The restriction of $\varphi$ to $H$ is the topological isomorphism we are looking for. □

The result just proved means that all topological groups $H$ containing a given topological group $G$ as a dense subgroup can be naturally identified with the subgroups $K$ of $\varrho G$ with $G \subset K \subset \varrho G$. This canonical identification will be used throughout the rest of the book.

The next result is a special case of Proposition 3.6.12. It establishes a basic property of Raĭkov's completion.

Corollary 3.6.17. *Every continuous homomorphism $f$ of a topological group $G$ to a topological group $H$ can be extended to a continuous homomorphism $f^*$ of $\varrho G$ to $\varrho H$.*

Here is another useful fact that complements Proposition 3.6.13:

Corollary 3.6.18. *Let $f: G \to H$ be a continuous homomorphism of topological groups, and $D$ a dense subgroup of $G$. If the restriction of $f$ to $D$ is a topological isomorphism of $D$ onto the subgroup $f(D)$ of $H$, then $f$ is a topological isomorphism of $G$ onto the subgroup $f(G)$ of $H$.*

PROOF. By Corollary 3.6.17, $f$ admits an extension to a continuous homomorphism $f^*: \varrho G \to \varrho H$. Put $E = f(D)$, and let $K$ be the closure of $E$ in $\varrho H$. Then $E$ is a dense subgroup of the Raĭkov complete group $K$, while $D$ is dense in the Raĭkov complete group $\varrho G$. Hence, according to Proposition 3.6.13, the restriction $g = f \restriction D$ admits an extension to a topological isomorphism $g^*: \varrho G \to K$. Since $g^*$ and $f^*$ coincide on the dense subgroup $D$ of $\varrho G$, we conclude that $g^* = f^*$. It follows that $f^*$ is a topological isomorphism of $\varrho G$ onto $K$ and, hence, $f$ is a topological isomorphism of $G$ onto $f(G)$. □

Corollary 3.6.18 fails to be true in the category of topological spaces. Indeed, consider the continuous mapping $g: I \to \mathbb{T}$ of the closed unit interval $I = [0, 1]$ onto the circle $\mathbb{T}$ given by $g(x) = e^{2\pi i x}$, for each $x \in I$. The restriction of $g$ to the dense open subspace $(0, 1)$ of $I$ is a homeomorphism of $(0, 1)$ onto the subspace $\mathbb{T} \setminus \{1\}$ of $\mathbb{T}$, but $g$ is clearly not a homeomorphism since $g(0) = g(1) = 1$.

One cannot expect that the homomorphic extension $f^*: \varrho G \to \varrho H$ in Corollary 3.6.17 be onto or quotient, even if the homomorphism $f: G \to H$ has both properties (see Problem 3.6.G). The next theorem shows, however, that if $f$ is quotient and $f^*$ is onto, then $f^*$ is quotient as well.

THEOREM 3.6.19. *Let $f: G \to H$ be a continuous onto homomorphism of topological groups. If $f$ is open, then the continuous homomorphic extension $f^*: \varrho G \to \varrho H$ of $f$ is open when considered as a mapping of $\varrho G$ onto its image $f^*(\varrho G)$, and the kernel of $f^*$ is the closure of* ker $f$ *in $\varrho G$. Therefore, $f^*: \varrho G \to \varrho H$ is open provided that $f$ is open and $f^*(\varrho G) = \varrho H$.*

PROOF. Let us prove the first part of the theorem. Suppose that $f$ is a quotient homomorphism. The kernel $N$ of $f$ is a closed invariant subgroup of $G$, and we claim that the closure of $N$ in $\varrho G$, say, $K$ is an invariant subgroup of $\varrho G$. If not, there exist elements $x \in \varrho G$ and $y \in K$ such that $xyx^{-1} \notin K$. Then $O = \varrho G \setminus K$ is an open neighbourhood of $xyx^{-1}$ in $\varrho G$, and the continuity of the multiplication in $\varrho G$ implies that we can find open sets $U$ and $V$ in $\varrho G$ such that $x \in U$, $y \in V$, and $UVU^{-1} \subset O$. Since $G$ is dense in $\varrho G$, there exist elements $a \in U \cap G$ and $b \in V \cap N$. It follows from the choice of $U$ and $V$ that $g = aba^{-1} \in O$, that is, $g \notin K$. Since $a \in G$ and $b \in N$, we conclude that $g \notin K \cap G = N$, which contradicts the invariance of $N$ in $G$.

Let $\pi: \varrho G \to \varrho G/K$ be the quotient homomorphism. It is clear that the kernel $P$ of $f^*$ is a closed subgroup of $\varrho G$ that contains $N$, whence it follows that $K \subset P$. Hence, there exists a homomorphism $\varphi: \varrho G/K \to \varrho H$ satisfying $f^* = \varphi \circ \pi$.

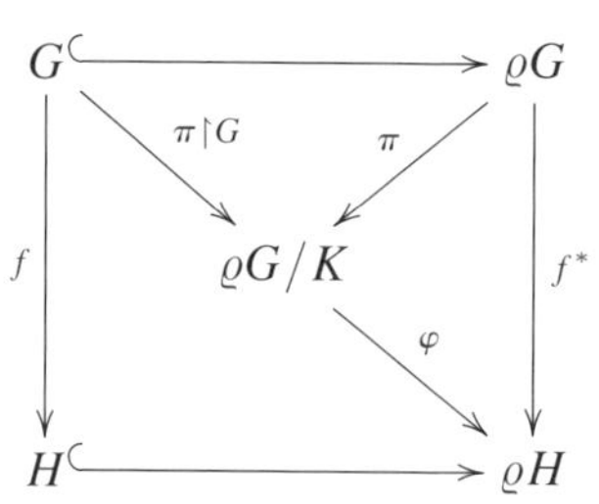


Since the homomorphism $\pi$ is quotient, $\varphi$ is continuous. Further, since $N = K \cap G$ is dense in $K$, it follows from Theorem 1.5.16 that the restriction of $\pi$ to $G$ is an open homomorphism of $G$ onto the subgroup $\pi(G)$ of $\varrho G/K$. We claim that the restriction of $\varphi$ to $\pi(G)$ is a topological isomorphism of $\pi(G)$ onto $H$.

Indeed, the kernel of the homomorphism $\pi\restriction G$ is the group $N = K \cap G$, and $\ker(\varphi\circ\pi)\restriction G = \ker f^*\restriction G = \ker f = N$. It follows that the kernels of the homomorphisms $\pi\restriction G$ and $(\varphi\circ\pi)\restriction G$ coincide, so that the restriction of $\varphi$ to $\pi(G)$ is a monomorphism. Further, the homomorphisms $\pi\restriction G$ and $f = (\varphi\circ\pi)\restriction G$ are open when considered as mappings onto their images $\pi(G)$ and $H$, respectively. Hence, the restriction of $\varphi$ to $\pi(G)$ is a homeomorphism of $\pi(G)$ onto $H$, which proves our claim.

Since $\pi(G)$ is dense in $\varrho G/K$, it follows from Corollary 3.6.18 that $\varphi$ is a topological isomorphism of $\varrho G/K$ onto the subgroup $\varphi(\varrho G/K) = f^*(\varrho G)$ of $\varrho H$. Hence, the equality $f^* = \varphi\circ\pi$ implies that $f^*$ is an open homomorphism of $\varrho G$ onto the subgroup $f^*(\varrho G)$ of $\varrho H$, and that $K$ is the kernel of $f^*$. This finishes the proof of the first part of the theorem.

Finally, if $f^*(\varrho G) = \varrho H$, then $f^*$ is open, by the first claim of the theorem. □

Now we are going to show that the class of metrizable topological groups is closed under taking Raĭkov completions. This result will be deduced from a more general fact given below.

PROPOSITION 3.6.20. *Let $H$ be a metrizable subgroup of a topological group $G$. Then the closure of $H$ in $G$ is also a metrizable subgroup of $G$.*

PROOF. Denote by $K$ the closure of $H$ in $G$. It follows from Corollary 1.4.14 that $K$ is a subgroup of $G$, while Proposition 1.4.16 implies that $K$ is also first-countable and, hence, metrizable by Theorem 3.3.12. □

COROLLARY 3.6.21. *For any metrizable topological group $G$, its Raĭkov completion $\varrho G$ is metrizable.*

The next theorem is analogous to a statement about complete uniform spaces (see [**165**, Theorem 8.3.9]). Its proof is close to that of the Tychonoff Compactness Theorem.

THEOREM 3.6.22. *Every topological product $G = \prod_{i\in I} G_i$ of Raĭkov complete groups is Raĭkov complete.*

PROOF. Let $\xi$ be a Cauchy filter on $G$. Then there exists an ultrafilter $\eta$ on $G$ with $\xi \subset \eta$. Clearly, $\eta$ is a Cauchy filter on $G$ as well. For every $i \in I$, let $\pi_i$ be the projection of $G$ to the factor $G_i$. By Lemma 3.6.11, the family $\eta_i = \{\pi_i(F) : F \in \eta\}$ is a Cauchy filter on $G_i$. Since the group $G_i$ is Raĭkov complete, $\eta_i$ converges to some point $b_i \in G_i$. Denote by $b$ the point in $G$ satisfying $\pi_i(b) = b_i$ for each $i \in I$. Since $\eta_i$ converges to $b_i$ for $i \in I$, we conclude that $\pi_i^{-1}(V) \cap F \neq \varnothing$ for every neighbourhood $V$ of $b_i$ in $G_i$ and for every $F \in \eta$. By our choice, $\eta$ is an ultrafilter, so that $\pi_i^{-1}(V) \in \eta$.

We claim that $\eta$ converges to $b$. Indeed, let $O$ be an arbitrary neighbourhood of $b$ in $G$. There exists a canonical open neighbourhood $U$ of $b$ in $G$ such that $U \subset O$. Then $U$ has the form $U = \pi_{i_1}^{-1}(U_1) \cap \ldots \cap \pi_{i_n}^{-1}(U_n)$, where $i_k \in I$ and $U_k \ni b_{i_k}$ is open in $G_{i_k}$ for each $k \leq n$. By the above observation, we have $O_k = \pi_{i_k}^{-1}(U_k) \in \eta$ for $k = 1, \ldots, n$. Since $\eta$ is a filter, this implies that $U = O_1 \cap \ldots \cap O_k \in \eta$. So, from $U \subset O$ it follows that $O \in \eta$. This and the inclusion $\xi \subset \eta$ together imply that every neighbourhood $O$ of $b$ intersects all elements of the filter $\xi$ or, in other words, $b$ is an accumulation point of $\xi$. But $\xi$, being a Cauchy filter, converges to the point $b$. This finishes the proof. □

Theorem 3.6.22 implies that the operations of taking topological products and Raĭkov completions of topological groups commute:

COROLLARY 3.6.23. *Let $G = \prod_{i\in I} G_i$ be a product of topological groups. Then $\varrho G$ is topologically isomorphic to the product group $\prod_{i\in I} \varrho G_i$.*

PROOF. It is clear that $G$ is a dense subgroup of the group $H = \prod_{i\in I} \varrho G_i$. Since the group $H$ is Raĭkov complete, by Theorem 3.6.22, and there exists a unique (up to a topological isomorphism) Raĭkov complete topological group containing $G$ as a dense subgroup (Theorem 3.6.14), we conclude that $\varrho G \cong H$. □

The following simple statement is good to keep in mind. It will be considerably generalized in Theorem 4.3.7.

THEOREM 3.6.24. *Every locally compact topological group $G$ is Raĭkov complete.*

PROOF. The group $G$ is an open subgroup of its Raĭkov completion $\varrho G$, since $G$ is locally compact and dense in $\varrho G$. However, every open subgroup of a topological group is closed in it. Therefore, $G = \varrho G$, that is, $G$ is Raĭkov complete. □

There exists a natural relation between Raĭkov complete groups and complete uniform spaces. To establish this relation we briefly recall the necessary concepts from the theory of uniform spaces.

Let $(X, \mathcal{U})$ be a uniform space. A filter $\xi$ of subsets of $X$ is called *Cauchy* in $(X, \mathcal{U})$ if, for each $U \in \mathcal{U}$, there exists an element $F \in \xi$ such that $F \times F \subset U$. The space $(X, \mathcal{U})$ is *complete* if every Cauchy filter in $(X, \mathcal{U})$ converges to some point of $X$ (see [**165**, Section 8.3], where Cauchy filters are called *filters with arbitrarily small sets*).

THEOREM 3.6.25. *A topological group $G$ is Raĭkov complete if and only if the uniform space $(G, \mathcal{V})$ is complete, where $\mathcal{V}$ is the two-sided uniformity of the group $G$.*

PROOF. Suppose that the group $G$ is Raĭkov complete. We have to verify that every Cauchy filter $\xi$ in $(G, \mathcal{V})$ converges. Take an arbitrary open symmetric neighbourhood $V$ of the neutral element $e$ in $G$. Then

$$O_V = \{(x, y) \in G \times G : x^{-1}y \in V,\ xy^{-1} \in V\}$$

is an element of the uniformity $\mathcal{V}$ (see (1.8)–(1.10)). Since $\xi$ is Cauchy in $(G, \mathcal{V})$, there exists $F \in \xi$ such that $F \times F \subset O_V$. If $x \in F$, then $F \subset xV \cap Vx$, so that $\xi$ is a Cauchy filter in the group $G$. The group $G$ being Raĭkov complete, $\xi$ must converge to some point of $G$. This proves that the uniform space $(G, \mathcal{V})$ is complete.

Conversely, suppose that the uniform space $(G, \mathcal{V})$ is complete. To show that the group $G$ is Raĭkov complete, take an arbitrary Cauchy filter $\xi$ in $G$. We claim that $\xi$ is also a Cauchy filter in $(G, \mathcal{V})$. Indeed, if $U \in \mathcal{V}$, there exist symmetric open neighbourhoods $V$ and $W$ of $e$ in $G$ such that $O_V \subset U$ and $W^2 \subset V$. It follows from our choice of $\xi$ that $F \subset aW \cap Wb$, for some $F \in \xi$ and $a, b \in G$. Let $x, y \in F$ be arbitrary. Since $W$ is symmetric, we have that $a^{-1}x \in W$, $x^{-1}a \in W$, and $a^{-1}y \in W$. Hence, $x^{-1}y = (x^{-1}a)(a^{-1}y) \in W^2$. Similarly, from $xb^{-1} \in W$, $yb^{-1} \in W$, and $by^{-1} \in W$ it follows that $xy^{-1} = (xb^{-1})(by^{-1}) \in W^2$. Since $W^2 \subset V$, we conclude that $(x, y) \in O_V$. This proves that $F \times F \subset O_V$, so $\xi$ is a Cauchy filter in $(G, \mathcal{V})$. Since, by our assumption, the uniform space $(G, \mathcal{V})$ is complete, $\xi$ converges to some point of $G$. We get to the conclusion that the group $G$ is Raĭkov complete. □

The reader has surely noted that in the proof of Theorem 3.6.25 we established the fact that a filter $\xi$ of subsets of the group $G$ is a Cauchy filter in $G$ if and only if $\xi$ is Cauchy in the uniform space $(G, \mathcal{V})$.

## Exercises

3.6.a. Verify that if $\eta$ is an open filter base, then $o(\eta)$ is an open filter containing $\eta$.

3.6.b. Show that if $\eta$ is a shrinking family of open sets, then $o(\eta)$ is also a shrinking family of open sets.

3.6.c. Verify that if $\eta$ is a Cauchy family of sets, then $o(\eta)$ is also a Cauchy family of sets.

3.6.d. Verify that if $\eta$ is an open filter base which is both Cauchy and shrinking, then $o(\eta)$ is a canonical filter containing $\eta$.

3.6.e. Show that if $\eta$ and $\xi$ are open filter bases, then $[\eta\xi]$ is also an open filter base.

3.6.f. Verify that if $\eta$ and $\xi$ are open filter bases, then $o([o(\eta)o(\xi)]) = o([\eta\xi])$.

3.6.g. Verify that for any $\eta \in G^*$, the family $\eta^{-1} = \{U^{-1} : U \in \eta\}$ is also an element of $G^*$.

3.6.h. Show that $(B_x)^{-1} = B_{x^{-1}}$ and $B_x \circ B_y = B_{xy}$, for all $x, y \in G$.

3.6.i. Verify that $\circ$ is a group operation on the set $G^*$, and that the canonical filter $B_e$ plays the role of the neutral element in $G^*$.

3.6.j. Verify that the mapping $i$ defined on page 183, after Proposition 3.6.1, is an isomorphism of the group $G$ onto the subgroup $i(G) = \{B_x : x \in G\}$ of $G^*$.

3.6.k. Show that if $\eta$ is a Cauchy filter on $G$ and the filter $o(\eta)$ converges in $G$, then $\eta$ converges in $G$ as well.

3.6.l. Show that for any continuous homomorphism of a topological group $G$ to a topological group $H$, the image of each Cauchy filter on $G$ is a Cauchy filter on $H$.

3.6.m. Prove that every closed subgroup of a Raĭkov complete topological group is Raĭkov complete.

3.6.n. Show that the Raĭkov completion of a second-countable topological group is a second-countable group.

3.6.o. Let $G$ be a compact topological group such that every element $a \in G$ distinct from $e_G$ generates a dense subgroup of $G$. Show that $G$ is a finite cyclic group with a prime number of elements. Can one generalize this to precompact topological groups (see page 118)?

3.6.p. Give an example of an $\omega$-narrow topological group that cannot be embedded as a topological subgroup into any paracompact topological group.
*Hint.* Use Problem 1.7.A and Theorem 3.6.22.

3.6.q. Let $X$ be a Tychonoff space. Show that the topological group $C_p(X)$ is Raĭkov complete iff the space $X$ is discrete. Is the similar statement true for the compact-open topology on $C_p(X)$?

3.6.r. Give an example of a continuous onto homomorphism $f : G \to H$ of topological groups such that the continuous extension $f^* : \varrho G \to \varrho H$ is open and onto, but $f$ fails to be open (see Theorem 3.6.19).

## Problems

3.6.A. Let $M$ be a metric space. Is the group $Is(M)$ of isometries of $M$, with the topology of uniform convergence, Raĭkov complete?

3.6.B. Let $M$ be a complete metric space. Show that the group $Is(M)$ of isometries of $M$, with the topology of uniform convergence, is Raĭkov complete. What if $Is(M)$ is endowed with the topology of pointwise convergence?

3.6.C. Let $M$ be a metric space. Show that the Raĭkov completion of the group $Is(M)$ of isometries of $M$, with the topology of uniform convergence, is the group of isometries of the completion of $M$ with respect to the metric, also taken with the topology of uniform convergence.

3.6.D. Let $M$ be a metric space. Show that Raĭkov completion of the group $Is(M)$ of isometries of $M$, with the topology of pointwise convergence, is the group of isometries of the completion of $M$ with respect to the metric, also taken with the topology of pointwise convergence.

3.6.E. Suppose that $G$ is an Abelian Raĭkov complete topological group, and $N$ is a closed subgroup of $G$. Is the quotient group $G/N$ Raĭkov complete? What if $N$ is compact or locally compact?

3.6.F. Let $N$ be a closed invariant subgroup of a topological group $G$ and suppose that the groups $N$ and $G/N$ are Raĭkov complete. Prove that $G$ is also Raĭkov complete.

3.6.G. Give an example of a quotient homomorphism $f: G \to H$ of a topological group $G$ onto $H$ such that the continuous homomorphic extension $f^*: \varrho G \to \varrho H$ (see Corollary 3.6.17) is not onto.

3.6.H. (C. Hernández and M. G. Tkachenko [**228**]) For a topological group $H$, denote by $(H)_\omega$ the underlying group $H$ with the new topology whose base of open sets consists of $G_\delta$-sets in $H$. The topological group $(H)_\omega$ is called the *$G_\delta$-modification* of $H$. Verify that $(H)_\omega$ is also a topological group and prove that $(H)_\omega$ is Raĭkov complete whenever $H$ is Raĭkov complete.

3.6.I. Construct a non-discrete Raĭkov complete Hausdorff group topology on the group $\mathbb{Z}$.

*Hint.* Let $x_n = 2^n$, for each $n \in \mathbb{N}$. Verify that there exists the maximal topological group topology $\mathcal{T}$ on $\mathbb{Z}$ such that the sequence $\{x_n : n \in \mathbb{N}\}$ converges to zero in $(\mathbb{Z}, \mathcal{T})$. Show that the topology $\mathcal{T}$ is Hausdorff and the group $(\mathbb{Z}, \mathcal{T})$ is Raĭkov complete. For details, see [**547**].

3.6.J. Give an example of a Raĭkov complete $\omega$-narrow topological group which contains an uncountable discrete family of non-empty open sets.

*Hint.* The group constructed in [**485**] has all the required properties.

3.6.K. Give an example of a Raĭkov complete topological group $H$ such that every point in $H$ is a $G_\delta$-set, but $H$ does not admit a continuous isomorphism onto a metrizable topological group. (Compare with Corollaries 3.4.24, 3.4.26, and Problem 3.4.A.)

*Hint.* Verify that the group $K^{\omega_1}$ endowed with the box topology and considered in the hint to Problem 3.4.A, is Raĭkov complete.

3.6.L. A filter $\xi$ of subsets of a topological group $G$ is said to be a *left Cauchy filter* in $G$ if for every neighbourhood $U$ of the neutral element in $G$, there exist $F \in \xi$ and $a \in G$ such that $F \subset aU$. A topological group $G$ is called *Weil complete* if every left Cauchy filter in $G$ converges.

(a) Verify that a subgroup $H$ of a Weil complete group $G$ is Weil complete iff $H$ is closed in $G$.

(b) Prove that the product of any family of Weil complete groups is Weil complete.

(c) Show that every Weil complete group is Raĭkov complete.

(d) Give an example of a Raĭkov complete group which fails to be Weil complete.

(e) Prove that a Raĭkov complete group $G$ is Weil complete if and only if $\xi^{-1} = \{F^{-1} : F \in \xi\}$ is a left Cauchy filter, for each left Cauchy filter $\xi$ in $G$. Deduce that Raĭkov completeness and Weil completeness coincide for Abelian topological groups.

(f) Use (e) to show that all locally compact topological groups are Weil complete.

(g) Give an example of a topological group that cannot be embedded into a Weil complete group as a topological subgroup. Conclude that Theorem 3.6.10 cannot be extended to Weil complete groups.

(h) Find out whether the assertions in Problems 3.6.F and 3.6.H remain valid for Weil complete groups.

## Open Problems

3.6.1. Suppose that $G$ is a Raĭkov complete Fréchet–Urysohn topological group. Is $G$ metrizable? What if, in addition, $G$ is countable?

3.6.2. Does there exist in $ZFC$ a Raĭkov complete $\omega$-narrow topological group which contains a discrete family $\gamma$ of open sets with $|\gamma| = 2^{\omega}$? (See also Problem 3.6.J.)

3.6.3. Suppose that $G$ and $H$ are homeomorphic topological groups such that $G$ is Raĭkov complete. Must $H$ be Raĭkov complete?

3.6.4. Suppose that $X$ is a Tychonoff space such that $C_p(X)$ is homeomorphic to a Raĭkov complete topological group. Is $X$ discrete?

3.6.5. Must the Sorgenfrey line $S$ be closed in every (Hausdorff, regular) paratopological group containing it as a paratopological subgroup?

3.6.6. Characterize Hausdorff (regular) paratopological groups that are closed in every Hausdorff (regular) paratopological group containing them as a paratopological subgroup.

## 3.7. Precompact groups and precompact sets

A left topological group $G$ is called *precompact* if, for every open neighbourhood $V$ of the neutral element in $G$, there exists a finite subset $A$ of $G$ such that $AV = G$. Similarly, a right topological group $G$ is called *precompact* if, given an open neighbourhood $V$ of the neutral element in $G$, one can find a finite subset $B$ of $G$ such that $VB = G$ (see also page 118).

For a quasitopological group $G$, the existence of a finite set $A \subset G$ with $G = AV$ in the above definition is equivalent to the existence of a finite set $B \subset G$ satisfying $G = VB$ (compare with Proposition 3.4.1). Therefore, for quasitopological groups and, in particular, for topological groups one can introduce the concept of precompactness in terms of left translations or, equivalently, right translations. Precompact topological groups are also said to be *totally bounded*, by analogy with metric spaces. Obviously, every compact topological group is precompact and precompact topological groups are $\omega$-narrow. It is also clear that a discrete group is precompact if and only if it is finite.

The proof of the next stability result is straightforward and is omitted.

PROPOSITION 3.7.1. *If $f$ is a continuous homomorphism of a precompact topological group $G$ onto a topological group $H$, then the group $H$ is also precompact.*

It is clear that every compact group is precompact. The following fact is a bit less obvious.

THEOREM 3.7.2. *Every pseudocompact topological group is precompact.*

PROOF. Let $V$ be an arbitrary symmetric open neighbourhood of the neutral element $e$ in $G$. As in the proof of Proposition 3.4.31, consider a maximal $V$-disjoint subset $A$ of $G$. If $U$ is a symmetric open neighbourhood of $e$ with $U^4 \subset V$, then the family of open sets $\{xU : x \in A\}$ is discrete in $G$, so $A$ must be finite since $G$ is pseudocompact. It follows from the maximality of $A$ that $G = AV$, thus implying the conclusion of the theorem. □

The concept of precompactness admits a natural extension to subsets of semitopological groups as follows. A subset $B$ of a semitopological group $G$ is called *precompact* in $G$ if, for every neighbourhood $U$ of the identity in $G$, there exists a finite set $F \subset G$ such that $B \subset FU$ and $B \subset UF$.

In general, one cannot weaken the requirement on the set $B$ in the above definition to the single condition $B \subset FU$ (or $B \subset UF$), not even if $G$ is a topological group (see Problem 3.7.B). If, however, $B$ is a symmetric subset of a quasitopological group $G$, the three different definitions coincide, by the argument given in the proof of Proposition 3.4.1.

It is clear that every compact subset of a topological group is precompact. The converse is obviously false — every pseudocompact non-compact group is a counterexample, by Theorem 3.7.2. Another way to see this is to take a countable dense subgroup of the circle group $\mathbb{T}$ and apply Proposition 3.7.4 below.

Let us show that a finite subset $F$ of $G$ in the definition of precompact sets in topological groups can always be chosen inside $B$.

LEMMA 3.7.3. *Let $B$ be a precompact subset of a topological group $G$ and $S$ dense in $B$. Then, for every neighbourhood $U$ of the identity in $G$, one can find a finite set $K \subset S$ such that $B \subset KU$ and $B \subset UK$.*

PROOF. Let $U$ be a neighbourhood of the identity $e$ in $G$. Choose a symmetric neighbourhood $V$ of $e$ in $G$ such that $V^2 \subset U$. Since $B$ is precompact in $G$, there exists a finite set $F$ in $G$ such that $B \subset FV$ and $B \subset VF$. If $x \in F$ and $B \cap xV \neq \varnothing$, then $S \cap xV \neq \varnothing$ and we pick a point $y_x \in S \cap xV$. Then the finite set

$$K_1 = \{y_x : x \in F \text{ and } B \cap xV \neq \varnothing\}$$

is contained in $S$ and satisfies $B \subset K_1U$. Indeed, if $b \in B$, then there exists $x \in F$ such that $b \in xV$. Clearly, $b \in B \cap xV \neq \varnothing$, so $y_x \in xV$ and $y_x^{-1}x \in V^{-1} = V$. Therefore, we have

$$b \in xV = y_x(y_x^{-1}x)V \subset y_xV^2 \subset y_xU \subset K_1U.$$

This implies the inclusion $B \subset K_1U$. Similarly, define a finite subset $K_2$ of $S$ choosing points $z_x \in S \cap Vx$, with $x \in F$, and apply the same argument to show that $B \subset UK_2$. Therefore, the finite set $K = K_1 \cup K_2 \subset S$ is as required. □

We also have the following fact analogous to Theorem 3.4.4:

PROPOSITION 3.7.4. *Every subgroup $H$ of a precompact topological group $G$ is a precompact topological group.*

PROOF. Let $e$ be the neutral element of $G$. Take an arbitrary open neighbourhood $U$ of $e$ in $H$ and choose an open neighbourhood $V$ of $e$ in $G$ such that $V \cap H = U$. Clearly, $H$ is a precompact subset of $G$, so we can apply Lemma 3.7.3 to find a finite set $F \subset H$ such that $H \subset FV$ and $H \subset VF$. Let us verify that $H \subset FU$ and $H \subset UF$. Indeed, for every $x \in H$, there are $f \in F$ and $y \in V$ such that $x = fy$. Since $H$ is a subgroup of $G$, we have that $y = f^{-1}x \in V \cap H = U$, which in its turn implies that $x \in FU$. Thus, $H \subset FU$ and, similarly, $H \subset UF$. It follows that the topological group $H$ is precompact. □

It is easy to see that if $C \subset B \subset G$ and $B$ is precompact in a topological group $G$, then $C$ is also precompact in $G$. Conversely, sometimes the precompactness of $C$ implies the same conclusion with respect to $B$.

LEMMA 3.7.5. *If a set $B$ in a topological group $G$ contains a dense precompact subset, then $B$ is also precompact in $G$. Hence, the closure of a precompact subset of $G$ is precompact in $G$.*

PROOF. Let $S$ be a dense precompact subset of $B$. Consider an arbitrary neighbourhood $U$ of the identity $e$ in $G$ and choose an open symmetric neighbourhood $V$ of $e$ in $G$ such that $V^2 \subset U$. There exists a finite subset $F$ of $G$ such that $S \subset FV$ and $S \subset VF$. We claim that $B \subset FU \cap UF$. Indeed, if $b \in B$, then $S \cap bV \neq \varnothing$, so we can choose a point $y \in S \cap bV$. Then $y \in xV$ for some $x \in F$, whence $b \in yV^{-1} \subset xVV^{-1} \subset xU$. This implies the inclusion $B \subset FU$, and a similar argument applies to show that $B \subset UF$. □

COROLLARY 3.7.6. *If a topological group $G$ contains a dense precompact subgroup, then $G$ is also precompact.*

One more important property of precompact subsets of topological groups is presented in the next lemma that generalizes Proposition 1.4.32.

LEMMA 3.7.7. *Let $B$ be a precompact subset of a topological group $G$. Then, for every neighbourhood $U$ of the identity $e$ in $G$, there exists a neighbourhood $V$ of $e$ in $G$ such that $bVb^{-1} \subset U$, for each $b \in B$.*

PROOF. Let $U$ be a neighbourhood of $e$ in $G$. Choose an open symmetric neighbourhood $W$ of $e$ in $G$ such that $W^3 \subset U$. Since $B$ is precompact, we can apply Lemma 3.7.3 to find a finite set $F \subset B$ such that $B \subset WF$. Then $V = \bigcap_{x\in F} x^{-1}Wx$. is an open neighbourhood of $e$ in $G$. If $b \in B$, then $b = wx$ for some $w \in W$ and $x \in F$. Therefore,

$$bVb^{-1} = wxVx^{-1}w^{-1} \subset wWw^{-1} \subset W^3 \subset U.$$

This proves the lemma. □

Taking $B = G$ in the above lemma, and applying Lemma 1.8.6, we obtain the following conclusion:

COROLLARY 3.7.8. *Every precompact topological group $G$ is balanced and, therefore, the invariance number of $G$ is countable.*

The elementary theory of precompact groups could be built in parallel to the elementary theory of $\omega$-narrow groups exhibited in Section 3.4. For example, we could show directly that the product of any family of precompact topological groups is a precompact topological group. However, we take another approach, discovering and using connections between precompact groups and some other important classes of groups, connections, which, unfortunately, do not admit a generalization to the class of $\omega$-narrow groups.

First we show that precompact subsets of topological groups admit a nice characterization via the Raĭkov completion of groups.

PROPOSITION 3.7.9. *If $B$ is a closed precompact subset of a Raĭkov complete topological group $G$, then the space $B$ is compact.*

PROOF. Let $\xi$ be any ultrafilter on the set $B$. We have to show that $\xi$ converges to some point in $B$. First, take an ultrafilter $\eta$ on $G$ such that $\xi \subset \eta$. Fix an open neighbourhood $U$ of the neutral element $e$ of $G$. Since $B$ is precompact, we can find elements $a_1, \ldots, a_n \in G$ such that $B \subset \bigcup_{i=1}^{n} a_iU$. Since $B \in \eta$ and $\eta$ is an ultrafilter on $G$, it follows that $a_iU \in \eta$, for some $i \leq n$. Therefore, $\eta$ is a Cauchy filter on $G$. Since the topological group $G$ is Raĭkov complete, it follows that $\eta$ converges to some element $x$ in the space $G$. However, the set $B$ is closed in $G$ and $B \in \eta$, so that $x \in B$. Hence, $\xi$ also converges to $x$, and $B$ is compact. □

THEOREM 3.7.10. *A subset $B$ of a topological group $G$ is precompact in $G$ iff the closure of $B$ in the Raĭkov completion $\varrho G$ of $G$ is compact.*

PROOF. Suppose that the closure $\overline{B}$ of $B$ in $\varrho G$ is compact. Let $U$ be an open neighbourhood of the identity $e$ in $G$. Choose an open neighbourhood $V$ of $e$ in $\varrho G$ such that $V \cap G = U$. Since $\overline{B}$ is compact (hence, precompact in $\varrho G$) and $B$ is dense in $\overline{B}$, we can apply Lemma 3.7.3 to find a finite subset $F$ of $B$ such that $\overline{B} \subset FV \cap VF$. This implies immediately that $B \subset FU \cap UF$. Indeed, for every $b \in B$, there are $x \in F$ and $v \in V$ such that $b = xv$. Then $v = x^{-1}b \in B^{-1}B \subset G$, whence $v \in G \cap V = U$. So, $b = xv \in FU$ and, hence, $B \subset FU$. Similarly, one shows that $B \subset UF$. This proves that $B$ is precompact in $G$.

Conversely, suppose that $B$ is precompact in $G$. Then $B$ is precompact in $\varrho G$, so Lemma 3.7.5 implies that $\overline{B}$ is also precompact in $\varrho G$. Therefore, $\overline{B}$ is compact, by Proposition 3.7.9. □

COROLLARY 3.7.11. *Let $A$ and $B$ be precompact subsets of a topological group $G$. Then the sets $A^{-1}$, $B^{-1}$ and $AB$ are precompact in $G$.*

PROOF. Let $\varrho G$ be the Raĭkov completion of the group $G$. Denote the closures of $A$ and $B$ in $\varrho G$ by $\overline{A}$ and $\overline{B}$, respectively. By Theorem 3.7.10, the sets $\overline{A}$ and $\overline{B}$ are compact. Therefore, $\overline{A}^{-1}$, $\overline{B}^{-1}$, and $\overline{A}\,\overline{B}$ are also compact. Since the sets $A^{-1}$, $B^{-1}$ and $AB$ are dense in $\overline{A}^{-1}$, $\overline{B}^{-1}$, and $\overline{A}\,\overline{B}$, respectively, we can apply Theorem 3.7.10 once again to conclude that the sets $A^{-1}$, $B^{-1}$, and $AB$ are precompact in $G$. □

COROLLARY 3.7.12. *Suppose that a topological group $H$ is algebraically generated by a precompact set $B \subset H$. Then $H$ is topologically isomorphic to a dense subgroup of a $\sigma$-compact topological group and, in particular, the group $H$ is $\omega$-narrow.*

PROOF. Let $\varrho H$ be the Raĭkov completion of $H$ and $K$ be the closure of $B$ in $\varrho H$. Then $K$ is compact by Theorem 3.7.10, so the subgroup $G = \langle K \rangle$ of $\varrho H$ is $\sigma$-compact and contains $H$ as a dense subgroup. Since every $\sigma$-compact group is $\omega$-narrow, we can apply Theorem 3.4.4 to conclude that the subgroup $H$ of $G$ is also $\omega$-narrow. □

According to the Tychonoff Compactness Theorem, compactness is a productive property in the class of topological spaces. Let us show that a similar assertion is valid for precompactness in the class of topological groups.

PROPOSITION 3.7.13. *Let $B_i$ be a precompact subset of a topological group $G_i$, for each $i \in I$. Then the set $B = \prod_{i\in I} B_i$ is precompact in the topological product $G = \prod_{i\in I} G_i$.*

PROOF. Denote by $\varrho G$ the Raĭkov completion of the group $G$. By Corollary 3.6.23, we can identify $\varrho G$ with the product $\prod_{i\in I} \varrho G_i$, where $\varrho G_i$ is the Raĭkov completion of $G_i$, for each $i \in I$. Then the closure of $B$ in $\varrho G$, say, $\overline{B}$ coincides with the product $\prod_{i\in I} \overline{B}_i$, where $\overline{B}_i = cl_{\varrho G_i} B_i$, for each $i \in I$. By Theorem 3.7.10, the sets $\overline{B}_i$ are compact and, therefore, so is $\overline{B}$. Since $B$ is dense in $\overline{B}$, the same theorem implies that $B$ is precompact in $G$. □

The next important statement is now immediate.

COROLLARY 3.7.14. *The product of a family of precompact topological groups is a precompact topological group.*

Here comes a series of basic facts on precompact topological groups.

THEOREM 3.7.15. *A topological group $G$ is compact if and only if it is precompact and Raĭkov complete.*

PROOF. This follows from Proposition 3.7.9 and the obvious fact that every compact topological group is precompact and Raĭkov complete. □

THEOREM 3.7.16. *A topological group $G$ is precompact if and only if its Raĭkov completion is compact.*

PROOF. Suppose that $G$ is precompact. Then, by Corollary 3.7.6, the Raĭkov completion $\varrho G$ is also precompact. Now it follows from Theorem 3.7.15 that $\varrho G$ is compact.

To prove the converse, it suffices to refer to Proposition 3.7.4 and to the obvious fact that every compact topological group is precompact. □

COROLLARY 3.7.17. *A topological group $G$ is precompact if and only if it is topologically isomorphic to a subgroup of a compact group.*

PROOF. Apply Theorem 3.7.16 and Proposition 3.7.4. □

COROLLARY 3.7.18. *The Raĭkov completion of every pseudocompact topological group is compact.*

PROOF. This follows from Theorems 3.7.16 and 3.7.2. □

A continuous extension $f^*\colon \varrho G \to \varrho H$ of a continuous onto homomorphism $f\colon G \to H$ need not be onto, even if $f$ is open (see Problem 3.6.G). The situation improves, however, if the kernel of $f$ is precompact:

PROPOSITION 3.7.19. *Let $f\colon G \to H$ be a quotient homomorphism of a topological group $G$ onto $H$. If the kernel $K$ of $f$ is precompact, then the continuous homomorphic extension $f^*\colon \varrho G \to \varrho H$ of the homomorphism $f$ is quotient and onto.*

PROOF. According to Theorem 3.6.19, it suffices to verify that the homomorphism $f^*$ is onto. The forementioned theorem also implies that the kernel $N$ of $f^*$ is the closure of $K$ in $\varrho G$, and that $f^*$ is open when considered as a mapping onto $f^*(\varrho G)$. Since $K$ is precompact, it follows from Theorem 3.7.10 that $N$ is a compact group. It is clear that the group $f^*(\varrho G)$ is dense in $\varrho H$, so the equality $\varrho H = f^*(\varrho G)$ will follow if we show that the image $T = f^*(\varrho G)$ is a Raĭkov complete group.

Let $\xi$ be a Cauchy filter in $T$. Since the family $\gamma = \{(f^*)^{-1}(P) : P \in \xi\}$ is a filter base in $\varrho G$, there exists an ultrafilter $\eta$ in $\varrho G$ such that $\gamma \subset \eta$. We claim that $\eta$ is a Cauchy filter in $\varrho G$. Indeed, let $U$ be an arbitrary neighbourhood of the neutral element $e$ in $\varrho G$. According to Proposition 1.4.32, there exists an open neighbourhood $V$ of $e$ such that $x^{-1}Vx \subset U$, for each $x \in N$. Take an open neighbourhood $W$ of $e$ in $\varrho G$ such that $W^2 \subset V$. Since $\xi$ is a Cauchy filter in $T$ and the homomorphism $f^*\colon \varrho G \to T$ is open, we can find $x_0 \in \varrho G$ and $P \in \xi$ such that $P \subset f^*(x_0W)$. Hence, $(f^*)^{-1}(P) \subset x_0WN$. By the compactness of $N$, there exist elements $x_1, \ldots, x_n \in N$ such that $N \subset \bigcup_{i=1}^n Wx_i$. Hence, our choice of the sets $V$ and $W$ implies that

$$(f^*)^{-1}(P) \subset \bigcup_{i=1}^n x_0 WWx_i \subset \bigcup_{i=1}^n x_0 Vx_i \subset \bigcup_{i=1}^n x_0 x_i U.$$

Since $(f^*)^{-1}(P)$ is an element of the ultrafilter $\eta$, we must have $(f^*)^{-1}(P) \cap x_0 x_i U \in \eta$, for some positive integer $i \leq n$. We have thus proved that for every neighbourhood $U$ of $e$ in $\varrho G$, there exist $y \in \varrho G$ and $Q \in \eta$ such that $Q \subset yU$. This implies that $\eta$ is a Cauchy filter in $\varrho G$.

Since the group $\varrho G$ is Raĭkov complete, $\eta$ converges to a point $a \in \varrho G$. Therefore, by the continuity of $f^*$, the filter $f^*(\eta) = \{f^*(Q) : Q \in \eta\}$ converges to the point $b = f^*(a) \in T$. Since $\xi$ is a Cauchy filter in $T$ and $\xi \subset f^*(\eta)$, it follows that $\xi$ also converges to $b$. This proves that the group $T = f^*(\varrho G)$ is Raĭkov complete and, hence, $\varrho H = f^*(\varrho G)$. □

We now need one simple topological fact.

PROPOSITION 3.7.20. *Let $Y$ be a dense pseudocompact subspace of a Tychonoff space $X$. Then $Y$ is $G_\delta$-dense in $X$.*

PROOF. Suppose to the contrary that there exists a non-empty $G_\delta$-set $P$ in $X$ disjoint from $Y$. Take a family $\gamma = \{U_n : n \in \omega\}$ of open sets in $X$ such that $P = \bigcap_{n\in\omega} U_n$. Choose a point $b \in P$ and define by induction a sequence $\{V_n : n \in \omega\}$ of open sets in $X$ such that $b \in V_{n+1} \subset \overline{V_{n+1}} \subset V_n \subset U_n$ for each $n \in \omega$. Since $X$ is Tychonoff, there exists a sequence $\{f_n : n \in \omega\}$ of continuous real-valued functions on $X$ such that $0 \leq f_n \leq 2^{-n}$, $f_n(b) = 0$, and $f_n(x) = 2^{-n}$ for each $x \in X \setminus V_n$, where $n \in \omega$. Then $f = \sum_{n=0}^\infty f_n$ is a continuous real-valued function on $X$ satisfying $f(b) = 0$ and $f(x) \geq 2^{-n}$ whenever $x \in X \setminus V_n$ and $n \in \omega$. Finally, the function $g$ on $Y$ defined by $g(x) = 1/f(x)$, for each $x \in Y$, is continuous and unbounded, thus contradicting the pseudocompactness of $Y$. □

COROLLARY 3.7.21. *Let $H$ be a pseudocompact topological group. Then $H$ is $G_\delta$-dense in the compact group $\varrho H$, the Raĭkov completion of $H$.*

The following statement follows directly from the definition of precompact left topological groups.

THEOREM 3.7.22. *Every precompact locally compact left topological group is compact.*

Here are two more elementary facts about precompact groups.

THEOREM 3.7.23. *If $G$ is a precompact topological group such that every point in $G$ is a $G_\delta$-set, then $G$ can be mapped by a continuous isomorphism onto a second-countable topological group.*

PROOF. Since $G$ is $\omega$-narrow and the points in $G$ are $G_\delta$-sets, there exists a continuous isomorphism of $G$ onto a metrizable topological group $M$, by Corollary 3.4.25. Then $M$ is also $\omega$-narrow, and Proposition 3.4.5 implies that $M$ is separable. □

THEOREM 3.7.24. *Every compact topological group is topologically isomorphic to a closed subgroup of the product of some family of second-countable topological groups.*

Proof. Since every compact topological group is $\omega$-narrow, the conclusion follows from Theorem 3.4.23. □

A characterization of $\omega$-narrow topological groups that admit closed embeddings into topological products of second-countable groups is given in Problem 5.1.D.

We conclude this introduction to the properties of precompact groups with a subtle result on the existence of countable non-closed sets in infinite precompact groups which generalizes Corollary 1.4.24. It is almost immediate that if all countable subsets of a precompact group $G$ are closed in $G$, then $G$ is finite. Indeed, otherwise we take a countable infinite subgroup $H$ of $G$ and, by the assumption, all subsets of $H$ are closed in $G$ and in $H$. Therefore, $H$ is discrete, thus contradicting Proposition 3.7.4. This implies that an infinite precompact group always contains a non-closed countable subset. This result can, however, be considerably strengthened. First we present two lemmas.

Lemma 3.7.25. *Let $A$ be a subset of a topological group $G$ with identity $e$ such that $e \in \overline{A} \setminus A$. Then $A$ contains an infinite discrete subset $\{x_n : n \in \omega\}$ such that the set $\{x_i x_j^{-1} : i, j \in \omega,\ i < j\}$ is also discrete. In addition, if $i < j$ and $k < l$, then $x_i x_j^{-1} = x_k x_l^{-1}$ only if $(i, j) = (k, l)$.*

Proof. We will construct a set $\{x_n : n \in \omega\} \subset A$ and two sequences $\{U_n : n \in \omega\}$ and $\{U_{i,j} : i, j \in \omega,\ i < j\}$ of open subsets of $G$ satisfying the following conditions for all $n, m, i, j \in \omega$:

(i) $x_n \in U_n$ and $x_i x_j^{-1} \in U_{i,j}$;
(ii) $\overline{U_n} \cap \overline{U_m} = \varnothing$ for distinct $n, m$;
(iii) $U_{i,j} \subset U_i$ for each $j > i$;
(iv) $U_{i,j} \cap U_{i,k} = \varnothing$ if $j \neq k$;
(v) $e \notin \overline{U_n}$ and $x_n \notin \overline{U_{n,j}}$ for each $j > n$.

Take an arbitrary element $x_0 \in A$ and choose an open neighbourhood $U_0$ of $x_0$ such that $e \notin \overline{U_0}$. Suppose that we have defined points $x_0, \ldots, x_n \in A$ and open sets $\{U_i : i \leq n\}$ and $\{U_{i,j} : i < j \leq n\}$ satisfying (i)–(v). By (i) and (v), there exists an open symmetric neighbourhood $V$ of $e$ in $G$ such that

(vi) $\overline{V} \cap \overline{U_i} = \varnothing$ and $x_i V \subset U_i$ for each $i \leq n$, and $x_i V \cap \overline{U_{i,j}} = \varnothing$ whenever $i < j \leq n$.

Since $e \in \overline{A}$, we can choose an element $x_{n+1} \in A \cap V$. Then $x_{n+1} \neq e$, so there exists an open neighbourhood $U_{n+1}$ of $x_{n+1}$ in $G$ such that $e \notin \overline{U_{n+1}} \subset V$. Let $i \leq n$ be arbitrary. Since $x_i x_{n+1}^{-1} \in x_i V \subset U_i$ and $x_i V \cap \overline{U_{i,j}} = \varnothing$ if $i < j \leq n$, there exists an open neighbourhood $U_{i,n+1}$ of $x_i x_{n+1}^{-1}$ in $G$ such that $x_i \notin \overline{U_{i,n+1}}$ and $\overline{U_{i,n+1}} \subset x_i V$. Hence, (vi) implies that $\overline{U_{i,n+1}} \cap \overline{U_{i,j}} = \varnothing$ for each $j \leq n$. It is easy to see that the points $x_0, \ldots, x_{n+1}$ and the sequences $\{U_i : i \leq n+1\}$, $\{U_{i,j} : i < j \leq n+1\}$ satisfy (i)–(v). This finishes our construction.

It remains to note that the set $\{x_n : n \in \omega\}$ is discrete by (i) and (ii), while the same property of the set $\{x_i x_j^{-1} : i, j \in \omega,\ i < j\}$ follows from (i)–(iv). The last assertion of the lemma follows from (i), (ii), and (iv). □

Lemma 3.7.26. *Let $A$ be a precompact subset of a topological group $G$ with neutral element $e$. If $\{x_n : n \in \omega\} \subset A$, then $e$ belongs to the closure of the set $B = \{x_i x_j^{-1} : i < j,\ i, j \in \omega\}$.*

PROOF. If $x_i = x_j$ for some distinct $i, j \in \omega$, then $e \in B \subset \overline{B}$. We can assume, therefore, that the elements of the sequence $X = \{x_n : n \in \omega\}$ are pairwise distinct. In particular, the set $X$ is infinite. Let $U$ be an open neighbourhood of $e$ in $G$. Choose an open neighbourhood $V$ of $e$ in $G$ such that $VV^{-1} \subset U$. Since the set $A$ is precompact in $G$, there exists a finite set $F \subset G$ such that $A \subset VF$. Then the intersection $X \cap Vy$ is infinite, for some $y \in F$. Choose distinct $i, j \in \omega$ such that $\{x_i, x_j\} \subset Vy$. If $i < j$, then $x_i x_j^{-1} \in (Vy)(Vy)^{-1} = VV^{-1} \subset U$, whence it follows that $U \cap B \neq \varnothing$. This finishes the proof. □

THEOREM 3.7.27. **[I. V. Protasov]** *Let $A$ be an infinite precompact subset of a topological group $G$ with identity $e$. Then $AA^{-1}$ contains a countable discrete subset $B$ such that $e \in \overline{B} \setminus B$.*

PROOF. Denote by $\varrho G$ the Raĭkov completion of the group $G$. Since $A$ is precompact, Theorem 3.7.10 implies that the set $K = cl_{\varrho G} A$ is compact. Choose an infinite discrete subset $D$ of $A$. Then $D$ has an accumulation point $g$ in $K$ and, clearly, $g \notin D$. The sets $D$ and $Dg^{-1}$ are precompact in $\varrho G$ and the identity $e$ of $\varrho G$ is an accumulation point of $Dg^{-1}$.

Since $D$ is infinite, Lemma 3.7.25 implies that the set $Dg^{-1}$ contains a sequence $\{x_n : n \in \omega\}$ of pairwise distinct elements such that $B = \{x_i x_j^{-1} : i < j,\ i, j \in \omega\}$ is a discrete subspace of $\varrho G$. It follows from the inclusions

$$B \subset Dg^{-1}(Dg^{-1})^{-1} = DD^{-1} \subset AA^{-1} \subset G$$

that $B$ is, in fact, a subspace of $G$. By Lemma 3.7.26, $e$ is an accumulation point of $B$ in $G$. This finishes the proof. □

Here is an application of the above results about discrete subsets of topological groups. Informally, it says that all infinite subsets of an extremally disconnected topological group are "big".

As usual, we say that a subspace $Y$ of a space $X$ is *$C^*$-embedded in $X$* if every bounded continuous real-valued function on $Y$ admits an extension to a continuous function on $X$ (see [**191**]).

THEOREM 3.7.28. *Every precompact subset of an extremally disconnected topological group is finite.*

PROOF. Suppose to the contrary that an extremally disconnected topological group $G$ contains an infinite precompact subset $A$, and let $e$ be the neutral element of $G$. Then $AA^{-1}$ and $A^* = AA^{-1} \setminus \{e\}$ are also precompact subsets of $G$, by Corollary 3.7.11. It follows from Lemma 3.7.26 that $e$ is in the closure of $A^*$, so we can assume without loss of generality that $e \in \overline{A} \setminus A$.

According to Lemma 3.7.25, $A$ contains a sequence $\{x_n : n \in \omega\}$ of pairwise distinct elements such that the set $B = \{x_i x_j^{-1} : i < j,\ i, j \in \omega\}$ is discrete and two elements $x_i x_j^{-1}$ and $x_k x_l^{-1}$ of $B$ coincide iff $(i, j) = (k, l)$. Let

$$C = \{x_{2i} x_{2j}^{-1} : i < j,\ i, j \in \omega\} \text{ and } D = \{x_{2i+1} x_{2j+1}^{-1} : i < j,\ i, j \in \omega\}.$$

Then $C$ and $D$ are disjoint subsets of the discrete set $B$. Applying Lemma 3.7.26 once again, we conclude that $e \in (\overline{C} \setminus C) \cap (\overline{D} \setminus D)$. Hence, the function $f$ on $B$ which is equal to 0 on $C$ and is equal to 1 on $D$ cannot be extended to a continuous function on $G$. This contradicts the fact that every countable discrete subset of a regular extremally disconnected space is $C^*$-embedded (see [**191**, 9.H] or [**60**, Ch. 6, Ex. 164]). □

COROLLARY 3.7.29. *Every precompact extremally disconnected topological group is finite.*

## Exercises

3.7.a. Prove Propositions 3.7.1 and 3.7.4.

3.7.b. Suppose that a topological group $G$ is homeomorphic to a precompact topological group. Must $G$ be precompact?

3.7.c. Show that subgroups of a precompact paratopological group may fail to be precompact (or even $\omega$-narrow).
*Hint.* Consider the circle group $\mathbb{T}$ endowed with the Sorgenfrey topology (in other words, consider the quotient paratopological group $\mathbb{R}/\mathbb{Z}$, where $\mathbb{R}$ carries the Sorgenfrey topology). Then the square of $\mathbb{T}$ contains an uncountable discrete subgroup.

3.7.d. Suppose that $B$ is a compact subset of a paratopological group $G$. Is the set $B^{-1}$ precompact in $G$?

3.7.e. Is every precompact Dieudonné complete topological group compact?

3.7.f. Give an example of an infinite precompact Abelian topological group $G$ such that all proper subgroups of $G$ are finite. (See also Problem 1.4.E.)
*Hint.* For a given prime $p$, consider the subgroup $G$ of $\mathbb{T}$ which consists of all elements $x \in \mathbb{T}$ satisfying $x^{p^n} = 1$, for some $n \in \mathbb{N}$.

3.7.g. Let $G$ be a compact Abelian group and $a \in G$ be an arbitrary element. Does there exist a (not necessarily closed) subgroup $H$ of $G$ such that $G = H \oplus \langle a \rangle$?
*Hint.* Consider the circle group $\mathbb{T}$, take an element $a \neq 1$ in $\mathbb{T}$ of order 2, and apply 3.7.f.

3.7.h. Give an example of an infinite precompact topological group $G$ such that every algebraic isomorphism of $G$ onto itself is a homeomorphism.
*Hint.* Let $G$ be an infinite Boolean group and $\mathcal{F}$ a family of all subgroups $H$ of $G$ such that the index $|G : H|$ of $H$ in $G$ is finite. Verify that $\mathcal{F}$ is a local base at the zero element for a Hausdorff topological group topology $\tau$ on $G$, and that the group $(G, \tau)$ is as required.

3.7.i. Give an example of an infinite precompact metrizable topological Abelian group $G$ such that *every* homomorphism of $G$ to $G$ is continuous.
*Hint.* Consider the $\sigma$-product of the cyclic groups $\mathbb{Z}(p)$, where $p$ runs through all primes numbers $\mathbb{P}$, taken with the topology inherited from the product group $\prod_{p\in\mathbb{P}} \mathbb{Z}(p)$.

## Problems

3.7.A. (W. W. Comfort and L. C. Robertson [**120**]) Let $H$ be a closed invariant subgroup of a topological group $G$. Prove that if both groups $H$ and $G/H$ are precompact, so is $G$.

3.7.B. A subset $X$ of a topological group $G$ is called *left-precompact* (*right-precompact*) if, for every neighbourhood $U$ of the neutral element of $G$, there exists a finite subset $F$ of $G$ such that $AU = G$ ($UA = G$). Give an example of a topological group $G$ and a left-precompact subset $X$ of $G$ which fails to be right-precompact.

3.7.C. Show that not every precompact paratopological group is topologically isomorphic to a subgroup of a compact paratopological group.

3.7.D. Give an example of a first-countable precompact paratopological group that is neither metrizable, nor compact.

3.7.E. Give an example of a countable precompact topological group $G$ without non-trivial convergent sequences.

3.7.F. Give an example of a countable precompact topological group $G$ such that $|\varrho G| = 2^{\mathfrak{c}}$, where $\mathfrak{c} = 2^{\omega}$.

3.7.G. Prove that every infinite precompact group is resolvable (see Exercise 1.4.1 and Problem 3.4.D).

3.7.H. Let $p$ be a prime number. We define a metric $\varrho_p$ on $\mathbb{Z}$ by the rule $\varrho_p(n, n) = 0$ for each $n \in \mathbb{Z}$, and $\varrho_p(m, n) = 2^{-k}$ for distinct $m, n \in \mathbb{Z}$, where $k$ is the biggest natural number such that $p^k$ divides $m - n$. Prove that the metric $\varrho_p$ generates a precompact group topology $\tau_p$ on the group $\mathbb{Z}$ and that the Raĭkov completion of the group $(\mathbb{Z}, \tau_p)$ is topologically isomorphic to the group of $p$-adic integers $\mathbb{Z}_p$ (see Example 1.3.16). This explains, in part, the choice of the symbol $\mathbb{Z}_p$ to denote the latter group.

3.7.I. A topological group $G$ is called *locally precompact* if there exists a neighbourhood $O$ of the neutral element $e$ in $G$ such that $O$ can be covered by finitely many left and right translates of each neighbourhood of $e$ in $G$. Prove that the Raĭkov completion $\varrho G$ of every locally precompact topological group $G$ is locally compact.

3.7.J. Let $H$ be an arbitrary locally pseudocompact topological group (see Exercise 2.4.5). Prove that $H$ is $G_\delta$-dense in $\varrho H$.

3.7.K. A topological group $G$ is said to be *$\sigma$-precompact* if it is the union of a countable family of precompact subsets. Suppose that $X$ is a Tychonoff space such that the group $C_p(X)$ is $\sigma$-precompact. Must $X$ be discrete?
*Hint.* See [**32**].

## Open Problems

3.7.1. Let $G$ be a topological group homeomorphic to the product group $\mathbb{R}^{\tau}$, for some cardinal $\tau > \omega$.
(a) Can $G$ be precompact?
(b) Is $G$ Raĭkov complete?

3.7.2. Suppose that a precompact topological group $G$ is homeomorphic to a Raĭkov complete topological group. Is $G$ compact?

3.7.3. Does every infinite precompact paratopological group contain a countable non-closed subset?

3.7.4. Suppose that a paratopological group $G$ is homeomorphic to one of the groups $\mathbb{Z}^{\tau}$ or $\mathbb{R}^{\tau}$, for some cardinal $\tau \geq \omega$. Is $G$ a topological group? (The answer is "yes" for $\tau \leq \omega$, see Corollary 2.4.3).

## 3.8. Embeddings into connected, locally connected groups

Our aim in this section is to show that every topological group $G$ is topologically isomorphic to a closed subgroup of a connected, locally connected topological group $G^{\bullet}$, and that the groups $G$ and $G^{\bullet}$ share many properties, such as metrizability, separability, $\omega$-narrowness, *etc.*

Evidently, taking the connected cone of a group does not help here. There is, however, a perfect substitute of connected cones in the theory of topological groups which is presented below.

CONSTRUCTION 3.8.1. [**S. Hartman and J. Mycielski**] Let $G$ be a topological group with identity $e$ and multiplication written multiplicatively. Consider the set $G^{\bullet}$ of all functions $f$ on $J = [0, 1)$ with values in $G$ such that, for some sequence $0 = a_0 < a_1 < \cdots < a_n = 1$, the function $f$ is constant on $[a_k, a_{k+1})$ for each $k = 0, \ldots, n-1$. Let us define a binary operation $*$ on $G^{\bullet}$ by $(f * g)(x) = f(x) \cdot g(x)$, for all $f, g \in G^{\bullet}$

and $x \in J$. Then every element $f \in G^\bullet$ has a unique inverse $f^{-1} \in G^\bullet$ defined by $(f^{-1})(r) = (f(r))^{-1}$, for each $r \in J$ (the inverse on the left and on the right side of the equality is taken in $G^\bullet$ and $G$, respectively). It is easy to see that $(G^\bullet, *)$ is a group with identity $e^\bullet$, where $e^\bullet(r) = e$ for each $r \in J$. The elements of $G^\bullet$ are called *step functions.*

To introduce a topology on $G^\bullet$, we take an open neighbourhood $V$ of $e$ in $G$ and a real number $\varepsilon > 0$, and define a subset $O(V, \varepsilon)$ of $G^\bullet$ by

$$O(V, \varepsilon) = \{f \in G^\bullet : \mu(\{r \in J : f(r) \notin V\}) < \varepsilon\},$$

where $\mu$ is the usual Lebesgue measure on $J$. Let us verify that the sets $O(V, \varepsilon)$ form a base of a Hausdorff topological group topology at the identity of $G^\bullet$, thus making $G^\bullet$ into a topological group. To this end, it suffices to verify that the family

$$\mathcal{N}(e^\bullet) = \{O(V, \varepsilon) : V \in \mathcal{N}(e),\ \varepsilon > 0\}$$

satisfies the six conditions i)–vi) of Theorem 1.3.12, where $\mathcal{N}(e)$ is a base for $G$ at $e$. We do it step by step.

i) Take an arbitrary $V \in \mathcal{N}(e)$ and fix $\varepsilon > 0$. Choose $U \in \mathcal{N}(e)$ with $U^2 \subset V$ and take $f, g \in O(U, \varepsilon/2)$. Then $\mu(\{r \in J : f(r) \cdot g(r) \notin V\}) < \varepsilon$, whence it follows that the square of $O(U, \varepsilon/2)$ is contained in $O(V, \varepsilon)$.

ii) Let $O(V, \varepsilon) \in \mathcal{N}(e^\bullet)$ be arbitrary. If $U \in \mathcal{N}(e)$ is symmetric and satisfies $U \subset V$, then $O(U, \varepsilon)$ is a symmetric set in $G^\bullet$ and, clearly, $e^\bullet \in O(U, \varepsilon) \subset O(V, \varepsilon)$.

iii) Suppose that $O(V, \varepsilon) \in \mathcal{N}(e^\bullet)$ and that $f \in O(V, \varepsilon)$. Then there exist real numbers $0 = a_0 < a_1 < \cdots < a_n = 1$ such that for each $k = 0, 1, \ldots, n-1$, $f$ is constant on $J_k = [a_k, a_{k+1})$, and $f$ takes a value $x_k \in G$ on $J_k$. Since $f \in O(V, \varepsilon)$, the number $\delta = \varepsilon - \mu(\{r \in J : f(r) \notin V\})$ is positive. Choose $U \in \mathcal{N}(e)$ such that if $0 \le k < n$ and $x_k \in V$, then $Ux_k \subset V$. A simple calculation shows that $O(U, \delta)f \subset O(V, \varepsilon)$.

iv) Let $O(V, \varepsilon) \in \mathcal{N}(e^\bullet)$ and $f \in G^\bullet$ be arbitrary. Then $f$, considered as a function from $J$ to $G$, takes only finitely many distinct values, say, $x_1, \ldots, x_m$. Choose an element $U \in \mathcal{N}(e)$ such that $x_i U x_i^{-1} \subset V$, for each $i = 1, \ldots, m$. A direct verification shows that $fO(U, \varepsilon)f^{-1} \subset O(V, \varepsilon)$.

v) Given two elements $O(V_1, \varepsilon_1)$ and $O(V_2, \varepsilon_2)$ of $\mathcal{N}(e^\bullet)$, put $U = V_1 \cap V_2$ and $\varepsilon = \min\{\varepsilon_1, \varepsilon_2\}$. Then, evidently, $O(U, \varepsilon) \subset O(V_1, \varepsilon_1) \cap O(V_2, \varepsilon_2)$.

vi) If $f \in G^\bullet$ and $f \neq e^\bullet$, then the number $\varepsilon = \mu(\{r \in J : f(r) \neq e\})$ is positive. Choose an element $V \in \mathcal{N}(e)$ which does not contain any value of $f$ distinct from $e$. Then $f \notin O(V, \varepsilon)$ and, therefore, $\bigcap \mathcal{N}(e^\bullet) = \{e^\bullet\}$.

We have thus proved that $G^\bullet$ admits a Hausdorff topological group topology with $\mathcal{N}(e^\bullet)$ being a local base at the identity of $G^\bullet$. □

Let us establish two of the most important properties of the group $G^\bullet$.

PROPOSITION 3.8.2. *The topological group $G^\bullet$ constructed in 3.8.1 is pathwise connected and locally pathwise connected.*

PROOF. Since every topological group is a homogeneous space, the local pathwise connectedness of $G^\bullet$ will follow if we show that each set $O(V, \varepsilon)$ defined in 3.8.1 is pathwise connected. Let $f \in O(V, \varepsilon)$ be arbitrary. We claim that there exists a continuous mapping $\varphi\colon [0, 1] \to O(V, \varepsilon)$ such that $\varphi(0) = e^\bullet$ and $\varphi(1) = f$. Indeed, by the definition of $G^\bullet$, there exist real numbers $a_0, a_1, \dots, a_n$ with $0 = a_0 < a_1 < \cdots < a_n = 1$ such that for each $k = 0, 1, \dots, n-1$, $f$ is constant on $[a_k, a_{k+1})$. For every $t \in [0,1]$ and for every non-negative $k < n$, put $b_{k,t} = a_k + t(a_{k+1} - a_k)$. Then $b_{k,0} = a_k$, $b_{k,1} = a_{k+1}$ and $a_k < b_{k,t} < a_{k+1}$ if $0 < t < 1$, for each $k = 0, 1, \dots, n-1$. Let us define a mapping $\varphi\colon [0,1] \to G^\bullet$ by $\varphi(0) = e^\bullet$, $\varphi(1) = f$ and, for $0 < t < 1$ and $0 \le r < 1$,

$$\varphi(t)(r) = \begin{cases} f(r) & \text{if } a_k \le r < b_{k,t}; \\ e & \text{if } b_{k,t} \le r < a_{k+1}. \end{cases}$$

Evidently, $f_t = \varphi(t) \in O(V, \varepsilon)$ for each $t \in [0,1]$. It follows from our definition of $\varphi$ that

$$\mu(\{r \in J : f_t(r) \ne f_s(r)\}) \le |t - s|$$

for all $s, t \in [0,1]$. This inequality, and the definition of the topology of the group $G^\bullet$ given in 3.8.1, together imply that the mapping $\varphi$ is continuous. Thus, every element $f \in O(V, \varepsilon)$ can be connected with the identity $e^\bullet$ of $G^\bullet$ by a continuous path lying in $O(V, \varepsilon)$. This implies immediately that every two element of $O(V, \varepsilon)$ can also be connected by a continuous path inside of $O(V, \varepsilon)$, so that the set $O(V, \varepsilon)$ is pathwise connected.

The same argument applied to whole group $G^\bullet$ in place of $O(V, \varepsilon)$ implies the pathwise connectedness of $G^\bullet$. □

THEOREM 3.8.3. *For every topological group $G$, there exists a natural topological isomorphism $i_G\colon G \to G^\bullet$ of $G$ onto a closed subgroup of the pathwise connected, locally pathwise connected topological group $G^\bullet$.*

PROOF. We assign to each $x \in G$ the element $x^\bullet$ of $G^\bullet$ defined by $x^\bullet(r) = x$ for all $r \in J$. It is easy to see that the function $i_G\colon G \to G^\bullet$, where $i(x) = x^\bullet$ for each $x \in G$, is a topological monomorphism of $G$ to $G^\bullet$, and the latter group is pathwise connected and locally pathwise connected by Proposition 3.8.2.

It remains to verify that $i_G(G)$ is a closed subgroup of $G^\bullet$. Take an arbitrary $f \in G^\bullet \setminus i_G(G)$. Then $f$ cannot be constant as a function from $J$ to $G$. Therefore, we can find real numbers $a_1, a_2, a_3, a_4$ satisfying $0 \le a_1 < a_2 \le a_3 < a_4 \le 1$ and two distinct elements $x_1, x_2 \in G$ such that $f$ is equal to $x_1$ on $[a_1, a_2)$ and $f$ is equal to $x_2$ on $[a_3, a_4)$. Choose an open symmetric neighbourhood $V$ of the identity in $G$ such that $x_1V \cap x_2V = \varnothing$, and put $\varepsilon = \min\{a_2 - a_1, a_4 - a_3\}$. We leave to the reader a simple verification of the fact that $i_G(G) \cap O(V, \varepsilon)f = \varnothing$. Thus, the complement $G^\bullet \setminus i_G(G)$ is open in $G^\bullet$ and, hence, $i_G(G)$ is closed in $G^\bullet$. □

In what follows we identify a topological group $G$ with its image $i_G(G) \subset G^\bullet$ defined in Theorem 3.8.3. Let us show that the group $G$ is placed in $G^\bullet$ in a very special way, permitting an extension of continuous bounded pseudometrics from $G$ over $G^\bullet$.

THEOREM 3.8.4. *Let $d$ be a continuous bounded pseudometric on a topological group $G$. Then $d$ admits an extension to a continuous bounded pseudometric $d^\bullet$ over the group $G^\bullet$. In addition, if $d$ is (left-) invariant on $G$, then $d^\bullet$ is (left-) invariant on $G^\bullet$, and if $d$ is a metric on $G$ generating the topology of $G$, then $d^\bullet$ is also a metric on $G^\bullet$ generating the topology of $G^\bullet$.*

PROOF. One can assume without loss of generality that $d$ is bounded by 1. Take arbitrary elements $f, g \in G^\bullet$ and a partition $0 = a_0 < a_1 < \cdots < a_n = 1$ of $J$ such that both $f$ and

$g$ are constant on each interval $J_k = [a_k, a_{k+1})$ and are equal to $x_k$ and $y_k$ on this interval, respectively. We define a distance $d^\bullet(f, g)$ by the formula

$$d^\bullet(f, g) = \sum_{k=0}^{n-1} (a_{k+1} - a_k) \cdot d(x_k, y_k).$$

It is easy to verify that the number $d^\bullet(f, g)$ is non-negative and does not depend on the choice of the partition $a_0, a_1, \ldots, a_n$ of $J$ which keeps $f$ and $g$ constant on each $[a_k, a_{k+1})$. Clearly, that the function $d^\bullet$ is symmetric and satisfies the triangle inequality, that is, $d^\bullet$ is a pseudometric on $G^\bullet$.

Let us show that $d^\bullet$ is continuous. Take an element $f \in G^\bullet$ and a number $\varepsilon > 0$. Choose a partition $0 = a_0 < a_1 < \cdots < a_n = 1$ of $J$ such that $f$ has a constant value $x_k$ on each $J_k = [a_k, a_{k+1})$. Since $d$ is continuous on $G$, there exists an open neighbourhood $V$ of the identity in $G$ such that $d(x_k, x_k y) < \varepsilon/2$ for each $y \in V$, where $k = 0, 1, \ldots, n-1$. It suffices to verify that $d^\bullet(f, g) < \varepsilon$ for each $g \in fO(V, \varepsilon/2)$. Suppose that $g \in fO(V, \varepsilon/2)$; we can assume without loss of generality that $g$ is constant on each $J_k$ and takes a value $y_k$ on this interval. Denote by $L$ the set of all integers $k \leq n-1$ such that $y_k \in x_k V$, and let $M = \{0, 1, \ldots, n-1\} \setminus L$. It follows from the choice of $d$ and $g$ that $d(x_k, y_k) < \varepsilon/2$ for each $k \in L$, and that $\sum_{k \in M}(a_{k+1} - a_k) < \varepsilon/2$. Since $d$ is bounded by 1, and $\sum_{k \in L}(a_{k+1} - a_k) \leq 1$, the definition of $d^\bullet$ implies that

$$\begin{aligned} d^\bullet(f, g) &\leq \sum_{k \in L} (a_{k+1} - a_k) d(x_k, y_k) + \sum_{k \in M} (a_{k+1} - a_k) \\ &< \max_{k \in L} d(x_k, y_k) + \varepsilon/2 < \varepsilon/2 + \varepsilon/2 = \varepsilon. \end{aligned}$$

This proves the continuity of $d^\bullet$ on $G^\bullet$.

Clearly, if $d$ is (left-) invariant on $G$, then $d^\bullet$ is (left-) invariant on $G^\bullet$. Finally, suppose that $d$ is a metric on $G$ generating the topology of $G$. Then $d^\bullet(f, g) > 0$ for any distinct $f, g \in G^\bullet$, so that $d^\bullet$ is a metric on the set $G^\bullet$. Take an arbitrary element $f \in G^\bullet$ and a basic open neighbourhood $fO(V, \varepsilon)$ of $f$ in $G^\bullet$. Let $0 = a_0 < a_1 < \cdots < a_n = 1$ be a partition of $J$ such that $f$ takes a constant value $u_k$ on each $[a_k, a_{k+1})$. There exists $\delta > 0$ such that $\{u \in G : d(u_k, u) < \delta\} \subset u_k V$, for each $k = 0, 1, \ldots, n-1$. Therefore, if $k < n$ and $y \in G \setminus u_k V$, then $d(u_k, y) \geq \delta$. Put $\delta_0 = \varepsilon\delta$. We claim that

$$\{g \in G^\bullet : d^\bullet(f, g) < \delta_0\} \subset fO(V, \varepsilon). \tag{3.8}$$

Indeed, suppose that an element $g \in G^\bullet$ satisfies $d^\bullet(f, g) < \delta_0$. There exists a partition $0 = b_0 < b_1 < \cdots < b_N = 1$ of $J$ refining the partition $0 = a_0 < a_1 < \cdots < a_n = 1$ such that $g$ has a constant value $y_i$ on each $[b_i, b_{i+1})$, for $i = 0, \ldots, N-1$. Then $f$ is also constant on each $[b_i, b_{i+1})$, and let $f(b_i) = x_i$. Denote by $P$ the set of all non-negative integers $i < N$ such that $y_i \notin x_i V$. Clearly,

$$\sum_{i \in P} (b_{i+1} - b_i) d(x_i, y_i) \leq \sum_{0 \leq i < N} (b_{i+1} - b_i) d(x_i, y_i) = d^\bullet(f, g) < \delta_0.$$

Since $d(x_i, y_i) \geq \delta$ if $y_i \in G \setminus x_i V$, we have that

$$\delta \cdot \sum_{i \in P} (b_{i+1} - b_i) \leq \sum_{i \in P} (b_{i+1} - b_i) d(x_k, y_i) < \delta_0.$$

It follows that

$$\sum_{i \in P}(b_{i+1} - b_i) < \delta_0/\delta = \varepsilon. \tag{3.9}$$

The definition of $P$ implies the equality

$$D = \{r \in J : g(r) \notin f(r)V\} = \bigcup\{[b_i, b_{i+1}) : i \in P\}.$$

Therefore, $\mu(D) < \varepsilon$ according to (3.9), whence $f^{-1}g \in O(V, \varepsilon)$ or, equivalently, $g \in fO(V, \varepsilon)$. This proves the inclusion (3.8). Since the sets $fO(V, \varepsilon)$ form a base of $G^\bullet$ at $f$ and the metric $d^\bullet$ is continuous, it follows that $d^\bullet$ generates the topology of the group $G^\bullet$. This finishes the proof. □

COROLLARY 3.8.5. *Let $f$ be a continuous real-valued bounded function on a topological group $G$. Then $f$ admits an extension to a bounded continuous function on the group $G^\bullet$. In other words, $G$ is $C^*$-embedded in $G^\bullet$.*

The construction of the group $G^\bullet$ in 3.8.1 has another important property, apart from the (local) path connectedness of $G^\bullet$ — it permits to extend continuous homomorphisms.

PROPOSITION 3.8.6. *Let $\varphi\colon G \to H$ be a continuous homomorphism of topological groups. Then $\varphi$ admits a natural extension to a continuous homomorphism $\varphi^\bullet : G^\bullet \to H^\bullet$. In addition, if $\varphi$ is open and onto, then so is $\varphi^\bullet$.*

PROOF. For an arbitrary $f \in G^\bullet$, define an element $\varphi^\bullet(f) \in H^\bullet$ by $\varphi^\bullet(f)(r) = \varphi(f(r))$, for each $r \in J = [0, 1)$. If $f, g \in G^\bullet$ and $r \in J$, then

$$\varphi^\bullet(f * g)(r) = \varphi(f(r) \cdot g(r)) = \varphi(f(r)) \cdot \varphi(g(r)) = [\varphi^\bullet(f) * \varphi^\bullet(g)](r).$$

Hence, $\varphi^\bullet(f*g) = \varphi^\bullet(f)*\varphi^\bullet(g)$, and we conclude that $\varphi^\bullet : G^\bullet \to H^\bullet$ is a homomorphism.

To show that $\varphi^\bullet$ is continuous, take an open neighbourhood $V$ of the identity in $H$ and a real number $\varepsilon > 0$. By the continuity of $\varphi$, there exists an open neighbourhood $U$ of the identity in $G$ such that $\varphi(U) \subset V$. Then the definition of $\varphi^\bullet$ implies immediately that $\varphi^\bullet(O(U, \varepsilon)) \subset O(V, \varepsilon)$, which proves that $\varphi^\bullet$ is continuous. It is also clear that $\varphi^\bullet \circ i_G = i_H \circ \varphi$, where $i_G\colon G \to G^\bullet$ and $i_H\colon H \to H^\bullet$ are natural topological monomorphisms defined in Theorem 3.8.3. In particular, identifying $G$ with $i_G(G)$ and $H$ with $i_H(H)$, the above equality takes the form $\varphi^\bullet \restriction G = \varphi$, i.e., $\varphi^\bullet$ is a continuous extension of $\varphi$ over $G^\bullet$.

Finally, suppose that the homomorphism $\varphi$ is open and $\varphi(G) = H$. We leave to the reader a simple verification of the equalities $\varphi^\bullet(G^\bullet) = H^\bullet$ and $\varphi^\bullet(O(V, \varepsilon)) = O(W, \varepsilon)$ for all $V \in \mathcal{N}_G(e)$ and $\varepsilon > 0$, where $W = \varphi(V)$ is an open neighbourhood of the identity in $H$. Therefore, $\varphi^\bullet$ is open and onto. □

The correspondences $G \mapsto G^\bullet$ and $\varphi \mapsto \varphi^\bullet$ are functorial since the equality $(\psi \circ \varphi)^\bullet = \psi^\bullet \circ \varphi^\bullet$ holds for any continuous homomorphisms $\varphi\colon G \to H$ and $\psi\colon H \to K$. This defines the covariant functor $^\bullet$ in the category of topological groups and continuous homomorphisms. It turns out that this functor preserves subgroups and quotient groups:

PROPOSITION 3.8.7. *Let $G$ be a topological group and $H$ a subgroup of $G$.*

a) *If $\varphi$ is the identity embedding of $H$ to $G$, then the natural homomorphic extension $\varphi^\bullet : H^\bullet \to G^\bullet$ is a topological monomorphism.*

b) *If $H$ is closed or invariant in $G$, so is $H^\bullet$ in $G^\bullet$ (we identify $H^\bullet$ with the corresponding subgroup $\varphi^\bullet(H^\bullet)$ of $G^\bullet$).*

c) *If $H$ is a closed invariant subgroup of $G$, then the groups $(G/H)^\bullet$ and $G^\bullet/H^\bullet$ are naturally topologically isomorphic.*

PROOF. a) It is clear that $\varphi^\bullet$ is a monomorphism. Let $V$ be an arbitrary open neighbourhood of the neutral element in $G$. Put $W = H \cap V$. For any $\varepsilon > 0$, we have that $\varphi^{-1}(O(V, \varepsilon)) = O(W, \varepsilon)$, whence item a) of the proposition follows.

b) A simple verification shows that $H^\bullet$ is invariant in $G^\bullet$ if and only if $H$ is invariant in $G$. Suppose that $H$ is closed in $G$, and take an arbitrary element $f \in G^\bullet \setminus H^\bullet$. There exists a partition $0 = a_0 < a_1 < \cdots < a_n = 1$ of $J$ such that $f$ is constant on each interval $[a_k, a_{k+1})$; let $x_k = f(a_k)$, for $k = 0, 1, \ldots, n-1$. It follows from the choice of $f$ that $x_k \in G \setminus H$, for some $k < n$. Since $H$ is closed in $G$, we can choose an open neighbourhood $V$ of the neutral element in $G$ such that $x_k V \cap H = \emptyset$. It is easy to see that $fO(V, \varepsilon) \cap H^\bullet = \emptyset$, where $\varepsilon = a_{k+1} - a_k$. Hence, $H^\bullet$ is closed in $G^\bullet$.

c) Suppose that $H$ is a closed invariant subgroup of $G$. Let $\pi \colon G \to G/H$ be the canonical homomorphism. Then the natural extension $\pi^\bullet \colon G^\bullet \to (G/H)^\bullet$ of $\pi$ is an open continuous onto homomorphism, by Proposition 3.8.6. Therefore, according to the first isomorphism theorem (see Theorem 1.5.13), it suffices to verify that the kernel of $\pi^\bullet$ coincides with $H^\bullet$. The latter follows, however, from the equality $\pi^\bullet(f)(r) = f(\pi(r))$ which holds for all $f \in G^\bullet$ and $r \in J$. □

The group $G^\bullet$ inherits many properties of $G$. For instance, if $G$ is Abelian, divisible, torsion, or torsion-free, so is $G^\bullet$. We show below that a similar assertion is valid for many topological properties, including metrizability and separability.

THEOREM 3.8.8. *Let $\kappa$ be an infinite cardinal number, and let $G$ be a topological group having one of the following properties:*

a) *$G$ is metrizable;*
b) *$G$ has a base of cardinality $\leq \kappa$;*
c) *$G$ has a local base at the identity of cardinality $\leq \kappa$;*
d) *$G$ has a network of cardinality $\leq \kappa$;*
e) *$G$ has a dense subset of cardinality $\leq \kappa$;*
f) *$G$ is $\kappa$-narrow.*

*Then the group $G^\bullet$ has the same property.*

PROOF. Since, by Theorem 3.3.12, the metrizability of a topological group is equivalent to having a countable base at the identity, item a) follows from c) with $\kappa = \omega$. Thus, suppose that $G$ has a local base $\mathcal{N}$ at the identity $e$ satisfying $|\mathcal{N}| \leq \kappa$. Then, by the definition of the topology of $G^\bullet$, the family $\mathcal{N}^\bullet = \{O(V, 1/n) : V \in \mathcal{N},\ n \in \mathbb{N}\}$ is a local base at the identity $e^\bullet$ of $G^\bullet$, and $|\mathcal{N}^\bullet| \leq |\mathcal{N}| \cdot \omega \leq \kappa$. This implies the assertion of the theorem for c) and a).

For e), take a dense set $D \subset G$ with $|D| \leq \kappa$. Denote by $S$ the set of all $f \in G^\bullet$ for which there exist rational numbers $b_0, b_1, \ldots, b_m$ with $0 = b_0 < b_1 < \cdots < b_m = 1$ such that $f$ is constant on each semi-open interval $J_k = [b_k, b_{k+1})$ and takes a value $x_k \in D$ on $J_k$. It is clear that $|S| \leq |D| \cdot \omega \leq \kappa$, and we claim that $S$ is dense in $G^\bullet$. Indeed, let $fO(V, \varepsilon)$ be a basic open neighbourhood of $f \in G^\bullet$, where $V$ is an open neighbourhood of $e$ in $G$ and $\varepsilon > 0$. Then there exist numbers $0 = a_0 < a_1 < \cdots < a_n = 1$ such that the function $f$ is constant on $[a_k, a_{k+1})$ for each $k < n$. Choose rationals $b_1, \ldots, b_{n-1}$ in $J$ such that $a_k \le b_k < a_{k+1}$ for each $k < n$ and $\sum_{k=1}^{n-1}(b_k - a_k) < \varepsilon$. Also, put $b_0 = 0$ and $b_n = 1$. For every $k < n$, choose a point $y_k \in D \cap x_k V$, where $x_k = f(a_k)$, and define an element $g \in S$ by letting $g(r) = y_k$ for each $r \in [b_k, b_{k+1})$; $k = 0, \ldots, n-1$. It follows that $g \in fO(V, \varepsilon)$, so $S$ is dense in $G^\bullet$ and $|S| \le \kappa$.

Suppose now that $G$ has a base of cardinality less than or equal to $\kappa$. Then, evidently, $G$ has properties c) and e). Therefore, as we have just proved above, the group $G^\bullet$ has a local base $\mathcal{N}^\bullet$ at the identity $e^\bullet$ with $|\mathcal{N}^\bullet| \le \kappa$ and it also contains a dense subset $D^\bullet$ with $|D^\bullet| \le \kappa$. As in Proposition 3.4.5, it is easy to see that the family

$$\mathcal{B}^\bullet = \{gV : g \in D^\bullet,\ V \in \mathcal{N}^\bullet\}$$

is a base for $G^\bullet$. Indeed, since $D^\bullet$ is dense in $G^\bullet$, we have $G^\bullet = D^\bullet U$ for each $U \in \mathcal{N}^\bullet$. Let $O$ be a neighbourhood of a point $f \in G^\bullet$. One can find $U, V \in \mathcal{N}^\bullet$ such that $fU \subset O$ and $V^{-1}V \subset U$. Take $g \in D^\bullet$ such that $f \in gV$. Then $g \in fV^{-1}$ and, hence,

$$gV \subset (fV^{-1})V = f(V^{-1}V) \subset fU \subset O,$$

that is, $gV \in \mathcal{B}^\bullet$ is an open neighbourhood of $f$ in $G^\bullet$, and $gV \subset O$. This proves that $\mathcal{B}^\bullet$ is a base for $G^\bullet$. The inequality $|\mathcal{B}^\bullet| \le |D^\bullet| \cdot |\mathcal{N}^\bullet| \le \kappa$ follows from the definition of $\mathcal{B}^\bullet$. Thus, $w(G^\bullet) \le \kappa$, which implies the necessary conclusion when b) holds.

In the case of d), take a network $\mathcal{P}$ for $G$ with $|\mathcal{P}| \le \kappa$. For every $m \in \mathbb{N}$, denote by $J(m)$ the set of all $m$-tuples $(b_1, \ldots, b_m)$ of rationals such that $0 < b_1 < \cdots < b_m = 1$. Given $m, n \in \mathbb{N}$, an element $\vec{b} = (b_1, \ldots, b_m) \in J(m)$ and $\vec{P} = (P_1, \ldots, P_m) \in \mathcal{P}^m$, we define a subset $Q(m, n, \vec{b}, \vec{P})$ of $G^\bullet$ as the set of all $g \in G^\bullet$ such that the measure (with respect to the Lebesgue measure $\mu$ on $J$) of the set of all $r \in J$ satisfying $b_k \le r < b_{k+1}$ and $g(r) \notin P_{k+1}$, for some $k = 0, 1, \ldots, m-1$, is less than $1/n$ (we always put $b_0 = 0$). Then the family $\mathcal{Q}$ of all sets $Q(m, n, \vec{b}, \vec{P})$ with $m, n \in \mathbb{N}$, $\vec{b} \in J(m)$ and $\vec{P} \in \mathcal{P}^m$ has the cardinality less than or equal to $\kappa$, and we claim that $\mathcal{Q}$ is a network for $G^\bullet$.

Indeed, take arbitrary $f \in G^\bullet$ and $O(V, \varepsilon) \ni e^\bullet$. There exist numbers $0 = a_0 < a_1 < \cdots < a_m = 1$ such that $f$ is constant on each semi-open interval $[a_k, a_{k+1})$. For every $k = 0, 1, \ldots, m-1$, put $x_k = f(a_k)$ and choose an element $P_{k+1} \in \mathcal{P}$ such that $x_k \in P_{k+1} \subset x_k V$. Then take $n \in \mathbb{N}$ with $1/n < \varepsilon$ and choose $\vec{b} = (b_1, \ldots, b_m) \in J(m)$ such that $a_k \le b_k < a_{k+1}$ for each $k = 1, \ldots, m-1$, and $\sum_{k=1}^{m-1}(b_k - a_k) < 1/(2n)$. All we need to verify is that $f \in Q(m, 2n, \vec{b}, \vec{P}) \subset fO(V, \varepsilon)$, where $\vec{P} = (P_1, \ldots, P_m) \in \mathcal{P}^m$. Let $b_0 = 0$ and $b_m = 1$.

It follows from the choice of $\vec{b}$ and $\vec{P}$ that $f(r) = x_k \in P_{k+1}$ for each $r \in [b_k, a_{k+1})$, where $k = 0, 1, \ldots, m-1$. Therefore, if $b_k \le r < b_{k+1}$ and $f(r) \notin P_{k+1}$ for some $k < m$, then $r \in [a_{k+1}, b_{k+1})$ and $k + 1 \ne m$. Since $\sum_{k=1}^{m-1}(b_k - a_k) < 1/(2n)$, we conclude that $f \in Q(m, 2n, \vec{b}, \vec{P})$. To show that $Q(m, 2n, \vec{b}, \vec{P}) \subset fO(V, \varepsilon)$, it suffices to verify that $f^{-1}g \in O(V, \varepsilon)$ for each $g \in Q(m, 2n, \vec{b}, \vec{P})$. Let $g \in Q(m, 2n, \vec{b}, \vec{P})$ be arbitrary. It follows from the definition of $Q(m, 2n, \vec{b}, \vec{P})$ that the set

$$L = \{r \in J : b_k \le r < b_{k+1} \text{ and } g(r) \notin P_{k+1} \text{ for some } k < m\}$$

satisfies $\mu(L) < 1/(2n)$. If $r \in J \setminus L$ and $b_k \le r < a_{k+1}$ for some $k < m$, then $g(r) \in P_{k+1}$ and $f(r) = x_k$, whence it follows that $(f^{-1} * g)(r) = x_k^{-1} g(r) \in x_k^{-1} P_{k+1} \subset V$. In its turn, this implies that

$$M = \{r \in J : (f^{-1} * g)(r) \notin V\} \subset L \cup \bigcup_{k=1}^{m-1} [a_k, b_k),$$

so that $\mu(M) < 1/(2n) + 1/(2n) = 1/n$. Hence, $f^{-1}g \in O(V, \varepsilon)$ and $g \in fO(V, \varepsilon)$, as claimed. This proves that $\mathfrak{Q}$ is a network for the group $G^\bullet$.

Finally, suppose that the group $G$ is $\kappa$-narrow. Let $O = O(V, \varepsilon)$ be a basic neighbourhood of $e^\bullet$ in $G^\bullet$, where $V$ is an open neighbourhood of $e$ in $G$ and $\varepsilon$ is a positive real number. Then there exists a set $D \subset G$ such that $G = DV$ and $|D| \leq \kappa$. Denote by $S$ the set of elements $g \in G^\bullet$ such that, for some rationals $b_0, b_1, \ldots, b_n$ with $0 = b_0 < b_1 < \cdots < b_n = 1$, the function $g$ is constant on each $J_k = [b_k, b_{k+1})$, and the value of $g$ on $J_k$ is an element of $D$. It is clear that $|S| \leq \kappa$. Let us verify that $G^\bullet = SO$. Take an arbitrary $f \in G^\bullet$. There exist $a_0, a_1, \ldots, a_n$ with $0 = a_0 < a_1 < \cdots < a_n = 1$ such that $f$ is constant on each $[a_k, a_{k+1})$. Similarly to the construction in e), choose rationals $b_0, b_1, \ldots, b_n$ such that $b_0 = 0$, $b_n = 1$, $a_k \leq b_k < a_{k+1}$, for $k = 1, \ldots, n-1$, and $\sum_{k=1}^{n-1}(b_k - a_k) < \varepsilon$. For every $k \in \{0, 1, \ldots, n-1\}$, pick an element $x_k \in D$ such that $f(a_k) \in x_k V$. Denote by $g$ the element of $G^\bullet$ which is constant on each $J_k = [b_k, b_{k+1})$ and takes the value $x_k$ on $J_k$. Then $g \in S$, and a simple verification shows that $f \in gO$. This proves the equality $G^\bullet = SO$ and implies that the group $G^\bullet$ is $\kappa$-narrow. The theorem is proved. □

Some properties the group $G^\bullet$ can never have, except for trivial cases. For example, $G^\bullet$ is neither pseudocompact nor precompact unless $|G| = 1$ (see Exercise 3.8.c). However, one can add $\sigma$-compactness to the list of properties given in Theorem 3.8.8:

THEOREM 3.8.9. *The group $G^\bullet$ is $\sigma$-compact if and only if $G$ is $\sigma$-compact.*

PROOF. The condition is necessary since $G$ is topologically isomorphic to a closed subgroup of $G^\bullet$, by Theorem 3.8.3. Conversely, let $G$ be the union of compact sets $K_i$, where $i \in \omega$. We can assume that $K_i \subset K_{i+1}$ for each $i \in \omega$. Let $I$ be the closed unit segment with usual interval topology. For every $n, m \in \mathbb{N}$, let

$$A_n = \{(a_1, \ldots, a_n) \in I^n : 0 < a_1 < \cdots < a_n < 1\}$$

and

$$A_{n,m} = \{(a_1, \ldots, a_n) \in A_n : a_{k+1} - a_k \geq 1/m \text{ for each } k \leq n\},$$

where, as usual, $a_0 = 0$ and $a_{n+1} = 1$. It is clear that $A_n = \bigcup_{m=1}^{\infty} A_{n,m}$ and that each $A_{n,m}$ is closed in $I^n$. In particular, the sets $A_{n,m}$ are compact.

Given $n \in \mathbb{N}$, we define a mapping $\varphi_n \colon G^{n+1} \times A_n \to G^\bullet$ by the rule $\varphi_n(x_0, \ldots, x_n, a_1, \ldots, a_n) = f$, where the function $f \colon J \to G$ takes the constant value $x_k$ on $[a_k, a_{k+1})$ for each $k \leq n$. We claim that the restriction of $\varphi_n$ to $G^{n+1} \times A_{n,m}$ is continuous for each $m \in \mathbb{N}$. Indeed, take $p = (x_0, \ldots, x_n, a_1, \ldots, a_n) \in G^{n+1} \times A_{n,m}$ and put $f = \varphi_n(p)$. Consider a basic open neighbourhood $fO(V, \varepsilon)$ of $f$ in $G^\bullet$, where $V$ is an open neighbourhood of the identity in $G$ and $\varepsilon > 0$. Choose a positive number $\delta < \min\{\varepsilon/(2n), 1/(2m)\}$, and define a neighbourhood $W$ of $p$ in $G^{n+1} \times \mathbb{R}^n$ by

$$W = x_0 V \times \cdots \times x_n V \times (a_1 - \delta, a_1 + \delta) \times \cdots \times (a_n - \delta, a_n + \delta).$$

Let us show that $\varphi_n(q) \in fO(V,\varepsilon)$, for each $q \in W \cap (G^{n+1} \times A_{n,m})$. Clearly, $q = (y_0, \dots, y_n, b_1, \dots, b_n)$, where $(y_0, \dots, y_n) \in G^{n+1}$ and $(b_1, \dots, b_n) \in A_{n,m}$. Set $g = \varphi_n(q)$. Clearly, $x_k^{-1} y_k \in V$ for each $k \le n$, and if $r \in J \setminus \bigcup_{k=1}^n (a_k - \delta, a_k + \delta)$ and $a_k \le r < a_{k+1}$ for some $k \le n$, then $b_k \le r < b_{k+1}$ and $g(r) = y_k$ (again, we put $b_0 = 0$ and $b_{n+1} = 1$). Hence, $f(r)^{-1} g(r) = x_k^{-1} y_k \in V$. This implies that

$$L = \{r \in J : (f^{-1} * g)(r) \notin V\} \subset \bigcup_{k=1}^{n} (a_k - \delta, a_k + \delta),$$

so that $\mu(L) \le 2n\delta < \varepsilon$. Therefore, $f^{-1} * g \in O(V,\varepsilon)$. We conclude that $g = \varphi_n(q)$ is an element of $fO(V,\varepsilon)$, that is, $\varphi_n$ is continuous on $G^{n+1} \times A_{n,m}$.

It is easy to see that $G^\bullet = \bigcup_{i,n,m=1}^{\infty} \varphi_n(K_i^{n+1} \times A_{n,m})$, where each image $\varphi_n(K_i^{n+1} \times A_{n,m})$ is a compact subset of $G^\bullet$, by the continuity of $\varphi_n$ on the product space $G^{n+1} \times A_{n,m}$. This proves that the group $G^\bullet$ is $\sigma$-compact. □

Combining Theorems 3.8.3 and 3.8.9, we deduce the following:

COROLLARY 3.8.10. *Every $\sigma$-compact group is topologically isomorphic to a closed subgroup of a $\sigma$-compact, pathwise connected, locally pathwise connected group.*

## Exercises

3.8.a. Fill in the details in the proof of Theorem 3.8.3. Show, in particular, that the function $i_G \colon G \to G^\bullet$, where $i(x) = x^\bullet$ for each $x \in G$, is a topological monomorphism of $G$ to $G^\bullet$.

3.8.b. Verify that the number $d^\bullet(f,g)$, defined in the proof of Theorem 3.8.4, is non-negative and does not depend on the choice of a partition $a_0, a_1, \dots, a_n$ of $J$ which keeps $f$ and $g$ constant on each $[a_k, a_{k+1})$. Show also that the function $d^\bullet$ is symmetric and satisfies the triangle inequality, that is, $d^\bullet$ is a pseudometric on $G^\bullet$.

3.8.c. Verify that the group $G^\bullet$ is never precompact, except for the trivial case when $|G| = 1$.

3.8.d. Is the homomorphism $\varphi^\bullet \colon G^\bullet \to H^\bullet$ in Proposition 3.8.6 a unique continuous one extending a given continuous homomorphism $\varphi \colon G \to H$? What if $\varphi(G) = H$?

3.8.e. Fill in the details omitted in the end of the proof of Proposition 3.8.6.

3.8.f. Let $H$ be a subgroup of a group $G$. Verify that the subgroup $H^\bullet$ of $G^\bullet$ is central in $G^\bullet$ iff $H$ is central in $G$.

3.8.g. Let $G$ be an arbitrary topological group. Show that, for every integer $n \ge 1$, the groups $G^\bullet$ and $(G^\bullet)^n$ are topologically isomorphic.

3.8.h. Verify that $G^\bullet$ is $\omega$-balanced iff $G$ is $\omega$-balanced.

3.8.i. A space $X$ is called *$\omega$-bounded* if the closure of every countable subset of $X$ is compact. Prove that if a topological group $G$ is the union of countably many $\omega$-bounded subspaces, then so is the group $G^\bullet$. Is a similar assertion valid for countable compactness in place of $\omega$-boundedness?

3.8.j. Verify that the construction of the embedding $G \hookrightarrow G^\bullet$ in this section can be extended to paratopological groups, to semitopological groups, to quasitopological groups, and to left topological groups.

## Problems

3.8.A. Can the group $G^\bullet$ be locally compact but not compact?

3.8.B. When is the group $G^\bullet$ Raĭkov complete?

3.8.C. Suppose that $G$ is an arbitrary subgroup of a compact connected topological group. Prove that $G$ is topologically isomorphic to a closed subgroup of a pseudocompact connected topological group. (In fact, it was shown by Ursul in [**507**] that every precompact topological group is topologically isomorphic to a closed subgroup of a connected pseudocompact topological group).

*Hint.* Let $G$ be a subgroup of a connected compact topological group $C$. We can assume that $|C| > 1$. Identify $G$ with the "diagonal" subgroup $\overline{G} = \{\overline{g} : g \in G\}$ of the product group $C^{\omega_1}$, where $\overline{g}_\alpha = g$ for each $\alpha < \omega_1$. Let $H = \Sigma C^{\omega_1}$ be the $\Sigma$-product of $\omega_1$ copies of the group $C$, that is, a dense subgroup of the group $C^{\omega_1}$ defined in Section 1.6. Then $p_A(H) = C^A$, for each countable set $A \subset \omega_1$, where $p_A : C^{\omega_1} \to C^A$ is the natural projection. Show that the subgroup $P$ of $C^{\omega_1}$ generated by the set $H \cup \overline{G}$ is connected, pseudocompact, and contains $\overline{G} \cong G$ as a closed subgroup.

3.8.D. Let $G$ be a topological group having one of the following properties:

(a) the cellularity of $G^n$ is countable, for each $n \in \mathbb{N}$;

(b) $G^n$ is pseudo-$\aleph_1$-compact, for each $n \in \mathbb{N}$.

Prove that the group $G^\bullet$ has the same property.

3.8.E. Prove that for every topological group $G$ with $|G| > 1$, the group $G^\bullet$ is resolvable.

3.8.F. Generalize Corollary 3.8.5 by showing that every topological group $G$ is $C$-embedded in the group $G^\bullet$. To this end, prove that there exists a mapping $B : C(G) \to C(G^\bullet)$ satisfying the following conditions for each $f \in C(G)$:

(i) $B(f)\restriction G = f$;

(ii) $B(r_G) = r_{G^\bullet}$ for each $r \in \mathbb{R}$, where $r_G$ and $r_{G^\bullet}$ are the constant functions on $G$ and $G^\bullet$, respectively, with value $r$;

(iii) $||B(f)|| = ||f||$ for each bounded function $f \in C(G)$, where $||\cdot||$ is the sup-norm.

Here $C(G)$ and $C(G^\bullet)$ are the spaces of all continuous real-valued functions on $G$ and $G^\bullet$, respectively. [Note that (i) means that $B$ is an extension operator, (ii) means that $B$ preserves constants, while (iii) tells us that $B$ preserves the norm of bounded functions.]

*Hint.* First, suppose that $f \in C(G)$ is bounded. Given an element $h \in G^\bullet$, take a partition $0 = a_0 < a_1 < \cdots < a_n = 1$ of $J = [0, 1)$ such that $h$ is constant on each segment $[a_k, a_{k+1})$. Then define a function $F(f) \in C(G^\bullet)$ by

$$F(f)(h) = \sum_{k=0}^{n-1} (a_{k+1} - a_k) f(h(a_k)).$$

Denote by $C^*(G)$ and $C^*(G^\bullet)$ the spaces of bounded continuous real-valued functions on $G$ and $G^\bullet$, respectively. Show that the mapping $F : C^*(G) \to C^*(G^\bullet)$ satisfies the same conditions (i)–(iii) and, in addition, if $f(G) \subset (-1, 1)$ for some $f \in C^*(G)$, then $F(f)(G^\bullet) \subset (-1, 1)$. The latter property of $F$ permits us to define a mapping $B : C(G) \to C(G^\bullet)$ as follows. Consider the homeomorphism $\varphi$ of $\mathbb{R}$ to $(-1, 1)$ defined by $\varphi(x) = 2\arctan(x)/\pi$, for each $x \in \mathbb{R}$. For every $f \in C(G)$, put

$$B(f) = \varphi^{-1} \circ F(\varphi \circ f).$$

This definition is correct, since $\varphi \circ f \in C^*(G)$ and all values of the function $\varphi \circ f$ lie in $(-1, 1)$, for each $f \in C(G)$. Verify that $B$ satisfies (i)–(iii).

## Open Problems

3.8.1. Characterize the topological groups $G$ such that the group $G^\bullet$ is realcompact or Dieudonné complete.

3.8.2. When is the group $G^\bullet$:
  (a) Lindelöf?
  (b) Čech-complete?
  (c) feathered (that is, contains a non-empty compact subset with countable neighbourhood base, see Section 4.3)?
  (d) paracompact?
  (e) normal?

## 3.9. Historical comments to Chapter 3

Quite a few results in Section 3.1 have their roots in [**291**] and [**128**] and in some other older papers. Theorem 3.1.1 generalizes the well-known theorem that every locally compact topological group is paracompact (Corollary 3.1.4). This corollary already appears in [**276**]; it can be also found in [**236**], where a reference to E. A. Michael is given. It is quite remarkable that paracompactness appears "automatically" in the presence of the algebraic structure. Strong paracompactness of locally compact groups plays an important role in Pasynkov's proof that all three classical dimensions for locally compact groups coincide [**363**]. Corollary 3.1.5, Theorem 3.1.6, and Proposition 3.1.8 are also very old results (see, for example, [**387, 325, 236**], or [**80, 81**]). In connection with Theorems 3.1.9, 3.1.11, and Corollary 3.1.12 see [**236**]. Theorem 3.1.15 appeared in [**105**]. The theory of dyadic compacta is a well-developed and interesting part of General Topology; see [**165**] in this connection. Regarding *NSS*-groups, see [**327**]. Theorem 3.1.27 and Corollary 3.1.28 are from [**387**].

Proposition 3.2.1, Theorems 3.2.2, 3.2.4, and Corollary 3.2.5 are recent results from [**47**]. Item 1) of Corollary 3.2.6 is a result of J. P. Serre [**425**]. It appears also in [**325**] and [**236**]. The proofs there were based on a careful analysis of the behaviour of compact sets under taking quotients. The new approach developed in [**47**], in particular, Theorem 3.2.2 obtained there permitted us to establish a series of new results on invariance of basic topological properties under quotients with respect to locally compact subgroups (see Corollary 3.2.5 and items 2)–8) of Corollary 3.2.6).

Prenorms on topological groups were introduced by A. A. Markov in [**308**] under the name of *pseudonorms*. He immediately recognized that continuous prenorms are destined to play the same role in the study of topological groups as continuous real-valued functions in the theory of Tychonoff spaces. Elementary Propositions 3.3.1, 3.3.2, 3.3.3, 3.3.4, 3.3.5, 3.3.7, 3.3.8, and Lemma 3.3.6 all originated in [**308**]. Lemma 3.3.10, on which the proof of Theorem 3.3.9 is based, comes from [**202**]. Note that a version of this lemma, that dealt with invariant pseudometrics, is found in [**264**]. The notion of a uniformly Tychonoff space is new, as well as, formally, Theorem 3.3.11, while L. S. Pontryagin was the first to establish in [**387**] that every topological group is a Tychonoff space. Theorem 3.3.12 is known as the Birkhoff–Kakutani theorem, it was proved in [**75**] and [**264**]. D. van Dantzig considered invariant metrics on groups in [**128**]. In connection with Corollary 3.3.13 see [**264, 75**], where this result was obtained (G. Birkhoff did not mention that the metric he constructed was left-invariant). For Corollary 3.3.14 see [**264, 236**] and [**279**].

As for Theorems 3.3.15 and 3.3.16, see [**202**] and [**24**]. The latter article surveys problems and concepts arising in this direction. Corollary 3.3.20 is a result of M. I. Graev [**202**], as well as Corollary 3.3.21. One finds Corollary 3.3.17 in [**22**]. It seems that Lemma 3.3.22 and Theorem 3.3.24 are new results.

The class of $\omega$-narrow topological groups was introduced by I. I. Guran in [**208**], where the main properties of $\omega$-narrow topological groups were established. Such groups were called there $\aleph_0$-*bounded*. Since there is also an older concept of an $\omega$-bounded topological space which has a quite different meaning, we have decided to change the terminology and call the groups in question $\omega$-*narrow*. The general notion of a $\tau$-narrow topological group was also introduced in [**208**]. An obvious prototype of these parallel notions is the concept of a totally bounded topological group.

Propositions 3.4.1, 3.4.2, 3.4.3, Theorem 3.4.4, Propositions 3.4.5, 3.4.6, and Theorem 3.4.7 were obtained in [**208**]. These results show that the class of $\omega$-narrow topological groups behaves quite nicely from the categorical point of view — it is closed under products, taking subgroups and continuous homomorphic images. This shows that this class of topological groups is analogous or parallel to the class of Tychonoff spaces. Indeed, the smallest class of topological spaces that contains all separable metrizable spaces and is closed under taking subspaces and arbitrary products, is precisely the class of Tychonoff spaces. This fact justifies the introduction of the class of $\omega$-narrow topological groups. Historically, there was another reason for that as well — it was an attempt to characterize the class of topological subgroups of Lindelöf topological groups. Though every such a subgroup is $\omega$-narrow, Theorem 3.4.7 shows that the class of $\omega$-narrow groups is considerably wider. Notice that Corollary 3.4.8 and Theorem 3.4.9 reflect very specific properties of topological groups. Corollary 3.4.19 was obtained by I. I. Guran in [**208**] and was a crucial step towards his proof of Theorem 3.4.23. It also implies Corollary 3.4.25, another interesting result from [**208**]. Corollary 3.4.27 was obtained in [**21**] (see also [**22**]). Corollaries 3.4.28 and 3.4.29 are, probably, new results. In [**24**], $\omega$-narrow groups were characterized as subgroups of topological groups of countable cellularity (see Theorem 5.1.11 in Chapter 5).

To a great extent, Theorem 3.4.23 is modeled on an important result of G. I. Katz in [**272**] characterizing subgroups of products of metrizable topological groups. These are precisely the groups we call $\omega$-*balanced*. In particular, Theorem 3.4.21 has its prototype for metrizable groups in [**272**]. Corollary 3.4.24 can be considered as a part of the folklore (see the discussion on this subject in [**24**]). It shows that in the class of $\omega$-balanced topological groups, some main questions are solved in a different way than in the general case of arbitrary topological groups.

Groups of isometries and groups of homeomorphisms are among the basic mathematical objects that served as a source for topological algebra. They possess two natural structures, a topology and a binary operation, blended in a whole. In particular, one finds a discussion of them in Bourbaki's tract [**81**], in [**387**], [**325**], and [**12**], where various natural topologies on the group of homeomorphisms are discussed. J. de Groot showed in [**204**] that every group is algebraically isomorphic to the homeomorphism group of some topological space. Some further results in this direction were obtained by J. van Mill in [**321**]. In connection with Theorem 3.5.1 see [**519**]. Theorem 3.5.2 is very close to some results in [**81**]. In connection with Corollary 3.5.3 and Theorem 3.5.4 see [**325**] and [**12**]; however, such results were surely known to L. S. Pontryagin. One finds Theorem 3.5.5 and Example 3.5.6

in [**81**]. The notion of $p$-homogeneity is new, so that Proposition 3.5.8 and Corollary 3.5.9 are also new. Theorem 3.5.10 and its proof are taken from [**515**] where further references are given. Theorem 3.5.11 is new. In connection with Propositions 3.5.12 and 3.5.13, see [**173**] and [**81**]. The concept of acceptable topology is from [**12**]. Theorem 3.5.15 can be found in [**81**]. Groups of uniformly continuous homeomorphisms of uniform spaces, with the topology of uniform convergence, were considered by J. Dieudonné [**132**]. See also [**171**] in this connection. For deep results on groups of homeomorphisms of manifolds, see [**86**] and [**98**].

The introduction of the concept of a uniform space by A. Weil, the theory of uniform spaces developed in [**532**], made it possible to treat topological groups from these positions, since every topological group has several natural uniform structures affiliated with it. In particular, it was studied in [**533, 80**] and in [**131**] when the uniform completion of a topological group $G$, with respect to one of these uniformities, is a topological group again, under the naturally extended operation. It turned out that this is always possible if the two-sided uniformity is used. The construction in Section 3.6 shows it. The resulting completion is the Raĭkov completion of a topological group. If the completion with respect to the left uniformity (called the Weil completion) is a group, then it automatically coincides with the Raĭkov completion. This is important, since the Raĭkov completion is always a group, while this need not be the case for the Weil completion.

Our approach to the construction of the Raĭkov completion follows, with some modifications, the original approach in [**396**]. Of course, there are also similarities with the less general construction in [**80**]. Raĭkov's construction was also described, with some modifications, in [**202**]. In particular, one finds statements similar to Propositions 3.6.1, 3.6.3, 3.6.6, Lemma 3.6.8, and Proposition 3.6.9 in [**396, 202**]. Theorem 3.6.10 is from [**396**]. For Proposition 3.6.12, Theorem 3.6.14, and Corollaries 3.6.17, 3.6.21 see also [**396**] and [**202**]. Theorem 3.6.19 is apparently new. Theorem 3.6.24 is from [**80**], where one also finds results on the Weil completion of a topological group in the case when the completion is a group. Raĭkov also established in [**396**] that Raĭkov complete groups are precisely *absolutely closed* topological groups in the sense of A. D. Alexandroff [**6**].

Precompact uniform spaces and precompact topological groups (called also *totally bounded*) were introduced and studied in [**532**]. Their definition, given in Section 3.7, provides a transparent internal characterization of subgroups of compact groups. The concept of precompactness plays an important role not only in topological algebra. In the context of metric spaces, it permits to analyze the basic notion of compactness as a straightforward combination of precompactness and metric completeness.

Pseudocompact groups were extensively studied in quite a big number of publications, we only mention here [**122, 121, 124**], and [**114**]. Such results as Propositions 3.7.1, 3.7.4, Corollaries 3.7.6 and 3.7.11 are either contained in [**532**], or were known to A. Weil and N. Bourbaki [**80**]. Theorems 3.7.15 and 3.7.16 should be also attributed to A. Weil [**532**], even thought the notion of Raĭkov completeness was not yet introduced at that time. For a precompact group, the Weil completion is also a group and coincides with the Raĭkov completion [**532**]. Corollary 3.7.21 is in [**122**]. Theorem 3.7.24 was definitely known to Pontryagin (see [**387**]). Lemma 3.7.25 and Theorem 3.7.27 are from [**391**]. Theorem 3.7.28 is new. It generalizes a theorem on compact subsets of extremally disconnected topological groups proved by A. V. Arhangel'skii in [**17**]. In connection with Corollaries 3.7.8 and 3.7.12, in particular, for further references, see [**484**].

Construction 3.8.1 is taken from [**220**], where Proposition 3.8.2 and Theorem 3.8.3, as well as several other relevant facts are established (see Theorem 3.8.8). In Section 3.8 some further properties of the Hartman–Mycielski construction are established, important for applications in later chapters (Theorem 3.8.4, Corollary 3.8.5, see also item f) of Theorem 3.8.8). Theorem 3.8.9 is new.

# Chapter 4

# Some Special Classes of Topological Groups

Chapter 4 is one of the main in the book. It develops, in various directions, the central theme that we have already touched upon in previous chapters — how the presence of a synchronized algebraic structure influences properties of a topology. We consider below, from this point of view, a series of most important topological properties such as compactness (Sections 4.1 and 4.2), Čech-completeness and featheriness (Section 4.3), the $P$-property, extremal disconnectedness (Sections 4.4 and 4.5), the Fréchet–Urysohn property and weak first countability (Section 4.7). The choice of the above mentioned properties is justified not only by their remarkable role in General Topology, but by the fact that each of them has already been a subject of deep and unexpected results in topological algebra.

In Section 4.1 we establish, using only techniques of General Topology, that every compact topological group is a dyadic compactum, and that the cellularity of every compact group is countable. In Section 4.2 we show, by a direct elementary construction, that every non-metrizable compact group contains a topological copy of the generalized Cantor discontinuum $D^{\omega_1}$. These statements have many corollaries for the topological structure of compact groups.

As it is well known, completeness type properties play the fundamental role in General Topology and its applications. In Section 4.3 we consider one of these properties, Čech-completeness, in its connections with topological groups. It turns out that, unlike the case of general topological spaces, Čech-completeness of topological groups implies paracompactness and Raĭkov completeness of the groups. We also show that Čech-complete groups and feathered groups are naturally related by open perfect homomorphisms to metrizable groups.

Section 4.4 is devoted to $P$-groups, that is, to topological groups in which every $G_\delta$-subset is open. Lindelöf $P$-spaces, in many respects, behave as compact Hausdorff spaces. We establish that every Lindelöf $P$-group is Raĭkov complete, and that every continuous homomorphism of a Lindelöf $P$-group onto another Lindelöf $P$-group is open.

In Section 4.5, devoted to extremally disconnected topological groups, we give an elementary proof of a well-known theorem on algebraic structure of extremally disconnected topological groups saying that every such group contains an open subgroup consisting of elements of order $\leq 2$. It is established that every extremally disconnected topological skew field is discrete. We also discuss the almost 40 years old problem: Is there in $ZFC$ alone a non-discrete extremally disconnected topological group? At the end of the section we present an important example of a non-discrete maximal (hence, extremally disconnected) topological group whose construction requires a weak form of Martin's Axiom abbreviated to $\mathfrak{p} = \mathfrak{c}$.

In Section 4.6 we have a look at the role of perfect mappings in the theory of topological groups from another angle. This allows to obtain original results on connections between some properties of subspaces of topological groups such as Čech-completeness, featheriness, paracompactness, metrizability, and similar properties of the group itself.

Section 4.7 treats certain delicate convergence properties in topological groups: Fréchet–Urysohn property, weak first countability, and bisequentiality. It turns out that in topological groups these properties are transformed greatly, and connections between them are considerably strengthened. We prove, in particular, that every Fréchet–Urysohn topological group is strongly Fréchet–Urysohn, that the product of a Fréchet–Urysohn topological group with a first-countable space is a Fréchet–Urysohn space, and that every weakly first-countable topological group, as well as every bisequential topological group, is metrizable.

In the chapter we formulate a series of open problems. The mastership of the material of this chapter can open good perspectives for research in various main stream directions of topological algebra.

## 4.1. Ivanovskij–Kuz'minov Theorem

In this section we prove the celebrated theorem of Ivanovskij[**259**] and Kuz'minov[**287**] that every compact topological group is a dyadic compactum. Recall that a *dyadic compactum* is a compact Hausdorff space that can be represented as an image of the generalized Cantor discontinuum $D^\tau$ under a continuous mapping, where $D = \{0, 1\}$ is the two-point discrete space and $\tau$ is a cardinal. It is well known that every metrizable compact space is a continuous image of the Cantor set $D^\omega$[**165**, 3.2.B]. Thus, all metrizable compacta are dyadic. However, not all compact spaces are dyadic, as we will see below. The proof of Ivanovskij–Kuz'minov's theorem is based on an important theorem of E. A. Michael on selections. We start with a proof of the later theorem, and then present a certain technique involving well-ordered inverse spectra of compact spaces, playing a crucial role in the proof of the main theorem.

If $M$ is a space, then $Exp(M)$ stands for the set of all closed non-empty subsets of $M$, and $\mathscr{E}(M)$ is the set of all non-empty subsets of $M$. A mapping $q$ of a topological space $X$ into the set $\mathscr{E}(M)$ is called *lower semicontinuous* if, for each open subset $V$ of $M$, the set $V_q$ of all $x \in X$ such that $q(x) \cap V \neq \varnothing$ is open in $X$. Particularly important is the case when $q(x) \in Exp(M)$ for each $x \in X$, that is, when each $q(x)$ is a non-empty closed subset of $M$. Then we, of course, say that $q$ is a lower semicontinuous mapping of $X$ into $Exp(M)$.

Let $X$ and $M$ be some spaces and $q$ a mapping of $X$ into $\mathscr{E}(M)$. A mapping $f$ of $X$ to $M$ is called a *selection* for $q$ if $f(x) \in q(x)$, for each $x \in X$. If, in addition, the mapping $f$ of $X$ into $M$ is continuous, $f$ is said to be a *continuous selection* for the mapping $q$.

We call a mapping $f$ *locally constant* if, for every point $x \in X$, there exists an open neighbourhood $W$ of $x$ such that $f(y) = f(x)$, for each $y \in W$. Of course, a locally constant function is continuous.

Now we are ready to formulate a version (not the strongest one) of Michael's Selection Theorem:

THEOREM 4.1.1. [**E. A. Michael**] *Let $M$ be a space metrizable by a complete metric, $X$ a zero-dimensional compact Hausdorff space, and $q$ any lower semicontinuous mapping of $X$ to the set $Exp(M)$ of all non-empty closed subsets of $M$. Then there exists a continuous selection for $q$.*

PROOF. There exists a complete bounded metric on $M$ generating the topology of $M$. Let $d$ be such a metric. Fix a positive number $\varepsilon$. A mapping $f$ of $X$ to $M$ will be called an *$\varepsilon$-selection* for $q$ if $d(f(x), q(x)) < \varepsilon$, for each $x \in X$.

**Claim.** *Let $\varepsilon$ and $\delta$ be any positive numbers, and $f$ a locally constant $\varepsilon$-selection for the mapping $q$. Then there exists a locally constant $\delta$-selection $g$ for $q$ such that $d(f(x), g(x)) < \varepsilon$, for every $x \in X$.*

Indeed, for every $x \in X$ choose a point $m(x) \in q(x)$ satisfying $d(f(x), m(x)) < \varepsilon$, and an open spherical neighbourhood $W = O_\delta(m(x))$ of radius $\delta$ of the point $m(x)$. Since $W$ is open in $M$, and $q$ is lower semicontinuous, there exists an open neighbourhood $V(x)$ of $x$ in $X$ such that $q(y) \cap W$ is not empty, for each $y \in V(x)$. Since $f$ is locally constant, we can also assume that $f$ is constant on $V(x)$.

The open covering $\{V(x) : x \in X\}$ contains a finite subcovering $\{V(x_i) : 1 \leq i \leq n\}$. Replacing $V(x_i)$, if necessary, by a smaller open and closed set $W(x_i)$ (which may be empty), we obtain a disjoint open covering $\{W(x_i) : 1 \leq i \leq n\}$ of $X$. Now let us define a locally constant mapping $g$ of $X$ into $M$ as follows: $g(x) = m(x_i)$, for each $x \in W(x_i)$. Clearly, $g$ satisfies the condition $d(f(x), g(x)) < \varepsilon$, for every $x \in X$, since $f(x) = f(x_i)$ and $g(x) = m(x_i)$, for each $x \in W(x_i)$, and $d(f(x_i), m(x_i)) < \varepsilon$, by the choice of $m(x_i)$.

Let $x \in W(x_i)$. Then $x \in V(x_i)$ and, therefore, by the choice of $V(x_i)$, $d(m(x_i), q(x)) < \delta$. Since $g(x) = m(x_i)$, by the definition of $g$, it follows that $d(g(x), q(x)) < \delta$. Thus, $g$ is a $\delta$-selection for $q$. Our Claim is proved.

We continue the proof of Michael's Selection Theorem. It is clear from the above Claim that we can construct recursively a sequence $\eta = \{f_n : n \in \omega\}$ of locally constant mappings of the space $X$ to the space $M$ such that the following two conditions are satisfied, for each $n \in \omega$:

(1) $f_n$ is a $1/2^n$-selection for $q$;
(2) $|f_n(x) - f_{n+1}(x)| < 1/2^n$, for each $x \in X$.

Since every locally constant mapping is continuous, all elements of $\eta$ are continuous mappings of $X$ to the metric space $M$. From conditions (1) and (2) and completeness of the metric space $M$ it follows that $\eta$ converges uniformly to a continuous mapping $f$ of $X$ to $M$ such that $d(f(x), q(x)) = 0$, for each $x \in X$. Since $q(x)$ is closed in $M$, we conclude that $f(x) \in q(x)$, for every $x \in X$. Thus, $f$ is a continuous selection for $q$. □

It is clear from the proof of Michael's theorem that it remains true if we drop the assumption of compactness of $X$ and replace it by the following condition: *Every open covering of $X$ can be refined by a disjoint open covering.* In other words, it suffices to assume that the space $X$ is *strongly zero-dimensional.*

It is easy to see that if $f$ is an open continuous mapping of a space $X$ onto a space $Y$, then the inverse mapping $f^{-1}$ of the space $Y$ into the set $Exp(X)$ of all non-empty closed subsets of $X$ is lower semicontinuous.

For the application of Michael's theorem we have in mind, we need a slightly more technical result:

LEMMA 4.1.2. *Let $f\colon X \to Z$ be a continuous mapping, $A$ a closed subset of $X$, and $p$ an open continuous mapping of a space $Y$ onto $Z$. Further, let $h\colon A \to Y$ be a continuous mapping such that $p(h(a)) = f(a)$, for each $a \in A$. Then the mapping $g\colon X \to Exp(Y)$ defined by the rule $g(x) = p^{-1}(f(x))$, for each $x \in X \setminus A$, and $g(x) = \{h(x)\}$, for each $x \in A$, is lower semicontinuous.*

PROOF. Notice that in any case $p(g(x)) = \{f(x)\}$. Let $U$ be a non-empty open subset of $Y$, and $x_0$ a point of $X$ such that $g(x_0) \cap U$ is not empty. We have to find an open neighbourhood $Ox_0$ of $x_0$ in $X$ such that $g(x) \cap U$ is not empty for each $x \in Ox_0$.

Put $z_0 = f(x_0)$ and $V = p(U)$. Then, since $z_0 = f(x_0) \in p(g(x_0) \cap U) \subset p(U) = V$ and the mapping $p$ is open, $V$ is an open neighbourhood of $z_0$ in $Z$. Now we have to distinguish two cases.

*Case 1.* $x_0 \in X \setminus A$. Then $g(x_0) = p^{-1}(z_0)$. Since $f$ is continuous, there exists an open neighbourhood $Ox_0$ of $x_0$ in $X$ such that $f(Ox_0) \subset V$. Since $A$ is closed in $X$ and $x_0$ is not in $A$, we can, in addition, assume that $Ox_0 \cap A = \varnothing$.

Let $x \in Ox_0$. Then $g(x) = p^{-1}(f(x))$ and $f(x) \in V = p(U)$. Therefore, there exists $y \in U$ such that $f(x) = p(y)$. Then $y \in p^{-1}(f(x)) \cap U = g(x) \cap U$, which implies that $g(x) \cap U$ is not empty, for each $x \in Ox_0$.

*Case 2.* $x_0 \in A$. Then $g(x_0) = \{h(x_0)\}$. Since, by the assumption, $g(x_0) \cap U$ is not empty, it follows that $h(x_0) \in U$. By the continuity of $h$, there exists an open neighbourhood $Ox_0$ of $x_0$ in $X$ such that $g(x) = \{h(x)\} \subset U$, for each $x$ in $Ox_0 \cap A$. Since $V = p(U)$ is an open neighbourhood of $z_0$ in $Z$, and $f$ is continuous, there exists an open neighbourhood $W$ of $x_0$ in $X$ such that $f(W) \subset V$ and $W \subset Ox_0$.

Take any $x \in W$. Let us show that $g(x) \cap U$ is not empty.

If $x \in A$, then this is so, since $g(x) = \{h(x)\} \subset U$, by the choice of $Ox_0$. Assume now that $x \in W \setminus A$. Then $g(x) = p^{-1}(f(x))$. Since $x \in W$, $f(x) \in V = p(U)$. Therefore, $f(x) = p(y)$, for some $y \in U$. Then $y \in p^{-1}(f(x)) \cap U = g(x) \cap U$, which implies that $g(x) \cap U$ is not empty. □

This lemma will be applied in combination with the next obvious assertion:

PROPOSITION 4.1.3. *Let $X$, $Y$, $M$ be some spaces, $g$ a lower semicontinuous mapping of $X$ into $\mathscr{E}(Y)$, and $\pi\colon Y \to M$ a continuous mapping. Then the mapping $q\colon X \to \mathscr{E}(M)$ defined by the formula $q(x) = \pi(g(x))$, for each $x \in X$, is lower semicontinuous.*

Now comes a technical statement, which follows easily from the above results, and which fits our purposes well.

LEMMA 4.1.4. *Let $Z$ be a space, $M$ a metrizable compact space, $Y$ a closed subspace of $Z \times M$, and $p$ the projection mapping of $Y$ to $Z$, that is, $p(z, m) = z$, for each $(z, m) \in Y$, which we assume to be open and onto $Z$. Further, let $f$ be a continuous mapping of a space $X$ onto $Z$, $A$ a closed subset of $X$, and $h$ a continuous mapping of $A$ to $Y$ such that $p(h(a)) = f(a)$, for each $a \in A$.*

*Suppose that a mapping $q$ of the space $X$ into $Exp(M)$ is defined in the following way. Take any $x \in X$. If $x \in A$, then $h(x) = (z, m)$ and we put $q(x) = \{m\}$. If $x \in X \setminus A$, we put $q(x) = M_x$, where $M_x = \{m \in M : (f(x), m) \in Y\}$ for each $x \in X \setminus A$. Then:*

a) *$q$ is lower semicontinuous;*
b) *if $X$ is, in addition, compact, zero-dimensional and Hausdorff, then there exists a continuous mapping $h_X$ of $X$ to $Y$ such that $ph_X(x) = f(x)$, for each $x \in X$, and $h_X$ is a continuous extension over $X$ of the mapping $h\colon A \to Y$.*

PROOF. The mapping $q$ is lower semicontinuous by Lemma 4.1.2 and Proposition 4.1.3. Thus, a) is verified. To prove b), we apply Theorem 4.1.1 to obtain a continuous selection $s$ for $q$. Now, let us define a mapping $h_X$ of $X$ to $Z \times M$ as follows: $h_X(x) = (f(x), s(x))$, for each $x \in X$. Clearly, $h_X$ is continuous, as the diagonal product of two continuous mappings. From the construction it is also clear that $ph_X(x) = f(x)$, for each $x$ in $X$, $h_X(X) \subset Y$, and $h_X(x) = h(x)$, for each $x \in A$. □

We also need the following fact:

THEOREM 4.1.5. *Every compact Hausdorff space $X$ can be represented as a continuous image of a closed subspace of the space $D^\tau$, where $D = \{0, 1\}$ is the two-point discrete space and $\tau$ is the weight of $X$.*

PROOF. We can assume without loss of generality that $X$ is infinite. Fix a base $\mathcal{B}$ of $X$ such that $|\mathcal{B}| = \tau = w(X)$, and for each $V \in \mathcal{B}$, let $F_V = \overline{V}$, $P_V = X \setminus V$, and $g_V(0) = F_V$, $g_V(1) = P_V$. Clearly, $X = g_V(0) \cup g_V(1)$ for each $V \in \mathcal{B}$. With an arbitrary point $z = (z_V)_{V \in \mathcal{B}}$ of the space $D^{\mathcal{B}}$ we associate the family $\xi_z = \{g_V(z_V) : V \in \mathcal{B}\}$ of closed subsets of $X$.

Let $Z$ be the set of all elements $z$ of $D^{\mathcal{B}}$ such that the family $\xi_z$ is centered. From the definition of topology in $D^{\mathcal{B}}$ it follows that $Z$ is closed in $D^{\mathcal{B}}$. Therefore, $Z$ is a compact subspace of $D^{\mathcal{B}}$. It is also obvious that both $D^{\mathcal{B}}$ and $Z$ are zero-dimensional spaces.

We claim that $X$ is a continuous image of $Z$. Take any $z \in Z$. Then, by the definition of $Z$, $\bigcap \xi_z$ is not empty. Pick up a point $x_z \in \bigcap \xi_z$. Let us show that, in fact, $\{x_z\} = \bigcap \xi_z$. We will prove slightly more:

**Claim.** *For every open neighbourhood $U$ of $x_z$ in $X$, there exists $B \in \xi_z$ such that $B \subset U$.*

Indeed, there exists $V \in \mathcal{B}$ such that $x_z \in V \subset \overline{V} \subset U$. Then $x_z \in F_V$ and $x_z$ is not in $P_V$. Since $x_z \in g_V(z_V)$, it follows that $g_V(z_V) = F_V$, which implies that $F_V \in \xi_z$. Since $F_V = \overline{V} \subset U$, our Claim is proved.

Now we define a mapping $f$ of $Z$ into $X$ by the rule: $f(z) = x_z$, where $\{x_z\} = \bigcap \xi_z$. Notice that from the claim and the definition of $\xi_z$ it follows immediately that $f$ is continuous: any point $z^*$ with the same $V$-th coordinate as $z$ is taken by $f$ into $U$, since then $F_V$ belongs to $\xi_{z^*}$ as well.

It remains to show that $f(Z) = X$. Fix $x \in X$, and for each $V \in \mathcal{B}$ choose $z_V \in D$ in such a way that $x \in g_V(z_V)$. Then, clearly, $z = (z_V)_{V \in \mathcal{B}}$ is an element of $Z$ and $f(z) = x$. □

Another proof of Theorem 4.1.5 can be based on the next two well-known facts (see Exercise 3.2.B and Theorem 3.2.5 of [**165**], respectively):

**Fact 1.** *The closed unit interval $I = [0, 1]$ is a continuous image of the Cantor set $D^\omega$.*

**Fact 2.** *Every compact Hausdorff space $X$ of weight $\le \tau$ can be topologically embedded into the Tychonoff cube $I^\tau$.*

Indeed, by Fact 2, a compact Hausdorff space $X$ is homeomorphic to a closed subspace of $I^\tau$, where $\tau = w(X)$. It follows from Fact 1 that the Tychonoff cube $I^\tau$ is a continuous image of the Cantor cube $D^\tau$ under a continuous mapping $f$. Then $H = f^{-1}(X)$ is a closed subspace of $D^\tau$, and the restriction of $f$ to $H$ is a continuous mapping of $H$ onto $X$.

Now we need some elementary notions and constructions involving inverse spectra of spaces. Below, whenever we consider a space $X_\alpha$, the symbol $\mathcal{T}_\alpha$ denotes the topology of $X_\alpha$. All spaces considered further in this section are assumed to be Hausdorff.

Let $X$ be a space, $(A, <)$ a well-ordered set, and let $f_\alpha$ be a quotient mapping of $X$ onto a space $X_\alpha$, for each $\alpha \in A$, such that the following two conditions are satisfied:

(S1) If $x, y \in X$, $\alpha, \beta \in A$ and $\alpha < \beta$, then $f_\beta(x) = f_\beta(y)$ implies $f_\alpha(x) = f_\alpha(y)$;
(S2) If $x, y \in X$ are distinct, then $f_\alpha(x) \neq f_\alpha(y)$, for some $\alpha \in A$.

Then we will say that $\mathcal{F} = \{f_\alpha : \alpha \in A\}$ is a *(well-ordered) spectral representation* of the space $X$, along the well-ordered set $(A, <)$.

It is easy to see that a space $X$ may have many different spectral representations. Let $\mathcal{F} = \{f_\alpha : \alpha \in A\}$ be a spectral representation of $X$. Then for any $\alpha, \beta \in A$ such that $\alpha < \beta$, we can define a mapping $p^\beta_\alpha \colon X_\beta \to X_\alpha$ by the formula $p^\beta_\alpha(x_\beta) = f_\alpha(f_\beta^{-1}(x_\beta))$, for each $x_\beta \in X_\beta$. The definition is correct in the sense that we obtain a single valued mapping $p^\beta_\alpha$, by condition (S1). Since $f_\beta$ is a quotient mapping onto, and $f_\alpha$ is a continuous mapping, it follows that $p^\beta_\alpha$ is a continuous mapping, for any $\alpha, \beta \in A$ with $\alpha < \beta$.

The mappings $p^\beta_\alpha$ will be called *connecting mappings* of the spectral representation, for an obvious reason. A spectral representation of $X$, together with the connecting mappings $p^\beta_\alpha$ will be called a *spectrum* of $X$. Notice that from condition (S1) it easily follows that *all connecting mappings are quotient mappings onto*. From the definition of the mappings $p^\beta_\alpha$ it is also clear that if $\alpha < \beta < \delta$, where $\alpha$, $\beta$, and $\delta$ are in $A$, then $p^\delta_\alpha = p^\beta_\alpha \circ p^\delta_\beta$.

Now, let $\mathcal{F}$ be a spectrum of $X$, and $Z$ some other space. We are going to introduce the notion of a spectral mapping of $Z$ into the spectrum $\mathcal{F}$. Here is the definition.

A *spectral mapping* $\mathcal{G}$ of a space $Z$ into a spectrum $\mathcal{F} = \{f_\alpha : \alpha \in A\}$ of a space $X$ is a family $\{g_\alpha : \alpha \in A\}$ of continuous mappings $g_\alpha$ of $Z$ to $X_\alpha$ such that the next condition is satisfied:

(S3) $p^\beta_\alpha \circ g_\beta = g_\alpha$ whenever elements $\alpha, \beta \in A$ satisfy $\alpha < \beta$.

We introduce some further notation. If $\mathcal{F} = \{f_\alpha : \alpha \in A\}$ is a spectrum of a space $X$, then for each $x \in X$ and for each $\alpha \in A$ we put $x_\alpha = f(x_\alpha)$. Then $(x_\alpha)_{\alpha \in A}$ is a point of the product space $P = \Pi\{X_\alpha : \alpha \in A\}$. We denote this point by $s(x)$ or by $x^*$. Thus, we have a canonical mapping $s$ of the space $X$ onto a subspace $s(X)$ of the product space $P$. We put $X^* = s(X)$ and restrict the range of $s$ to $s(X)$. From the continuity of the mappings $f_\alpha$ and condition (S2) it easily follows that $s$ is a continuous one-to-one mapping of the space $X$ onto the subspace $s(X)$ of $P$.

THEOREM 4.1.6. *Let $X$ be a compact Hausdorff space and $\mathcal{F} = \{f_\alpha : \alpha \in A\}$ a spectrum of $X$. Let, further, $\mathcal{G} = \{g_\alpha : \alpha \in A\}$ be a spectral mapping of a space $Z$ into $\mathcal{F}$. Then there exists a continuous mapping $g$ of the space $Z$ to the space $X$ such that the spectral mapping $\mathcal{G}$ is generated by $g$, that is, $g_\alpha = f_\alpha \circ g$ for each $\alpha \in A$.*

PROOF. Take any $z \in Z$, and put $F_\alpha(z) = f_\alpha^{-1}(g_\alpha(z))$, and $\eta_z = \{F_\alpha(z) : \alpha \in A\}$. Then, clearly, $\eta_z$ is a centered family of non-empty closed sets in $X$. Actually, it is a chain, that is, $F_\beta(z) \subset F_\alpha(z)$ if $\alpha < \beta$. Therefore, its intersection $P_z = \bigcap_{\alpha \in A} F_\alpha(z)$ is not empty.

Let us show that $P_z$ consists of exactly one point, for each $z \in Z$. This follows from condition (S2). Indeed, if $x \in P_z$ and $y \in P_z$, then $f_\alpha(x) = g_\alpha(z) = f_\alpha(y)$ for each $\alpha \in A$, and (S2) implies that $x = y$. Thus, $P_z = \{x_z\}$ for some $x_z \in X$.

Now we define a mapping $g\colon Z \to X$ by the rule $g(z) = x_z$. Since $x_z \in F_\alpha(z) = f_\alpha^{-1}(g_\alpha(z))$, we have $f_\alpha(g(z)) = f_\alpha(x_z) = g_\alpha(z)$ for each $z \in Z$, that is, $f_\alpha \circ g = g_\alpha$, for each $\alpha \in A$.

Let us show that $g$ is continuous. Fix $z \in Z$ and put $x = g(z) = x_z$ and $x_\alpha = f_\alpha(x) = g_\alpha(z)$ for each $\alpha \in A$. Let $Ox$ be any open neighbourhood of $x$ in $X$. Since $\bigcap \eta_z$ is contained in $Ox$, and $\eta_z$ is a chain consisting of closed sets, it follows that $F_\alpha(z) \subset Ox$, for some $\alpha \in A$.

Since $X$ is compact and $X_\alpha$ is Hausdorff, the mapping $f_\alpha$ is closed. Since $f_\alpha^{-1}(x_\alpha) = F_\alpha(z) \subset Ox$, it follows that there exists an open neighbourhood $V_\alpha$ of $x_\alpha$ in $X_\alpha$ such that $f_\alpha^{-1}(V_\alpha) \subset Ox$. Put $U = g_\alpha^{-1}(V_\alpha)$. Clearly, $U$ is an open neighbourhood of $z$ in $Z$, and $g(U) \subset Ox$, since $f_\alpha(g(t)) = g_\alpha(t) \in V_\alpha$, for each $t \in U$. Notice that we do not claim that $g$ is necessarily onto $X$. □

The following question is quite natural: Is every compact Hausdorff space $X$ a continuous image of $D^\tau$, for some $\tau$, where $D = \{0, 1\}$ is the two-point discrete space? The answer is "no", since the cellularity of $D^\tau$ is always countable and the Souslin number never increases under continuous onto mappings. Thus, not every compact space is dyadic.

The theory of dyadic compacta is well-developed and rich in results (see [**165**]), and one of the most remarkable results in this theory is Ivanovskij–Kuz'minov's theorem, which we are going to prove now. In the proof, we apply Theorem 3.7.24 stating that every compact topological group $G$ is topologically isomorphic to a closed subgroup of the product of some family of second-countable topological groups.

THEOREM 4.1.7. [**L. N. Ivanovskij**, **V. I. Kuz'minov**] *Every compact topological group $G$ is a dyadic compactum.*

PROOF. By Theorem 3.7.24, we can assume that $G$ is a subgroup of the product $M = \Pi_{\alpha \in A} M_\alpha$ of separable metrizable topological groups $M_\alpha$. Taking projections of $G$ to the factors, we may assume that each $M_\alpha$ is compact. Let us also fix a well-ordering $<$ of the index set $A$. Now a spectrum, naturally associated with $G$ as a subgroup of the product group $M$, is defined as follows.

For each $\alpha \in A$, let $\pi_\alpha$ be the natural projection of $M$ onto the product $L_\alpha = \Pi_{\nu < \alpha} M_\nu$, $G_\alpha = \pi_\alpha(G)$, and $f_\alpha$ be the restriction of $\pi_\alpha$ to $G \subset M$. Since $G$ is compact, and $G_\alpha$ is Hausdorff, the mapping $f_\alpha$ is closed for each $\alpha \in A$. Clearly, conditions (S1) and (S2) are also satisfied for $\mathscr{F} = \{f_\alpha : \alpha \in A\}$. Therefore, $\mathscr{F}$ is a spectrum of $G$.

**Claim 1.** *Each $f_\alpha$ is an open homomorphism of $G$ onto $G_\alpha$.*

Indeed, it is clear that $f_\alpha$ is a continuous homomorphism, and every continuous homomorphism of one compact group onto another compact group is open, by Theorem 3.1.27.

For $\alpha \in A$, let $\alpha + 1$ stand for the successor of $\alpha$ in the well-ordered set $(A, <)$, and we write $p_\alpha$ for the natural projection $p_\alpha^{\alpha+1}$ of $G_{\alpha+1}$ onto $G_\alpha$. Notice that $f_\alpha = p_\alpha \circ f_{\alpha+1}$, that is, $p_\alpha$ is one of the connecting mappings of the spectrum $\mathcal{F}$. The next fact is immediate.

**Claim 2.** *The topological group $G_{\alpha+1}$ is a subgroup of the product group $G_\alpha \times M_\alpha$, and $p_\alpha$ is the restriction of the projection of $G_\alpha \times M_\alpha$ onto $G_\alpha$. As a continuous homomorphism of one compact group onto another, $p_\alpha$ is closed and open.*

Extending the above notation, we denote by $p_\beta^\alpha$ the natural projection of $G_\alpha$ to $G_\beta$ provided that $\alpha, \beta \in A$ and $\alpha < \beta$.

**Claim 3.** *For every $\alpha \in A$ which does not have an immediate predecessor, the family $\mathcal{F}_\alpha = \{p_\beta^\alpha : \beta < \alpha\}$ is a spectrum of the space $G_\alpha$.*

This is obvious since all mappings $p_\beta^\alpha$ are continuous surjective homomorphisms of compact groups and are, therefore, open.

According to Theorem 4.1.5, there exist a cardinal number $\tau$, a closed subspace $H$ of $D^\tau$, and a continuous mapping $h$ of $H$ onto $G$. We put $h_\alpha = f_\alpha \circ h$, for $\alpha \in A$. Obviously, $D^\tau$ is a zero-dimensional compact space, which opens the door for applications of Michael's theorem.

We will construct a continuous mapping $g$ of $D^\tau$ into $G$ that will be is extension of the mapping $h$. Because of that, the mapping $g$ will be onto $G$. And to define $g$, we will first define a spectral mapping of $D^\tau$ into the spectrum $\mathcal{F}$ of $G$. A spectral mapping $\mathcal{G} = \{g_\alpha : \alpha \in A\}$ will be constructed by transfinite recursion along the well-ordered set $A$ in such a way that each $g_\alpha$ will be a continuous extension of the mapping $h_\alpha \colon H \to G_\alpha$. Here is how the construction goes.

Clearly, we may assume that $G_0$ consists only of one element $e_0$. Then the mapping $g_0$ is trivially defined by $g_0(z) = e_0$ for each $z \in D^\tau$, and $g_0$ extends $h_0$.

Assume now that, for some $\alpha \in A$, a continuous mapping $g_\beta$ of $D^\tau$ to $G_\beta$ is defined for every $\beta < \alpha$, and that the following conditions are satisfied:

$(1_\alpha)$ $g_\beta$ is an extension of $h_\beta$ for each $\beta < \alpha$;

$(2_\alpha)$ if $\delta < \beta < \alpha$, then $g_\delta = p_\delta^\beta \circ g_\beta$.

We want to define a continuous mapping $g_\alpha$ of $D^\tau$ to $G_\alpha$ extending the mapping $h_\alpha$ and such that $g_\delta = p_\delta^\alpha \circ g_\alpha$, for every $\delta < \alpha$.

*Case 1.* The element $\alpha$ does not have an immediate predecessor in $(A, <)$.

Then $\mathcal{F}_\alpha = \{p_\beta^\alpha : \beta < \alpha\}$ is, obviously, a spectrum of the space $G_\alpha$, and $\mathcal{G}_\alpha = \{g_\beta : \beta < \alpha\}$ is a spectral mapping of the space $D^\tau$ into the spectrum $\mathcal{F}_\alpha$, in view of conditions $(1_\alpha)$ and $(2_\alpha)$. Since $G_\alpha$ is compact, we can apply Theorem 4.1.5, which implies that the spectrum $\mathcal{F}_\alpha$ is generated by a continuous mapping $g_\alpha$ of $D^\tau$ to $G_\alpha$. Clearly, $g_\alpha$ satisfies the composition condition $g_\beta = p_\beta^\alpha \circ g_\alpha$ for each $\beta < \alpha$ (see Theorem 4.1.5). It remains to check that $g_\alpha$ is an extension of $h_\alpha$.

Take any $z \in H$. Then $h_\alpha(z) = f_\alpha(h(z))$, by the definition of $h_\alpha$. Assume that $g_\alpha(z) \neq h_\alpha(z)$. Then, for some $\beta < \alpha$, $p_\beta^\alpha(g_\alpha(z)) \neq p_\beta^\alpha(h_\alpha(z))$. On the other hand, it follows from the definitions that $p_\beta^\alpha \circ g_\alpha = g_\beta$ and $p_\beta^\alpha \circ h_\alpha = h_\beta$. Therefore, since $g_\beta$, by our assumptions, is an extension of $h_\beta$, we have $g_\beta(z) = h_\beta(z)$, that is, $p_\beta^\alpha(g_\alpha(z)) = p_\beta^\alpha(h_\alpha(z))$, which is a contradiction. Hence, $g_\alpha$ is an extension of $h_\alpha$.

*Case 2.* $\alpha = \beta + 1$, that is, there exists an immediate predecessor $\beta$ of $\alpha$ in $(A, <)$.

In this case the result is achieved by a straightforward application of Lemma 4.1.4, where $Z = G_\beta$, $M = M_\beta$, $Y = G_\alpha$, $p = p^\alpha_\beta$, $f = g_\beta$, $X = D^\tau$, $A = H$, and $h = h_\alpha$. Notice that $G_\alpha$ is a subspace of $G_\beta \times M_\beta$, and $M_\beta$ is compact and, therefore, is a complete metric space.

Our definition of $g_\alpha$ is complete, and it is clear that conditions $(1_{\alpha+1})$ and $(2_{\alpha+1})$ are satisfied. In this way, a spectral mapping $\mathcal{G} = \{g_\alpha : \alpha \in A\}$ of $D^\tau$ into the spectrum $\mathcal{F}$ of $G$ is defined such that $g_\alpha$ is a continuous extension of $h_\alpha$ for each $\alpha \in A$.

Now we apply Theorem 4.1.5 once again; according to it, there exists a continuous mapping $g$ of $D^\tau$ to $G$ such that $g_\alpha = f_\alpha \circ g$, for each $\alpha \in A$. Arguing exactly as in Case 1 (when we proved that $g_\alpha$ extended $h_\alpha$), we establish that $g$ is an extension of $h$. Since $h(H) = G$, it follows that $g(D^\tau) = G$. □

Theorem 4.1.7 has many interesting applications. The first of them is almost immediate.

COROLLARY 4.1.8. *The cellularity of any precompact topological group is countable. In particular, the cellularity of any compact topological group is countable.*

PROOF. Let $G$ be a compact topological group. Then $G$ is a continuous image of $D^\tau$, for some $\tau$, by Theorem 4.1.7. The Souslin number of the product of any family of separable metrizable spaces is countable by [**165**, Coro. 2.3.18]; therefore, the Souslin number of $D^\tau$ is countable. It remains to note that under continuous onto mappings, the Souslin number does not increase.

If $G$ is a precompact topological group, then Theorem 3.7.16 implies that the Raĭkov completion $\varrho G$ of $G$ is a compact topological group. Hence the cellularity of $\varrho G$ is countable. Since $G$ is dense in $\varrho G$, any disjoint family of open sets in $G$ is also countable. □

The above result will be considerably generalized in Section 5.3 (see Corollary 5.3.22).

## Exercises

4.1.a. Prove that the cellularity of any pseudocompact topological group is countable.

4.1.b. A space $X$ is called $\tau$-*monolithic*, where $\tau$ is a cardinal number, if, whenever the cardinality of a subset $A$ of $X$ does not exceed $\tau$, the closure of $A$ has a network $S$ such that $|S| \leq \tau$. Prove that every compact topological group $G$ contains a dense countably compact $\aleph_0$-monolithic subspace.

4.1.c. Suppose that in a compact topological group $G$, every countably compact subspace is compact. Prove that $G$ is metrizable.

4.1.d. Let $G$ be a compact topological group, and let $\gamma$ be an uncountable family of non-empty open subsets of $G$. Prove that there exists an uncountable centered subfamily $\xi$ of $\gamma$.

4.1.e. Show that not every compact topological group can be represented as an image of $D^\tau$, for some $\tau$, under an open continuous mapping.

*Hint.* Consider the behaviour of dimension under open continuous mappings.

4.1.f. Let $G$ be an infinite compact topological group. Show that for every point $x \in G$, there exists a sequence $\{x_n : n \in \omega\} \subset G \setminus \{x\}$ converging to $x$.

*Hint.* Consider a continuous onto mapping $f\colon D^\tau \to G$ which exists by Theorem 4.1.7. Since $D^\tau$ is a compact space, the mapping $f$ is closed. This implies that the fiber $f^{-1}(x)$ cannot be open in $D^\tau$. Take a point $a \in f^{-1}(x)$ lying in the closure of $D^\tau \setminus f^{-1}(x)$ and define a sequence $\{a_n : n \in \omega\} \subset D^\tau \setminus f^{-1}(x)$ converging to $a$ (see [**165**]).

## Problems

4.1.A. Every compact $G_\delta$-subset of an arbitrary topological group $G$ is a dyadic compactum.
*Remark.* This far reaching generalization of the Ivanovskij–Kuz'minov theorem was established by M. M. Choban in [**101**]. The proof is again based on Michael's selection theorem and uses the method of inverse spectra. Even more general results will be presented in Section 10.3.

4.1.B. Recall that a space $X$ is said to be homogeneous if, for any points $x, y \in X$, there exists a homeomorphism $f$ of $X$ onto itself such that $f(x) = y$. Is every homogeneous compact Hausdorff space homeomorphic to a compact topological group?
*Hint.* See the next problem.

4.1.C. Is every homogeneous compact Hausdorff space dyadic?
*Hint.* See Problem 4.1.D.

4.1.D. Construct a homogeneous first-countable compact Hausdorff space $X$ such that the cellularity $c(X)$ of $X$ is uncountable.
*Hint.* Let $X$ be the *Alexandroff duplicate* of the Cantor set (see [**165**, Example 3.1.26]). Then $X$ is a first-countable non-metrizable compact Hausdorff space with an open discrete subspace of cardinality $\mathfrak{c} = 2^\omega$. Thus, the Souslin number of $X$ is uncountable. The space $X$ is also zero-dimensional. Therefore, by a theorem of D. B. Motorov in [**332**], the space $X^\omega$ is homogeneous. Clearly, the Souslin number of $X$ is equal to $\mathfrak{c}$.

4.1.E. Let $G$ be a compact topological group such that every subgroup of $G$ is a normal space. Prove that under the assumption $2^{\aleph_0} < 2^{\aleph_1}$, the group $G$ is metrizable. (The cardinality assumption is, in fact, not necessary but, to avoid it, one has to use more subtle methods from [**120**].)
*Hint.* Suppose that $G$ is not metrizable, and use Theorem 4.1.7 to find a closed separable non-metrizable subgroup $H$ of $G$. Then the character of $H$ is uncountable, so $|H| \geq 2^{\aleph_1}$ by the Čech–Pospíšil theorem, see [**165**, 3.12.11(a)]. Choose an arbitrary point $y_0 \in G$ distinct from the neutral element of $G$ and apply 4.1.f to choose a sequence $S = \{x_n : n \in \omega\}$ in $G \setminus \{y_0\}$ converging to $y_0$. Show that $S$ can be additionally chosen to satisfy $y_0 \notin \langle S \rangle$. Then use the assumption $2^{\aleph_0} < 2^{\aleph_1}$ to define by recursion a dense pseudocompact subgroup $P$ of $H$ such that $S \subset P$ and $y_0 \notin P$. Finally, verify that the space $P$ is not normal.

4.1.F. Prove that the $\delta$-tightness and $G_\delta$-tightness of every compact topological group are countable.
*Hint.* Verify that continuous onto mappings of compact spaces do not increase $\delta$-tightness. Then apply Theorems 4.1.7 and 1.6.11.

## Open Problems

4.1.1. Let $G$ be a compact group. Is there a dense subspace $X$ of $G$ such that $t(X) \leq \omega$?

4.1.2. Let $G$ be a compact group. Is there a dense sequential subspace of $G$?

4.1.3. Let $G$ be a compact group. Is there a dense countably compact $\aleph_0$-monolithic subspace $X$ of $G$ such that the tightness of $X$ is countable?

4.1.4. Let $G$ be a compact group. Is there a dense subgroup of $G$ satisfying the conditions in one of the above three problems?

4.1.5. Is every compact $G_\delta$-subset of a (Tychonoff) paratopological group dyadic?

4.1.6. Is every compact $G_\delta$-subset of a Hausdorff semitopological group dyadic?

4.1.7. Is it true that every homogeneous compact Hausdorff space contains a dense subspace of countable tightness?

## 4.2. Embedding $D^{\omega_1}$ in a non-metrizable compact group

We start with the proof of the following important result:

THEOREM 4.2.1. [**R. Engelking**] *If $G$ is a non-metrizable compact topological group of weight $\tau$, then the space $D^\tau$ is homeomorphic to a subspace of $G$.*

PROOF. Let $G$ be a compact group of weight $\tau > \omega$. Since $G$ is compact, the character and pseudocharacter of $G$ at the neutral element $e$ coincide, according to [**165**, 3.1.F. (a)]. Hence, by the Birkhoff–Kakutani theorem, the singleton $\{e\}$ cannot be a $G_\delta$-set in $G$. For each open neighbourhood $U$ of $e$ in $G$, there exists a closed invariant subgroup $H$ of $G$ such that $H \subset U$ and $H$ is a $G_\delta$-set in $G$ (see Corollary 3.4.19 and Proposition 3.4.6). Therefore, there exists a family $\mathcal{H} = \{H_\alpha : \alpha < \tau\}$ of closed invariant subgroups of $G$ such that the following three conditions are satisfied for each $\alpha < \tau$:

1) $H_\alpha$ is a $G_\delta$-set in $G$;
2) $\bigcap \mathcal{H} = \{e\}$;
3) $\bigcap_{\beta<\alpha} H_\beta \neq \{e\}$.

We can also assume that $H_0 \neq G$. Let $A$ be the set of all $\alpha < \tau$ such that $\bigcap_{\beta<\alpha} H_\beta$ is not contained in $H_\alpha$. If we replace $\tau$ with the naturally well-ordered set $A$, then conditions 1)–3) will be still satisfied, and, in addition, the next condition will hold:

4) $(\bigcap_{\beta<\alpha} H_\beta) \setminus H_\alpha$ is not empty, for each $\alpha \in A$.

So without loss in generality we can assume that the family $\mathcal{H} = \{H_\alpha : \alpha < \tau\}$ satisfies all four conditions 1)–4) for each $\alpha < \tau$.

Now, for each $\alpha < \tau$, consider the quotient homomorphism of $G$ onto the compact metrizable group $M_\alpha = G/H_\alpha$. Clearly, the diagonal product of these quotient homomorphisms is a continuous monomorphism of $G$ to the product of separable metrizable groups $M_\alpha$. Therefore, $G$ is represented as a topological subgroup of the product of compact metrizable groups $M_\alpha$, that is, $G \subset \prod_{\alpha<\tau} M_\alpha$. Let $\mathcal{P} = \{p_\alpha : \alpha < \tau\}$ be the natural spectrum of $G$ associated with the given embedding of $G$ into the product group $M = \prod_{\alpha<\tau} M_\alpha$ and with the natural well-ordering on $\tau = \{\alpha : \alpha < \tau\}$. In other words, for every $\alpha < \tau$, let $\pi_\alpha$ be the natural projection of $M$ onto the subproduct $P_\alpha = \prod_{\beta<\alpha} M_\beta$ and $p_\alpha$ be the restriction of $\pi_\alpha$ to $G$. We also put $G_\alpha = p_\alpha(G)$ for each $\alpha < \tau$. As in Section 4.1, for all $\alpha, \beta$ satisfying $\beta < \alpha < \tau$, there exists a quotient mapping $p^\alpha_\beta\colon G_\alpha \to G_\beta$ such that $p_\beta = p^\alpha_\beta \circ p_\alpha$.

**Claim 1.** *For all elements $\alpha, \beta$ satisfying $\beta < \alpha < \tau$, the projection $p^\alpha_\beta$ of $G_\alpha$ to $G_\beta$ is an open continuous homomorphism of the group $G_\alpha$ onto the group $G_\beta$ which is not one-to-one.*

Indeed, $p^\alpha_\beta$ is not one-to-one by condition 4) above. The validity of the remaining part of the claim is clear.

Let $D_\alpha = \{0, 1\}$ be the two-point discrete space for each $\alpha < \tau$, and put $F_\alpha = \prod_{\beta<\alpha} D_\beta$, $\mathcal{F} = \{f_\alpha : \alpha < \tau\}$, where $f_\alpha$ is the natural projection of $D^\tau = \prod_{\nu<\tau} D_\nu$ onto $F_\alpha$. Clearly, each $f_\alpha$ is a continuous homomorphism, and $F_{\alpha+1} = F_\alpha \times D_\alpha$. The natural projection of $F_\alpha$ onto $F_\beta$, where $\beta < \alpha$, is denoted by $q^\alpha_\beta$.

For every $\alpha < \tau$, we are going to construct a homeomorphism $t_\alpha$ of $F_\alpha$ onto a subspace $B_\alpha$ of $G_\alpha$ such that

(5) $p^\alpha_\beta \circ t_\alpha = t_\beta \circ q^\alpha_\beta$ whenever $0 < \beta < \alpha < \tau$.

Clearly, $F_1 = D_0 = \{0, 1\}$. There are two distinct points $a$ and $b$ in $G_1 = M_0 = G/H_0$; we fix them and put $t_1(0) = a$ and $t_1(1) = b$. That is how we start.

Now the inductive step follows. Assume that for some $\alpha < \tau$ the homeomorphisms $t_\beta$ are already defined for all $\beta$ with $0 < \beta < \alpha$, satisfying the version of condition (5):

($6_\alpha$) $p^\delta_\beta \circ t_\delta = t_\beta \circ q^\delta_\beta$ whenever $0 < \beta < \delta < \alpha$.

*Case 1.* The ordinal $\alpha$ is limit. Put $s^\alpha_\beta = t_\beta \circ q^\alpha_\beta$, for each $\beta < \alpha$.

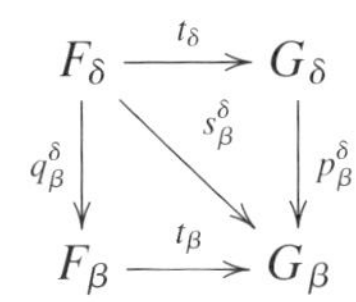


Then $\mathcal{S}_\alpha = \{s^\alpha_\beta : 0 < \beta < \alpha\}$ is a spectral mapping of $F_\alpha$ into the spectrum $\mathcal{P}_\alpha = \{p^\alpha_\beta : 0 < \beta < \alpha\}$ of $G_\alpha$. Since $G_\alpha$ is compact, there exists a continuous mapping $t_\alpha$ of $F_\alpha$ to $G_\alpha$ which generates the spectral mapping $\mathcal{S}_\alpha$. From condition (S2) for the spectrum $\mathcal{P}_\alpha$ (see page 221) and ($6_\alpha$) it is clear that $t_\alpha$ is one-to-one. Since $F_\alpha$ is compact and $G_\alpha$ is Hausdorff, it follows that $t_\alpha$ is a homeomorphism of $F_\alpha$ onto the subspace $t_\alpha(F_\alpha)$ of $G_\alpha$. From the definition of $t_\alpha$ it also follows that condition ($6_{\alpha+1}$) is satisfied.

*Case 2.* $\alpha = \beta+1$. Then $B_\beta = t_\beta(F_\beta)$ is a zero-dimensional compact space (homeomorphic to $D^\beta$) lying in $G_\beta$. Put $C_\alpha = (p^\alpha_\beta)^{-1}(B_\beta)$, where $p^\alpha_\beta$ is the natural projection of $G_\alpha$ onto $G_\beta$. Since $p^\alpha_\beta$ is open and continuous, the restriction $u$ of $p^\alpha_\beta$ to $C_\alpha$ is an open continuous mapping of $C_\alpha$ onto $B_\beta$. Now $C_\alpha$ is a closed subspace of $B_\beta \times M_\beta$, and $M_\beta$ is a compact metric space. Let $\pi$ be the natural projection of $B_\beta \times M_\beta$ onto $M_\beta$. Then the mapping $\psi$ of $B_\beta$ to $Exp(M_\beta)$ defined by the rule $\psi(b) = \pi(u^{-1}(b))$, for each $b \in B_\beta$, is lower semicontinuous. Therefore, by Theorem 4.1.1, there exists a continuous selection $m\colon B_\beta \to M_\beta$ for $\psi$. Put $v(b) = (b, m(b))$. Then, clearly, $v$ is a continuous mapping of $B_\beta$ to $C_\alpha$, and $v$ is a continuous selection for $u^{-1}$. It follows that $v$ is a topological embedding of $B_\beta$ into $C_\alpha \subset G_\alpha$, and $p^\alpha_\beta \circ v$ is the identity mapping of $B_\beta$ onto itself.

Take any $x \in v(B_\beta)$. Since $p^\alpha_\beta$ is a homomorphism and, by Claim 1, it is not one-to-one, there exists $y \in G_\alpha$ such that $x \neq y$ and $p^\alpha_\beta(x) = p^\alpha_\beta(y)$. We fix such a point $y$. Since $u = p^\alpha_\beta \restriction C_\alpha$, for any $a \in v(B_\beta)$ we have $u(yx^{-1}a) = u(y)u(x)^{-1}u(a) = u(a)$, by the choice of $y$. Since the restriction of $u$ to $v(B_\beta)$ is one-to-one, it follows that $yx^{-1}a$ is not in $v(B_\beta)$, for any $a \in v(B_\beta)$. Therefore, the sets $yx^{-1}v(B_\beta)$ and $v(B_\beta)$ are disjoint.

Consider the disjoint union $F_\alpha = (F_\beta \times \{0\}) \cup (F_\beta \times \{1\})$. Let us put $t_\alpha(z) = v(s^\alpha_\beta(z))$, for each $z \in F_\beta \times \{0\}$, and $t_\alpha(z) = yx^{-1}v(s^\alpha_\beta(z))$, for each $z \in F_\beta \times \{1\}$. Obviously, $t_\alpha$ is a homeomorphic embedding of $F_\alpha$ to $G_\alpha$, while ($6_\alpha$) and the definition of $t_\alpha$ imply the validity of ($6_{\alpha+1}$).

The recursive construction of the family $\{t_\alpha : \alpha < \tau\}$ is complete. Now we put $\phi_\alpha = t_\alpha \circ f_\alpha$ for each $\alpha < \tau$. Clearly, $\{\phi_\alpha : \alpha < \tau\}$ is a spectral mapping of $D^\tau$ into the spectrum $\mathcal{P} = \{p_\alpha : \alpha < \tau\}$ of $G$. Since $G$ is compact, this spectral mapping is generated by a continuous mapping $t$ of $D^\tau$ to $G$. From (S2) and the fact that each $t_\alpha$ is a homeomorphism it obviously follows that $t$ is one-to-one. Since $D^\tau$ is compact and $G$ is Hausdorff, we conclude that $t$ is a homeomorphism of $D^\tau$ onto the subspace $t(D^\tau)$ of $G$. □

Recall that the *tightness* of a space $X$ is countable if for every $A \subset X$ and every $x \in \overline{A}$, there exists a countable subset $B$ of $A$ such that $x \in \overline{B}$. It is easy to construct compact spaces of countable tightness that are not first-countable — take, for example, the one-point compactification of an uncountable discrete space. However, for compact groups we have the following remarkable fact:

COROLLARY 4.2.2. *Every compact topological group $G$ of countable tightness is metrizable.*

PROOF. Assume the contrary. Then, by virtue of Theorem 4.2.1, $G$ contains a topological copy of the space $D^{\omega_1}$. It is easy to verify that the tightness of $D^{\omega_1}$ is uncountable (take $A$ to be the $\Sigma$-product with center at an arbitrary point of $D^{\omega_1}$ and as $x$ any point in the complement of $A$). Therefore, the tightness of $G$ is uncountable, a contradiction. □

The following result is a part of topological folklore, but we prefer to supply the reader with its proof.

PROPOSITION 4.2.3. *A compact Hausdorff space $X$ without isolated points admits a continuous mapping onto the closed unit interval.*

PROOF. First we show that the Cantor set $C = \{0, 1\}^\omega$ is a continuous image of some closed subspace of $X$. In what follows we identify $C$ with the family of all functions from $\omega$ to $2 = \{0, 1\}$.

For every $n \in \mathbb{N}$, let $2^n$ be the family of all functions from $n = \{0, 1, \dots, n-1\}$ to 2. Put $\mathscr{C} = \bigcup_{n \in \mathbb{N}} 2^n$. We are going to construct, for each $f \in \mathscr{C}$, a non-empty open subset $U_f$ of $X$ such that if $f, g \in \mathscr{C}$ and $g$ is a proper extension of $f$, then $\overline{U_g} \subset U_f$.

Choose non-empty open subsets $U_0$ and $U_1$ of $X$ such that $\overline{U_0} \cap \overline{U_1} = \varnothing$ (here we identify 0 and 1 with the functions $(0, 0)$ and $(0, 1)$, respectively). Suppose that for some $n \in \mathbb{N}$, we have defined the sets $U_f$ for all $f \in 2^n$. For every $f \in 2^n$ and every $i = 0, 1$, let $f^\frown i$ denote the function $g \in 2^{n+1}$ such that $g{\restriction}n = f$ and $g(n) = i$. Since $X$ has no isolated points, we can choose non-empty open sets $U_{f^\frown 0}$ and $U_{f^\frown 1}$ in $X$ such that $\overline{U_{f^\frown 0}} \cap \overline{U_{f^\frown 1}} = \varnothing$ and $\overline{U_{f^\frown 0}} \cup \overline{U_{f^\frown 1}} \subset U_f$. The construction of the family $\{U_f : f \in \mathscr{C}\}$ is complete.

For every function $h \in 2^\omega = C$, put $F_h = \bigcap_{n \in \mathbb{N}} \overline{U_{h{\restriction}n}}$. It follows from the choice of the sets $U_f$ that $F_h$ is a non-empty closed subset of $X$, for each $h$. It also follows that the sets $F_h$ and $F_{h'}$ are disjoint whenever $h \neq h'$. Let

$$Y = \bigcap_{n \in \mathbb{N}} \bigcup_{f \in 2^n} \overline{U_f}.$$

Since the family $2^n$ is finite for each $n \in \mathbb{N}$, the set $Y$ is closed in $X$. Notice that $Y = \bigcup_{h \in C} F_h$. We define a mapping $\varphi\colon Y \to C$ by $\varphi(y) = h$ if $y \in F_h$. Since the sets $F_h$ are disjoint, our definition of $\varphi$ is correct. It is clear that $\varphi$ maps $Y$ onto $C$. It remains to verify the continuity of $\varphi$.

Suppose that $y \in Y$, and let $O$ be a neighbourhood of the element $h = \varphi(y)$ in $C$. It follows from the definition of the product topology in $C = 2^\omega$ that $V = p_n^{-1}(h{\restriction}n) \subset O$ for some $n \in \mathbb{N}$, where $p_n$ is the projection of $C$ onto $2^n$. Clearly, $h \in V$. Put $f = h{\restriction}n$. Then $y \in F_h \subset U_f$, and it follows from the definition of the mapping $\varphi$ that $\varphi(U_f \cap Y) \subset V$. Indeed, if $z \in U_f \cap Y$, then the image $h' = \varphi(z)$ satisfies $h'{\restriction}n = h{\restriction}n$; hence, $h' \in V$. Thus, $\varphi(U_f \cap Y) \subset V \subset O$, and the continuity of $\varphi$ follows.

By [**165**, 3.2.B], there exists a continuous mapping $\psi$ of the Cantor set $C$ onto $I = [0, 1]$. Then the composition $\psi \circ \varphi$ is a continuous mapping of $Y$ onto $I$. Since $Y$ is a closed subset of the compact (hence, normal) space $X$, one can extend $\psi \circ \varphi$ to a continuous mapping $f : X \to I$, by the Tietze–Urysohn theorem (see [**165**, Theorem 2.1.8]). Clearly, $I = f(Y) \subset f(X) \subset I$; hence, $f(X) = I$. □

THEOREM 4.2.4. *Let $G$ be an infinite compact topological group of weight $\tau$. Then $G$ admits a continuous mapping onto the Tychonoff cube $I^\tau$.*

PROOF. We start with a simple observation. By Fact 1 on page 220, the Cantor set $C = D^\omega$ admits a continuous mapping $f$ onto the closed unit interval $I$. Hence the product mapping $f^\omega$ maps $C^\omega$ onto $I^\omega$. Since $C^\omega = (D^\omega)^\omega \cong C$, we conclude that $C$ admits a continuous mapping $g$ onto $I^\omega$. Further, since $C$ is homeomorphic to a closed subspace of $I$, Urysohn's lemma implies that $g$ admits an extension to a continuous mapping of $I$ onto $I^\omega$.

Clearly, an infinite compact group has no isolated points. Hence, in the case $\tau = \omega$ the conclusion follows from Proposition 4.2.3 and the above observation. Suppose, therefore, that $\tau > \omega$. By Theorem 4.2.1, $G$ contains a topological copy of $D^\tau$. Again, there exists a continuous mapping of $D^\tau$ onto $I^\tau$. Since $G$ is normal, we can use Urysohn's lemma to extend this mapping to a continuous mapping of $G$ onto $I^\tau$. □

If the group $G$ is Abelian, the conclusion in Theorem 4.2.4 can be considerably strengthened (see Problem 9.6.B).

COROLLARY 4.2.5. *For every non-metrizable compact group $G$, there exists a continuous mapping of $G$ onto the Tychonoff cube $I^{\omega_1}$.*

## Exercises

4.2.a. Prove that a compact group $G$ is $\aleph_0$-monolithic (see Exercise 4.1.b) if and only if $G$ is metrizable.

4.2.b. Show that an $\aleph_0$-monolithic countably compact group need not be metrizable.

4.2.c. Apply Corollary 4.2.2 to show that every hereditarily separable compact group is metrizable.

4.2.d. Give an example of a countably compact non-metrizable group of countable tightness. Deduce that Theorem 4.2.1 cannot be extended to countably compact groups.

4.2.e. Show that every non-metrizable compact group contains an uncountable discrete subspace.

4.2.f. Prove that a compact group $G$ is metrizable if and only if every subspace of $G$ is normal (see also Problem 4.1.E).

4.2.g. Give an alternative solution to Problem 4.1.C, using the material in this section — show that not every homogeneous compactum is dyadic, and therefore, not every homogeneous compactum is homeomorphic to a compact topological group.

*Hint.* Take a first-countable non-metrizable homogeneous compact space $X$ (for example, we can take $X$ to be the *two arrows space* [**165**, 3.10.C]). Now use the following generalization of Corollary 4.2.2: Every dyadic compact space of countable tightness is metrizable (see [**165**, 3.12.12(h)]).

## Problems

4.2.A. Let $X$ be a closed subset of a compact topological group $G$. Suppose that $X$ has countable tightness and that $X$ algebraically generates the group $G$. Prove that $G$ is metrizable.

4.2.B. A space $X$ is called *scattered* if every non-empty subset of $X$ contains an isolated point. Prove that if $G$ is a non-discrete compact group, then no scattered subspace of $G$ algebraically generates it.

4.2.C. Let us call a space $X$ *left-separated* if there exists a well-ordering $\prec$ of $X$ such that the set $X_{<x} = \{y \in X : y \prec x\}$ is closed in $X$, for each $x \in X$. Prove that if $G$ is a non-discrete compact group, then no left-separated subspace of $G$ algebraically generates it.

4.2.D. Show that every left-separated countably compact topological group is finite.

4.2.E. (M. G. Tkachenko [**467**]) Present an example of an infinite pseudocompact left-separated topological group.

4.2.F. A space $X$ is called *$C$-closed* if every countably compact subspace of $X$ is closed in $X$. Prove that every $C$-closed compact topological group is metrizable.

4.2.G. It is not possible to construct in $ZFC$ a hereditarily separable countably compact topological group which is not compact.

*Remark.* Under the Continuum Hypothesis, A. Hajnal and I. Juhász constructed in [**215**] a countably compact hereditarily separable hereditarily normal topological group $G$ which was not compact. Their group $G$ is a dense subgroup of $\mathbb{Z}(2)^{\omega_1}$.

4.2.H. It is not possible to construct in $ZFC$ a hereditarily separable pseudocompact group $G$ which is not compact.

*Remark.* S. Todorčević proved in [**495**] that the following statement is consistent with $ZFC$:

(S) Every regular hereditarily separable space is (hereditarily) Lindelöf.

It follows from (S) that every hereditarily separable pseudocompact topological group is compact and first-countable and, therefore, is metrizable.

## Open Problems

4.2.1. Is it possible to construct in $ZFC$ a hereditarily normal countably compact non-compact topological group? (See also Problem 6.4.H and the remark after Problem 4.2.G).

4.2.2. Is every monolithic hereditarily normal countably compact group compact?

4.2.3. Is it true that every monolithic non-metrizable countably compact group contains an uncountable discrete subspace?

4.2.4. Does every non-metrizable sequentially compact topological group contain an uncountable discrete subspace?

4.2.5. Is it true that every countably compact group contains a dense subspace of countable tightness? Contains a dense sequential subspace?

4.2.6. Suppose that $G$ is an infinite countably compact topological group. Is there a non-trivial convergent sequence in $G$? (Assuming Martin's Axiom, E. van Douwen constructed in [**149**] an infinite countably compact topological group without non-trivial convergent sequences.)

## 4.3. Čech-complete and feathered topological groups

A Tychonoff space $X$ is called *Čech-complete* if $X$ is homeomorphic to a $G_\delta$-set in a compact space. The class of Čech-complete spaces is stable with respect to taking closed subspaces, countable products and perfect images (see Theorems 3.9.6, 3.9.8, and 3.9.10 of [**165**]). In addition, for metrizable spaces, the properties of being Čech-complete and admitting a complete metric coincide, by the Alexandroff–Hausdorff theorem [**165**, Theorem 4.3.26].

Čech-complete spaces need not be *topologically complete* (or, equivalently, *Dieudonné complete*, see Section 6.5). Indeed, let $x$ be an arbitrary point of the Tychonoff cube $I^\tau$ of an uncountable weight $\tau$ and let $X = I^\tau \setminus \{x\}$. Then $X$ is an open subspace of the compact space $I^\tau$ and, hence, is Čech-complete. However, every continuous real-valued function on $X$ can be extended to a continuous real-valued function on $I^\tau$. This implies (see [**165**, 8.5.13]) that the *Dieudonné completion* $\mu X$ of $X$ is homeomorphic to $I^\tau$. Hence, $X \neq \mu X$.

The space $\omega_1$ of all countable ordinals with the order topology is another example of a locally compact space that is not Dieudonné complete, since every continuous real-valued function on $\omega_1$ admits an extension to a continuous function on $\omega_1+1$ [**165**, Example 3.1.27]. The space $\omega_1$ is even locally metrizable. We shall see in this section that for topological groups the relationship between Čech-completeness and Dieudonné completeness is quite different.

A *Čech-complete group* is a topological group whose underlying space is Čech-complete. It is not clear from the definition whether there exists a connection between Čech-complete groups and Raĭkov complete groups (see Section 3.6). Theorem 4.3.7 states that all Čech-complete groups are Raĭkov complete (hence, Dieudonné complete), so that Čech-complete groups form a subclass of the class of Raĭkov complete groups. Example 4.3.9 shows that this subclass is proper.

The next result follows immediately from Theorems 3.9.6 and 3.9.8 of [**165**].

PROPOSITION 4.3.1. *The class of Čech-complete groups is closed under taking closed subgroups and countable products.*

A family $\gamma$ of open sets in a space $X$ is called a *base for $X$ at a set $F \subset X$* if all elements of $\gamma$ contain $F$ and, for every open set $V$ that contains $F$, there exists $U \in \gamma$ such that $U \subset V$. The *character of $X$ at a set $F \subset X$* or, equivalently, the *character of $F$ in $X$* is the smallest cardinality of a base for $X$ at $F$. The character of $X$ at $F$ is denoted by $\chi(F, X)$ (see [**165**, 2.1.C (b)]). We start with several facts from general topology that involve the character of spaces at certain sets.

LEMMA 4.3.2. *Let $\gamma = \{V_n : n \in \omega\}$ be a sequence of non-empty open sets in a countably compact space $X$ such that $\overline{V}_{n+1} \subset V_n$ for each $n \in \omega$. Then $\gamma$ is a base for $X$ at the set $F = \bigcap_{n\in\omega} V_n$.*

PROOF. First, note that $F = \bigcap_{n\in\omega} V_n = \bigcap_{n\in\omega} \overline{V}_n$, so $F$ is closed in $X$. Let $O$ be an arbitrary open set in $X$ which contains $F$. If $K_n = \overline{V}_n \cap (X \setminus O) \neq \varnothing$ for each $n \in \omega$, then the decreasing sequence $\{K_n : n \in \omega\}$ of closed non-empty sets in $X$ has empty intersection, which contradicts the countable compactness of $X$. Therefore, $V_n \cap (X \setminus O) = \varnothing$ for some $n \in \omega$ or, equivalently, $V_n \subset O$. This proves that $\gamma$ is a base for $X$ at $F$. □

The following result generalizes Lemma 1.4.15.

LEMMA 4.3.3. *Let $Y$ be a dense subspace of a regular space $X$, and $K$ be a compact subset of $Y$. Then $\chi(K, Y) = \chi(K, X)$.*

PROOF. Let $\mathcal{B}$ be a base for $X$ at $K$ with $|\mathcal{B}| = \chi(K, X)$. We claim that the family $\mathcal{C} = \{U \cap Y : U \in \mathcal{B}\}$ is a base for $Y$ at $K$. Indeed, let $W$ be an open neighbourhood of

$K$ in $Y$. Then the set $F = Y \setminus W$ is closed in $Y$ and the closure $\overline{F}$ of $F$ in $X$ is disjoint from $K$. Hence there exists $U \in \mathcal{B}$ with $U \subset X \setminus \overline{F}$. It is clear that $V = U \cap Y \in \mathcal{C}$ and $K \subset V \subset W$. So, $\mathcal{C}$ is a base for $Y$ at $K$ and $|\mathcal{C}| \leq |\mathcal{B}| = \chi(K, X)$. This proves the inequality $\chi(K, Y) \leq \chi(K, X)$.

Conversely, let $\mathcal{C}$ be a base for $Y$ at $K$ satisfying $|\mathcal{C}| = \chi(K, Y)$. For every $V \in \mathcal{C}$, choose an open set $U_V$ in $X$ with $U_V \cap Y = V$. It remains to verify that the family $\mathcal{B} = \{U_V : V \in \mathcal{C}\}$ is a base for $X$ at $K$. Suppose that $O$ is a neighbourhood of $K$ in $X$. Since $X$ is regular and $K$ is compact, for every $V \in \mathcal{C}$ there exists an open set $W$ in $X$ such that $K \subset W \subset \overline{W} \subset O$. Choose $V \in \mathcal{C}$ with $V \subset W$. Since $U_V \cap Y = V$ and $Y$ is dense in $X$, we have $U_V \subset \overline{U}_V = \overline{V} \subset \overline{W} \subset O$. So, $\mathcal{B}$ is a base for $X$ at $K$ and, clearly, $|\mathcal{B}| \leq |\mathcal{C}| = \chi(K, Y)$. Therefore, $\chi(K, X) \leq \chi(K, Y)$. This proves the required equality. □

LEMMA 4.3.4. *Let $X$ be a Čech-complete space. Then, for every point $x \in X$ and every neighbourhood $V$ of $x$ in $X$, there exists a compact set $F$ of countable character in $X$ such that $x \in F \subset V$. In other words, every Čech-complete space has pointwise countable type.*

PROOF. According to the definition of Čech-complete spaces, we can identify $X$ with a $G_\delta$-subset of a compact space $Y$. Replacing $Y$ by the closure of $X$, if necessary, and using the fact that closed subspaces of a Čech-complete space are also Čech-complete, we can assume that $X$ is dense in $Y$. Let $X = \bigcap_{n\in\omega} U_n$, where each $U_n$ is open in $Y$. Take an arbitrary point $x \in X$ and a neighbourhood $V$ of $x$ in $X$. By recursion, one can define a sequence $\{V_n : n \in \omega\}$ of open sets in $Y$ such that $V_0 \cap X \subset V$ and $x \in V_{n+1} \subset \overline{V}_{n+1} \subset V_n \cap U_n$ for each $n \in \omega$. Then the set $F = \bigcap_{n\in\omega} V_n = \bigcap_{n\in\omega} \overline{V}_n$ is closed in $Y$ (hence compact), it contains the point $x$ and has countable character in $Y$ — the family $\{V_n : n \in \omega\}$ is a base for $Y$ at $F$ by Lemma 4.3.2. Note that $F \subset \bigcap_{n\in\omega} U_n = X$, so $F$ has countable character in $X$ by Lemma 4.3.3. □

The next corollary of Lemma 4.3.4 is immediate; a stronger (and more general) result will be proved in Proposition 4.3.11.

COROLLARY 4.3.5. *Every Čech-complete group $G$ contains a non-empty compact set of countable character in $G$.*

The existence of compact sets of countable character has an unexpectedly strong impact on the properties of topological groups, as we shall see in Theorem 4.3.15, Proposition 4.3.17, and Corollary 4.3.21.

A subset $A$ of a space $X$ is said to be *meager* (equivalently, *of the first category*) in $X$ if $A$ is the union of a countable family of nowhere dense sets in $X$. Let us call a set $A \subset X$ *almost open* in $X$ if there are meager sets $B$ and $C$ in $X$ such that the set $(A \setminus B) \cup C$ is open in $X$. The next general result has several important application to Čech-complete groups.

PROPOSITION 4.3.6. *If $A$ is an almost open non-meager subset of a semitopological group $G$, then $AA^{-1}$ and $A^{-1}A$ are neighbourhoods of the identity in $G$.*

PROOF. If $A$ is not meager in $G$, then so is $G$. We claim that every non-empty open subset of $G$ is not meager either. Suppose to the contrary that $G$ contains a non-empty open meager set $U$. Since $G$ is homogeneous, we can assume that $U$ contains the identity $e$ of $G$. Denote by $\gamma$ a maximal disjoint family of open subsets of $G$ of the form $xV$, where

$e \in V \subset U$ and $x \in G$. Since left translations in $G$ are homeomorphisms, all elements of $\gamma$ are meager in $G$. In addition, the maximality of $\gamma$ implies that the open set $O = \bigcup \gamma$ is dense in $G$, so that $F = G \setminus O$ is nowhere dense in $G$. Let $\gamma = \{W_i : i \in I\}$. For every $i \in I$, there exists a family $\{B_{i,n} : n \in \omega\}$ of meager subsets of $G$ such that $W_i = \bigcup_{n\in\omega} B_{i,n}$. Then the set $C_n = \bigcup_{i\in I} B_{i,n}$ is nowhere dense in $G$ for each $n \in \omega$ and, clearly, $G = F \cup \bigcup_{n\in\omega} C_n$. This contradicts the fact that $G$ is not meager in itself, thus proving the claim.

For every almost open subset $B$ of $G$, denote by $B^*$ the union of all open sets $U$ in $G$ for which the complement $U \setminus B$ is meager in $G$. It is clear that if $B$ and $C$ are almost open sets in $G$ and $B \subset C$, then $B^* \subset C^*$.

**Claim.** *Let $B$ and $C$ be almost open subsets of $G$ and $x \in G$. Then $(xB)^* = xB^*$ and $(B \cap C)^* = B^* \cap C^*$.*

Indeed, the first equality $(xB)^* = xB^*$ is evident, so we only verify the second one. Since the inclusion $(B\cap C)^* \subset B^*\cap C^*$ is trivial, it suffices to check that $B^*\cap C^* \subset (B\cap C)^*$. If $y \in B^* \cap C^*$, one can find open sets $U$ and $V$ in $G$ such that $y \in U \cap V$ and the sets $U \setminus B$ and $V \setminus C$ are meager in $G$. Since the complement $(U \cap V) \setminus (B \cap C) \subset (U \setminus B) \cup (V \setminus C)$ is meager in $G$, we conclude that $y \in U \cap V \subset (B \cap C)^*$. Therefore, $B^* \cap C^* \subset (B \cap C)^*$, and the claim is proved.

Let $x$ be an arbitrary element of $G$. By the above claim, we have $xA^* \cap A^* = (xA \cap A)^*$, so $xA^* \cap A^* \neq \varnothing$ implies that $xA \cap A \neq \varnothing$. Therefore, we have

$$A^*(A^*)^{-1} = \{x \in G : xA^* \cap A^* \neq \varnothing\} \subset \{x \in G : xA \cap A \neq \varnothing\} = AA^{-1}.$$

In other words, the set $AA^{-1}$ contains the open neighbourhood $A^*(A^*)^{-1}$ of the identity in $G$ (notice that $A^* \neq \varnothing$ since $A$ is not meager in $G$). A similar argument shows that $A^{-1}A$ is also a neighbourhood of the identity in $G$. □

In the next theorem we establish a clear relation between Čech-complete and Raĭkov complete topological groups.

THEOREM 4.3.7. *If a topological group $G$ contains a non-empty open Čech-complete subset, then $G$ is Raĭkov complete. In particular, every Čech-complete group is Raĭkov complete.*

PROOF. Let $U$ be a non-empty open Čech-complete subset of $G$. Then $U$ is not meager in itself by the Baire category theorem [**165**, Theorem 3.9.3]. Denote by $\varrho G$ the Raĭkov completion of $G$, and let $X$ be the Čech–Stone compactification of the space $\varrho G$. Consider the closure $C = \overline{U}$ of $U$ in $X$. Since $C$ is compact and $U$ is a dense Čech-complete subspace of $C$, [**165**, Theorem 3.9.1] implies that the complement $C \setminus U$ is an $F_\sigma$-set in $C$. In particular, $cl_{\varrho G}(U) \setminus U$ is an $F_\sigma$-set in $\varrho G$. Since $U$ is dense in $cl_{\varrho G}(U)$, we conclude that the complement $cl_{\varrho G}(U) \setminus U$ is meager in $cl_{\varrho G}(U)$ and in $\varrho G$. By the same reason, $U$ is not meager either in $C$, $cl_{\varrho G} U$ or in $\varrho G$. Observe that the set $cl_{\varrho G}(U)$ contains the open subset $V = \varrho G \setminus cl_{\varrho G}(G \setminus U)$ of $\varrho G$ and $U \subset V$, so $U$ is an almost open non-meager subset of $\varrho G$.

By Proposition 4.3.6, the set $UU^{-1} \subset G$ contains a neighbourhood of the identity in $\varrho G$. Therefore, $G$ is an open and closed subgroup of $\varrho G$ and, hence, $G = \varrho G$. This means that the group $G$ is Raĭkov complete. □

Under some additional conditions, Raĭkov complete groups turn out to be Čech-complete, as is shown in the next proposition.

PROPOSITION 4.3.8. *If a metrizable topological group $G$ is Raĭkov complete, then it is Čech-complete or, equivalently, $G$ is metrizable by a complete metric.*

PROOF. Let $\{U_n : n \in \omega\}$ be a base at the identity $e$ of the group $G$ such that $U_n^{-1} = U_n$ and $U_{n+1}^2 \subset U_n$, for each $n \in \omega$. By Lemma 3.3.10, there exists a continuous prenorm $N$ on $G$ such that

$$\{x \in G : N(x) < 1/2^n\} \subset U_n \subset \{x \in G : N(x) \le 2/2^n\}$$

for all $n \in \omega$. It follows from the choice of the sets $U_n$ that $N(x) = 0$ iff $x = e$. Therefore, the function $d$ defined by $d(x, y) = N(x^{-1}y) + N(xy^{-1})$ for $x, y \in G$ is a metric that generates the original topology of $G$.

We claim that the metric $d$ is complete and, hence, $G$ is Čech-complete by [**165**, Theorem 4.3.26]. Indeed, let $\xi = \{x_n : n \in \omega\}$ be a Cauchy sequence in $G$ with respect to the metric $d$. For every $n \in \omega$, let $F_n = \{x_k : k \ge n\}$. It is easy to see that the family $\{F_n : n \in \omega\}$ is a base of a Cauchy filter in the group $G$. To verify this fact, take an arbitrary neighbourhood $W$ of $e$ in $G$. There exists $m \in \omega$ such that $U_m \subset W$. Since $\xi$ is a Cauchy sequence with respect to $d$, we can find an integer $n$ such that $d(x_k, x_l) < 1/2^{n+1}$ for all $k, l \ge n$. Our choice of $N$ and $d$ implies that $x_k \in x_n U_n$ and $x_k \in U_n x_n$ for each $k \ge n$, whence it follows that $F_n \subset x_n U_n \subset x_n W$ and $F_n \subset U_n x_n \subset W x_n$. Thus, $\{F_n : n \in \omega\}$ is a base of a Cauchy filter. Since the group $G$ is Raĭkov complete, this filter converges to an element $x \in G$. The latter implies immediately that the sequence $\{x_n : n \in \omega\}$ converges to $x$ in $G$, which completes the proof. □

Combining Theorem 4.3.7 and Proposition 4.3.8, we conclude that Čech-completeness and Raĭkov completeness coincide for metrizable groups. We shall extend this result to a wider class of topological groups in Theorem 4.3.15. In general, however, Raĭkov complete groups need not be Čech-complete, not even in the class of Abelian $\omega$-narrow groups. According to Proposition 4.3.1, the class of Čech-complete groups is countably productive, while arbitrary products of Raĭkov complete groups are Raĭkov complete. These facts suggest a way of constructing a Raĭkov complete group that fails to be Čech-complete.

EXAMPLE 4.3.9. Let $\mathbb{Z}$ be the additive group of integers with the discrete topology. The group $\mathbb{Z}^\tau$ is Raĭkov complete but it fails to be Čech-complete, for any cardinal $\tau > \aleph_0$. Indeed, by Theorem 3.6.22, the class of Raĭkov complete groups contains arbitrary topological products, so the group $\mathbb{Z}^\tau$ is Raĭkov complete. Suppose to the contrary that $\mathbb{Z}^\tau$ is Čech-complete for some $\tau > \aleph_0$. Then, by Corollary 4.3.5, there exists a non-empty compact set $F$ of countable character in the product space $\mathbb{Z}^\tau$. Choose a countable family $\{U_n : n \in \omega\}$ of open sets in $\mathbb{Z}^\tau$ such that $F = \bigcap_{n\in\omega} U_n$. Let $x$ be an arbitrary point of $F$. It is easy to define a sequence $\{V_n : n \in \omega\}$ of canonical open neighbourhoods of $x$ in $\mathbb{Z}^\tau$ such that $V_{n+1} \subset \overline{V}_{n+1} \subset V_n \cap U_n$ for each $n \in \omega$. Then $K = \bigcap_{n\in\omega} V_n = \bigcap_{n\in\omega} \overline{V}_n$ is closed in $\mathbb{Z}^\tau$, and $x \in K \subset F$. Hence $K$ is compact.

Given a non-empty set $A \subset \tau$, denote by $\pi_A$ the projection of $\mathbb{Z}^\tau$ onto $\mathbb{Z}^A$. For every $n \in \omega$, there exists a finite set $A_n \subset \tau$ such that $V_n = \pi_{A_n}^{-1}\pi_{A_n}(V_n)$. Then $A = \bigcup_{n\in\omega} A_n$ is a countable subset of $\tau$ and $V_n = \pi_A^{-1}\pi_A(V_n)$ for each $n \in \omega$. This implies immediately that $K = \pi_A^{-1}\pi_A(K)$. In particular, $\pi_\alpha(K) = \mathbb{Z}$ for each $\alpha \in \tau \setminus A$, which contradicts the compactness of $K$. Therefore, the group $\mathbb{Z}^\tau$ is not Čech-complete. □

A topological group $G$ is *feathered* if it contains a non-empty compact set $K$ of countable character in $G$. Therefore, all Čech-complete groups are feathered. It is also clear that all metrizable topological groups are feathered. These simple facts show that the class of feathered groups is very wide.

Let us show that the compact set $K$ in the above definition can always be chosen to be a *subgroup* of $G$. The proof of this fact requires a technical lemma which will be applied several times in the sequel (see Theorems 3.1.25 and 3.1.26 in this connection).

LEMMA 4.3.10. *Let $G$ be a topological group with neutral element $e$ and $F$ be a compact subset of $G$ containing $e$ and having a countable base $\{U_n : n \in \omega\}$ in $G$. Suppose that a sequence $\gamma = \{V_n : n \in \omega\}$ of open symmetric neighbourhoods of $e$ in $G$ satisfies $V_{n+1}^2 \subset V_n \cap U_n$ for each $n \in \omega$. Then $H = \bigcap_{n\in\omega} V_n$ is a compact subgroup of $G$, $H \subset F$, and $\gamma$ is a base for $G$ at $H$.*

PROOF. Since $V_{n+1}^2 \subset V_n$, we have $\overline{V}_{n+1} \subset V_n$ for each $n \in \omega$, and from the definition of $H$ it follows immediately that $H = \bigcap_{n\in\omega} V_n = \bigcap_{n\in\omega} \overline{V}_{n+1}$ is a closed subgroup of $G$. It is also clear that $H \subset \bigcap_{n\in\omega} U_n = F$, so the group $H$ is compact.

Let $W$ be an open neighbourhood of $H$ in $G$. Then $K = F \setminus W$ is a compact subset of $G$ disjoint from $H$. If $\overline{V}_n \cap K \neq \emptyset$ for each $n \in \omega$, then, by the compactness of $K$, the set $K \cap \bigcap_{n\in\omega} \overline{V}_n = K \cap H$ is not empty, which is a contradiction. So, $K \cap V_n = \emptyset$ for some $n \in \omega$. It is clear that $F \subset W \cup K \subset W \cup KV_{n+1}$. Since the set $W \cup KV_{n+1}$ is open in $G$, there exists an integer $m$ such that $U_m \subset W \cup KV_{n+1}$. Let $k = \max\{m, n\}$. Observe that $K \cap V_{k+1}V_{n+1} \subset K \cap V_n = \emptyset$, whence it follows that $KV_{n+1} \cap V_{k+1} = \emptyset$. Therefore,

$$V_{k+1} = V_{k+1} \cap U_m \subset V_{k+1} \cap (W \cup KV_{n+1}) = V_{k+1} \cap W \subset W,$$

that is, $V_{k+1} \subset W$. This proves that $\gamma$ is a base for $G$ at $H$. □

PROPOSITION 4.3.11. *Let $G$ be a feathered group and $O$ be a neighbourhood of the neutral element in $G$. Then there exists a compact subgroup $H$ of countable character in $G$ satisfying $H \subset O$.*

PROOF. Since the group $G$ is feathered, it contains a non-empty compact set $F$ with $\chi(F, G) \leq \omega$. By the homogeneity of $G$, we can assume that $F$ contains the neutral element $e$ of $G$. Let $\{U_n : n \in \omega\}$ be a countable base for $G$ at $F$. We define by induction a sequence $\{V_n : n \in \omega\}$ of symmetric open neighbourhoods of $e$ in $G$ satisfying the following conditions:

(i) $V_0 \subset O$;

(ii) $V_{n+1}^2 \subset U_n \cap V_n$ for each $n \in \omega$.

Put $H = \bigcap_{n\in\omega} V_n$. Then $H \subset O$, by (i). Lemma 4.3.10 implies that $H$ is a compact subgroup of $G$ and that $\{V_n : n \in \omega\}$ is a base for $G$ at $H$. □

COROLLARY 4.3.12. *Let $G$ be a feathered group and $\mathcal{H}$ be the family of all compact subgroups of $G$ which have countable character in $G$. Denote by $\mathcal{B}$ the family of all sets of the form $\pi_H^{-1}(V)$ in $G$, where $H \in \mathcal{H}$, $\pi_H\colon G \to G/H$ is the quotient mapping onto the left coset space $G/H$ and $V$ is open in $G/H$. Then $\mathcal{B}$ is a base of $G$.*

PROOF. It is easy to see that the family $\mathcal{B}$ is closed under taking left and right translates in $G$. Therefore, it suffices to verify that the subfamily $\mathcal{B}(e) = \{U \in \mathcal{B} : e \in U\}$ is a base at the identity $e$ of $G$. Let $W$ be a neighbourhood of $e$ in $G$. Choose a neighbourhood

$O$ of $e$ such that $O^2 \subset W$. By Proposition 4.3.11, there exists a compact subgroup $H$ of $G$ such that $H \in \mathcal{H}$ and $H \subset O$. The set $V = \pi_H(O)$ is open in $G/H$ since $\pi_H^{-1}(V) = \pi_H^{-1}\pi_H(O) = OH$ is open in $G$. In addition, $OH \in \mathcal{B}(e)$ and $OH \subset O^2 \subset W$, which finishes the proof. □

Similarly to metrizable topological groups, the class of feathered groups is countably productive. As we shall see in the proof that follows, this fact is topological in nature.

PROPOSITION 4.3.13. *The product $G = \prod_{n\in\omega} G_n$ of countably many feathered groups is a feathered group.*

PROOF. For every $n \in \omega$, take a compact set $K_n \subset G_n$ of countable character in $G_n$ containing the neutral element of $G_n$. Consider the compact set $K = \prod_{n\in\omega} K_n$ in $G$. Let $\gamma_n$ be a countable base for $G_n$ at $K_n$, $n \in \omega$. We claim that the family

$$\mathcal{B} = \{\pi_0^{-1}(U_0) \cap \ldots \cap \pi_k^{-1}(U_k) : U_0 \in \gamma_0, \ldots, U_k \in \gamma_k,\ k \in \omega\}$$

is a base for $G$ at $K$, where $\pi_i \colon G \to G_i$ is the projection for each $i \in \omega$. Indeed, let $W$ be a neighbourhood of $K$ in $G$. By Wallace's theorem (see [**165**, Theorem 3.2.10]), one can find open sets $W_n \subset G_n$ such that $W_n \neq G_n$ for only finitely many $n \in \omega$ and $K \subset \prod_{n\in\omega} W_n \subset W$. Choose $k \in \omega$ such that $W_n = G_n$ for all $n > k$ and, for every $i \leq k$, take an element $U_i \in \gamma_i$ satisfying $U_i \subset W_i$. Clearly, the set $U = \pi_0^{-1}(U_0) \cap \ldots \cap \pi_k^{-1}(U_k)$ belongs to the family $\mathcal{B}$ and satisfies $K \subset U \subset W$. Therefore, $\chi(K, G) \leq |\mathcal{B}| \leq \omega$, as required. □

By Theorem 4.3.7, Raĭkov completeness is a necessary condition for a topological group to be Čech-complete. It turns out that this condition is sufficient in the class of feathered groups, as Theorem 4.3.15 below shows. To prove this important fact, we need a lemma.

LEMMA 4.3.14. *Let a compact subset $K$ of a topological group $G$ contain the identity of $G$, and suppose that $\{V_n : n \in \omega\}$ is a base for $G$ at $K$. Then, for every element $x \in G$, the family $\{xV_n \cap V_n x : n \in \omega\}$ is a base for $G$ at the set $xK \cap Kx$.*

PROOF. We can assume without loss of generality that the sequence $\gamma = \{V_n : n \in \omega\}$ is decreasing. Let $O$ be an open neighbourhood of the set $xK \cap Kx$ in $G$. Suppose to the contrary that $(xV_n \cap V_n x) \setminus O \neq \varnothing$ for all $n \in \omega$. Put $W = Ox^{-1}$. Then $xKx^{-1} \cap K \subset W$ and the sets $(xV_n x^{-1} \cap V_n) \setminus W$ are not empty, so we can choose a sequence $\{a_n : n \in \omega\}$ of points in $G$ satisfying the following conditions for all $n \in \omega$:

(i) $a_n \in V_n$;
(ii) $b_n = xa_n x^{-1} \in V_n \setminus W$.

Since $\gamma$ is a base for $G$ at $K$, (i) implies that the set $\{a_n : n \in \omega\}$ has an accumulation point $y \in K$. In its turn, (ii) implies that the element $z = xyx^{-1}$ is an accumulation point of the set $B = \{b_n : n \in \omega\} \subset G \setminus W$. Since $W$ is open in $G$, we conclude that $z \notin W$. In addition, since $b_n \in V_n$ and $V_{n+1} \subset V_n$ for each $n \in \omega$, all accumulation points of $B$ lie in $K$. Therefore, $z \in K \setminus W$. However, $z = xyx^{-1} \in xKx^{-1} \cap K \subset W$, which is a contradiction. □

THEOREM 4.3.15. [**M. M. Choban**] *A feathered group is Čech-complete if and only if it is Raĭkov complete.*

Proof. Let $G$ be a feathered group. If $G$ is Čech-complete, then Theorem 4.3.7 implies that $G$ is Raĭkov complete. Conversely, suppose that $G$ is Raĭkov complete. Since $G$ is feathered, there exists a compact set $F$ of countable character in $G$ which contains the identity $e$ of $G$. Let $\{U_n : n \in \omega\}$ be a base for $G$ at $F$. Choose a sequence $\{V_n : n \in \omega\}$ of open symmetric neighbourhoods of $e$ in $G$ satisfying $V_{n+1}^2 \subset V_n \cap U_n$ for each $n \in \omega$. Then, by Lemma 4.3.10, $H = \bigcap_{n\in\omega} V_n$ is a compact subgroup of $G$, $H \subset F$ and $\{V_n : n \in \omega\}$ is a base for $G$ at $H$.

Apply Lemma 3.3.10 to define a continuous prenorm $N$ on the group $G$ which satisfies

$$\{x \in G : N(x) < 1/2^n\} \subset V_n \subset \{x \in G : N(x) \leq 2/2^n\} \tag{4.1}$$

for each integer $n \geq 0$. From our choice of $N$ it follows that

$$H = \{x \in G : N(x) = 0\}. \tag{4.2}$$

Let $\varrho$ be a pseudometric on $G$ defined by

$$\varrho(x, y) = N(xy^{-1}) + N(x^{-1}y) \tag{4.3}$$

for all $x, y \in G$. Then $\varrho$ is continuous and $\varrho(x, y) = 0$ iff $xy^{-1} \in H$ and $x^{-1}y \in H$.

Consider the equivalence relation $\sim$ on $G$ defined by $x \sim y$ iff $y \in xH \cap Hx$ or, equivalently, if $x^{-1}y \in H$ and $yx^{-1} \in H$. Denote by $X$ the quotient space $G/{\sim}$ and let $\pi\colon G \to X$ be the quotient map. Let also $\varrho^*$ be the function on $X \times X$ defined by

$$\varrho^*(\pi(x), \pi(y)) = \varrho(x, y)$$

for all $x, y \in G$. From (4.2) and (4.3) it follows that $\varrho^*$ is correctly defined, since $\varrho(x, y) = \varrho(x', y')$ whenever $\pi(x) = \pi(x')$ and $\pi(y) = \pi(y')$. In addition, our definition of $\varrho^*$ implies that $\varrho^*(z, t) = 0$ if and only if $z = t$, that is, $\varrho^*$ is a metric on $X$. For any $x \in G$, $z \in X$ and $\varepsilon > 0$, we put

$$B(x, \varepsilon) = \{x' \in G : \varrho(x', x) < \varepsilon\}, \; B^*(z, \varepsilon) = \{z' \in X : \varrho^*(z', z) < \varepsilon\}.$$

A simple verification, with the use of (4.2) and (4.3), shows that

$$B(x, \varepsilon) = \pi^{-1}(B^*(\pi(x), \varepsilon)) \text{ for all } x \in G \text{ and } \varepsilon > 0. \tag{4.4}$$

Therefore, the metric $\varrho^*$ is continuous on the quotient space $X$. We claim that $\varrho^*$ generates the quotient topology of the space $X = G/{\sim}$. Indeed, suppose that the preimage $O = \pi^{-1}(W)$ is open in $G$, where $W$ is a non-empty subset of $X$. Take an arbitrary point $z \in W$. Then $\pi^{-1}(z) = xH \cap Hx \subset O$, where $x$ is an arbitrary point of the fiber $\pi^{-1}(z)$. By Lemma 4.3.14, there exists $n \in \omega$ such that $xV_n \cap V_n x \subset O$. Let $\varepsilon = 2^{-n}$. Then (4.1) and (4.3) imply that $B(x, \varepsilon) \subset xV_n \cap V_n x$. So, from (4.4) it follows that

$$\pi^{-1}(B^*(z, \varepsilon)) = B(x, \varepsilon) \subset xV_n \cap V_n x \subset O.$$

The above inclusion means that $B(z, \varepsilon) \subset W$, so the set $W$ is the union of a family of open balls in $(X, \varrho^*)$. Hence $W$ is open in $(X, \varrho^*)$, which proves the claim.

A similar argument shows that the mapping $\pi\colon G \to X$ is closed. Indeed, let $F$ be a closed subset of $G$ and $z \in X \setminus \pi(F)$ be an arbitrary point. Choose a point $x \in G$ with $\pi(x) = z$. Then $C = \pi^{-1}(z) = xH \cap Hx$ is a compact subset of $G$ disjoint from $F$, so one can apply Theorem 1.4.29 to find an open neighbourhood $V$ of $e$ in $G$ such that $CV \cap F = \varnothing$. By Lemma 4.3.14, there exists $n \in \omega$ such that $xV_n \cap V_n x \subset CV$. The above argument implies that $\pi^{-1}(B^*(z, 2^{-n})) \subset xV_n \cap V_n x \subset CV$, which in its turn gives

$B^*(z, 2^{-n}) \cap \pi(F) = \varnothing$. Thus, $\pi(F)$ is closed in $X$ and, hence, the mapping $\pi$ is closed. Observe that all fibers of $\pi$ are compact, so the mapping $\pi$ is, in fact, perfect.

Let us prove that the metric space $(X, \varrho^*)$ is complete. Suppose not. Then $X$ contains a sequence of open balls $\{B^*(z_n, 2^{-k_n}) : n \in \omega\}$ satisfying the following two conditions:

(i) $k_n \geq n$ and $B^*(z_{n+1}, 2^{-k_{n+1}}) \subset B^*(z_n, 2^{-k_n})$ for each $n \in \omega$;

(ii) $\bigcap_{n=0}^{\infty} \overline{B_n^*(z_n, 2^{-k_n})} = \varnothing$, the closures are taken in $(X, \varrho^*)$.

For every $n \in \omega$, set $U_n = \pi^{-1}(B^*(z_n, 2^{-k_n}))$ and pick an arbitrary point $x_n \in \pi^{-1}(z_n)$. Consider the set $W_n = \{x \in G : N(x) < 2^{-k_n}\}$. From (4.1), (4.3), and (4.4) it follows that $U_n = B(x_n, 2^{-k_n}) \subset x_n W_n \cap W_n x_n$. Since the sequence $\{U_n : n \in \omega\}$ is decreasing, it is contained in a maximal open filter $\xi$ on $G$. Let $V$ be an arbitrary open neighbourhood of $e$ in $G$. Observe that, by (4.1), the family $\{W_n : n \in \omega\}$ is a base for $G$ at $H$. Hence one can find elements $h_1, \dots, h_m \in H$ and an integer $n_0 \in \omega$ such that

$$H \subset W_{n_0} \subset (\bigcup_{i=0}^{m} h_i V) \cap (\bigcup_{i=0}^{m} V h_i).$$

Since $U_{n_0} \in \xi$ and $U_{n_0} \subset (\bigcup_{i=0}^{m} x_{n_0} h_i V) \cap (\bigcup_{i=0}^{m} V h_i x_{n_0})$, there exist integers $i, j \leq m$ such that $x_{n_0} h_i V \in \xi$ and $V h_j x_{n_0} \in \xi$. Therefore, $\xi$ is a Cauchy filter on $G$. Further, from (ii) and our definition of the sets $U_n$ it follows that $\bigcap_{U \in \xi} \overline{U} \subset \bigcap_{n=0}^{\infty} \overline{U}_n = \varnothing$. This contradicts our assumption that the group $G$ is Raĭkov complete, thus proving the completeness of the metric space $(X, \varrho^*)$.

Finally, $X$ is Čech-complete as is every complete metric space (see [**165**, Theorem 4.3.26]), and so is the perfect preimage $G$ of $X$ by [**165**, Theorem 3.9.10]. □

Feathered groups (even metrizable groups) need not be Čech-complete — take the group of rational numbers, for example. It turns out, however, that feathered groups are subgroups of Čech-complete groups.

THEOREM 4.3.16. [**M. M. Choban**] *Every feathered group can be embedded as a subgroup in a Čech-complete topological group.*

PROOF. Let $G$ be a feathered group. Then $G$ is a dense subgroup of $\varrho G$, the Raĭkov completion of $G$. Let $K$ be a non-empty compact subset of countable character in $G$. Then Lemma 4.3.3 implies that $K$ has countable character in $\varrho G$, so that the complete group $\varrho G$ is feathered as well. Hence from Theorem 4.3.15 it follows that $\varrho G$ is a Čech-complete group. □

In general, a local property does not imply the corresponding global property. For example, a locally connected space is not necessarily connected, and a locally compact space can easily fail to be compact. The same happens in the class of topological groups. For example, the group $\mathbb{R} \times \mathbb{Z}$ is both locally connected and locally compact, while it is neither connected nor compact. Let us show that, for topological groups, local Čech-completeness and Čech-completeness coincide.

PROPOSITION 4.3.17. *Every locally Čech-complete topological group is Čech-complete.*

PROOF. Let $U$ be a Čech-complete neighbourhood of the identity $e$ in a topological group $G$. Since Čech-completeness is hereditary with respect to taking open subspaces, we can assume that $U$ is open in $G$. By Lemma 4.3.4, there exists a compact set $K$ in $G$ such that $e \in K \subset U$ and $\chi(K, U) \leq \omega$. Since $U$ is open in $G$, we have $\chi(K, G) = \chi(K, U) \leq \omega$. Therefore, the group $G$ is feathered. In addition, $G$ is Raĭkov complete by Theorem 4.3.7. The conclusion now follows from Theorem 4.3.15. □

COROLLARY 4.3.18. *Suppose that $G$ is a topological group, and let $H$ be a locally compact subgroup of $G$ such that the quotient space $G/H$ is locally Čech-complete. Then $G$ is also Čech-complete.*

PROOF. From item 7) of Corollary 3.2.6 it follows that $G$ is locally Čech-complete. It remains to refer to Proposition 4.3.17, according to which any locally Čech-complete topological group is Čech-complete. □

Let us say that a topological group $G$ is *locally feathered* if $G$ contains an open set $U$ and a non-empty compact set $K$ such that $K \subset U$ and $\chi(K, U) \leq \omega$. Then, as in the proof of Proposition 4.3.17, we have $\chi(K, G) = \chi(K, U) \leq \omega$, whence it follows that the group $G$ is feathered. Therefore, the classes of locally feathered groups and feathered groups trivially coincide.

Čech-complete and feathered groups admit useful characterizations in terms of quotient spaces given in Theorem 4.3.20 below. Its proof is based on the following lemma.

LEMMA 4.3.19. *Let $G$ be a topological group and $H$ be a compact subgroup of $G$. If $H$ has countable character in $G$, then the quotient space $G/H$ is metrizable.*

PROOF. By Theorem 1.5.7, the mapping $\pi_H$ is perfect. Suppose now that $\{U_n : n \in \omega\}$ is a countable base for $G$ at $H$. We define by induction a sequence $\{V_n : n \in \omega\}$ of open symmetric neighbourhoods of the identity $e$ in $G$ such that $V_{n+1}^2 \subset V_n \cap U_n$ for each $n \in \omega$. Put $P = \bigcap_{n\in\omega} V_n$. Then, by Lemma 4.3.10, $P$ is a compact subgroup of $G$, $P \subset H$, and $\{V_n : n \in \omega\}$ is a countable base for $G$ at $P$. Let $\pi_P \colon G \to G/P$ be the quotient mapping of $G$ onto the left coset space $G/P$.

Apply Lemma 3.3.10 to choose a continuous prenorm $N$ on $G$ which satisfies

$$\{x \in G : N(x) < 1/2^n\} \subset V_n \subset \{x \in G : N(x) \leq 2/2^n\}$$

for each integer $n \geq 0$. It is clear that $N(x) = 0$ if and only if $x \in P$. Define a continuous pseudometric $d$ on $G$ by $d(x, y) = N(x^{-1}y)$ for all $x, y \in G$. Observe that if $x' \in xP$ and $y' \in yP$ for some $x, y \in G$, then $d(x', y') = d(x, y)$. This enables us to define a function $\varrho$ on $G/P \times G/P$ by

$$\varrho(\pi_P(x), \pi_P(y)) = d(x, y)$$

for all $x, y \in G$. Since the mapping $\pi_P$ is quotient, $\varrho$ is a continuous metric on $Y = G/P$. Let us verify that $\varrho$ generates the quotient topology of the space $Y$. Given points $x \in G$, $y \in Y$ and a positive number $\varepsilon$, we define open balls

$$B(x, \varepsilon) = \{x' \in G : d(x', x) < \varepsilon\}$$

and

$$B^*(y, \varepsilon) = \{y' \in G/P : \varrho(y', y) < \varepsilon\}$$

in $G$ and $Y$, respectively. From our definition of $\varrho$ it follows that if $x \in G$ and $y = \pi_P(x)$, then $B(x, \varepsilon) = \pi_P^{-1}(B^*(y, \varepsilon))$. Suppose that the preimage $O = \pi_P^{-1}(W)$ is open in $G$, where $W$ is a non-empty subset of $Y$. Take an arbitrary point $y \in W$. Then $\pi_P^{-1}(y) = xP \subset O$, where $x$ is an arbitrary point of the fiber $\pi_P^{-1}(y)$. Since $\{V_n : n \in \omega\}$ is a base for $G$ at $P$, there exists $n \in \omega$ such that $xV_n \subset O$. Let $\varepsilon = 2^{-n}$. Then our choice of $N$ and $d$ implies that $B(x, \varepsilon) \subset xV_n$. Therefore, we have

$$\pi_P^{-1}(B^*(y, \varepsilon)) = B(x, \varepsilon) \subset xV_n \subset O.$$

It follows that $B^*(y, \varepsilon) \subset W$, so the set $W$ is the union of a family of open balls in $(Y, \varrho)$. Hence $W$ is open in $(Y, \varrho)$, which proves that the metric and quotient topologies on $Y = G/P$ coincide.

Since $P \subset H$, we can define the natural mapping $\varphi\colon G/P \to G/H$ by $\varphi(xP) = xH$ for every $x \in G$. Then $\varphi$ satisfies the equality $\pi_H = \varphi \circ \pi_P$. Since $\pi_H$ is a perfect map, so is $\varphi$ (see [**165**, Proposition 3.7.10]). Therefore, the perfect image $G/H$ of the metrizable space $G/P$ is also metrizable by [**165**, Theorem 4.4.15]. □

THEOREM 4.3.20. [**B. A. Pasynkov**] *A topological group $G$ is feathered if and only if it contains a compact subgroup $H$ such that the left quotient space $G/H$ is metrizable. Similarly, the group $G$ is Čech-complete if and only if it contains a compact subgroup $H$ such that the left quotient space $G/H$ is metrizable by a complete metric.*

PROOF. Suppose that $G$ contains a compact subgroup $H$ such that the left quotient space $X = G/H$ is metrizable. Then, by Theorem 1.5.7, the mapping $\pi\colon G \to X$ is perfect. We now claim that the subgroup $H \subset G$ has countable character in $G$. Indeed, let $\{U_n : n \in \omega\}$ be a countable base for $X$ at the point $\pi(e)$, where $e$ is the neutral element of $G$. Then the family $\{V_n : n \in \omega\}$ is a base for $G$ at $H$, where $V_n = \pi^{-1}(U_n)$ for each $n \in \omega$. To verify this fact, choose an arbitrary open neighbourhood $O$ of $H$ in $G$. Then $F = G \setminus O$ is closed in $G$ and the image $\pi(F)$ is a closed subset of $X$ which does not contain $\pi(e)$. Hence $U_n \cap \pi(F) = \varnothing$ for some $n \in \omega$, which in its turn implies that $V_n \cap F = \varnothing$. Therefore, $H \subset V_n \subset O$, thus proving that $\chi(H, G) \leq \omega$. Since $H$ is compact, the group $G$ is feathered.

Conversely, if the group $G$ is feathered, it contains a compact subgroup $H$ of countable character in $G$ by Proposition 4.3.11. Then Lemma 4.3.19 implies that the left quotient space $X = G/H$ is metrizable. This proves the first part of the theorem.

Now, suppose that the group $G$ is Čech-complete. Then $G$ is feathered by Corollary 4.3.5, so Proposition 4.3.11 implies that $G$ contains a compact subgroup $H$ of countable character in $G$. In its turn, it follows from Theorem 1.5.7 and Lemma 4.3.19 that the quotient mapping $\pi\colon G \to G/H$ is perfect and the left coset space $G/H$ is metrizable. Since Čech-completeness is an invariant of perfect mappings (see [**165**, Theorem 3.9.10]), the space $G/H$ is Čech-complete. It remains to note that every metrizable Čech-complete space is metrizable by a complete metric [**165**, Theorem 4.3.26].

Conversely, if the group $G$ contains a compact subgroup $H$ such that the left coset space $G/H$ is metrizable by a complete metric, then the space $G/H$ is Čech-complete by [**165**, Theorem 4.3.26]. Since, by Theorem 1.5.7, the quotient mapping $\pi\colon G \to G/H$ is perfect, and Čech-completeness is an inverse invariant of perfect mappings (see [**165**, Theorem 3.9.10]), we conclude that the space $G$ is Čech-complete. □

By Corollary 3.1.4, locally compact groups are strongly paracompact. We apply the above theorem to establish that feathered groups (which form a considerably wider class than locally compact groups) have a slightly weaker property.

COROLLARY 4.3.21. *Every feathered group is paracompact.*

PROOF. Let $G$ be a feathered group. By Theorem 4.3.20, $G$ contains a compact subgroup $H$ such that the left coset space $G/H$ is metrizable. Then the quotient mapping $\pi\colon G \to G/H$ is perfect by Theorem 1.5.7. Hence, the space $G$ is paracompact according to [**165**, Theorem 5.1.35]. □

Since Čech-complete groups are feathered, we obtain the next result:

COROLLARY 4.3.22. *Every Čech-complete group is paracompact.*

Corollary 4.3.22 is not valid for Čech-complete spaces. Indeed, let $W$ be the space of ordinal numbers less than or equal to $\omega_1$, with the topology generated by the natural order, and let $W'$ be the subspace of $W$ consisting of ordinals $\leq \omega$. Then $X = W \times W'$ is a compact space and $T = W \times W' \setminus \{(\omega_1, \omega)\}$ is an open subspace of $X$ known as the *Tychonoff plank.* It is clear that $T$ is locally compact and, hence, Čech-complete. However, $T$ fails to be normal by [**165**, 3.12.19 (a)]. Thus, $T$ is a Čech-complete non-paracompact space.

Our aim now is to study quotient spaces of feathered and Čech-complete groups. We will show that every quotient space of a feathered group is paracompact.

THEOREM 4.3.23. *Suppose that $H$ is a closed subgroup of a feathered group $G$. Then the quotient space $G/H$ is paracompact.*

PROOF. By Theorem 4.3.20, $G$ contains a compact subgroup $K$ such that the left coset space $G/K$ is metrizable. Since the mapping $i\colon G/K \to K\backslash G$ defined by $i(xK) = Kx$ for each $x \in G$, is a homeomorphism between $G/K$ and the right coset space $K\backslash G$, the latter space is also metrizable. Let $\pi_K\colon G \to K\backslash G$ be the quotient mapping. We also consider the quotient mapping $\pi_H\colon G \to G/H$ onto the left coset space $G/H$. Denote by $Z$ the double coset space $K\backslash G/H$ endowed with the quotient topology with respect to the canonical mapping $\pi\colon G \to Z$ (see Proposition 1.8.17). Then we define the natural mappings $q_H\colon K\backslash G \to K\backslash G/H$ and $q_K\colon G/H \to K\backslash G/H$ by $q_H(Kx) = KxH$ and $q_K(xH) = KxH$, for each $x \in G$. It is clear that $\pi = q_K \circ \pi_H = q_H \circ \pi_K$.

$$
\begin{array}{ccc}
G & \xrightarrow{\pi_H} & G/H \\
{\scriptstyle \pi_K}\downarrow & \searrow^{\pi} & \downarrow{\scriptstyle q_K} \\
K\backslash G & \xrightarrow{q_H} & K\backslash G/H
\end{array}
$$

The space $Z = K\backslash G/H$ is metrizable. Indeed, consider the right uniformity $\mathscr{E}^r_Z$ on $Z$ (see Theorem 1.8.21). Since the quotient space $K\backslash G$ is metrizable and the mapping $\pi_K$ of $G$ onto $K\backslash G$ is perfect by Theorem 1.5.7, the group $G$ has a countable neighbourhood base $\{U_n : n \in \omega\}$ at $K$. For every $n \in \omega$, let

$$O_n = \{(\pi(x), \pi(y)) : y \in U_n x\}.$$

It follows from the definition of $\mathscr{E}^r_Z$ that $O_n \in \mathscr{E}^r_Z$ for all $n \in \omega$, and we will verify that the family $\{O_n : n \in \omega\}$ is a base for the uniformity $\mathscr{E}^r_Z$. Every element of $\mathscr{E}^r_Z$ contains the set

$$E_V = \{(\pi(x), \pi(y)) : y \in Vx\},$$

for some open neighbourhood $V$ of the identity in $G$. Since the set $KV$ is open in $G$ and contains $K$, there exists $n \in \omega$ with $U_n \subset KV$. It is clear that $E_W \subset E_V$ whenever $W \subset V$

and that $E_{KV} = E_V$. Therefore, $O_n = E_{U_n} \subset E_{KV} = E_V$. Thus, the uniform space $(Z, \mathscr{E}^r_Z)$ has a countable base, so the space $Z$ is metrizable.

It follows from Proposition 1.8.18 that the mapping $q_K$ is perfect. By [**165**, Theorem 5.1.3], every metrizable space is paracompact. Therefore, the quotient space $G/H$ is paracompact as a perfect preimage of the metrizable space $K\backslash G/H$ according to [**165**, Theorem 5.1.35]. The proof is complete. □

The next result can be deduced from Theorem 4.3.23, but we prefer to supply it with a direct proof.

COROLLARY 4.3.24. *Let $f\colon G \to H$ be an open continuous homomorphism of a feathered group $G$ onto a group $H$. Then $H$ is feathered.*

PROOF. The group $G$ contains a non-empty compact set $K$ which has a countable base in $G$. Let $\{U_n : n \in \omega\}$ be such a base for $G$ at $K$. Then the family $\{f(U_n) : n \in \omega\}$ is a countable base for $H$ at the compact set $f(K)$, so the group $H$ is feathered. □

Similarly to feathered groups, Čech-complete groups are stable under taking quotient spaces and quotient groups. Our proof of this fact requires an important result from general topology given below.

PROPOSITION 4.3.25. *Let $f\colon X \to Y$ be an open continuous mapping of a Čech-complete space $X$ onto a paracompact space $Y$. Then $Y$ is Čech-complete.*

PROOF. Extend $f$ to a continuous mapping $g\colon \beta X \to \beta Y$, where $\beta X$ and $\beta Y$ are Čech–Stone compactifications of $X$ and $Y$, respectively. Since the mapping $g$ is perfect, so is the restriction of $g$ to the inverse image $Z = g^{-1}(X)$ (see [**165**, Prop. 3.7.4]). First, we prove the following:

**Claim.** *For every open set $U$ in $Z$ satisfying $f(U \cap X) = Y$, there exists an open set $V$ in $Z$ such that $cl_Z V \subset U$ and $f(V \cap X) = Y$.*

For every $y \in Y$, we choose an open set $O_y$ in $Z$ such that $cl_Z O_y \subset U$ and $O_y \cap f^{-1}(y) \neq \varnothing$. Then $f$ being open, the family $\{f(O_y \cap X) : y \in Y\}$ is an open covering of the space $Y$. Since $Y$ is paracompact, this covering has a locally finite open refinement $\{U_i : i \in I\}$. For every $i \in I$, choose a point $y_i \in Y$ such that $U_i \subset f(O_{y_i} \cap X)$. Then the set

$$V = \bigcup_{i\in I} O_{y_i} \cap g^{-1}(U_i)$$

is as required. Indeed, $V$ is open in $Z$ and $V \subset U$. Take an arbitrary point $x \in cl_Z V$. Since the family $\{U_i : i \in I\}$ is locally finite, we can find an open neighbourhood $W$ of $x$ in $Z$ and a finite subset $F$ of $I$ such that $f(W) \cap U_i = \varnothing$ for each $i \in I \setminus F$. This implies that $W \cap g^{-1}(U_i) = \varnothing$ for each $i \in I \setminus F$, so that

$$x \in cl_Z \left( \bigcup_{i\in F} O_{y_i} \cap g^{-1}(U_i) \right) \subset \bigcup_{i\in F} cl_Z O_{y_i} \subset U.$$

Therefore, $cl_Z V \subset U$.

To show that $f(V \cap X) = Y$, take an arbitrary point $y \in Y$. Then $y \in U_i$ for some $i \in I$, so we have:

$$y \in f(O_{y_i} \cap X) \cap f(f^{-1}(U_i)) = f(O_{y_i} \cap f^{-1}(U_i)) \subset f(V \cap X).$$

This finishes the proof of the Claim.

Since $X$ is Čech-complete, there exists a family $\{U_n : n \in \omega\}$ of open sets in $\beta X$ such that $X = \bigcap_{n\in\omega} U_n$. Apply Claim to construct by induction a sequence $\{V_n : n \in \omega\}$ of open sets in $Z$ such that $cl_Z V_0 \subset U_0$, $cl_Z(V_{n+1}) \subset V_n \cap U_{n+1}$, and $f(V_n \cap X) = Y$, for each $n \in \omega$. Then $P = \bigcap_{n\in\omega} V_n$ is a $G_\delta$-set in $Z$ and $P \subset X$. It also follows that $P = \bigcap_{n\in\omega} V_n = \bigcap_{n\in\omega} cl_Z V_n$ is closed in $Z$.

Let us show that $f(P) = Y$. For every $y \in Y$, the fiber $g^{-1}(y)$ is a compact subset of $Z$. It follows from $f(V_n \cap X) = Y$ that $g^{-1}(y) \cap cl_Z V_n \neq \varnothing$ for each $n \in \omega$. Since $cl_Z V_{n+1} \subset cl_Z V_n$ for each $n \in \omega$, the intersection

$$g^{-1}(y) \cap \bigcap_{n=0}^{\infty} cl_Z V_n = g^{-1}(y) \cap \bigcap_{n=0}^{\infty} V_n = g^{-1}(y) \cap P$$

is not empty. Therefore, $f(x) = y$ for each $x \in g^{-1}(y) \cap P$, whence it follows that $f(P) = Y$.

To complete the proof, we note that $P$ is Čech-complete as a closed subspace of the Čech-complete space $Z$. Hence the restriction of $g$ to $P$ is a perfect mapping of $P$ onto $X$ (see [**165**, Prop. 3.7.4]) and, consequently, $X$ is Čech-complete by [**165**, Theorem 3.9.10]. □

THEOREM 4.3.26. *If $N$ is a closed subgroup of a Čech-complete group $G$, then the quotient space $G/N$ is also Čech-complete.*

PROOF. Since every Čech-complete group is feathered, the space $G/N$ is paracompact by Theorem 4.3.23. Now the conclusion follows from Proposition 4.3.25. □

Our aim now is to establish that, under some mild addition conditions, every continuous onto homomorphism of Čech-complete groups is open. First, we introduce the necessary concepts and then prove a theorem from general topology which is interesting in itself.

Let us call a subset $U$ of a space $X$ *nearly open* if $U$ is in the interior of its closure. A mapping $f: X \to Y$ is said to be *nearly continuous* if the preimage $f^{-1}(U)$ is nearly open in $X$ for each open set $U \subset Y$. Similarly, the mapping $f$ is called *nearly open* if $f(V)$ is nearly open in $Y$, for each open set $V \subset X$. Notice that for a bijection $f: X \to Y$, $f$ is nearly open if and only if $f^{-1}: Y \to X$ is nearly continuous.

We also say that a mapping $f: X \to Y$ *has closed graph* if the graph of $f$,

$$Gr(f) = \{(x, y) \in X \times Y : x \in X,\ y = f(x)\},$$

is a closed subset of the product space $X \times Y$.

THEOREM 4.3.27. *Let $f: X \to Y$ be a nearly continuous mapping of a Hausdorff space $X$ to a Čech-complete space $Y$. If $f$ has the closed graph and $f^{-1}(C)$ is compact for every compact set $C \subset Y$, then $f$ is continuous.*

PROOF. Given an open covering $\mathcal{U}$ of the space $Y$ and a set $A \subset Y$, we will say that *$A$ has the diameter less than* $\mathcal{U}$ provided that $A \subset U$ for some $U \in \mathcal{U}$. A sequence $\{\mathcal{U}_n : n \in \omega\}$ of open coverings of $Y$ is called *complete* if, for every family $\mathcal{F}$ of non-empty closed sets in $Y$ which has the finite intersection property and contains elements of diameter less than $\mathcal{U}_n$ for each $n \in \omega$, the intersection $\bigcap \mathcal{F}$ is non-empty and compact. According to [**165**, Theorem 3.9.2], the Čech-complete space $Y$ has a complete sequence $\{\mathcal{U}_n : n \in \omega\}$ of open coverings.

First, we prove the following auxiliary fact:

**Claim.** *Let $V$ and $W$ be open subsets of $Y$ such that $\overline{f^{-1}(V)} \cap f^{-1}(W) \neq \varnothing$. Then for every $n \in \omega$, there exist open sets $V_n$ and $W_n$ in $Y$ of diameter less than $\mathcal{U}_n$ such that $\overline{V_n} \subset V$, $\overline{W_n} \subset W$, and $\overline{f^{-1}(V_n)} \cap f^{-1}(W_n) \neq \varnothing$.*

Indeed, pick a point $x \in \overline{f^{-1}(V)} \cap f^{-1}(W)$ and choose an open neighbourhood $W_n$ of $f(x)$ in $Y$ of diameter less than $\mathcal{U}_n$ and such that $\overline{W_n} \subset W$. Then $\overline{f^{-1}(W_n)}$ is a neighbourhood of $x$ in $X$, so there is $y \in f^{-1}(V) \cap \overline{f^{-1}(W_n)}$. Let $V_n$ be an open neighbourhood of $f(y)$ in $Y$ of diameter less than $\mathcal{U}_n$ and such that $\overline{V_n} \subset V$. Since $\overline{f^{-1}(V_n)}$ is a neighbourhood of $y$, we have that $\overline{f^{-1}(V_n)} \cap f^{-1}(W_n) \neq \varnothing$. This proves our Claim.

We turn to the proof of the theorem. To deduce the continuity of $f$, it suffices to verify that the inclusion

$$\overline{f^{-1}(G)} \subset f^{-1}(\overline{G}) \tag{4.5}$$

holds for every open set $G \subset Y$. Indeed, for an arbitrary point $x \in X$, let $O$ be a neighbourhood of $y = f(x)$ in $Y$. Since $Y$ is completely regular (hence, regular), there exists an open neighbourhood $W$ of $y$ such that $\overline{W} \subset O$. Then $P = \overline{f^{-1}(W)}$ is a neighbourhood of $x$ in $X$ and (4.5) implies that $f(P) \subset f(f^{-1}(\overline{W})) \subset \overline{W} \subset O$. Therefore, $f$ is continuous.

Suppose to the contrary that (4.5) fails to hold for some open set $G \subset Y$. Then $\overline{f^{-1}(G)} \cap f^{-1}(H) \neq \varnothing$, where $H = Y \setminus \overline{G}$. Let us put $V_0 = G$ and $W_0 = H$. Apply the above Claim to define by induction sequences $\{V_n : n \in \omega\}$ and $\{W_n : n \in \omega\}$ of open sets in $Y$ satisfying the following conditions for each $n \in \omega$:

(i) $\overline{V_{n+1}} \subset V_n$, $\overline{W_{n+1}} \subset W_n$;
(ii) both sets $V_n$ and $W_n$ have the diameter less than $\mathcal{U}_n$;
(iii) $\overline{f^{-1}(V_n)} \cap f^{-1}(W_n) \neq \varnothing$.

Consider the sets $A = \bigcap_{n \in \omega} \overline{V_n}$ and $B = \bigcap_{n \in \omega} \overline{W_n}$. It follows from the completeness of the sequence $\{\mathcal{U}_n : n \in \omega\}$ and (ii) that $A$ and $B$ are compact and non-empty. In addition, we have that

(a) the families $\{V_n : n \in \omega\}$ and $\{W_n : n \in \omega\}$ are bases for $Y$ at $A$ and $B$, respectively.

Suppose to the contrary that there exists on open neighbourhood $O$ of $A$ in $Y$ such that $V_n \setminus O \neq \varnothing$ for each $n \in \omega$. Put $F_n = \overline{V_n} \setminus O$, where $n \in \omega$. Then $\{F_n : n \in \omega\}$ is a decreasing sequence of non-empty closed sets in $Y$. It follows from (i) and (ii) that for every $n \in \omega$, the set $F_{n+1}$ has the diameter less than $\mathcal{U}_n$. We infer, by the completeness of $\{\mathcal{U}_n : n \in \omega\}$, that the set $F = \bigcap_{n=0}^{\infty} F_n$ is non-empty, which contradicts the fact that

$$F = \bigcap_{n=0}^{\infty} F_n \subset \bigcap_{n=0}^{\infty} (\overline{V_n} \setminus O) = A \setminus O = \varnothing.$$

The same argument applies to $B$ and $\{W_n : n \in \omega\}$, thus implying (a).

Apply (iii) to choose, for every $n \in \omega$, a point $y_n \in Y$ such that

(iv) $y_n \in W_n$ and $\overline{f^{-1}(V_n)} \cap f^{-1}(y_n) \neq \varnothing$.

It follows from (a) that the subset $C = \{y_n : n \in \omega\} \cup B$ of the space $Y$ is compact. Let $Gr(f) \subset X \times Y$ be the graph of $f$. Since the sets $A$ and $C$ are disjoint, we have that $(f^{-1}(C) \times A) \cap Gr(f) = \varnothing$. By assumptions of the theorem, the set $f^{-1}(C)$ is compact. Therefore, $f^{-1}(C) \times A$ is a compact rectangular subset of the space $X \times Y$ disjoint from the closed set $Gr(f)$. By [**165**, Theorem 3.2.8], there exist open sets $U \subset X$ and $W \subset Y$ such that

(b) $f^{-1}(C) \times A \subset U \times W \subset (X \times Y) \setminus Gr(f)$.

By (a), there exists $n \in \omega$ such that $V_n \subset W$. It follows from (iv) and (b) that

$$\varnothing \neq \overline{f^{-1}(V_n)} \cap f^{-1}(C) \subset \overline{f^{-1}(W)} \cap U,$$

which contradicts the equality $f^{-1}(W) \cap U = \varnothing$. □

COROLLARY 4.3.28. *If $g\colon X \to Y$ is a nearly open continuous bijection of a Čech-complete space $X$ onto a Hausdorff space $Y$, then $g$ is a homeomorphism.*

PROOF. The inverse mapping $f = g^{-1}$ of $Y$ to $X$ is nearly continuous. In addition, $f$ has the closed graph. Indeed, let $Gr(f)$ and $Gr(g)$ be the graphs of $f$ and $g$, respectively. Denote by $T$ the natural homeomorphism of $X \times Y$ onto $Y \times X$, $T(x, y) = (y, x)$ for all $x \in X$ and $y \in Y$. The graph of $g$ is closed in $X \times Y$ since $g$ is continuous, and the obvious equality $Gr(f) = T(Gr(g))$ implies that $Gr(f)$ is closed in $Y \times X$.

It is also clear that if $C$ is a compact subset of $X$, then $f^{-1}(C) = g(C)$ is a compact set in $Y$, by the continuity of $g$. Therefore, we can apply Theorem 4.3.27 to conclude that $f$ is continuous. Hence $g$ is a homeomorphism. □

We now need two auxiliary results about nearly open homomorphisms. The first of them is close to Proposition 1.5.15 while the second one is, in fact, a part of Theorem 4.3.30.

LEMMA 4.3.29. *Let $f\colon G \to H$ be a continuous homomorphism of topological groups. Then the following are equivalent:*

a) *the interior of $\overline{f(U)}$ is not empty, for each neighbourhood $U$ of the identity in $G$;*
b) *the homomorphism $f$ is nearly open.*

PROOF. Evidently, it suffices to show that a) implies b). Suppose a) holds. Denote by $\mathcal{N}(e_G)$ and $\mathcal{N}(e_H)$ the families of neighbourhoods at the neutral elements $e_G$ and $e_H$ of $G$ and $H$, respectively. We claim that $\overline{f(U)} \in \mathcal{N}(e_H)$, for each $U \in \mathcal{N}(e_G)$. Indeed, given an $U \in \mathcal{N}(e_G)$, choose $V \in \mathcal{N}(e_G)$ with $V^{-1}V \subset U$. By our assumption, the set $\overline{f(V)}$ contains a non-empty open set $W$. Clearly, $W^{-1}W$ is an open neighbourhood of $e_H$ and we have that

$$W^{-1}W \subset \left(\overline{f(V)}\right)^{-1}\overline{f(V)} = \overline{f(V)^{-1}f(V)} = \overline{f(V^{-1}V)} \subset \overline{f(U)}.$$

This proves our claim.

To finish the proof, consider an arbitrary open neighbourhood $U$ of $e_G$, take a point $x \in U$ and put $y = f(x)$. Then choose $V \in \mathcal{N}(e_G)$ such that $xV \subset U$. It follows from the claim that $\overline{f(V)}$ contains an open neighbourhood $O$ of $e_H$ in $H$. Therefore, the open set $yO \subset H$ contains $y$ and satisfies $yO \subset y \cdot \overline{f(V)} = \overline{f(xV)} \subset \overline{f(U)}$. We have thus proved that every point $y \in f(U)$ is contained in the interior of $\overline{f(U)}$, so the homomorphism $f$ is nearly open. □

THEOREM 4.3.30. *Every continuous nearly open homomorphism $f\colon G \to H$ of Čech-complete groups is open.*

PROOF. Let $N$ be the kernel of $f$ and $K = G/N$ be the corresponding quotient group. Let also $\pi\colon G \to G/N$ be the quotient homomorphism. Clearly, there exists a monomorphism $g\colon G/N \to H$ such that $g \circ \pi = f$. Then $g$ is continuous since the homomorphism $\pi$ is open. In addition, $g$ is a nearly open mapping since $f$ is.

By Theorem 4.3.26, the group $G/N$ is Čech-complete. Therefore, $g\colon G/N \to H$ is a continuous nearly open monomorphism of Čech-complete groups. Therefore, the subgroup $P = f(G)$ of $H$ is dense in some open subset $U$ of $H$. We can now apply Corollary 4.3.28 to conclude that $g\colon G/N \to P$ is a homeomorphism, i.e., $g$ is a topological isomorphism. Hence both groups $G/N$ and $P$ are Raĭkov complete by Theorem 4.3.7. Denote by $K$ the closure of $P$ in $H$. Then $K$ is a closed subgroup of $H$ and $K$ contains $U$. Therefore, the group $K$ is open in $H$. Since $P$ is a dense Raĭkov complete subgroup of $K$, we must have $K = P$. Therefore, $g$ is a topological isomorphism between $G/N$ and the open subgroup $K$ of $H$ and, hence, the homomorphism $f = g \circ \pi$ is open. □

The next example shows that all assumptions in Theorem 4.3.30 are essential, so the result we just proved is, in a sense, best possible.

Example 4.3.31. There exists a continuous non-open isomorphism $f\colon G \to H$ of topological groups, where $H$ is compact metrizable and $G$, $f$ satisfy one of the following conditions:

a) $G$ is locally compact (hence, Čech-complete);
b) $G$ is precompact and $f$ is nearly open.

For a), let $G$ be the group $\mathbb{T}$ with the discrete topology and $H$ be the same group $\mathbb{T}$ with the usual compact topology. For b), take an arbitrary discontinuous homomorphism $g\colon \mathbb{T} \to \mathbb{T}$ and let $G$ be the graph of $g$. Then $G$ is a subgroup of the product group $\mathbb{T} \times \mathbb{T}$ and we endow $G$ with the subspace topology. Again, let $H$ be the same as in a). Then the restriction to $G$ of the projection of $\mathbb{T} \times \mathbb{T}$ onto the first factor is a continuous isomorphism $f$ of $G$ onto $H$. The fact that $f$ is nearly open follows from the proposition given below. □

Proposition 4.3.32. *A continuous homomorphism $f\colon G \to H$ of an $\omega$-narrow group $G$ onto a group $H$ with the Baire property is nearly open.*

Proof. According to Lemma 4.3.29, it suffices to show that the interior of the set $\overline{f(U)}$ is not empty, for every open neighbourhood $U$ of the neutral element in $G$. Since the group $G$ is $\omega$-narrow, there exists a countable set $C \subset G$ such that $G = \bigcup_{x \in C} xU$. Then the sets $f(xU)$, with $x \in C$, cover the group $H$. By the Baire property of $H$, there must exist $x \in C$ such that the closure of $f(xU)$ has a non-empty interior. Since the translations in $H$ are homeomorphisms, the interior of $\overline{f(U)}$ is not empty either, as required. □

In some special cases, one can omit the assumption of nearly openness of the homomorphism $f$ in Theorem 4.3.30. For example, combining Theorem 4.3.30 and Proposition 4.3.32, we obtain the following:

Corollary 4.3.33. *Every continuous onto homomorphism of $\omega$-narrow Čech-complete groups is open.*

Since every space metrizable by a complete metric is Čech-complete (see [**165**, Theorem 4.3.26]), we also deduce the following result:

Corollary 4.3.34. *Every continuous onto homomorphism of separable completely metrizable groups is open.*

It follows from Theorem 4.3.20 that every feathered topological group is a preimage of a metrizable space under a perfect mapping. In the general setting, preimages of metrizable spaces under perfect mappings were characterized in [**16**] (see also [**60**, Ch. 5, Problem 228]) as paracompact $p$-spaces. A space $X$ is said to be a *p-space* or *feathered space* if it is Tychonoff and there exists a countable collection $\mathscr{E} = \{\gamma_n : n \in \omega\}$ of families $\gamma_n$ of open sets in the Čech–Stone compactification $\beta X$ of $X$ such that $\bigcap\{St(x, \gamma_n) : n \in \omega\} \subset X$, for every $x \in X$. Here $St(x, \gamma_n)$ is the union of all elements of the family $\gamma_n$ that contain $x$. It is easy to see that every metrizable space is a $p$-space, and every Čech-complete space is a $p$-space. Thus, each locally compact space is a $p$-space. On the other hand, every non-empty $p$-space $X$, evidently, contains a non-empty compact subset of countable character in $X$. It was established in [**16**] that a Tychonoff space $X$ admits a perfect mapping onto a metrizable space if and only if $X$ is a paracompact $p$-space. Hence, applying Theorem 4.3.20, we obtain the following:

THEOREM 4.3.35. *A topological group is feathered iff it is a p-space, and iff it is a paracompact p-space.*

We have established in Section 3.2 that when $H$ is a locally compact subgroup of an arbitrary topological group $G$, then the natural quotient mapping $\pi$ of $G$ onto the quotient space $G/H$ has some nice properties. Now we will apply these results to the class of feathered topological groups.

Čech-completeness is preserved by perfect preimages [**165**, Theorem 3.9.10]. Similarly, we have the following preservation result.

PROPOSITION 4.3.36. *The following are valid:*

a) *A closed subspace of a feathered space is feathered.*
b) *If $f: X \to Y$ is a perfect onto mapping of Tychonoff spaces and the space $Y$ is feathered, then so is $X$.*

PROOF. Item a) is trivial, so we only verify b). If $Y$ is feathered, then there exists a countable collection $\mathscr{E} = \{\gamma_n : n \in \omega\}$ of open coverings of $Y$ in $\beta Y$ such that $S(y) = \bigcap_{n\in\omega} St(y, \gamma_n) \subset Y$, for each $y \in Y$. Extend $f$ to a continuous mapping $g: \beta X \to \beta Y$. For every $n \in \omega$, let $\lambda_n = \{g^{-1}(U) : U \in \gamma_n\}$. Then $\lambda_n$ is an open covering of $X$ in $\beta X$. Take an arbitrary point $x \in X$ and put $y = f(x) = g(x)$. It follows from the definition of the coverings $\lambda_n$ that $St(x, \lambda_n) = g^{-1}(St(y, \gamma_n))$ for each $n \in \omega$, so that

$$T(x) = \bigcap_{n\in\omega} St(x, \lambda_n) = g^{-1}\left(\bigcap_{n\in\omega} St(y, \gamma_n)\right) = g^{-1}(S(y)).$$

Since the mapping $f$ is perfect, we have that $g(\beta X \setminus X) \subset \beta Y \setminus Y$ or, equivalently, $X = g^{-1}(Y)$ (see [**165**, Theorem 3.7.15]). Therefore, the inclusion $S(y) \subset Y$ implies that $T(x) = g^{-1}(S(y)) \subset X$, and we conclude that the space $X$ is feathered. □

The above proposition leads to the following two results that complement Corollary 3.2.6:

THEOREM 4.3.37. *Suppose that $G$ is a topological group, and $H$ a locally compact subgroup of $G$ such that the quotient space $G/H$ is a feathered space. Then $G$ is a paracompact feathered space.*

PROOF. From Theorem 3.2.2 it follows that there exists an open neighbourhood $U$ of the neutral element $e$ in $G$ such that $\overline{U}$ is a preimage of a closed subset of $G/H$ under a perfect mapping. The class of feathered spaces is closed under taking closed subspaces, by a) of Proposition 4.3.36, so $\overline{U}$ is a feathered space itself by b) of the same proposition. It follows that $U$ contains a non-empty compact subspace $F$ with a countable base of neighbourhoods in $G$ which, by Corollary 4.3.21, implies that $G$ is a paracompact feathered space. □

Since every metrizable space is feathered, we have:

COROLLARY 4.3.38. *Suppose that $G$ is a topological group, and $H$ is a locally compact subgroup of $G$ such that the quotient space $G/H$ is metrizable. Then the group $G$ is feathered and paracompact.*

## Exercises

4.3.a. Show that every precompact Čech-complete group is compact.

4.3.b. Verify that every $\sigma$-compact Čech-complete group is locally compact.

4.3.c. Let $G$ be a topological group, and $H$ a locally compact subgroup of $G$ such that the quotient space $G/H$ has a countable network. Then $G$ contains an open subgroup $M$ which is a Lindelöf space, and, therefore, $G$ is a free topological sum of Lindelöf subspaces.

4.3.d. Formulate and prove a generalization of Lemma 4.3.14 for an arbitrary compact subset $K$ of a topological group $G$ (without assuming that $K$ has countable character in $G$).

4.3.e. Give an example of a countable non-feathered topological group.

4.3.f. Let us call a topological group $G$ *weakly feathered* if $G$ contains a non-empty compact $G_\delta$-set.
   (a) Give an example of a weakly feathered group which is not feathered.
   (b) Give an example of a weakly feathered group which fails to be normal (observe that, by Corollary 4.3.21, such a group cannot be feathered).
   (c) Prove that $G$ is weakly feathered if and only if $G$ contains a compact subgroup $H$ such that the left coset space $G/H$ is *submetrizable*, that is, admits a coarser metrizable topology (cf. Theorem 4.3.20).

4.3.g. Give an example of a first-countable non-feathered paratopological group.

4.3.h. Give an example of a first-countable regular non-paracompact paratopological group.

4.3.i. Show that the quotient space $G/H$ in Theorem 4.3.23 is a paracompact $p$-space.

4.3.j. Is the continuity of the homomorphism in Proposition 4.3.32 essential?

## Problems

4.3.A. Prove that every $\omega$-narrow feathered group is Lindelöf. Give an example of a Lindelöf topological group which is not feathered.

4.3.B. (A. Bouziad [**84**]) Prove that if a semitopological group $G$ is a Baire feathered space, then $G$ is a paratopological group.

4.3.C. Let $G$ be a topological group, and $bG$ a Hausdorff compactification of the space $G$ such that the remainder $bG \setminus G$ is Lindelöf. Prove that $G$ is a paracompact $p$-space.

4.3.D. Show that the conclusion in Problem 4.3.C cannot be extended to Tychonoff paratopological groups.

4.3.E. Construct a topological group $G$ and a closed subgroup $N$ of $G$ such that the quotient space $G/N$ is locally compact but fails to be paracompact.
*Hint.* See [**410**, Example 5.8 (a)].

4.3.F. Let $N$ be a closed invariant subgroup of a topological group $G$. Prove that if both groups $N$ and $G/N$ are Čech-complete, then so is $G$.

4.3.G. (A. V. Arhangel'skii [**49**]) Let $G$ be a topological group with a dense Čech-complete subspace. Prove that $G$ is Čech-complete.

4.3.H. Prove that the usual topological group $\mathbb{Q}$ of rational numbers is not homeomorphic to any Raĭkov complete topological group.

4.3.I. Suppose that $G$ is a metrizable non-discrete Raĭkov complete topological group. Prove that the cardinality of $G$ is not less than $2^{\omega}$.

## Open Problems

4.3.1. Let $G$ be a regular first-countable paratopological group. Is the space $G$ Tychonoff? (See also Problem 1.3.1.)

4.3.2. Let $G$ be a first-countable Hausdorff semitopological (paratopological) group. Is the space $G$ subparacompact? (See Exercise 3.2.a.)

4.3.3. Let $G$ be a first-countable regular (Tychonoff) paratopological group. Is the space $G$ subparacompact?

## 4.4. *P*-groups

We call $X$ a *P-space* if every $G_\delta$-set in $X$ is open. Similarly, a *P-group* is a topological group whose underlying space is a $P$-space. It is easy to see that every regular $P$-space is zero-dimensional. Therefore, all $P$-groups are zero-dimensional. The classes of $P$-spaces and $P$-groups are peculiar in many respects; they may serve as a source of examples and counterexamples of topological groups with unusual combinations of properties.

We start with some general results on $P$-groups.

LEMMA 4.4.1. *Suppose that $G$ is a $P$-group. Then:*

a) *$G$ has a base at the identity consisting of open subgroups, so $G$ is zero-dimensional.*
b) *If $G$ is $\omega$-narrow, then it has a base at the identity which consists of open invariant subgroups.*
c) *Every (topological) quotient group of $G$ is also a $P$-group.*
d) *If $G$ is a dense subgroup of a topological group $H$, then $H$ is a $P$-group.*

PROOF. a) For a neighbourhood $U$ of the identity $e$ in $G$, there exists a sequence $\{U_n : n \in \omega\}$ of open symmetric neighbourhoods of $e$ in $G$ such that $U_0 \subset U$ and $U_{n+1}^2 \subset U_n$ for each $n \in \omega$. Then $N = \bigcap_{n=0}^{\infty} U_n$ is a subgroup of $G$ lying in $U$. Since $G$ is a $P$-group, $N$ is open in $G$. Every open subgroup of $G$ is closed, so $G$ has a base of closed and open sets at $e$. Now the homogeneity of $G$ implies the conclusion.

b) Let $U$ be a neighbourhood of the identity in the $\omega$-narrow $P$-group $G$. By a), there exists an open subgroup $N$ of $G$ such that $N \subset U$. Consider the invariant subgroup $P = \bigcap_{x\in G} xNx^{-1}$ of $G$. It is clear that $P \subset N \subset U$ and we claim that $P$ is open in $G$. Indeed, since $G$ is $\omega$-narrow, there exists a countable set $F$ in $G$ such that $F \cdot N = G$. Note that if $x, y \in G$ and $y^{-1}x \in N$, then $xNx^{-1} = yNy^{-1}$. Since every element $x \in G$ belongs to $yN$ for some $y \in F$, we conclude that $P = \bigcap_{x\in G} xNx^{-1} = \bigcap_{y\in F} yNy^{-1}$ is open in $G$.

c) Suppose that $\pi\colon G \to H$ is a continuous open homomorphism of $G$ onto a group $H$. If $Q$ is a $G_\delta$-set in $H$, then $\pi^{-1}(Q)$ is a $G_\delta$-set in the $P$-group $G$, so that $\pi^{-1}(Q)$ is open in $G$. Therefore, $Q = \pi\pi^{-1}(Q)$ is open in $H$. This proves that $H$ is a $P$-group.

d) Let $\{U_n : n \in \omega\}$ be a sequence of open neighbourhoods of the identity $e$ in $H$. There exists a sequence $\{V_n : n \in \omega\}$ of open symmetric neighbourhoods of $e$ in $H$ such that $V_{n+1}^2 \subset V_n \subset U_n$ for each $n \in \omega$. Then $N = \bigcap_{n=0}^{\infty} V_n$ is a closed subgroup of $H$. Since $G$ is a $P$-group, $P = N \cap G = \bigcap_{n=0}^{\infty}(V_n \cap G)$ is an open subgroup of $G$. Therefore, we can take an open set $W$ in $H$ such that $W \cap G = P$. It is clear that $e \in W$, and the density of $G$ in $H$ implies that $cl_H W = cl_H P \subset N$. Since $N \subset \bigcap_{n=0}^{\infty} U_n$, we conclude that the intersection $\bigcap_{n=0}^{\infty} U_n$ contains the open neighbourhood $W$ of $e$ in $H$. Therefore, $H$ is a $P$-group. □

The following simple property of $\omega$-narrow $P$-groups will be used in Sections 5.6 and 8.6.

LEMMA 4.4.2. *Let $G$ be an $\omega$-narrow $P$-group. Then every homomorphic continuous image $K$ of $G$ with $\psi(K) \leq \omega$ is countable.*

PROOF. Consider a continuous homomorphism $\pi \colon G \to K$ onto a group $K$ of countable pseudocharacter. Since $G$ is a $P$-group, the kernel $N$ of $\pi$ is an open invariant subgroup of $G$. By assumption, the group $G$ is $\omega$-narrow, so it can be covered by countably many translates of $N$. Therefore, the quotient group $G/N$ is countable. Finally, the groups $K$ and $G/N$ are algebraically isomorphic, whence $|K| \leq \omega$. □

We shall show in Chapter 8 that $\omega$-narrow $P$-groups need not be complete (see Example 8.2.1). The situation changes if one replaces $\omega$-narrowness by the stronger Lindelöf property: every Lindelöf $P$-group is Raĭkov complete. To deduce this result, we recall the following well-known topological fact.

LEMMA 4.4.3. *A Lindelöf subset $Y$ of a Hausdorff $P$-space $X$ is closed in $X$.*

PROOF. Suppose that $Y$ is a proper Lindelöf subset of $X$. Let $a \in X \setminus Y$ be arbitrary. For every $y \in Y$, there exist disjoint open sets $U_y$ and $V_y$ in $X$ such that $y \in U_y$ and $a \in V_y$. Since $Y$ is Lindelöf, the family $\{U_y : y \in Y\}$ contains a countable subfamily $\{U_y : y \in C\}$ such that $Y \subset \bigcup_{y \in C} U_y$. Put $U = \bigcup_{y \in C} U_y$ and $V = \bigcap_{y \in C} V_y$. Then $Y \subset U$, $V$ is an open neighbourhood of $a$ in $X$ and $U \cap V = \varnothing$. In particular, $Y \cap V = \varnothing$. This implies that the complement $X \setminus Y$ is open in $X$, so that $Y$ is closed in $X$. □

COROLLARY 4.4.4. *Let $f \colon X \to Y$ be a continuous mapping of $X$ to a Hausdorff space $Y$. If $X$ is Lindelöf and $Y$ is a $P$-space, then the mapping $f$ is closed.*

PROOF. If $F$ is a closed subset of $X$, then both $F$ and $f(F)$ are Lindelöf, so Lemma 4.4.3 implies that the image $f(F)$ is closed in $Y$. □

PROPOSITION 4.4.5. *Every Lindelöf $P$-group $G$ is Raĭkov complete.*

PROOF. Suppose that $G$ is a dense subgroup of a topological group $H$. By d) of Lemma 4.4.1, $H$ is also a $P$-group. Then Lemma 4.4.3 implies that $G$ is closed in $H$, whence it follows that $G = H$. Therefore, the group $G$ is Raĭkov complete. □

The next result shows that Lindelöf $P$-groups behave, in a sense, similarly to locally compact $\sigma$-compact groups or to Čech-complete $\omega$-narrow groups (see Theorem 3.1.27 and Corollary 4.3.33).

LEMMA 4.4.6. *Let $\pi \colon G \to H$ be a continuous onto homomorphism of Lindelöf $P$-groups. Then $\pi$ is open.*

PROOF. By Lemma 4.4.3, the homomorphism $\pi$ is closed and, hence, quotient. Let $U$ be an open subset of $G$. Then $\pi^{-1}\pi(U) = U \cdot K$ is open in $G$, where $K$ is the kernel of $\pi$. Since $\pi$ is a quotient mapping, the set $\pi(U)$ has to be open in $H$. So $\pi$ is an open homomorphism. □

It is natural to ask, whether there exist Lindelöf $P$-groups of arbitrarily large cardinality. We will show below, in Example 4.4.11, that the answer is "yes". Let us first present several results on Lindelöf $P$-spaces and their products that form a part of the general topology background.

LEMMA 4.4.7. *Let $P = \prod_{i\in\omega} X_i$ be a product space. Suppose that for every $n \in \omega$, the projection $\pi_n^{n+1}\colon P_{n+1} \to P_n$ is closed, where $P_k = \prod_{i\leq k} X_i$ for $k \in \omega$. Then the projection $p_n\colon P \to P_n$ is closed for each $n \in \omega$.*

PROOF. Let $n \in \omega$ be arbitrary, and suppose that $p_n(F)$ is not closed in $P_n$ for a subset $F$ of $P$. Choose a point $x_n$ in $\overline{C_n} \setminus C_n$, where $C_n = p_n(F)$. Suppose that for some $m \geq n$, we have defined points $x_n, \ldots, x_m$ satisfying the following conditions for each $i$ with $n \leq i \leq m$:

(1) $x_i \in \overline{p_i(F)}$;
(2) $\pi_{i-1}^i(x_i) = x_{i-1}$ if $i > n$.

Let $C_k = p_k(F)$ for every $k \geq n$. Since the mapping $\pi_m^{m+1}$ is closed, we have $\pi_m^{m+1}(\overline{C_{m+1}}) = \overline{C_m}$. Hence from $x_m \in \overline{C_m}$ (see (1)) it follows that $(\pi_m^{m+1})^{-1}(x_m) \cap \overline{C_{m+1}} \neq \emptyset$, so we can choose a point $x_{m+1} \in (\pi_m^{m+1})^{-1}(x_m) \cap \overline{C_{m+1}}$. It is clear that the points $x_n, \ldots, x_m, x_{m+1}$ satisfy (1) and (2) for each $i = n, \ldots, m, m+1$. This finishes our construction of the sequence $\{x_k : k \geq n\}$.

By (2), there exists a point $x \in P$ such that $p_k(x) = x_k$ for all $k \geq n$. Then (1) implies that $x \in \overline{F}$. Since $x_n = p_n(x) \notin p_n(F)$ by our choice of $x_n$, we conclude that $x \notin F$. Hence $x \in \overline{F} \setminus F$, so that the set $F$ is not closed in $P$. This proves that the projection $p_n\colon P \to P_n$ is closed. □

LEMMA 4.4.8. *Let $X$ be a Lindelöf space and $Y$ be a Hausdorff $P$-space. Then the projection $\pi\colon X \times Y \to Y$ is a closed mapping.*

PROOF. Suppose that a set $F \subset X \times Y$ is closed, and take an arbitrary point $y \in Y \setminus \pi(F)$. Then $(X \times \{y\}) \cap F = \emptyset$. For every point $x \in X$, choose open sets $U_x$ and $V_x$ in $X$ and $Y$, respectively, such that $x \in U_x$, $y \in V_x$ and $(U_x \times V_x) \cap F = \emptyset$. Since $X$ is Lindelöf, there exists a countable set $C \subset X$ such that $X = \bigcup_{y\in C} U_x$. Then the set $V = \bigcap_{x\in C} V_x$ is open in $Y$ and contains $y$. Clearly, $V \subset V_x$ for each $y \in C$, so $(X \times V) \cap F = \emptyset$. This implies immediately that $V \cap \pi(F) = \emptyset$ and, hence, $y \notin \overline{\pi(F)}$. We conclude, therefore, that the set $\pi(F)$ is closed in $Y$. Hence $\pi$ is a closed mapping. □

PROPOSITION 4.4.9. *Let $X$ and $Y$ be Lindelöf $P$-spaces. Then the product $X \times Y$ is Lindelöf and the projection of $X \times Y$ to $X$ is closed.*

PROOF. The projection $p\colon X \times Y \to X$ is closed, by Lemma 4.4.8. Since $X$ is Lindelöf and the fibers $p^{-1}(x) \cong Y$ of the mapping $p$ are also Lindelöf, it follows from [**165**, Th. 3.8.8] that the product $X \times Y$ is Lindelöf. □

THEOREM 4.4.10. *The product of countably many Lindelöf $P$-spaces is Lindelöf.*

Proof. Let $\{X_i : i \in \omega\}$ be a family of Lindelöf $P$-spaces. Put $P = \prod_{i\in\omega} X_i$ and $P_k = \prod_{i\leq k} X_i$, for $k \in \omega$. As in Lemma 4.4.7, let $p_n\colon P \to P_n$ and $\pi^{n+1}_n\colon P_{n+1} \to P_n$ be natural projections. It is clear that each $X_n$ is a $P$-space. Apply induction on $n \in \omega$ along with Proposition 4.4.9 to show that the space $X_n$ is Lindelöf and the projection $\pi^{n+1}_n$ is closed for each $n \in \omega$. Hence Lemma 4.4.7 implies that the projections $p_n\colon P \to P_n$ are also closed.

Let $\{F_\alpha : \alpha < \tau\}$ be a decreasing sequence of non-empty closed sets in $P$, where $\tau > \omega$ is a regular cardinal. For every $n \in \omega$, put $K_n = \bigcap_{\alpha<\tau} p_n(F_\alpha)$. Since $\{p_n(F_\alpha) : \alpha < \tau\}$ is a decreasing sequence of non-empty closed sets in the Lindelöf space $P_n$, the closed set $K_n$ is not empty. It is also clear that $\pi^{n+1}_n(K_{n+1}) \subset K_n$ for each $n \in \omega$. In fact, this inclusion is equality. Indeed, suppose that $x \in K_n$ for some $n \in \omega$. From $p_n = \pi^{n+1}_n \circ p_{n+1}$ it follows that $(\pi^{n+1}_n)^{-1}(x) \cap p_{n+1}(F_\alpha) \neq \varnothing$ for each $\alpha < \tau$. Since the fiber $(\pi^{n+1}_n)^{-1}(x) \cong X_{n+1}$ is Lindelöf and the sets $p_{n+1}(F_\alpha)$ are closed, the intersection $(\pi^{n+1}_n)^{-1}(x) \cap K_{n+1}$ must be non-empty. Hence $x \in \pi^{n+1}_n(K_{n+1})$, whence the equality $\pi^{n+1}_n(K_{n+1}) = K_n$ follows.

The above equality enables us to define by induction a sequence $\{x_n : n \in \omega\}$ such that $x_n \in K_n$ and $\pi^{n+1}_n(x_{n+1}) = x_n$ for each $n \in \omega$. As in the proof of Lemma 4.4.7, choose a point $x \in P$ such that $p_n(x) = x_n$ for all $n \in \omega$. Fix an arbitrary $\alpha < \tau$. Our choice of $x$ implies that $p_n(x) \in p_n(F_\alpha)$ for every $n \in \omega$, whence it follows that $x \in F_\alpha$. Therefore, $x \in \bigcap_{\alpha<\tau} F_\alpha \neq \varnothing$. This proves that the product space $P$ is Lindelöf. □

The construction that follows makes use of $\sigma$-products introduced in Section 1.6.

Example 4.4.11. For every infinite cardinal $\tau$, there exists an Abelian Lindelöf $P$-group $G_\tau$ of cardinality $\tau$ such that $(G_\tau)^k$ is topologically isomorphic to $G_\tau$, for each integer $k \geq 1$. In addition, for every $\tau < \aleph_\omega$, the group $G_\tau$ may be chosen to satisfy $w(G_\tau) = \tau$.

In the case $\tau = \aleph_0$ one can take $G_\tau$ to be a countable infinite Boolean group and endow $G_\tau$ with the discrete topology. Now suppose that $A$ is a set of cardinality $\tau > \aleph_0$ and let $K$ be a non-trivial countable discrete Abelian group with identity $e$. In the product group $\Pi = K^A$, consider the subgroup

$$G_\tau = \{x \in \Pi : |\operatorname{supp}(x)| < \omega\},$$

where $\operatorname{supp}(x)$ denotes the set $\{\alpha \in A : x(\alpha) \neq e\}$. Clearly, the group $G_\tau$ is the *$\sigma$-product* of $\tau$ copies of the group $K$ (see Section 1.6). Since $|A| = \tau$ and $K$ is countable, we have the equality $|G_\tau| = \tau$. For every $B \subset A$, let

$$U_B = \{x \in G_\tau : \operatorname{supp}(x) \cap B = \varnothing\}.$$

Then the family $\mathcal{N} = \{U_B : B \subset A,\ |B| \leq \omega\}$ consists of invariant subgroups of countable index in the group $G_\tau$ and forms a base at the identity for a Hausdorff group topology $\mathcal{T}$ on $G_\tau$. Sometimes $\mathcal{T}$ is called the *$\omega$-box topology*. We claim that the group $G_\tau$ endowed with topology $\mathcal{T}$ is as required.

Clearly, the family $\mathcal{N}$ is closed under countable intersections, so $G_\tau$ is a $P$-group. The space $G_\tau$ is Lindelöf, by Proposition 1.6.44.

Let us show that the group $G_\tau$ is topologically isomorphic with $(G_\tau)^k$ for each integer $k \geq 2$. There exists a partition $A = A_1 \cup A_2 \cup \ldots \cup A_k$ of the set $A$ such that $|A_i| = \tau$ for each $i = 1, \ldots, k$. Then

$$N_i = \{x \in G_\tau : \operatorname{supp}(x) \subset A_i\}$$

is a subgroup of $G_\tau$ which is topologically isomorphic to $G_\tau$, $i \le k$. Note that the group $G_\tau$ is topologically isomorphic to the product group $N_1 \times \cdots \times N_k$, so that $G_\tau \cong (G_\tau)^k$.

It remains to show that $w(G_\tau) = \tau$ whenever $\tau < \aleph_\omega$. Denote by $([A]^{\le\omega}, \subset)$ the family of all countable subsets of $A$ ordered by inclusion. We say that a subfamily $\gamma$ of $[A]^{\le\omega}$ is *dominating* if every element of $[A]^{\le\omega}$ is contained in some element of $\gamma$. If $|A| = \tau$, the minimal size of a dominating subfamily of $([A]^{\le\omega}, \subset)$ is denoted by $D(\tau)$. First, we prove the following:

**Claim.** *If $\tau = \aleph_n$ for some integer $n \ge 1$, then $D(\tau) = \tau$.*

Clearly, $\tau \le D(\tau)$ for every $\tau > \omega$, because the set $A$ can be partitioned into $\tau$ disjoint countably infinite subsets, and these cannot be covered by any collection of fewer than $\tau$ countable subsets. Hence, it suffices to show that $D(\tau) \le \tau$. Since $|A| = \aleph_n$, we can identify $A$ with the set $\aleph_n$. If $n = 1$, then the required dominating family in $([\aleph_1]^{\le\omega}, \subset)$ is $\{\alpha : \alpha < \omega_1\}$. Suppose that the equality $D(\aleph_k) = \aleph_k$ holds for each $k \le n$, where $n \ge 1$. By the assumption, for every uncountable ordinal $\alpha < \aleph_{n+1}$ there exists a dominating family $\gamma_\alpha$ in $([\alpha]^{\le\omega}, \subset)$ satisfying $|\gamma_\alpha| = |\alpha| \le \aleph_n$. Put $\gamma = \bigcup\{\gamma_\alpha : \omega_1 \le \alpha < \aleph_{n+1}\}$. Then $|\gamma| \le \aleph_{n+1}$, and it is easy to see that $\gamma$ is dominating in $([\aleph_{n+1}]^{\le\omega}, \subset)$. Indeed, if $A$ is a countable subset of $\aleph_{n+1}$, then $A \subset \alpha$ for some uncountable $\alpha < \aleph_{n+1}$, so there exists $B \in \gamma_\alpha$ with $A \subset B$. Since $\gamma_\alpha \subset \gamma$, this proves that $\gamma$ is dominating in $([\aleph_{n+1}]^{\le\omega}, \subset)$. Therefore, $D(\tau) \le \aleph_{n+1}$, whence our Claim follows.

To calculate the weight of the group $G_\tau$, we argue as follows. Let $\mathcal{B}$ be a local base at the neutral element of $G_\tau$, and suppose that $\mathcal{B}$ has the minimal possible cardinality. It is clear that $|\mathcal{B}| \le w(G_\tau)$. We can assume without loss of generality that $\mathcal{B} \subset \mathcal{N}$, so each element of $\mathcal{B}$ has the form $U_B$, for some countable subset $B$ of the index set $A$. Suppose that $\tau = \aleph_n$, for some integer $n \ge 1$. Then, clearly, $|A| = \aleph_n$. Consider the subfamily $\gamma$ of $[A]^{\le\omega}$ defined by

$$\gamma = \{B \subset A : |B| \le \omega,\ U_B \in \mathcal{B}\}.$$

It is clear that $|\gamma| = |\mathcal{B}|$. Since $\mathcal{B}$ is a local base at the neutral element of $G_\tau$, the family $\gamma$ is dominating in $([A]^{\le\omega}, \subset)$. Hence, the above Claim implies that $|\gamma| \ge |A| = \aleph_n$. It follows that $|\mathcal{B}| \ge \aleph_n = \tau$. On the other hand, take a dominating family $\lambda$ in $([A]^{\le\omega}, \subset)$ satisfying $|\lambda| = \tau$. Then $\mathcal{U} = \{U_B : B \in \lambda\}$ is a local base at the neutral element of $G_\tau$ with $|\mathcal{U}| = \tau$. Clearly, $\mathcal{U}^* = \{x + U_B : x \in G_\tau,\ B \in \lambda\}$ is a base for $G_\tau$, and $|\mathcal{U}^*| \le |\lambda| \cdot |G_\tau| = \tau$. We conclude, therefore, that $w(G_\tau) = \tau$. □

## Exercises

4.4.a. Show that every precompact $P$-group is finite (and discrete).

4.4.b. Suppose that every non-empty $G_\delta$-set in a topological group $H$ has a non-empty interior. Show that $H$ is a $P$-group.

4.4.c. Let $H$ be a closed subgroup of a topological group $G$, and suppose that both $H$ and the quotient space $G/H$ are $P$-spaces. Prove that $G$ is a $P$-group.

4.4.d. Let $H$ be a closed subgroup of a topological group $G$. Prove that if both $H$ and $G/H$ are Lindelöf $P$-spaces, then so is $G$.

4.4.e. Give an example of a $P$-group which is not Raĭkov complete and does not have a base at the identity consisting of open invariant subgroups.

4.4.f. Show that continuous onto homomorphisms of $\omega$-narrow $P$-groups need not be open.

4.4.g. Prove that every regular $P$-space $X$ is homeomorphic to a closed subspace of a $P$-group $G$. In addition, if $X$ is Lindelöf, then $G$ can also be chosen to be Lindelöf.

4.4.h. Show that every continuous real-valued function defined on a Lindelöf $P$-group is uniformly continuous. Can this assertion be extended to $\omega$-narrow $P$-groups?

4.4.i. Verify that for every $\tau > \aleph_0$ and every non-trivial countable discrete group $K$, the group $G_\tau$ in Example 4.4.11 contains a proper dense subgroup.

## Problems

4.4.A. Let $G$ be a topological group, and $H$ be a closed Lindelöf subgroup of $G$ such that $G/H$ is a $P$-space. Show that the natural projection $\pi\colon G \to G/H$ is closed.

4.4.B. Is it possible to embed an arbitrary regular $P$-space in a Lindelöf $P$-group?

4.4.C. Is every regular $P$-space homeomorphic to a closed subspace of an $\omega$-narrow $P$-group?

4.4.D. Characterize topological groups that can be represented as isomorphic continuous images of Lindelöf $P$-groups.

4.4.E. Let $\{G_i : i \in I\}$ be a family of Lindelöf $P$-groups, and $\sigma\Pi \subset \prod_{i\in I} G_i$ be the $\sigma$-product of this family, considered as a topological subgroup of the product group $\prod_{i\in I} G_i$. Prove that the topological group $(\sigma\Pi)_\omega$ (see Exercise 3.6.H) is a Lindelöf $P$-space.

4.4.F. Suppose that $G$ is an $\omega$-narrow $P$-group. Is the Raĭkov completion $\varrho G$ of $G$ a Lindelöf group?

4.4.G. Suppose that $G$ is a topological group such every bounded continuous real-valued function on $G$ is uniformly continuous, i.e., $G$ is $b$-fine (see Problem 1.8.C). Prove the following:
(a) If $G$ is a $P$-space, then $G$ is fine.
(b) $G$ is pseudocompact or is a $P$-group.
*Hint.* For (b), assume that $G$ is not a $P$-space and show that $G$ must be precompact. Then apply Problem 1.8.C to deduce that $G$ is pseudocompact.

4.4.H. Combine the conclusions in Problems 4.4.G and 1.8.C to deduce that every $b$-fine topological group is fine. Therefore, the classes of fine and $b$-fine topological groups coincide.

4.4.I. (O. Alas [**3**]) Prove that a $P$-group $G$ is fine iff for every clopen set $U$ in $G$, there exists an open neighbourhood $V$ of the neutral element in $G$ such that $U = VU$.

4.4.J. Let $\tau$ be an uncountable cardinal number. Give an example of two Lindelöf $P$-groups of the weight (exactly) $\tau$ which are not homeomorphic.

4.4.K. Does there exist a Lindelöf $P$-group of weight $\aleph_\omega$?

4.4.L. Let $X$ be a compact Hausdorff space, and $G$ a topological subgroup of the additive group $C_p(X)$. Show that $G$ is a $P$-group if and only if $G$ is discrete.

4.4.M. Give an example of a Tychonoff space $X$, and a topological subgroup $G$ of $C_p(X)$ such that $G$ is a non-discrete $P$-space.

## Open Problems

4.4.1. Characterize the algebraic structure of Abelian groups admitting a Hausdorff topological group topology that makes them into Lindelöf $P$-groups.

4.4.2. Does every Abelian group admit a Hausdorff topology that makes it into a Lindelöf topological group?

4.4.3. A space is called *linearly Lindelöf* if every uncountable subset of regular cardinality has a point of complete accumulation in $X$. Let $G$ be a linearly Lindelöf $P$-group. Is $G$ Lindelöf?

4.4.4. Suppose that $G$ is a regular paratopological (semitopological) group such that $G$ is a Lindelöf $P$-space. Is $G$ a topological group?

4.4.5. Is there a non-discrete Lindelöf $P$-group such that every stronger (topological group) topology on $G$ has at least one isolated point?

4.4.6. Let $\tau$ be an uncountable cardinal. Is there a Lindelöf $P$-group $U$ such that every Lindelöf $P$-group $G$ of weight $\leq \tau$ can be represented as a continuous image of the space $U$?

4.4.7. Let $G$ and $H$ be Lindelöf $P$-groups of weight $\aleph_1$. Must $G$ and $H$ be locally homeomorphic?

4.4.8. Let $G$ be a non-discrete Lindelöf $P$-group. Does $G$ contain a Lindelöf subgroup of cardinality $\aleph_1$? Does $G$ contain a (closed) subgroup $H$ such that $|G/H| = \aleph_1$?

4.4.9. Characterize the Tychonoff spaces $X$ such that every $P$-subgroup of $C_p(X)$ is discrete.

4.4.10. Find conditions on Tychonoff spaces $X$ and $Y$ under which the $G_\delta$-modifications of $C_p(X)$ and $C_p(Y)$ are topologically isomorphic as topological groups.

4.4.11. Find conditions on Tychonoff spaces $X$ and $Y$ under which the $G_\delta$-modifications of $C_p(X)$ and $C_p(Y)$ are homeomorphic.

## 4.5. Extremally disconnected topological and quasitopological groups

Starting with a very simple proof of Frolík's theorem on homeomorphisms of extremally disconnected spaces, we show how this theorem implies a well-known result of Malykhin's [**298**]: Every extremally disconnected topological group contains an open and closed subgroup consisting of elements of order 2. We also apply Frolík's theorem to obtain some further theorems on the structure of extremally disconnected topological groups and of quasitopological groups. In particular, it turns out that every Lindelöf extremally disconnected quasitopological group with square roots is countable, and every extremally disconnected topological field is discrete.

THEOREM 4.5.1. **[Z. Frolík]** *Let $X$ be an extremally disconnected Hausdorff space, and $h$ a homeomorphism of $X$ onto itself. Then the set $M = \{x \in X : h(x) = x\}$ of all fixed points of $h$ is an open and closed subset of $X$.*

PROOF. A subset $A$ of $X$ will be called *$h$-simple* if $h(A) \cap A = \varnothing$. Let (using Zorn's Lemma) $C$ be a maximal chain of $h$-simple open subsets of $X$. Put $U = \bigcup C$. Then, by an obvious standard argument, $U$ is $h$-simple. Thus, the sets $U$ and $h(U)$ are disjoint. Therefore, since $h$ is a homeomorphism, $h^{-1}(U)$ and $h(U)$ are also $h$-simple open sets. Since $X$ is extremally disconnected, it follows that the closures of $U$ and $h(U)$ are disjoint open sets as well. Thus, $\overline{U}$ is $h$-simple. Notice that the maximality of the chain $C$ and the definition of $U$ imply that $U$ is a maximal $h$-simple open subset of $X$. Therefore, $\overline{U}$ coincides with $U$, that is, $U$ is closed. It follows that the sets $h(U)$ and $h^{-1}(U)$ are also closed. Hence, the set $F = U \cup h(U) \cup h^{-1}(U)$ is closed.

Obviously, the intersection of $M$ with any $h$-simple set is empty. Since $F$ is the union of three $h$-simple sets, it follows that $M \cap F = \varnothing$. Therefore, $X \setminus F$ is an open set containing $M$. Let us show that $M = X \setminus F$ (which will obviously make the proof of Theorem 4.5.1 complete).

Assume the contrary. Then there exists $a \in X \setminus F$ such that $h(a) \neq a$. Since $X$ is Hausdorff and $h$ is continuous, there exists an open neighbourhood $W$ of $a$ such that $h(W) \cap W = \varnothing$ and $W \cap F = \varnothing$. Then $W$ is $h$-simple, and $W \cap U = \varnothing$, $W \cap h(U) = \varnothing$, $W \cap h^{-1}(U) = \varnothing$, from which it follows that $U \cup W$ is an $h$-simple open set that properly contains $U$. On the other hand, by the maximality of $U$ this is impossible, a contradiction. □

The sets $U$ and $h(U)$ are disjoint, as well as the sets $U$ and $h^{-1}(U)$, while the sets $h(U)$ and $h^{-1}(U)$ may have a non-empty intersection. If we wish to have a disjoint covering of the complement of $M$ by open and closed $h$-simple subsets of $X$, we only have to replace $h^{-1}(U)$ by the set $h^{-1}(U) \setminus h(U)$.

We recall that a group $G$ is *Boolean* if every element of $G$ is of order 2 (see Example 1.2.9). It is easy to see that every Boolean group is Abelian. Indeed, if $a, b \in G$ then $abab = e$ and, multiplying this equality by $a$ from the left and by $b$ from the right, we obtain $ba = ab$.

Extremally disconnected groups need not be Boolean — the product of any extremally disconnected group $G$ with a discrete group $H$ is extremally disconnected, while $H$ may have no elements $x$ satisfying $x^2 = e_H$, except for the identity $e_H$. However, every extremally disconnected group must contain a relatively big Boolean subgroup:

THEOREM 4.5.2. [**V. I. Malykhin**] *Let $G$ be an extremally disconnected topological group. Then there exists an open and closed Abelian subgroup $H$ of $G$ such that $a^2 = e$, for each $a \in H$.*

PROOF. The mapping $h\colon G \to G$ defined by $h(a) = a^{-1}$ for each $a \in G$ is a homeomorphism of $G$. By Theorem 4.5.1, the set $U = \{a \in G : a^2 = e\}$ is an open neighbourhood of the neutral element $e$. Since $G$ is a topological group, there exists an open neighbourhood $V$ of $e$ such that $V^2 \subset U$. Every two elements $a$ and $b$ of $V$ commute. Indeed, $abab = e$, since $ab \in U$. Now from $a^2 = e$ and $b^2 = e$ it follows that $ab = ba$. Therefore, the subgroup $H$ of $G$ generated by $V$ is Abelian. Since $V$ is open, the subgroup $H$ is also open and, therefore, closed in $X$. Finally, since the Abelian group $H$ is generated by $V$, and all elements of $V$ are of order 2, it follows that $a^2 = e$, for every $a \in H$. □

The above proof depends heavily on the assumption that $G$ is a topological group, in particular, on the joint continuity of multiplication in $G$. If we replace this assumption with the weaker one that the multiplication is separately continuous, we are no longer able to derive a conclusion as strong as that in Theorem 4.5.2, but we can still obtain some interesting information on the topological and algebraic structure of $G$.

THEOREM 4.5.3. *Let $G$ be an extremally disconnected quasitopological group. Then the set $W = \{a \in G : a^2 = e\}$ is an open and closed neighbourhood of the neutral element $e$ of $G$.*

PROOF. The inverse mapping of $G$ onto itself is a homeomorphism, and $e$ is a fixed point of this mapping. It remains to apply Theorem 4.5.1. □

It is well known that elements of order 2 in a group need not constitute a subgroup. This happens because they do not have to commute. In this light, the next result is of some interest.

PROPOSITION 4.5.4. *Let $G$ be an extremally disconnected quasitopological group. Then, for every element $a$ of $G$ of order 2, there exists an open neighbourhood $V$ of the neutral element $e$ such that $a$ commutes with every element of $V \cup aV$.*

PROOF. By Theorem 4.5.3, the set $U$ of all elements of $G$ of order 2 is open in $G$. Since $G$ is semitopological and $a \in U$, there exists an open neighbourhood $V$ of the neutral element $e$ such that $V \subset U$ and $aV \subset U$. Let $b \in V$. Then $ab \in U$ and, therefore, $abab = e$. Since $a^2 = e$ and $b^2 = e$, it follows that $ab = ba$. Thus, $a$ commutes with every element of $V$.

Now, let $c \in aV$. Then $c = ab$, for some $b \in V$, and $ac = aab$, $ca = aba = aab$. Therefore, $ac = ca$. □

In the proof of Theorem 4.5.6 below we apply Proposition 4.5.4. However, the next stronger statement is proved by a slightly more elaborate argument. If $G$ is a group and $a \in G$, we denote by $C_a$ the set of all $b \in G$ which commute with $a$ (that is, satisfy the condition $ab = ba$). It is clear that $C_a$ is a subgroup of $G$, and if $G$ is a topological group, then $C_a$ is closed in $G$.

THEOREM 4.5.5. *Let $G$ be an extremally disconnected quasitopological group. Then, for any $a \in G$, the set $C_a$ of all $b \in G$ that commute with $a$ is an open and closed subgroup of $G$ (containing $a$).*

PROOF. Let $\phi$ be the mapping of $G$ to $G$ given by the rule $\phi(x) = a^{-1}xa$, for each $x \in G$. Clearly, $\phi$ is a homeomorphism of the space $G$ onto itself. Therefore, by Theorem 4.5.1, the set $F$ of all fixed points under $\phi$ is open and closed. Since $C_a$ is a subgroup of $G$, it remains to check that $C_a = F$. We have: $\phi(x) = x$ if and only if $a^{-1}xa = x$ if and only if $ax = xa$ if and only if $x \in C_a$. □

REMARK. Theorem 4.5.5 allows us to strengthen Theorem 4.5.2 in the following way. Let $G$ be an extremally disconnected topological group. Then, for any $a \in G$, there exists an open Boolean subgroup $H$ of $G$ such that $ab = ba$, for every element $b$ of $H$.

THEOREM 4.5.6. *Let $G$ be an extremally disconnected quasitopological group such that $G$ is generated by every open neighbourhood of the neutral element $e$. Then the group is $G$ is Boolean.*

PROOF. By Theorem 4.5.3, $U = \{a \in G : a^2 = e\}$ is an open neighbourhood of $e$. Take any $a \in U$. It follows from Proposition 4.5.4 that there exists an open neighbourhood $V$ of $e$ such that $a$ commutes with every element of $V$. Then, obviously, $a$ commutes with every element of the subgroup $H$ algebraically generated by $V$. However, $H$ coincides with $G$ by the assumption. Hence, $a$ commutes with every element of $G$. It follows, in particular, that any two elements of $U$ commute. By the assumption, $U$ generates $G$. In addition, if $a \in U$ and $b \in U$, then $a^{-1} = a \in U$, and $ab \in U$, since $abab = abba = aea = a^2 = e$. Therefore, $U$ is a subgroup of $G$. It follows that $G = U$. □

COROLLARY 4.5.7. *Let $h$ be a homeomorphism of the Čech–Stone compactification $\beta\mathbb{N}$ of the infinite discrete space $\mathbb{N}$ onto $\beta\mathbb{N}$ such that there are no fixed points of $h$ in $\mathbb{N}$. Then no point of $\beta\mathbb{N}$ is fixed under $h$.*

PROOF. Since the space $\beta\mathbb{N}$ is extremally disconnected and $\mathbb{N}$ is dense in $\beta\mathbb{N}$, the conclusion follows from Theorem 4.5.1. □

It may be useful to mention that the following version of Corollary 4.5.7 is valid: If the homomorphism $h$ has only finitely many fixed points in $\mathbb{N}$, then $h$ has no fixed points in $\beta\mathbb{N} \setminus \mathbb{N}$.

THEOREM 4.5.8. *Let $G$ be a separable extremally disconnected quasitopological group. Then there exists an Abelian subgroup $H$ of $G$ such that $H$ is a closed $G_\delta$-set in $G$. Moreover, $H$ can be chosen so that every element of $H$ commutes with every element of $G$.*

PROOF. Fix a countable dense subset $A$ of $G$. By Theorem 4.5.5, for each $a \in A$ there exists an open and closed subgroup $H_a$ of $G$ such that every element of $H_a$ commutes with $a$. Put $H = \bigcap\{H_a : a \in A\}$. Then $H$ is a closed subgroup of $G$ and a $G_\delta$-set in $G$; it is also clear that every $x \in H$ commutes with every element of $A$. Since $A$ is dense in $G$, and left and right translations are continuous, it follows that every $x \in H$ commutes with each element of $G$. In particular, $H$ is Abelian. □

THEOREM 4.5.9. *Let $G$ be an extremally disconnected quasitopological group, and let $b$ be any element of $G$. Then the set $M_b = \{x \in G : x^2 = b\}$ is open and closed in $G$.*

PROOF. Let $h_b$ be the mapping of $G$ into itself given by the rule: $h_b(x) = x^{-1}b$, for each $x \in G$. Obviously, $h_b$ is a homeomorphism of the space $G$ onto itself. Therefore, the set $F$ of all fixed points under $h_b$ is an open and closed subset of $G$ by Theorem 4.5.1. Now, $F$ coincides with $M_b$. Indeed, for $a \in G$, $h_b(a) = a$ if and only if $a = a^{-1}b$ if and only if $a^2 = b$. □

COROLLARY 4.5.10. *Let $G$ be an extremally disconnected quasitopological group, and let $S_2(G) = \{M_b : b \in G\}$, where $M_b = \{x \in G : x^2 = b\}$. Then $S_2(G)$ is a disjoint open covering of the space $G$.*

We recall that a space $X$ is *pseudo-$\aleph_1$-compact* if every discrete (equivalently, locally finite) in $X$ family of non-empty open subsets of $X$ is countable.

PROPOSITION 4.5.11. *Let $G$ be an extremally disconnected quasitopological group. If $G$ is pseudo-$\aleph_1$-compact, then the set $\{a^2 : a \in G\}$ is countable.*

PROOF. This follows from Corollary 4.5.10 which guarantees that, under the restrictions of the proposition, the disjoint open covering $S_2(G)$ of $G$ is countable. □

We will call a group $G$ *group with square roots* if for each $b \in G$, there exists $a \in G$ such that $a^2 = b$. From Proposition 4.5.11 we obtain the next result immediately:

THEOREM 4.5.12. *Suppose that $G$ is an extremally disconnected quasitopological group with square roots. If $G$ is pseudo-$\aleph_1$-compact, then it is countable.*

COROLLARY 4.5.13. *Suppose that $G$ is a pseudocompact extremally disconnected quasitopological group with square roots. Then $G$ is finite.*

PROOF. By Theorem 4.5.12, $G$ is countable. Therefore, $G$ is compact as every countable pseudocompact space. Since every compact, countable Hausdorff space has an isolated point, $G$ must be discrete. Therefore, $G$ is finite. □

COROLLARY 4.5.14. *Suppose that $G$ is a Lindelöf extremally disconnected quasitopological group with square roots. Then $G$ is countable.*

COROLLARY 4.5.15. *Suppose that $G$ is an extremally disconnected quasitopological group with square roots such that the cellularity of $G$ is countable. Then $G$ is countable.*

THEOREM 4.5.16. *Let $G$ be an extremally disconnected quasitopological group with square roots such that $G$ is pseudo-$\aleph_1$-compact and has the Baire property. Then $G$ is countable and discrete.*

PROOF. This assertion follows from Theorem 4.5.12, since every countable $T_1$-space with the Baire property has an isolated point. Indeed, then the space $G$, being homogeneous, must be discrete. □

It should be noted that if $G$ is an extremally disconnected topological group, then the set $L = \{x \in G : x^3 = e\}$ need not be open in $G$. Indeed, if $L$ is open, then $L$ is a neighbourhood of $e$; therefore, $L \cap M_e$ is also an open neighbourhood of the neutral element $e$ in $G$ (see Corollary 4.5.10). On the other hand, it is clear that $M_e \cap L = \{e\}$; therefore, $e$ is isolated in $G$, which implies that $G$ is discrete.

A *topological skew field* $F$ is a skew field in which addition and multiplication are continuous and the set $G = F \setminus \{0\}$ of all non-zero elements is a topological group under multiplication.

THEOREM 4.5.17. *If a topological skew field $F$ is extremally disconnected, then it is discrete.*

PROOF. Suppose to the contrary that $F$ is not discrete. Let 0 and 1 denote the zero element and the unit element of $F$. Notice that $G = F \setminus \{0\}$ is dense in $F$ and, therefore, the space $G$ is also extremally disconnected.

Since $F$ is an extremally disconnected topological group with respect to addition, there exists an open neighbourhood $V$ of 0 such that $a + a = 0$, for each $a \in V$. Since $G$ is an extremally disconnected topological group with respect to multiplication, there exists an open neighbourhood $W$ of 1 in $G$ such that $b^2 = 1$, for each $b \in W$. Clearly, $W$ is open in $F$ since $G$ is open in $F$.

Since $F$ is a topological group with respect to addition, there exits an open neighbourhood $U$ of 0 such that $U \subset V$ and $1 + U \subset W$. Then for any $a \in U$, we have that $(1+a)(1+a) = 1 + (a+a) + a^2 = 1 + 0 + a^2 = 1 + a^2$, since $a \in U \subset V$. On the other hand, $(1+a)^2 = 1$, since $1 + a \in W$. Therefore, $1 = 1 + a^2$ which implies that $a^2 = 0$. Since all non-zero elements of $F$ are invertible, it follows that $a = 0$. Therefore, $U = \{0\}$, and, hence, $F$ is discrete. This contradiction completes the proof. □

Theorem 4.5.17, as it is clear from its proof, remains valid if we only assume that $F$ is an extremally disconnected semitopological skew field, that is, both $F$ and $G$ are quasitopological groups.

We conclude this section with a subtle and very important construction of a non-discrete extremally disconnected topological group $G$. Since no examples of such a group are known to exist in $ZFC$, we use a weak form of Martin's Axiom (see [**413**, Chapter III] or [**263**, Chapter 19]) to construct the group $G$.

Let $X$ be a Hausdorff space without isolated points, and let $\mathcal{T}$ be the topology of $X$. The space $X$ is called *maximal* if every topology $\mathcal{T}'$ on $X$ strictly finer than $\mathcal{T}$ has isolated points. It is easy to see that every Hausdorff space $(X, \mathcal{T})$ without isolated points admits a finer (hence, Hausdorff) topology $\mathcal{T}^*$ such that the space $(X, \mathcal{T}^*)$ is maximal — it suffices to apply Zorn's lemma to the family of all topologies on $X$ which are finer than $\mathcal{T}$ and have no isolated points.

One of important properties of the maximal spaces is that they all are extremally disconnected. In fact, we prove a bit more in the next lemma.

LEMMA 4.5.18. *Let $X$ be a maximal space. Then:*

a) *If $U$ is open in $X$ and $x \in \overline{U}$, then $U \cup \{x\}$ is open in $X$ and, in particular, $X$ is extremally disconnected;*
b) *every open subspace of $X$ is maximal;*
c) *every dense subspace of $X$ is open;*
d) *every nowhere dense subset of $X$ is closed and discrete in $X$.*

PROOF. a) Suppose that $U$ is a non-empty open subset of $X$ and that $x \in \overline{U}$. Put $V = U \cup \{x\}$ and consider the topology $\mathcal{T}'$ on $X$ with the subbase $\mathcal{T} \cup \{V\}$, where $\mathcal{T}$ is the original topology of $X$. Then $\mathcal{T}'$ is finer than $\mathcal{T}$ and has no isolated points. Since the space $(X, \mathcal{T})$ is maximal, this implies that $\mathcal{T}' = \mathcal{T}$ or, equivalently, $V \in \mathcal{T}$. Therefore, $\{x\} \cup U$ is open in $X$ for each $x \in \overline{U}$ and, hence, the closure of $U$ is open in $X$. This proves a).

b) Consider an arbitrary non-empty open subset $U$ of $X$. Clearly, $U$ has no isolated points. Denote by $\mathcal{T}_U$ the topology $U$ inherits from $X$. Suppose that $\mathcal{T}'_U$ is a topology on $U$ without isolated points and finer than $\mathcal{T}_U$. Denote by $\mathcal{T}'$ the family of all sets $V$ of the form $V \cup W$, where $V \in \mathcal{T}$ and $W \in \mathcal{T}'_U$. Then $\mathcal{T}'$ is a topology on $X$ finer than its original topology $\mathcal{T}$, and the space $(X, \mathcal{T}')$ has no isolated points. The maximality of $(X, \mathcal{T})$ implies that $\mathcal{T}' = \mathcal{T}$, so that $\mathcal{T}'_U = \mathcal{T}_U$ and, hence, the subspace $U$ of $X$ is maximal.

c) If $S$ is a dense subset of $X$, consider the topology $\mathcal{T}'$ on $X$ generated by the family $\mathcal{T} \cup \{S\}$, where $\mathcal{T}$ is the original topology of $X$. Then $\mathcal{T}'$ is finer than $\mathcal{T}$ and the space $(X, \mathcal{T}')$ has no isolated points. As above, we conclude that $\mathcal{T}' = \mathcal{T}$ and, hence, $S$ is open in $X$.

d) Suppose that a set $A \subset X$ is nowhere dense in $X$. Then $F = \overline{A}$ is also nowhere dense in $X$, and the complement $O = X \setminus F$ is an open dense subset of $X$. Take $x \in F$ and put $V_x = O \cup \{x\}$. Since $x \in \overline{O}$, item a) implies that the set $V_x$ is open in $X$ for each $x \in F$. Since $V_x \cap F = \{x\}$, every point of $F$ is isolated in $F$. Thus, $F$ is a closed discrete subset of $X$ and it follows from $F = \overline{A}$ that $A = F$. □

In what follows we will use the next characterization of maximal spaces:

PROPOSITION 4.5.19. *Let $X$ be a Hausdorff space without isolated points. Then $X$ is maximal if and only if for every $x \in X$ and every disjoint subsets $A$ and $B$ of $X \setminus \{x\}$, the point $x$ belongs to at most one of the sets $\overline{A}$, $\overline{B}$.*

PROOF. Suppose that the space $X$ is maximal. We claim that every subset $A$ of $X$ has the form $U \cup D$, where $U \subset X$ is open and $D$ is a closed discrete subset of $X$. Indeed, let $F$ be the closure of $A$ and $O$ be the interior of $F$ in $X$. Then $F \setminus O$ is a nowhere dense set in $X$, so d) of Lemma 4.5.18 implies that $D = A \setminus O \subset F \setminus O$ is a closed discrete subset of $X$. Since $A \cap O$ is dense in the open set $O$, the set $U = A \cap O$ is open in $X$, by b) and c) of Lemma 4.5.18. Clearly, we have the equality $A = U \cup D$, as required.

It easily follows from the above claim that if $x \in X$ and $x \in \overline{A}$ for some $A \subset X \setminus \{x\}$, then $A \cup \{x\}$ is a neighbourhood of $x$ in $X$. Indeed, by the claim, the set $A$ is the union of an open set $U \subset X$ and a closed discrete set $D$ in $X$. Since $x \in \overline{A} \setminus A$, we must have $x \in \overline{U}$. Then a) of Lemma 4.5.18 implies that $V = U \cup \{x\}$ is open in $X$, so $A \cup \{x\}$ contains the open neighbourhood $V$ of $x$ in $X$.

Suppose now that $x \in X$ and that $A$ and $B$ are disjoint subsets of $X \setminus \{x\}$. If $x \in \overline{A}$, then $A \cup \{x\}$ is a neighbourhood of $x$ in $X$, so there exists an open set $V$ in $X$ such that $x \in V \subset A \cup \{x\}$. Since $A$ and $B$ are disjoint subsets of $X \setminus \{x\}$, we have that $B \subset X \setminus V$ and, hence, $\overline{B} \subset X \setminus V$. This implies that $x \notin \overline{B}$, as required.

Conversely, suppose that a Hausdorff space $(X, \mathcal{T})$ has no isolated points and is not maximal. Take a topology $\mathcal{T}'$ on $X$ strictly finer than $\mathcal{T}$ and such that the space $(X, \mathcal{T}')$ has no isolated points. Then there exists an element $V \in \mathcal{T}' \setminus \mathcal{T}$. Clearly, the set $V$ is not empty and, considered as a subspace of $(X, \mathcal{T}')$, the space $V$ has no isolated points either. Put $B = X \setminus V$. It follows from $V \notin \mathcal{T}$ that there exists a point $x \in V \cap \overline{B}$, where the closure of $B$ is taken in the space $(X, \mathcal{T})$. Since $V$ has no isolated points, $x$ is an accumulation point of the set $A = V \setminus \{x\}$ in both spaces $(X, \mathcal{T}')$ and $(X, \mathcal{T})$. Thus, $A$ and $B$ are disjoint subsets of $X \setminus \{x\}$ and the closures of both sets in $(X, \mathcal{T})$ contain the point $x$. This implies the equivalence of the two conditions for $X$. □

Let $\gamma$ be a family of infinite subsets of $\omega$ and suppose that the intersection of every finite subfamily of $\gamma$ is infinite. We say that such a family $\gamma$ has the *strong intersection property.* An infinite subset $A$ of $\omega$ is called a *pseudointersection* of $\gamma$ if the complement $A \setminus B$ is finite, for each $B \in \gamma$. In other words, the set $A$ is *almost contained* in every element $B \in \gamma$. It is clear that every countable family $\gamma$ with the strong intersection property has a pseudointersection. Indeed, let $\{B_n : n \in \omega\}$ be an enumeration of $\gamma$. For every $n \in \omega$, choose an integer $a_n \in \bigcap_{i \leq n} B_i$ distinct from each $a_i$ with $i < n$ — this is possible since the intersection $\bigcap_{i \leq n} B_i$ is infinite. Then the set $A = \{a_n : n \in \omega\}$ is infinite and $|A \setminus B_n| \leq n$ for each $n \in \omega$, that is, $A$ is a pseudointersection of the family $\gamma$.

Denote by $\mathfrak{p}$ the minimum cardinality of the families $\gamma$ of subsets of $\omega$ with the strong intersection property such that $\gamma$ has no pseudointersection. It follows from the above observation that $\aleph_1 \leq \mathfrak{p}$ and, clearly, $\mathfrak{p} \leq 2^\omega$. Therefore, the Continuum Hypothesis (i.e., the assumption that $\aleph_1 = 2^\omega$) implies the equality $\mathfrak{p} = 2^\omega$. In fact, the equality $\mathfrak{p} = 2^\omega$ is equivalent to a weak form of Martin's Axiom (namely, Martin's Axiom for $\sigma$-centered partially ordered sets, see [**176**]). It is shown in [**285**, Chapter 8] that Martin's Axiom is consistent with the negation of the Continuum Hypothesis, so we conclude that the combination $\aleph_1 < \mathfrak{p} = 2^\omega$ is consistent with $ZFC$.

The maximal topological group we have in mind will be constructed under the assumption that $\mathfrak{p} = 2^\omega$. The construction has two main ingredients; one of them is Corollary 2.2.27 and the other is the characterization of maximal spaces given in Proposition 4.5.19.

From now on, we fix a countable infinite Boolean group $G$, without topology. One can take $G$ to be the direct sum of $\omega$ copies of the group $\mathbb{Z}(2)$. Let us say that a topological group topology $\mathcal{T}$ on $G$ is *linear* if the topological group $(G, \mathcal{T})$ has a local base at the neutral element 0 consisting of open subgroups. It is clear that every linear topology on $G$ makes $G$ into a zero-dimensional topological group.

We will use the assumption $\mathfrak{p} = 2^\omega$ for a recursive construction of length $\mathfrak{c} = 2^\omega$ of a maximal linear group topology on $G$. Let us start with two lemmas that explain the strategy of the construction.

Lemma 4.5.20. *Suppose that $\mathcal{T}$ is a non-discrete linear group topology on $G$ such that the group $(G, \mathcal{T})$ has a local base at zero of cardinality less than $\mathfrak{c}$. Then, under $\mathfrak{p} = \mathfrak{c}$, $G$ admits a finer non-discrete linear topological group topology $\mathcal{T}'$ with a countable base.*

Proof. Since $\mathcal{T}$ is a linear group topology on $G$, there exists a local base $\mathcal{B}$ at zero of the group $(G, \mathcal{T})$ consisting of open subgroups and satisfying $|\mathcal{B}| < \mathfrak{c}$. The group $G$ being countable and infinite, there exists a bijection $f \colon G \setminus \{0\} \to \omega$. For every $U \in \mathcal{B}$, put

$U^* = U \setminus \{0\}$ and consider the family

$$\gamma = \{f(U^*) : U \in \mathcal{B}\}$$

of infinite subsets of $\omega$. Since the group $(G, \mathcal{T})$ is non-discrete, the family $\gamma$ has the strong intersection property. Therefore, it follows from $|\gamma| < \mathfrak{p} = \mathfrak{c}$ that the family $\gamma$ has a pseudointersection $A$. Choose a sequence $X = \{x_n : n \in \omega\}$ of pairwise distinct elements of the infinite set $f^{-1}(A)$. It follows from the choice of $A$ and $X$ that the complement $X \setminus U$ is finite for each $U \in \mathcal{B}$.

For every $n \in \omega$, denote by $H_n$ the subgroup of $G$ generated by the set $\{x_k : n \leq k \in \omega\}$. We claim that every $U \in \mathcal{B}$ contains $H_n$ for some $n \in \omega$. Indeed, since the set $X \setminus U$ is finite, there exists $n \in \omega$ such that $\{x_k : n \leq k \in \omega\} \subset U$. The fact that $U$ is a subgroup of $G$ implies the inclusion $H_n \subset U$.

Finally, let $\mathcal{T}'$ be the topology on $G$ whose base $\mathcal{B}'$ consists of the sets $x + H_n$, with $x \in G$ and $n \in \omega$. Clearly, the base $\mathcal{B}'$ is countable, so that $\mathcal{T}'$ is a non-discrete, second-countable linear group topology on $G$ which is finer than $\mathcal{T}$. □

LEMMA 4.5.21. *Let $\mathcal{T}$ be a non-discrete, second-countable linear topological group topology on $G$, and suppose that $G \setminus \{0\} = P_1 \cup P_2$, where $P_1$ and $P_2$ are disjoint. Then there exists a non-discrete second-countable linear group topology $\mathcal{T}'$ on $G$ finer than $\mathcal{T}$ such that at most one of the sets $P_1$, $P_2$ accumulates at zero in $(G, \mathcal{T}')$.*

PROOF. By the assumptions of the lemma, there exists a countable decreasing local base $\{U_n : n \in \omega\}$ at zero of $G$ consisting of open subgroups. Choose a sequence $X = \{x_n : n \in \omega\}$ of pairwise distinct non-zero elements of $G$ satisfying the following conditions for each $n \in \omega$:

(i) $x_n \in U_n$;
(ii) $x_{n+1}$ is not in the subgroup generated by the elements $x_0, \ldots, x_n$.

Notice that every finitely generated subgroup of $G$ is finite, so the construction of the sequence $X = \{x_n : n \in \omega\}$ with properties (i) and (ii) presents no difficulties. Clearly, this sequence converges to zero in $(G, \mathcal{T})$. Notice also that by (ii), the set $X$ is *independent* in the sense that all sums $x_{n_1} + \cdots + x_{n_k}$, with $n_1 < \cdots < n_k$, are distinct from zero.

Let $H$ be the subgroup of $G$ generated by the set $X$. The disjoint sets $P_1$ and $P_2$ cover $H \setminus \{0\}$, so we can apply Corollary 2.2.27 to find an infinite subgroup $K$ of $H$ contained in one of the sets $P_1 \cup \{0\}$ or $P_2 \cup \{0\}$. We can assume without loss of generality that $K \subset P_1 \cup \{0\}$.

Let $\mathcal{B}' = \mathcal{B} \cup \{K\}$. Then $\mathcal{B}'$ is a base for a linear group topology $\mathcal{T}'$ on $G$ finer than the group topology $\mathcal{T}$. Since $K \cap P_2 = \emptyset$, the set $P_2$ does not accumulate at zero in $(G, \mathcal{T}')$. It remains to verify that the topology $\mathcal{T}'$ is non-discrete or, equivalently, $K \cap U_n$ is infinite for each $n \in \omega$. Suppose to the contrary that $K \cap U_m$ is finite for some $m \in \omega$. Let $H_m$ be the subgroup of $H$ generated by the set $\{x_n : m < n \in \omega\}$. It follows from (i) that $H_m \subset U_m$, so the intersection $K \cap H_m$ is finite as well. Denote by $\pi$ the canonical homomorphism of $H$ onto the quotient group $H/H_m$. Clearly, that the group $H/H_m$ is generated by the elements $\pi(x_0), \ldots, \pi(x_m)$. Hence the group $H/H_m$ is finite, since it is Boolean. The restriction $\varphi = \pi \restriction K$ is a homomorphism of $K$ onto a subgroup of the finite group $H/H_m$, and the kernel of $\varphi$ is the finite group $K \cap H_m$. Therefore, $K$ is finite, which is a contradiction. We have proved that the linear group topology $\mathcal{T}'$ on $G$ is non-discrete, as claimed. □

Now we are in position to show that the existence of a non-discrete maximal topological group is consistent with $ZFC$.

THEOREM 4.5.22. [**V. I. Malykhin**] *Under the assumption* $\mathfrak{p} = \mathfrak{c}$, *the countable infinite Boolean group* $G$ *admits a non-discrete Hausdorff maximal linear group topology.*

PROOF. Let $\mathcal{P} = \{(P_{\alpha,1}, P_{\alpha,2}) : \alpha < \mathfrak{c}\}$ be an enumeration of all pairs $(P_1, P_2)$ such that $P_1 \cap P_2 = \varnothing$ and both $P_1$ and $P_2$ are subsets of $G \setminus \{0\}$. Such an enumeration exists since the group $G$ is countable. By recursion of length $\mathfrak{c}$ we will construct a family $\{\mathcal{T}_\alpha : \alpha < \mathfrak{c}\}$ of non-discrete, second-countable linear group topologies on $G$ satisfying the following conditions for all $\alpha, \beta < \mathfrak{c}$:

(i) $\mathcal{T}_\alpha \subset \mathcal{T}_\beta$ if $\alpha < \beta$;

(ii) the neutral element $0_G$ of $G$ belongs to the closure in $(G, \mathcal{T}_\alpha)$ of at most one of the sets $P_{\alpha,1}, P_{\alpha,2}$.

Let $\mathcal{T}_0$ be an arbitrary non-discrete, Hausdorff, linear group topology on $G$. For example, one can identify $G$ with the $\sigma$-product of $\omega$ copies of the discrete group $\mathbb{Z}(2)$ and consider it with the topology inherited from the compact group $\mathbb{Z}(2)^\omega$.

Suppose that for some $\alpha < \mathfrak{c}$, we have defined a sequence $\{\mathcal{T}_\nu : \nu < \alpha\}$ of non-discrete, second-countable linear group topologies on $G$ satisfying (i) and (ii). It follows from (i) that the linear group topology $\gamma$ on $G$ with the base $\bigcup_{\nu<\alpha} \mathcal{T}_\nu$ is non-discrete. Since each topology $\mathcal{T}_\nu$ with $\nu < \alpha$ has a countable local base at zero of $G$, the group $(G, \gamma)$ has a local base at zero of cardinality less than $\mathfrak{c}$. Therefore, we can apply Lemma 4.5.20 to find a non-discrete, second-countable linear group topology $\gamma_1$ on $G$ finer than $\gamma$. Then, by Lemma 4.5.21, there exists a non-discrete, second-countable linear group topology $\mathcal{T}_\alpha$ on $G$ such that $\mathcal{T}_\alpha$ is finer than $\gamma_1$ and the neutral element of $G$ belongs to the closure in $(G, \mathcal{T}_\alpha)$ of at most one of the sets $P_{\alpha,1}$ or $P_{\alpha,2}$. Clearly, the family $\{\mathcal{T}_\nu : \nu \le \alpha\}$ satisfies (i) and (ii) at the stage $\alpha$. This finishes our recursive construction.

Let $\mathcal{T}$ be the linear group topology on $G$ with the base $\bigcup_{\alpha<\mathfrak{c}} \mathcal{T}_\alpha$. Since the group topologies $\mathcal{T}_\alpha$ are non-discrete, it follows from (i) that so is $\mathcal{T}$. Suppose that $P_1$ and $P_2$ are disjoint subsets of $G \setminus \{0_G\}$. Then there exists $\alpha < \mathfrak{c}$ such that $(P_1, P_2) = (P_{\alpha,1}, P_{\alpha,2})$, and (ii) implies that the neutral element of $G$ lies in the closure in $(G, \mathcal{T}_\alpha)$ of at most one of the sets $P_1$ or $P_2$. Since the topology $\mathcal{T}$ is finer than $\mathcal{T}_\alpha$, the same remains valid for the closures of $P_1$ and $P_2$ in the group $(G, \mathcal{T})$. Therefore, the group $(G, \mathcal{T})$ is a maximal space, by Proposition 4.5.19 and the homogeneity argument. □

## Exercises

4.5.a. Let $G$ be an abstract group with identity $e$ such that the set of all elements of order 2 in $G$ is finite, and $\mathcal{T}$ be a topology on $G$ such that the inverse operation is continuous and the one-point set $\{e\}$ does not belong to $\mathcal{T}$. Then $\mathcal{T}$ is not extremally disconnected.

4.5.b. Let $G$ be a group such that the set of elements of order 2 in $G$ is finite, and let $\mathcal{T}$ be a topology on $G$ such that the inverse operation is continuous and the space $(G, \mathcal{T})$ is homogeneous. Then the space $(G, \mathcal{T})$ is extremally disconnected if and only if it is discrete.

4.5.c. Show that the assumption in 4.5.b that the space $(G, \mathcal{T})$ is homogeneous cannot be dropped. *Hint.* Fix an extremally disconnected topology $\mathcal{T}$ on the set $P$ of all positive real numbers such that $(P, \mathcal{T})$ is dense in itself. For $V \subset P$, put $-V = \{-x : x \in V\}$. The family

$\mathcal{B} = \mathcal{T} \cup \{-V : V \in \mathcal{T}\} \cup \{\{0\}\}$ is a base of a non-discrete extremally disconnected topology $\mathcal{T}^*$ on the additive group $\mathbb{R}$ such that the inverse mapping is continuous.

4.5.d. Let $G$ be the maximal topological group constructed in Theorem 4.5.22. Denote by $\mathcal{F}$ the family of all subset $A$ of $G \setminus \{0\}$ such that $0 \in \overline{A}$. Verify that $\mathcal{F}$ is a free ultrafilter on the set $G \setminus \{0\}$.

4.5.e. Give an example of a Tychonoff space $X$ such that $C_p(X)$ contains an infinite extremally disconnected compact subspace.

## Problems

4.5.A. Prove that the product of two non-discrete extremally disconnected topological groups is never extremally disconnected.

4.5.B. Show that every non-discrete subgroup $H$ of a maximal topological group $G$ is open in $G$ and, hence, is maximal.

4.5.C. Let $H$ be a closed subgroup of a maximal topological group $G$. Prove that if the quotient space $G/H$ is not discrete, then it is maximal. Is an analogous assertion valid for extremal disconnectedness?

4.5.D. Let us call a group $G$ with a topology *sub-extremally disconnected* if $G$ is a subspace of some extremally disconnected space. Which results in this section remain true for sub-extremally disconnected topological (paratopological) groups?

4.5.E. Prove that the space $X$ in Exercise 4.5.e cannot be compact.

4.5.F. (V. I. Arnautov and E. G. Zelenyuk [**64**]) Let $G$ be a countable infinite Boolean group. Prove the following:

a) Under $\mathfrak{p} = 2^\omega$, there exists a Raĭkov complete maximal topological group topology on $G$;

b) Under $\mathfrak{p} = 2^\omega$, there exists a maximal topological group topology on $G$ that fails to be Raĭkov complete;

c) Under the Continuum Hypothesis, every infinite Abelian group admits a Raĭkov complete maximal topological group topology.

In particular, under $\mathfrak{p} = 2^\omega$, an extremally disconnected topological group may fail to be Raĭkov complete.

## Open Problems

4.5.1. Are all subgroups of an extremally disconnected topological group extremally disconnected?

4.5.2. Is every compact subspace of an extremally disconnected Tychonoff (regular) paratopological group finite?

4.5.3. Is every compact subspace of an extremally disconnected Tychonoff (regular) quasitopological group finite?

4.5.4. Is there in $ZFC$ a non-discrete extremally disconnected topological group?

4.5.5. Let $G$ be an extremally disconnected quasitopological group. Is it true that $G$ has an open and closed Abelian subgroup?

4.5.6. Characterize Tychonoff spaces $X$ such that every extremally disconnected subspace of $C_p(X)$ is discrete.

## 4.6. Perfect mappings and topological groups

In a topological group the product operation generates a variety of continuous mappings. Some of them are well known and play a fundamental role: the translations, for example.

However, it is natural to try to treat the subject in a more systematic way. In this section we are going to consider the following general questions. Suppose that $G$ is a topological group, $A$ and $B$ are subsets of $G$, and $f$ is the mapping of the product space $A \times B$ onto the subspace $AB$ of $G$ given by the product operation in $G$. Under certain natural restrictions on $A$ and $B$, what can be said about the properties of the mapping $f$? How the properties of the subspace $AB$ are related to the properties of the subspaces $A$ and $B$? We establish some results in this direction, in particular, we show that the theory of perfect mappings, a well-developed chapter of General Topology, can be effectively applied to answer some of these questions. Under this approach, the quotient groups also naturally enter the picture.

Let $G$ be a topological group, and $A$ and $B$ be subsets of $G$. Let us say that $A$ and $B$ are *cross-complementary* in $G$ if $G = AB = \{ab : a \in A,\ b \in B\}$. Suppose that $\mathcal{P}$ is a topological (algebraic, or a mixed nature) property, and $A$ is a subset of $G$. We will say that *$A$ has a $\mathcal{P}$-grasp* on $G$ if there exists a subset $B$ of $G$ such that $B$ is cross-complementary to $A$ and has the property $\mathcal{P}$. In particular, $A \subset G$ has a *compact grasp* on $G$ if there exists a compact subspace $B$ of $G$ such that $AB = G$. Similarly, $A$ has a *metrizable grasp* on $G$ if $G = AB$, for some metrizable subspace $B$ of $G$.

PROPOSITION 4.6.1. *Suppose that $G$ is a topological group, and $H$ a metrizable invariant subgroup of $G$ such that $H$ has a countably tight compact grasp on $G$. Then $G$ is metrizable.*

PROOF. The closure $K$ of $H$ in $G$ is first-countable by Proposition 1.4.16, so the Birkhoff–Kakutani theorem (see Theorem 3.3.12) implies that $K$ is metrizable. Also, $K$ is an invariant subgroup of $G$. Thus, we may assume that $H$ is closed in $G$. By the assumption, there exists a compact subspace $B$ of $G$ such that $HB = G$ and the tightness of $B$ is countable.

Let us consider now the quotient group $G/H$ and the quotient mapping $p\colon G \to G/H$. Since $HB = G$, we have $p(B) = G/H$. The restriction of the mapping $p$ to the compact subspace $B \subset G$ is closed and, in particular, quotient. Therefore, $G/H$ is a compactum of countable tightness by [**165**, 3.12.8(a)]. Since $G/H$ is also a topological group, and every compact group of countable tightness is metrizable by Corollary 4.2.2, we conclude that $G/H$ is metrizable. Since $H$ is metrizable as well, it follows from Corollary 1.5.21 that $G$ is also metrizable. □

The next simple fact can be formulated in a more general form, for continuous mappings of topological spaces. However, the formulation that follows completely suits for our purpose.

LEMMA 4.6.2. *Let $H$ be a closed subgroup of a topological group $G$. If both the group $H$ and the quotient space $G/H$ have countable pseudocharacter, then so has $G$.*

PROOF. Let $p\colon G \to G/H$ be the quotient mapping. Choose a countable family $\gamma$ of open sets in $G/H$ such that $\bigcap \gamma = \{p(e)\}$, where $e$ is the neutral element of $G$. Also, choose a countable family $\lambda$ of open sets in $G$ such that $H \cap \bigcap \lambda = \{e\}$. Then the family $\mathcal{F} = \lambda \cup \{p^{-1}(U) : U \in \gamma\}$ is countable, consists of open sets in $G$, and $\bigcap \mathcal{F} = \{e\}$. Therefore, the pseudocharacter of $G$ at $e$ and, by homogeneity, at every point $x \in G$ is countable. □

The argument in the proof of Proposition 4.6.1, after an obvious modification, can be turned into a proof of the following statement.

THEOREM 4.6.3. *Let $G$ be a topological group and $H$ a closed invariant subgroup of $G$ such that $H$ has a countably tight compact grasp on $G$ and the pseudocharacter of $H$ is countable. Then the pseudocharacter of $G$ is also countable.*

PROOF. As in the proof of Proposition 4.6.1, we conclude that the group $G/H$ is metrizable. Since the character and pseudocharacter of metrizable spaces are countable, the conclusion follows from Lemma 4.6.2. □

THEOREM 4.6.4. *Suppose that $G$ is a topological group, and let $H$ be a closed subgroup of $G$ such that the pseudocharacter of $H$ is countable and $H$ has a cosmic grasp on $G$. Then the pseudocharacter of $G$ is also countable.*

PROOF. There exists a subspace $B$ of $G$ such that $HB = G$ and $B$ has a countable network. Let us consider the quotient space $G/H$ and the corresponding quotient mapping $p\colon G \to G/H$. Since $HB = G$, we have $p(B) = G/H$. Therefore, $G/H$ has a countable network. It follows that the pseudocharacter of $G/H$ is countable. Hence, the pseudocharacter of $G$ is countable by Lemma 4.6.2. □

The following result has many applications.

PROPOSITION 4.6.5. *Suppose that $F$ is a compact subspace of a topological group $G$. Then the restriction $f$ of the product mapping $G \times G \to G$ to the subspace $F \times G$ is a perfect and open mapping of $F \times G$ onto $G$. The same is valid for the restriction of the product mapping to the subspace $G \times F$ of $G \times G$.*

PROOF. Consider a mapping $s\colon F \times G \to F \times G$ defined by $s(x, y) = (x, xy)$, for each $(x, y) \in F \times G$. Obviously, $s$ is continuous, one-to-one, and $s(F \times G) = F \times G$. Clearly, the inverse mapping $s^{-1}$ is described by the formula $s^{-1}(x, y) = (x, x^{-1}y)$. Therefore, $s^{-1}$ is also continuous. Hence, $s$ is a homeomorphism. Let $p\colon F \times G \to G$ be the natural projection mapping given by $p(x, z) = z$, for each $(x, z) \in F \times G$. Since $xy = p(x, xy) = ps(x, y)$ for all $x \in F$ and $y \in G$, we conclude that $f$ is the composition of $s$ and $p$, that is, $f = p \circ s$. However, since $F$ is compact, the mapping $p$ is closed by [**165**, Theorem 3.1.16]. Therefore, the mapping $f$, being a composition of a homeomorphism with a closed mapping, is itself closed. Clearly, $f$ has compact fibers, so it is perfect.

Let $U$ be an open subset of $F \times G$. Denote by $\pi$ the projection of $F \times G$ onto the first factor and put $O = \pi(U)$. For every $x \in O$, the set $U_x = \{y \in G : (x, y) \in U\}$ is open in $G$ as the projection of the open subset $U \cap \pi^{-1}(x)$ of $\{x\} \times G$ onto the second factor. Therefore, $f(U) = \bigcup_{x \in O} xU_x$ is open in $G$, which implies that $f$ is an open mapping.

Finally, since the mapping $i$ of $G \times G$ onto itself, defined by $i(x, y) = (y, x)$ for $x, y \in G$, is a homeomorphism, and $i(F \times G) = G \times F$, the rest of the proposition is immediate. □

The mapping $f$ in the above proposition remains open for an arbitrary subspace $F$ of $G$.

COROLLARY 4.6.6. *Suppose that $F$ is a compact subspace of a topological group $G$, and let $M$ be a closed subspace of $G$. Then the restriction $f$ of the product mapping $G \times G \to G$ to the subspace $M \times F$ is a perfect mapping of $M \times F$ onto a closed subspace of $G$.*

PROOF. To derive this corollary from Proposition 4.6.5, we only have to observe that the restriction of a perfect mapping to a closed subspace is again a perfect mapping and that the product of a compact set with a closed set in a topological group is closed by Theorem 1.4.30. □

Note that the perfect mapping of $M \times F$ onto $MF$ in Corollary 4.6.6 need not be open. In the next example we show that the assumption that $M$ be closed in $G$ cannot be dropped either in Corollary 4.6.6 or in Proposition 4.6.5.

EXAMPLE 4.6.7. Take any compact metrizable group $G$ with a proper dense subgroup $M$ (the group $\mathbb{T}$ and the torsion subgroup of $\mathbb{T}$ make the job). Then the natural mapping $f$ of the space $G \times M$ onto $G$ generated by the product operation is not closed. Indeed, each fiber of $f$ is nowhere dense in $G \times M$, so if $f$ were closed, it would have compact fibers, by Vaĭnšteĭn's lemma [**165**, Lemma 4.4.16], and be perfect. Since $G$ is compact, this would imply that $G \times M$ is compact, which is a contradiction. □

Corollary 4.6.6 is one of the main technical tools in this section. Of course, it works in combination with other results of the theory of perfect mappings, some of which are quite deep.

THEOREM 4.6.8. *Suppose that $F$ is a compact subspace of a topological group $G$, and that $M$ a closed subspace of $G$. Suppose also that both $M$ and $F$ have countable tightness. Then the tightness of $MF$ is also countable. If, in addition, $G = FM$, then the tightness of $G$ is countable.*

PROOF. Perfect mappings do not increase the tightness and the tightness of the product space $M \times F$ is countable by [**165**, 3.12.8(a), (f)]. Therefore, it follows from Corollary 4.6.6 that the tightness of $MF$ is also countable. □

THEOREM 4.6.9. *Suppose that $F$ is a compact metrizable subspace of a topological group $G$, and let $M$ be a closed metrizable subspace of $G$. Then $FM$ is closed in $G$ and metrizable.*

PROOF. By Theorem 1.4.30, the set $FM$ is closed in $G$. It remains to apply Corollary 4.6.6 and the well-known fact that perfect mappings preserve metrizability in the direction of the image (see [**165**, Theorem 4.4.15]). □

The above techniques can be also applied to the study of the topological structure of a topological group under certain weaker assumptions than in Theorem 4.6.9. For example, we have:

THEOREM 4.6.10. *Let $G$ be a topological group, and let $H$ be a subgroup of $G$ algebraically generated by a compact metrizable subspace $F$. Suppose further that $G = HM$, where $M$ is a closed metrizable subspace of $G$. Then $G$ is the union of a countable family of closed metrizable subspaces.*

PROOF. Clearly, we can assume that $F$ is symmetric, $F = F^{-1}$, and that $F$ contains the neutral element of $G$. Now we inductively define a sequence of subspaces $M_n$ of $G$ by the rule: $M_0 = M$, and $M_{n+1} = FM_n$, for each $n \in \omega$. Since $F$ generates $H$ and $G = HM$, we have the equality $G = \bigcup_{n=0}^{\infty} M_n$. Theorem 4.6.9 guarantees that each $M_n$ is a closed metrizable subspace of $G$. □

THEOREM 4.6.11. *Suppose that $F$ is a compact subspace of a topological group $G$, and let $M$ be a paracompact, closed subspace of $G$. Then $FM$ is also a paracompact space. In particular, if $G = FM$, then $G$ is paracompact.*

PROOF. Since the product of a paracompact space with a compact space is paracompact by [**165**, Theorem 5.1.36], the conclusion follows from Corollary 4.6.6 and the well-known fact that paracompactness is preserved by closed mappings (see [**165**, Theorem 5.1.33]). □

The next result is somewhat unexpected: it shows that certain closed subsets of a topological group with a compact grasp on this group must be paracompact.

THEOREM 4.6.12. *Suppose that $G$ is a topological group such that $G = FM$, where $F \subset G$ is compact and $M$ is a Čech-complete closed subspace of $G$. Then $G$ is also Čech-complete, and both $M$ and $G$ are paracompact.*

PROOF. The product of a Čech-complete space and a compact space is Čech-complete. Therefore, the space $M \times F$ is Čech-complete. Since Čech-completeness is preserved by perfect mappings, by [**165**, Theorem 3.9.10], it follows from Corollary 4.6.6 that $G$ is Čech-complete. Hence, $G$ is paracompact by Corollary 4.3.22. Since $M$ is a closed subspace of $G$, the space $M$ is also paracompact. □

Now we are going to generalize Proposition 4.6.5. A subset $A$ of a topological space $X$ is said to be *$G_\delta$-closed* in $X$ if, for each $x \in X \setminus A$, there exists a $G_\delta$-set $P$ in $X$ such that $x \in P \subset X \setminus A$. Of course, in a space of countable pseudocharacter every subset is $G_\delta$-closed. On the other hand, a Lindelöf subspace is $G_\delta$-closed in every larger Hausdorff space, by [**165**, 3.12.24 (a)].

A mapping $f$ of a space $X$ to a space $Y$ will be called *$G_\delta$-closed* if, for every closed subset $P$ of $X$, the image $f(P)$ is $G_\delta$-closed in $Y$. Clearly, every closed mapping is $G_\delta$-closed. We also have:

LEMMA 4.6.13. *Every continuous mapping $f$ of a Lindelöf space $X$ to a Hausdorff space $Y$ is $G_\delta$-closed.*

PROOF. This is so, since all closed subsets of a Lindelöf space are Lindelöf and every Lindelöf subspace of a Hausdorff space is $G_\delta$-closed in that space [**165**, 3.12.24]. □

The next result is a kind of Kuratowski's theorem (see [**165**, Theorem 3.1.16]). Its proof is close to that of Lemma 4.4.8:

PROPOSITION 4.6.14. *Let $p\colon X \times Y \to Y$ be the natural projection, where $X$ is a Lindelöf space. Then $p$ is a $G_\delta$-closed mapping.*

PROOF. Take a closed subset $F$ of $X \times Y$ and an arbitrary point $y \in Y \setminus p(F)$. We have to find a $G_\delta$-set $P$ in $Y$ such that $y \in P \subset Y \setminus p(F)$. It is clear that the fiber $p^{-1}(y) = X \times \{y\}$ is homeomorphic to $X$ and, hence, is Lindelöf. Since $p^{-1}(y)$ is disjoint from $F$, we can cover $p^{-1}(y)$ by open in $X \times Y$ rectangular sets of the form $U \times V$, where each of them is disjoint from $F$ and $y \in V$. The fiber $p^{-1}(y)$ being Lindelöf, there exists a countable covering $\gamma = \{U_n \times V_n : n \in \omega\}$ of $p^{-1}(y)$ by such sets. Then $P = \bigcap_{n=0}^{\infty} V_n$ is a $G_\delta$-set in $Y$, it contains the point $y$ and does not intersect $p(F)$. □

THEOREM 4.6.15. *Suppose that $F$ is a Lindelöf subspace of a topological group $G$. Then the restriction $f$ of the product mapping $G \times G \to G$ to the subspace $F \times G$ is a $G_\delta$-closed mapping of $F \times G$ onto $G$.*

PROOF. As in the proof of Proposition 4.6.5, consider the mapping $s\colon F \times G \to F \times G$ defined by $s(x, y) = (x, xy)$, for each $(x, y) \in F \times G$. Again, $s$ is a homeomorphism. Since $xy = p(x, xy) = ps(x, y)$, where $p\colon F \times G \to G$ is the projection onto the second factor, we conclude that $f = p \circ s$. However, since $F$ is Lindelöf, the mapping $p$ is $G_\delta$-closed by Proposition 4.6.14. Therefore, the mapping $f$ being a composition of a homeomorphism with a $G_\delta$-closed mapping, is itself $G_\delta$-closed. □

Here we give just one application of Theorem 4.6.15 which requires the following auxiliary fact:

LEMMA 4.6.16. *Let $X$ and $Y$ be regular $P$-spaces, where $X$ is Lindelöf and $Y$ is paracompact. Then the product space $X \times Y$ is also paracompact.*

PROOF. Evidently, $X \times Y$ is a regular $P$-space and, hence, is zero-dimensional. Consider an open covering $\gamma$ of $X \times Y$. Let $p\colon X \times Y \to Y$ be the natural projection. The fiber $p^{-1}(y) = X \times \{y\}$ is Lindelöf for each $y \in Y$, so there exists a countable subfamily $\gamma(y)$ of $\gamma$ that covers $p^{-1}(y)$. Since $X$ is zero-dimensional and Lindelöf, we can find a countable disjoint family $\gamma^*(y)$ of open rectangular sets in $X \times Y$ which covers $p^{-1}(y)$ and such that every element of $\gamma^*(y)$ intersects $p^{-1}(y)$ and is contained in some element of $\gamma(y)$.

By Lemma 4.4.8, the mapping $p$ is closed. Therefore, for every $y \in Y$, there exists an open neighbourhood $V_y$ of $y$ in $Y$ such that $p^{-1}(V_y) \subset \bigcup \gamma^*(y)$. Since $Y$ is paracompact, the open covering $\{V_y : y \in Y\}$ of $Y$ admits an open locally finite refinement $\lambda$. It follows from our choice of $\lambda$ that for every $V \in \lambda$, there exists a point $y_V \in Y$ such that $p^{-1}(V) \subset \bigcup \gamma^*(y_V)$. Then the family

$$\mu = \{p^{-1}(V) \cap U : V \in \lambda,\ U \in \gamma^*(y_V)\}$$

is an open covering of the space $X \times Y$ which refines $\gamma$. Since the family $\lambda$ is locally finite and $\gamma^*(y_V)$ is disjoint for each $V \in \lambda$, we conclude that the covering $\mu$ is locally finite. This implies that the space $X \times Y$ is paracompact. □

THEOREM 4.6.17. *Suppose that $G$ is a topological $P$-group, $F$ is a Lindelöf subspace of $G$ and let $M$ be a paracompact closed subspace of $G$. Then $FM$ is a paracompact closed subspace of $G$. If, in addition, $G = FM$ then $G$ is paracompact.*

PROOF. It follows from Lemma 4.6.16 that the product space $F \times M$ is paracompact. The mapping $f$ of $F \times G$ onto $G$ given by the product operation, is continuous and $G_\delta$-closed, by Theorem 4.6.15. Since $G$ is a $P$-space, it follows that $f$ is closed. Since $F \times M$ is closed in $F \times G$, the restriction of $f$ to $F \times M$ is a closed mapping of $F \times M$ to $G$. Hence, by a well-known theorem of E. A. Michael (see [**165**, Theorem 5.1.33]), $FM$ is paracompact and closed in $G$. □

Here is one more modification of Proposition 4.6.5. It can be proved by almost the same argument as Proposition 4.6.5 and Theorem 4.6.15, we just have to refer to [**165**, Theorem 3.10.7].

THEOREM 4.6.18. *Let $G$ be a sequential topological group, and let $F$ be a countably compact subspace of $G$. Then the restriction of the product mapping $G \times G \to G$ to the subspace $G \times F$ is a closed continuous mapping of $G \times F$ onto $G$.*

Recall that a mapping $f\colon X \to Y$ is *locally perfect* if, for each $x \in X$, there exists an open neighbourhood $U$ of $x$ such that the restriction of $f$ to the closure of $U$ is a perfect mapping of $\overline{U}$ to $Y$. The following generalization of Proposition 4.6.5 is sometimes useful.

THEOREM 4.6.19. *Let $Y$ be a locally compact subspace of a topological group $G$. Then the restriction of the multiplication mapping $G \times G \to G$ to the subspace $G \times Y$ is an open locally perfect mapping of $G \times Y$ onto $G$.*

PROOF. This follows from Proposition 4.6.5. □

After the results obtained above, and taking into account that every locally compact topological group is feathered, it is natural to pose the following question. Suppose that $G$ is a topological group, and $H$ a closed invariant subgroup of $G$ such that both $H$ and the quotient group $G/H$ are feathered. Must then $G$ be feathered as well? In Section 7.3 we show that the answer is negative. Notice that the similar question for Čech-completeness is answered in the affirmative in 4.3.F.

Another natural general question for consideration is the following one. Suppose that $H$ is a closed subgroup of a topological group $G$. When does $H$ have a compact grasp on $G$? Clearly, a necessary condition for this is compactness of the quotient space $G/H$. This condition is not sufficient (see Theorem 7.3.1). We are going to answer the above question completely in the case when the subgroup $H$ is locally compact.

A mapping $f\colon X \to Y$ is said to be *compact-covering* or *$k$-covering* if, for each compact subset $F$ of $Y$, there exists a compact subset $K$ of $X$ such that $f(K) = F$ (see [**165**, 5.5.11]). The relevance of this notion to our considerations is revealed by the following statement the proof of which is obvious.

PROPOSITION 4.6.20. *Suppose that $G$ is a topological group, and let $H$ be a closed subgroup of $G$. Then $H$ has a compact grasp on $G$ if and only if the quotient space $G/H$ is compact and the quotient mapping $\pi\colon G \to G/H$ is compact-covering.*

One can consider Proposition 4.6.20 as an answer to the question formulated above. However, then the next question comes: When is the quotient mapping $\pi\colon G \to G/H$ compact-covering? The following result helps us to answer it in the special case when the subgroup $H$ of $G$ is locally compact.

PROPOSITION 4.6.21. *Every open locally perfect mapping $f$ of a space $X$ onto a space $Y$ is compact-covering.*

PROOF. Let $F$ be any compact subspace of $Y$. For each $x \in X$, fix an open neighbourhood $V_x$ of $x$ in $X$ such that the restriction of $f$ to the closure of $V_x$ is a perfect mapping of $\overline{V_x}$ to $Y$.

Since $f$ is open and $f(X) = Y$, the family $\gamma = \{f(V_x) : x \in X\}$ is an open covering of $Y$. Since $F$ is compact and $F \subset \bigcup \gamma$, there exists a finite subset $K$ of $X$ such that $F \subset \bigcup\{f(V_x) : x \in K\}$. Put $F_x = F \cap f(\overline{V_x})$ for $x \in K$. Observe that $f(\overline{V_x})$ is closed in $Y$, since the restriction $f_x$ of $f$ to $\overline{V_x}$ is a perfect mapping to $Y$. Hence, $F_x$ is compact, for each $x \in K$. From the fact that $f_x$ is perfect it follows also that $P_x = \overline{V_x} \cap f^{-1}(F_x) = f_x^{-1}(F_x)$

is a compact subset of $X$. Obviously, $f(P_x) = F_x$. Put $P = \bigcup\{P_x : x \in K\}$. Then $P$ is compact, and $f(P) = F$. □

THEOREM 4.6.22. *Suppose that $G$ is a topological group, and let $H$ be a locally compact subgroup of $G$. Then the quotient mapping $\pi: G \to G/H$ is compact-covering.*

PROOF. By Theorem 3.2.2, the quotient mapping $\pi$ is locally perfect. Since the mapping $\pi$ is open, the required conclusion follows from Proposition 4.6.21. □

Let us call a mapping $f: X \to Y$ *Lindelöf-covering* or *l-covering* if, for each Lindelöf subspace $M$ of $Y$, there exists a Lindelöf subspace $L$ of $X$ such that $f(L) = M$. Introducing obvious changes in the proofs of the last two statements, we obtain the following results.

PROPOSITION 4.6.23. *Every open locally perfect mapping $f$ of a topological space $X$ onto a topological space $Y$ is Lindelöf-covering.*

THEOREM 4.6.24. *Suppose that $H$ is a locally compact subgroup of a topological group $G$. Then the quotient mapping $\pi: G \to G/H$ is Lindelöf-covering.*

Theorem 4.6.22 allows us to clarify completely when a locally compact subgroup $H$ of a topological group $G$ has a compact grasp on $G$:

THEOREM 4.6.25. *A locally compact subgroup $H$ of a topological group $G$ has a compact grasp on $G$ if and only if the quotient space $G/H$ is compact.*

PROOF. Assume that $G = FH$, where $F$ is compact. Then $G/H = \pi(G) = \pi(F)$, where $\pi$ is the quotient mapping of $G$ onto $G/H$. Therefore, by the continuity of $\pi$, the space $G/H$ is compact. The inverse statement follows from Theorem 4.6.22. □

Similarly, we have:

THEOREM 4.6.26. *A locally compact subgroup $H$ of a topological group $G$ has a Lindelöf grasp on $G$ if and only if the quotient space $G/H$ is Lindelöf.*

PROOF. The proof is the same as that of Theorem 4.6.25; we just refer to Theorem 4.6.24. □

In connection with Theorem 4.6.22, note that the quotient mapping of a topological group $G$ onto a quotient group $G/H$ need not be compact-covering, even under very strong restrictions on $G/H$ and $H$. Indeed, as we have already mentioned, there exists an Abelian topological group $G$ which is not a paracompact $p$-space but has a closed metrizable subgroup$H$ such that the quotient group $G/H$ is compact (see Theorem 7.3.1). In this case the natural quotient mapping $\pi: G \to G/H$ is not compact-covering.

## Exercises

4.6.a. Consider a separable topological group $G$ metrizable by a complete metric, and let $A$ be a discrete subspace of $G$. Show that $A$ does not have a locally compact grasp on $G$.

4.6.b. Let $A$ be any metrizable subspace of the topological group $G = \mathbb{R}^{\omega_1}$. Show that $A$ does not have a compact grasp on $G$. Verify that the same assertion remains valid if one replaces $G$ with any $\Sigma$-product lying in $\mathbb{R}^{\omega_1}$.

4.6.c. Suppose that a topological group $G$ contains a closed invariant metrizable subgroup $H$ such that $H$ has a closed discrete grasp on $G$. Is $G$ metrizable? What if $H$ is also second-countable?

4.6.d. Let $H$ be a closed, invariant, Čech-complete subgroup of a topological group $G$, and suppose that $H$ has a Čech-complete grasp on $G$. Is $G$ Čech-complete?

## Problems

4.6.A. Let $F$ be a compact subspace of a topological group $G$. Suppose further that $M$ is a closed metrizable subspace of $G$. Then $FM$ and $MF$ are paracompact $p$-spaces. In particular, if $G = FM$ or $G = MF$, then $G$ is feathered.
*Sketch of the proof.* The product $M \times F$ is a paracompact $p$-space by [**16**], and $G$ is an image of $M \times F$ under a perfect mapping, by Corollary 4.6.6. It follows from a theorem of V. V. Filippov in [**169**] that $MF$ is a paracompact $p$-space as well.

4.6.B. Suppose that $G$ is a topological group such that $G = FM$, where $F \subset G$ is a compact subspace of countable tightness, and $M$ is a closed metrizable subspace of $G$. Then $G$ is metrizable.

*Sketch of the proof.* By Problem 4.6.A, the group $G$ is feathered. It follows from Theorem 4.6.8 that the tightness of $G$ is countable. However, every feathered topological group of countable tightness is metrizable. Indeed, every such a group $G$ must contain a compact subgroup $H$ with a countable base of neighbourhoods. Then $H$ is metrizable, since every compact group of countable tightness is metrizable by Corollary 4.2.2. It follows that both spaces $H$ and $G/H$ are first-countable and, therefore, $G$ is metrizable by Corollary 1.5.21.

4.6.C. Let $G$ be a topological group and $H$ a closed subgroup of $G$ such that $w(H) \leq \omega$ and $nw(G/H) \leq \omega$. Prove that $nw(G) \leq \omega$.
*Hint.* Let $\{U_n : n \in \omega\}$ be a sequence of open symmetric neighbourhoods of the neutral element $e$ in $G$ such that $U_{n+1}^2 \subset U_n$ for each $n \in \omega$ and the family $\{U_n \cap H : n \in \omega\}$ is a local base for $H$ at $e$. Choose a countable network $\{P_k : k \in \omega\}$ for the quotient space $G/H$. Apply Theorem 1.5.23 to find a countable dense subset $\{b_m : m \in \omega\}$ of $G$. Then the family $\{\pi^{-1}(P_k) \cap b_m U_n : k, m, n \in \omega\}$ is a network for $G$, where $\pi\colon G \to G/H$ is the quotient mapping.

4.6.D. Let $H$ be a closed subgroup of an Abelian topological group $G$, and suppose that $nw(H) \leq \omega$ and $w(G/H) \leq \omega$. Is it true that $nw(G) \leq \omega$?
*Hint.* The answer is "no" (see [**511**]).

4.6.E. Let $G$ be a topological group, and $X$ a locally compact subspace of $G$ with a locally compact grasp on $G$. Must $G$ be locally compact? What if, in addition, $X$ is closed in $G$?

## Open Problems

4.6.1. Suppose that $G$ is a topological group such that $G = FM$, where $F$ is a compact set of countable tightness and that $M$ is a first-countable closed subspace of $G$. Is $G$ metrizable?

4.6.2. Is there a discrete subspace $A$ of the topological group $G = \mathbb{R}^{\omega_1}$ that has a countably compact grasp on $G$?

4.6.3. Is there a discrete subspace $A$ of the group $G = \mathbb{R}^{\omega_1}$ that has a discrete grasp on $G$?

4.6.4. Is there a discrete subspace $A$ of the group $G = \mathbb{R}^{\omega_1}$ that has a locally compact grasp on $G$?

4.6.5. Is there a discrete (metrizable, locally compact) subspace $A$ of the group $G = \mathbb{R}^{\omega_1}$ such that $A$ algebraically generates $G$?

4.6.6. Is there a locally compact (Čech-complete) subspace $A$ of the group $G = \mathbb{R}^{\omega_1}$ that has a locally compact (Čech-complete) grasp on $G$?

4.6.7. Is there a metrizable subspace $A$ of the group $G = \mathbb{R}^{\omega_1}$ that has a metrizable grasp on $G$?

4.6.8. Consider problems similar to the last three problems, where $G$ is the $\Sigma$-product of $\omega_1$ copies of $\mathbb{R}$.

4.6.9. Which theorems in this section remain true for paratopological groups? For semitopological (for quasitopological) groups?

## 4.7. Some convergence phenomena in topological groups

It is natural and quite plausible to expect that certain topological properties of spaces should become stronger in the presence of an appropriately related algebraic structure. The most obvious example is that of first countability which becomes equivalent to metrizability in topological groups. We are now going to strengthen this result.

Let $X$ be a topological space. Suppose that it is possible to assign to each point $x \in X$ a sequence $\{V_n(x) : n \in \omega\}$ of subsets of $X$ containing $x$ in such a way that $V_{n+1}(x) \subset V_n(x)$ and the following condition is satisfied: A subset $U$ of $X$ is open in $X$ if and only if for each $x \in U$, there exists $n \in \omega$ such that $V_n(x) \subset U$. Then we will say that $\{V_n(x) : n \in \omega,\ x \in X\}$ is a *weak $\omega$-assignment* on $X$. Notice that the elements $V_n(x)$ of a weak $\omega$-assignment on $X$ are not necessarily open in $X$.

A topological space $X$ is called *weakly first-countable* if there exists a weak $\omega$-assignment on $X$. It is easy to construct weakly first-countable Tychonoff spaces which are not first-countable (see Problem 113 in Chapter II of [**60**]). In addition, (weakly) first-countable compact spaces need not be metrizable, as the two arrows space shows. However, the situation in the class of topological groups is quite different. To show this, we start with a study of weak $\omega$-assignments on spaces.

LEMMA 4.7.1. *Suppose that $\{V_n(x) : n \in \omega,\ x \in X\}$ and $\{W_n(x) : n \in \omega,\ x \in X\}$ are two weak $\omega$-assignments on a Hausdorff space $X$. Then, for each $x \in X$ and each $n \in \omega$, there exists $m \in \omega$ such that $W_m(x) \subset V_n(x)$.*

PROOF. Assume the contrary. Then, for some $x \in X$ and some $k \in \omega$, we have $W_m(x) \setminus V_k(x) \neq \varnothing$ for each $m \in \omega$, and we fix $x_m \in W_m(x) \setminus V_k(x)$. Clearly, the sequence $\eta = \{x_m : m \in \omega\{\}$ converges to $x$. Since $X$ is Hausdorff, the set $P = \{x_m : m \in \omega\} \cup \{x\}$ is closed in $X$. Let $P_0 = P \setminus \{x\} = \{x_m : m \in \omega\}$. Since $\{V_n(x) : n \in \omega,\ x \in X\}$ is a weak $\omega$-assignment on $X$, for each $y \in X \setminus P$ there exists $n(y) \in \omega$ such that $V_{n(y)}(y) \cap P = V_{n(y)}(y) \cap P_0 = \varnothing$. We also have $P_0 \cap V_k(x) = \varnothing$. Since $\{V_n(x) : n \in \omega,\ x \in X\}$ is a weak $\omega$-assignment on $X$, it follows that the set $P_0$ is closed in $X$, a contradiction with $x \in \overline{P_0} \setminus P_0$. □

Sequences $\{A_n : n \in \omega\}$ and $\{B_n : n \in \omega\}$ of sets will be called *cofinal* if for each $n \in \omega$, there exist $m, k \in \omega$ such that $B_m \subset A_n$ and $A_k \subset B_n$. Though the next proposition sounds a little bit unexpected, it is an easy corollary of Lemma 4.7.1.

PROPOSITION 4.7.2. *Let $\{V_n(x) : n \in \omega,\ x \in X\}$ be a weak $\omega$-assignment on a homogeneous Hausdorff space $X$, and let $b$ be an element of $X$. Suppose further that, for each $x \in X$, $f_x$ is a homeomorphism of $X$ onto $X$ such that $f_x(b) = x$. Put $W_n(x) = f_x(V_n(b))$. Then $\{W_n(x) : n \in \omega,\ x \in X\}$ is a weak $\omega$-assignment on $X$.*

PROOF. Fix $x \in X$, and put $U_n(y) = f_x(V_n(f_x^{-1}(y))$, for each $y \in X$. Since $f_x$ is a homeomorphism of $X$ onto $X$, it is clear that $\{U_n(y) : n \in \omega,\ y \in X\}$ is a weak $\omega$-assignment on $X$. Therefore, the sequences $\{U_n(x) : n \in \omega\}$ and $\{V_n(x) : n \in \omega\}$ are cofinal. However, $U_n(x) = W_n(x)$, for each $n \in \omega$. Thus, by Lemma 4.7.1, the sequences

$\{W_n(x) : n \in \omega\}$ and $\{V_n(x) : n \in \omega\}$ are cofinal, for each $n \in \omega$. It follows that $\{W_n(x) : n \in \omega,\ x \in X\}$ is a weak $\omega$-assignment on $X$. □

The next lemma plays the key role in the proof of the main result of this section, Theorem 4.7.5.

LEMMA 4.7.3. *Suppose that $\{V_n(x) : n \in \omega,\ x \in G\}$ is a weak $\omega$-assignment on a paratopological group $G$. Put, for each $x \in G$ and each $n \in \omega$, $W_n(x) = x(V_n(e))^2$, where $e$ is the neutral element of $G$. Then $\{W_n(x) : n \in \omega,\ x \in X\}$ is a weak $\omega$-assignment on $G$.*

PROOF. We can assume that $V_n(x) = xV_n(e)$, for each $x \in G$ and each $n \in \omega$. Indeed, this follows from Proposition 4.7.2 since the left translation $l_x$ on $G$ given by the formula $l_x(y) = xy$, for each $y \in G$, is a homeomorphism of $G$ onto itself. Now the continuity of multiplication in $G$ implies that $\{W_n(x) : n \in \omega,\ x \in X\}$ is a weak $\omega$-assignment on $G$. □

PROPOSITION 4.7.4. *Suppose that $\{V_n(x) : n \in \omega,\ x \in G\}$ is a weak $\omega$-assignment on a paratopological group $G$. Then, for each $n \in \omega$, there exists $k \in \omega$ such that $(V_k(e))^2 \subset V_n(e)$.*

PROOF. This follows from Lemma 4.7.3 and Lemma 4.7.1. □

Now we are ready to prove the following basic fact:

THEOREM 4.7.5. [**P. J. Nyikos**] *Every weakly first-countable Hausdorff paratopological group is first-countable.*

PROOF. Let $\{V_n(x) : n \in \omega,\ x \in G\}$ be a weak $\omega$-assignment on a Hausdorff paratopological group $G$. By Proposition 4.7.2, we can assume that $V_n(x) = xV_n(e)$, for each $x \in G$ and each $n \in \omega$.

Let us show that, for each $n \in \omega$, $V_n(e)$ contains an open neighbourhood of $e$ in $G$. Let $U_n$ be the set of all points $x \in V_n(e)$ such that $xV_k(e) \subset V_n(e)$, for some $k \in \omega$. Clearly, $e \in U_n \subset V_n(e)$. We claim that the set $U_n$ is open. Indeed, take any $y \in U_n$. Then $yV_k(e) \subset V_n(e)$, for some $k \in \omega$. By Proposition 4.7.4, there is $m \in \omega$ such that $(V_m(e))^2 \subset V_k(e)$. Then $(yV_m(e))V_m(e) \subset yV_k(e) \subset V_n(e)$, which implies that $V_m(y) = yV_m(e) \subset U_n$. Since $y$ was an arbitrary point of $U_n$, it follows that $U_n$ is open in $G$. Now it is clear that $\{U_n : n \in \omega\}$ is a countable base of $G$ at $e$. □

COROLLARY 4.7.6. [**M. M. Choban and S. J. Nedev**] *Every weakly first-countable topological group $G$ is metrizable.*

PROOF. This follows from Theorem 4.7.5, since every first-countable topological group is metrizable, by the Birkhoff–Kakutani theorem (see Theorem 3.3.12). □

Our next aim is to prove an interesting fact, Theorem 4.7.8, that complements the results of Section 2.3. Recall that a topological space is said to be *symmetrizable* if its topology is generated by a *symmetric*, that is, by a distance function satisfying all the usual restrictions on a metric, except for the triangle inequality. In other words, if $d$ is a symmetric on a space $X$ that generates the topology of $X$, then a subset $U$ of $X$ is open iff for every $x \in U$, there exists $\varepsilon > 0$ such that the $\varepsilon$-ball $B_\varepsilon(x) = \{y \in X : d(x, y) < \varepsilon\}$ is contained in $U$. It is clear that every symmetrizable space is weakly first-countable, but not vice versa (the Sorgenfrey line is a counterexample).

PROPOSITION 4.7.7. *Let $d$ be a symmetric on a first-countable Hausdorff space $X$ that generates the topology of $X$. Then, for every $x \in X$ and every $\varepsilon > 0$, the point $x$ is in the interior of the set $B_\varepsilon(x) = \{y \in X : d(x, y) < \varepsilon\}$.*

PROOF. Given a point $x \in X$ and a non-empty set $A \subset X$, we put

$$d(x, A) = \inf\{d(x, y) : y \in A\}.$$

Let us establish the following.

**Claim 1.** *If a sequence $S = \{x_n : n \in \omega\}$ of pairwise distinct points of $X$ converges to a point $a \in X$, then $d(x, x_n)$ tends to zero when $n \to \infty$ and, therefore, $d(a, S) = 0$.*

Suppose to the contrary that there exist $\delta > 0$ and an infinite set $M \subset \omega$ such that $d(a, x_n) \geq \delta$, for each $n \in M$. Then $x_n \neq a$, for each $n \in M$. Hence, to obtain a contradiction, it suffices to show that the set $S' = \{x_n : n \in M\}$ is closed in $X$. Clearly, $T = S' \cup \{a\}$ is a compact subset of $X$, so that $T$ is closed in $X$. Hence, for every $x \in X \setminus T$, there exists $\varepsilon > 0$ such that $d(x, T) \geq \varepsilon$. In particular, $d(x, S') \geq \varepsilon$. We also have that $d(a, S') \geq \delta > 0$. Since $d$ generates the topology of $X$, this implies that the set $U = X \setminus S'$ is open, i.e., $S'$ is closed in $X$. Claim 1 is proved.

Assume that there are a point $x \in X$ and $\varepsilon > 0$ such that $x$ is not in the interior of $B_\varepsilon(x)$. Then $x \in \overline{X \setminus B_\varepsilon(x)}$. Since $X$ is first-countable, we can find a sequence $\{x_n : n \in \omega\}$ of pairwise distinct points of $X \setminus B_\varepsilon(x)$ converging to $x$. It follows from Claim 1 that $d(x, x_n) < \varepsilon$ for some $n \in \omega$ and, hence, $x_n \in B_\varepsilon(x)$. This contradiction completes the proof. □

THEOREM 4.7.8. *Every symmetrizable Hausdorff paratopological group $G$ with the Baire property is a metrizable topological group.*

PROOF. We fix a symmetric $d$ on the space $G$ generating the topology of $G$. Let $U$ be any open neighbourhood of the neutral element $e$ of $G$. For each positive integer $n$ and each $g \in G$, put $B_n(g) = \{x \in G : d(g, x) < 1/n\}$. Since every symmetrizable space is weakly first-countable and, by Theorem 4.7.5, every weakly first-countable Hausdorff paratopological group is first-countable, it follows from Proposition 4.7.7 that $g \in \operatorname{Int}(B_n(g))$, for each $n \in \mathbb{N}$ and each $g \in G$. Let $A_n = \{g \in G : B_n(g) \subset gU\}$. Clearly, $G = \bigcup_{n=1}^{\infty} A_n$ and $A_i \subset A_j$ whenever $i < j$. Since $G$ has the Baire property, there exist $k, n \in \mathbb{N}$ and $h \in G$ such that $k \leq n$ and the set $V = \overline{B_n(h)}$ is contained in $\overline{A_k}$. Then $h \in \overline{A_k}$ and, since $V$ is a neighbourhood of $h$, we have $h \in \overline{A_k \cap V}$. Take any $v \in V \cap A_k$. Then $B_n(v) \subset B_k(v) \subset vU$. However, $h \in B_n(v)$, since $v \in V = B_n(h)$. Hence, $h \in vU$, for each $v \in V \cap A_k$. It follows that $h^{-1}v \in U^{-1}$, for each $v \in V \cap A_k$, that is, $h^{-1}(V \cap A_k) \subset U^{-1}$. Let $V_0 = \operatorname{Int}(V)$. Clearly, $V_0 \cap A_k$ is dense in $V_0$, and the multiplication by $h^{-1}$ is continuous. It follows that $h^{-1}V_0 \subset \overline{U^{-1}}$. Since $h \in V_0$, we also have $e \in h^{-1}V_0$. Therefore, since $h^{-1}V_0$ is an open set, we obtain $e \in \operatorname{Int}(\overline{U^{-1}})$. Now it follows from Corollary 2.3.20 that $G$ is a topological group. □

A curious metamorphosis occurs to the Fréchet–Urysohn property in topological groups. A space $X$ is said to be *strongly Fréchet–Urysohn* if the following condition is satisfied:

(SFU) *For each $x \in X$ and every sequence $\xi = \{A_n : n \in \omega\}$ of subsets of $X$ such that $x \in \bigcap_{n\in\omega} \overline{A_n}$, there exists a sequence $\eta = \{b_n : n \in \omega\}$ in $X$ converging to $x$ and intersecting infinitely many members of $\xi$.*

Clearly, every strongly Fréchet–Urysohn space is Fréchet–Urysohn, as the name of the new property suggests. However, there are many Fréchet–Urysohn spaces which are not strongly Fréchet–Urysohn — the countable Fréchet–Urysohn fan obtained by identifying the limit points of countably many convergent sequences is a standard example of such a space.

THEOREM 4.7.9. *If a topological group $G$ is Fréchet–Urysohn, then it is strongly Fréchet–Urysohn.*

PROOF. We can assume that the group $G$ is non-discrete. It is enough to verify condition (SFU) for $x = e$. Suppose that $e \in \bigcap_{n\in\omega} \overline{A_n}$, where each $A_n$ is a subset of $G$. Fix a sequence $\{a_n : n \in \omega\}$ in $G \setminus \{e\}$ converging to $e$. For each $n \in \omega$, fix a symmetric open neighbourhood $V_n$ of $e$ such that $a_n \notin V^2$. Since $e \in \overline{A_n}$, we may assume that $A_n \subset V_n$, for each $n \in \omega$ (otherwise, replace $A_n$ with the intersection $A_n \cap V_n$). Put $C_n = a_n A_n$, for $n \in \omega$. From the choice of $V_n$ it is clear that $e \notin \overline{C_n}$, while $a_n \in \overline{C_n}$, for $n \in \omega$. The last condition and the fact that $\{a_n : n \in \omega\}$ converges to $e$ implies that $e \in \overline{C}$, where $C = \bigcup\{C_n : n \in \omega\}$.

Since the space $G$ is Fréchet-Urysohn, there exists a sequence $\eta = \{c_n : n \in \omega\}$ in $C$ converging to $e$. Since $e$ is not in the closure of $C_n$, the sequence $\eta$ must intersect $C_n$ for infinitely many values of $n$. For every $n \in \omega$, choose $k_n \in \omega$ such that $c_n \in C_{k_n}$, and put $b_n = (a_{k_n})^{-1} c_n$. Clearly, the sequence $\{b_n : n \in \omega\}$ converges to $e$ and intersects infinitely many $A_n$'s. Thus, condition (SFU) is satisfied, and the space $G$ is strongly Fréchet–Urysohn. □

The product of a Fréchet–Urysohn space with a metrizable space need not be Fréchet–Urysohn [**165**, 2.4.G (c)]. However, with the help of the preceding result, we can establish that the situation in the class of topological groups is different.

THEOREM 4.7.10. *The product of a Fréchet–Urysohn topological group $G$ with a first-countable space $M$ is Fréchet–Urysohn.*

PROOF. Take any subset $A$ of $G \times M$ and any point $(x, y) \in G \times M$ in the closure of $A$. Let $p$ be the natural projection of $G \times M$ onto $G$. Fix a decreasing countable base $\{U_n : n \in \omega\}$ of the space $M$ at the point $y$, and put $B_n = p((G \times U_n) \cap A)$. Clearly, $x \in \overline{B_n}$, for each $n \in \omega$. We also have $B_{n+1} \subset B_n$, since $U_{n+1} \subset U_n$. From Theorem 4.7.9 it follows that there exists a sequence $\{b_n : n \in \omega\}$ in $G$ converging to $x$ and intersecting $B_n$ for infinitely many $n \in \omega$. For each $k \in \omega$, there are $b_{n_k} \in B_k$ and $c_k \in U_k$ such that $(b_{n_k}, c_k) \in A$ and $n_k > k$. Then, clearly, the sequence $\{(b_{n_k}, c_k) : k \in \omega\}$ converges to the point $(x, y)$. □

Bisequential spaces constitute an important subclass of the class of strongly Fréchet–Urysohn spaces. Recall that a *prefilter* $\eta$ on a space $X$ is any family of non-empty subsets of $X$ such that whenever $P_1$ and $P_2$ are in $\eta$, there exists $P \in \eta$ such that $P \subset P_1 \cap P_2$. A prefilter $\eta$ is called *open* if all elements of $\eta$ are open sets. A prefilter $\eta$ on a space $X$ is said to *converge to a point* $x \in X$ if every open neighbourhood of $x$ contains an element of $\eta$. If $x \in X$ belongs to the closure of each element of a prefilter $\eta$ on $X$, we say that

$\eta$ *accumulates to* $x$ or that $x$ is a cluster point for $\eta$. Clearly, if $\eta$ converges to $x$, then $\eta$ accumulates to $x$. Two prefilters $\eta$ and $\xi$ on $X$ are said to be *synchronous* if, for any $P \in \xi$ and any $Q \in \eta$, $P \cap Q \neq \varnothing$. A space $X$ is called *bisequential* if, for every prefilter $\eta$ on $X$ accumulating to a point $x \in X$, there exists a countable prefilter $\eta$ on $X$ converging to the same point $x$ such that $\eta$ and $\xi$ are synchronous.

Clearly, every first-countable space is bisequential. On the other hand, the one-point compactification of an uncountable discrete space is an example of a bisequential space which is not first-countable at the single non-isolated point. It is an easy exercise to show that every bisequential space is strongly Fréchet–Urysohn. However, in the class of topological groups a much stronger statement holds, as Theorem 4.7.13 shows.

LEMMA 4.7.11. *For every regular bisequential space* $X$ *and each* $x \in X$, *there exists a countable open prefilter on* $X$ *converging to* $x$.

PROOF. Let $\eta$ be the family of all open dense subsets of $X$. Clearly, $\eta$ accumulates to every point of $X$. Now fix $x \in X$. Since $\eta$ accumulates to $x$, and $X$ is bisequential, there exists a countable prefilter $\xi$ on $X$ converging to $x$ and synchronous with $\eta$. Since $X$ is regular, we may assume that all elements of $\xi$ are closed in $X$. Take any $P \in \xi$. The interior of $P$ cannot be empty, since otherwise $X \setminus P$ is an open dense subset of $X$ and, therefore, $X \setminus P \in \eta$ which implies that $\xi$ and $\eta$ are not synchronous. Since $\xi$ is a prefilter, it follows that $\{\mathrm{Int}(P) : P \in \xi\}$ is a countable open prefilter converging to $x$. □

The next auxiliary fact will be considerably generalized in Chapter 5 (see Proposition 5.2.6).

LEMMA 4.7.12. *Let* $G$ *be a topological group with identity* $e$, *and suppose that there exists a countable open prefilter* $\xi$ *on* $G$ *converging to* $e$. *Then the space* $G$ *is first-countable.*

PROOF. Obviously, $\gamma = \{P^{-1}P : P \in \xi\}$ is a base at the identity of $G$. Indeed, all elements of $\gamma$ are open in $G$ and contain $e$. Take open neighbourhoods $U$ and $V$ of $e$ in $G$ such that $V^{-1}V \subset U$. Since $\xi$ converges to $e$, there exists $P \in \xi$ such that $P \subset V$. Therefore, $P^{-1}P \subset V^{-1}V \subset U$. Hence, $G$ is first-countable. □

Combining Lemma 4.7.11 and Lemma 4.7.12, we obtain the second important result of this section:

THEOREM 4.7.13. [**A. V. Arhangel'skii**] *Every bisequential topological group* $G$ *is metrizable.*

The statement and argument above can be extended to the wider class of biradial topological spaces and groups. A space $X$ is called *biradial* if for every prefilter $\xi$ on $X$ converging to a point $x \in X$, there exists a chain $\eta$ of subsets of $X$ converging to $x$ and synchronous with $\xi$. Obviously, every bisequential space is biradial. In addition, every linearly ordered topological space is also biradial. Thus, for example, the space $\omega_1 + 1$, in the usual order topology, is biradial but not bisequential (since it is not Fréchet–Urysohn). Clearly, every subspace of a biradial space is biradial.

PROPOSITION 4.7.14. *If* $X$ *is a regular biradial space then, for each* $x \in X$, *there exists a chain* $\eta$ *of non-empty open subsets of* $X$ *converging to* $x$.

PROOF. Let $\xi$ be the family of all open sets $W \subset X$ such that $x$ belongs to the interior of $\overline{W}$. Clearly, $\xi$ is a prefilter converging to $x$. Since $X$ is biradial, there exists a chain $\eta$ in $X$ converging to $x$ and synchronous with $\xi$. As $X$ is regular, we can assume that all elements of $\eta$ are closed in $X$.

Take any $P \in \eta$, and put $U_P = \mathrm{Int}(P)$. Then $U_P$ is not empty since, otherwise, the open set $V = X \setminus P$ is dense in $X$, which implies that $V \in \xi$. Since $V \cap P = \varnothing$, it follows that $\xi$ and $\eta$ are not synchronous, a contradiction.

Since $\eta$ is a chain converging to $x$, the family $\mu = \{U_P : P \in \eta\}$ is also a chain converging to $x$. Since all elements of $\mu$ are open and non-empty, $\mu$ is the chain we were looking for. □

A chain consisting of open sets will be called a *nest*. A space $X$ is said to be *nested* if for each $x \in X$, there exists a nest which is a base of $X$ at $x$.

THEOREM 4.7.15. *Every biradial topological group $G$ is nested.*

PROOF. Since the space $G$ is biradial and regular, it follows from Proposition 4.7.14 that there exists a nest $\xi$ in $G$ converging to the neutral element $e$ of $G$. Then, by the continuity of operations in $G$, $\eta = \{UU^{-1} : U \in \xi\}$ is again a nest converging to $e$. Since $e$ belongs to every element of $\eta$, it follows that $\eta$ is a base for $G$ at $e$. □

In a sense, Theorem 4.7.14 is parallel to the theorem on metrizability of bisequential topological groups.

We are now going to present a few results on the behaviour of the tightness and Fréchet–Urysohn property in topological groups under group extensions.

PROPOSITION 4.7.16. *Suppose that $f : X \to Y$ is a closed continuous mapping of a regular space $X$ onto a space $Y$ of countable tightness. Suppose further that the tightness of every fiber $f^{-1}(y)$, for $y \in Y$, is countable. Then the tightness of $X$ is also countable.*

PROOF. For a subset $A$ of $X$, we put $[A]_\omega = \bigcup\{\overline{B} : B \subset A,\ |B| \le \omega\}$. It is easy to see that $[[A]_\omega]_\omega = [A]_\omega$, for any $A \subset X$.

Take any $M \subset X$ and any $x \in \overline{M}$. Put $y = f(x)$, $F = f^{-1}(y)$, and $P = F \cap [M]_\omega$. Let us show that $x$ is in the closure of $P$. Take open neighbourhoods $V$ and $W$ of $x$ in $X$ such that $\overline{W} \subset V$. Then $x \in \overline{K}$, where $K = M \cap W$, and $y \in \overline{L}$, where $L = f(K)$. Since $t(Y) \le \omega$, there exists a countable subset $S$ of $L$ such that $y \in \overline{S}$. Take a countable subset $C$ of $K$ such that $f(C) = S$. Since the mapping $f$ is closed, we have $y \in \overline{S} = \overline{f(C)} \subset f(\overline{C})$. Hence, $\overline{C} \cap F \neq \varnothing$. Clearly, $\overline{C} \subset [M]_\omega \cap V$. It follows that $P \cap V \neq \varnothing$. Therefore, $x \in \overline{P}$. However, $P \subset F$, $F$ is closed in $X$, and $t(F) \le \omega$. Therefore, $\overline{P} = [P]_\omega$. Since $P = F \cap [M]_\omega$ and $[[M]_\omega]_\omega = [M]_\omega$, we conclude that $x \in [P]_\omega \subset [M]_\omega$. Hence, the tightness of $X$ is countable. □

THEOREM 4.7.17. *Suppose that $G$ is a topological group, and that $H$ is a locally compact metrizable subgroup of $G$ such that tightness of the quotient space $G/H$ is countable. Then the tightness of $G$ is also countable.*

PROOF. It follows from Theorem 3.2.2 that there exists an open neighbourhood $U$ of the neutral element $e$ in $G$ such that $\overline{U}$ is a preimage of a space of countable tightness under a perfect mapping with metrizable fibers. Hence, the tightness of $\overline{U}$ is also countable, by

Proposition 4.7.16. Since $U$ is a non-empty open subset of the homogeneous space $G$, we conclude that the tightness of $G$ is countable. □

Let us say that a space $X$ is *Fréchet–Urysohn at a point* $x \in X$ if, for every $A \subset X$ with $x \in \overline{A}$, there exists a sequence $\{x_n : n \in \omega\} \subset A$ converging to $x$. Evidently, $X$ is Fréchet–Urysohn iff it is Fréchet–Urysohn at every point $x \in X$.

PROPOSITION 4.7.18. *Suppose that $X$ is a regular space, and that $f: X \to Y$ is a closed continuous mapping. Suppose also that $b \in X$ is a $G_\delta$-point in the space $F = f^{-1}(f(b))$ and $F$ is Fréchet–Urysohn at $b$. If the space $Y$ is strongly Fréchet–Urysohn, then $X$ is Fréchet–Urysohn at $b$.*

PROOF. Using the regularity of $X$, we can construct in a standard way a sequence $\{U_n : n \in \omega\}$ of open neighbourhoods of $b$ in $X$ such that $\overline{U_{n+1}} \subset U_{n+1}$ for each $n \in \omega$, and

$$\{b\} = F \cap \bigcap_{n\in\omega} U_n.$$

Now take any subset $P$ of $X$ such that $b \in \overline{P}$. We have to find a sequence in $P$ converging to $b$. If $b \in \overline{P \cap F}$, then such a sequence exists, since $F$ is Fréchet–Urysohn at $b$. The remaining case, in view of regularity of $X$, obviously reduces to the case when $P \cap F = \varnothing$. So we make this assumption.

Put $B_n = P \cap U_n$ and $C_n = f(B_n)$, for $n \in \omega$. Clearly, $b \in \overline{B_n}$ and, therefore, the continuity of $f$ implies that $c = f(b) \in \overline{C_n}$, for $n \in \omega$. It is also clear that $C_{n+1} \subset C_n$, and $c \notin C_n$, since $P \cap F = \varnothing$. Since $Y$ is strongly Fréchet–Urysohn, and the sequence $\{C_n : n \in \omega\}$ is decreasing, we can select $y_n \in C_n$, for each $n \in \omega$, in such a way that the sequence $\{y_n : n \in \omega\}$ converges to $c$. For each $n \in \omega$, fix $x_n \in B_n$ with $f(x_n) = y_n$.

We claim that the sequence $\xi = \{x_n : n \in \omega\}$ converges to $b$. Note that $\xi$ is disjoint from $F$. However, the closure of every subsequence of $\xi$ must intersect $F$, since the mapping $f$ is closed and every subsequence of $\{y_n : n \in \omega\}$ converges to $c$. It follows that every subsequence of $\xi$ has an accumulation point in $F$. Take any point $z \in F$ distinct from $b$. There exists $k \in \omega$ such that $z \notin \overline{U_k}$. Since $\{x_n : n \in \omega,\ n > k\} \subset B_k \subset U_k$, it follows that $z$ cannot be an accumulation point for $\xi$ or for a subsequence of $\xi$. Thus, $b$ is the unique point of accumulation of every subsequence of $\xi$ which implies that $\xi$ converges to $b$. Since $\xi$ is contained in $P$, the space $X$ is Fréchet–Urysohn at $b$. □

Now we can prove a theorem on the behaviour of the Fréchet–Urysohn property under taking quotients that is similar to Theorem 4.7.17.

THEOREM 4.7.19. *Suppose that $G$ is a topological group, and $H$ a locally compact metrizable subgroup of $G$ such that the quotient space $G/H$ is Fréchet–Urysohn. Then the space $G$ is also Fréchet-Urysohn.*

PROOF. We argue as in the proof of Theorem 4.7.17 using, in addition to Corollary 3.2.6, the fact that every Fréchet–Urysohn topological group is a strong Fréchet–Urysohn space, by Theorem 4.7.9, and referring to Proposition 4.7.18. □

## Exercises

4.7.a. Prove that if $G$ is a biradial topological group of countable pseudocharacter, then $G$ is first-countable.

4.7.b. Give an example of a weakly first-countable, not first-countable, Hausdorff quasitopological group.
*Hint.* Consider the cross topology on the plane.

4.7.c. Show that not every countable Fréchet–Urysohn space can be topologically embedded in a Fréchet–Urysohn topological group.
*Hint.* Take the Fréchet–Urysohn fan $V(\omega)$. Verify that this space is not strongly Fréchet–Urysohn and apply Theorem 4.7.9.

4.7.d. Show that every quotient space of a Fréchet–Urysohn (biradial) topological group is again Fréchet–Urysohn (biradial).

4.7.e. Give an example of two biradial topological groups whose product is not biradial.

4.7.f. A topological space $X$ is called *linearly orderable* if there exists a linear order $\leq$ on $X$ such that the sets of the form $X_{<a} = \{x \in X : x < a\}$ and $X_{>a} = \{x \in X : a < x\}$, with $a \in X$, constitute a subbase for the topology of $X$. A topological group $G$ is called *topologically orderable* if it is orderable as a topological space. Prove the following:
  a) The Cantor set $C \cong \mathbb{Z}(2)^\omega$ is a topologically orderable compact topological group.
  b) The compact group $\mathbb{T}$ is not topologically orderable.
  c) (P. J. Nyikos and H. C. Reichel [**357**]) If $(G, <)$ is a topologically ordered group such that the neutral element $e$ of $G$ is not isolated either from above or from below, then the cofinality of the set $(G_{<e}, <)$ equals the cofinality of the set $(G_{>e}, >)$. Equivalently, the cofinality of $(G, <)$ *below* $e$ is equal to the cofinality of $(G, <)$ *above* $e$.

4.7.g. A group $G$ is *algebraically orderable* if there is a subset $P$ of $G$ (called the set of *positive elements*) such that $PP \subset P$ and $G$ is the disjoint union of $P$, $P^{-1}$, and the neutral element of $G$. Prove that if $G$ is an algebraically ordered group, then:
  a) every element of $G$ distinct from the neutral element is of infinite order;
  b) $G$ endowed with the topology induced by the linear order defined by letting $x < y$ if $yx^{-1} \in P$, is a topologically ordered group.

4.7.h. Let $\mathbb{Z}$ be the additive group of integers. Prove the following:
  a) The only algebraic orderings on the group $\mathbb{Z}$ are defined by taking $P$ to be the positive integers or the negative integers, both of which turn $\mathbb{Z}$ into a discrete space.
  b) There exist many distinct non-discrete topologies $\tau$ on $\mathbb{Z}$ compatible with the group structure of $\mathbb{Z}$ such that $(\mathbb{Z}, \tau)$ is a topologically orderable group (so, topologically orderable groups can fail to be algebraically orderable).

## Problems

4.7.A. Suppose that $G$ is a topological group such that $G = FM$, where $F$ is a first-countable compact space and $M$ is a first-countable closed subspace of $G$. Then $G$ is metrizable.
*Hint.* The space $G$ is an image of the first-countable space $M \times F$ under a perfect mapping. Therefore, the space $G$ is bisequential. However, every bisequential topological group is metrizable, by Theorem 4.7.13.

4.7.B. If a topological group $G$ has a Hausdorff compactification $bG$ of countable tightness, then $G$ is metrizable.
*Hint.* There is a countable $\pi$-base in $bG$ at any point of $G$, since the tightness of $bG$ is countable [**440**]. Take a countable $\pi$-base $\gamma$ at the neutral element $e$ of $G$ and show that the family $\{UU^{-1} : U \in \gamma\}$ is a local base at $e$ in $G$.

4.7.C. Show that a paratopological group $G$ that has a Hausdorff first-countable compactification need not be metrizable.
*Hint.* Consider the Sorgenfrey line.

4.7.D. Construct a precompact topological group $G$ and a closed invariant subgroup $H$ of $G$ such that $H$ is Fréchet–Urysohn and countably compact, the quotient group $G/H$ is compact and metrizable, but the tightness of $G$ is uncountable.
*Hint.* Take the group $K = D^{\omega}$ with the product topology, where $D = \{0, 1\}$ is the discrete two-element group. Clearly, $K$ is compact and metrizable. Let also $L = D^{\mathfrak{c}}$, where $\mathfrak{c} = 2^{\omega}$. Then the $\Sigma$-product in $L$ with center at the neutral element of $L$ is a dense countably compact subgroup of $L$ which is denoted by $\Sigma$. The space $\Sigma$ is Fréchet–Urysohn, by Corollary 1.6.25, and satisfies $|\Sigma| = \mathfrak{c}$. Define by recursion of length $\mathfrak{c}$ an algebraic homomorphism $\varphi\colon K \to L$ such that the intersection of the graph of $\varphi$, $P = \{(x, \varphi(x)) : x \in K\}$, with $K \times \Sigma$ is dense in $K \times \Sigma$ and the complement $P \setminus (K \times \Sigma)$ is not empty. Then $H = \{0_K\} \times \Sigma$ is a closed subgroup of the group $G = H + P$ and the quotient group $G/H$ is topologically isomorphic to $K$. Verify that $t(G) > \omega$.

4.7.E. Does Theorem 4.7.19 remain valid if one assumes that $H$ is an arbitrary closed metrizable subgroup of the group $G$?

4.7.F. Show that is consistent with $ZFC$ that a compact sequentially compact topological group need not be metrizable.
*Hint.* Under Martin's Axiom combined with negation of the Continuum Hypothesis, $D^{\omega_1}$ is the group we are looking for.

4.7.G. (M. Venkataraman, M. Rajagopalan, and T. Soundararajan [**527**]) Let $G$ be a topologically orderable topological group. Prove the following:
   a) If $G$ is not totally disconnected, then it contains an open invariant subgroup which is topologically isomorphic to the group of reals $\mathbb{R}$.
   b) If $G$ is infinite, locally compact, and totally disconnected, then it contains an open subgroup homeomorphic as a space to the Cantor set.

4.7.H. (M. Venkataraman, M. Rajagopalan, and T. Soundararajan [**527**]) A separable totally disconnected topological group is topologically orderable iff it is metrizable and zero-dimensional.

4.7.I. (M. Sanchis and A. Tamariz-Mascarúa [**417**]) Prove that a non-discrete topologically orderable group $G$ is metrizable if and only if $G$ contains an infinite precompact subset. (See also Exercise 4.7.f.)

4.7.J. (P. J. Nyikos and H. C. Reichel [**357**]) Let $G$ be a non-metrizable topological group. Prove that the following conditions are equivalent:
   a) $G$ is topologically orderable;
   b) the neutral element of $G$ has a linearly ordered (by inverse inclusion) local base;
   c) the neutral element of $G$ has a linearly ordered (by inverse inclusion) local base consisting of open subgroups.

## Open Problems

4.7.1. Can any first-countable Tychonoff space $X$ be topologically embedded in a Fréchet–Urysohn topological group? What if $X$ is also compact?

4.7.2. A space $X$ is called *radial* if for every $A \subset X$ and every point $x \in \overline{A} \setminus A$, there exists a subset $B \subset A$ such that $x \in \overline{B}$ and $|B \setminus U| < |B|$, for each neighbourhood $U$ of $x$ in $G$. Can Theorem 4.7.9 be generalized to radial topological groups? Of course, condition (SFU) should be also generalized to this case.

4.7.3. Is the product of a radial topological group with a nested topological group radial?

4.7.4. Is it consistent with $ZFC$ that every countable Fréchet–Urysohn topological group is metrizable? In other words, can one construct in $ZFC$ a non-metrizable countable topological group which is Fréchet–Urysohn?
*Remark.* There are countable Fréchet–Urysohn non-metrizable topological groups under Martin's Axiom combined with negation of the Continuum Hypothesis. One can simply take any countable dense subgroup of the compact group $D^{\omega_1}$, where $D = \{0, 1\}$ is the discrete two-element group.

4.7.5. Is there in $ZFC$ an example of a (countable) Fréchet–Urysohn topological group $G$ such that $G \times G$ is not Fréchet–Urysohn?
*Remark.* Consistent examples of such groups are known. They were constructed by A. Shibakov [**444**] (see also the survey [**436**]).

4.7.6. Is every countably compact sequential topological group $G$ Fréchet–Urysohn? What if $G$ is also a normal space?

4.7.7. Prove in $ZFC$ that if a topological group $G$ has a Hausdorff compactification $bG$ such that the tightness of $bG$ is countable at every $x \in G$, then $G$ is metrizable.

4.7.8. Let us say that the *dyadicity index* of a space $X$ is countable if there exists no continuous mapping of $X$ onto the Tychonoff cube $I^{\omega_1}$. Suppose that a topological group $G$ has a Hausdorff compactification $bX$ of countable dyadicity index. Is $G$ metrizable?

## 4.8. Historical comments to Chapter 4

The class of dyadic compacta was introduced by P. S. Alexandroff in [**7**] by means of the following question: Is it possible to represent an arbitrary compact Hausdorff space as a continuous image of some generalized Cantor discontinuum $D^\tau$? The negative answer was given by E. Marczewski in [**302**] (see [**165**] for further references). He noticed that the cellularity of every dyadic compactum is countable. The crucial Theorem 4.1.1 was proved by E. A. Michael in [**320**] who created the theory of continuous selections. This theory found many important applications (see, for example, [**74**]). Theorem 4.1.7 was independently proved by L. N. Ivanovskij [**259**] and V. I. Kuz'minov [**287**]; this is one of the major achievements in the topological theory of compact groups. The proof of Theorem 4.1.7 in this book follows the argument in [**517, 515, 516**] very closely. This concerns, in particular, Lemma 4.1.2, Proposition 4.1.3, Lemma 4.1.4, and Theorem 4.1.6. The argument is based on a method created by R. Haydon in [**221**] and developed further by E. V. Schepin in [**420, 421**]. In connection with Theorem 4.1.5 see [**7**]. Theorem 4.2.1 is an easy corollary from Theorem 4.1.7 and a general result of R. Engelking on dyadic compacta in [**164**]. Some strong partial results in this direction were obtained by B. A. Efimov in [**158**]. The proof of Theorem 4.2.1 given in the book is based on Shakhmatov's argument from [**434**]. Corollary 4.2.2 was obtained in [**59**]. L. B. Shapiro established in [**439**] that every homogeneous zero-dimensional dyadic compact space of weight $\aleph_1$ is homeomorphic to the topological group $D^{\omega_1}$. This result was also obtained by M. G. Bell in [**70**]. Thus, the assumption of dyadicity imposes very strong structural constraints on a compact space.

Proposition 4.3.1 and Lemmas 4.3.2 and 4.3.3 are among the standard tools of general topology. One finds analogous statements in [**276**], [**80**], and [**165**]. M. Henriksen and J. Isbell introduced spaces of countable type in [**223**]. Spaces of pointwise countable type (or of point-countable type) were defined later by A. V. Arhangel'skii in [**15**]. This notion brings to light a natural common extension of the classes of compact spaces and first-countable spaces. Lemma 4.3.4 and Corollary 4.3.5 go back to [**223**]. For almost open sets and Proposition 4.3.6 see [**249**]. The class of feathered spaces, or $p$-spaces, was introduced in [**14, 16**], by a different condition. B. A. Pasynkov established in [**364**] that a topological group $G$ is feathered if and only if it contains a non-empty compact subset of countable character in $G$, that is, if $G$ is a space of pointwise countable type. He called these groups *almost metrizable*. Lemma 4.3.10, Propositions 4.3.11, 4.3.13, and Corollary 4.3.12 came from [**364**], where some further interesting results were obtained. Theorems 4.3.15 and 4.3.16 are due to M. M. Choban [**100**]. Later, Theorem 4.3.15 was also proved by L. G. Brown in [**88**]. Corollary 4.3.18 is from [**47**]. Lemma 4.3.19 was known to L. S. Pontryagin (see [**387**]). Theorem 4.3.20 and Corollaries 4.3.21, 4.3.22, and 4.3.24 are from [**364**].

Proposition 4.3.25 and Theorem 4.3.26 are from [**365**]. Proposition 4.3.36 appeared in [**16**]. Theorem 4.3.37 and Corollary 4.3.38 were proved in [**47**].

In general topology $P$-spaces were introduced a long time ago (see [**350**], for example). They are especially interesting in the context of function spaces; that was shown in [**191**] and [**32**]. Clearly, every topological space gives rise to a $P$-space by means of the very natural operation of taking the $G_\delta$-modification of the original topology — a base of this topology consists of $G_\delta$-sets in the original space.

In topological algebra $P$-spaces made their appearance quite recently; one of them was in [**109**], in connection with the study of pseudocompact groups. They are still looked upon as exotic objects. However, there is a very important point about $P$-spaces: Lindelöf $P$-spaces behave very much similar to compacta! Thus, they can serve as objects of a nice theory, not only as a source of peculiar examples. Many elementary results in Section 4.4 could be, probably, viewed as a part of the folklore. The proofs of some of them may appear in print for the first time. Lemma 4.4.3 and Corollary 4.4.4 are of the purely topological nature, they were proved in [**350**] and [**351**], respectively. Proposition 4.4.5 was obtained by C. Hernández in [**225**]. For Lemmas 4.4.7 and 4.4.8, Proposition 4.4.9, and Theorem 4.4.10 see [**350**] and [**351**]. Example 4.4.11 goes back to W. W. Comfort and K. A. Ross' article [**122**] (where it was presented for the special case $\tau = \aleph_1$).

The first consistent example (under the Continuum Hypothesis) of a non-discrete extremally disconnected topological group was constructed by S. Sirota in [**455**]. The problem of finding such a group was formulated in [**17**], where it was shown that every compact subspace of any extremally disconnected topological group was discrete (in Theorem 3.7.28 this result is extended to precompact subsets of topological groups). The problem is still open in $ZFC$. In [**298**] V. I. Malykhin made an important step forward — he proved Theorem 4.5.2. Theorem 4.5.1 goes back to Z. Frolík [**180**] and M. G. Katetov [**270**]. Our treatment of the topic of extremal disconnectedness follows [**38**]. In particular, Theorems 4.5.3, 4.5.5, 4.5.6, 4.5.8, 4.5.9, and Proposition 4.5.4 are from [**38**]. In connection with Corollary 4.5.7 see [**397**]. Theorems 4.5.12, 4.5.16, 4.5.17, and Corollaries 4.5.14, 4.5.15 are from [**38**]. Lemma 4.5.18 and Proposition 4.5.19 can be found in [**152**] and [**52**], respectively. Theorem 4.5.22 is due to V. I. Malykhin (see [**298, 299**]).

The notions of grasp on a group and of cross-complementary subsets of a group were introduced in [**46**]. Proposition 4.6.1, Lemma 4.6.2, Theorems 4.6.3 and 4.6.4 are taken from [**46**]. Proposition 4.6.5 goes back to N. Bourbaki's [**81**]. Example 4.6.7, Theorems 4.6.8, 4.6.9, 4.6.10, and 4.6.11 are from [**46**]. The notion of a $G_\delta$-closed mapping was also introduced in [**46**]. Lemma 4.6.13, Proposition 4.6.14, Theorems 4.6.15 and 4.6.17 are from [**46**], though the first three of them have their prototypes in the theory of $P$-spaces (see, in particular, [**350**] and [**351**]). Theorems 4.6.22, 4.6.24, 4.6.25, and 4.6.26 also come from [**46**].

Weakly first-countable spaces were studied by S. J. Nedev in [**340**] under the name of *o-metrizable spaces.* Lemmas 4.7.1, 4.7.3, and Propositions 4.7.2 and 4.7.4 appear in print, probably, for the first time. Theorem 4.7.5 is P. Nyikos' result from [**355**]. Corollary 4.7.6 was obtained in [**341**] and later in [**355**]. Proposition 4.7.7 is from [**60**]. Theorem 4.7.8 was proved in [**62**]. Theorem 4.7.9 is due to P. Nyikos, see [**355, 356**]. Lemmas 4.7.11, 4.7.12, and Theorem 4.7.13 were established in [**33**]. Proposition 4.7.14 and Theorem 4.7.15 are taken from [**31**]. Theorems 4.7.17, 4.7.19, and Proposition 4.7.18 are quite recent results from [**47**].

# Chapter 5

# Cardinal Invariants of Topological Groups

The definition of the notions of topology and topological space, based on the axiomatic approach, is of necessity of a purely set-theoretic nature. Indeed, a topology is just a family of sets satisfying certain axioms. Not so many elementary and natural properties of sets can be formulated without recourse to special, more complicated, structures or tools of mathematical logic. The most important, and almost the only such property is the cardinality of a set. So no wonder that in General Topology cardinal invariants, that is, characteristics of spaces preserved by homeomorphisms and formulated in terms of cardinal numbers and of families of sets, play a central, almost universal, role. Cardinal invariants measure the size of the space in various ways, the local behaviour of the space, and, most importantly, they are used to bring to light specific features of the space. When we consider continuous mappings, it is important to know which cardinal invariants do not increase under certain natural restrictions on these mappings. When studying products of spaces, it is most useful, whatever our main interest may be, to know how certain cardinal invariants behave under the Tychonoff product operation. Some of the questions of this kind are quite deep and difficult, and the work on them has generated much of the progress not only in General Topology, but in Set Theory and in Mathematical Logic as well. To make the point, it is enough to mention the following question. Suppose that the Souslin number of a space $X$ is countable. Is the Souslin number of the square $X \times X$ of $X$ countable? This question, which is so easy to formulate and understand, is intimately related to the famous Souslin Conjecture and to Martin's Axiom, coined in Mathematical Logic, both of which have so many consequences in Topology and Analysis.

Cardinal invariants of a somewhat different kind play a fundamental role in algebra also; one can refer to the cardinality of a basis of a vector space, to the $p$-rank and torsion-free rank of an Abelian group (see Section 9.9), and to similar concepts. So we should expect that cardinal invariants must have a prominent role in topological algebra. Examples of that were seen in preceding chapters; it suffices to mention the Birkhoff–Kakutani theorem on the metrizability of first-countable topological groups, the theorem on the metrizability of compact groups of countable tightness, or the theorem stating that the cellularity of every compact topological group is countable.

This chapter is devoted to a deeper, and more systematic, study of cardinal invariants of topological groups, with detours into the realm of paratopological groups. We introduce a variety of cardinal invariants, some of them of mixed, topological and algebraic nature, and we study relationships between them. One of the main points of interest is to clarify how the presence of a "synchronous" algebraic structure influences the behaviour of purely topological cardinal invariants in the new ambient, how the relationship between them changes.

One of the ways to understand the essence and the role of a topological cardinal invariant is to consider the class of all objects (topological spaces, topological groups, and so on) satisfying certain restriction on the value of this invariant, and to study the categorical properties of the class so obtained, that is, to investigate whether the class is closed under the product operation, whether it is preserved by various classes of mappings, whether it is hereditary.

We study such questions below and, although the theory is obviously far from complete, we present, along with simple basic facts, certain deep and delicate results of a very general nature. We hope that some of these results will serve as corner stones for the emerging theory.

## 5.1. More on embeddings in products of topological groups

We already know when a topological group $G$ is topologically isomorphic to a subgroup of a product of second-countable groups — according to Guran's theorem, this happens if and only if $G$ is $\omega$-narrow (see Theorem 3.4.23). It is very natural to introduce more general classes of groups by taking subgroups of arbitrary products of some "nice" topological groups. Given a class $\mathcal{P}$ of topological groups, it is also natural to try to find an internal characterization of the subgroups of the groups in $\mathcal{P}$. This is still an open problem for the class $\mathcal{P}$ of Lindelöf topological groups. On the other hand, subgroups of compact topological groups are precisely the precompact groups, by Corollary 3.7.17. The following definition generalizes the concepts of precompact and $\omega$-narrow groups.

Let $\tau$ be an infinite cardinal. A left topological group $G$ is called *$\tau$-narrow* if, for every neighbourhood $U$ of the identity in $G$, there exists a subset $K \subset G$ with $|K| \leq \tau$ such that $KU = G$.

We collect several simple properties of $\tau$-narrow topological groups in the following proposition.

Proposition 5.1.1.

a) *Every subgroup of a $\tau$-narrow topological group is $\tau$-narrow.*
b) *If $\pi: G \to H$ is a continuous homomorphism of a $\tau$-narrow left topological group $G$ onto a left topological group $H$, then $H$ is $\tau$-narrow.*
c) *The topological product of arbitrarily many $\tau$-narrow left topological groups is $\tau$-narrow.*
d) *If $G$ is a dense $\tau$-narrow subgroup of a topological group $H$, then $H$ is also $\tau$-narrow.*

Let us mention that in the case $\tau = \omega$, item a) of Proposition 5.1.1 coincides with Theorem 3.4.4, item b) coincides with Proposition 3.4.2, item c) is exactly Proposition 3.4.3, and item d) is Theorem 3.4.9. Since the general case of $\tau$-narrow groups does not substantially differ from that of $\omega$-narrow groups, we leave Proposition 5.1.1 without proof.

The above proposition implies that $\mathbb{R}^\omega$ is an $\omega$-narrow group that fails to be $\sigma$-compact. In fact, many non-closed subgroups of the groups $\mathbb{R}$ and $\mathbb{T}$ are not $\sigma$-compact; all of them are $\omega$-narrow, by a) of Proposition 5.1.1. The next example shows even more.

EXAMPLE 5.1.2. The group $\mathbb{R}^\omega$ cannot be embedded as a subgroup into a $\sigma$-compact topological group.

Indeed, the group $\mathbb{R}^\omega$ is Raĭkov complete, by Theorem 3.6.22, and if it were topologically isomorphic to a subgroup $H$ of a $\sigma$-compact topological group $G$, $H$ would be closed in $G$ and $\sigma$-compact, which is impossible. □

Proposition 5.1.1 also implies that subgroups of topological products of $\sigma$-compact groups are $\omega$-narrow. However, we shall see in Theorem 5.1.24 that not every $\omega$-narrow group can be embedded as a subgroup into a topological product of $\sigma$-compact topological groups.

The following result provides us with more examples of $\tau$-narrow groups. We recall that $l(G)$ and $c(G)$ are the Lindelöf number and the cellularity of $G$, respectively.

PROPOSITION 5.1.3. *Let $G$ be a topological group.*

a) *If $G$ satisfies $l(G) \leq \tau$, then $G$ is $\tau$-narrow.*
b) *If $c(G) \leq \tau$, then $G$ is $\tau$-narrow.*

PROOF. Claim a) is almost obvious. Indeed, if $U$ is an open neighbourhood of the identity in $G$, then $\{xU : x \in G\}$ is an open covering of $G$. Since $l(G) \leq \tau$, there is a set $C \subset G$ with $|C| \leq \tau$ such that the family $\{xU : x \in C\}$ covers $G$ or, equivalently, $CU = G$. Hence, $G$ is $\tau$-narrow.

To deduce b), apply the argument in the proof of Theorem 3.4.7. □

Since a separable space has countable cellularity, item b) of Proposition 5.1.3 implies that every separable topological group is $\omega$-narrow. This suggests that the class of $\omega$-narrow groups is wider than the class of subgroups of Lindelöf topological groups. Indeed, the $\omega$-narrow group $\mathbb{Z}^{\omega_1}$, where the group $\mathbb{Z}$ carries the discrete topology, is Raĭkov complete and contains an uncountable closed discrete subset, by [**165**, 2.7.16]. So, $\mathbb{Z}^{\omega_1}$ cannot be embedded as a subgroup into a Lindelöf topological group.

By Theorem 3.4.23, $\omega$-narrow groups are precisely the subgroups of topological products of second-countable groups. Similarly, every $\tau$-narrow topological group can be embedded as a subgroup into a product of topological groups of weight at most $\tau$ (see Theorem 5.1.10). The argument in this case is close to that for $\omega$-narrow groups given in Section 3.4, so we only sketch the proof here.

Let $G$ be a topological group. We say that the *invariance number* of $G$ is less than or equal to $\tau$ or, in symbols, $inv(G) \leq \tau$ if for every neighbourhood $U$ of the identity $e$ in $G$, there exists a family $\gamma$ of open neighbourhoods of $e$ with $|\gamma| \leq \tau$ such that for each $x \in G$, one can find $V \in \gamma$ satisfying $x^{-1}Vx \subset U$. Following Section 3.4, we call such a family $\gamma$ *subordinated to $U$*. The next three facts are completely analogous to Propositions 3.4.10, 3.4.5, and Theorem 3.4.18, respectively, so we omit their proofs.

LEMMA 5.1.4. *If $G$ is a $\tau$-narrow topological group, then $inv(G) \leq \tau$.*

LEMMA 5.1.5. *Let $H$ be a $\tau$-narrow topological group of character less than or equal to $\tau$. Then $w(H) \leq \tau$.*

LEMMA 5.1.6. *Let $U$ be an open neighbourhood of the identity in a topological group $G$ with $inv(G) \leq \tau$. Then there exists a continuous homomorphism $\pi\colon G \to H$ of $G$ onto a topological group $H$ with $\chi(H) \leq \tau$ such that $\pi^{-1}(V) \subset U$, for some open neighbourhood $V$ of the identity in $H$.*

Combining Lemmas 5.1.4–5.1.6, we arrive at the following conclusion.

COROLLARY 5.1.7. *Let $U$ be an open neighbourhood of the identity in a $\tau$-narrow topological group $G$. Then there exists a continuous homomorphism $\pi\colon G \to H$ of $G$ onto a topological group $H$ with $w(H) \leq \tau$ such that $\pi^{-1}(V) \subset U$, for some open neighbourhood $V$ of the identity in $H$.*

COROLLARY 5.1.8. *Let $G$ be a $\tau$-narrow topological group with identity $e$ and let $P$ be a subset of $G$ such that $e \in P$ and $\psi(P, G) \leq \tau$. Then there exists a continuous homomorphism $p\colon G \to H$ onto a topological group $H$ with $w(H) \leq \tau$ such that* $\ker p \subset P$.

PROOF. Since $\psi(P, G) \leq \tau$, there exists a family $\{U_\alpha : \alpha < \tau\}$ of open neighbourhoods of $e$ in $G$ such that $P = \bigcap_{\alpha<\tau} U_\alpha$. Apply Corollary 5.1.7 to find, for every $\alpha \in A$, a continuous homomorphism $p_\alpha\colon G \to H_\alpha$ onto a topological group $H_\alpha$ with $w(H_\alpha) \leq \tau$ such that $p_\alpha^{-1}(V_\alpha) \subset U_\alpha$ for some open neighbourhood $V_\alpha$ of the identity in $H_\alpha$. In particular, $\ker p_\alpha \subset U_\alpha$. The diagonal product $p = \Delta_{\alpha<\tau} p_\alpha$ is a continuous homomorphism of $G$ to the group $H = \prod_{\alpha<\tau} H_\alpha$. It is clear that $w(H) \leq \tau$. In addition,

$$\ker p = \bigcap_{\alpha<\tau} \ker p_\alpha \subset \bigcap_{\alpha<\tau} U_\alpha = P. \qquad \square$$

Now we extend Theorem 3.4.22 to topological groups with invariance number less than or equal to an infinite cardinal number $\tau$.

THEOREM 5.1.9. *Every topological group $G$ with $inv(G) \leq \tau$ can be embedded as a subgroup into a topological product of topological groups of character $\leq \tau$.*

PROOF. Let $e$ be the identity of $G$. Consider the family $\mathcal{B} = \{U_i : i \in I\}$ of all open neighbourhoods of $e$ in $G$. By Lemma 5.1.6 we can find, for every $i \in I$, a continuous homomorphism $\pi_i\colon G \to H_i$ of $G$ onto a topological group $H_i$ with $\chi(H_i) \leq \tau$ such that $\pi_i^{-1}(V_i) \subset U_i$ for some open neighbourhood $V_i$ of the identity in $H_i$. Let $\Pi = \prod_{i\in I} H_i$ be the topological product of the groups $H_i$'s and let $\varphi\colon G \to \Pi$ be the diagonal product of the homomorphisms $\pi_i$, where $i \in I$. It is clear that $\varphi$ is a continuous homomorphism of $G$ to $\Pi$. It remains to show that $\varphi$ is a topological embedding.

Let $H = \varphi(G)$. Choose an arbitrary neighbourhood $U$ of $e$ in $G$. There exists $i \in I$ such that $U_i \subset U$, and then $\pi_i^{-1}(V_i) \subset U_i$ by the choice of the open neighbourhood $V_i$ of the identity in $H_i$. Denote by $p_i$ the projection of $\Pi$ onto the factor $H_i$. The set $W = p_i^{-1}(V_i)$ is an open neighbourhood of the identity in $\Pi$. Since $\varphi$ is the diagonal product of the mappings $\pi_j$, $j \in I$, we have $\pi_i = p_i \circ \varphi$. Therefore, $\varphi^{-1}(W) = \pi_i^{-1}(V_i) \subset U_i \subset U$. In particular, $O = W \cap H$ is an open neighbourhood of the identity in $H$ which satisfies $\varphi^{-1}(O) \subset U$.

So we have proved that for every open set $U$ in $G$ containing $e$, there exists an open set $O$ in $H$ containing the identity of $H$ such that $\varphi^{-1}(O) \subset U$. This implies immediately that $\varphi\colon G \to H$ is a continuous isomorphism and that its inverse $\varphi^{-1}$ is also continuous. Therefore, $\varphi\colon G \to H$ is a topological isomorphism. $\square$

Theorem 5.1.9 enables us to deduce the main result of this section, which characterizes $\tau$-narrow topological groups in terms of embeddings into topological products and implies Theorem 3.4.23 in the special case when $\tau = \omega$:

THEOREM 5.1.10. [**I. I. Guran**] *A topological group $G$ is $\tau$-narrow if and only if $G$ is topologically isomorphic to a subgroup of a topological product of topological groups of weight less than or equal to $\tau$.*

PROOF. Clearly, every topological group of weight $\leq \tau$ is $\tau$-narrow. Therefore, items a) and c) of Proposition 5.1.1 imply that every subgroup of a product $\prod_{i \in I} H_i$ of topological groups is $\tau$-narrow provided that $w(H_i) \leq \tau$, for each $i \in I$.

Conversely, let $G$ be a $\tau$-narrow topological group. By Lemma 5.1.5 and Theorem 5.1.9, we can identify $G$ with a subgroup of a product $\Pi = \prod_{i \in I} H_i$ of topological groups $H_i$ satisfying $\chi(H_i) \leq \tau$ for each $i \in I$. Let $\pi_i \colon \Pi \to H_i$ be the projection. Clearly, we can assume that $H_i = \pi_i(G)$, for all $i \in I$. Then each group $H_i$ is $\tau$-narrow by b) of Proposition 5.1.1, so Lemma 5.1.5 implies that $w(H_i) \leq \tau$. □

Below, we apply Theorem 5.1.10 to give another characterization of $\tau$-narrow topological groups that avoids mentioning topological products:

THEOREM 5.1.11. *A topological group $G$ is $\tau$-narrow iff it can be embedded as a subgroup into a topological group $H$ satisfying $c(H) \leq \tau$.*

PROOF. Every subgroup of a topological group $H$ with $c(H) \leq \tau$ is $\tau$-narrow, by b) of Proposition 5.1.3 and a) of Proposition 5.1.1. Conversely, by Theorem 5.1.10, a $\tau$-narrow topological group $G$ is topologically isomorphic to a subgroup of a product $H$ of topological groups $H_i$ with $w(H_i) \leq \tau$, and such a product satisfies $c(H) \leq \tau$, by [**165**, Theorem 2.3.17]. □

A topological space is said to be *k-separable* if it has a dense $\sigma$-compact subspace. We shall see in Corollary 5.3.22 below that every $\sigma$-compact topological group has countable cellularity. Since every $k$-separable topological group contains a dense $\sigma$-compact subgroup, all $k$-separable groups have countable cellularity. Hence, the next result is, in a sense, a refinement of Theorem 5.1.11 in the case when $\tau = \omega$.

THEOREM 5.1.12. [**V. G. Pestov**] *The class of $\omega$-narrow groups coincides with the class of subgroups of $k$-separable topological groups.*

PROOF. Every $\sigma$-compact topological group $H$ is Lindelöf and, hence, $\omega$-narrow. If $H$ is dense in a topological group $G$, then $G$ is $\omega$-narrow by d) of Proposition 5.1.1, so that all subgroups of $G$ are also $\omega$-narrow by a) of the same proposition.

Conversely, let $G$ be an arbitrary $\omega$-narrow topological group. It follows from Theorem 5.1.10 that we can identify $G$ with a subgroup of a topological product $\Pi = \prod_{i \in I} H_i$ of second-countable groups $H_i$'s. For every $i \in I$, fix a countable dense subgroup $D_i$ of $H_i$. Let $D$ be a subgroup of $\prod_{i \in I} D_i$ consisting of the points whose all coordinates except for finitely many are identities. Thus, $D$ is the $\sigma$-product of the family $\{D_i : i \in I\}$ (see Section 1.6). Then $D$ is dense in $\prod_{i \in I} D_i$ and in $\Pi$. In addition, from Proposition 1.6.41 it follows that $D$ is $\sigma$-compact. □

Continuous homomorphic images of $\tau$-narrow groups are $\tau$-narrow, by b) of Proposition 5.1.1. It turns out that, in the class of Abelian topological groups $G$, continuous homomorphic images $H$ with $\chi(H) \leq \omega$ of a given group $G$ determine whether $G$ is $\tau$-narrow or not.

PROPOSITION 5.1.13. *Let $G$ be an Abelian topological group and suppose that every continuous homomorphic image $H$ of $G$ with $\chi(H) \leq \omega$ is $\tau$-narrow. Then the group $G$ is also $\tau$-narrow.*

PROOF. Let $U$ be an open neighbourhood of the neutral element $e$ in $G$. There exists a sequence $\{U_n : n \in \omega\}$ of open symmetric neighbourhoods of $e$ in $G$ such that $U_0 \subset U$ and $U_{n+1}^2 \subset U_n$ for each $n \in \omega$. Then $N = \bigcap_{n\in\omega} U_n$ is a closed subgroup of $G$. Since $G$ is Abelian, we can consider the algebraic quotient group $G/N$. Let $\pi\colon G \to G/N$ be the natural homomorphism. Obviously, the family $\{\pi(U_n) : n \in \omega\}$ is a base for a Hausdorff group topology $\mathcal{T}$ on $G/N$ at the neutral element of this group. Let $H = (G/N, \mathcal{T})$. Clearly, $\chi(H) \leq \omega$ and the homomorphism $\pi\colon G \to H$ is continuous. Therefore, the group $H$ is $\tau$-narrow. For the open set $V = \pi(U_1)$, choose a set $K \subset H$ such that $KV = H$ and $|K| \leq \tau$. Let $F$ be any subset of $G$ such that $\pi(F) = K$ and $|F| \leq \tau$. We claim that $FU = G$. Indeed, take an arbitrary element $x \in G$. Then $\pi(x) \in bV$ for some $b \in K$. Choose an element $a \in F$ with $\pi(a) = b$. Clearly, $\pi(x) \in bV = \pi(aU_1)$, whence it follows that

$$x \in \pi^{-1}\pi(aU_1) = aU_1N \subset aU_1U_1 \subset aU_0 \subset aU \subset FU.$$

This implies that $FU = G$, so the group $G$ is $\tau$-narrow. □

Sometimes, the $\tau$-narrowness of a topological group $G$ can be deduced from the existence of a certain type of subspace of $G$ (see Theorem 5.1.19 below). The next definition extends the concept of $\tau$-narrowness to subsets of semitopological groups.

A subset $B$ of a semitopological group $G$ is called *$\tau$-narrow in $G$* if, for every neighbourhood $U$ of the identity in $G$, there exists a subset $F$ of $G$ such that $B \subset FU \cap UF$ and $|F| \leq \tau$. Clearly, $G$ is $\tau$-narrow iff $G$ is $\tau$-narrow in itself, and every subset of a $\tau$-narrow semitopological group is $\tau$-narrow in this group. In what follows we focus in subsets of topological groups.

The lemma below is analogous to Proposition 5.1.3, so its proof is omitted. In fact, we will prove below, in Proposition 5.1.15, a considerably more general result.

LEMMA 5.1.14. *A subset $B$ of a topological group $G$ is $\tau$-narrow in each of the following cases:*

a) $l(B) \leq \tau$;
b) $c(B) \leq \tau$.

The *discrete cellularity number* $dc(X)$ of a space $X$ is the smallest infinite cardinal number $\tau$ such that the cardinality of any discrete family of non-empty open subsets of $X$ is strictly less than $\tau$. Thus, $dc(X) = \aleph_0$ iff $X$ is pseudocompact, and $dc(X) \leq \aleph_1$ iff every discrete family of non-empty open sets in $X$ is countable, that is, $X$ is pseudo-$\aleph_1$-compact. It is clear that each of the conditions $l(B) \leq \tau$ or $c(B) \leq \tau$ implies that $dc(X) \leq \tau^+$, so Lemma 5.1.14 follows from the next result:

PROPOSITION 5.1.15. *Let $B$ be a subspace of a topological group $G$ satisfying $dc(B) \leq \tau^+$. Then $B$ is $\tau$-narrow in $G$.*

PROOF. Assume to the contrary that $B$ fails to be $\tau$-narrow in $G$. Then there exists a neighbourhood $U$ of the identity $e$ in $G$ such that either $B \setminus FU \neq \varnothing$ or $B \setminus UF \neq \varnothing$ for every $F \subset G$ with $|F| \leq \tau$. We can assume without loss of generality that the first case takes place. By Zorn's lemma, there exists a maximal subset $X$ of $B$ with the property that $X \cap xU = \{x\}$, for all $x \in X$. Our assumption implies that $|X| > \tau$. Choose an open symmetric neighbourhood $V$ of $e$ in $G$ such that $V^4 \subset U$. Then Lemma 1.4.22 implies that the family $\gamma = \{xV : x \in X\}$ is discrete in $G$. We conclude that the family $\theta = \{B \cap W : W \in \gamma\}$ of non-empty open sets in $B$ is discrete in $B$ and $|\theta| = |\gamma| > \tau$, which contradicts $dc(B) \leq \tau^+$. □

It turns out that the property of a subset $B$ of a topological group $G$ being $\tau$-narrow in $G$ depends only on the subgroup $\langle B \rangle$ of $G$ algebraically generated by $B$:

PROPOSITION 5.1.16. *The following properties are equivalent for a subset $B$ of a topological group $G$:*

a) *$B$ is $\tau$-narrow in $G$;*
b) *$B$ is $\tau$-narrow in some subgroup $H$ of $G$ containing $B$;*
c) *$B$ is $\tau$-narrow in every subgroup $H$ of $G$ containing $B$;*
d) *$H$ is $\tau$-narrow in the subgroup $\langle B \rangle$ of $G$.*

PROOF. If $H$ is a subgroup of $G$ with $B \subset H$, then clearly $\langle B \rangle \subset H$. Therefore, d) $\Rightarrow$ c) $\Rightarrow$ b) $\Rightarrow$ a). It remains to show that d) follows from a). Suppose that $B$ is $\tau$-narrow in $G$ and put $K = \langle B \rangle$. Let $U$ be an arbitrary open neighbourhood of the identity $e$ in $K$. Choose an open symmetric neighbourhood $V$ of $e$ in $G$ such that $V \cap K \subset U$. Let $X$ be a maximal subset of $B$ with the property that $X \cap xV = \{x\}$, for all $x \in X$. We claim that $|X| \leq \tau$. Indeed, if $|X| > \tau$, choose an open symmetric neighbourhood $W$ of $e$ in $G$ satisfying $W^2 \subset V$. Since $B$ is $\tau$-narrow in $G$, there exists a set $F \subset G$ such that $B \subset FW$ and $|F| \leq \tau$. It follows from $|F| < |X|$ and $X \subset FW$ that $|X \cap yW| \geq 2$ for some $y \in F$. Let $x_1$ and $x_2$ be distinct elements of the intersection $X \cap yW$. Then $x_1^{-1}x_2 \in (yW)^{-1}yW = W^2 \subset V$, whence $x_2 \in x_1V$. This contradicts our choice of the set $X$, so $|X| \leq \tau$.

The maximality of $X$ implies that $B \subset XV$, and we claim that $B \subset XU$. Indeed, if $b \in B$, then $b = xv$ for some $x \in X$ and $v \in V$, whence $v \in x^{-1}b \in \langle B \rangle = K$. Therefore, $v \in K \cap V \subset U$ which in its turn implies that $b = xv \in xU$. This proves that $B \subset XU$, where $|X| \leq \tau$. Similarly, one can choose a subset $Y$ of $B$ such that $B \subset UY$ and $|Y| \leq \tau$. Hence $B$ is $\tau$-narrow in $\langle B \rangle$. □

Since the sets $X$ and $Y$ in the proof of Proposition 5.1.16 were chosen as subsets of $B$, we have:

COROLLARY 5.1.17. *Let $B$ be a $\tau$-narrow subset of a topological group $G$. Then, for every neighbourhood $U$ of the identity in $G$, there exists a subset $F$ of $B$ with $|F| \leq \tau$ such that $B \subset FU \cap UF$.*

In the following lemma we establish that $\tau$-narrow subsets of a topological group are stable under the inverse and multiplication in the group.

LEMMA 5.1.18. *Let $A$ and $B$ be $\tau$-narrow subsets of a topological group $G$. Then the sets $A^{-1}$ and $AB$ are also $\tau$-narrow in $G$.*

PROOF. Let $U$ be a neighbourhood of the identity $e$ in $G$. We can find a neighbourhood $O$ of $e$ in $G$ and a set $F \subset G$ such that $O^{-1} \subset U$ and $A \subset FO \cap OF$. Put $K = F^{-1}$. Then, clearly, $|K| = |F| \le \tau$ and

$$A^{-1} \subset (FO)^{-1} \cap (OF)^{-1} = O^{-1}K \cap KO^{-1} \subset UK \cap KU.$$

Hence, the set $A^{-1}$ is $\tau$-narrow in $G$.

Let us show that $AB$ is also $\tau$-narrow in $G$. Choose an open neighbourhood $V$ of $e$ in $G$ with $V^2 \subset U$ and a subset $L$ of $G$ with $|L| \le \tau$ satisfying $B \subset LV$. For every $y \in L$, choose a neighbourhood $W_y$ of $e$ in $G$ such that $y^{-1}W_y y \subset V$. Since $A$ is $\tau$-narrow in $G$, for every $y \in L$ there exists a subset $K_y$ of $G$ such that $|K_y| \le \tau$ and $A \subset K_y W_y$. Put $K = \bigcup_{y \in L} K_y$ and $M = KL$. It is clear that $|M| \le \tau$. Let us verify that $AB \subset MU$. Suppose that $a \in A$ and $b \in B$. Choose $y \in L$ such that $b \in yV$. Then there exists $x \in K_y$ such that $a \in xW_y$. Therefore, we have

$$ab \in xW_y yV = xy(y^{-1}W_y y)V \subset xyVV \subset xyU,$$

that is, $ab \in MU$. This proves that $AB \subset MU$. One can prove in a similar way that there exists a subset $M'$ of $G$ such that $|M'| \le \tau$ and $AB \subset UM'$. Clearly, $AB \subset EU \cap UE$, where the set $E = M \cup M'$ satisfies $|E| \le \tau$. Therefore, the product $AB$ is $\tau$-narrow in $G$. □

THEOREM 5.1.19. *Let $X$ be a $\tau$-narrow subset of a topological group $G$ which algebraically generates $G$. Then the group $G$ is $\tau$-narrow.*

PROOF. By Lemma 5.1.18, the set $Y_0 = X \cup X^{-1}$ is $\tau$-narrow in $G$. Consider the sequence $\{Y_n : n \in \omega\}$ of subsets of $G$ defined by $Y_{n+1} = Y_n Y_0$ for all $n \in \omega$. Apply Lemma 5.1.18 to show, by induction on $n$, that $Y_{n+1}$ is $\tau$-narrow for each $n \in \omega$. Since $\tau \ge \omega$ and $\langle X \rangle = \bigcup_{n=0}^{\infty} Y_n$, the group $G$ is $\tau$-narrow as well. □

The following two corollaries to Theorem 5.1.19 are almost immediate.

COROLLARY 5.1.20. *If a topological group $G$ contains a dense subgroup algebraically generated by a Lindelöf subspace, then $G$ is $\omega$-narrow.*

PROOF. Let $B$ be a Lindelöf subspace of $G$ which generates a dense subgroup $H$ of $G$. Then $B$ is $\tau$-narrow in $G$ by Lemma 5.1.14, so the group $H$ is $\tau$-narrow by Theorem 5.1.19. Hence, item d) of Proposition 5.1.1 implies that $G$ is $\tau$-narrow as well. □

Similarly, we have:

COROLLARY 5.1.21. *If a topological group $G$ contains a dense subgroup algebraically generated by a subspace $B$ with $c(B) \le \tau$, then $G$ is $\tau$-narrow.*

The subgroups of $\sigma$-compact groups form a proper subclass of the $\omega$-narrow groups (see Example 5.1.2). However, this still leaves open the possibility that every $\omega$-narrow group could be embedded as a subgroup into a product of $\sigma$-compact topological groups. Let us show that this is not the case either. Our argument makes use of *topological vector spaces* (a more detailed discussion of this subject is presented in Section 9.2).

Suppose that $L$ is a vector space over a field $K$, where $K$ is $\mathbb{R}$ or $\mathbb{C}$. A non-negative real-valued function $\|\cdot\|$ on $L$ is called a *norm* if it satisfies the following conditions:

(N1) if $x \in L$ and $\|x\| = 0$, then $x = 0_L$;
(N2) $\|\lambda x\| = |\lambda| \cdot \|x\|$, for all $\lambda \in K$ and $x \in L$;
(N3) $\|x + y\| \le \|x\| + \|y\|$, for all $x, y \in L$.

A vector space $L$ with a norm $\|\cdot\|$ is called a *normed space*. The normed space $(L, \|\cdot\|)$ admits a natural metrizable group topology. Indeed, for every positive number $\varepsilon$, consider the "open ball"

$$B(\varepsilon) = \{x \in L : \|x\| < \varepsilon\}$$

with center $0_L$ in $L$. One can easily verify that the family $\{B(\varepsilon) : \varepsilon > 0\}$ is a base at the neutral element $0_L$ for a Hausdorff group topology $\tau$ on the Abelian group $L$. In addition, the metric $d$ on $L$ defined by $d(x, y) = \|x - y\|$ for all $x, y \in L$ is invariant and generates the same topology $\tau$. Therefore, the group topology $\tau$ is metrizable. If the metric space $(L, d)$ is complete, we call $(L, \|\cdot\|)$ a *Banach space*. In what follows we omit the symbol of norm and simply say that $L$ is a normed or Banach space. It is clear that if $L$ is a Banach space, then $(L, \tau)$ is a Raĭkov complete topological group.

Let us say that a topological group $G$ is *locally minimal* if there exists a neighbourhood $U$ of the identity in $G$ such that $G$ admits no strictly weaker Hausdorff topological group topology for which $U$ remains a neighbourhood of the identity. Evidently, every discrete topological group is locally minimal. More generally, every locally compact topological group is locally minimal simply because all translations of a topological group are homeomorphisms. The next result shows that locally minimal groups can differ substantially from locally compact groups.

Proposition 5.1.22. *Every normed vector space considered as an Abelian topological group is locally minimal.*

Proof. Let $L$ be a vector space with a norm $\|\cdot\|$. Consider the unit open ball $B(1) = \{x \in L : \|x\| < 1\}$ in $L$ and suppose that $\tau$ is a weaker Hausdorff group topology on $L$ such that $B(1)$ is a neighbourhood of $0_L$ in $(L, \tau)$. Choose an open neighbourhood $U$ of $0_L$ in $(L, \tau)$ with $U \subset B(1)$. To show that $\tau$ coincides with the original topology of $L$, it suffices to verify that for every integer $n \geq 1$, the set $B(1/n) = \{x \in L : \|x\| < 1/n\}$ is a neighbourhood of the neutral element $0_L$ in $(L, \tau)$. By the continuity of the sum operation in $(L, \tau)$, there exists an open neighbourhood $V$ of $0_L$ in $(L, \tau)$ such that $V + \cdots + V \subset U$, where $V$ is $n$ times a summand. Then $V \subset B(1/n)$. Indeed, otherwise there exists $x \in U$ with $\|x\| \geq 1/n$, whence $\|nx\| \geq 1$. The latter contradicts the fact that $nx \in U \subset B(1)$. □

We need one more useful concept. Recall from Chapter 3 that a topological group $G$ is a *group with no small subgroups* or *NSS-group* if there exists a neighbourhood of the identity in $G$ containing no subgroups except for the trivial one. Obviously, every normed vector space is an $NSS$-group when considered as an Abelian topological group. In the following proposition we establish an important property of locally minimal $NSS$-groups.

Proposition 5.1.23. *Let $G$ be a subgroup of a topological product $\Pi = \prod_{i\in I} G_i$ of topological groups $G_i$'s. If $G$ is a locally minimal $NSS$-group, then there exists a finite set $J \subset I$ such that the restriction of the projection $\pi_J : \Pi \to \prod_{i\in J} G_i$ to $G$ is a topological embedding.*

Proof. Choose a neighbourhood $U$ of the identity $e$ in $G$ which does not contains non-trivial subgroups. Let also $V$ be a neighbourhood of $e$ in $G$ such that $G$ admits no strictly weaker Hausdorff group topology for which $V$ remains a neighbourhood of $e$. Then there exists a finite set $J \subset I$ and an open neighbourhood $W$ of the identity $e'$ in $\Pi_J = \prod_{i\in J} G_i$

such that $\pi_J^{-1}(W) \subset U \cap V$. It remains to verify that the restriction of $\pi_J$ to $G$ is a topological isomorphism of $G$ onto $\pi_J(G)$.

By the choice of $U$, $G \cap \pi_J^{-1}(e')$ is a trivial subgroup of $G$, so $p = \pi_J \restriction G$ is a continuous monomorphism. Therefore,

$$\tau = \{G \cap \pi_J^{-1}(O) : O \text{ is open in } \Pi_J\}$$

is a Hausdorff group topology on $G$ weaker than the original topology $t$ of $G$ inherited from $\Pi$. Since $V$ is a neighbourhood of $e$ in $(G, \tau)$, we conclude that $\tau = t$. Hence, the mapping $p$ is a topological embedding. □

THEOREM 5.1.24. [**T. Banakh**] *Let $G$ be any infinite-dimensional Banach space considered as an Abelian topological group. Then $G$ does not admit an embedding as a topological subgroup into a topological product of $\sigma$-compact topological groups.*

PROOF. Indeed, assume the contrary. Then, by Proposition 5.1.23, $G$ is $\sigma$-compact. Let $\gamma$ be a countable family of compact subsets of $G$ such that $G = \bigcup \gamma$. Since $G$ is a complete metric space, it has the Baire property. It follows that some element $C \in \gamma$ has a non-empty interior in $G$, say, $U$. Take any $x \in U$. Then $V = U - x$ is an open neighbourhood of the neutral element in $G$, and the closure of $V$ is compact. Therefore, the space $G$ is locally compact. It is well known, however, that every locally compact Banach space is finite-dimensional (see [**81**] or [**157**]), which contradicts our assumptions about $G$. We conclude that $G$ cannot be embedded into any product of $\sigma$-compact topological groups. □

## Exercises

5.1.a. A subset $Y$ of a uniform space $(X, \mathcal{U})$ is called *$\tau$-narrow* if, for every $U \in \mathcal{U}$, there exists a set $F \subset X$ with $|F| \leq \tau$ such that $Y \subset \bigcup_{x \in F} B(x, U)$. Prove that a subset $B$ of a topological group $G$ is $\tau$-narrow in $G$ iff $B$ is a $\tau$-narrow subset of the uniform space $(G, \mathcal{V})$, where $\mathcal{V}$ is the two-sided uniformity of the group $G$.

5.1.b. For a subset $Y$ of a space $X$, the inequality $c(Y, X) \leq \tau$ means that every disjoint family of open sets in $X$ contains at most $\tau$ elements which intersect $Y$. Prove that every subset $B$ of a topological group $G$ satisfying $c(B, G) \leq \tau$ is $\tau$-narrow in $G$. (This generalizes Corollary 5.1.21.)

5.1.c. Let $\pi \colon G \to H$ be a continuous homomorphism of topological groups, $L$ a subgroup of $H$, and let $K = \pi^{-1}(L)$. Verify that $K$ is topologically isomorphic to a closed subgroup of the group $G \times L$.

5.1.d. Show that a subgroup of a compact group can fail to be Lindelöf. Define, for every cardinal $\tau \geq \omega$, a subgroup $H$ of $\mathbb{T}^\tau$ satisfying $l(H) = \tau$.

5.1.e. Let $G$ be a hereditarily Lindelöf topological group. Show that every pseudocompact subspace of $G$ is metrizable and compact.
*Hint.* Show that the space $G$ is submetrizable, that is, admits a weaker metrizable topology.

5.1.f. (R. D. Kopperman *et al.* [**281**], Y. Torres [**500**]) Show that a topological group $G$ is $\tau$-narrow iff every subgroup $H$ of $G$ with $|H| \leq \tau^+$ is $\tau$-narrow. Verify that if all countable subgroups of a topological group $G$ are precompact, then so is $G$.

5.1.g. Prove that the product of an arbitrary family of $\omega$-narrow paratopological (quasitopological, semitopological) groups is an $\omega$-narrow paratopological (quasitopological, semitopological) group.

5.1.h. Verify that every locally compact Banach space is finite-dimensional (this fact was used in the proof of Theorem 5.1.24).

## Problems

5.1.A. Give an example of a paracompact $\omega$-narrow topological group $G$ with an uncountable closed discrete subspace.
*Hint.* Take a proper dense subgroup of a Lindelöf $P$-group of weight $\aleph_1$.

5.1.B. Show that a closed subgroup of an $\omega$-narrow paratopological group can fail to be $\omega$-narrow.
*Hint.* Consider the second diagonal in the square of the Sorgenfrey line.

5.1.C. Prove that every dense subgroup of an $\omega$-narrow paratopological group is $\omega$-narrow.

5.1.D. (M. A. López and M. G. Tkachenko [**295**]; items (a) and (c) were also obtained by G. Lukács [**296**]) Let $H$ be an $\omega$-narrow topological group. Prove the following:

(a) $H$ is topologically isomorphic to a closed subgroup of a product of second-countable groups if and only if the set $\varrho H \setminus H$ is the union of a family of $G_\delta$-sets in the Raĭkov completion $\varrho H$ of the group $H$.

(b) Every Lindelöf topological group is topologically isomorphic to a closed subgroup of the product of some family of second-countable groups.

(c) If the neutral element of $H$ is a $G_\delta$-set in $H$, then the group $H$ is topologically isomorphic to a closed subgroup of the product of a family of second-countable groups.

(d) The class of closed subgroups of products of Lindelöf topological groups and the class of topological groups described in (a) coincide.

(e) Give an example of an $\omega$-narrow Fréchet–Urysohn topological group that cannot be embedded as a closed subgroup into a product of second-countable topological groups.

(f) Give an example of an $\omega$-narrow hereditarily paracompact topological group that cannot be embedded as a closed subgroup into a product of second-countable topological groups.

(See also Problems 5.1.E, 6.5.D, 6.5.E, 8.3.C, and Exercises 6.5.c and 8.3.a.)

*Hint.* For (a), consider the Raĭkov completion $\varrho H$ of the group $H$ and note that the group $\varrho H$ is $\omega$-narrow. Since every point $x \in \varrho H \setminus H$ is contained in a $G_\delta$-subset $P$ of $\varrho H$ such that $P \cap H = \varnothing$, one can apply Corollary 3.4.19 to define a family $\mathcal{H} = \{h_i : i \in I\}$ of continuous homomorphisms of $\varrho H$ to second-countable topological groups $H_i$ such that $H = \bigcap_{i\in I} h_i^{-1}(h_i(H))$. Let $h$ be the diagonal product of the family $\mathcal{H}$. By Exercise 5.1.c, the group $H$ is topologically isomorphic to a closed subgroup of the product group $\varrho H \times \prod_{i\in I} H_i$. Finally, apply Theorem 3.4.23 to embed $\varrho H$ into a product $\Pi$ of second-countable topological groups and note that $\varrho H$ is closed in $\Pi$. Hence $H$ is a closed subgroup of the product $\Pi \times \prod_{i\in I} H_i$.

Conversely, if $H$ is a closed subgroup of a product $G = \prod_{i\in I} G_i$ of second-countable topological groups, consider the product $\varrho G = \prod_{i\in I} \varrho G_i$ of Raĭkov completions of the groups $G_i$ (see Corollary 3.6.23) and note that the closed subgroup $\varrho H$ of $\varrho G$ satisfies $G \cap \varrho H = H$. Since the factors $G_i$ and $\varrho G_i$ are second-countable, the complement $\varrho G \setminus G$ is the union of a family of $G_\delta$-sets in $\varrho G$. This implies the required property of $H$ with respect to $\varrho H$.

Parts (b), (c), and (d) of the problem are now immediate. For (e), take the $\Sigma$-product of uncountably many copies of the discrete group $\mathbb{Z}(2)$ and apply Corollary 1.6.35 along with (a) of the problem. For (f), consider the group $G_\tau$ with $\tau = \aleph_1$ defined in Example 4.4.11 and apply Exercise 4.4.i.

5.1.E. (M. A. López and M. G. Tkachenko [**295**]; G. Lukács [**296**]) Prove that a topological group $H$ is topologically isomorphic to a closed subgroup of a product of metrizable topological groups if and only if the group $H$ is $\omega$-balanced and the complement $\varrho H \setminus H$ is the union of a family of $G_\delta$-sets in $\varrho H$.

5.1.F. (M. A. López and M. G. Tkachenko [**295**]) Apply the conclusion of Problem 5.1.E to show that every Abelian topological group of countable pseudocharacter is topologically isomorphic to a closed subgroup of a product of metrizable topological groups.

## Open Problems

5.1.1. When is a topological group $G$ topologically isomorphic to a subgroup of a $\sigma$-product of second-countable groups?

5.1.2. Characterize the topological subgroups of Lindelöf topological groups.

5.1.3. Characterize the topological subgroups of paracompact topological groups. (See also Exercise 3.6.p.)

5.1.4. When is the Raĭkov completion of a topological group $G$ Lindelöf?

5.1.5. Let $G_1$ and $G_2$ be Lindelöf topological groups. Is it possible to find a Lindelöf topological group $G$ such that $G_1$ and $G_2$ are topologically isomorphic to subgroups of $G$?

5.1.6. Let $G_1$ and $G_2$ be Lindelöf topological groups. Does there exist a Lindelöf topological group $G$ such that $G_1$ and $G_2$ are topologically isomorphic to closed subgroups of $G$?

5.1.7. Given a countable family $\mathcal{P}$ of Lindelöf topological groups, is there a Lindelöf topological group $G$ such that every group $H \in \mathcal{P}$ is topologically isomorphic to a (closed) subgroup of $G$?

5.1.8. Is it true that every $\omega$-narrow paratopological group is topologically isomorphic to a subgroup of the product of some family of Lindelöf paratopological groups?

5.1.9. Let $A$ and $B$ be $\omega$-narrow subsets of a paratopological group $G$. Is the set $AB$ necessarily $\omega$-narrow in $G$?

5.1.10. Let $G$ be a Lindelöf topological group and $f\colon G \to H$ a continuous homomorphism of $G$ onto a metrizable group $H$. Let $M$ be an arbitrary subgroup of $H$. Is $f^{-1}(M)$ Lindelöf?

5.1.11. Let $G$ be a paracompact $\omega$-narrow topological group such that the space $G$ is the union of a countable family of closed discrete subspaces of $G$ (that is, $G$ is *strongly $\sigma$-discrete*). Is $G$ Lindelöf?

5.1.12. (I. I. Guran) Suppose that $G$ is a topological group such that for each open neighbourhood $U$ of the neutral element in $G$, there exists a countable subset $M$ of $G$ satisfying $UMU = G$. Is $G$ $\omega$-narrow?

## 5.2. Some basic cardinal invariants of topological groups

The results of this section show that cardinal functions behave much better on topological groups than on Tychonoff spaces. In particular, we establish that some cardinal functions coincide on the class of topological groups, while they are distinct, even for compact Hausdorff spaces. An example of this kind is the equality $\chi(G) = \pi\chi(G)$ which holds for every topological group $G$ (see Proposition 5.2.6).

Below $w(X)$, $nw(X)$, $d(X)$, $\chi(X)$, $\psi(X)$, $l(X)$, and $c(X)$ denote the weight, network weight, density, character, pseudocharacter, Lindelöf number and cellularity of a space $X$, respectively. The tightness, $\pi$-weight, and $\pi$-character of $X$ are denoted by $t(X)$, $\pi w(X)$, and $\pi\chi(X)$. The minimal number of compact subsets of $X$ required to cover $X$ is denoted by $k(X)$ and is called the *compact-covering number* of $X$. Needless to say all these cardinal functions are defined for an arbitrary topological or semitopological group $G$ as well.

An important cardinal function is the *index of narrowness* of a topological group $G$ denoted by $ib(G)$. By definition, $ib(G)$ is the minimal cardinal $\tau \geq \omega$ such that $G$ is $\tau$-narrow. The following result is a reformulation of Proposition 5.1.3.

PROPOSITION 5.2.1. *Every topological group $G$ satisfies the inequalities $ib(G) \leq l(G)$ and $ib(G) \leq c(G)$.*

The first of the inequalities in Proposition 5.2.1 can be improved as follows. Given a space $X$, we denote by $e(X)$ the supremum of cardinalities of closed discrete subsets of $X$. The cardinal invariant $e(X)$ is called the *extent* of $X$. Clearly, the extent of every Lindelöf space is countable.

PROPOSITION 5.2.2. *$ib(G) \leq e(G)$, for every topological group $G$.*

PROOF. Let $\kappa = e(G)$. It suffices to show that the group $G$ can be covered by at most $\kappa$ translates of every open symmetric neighbourhood $U$ of the neutral element $e$ of $G$. If this fails to be true for such a $U$, we can define by recursion a transfinite sequence $X = \{x_\alpha : \alpha < \kappa^+\}$ of elements of $G$ such that $x_\beta \notin x_\alpha U$ whenever $\alpha < \beta < \kappa^+$. Clearly, the set $X$ is $U$-disjoint. Take a symmetric open neighbourhood $V$ of $e$ in $G$ such that $V^4 \subset U$. By Lemma 1.4.22, the family $\{x_\alpha V : \alpha < \kappa^+\}$ of open sets is discrete in $G$, whence if follows that $X$ is a closed discrete subset of $G$. Since $|X| = \kappa^+$, this contradicts the definition of $\kappa$. □

The weight of a topological group can be expressed in terms of its character and index of narrowness.

PROPOSITION 5.2.3. *The equality $w(G) = ib(G) \cdot \chi(G)$ is valid for every topological group $G$.*

PROOF. The inequalities $ib(G) \leq l(G) \leq w(G)$ and $\chi(G) \leq w(G)$ are clear, so it suffices to verify that $w(G) \leq ib(G) \cdot \chi(G)$.

Let $\tau = ib(G) \cdot \chi(G)$. Denote by $\mathcal{H}$ a base at the identity $e$ of $G$ satisfying $|\mathcal{H}| \leq \tau$. Since $G$ is $\tau$-narrow, we can find, for every $U \in \mathcal{H}$, a set $S_U \subset G$ with $|S_U| \leq \tau$ such that $S_U U = G$. For every $U \in \mathcal{H}$, put $\mathcal{B}_U = \{xU : x \in S_U\}$. The family $\mathcal{B} = \bigcup\{\mathcal{B}_U : U \in \mathcal{H}\}$ satisfies $|\mathcal{B}| \leq \tau$, and we claim that $\mathcal{B}$ is a base of $G$.

Indeed, let $O$ be a neighbourhood of a point $a \in G$. One can find $U, V \in \mathcal{H}$ such that $aU \subset O$ and $V^{-1}V \subset U$. There exists $x \in S_V$ such that $a \in xV$, whence $x \in aV^{-1}$. We have

$$xV \subset (aV^{-1})V = a(V^{-1}V) \subset aU \subset O,$$

that is, $xV$ is an open neighbourhood of $a$ and $xV \subset O$. It remains to note that $xV \in \mathcal{B}$. □

COROLLARY 5.2.4. *If $G$ is a non-discrete $\omega$-narrow topological group, then $w(G) = \chi(G)$. In particular, every infinite precompact group $G$ satisfies $w(G) = \chi(G)$.*

Proposition 5.2.3 can be used to establish several non-trivial relations between well-known cardinal functions in topological groups. Since $ib(G) \leq c(G) \leq d(G)$ and $ib(G) \leq l(G) \leq k(G)$ for any topological group $G$, we have the following three inequalities.

THEOREM 5.2.5. *Every topological group $G$ satisfies:*

a) $w(G) \leq d(G) \cdot \chi(G)$;

b) $w(G) \leq k(G) \cdot \chi(G)$;
c) $w(G) \leq l(G) \cdot \chi(G)$.

Theorem 5.2.5 fails to be valid for compact Hausdorff spaces — the two arrows space $Z$ is compact, first-countable, and separable, but $w(Z) = \mathfrak{c}$. Therefore, neither a), b), nor c) of Theorem 5.2.5 can be extended to compact Hausdorff spaces.

The following result shows that the difference between several cardinal functions disappears in the realm of topological groups.

PROPOSITION 5.2.6. *Let $G$ be a topological group. Then:*

a) $\chi(G) = \pi\chi(G)$;
b) $w(G) = \pi w(G)$.

PROOF. a) It suffices to show that $\chi(G) \leq \pi\chi(G)$. Let $\gamma$ be a $\pi$-base at the identity $e$ of $G$ such that $|\gamma| = \pi\chi(G)$. Then the family $\mu = \{UU^{-1} : U \in \gamma\}$ is a local base at $e$. Indeed, if $O$ is a neighbourhood of $e$ in $G$, then there exists a neighbourhood $V$ of $e$ such that $VV^{-1} \subset O$. Since $\gamma$ is a $\pi$-base at $e$, one can find $U \in \gamma$ with $U \subset V$. Then $W = UU^{-1} \in \mu$ and $e \in W \subset O$. This proves that $\mu$ is a local base at the identity of $G$. Since $|\mu| \leq |\gamma| = \pi\chi(G)$, we conclude that $\chi(G) \leq \pi\chi(G)$. This proves a).

b) Note that $d(G) \leq \pi w(G)$ and $\pi\chi(G) \leq \pi w(G)$. Therefore, according to the previous item a) and Theorem 5.2.5 a), we have that

$$w(G) \leq d(G) \cdot \chi(G) \leq d(G) \cdot \pi\chi(G) \leq \pi w(G).$$

Since every base is a $\pi$-base, we conclude that $\pi w(G) \leq w(G)$. □

Again, none of the equalities in Proposition 5.2.6 is valid for compact Hausdorff spaces. Indeed, the one-point compactification $\alpha D$ of an uncountable discrete space $D$ satisfies $\aleph_0 = \pi\chi(\alpha D) < \chi(\alpha D) = |D|$. In addition, the two arrows space $Z$ is compact and satisfies $\aleph_0 = \pi w(Z) < w(Z) = \mathfrak{c}$.

In the case of compact topological groups, the relations between cardinal characteristics become even stronger. Given an infinite cardinal $\tau$, we denote by $\operatorname{Ln} \tau$ the minimal cardinal $\lambda$ such that $2^\lambda \geq \tau$.

COROLLARY 5.2.7. *Let $G$ be an infinite compact topological group. Then:*

a) $\pi\chi(G) = \chi(G) = w(G)$;
b) $|G| = 2^{w(G)}$;
c) $d(G) = \operatorname{Ln} w(G)$.

PROOF. a) The equality $\pi\chi(G) = \chi(G)$ follows from item a) of Proposition 5.2.6. Since every compact is Lindelöf, c) of Theorem 5.2.5 implies the inequality $w(G) \leq \chi(G)$. Hence $\pi\chi(G) = \chi(G) = w(G)$.

b) Let $\tau = w(G)$. Since the space $G$ is homogeneous and non-discrete, the character of $G$ at each point is equal to $\tau$, by a). Therefore, Čech–Pospíšil's theorem (see [**165**, 3.2.11]) implies that $|G| \geq 2^\tau$. Since every $T_1$ space $X$ satisfies $|X| \leq 2^{w(X)}$, we conclude that $|G| = 2^\tau$. Another way to obtain this equality is to apply Theorem 4.2.1.

c) The group $G$ is dyadic by Theorem 4.1.7, so for some cardinal $\kappa$, there exists a continuous mapping $f\colon D^\kappa \to G$ of a Cantor cube $D^\kappa$ onto the space $G$. We can assume, by Corollary 1.7.4, that $\kappa = w(G)$. Since $d(D^{\kappa}) = \operatorname{Ln}\kappa$ by the Hewitt–Marczewski–Pondiczery theorem, we conclude that $d(G) \le d(D^{w(G)}) = \operatorname{Ln} w(G)$. The inverse inequality follows from the fact that $w(G) \le 2^{d(G)}$ (see [**165**, Th. 1.5.6]). □

Recall that the *weak Lindelöf number* $wl(X)$ of a space $X$ is the least cardinal $\tau \ge \aleph_0$ such that every open covering of $X$ contains a subfamily of cardinality $\le \tau$ whose union is dense in $X$. It is easy to see that $wl(X) \le l(X)$ and $wl(X) \le c(X)$ for every space $X$. The spaces $X$ with $wl(X) \le \aleph_0$ are called *weakly Lindelöf*. The following result improves both the inequalities $ib(G) \le l(G)$ and $ib(G) \le c(G)$ from Proposition 5.2.1.

PROPOSITION 5.2.8. *Every topological group $G$ satisfies the inequality $ib(G) \le wl(G)$.*

PROOF. Let $\tau = wl(G)$. We have to show that $G$ can be covered by at most $\tau$ translates of any neighbourhood $U$ of the identity in $G$. Choose an open symmetric neighbourhood $V$ of the identity such that $V^2 \subset U$ and consider the open covering $\gamma = \{xV : x \in G\}$ of $G$. By definition of $\tau$, there exists a subfamily $\mu$ of $\gamma$ such that $\bigcup \mu$ is dense in $G$ and $|\mu| \le \tau$. Then there exists a subset $K$ of $G$ with $|K| \le \tau$ such that $\bigcup \mu = KV$. It remains to show that $KU = G$. Let $a \in G$ be arbitrary. Since $KV$ is dense in $G$, there exists $x \in K$ such that $aV \cap xV \ne \emptyset$. Therefore, $a \in xVV^{-1} = xV^2 \subset xU$. This implies that $KU = G$, whence $ib(G) \le \tau$. □

COROLLARY 5.2.9. *Every weakly Lindelöf topological group is $\omega$-narrow.*

The *$i$-weight* of a Tychonoff space $X$ denoted by $iw(X)$ is defined as the minimal cardinal $\tau$ such that there exists a continuous bijection of $X$ onto a Tychonoff space of weight $\tau$. Equivalently, $iw(X) \le \tau$ iff $X$ admits a weaker Tychonoff topology of weight less than or equal to $\tau$. It is clear that $iw(X) \le w(X)$. We refine this inequality in the next lemma which extends [**165**, Lemma 3.1.18] to Tychonoff spaces.

LEMMA 5.2.10. *The inequality $iw(X) \le nw(X)$ holds for every Tychonoff space $X$.*

PROOF. Let $\gamma$ be a network for $X$ satisfying $|\gamma| = \tau$, where $\tau = nw(X)$. We call a pair $P = (K, L) \in \gamma \times \gamma$ *separated* if there exists a continuous real-valued function $f_P\colon X \to \mathbb{R}$ such that $f_P(K) \cap f_P(L) = \emptyset$. Denote by $\mathcal{P}$ the family of all separated pairs in $\gamma \times \gamma$. It is clear that $\kappa = |\mathcal{P}| \le \tau$. Let $f$ be the diagonal product of the functions $f_P$ with $P \in \mathcal{P}$. Then $f$ is a continuous mapping of $X$ to $\mathbb{R}^{\mathcal{P}} \cong \mathbb{R}^{\kappa}$ and the subspace $Y = f(X)$ of $\mathbb{R}^{\kappa}$ satisfies $w(Y) \le \kappa \le \tau$. It remains to verify that $f\colon X \to Y$ is a bijection. If $x, y \in X$ and $x \ne y$, then there exists a continuous real-valued function $g$ on $X$ such that $g(x) = 0$ and $g(y) = 1$. Choose open neighbourhoods $U$ and $V$ of $x$ and $y$, respectively, such that $|g(y)| < 1/3$ and $|g(z) - 1| < 1/3$ for all $y \in U$ and $z \in V$. Since $\gamma$ is a network for $X$, we can find $K, L \in \gamma$ such that $x \in K \subset U$ and $y \in L \subset V$. It is clear that $g(K) \cap g(L) \subset g(U) \cap g(V) = \emptyset$, so the pair $P = (K, L)$ is separated. Therefore, $f_P(K) \cap f_P(L) = \emptyset$, whence it follows that $f_P(x) \ne f_P(y)$ and $f(x) \ne f(y)$.

We conclude that $f\colon X \to Y$ is a continuous bijection onto the Tychonoff space $Y$ satisfying $w(Y) \le \tau$, so $iw(X) \le \tau = nw(X)$. □

The index of narrowness is instrumental in estimating the $i$-weight of topological groups.

PROPOSITION 5.2.11. *Let $G$ be a topological group. Then there exists a continuous isomorphism $\varphi\colon G \to H$ of $G$ onto a topological group $H$ satisfying $w(H) \le ib(G) \cdot \psi(G)$. Thus, $iw(G) \le ib(G) \cdot \psi(G)$.*

Proof. Let $\tau = ib(G) \cdot \psi(G)$. Then there exists a family $\gamma = \{U_\alpha : \alpha < \tau\}$ of open neighbourhoods of the identity $e$ in $G$ such that $\bigcap \gamma = \{e\}$. Apply Lemma 5.1.7 to find, for every $\alpha < \tau$, a continuous homomorphism $\pi_\alpha \colon G \to H_\alpha$ onto a topological group $H_\alpha$ with $w(H_\alpha) \leq \tau$ such that $\pi_\alpha^{-1}(V_\alpha) \subset U_\alpha$, for some open neighbourhood $V_\alpha$ of the identity in $H_\alpha$.

Let $\varphi \colon G \to \prod_{\alpha<\tau} H_\alpha$ be the diagonal product of the homomorphisms $\pi_\alpha$ with $\alpha < \tau$. Put $H = \varphi(G)$. It is clear that $w(H) \leq \tau$, and from the choice of $\gamma$ and $\varphi$ it follows that the kernel of $\varphi$ is trivial, so $\varphi \colon G \to H$ is a continuous isomorphism. Therefore, $iw(G) \leq \tau$. □

As a special case of Proposition 5.2.11, we obtain:

Corollary 5.2.12. *An $\omega$-narrow topological group of countable pseudocharacter admits a continuous isomorphism onto a second-countable topological group.*

Theorem 5.2.13. *Let $G$ be a topological group such that $nw(G) \leq \tau$. Then there exists a continuous isomorphism of $G$ onto a topological group $H$ with $w(H) \leq \tau$.*

Proof. This follows from Proposition 5.2.11, since both the index of narrowness of $G$ and the pseudocharacter of $G$ do not exceed the network weight of $G$. □

Corollary 5.2.14. *Every topological group with a countable network admits a continuous isomorphism onto a second-countable topological group.*

Proposition 5.2.11 provides strong restraints for the cardinality of topological groups.

Theorem 5.2.15. *Every topological group $G$ satisfies $|G| \leq 2^{ib(G)\cdot\psi(G)}$. Therefore, $|G| \leq 2^{l(G)\cdot\psi(G)}$ and $|G| \leq 2^{c(G)\cdot\psi(G)}$.*

Proof. Any Tychonoff space $X$ satisfies $|X| \leq 2^{w(X)}$ (see [**165**, Theorem 1.5.1]). Since bijections do not change the size of spaces, we have the stronger inequality $|X| \leq 2^{iw(X)}$. Therefore, the inequality $|G| \leq 2^{ib(G)\cdot\psi(G)}$ follows directly from Proposition 5.2.11. By Proposition 5.2.1, $ib(G) \leq l(G)$ and $ib(G) \leq c(G)$, so the rest of the theorem is immediate. □

Corollary 5.2.16. *Every $\omega$-narrow topological group $G$ satisfies $|G| \leq 2^{\psi(G)}$.*

None of the inequalities in Theorem 5.2.15 is valid for Tychonoff spaces. Indeed, it is consistent with $ZFC$ that there exists a regular Lindelöf space $X$ of countable pseudocharacter with $|X| > \mathfrak{c}$ (see [**441**]), so we have $2^{l(X)\cdot\psi(X)} = \mathfrak{c} < |X|$. As for the second inequality, there exist in $ZFC$ a Tychonoff space $X$ as large as we wish satisfying the conditions $c(X) \cdot \psi(X) \leq \aleph_0$ (see Exercise 5.2.f).

In the next result we will see how the compact-covering number $k(G)$ of a topological group $G$ can be used to bound the network weight.

Proposition 5.2.17. *Let $G$ be a topological group. Then $nw(G) \leq k(G) \cdot \psi(G)$.*

Proof. Put $\tau = k(G) \cdot \psi(G)$. Since $ib(G) \leq l(G) \leq k(G)$, Proposition 5.2.11 implies that there exists a continuous isomorphism $\varphi \colon G \to H$ of $G$ onto a topological group $H$ with $w(H) \leq \tau$. Let $\mathcal{K}$ be a family of compact sets in $G$ such that $\bigcup \mathcal{K} = G$ and $|\mathcal{K}| = k(G) \leq \tau$. The restriction of $\varphi$ to every $K \in \mathcal{K}$ is a homeomorphism, so that $w(K) \leq \tau$ for each $K \in \mathcal{K}$. Since $G = \bigcup \mathcal{K}$, we conclude that $nw(G) \leq \tau \cdot \tau = \tau$. □

The two arrows space $Z$ is compact, first-countable and $nw(Z) = w(Z) = \mathfrak{c}$, whence it follows that $\aleph_0 = k(Z) \cdot \psi(Z) < nw(Z) = \mathfrak{c}$. Hence, the inequality in Proposition 5.2.17 fails to hold for compact Hausdorff spaces. In Corollary 5.3.25 we will give an interesting generalization of Proposition 5.2.17.

Clearly, many important cardinal invariants are not productive when taking products of arbitrary families of topological groups. For example, the Lindelöf number of the product of a sufficiently large family of Lindelöf topological groups can be as large as we wish (consider the topological groups $R^\tau$), the same holds for the density and for the pseudocharacter. It is for this reason that we should appreciate the index of narrowness — it is not increased, as we saw above, either by arbitrary products, or by passing to subgroups. However, when we take finite powers of topological groups, the situation is not so clear regarding many natural properties, and it becomes much harder to find counterexamples to certain tempting conjectures. We discuss some concrete questions arising here in the problem section below.

There are many reasons why we should pay particular attention to the behaviour of cardinal invariants and of other topological properties under products of topological groups. Given a class $\mathcal{P}$ of topological spaces and two spaces $X$ and $Y$ from $\mathcal{P}$, we might be willing, for various reasons, to represent both $X$ and $Y$ as closed subspaces of some third element $Z$ of the class $\mathcal{P}$. In almost all classes $\mathcal{P}$ of topological spaces this problem is solved trivially: most often, by taking the disjoint topological union of $X$ and $Y$. The situation changes drastically if $\mathcal{P}$ is a class of topological groups. There is no operation on topological groups which could be claimed to be analogous to the operation of free topological sum. The simplest way to represent two topological groups as closed subgroups of another topological group is to take their topological product. This, evidently, brings us to the problem of preservation of properties of topological groups under finite products. We will encounter concrete problems of this kind in many sections of this book.

## Exercises

5.2.a. Generalize Corollary 5.1.20 and show that if a subset $B$ of a topological group $G$ satisfies $wl(B) \leq \tau$ and generates a dense subgroup of $G$, then $G$ is $\tau$-narrow.

5.2.b. Give an example of a Tychonoff $\omega$-narrow quasitopological group $G$ and a closed subgroup $H$ of $G$ such that $H$ fails to be $\omega$-narrow.

*Hint.* Verify that the additive group $\mathbb{R}^2$ endowed with topology $\mathcal{T}$ defined in the hint to Exercise 1.4.d is a Tychonoff quasitopological group, and that the $x$-axis is a closed discrete subgroup of $(\mathbb{R}^2, \mathcal{T})$.

5.2.c. Construct an infinite countably compact topological Abelian group $G$ satisfying $|G| < w(G)$.

5.2.d. Let $G$ be an $\omega$-narrow topological group, and let $A$ be a subspace of $G$ with a countable network. Then there exists a continuous homomorphism $f$ of $G$ onto a topological group $H$ with a countable base such that $f$ restricted to $A$ is a one-to-one mapping.

5.2.e. Give an example of a regular hereditarily Lindelöf hereditarily separable first-countable paratopological group $G$ such that $G \times G$ is neither normal nor hereditarily separable.

5.2.f. (D. B. Shakhmatov [**428**]) Show that the inequality $|G| \leq 2^{c(G)\cdot\psi(G)}$ in Theorem 5.2.15 is not valid for Tychonoff spaces. Construct, for every cardinal $\tau \geq \omega$, a dense subspace $X_\tau$ of $I^\tau$ such that $c(X_\tau) \cdot \psi(X_\tau) \leq \omega$ and $|X_\tau| \geq \tau$.

## Problems

5.2.A. (For the Hausdorff case, D. B. Shakhmatov [**427**]; for the regular and Tychonoff cases, C. Hernández [**226**])

a) Every Hausdorff (regular, Tychonoff) paratopological group with a countable network can be mapped onto a Hausdorff (regular, Tychonoff) paratopological group with a countable base by a continuous isomorphism.

b) Every Hausdorff semitopological (quasitopological) group $G$ with a countable network admits a continuous isomorphism onto a Hausdorff semitopological (quasitopological) group with a countable base. The same assertion remains valid for regular and Tychonoff semitopological (quasitopological) groups.

c) Let $f$ be a continuous real-valued function on a Hausdorff (regular, Tychonoff) paratopological group $G$ with a countable network. Then there exists a weaker second-countable Hausdorff (regular, Tychonoff) paratopological group topology on $G$ that keeps $f$ continuous. A similar assertion is valid for Hausdorff, regular, and Tychonoff semitopological (quasitopological) groups.

5.2.B. (V. V. Uspenskij [**514**]) There exists a second-countable topological group $G$ such that every second-countable topological group $H$ is topologically isomorphic to a subgroup of $G$.

5.2.C. (S. A. Shkarin [**447**]) There exists a second-countable Abelian topological group $G$ such that every second-countable Abelian topological group is topologically isomorphic to a subgroup of $G$.

5.2.D. (V. I. Malykhin [**300**]) Under the Continuum Hypothesis, there exists a topological group $G$ of countable cellularity such that the cellularity of $G \times G$ is uncountable.

5.2.E. (A. V. Arhangel'skii and D. K. Burke [**51**]) Show that there exists a regular paratopological group $G$ with a countable $\pi$-base that fails to be first-countable.

5.2.F. (A. V. Arhangel'skii and A. Bella [**50**]) Let $G$ be a separable semitopological group. Show that $G$ is $\omega$-narrow.

5.2.G. (V. I. Malykhin [**300**]) It is consistent with $ZFC$ that there is a hereditarily separable countably compact topological group $G$ such that the space $G \times G$ has uncountable tightness and, hence, is not hereditarily separable.

## Open Problems

5.2.1. Suppose that a Hausdorff (regular, Tychonoff) paratopological group $G$ contains a dense $\tau$-narrow subgroup. Is $G$ $\tau$-narrow?

5.2.2. Suppose that a Hausdorff (regular, Tychonoff) quasitopological group $G$ contains a dense $\tau$-narrow subgroup. Is $G$ then $\tau$-narrow?

5.2.3. Let $G$ be a $\tau$-narrow Hausdorff (regular, Tychonoff) paratopological group such that $\psi(G) \leq \tau$. Is the cardinality of $G$ not greater than $2^\tau$? (See also Problem 3.4.F).

5.2.4. Let $G$ be a Hausdorff (regular, Tychonoff) $\tau$-narrow quasitopological group such that $\psi(G) \leq \tau$. Is the cardinality of $G$ not greater than $2^\tau$? (See also Problems 5.2.3 and 3.4.F).

5.2.5. Let $G$ be a Lindelöf regular paratopological group of countable pseudocharacter. Does $G$ admit a continuous isomorphism onto a regular second-countable paratopological group? (See also Problems 3.4.3, 5.7.P, and the article [**51**]).

5.2.6. Let $G$ be a Lindelöf paratopological (semitopological) group. Does $G$ satisfy $inv(G) \leq \omega$?

5.2.7. Is there in $ZFC$ a topological group $G$ of countable tightness such that the tightness of $G \times G$ is uncountable?

5.2.8. Is there in $ZFC$ a paratopological group $G$ of countable tightness such that the tightness of $G \times G$ is uncountable?

5.2.9. Is it true that, for every infinite cardinal $\tau$, there exists a topological group $G_\tau$ of weight $\leq \tau$ such that every topological group $G$ of weight $\leq \tau$ is topologically isomorphic to a subgroup of $G_\tau$? (See also Problem 5.2.B.)

5.2.10. Is it true that, for every infinite cardinal $\tau$, there exists an Abelian topological group $H_\tau$ such that every Abelian topological group $H$ of weight $\leq \tau$ is topologically isomorphic to a subgroup of $H_\tau$? (See Problem 5.2.C.)

5.2.11. Is there a regular second-countable paratopological group $G$ such that every regular second-countable paratopological group $H$ is topologically isomorphic to a subgroup of $G$?

5.2.12. Is there a regular second-countable Abelian paratopological group $G$ such that every regular second-countable Abelian paratopological group $H$ is topologically isomorphic to a subgroup of $G$?

## 5.3. Lindelöf Σ-groups and Nagami number

In this section, we introduce a cardinal function on the class of Tychonoff topological spaces called the *Nagami number.* The spaces with countable Nagami number are called *Lindelöf Σ-spaces*, while topological groups with the same property are said to be *Lindelöf Σ-groups.* We will see that $\sigma$-compact spaces form a proper subclass of Lindelöf Σ-spaces; however, Lindelöf Σ-spaces and $\sigma$-compact spaces share many topological properties.

Lindelöf Σ-groups are remarkable in many respects. First, this class of groups is stable with respect to taking countable products, closed subgroups and continuous homomorphic images. Second, every family $\gamma$ of $G_\delta$-sets in a Lindelöf Σ-group contains a countable subfamily $\mu$ such that $\bigcup \mu$ is dense in $\bigcup \gamma$ (see Corollary 5.3.19). In particular, every subgroup of a Lindelöf Σ-group has countable cellularity (Corollary 5.3.22).

Suppose that $X$ is a subset of $Y$ and that $\gamma$ is a family of subsets of $Y$. We say that $\gamma$ *separates $X$ from $Y \setminus X$* if for every $x \in X$ and every $y \in Y \setminus X$, there exists $F \in \gamma$ such that $x \in F$ and $y \notin F$.

Let $\beta X$ be the Čech–Stone compactification of a Tychonoff space $X$ and let $\mathcal{F}$ be the family of all closed subsets of $\beta X$. We define the *Nagami number* $Nag(X)$ of $X$ as follows:

$$Nag(X) = \min\{|\mathcal{P}| : \mathcal{P} \subset \mathcal{F},\ \mathcal{P} \text{ separates } X \text{ from } \beta X \setminus X\}.$$

It is immediate that $Nag(X) \leq k(X)$ for every Tychonoff space $X$. In particular, every $\sigma$-compact regular space $X$ satisfies $Nag(X) \leq \aleph_0$. A Tychonoff space $X$ such that $Nag(X) \leq \aleph_0$ is called a *Lindelöf Σ-space.* We establish below some properties of the cardinal function $Nag$.

First, we show that the Čech–Stone compactification $\beta X$ of $X$ in the definition of $Nag(X)$ can be replaced by any compactification $bX$ of the space $X$.

LEMMA 5.3.1. *The following conditions are equivalent for a Tychonoff space $X$ and an infinite cardinal $\tau$:*

a) $Nag(X) \leq \tau$;

b) *there exist a Hausdorff compactification $bX$ of $X$ and a family $\mathcal{F}$ of closed subsets of $bX$ separating $X$ from $bX \setminus X$ such that $|\mathcal{F}| \leq \tau$;*

c) *for every Hausdorff compactification $bX$ of $X$, there exists a family $\mathcal{F}$ of closed subsets of $bX$ separating $X$ from $bX \setminus X$ such that $|\mathcal{F}| \leq \tau$.*

PROOF. Clearly, b) follows from a) and c) implies a). Let us verify that a) $\Rightarrow$ c) and that b) $\Rightarrow$ a).

a) $\Rightarrow$ c). If $Nag(X) \leq \tau$, we can find a family $\mathcal{P}$ of closed subsets of $\beta X$ separating $X$ from $\beta X \setminus X$ and satisfying $|\mathcal{P}| \leq \tau$. We can also assume that the family $\mathcal{P}$ is closed under finite intersections. Let $bX$ be an arbitrary Hausdorff compactification of $X$. Extend the identity mapping $id_X$ of $X$ to a continuous mapping $f \colon \beta X \to bX$ and consider the family

$$\mathcal{F} = \{f(P) : P \in \mathcal{P}\}$$

of closed subsets of $bX$. We claim that $\mathcal{F}$ separates $X$ from $bX \setminus X$. Indeed, take arbitrary points $x \in X$ and $y \in bX \setminus X$. Clearly, the compact set $K = f^{-1}(y)$ is disjoint from $X$. Note that the family $\mathcal{P}(x) = \{P \in \mathcal{P} : x \in P\}$ is closed under finite intersections and that $C = \bigcap \mathcal{P}(x) \subset X$. Then $U = \beta X \setminus K$ is an open neighbourhood of $C$ in $\beta X$ and, hence, there exists a finite subfamily $\gamma$ of $\mathcal{P}(x)$ such that $P = \bigcap \gamma \subset U$. Therefore, $P \in \mathcal{P}(x) \subset \mathcal{P}$, $x \in P$ and $P \cap K = \varnothing$. This implies immediately that $F = f(P) \in \mathcal{F}$, $x \in F$ and $y \notin F$. So, $\mathcal{F}$ separates $X$ from $bX \setminus X$. Clearly, $|\mathcal{F}| \leq |\mathcal{P}| \leq \tau$.

b) $\Rightarrow$ a). Let $bX$ be a Hausdorff compactification of $X$ and let $\mathcal{F}$ be a family of closed subsets of $bX$ separating $X$ from $bX \setminus X$ and satisfying $|\mathcal{F}| \leq \tau$. The identity mapping $id_X$ can be extended to a continuous mapping $g \colon \beta X \to bX$. We claim that the family

$$\mathcal{P} = \{f^{-1}(F) : F \in \mathcal{F}\}$$

of closed subsets of $\beta X$ separates $X$ from $\beta X \setminus X$. Indeed, let $x \in X$ and $y \in \beta X \setminus X$ be arbitrary points. Since $id_X$ is a perfect mapping, its extension $g$ satisfies $g(\beta X \setminus X) \subset bX \setminus X$ by [**165**, Theorem 3.7.15]. Hence, $z = g(y) \in bX \setminus X$. By the choice of $\mathcal{F}$, there exists $F \in \mathcal{F}$ such that $x \in F \not\ni z$. Then $P = g^{-1}(F)$ belongs to $\mathcal{P}$, $x \in P$ and $y \notin P$. This proves our claim. It is also clear that $|\mathcal{P}| \leq |\mathcal{F}| \leq \tau$. □

Corollary 5.3.2. *If $K$ is a closed subspace of a space $X$, then $Nag(K) \leq Nag(X)$. In particular, the class of Lindelöf $\Sigma$-spaces is hereditary with respect to closed subspaces.*

Proof. Denote by $\mathcal{F}$ a family of closed subsets of $\beta X$ which separates $X$ from $\beta X \setminus X$ and satisfies $|\mathcal{F}| = Nag(X)$. Denote by $bK$ the closure of $K$ in $\beta X$. It is clear that $bK$ is a compactification of $K$. For every $F \in \mathcal{F}$, denote by $C_F$ the closure of $F \cap K$ in $\beta X$. Then $C_F \subset bK$ for each $F \in \mathcal{F}$. Since $bK \setminus K \subset \beta X \setminus X$, the family

$$\mathcal{C} = \{C_F : F \in \mathcal{F}\}$$

separates $K$ from $bK \setminus K$ and satisfies $|\mathcal{C}| \leq |\mathcal{F}|$. Therefore, by Lemma 5.3.1, we have $Nag(K) \leq |\mathcal{C}| \leq |\mathcal{F}| = Nag(X)$. □

The next result clarifies the relations between the Lindelöf number, Nagami number, and network weight.

Proposition 5.3.3. *Every Tychonoff space $X$ satisfies $l(X) \leq Nag(X) \leq nw(X)$.*

Proof. First, we show that $Nag(X) \leq nw(X)$. Let $\gamma$ be a network for $X$ with $|\gamma| = nw(X)$. Consider the family $\mathcal{F} = \{cl_{\beta X} K : K \in \gamma\}$. The elements of $\mathcal{F}$ are closed in the Čech–Stone compactification $\beta X$ of $X$ and $|\mathcal{F}| \leq |\gamma|$. Let $x \in X$ and $y \in \beta X \setminus X$ be arbitrary points. In $\beta X$, choose disjoint open neighbourhoods $U$ and $V$ of $x$ and $y$, respectively. Since $\gamma$ is a network in $X$, there exists $K \in \gamma$ such that $x \in K \subset U$. Put $F = cl_{\beta X} K$. Then $F \in \mathcal{F}$ and $x \in F$. Since $U$ and $V$ are disjoint, we have $F \cap V = \varnothing$, whence $y \notin F$. Therefore, the family $\mathcal{F}$ separates $X$ from $\beta X \setminus X$ and, hence, $Nag(X) \leq |\mathcal{F}| \leq |\gamma| = nw(X)$.

It remains to verify that $l(X) \leq Nag(X)$. Choose a family $\mathcal{P}$ of closed subsets of $\beta X$ with $|\mathcal{P}| = Nag(X)$. We can assume that $\mathcal{P}$ is closed under finite intersections. Suppose that $\mu$ is a covering of $X$ by open sets in $\beta X$. All we need to show is that $\mu$ contains a subfamily $\nu$ which covers $X$ and satisfies $|\nu| \leq Nag(X)$.

Let us call an element $P \in \mathcal{P}$ *small* if there exists a finite subfamily of $\mu$ covering $P$. Denote by $\mathcal{P}_s$ the subfamily of $\mathcal{P}$ consisting of all small elements. We claim that $X \subset \bigcup \mathcal{P}_s$. Indeed, let $x$ be an arbitrary point of $X$. Since $\mathcal{P}$ separates $X$ from $\beta X \setminus X$, the intersection $C(x)$ of the family $\mathcal{P}(x) = \{P \in \mathcal{P} : x \in P\}$ is a compact subset of $X$. Hence, there exists a finite subfamily $\nu(x)$ of $\mu$ which covers $C(x)$. Let $U(x) = \bigcup \nu(x)$. Then $C(x) \subset U(x)$. Since $\beta X$ is compact and the family $\mathcal{P}(x)$ is closed under finite intersections, we can find $P \in \mu$ such that $x \in P \subset U(x)$. This proves that $x \in P \in \mathcal{P}_s$, whence our claim follows.

For every $P \in \mathcal{P}_s$, choose a finite subfamily $\mu_P$ of $\mu$ which covers $P$. Let also $\nu = \bigcup\{\mu_P : P \in \mathcal{P}_s\}$. Then $X \subset \bigcup \mathcal{P}_s \subset \bigcup \nu$ and $|\nu| \leq |\mathcal{P}| \cdot \omega = Nag(X)$. □

COROLLARY 5.3.4. *Every Tychonoff space $X$ satisfies $Nag(X) \leq w(X)$. In particular, all regular second-countable spaces are Lindelöf $\Sigma$-spaces.*

It turns out that many results valid for $\sigma$-compact spaces remain valid for Lindelöf $\Sigma$-spaces. In particular, continuous mappings preserve the class of Lindelöf $\Sigma$-spaces.

PROPOSITION 5.3.5. *Let $f : X \to Y$ be a continuous onto mapping of Tychonoff spaces. Then $Nag(Y) \leq Nag(X)$. In particular, if $X$ is a Lindelöf $\Sigma$-space, so is $Y$.*

PROOF. Let $g : \beta X \to \beta Y$ be a continuous extension of $f$. Denote by $\mathcal{F}$ a family of closed subsets of $\beta X$ which separates $X$ from $\beta X \setminus X$ and satisfies $|\mathcal{F}| = Nag(X)$. We can assume that $\mathcal{F}$ is closed under finite intersections. Let us verify that the family $\mathcal{G} = \{g(F) : F \in \mathcal{F}\}$ separates $Y$ from $\beta Y \setminus Y$.

Take arbitrary points $y \in Y$, $z \in \beta Y \setminus Y$ and put $K = g^{-1}(z)$. Then $K$ is a compact subset of $\beta X$ and $K \cap X = \varnothing$. Pick a point $x \in X$ with $f(x) = y$. Clearly, the family $\mathcal{F}(x) = \{F \in \mathcal{F} : x \in F\}$ is closed under finite intersections and $C_x = \bigcap \mathcal{F}(x)$ is a subset of $X$. Since the elements of $\mathcal{F}$ are closed in $\beta X$ and $U = \beta X \setminus K$ is an open neighbourhood of $C_x$, there exists $F \in \mathcal{F}(x)$ disjoint from $K$. Therefore, $G = f(F)$ is an element of $\mathcal{G}$ which does not contain the point $z$. We conclude that $x \in G \not\ni z$, so $\mathcal{G}$ separates $Y$ from $\beta Y \setminus Y$.

Finally, $|\mathcal{G}| \leq |\mathcal{F}| = Nag(X)$; hence, $Nag(Y) \leq |\mathcal{G}| \leq Nag(X)$. □

PROPOSITION 5.3.6. *Let $f : X \to Y$ be a perfect onto mapping of Tychonoff spaces. Then $Nag(X) = Nag(Y)$.*

PROOF. By Proposition 5.3.5, $Nag(Y) \leq Nag(X)$. Therefore, it suffices to verify that $Nag(X) \leq Nag(Y)$. Let $\mathcal{P}$ be a family of closed subsets of $\beta Y$ separating $Y$ from $\beta Y \setminus Y$ and satisfying $|\mathcal{P}| = Nag(Y)$. Extend $f$ to a continuous mapping $g : \beta X \to \beta Y$. We claim that the family $\mathcal{F} = \{g^{-1}(P) : P \in \mathcal{P}\}$ of closed subsets of $\beta X$ separates $X$ from $\beta X \setminus X$. Indeed, take arbitrary points $x \in X$ and $y \in \beta X \setminus X$. Since $f$ is perfect, we have $g(\beta X \setminus X) \subset \beta Y \setminus Y$ by [**165**, Theorem 3.7.15], whence $y' = g(y) \in \beta Y \setminus Y$. Clearly, $x' = g(x) = f(x) \in Y$, so there exists $P \in \mathcal{P}$ such that $x' \in P$ and $y' \notin P$. Then $F = g^{-1}(P) \in \mathcal{F}$, $x \in F$ and $y \notin F$. This proves the claim.

Therefore, $Nag(X) \leq |\mathcal{F}| \leq |\mathcal{P}| = Nag(Y)$. □

COROLLARY 5.3.7. *Let $X$ be a Tychonoff space such that $w(X) \leq \tau$ and let $K$ be a compact space. Then $Nag(Y) \leq \tau$, for every closed subspace $Y$ of the product space $X \times K$.*

PROOF. Denote by $\pi$ the projection of $X \times K$ to $X$. Since $X$ is compact, the mapping $\pi$ is perfect. Applying Propositions 5.3.6 and 5.3.3, we deduce that

$$Nag(X \times K) = Nag(X) \leq nw(X) \leq w(X) \leq \tau.$$

Hence, by Corollary 5.3.2, $Nag(Y) \leq Nag(X \times K) \leq \tau$ for every closed subspace $Y$ of $X \times K$. □

The analogy between $\sigma$-compact and Lindelöf $\Sigma$-spaces can be extended further. Clearly, if $\gamma$ is countable family of $\sigma$-compact subsets of a space $X$, then $Y = \bigcup \gamma \subset X$ is also $\sigma$-compact. Similarly, we have the following result expressed in terms of the cardinal function $Nag$.

PROPOSITION 5.3.8. *Let $\tau$ be an infinite cardinal and let $\gamma$ be a family of subspaces of a Tychonoff space $X$ such that $Nag(P) \leq \tau$ for each $P \in \gamma$ and $|\gamma| \leq \tau$. Then the subspace $Y = \bigcup \gamma$ of $X$ satisfies $Nag(Y) \leq \tau$. Therefore, if $X$ is covered by a countable family of its Lindelöf $\Sigma$-subspaces, then $X$ is also a Lindelöf $\Sigma$-space.*

PROOF. Let $\gamma = \{P_i : i \in I\}$, where $|I| \leq \tau$. Denote by $S$ the free topological sum of the elements of $\gamma$, that is, $S = \bigoplus_{i \in I} P_i$. Consider the natural mapping $f$ of $S$ onto $Y$ whose restriction to each $P_i$ coincides with the identity embedding of $P_i$ to $X$. Clearly, $f$ is continuous. By Proposition 5.3.5, it suffices to verify that $Nag(S) \leq \tau$.

Let $\beta S$ be the Čech–Stone compactification of $S$. Since every $P_i$ is closed and open in $S$, the closure $K_i$ of $P_i$ in $\beta S$ is naturally homeomorphic to $\beta P_i$. Let $\mathcal{F}_i$ be a family of closed subsets of $K_i$ which separates $P_i$ from $K_i \setminus P_i$ and satisfies $|\mathcal{F}_i| \leq \tau$. Then the family

$$\mathcal{F} = \{K_i : i \in I\} \cup \bigcup_{i \in I} \mathcal{F}_i$$

consists of closed subsets of $\beta S$ and satisfies $|\mathcal{F}| \leq \tau \cdot \tau = \tau$. We claim that $\mathcal{F}$ separates $S$ from $\beta S \setminus S$. Indeed, let $x \in S$ and $y \in \beta S \setminus S$ be arbitrary points. Then $x \in P_i$ for some $i \in I$. If $y \notin K_i$, then $x \in K_i \not\ni y$. Otherwise, by the choice of $\mathcal{F}_i$, there exists $F \in \mathcal{F}_i$ with $x \in F \not\ni y$. This proves our claim. Therefore, $Nag(S) \leq |\mathcal{F}| \leq \tau$.

The second part of the proposition is immediate. □

The next result shows, in particular, that the class of Lindelöf $\Sigma$-spaces is countably productive. Note that an analogous assertion for $\sigma$-compact spaces is false: $\mathbb{Z}^\omega$ and $\mathbb{R}^\omega$ are counterexamples.

PROPOSITION 5.3.9. *Let $X = \prod_{i \in I} X_i$, where $Nag(X_i) \leq \tau$ for each $i \in I$ and $|I| \leq \tau$. Then $Nag(X) \leq \tau$.*

PROOF. For every $i \in I$, let $\beta X_i$ be the Čech–Stone compactification of $X_i$. By the assumptions, there exists a family $\mathcal{F}_i$ of closed subsets of $\beta X_i$ which separates $X_i$ from $\beta X_i \setminus X_i$ and satisfies $|\mathcal{F}_i| \leq \tau$, $i \in I$. Consider the product space $K = \prod_{i \in I} \beta X_i$. For every $i \in I$, let $\pi_i \colon K \to \beta X_i$ be the projection. The family

$$\mathcal{F} = \{\pi_i^{-1}(F) : i \in I,\ F \in \mathcal{F}_i\}$$

consists of closed subsets of $K$ and satisfies $|\mathcal{F}| \le \tau$. Let us verify that $\mathcal{F}$ separates $X$ from $K \setminus X$. Let $x \in X$ and $y \in K \setminus X$ be arbitrary points. Suppose that $x = \{x_i\}_{i \in I}$ and $y = \{y_i\}_{i \in I}$. Since $y \in K \setminus X$, there exists $i \in I$ such that $y_i \notin X_i$. Then, by the choice of $\mathcal{F}_i$, we can find $F_i \in \mathcal{F}_i$ such that $x_i \in F_i$ and $y_i \notin F_i$. Put $F = \pi_i^{-1}(F_i)$. Then $F \in \mathcal{F}$, $x \in F$ and $y \notin F$.

Finally, $K$ is a compactification of $X$, so Lemma 5.3.1 implies that $Nag(X) \le |\mathcal{F}| \le \tau$. □

PROPOSITION 5.3.10. *A topological group generated by a Lindelöf Σ-space is also a Lindelöf Σ-space. More generally, if $X$ a subspace of a topological group $G$ and $G = \langle X \rangle$, then $Nag(G) \le Nag(X)$.*

PROOF. Let $\tau$ be such that $Nag(X) = \tau$. It suffices to prove the second assertion. Consider the subspace $Z = X \cup \{e\} \cup X^{-1}$ of $G$, where $e$ is the identity of $G$. By Proposition 5.3.8, $Nag(Z) \le \tau$. For every $n \in \mathbb{N}$, consider the natural mapping $i_n \colon Z^n \to G$ which assigns to a point $z = (z_1, \ldots, z_n) \in Z^n$ the element $i_n(z) z_1 \cdots z_n$ of $G$. Since the multiplication mapping of $G^n$ to $G$ is continuous and $Z^n$ is a subspace of $G^n$, $i_n$ is also continuous. Hence, from Propositions 5.3.9 and 5.3.5 it follows that $Nag(Z^n) \le \tau$ and $Nag(Y_n) \le \tau$, where $Y_n = i_n(Z^n)$. Clearly, $G = \bigcup_{n=1}^{\infty} Y_n$; hence, by Proposition 5.3.8, $Nag(G) \le \tau$. □

By Proposition 5.3.5, continuous homomorphisms preserve the class of Lindelöf Σ-groups. It turns out that a similar assertion is valid for subgroups of Lindelöf Σ-groups.

PROPOSITION 5.3.11. *The class of subgroups of Lindelöf Σ-groups is closed under taking continuous homomorphic images.*

PROOF. Let $G$ be a subgroup of a Lindelöf Σ-group $G^*$. Consider a continuous homomorphism $\pi \colon G \to H$ onto a topological group $H$. Denote by $K$ the closure of $G$ in $G^*$. Then $K$ is a closed subgroup of $G^*$, so $Nag(K) \le Nag(G^*) \le \aleph_0$ by Corollary 5.3.2. Let $\varrho K$ and $\varrho H$ be the Raĭkov completions of the groups $K$ and $H$, respectively. Since $G$ is dense in $\varrho K$, we can extend $\pi$ to a continuous homomorphism $\varrho\pi \colon \varrho K \to \varrho H$. Then the subgroup $H^* = \varrho\pi(K)$ of $\varrho H$ contains $H$ and Proposition 5.3.5 implies that $Nag(H^*) \le Nag(K) \le \aleph_0$. Therefore, $H$ is a subgroup of the Lindelöf Σ-group $H^*$. □

The study of cellularity type properties in topological groups requires a more profound knowledge of Lindelöf Σ-spaces and, more generally, of the spaces $X$ satisfying $Nag(X) \le \tau$ for an infinite cardinal $\tau$. In Theorems 5.3.12 and 5.3.13, we give two useful characterizations of these classes of spaces.

THEOREM 5.3.12. *A Tychonoff space $X$ satisfies $Nag(X) \le \tau$ iff there exist a Tychonoff space $M$ with $w(M) \le \tau$ and a compact Hausdorff space $K$ such that $X$ is a continuous image of a closed subspace of $M \times K$.*

PROOF. The sufficiency of the condition follows directly from Corollary 5.3.7 and Proposition 5.3.5. Conversely, suppose that $X$ satisfies $Nag(X) \le \tau$. Then there exists a family $\mathcal{P}$ of closed subsets of $\beta X$ which separates $X$ from $\beta X \setminus X$ and satisfies $|\mathcal{P}| \le \tau$. Let $\mathcal{P} = \{P_\alpha : \alpha < \tau\}$. Consider the space $\tau^\tau$, where $\tau$ carries the discrete topology. We define a subspace $M$ of $\tau^\tau$ by

$$M = \{\varphi \in \tau^\tau : \bigcap_{\alpha\in\tau} P_{\varphi(\alpha)} \subset X\}.$$

It is clear that $w(M) \leq w(\tau^\tau) = \tau$. Now we set $K = \beta X$ and define a subspace $Y$ of $M \times K$ by

$$Y = \{(\varphi, x) \in M \times K : x \in P_{\varphi(\alpha)} \text{ for each } \alpha \in \tau\}.$$

We claim that $Y$ is closed in $M \times K$. Indeed, let $(\varphi, x) \in (M \times K) \setminus Y$ be arbitrary. Our definition of $Y$ implies that $x \notin P_{\varphi(\alpha)}$ for some $\alpha \in \tau$. Then the set $W = U \times (K \setminus P_{\varphi(\alpha)})$ is open in $M \times K$, where $U = \{\psi \in \tau^\tau : \psi(\alpha) = \varphi(\alpha)\}$. Obviously, $(\varphi, x) \in W$. If $(\psi, y) \in W$, then $\psi(\alpha) = \varphi(\alpha)$ and $y \notin P_{\varphi(\alpha)} = P_{\psi(\alpha)}$, whence $(\psi, y) \notin Y$. Therefore, $W \cap Y = \varnothing$, which proves that $Y$ is closed in $M \times K$.

It remains to verify that $X$ is a continuous image of $Y$. Denote by $\pi$ the projection of $M \times K$ to the second factor. For every $\varphi \in \tau^\tau$, set $P_\varphi = \bigcap_{\alpha\in\tau} P_{\varphi(\alpha)}$. From our definition of $Y$ it follows that

$$\pi(Y) = \bigcup\{P_\varphi : \varphi \in M\}.$$

Applying the definition of $M$, we conclude that $P_\varphi \subset X$ for each $\varphi \in M$, so that $\pi(Y) \subset X$. On the other hand, given an arbitrary point $x \in X$, we can find $\varphi \in M$ such that $x \in P_\varphi$, whence it follows that $X \subset \pi(Y)$. Hence, $X = \pi(Y)$. □

Combining the facts established above, we obtain the next important result characterizing the spaces $X$ with $Nag(X) \leq \tau$ in terms of continuous mappings.

THEOREM 5.3.13. *A space $X$ satisfies $Nag(X) \leq \tau$ iff there exist a space $N$ with $w(N) \leq \tau$, a space $Z$ and continuous onto mappings $f\colon Z \to X$ and $p\colon Z \to N$, where $p$ is perfect.*

PROOF. Suppose that $Nag(X) \leq \tau$. By Theorem 5.3.12, one can find a space $M$ with $w(M) \leq \tau$, a compact space $K$, a closed subspace $Z$ of $M \times K$, and a continuous onto mapping $f\colon Z \to X$. Clearly, the projection $\pi_M\colon M \times K \to M$ is perfect and so is its restriction $p = \pi_M \restriction Z$ to the closed subspace $Z$ of $M \times K$. Then the spaces $Z$, $N = p(Z) \subset M$ and the mappings $f$, $p$ are as required.

Conversely, suppose that $f\colon Z \to X$ and $p\colon Z \to N$ satisfy the conditions of the theorem. Then $Nag(N) \leq w(N) \leq \tau$ by Corollary 5.3.4. It remains to apply Propositions 5.3.5 and 5.3.6 to conclude that $Nag(X) \leq Nag(Z) = Nag(N) \leq \tau$. □

Recall that a *Lindelöf p-space* is a Tychonoff space that admits a perfect mapping onto a regular second-countable space.

COROLLARY 5.3.14. *A Tychonoff space $Y$ is a Lindelöf $\Sigma$-space if and only if $Y$ is a continuous image of a Lindelöf p-space.*

With the help of Theorem 5.3.13 we establish an important formula relating three basic cardinal functions.

PROPOSITION 5.3.15. *Every Tychonoff space $X$ satisfies the equality $nw(X) = Nag(X) \cdot iw(X)$.*

PROOF. By Lemma 5.2.10 and Proposition 5.3.3, we have $Nag(X) \cdot iw(X) \leq nw(X)$. So, it suffices to verify the inverse inequality.

Set $\tau = Nag(X) \cdot iw(X)$. Then there exists a continuous bijection $i\colon X \to Y$ of $X$ onto a Tychonoff space $Y$ satisfying $w(Y) \leq \tau$. By Theorem 5.3.13, we can find Tychonoff spaces $N$ and $Z$, a perfect onto mapping $p\colon Z \to N$ and a continuous onto mapping $f\colon Z \to X$, where $w(N) \leq \tau$.

It is well known that the diagonal product of a perfect mapping and a continuous mapping is perfect (see [**165**, Theorem 3.7.9]). Therefore, the diagonal product $\varphi = p\Delta f\colon Z \to N \times X$ is a perfect mapping. Denote by $\pi_N$ and $\pi_X$ the projections of $N \times X$ to the first and the second factor, respectively. Then $f = \pi_X \circ \varphi$ and $p = \pi_N \circ \varphi$. Consider the mapping $g = id_N \times i\colon N \times X \to N \times Y$ defined by $g(n, x) = (n, i(x))$ for each $(n, x) \in N \times X$, where $id_N$ is the identity mapping of $N$. Evidently, $g$ is a continuous bijection. We also consider the diagonal product $h = \pi_N \Delta g\colon N \times X \to N \times N \times Y$ which is clearly a continuous one-to-one mapping. Finally, we denote by $pr_N$ and $pr_{N\times Y}$ the projections of $N \times (N \times Y)$ to $N$ and $N \times Y$, respectively, and obtain the following commutative diagram:

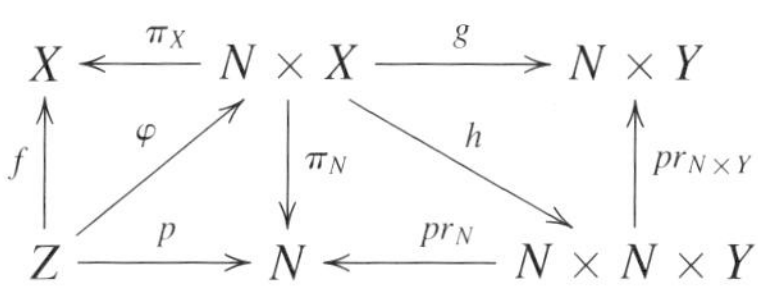


Let $R = \varphi(Z)$ and $\pi = \pi_N \restriction R$. Then $\pi \circ \varphi = p$. Since $p$ is a perfect mapping, so are $\pi$ and $\varphi$ (see [**165**, Theorem 3.7.10]). Let also $T = g(R)$ and $g_1 = g\restriction R$. Then $T \subset N \times Y$, whence $w(T) \leq \tau$. Note that the mapping $h_1 = h\restriction R$ satisfies the equality $\pi = pr_N \circ h_1$. Again, since $\pi$ is perfect, so is $h_1$. Hence, the perfect bijection $h_1\colon R \to h_1(R)$ is a homeomorphism. Clearly, $h_1(R)$ is a subspace of $N \times T$, whence it follows that $w(R) \leq w(N \times T) \leq \tau$.

Finally, since the mapping $\pi_X$ is continuous and $\pi_X(R) = X$, the images under $\pi_X$ of elements of a base for $R$ form a network for $X$. Hence, we conclude that $nw(X) \leq w(R) \leq \tau$. □

Recall that a space $X$ is said to be *$\tau$-cellular* if every family $\gamma$ of $G_\tau$-sets in $X$ contains a subfamily $\eta$ such that $|\eta| \leq \tau$ and $\overline{\bigcup \eta} = \overline{\bigcup \gamma}$. Our goal now is to show that every topological group $G$ with $Nag(G) \leq \tau$ is $\tau$-cellular. We start with two auxiliary results.

LEMMA 5.3.16. *Let $\{x_\alpha : \alpha < \tau^+\}$ be a sequence of points in a Tychonoff space $X$, $\{U_\alpha : \alpha < \tau^+\}$ a sequence of open neighbourhoods of the diagonal $\Delta_X$ in $X^2$, and let $\{\varphi_\alpha : \alpha < \tau^+\}$ be a family of continuous mappings of $X$ to Tychonoff spaces of network weight $\leq \tau$. If $Nag(X) \leq \tau$, then there exist $\alpha, \beta < \tau^+$, $\alpha < \beta$, and a point $x \in X$ such that $(x, x_\alpha) \in U_\beta$ and $\varphi_\alpha(x) = \varphi_\alpha(x_\beta)$.*

PROOF. Since $Nag(X) \leq \tau$, we can apply Theorem 5.3.13 to find two spaces $Y$ and $Z$ with $w(Z) \leq \tau$, a perfect mapping $\varphi^*\colon Y \to Z$ and a continuous mapping $\psi\colon Y \to X$ such that $\psi(Y) = X$ and $\varphi^*(Y) = Z$. Taking preimages, we can replace the sequences in $X$ and $X \times X$ with corresponding sequences in $Y$ and $Y \times Y$. Hence, we can assume that $X = Y$ and that the mapping $\varphi^*\colon X \to Z$ is perfect. For every $\alpha < \tau^+$, let $f_\alpha\colon X \to Z \times \prod_{\beta<\alpha} X_\beta$ be the diagonal product of the mappings $\{\varphi^*\} \cup \{\varphi_\beta : \beta < \alpha\}$, where $X_\beta = \varphi_\beta(X)$ for every $\beta < \tau^+$. Since $\varphi^*$ is perfect, so are the mappings $f_\alpha$'s (see [**165**, Theorem 3.7.9]). Let $P = \{x_\alpha : \alpha < \tau^+\}$ and $P_\alpha = \{x_\beta : \beta < \alpha\}$, where $\alpha < \tau^+$. Since $f_\alpha(X)$ is a subspace of $Z \times \prod_{\beta<\alpha} X_\beta$, we have $nw(f_\alpha(X)) \leq \tau$ for each $\alpha < \tau^+$. Therefore, for every $\alpha < \tau^+$, there exists $\alpha' < \tau^+$ such that $\alpha < \alpha'$ and $f_\alpha(P_{\alpha'})$ is dense in $f_\alpha(P)$. Let us define a sequence $\{\beta_n : n \in \omega\} \subset \tau^+$ by the rule $\beta_0 = 0$ and $\beta_{n+1} = (\beta_n)'$. If $\beta$ is the limit of this sequence, then $\beta < \tau^+$ and $f_\beta(P_\beta)$ is dense in $f_\beta(P)$. This implies, in particular, that $f_\beta(x_\beta) \in \overline{f_\beta(P_\beta)} = f_\beta(\overline{P_\beta})$. Choose a point $x \in \overline{P_\beta}$ such that $f_\beta(x) = f_\beta(x_\beta)$. Then $\varphi_\alpha(x) = \varphi_\alpha(x_\beta)$ for each $\alpha < \beta$, and from $x \in \overline{P_\beta}$ it follows that $(x, x_\alpha) \in U_\beta$, for some $\alpha < \beta$. □

The following lemma enables us to replace certain families of $G_\tau$-sets in $X^2$ by a single continuous mapping of $X$ to $\mathbb{R}^\tau$. As usual, $\Delta_X = \{(x, x) : x \in X\}$ denotes the diagonal in $X^2$.

LEMMA 5.3.17. *Let $X$ be a space with $l(X) \leq \tau$. If $\Delta_X \subset K \subset X^2$ and $K$ is of type $G_\tau$ in $X^2$, then there exists a continuous mapping $\varphi$ of $X$ to $\mathbb{R}^\tau$ such that for all $x, y \in X$, $\varphi(x) = \varphi(y)$ implies $(x, y) \in K$.*

PROOF. By the assumptions in the lemma, there exists a family $\{O_\alpha : \alpha < \tau\}$ of open sets in $X^2$ such that $K = \bigcap_{\alpha<\tau} O_\alpha$. For every $x \in X$ and $\alpha < \tau$, choose a cozero-set $U_\alpha(x)$ in $X$ such that $x \in U_\alpha(x)$ and $U_\alpha(x) \times U_\alpha(x) \subset O_\alpha$. Since $l(X) \leq \tau$, we can find, for every $\alpha < \tau$, a subset $Y_\alpha$ of $X$ with $|Y_\alpha| \leq \tau$ such that

$$X = \bigcup\{U_\alpha(x) : x \in Y_\alpha\}.$$

For $\alpha < \tau$ and $x \in Y_\alpha$, choose a continuous function $f_{\alpha,x}\colon X \to \mathbb{R}$ such that $X \setminus U_\alpha(x) = f_{\alpha,x}^{-1}(0)$. Note that the family $\{f_{\alpha,x} : \alpha < \tau,\ x \in Y_\alpha\}$ is of cardinality $\leq \tau$, so the diagonal product $\varphi$ of this family maps $X$ to a subspace of $\mathbb{R}^\tau$. We claim that the mapping $\varphi$ is as required.

Indeed, suppose that $\varphi(a) = \varphi(b)$ for some $a, b \in X$. Then $f_{\alpha,x}(a) = f_{\alpha,x}(b)$ for all $\alpha < \tau$ and $x \in Y_\alpha$. Let $\alpha < \tau$ be arbitrary. Then $a \in U_\alpha(x)$ for some $x \in Y_\alpha$, so our choice of the function $f_{\alpha,x}$ implies that $f_{\alpha,x}(a) \neq 0$. Consequently, $f_{\alpha,x}(b) = f_{\alpha,x}(a) \neq 0$ and $b \in U_\alpha(x)$. This implies that $(a, b) \in U_\alpha(x) \times U_\alpha(x) \subset O_\alpha$. Since the latter is valid for each $\alpha < \tau$, we conclude that $(a, b) \in \bigcap_{\alpha<\tau} O_\alpha = K$. □

The next especially important result has a number of applications.

THEOREM 5.3.18. *Every topological group $H$ with $Nag(H) \leq \tau$ is $\tau$-cellular or, equivalently, $cel_\tau(H) \leq \tau$.*

PROOF. Assume the contrary. Then there exist a family $\{F_\alpha : \alpha < \tau^+\}$ of non-empty $G_\tau$-sets in $H$ and a family $\{O_\alpha : \alpha < \tau^+\}$ of open subsets of $H$ such that $F_\alpha \subset O_\alpha$ and $F_\alpha \cap O_\beta = \varnothing$ whenever $\alpha < \beta < \tau^+$. For every $\alpha < \tau$, pick a point $x_\alpha \in F_\alpha$. Let $f\colon H^3 \to H$ be the mapping defined by $f(x, y, z) = xy^{-1}z$. Then $f$ is continuous, so $U_\alpha = \{(x, z) \in H^2 : f(x, z, x_\alpha) \in O_\alpha\}$ is an open neighbourhood of the diagonal in $H^2$ for each $\alpha < \tau^+$. The set $K_\alpha = \{(x, z) \in H^2 : f(x_\alpha, x, z) \in F_\alpha\}$ contains the diagonal of $H^2$ and is of type $G_\tau$ in $H^2$. By Lemma 5.3.17, there exists a continuous mapping $\varphi_\alpha\colon H \to \mathbb{R}^\tau$ such that

$$\{(x, y) \in H \times H : \varphi(x) = \varphi(y)\} \subset K_\alpha.$$

We can now apply Lemma 5.3.16 to the sequences $\{x_\alpha : \alpha < \tau^+\}$, $\{U_\alpha : \alpha < \tau^+\}$ and $\{\varphi_\alpha : \alpha < \tau^+\}$ to find $\alpha, \beta$ with $\alpha < \beta < \tau^+$ and a point $x \in H$ such that $\varphi_\alpha(x) = \varphi_\alpha(x_\beta)$ and $(x_\alpha, x) \in U_\beta$. Then $(x, x_\beta) \in K_\alpha$ and $(x_\alpha, x) \in U_\beta$, whence $f(x_\alpha, x, x_\beta) \in F_\alpha \cap O_\beta$. This contradiction completes the proof. □

Theorem 5.3.18 implies the following two corollaries immediately.

COROLLARY 5.3.19. *Every family $\gamma$ of $G_\delta$-sets in a Lindelöf $\Sigma$-group contains a countable subfamily $\mu$ such that $\overline{\bigcup \mu} = \overline{\bigcup \gamma}$.*

COROLLARY 5.3.20. *Let $H$ be a $\sigma$-compact topological group. Then every family $\gamma$ of $G_\delta$-sets in $H$ contains a countable subfamily $\eta$ such that $\bigcup \eta$ is dense in $\bigcup \gamma$.*

Subgroups of Lindelöf $\Sigma$-groups need not be Lindelöf (see Exercise 5.1.d). Nevertheless, we have the following:

COROLLARY 5.3.21. *Any subgroup of a Lindelöf $\Sigma$-group has countable cellularity.*

PROOF. Let $H$ be a subgroup of a Lindelöf $\Sigma$-group $G$. Denote by $H^*$ the closure of $H$ in $G$. Then $H^*$ is a closed subgroup of $G$, so $Nag(H^*) \leq Nag(G) \leq \omega$ by Corollary 5.3.2. Corollary 5.3.19 implies that $c(H^*) \leq cel_\omega(H^*) \leq \omega$. Since $H$ is dense in $H^*$, we conclude that $c(H) = c(H^*) \leq \omega$. □

Since the class of Lindelöf $\Sigma$-spaces contains $\sigma$-compact spaces, the next result is a special case of Corollary 5.3.21.

COROLLARY 5.3.22. [**M. G. Tkachenko**] *Every $\sigma$-compact topological group has countable cellularity.*

We can extend the analogy between Cantor cubes $\{0, 1\}^\kappa$ and Lindelöf $\Sigma$-groups as follows. By Theorem 1.6.18, for any family $\gamma$ of $G_\delta$-sets in $\{0, 1\}^\kappa$, the closure of $\bigcup \gamma$ is also a $G_\delta$-set or, equivalently, every Cantor cube is an Efimov space (see Section 1.6). It turns out that Lindelöf $\Sigma$-groups have the same property. To prove this fact, we need two lemmas.

LEMMA 5.3.23. *Let $H$ be a topological group such that $l(H) \leq \tau$ and let $N$ be a closed invariant subgroup of type $G_\tau$ in $H$. Then the quotient group $K = H/N$ satisfies $\psi(K) \leq \tau$.*

PROOF. Suppose that $\{O_\alpha : \alpha < \tau\}$ is a family of open sets in $H$ such that $N = \bigcap_{\alpha<\tau} O_\alpha$. If $\alpha < \tau$ and $x \in H \setminus O_\alpha$, choose open symmetric neighbourhoods $U_\alpha(x)$ and $V_\alpha(x)$ of the identity $e$ in $H$ such that $xU_\alpha(x) \cap N = \varnothing$ and $V_\alpha^2(x) \subset U_\alpha(x)$. The family $\{xV_\alpha(x) : x \in H \setminus O_\alpha\}$ covers the closed set $H \setminus O_\alpha$, so there exists a subset $S_\alpha$ of $H \setminus O_\alpha$ such that $H \setminus O_\alpha \subset \bigcup\{xV_\alpha(x) : x \in S_\alpha\}$ and $|S_\alpha| \leq \tau$. Put $\mu = \{V_\alpha(x) : x \in S_\alpha,\ \alpha < \tau\}$. Clearly, $|\mu| \leq \tau$.

Denote by $\pi$ the quotient homomorphism of $H$ onto $H/N$. We claim that the set $\bigcap\{\pi(V) : V \in \mu\}$ contains only the identity $e$ of $H/N$. Indeed, let $z \in H/N$ be arbitrary, $z \neq e$. Choose $y \in H$ with $\pi(y) = z$. Then $y \notin N$ and, hence, $y \in H \setminus O_\alpha$ for some $\alpha < \tau$. By the definition of $\mu$, we can find $x \in S_\alpha$ such that $y \in xV_\alpha(x)$. Since $yV_\alpha(x) \subset xV_\alpha^2(x) \subset xU_\alpha(x)$ and $xU_\alpha(x) \cap N = \varnothing$, we conclude that $yV_\alpha(x) \cap N = \varnothing$, whence $y \notin NV_\alpha(x)$. Therefore, $z = \pi(y) \notin \pi(V_\alpha(x))$. Since the cardinality of the family $\{\pi(V) : V \in \mu\}$ is not greater than $\tau$, we conclude that $\psi(H/N) \leq \tau$. □

In the next lemma, we present three important properties of topological groups $H$ with $Nag(H) \leq \tau$.

LEMMA 5.3.24. *Let $H$ be a topological group such that $Nag(H) \leq \tau$. Then:*

a) *for every closed invariant subgroup $N$ of type $G_\tau$ in $H$, the quotient group $H/N$ satisfies $nw(H/N) \leq \tau$;*
b) *the sets of the form $\pi^{-1}(V)$ form a base of $H$, where $\pi\colon H \to K$ is an open continuous homomorphism onto a topological group $K$ with $nw(K) \leq \tau$ and $V$ open in $K$;*
c) *for every closed $G_\tau$-set $P$ in $H$, there exist a continuous open homomorphism $\pi\colon H \to K$ onto a topological group $K$ with $nw(K) \leq \tau$ and a closed subset $F \subset K$ such that $P = \pi^{-1}(F)$.*

PROOF. a) Let $K = H/N$, where $N$ is a closed invariant subgroup of type $G_\tau$ in $H$. From Proposition 5.3.3 it follows that $l(H) \leq Nag(H) \leq \tau$, so Lemma 5.3.23 implies that $\psi(K) \leq \tau$. In addition, we have $ib(K) \leq ib(H) \leq l(H) \leq \tau$. Therefore, by Proposition 5.2.11, there exists a continuous isomorphism of $K$ onto a topological group $L$ with $w(L) \leq \tau$. In particular, $iw(K) \leq \tau$. Since $K$ is a continuous image of $H$, we also have $Nag(K) \leq Nag(H) \leq \tau$. Then Proposition 5.3.15 implies that $nw(K) = Nag(K) \cdot iw(K) \leq \tau$.

b) Let $U$ be a neighbourhood of the identity in $H$. By Lemma 5.1.6, one can find a continuous homomorphism $p\colon H \to L$ of $H$ onto a topological group $L$ with $w(L) \leq \tau$ and an open neighbourhood $W$ of the identity in $L$ such that $p^{-1}(W) \subset U$. Denote by $N$ the kernel of $p$ and consider the quotient group $K = H/N$. Let $\pi\colon H \to H/N$ be the quotient homomorphism. Clearly, $N$ is of type $G_\tau$ in $H$, so $nw(K) \leq \tau$, by a). The groups $K$ and $L$ are algebraically isomorphic, so there exists an isomorphism $\varphi\colon K \to L$ such that $\varphi \circ \pi = p$. Since $\pi$ is an open mapping, the isomorphism $\varphi$ is continuous. Hence, $V = \varphi^{-1}(W)$ is an open neighbourhood of the identity in $K$ and $\pi^{-1}(V) = \pi^{-1}(\varphi^{-1}(W)) = p^{-1}(W) \subset U$.

c) Let $P$ be a closed $G_\tau$-set in $H$ and let $\gamma$ be a family of open sets in $H$ such that $P = \bigcap \gamma$ and $|\gamma| \leq \tau$. Fix an arbitrary element $W \in \gamma$. There is an open covering $\eta$ of $P$ by sets of the form $\pi^{-1}(V)$ satisfying $\pi^{-1}(V) \subset W$, where $\pi$ and $V$ are as in b). By Proposition 5.3.3, $l(P) \leq l(H) \leq Nag(H) \leq \tau$, so the covering $\eta$ of $P$ contains a subcovering $\mu_W$ of cardinality $\leq \tau$. Let $\mu_W = \{\pi_\alpha^{-1}(V_\alpha) : \alpha < \tau\}$. Then the diagonal product $\varphi_W$ of the family $\{\pi_\alpha : \alpha < \tau\}$ is a continuous homomorphism of $H$ to the product of $\tau$ many topological groups of network weight $\leq \tau$ and hence, the group $H_W = \varphi_W(H)$ satisfies $nw(H_W) \leq \tau$. One easily verifies that the set $O_W = \bigcup \mu_W$ satisfies $P \subset O_W = \varphi_W^{-1}\varphi_W(O_W) \subset W$ and that, in addition, $\varphi_W(O_W)$ is open in $H_W$ (the mapping $\varphi_W$ is not necessarily open).

Let $p$ be the diagonal product of the family $\{\varphi_W : W \in \gamma\}$. Since $|\gamma| \leq \tau$, the group $L = p(H)$ satisfies $nw(L) \leq \tau$. From the definition of $p$ it follows that $O_W = p^{-1}p(O_W) \subset W$ for each $W \in \gamma$, so the equality $P = \bigcap \gamma$ implies that $P = p^{-1}p(P)$. As in b), consider the kernel $N$ of the homomorphism $p$ and the quotient group $K = H/N$. Let $\pi\colon H \to H/N$ be the quotient homomorphism. Then there exists a continuous homomorphism $\varphi\colon H/N \to L$ such that $\varphi \circ \pi = p$. Note that $N$ is of type $G_\delta$ in $H$, so a) implies that $nw(K) \leq \tau$. Since the homomorphism $\pi$ is quotient and $P = \pi^{-1}\pi(P)$, the set $F = \pi(P)$ is closed in $K$. □

The next corollary shows that the $i$-weight in Proposition 5.3.15 can be replaced by the pseudocharacter in the case of topological groups.

Corollary 5.3.25. *Every topological group $H$ satisfies $nw(H) = Nag(H) \cdot \psi(H)$.*

Proof. Clearly, $\psi(H) \leq nw(H)$. Proposition 5.3.3 implies that $Nag(H) \cdot \psi(H) \leq nw(H)$ (this inequality holds for every Tychonoff space). Conversely, if $\tau = Nag(H) \cdot \psi(H)$, then the identity $e$ of $H$ is a closed $G_\tau$-set in $H$, and c) of Lemma 5.3.24 implies that there exists an open continuous homomorphism $\pi \colon H \to K$ onto a topological group $K$ with $nw(K) \leq \tau$ such that $\{e\} = \pi^{-1}\pi(e)$. Therefore, $\pi$ is a topological isomorphism of $H$ onto $K$ and $nw(H) \leq \tau$. □

The following result complements Theorem 5.3.18.

Theorem 5.3.26. *Let $\gamma$ be a family of $G_\tau$-sets in a topological group $H$ such that $Nag(H) \leq \tau$. Then $\overline{\bigcup \gamma}$ is also a $G_\tau$-set in $H$.*

Proof. Without loss of generality, we can assume that all the elements of $\gamma$ are closed in $H$. Let $F = \overline{\bigcup \gamma}$. By Theorem 5.3.18, there exists a subfamily $\mu$ of $\gamma$ with $|\mu| \leq \tau$ such that $\overline{\bigcup \mu} = F$. Apply (c) of Lemma 5.3.24 to find, for every $P \in \mu$, a continuous homomorphism $\pi_P \colon H \to H_P$ of $H$ to a topological group $H_P$ with $nw(H_P) \leq \tau$ such that $P = \pi_P^{-1}\pi_P(P)$. Let $p$ be the diagonal product of the family $\{\pi_P : P \in \mu\}$. The group $L = p(H)$ satisfies $nw(L) \leq \tau$, so we can define (as in (b) of Lemma 5.3.24) an open continuous homomorphism $\pi \colon H \to K$ onto a topological group $K$ with $nw(K) \leq \tau$ and a continuous isomorphism $\varphi \colon K \to L$ such that $\varphi \circ \pi = p$. It is clear that $P = \pi^{-1}\pi(P)$ for each $P \in \mu$. Put $D = \bigcup \mu$. Then $D = \pi^{-1}\pi(D)$ and, since the homomorphism $\pi$ is open, we have

$$F = \overline{\bigcup \mu} = \overline{D} = \overline{\pi^{-1}(\pi(D))} = \pi^{-1}(\overline{\pi(D)}).$$

It follows from $nw(K) \leq \tau$ that every closed subset of $K$ is of the type $G_\tau$ in $K$. Hence, $F = \pi^{-1}(E)$ is of the type $G_\tau$ in $H$, where $E = \overline{\pi(D)}$. □

Corollary 5.3.27. *Every Lindelöf Σ-group is an Efimov space.*

Corollary 5.3.28. *Every $\sigma$-compact topological group $H$ is an Efimov space.*

Corollary 5.3.29. *Let $H$ be an arbitrary subgroup of a $\sigma$-compact topological group. Then every regular closed subset of $H$ is a zero-set in $H$.*

Proof. Suppose that $H$ is a topological subgroup of a $\sigma$-compact group $G$. Clearly, we can assume that $H$ is dense in $G$. Let $U$ be an open subset of $H$. We have to verify that $F = cl_H U$ is a zero-set in $H$. Take an open set $V$ in $G$ such that $U = H \cap V$ and put $K = cl_G V$. Then $K$ is of the type $G_\delta$ in $G$ and, since $G$ is $\sigma$-compact, $K$ is a zero-set in $G$. Therefore, $F = K \cap H$ is a zero-set in $G$. □

We can extend Theorems 5.3.18 and 5.3.26 to arbitrary products of topological groups $G_i$ satisfying $Nag(G_i) \leq \tau$ as follows:

Theorem 5.3.30. *Let $\Pi = \prod_{i \in I} G_i$ be a product of topological groups satisfying $Nag(G_i) \leq \tau$ for each $i \in I$. Then we have:*

a) *$cel_\tau(\Pi) \leq \tau$;*
b) *if $\mathcal{F}$ is a family of $G_\tau$-sets in $\Pi$, then $\overline{\bigcup \mathcal{F}}$ is of type $G_\tau$ in $\Pi$;*
c) *for every closed $G_\tau$-set $P$ in $\Pi$, there exists $J \subset I$ with $|J| \leq \tau$ such that $P = \pi_J^{-1}\pi_J(P)$, where $\pi_J \colon \Pi \to \Pi_J = \prod_{i \in J} G_i$ is the projection.*

PROOF. Let us say that a closed subset $F$ of $\Pi$ is a $\tau$-*cube* if it has the form $F = \pi_K^{-1}(F_K)$, where $K \subset I$, $|K| \leq \tau$, $F_K = \prod_{i\in K} F_i$, and every $F_i$ is a closed $G_\tau$-set in $G_i$. The set $K$ is called the *core* of $F$. Note that if $F$ is a $\tau$-cube in $\Pi$, then $\pi_J(F)$ is of type $G_\tau$ in $\Pi_J$ for each $J \subset I$.

a) Let $\mathcal{F}$ be a family of $G_\tau$-sets in $\Pi$. Since every $G_\tau$-set in $\Pi$ is a union of $\tau$-cubes, we can assume that all elements of $\mathcal{F}$ are $\tau$-cubes. For every $J \subset I$ with $|J| \leq \tau$, the group $\Pi_J$ satisfies $Nag(\Pi_J) \leq \tau$ by Proposition 5.3.9, so $cel_\tau(\Pi_J) \leq \tau$ by Theorem 5.3.18. As in the proof of Theorem 1.6.18, one can define a sequence $\{J(n) : n \in \omega\}$ of subsets of $I$ and a sequence $\{\mathcal{F}_n : n \in \omega\}$ of subfamilies of $\mathcal{F}$ satisfying the following conditions for each $n \in \omega$:

(i) $|J(n)| \leq \tau$ and $|\mathcal{F}_n| \leq \tau$;
(ii) $J(n) \subset J(n+1)$ and $\mathcal{F}_n \subset \mathcal{F}_{n+1}$;
(iii) $\pi_{J(n)}(\bigcup \mathcal{F}_n)$ is dense in $\pi_{J(n)}(\bigcup \mathcal{F})$;
(iv) $J_{n+1}$ contains the core of every element of $\mathcal{F}_n$.

Put $J = \bigcup_{n\in\omega} J(n)$, $\mathcal{G} = \bigcup_{n\in\omega} \mathcal{F}_n$ and $D = \bigcup \mathcal{G}$. It is clear that $|\mathcal{G}| \leq \tau$, $D = \pi_J^{-1}\pi_J(D)$, and that $\pi_J(D)$ is dense in $\pi_J(\bigcup \mathcal{F})$. Consequently,

$$\bigcup \mathcal{F} \subset \pi_J^{-1}\pi_J(\bigcup \mathcal{F}) \subset \pi_J^{-1}\overline{\pi_J(D)} = \overline{\pi_J^{-1}\pi_J(D)} = \overline{D}.$$

Hence, $\bigcup \mathcal{G}$ is dense in $\bigcup \mathcal{F}$. Since $|\mathcal{G}| \leq \tau$, we have $cel_\tau(\Pi) \leq \tau$.

b) Let $\mathcal{F}$ be a family of $G_\tau$-sets in $\Pi$. Again, we can assume that all the elements of $\mathcal{F}$ are $\tau$-cubes. Since $cel_\tau(\Pi) \leq \tau$ by a), one can find a subfamily $\mathcal{G} \subset \mathcal{F}$ such that $|\mathcal{G}| \leq \tau$ and $\bigcup \mathcal{G}$ is dense in $\bigcup \mathcal{F}$. Denote by $J$ the union of the cores of the elements of $\mathcal{G}$. Then $|J| \leq \tau$ and the equality $F = \pi_J^{-1}\pi_J(F)$ holds for each $F \in \mathcal{G}$. Therefore, the set $E = \bigcup \mathcal{G}$ satisfies $E = \pi_J^{-1}\pi_J(E)$. Since the mapping $\pi_J$ is open, we have

$$\overline{\bigcup \mathcal{F}} = \overline{E} = \overline{\pi_J^{-1}\pi_J(E)}\pi_J^{-1}\overline{\pi_J(E)}. \tag{5.1}$$

It follows from the choice of $\mathcal{G}$ that the family $\mathcal{G}_J = \{\pi_J(F) : F \in \mathcal{G}\}$ consists of $G_\tau$-sets in $\Pi_J$. In addition, $Nag(\Pi_J) \leq \tau$ because $|J| \leq \tau$. Thus, the closure of the set $\pi_J(E) = \bigcup \mathcal{G}_J$ is of type $G_\tau$ in $\Pi_J$ by Theorem 5.3.26. By (5.1), we have that $\overline{\bigcup \mathcal{F}} = \pi_J^{-1}\overline{\pi_J(E)}$ and hence, $\overline{\bigcup \mathcal{F}}$ is a $G_\tau$-set in $\Pi$.

c) A closed $G_\tau$-subset $P$ of $\Pi$ is a union of $\tau$-cubes in $\Pi$, say $P = \bigcup \gamma$. By a), $\gamma$ contains a subfamily $\mu$ with $|\mu| \leq \tau$ such that $\bigcup \mu$ is dense in $P$. Since the cardinality of the core of every element $F \in \mu$ is at most $\tau$, there exists a set $J \subset I$ with $|J| \leq \tau$ such that $F = \pi_J^{-1}\pi_J(F)$ for each $F \in \mu$. Then $P = \pi_J^{-1}\pi_J(P)$. □

COROLLARY 5.3.31. *The product of any family of Lindelöf $\Sigma$-groups is an $\omega$-cellular Efimov space.*

## Exercises

5.3.a. Show that every non-empty topological space can be represented as an image of a non-discrete Abelian topological group under a continuous mapping.

5.3.b. Verify that every non-discrete countable topological group is an image under a continuous homomorphism of a non-discrete countable metrizable group.

5.3.c. Let $G$ be a Lindelöf $\Sigma$-group with a $\sigma$-disjoint $\pi$-base. Show that $G$ is metrizable.

5.3.d. Show that if $G$ is a topological group, and $H$ is a locally compact subgroup of $G$ such that the quotient space $G/H$ is a Lindelöf $\Sigma$-space, then $G$ contains an open subgroup $M$ which is a Lindelöf $\Sigma$-space (which implies that $G$ is a free topological sum of Lindelöf $\Sigma$-spaces).
5.3.e. Give an example of a precompact Lindelöf $\Sigma$-group $G$ which is neither compact nor Raĭkov complete.
5.3.f. Let $G$ be a Lindelöf $\Sigma$-group. Is the Raĭkov completion of $G$ a Lindelöf $\Sigma$-group?
5.3.g. Show that every Lindelöf $\Sigma$-group with no small subgroups is cosmic.
5.3.h. Verify that the $G_\delta$-tightness of every Lindelöf $\Sigma$-group is countable.
5.3.i. Let us say that $X$ is a *Mal'tsev space* if there exists a continuous mapping $f\colon X^3 \to X$ such that $f(x, y, y) = f(y, y, x) = x$, for all $x, y \in X$. Show that the following hold:
   a) Every topological group $G$ is a Mal'tsev space.
   b) Every retract of a topological group is a Mal'tsev space.
   c) Verify that Theorems5.3.18, 5.3.26, 5.3.30, 5.4.7 as well as Corollaries 5.3.21, 5.3.19, 5.3.28, and 5.4.9 are valid for Mal'tsev spaces.

## Problems

5.3.A. Suppose that $G$ is a topological group which is a Lindelöf $p$-space. Prove that $G$ contains a dense $\sigma$-compact subgroup and hence, $G$ is $k$-separable.
5.3.B. Suppose that $G$ is a Lindelöf $\Sigma$-group. Is $G$ a continuous homomorphic image of a topological group $H$ which is a Lindelöf $p$-space?
*Hint.* The answer is "no". Take a Lindelöf $\Sigma$-group $G$ that is not $k$-separable. For example, put $G = C_p(X)$, where $X$ is a Gul'ko compactum that is not Eberlein, see [**462**] (every Gul'ko compactum is a Corson compactum, by a theorem in [**207**]).
5.3.C. Let $G$ be a topological group such that $G$ is a Lindelöf $p$-space all compact subspaces of which are metrizable. Prove that $G$ is metrizable.
5.3.D. Give an example of a non-metrizable Lindelöf $\Sigma$-group $G$ such that every compact subspace of $G$ is metrizable.
5.3.E. Suppose that $G$ is a Lindelöf $\Sigma$-group which is a $P$-space. Prove that $G$ is discrete.
5.3.F. Show that a Lindelöf subgroup of a Lindelöf $\Sigma$-group need not be a Lindelöf $\Sigma$-group.
5.3.G. Show that a subgroup of type $G_\delta$ of a Lindelöf $\Sigma$-group need not be Lindelöf.
5.3.H. A topological group $H$ is said to be a *strong Lindelöf $\Sigma$-group* if there exists a continuous homomorphism of a Lindelöf $p$-group onto $H$. Prove that a $G_\delta$-subgroup of a strong Lindelöf $\Sigma$-group is again a strong Lindelöf $\Sigma$-group.
5.3.I. (V. G. Pestov and D. B. Shakhmatov [**380**]) Prove that not every cosmic topological group can be represented as an image of a second-countable topological group under a continuous homomorphism.
5.3.J. Let $G$ be a Lindelöf $\Sigma$-group that is hereditarily Lindelöf. Prove that $G$ is cosmic.
5.3.K. Let $G$ be a Lindelöf $\Sigma$-group that is hereditarily separable. Is it true in $ZFC$ that $G$ is cosmic?
5.3.L. Suppose that $f\colon G \to H$ is a continuous homomorphism of a Lindelöf $\Sigma$-group $G$ onto a metrizable group $H$. Suppose further that $M$ is a subgroup of $H$. Prove that $f^{-1}(M)$ is a Lindelöf $\Sigma$-group.
5.3.M. (J. van Mill [**323**]) Show that every regular second-countable space is an image of a second-countable group under a closed continuous mapping.

## Open Problems

5.3.1. Characterize the topological subgroups of Lindelöf $\Sigma$-groups.
5.3.2. Does every $\sigma$-compact (Lindelöf $\Sigma$-) topological group have countable $\delta$-tightness? (See also Exercise 5.3.h and Problems 4.1.F and 6.6.1.)

5.3.3. Does every countably compact topological group have countable $\delta$-tightness?
5.3.4. Is every Lindelöf $\Sigma$-group $G$ of countable spread (which means that every discrete subspace of $G$ is countable) cosmic?
5.3.5. Suppose that $G$ is a (regular) paratopological group and $F$ is a compact subspace of $G$. Suppose further that $S$ is the minimal subsemigroup of $G$ with $F \subset S$. Is the Souslin number of the subspace $S$ countable?
5.3.6. Is it possible to represent an arbitrary Tychonoff (regular, Hausdorff) space as an image of a topological group under a closed continuous mapping?
5.3.7. Is it possible to represent an arbitrary Tychonoff (regular, Hausdorff) space as an image of a topological group under an open continuous mapping?

## 5.4. Cellularity and weak precalibres

Here we show that subgroups of Lindelöf $\Sigma$-groups satisfy a certain restriction on linked families of open sets. This condition also implies that the cellularity of such groups is countable (which was already established in Corollary 5.3.21). Then we prove in Theorem 5.4.10 that, unlike the case of topological spaces, the cellularity of each subgroup of a given topological group $G$ is bounded by $2^{c(G)}$.

A cardinal $\tau > \omega$ is said to be a *precalibre* of a space $X$ if every family $\gamma$ of (non-empty) open subsets of $X$ with $|\gamma| = \tau$ contains a subfamily $\lambda$ of the same cardinality $\tau$ with the finite intersection property. If $\tau$ is a precalibre of $X$, then the Souslin number of $X$ is, obviously, strictly less than $\tau$.

It is quite a delicate question whether $\aleph_1$ is a precalibre of every $\sigma$-compact topological group (see Problem 5.4.H). In fact, the question is undecidable in $ZFC$. However, there exists a slightly weaker property of a similar nature enjoyed by all $\sigma$-compact and, more generally, by all Lindelöf $\Sigma$-groups. Here is the definition.

Let $\tau$ be an infinite cardinal and $n \geq 2$ an integer. A family $\gamma$ of sets is called *$n$-linked* if $U_1 \cap \cdots \cap U_n \neq \varnothing$, for all $U_1, \ldots, U_n \in \gamma$. A pair $(\tau, n)$ is said to be a *weak precalibre* of a space $X$ if every family $\gamma$ of open sets in $X$ with $|\gamma| \geq \tau$ contains an $n$-linked subfamily of cardinality $\tau$.

If $\tau$ is an infinite cardinal and $2 \leq m < n$, then we obviously have the following implications for every space $X$:

$$\begin{aligned} \tau \text{ is a precalibre } &\implies (\tau, n) \text{ is a weak precalibre} \\ &\implies (\tau, m) \text{ is a weak precalibre} \end{aligned}$$

Thus, for a given cardinal $\tau$, the property "$(\tau, 2)$ is a weak precalibre" is the weakest in the above sequence. If $(\aleph_1, 2)$ is a weak precalibre of a space $X$, it is also said that $X$ has the *Knaster property*. It is clear from the definitions that every separable space has the Knaster property, and all spaces with the Knaster property have countable cellularity.

In contrast with the property of having countable cellularity, precalibres and weak precalibres are stable with respect to taking products of spaces:

THEOREM 5.4.1. *Let $X = \prod_{i \in I} X_i$ be a product space and suppose that $(\tau, n)$ is a weak precalibre for each factor $X_i$, where $\tau$ is a regular uncountable cardinal. Then $(\tau, n)$ is a weak precalibre for $X$. Similarly, if $\tau$ is a precalibre for each factor $X_i$, then $\tau$ is a precalibre of $X$.*

PROOF. We prove only the first part of the theorem, for weak precalibres, leaving the rest to the reader.

If $(\tau, n)$ is a weak precalibre for spaces $Y$ and $Z$, then $(\tau, n)$ is a weak precalibre for the product $Y \times Z$. Indeed, take a family $\gamma$ of rectangular open sets in $Y \times Z$ and suppose that $|\gamma| = \tau$. Since the projection $\pi_Y$ of $Y \times Z$ to $Y$ is open, there exists a subfamily $\eta$ of $\gamma$ of cardinality $\tau$ such that the family $\eta_Y = \{\pi_Y(U) : U \in \eta\}$ is $n$-linked. Similarly, $\eta$ contains a subfamily $\lambda$ of cardinality $\tau$ such that the family $\lambda_Z = \{\pi_Z(V) : V \in \lambda\}$ is $n$-linked, where $\pi_Z \colon Y \times Z \to Z$ is the projection. Since the elements of $\lambda$ are rectangular, it follows that the family $\lambda$ itself is $n$-linked. Hence, $(\tau, n)$ is a weak precalibre of $Y \times Z$.

We can conclude, therefore, that $(\tau, n)$ is a weak precalibre for each subproduct $X_K = \prod_{i \in K} X_i$, with $K$ a finite subset of $I$.

Let $\gamma$ be a family of non-empty open sets in $X$ with $|\gamma| = \tau$. Without loss of generality we can assume that $\gamma$ consists of canonical open sets. For every non-empty finite set $K \subset I$, denote by $p_K$ the projection of $X$ onto $X_K$. Let $\{U_\alpha : \alpha < \tau\}$ be a faithful enumeration of the elements of $\gamma$. Take $A_\alpha$ to be a finite subset of $I$ such that $U_\alpha = p_{A_\alpha}^{-1} p_{A_\alpha}(U_\alpha)$, where $\alpha < \tau$. According to Theorem 1.6.20, there exists a set $B \subset \tau$ of cardinality $\tau$ and a finite subset $R$ of $I$ such that $A_\alpha \cap A_\beta = R$, for all distinct $\alpha, \beta \in B$. Since $(\tau, n)$ is a weak precalibre of $X_R$ and the projection $p_R$ is open, the exists a set $C \subset B$ with $|C| = \tau$ such that the family $\{p_R(U_\alpha) : \alpha \in C\}$ is $n$-linked. Then the family $\lambda = \{U_\alpha : \alpha \in C\}$ is $n$-linked as well. Indeed, take distinct elements $U_{\alpha_1}, \ldots, U_{\alpha_n} \in \lambda$ and choose a point $y \in p_R(U_{\alpha_1}) \cap \cdots \cap p_R(U_{\alpha_n})$. Since the family $\{A_{\alpha_i} \setminus R : i = 1, \ldots, n\}$ is disjoint, we can take a point $x \in X$ such that $x_\alpha = y_\alpha$ for each $\alpha \in R$, and $x_\alpha \in p_\alpha(U_i)$ whenever $\alpha \in A_{\alpha_i} \setminus R$, for $1 \leq i \leq n$. Then $x \in U_{\alpha_1} \cap \cdots \cap U_{\alpha_n} \neq \varnothing$. This proves that $(\tau, n)$ is a weak precalibre of $X$. □

COROLLARY 5.4.2. *The product of an arbitrary family of spaces with the Knaster property also has the Knaster property.*

COROLLARY 5.4.3. *Let $X$ be a dyadic compactum. Then every regular cardinal $\tau > \omega$ is a precalibre of $X$.*

PROOF. Clearly, $X$ is a continuous image of $D^\kappa$, for some cardinal $\kappa$, where $D = \{0, 1\}$ is discrete. It follows from Theorem 5.4.1 that every regular cardinal $\tau > \omega$ is a precalibre of $D^\kappa$. It remains to note that continuous onto mappings preserve the latter property. □

In general, the Souslin number can increase when taking products of spaces or, even worse, of topological groups (see Problem 5.4.G). Theorem 5.4.1 and Corollary 5.4.2 show that weak precalibres serve as a good substitute for the cellularity (see, for example, Theorem 6.4.21).

Let us show that $(\tau, 2)$ is a weak precalibre of every Lindelöf $\Sigma$-group provided that $\tau$ is regular and uncountable. This requires some preliminary work.

If $X$ is a set and $n \in \mathbb{N}$, then we denote by $[X]^n$ the family of all subsets of $X$ of cardinality $n$. Recall that for cardinal numbers $\tau, \lambda, \kappa$ and an integer $n \geq 1$, the formula $\tau \to (\lambda)^n_\kappa$ means that for every partition $\{B_\alpha : \alpha < \kappa\}$ of the set $[\tau]^n$ into $\kappa$ (not necessarily disjoint) subsets, there exist $\alpha < \kappa$ and a subset $A$ of $\tau$ with $|A| = \lambda$ such that $[A]^n \subset B_\alpha$. Similarly, $\tau \to (\lambda_1, \lambda_2)^n$ means that for every partition $\{B_1, B_2\}$ of $[\tau]^n$, there exist $i \in \{1, 2\}$ and a set $A \subset \tau$ such that $|A| = \lambda_i$ and $[A]^n \subset B_i$. We use the following well-known combinatorial results (see [**262, 413**]):

THEOREM 5.4.4. *The following partition relations hold for every integer $n \geq 1$ and a for each cardinal $\tau \geq \omega$:*

a) $\omega \to (\omega)^2_n$;
b) $(2^\tau)^+ \to (\tau^+)^2_\tau$;
c) $\tau^+ \to (\tau^+, \omega)^2$

One more definition will be helpful. Given a set $X$, a point $x \in X$, and a covering $\mathcal{C}$ of $X$, we put $St(x, \mathcal{C}) = \bigcup\{K \in \mathcal{C} : x \in K\}$.

LEMMA 5.4.5. *Let $X$ be a set, let $\tau, \lambda, \kappa, \mu$ be cardinals with $\mu^2 \leq \kappa$, let $\{x_\alpha : \alpha < \tau\}$ be a sequence of points of $X$, and let $\{\mathcal{C}_\alpha : \alpha < \tau\}$ be a sequence of coverings of $X$ such that $|\mathcal{C}_\alpha| \leq \mu$ for each $\alpha < \tau$. If $\tau \to (\lambda)^2_\kappa$ holds, then there exists a set $A \subset \tau$ with $|A| = \lambda$ such that $x_\beta \in St(x_\alpha, \mathcal{C}_\gamma) \cap St(x_\gamma, \mathcal{C}_\alpha)$ for any $\alpha, \beta, \gamma \in A$ with $\alpha < \beta < \gamma$.*

PROOF. For every $\alpha < \tau$, let $\mathcal{C}_\alpha = \{K_{\alpha,i} : i \in \mu\}$ be an enumeration of $\mathcal{C}_\alpha$. Given any $\{\alpha, \beta\} \in [\tau]^2$ with $\alpha < \beta$, we define $f(\{\alpha, \beta\}) = (i, j) \in \kappa \times \kappa$ in such a way that $x_\alpha \in K_{\beta,i}$ and $x_\beta \in K_{\alpha,j}$. The function $f \colon [\tau]^2 \to \mu \times \mu$ partitions $[\tau]^2$ into at most $|\mu \times \mu| \leq \kappa$ parts, so there exist $(i, j) \in \mu \times \mu$ and $A \subset \tau$ with $|A| = \lambda$ such that $f(\{\alpha, \beta\}) = (i, j)$ for all distinct $\alpha, \beta \in A$. Let $\alpha < \beta < \gamma$ be arbitrary elements of $A$. From $f(\{\alpha, \beta\}) = (i, j)$ it follows that $x_\beta \in K_{\alpha,j}$; $f(\{\beta, \gamma\}) = (i, j)$ implies $x_\beta \in K_{\gamma,i}$, and $f(\{\alpha, \gamma\}) = (i, j)$ implies that $x_\alpha \in K_{\gamma,i}$ and $x_\gamma \in K_{\alpha,j}$. Therefore, we have

$$x_\beta \in K_{\gamma,i} \subset St(x_\alpha, \mathcal{C}_\gamma) \text{ and } x_\beta \in K_{\alpha,j} \subset St(x_\gamma, \mathcal{C}_\alpha).$$

□

LEMMA 5.4.6. *Let $X$ be a Tychonoff space such that $Nag(X) \leq \tau$, let $\{x_\alpha : \alpha < \tau^+\}$ be a sequence of points of $X$, and let $\{\mathcal{C}_\alpha : \alpha < \tau^+\}$ be a sequence of open coverings of $X$. Then there exists a set $A \subset \tau^+$ with $|A| = \tau^+$ such that $St(x_\alpha, \mathcal{C}_\beta) \cap St(x_\beta, \mathcal{C}_\alpha) \neq \emptyset$, for any distinct $\alpha, \beta \in A$.*

PROOF. Since $Nag(X) \leq \tau$, there exists a family $\mathcal{P}$ of closed subsets of $\beta X$ with $|\mathcal{P}| \leq \tau$ which separates $X$ from $\beta X \setminus X$. We can assume that $\mathcal{P}$ is closed under finite intersections. Note that $\bigcap\{F \in \mathcal{P} : x \in F\} \subset X$ for each $x \in X$. For every $\alpha < \tau$, choose a family $\mathcal{D}_\alpha$ of open sets in $\beta X$ such that $\mathcal{C}_\alpha = \{X \cap U : U \in \mathcal{D}_\alpha\}$. Since the family $\mathcal{P}$ is closed under finite intersections, we can find, for every $\alpha < \tau$, an element $F_\alpha \in \mathcal{P}$ such that $x_\alpha \in F_\alpha \subset \bigcup \mathcal{D}_\alpha$. From $|\mathcal{P}| \leq \tau$ it follows that there are $F \in \mathcal{P}$ and $B \subset \tau^+$ with $|B| = \tau^+$ such that $F_\alpha = F$ for each $\alpha \in B$. We can assume without loss of generality that $B = \tau^+$. As $F$ is compact, for every $\alpha < \tau^+$, there exists a finite subfamily $\mathcal{E}_\alpha$ of $\mathcal{D}_\alpha$ which covers $F$. We can additionally assume that there exists an integer $n$ such that $|\mathcal{E}_\alpha| \leq n$ for each $\alpha < \tau^+$. The rest of the proof is a mere application of the partition calculus technique.

Let $k = n^2$. Since $\omega \to (\omega)^2_k$ by item a) of Theorem 5.4.4 a), Lemma 5.4.5 applied to the sequences $\{x_\alpha : \alpha < \tau^+\}$ and $\{\mathcal{E}_\alpha : \alpha < \tau^+\}$ yields the following:

**Claim.** *Every infinite set $I \subset \tau^+$ contains an infinite subset $J$ such that $x_\beta \in St(x_\alpha, \mathcal{E}_\gamma) \cap St(x_\gamma, \mathcal{E}_\alpha)$ for any $\alpha, \beta, \gamma \in J$ with $\alpha < \beta < \gamma$.*

Let us define a mapping $h \colon [\tau^+]^2 \to \{0, 1\}$ as follows:

$$h(\{\alpha, \beta\}) = 0 \text{ if } St(x_\alpha, \mathcal{E}_\beta) \cap St(x_\beta, \mathcal{E}_\alpha) = \emptyset.$$

Since $\tau^+ \to (\tau^+, \omega)^2$ by c) of Theorem 5.4.4, either there exists a subset $A$ of $\tau^+$ of cardinality $\tau^+$ with $[A]^2 \subset h^{-1}(1)$ or there exists an infinite set $R \subset \tau^+$ such that $[R]^2 \subset h^{-1}(0)$. The latter, however, is impossible by the above Claim. Therefore, for an appropriate subset $A$ of $\tau^+$, we have $St(x_\alpha, \mathscr{E}_\beta) \cap St(x_\beta, \mathscr{E}_\alpha) \neq \varnothing$ for all $\alpha, \beta \in A$. Since $\mathscr{E}_\alpha \subset \mathscr{D}_\alpha$ and the trace of the family $\mathscr{D}_\alpha$ on $X$ coincides with $\mathscr{C}_\alpha$ for each $\alpha < \tau^+$, the conclusion of the lemma is immediate. □

Lemma 5.4.6 enables us to present a short proof of the following result which complements Corollary 5.3.21.

THEOREM 5.4.7. *Let $H$ be an arbitrary subgroup of a topological group $G$ such that $Nag(G) \leq \tau$. Then $(\tau^+, 2)$ is a weak precalibre of $H$.*

PROOF. To have $(\tau^+, 2)$ as a weak precalibre is a hereditary property with respect to passing to dense subspaces. Since the closure $H^*$ of $H$ in $G$ is a closed subgroup of $G$ such that $Nag(H^*) \leq Nag(G) \leq \tau$, we can assume without loss of generality that $H = G$.

Let $\{O_\alpha : \alpha < \tau^+\}$ be a family of non-void open sets in $G$. Consider the continuous mapping $f\colon G^3 \to G$ defined by $f(x, y, z) = xy^{-1}z$. For every $\alpha < \tau^+$, pick a point $x_\alpha \in O_\alpha$ and choose an open covering $\mathscr{C}_\alpha$ of $G$ such that if $y, z \in U$ for some $U \in \mathscr{C}_\alpha$, then $f(x_\alpha, y, z) \in O_\alpha$ and $f(y, z, x_\alpha) \in O_\alpha$. By Lemma 5.4.6, one can find a set $A \subset \tau^+$ of cardinality $\tau^+$ such that $St(x_\alpha, \mathscr{C}_\beta) \cap St(x_\beta, \mathscr{C}_\alpha) \neq \varnothing$ for all distinct $\alpha, \beta \in A$. If $x \in St(x_\alpha, \mathscr{C}_\beta) \cap St(x_\beta, \mathscr{C}_\alpha)$, then there exist $U \in \mathscr{C}_\alpha$ and $V \in \mathscr{C}_\beta$ such that $x, x_\beta \in U$ and $x_\alpha, x \in V$. Therefore, $f(x_\alpha, x, x_\beta) \in O_\alpha \cap O_\beta$, by the choice of the coverings $\mathscr{C}_\alpha$ and $\mathscr{C}_\beta$. We have thus proved that $O_\alpha \cap O_\beta \neq \varnothing$ for all $\alpha, \beta \in A$. □

COROLLARY 5.4.8. *Every subgroup of a $\sigma$-compact topological group has the Knaster property.*

In general, the cellularity is not productive, even in topological groups. It is known, for example, that there exists in $ZFC$ a topological group $G$ such that $c(G \times G) > c(G)$ (see Problem 5.4.G). Nevertheless, a weak form of compactness in topological groups improves the situation.

COROLLARY 5.4.9. *Let $G$ be a topological group with $Nag(G) \leq \tau$. Then $c(G \times X) \leq \tau$, for every space $X$ satisfying $c(X) \leq \tau$.*

PROOF. Let $\gamma = \{O_\alpha : \alpha < \tau^+\}$ be a family of non-void open sets in $G \times X$, where the space $X$ satisfies $c(X) \leq \tau$. We can assume without loss of generality that every $O_\alpha$ has the form $U_\alpha \times V_\alpha$. By Theorem 5.4.7, $(\tau^+, 2)$ is a weak precalibre of $G$, so there exists a subset $A$ of $\tau^+$ with $|A| = \tau^+$ such that $U_\alpha \cap U_\beta \neq \varnothing$ for all $\alpha, \beta \in A$. Since $c(X) \leq \tau$, one can find distinct $\alpha, \beta \in A$ such that $V_\alpha \cap V_\beta \neq \varnothing$. Then $O_\alpha \cap O_\beta \neq \varnothing$, whence the conclusion follows. □

Since every Tychonoff space can be topologically embedded into the Tychonoff cube $I^\tau$ or, equivalently, into the compact topological group $\mathbb{T}^\tau$ for some $\tau$, there is no upper bound for the cellularity of subspaces of $\omega$-cellular spaces. The situation changes radically if we consider subgroups of topological groups. First, we prove a general theorem that relates the cellularity and the index of narrowness in topological groups.

THEOREM 5.4.10. $c(G) \leq 2^{ib(G)}$, *for every topological group $G$.*

PROOF. The proof is very much like that of Theorem 5.4.7. Set $\kappa = ib(G)$ and $\tau = 2^\kappa$. Let $\{O_\alpha : \alpha < \tau^+\}$ be a family of non-empty open subsets of $G$. For every $\alpha < \tau^+$, choose a point $x_\alpha \in O_\alpha$ and open symmetric neighbourhoods $U_\alpha$ and $V_\alpha$ of the identity in $G$ such that $x_\alpha U_\alpha \subset O_\alpha$, $U_\alpha x_\alpha \subset O_\alpha$ and $V_\alpha^2 \subset U_\alpha$. Since the group $G$ is $\kappa$-narrow, there exists $K_\alpha \subset G$ with $|K_\alpha| \leq \kappa$ such that $K_\alpha V_\alpha = G$ and $V_\alpha K_\alpha = G$. Denote by $\mathscr{C}_\alpha$ an open covering of $G$ refining the coverings $\{xV_\alpha : x \in K_\alpha\}$ and $\{V_\alpha x : x \in K_\alpha\}$ and satisfying $|\mathscr{C}_\alpha| \leq \kappa$. Since $(2^\kappa)^+ \to (\kappa^+)^2_\kappa$, by b) of Theorem 5.4.4, we can apply Lemma 5.4.5 to the sequences $\{x_\alpha : \alpha < \tau^+\}$, $\{\mathscr{C}_\alpha : \alpha < \tau^+\}$ and find $\alpha, \beta, \gamma < \tau^+$ with $\alpha < \beta < \gamma$ such that $x_\beta \in St(x_\alpha, \mathscr{C}_\gamma) \cap St(x_\gamma, \mathscr{C}_\alpha)$. Therefore, there exist $a, b \in G$ such that $x_\alpha, x_\beta \in V_\gamma a$ and $x_\beta, x_\gamma \in bV_\alpha$, whence $x_\alpha x_\beta^{-1} \in V_\gamma^2$ and $x_\beta^{-1} x_\gamma \in V_\alpha^2$. We conclude that

$$x_\alpha x_\beta^{-1} x_\gamma \in x_\alpha V_\alpha^2 \cap V_\gamma^2 x_\gamma \subset x_\alpha U_\alpha \cap U_\gamma x_\gamma \subset O_\alpha \cap O_\gamma.$$

This proves that the family $\{O_\alpha : \alpha < \tau^+\}$ is not disjoint, so that $c(G) \leq 2^\kappa$. □

THEOREM 5.4.11. *Every subgroup $H$ of a topological group $G$ satisfies $c(H) \leq 2^{c(G)}$.*

PROOF. If $H$ is a subgroup of $G$, then $ib(H) \leq ib(G) \leq c(G)$ by Propositions 5.1.1 a) and 5.1.3 b), so Theorem 5.4.10 implies that $c(H) \leq 2^{ib(H)} \leq 2^{c(G)}$. □

Since topological products of second-countable groups have countable cellularity by [**165**, Coro. 2.3.18], Theorem 5.4.11 implies the following useful result.

COROLLARY 5.4.12. *Every subgroup $H$ of a product of second-countable groups satisfies $c(H) \leq \mathfrak{c}$.*

In the next example we show that the upper bound $\mathfrak{c}$ for the cellularity of $\omega$-narrow groups in Corollary 5.4.12 is exact.

EXAMPLE 5.4.13. Let $H = A(\omega)$ be the free Abelian group over the set $\omega$, that is, the direct sum of $\omega$ copies of the group $\mathbb{Z}$. We consider $H$ with the discrete topology. Clearly, $H$ is countable. Our aim is to show that the group $H^\mathfrak{c}$, taken with the product topology, contains a subgroup $G$ satisfying $c(G) = \mathfrak{c}$. Denote by $I$ the set of all mappings $g \colon \omega \setminus \{0\} \to \{0, 1\}$. Since $|I| = \mathfrak{c}$, we will construct $G$ as a subgroup of $H^I$.

Let $[I]^2$ be the family of all unordered pairs of $I$. For any distinct $g, h \in I$, set

$$\varphi(\{g, h\}) = \min\{n \in \omega \setminus \{0\} : g(n) \neq h(n)\}.$$

It is easy to see that the mapping $\varphi \colon [I]^2 \to \{0, 1\}$ has the following property:

**Fact 1.** *If $f, g, h$ are distinct elements of $I$, then two of the three numbers $\varphi(\{f, g\})$, $\varphi(\{g, h\})$ and $\varphi(\{h, f\})$ are equal and strictly less than the third one.*

For every $g \in I$, we define a mapping $x_g \colon I \to H$ by $x_g(h) = \varphi(\{g, h\})$ if $g \neq h$ and $x_g(h) = 0$ if $g = h$. Then $x_g \in H^I$. Let $G$ be the subgroup of $H^I$ generated by the set $X = \{x_g : g \in I\}$. Since the weight of the group $H^I$ is equal to the cardinality of $I$, we have $c(G) \leq w(G) \leq w(H^I) \leq \mathfrak{c}$. It remains to show that $c(G) \geq \mathfrak{c}$.

For every $g \in I$, consider the open subset

$$V_g = \{y \in G : y(g) = 0\}$$

of $G$. Since $V_g$ is not empty for each $g \in I$, we only need to prove the following fact.

**Fact 2.** *The family $\{V_g : g \in I\}$ is disjoint.*

Assume the contrary. Then $V_g \cap V_h \neq \varnothing$ for some distinct $g, h \in I$. Let $n = \varphi(\{g, h\})$. Then $x_g(h) = x_h(g) = n > 0$. Denote by $\pi$ the projection of $H^I$ to $H^K$, where $K = \{g, h\}$. Fact 1 implies that the group $\pi(G)$ is generated by the set

$$X_K = \{(0, n)\} \cup \{(n, 0)\} \cup \{(i, j) : 0 < i = j \text{ or } i = n < j \text{ or } j = n < i\}.$$

Since $V_g \cap V_h \neq \varnothing$, the element $(0, 0)$ belongs to $\pi(G)$. Therefore, we can find distinct elements $(u_1, v_1), \dots, (u_p, v_k)$ of $X_K$ and non-zero integers $k_1, \dots, k_p$ such that

$$(0, 0) = \sum_{i=1}^{p} k_i \cdot (u_i, v_i). \tag{5.2}$$

Hence, both pairs $(0, n)$ and $(n, 0)$ appear among $(u_1, v_1), \dots, (u_p, v_k)$. In addition, if $(0, n) = (u_i, v_i)$ and $(n, 0) = (u_j, v_j)$ for some $i, j \leq p$, then $i \neq j$ and $k_i = k_j = 1$. We can assume that $i = p - 1$ and $j = p$. Moving $(u_i, v_i)$ and $(u_j, v_j)$ to the left part of (5.2), we obtain the following equality after obvious cancellations:

$$-(n, n) = k_1 \cdot (u_1, v_1) + \cdots + k_q \cdot (u_q, v_q), \tag{5.3}$$

where $q = p - 2$ and all $u_1, v_1, \dots, u_q, v_q$ are distinct from 0. Since each $(u_m, v_m)$ is in $X_K$, we conclude that either $u_m = v_m = i$ for some $i < n$, or there exists $j > n$ such that $(u_m, v_m) = (n, j)$ or $(u_m, v_m) = (j, n)$.

Let $k = \max\{u_1, v_1, \dots, u_q, v_q\}$. It is clear that $k > n$. We can assume that $(k, n) = (u_m, v_m)$ for some $m \leq q$. Then $u_i \neq k$ for each $i \neq m$. Rewriting the vector equality (5.3) for the first coordinate, we see that the summand $k$ cannot be cancelled, which is a contradiction. This proves Fact 2 and implies the equality $c(G) = \mathfrak{c}$. □

Example 5.4.13 also shows that the cellularity is not monotonous when taking subgroups of (Abelian) topological groups. On the other hand, if $H$ is a dense or open subgroup of a topological group $H^*$, then $c(H) \leq c(H^*)$. Finally, it is worth mentioning that there exist an Abelian topological group $H^*$ and a closed subgroup $H$ of type $G_\delta$ in $H^*$ such that $\omega = c(H^*) < c(H) = \mathfrak{c}$ (see Exercise 5.4.a).

## Exercises

5.4.a. Use the group constructed in Example 5.4.13 to find a topological Abelian group $H^*$ and a closed subgroup $H$ of type $G_\delta$ in $H^*$ such that $\omega = c(H^*) < c(H) = \mathfrak{c}$.

5.4.b. Let $H$ be an arbitrary subgroup of the $\sigma$-product of a family of second-countable topological groups. Prove that the cellularity of $H$ is countable. Is $(\omega_1, 2)$ a weak precalibre of $H$?

5.4.c. Show that if $G$ is the product of a family of $\sigma$-compact topological groups, then $(\tau, 2)$ is a weak precalibre of $G$, for every regular cardinal $\tau > \omega$.

5.4.d. Let $S$ be the Sorgenfrey line considered as a paratopological group, $\kappa$ a cardinal, and $X \subset S^\kappa$ be a compact subspace of $S^\kappa$. Denote by $H$ the minimal subsemigroup of $S^\kappa$ containing $X$, that is, $x + y \in H$, for all $x, y \in H$. Is the cellularity of $H$ countable?

5.4.e. Show that Theorem 5.4.7 and Corollary 5.4.9 are valid for Mal'tsev spaces (see Exercise 5.3.i).

## Problems

5.4.A. Prove that there exists a $\sigma$-compact topological group which is not a continuous image of the space $\mathbb{N} \times D^\tau$ for any cardinal $\tau$, where the factors $\mathbb{N}$ and $D = \{0, 1\}$ carry the discrete topology.

5.4.B. Give an example of an infinite topological group $G$ such that the space $G \times G$ cannot be represented as a continuous image of the space $G$.

5.4.C. Give an example of an infinite topological group $G$ such that $G$ is not homeomorphic to any dense subspace of any Tychonoff cube $I^\tau$ and still the cellularity of $G$ is countable.

5.4.D. (V. V. Uspenskij [**521**]) Construct a subgroup $H$ of the topological group $\mathbb{Z}^{\mathfrak{c}}$ such that $c(H) = \mathfrak{c}$.
*Hint.* Apply the Chinese remainder theorem to construct a subgroup of $\mathbb{Z}^{\mathfrak{c}}$ with the required property.

5.4.E. (P. Gartside *et al.* [**187**]) Show that there exists a Lindelöf topological group $G$ satisfying $c(G) = \mathfrak{c}$.

5.4.F. (V. V. Uspenskij [**512**]) Prove that the cellularity of every Lindelöf $P$-group does not exceed $\aleph_1$.
*Hint.* The proof of this fact can be found in Section 8.6.

5.4.G. (S. Todorčević [**494**]) Prove in $ZFC$ that there exists a topological group $G$ such that $c(G) < c(G \times G)$.

5.4.H. (D. B. Shakhmatov [**429**]) It is consistent with and independent of $ZFC$ that $\aleph_1$ is a precalibre of every $\sigma$-compact topological group.

5.4.I. Let $G = \prod_{i \in I} G_i$ be the product of a family of topological groups, where $\tau$ is an infinite cardinal and $c(G_i) \leq \tau$, for each $i \in I$. Then $c(H) \leq 2^\tau$, for every subgroup $H$ of $G$.

5.4.J. Suppose that $H = \Sigma \prod \xi$ is the $\Sigma$-product of a family $\xi$ of topological groups, where each $K \in \xi$ is either discrete or satisfies $|K| \leq \aleph_1$. Let $G$ be the $G_\delta$-modification of $H$, where $H$ carries the usual subspace topology inherited from $\prod \xi$. Prove that $t(G) \leq \aleph_1$.

5.4.K. Let $G$ be a $\sigma$-compact topological group such that the $G_\delta$-modification of $G$ is Lindelöf. Prove that $G$ is countable.

5.4.L. Let $G$ be a Lindelöf $\Sigma$-group such that the $G_\delta$-modification of $G$ is Lindelöf. Must $G$ be countable?

5.4.M. (I. V. Protasov [**389**]) Let $G$ be a topological group. A sequence $\{x_n : n \in \omega\}$ of elements of $G$ is called a *left Cauchy sequence* if, for every neighbourhood $U$ of the neutral element in $G$, there exists an integer $N$ such that $x_n^{-1} x_m \in U$ for all $m, n > N$. Suppose that a sequence $\{x_n : n \in \omega\}$ of elements of a metrizable topological group $G$ does not contain a left Cauchy subsequence. Prove that one can find an infinite set $A \subset \omega$ and a neighbourhood $U$ of the neutral element in $G$ such that $x_i U \cap x_j U = \varnothing$, for all distinct $i, j \in A$.
*Hint.* Assume the contrary and apply Ramsey's theorem (see item a) of Theorem 5.4.4) to construct by induction an infinite set $A \subset \omega$ such that the subsequence $\{x_i : i \in A\}$ of $\{x_n : n \in \omega\}$ is left Cauchy.

5.4.N. (I. V. Protasov [**389**]) Prove that if a metrizable topological group $G$ is functionally balanced (see Problem 1.8.B), then $G$ is balanced. Extend the result to the class of feathered topological groups.
*Hint.* For a metrizable group $G$, apply the conclusion of Problem 5.4.M along with (b) of Problem 1.8.B. For a feathered group $G$, use Proposition 4.3.11.

## Open Problems

5.4.1. Let $G$ be an infinite topological group. Is it always possible to represent $G \times G$ as a continuous image of some subspace of $G$?

5.4.2. Let $G$ be a Lindelöf topological group of countable tightness. Is the cellularity of $G$ countable?

5.4.3. Let $G$ be an $\omega$-narrow topological group such that the space $G$ is normal and of countable tightness. Is the Souslin number of $G$ countable?

5.4.4. Let $G$ be a Lindelöf topological group such that $C_p(G)$ is Lindelöf. Is the cellularity of $G$ countable?

5.4.5. Can one refine the conclusion of Lemma 5.4.6 to claim the existence of a set $A \subset \tau^+$ such that $|A| = \tau^+$ and $x_\beta \in St(x_\alpha, \mathscr{C}_\gamma) \cap St(x_\gamma, \mathscr{C}_\alpha)$ for all $\alpha, \beta, \gamma \in A$ with $\alpha < \beta < \gamma$?

5.4.6. Let $G$ be a topological group with $Nag(G) \leq \tau$. Is $(\tau^+, n)$ a weak precalibre of $G$ for $n = 3$? Is this true for all integers $n \geq 2$? What is the answer in the case when $G$ is $\sigma$-compact and $\tau = \aleph_1$?

5.4.7. Let $G$ be the $\Sigma$-product of $\tau$ copies of the discrete group $\mathbb{Z}$. Does every subgroup of $G$ have countable cellularity?

## 5.5. *o*-tightness in topological groups

A space $X$ satisfies $c(X) \leq \tau$ if and only if every family $\gamma$ of open sets in $X$ contains a subfamily $\eta$ such that $|\eta| \leq \tau$ and $\overline{\bigcup \eta} = \overline{\bigcup \gamma}$. Here we consider a pointwise version of the Souslin number known as the *o-tightness.*

The *o-tightness* of a space $X$ is the smallest infinite cardinal number $\tau$ such that whenever a point $a \in X$ belongs to the closure of $\bigcup \gamma$, where $\gamma$ is any family of open sets in $X$, there exists a subfamily $\eta$ of $\gamma$ such that $|\eta| \leq \tau$ and $a$ is in the closure of $\bigcup \eta$. If the $o$-tightness of $X$ is equal to $\tau$, we write $ot(X) = \tau$.

The $o$-tightness of $X$ is not greater than the tightness of $X$ and the cellularity of $X$. It also clear that the $o$-tightness of $X$ does not exceed the $G_\delta$-tightness of $X$, that is, $ot(X) \leq get(X)$.

Obviously, an open subspace $U$ of a space $X$ satisfies $ot(U) \leq ot(X)$. It is worth mentioning the following simple property of the $o$-tightness:

PROPOSITION 5.5.1. *If $Y$ is a dense subspace of a space $X$, then $ot(Y) \leq ot(X)$.*

PROOF. Let $\gamma$ be a family of open sets in $Y$, and suppose that $y \in Y$ is in the closure of the set $\bigcup \gamma$. Let $\eta$ be a family of open sets in $X$ such that $\gamma = \{U \cap Y : U \in \eta\}$. Evidently, $y \in \overline{\bigcup \eta}$ and, since $ot(X) \leq \omega$, there exists a countable subfamily $\eta'$ of $\eta$ such that $y \in \overline{\bigcup \eta'}$. Then $\gamma' = \{U \cap Y : U \in \eta'\}$ is a countable subfamily of $\gamma$. Since $Y$ is dense in $X$, it follows that $y \in \overline{\bigcup \gamma'}$, so we conclude that the $o$-tightness of $Y$ is countable. □

There is no upper bound for the $o$-tightness of compact spaces. Indeed, let $\tau$ be an infinite regular cardinal. Denote by $\tau + 1$ the space $\tau \cup \{\tau\}$ endowed with the order topology. Then $\tau + 1$ is a compact space, but the end point $\{\tau\}$ is not in the closure of any set $K \subset \tau$ with $|K| < \tau$. Since the set $D$ of all isolated points of $\tau$ has the cardinality $\tau$, we conclude that $ot(\tau + 1) = \tau$.

Since the space $\tau + 1$ is homeomorphic to a subspace of the Tychonoff cube $I^\tau$ and the $o$-tightness of $I^\tau$ is countable (note that the cellularity of $I^\tau$ is countable), it follows that the $o$-tightness can increase when passing to a (closed) subspace. In particular, one cannot drop the condition of $Y$ being dense in $X$ in Proposition 5.5.1. The same phenomenon occurs in topological groups (see Exercise 5.5.c).

On the other hand, we show in Theorem 5.5.4 below that a mild compactness type condition imposed on a topological group $G$ implies that $ot(G) \leq \omega$. This result requires the notion of an *admissible subgroup.*

A subgroup $H$ of a topological group $G$ is called *admissible* if there exists a sequence $\{U_n : n \in \omega\}$ of open symmetric neighbourhoods of the identity in $G$ such that $U_{n+1}^3 \subset U_n$ for each $n \in \omega$ and $H = \bigcap_{n \in \omega} U_n$.

In the next lemma, we establish three simple properties of admissible subgroups.

LEMMA 5.5.2. *Let $G$ be a topological group. Then:*

a) *every admissible subgroup $H$ of $G$ is closed in $G$ and the quotient space $G/H$ has countable pseudocharacter;*
b) *every neighbourhood of the identity $e$ in $G$ contains an admissible subgroup;*
c) *the intersection of countably many admissible subgroups of $G$ is again an admissible subgroup of $G$.*

PROOF. a) Suppose that $H$ is an admissible subgroup of $G$. Then there exists a sequence $\{U_n : n \in \omega\}$ of open symmetric neighbourhoods of $e$ in $G$ such that $U_{n+1}^3 \subset U_n$ for each $n \in \omega$ and $H = \bigcap_{n\in\omega} U_n$. From $U_{n+1}^3 \subset U_n$ it follows that $U_n$ contains the closure of $U_{n+1}$ in $G$ for each $n \in \omega$, so that the intersection of the sets $U_n$ coincides with the intersection of their closures. Hence, $H$ is closed in $G$.

Let $\pi\colon G \to G/H$ be the quotient mapping of $G$ onto the left coset space $G/H$. For every $n \in \omega$, we have

$$\pi^{-1}\pi(U_{n+1}) = U_{n+1}H \subset U_{n+1}^2 \subset U_n.$$

Therefore, the set $P = \bigcap_{n\in\omega} \pi(U_n)$ satisfies

$$\pi^{-1}(P) = \bigcap_{n\in\omega} \pi^{-1}\pi(U_{n+1}) \subset \bigcap_{n\in\omega} U_n = H.$$

This proves that $P = \{\pi(e)\}$, i.e., the point $\pi(e)$ has countable pseudocharacter in $G/H$. To finish the proof, it remains to note that the quotient space $G/H$ is homogeneous.

b) Let $U$ be an arbitrary neighbourhood of $e$ in $G$. Define a sequence $\{U_n : n \in \omega\}$ of open symmetric neighbourhoods of $e$ in $G$ such that $U_0 \subset U$ and $U_{n+1}^3 \subset U_n$ for each $n \in \omega$. Then $H = \bigcap_{n\in\omega} U_n$ is an admissible subgroup of $G$ and $H \subset U_0 \subset U$.

c) Let $\{H_n : n \in \omega\}$ be a sequence of admissible subgroups of $G$. For every $n \in \omega$, we can find a sequence $\{U_{n,k} : k \in \omega\}$ of open symmetric neighbourhoods of $e$ in $G$ such that $(U_{n,k+1})^3 \subset U_{n,k}$ for each $k \in \omega$ and $H_n = \bigcap_{k\in\omega} U_{n,k}$. Consider the sequence $\{V_n : n \in \omega\}$, where $V_n = \bigcap_{i=0}^n U_{i,n}$ for each $n \in \omega$. Clearly, every $V_n$ is an open symmetric neighbourhood of $e$ in $G$. In addition, $V_{n+1}^3 \subset U_{i,n+1}^3 \subset U_{i,n}$ whenever $i \le n$, so $V_{n+1}^3 \subset \bigcap_{i=0}^n U_{i,n} = V_n$ for all $n \in \omega$. Therefore, $H = \bigcap_{n\in\omega} V_n$ is an admissible subgroup of $G$, and our definition of the sets $V_n$ implies that

$$H = \bigcap_{n\in\omega} V_n = \bigcap_{n\in\omega} \bigcap_{i=0}^{n} U_{i,n} = \bigcap_{i\in\omega} \bigcap_{n=i}^{\infty} U_{i,n} = \bigcap_{i\in\omega} H_i.$$

So, the group $\bigcap_{i\in\omega} H_i$ is admissible. □

One more technical notion will be useful in the sequel. Let us call a subset $F$ of a topological group $H$ *standard* if one can find an admissible subgroup $N$ of $H$ and a $G_\delta$-set $K$ in the quotient space $H/N$ such that $F = \pi_N^{-1}(K)$, where $\pi_N\colon H \to H/N$ is the quotient mapping. Observe that every standard set in $H$ is of type $G_\delta$ in $H$.

LEMMA 5.5.3. *Let $H$ be any topological group. Then:*

a) *Every $G_\delta$-set in $H$ is the union of a family of standard sets.*

b) *If $\eta$ is a countable family of standard subsets of $H$, then $\overline{\bigcup \eta}$ is the union of standard sets in $H$.*

PROOF. Clearly, a) follows from b) and c) of Lemma 5.5.2. Let us prove b). Suppose that $\eta$ is a countable family of standard sets in $H$. For every $F \in \eta$, choose an admissible subgroup $N_F$ of $H$ and a $G_\delta$-set $K_F$ in $H/N_F$ satisfying $F = \pi_{N_F}^{-1}(K_F)$, where $\pi_{N_F}\colon H \to H/N_F$ is the natural quotient mapping. Then $N = \bigcap_{F\in\eta} N_F$ is an admissible subgroup of $H$, by c) of Lemma 5.5.2. It is clear that the equality $F = \pi^{-1}\pi(F)$ holds for every $F \in \eta$, where $\pi\colon H \to H/N$ is the coset mapping. By a) of Lemma 5.5.2, the quotient space $H/N$ has countable pseudocharacter, so every fiber $\pi^{-1}(y)$ is a standard set in $H$. Put $F_0 = \bigcup \eta$ and $K_0 = \pi(K_0)$. Then $F_0 = \pi^{-1}(K_0)$. Since $\pi$ is an open mapping, we have $\overline{F}_0 = \pi^{-1}(\overline{K}_0)$, which implies that $\overline{F}_0$ is a union of standard sets in $H$. □

Let $X$ be a Hausdorff space such that a subset $F$ of $X$ is closed if and only if the intersection $F \cap K$ is closed in $K$, for each compact set $K$ in $X$. Such a space $X$ is called a *$k$-space* (see [**165**, Section 3.3]). All $G_\delta$-sets and their unions in a compact space are $k$-spaces. In addition, all Fréchet–Urysohn spaces and all sequential spaces are $k$-spaces as well [**165**, Th. 3.3.20]. If $G$ is a topological group, we will call $G$ a *$k$-group* if $G$ is a $k$-space as a topological space.

THEOREM 5.5.4. [**M. G. Tkachenko**] *Let $H$ be a $k$-group. If $\gamma$ is a family of $G_\delta$-sets in $H$ and $x \in H$ is a cluster point of $\gamma$, then there exists a countable subfamily $\eta$ of $\gamma$ such that $x \in \overline{\bigcup \eta}$. Therefore, both the $G_\delta$-tightness and $o$-tightness of every $k$-group $H$ are countable.*

PROOF. Suppose that $\gamma$ is a family of $G_\delta$-sets in $H$. To prove the first assertion of the theorem, it suffices to show that the set

$$X = \bigcup\{\overline{\bigcup \eta} : \eta \subset \gamma,\ |\eta| \le \omega\}$$

is closed in $H$. We may assume without loss of generality that all elements of $\gamma$ are standard sets in $H$. Let $B$ be any compact subset of $H$ and let $\langle B\rangle$ be the subgroup of $H$ generated by $B$. Then $\langle B\rangle$ is a $\sigma$-compact subgroup of $H$ and, by Lemma 5.5.3, there exists a family $\mathcal{P}$ of $G_\delta$-sets in $\langle B\rangle$ such that $X \cap \langle B\rangle = \bigcup \mathcal{P}$ and every element $P \in \mathcal{P}$ is contained in the closure of the union of some countable subfamily $\mu_P$ of $\gamma$. By Corollary 5.3.20, $\mathcal{P}$ contains a countable subfamily $\lambda$ such that $\bigcup \lambda$ is dense in $\bigcup \mathcal{P}$. Put $\mu_0 = \bigcup_{P\in\lambda} \mu_P$. Then $\mu_0$ is countable, so we have $\overline{\bigcup \mathcal{P}} = \overline{\bigcup \lambda} \subset \overline{\bigcup \mu_0} \subset X$. On the other hand, the equality $X \cap \langle B\rangle = \bigcup \mathcal{P}$ implies that $X \cap \langle B\rangle = \overline{\bigcup \mu_0} \cap \langle B\rangle$. Thus, $X \cap \langle B\rangle$ is closed in $\langle B\rangle$ and $X \cap B$ is closed in $B$. Since $H$ is a $k$-group and $B$ is an arbitrary compact subset of $H$, the set $X$ is closed in $H$.

The inequality $ot(H) \le get(H) \le \omega$ is now immediate, and the theorem is proved. □

Recall that a space $X$ is $G_\delta$-preserving if for every family $\gamma$ of $G_\delta$-sets in $X$, the set $\overline{\bigcup \gamma}$ is again the union of some family $G_\delta$-sets in $X$ (see Section 1.6). The next result gives a useful sufficient condition for a topological group to be $G_\delta$-preserving.

PROPOSITION 5.5.5. *Every topological group $H$ of countable $G_\delta$-tightness is $G_\delta$-preserving.*

PROOF. Let $\gamma$ be a family of $G_\delta$-sets in the group $H$. Since every $G_\delta$-set in $H$ is the union of standard sets by (a) of Lemma 5.5.3, we may assume that all elements of $\gamma$ are standard sets. Put $P = \bigcup \gamma$. Since $get(H) \leq \omega$, it follows that

$$\overline{P} = \bigcup \{\overline{\bigcup \eta} : \eta \subset \gamma,\ |\eta| \leq \omega\}.$$

By (b) of Lemma 5.5.3, the set $\overline{\bigcup \eta}$ is the union of a family of standard subsets of $H$ for every countable subfamily $\eta$ of $\gamma$, and each standard set is of type $G_\delta$ in $H$. This implies the required conclusion. □

Combining Theorem 5.5.4 and Proposition 5.5.5, we deduce the following fact.

COROLLARY 5.5.6. *Let $\gamma$ be a family of $G_\delta$-sets in a $k$-group $H$. Then $\overline{\bigcup \gamma}$ is the union of a family of $G_\delta$-sets in $H$.*

Since every feathered topological group is a $k$-group, from Theorem 5.5.4 it follows:

COROLLARY 5.5.7. *The $G_\delta$-tightness and the o-tightness of every feathered topological group are countable.*

It is not known whether the $o$-tightness is productive in the realm of topological groups (see Problem 5.5.5). However, the $o$-tightness becomes productive under certain restrictions on factors by Theorem 5.5.9 below. We start with a lemma.

LEMMA 5.5.8. *The product $X \times Y$ of a space $X$ with $ot(X) \leq \omega$, and a first-countable space $Y$ has countable o-tightness.*

PROOF. Let $\gamma$ be a family of open sets in $X \times Y$ and suppose that a point $(x, y) \in X \times Y$ belongs to the closure of $\bigcup \gamma$. We may assume that every element of $\gamma$ has the rectangular form $U \times V$. Choose a countable base $\{V_n : n \in \omega\}$ at the point $y$ in $Y$ and, for every $n \in \omega$, put

$$\gamma_n = \{U \times V \in \gamma : V \cap V_n \neq \varnothing\}.$$

It is clear that $(x, y) \in \overline{\bigcup \gamma_n}$. Let $\pi\colon X \times Y \to X$ be the projection. From $ot(X) \leq \omega$ it follows that there exists a countable subfamily $\mu_n \subset \gamma_n$ such that $x \in \overline{\pi(\bigcup \mu_n)}$, $n \in \omega$. Put $\mu = \bigcup_{n\in\omega} \mu_n$. Then the family $\mu$ is countable, so it remains to verify that $(x, y)$ is in the closure of $\bigcup \mu$.

Take an arbitrary rectangular neighbourhood $U \times V$ of $(x, y)$ in $X \times Y$. Choose $n \in \omega$ such that $y \in V_n \subset V$. Since $x \in \overline{\pi(\bigcup \mu_n)}$, there exists an element $U' \times V'$ of $\mu_n$ such that $U \cap U' \neq \varnothing$. From our definition of $\mu_n$ it follows that $V \cap V' \neq \varnothing$, so that $(U \times V) \cap (U' \times V') \neq \varnothing$. Since $U' \times V' \in \mu_n \subset \mu$, we conclude that $(x, y) \in \overline{\bigcup \mu}$. □

THEOREM 5.5.9. *Let $G$ be a topological group such that $ot(G) \leq \omega$. Then $ot(G \times H) \leq \omega$, for every feathered topological group $H$.*

PROOF. Denote by $\mathcal{K}$ the family of all compact subgroups $K$ of $H$ satisfying $\chi(K, H) \leq \omega$. Let $\mathcal{B}$ be the family of all sets of the form $\pi_K^{-1}(O)$ in $H$, where $K \in \mathcal{K}$, $\pi_K\colon H \to H/K$ is the quotient mapping onto the left coset space $H/K$ and $O$ is open in $H/K$. By Corollary 4.3.12, $\mathcal{B}$ is a base for $H$. Suppose that $(x, y) \in G \times H$ is an accumulation point of a family $\gamma$ of open sets in $G \times H$. Since topological groups are homogeneous, we may assume that $x = e_G$ and $y = e_H$. In addition, we may assume that every element of $\gamma$ has the form $U \times V$, where $U$ and $V$ are open in $G$ and $H$, respectively, and $V \in \mathcal{B}$.

Let $i_G$ be the identity isomorphism of $G$ and let $p\colon G \times H \to H$ be the projection. We will define by recursion on $n \in \omega$ two sequences $\{K_n : n \in \omega\}$ and $\{\gamma_n : n \in \omega\}$ satisfying the following conditions for each $n \in \omega$:

(i) $K_n$ is a compact subgroup of $H$ and $\chi(K_n, H) \leq \omega$;
(ii) $K_{n+1} \subset K_n$;
(iii) $\gamma_n \subset \gamma_{n+1} \subset \gamma$ and $|\gamma_n| \leq \omega$;
(iv) $p(W) = \pi_{n+1}^{-1}\pi_{n+1}(p(W))$ for each $W \in \gamma_n$, where $\pi_{n+1}$ is the quotient mapping of $H$ onto the left coset space $H/K_{n+1}$;
(v) $(e_G, \pi_n(e_H))$ is in the closure of $\varphi_n(\bigcup \gamma_n)$, where $\varphi_n = i_G \times \pi_n$.

By Proposition 4.3.11, there is $K_0 \in \mathcal{K}$. Then the quotient space $H_0 = H/K_0$ is metrizable by Lemma 4.3.19. Let $\pi_0\colon H \to H_0$ be the quotient mapping and $\varphi = i_G \times \pi_0$. Then $\varphi$ is a continuous mapping of $G \times H$ onto $G \times H_0$. Clearly, $\varphi$ is open as the product of open mappings $i_G$ and $\pi_0$. Since the group $H_0$ is metrizable, Lemma 5.5.8 implies that $ot(G \times H_0) \leq \omega$. Hence we can find a countable family $\gamma_0 \subset \gamma$ such that $(e_G, \pi_0(e_H)) \in \overline{\varphi_0(\bigcup \gamma_0)}$.

Suppose that, for some $n \in \omega$, we have already defined two sequences $\{K_i : i \leq n\}$ and $\{\gamma_i : i \leq n\}$ satisfying (i)–(v). The family $\gamma_n$ is countable by (iii), so we can write $\gamma_n = \{U_k \times V_k : k \in \omega\}$. Note that $\gamma_n \subset \gamma$ and, hence, $V_k \in \mathcal{B}$ for every $k \in \omega$. So, there exists $L_k \in \mathcal{K}$ such that $V_k = \pi_{L_k}^{-1}\pi_{L_k}(V_n)$. Put $K_{n+1} = K_n \cap \bigcap_{k \in \omega} L_k$. It is easy to see that $K_{n+1} \in \mathcal{K}$; obviously, $K_{n+1} \subset K_n$. In addition, if $k \in \omega$, then $K_{n+1} \subset L_k$, so that $V_k = \pi_{n+1}^{-1}\pi_{n+1}(V_k)$, where $\pi_{n+1}\colon H \to H/K_{n+1} = H_{n+1}$ is the quotient mapping. Again, the group $H_{n+1}$ is metrizable and the mapping $\varphi_{n+1} = i_G \times \pi_{n+1}$ of $G \times H$ onto $G \times H_{n+1}$ is continuous and open. Hence $ot(G \times H_{n+1}) \leq \omega$ and we can find a countable family $\gamma_{n+1} \subset \gamma$ such that $\gamma_n \subset \gamma_{n+1}$ and the identity of $G \times H_{n+1}$ is in the closure of the set $\varphi_{n+1}(\bigcup \gamma_{n+1})$. Therefore, the families $\{K_i : i \leq n+1\}$ and $\{\gamma_i : i \leq n+1\}$ satisfy (i)–(v) at the stage $n+1$.

Put $K = \bigcap_{n \in \omega} K_n$ and $\mu = \bigcup_{n \in \omega} \gamma_n$. Then $K \in \mathcal{K}$ and $\mu \subset \gamma$, $|\mu| \leq \omega$. Let $\pi\colon H \to H/K = M$ be the quotient mapping. Since $K \subset K_{n+1}$ for all $n \in \omega$, from (iv) it follows that $p(W) = \pi^{-1}\pi(p(W))$ for each $W \in \mu$. In its turn, this implies that the continuous open mapping $\varphi = i_G \times \pi$ of $G \times H$ onto $G \times M$ satisfies $W = \varphi^{-1}\varphi(W)$ for each $W \in \mu$. Hence $\bigcup \mu = \varphi^{-1}\varphi(\bigcup \mu)$. We claim that $(e_G, e_H) \in \overline{\bigcup \mu}$.

Assume that $(e_G, e_H) \notin \overline{\bigcup \mu}$ and choose an open rectangular neighbourhood $U \times O$ of the point $(e_G, e_H)$ in $G \times H$ disjoint from $\bigcup \mu$. Since $\bigcup \mu = \varphi^{-1}\varphi(\bigcup \mu)$, the set $\varphi^{-1}\varphi(U \times O) = U \times \pi^{-1}\pi(O)$ is an open neighbourhood of the identity in $G \times H$ disjoint from $\bigcup \mu$. It is clear that $K = \pi^{-1}\pi(e_H) \subset \pi^{-1}\pi(O)$. Since $K = \bigcap_{n \in \omega} K_n$ is the intersection of a decreasing sequence of compact sets, we have $K_n \subset \pi^{-1}\pi(O)$ for some $n \in \omega$. Apply Theorem 1.4.29 to find an open neighbourhood $V$ of $e_H$ such that $V' = K_n V \subset \pi^{-1}\pi(O)$. Then $V'$ is an open neighbourhood of $e_H$ and $(U \times V') \cap \bigcup \mu = \varnothing$. Observe that $V' = \pi_n^{-1}\pi_n(V')$ and $U \times V' = \varphi_n^{-1}\varphi_n(V')$. Hence, the open neighbourhood $\varphi_n(U \times V')$ of $(e_G, \pi_n(e_H))$ in $G \times H_n$ is disjoint from $\varphi_n(\bigcup \mu) \supseteq \varphi_n(\bigcup \mu_n)$. This contradicts (v).

Thus, the identity of $G \times H$ is in the closure of the set $\bigcup \mu$, so that $ot(G \times H) \leq \omega$. □

COROLLARY 5.5.10. *The product $G \times H$ of an arbitrary topological group $G$ satisfying $ot(G) \leq \omega$ and a precompact topological group $H$ has countable o-tightness.*

PROOF. Let $\varrho H$ be the Raĭkov completion of $H$. Then the group $\varrho H$ is compact and, hence, feathered. By Theorem 5.5.9, the product group $G \times \varrho H$ has countable $o$-tightness. Since $G \times H$ is dense in $G \times \varrho H$, we conclude that $ot(G \times H) \leq \omega$. □

A simple modification of the proof of Lemma 5.5.8 yields the following:

LEMMA 5.5.11. *The product $X \times Y$ of a space $X$ of countable $G_\delta$-tightness and a first-countable space $Y$ also has countable $G_\delta$-tightness.*

One can deduce the next result following the scheme in the proof of Theorem 5.5.9 and applying Lemma 5.5.11:

THEOREM 5.5.12. *Let $G$ be a topological group of countable $G_\delta$-tightness. Then the $G_\delta$-tightness of the product group $G \times H$ is countable, for each feathered topological group $H$.*

Again, we use Theorem 5.5.12 to establish one more result on the preservation of $G_\delta$-tightness:

COROLLARY 5.5.13. *The product of a topological group $G$ of countable $G_\delta$-tightness and a pseudocompact topological group $H$ has countable $G_\delta$-tightness.*

PROOF. The Raĭkov completion $\varrho H$ of the pseudocompact group $H$ is compact, by Corollary 3.7.18. In addition, Corollary 3.7.21 implies that $H$ is $G_\delta$-dense in $\varrho H$. Therefore, $G \times H$ is $G_\delta$-dense in $G \times \varrho H$. According to Theorem 5.5.12, the product group $G \times \varrho H$ has countable $G_\delta$-tightness. Since $G \times H$ is $G_\delta$-dense in $G \times \varrho H$, it follows that the $G_\delta$-tightness of the group $G \times H$ is countable as well (see also Exercise 1.6.e). □

## Exercises

5.5.a. Present detailed proofs of Lemma 5.5.11 and Theorem 5.5.12.

5.5.b. Give an example of a Lindelöf topological group of uncountable $o$-tightness.

5.5.c. Verify that the $o$-tightness of the product of any family of separable spaces is countable. Give an example of a subgroup $G$ of a topological product of countable discrete groups such that $ot(G) > \omega$. Deduce that the $o$-tightness is not monotonous with respect to taking subgroups.

## Problems

5.5.A. Is the $o$-tightness of every subgroup of the topological group $\mathbb{Z}^{\mathfrak{c}}$ countable?

5.5.B. Let $G$ be the $\sigma$-product of a family of metrizable topological groups. Does every subgroup of $G$ have countable $o$-tightness?

## Open Problems

5.5.1. Let $G$ be a topological group of countable $o$-tightness ($G_\delta$-tightness), and $H$ be a $\sigma$-compact topological group. Is the $o$-tightness (the $G_\delta$-tightness) of $G \times H$ countable?

5.5.2. Let $G$ be a topological group with $ot(G) \leq \omega$. Is it true that $ot(G \times H) \leq \omega$ for every Lindelöf $\Sigma$-group $H$?

5.5.3. Suppose that $G$ is a topological group of countable $o$-tightness. Is $ot(G \times H) \leq \omega$, for every $k$-group $H$?

5.5.4. Is it true that the product of an arbitrary family of $k$-groups has countable $o$-tightness?

5.5.5. Does the equality $ot(G \times H) = ot(G) \cdot ot(H)$ hold for all topological groups $G$ and $H$?

## 5.6. Steady and stable topological groups

Let $\tau$ be an infinite cardinal. A topological group $G$ is called $\tau$-*steady* if every continuous homomorphic image $H$ of $G$ with $\psi(H) \leq \tau$ satisfies $nw(H) \leq \tau$. If the group $G$ is $\tau$-steady for every infinite cardinal $\tau$, then $G$ is called *steady*. By a) of Lemma 5.3.24, every topological group $G$ with $Nag(G) \leq \tau$ is $\tau$-steady. In particular, Lindelöf $\Sigma$-groups and $\sigma$-compact groups are steady. However, the class of $\tau$-steady groups is considerably wider than the class of topological groups $G$ satisfying $Nag(G) \leq \tau$, since topological products of $\tau$-steady groups are $\tau$-steady (see Theorem 5.6.4). Our proof of this fact leans on two technical results given below.

LEMMA 5.6.1. *Let $\varphi\colon G = \prod_{\alpha\in A} G_\alpha \to K$ be a continuous homomorphism of the product group $G$ to a topological group $K$. For every $\alpha \in A$, denote by $i_\alpha$ be the canonical embedding of $G_\alpha$ to $G$, and let $N_\alpha = i_\alpha^{-1}(\ker\varphi)$. Then $N = \prod_{\alpha\in A} N_\alpha \subset \ker\varphi$.*

PROOF. For every finite set $B \subset A$, put

$$N_B = \{x \in N : \pi_\alpha(x) = e_\alpha \text{ for each } \alpha \in A \setminus B\},$$

where $e_\alpha$ is the neutral element of the group $G_\alpha$ and $\pi_\alpha$ is the projection of $G$ to the factor $G_\alpha$, $\alpha \in A$. Consider the product homomorphism $i_B = \prod_{\alpha\in B} i_\alpha$ of $G_B = \prod_{\alpha\in B} G_\alpha$ to $G$. It is clear that the kernel of the homomorphism $\varphi\circ i_B\colon G_B \to K$ contains the product $N_B^* = \prod_{\alpha\in B} N_\alpha$. Therefore, $N_B = i_B(N_B^*) \subset \ker\varphi$ for each finite set $B \subset A$. Observe that the set

$$S = \bigcup\{N_B : B \subset A,\ |B| < \omega\}$$

is dense in $N$, $S \subset \ker\varphi$ and $\ker\varphi$ is closed in $G$. This implies that $N = \overline{S} \subset \ker\varphi$. □

LEMMA 5.6.2. *Let $G = \prod_{\alpha\in A} G_\alpha$ be a product of topological groups and let $\varphi\colon G \to K$ be a continuous homomorphism of $G$ to a topological group $K$ with $\psi(K) \leq \tau$. Then one can find a set $B \subset A$ with $|B| \leq \tau$ and, for every $\alpha \in B$, a continuous open homomorphism $p_\alpha\colon G_\alpha \to H_\alpha$ onto a topological group $H_\alpha$ with $\psi(H_\alpha) \leq \tau$, and a continuous homomorphism $h\colon \prod_{\alpha\in B} H_\alpha \to K$ such that $\varphi = h\circ p\circ \pi_B$, where $\pi_B\colon G \to G_B = \prod_{\alpha\in B} G_\alpha$ is the projection and $p\colon G \to \prod_{\alpha\in A} H_\alpha$ is the product of the homomorphisms $p_\alpha$.*

PROOF. Let $\{U_\nu : \nu < \tau\}$ be a pseudobase at the identity $e_K$ of the group $K$. Since the homomorphism $\varphi$ of $G$ to $K$ is continuous, we can choose, for every $\nu < \tau$, a canonical open neighbourhood $V_\nu = \prod_{\alpha\in A} V_{\nu,\alpha}$ of the identity $e$ in $G$ such that $\varphi(V_\alpha) \subset U_\nu$. Then the set $B_\nu = \{\alpha \in A : V_{\nu,\alpha} \neq G_\alpha\}$ is finite for each $\nu < \tau$, so the set $B = \bigcup_{\nu<\tau} B_\nu$ has cardinality less than or equal to $\tau$. Clearly, $\ker\pi_B \subset \ker\varphi$, where $\pi_B\colon G \to G_B\prod_{\alpha\in B} G_\alpha$ is the projection. Hence there exists a homomorphism $\varphi_B\colon G_B \to K$ satisfying $\varphi = \varphi_B\circ\pi_B$. Since the homomorphism $\pi_B$ is open, $\varphi_B$ is continuous.

Let $P = \ker\varphi_B$. For every $\alpha \in B$, denote by $i_\alpha$ the canonical embedding of $G_\alpha$ into $G_B$ and put $N_\alpha = i_\alpha^{-1}(P)$. Since $\varphi_B$ and $i_\alpha$ are continuous homomorphisms, we conclude that $N_\alpha$ is a closed invariant subgroup of $G_\alpha$. Let $p_\alpha\colon G_\alpha \to G_\alpha/N_\alpha$ be the quotient homomorphism. The groups $K_\alpha = \varphi_B(i_\alpha(G_\alpha)) \subset K$ and $H_\alpha = G_\alpha/N_\alpha$ are algebraically isomorphic and the natural isomorphism $j_\alpha\colon H_\alpha \to K_\alpha$ satisfies the equality $\varphi_B\circ i_\alpha = j_\alpha\circ p_\alpha$. The isomorphism $j_\alpha$ is continuous because $p_\alpha$ is open. Therefore, $\psi(H_\alpha) \leq \psi(K_\alpha) \leq \psi(K) \leq \tau$ for each $\alpha \in B$.

We now set $H = \prod_{\alpha\in B} H_\alpha$. Then the product homomorphism $p = \prod_{\alpha\in B} p_\alpha$ of $G_B$ onto $H$ is continuous and open. Since $N_\alpha = \ker p_\alpha$ for each $\alpha \in B$, we have $\ker p = \prod_{\alpha\in B} N_\alpha$. Therefore, Lemma 5.6.1 implies that $\ker p \subset \ker \varphi_B$. Hence there exists a homomorphism $h\colon H \to K$ such that $\varphi_B = h \circ p$.

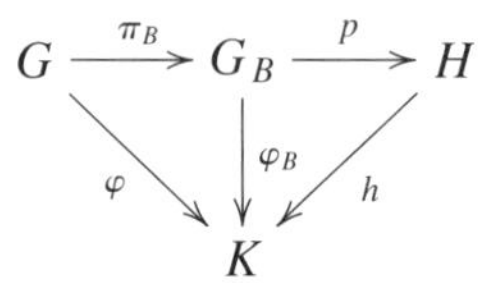


Since $p$ is open, the homomorphism $h$ is continuous. In addition, $\varphi = \varphi_B \circ \pi_B = h \circ p \circ \pi_B$. □

Proposition 5.6.2 admits a simplified form given below.

Corollary 5.6.3. *Let $G = \prod_{i\in I} G_i$ be an arbitrary product of topological groups. Then for every continuous homomorphism $p\colon G \to K$ to a topological group $K$ of countable pseudocharacter, there exist a countable subset $J \subset I$ and a continuous homomorphism $\psi\colon G_J \to K$ such that $p = \psi \circ \pi_J$, where $\pi_J\colon G \to G_J = \prod_{i\in J} G_i$ is the projection.*

Theorem 5.6.4. *The product of an arbitrary family of $\tau$-steady topological groups is a $\tau$-steady topological group.*

Proof. Let $G = \prod_{\alpha\in A} G_\alpha$ be a product of $\tau$-steady topological groups and let $\varphi\colon G \to K$ be a continuous homomorphism of $G$ onto a topological group $K$ satisfying $\psi(K) \le \tau$. Using Lemma 5.6.2 (and notation introduced there) we find a set $B \subset A$ with $|B| \le \tau$, open continuous homomorphisms $p_\alpha\colon G_\alpha \to H_\alpha$ onto topological groups $H_\alpha$ with $\psi(H_\alpha) \le \tau$, for $\alpha \in B$, and a continuous homomorphism $h$ of $H = \prod_{\alpha\in B} H_\alpha$ to $K$ such that $\varphi = h \circ \pi_B \circ p$, where $p = \prod_{\alpha\in B} p_\alpha\colon G_B \to H$. Then $h(H) = \varphi(G) = K$, that is, the homomorphism $h$ is surjective. By assumption, each group $G_\alpha$ is $\tau$-steady, so $nw(H_\alpha) \le \tau$. This implies that $nw(H) \le |B| \cdot \tau = \tau$. Therefore, the continuous image $K = h(H)$ satisfies the same inequality $nw(K) \le \tau$. This proves that the product group $G$ is $\tau$-steady. □

Corollary 5.6.5. *The product of any family of steady topological groups is a steady topological group.*

Since every Lindelöf $\Sigma$-group is steady by Lemma 5.3.24, the following corollary is immediate.

Corollary 5.6.6. *The topological product of any family of Lindelöf $\Sigma$-groups is a steady group.*

The above result will be given a more general form in Corollary 5.6.17. The next statement is obvious.

Theorem 5.6.7. *A continuous homomorphic image of a $\tau$-steady topological group is $\tau$-steady.*

Suppose that $X$ is a Tychonoff space and that $\tau$ is an infinite cardinal. The space $X$ is called *$\tau$-stable* if every Tychonoff continuous image $Y$ of $X$ with $iw(Y) \le \tau$ satisfies $nw(Y) \le \tau$. Equivalently, a Tychonoff space $X$ is $\tau$-stable provided that for any continuous onto mappings $f\colon X \to Y$ and any $i\colon Y \to Z$, where $Y$ and $Z$ are Tychonoff spaces, if $i$ is one-to-one and $w(Z) \leq \tau$, then the space $Y$ satisfies $nw(Y) \leq \tau$. If $X$ is $\tau$-stable for every $\tau \geq \omega$, then $X$ is called *stable*. We call a topological group $G$ $\tau$-stable if $G$ is $\tau$-stable as a topological space.

The study of $\tau$-stable topological groups requires several results concerning topological spaces in general. Let $\tau$ be an infinite cardinal. We recall that a space $X$ is said to be *pseudo-$\tau$-compact* if every discrete (equivalently, locally finite) family of open sets in $X$ has cardinality strictly less than $\tau$ (see page 54). The next result relates the notions of $\tau$-stability and pseudo-$\tau$-compactness.

PROPOSITION 5.6.8. *Every $\tau$-stable Tychonoff space is pseudo-$\tau^+$-compact.*

PROOF. Suppose that $X$ contains a discrete family $\{U_\alpha : \alpha < \kappa\}$ of non-empty open sets, where $\kappa = \tau^+$. For every $\alpha < \kappa$, pick a point $x_\alpha \in U_\alpha$ and define a continuous real-valued function $f_\alpha$ on $X$ with values in $[0, 1]$ such that $f_\alpha(x_\alpha) = 1$ and $f_\alpha(X \setminus U_\alpha) \subset \{0\}$. Denote by $C_p(X)$ the space of all continuous real-valued functions on $X$ endowed with the pointwise convergence topology (see Section 1.2). Let us define a mapping $\Phi\colon [0, 1]^\kappa \to C_p(X)$ by the formula

$$\Phi(t)(x) = \sum_{\alpha<\kappa} t(\alpha) \cdot f_\alpha(x),$$

where $t \in [0, 1]^\kappa$ and $x \in X$. Since the family $\{U_\alpha : \alpha < \kappa\}$ is discrete, our choice of the functions $f_\alpha$'s guarantees that each $\Phi(t)$ is continuous on $X$. It is easy to see that the mapping $\Phi$ is injective. Indeed, if $t_1, t_2$ are distinct points of $[0, 1]^\kappa$, then $t_1(\alpha) \neq t_2(\alpha)$ for some $\alpha < \kappa$ and, hence,

$$\Phi(t_1)(x_\alpha) = t_1(\alpha) \neq t_2(\alpha) = \Phi(t_2)(x_\alpha).$$

So, $\Phi(t_1) \neq \Phi(t_2)$.

We claim that the mapping $\Phi$ is continuous. Indeed, let $t_0 \in [0, 1]^\kappa$ be arbitrary and let $V$ be a neighbourhood of $g_0 = \Phi(t_0)$ in $C_p(X)$. Then there exist points $x_1, \ldots, x_n \in X$ and a number $\varepsilon > 0$ such that the basic open set

$$W = \{g \in C_p(X) : |g(x_i) - g_0(x_i)| < \varepsilon \text{ for each } i = 1, \ldots, n\}$$

satisfies $g_0 \in W \subset V$. Clearly, there are at most $n$ distinct ordinals $\alpha < \kappa$ such that $x_i \in U_\alpha$ for some $i \leq n$. Let $\alpha_1, \ldots, \alpha_k$ be the list of such $\alpha$'s. Put

$$O = \{t \in [0, 1]^\kappa : |t(\alpha_j) - t_0(\alpha_j)| < \varepsilon \text{ for each } j = 1, \ldots, k\}.$$

Then $O$ is an open neighbourhood of $t_0$ in $[0, 1]^\kappa$. Let $t \in O$ be arbitrary. If $i \leq n$ and $x_i \notin \bigcup_{\alpha<\kappa} U_\alpha$, then $\Phi(t)(x_i) = 0 = g_0(x_i)$. If $x_i \in U_{\alpha_j}$ for some $i \leq n$ and $j \leq k$, then we have (with $\alpha = \alpha_j$):

$$|\Phi(t)(x_i) - g_0(x_i)| = |\sum_{\beta<\kappa} t(\beta)\, f_\beta(x_i) - \sum_{\beta<\kappa} t_0(\beta)\, f_\beta(x_i)| =$$

$$|t(\alpha)\, f_\alpha(x_i) - t_0(\alpha)\, f_\alpha(x_i)| = f_\alpha(x_i) \cdot |t(\alpha) - t_0(\alpha)| < 1 \cdot \varepsilon = \varepsilon.$$

We conclude that $\Phi(t) \in W$ for each $t \in O$ and, hence, $\Phi(O) \subset W \subset V$. This proves the continuity of $\Phi$.

Since the mapping $\Phi\colon [0, 1]^\kappa \to C_p(X)$ is one-to-one and continuous, it must be a topological embedding. Hence, $C_p(X)$ contains a copy of the Tychonoff cube $[0, 1]^\kappa$. Now we apply [**27**, Theorem 21]: If $X$ is $\tau$-stable, then every subset $C$ of $C_p(X)$ with

$d(C) \leq \tau$ satisfies $nw(C) \leq \tau$. Since the Tychonoff cube $C = [0, 1]^\kappa$ satisfies $d(C) \leq \tau$ and $nw(C) = \kappa > \tau$, we conclude that $X$ is not $\tau$-stable. □

The converse to Proposition 5.6.8 is false. Indeed, the Niemytzki plane $L$ is a Tychonoff separable (hence, pseudo-$\aleph_1$-compact) space that admits a continuous bijection onto a second-countable space (the upper half-plane in $\mathbb{R}^2$ with the usual Euclidean topology), but $L$ contains a closed discrete subset of cardinality $2^\omega$, whence it follows that $nw(L) = 2^\omega$. Therefore, $L$ is not $\omega$-stable.

In the case of $P$-spaces, we have the following partial converse to Proposition 5.6.8:

PROPOSITION 5.6.9. *Every pseudo-$\aleph_1$-compact $P$-space $X$ is $\omega$-stable.*

PROOF. Let $f: X \to Y$ be a continuous mapping onto a space $Y$ that admits a continuous one-to-one mapping onto a second-countable space. Then every point $y \in Y$ is a $G_\delta$-set in $Y$, so that each fiber $f^{-1}(y)$, being a $G_\delta$-set in the $P$-space $X$, is open in $X$. It follows that $\gamma = \{f^{-1}(y) : y \in Y\}$ is a disjoint open covering of the pseudo-$\aleph_1$-compact space $X$, and we conclude that $\gamma$ is countable, that is, $|Y| \leq \omega$. In particular, $nw(Y) \leq \omega$. □

COROLLARY 5.6.10. *Every Lindelöf $P$-space $X$ is $\omega$-stable.*

We now turn back to topological groups. Combining Propositions 3.4.31 and 5.6.8, one obtains the following statement:

COROLLARY 5.6.11. *Every $\tau$-stable topological group is $\tau$-narrow.*

For Abelian topological groups we can weaken the assumption in Corollary 5.6.11 as follows:

PROPOSITION 5.6.12. *Every $\tau$-steady Abelian topological group is $\tau$-narrow.*

PROOF. Let $p: G \to H$ be a continuous homomorphism of $G$ onto a topological group $H$ of countable character. Since $G$ is $\tau$-steady, the group $H$ satisfies $l(H) \leq nw(H) \leq \tau$. This implies immediately that $H$ is $\tau$-narrow (see Proposition 5.1.3). Hence, the group $G$ is $\tau$-narrow by Proposition 5.1.13. □

The analogy between the definitions of $\tau$-steady groups and $\tau$-stable spaces is obvious. By the next result, we establish the exact relationship between the two properties in the class of topological groups.

PROPOSITION 5.6.13. *Every $\tau$-stable topological group is $\tau$-steady.*

PROOF. Suppose that $p: G \to H$ is a continuous homomorphism of a $\tau$-stable topological group $G$ onto a topological group $H$ satisfying $\psi(H) \leq \tau$. The group $G$ is $\tau$-narrow by Corollary 5.6.11, and so is the continuous homomorphic image $H$ of $G$. Therefore, by Proposition 5.2.11, there exists a continuous isomorphism $i: H \to K$ onto a topological group $K$ satisfying $w(K) \leq \tau$. Since $G$ is $\tau$-stable, we conclude that $nw(H) \leq \tau$. This proves that the group $G$ is $\tau$-steady. □

The next example shows that the implications in Proposition 3.4.31 and in Corollary 5.6.11 are not invertible for $\tau = \omega$ (in Exercise 5.6.b, this is generalized for an arbitrary cardinal $\tau \geq \omega$). In addition, it also shows that $\omega$-steady groups need not be $\omega$-stable, not even for Abelian topological groups.

EXAMPLE 5.6.14. There exists an $\omega$-steady Abelian $P$-group $H$ which contains an uncountable discrete family of open sets. Therefore, $H$ is neither pseudo-$\aleph_1$-compact nor $\omega$-stable.

Our construction of such a group $H$ makes use of Example 4.4.11. Let $K = \{0, 1\}$ be the discrete two-element group with the usual addition, and let $\Pi = K^{\omega_1}$ be the product group endowed with the $\omega$-box topology. Then $\Pi$ is an Abelian topological $P$-group. For every $x \in \Pi$, denote by $\mathrm{supp}(x)$ the set $\{\alpha \in \omega_1 : x(\alpha) = 1\}$. Then

$$G = \{x \in \Pi : \mathrm{supp}(x) \text{ is finite}\}$$

is a subgroup of $\Pi$. With the topology inherited from $\Pi$, $G$ becomes a Lindelöf $P$-group (see Example 4.4.11). Finally, we define $H$ by

$$H = \{x \in G : |\mathrm{supp}(x)| \text{ is even}\}.$$

It is clear that $H$ is a proper subgroup of $G$. For every $A \subset \omega_1$, denote by $\pi_A$ the projection of $\Pi$ onto $K^A$. Let us verify the following:

**Claim.** $\pi_A(H) = \pi_A(G)$ *for every countable set* $A \subset \omega_1$.

Indeed, suppose that $A \subset \omega_1$, $|A| \leq \omega$ and $x \in G$. If $|\mathrm{supp}(x)|$ is even, there is nothing to prove, so we can assume that $|\mathrm{supp}(x)|$ is odd. Choose an arbitrary $\alpha \in \omega_1 \setminus (A \cup \mathrm{supp}(x))$ and define an element $y \in H$ by $y(\beta) = 1$ if and only if $\beta \in \mathrm{supp}(x) \cup \{\alpha\}$. Then $\pi_A(y) = \pi_A(x)$, as required.

The above Claim implies, in particular, that $H$ is a proper dense subgroup of $G$.

Let $p\colon H \to L$ be a continuous homomorphism onto a topological group $L$ of countable pseudocharacter. Then $L$ is countable by Lemma 4.4.2, which implies that $nw(L) \leq \omega$. This proves that $H$ is $\omega$-steady.

For every $\alpha \in \omega_1$, let

$$N_\alpha = \{x \in G : x(\beta) = 0 \text{ for each } \beta \leq \alpha\}.$$

Then $N_\alpha$ is an open subgroup of $G$ and the family $\{N_\alpha : \alpha < \omega_1\}$ forms a decreasing base of $G$ at the neutral element. Since $G$ is a non-discrete $P$-group, this implies that $\chi(G) = \omega_1$. Since each subgroup $N_\alpha$ is closed in $G$, the group $G$ is zero-dimensional.

Take an arbitrary element $g \in G \setminus H$ and fix a strictly decreasing base $\{U_\alpha : \alpha < \omega_1\}$ of $G$ at $g$. Since $G$ is zero-dimensional, we can assume that the sets $U_\alpha$ are closed in $G$. Then the non-empty set $V_\alpha = U_\alpha \setminus U_{\alpha+1}$ is open and closed in $G$ for each $\alpha < \omega_1$. Clearly, $V_\alpha \cap V_\beta = \varnothing$ if $\alpha \neq \beta$. Therefore, the only accumulation point of the disjoint family $\mathcal{V} = \{V_\alpha : \alpha < \omega_1\}$ in $G$ is the point $g$. Indeed, let $x \in G$, $x \neq g$. Choose disjoint open neighbourhoods $O_x$ and $O_g$ of $x$ and $g$ in $G$, respectively. Then there exists $\alpha < \omega_1$ such that $N_\alpha \subset O_g$. This implies that $V_\beta \subset N_\alpha \subset O_g$ and, hence, $O_x \cap V_\beta = \varnothing$ for each $\beta \geq \alpha$. Since $G$ is a $P$-group, the countable disjoint family $\{V_\gamma : \gamma \leq \alpha\}$ has no accumulation points in $G$. Therefore, $x$ cannot be an accumulation point of $\mathcal{V}$.

For every $\alpha < \omega_1$, put $W_\alpha = V_\alpha \cap H$. Then the disjoint family $\{W_\alpha : \alpha < \omega_1\}$ of non-empty open sets in $H$ has no accumulations points in $H$. This means that the family $\{W_\alpha : \alpha < \omega_1\}$ is discrete in $H$, so that $H$ is not pseudo-$\aleph_1$-compact. Finally, according to Proposition 5.6.8, every $\omega$-stable space is pseudo-$\aleph_1$-compact. Therefore, $H$ fails to be $\omega$-stable. □

Using results of Section 5.3, one can easily verify that every Lindelöf $\Sigma$-space is stable. In Proposition 5.6.16 below we prove a considerably more general fact. Our argument makes use of the following lemma that complements the results of Section 1.7.

LEMMA 5.6.15. *Let $X = \prod_{i\in I} X_i$ be the product of spaces satisfying $Nag(X_i) \leq \tau$ for each $i \in I$, and $f\colon X \to Z$ be a continuous mapping to a Tychonoff space $Z$ with $w(Z) \leq \tau$. Then there exist a set $J \subset I$ with $|J| \leq \tau$ and a continuous mapping $h\colon X_J \to Z$ such that $f = h \circ \pi_J$, where $X_J = \prod_{i\in J} X_i$ and $\pi_J\colon X \to X_J$ is the projection.*

PROOF. By Propositions 5.3.3 and 5.3.9, the subproduct $X_K = \prod_{i\in K} X_i$ satisfies $l(X_K) \leq Nag(X_K) \leq \tau$ for every finite set $K \subset I$. Choose a point $p \in X$ and denote by $\sigma(p) \subset X$ the $\sigma$-product of the spaces $X_i$ with center at $p$. Then $l(\sigma(p)) \leq \tau$, according to Corollary 1.6.45. Since $\sigma(p)$ is dense in $X$, the latter inequality implies that the space $X$ is pseudo-$\tau^+$-compact. Hence the existence of the required set $J \subset I$ and mapping $h\colon X_J \to Z$ follows from Theorem 1.7.3. □

PROPOSITION 5.6.16. *Let $\{X_i : i \in I\}$ be a family of spaces such that $Nag(X_i) \leq \tau$, for each $i \in I$. Then the product space $X = \prod_{i\in I} X_i$ is $\tau$-stable. In particular, an arbitrary product of Lindelöf $\Sigma$-spaces is stable.*

PROOF. Let $f\colon X \to Y$ and $g\colon Y \to Z$ be continuous onto mappings, where $g$ is one-to-one and $w(Z) \leq \tau$. We have to show that $nw(Y) \leq \tau$. The composition $h = g \circ f$ is a continuous mapping of $X$ onto $Z$, so we can apply Lemma 5.6.15 to find a set $J \subset I$ with $|J| \leq \tau$ and a continuous mapping $h\colon X_J \to Z$ such that $f = h \circ \pi_J$, where $X_B = \prod_{i\in J} X_i$ and $\pi_B\colon X \to X_B$ is the projection. Since $g\colon Y \to Z$ is one-to-one, we can consider the mapping $p = g^{-1} \circ h$ of $X_J$ to $Y$. It is clear that $h = g \circ p$. Hence, $f = g^{-1} \circ h \circ \pi_J = p \circ \pi_J$. Note that $\pi_J$ is an open mapping, so $p$ is continuous.

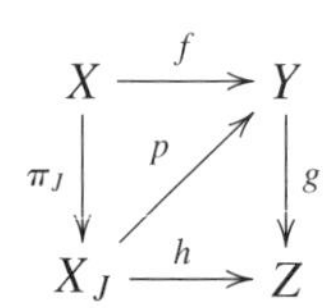

Since $|J| \leq \tau$, Proposition 5.3.9 implies that $Nag(X_J) \leq \tau$. Therefore, the continuous image $Y = p(X_J)$ satisfies $Nag(Y) \leq Nag(X_J) \leq \tau$ by Proposition 5.3.5. It is also clear that $iw(Y) \leq \tau$, since the mapping $g\colon Y \to Z$ is one-to-one and $w(Z) \leq \tau$. It remains to apply Proposition 5.3.15 to conclude that $nw(Y) \leq iw(Y) \cdot Nag(Y) \leq \tau$. So $Y$ is $\tau$-stable. The stability of products of Lindelöf $\Sigma$-spaces is now obvious. □

COROLLARY 5.6.17. *The product of an arbitrary family of Lindelöf $\Sigma$-groups is stable.*

A space $X$ is called *perfectly $\kappa$-normal* if the closure of any open set (that is, every *regular closed set*) is a zero-set. Recall that a subset $P$ of a space $X$ is said to be a *zero-set* if there exists a continuous real-valued function $f$ on $X$ such that $P = \{x \in X : f(x) = 0\}$. Clearly, every metrizable space is perfectly normal and, hence, perfectly $\kappa$-normal. Furthermore, the product of any family of metrizable spaces is perfectly $\kappa$-normal, by a theorem in [**420**].

THEOREM 5.6.18. *Every $\omega$-steady topological group of countable cellularity is perfectly $\kappa$-normal.*

Proof. Let $O$ be a non-empty open subset of an $\omega$-steady topological group $G$ satisfying $c(G) \leq \omega$. By Theorem 3.4.7, the group $G$ is $\omega$-narrow. Denote by $\mathcal{B}$ the family of all subsets $U$ of $G$ which have the form $U = p^{-1}(V)$, where $p\colon G \to H$ is a continuous homomorphism onto a second-countable topological group $H$ and $V$ is an open set in $H$. From Corollary 3.4.19 it follows that $\mathcal{B}$ is a base for $G$. Therefore, there exists a subfamily $\gamma$ of $\mathcal{B}$ such that $O = \bigcup \gamma$. Since the cellularity of $G$ is countable, we can find a countable subfamily $\mu$ of $\gamma$ such that $\bigcup \mu$ is dense in $O$. Let $\mu = \{U_n : n \in \omega\}$. For every $n \in \omega$, take a continuous homomorphism $p_n\colon G \to H_n$ onto a second-countable topological group $H_n$ and an open set $V_n \subset H_n$ such that $U_n = p_n^{-1}(V_n)$. Then the diagonal product $p = \Delta_{n\in\omega} p_n$ is a continuous homomorphism of $G$ to the topological product $P = \prod_{n\in\omega} H_n$. Clearly, the image $H = p(G) \subset P$ satisfies $w(H) \leq w(P) \leq \omega$. Denote by $\pi_n$ the projection of $P$ to $H_n$, $n \in \omega$. Then $W_n = \pi_n^{-1}(V_n) \cap H$ is an open subset of $H$, and from $p_n = \pi_n \circ p$ it follows that $U_n = p^{-1}(W_n)$ for each $n \in \omega$. The set $W = \bigcup_{n\in\omega} W_n$ is open in $H$ and $p^{-1}(W) = \bigcup_{n\in\omega} U_n$ is dense in $O$.

Let $N$ be the kernel of $p$ and let $\varphi\colon G \to G/N$ be the quotient homomorphism. Then the natural isomorphism $i\colon G/N \to H$ satisfies $p = i \circ \varphi$, so $i$ is continuous.

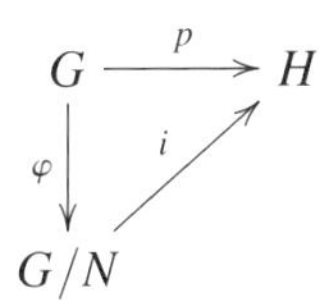


In particular, the quotient group $G/N$ has countable pseudocharacter. Therefore, since $G$ is $\omega$-steady, we have $nw(G/N) \leq \omega$. Obviously, the set $W_0 = i^{-1}(W)$ is open in $G/N$ and $\varphi^{-1}(W_0) = p^{-1}(W)$ is dense in $O$. Since the homomorphism $\varphi$ is open, we have that

$$cl_G(O) = cl_G(\varphi^{-1}(W_0)) = \varphi^{-1}(cl_{G/N}(W_0)).$$

Every closed subset of $G/N$ is a zero-set, since $G/N$ has a countable network. Therefore, $cl_{G/N}(W_0)$ is a zero-set in $G/N$ and its inverse image under $\varphi$, $cl_G(O)$, is a zero-set in $G$. This implies that the group $G$ is perfectly $\kappa$-normal. □

By Proposition 5.6.13, every $\omega$-stable topological group is $\omega$-steady. Therefore, we have:

Corollary 5.6.19. *Every $\omega$-stable topological group of countable cellularity is perfectly $\kappa$-normal.*

Corollary 5.6.20. *Every topological group, which is a Lindelöf $\Sigma$-space, is perfectly $\kappa$-normal.*

## Exercises

5.6.a. Find an example of a Lindelöf $\omega$-stable topological group $G$ such that $G$ is not perfectly $\kappa$-normal and $c(G) > \omega$.

5.6.b. For every regular cardinal $\tau \geq \omega$, give an example of a $\tau$-steady Abelian topological group $G$ which fails to be pseudo-$\tau^+$-compact.

5.6.c. Give an example of a precompact topological group which is not $\omega$-steady.
*Hint.* Embed the Niemytzki plane $L$ in the product group $\mathbb{T}^{\mathfrak{c}}$ and consider the subgroup $G$ of $\mathbb{T}^{\mathfrak{c}}$ generated by $L$. Use the fact that $L$ admits a continuous bijection onto a regular second-countable space.

5.6.d. Prove that every $\omega$-stable topological group with a $\sigma$-discrete $\pi$-base is metrizable.

## Problems

5.6.A. Show that a closed subgroup of a $\tau$-stable topological group need not be $\tau$-stable.
5.6.B. Give an example of a perfectly $\kappa$-normal topological group that is not $\omega$-steady.
5.6.C. Must the Raĭkov completion of any $\omega$-stable topological group be an $\omega$-stable group?
5.6.D. Must the Raĭkov completion of any $\omega$-steady topological group be $\omega$-steady?
5.6.E. Is it true that if an Abelian topological group $G$ of countable tightness is $\omega$-steady, then $G$ is $\omega$-stable?

## Open Problems

5.6.1. Is the product of two stable topological groups $\omega$-stable?
5.6.2. Is the product of a Lindelöf $\Sigma$-group and a Lindelöf $P$-group $\omega$-stable?

## 5.7. Cardinal invariants in paratopological and semitopological groups

In this section, we establish certain new connections between cardinal invariants in paratopological groups and in semitopological groups. In particular, it is proved that if $G$ is a bisequential paratopological group such that the space $G \times G$ is Lindelöf, then $G$ is cosmic. Under the Continuum Hypothesis (abbreviated to $CH$) we prove that if $G$ is a separable first-countable paratopological group such that $G \times G$ is normal, then $G$ has a countable base. This sheds a new light on why the square of the Sorgenfrey line is not normal.

There exists a first-countable, non-normal (therefore, non-paracompact) paratopological group — it suffices to take the square of the Sorgenfrey line. However, first countability has a strong impact on the properties of semitopological groups. The following concepts are instrumental in this direction.

Let $G$ be an abstract group. A family $\mathscr{E}$ of subsets of $G$ will be called *discerning* or a *Hausdorff discerner on* $X$ if all the elements of $\mathscr{E}$ are non-empty and, for every $z \in G$ distinct from the neutral element $e$, there exists $P \in \mathscr{E}$ such that $zP \cap P = \varnothing$.

If $G$ is a Hausdorff semitopological group and $\mathscr{B}$ is a base of $G$ at some $a \in G$, then $\mathscr{B}$ is a Hausdorff discerner on $G$. We obtain a slightly less trivial and much more useful example of a Hausdorff discerner when we take an arbitrary $\pi$-network of $G$ at $e$. Recall that a family $\mathscr{E}$ of subsets of a topological space $X$ is said to be a $\pi$-*network* of $X$ at a point $a \in X$ if all the elements of $\mathscr{E}$ are non-empty and every open neighbourhood of $a$ in $X$ contains an element of $\mathscr{E}$. If $\mathscr{E}$ is a $\pi$-network of $X$ at $a \in X$ and all the elements of $\mathscr{E}$ are open, we call $\mathscr{E}$ a $\pi$-*base* of $X$ at $a$. The next statement is obvious:

PROPOSITION 5.7.1. *Suppose that $G$ is a Hausdorff semitopological group. Then every $\pi$-network $\mathscr{E}$ of $G$ at $e$ is a Hausdorff discerner on $G$.*

Here is one of the main technical results on Hausdorff discerners.

PROPOSITION 5.7.2. *Suppose that $G$ is a group and $\mathscr{E}$ a Hausdorff discerner on $G$. Then $\bigcap\{PP^{-1} : P \in \mathscr{E}\} = \{e\}$.*

PROOF. Put $F = \bigcap\{PP^{-1} : P \in \mathscr{E}\}$. Clearly, $e \in F$. Now take any $z \in F$. We have to show that $z = e$. Assume the contrary. Then we can fix $P \in \mathscr{E}$ such that $zP \cap P = \varnothing$. Now we have $z \notin PP^{-1}$. Indeed, otherwise $z = ab^{-1}$, for some $a, b \in P$, and $zb = a \in zP \cap P$, a contradiction. It follows that $z \notin F$, a contradiction. Hence, $F = \{e\}$. □

Let $G$ be a semitopological group. A *topological discerner $\mathscr{E}$ on $G$* is a Hausdorff discerner on $G$ such that the interior of $PP^{-1}$ contains $e$, for each $P \in \mathscr{E}$. A discerner is called *open* if all its elements are open sets. Finally, a discerner $\mathscr{E}$ is said to be *coopen* if $P^{-1}$ is open, for every $P \in \mathscr{E}$. It is clear that open discerners and coopen discerners are topological discerners.

PROPOSITION 5.7.3. *Suppose that $G$ is a left topological group with a countable topological discerner. Then $G$ has countable pseudocharacter.*

PROOF. Since every left topological group is a homogeneous space, the conclusion follows from the definition of a topological discerner and Proposition 5.7.2. □

In the special case of open discerners, Proposition 5.7.3 can be considerably strengthened as follows.

THEOREM 5.7.4. *Suppose that $G$ is a semitopological group with a countable open Hausdorff discerner $\mathscr{E}$. Then the diagonal $\Delta$ in $G \times G$ is a $G_\delta$-set.*

PROOF. By Proposition 5.7.2, $\bigcap\{VV^{-1} : P \in \mathscr{E}\} = \{e\}$. For each $V \in \mathscr{E}$, put $U_V = \bigcup\{Vx \times Vx : x \in G\}$. Since every $V \in \mathscr{E}$ is an open set, the set $U_V$ is an open neighbourhood of the diagonal $\Delta$ in $G \times G$.

Let us show that $\Delta = \bigcap\{U_V : V \in \mathscr{E}\}$. Assume the contrary. Then there exist distinct points $y$ and $z$ in $G$ such that $(y, z) \in U_V$ for every $V \in \mathscr{E}$. Put $b = yz^{-1}$. Then $b \neq e$ and, for each $V \in \mathscr{E}$, there exists $x \in G$ such that $y \in Vx$ and $z \in Vx$. It follows that $b = yz^{-1} \in Vx(Vx)^{-1} = Vxx^{-1}V^{-1} = VV^{-1}$. Hence $b \in \bigcap\{VV^{-1} : P \in \mathscr{E}\} = \{e\}$, a contradiction with $b \neq e$. □

COROLLARY 5.7.5. *If $G$ is a Hausdorff semitopological group of countable $\pi$-character, then the diagonal $\Delta$ in $G \times G$ is a $G_\delta$-set.*

PROOF. Every countable $\pi$-base of the space $G$ at the identity $e$ is a countable open Hausdorff discerner on $G$. It remains to apply Theorem 5.7.4. □

THEOREM 5.7.6. *For every paracompact Hausdorff semitopological group $G$ of countable $\pi$-character, there exists a continuous one-to-one mapping of $G$ onto a metrizable space.*

PROOF. By Corollary 5.7.5, $G$ is a space with a $G_\delta$-diagonal. Since $G$ is paracompact, it follows from [**165**, 5.5.7] that the topology of $G$ contains a metrizable topology. □

Our next result about semitopological groups, Theorem 5.7.8, requires the following auxiliary fact:

LEMMA 5.7.7. *Suppose that a space $X$ admits a perfect mapping $f$ onto a Hausdorff space $Y$ and a one-to-one continuous mapping $g$ onto a Hausdorff space $Z$. Then $X$ is homeomorphic to a closed subspace of $Y \times Z$.*

PROOF. Denote by $h$ the diagonal product of the mappings $f$ and $g$. It follows from [**165**, Theorem 3.7.9] that $h$ is perfect. Since $g$ is one-to-one, so is $h$. Therefore, $h$ is a perfect one-to-one mapping of $X$ to $Y \times Z$ or, in other words, $h$ is a homeomorphic embedding of $X$ into $Y \times Z$, and the image $h(X)$ is closed in $Y \times Z$. □

THEOREM 5.7.8. *Suppose that $G$ is a paracompact feathered semitopological group of countable $\pi$-character. Then $G$ is metrizable.*

PROOF. The group $G$, being feathered, is a $p$-space. Since, in addition, $G$ is paracompact, it admits a perfect mapping onto a metrizable space (see [**60**, Ch. 5, Problem 228]). It also follows from Theorem 5.7.6 that $G$ admits a continuous one-to-one mapping onto a metrizable space. Hence, Lemma 5.7.7 implies the required conclusion. □

The above results are applicable, in particular, to Čech-complete and to first-countable spaces. In particular, we have:

COROLLARY 5.7.9. *Every first-countable paracompact $p$-space homeomorphic to a semitopological group is metrizable.*

Let us now have a look at the Sorgenfrey line $S$. We know that $S$ is a first-countable Lindelöf paratopological group with the Baire property. On the other hand, $S$ is not Čech-complete, not metrizable, and not even a $p$-space (this follows from Corollary 5.7.9). The square $X = S \times S$ is again a first-countable paratopological group. However, $X$ is no longer paracompact, but is subparacompact in the sense that every open covering of $X$ can be refined by a $\sigma$-discrete covering (see [**93**]). This combination of properties of the Sorgenfrey line $S$ is not just an individual feature of $S$, it is typical for non-metrizable first-countable paratopological groups.

It turns out that a number of results established in Sections 5.3 and 5.4 can be extended to paratopological groups. We will only generalize Theorem 5.3.18 and Corollaries 5.3.22 and 5.4.8 here. This will be done with the help of the following corollary from Proposition 2.3.24, which is especially convenient for applications.

LEMMA 5.7.10. *For every Hausdorff paratopological group $G$, there exists a topological group $T$ homeomorphic to a closed subspace of $G \times G$ such that $T$ can be mapped by a continuous isomorphism onto $G$.*

THEOREM 5.7.11. *Let $G$ be a paratopological group satisfying $Nag(G) \leq \tau$. Then $G$ is $\tau$-cellular and $(\tau^+, 2)$ is a weak precalibre of $G$. Hence, $c(G) \leq \tau$.*

PROOF. Take a topological group $T$ as in Lemma 5.7.10. Since $T$ is homeomorphic to a closed subspace of $G \times G$, it follows that $Nag(T) \leq Nag(G \times G) \leq \tau$. Hence, according to Theorems 5.3.18 and 5.4.7, the space $T$ is $\tau$-cellular and $(\tau^+, 2)$ is a weak precalibre of $T$. In particular, $c(T) \leq \tau$. Since continuous mappings preserve each of these properties, and $G$ is a continuous isomorphic image of $T$, the conclusion of the theorem is immediate. □

COROLLARY 5.7.12. *If $G$ is a $\sigma$-compact paratopological group, then $(\tau, 2)$ is a weak precalibre of $G$, for each regular cardinal $\tau > \omega$. In particular, the cellularity of $G$ is countable.*

The proof of the next statement is another typical application of Lemma 5.7.10.

THEOREM 5.7.13. *Suppose that $G$ is a Hausdorff paratopological group with the Baire property such that the extent of $G \times G$ is countable. Then $G$ is a topological group.*

PROOF. Let $H$ be a topological group as in Lemma 5.7.10. Then $H$ is homeomorphic to a closed subset of $G \times G$ and, hence, the extent of $H$ is countable. It follows from Proposition 5.2.2 that the topological group $H$ is $\omega$-narrow. Now Theorem 2.3.23 implies that $G$ is a topological group. □

We also need the following fact:

PROPOSITION 5.7.14. [**O. V. Ravsky**] *Every first-countable cosmic paratopological group $G$ has a countable base.*

PROOF. Fix a countable base $\mathcal{B}$ at the neutral element $e$ of $G$, and let $\mathcal{S}$ be a countable network of $G$. By the continuity of the multiplication in $G$, the countable family $\{VP : V \in \mathcal{B},\ P \in \mathcal{S}\}$ is a base of the space $G$. □

Clearly, a hereditarily separable first-countable paratopological group need not have a countable base, as the example of the Sorgenfrey line shows.

THEOREM 5.7.15. *Suppose that $G$ is a Hausdorff first-countable paratopological group. Then the following three conditions are equivalent:*

1) *$G \times G$ is Lindelöf;*
2) *$e(G \times G) \leq \omega$;*
3) *$G$ has a countable base.*

PROOF. Clearly, 3) implies 1), and 1) implies 2). Assume now that 2) holds. By Lemma 5.7.10, we can fix a topological group $H$ homeomorphic to a closed subspace of $G \times G$ and a continuous mapping $j$ of $H$ onto $G$. Then $e(H) \leq \omega$, since $H$ is closed in $G \times G$. Since $G$ is first-countable, the spaces $G \times G$ and $H$ are also first-countable. Therefore, the topological group $H$ is metrizable. Since $e(H) \leq \omega$, the space $H$ is separable and has a countable base. It follows that $G$ has a countable network, since $G$ is a continuous image of $H$. It remains to apply Proposition 5.7.14. □

For any topological group $G$, the condition that $G$ has pointwise countable type (that is, $G$ contains a non-empty compact subset with a countable base of neighbourhoods in $G$) is equivalent to the condition that $G$ is a paracompact $p$-space (see Theorem 4.3.35). This result does not generalize to paratopological groups, as the Sorgenfrey line shows. However, we have the next result about paratopological groups of pointwise countable type which is parallel to Theorem 5.7.15:

THEOREM 5.7.16. *Suppose that $G$ is a Hausdorff paratopological group of pointwise countable type. Then the following three conditions are equivalent:*

1) *$G \times G$ is Lindelöf;*
2) *$e(G \times G) \leq \omega$;*
3) *$G$ is a Lindelöf $\Sigma$-space.*

PROOF. The implications 3) $\Rightarrow$ 1) and 1) $\Rightarrow$ 2) are evident. Suppose that 2) holds. According to Lemma 5.7.10, we can fix a topological group $H$ homeomorphic to a closed subspace of $G \times G$ and a continuous mapping $j$ of $H$ onto $G$. Since $G$ is of pointwise countable type, so are the spaces $G \times G$ and $H$. Since $H$ is a topological group, it follows from Theorem 4.3.35 that $H$ is a paracompact $p$-space. We know that $e(G \times G) \leq \omega$ and $H$ is closed in $G \times G$; hence, $e(H) \leq \omega$. Since $H$ is also paracompact, it follows that $H$ is Lindelöf. The space $G$ being a continuous image of $H$, it remains to apply Corollary 5.3.14 to conclude that $G$ is a Lindelöf $\Sigma$-space. □

Now we are going to present two results which depend on the additional assumption that $2^{\aleph_0} < 2^{\aleph_1}$.

THEOREM 5.7.17. *Suppose that $2^{\aleph_0} < 2^{\aleph_1}$. Let $G$ be a separable paratopological group such that $G \times G$ is normal. If $G$ is of pointwise countable type, then $G$ is a Lindelöf $\Sigma$-space.*

PROOF. Take the same subspace $H$ of $G \times G$ as in Lemma 5.7.10. Then, arguing as in the proof of Theorem 5.7.16, we conclude that $H$ is a paracompact $p$-space. Since $G \times G$ is separable and normal and $H$ is closed in $G \times G$, it follows from the assumption $2^{\aleph_0} < 2^{\aleph_1}$ that every closed discrete subspace of $H$ is countable (argue as in [**165**, Example 1.5.9]). Since $H$ is paracompact, this implies that $H$ is Lindelöf. Hence, Corollary 5.3.14 implies that $G$ is a Lindelöf $\Sigma$-space, as a continuous image of the Lindelöf $p$-space $H$. □

Combining Theorems 5.7.17 and 5.7.15 and taking into account that the class of Lindelöf $\Sigma$-spaces is finitely productive, we obtain the following statement:

THEOREM 5.7.18. *Suppose that $2^{\aleph_0} < 2^{\aleph_1}$. Let $G$ be a separable first-countable paratopological group such that $G \times G$ is normal. Then $G$ has a countable base.*

The last five results clarify from the point of view of topological algebra why the square of the Sorgenfrey line is neither Lindelöf nor normal.

Recall that a Tychonoff space $X$ is said to be *weakly pseudocompact* if there exists a Hausdorff compactification $bX$ of $X$ such that $X$ is $G_\delta$-dense in $bX$, that is, every non-empty $G_\delta$-set in $bX$ intersects $X$. Of course, every pseudocompact space is weakly pseudocompact. The converse is false, since every uncountable discrete space $D$ is also weakly pseudocompact — it suffices to take the one-point compactification of $D$.

Below, after a series of more special results, we are going to establish a simple sufficient condition for a weakly pseudocompact semitopological group to be metrizable. First, we need the following statement.

PROPOSITION 5.7.19. *Every weakly pseudocompact Tychonoff space $X$ with a $G_\delta$-diagonal is Čech-complete.*

PROOF. Let $bX$ be a Hausdorff compactification of $X$ such that $X$ is $G_\delta$-dense in $bX$. We are going to show that $X$ is a $G_\delta$-subset of $bX$. Since $X$ has a $G_\delta$-diagonal, there is a sequence $\{\gamma_n : n \in \omega\}$ of coverings of $X$ by open sets in $\beta X$ satisfying the following condition:

(•) For any distinct $x$ and $y$ in $X$, there exists $n \in \omega$ such that no element of $\gamma_n$ contains both $x$ and $y$.

Put $W_n = \bigcup \gamma_n$, for each $n \in \omega$, and let $P = \bigcap_{n\in\omega} W_n$. Clearly, $X \subset P$. It remains to show that $P \subset X$. Assume the contrary, and take any $y \in P \setminus X$. For each $n \in \omega$, fix $V_n \in \gamma_n$ such that $y \in V_n$. Put $G = \bigcap_{n\in\omega} V_n$. Since $X$ is $G_\delta$-dense in $bX$, the set $G \cap X$ is dense in $G$ and, hence, contains more than one point. Taking two distinct points $x, z \in G \cap X$, we arrive at a contradiction with $(\bullet)$ : $\{x, z\} \subset V_n \in \gamma_n$, for each $n \in \omega$. □

Corollary 5.7.20. *Every pseudocompact space with a $G_\delta$-diagonal is Čech-complete.*

Here is another general fact we need:

Proposition 5.7.21. *Suppose that $X$ is a weakly pseudocompact space and $K$ is a compact $G_\delta$-subset of $X$. Then $K$ has a countable base of neighbourhoods in $X$.*

Proof. Take a Hausdorff compactification $bX$ of $X$ such that $X$ is $G_\delta$-dense in $bX$. Since $K$ is a $G_\delta$-set in $X$, there exists a countable family $\gamma$ of open sets in $bX$ such that $X \cap \bigcap \gamma = K$. Put $\eta = \gamma \cup \{bX \setminus K\}$. Then $\eta$ is a countable family of open sets in $bX$ (note that $K$ is closed in $bX$). Clearly, $X \cap \bigcap \eta = \varnothing$. Since $\bigcap \eta$ is a $G_\delta$-set in $bX$, and $X$ is $G_\delta$-dense in $bX$, it follows that $\bigcap \eta = \varnothing$. Therefore, $\bigcap \gamma = K$, that is, $K$ is a $G_\delta$-set in $bX$. Since $K$ and $bX$ are compact, it follows that $K$ has a countable base of open neighbourhoods in $bX$. Hence, $X$ has also a countable base of open neighbourhoods in $X$. □

Corollary 5.7.22. *If $X$ is a weakly pseudocompact space and $x$ is $G_\delta$-point in $X$, then $X$ has a countable local base at $x$.*

The next result follows directly from Proposition 5.7.3 and Corollary 5.7.22:

Theorem 5.7.23. *If $G$ is a weakly pseudocompact left topological group with a countable topological discerner, then $G$ is first-countable.*

Here is the promised result concerning metrizability of semitopological groups.

Theorem 5.7.24. *Every weakly pseudocompact semitopological group $G$ with a countable topological discerner is a topological group metrizable by a complete metric.*

Proof. By Theorem 5.7.23, the space $G$ is first-countable. Therefore, it follows from Corollary 5.7.5 that $G$ is a space with a $G_\delta$-diagonal. Now Proposition 5.7.19 implies that $G$ is Čech-complete. By Theorem 2.4.12, every Čech-complete semitopological group is a topological group. Therefore, $G$ is a topological group. Since $G$ is first-countable, it follows that $G$ is metrizable. It remains to observe that every metrizable Čech-complete space is metrizable by a complete metric [**165**, Theorem 4.3.26]. □

The three facts below are special cases of Theorem 5.7.24:

Corollary 5.7.25. *Every first-countable weakly pseudocompact semitopological group is a topological group metrizable by a complete metric.*

Corollary 5.7.26. *Every weakly pseudocompact semitopological group of countable $\pi$-character is a topological group metrizable by a complete metric.*

Corollary 5.7.27. *Every pseudocompact semitopological group $G$ of countable $\pi$-character is a compact metrizable topological group.*

## Exercises

5.7.a. (I. I. Guran, cited in [**68**]) A paratopological group $G$ is *saturated* if the inverse $U^{-1}$ of every non-empty open subset of $G$ contains a non-empty open subset of $G$. Show that every subgroup of the Sorgenfrey line is a saturated paratopological group.

5.7.b. A space $X$ is said to be *rectifiable* if there exists a homeomorphism $f$ of the space $X \times X$ onto itself satisfying the following two conditions:
   (i) For any $z \in X \times X$, the first coordinates of $z$ and $f(z)$ coincide;
   (ii) The image $f(\Delta_X)$ of the diagonal $\Delta_X = \{(x, x) : x \in X\}$ is the set $Z_e = \{(x, e) : x \in X\}$, for some $e \in X$.
Show that every topological group $G$ is a rectifiable space.

5.7.c. Prove that every rectifiable space is regular and homogeneous.

5.7.d. Prove that the product of an arbitrary family of rectifiable spaces is (naturally) rectifiable.

## Problems

5.7.A. (G. Itzkowitz and V. V. Tkachuk [**258**]) For a non-discrete topological group $G$, let $p(G)$ be the least number of open sets in $G$ whose intersection is not open. Prove that if a topological group $G$ is pseudo-$\tau$-compact and $p(G) = \tau$, then $G$ is fine (see Problems 1.8.C, 4.4.G, 4.4.H, and 4.4.I).

5.7.B. (A. V. Arhangel'skii and D. K. Burke [**51**]) Let $G$ be a regular first-countable semitopological group of countable extent. Prove that $|G| \leq 2^{\omega}$.

5.7.C. (A. V. Arhangel'skii and D. K. Burke [**51**]) Suppose that $G$ is a Tychonoff separable semitopological group of countable pseudocharacter. Prove that there exists a weaker regular second-countable topology on $G$.

5.7.D. Give an example of a regular quasitopological group $G$ with a countable $\pi$-base which is not first-countable.

5.7.E. (A. S. Gul'ko [**206**]) Prove that every first-countable rectifiable space is metrizable.

5.7.F. Show that the Sorgenfrey line is not rectifiable.

5.7.G. Show that every pseudocompact subspace of a rectifiable space of countable pseudocharacter is compact and metrizable.

5.7.H. Show that a regular first-countable $\omega$-narrow paratopological group need not be Lindelöf.

5.7.I. (O. V. Ravsky [**400**]) Show that every totally bounded paratopological group $G$ is saturated.
*Hint.* Take a non-empty open subset $U$ of $G$. There exists a finite subset $F$ of $G$ such that $\overline{FU^{-1}} = G$. Therefore, $aU^{-1}$ is not nowhere dense in $G$, for some $a \in F$, that is, $V \subset \overline{aU^{-1}}$, for some non-empty open subset $V$ of $G$. Since $G$ is a paratopological group, it follows that $V \subset aU^{-1}U^{-1}$. Hence, $(U^2)^{-1}$ has a non-empty open interior. Clearly, in a paratopological group this implies that the interior of $U^{-1}$ is non-empty.

5.7.J. (T. Banakh and O. V. Ravsky [**68**]) Show that a Hausdorff paratopological group $H$ is topologically isomorphic to a paratopological subgroup of a saturated Hausdorff paratopological group if an and only if there exists a continuous isomorphism of $H$ onto a topological group.

5.7.K. (T. Banakh and O. V. Ravsky [**68**]) Show that a Hausdorff paratopological group $H$ is topologically isomorphic to a closed paratopological subgroup of a Hausdorff totally bounded paratopological group if and only if there exists a continuous isomorphism of $H$ onto a totally bounded topological group.

5.7.L. (O. V. Ravsky [**399**]) Show that a regular paratopological group with the Baire property need not be saturated.
*Hint.* Take the product $G = \mathbb{R}^{\omega}$ as a group, but not with the natural topology. For $n \in \omega$, denote by $U_n$ the set of all elements $(x_i)_{i \in \omega}$ of $G$ such that $x_i = 0$, for each $i \leq n$, and $x_i \geq 0$, for each $i \geq n$. The sets $U_n$, where $n \in \omega$, constitute a local base at the neutral element for

a topology on $G$ that turns $G$ into a paratopological group. Show that this paratopological group has the Baire property, but is not saturated.

5.7.M. (M. Fernández [**168**]) Prove that any subgroup of an arbitrary power of the Sorgenfrey line is saturated.

5.7.N. (A. V. Arhangel'skii and A. Bella [**50**]) Suppose that $G$ is a first-countable $\omega$-narrow Tychonoff paratopological group. Show that $G$ is submetrizable.

5.7.O. (A. V. Arhangel'skii and D. K. Burke [**51**]) Give an example of a countable regular paratopological group with a countable $\pi$-base that fails to be first-countable. (Compare with Theorems 5.7.6 and 5.7.8.)

5.7.P. (M. Sanchis and M. G. Tkachenko [**418**]) Let $H$ be a regular paratopological group with identity $e$ such that the space $H^2$ is Lindelöf. Prove the following:

a) If $\gamma$ is a countable family of open neighbourhoods of $e$ in $H$, then one can find a continuous homomorphism $p\colon H \to K$ onto a regular second-countable paratopological group $K$ such that $\ker p \subset \bigcap \gamma$ and $p(U)$ is a neighbourhood of the neutral element of $K$, for each $U \in \gamma$.

b) If $H$ has countable pseudocharacter, then $H$ admits a continuous isomorphism onto a regular second-countable paratopological group.

5.7.Q. (M. Sanchis and M. G. Tkachenko [**418**]) Let $G$ be a paratopological group such that $G$ is a Lindelöf $\Sigma$-space. Prove that $G$ is an Efimov space. Hence, every regular closed subset of $G$ is a zero-set. (Compare with Corollaries 5.3.28 and 5.3.29.)

### Open Problems

5.7.1. Is every regular paracompact $\omega$-narrow paratopological group $G$ of countable $o$-tightness Lindelöf?

5.7.2. Suppose that $G$ is a first-countable $\omega$-narrow regular paratopological group. Is $G$ submetrizable?

5.7.3. Suppose that $G$ is a Tychonoff semitopological (paratopological) group of countable pseudocharacter such that the cellularity of $G$ is countable. Is there a weaker regular second-countable topology on $G$?

5.7.4. Suppose that $G$ is a first-countable semitopological (paratopological) group which is a $p$-space. Is $G$ a Moore space?

5.7.5. Is every paratopological group, which is a Moore space, metrizable?

5.7.6. Suppose that $G$ is a (regular, Tychonoff) paratopological group which is also a rectifiable space. Is $G$ homeomorphic to a topological group?

5.7.7. Is every regular rectifiable space Tychonoff?

5.7.8. Is every regular rectifiable space of countable pseudocharacter submetrizable? Is it Tychonoff?

## 5.8. Historical comments to Chapter 5

The results of Section 5.1 concerning $\tau$-narrow topological groups are essentially due to I. I. Guran [**208**]. In particular, Propositions 5.1.1, 5.1.3, Lemmas 5.1.4, 5.1.6, Corollary 5.1.7, and Theorem 5.1.10 all originated in [**208**] and [**210**]. A prototype of Lemma 5.1.6 can be found in [**202**]. However, Theorem 5.1.10 appeared without proof in [**208**], and the first published proof of the theorem (apart from that in Guran's PhD thesis [**209**]) was given by V. V. Uspenskij in [**516**]. The alternative characterization of $\omega$-narrow groups as subgroups of $k$-separable groups in Theorem 5.1.12 was found by V. G. Pestov

[**375**]. Theorem 5.1.11 was obtained in [**24**]. A common prototype of Theorems 5.1.10 and 5.1.19 was obtained by V. K. Bel'nov in [**72**] — he proved that if a topological group $G$ is algebraically generated by a Lindelöf subspace, then $G$ is topologically isomorphic to a subgroup of the product of some family of second-countable groups. In the present form, Theorem 5.1.19 appeared in [**142**]. Proposition 5.2.6 is from [**22**]. Lemma 5.2.10 originated in [**60**] (see also [**21**]). Corollary 5.2.14 was proved in [**21**]. Theorems 5.2.13 and 5.2.15 are taken from [**21**].

Theorem 5.1.24 and Proposition 5.1.22 were proved by T. Banakh in [**67**].

Lindelöf $\Sigma$-spaces were introduced and studied by K. Nagami in [**335**] as a subclass of a wider class of $\Sigma$-*spaces*. It is remarkable that this class of spaces contains all $\sigma$-compact spaces, all second-countable spaces and all cosmic spaces, and is closed under countable products and continuous images. The countable versions of Corollaries 5.3.2, 5.3.4, Propositions 5.3.3, 5.3.5, 5.3.8, 5.3.9, and of some other results in this book were established in [**335**]. An important corollary from these results for the theory of topological groups is Proposition 5.3.10.

The majority of other results in Section 5.3 on Lindelöf $\Sigma$-spaces, Lindelöf $\Sigma$-groups, and the Souslin number of topological groups are due to A. V. Arhangel'skii, M. G. Tkachenko, and V. V. Uspenskij. Corollary 5.3.22 on the cellularity of $\sigma$-compact groups appeared originally in [**469**], and then V. V. Uspenskij extended the result to Lindelöf $\Sigma$-groups in [**508**]. The Nagami number of a Tychonoff space appeared in [**25**], where Theorems 5.3.12 and 5.3.13 were proved. For the special case $\tau = \aleph_0$, Theorem 5.3.18 was proved by Uspenskij in [**512**], while the present version of the theorem was given in [**481**]. The notions of a Mal'tsev operation and a Mal'tsev space were introduced by V. V. Uspenskij in [**512**] (see also [**518**], [**187**], and [**449**]). In this connection, see also the article [**206**] of A. S. Gul'ko, where rectifiable spaces are studied.

The articles [**469**] and [**471**] by M. G. Tkachenko cover a good part of the material of Section 5.4. Corollary 5.4.3 is a famous result of N. A. Shanin [**438**]. In connection with this result and Theorem 5.4.1 see also [**437**]. Example 5.4.13 is due to Uspenskij, see [**508**]. Theorem 5.5.4 and Corollary 5.5.10 appeared in [**479**].

The concepts of $\tau$-stability and stability were introduced in [**25**] by A. V. Arhangel'skii, where Proposition 5.6.16 and Corollary 5.6.17 were proved. These concepts turned out to be useful in $C_p$-theory (see [**32**]). In this connection, see also [**61**]. Our Theorem 5.6.18 extends Uspenskij's result (obtained in [**512**] for $\omega$-stable groups) to $\omega$-steady groups.

Almost all results of Section 5.7 are due to Arhangel'skii and Reznichenko [**62**]. Proposition 5.7.14 was proved by O. V. Ravsky. Some properties of topological groups find a non-trivial reflection in the properties of remainders of the groups in their Hausdorff compactifications. See [**48**] about these matters, which we left almost untouched in this book.

# Chapter 6

# Moscow Topological Groups and Completions of Groups

It was established in Section 3.6 that every topological group $G$ is a dense subgroup of a Raĭkov complete topological group denoted by $\varrho G$ and called the Raĭkov completion of $G$. In the class of Tychonoff spaces, there are at least three distinct ways to "complete" a given space $X$, by taking the Čech–Stone compactification $\beta X$, Hewitt–Nachbin completion $\upsilon X$, and Dieudonné completion $\mu X$ of $X$, respectively. The latter three extensions are related by the inclusions $\mu X \subset \upsilon X \subset \beta X$ and, similarly to the Raĭkov completion, have natural functorial properties (permitting extension of continuous mappings of spaces over corresponding completions).

We will see in Corollary 6.6.4 that the Čech–Stone compactification $\beta G$ of a pseudocompact topological group $G$ is again a topological group containing $G$ as a dense topological subgroup. However, for very simple topological groups $G$, the Čech–Stone compactification $\beta G$ may fail to be a topological group. For example, neither $\beta\mathbb{Z}$ nor $\beta\mathbb{R}$ is a topological group, where $\mathbb{Z}$ is the discrete group of integers and $\mathbb{R}$ is the group of reals with the usual topology and sum operation. In fact, the spaces $\beta\mathbb{Z}$ and $\beta\mathbb{R}$ are not even homogeneous.

If, however, one considers the Dieudonné (or Hewitt–Nachbin) completion of a topological group $G$, it surprisingly often turns out that the group operations of $G$ can be extended over the space $\mu G$ (or $\upsilon G$), thus making $\mu G$ into a topological group containing $G$ as a dense topological subgroup. If this happens, we say that $G$ is a *PT-group*. The reader will find in Theorem 6.5.24 a long (but far from complete) list of properties of a group $G$ each of which guarantees that $\mu G$ is a topological group.

The basic concept in this chapter is that of a Moscow space. The class of Moscow spaces is very large; it contains, on one hand, all extremally disconnected spaces and, on the other, all spaces of countable pseudocharacter. In the first three sections we study in detail this class of spaces and establish some permanence properties of Moscow spaces. The results in Section 6.4 show that the class of Moscow topological (and paratopological) groups is even wider compared to the class of Moscow spaces. In fact, very mild compactness type conditions imposed on a topological group $G$ imply that the space $G$ is Moscow.

The principal result of Section 6.5, Theorem 6.5.1, says that if $G$ is a topological group and the space $G$ is Moscow, then $\mu G$ is also a topological group that contains $G$ as a dense topological subgroup. This important result has a number of corollaries, sometimes quite unexpected. For example, we show in Section 6.6 that a pseudocompact topological group $G$ is $C$-embedded in its Raĭkov completion $\varrho G$, so that $\beta G = \varrho G$ in this case, and the product of an arbitrary family of pseudocompact topological groups is pseudocompact (both results due to W. W. Comfort and K. A. Ross). The commutativity of the Hewitt–Nachbin completion and the product operation is studied in Section 6.7, where we show that if the product group $G = \prod_{i \in I} G_i$ is Moscow and has Ulam non-measurable cardinality, then $\upsilon G = \prod_{i \in I} \upsilon G_i$. In the case of two factors, one can go further and characterize the groups $G$ and $H$ satisfying the formula $\upsilon(G \times H) = \upsilon G \times \upsilon H$. Roughly speaking, the equality holds if and only if the product $G \times H$ is a $PT$-group (see Theorem 6.7.10).

Subgroups of Moscow topological groups are considered in Section 6.8. It is shown that every Abelian topological group is topologically isomorphic to a closed subgroup of a Moscow topological group. Therefore, closed subgroups of Moscow groups may fail to be Moscow. We also give an example of a closed $C$-embedded subgroup $H$ of a Moscow topological group such that $H$ is not Moscow.

Pointwise pseudocompact topological groups are studied in Section 6.9. One of the main results of the section, Theorem 6.9.14, says that a topological group $G$ is pointwise pseudocompact if and only if its Dieudonné completion $\mu G$ is a feathered topological group. The class of pointwise pseudocompact groups has nice properties. For example, it is countably productive, and every group from this class has countable $o$-tightness. Furthermore, the product of an arbitrary family of pointwise pseudocompact groups is a Moscow space.

The last section of this chapter is dedicated to bounded and $C$-compact subsets of topological groups.

## 6.1. Moscow spaces and $C$-embeddings

In this section we introduce the class of Moscow spaces, and establish some of its basic properties. Moscow spaces work especially well in combination with homogeneity; this explains why this notion turned out to be a powerful tool in the theory of topological groups. Applications of Moscow spaces involve Dieudonné completions of topological groups and $C$-embeddings.

A Hausdorff space $X$ is called *Moscow* if for every open subset $U$ of $X$, the closure of $U$ in $X$ is the union of a family of $G_\delta$-sets in $X$, that is, for each $x \in \overline{U}$ there exists a $G_\delta$-set $P$ in $X$ such that $x \in P \subset \overline{U}$.

Clearly, the notion of a Moscow space generalizes the notion of a perfectly $\kappa$-normal space. The class of Moscow spaces is much wider than the class of perfectly $\kappa$-normal spaces, since every first-countable space, and even every space of countable pseudocharacter is a Moscow space, while not every first-countable compact space is perfectly $\kappa$-normal. For example, the square of the two arrows space $Z$ (see [**165**, 3.10.C]) is a first-countable compact space which is not perfectly $\kappa$-normal. Indeed, assume the contrary. Then the diagonal $\Delta_Z$ in $Z \times Z$ can be represented as the intersection of two regular closed sets, since $Z$ is linearly ordered. Therefore, $\Delta_Z$ is a $G_\delta$-set in $Z \times Z$. It remains to refer to the fact that every compact space with the diagonal of type $G_\delta$ is metrizable, by [**242**, Corollary 7.6]. This contradiction shows that $Z \times Z$ is not perfectly $\kappa$-normal.

The notion of a Moscow space can be also considered as a straightforward generalization of extremal disconnectedness (see Sections 2.2 and 4.5).

We sum up the above observations in the next statement.

PROPOSITION 6.1.1. *Suppose that a space $X$ has one of the following properties:*

a) *perfect $\kappa$-normality;*
b) *extremal disconnectedness;*
c) *countable pseudocharacter.*

*Then $X$ is Moscow.*

We will see in Corollary 6.3.15 that the product of an arbitrary family of first-countable spaces is Moscow. It follows that every Tychonoff cube $I^\tau$, as well as every generalized Cantor discontinuum $D^\tau$, is Moscow. The space $\mathbb{R}^\tau$ is also Moscow. The next statement is obviously true:

PROPOSITION 6.1.2. *Every dense subspace of a Moscow space is Moscow.*

However, we cannot claim that every closed subspace of a Moscow space is Moscow.

EXAMPLE 6.1.3. Let $D(\tau)$ be an uncountable discrete space and $\alpha D(\tau) = D(\tau) \cup \{*\}$ be the one-point compactification of $D(\tau)$. Then $D(\tau)$ is a Moscow space, and $D(\tau)$ is $G_\delta$-dense in $\alpha D(\tau)$, while $\alpha D(\tau)$ is not a Moscow space. Indeed, let $U$ be any countable infinite subset of $D(\tau)$. Then $U$ is open in $\alpha D(\tau)$, and $\overline{U} = U \cup \{*\}$, where $*$ is the only non-isolated point in $\alpha D(\tau)$. Every $G_\delta$-set in $\alpha D(\tau)$ containing the point $*$ is uncountable; therefore, $\overline{U}$ is not the union of any family of $G_\delta$-sets in $\alpha D(\tau)$. Since $\alpha D(\tau)$ is a closed subspace of the Tychonoff cube $I^\tau$, we conclude that the class of Moscow spaces is not closed hereditary. □

The notion of a Moscow space plays a major role in the theory of $C$-embeddings. Recall that a subspace $Y$ of a space $X$ is said to be *$C$-embedded* in $X$ if every continuous real-valued function on $Y$ can be extended to a continuous real-valued function on $X$. When is a subspace $Y$ $C$-embedded in the whole space $X$? Of course, the normality of $X$ guarantees that every closed subspace of $X$ is $C$-embedded in $X$. Let us now consider an "orthogonal" question: When is a dense subspace $Y$ of $X$ $C$-embedded in $X$? The next important general result provides us with a necessary condition.

THEOREM 6.1.4. [**L. Gillman and M. Jerison**] *If a dense subspace $Y$ of a Tychonoff space $X$ is $C$-embedded in $X$, then $Y$ is $G_\delta$-dense in $X$.*

PROOF. Assume that $Y$ is not $G_\delta$-dense in $X$. Then, since $X$ is Tychonoff, there exists a non-empty zero-set $P$ in $X$ contained in $X \setminus Y$. Fix a continuous real-valued function $f$ on $X$ such that $P = \{x \in X : f(x) = 0\}$. Put $g(y) = 1/f(y)$, for every $y \in Y$. Obviously, $g$ is continuous on $Y$.

It is also clear from the choice of $f$ that $g$ cannot be continuously extended over $X$, since the value of any such extension at any point of the set $P$ would have been infinite. □

The converse to Theorem 6.1.4 is not true. Indeed, in Example 6.1.3, the subspace $D(\tau)$ is $G_\delta$-dense in $\alpha D(\tau)$ while, obviously, $D(\tau)$ is not $C$-embedded in $\alpha D(\tau)$.

To present one of the key properties of Moscow spaces in Theorem 6.1.7 below, we recall the *Urysohn Extension Theorem* characterizing $C^*$-embedded subsets of a space. Its proof is almost the same as that of [**165**, Theorem 2.1.8] (a complete argument is given in [**191**, 1.17]), so we omit it. As usual, subsets $A$ and $B$ of a space $Y$ are called *completely separated* in $Y$ if there exists a continuous function $f$ on $Y$ with values in the unit interval $[0, 1]$ such that $f(A) \subset \{0\}$ and $f(B) \subset \{1\}$.

THEOREM 6.1.5. [**P. S. Urysohn**] *A subspace $Y$ of a space $X$ is $C^*$-embedded in $X$ if and only if any two completely separated sets in $Y$ are completely separated in $X$.*

The next result shows that, for $G_\delta$-dense subsets of a space, the properties of being $C^*$-embedded and $C$-embedded coincide (a more general fact will be established in Lemma 9.9.35).

PROPOSITION 6.1.6. *Let $Y$ be a $G_\delta$-dense subspace of a space $X$. Then $Y$ is $C$-embedded in $X$ iff it is $C^*$-embedded in $X$.*

PROOF. It suffices to verify that '$C^*$-embedded' implies '$C$-embedded'. Consider a continuous real-valued function $f$ on $Y$. Denote by $h$ a homeomorphism of the real line $\mathbb{R}$ onto the open interval $(0, 1)$. Since $Y$ is $C^*$-embedded in $X$, the function $h \circ f$ on $Y$ can be extended to a continuous function $g\colon X \to [0, 1]$. Then the image $g(X)$ does not contain either 0 or 1.

Indeed, suppose to the contrary that $0 \in g(X)$ (the case $1 \in g(X)$ is similar). Then $F = g^{-1}(0)$ is a non-empty $G_\delta$-set in $X$ and, since $Y$ is $G_\delta$-dense in $X$, the intersection $Y \cap F$ is not empty. If $x \in Y \cap F$, then $g(x) = 0$. On the other hand, $g(x) = h(f(x)) \neq 0$. This contradiction shows that $g(X) \cap \{0, 1\} = \varnothing$. Therefore, $\tilde{f} = h^{-1} \circ g$ is the required continuous extension of $f$ over $X$. □

Here is the promised key property of Moscow spaces.

THEOREM 6.1.7. [**V. V. Uspenskij, M. G. Tkachenko**] *For a Moscow space $X$, every $G_\delta$-dense subset $Y$ of $X$ is $C$-embedded in $X$.*

PROOF. Assume that $Y$ is not $C$-embedded in $X$. By Proposition 6.1.6, $Y$ is not $C^*$-embedded in $X$ either, so Theorem 6.1.5 implies that $Y$ contains two completely separated subsets $A$ and $B$ whose closures in $X$ intersect. Take open subsets $V_1$ and $V_2$ in $Y$ such that $A \subset V_1$, $B \subset V_2$ and the closures of $V_1$ and $V_2$ in $Y$ are disjoint. Clearly, the intersection of the closures of $V_1$ and $V_2$ in $X$ is not empty. Fix a point $x$ in $\overline{V_1} \cap \overline{V_2}$, and let $U_i$ be the interior of the closure of $V_i$ in $X$, $i = 1, 2$. Obviously, $V_i \subset U_i$ and, therefore, $U_i$ is not empty and $x \in \overline{U_1} \cap \overline{U_2}$.

Since $X$ is a Moscow space, we can find $G_\delta$-sets $P_i$ in $X$ such that $x \in P_i \subset \overline{U_i}$, for $i = 1, 2$. Then $P = P_1 \cap P_2$ is a $G_\delta$-subset of $X$ and $x \in P$; therefore, $P \cap Y$ is not empty. Clearly, every point of $P \cap Y$ belongs to the intersection of the closures of the sets $V_1$ and $V_2$ in $Y$, which is impossible, since this intersection is empty, by the choice of $V_1$ and $V_2$. □

Example 6.1.3 shows that it is not enough to assume $Y$ to be a Moscow space in the next theorem. However, the homogeneity of the enveloping space $X$ suffices. Much of what follows depends upon this result.

THEOREM 6.1.8. [**A. V. Arhangel'skii**] *If a Moscow space $Y$ is a $G_\delta$-dense subspace of a homogeneous space $X$, then $X$ is also a Moscow space and $Y$ is $C$-embedded in $X$.*

PROOF. Let $U$ be an open subset of $X$ and $x$ a point in the closure of $U$. We have to show that there exists a $G_\delta$-set $P$ in $X$ such that $x \in P \subset \overline{U}$.

Since $X$ is homogeneous, we may assume that $x \in Y$. Then $x \in \overline{U \cap Y}$ and, since $Y$ is a Moscow space, there exists a $G_\delta$-set $Q$ in the space $Y$ such that $x \in Q \subset \overline{U \cap Y}$.

Thus, there exists a countable family $\{U_n : n \in \omega\}$ of open subsets of $X$ such that their intersection $P = \bigcap_{n\in\omega} U_n$ satisfies the condition:

$$x \in P \cap Y \subset \overline{U}.$$

We claim that $P \subset \overline{U}$. Indeed, assume the contrary. Then $P \setminus \overline{U}$ is a non-empty $G_\delta$-set in $X$, and since $Y$ is $G_\delta$-dense in $X$, it follows that $(P \setminus \overline{U}) \cap Y$ is not empty. On the other hand, $(P \setminus \overline{U}) \cap Y = (P \cap Y) \setminus \overline{U} = \varnothing$, by the above formula. This contradiction shows that $x \in P \subset \overline{U}$. Thus, $X$ is a Moscow space. To conclude that $Y$ is $C$-embedded in $X$ it remains to apply Theorem 6.1.7. □

From Theorem 6.1.8 and Proposition 6.1.2 we obtain immediately the following statement:

COROLLARY 6.1.9. *Let $X$ be a homogeneous space and $Y$ a $G_\delta$-dense subspace of $X$. Then $X$ is a Moscow space if and only if $Y$ is a Moscow space.*

We can sum up the information obtained so far as follows:

COROLLARY 6.1.10. *Let $Y$ be a dense subspace of a homogeneous space $X$. Then the following conditions are equivalent:*

1) *$X$ is a Moscow space and $Y$ is $G_\delta$-dense in $X$;*
2) *$X$ is a Moscow space and $Y$ is $C$-embedded in $X$;*
3) *$Y$ is a Moscow space and $Y$ is $G_\delta$-dense in $X$;*
4) *$Y$ is a Moscow space and $Y$ is $C$-embedded in $X$.*

The next characterization of Moscow spaces shows that the relationship of this class of spaces to $C$-embeddings is even deeper than one might presume. To formulate the result, we recall that the *$G_\delta$-closure* of a set $Y \subset X$ in a space $X$ is defined as the set of all points $x \in X$ such that every $G_\delta$-set $P$ in $X$ containing $x$ intersects $Y$.

THEOREM 6.1.11. *A space $X$ is Moscow if and only if every dense subspace $Y$ of $X$ is $C$-embedded in the $G_\delta$-closure of $Y$ in $X$.*

PROOF. The necessity of the condition follows from Theorem 6.1.7 and the fact that every dense subspace of a Moscow space is Moscow. Conversely, suppose that every dense subset $Y$ of $X$ is $C$-embedded in its $G_\delta$-closure $Z$. Suppose that $U$ is a non-empty open subset of $X$, and let $V = X \setminus \overline{U}$. Then the function $f$ on $Y = U \cup V$ which equals 0 on $U$ and 1 on $V$ is continuous. Hence, $f$ admits an extension to a continuous function $g$ on the $G_\delta$-closure $Z$ of $Y$. Notice that $Z \cap \overline{U} \cap \overline{V} = \varnothing$, since no extension of $f$ is continuous at the points of the set $\overline{U} \cap \overline{V}$. Therefore, for every $x \in \overline{U}$, either $x \notin \overline{V}$ or $x \notin Z$. In the first case, $O = X \setminus \overline{V}$ is an open neighbourhood of $x$ in $X$ satisfying $O \subset \overline{U}$. In the second case, there exists a $G_\delta$-set $P$ in $X$ such that $x \in P$ and $P \cap (U \cup V) = \varnothing$, whence it follows that $P \subset \overline{U}$. This implies that $\overline{U}$ is the union of a family of $G_\delta$-sets in $X$, so that $X$ is a Moscow space. □

The next example shows that the conditions on the space $X$ in Theorem 6.1.11 cannot be weakened.

EXAMPLE 6.1.12. There exists a non-Moscow space $X$ such that every $G_\delta$-dense subspace $Y$ of $X$ is $C$-embedded in $X$.

It turns out that $\omega_1+1$ with the order topology is such a space. Indeed, the subspace $\omega_1$ of $\omega_1+1$ is countably compact and, hence, pseudocompact. Since every $G_\delta$-dense subspace $Y$ of $\omega_1+1$ is either $\omega_1$ or $\omega_1+1$, and $\omega_1$ is $C^*$-embedded in $\omega_1+1$ [**165**, Example 3.1.27], it follows that every $G_\delta$-dense subspace of $\omega_1+1$ is $C$-embedded in $\omega_1+1$. To see that $\omega_1+1$ is not Moscow, take two disjoint uncountable sets $U$ and $V$ consisting of non-limit ordinals. Then $U$ and $V$ are open and the end point $\omega_1$ of $\omega_1+1$ is the only point in the intersection of their closures. Therefore, $\omega_1+1$ is not Moscow. □

Here are some curious applications of Theorems 6.1.7 and 6.1.8.

COROLLARY 6.1.13. *If $X$ is a pseudocompact Moscow space and $bX$ is a homogeneous compactification of $X$, then $bX$ is the Čech–Stone compactification of $X$.*

PROOF. Since $X$ is pseudocompact, $X$ is $G_\delta$-dense in $bX$. By Theorem 6.1.8, $X$ is $C$-embedded in $bX$. Therefore, $bX=\beta X$, according to [**165**, Coro. 3.6.3]. □

COROLLARY 6.1.14. *Every $G_\delta$-dense subspace $Y$ of a compact Moscow space $X$ is pseudocompact, and $X$ is the Čech–Stone compactification of $Y$.*

PROOF. By Theorem 6.1.7, $Y$ is $C$-embedded in $X$. Since $X$ is compact, it follows that $Y$ is pseudocompact, and that $X$ is the Čech–Stone compactification of $Y$. □

## Exercises

6.1.a. Let $U$ be a regular open subset of a Moscow space $X$. Show that $\overline{U}\setminus U$ is the union of a family of $G_\delta$-sets in $X$.

6.1.b. A subset $Y$ of a space $X$ is said to be *residually Moscow* in $X$ if, for each open subset $U$ of $X$ and each $z\in\overline{U}\setminus Y$, there exists a $G_\delta$-set $P$ in $X$ such that $z\in P$ and $P\cap Y\subset\overline{U}$. Prove that if $Y$ is $G_\delta$-dense in $X$, and $Y$ is residually Moscow in $X$, then $Y$ is $C$-embedded in $X$.

6.1.c. Let $Y$ be a subspace of a space $X$. We say that $Y$ is *$h$-dense* in $X$ if $Y$ is dense in $X$ and, for each $x\in X$, there exists a homeomorphism $h$ of $X$ onto itself such that $h(x)\in Y$. We also say in this case that $X$ is *$Y$-homogeneous*. Prove that if a Moscow space $Y$ is a $G_\delta$-dense subspace of a $Y$-homogeneous space $X$, then $X$ is also a Moscow space. Show that the assumption that $X$ is $Y$-homogeneous cannot be dropped (consider the spaces $D(\tau)$ and $\alpha D(\tau)$ in Example 6.1.3).

## Problems

6.1.A. Let $X$ be a pseudocompact space, and suppose that the Čech–Stone compactification $\beta X$ of $X$ is a Moscow space. Is $X$ perfectly $\kappa$-normal?

6.1.B. Is the class of compact Moscow spaces finitely productive?

6.1.C. Suppose that the Čech–Stone compactification of a Tychonoff space $X$ admits the structure of a topological group. Prove that $cel_\omega(X)\le\omega$.
*Hint.* According to [**147**] or [**126**], the assumptions about $X$ imply that $X$ is pseudocompact. Then $X$ is $G_\delta$-dense in $\beta X$, so one can apply Corollary 5.3.20.

## Open Problems

6.1.1. When does the Čech–Stone compactification $\beta X$ of a Tychonoff space $X$ turn out to be a Moscow space?

6.1.2. Given a Tychonoff space $X$, when does there exist a Hausdorff compactification of $X$ which is a Moscow space?

## 6.2. Moscow spaces, $P$-spaces, and extremal disconnectedness

In this section we consider the notions of extremal disconnectedness and of $P$-space in the context of Moscow spaces. It turns out that the notions of a $P$-space and of a Moscow space are almost incompatible. The next statement is obvious.

PROPOSITION 6.2.1. *A $P$-space $X$ is a Moscow space if and only if it is extremally disconnected.*

A point $x$ of a space $X$ is called a *point of extremal disconnectedness* of $X$ or, briefly, an *ed-point* of $X$ if, for any disjoint open sets $U$ and $V$ in $X$, $x$ is not in $\overline{U} \cap \overline{V}$. Clearly, a space $X$ is extremally disconnected iff every point of $X$ is an *ed*-point.

A point $x \in X$ is a *$P$-point* of $X$ if every $G_\delta$-set in $X$ containing $x$ also contains an open neighbourhood of $x$.

An infinite cardinal number $\tau$ is said to be *Ulam measurable* if there exists a countably closed free ultrafilter on a set $A$ of cardinality $\tau$. 'Countably closed' means that the intersection of every countable subfamily of $\xi$ belongs to $\xi$. Clearly, a cardinal $\tau$ is *Ulam non-measurable* if every countably closed ultrafilter $\xi$ on a set $A$ of cardinality $\tau$ is fixed, that is, has a non-empty intersection. It is immediate from the definition that the cardinal $\omega$ is Ulam non-measurable. In addition, if $\tau$, $\lambda$ are infinite cardinals, $\tau$ is Ulam measurable and $\lambda > \tau$, then $\lambda$ is Ulam measurable as well.

The relation of Ulam measurable cardinals with measures is quite clear. It is easy to verify that an infinite cardinal $\tau$ is Ulam measurable if and only if there exists a two-valued measure (with values 0 and 1) on the set $\mathcal{P}(A)$ of all subsets of a set $A$ of cardinality $\tau$ which vanishes at singletons, is monotone and countably additive. Indeed, given a countably closed free ultrafilter $\xi$ on $A$, define a mapping $m\colon \mathcal{P}(A) \to \{0, 1\}$ by $m(B) = 1$ iff $B \in \xi$. Evidently, $m$ is the required measure on $A$. Conversely, if $m$ is a two-valued countably additive measure on $A$ vanishing at singletons, put $\xi = \{B \subset A : m(B) = 1\}$. Then $\xi$ is a countably closed free ultrafilter on the set $A$ (see [**263**, Section 23]).

In fact, Ulam measurable cardinals are extremely large. To see it, we present the following result that combines Theorems 23.12 and 23.14 of [**263**] and will be used shortly afterwards.

THEOREM 6.2.2. *Suppose that a cardinal $\tau \geq \omega$ is Ulam non-measurable. Then:*

a) *the cardinal $2^\tau$ is Ulam non-measurable;*
b) *if $\{\lambda_\alpha : \alpha < \tau\}$ is a sequence of Ulam non-measurable cardinals, then the cardinal $\sup_{\alpha<\tau} \lambda_\alpha$ is Ulam non-measurable as well.*

It follows from b) of Theorem 6.2.2 that the first Ulam measurable cardinal $\mathfrak{m}_0$ is regular, while a) and b) together imply that $\mathfrak{m}_0$ is *strongly inaccessible.* In particular, the cardinal numbers $\mathfrak{c} = 2^\omega, 2^\mathfrak{c}, 2^{2^\mathfrak{c}}$, *etc.*, are all Ulam non-measurable.

We call $x \in X$ an *Ulam $P$-point* of $X$ or, briefly, a *UP-point* of $X$ if, for each family $\gamma$ of open sets in $X$ such that $x \in \bigcap \gamma$ and $|\gamma|$ is an Ulam non-measurable cardinal, $\bigcap \gamma$ contains an open neighbourhood of $x$. If every point of $X$ is an *UP*-point, we say that $X$ is a *UP-space.*

THEOREM 6.2.3. *Let $X$ be a regular space, and $b \in X$ an ed-point of $X$. Then either $b$ is not a $P$-point of $X$, or $b$ is a UP-point of $X$.*

Proof. Assume to the contrary that $b$ is a $P$-point and is not a $UP$-point. Then there exists a family $\gamma$ of open sets in $X$ such that $\tau = |\gamma|$ is Ulam non-measurable, $b \in \bigcap \gamma$, and $b \in \overline{X \setminus \bigcap \gamma}$. We can also choose $\gamma$ to be of the smallest possible cardinality. Then $\tau > \omega$, since $b$ is a $P$-point, and $\tau$ is a regular cardinal.

Since $X$ is regular, we may assume that $b \in \overline{W}$, where $W = X \setminus \bigcap\{\overline{U} : U \in \gamma\}$. Clearly, $W$ is open, and $b \notin W$. Let $\mathcal{F} = \{V_\alpha : \alpha < \tau\}$ be the family of all sets $X \setminus \overline{U}$, where $U \in \gamma$. Clearly, $V_\alpha$ is an open subset of $X$, $b$ is not in the closure of $V_\alpha$, and $b \in \overline{\bigcup\{V_\alpha : \alpha < \tau\}}$. Put $W_\alpha = V_\alpha \setminus \overline{\bigcup\{V_\beta : \beta < \alpha\}}$, for each $\alpha < \tau$. Obviously, $\xi = \{W_\alpha : \alpha < \tau\}$ is a disjoint family of open subsets of $X$, $b$ is in the closure of $\bigcup \xi$, but $b \notin \overline{W_\alpha}$, for any $\alpha < \tau$.

Let $\mathcal{P}$ be the family of all non-empty open subsets $V$ of $X$ contained in at least one element of $\xi$. By Zorn's Lemma, there exists a maximal disjoint subfamily $\mathcal{M}$ of $\mathcal{P}$. Obviously, $b$ is in the closure of $\bigcup \mathcal{M}$.

Denote by $\eta$ the collection of all subfamilies $\mathcal{A}$ of $\mathcal{M}$ such that $b$ is in the closure of $\bigcup \mathcal{A}$. Clearly, $\mathcal{M} \in \eta$, and every $\mathcal{A} \in \eta$ is non-empty. It is also clear that, for any $\mathcal{A} \subset \mathcal{M}$, either $\mathcal{A} \in \eta$, or the complement $\mathcal{M} \setminus \mathcal{A}$ is in $\eta$. Since $b$ is an *ed*-points of $X$, it follows that $\mathcal{A} \in \eta$ if and only if the complement of $\mathcal{A}$ in $\mathcal{M}$ is not in $\eta$.

Let us show that $\eta$ is $\omega$-centered, that is, $\bigcap \mu \in \eta$, for any countable subfamily $\mu$ of $\eta$. Indeed, if $\mathcal{A} \in \mu$, then $\mathcal{M} \setminus \mathcal{A}$ is not in $\mu$, that is, $b$ is not in the closure of $\bigcup(\mathcal{M} \setminus \mathcal{A})$. Since $b$ is a $P$-point, it follows that $b$ is not in the closure of $\bigcup\{\bigcup(\mathcal{M} \setminus \mathcal{A}) : \mathcal{A} \in \mu\}$. By passing to the complement, we conclude that $b$ is in the closure of $\bigcup \bigcap \mu$, that is, $\bigcap \mu \in \eta$, and therefore, $\bigcap \mu$ is non-empty. Hence, $\eta$ is a countably closed ultrafilter on the set $\mathcal{M}$. Since $\tau$ is Ulam non-measurable, for any subfamily $\nu$ of $\eta$ with $|\nu| \leq \tau$, we have that $\bigcap \nu \neq \varnothing$.

Now, let $\mathcal{A}_\alpha$ be the family of all $V \in \mathcal{M}$ such that $V \cap W_\alpha = \varnothing$, and $\nu = \{\mathcal{A}_\alpha : \alpha < \tau\}$. Clearly, $\mathcal{A}_\alpha \in \eta$, $|\nu| \leq \tau$, and $\bigcap \nu = \varnothing$. This contradiction completes the proof. □

Theorem 6.2.4. *Let $X$ be a regular extremally disconnected $P$-space and $b \in X$ a point such that $\chi(b, X) \leq \mathfrak{m}_0$, where $\mathfrak{m}_0$ is the first Ulam measurable cardinal. Then the point $b$ is isolated in $X$.*

Proof. Assume the contrary. By an obvious transfinite recursion we can define, making use of Theorem 6.2.3, a transfinite sequence $\xi = \{V_\alpha : \alpha < \tau^*\}$ of non-empty disjoint open subsets of $X$ such that the next two conditions are satisfied:

1) For each open neighbourhood $U$ of $b$, there exists $\alpha < \tau^*$ such that $V_\beta \subset U$ whenever $\alpha \leq \beta < \tau^*$;
2) For any $\alpha < \tau^*$, $b$ is not in $\overline{\bigcup\{V_\beta : \beta < \alpha\}}$.

Let $W_1$ be the union of all $V_\alpha$'s such that $\alpha$ is a limit ordinal, and $W_2$ the union of all $V_\alpha$'s such that $\alpha$ is a successor ordinal. Then $W_1$ and $W_2$ are disjoint open sets and, by 1), $b$ is in the closure of both of them. This contradicts the extremal disconnectedness of $X$. □

Theorem 6.2.5. *Let $X$ be a regular extremally disconnected non-discrete $P$-space. Then:*

1) *There exists a disjoint family $\gamma$ of non-empty open and closed subsets of $X$ such that $\bigcup \gamma$ is dense in $X$ and $|\gamma|$ is an Ulam measurable cardinal;*
2) *Every disjoint family $\eta$ of open and closed subsets of $X$ such that $|\eta|$ is Ulam non-measurable, is discrete in $X$;*
3) *$dc(X)$ is not less than the first Ulam measurable cardinal.*

PROOF. The family of all open and closed subsets of $X$ is a base of $X$, since $X$ is a regular $P$-space. Fix a non-isolated point $b$ in $X$, and let $\gamma$ be a maximal disjoint family of non-empty open and closed subsets $U$ of $X$ such that $b$ is not in $U$ (by Zorn's Lemma, such a family $\gamma$ exists). Then $\bigcup \gamma$ is dense in $X$ and $b \in A$, where $A = \bigcap\{X \setminus U : U \in \gamma\}$. Clearly, $A$ does not contain any open neighbourhood of $b$ in $X$. By Theorem 6.2.3, $b$ is a $UP$-point in $X$. It follows that $|\gamma| \geq \mathfrak{m}_0$, where $\mathfrak{m}_0$ is the first Ulam measurable cardinal. Therefore, $|\gamma|$ is also Ulam measurable. Thus, 1) is proved.

To prove 2), assume that $\eta$ is not discrete in $X$, and take an accumulation point $y$ of $\eta$. Then $y \in B$, where $B = \bigcap\{X \setminus U : U \in \eta\}$, and $B$ does not contain any open neighbourhood of $y$ in $X$. Therefore, since $|\eta|$ is Ulam non-measurable, $y$ is not a $UP$-point in $X$, thus contradicting Theorem 6.2.3. Hence, 2) holds.

Assertion 3) follows from 1) and 2). Indeed, take a disjoint family $\gamma$ of non-empty open and closed subsets of $X$ such that $|\gamma|$ is an Ulam measurable cardinal. Obviously, we may assume that $|\gamma|$ is the first Ulam measurable cardinal $\mathfrak{m}_0$.

Now let $\tau$ be any cardinal number less than $\mathfrak{m}_0$. There exists a subfamily $\eta$ of $\gamma$ such that $|\eta| = \tau$. Then, by 2), $\eta$ is discrete in $X$, which implies that $dc(X) > \tau$. Since $\tau < \mathfrak{m}_0$, we conclude that $dc(X) \geq \tau^*$. □

Clearly, if the cellularity or the Lindelöf number of $X$ is Ulam non-measurable, then the discrete cellularity number $dc(X)$ of $X$, defined in Section 5.1, is Ulam non-measurable as well. Therefore, the next four results follow from Theorem 6.2.5.

COROLLARY 6.2.6. *If the cellularity of a regular extremally disconnected $P$-space $X$ is Ulam non-measurable, then $X$ is discrete.*

COROLLARY 6.2.7. *If the Lindelöf number of a regular extremally disconnected $P$-space $X$ is Ulam non-measurable, then $X$ is discrete.*

By Proposition 6.2.1, every Moscow $P$-space is extremally disconnected. Therefore, we have the following:

COROLLARY 6.2.8. *A regular Moscow space of Ulam non-measurable cardinality is a $P$-space if and only if it is discrete.*

COROLLARY 6.2.9. *A Lindelöf Moscow space $X$ is a $P$-space if and only if it is discrete and countable.*

PROOF. By our definition, all Moscow spaces are Hausdorff. To apply Corollary 6.2.7, it suffices to verify that the Lindelöf $P$-space $X$ is regular. Let $x \in X$ be a point and $U$ an open set in $X$ containing $x$. Then $F = X \setminus U$ is closed in $X$. For each $y \in F$, choose disjoint open sets $V_y$ and $W_y$ in $X$ such that $x \in V_y$ and $y \in W_y$. Since $X$ is Lindelöf, the open covering $\{W_y : y \in F\}$ of the set $F$ contains a subcovering $\{W_y : y \in C\}$, where $C$ is a countable subset of $F$. Then $V = \bigcap_{y \in C} V_y$ is an open neighbourhood of $x$ disjoint from the open set $W = \bigcup_{y \in C} W_y$. Since $X \setminus U = F \subset W$, we conclude that the closure of $V$ is contained in $U$, so that the space $X$ is regular. □

It is well known that every regular space $X$ with a topology $\mathcal{T}$ can be turned into a regular $P$-space by means of the following simple construction. Take the *Baire topology* $\mathcal{T}_{\mathcal{B}}$ on $X$ such that the family of all $G_\delta$-sets in the original space $X$ is a base of $\mathcal{T}_{\mathcal{B}}$. Clearly, $A$ is in $\mathcal{T}_{\mathcal{B}}$ if and only if $A$ is the union of a family of $G_\delta$-subsets of $X$. Let $(X)_\omega$ be the space obtained when we endow the set $X$ with the topology $\mathcal{T}_{\mathcal{B}}$. We call this space the $G_\delta$-*modification* of the space $X$. Obviously, $(X)_\omega$ is a regular $P$-space. Assume now that the cardinality of $X$ is Ulam non-measurable. Then, by Corollary 6.2.8, $(X)_\omega$ is a Moscow space if and only if $X^*$ is discrete. This result can be also reformulated in terms of the original space $X$:

THEOREM 6.2.10. *Let $X$ be a regular space of Ulam non-measurable cardinality. Then the following conditions are equivalent:*

1) *Every point of $X$ is a $G_\delta$-set;*
2) *For every family $\gamma$ of $G_\delta$-sets in $X$, the $G_\delta$-closure of the union of $\gamma$ is also the union of a family of $G_\delta$-sets in $X$;*
3) *The $G_\delta$-modification of $X$ is a Moscow space;*
4) *The $G_\delta$-modification of $X$ is discrete.*

PROOF. By Corollary 6.2.8, the conditions in items 1), 2), and 4) are equivalent, while 3) says that the space $(X)_\omega$ is Moscow. □

COROLLARY 6.2.11. *Let $X$ be a regular space such that the cardinality of $X$ is Ulam non-measurable, and at least one point of $X$ is not a $G_\delta$-set in $X$. Then the $G_\delta$-modification of the space $X$ is not a Moscow space.*

In connection with the above results, we should mention that one cannot get rid of Ulam measurable cardinals in their formulations since, under the assumption that such a cardinal exists, V. I. Malykhin constructed a non-discrete extremally disconnected Abelian $P$-group (see [**299**]).

## Exercises

6.2.a. Show that the remainder $\beta\omega \setminus \omega$ of the Čech–Stone compactification of the discrete group space $\omega$ is not a Moscow space.

6.2.b. Let us call a Tychonoff space $X$ *weakly Moscow* if the closure of each cozero-set in $X$ is the union of a family of $G_\delta$-sets in $X$. Give an example of a weakly Moscow space which is not Moscow.

6.2.c. Verify that a weakly Moscow space of countable $o$-tightness is Moscow.

6.2.d. Find out whether every compact weakly Moscow space is Moscow.

## Problems

6.2.A. Is every Hausdorff extremally disconnected space zero-dimensional?

6.2.B. Can the regularity of a Moscow space in Corollary 6.2.8 be weakened to the Hausdorff property?

6.2.C. Assuming $CH$, prove that the compact space $\beta\omega \setminus \omega$ contains a dense subspace $X$ which is a $P$-space.

## 6.3. Products and mappings of Moscow spaces

In this section we present some results on the behaviour of the class of Moscow spaces with respect to products and continuous mappings which we later apply to topological groups.

We recall that a mapping $f\colon X \to Y$ is called *compact* if the fiber $f^{-1}(y)$ is compact, for each $y \in Y$.

THEOREM 6.3.1. *If $f$ is an open continuous compact mapping of a Moscow space $X$ onto a space $Y$, then $Y$ is also a Moscow space.*

PROOF. Let $V$ be an open subset of $Y$ and $U = f^{-1}(V)$. Put $F = \overline{U}$. Since $f$ is open and continuous, we have $F = f^{-1}(\overline{V})$ and, hence, $f(F) = \overline{V}$. Take any $y \in \overline{V}$, and fix $x \in F$ such that $f(x) = y$. Since $F$ is a regular closed set the Moscow space $X$, there exist open sets $W_n$ in $X$, where $n \in \omega$, such that $x \in W_n$ and $\bigcap_{n=0}^{\infty} W_n \subset F$. Since $X$ is regular, we may also assume that $x \in \overline{W_{n+1}} \subset W_n$, for each $n \in \omega$. Put $O_n = f(W_n)$. Then, obviously, $y = f(x) \in O_n$ and each $O_n$ is open in $Y$. Therefore, $y \in P$, where $P = \bigcap_{n=0}^{\infty} O_n$ is a $G_\delta$-set in $Y$.

Let us show that $P \subset \overline{V}$. Take any $y_0 \in Y \setminus \overline{V}$, and put $F_0 = f^{-1}(y_0)$. Since $f(F) = \overline{V}$, it follows that $F \cap F_0 = \varnothing$. For every $n \in \omega$, put $M_n = \overline{W_n} \cap F_0$ and consider the family $\eta = \{M_n : n \in \omega\}$. Since $F_0$ is compact, $\eta$ is a decreasing sequence of compact subsets of $X$. Clearly, the intersection of $\eta$ is empty, since it is contained in $F_0 \cap F = \varnothing$. Therefore, some element $M_k$ of $\eta$ is empty. Then $W_k \cap F_0 = \varnothing$, which implies that $y_0$ is not $f(W_k) = O_k$. It follows that $y \in P \subset \overline{V}$. Hence, $Y$ is a Moscow space. □

Let us say that a subspace $Y$ of a space $X$ is *canonically embedded* in $X$ if, for every open subset $V$ of $Y$, there exists an open subset $U$ of $X$ such that the closure of $V$ in $Y$ is the set $\overline{U} \cap Y$.

It is clear that every retract $Y$ of a space $X$ is canonically embedded in $X$. Indeed, let $r\colon X \to X$ be a continuous retraction such that $Y = r(X)$. Clearly, $Y$ is closed in $X$. Given an open subset $V$ of $Y$, put $U = r^{-1}(V)$. Since $r$ is a retraction, we have that $U \cap Y = V$. Hence, $\overline{V} \subset \overline{U} \cap Y$. It also follows from the continuity of $r$ that $r(\overline{U}) \subset \overline{V}$, so that $\overline{U} \cap Y \subset \overline{V}$. This implies the equality $\overline{V} = \overline{U} \cap Y$, so $Y$ is canonically embedded in $X$.

The next statement is obvious:

PROPOSITION 6.3.2. *If a subspace $Y$ of a Moscow space $X$ is canonically embedded in $X$, then $Y$ is also a Moscow space.*

Here is another preservation result for Moscow spaces which follows from Proposition 6.3.2 and the observation made after the definition of canonical embeddings.

THEOREM 6.3.3. *Every retract of a Moscow space is a Moscow space.*

COROLLARY 6.3.4. *If a Moscow space $Z$ is the topological product of some family $\mathcal{F}$ of spaces, then every element of $\mathcal{F}$ is a Moscow space.*

PROOF. Clearly, we can assume that $\mathcal{F}$ consists of two spaces $X$ and $Y$, so $Z = X \times Y$. Now it is easy to see that the conclusion follows from Proposition 6.3.2, since every factor is embedded canonically into the product space. In fact, every factor is homeomorphic to a retract of $Z$. □

Unfortunately, it is not true in general that the product of two Moscow spaces is a Moscow space.

EXAMPLE 6.3.5. There exists a separable compact Moscow space $B$ such that $B \times B$ is not Moscow. Indeed, let $B$ be $\beta\omega$, the Čech–Stone compactification of the discrete space

$\omega$. Then $B$ is extremally disconnected. Therefore, $B$ is a Moscow space. Besides, $B$ is compact and separable.

Let us show that $B \times B$ is not a Moscow space. Fix any point $p \in \beta\omega \setminus \omega$, and put $X = \omega \cup \{p\}$ and $Y = B \setminus \{p\}$. The set $Y$ is $G_\delta$-dense in $B$, since $B$ is not first-countable at $p$. Therefore, $X \times Y$ is $G_\delta$-dense in $X \times B$. Put $Z = (X \times Y) \cup \{(p, p)\}$. Since $(p, p) \in X \times B$, it follows that $X \times Y$ is $G_\delta$-dense in the space $Z$.

Assume now that $B \times B$ is Moscow. Since $Z$ is dense in $B \times B$, it follows that $Z$ is also a Moscow space. Put $U = \{(n, n) : n \in \omega\}$. Then $U$ is a closed subset of $X \times Y$, since $\omega = X \cap Y$. Therefore, the set $V = (X \times Y) \setminus U$ is open in $X \times Y$. The set $U$ is also open in $X \times Y$, since each $n \in \omega$ is isolated in $X$ and in $Y$. Since $X \times Y$ is obviously open in $Z$, it follows that the sets $U$ and $V$ are open in $Z$. Clearly, $U$ and $V$ are disjoint, and $(p, p) \in \overline{U} \cap \overline{V}$. Since $U \cup V = X \times Y$, and $U$ and $V$ are disjoint open sets, it follows that $\{(p, p)\} = \overline{U} \cap \overline{V}$ (the closures are taken in the space $Z$). Therefore, $(p, p)$ is a $G_\delta$-point in $Z$. This contradicts the earlier established fact that $X \times Y$ is $G_\delta$-dense in the space $Z$. □

Though the class of Moscow spaces is not productive, as it is witnessed by Example 6.3.5, there are quite a few large classes of Moscow spaces each of which is closed under the product operation or *productive*, as we say below.

We will now establish a general theorem regarding products of Moscow spaces. Let us start with several technical facts that have a close connection with results of Section 1.6.

LEMMA 6.3.6. *Let $\{X_n : n \in \omega\}$ be a countable family of spaces such that the space $X_K = \prod_{n\in K} X_n$ is Moscow, for every finite subset $K$ of $\omega$. Then the product space $X = \prod_{n\in\omega} X_n$ is also Moscow.*

PROOF. Let $U$ be an open set in $X$, and $x$ a point in the closure of $U$. Take any finite subset $K$ of $\omega$, and denote by $p_K$ the projection of $X$ onto $X_K$. Since $X_K = \prod_{n\in K} X_n$ is Moscow, and the set $p_K(U)$ is open in $X_K$, there exists a $G_\delta$-subset $P_K$ of $X_K$ such that $p_K(x) \in P_K$ and $P_K$ is contained in the closure of the set $p_K(U)$ in the space $X_K$. Put $P = \bigcap\{p_K^{-1}(P_K) : K \in \mathcal{K}\}$, where $\mathcal{K}$ is the family of all finite subsets of $\omega$. Then, clearly, $x \in P$, and $P$ is a $G_\delta$-set in $X$.

Let us show that $P \subset \overline{U}$, which will complete the proof. Take any $y \in P$ and any standard open neighbourhood $O$ of $y$ in $X$. Then $O = p_F^{-1} p_F(O)$, for some finite $F \subset \omega$. We have $p_F(y) \in P_F$, since $y \in P$. Therefore, $p_F(y) \in \overline{p_F(U)}$. Since, obviously, $p_F(O)$ is an open neighbourhood of $p_F(y)$ in the space $X_F$, the set $p_F(O) \cap p_F(U)$ is not empty. Since $O = p_F^{-1} p_F(O)$, it follows that $O \cap U \neq \varnothing$. Hence, $y \in \overline{U}$ and $P \subset \overline{U}$. □

In the next lemma we show that the $o$-tightness of a product space is defined by the $o$-tightness of finite subproducts of the product (this cardinal function was defined in Section 5.5).

LEMMA 6.3.7. *Let $\{X_n : n \in \omega\}$ be a countable family of spaces such that the $o$-tightness of $X_K = \prod_{n\in K} X_n$ is countable, for every finite subset $K$ of $\omega$. Then the $o$-tightness of the product $X = \prod_{n\in\omega} X_n$ is countable.*

PROOF. Let $\gamma$ be a family of open sets in $X$ and $x$ a point in the closure of the set $U = \bigcup\gamma$. Take any finite subset $K$ of $\omega$. Since $ot(X_K) \leq \omega$, there exist a countable subfamily $\gamma_K$ of $\gamma$ such that $p_K(x)$ is in the closure of the set $\bigcup\{p_K(V) : V \in \gamma_K\}$ in the space $X_K$, where $p_K \colon X \to X_K$ is the projection.

Put $\eta = \bigcup\{\gamma_K : K \in \mathcal{K}\}$, where $\mathcal{K}$ is the family of all finite subsets of $\omega$. Then, clearly, $\eta$ is a countable subfamily of $\gamma$, and $x \in \overline{\bigcup \eta}$. □

PROPOSITION 6.3.8. *Let $\{X_a : a \in A\}$ be a family of topological spaces such that the o-tightness of $X_K = \prod_{a \in K} X_a$ is countable, for every finite subset $K$ of $A$, and let $X = \prod_{a \in A} X_a$. Then, for any family $\gamma$ of open sets in $X$ and for any point $x$ in the closure of the set $U = \bigcup \gamma$, there exist a countable subfamily $\eta$ of $\gamma$ and an $\omega$-cube $B$ such that $x \in B \subset \overline{\bigcup \eta}$. In particular, $ot(X) \leq \omega$.*

PROOF. Without loss in generality we may assume that $\gamma$ consists of open $\omega$-cubes in $X$ with finite core. For every $V \in \gamma$, let $A_V$ be the finite core of $V$ (see Section 1.6). Then clearly $V = p_{A_V}^{-1} p_{A_V}(V)$, where $p_K$ denotes the projection of $X$ onto $X_K$, for each non-empty $K \subset A$. We are going to define, by induction, an increasing sequence of countable subsets $A_n$ of $A$ and a sequence of countable subfamilies $\gamma_n$ of $\gamma$ in the following way.

We take $A_0$ to be any non-empty countable subset of $A$. Assume that a countable set $A_n \subset A$ is already defined, and put $K = A_n$. By Lemma 6.3.7, $ot(X_K) \leq \omega$. Therefore, there exists a countable subfamily $\gamma_n$ of $\gamma$ such that $p_K(x)$ is in the closure of the set $\bigcup\{p_K(V) : V \in \gamma_n\}$ in the space $X_K$. Now put $A_{n+1} = A_n \cup \bigcup\{A_V : V \in \gamma_n\}$. The inductive step is complete.

Put $M = \bigcup\{A_n : n \in \omega\}$ and $\eta = \bigcup\{\gamma_n : n \in \omega\}$. Clearly, $\eta$ is a countable subfamily of $\gamma$. Let $H$ be the closure of $\bigcup \eta$, and let $y$ be any point in $H$. Since $\eta$ consists of $\omega$-cubes, there exists an $\omega$-cube $B$ in $X$ such that $y \in B \subset H$. Indeed, let $B$ be the set of all $z \in X$ such that $z_a = y_a$, for each $a \in M$. Since $y \in H$, we clearly have $B \subset H$.

It remains to show that $x \in H$. Take any standard open neighbourhood $O$ of $x$ in $X$. Then $O = p_F^{-1} p_F(O)$, for some finite $F \subset A$. We claim that $O \cap \bigcup \eta \neq \varnothing$. Since $p_M^{-1} p_M(H) = H$, it suffices to consider the case when $F \subset M$. Since the sequence $\{A_n : n \in \omega\}$ is increasing, there exists $n \in \omega$ such that $F \subset A_n$. Then, by the choice of $\gamma_n$, $p_F(x)$ is in the closure of the set $\bigcup\{p_F(V) : V \in \gamma_n\}$ in the space $X_F$. Therefore, there exists a point $z \in \bigcup \eta$ such that $p_F(z) \in p_F(O)$. It follows that $z$ is in $O \cap \bigcup \eta$. Hence, $x \in H$. □

COROLLARY 6.3.9. [**M. G. Tkachenko**] *The product of an arbitrary family of first-countable spaces has countable o-tightness.*

Now we need an obvious lemma.

LEMMA 6.3.10. *Let $\{X_a : a \in A\}$ be a family of topological spaces such that the pseudocharacter of $X_a$ is countable, for each $a \in A$. Then every $\omega$-cube $B$ in the product space $X = \prod_{a \in A} X_a$ is the union of a family of $G_\delta$-sets in $X$.*

Finally, we prove the main assertion:

THEOREM 6.3.11. *Let $\{X_a : a \in A\}$ be a family of topological spaces such that the o-tightness of $X_K = \prod_{a \in K} X_a$ is countable, for every finite subset $K$ of $A$, and the pseudocharacter of $X_a$ is countable, for each $a \in A$. Then the product space $X = \prod_{a \in A} X_a$ is Moscow.*

PROOF. Take any open set $U$ in $X$, and let $x$ be any point in the closure of $U$. Put $\gamma = \{U\}$. From Proposition 6.3.8 it follows that there exists an $\omega$-cube $B$ in $X$ such that

$x \in B \subset \overline{U}$. By Lemma 6.3.10, there exists a $G_\delta$-set $P$ in $X$ such that $x \in P \subset B$. Now we have $x \in P \subset B \subset \overline{U}$. Hence, $X$ is Moscow. □

The last theorem can be generalized as follows:

THEOREM 6.3.12. *Let $\{X_a : a \in A\}$ be a family of topological spaces such that $X_K = \prod_{a\in K} X_a$ is a Moscow space of countable o-tightness, for every finite subset K of A. Then the product space $X = \prod_{a\in A} X_a$ is Moscow.*

PROOF. We have to make a few minor changes in the proof of Proposition 6.3.8. We use notation from that proof. By Lemma 6.3.6, $X_{A_n}$ is a Moscow space, where $A_n$ is a countable subset of $A$. Hence, there exists a $G_\delta$-set $F_n$ in the space $X_{A_n}$ such that $p_{A_n}(x) \in F_n$ and $F_n$ is contained in the closure of $\bigcup\{p_{A_n}(V) : V \in \gamma_n\}$. Now put $P_n = p_{A_n}^{-1}(F_n)$, for each $n \in \omega$, and $P = \bigcap_{n\in\omega} P_n$. Clearly, $x \in P$ and $P$ is a $G_\delta$-set in $X$. Since the sets $\bigcup \eta$ and $P$ do not depend on coordinates in $A \setminus M$, and the sequence $\{A_n : n \in \omega\}$ is increasing, it follows that $P$ is contained in the closure of $U$. Hence, $X$ is Moscow. □

The condition in Lemma 6.3.6 and in Theorem 6.3.12 that $X_K$ is Moscow for each finite set $K$ of coordinates cannot be replaced by the weaker assumption that each factor is Moscow. Indeed, according to Example 6.3.5, $\beta\omega$ is a separable extremally disconnected compact space such that $\beta\omega \times \beta\omega$ is not Moscow. However, the space $\beta\omega \times \beta\omega$ is separable and, therefore, the $o$-tightness of it is countable.

COROLLARY 6.3.13. *Let $\{X_a : a \in A\}$ be a family of compact spaces of countable tightness such that $X_K$ is a Moscow space, for every finite subset K of A. Then the product space $X = \prod_{a\in a} X_a$ is Moscow.*

PROOF. Since the product of a finite family of compact spaces of countable tightness has countable tightness [**165**, 3.12.8 (e)], it suffices to apply Theorem 6.3.12. □

Here are three more corollaries from Theorem 6.3.11.

COROLLARY 6.3.14. *Let X be the product of some family of cosmic spaces. Then X is a Moscow space.*

COROLLARY 6.3.15. *The product of any family of first-countable spaces is a Moscow space.*

COROLLARY 6.3.16. *Every dense subspace of the product of any family of metrizable spaces is a Moscow space.*

## Exercises

6.3.a. Show that every compact space is a continuous image of a compact Moscow space.

6.3.b. Prove that every Tychonoff space is an image of a Moscow space under a continuous open mapping.

6.3.c. Give an example of a family $\{X_i : i \in I\}$ of Tychonoff spaces, each of countable pseudocharacter, such that the product $\prod_{i\in I} X_i$ fails to be a Moscow space.

## Problems

6.3.A. Let $X$ and $Y$ be compact spaces, where $X$ is first-countable and $Y$ is extremally disconnected. Is the product space $X \times Y$ Moscow?

6.3.B. Let $X = \prod_{i \in I} X_i$ be the product of a family of compact spaces such that, for every finite set $J \subset I$, the subproduct $X_J = \prod_{i \in J} X_i$ is a Moscow space. Is $X$ a Moscow space?

## Open Problems

6.3.1. Suppose that $X$ and $Y$ are compact homogeneous Moscow spaces. Is the product $X \times Y$ a Moscow space?

6.3.2. Let $X$ be a compact space satisfying $cel_\omega(X) \leq \omega$. Is $X$ a Moscow space?

## 6.4. The breadth of the class of Moscow groups

In this section we demonstrate that the class of Moscow groups contains many other important classes of groups. In particular, we show that a topological group is much more often a Moscow space than a topological space in general. To be more precise, there is a long list of cardinal invariants such that the countability restriction on any one of them guarantees that a topological group is a Moscow space. This phenomenon does not take place in the class of general topological spaces.

Recall that an open subset $U$ of a space $X$ is said to be *regular open* in $X$ if $U$ is the interior of its closure.

A point $x \in X$ is said to be a *point of canonical weak pseudocompactness* or a *cwp-point* of $X$, for brevity, if the following condition is satisfied:

(cwp) For every regular open subset $U$ of $X$ such that $x \in \overline{U}$, there exists a sequence $\{A_n : n \in \omega\}$ of subsets of $U$ such that $x \in \overline{A_n}$, for each $n \in \omega$, and for every indexed family $\eta = \{O_n : n \in \omega\}$ of open subsets of $X$ satisfying $O_n \cap A_n \neq \varnothing$ for all $n \in \omega$, the family $\eta$ has an accumulation point in $X$.

We say that a space $X$ is *canonically weakly Fréchet–Urysohn at a point* $x \in X$, or *$\kappa$-Fréchet–Urysohn at* $x$ if whenever $x \in \overline{U}$, where $U$ is a regular open subset of $G$, some sequence of points of $U$ converges to $x$. If $X$ is $\kappa$-Fréchet–Urysohn at every point of $X$, we say that $X$ is *$\kappa$-Fréchet–Urysohn*. Obviously, if a space $X$ is $\kappa$-Fréchet–Urysohn at a point $x \in X$, then $x$ is also a point of canonical weak pseudocompactness of $X$.

A space $X$ is *pointwise canonically weakly pseudocompact* if each point of $X$ is a point of canonical weak pseudocompactness. All $\kappa$-Fréchet–Urysohn spaces are pointwise canonically weakly pseudocompact.

THEOREM 6.4.1. *If a topological group $G$ is pointwise canonically weakly pseudocompact, then $G$ is a Moscow space.*

PROOF. Let $U$ be a regular open subset of $G$. Clearly, it suffices to show that if $e \in \overline{U}$, then there exists a $G_\delta$-set $P$ in $G$ such that $e \in P \subset \overline{U}$. So let us assume that $e \in \overline{U}$ and fix subsets $A_n$ of $U$ such as in condition (cpw), where $x = e$.

We are going to define a sequence $\{V_n : n \in \omega\}$ of open neighbourhoods of $e$, and a sequence $\{a_n : n \in \omega\}$ of points in $U$ such that $a_n \in A_n$, for each $n \in \omega$. First, choose $a_0 \in A_0$, and let $V_0$ be an open neighbourhood of $e$ such that $a_0 V_0 \subset U$. Assume now that an open neighbourhood $V_k$ of $e$ is already defined, for some $k \in \omega$. Then we let $a_{k+1}$ to be any point of $A_{k+1} \cap V_k$. Let $V_{k+1}$ be a symmetric open neighbourhood of $e$ such that $V_{k+1}^2 \subset V_k$ and $a_{k+1}V_{k+1} \subset U$. The recursive definition is complete.

By condition (cpw), the indexed family $\eta = \{a_n V_{n+1} : n \in \omega\}$ has a point of accumulation in $G$. We denote by $H$ the set of all accumulation points of $\eta$ in $G$. Thus, $H$ is not empty. Put $P = \bigcap_{n\in\omega} V_n$. From the construction it is clear that $P$ is a closed subgroup of $G$.

**Claim 1.** $H \subset P$.

Indeed, take any $a \in H$, and fix $m \in \omega$. Put $k = m + 2$. There exists $n > k$ such that $aV_k \cap a_n V_{n+1} \neq \varnothing$. Then $ac = a_n b$, for some $b \in V_{n+1}$ and $c \in V_k$. Hence, $a = a_n bc^{-1} \in V_{n-1}V_{n+1}V_k \subset V_{m+2}^3 \subset V_{m+1}^2 \subset V_m$. Therefore, $a \in V_m$, for each $m \in \omega$. It follows that $a \in P$, whence Claim 1 follows.

**Claim 2.** $aP = P$, *for any* $a \in H$.

Indeed, this is so, since $H \subset P$ and $P$ is a subgroup of $G$.

Fix $a \in H$. Then, by Claim 2, $P = aP \subset \overline{\bigcup_{n\in\omega} a_n V_{n+1} P} \subset \overline{\bigcup_{n\in\omega} a_n V_n}$. Since $P$ is a subgroup of $G$, we have that $e \in P$. Therefore,

$$e \in P \subset \overline{\bigcup_{n\in\omega} a_n V_n} \subset \overline{U}.$$

Since $P$ is a $G_\delta$-set, the proof is complete. □

THEOREM 6.4.2. *Every dense subspace of a pointwise canonically weakly pseudocompact topological group is a Moscow space. In particular, if a topological group G satisfies at least one of the following conditions, then it is a Moscow space:*

1) *G is locally pseudocompact;*
2) *G is precompact;*
3) *G is a dense subgroup of a κ-Fréchet–Urysohn group.*

PROOF. Evidently, a locally pseudocompact group is canonically weakly pseudocompact. Also, every totally bounded group is topologically isomorphic to a dense subgroup of a compact topological group (see Corollary 3.7.16). Therefore, in each of the cases 1), 2), or 3), the required conclusion follows from Theorem 6.4.1 and Proposition 6.1.2. □

The following cardinal invariant of the tightness type serves to push farther away the limits of the class of Moscow groups.

Let $G$ be a left topological group, and $U \subset G$. A subset $A$ of $G$ is called *ω-deep* in $U$ if there exists a $G_\delta$-set $P$ in $G$ such that $e \in P$ and $AP \subset U$. We say that the *g-tightness* $t_g(G)$ of $G$ is countable (and write $t_g(G) \leq \omega$) if, for each regular open subset $U$ of $G$ and each $x \in \overline{U}$, there exists an $\omega$-deep subset $A$ of $U$ such that $x \in \overline{A}$.

Before we state the main result on semitopological groups of countable $g$-tightness, Theorem 6.4.9, we present several almost obvious statements showing how large is the class of these objects.

PROPOSITION 6.4.3. *The union of any countable family of ω-deep subsets of U is an ω-deep subset of U, for any set U in a left topological group G.*

PROPOSITION 6.4.4. *If $G$ is a left topological group of countable tightness, then the g-tightness of $G$ is also countable.*

PROOF. Clearly, if $U$ is an open subset of $G$, then every one-point subset of $U$ is $\omega$-deep in $U$. It remains to apply Proposition 6.4.3. □

PROPOSITION 6.4.5. *If $G$ is a paratopological group of countable o-tightness, then the g-tightness of $G$ is also countable.*

PROOF. Let $U$ be a regular open subset of $G$, and suppose that $x \in \overline{U}$. Denote by $\gamma$ the family of all non-empty $\omega$-deep open subsets of $U$. Since $G$ is a paratopological group, we have the equality $U = \bigcup \gamma$. Since the $o$-tightness of $G$ is countable, there exists a countable subfamily $\xi$ of $\gamma$ such that $x \in \overline{\bigcup \xi}$. The family $\xi$ being countable, the set $O = \bigcup \xi$ is an $\omega$-deep subset of $U$, by Proposition 6.4.3. Therefore, $t_g(G) \leq \omega$. □

Since the $o$-tightness of a space $X$ is less than or equal to the cellularity of $X$, the next corollary to Proposition 6.4.5 is immediate.

COROLLARY 6.4.6. *Let $G$ be a paratopological group of countable cellularity. Then the g-tightness of $G$ is countable.*

The next two statements are obvious.

PROPOSITION 6.4.7. *Let $G$ be an extremally disconnected semitopological group. Then the g-tightness of $G$ is countable.*

PROPOSITION 6.4.8. *Let $G$ be a left topological group of countable pseudocharacter. Then the g-tightness of $G$ is countable.*

Here is an important fact (with a simple proof) which has a number of applications.

THEOREM 6.4.9. *Every semitopological group $G$ of countable g-tightness is a Moscow space.*

PROOF. Take any regular open subset $U$ of $G$, and any point $x \in \overline{U}$. Since $t_g(G) \leq \omega$, there exists an $\omega$-deep subset $A$ of $U$ such that $x \in \overline{A}$. Now we can fix a $G_\delta$-subset $P$ of $G$ such that $e \in P$ and $AP \subset U$. Then, since the right and left translations in $G$ are homeomorphisms, we have that $x \in xP \subset \overline{AP} \subset \overline{U}$, and $xP$ is a $G_\delta$-set in $G$. Thus, $G$ is a Moscow space. □

Theorems 6.4.1 and 6.4.9 together cover very large classes of topological groups. It is really amazing how many topological conditions, which are innocently weak in the general case of arbitrary topological spaces, turn out to be sufficient for a topological or paratopological group to be a Moscow space. Here is one more result of this nature.

THEOREM 6.4.10. *If $G$ is a dense subgroup of the product of a family of first-countable paratopological groups, then the g-tightness of $G$ is countable, and $G$ is a Moscow space.*

PROOF. Since the $o$-tightness does not increase when passing to dense subspaces, the $o$-tightness of $G$ is countable, by Corollary 6.3.9. The conclusions now follow from Proposition 6.4.5 and Theorem 6.4.9. □

Let us say that the *$\kappa$-tightness* of a space $X$ does not exceed the cardinal number $\tau$ (notation: $\kappa t(X) \leq \tau$) if, for every regular open set $U$ in $X$ and each point $x$ in the closure of $U$, there exists a subset $B$ of $U$ such that $x \in \overline{B}$ and $|B| \leq \tau$. Clearly, for every $\kappa$-Fréchet–Urysohn space $X$, the $\kappa$-tightness of $X$ is countable. It is also clear that $\kappa t(X) \leq t(X)$, for every space $X$, and $t_g(G) \leq \omega$ whenever $G$ is a left topological group of countable $\kappa$-tightness.

We sum up the results of this section in the following corollary.

COROLLARY 6.4.11. *Every dense subspace of a paratopological group $G$ of countable $g$-tightness is a Moscow space. In particular, if a paratopological group $G$ satisfies at least one of the following conditions, then it is Moscow:*

1) *the pseudocharacter of $G$ is countable, that is, each point in $G$ is a $G_\delta$-set;*
2) *$G$ is extremally disconnected;*
3) *the tightness of $G$ is countable;*
4) *the $o$-tightness of $G$ is countable;*
5) *the cellularity of $G$ is countable;*
6) *$G$ is a dense subgroup of a paratopological group $F$ of countable $\kappa$-tightness;*
7) *$G$ is a subgroup of a topological group $F$ such that $F$ is a $k$-space.*

PROOF. The main assertion of the corollary follows from Theorem 6.4.9 and Proposition 5.5.1. Items 1) and 2) are evident. Since $ot(G) \leq t(G)$ and $ot(G) \leq c(G)$, items 3), 4), and 5) follow from Proposition 6.4.5 and Theorem 6.4.9. Since every paratopological group $H$ of countable $\kappa$-tightness satisfies $t_g(H) \leq \omega$, item 6) follows from Theorem 6.4.9 (and Proposition 5.5.1). Finally, every $k$-group has countable $o$-tightness by Theorem 5.5.4; hence, the fact that a closed subgroup of a $k$-group is again a $k$-group and item 4) (along with Proposition 5.5.1) imply item 7). □

The $G_\delta$-closure of a topological group $G$ in its Raĭkov completion $\varrho G$ (see Section 3.5) will be denoted by $\varrho_\omega G$. The $G_\delta$-closure of a subgroup in a topological group $H$ is, obviously, a subgroup of $H$. Therefore, $\varrho_\omega G$ is a subgroup of $\varrho G$ containing $G$.

Another device allowing to build new examples of Moscow groups from the Moscow groups already in existence is provided by the following statement, which follows from Proposition 6.1.2, Corollary 6.4.11, and Theorem 6.1.8.

THEOREM 6.4.12. *Let $G$ be a topological group satisfying at least one of the following conditions:*

1) *the tightness of $G$ is countable;*
2) *$G$ is a dense subgroup of a topological group $H$ such that the $\kappa$-tightness of $H$ is countable;*
3) *$G$ is a $k$-space.*

*Then $\varrho_\omega G$, as well as any subgroup between $G$ and $\varrho_\omega G$, is a Moscow group.*

In general, the product of two topological groups of countable $g$-tightness need not be a group of countable $g$-tightness (see Problem 6.4.A). However, there are very large productive classes of Moscow groups, as we will presently see.

Note that every precompact topological group is a dense subgroup of a compact group. Therefore, we have the next corollary to Theorem 6.4.2:

COROLLARY 6.4.13. *The product of any family of dense subspaces of precompact topological groups is a Moscow space.*

Whether a similar assertion holds for topological groups with countable cellularity is much less clear. Here we have only a consistency result.

COROLLARY 6.4.14. *Assume Martin's Axiom and the negation of the Continuum Hypothesis, $MA + \neg CH$. Then the product of any family of paratopological groups of countable cellularity is a Moscow space.*

PROOF. Under $MA + \neg CH$, the product of any family of topological spaces with countable Souslin number is a space with countable Souslin number (see [**285**, Theorem 2.24]). It remains to refer to the fact that every paratopological group of countable cellularity is a Moscow space, by item 5) of Corollary 6.4.11. □

We can extend further the reach of the above theorems using the following approach. A space $X$ is called a *groupy space* if it can be represented as a dense subspace of a paratopological group. First, we need a lemma.

LEMMA 6.4.15. *If $X$ is a dense subspace of a homogeneous space $Z$ and the o-tightness of $X$ is countable, then the o-tightness of $Z$ is also countable.*

PROOF. Let $\gamma$ be a family of open sets in $Z$ and $z \in Z$, $z \in \overline{\bigcup \gamma}$. Since $Z$ is homogeneous, we may assume that $z \in X$. Since $X$ is dense in $Z$, we have that $z \in \overline{\bigcup \eta}$, where $\eta = \{U \cap X : U \in \gamma\}$. From $ot(X) \le \omega$ it follows that there exists a countable subfamily $\xi$ of $\eta$ such that $z \in \overline{\bigcup \xi}$. Take a countable subfamily $\lambda$ of $\gamma$ such that $\xi = \{U \cap Y : U \in \lambda\}$. Then clearly $z \in \overline{\bigcup \lambda}$ and, hence, $ot(Z) \le \omega$. □

THEOREM 6.4.16. *If the o-tightness of a groupy space $X$ is countable, then $X$ is a Moscow space.*

PROOF. Let $G$ be a paratopological group such that $X$ is a dense subspace of $G$. Then the space $G$ is homogeneous and $ot(G) \le \omega$, by Lemma 6.4.15. According to item 4) of Corollary 6.4.11, $G$ is a Moscow space. Therefore $X$, as a dense subspace of $G$, is also a Moscow space (see Proposition 6.1.2). □

Since every space with the Souslin property has countable $o$-tightness, the next corollary to Theorem 6.4.16 is immediate.

COROLLARY 6.4.17. *Every groupy space of countable cellularity is Moscow.*

Clearly, if a dense subspace of a given space has countable cellularity, then the space itself has countable cellularity. Therefore, making use of Corollaries 6.4.14 and 6.4.17, we come to the following conclusion:

COROLLARY 6.4.18. *Assume that $MA + \neg CH$ holds. Then the product of any family of groupy spaces of countable cellularity is a Moscow space.*

It is not clear whether it is possible to drop the assumption $MA + \neg CH$ in Corollary 6.4.18. However, we have the next result in $ZFC$:

THEOREM 6.4.19. *The product of any family of separable groupy spaces is a Moscow space.*

PROOF. The product of any family of separable spaces is a space of countable cellularity [**165**, Coro. 2.3.18]. It remains to observe that, by 5) of Corollary 6.4.11, a paratopological group of countable cellularity is a Moscow space, and every dense subspace of a Moscow space is a Moscow space according to Proposition 6.1.2. □

In Theorem 6.4.20 below the separability of the factors is replaced by a considerably weaker condition. However, we strengthen another requirement on the factors — they will be subspaces of topological groups.

THEOREM 6.4.20. *Let $\{X_i : i \in I\}$ be a family of k-separable spaces, where each $X_i$ is a dense subspace of a topological group. Then the product space $X = \prod_{i \in I} X_i$ is Moscow.*

PROOF. By the assumptions of the theorem, every space $X_i$ contains a dense $\sigma$-compact subspace $Y_i$. Let $Y = \prod_{i \in I} Y_i$ be the product space and $\sigma Y \subset Y$ the corresponding $\sigma$-product of the spaces $Y_i$, with $i \in I$. Then $\sigma Y$ is dense in $Y$ and $Y$ is dense in $X$. It follows from Proposition 1.6.41 that $\sigma Y$ is $\sigma$-compact, so $X$ is $k$-separable.

Every $\sigma$-compact subspace of a topological group is contained in a $\sigma$-compact subgroup of the group, by Lemma 1.9.4. According to Corollary 5.3.22, the cellularity of every $\sigma$-compact topological group is countable, so the cellularity of every $k$-separable topological group is countable as well. Since, by Corollary 5.3.22, every topological group of countable cellularity is a Moscow space, and every dense subspace of a Moscow space is a Moscow space, the conclusion follows. □

A large productive class of Moscow spaces can be described as follows. Recall that $\aleph_1$ is said to be a *precalibre* of a space $X$ if every uncountable family of open subsets of $X$ contains an uncountable subfamily with the finite intersection property. If $\aleph_1$ is a precalibre of $X$, then the cellularity of $X$ is, obviously, countable. The converse is true if we assume $MA + \neg CH$ [**285**, Theorem 2.24]. Theorem 5.4.1 implies that if $\aleph_1$ is a precalibre of $X_\alpha$, for each $\alpha \in A$, then $\aleph_1$ is a precalibre of the product of the spaces $X_\alpha$. If $X$ is a dense subspace of the product $\Pi$ of a family of second-countable spaces, then $\aleph_1$ is a precalibre of $X$. Clearly, this follows from Theorem 5.4.1, since the dense subspace $X$ of $\Pi$ shares the same precalibres with $\Pi$.

From the above fact and 5) of Corollary 6.4.11, we obtain immediately the following result:

THEOREM 6.4.21. *If $\{X_\alpha : \alpha \in A\}$ is a family of groupy spaces, and $\aleph_1$ is a precalibre of each $X_\alpha$, then the product space $\prod_{\alpha \in A} X_\alpha$ is Moscow.*

In conclusion, we formulate one more general theorem which easily follows from the results already obtained.

THEOREM 6.4.22. *Suppose that $\{G_a : a \in A\}$ is a family of paratopological group such that the o-tightness of the space $G_K = \prod_{a \in K} G_a$ is countable, for every finite subset $K$ of $A$, and let $G = \prod_{a \in A} G_a$ be their topological product. Then the space $G$ is Moscow.*

PROOF. By Proposition 6.3.8, the $o$-tightness of the space $G$ is countable. Since $G$ is a paratopological group, it follows from item 4) of Corollary 6.4.11 that the space $G$ is Moscow. □

## Exercises

6.4.a. Give an example of a compact Fréchet–Urysohn space which is not a Moscow space (compare with Theorem 6.4.1).

6.4.b. Show that "precompact" in item 2) of Theorem 6.4.2 can be weakened to "locally precompact" (see Problem 3.7.I).

6.4.c. Show that there exists a compact right topological semigroup without isolated points that has uncountable $o$-tightness.

6.4.d. Verify that if a paratopological group $G$ contains a dense subgroup of countable tightness, then $G$ is a Moscow space.

## Problems

6.4.A. Give an example of two topological groups $G$ and $H$, both of countable $g$-tightness, such that the product group $G \times H$ has uncountable $g$-tightness.

6.4.B. Show that a pseudocompact quasitopological group of countable cellularity need not be a Moscow space.

6.4.C. Show that pseudocompact quasitopological groups can have uncountable $o$-tightness.

6.4.D. Is it true that every pointwise canonically weakly pseudocompact paratopological group is a Moscow space (see Theorem 6.4.1)?

6.4.E. Let $G$ be a left topological group.
(a) If $G$ has countable tightness (or $o$-tightness), is then the $g$-tightness of $G$ countable?
(b) Under the same assumptions as in (a), is $G$ a Moscow space?

6.4.F. Suppose that a topological (paratopological) group $G$ contains a dense subgroup of countable $g$-tightness. Is $G$ a Moscow space?

6.4.G. Does there exist a topological group $G$ that contains a dense Moscow subgroup, but still fails to be Moscow itself? (See Problem 6.8.B.)

6.4.H. Prove that if a countably compact non-compact topological group $G$ is a hereditarily normal space, then $G$ does not contain non-trivial convergent sequences.

## Open Problems

6.4.1. Is it true that every Moscow topological group has countable $g$-tightness?

6.4.2. Let $G$ be a topological group of countable tightness. Is the $o$-tightness of $G \times G$ countable? Is the $g$-tightness of $G \times G$ countable?

6.4.3. Let $G$ be a topological group of countable tightness. Is $G \times G$ a Moscow group? (We still do not have an example in $ZFC$ of a topological group $G$ of countable tightness such that the tightness of $G \times G$ is uncountable.)

6.4.4. Let $G$ be an extremally disconnected topological group. Is $G \times G$ a Moscow group?

6.4.5. Let $G$ be an extremally disconnected topological group and $B$ a compact group. Is $G \times B$ a Moscow group?

6.4.6. Can we drop the assumption $MA + \neg CH$ in Corollary 6.4.18?

6.4.7. Let $X_i$ be a $k$-separable groupy space, for each $i \in I$. Is the product space $X = \prod_{i \in I} X_i$ Moscow? (Compare with Theorem 6.4.20.)

6.4.8. Suppose that a paratopological group $G$ contains a non-empty compact set of countable character in $G$. Does there exist a perfect mapping of $G$ onto a metrizable space? Is the space $G$ Moscow?

6.4.9. Is it true that every quasitopological group $G$ of countable $o$-tightness has countable $g$-tightness? What if the cellularity of $G$ is countable?

## 6.5. When the Dieudonné completion of a topological group is a group?

The techniques based on the notion of Moscow space play a vital role in the solution of the next question:

Let $G$ be a topological group, and $\mu G$ the Dieudonné completion of the space $G$. Can the operations on $G$ be extended to $\mu G$ in such a way that $\mu G$ becomes a topological group containing $G$ as a topological subgroup?

Recall that the Dieudonné completion $\mu X$ of a space $X$ is the completion of $X$ with respect to the maximal uniformity $\mathcal{U}_X$ on $X$ compatible with the topology of $X$ [**165**, 8.5.13]. In particular, the space $X$ is Dieudonné complete iff the uniformity $\mathcal{U}_X$ is complete. It is well known that the Dieudonné completion of a topological space $X$ is always contained in the Hewitt–Nachbin completion $\upsilon X$ of $X$, so that $\mu X \subset \upsilon X \subset \beta X$ (this also follows from [**165**, 8.5.13]). In fact, $\mu X$ is the smallest Dieudonné complete subspace of $\beta X$ containing $X$. Moreover, if there are no Ulam measurable cardinals, then $\upsilon X$ and $\mu X$ coincide, by [**165**, 8.5.13 (h)]. Therefore, the next question is almost equivalent to the one formulated above: For a topological group $G$, can the operations on $G$ be extended to $\upsilon G$ in such a way that $\upsilon G$ becomes a topological group containing $G$ as a topological subgroup?

Clearly, if there exists an Ulam measurable cardinal $\tau$, then, for any discrete group $G$ of cardinality $\tau$, the answer to the last question is in the negative (since in this case the Hewitt–Nachbin completion $\upsilon G$ is a non-discrete non-homogeneous space).

In this section we apply the results of Section 6.4 to topological groups and their Raĭkov completions, and describe large classes of topological groups $G$ for which $\mu G$ is a topological group. One of the most important facts going in this direction is the following:

THEOREM 6.5.1. [**A. V. Arhangel'skii**] *Let $G$ be a Moscow topological group. Then the operations on $G$ can be extended to the Dieudonné completion $\mu G$ of $G$ in such a way that $\mu G$ becomes a topological group containing $G$ as a topological subgroup.*

To prove Theorem 6.5.1 we need several preliminary results. First we recall one concept introduced in Section 4.6. A subset $Y$ of a space $X$ is *$G_\delta$-closed in $X$* if, for every point $x \in X \setminus Y$, there is a $G_\delta$-set $P_x$ in $X$ such that $x \in P_x$ and $P_x \cap Y = \varnothing$.

PROPOSITION 6.5.2. *Let $G$ be a topological group. Then $\varrho_\omega G$ is a Dieudonné complete topological group. If, in addition, $G$ is Moscow, then $\varrho_\omega G$ is also Moscow, and $G$ is $C$-embedded in $\varrho_\omega G$.*

PROOF. It follows from Theorem 3.6.25 that the uniform space $(\varrho G, \mathcal{V})$ is complete, where $\mathcal{V}$ is the two-sided uniformity of the group $\varrho G$. Therefore, the space $\varrho G$ is Dieudonné complete, as every space that admits a complete uniformity [**165**, 8.5.13]. Every $G_\delta$-closed subspace of a Dieudonné complete space is also Dieudonné complete [**165**, 8.5.13 (f)]. Of course, $\varrho_\omega G$ is $G_\delta$-closed in $\varrho G$. Hence, $\varrho_\omega G$ is Dieudonné complete. Since $\varrho_\omega G$ is homogeneous, Theorem 6.1.8 implies that $\varrho_\omega G$ is Moscow and $G$ is $C$-embedded in $\varrho_\omega G$. □

A topological property $\mathcal{P}$ will be called *invariant under intersections* if, for every family $\gamma$ of subspaces of a topological space $X$ such that every $Y \in \gamma$ has $\mathcal{P}$, the subspace $Z = \bigcap \gamma$ of $X$ also has the property $\mathcal{P}$.

PROPOSITION 6.5.3. *Hewitt–Nachbin completeness and Dieudonné completeness are topological properties invariant under intersections.*

PROOF. Indeed, both properties are multiplicative, by Theorem 3.11.5 and Problem 8.5.13 (a) of [**165**], respectively. They are also closed hereditary, according to Theorem 3.11.4 and Problem 8.5.13 (a) of [**165**]. Suppose that $\gamma = \{Y_i : i \in I\}$ is a family of subspaces of a Tychonoff space $X$ such that each $Y_i$ is Dieudonné complete, and let $Y = \bigcap \gamma$. Consider the product $P = \prod_{i \in I} Y_i$ of the family $\gamma$ and the diagonal $\Delta$ of $P$ which consists of the points all whose coordinates coincide. It is easy to see that $\Delta$ is closed in $P$ and $Y$ is naturally homeomorphic to $\Delta$. Therefore, the spaces $P$, $\Delta$, and $Y$ are Dieudonné complete. The argument for Hewitt–Nachbin completeness is the same. □

Recall that a *quasitopological group* is a group with a topology on it such that the multiplication is separately continuous and the inverse operation is a homeomorphism (see Section 1.2).

PROPOSITION 6.5.4. *Let $H$ be a subgroup of a quasitopological group $G$, and $\mathcal{P}$ be a topological property invariant under intersections. Let $\gamma_{\mathcal{P}}$ be the family of all subspaces $X$ of $G$ such that $X$ has the property $\mathcal{P}$ and $H \subset X$. Then either $\gamma_{\mathcal{P}}$ is empty, or there exists the smallest (by inclusion) element $M$ in $\gamma_{\mathcal{P}}$, and $M$ is a subgroup of $G$ containing $H$.*

PROOF. Assume that $\gamma_{\mathcal{P}}$ is not empty, and let $M$ be the intersection of the family $\gamma_{\mathcal{P}}$. Clearly, $H \subset M$. Since $\mathcal{P}$ is invariant under intersections, $M$ also has the property $\mathcal{P}$. Therefore, $M \in \gamma_{\mathcal{P}}$, and $M$ is the smallest element of $\gamma_{\mathcal{P}}$.

It remains to show that $M$ is a subgroup of $G$. Note that $M^{-1}$ is homeomorphic to $M$; therefore, $M^{-1}$ also has the property $\mathcal{P}$. Since $H = H^{-1} \subset M^{-1} \subset G$, it follows that $M^{-1} \in \gamma_{\mathcal{P}}$ and, therefore, $M \subset M^{-1}$. Hence, $M^{-1} \subset (M^{-1})^{-1} = M$ and, finally, $M = M^{-1}$.

For every $a \in H$, we have that $H = aH \subset aM \subset aG = G$, which implies that $M \subset aM$, since $aM$ is homeomorphic to $M$ and, therefore, has the property $\mathcal{P}$. It follows that $a^{-1}M \subset M$. Since $H = H^{-1}$, this implies that $aM \subset M$, for each $a \in H$. Therefore, $HM \subset M$. It follows that $Hb \subset M$ or, equivalently, $H \subset Mb^{-1}$, for any $b \in M$. Since $M^{-1} = M$, we conclude that $H \subset Mb$, for any $b \in M$. Since $Mb$ is homeomorphic to $M$, it follows that $Mb$ is in $\gamma_{\mathcal{P}}$ and $M \subset Mb$. Hence, $Mb^{-1} \subset M$. Since $M = M^{-1}$, it follows that $Mb \subset M$, for each $b \in M$. Now it is clear that $M$ is closed under multiplication. Hence, $M$ is a subgroup of $G$. □

A space $X$ is called a *minimal Dieudonné extension* of $Y$ if $X$ is Dieudonné complete, $Y$ is dense in $X$, and every Dieudonné complete subspace of $X$ containing $Y$ coincides with $X$. It is now clear how to define a *minimal Hewitt–Nachbin extension* of $Y$.

The next statement is obvious.

PROPOSITION 6.5.5. *The Dieudonné completion $\mu X$ of a space $X$ is a minimal Dieudonné extension of $X$ in which $X$ is $C$-embedded.*

The converse of Proposition 6.5.5 easily follows from the definitions, but we supply the reader with its proof because of the importance of the fact.

PROPOSITION 6.5.6. *If a space $X$ is a minimal Dieudonné extension of a subspace $Y$, and $Y$ is $C$-embedded in $X$, then $X = \mu Y$, that is, $X$ is the Dieudonné completion of $Y$.*

PROOF. Let $\beta X$ be the Čech–Stone compactification of the space $X$. Since $Y$ is a dense $C$-embedded subspace of $X$, it follows that $Y$ is $C^*$-embedded in $\beta X$ and, therefore, we can identify topologically $\beta Y$ with $\beta X$, by [**165**, Coro. 3.6.3]. In other words, we have the inclusions $Y \subset X \subset \beta Y$. Since $\mu Y$ is a subspace of $\beta Y$ containing $Y$ and, by our assumption, $X$ is Dieudonné complete, it follows that $Z = X \cap \mu Y$ is a Dieudonné complete subspace of $X$ containing $Y$. We conclude, by the minimality of $X$, that $Z = X$. Therefore, $X \subset \mu Y$. According to Proposition 6.5.5, $\mu Y$ is a minimal Dieudonné complete subspace of $\beta Y$ containing $Y$, whence it follows that $X = \mu Y$. □

Propositions 6.5.3 and 6.5.4 imply the next result:

PROPOSITION 6.5.7. *If $H$ is a subgroup of a quasitopological group $G$, and there exists a Dieudonné complete subspace $X$ of $G$ such that $H \subset X$, then there exists a subgroup $M$ of $G$ such that the space $M$ is a minimal Dieudonné extension of $H$ and $M \subset X$. Similarly, if there exists a realcompact space $Y$ with $H \subset Y \subset G$, then there also exists a subgroup $N$ of $G$ such that the space $N$ is a minimal Hewitt–Nachbin extension of $H$ and $N \subset Y$.*

Here is a crucial fact on Dieudonné completeness in quasitopological groups.

THEOREM 6.5.8. *Let $H$ be a subgroup of a quasitopological group $G$, and $X$ a subspace of $G$ containing $H$ such that $X$ is the Dieudonné completion of the space $H$. Then $X$ is a subgroup of $G$.*

PROOF. By Proposition 6.5.7, there is a subgroup $M$ of $G$ contained in $X$ such that $H \subset M$ and the space $M$ is Dieudonné complete. However, $X$ is a minimal Dieudonné extension of the space $H$. Therefore, $M = X$, and $X$ is a subgroup of $G$. □

We now have all necessary tools for the proof of the result announced at the beginning of the section.

*Proof of Theorem 6.5.1.* Let $G$ be a topological group such that $G$ is a Moscow space. The Raĭkov completion $\varrho G$ of $G$ and the subgroup $\varrho_\omega G$ of $\varrho G$ are Dieudonné complete spaces, by Proposition 6.5.2. Hence, we can apply Proposition 6.5.7 to find a subgroup $M$ of $\varrho_\omega G$ such that $G \subset M$ and $M$ is a minimal Dieudonné extension of $G$. Obviously, $G$ is $G_\delta$-dense in $\varrho_\omega G$ and in $M$, so Theorem 6.1.8 implies that $G$ is $C$-embedded in $M$. Since $M$ is the minimal Dieudonné extension of $G$, it follows from Proposition 6.5.6 that $M = \mu G$. □

A topological group $G$ is called a *PT-group* if the operations on $G$ can be extended to the Dieudonné completion $\mu G$ in such a way that $G$ becomes a topological subgroup of the topological group $\mu G$. Combining Corollary 3.6.16 and Theorem 6.5.8, we obtain the following conclusion:

COROLLARY 6.5.9. *Let $G$ be a topological group. Then $\mu G$ is a topological group (containing $G$ as a subgroup) iff there exists a subspace $X$ of the Raĭkov completion $\varrho G$ of $G$ such that $G \subset X$ and $X$ is the Dieudonné completion of $G$. That is, $G$ is a PT-group iff $\varrho G$ naturally contains the Dieudonné completion of $G$.*

The next result is a version of Corollary 6.5.9 for the Hewitt–Nachbin completion of a topological group. The proof of it is similar to the proof of Theorem 6.5.8.

Corollary 6.5.10. *Suppose that $G$ is a topological group and $X$ a subspace of the Raĭkov completion $\varrho G$ of $G$ such that $G \subset X$ and $X$ is the Hewitt–Nachbin completion of $G$. Then $X$ is a subgroup of $\varrho_\omega G$.*

One of the main questions considered in this section can be reformulated as follows: *Is every topological group a $PT$-group?* One can restate Theorem 6.5.1 by saying that every Moscow topological group is a $PT$-group. We will strengthen this result in Theorem 6.5.13 below.

In general, $PT$-groups admit the following characterization:

Theorem 6.5.11. *A topological group $G$ is a $PT$-group if and only if it is $C$-embedded in some Dieudonné complete topological group as a topological subgroup.*

Proof. If $G$ is a $PT$-group, then $\mu G$ is a topological group containing $G$ as a dense subgroup, so Corollary 3.6.16 implies that $\mu G$ is a topological subgroup of $\varrho G$ containing $G$.

Conversely, suppose that $G$ is a $C$-embedded subgroup of a Dieudonné complete topological group $G^*$. Then $G$ is $C$-embedded in its closure $H$ in $G^*$. Obviously, $H$ is Dieudonné complete and $G$ is dense in $H$. Apply Proposition 6.5.7 to find a subgroup $M$ of $G^*$ such that $M$ is a minimal Dieudonné extension of $G$ and $G \subset M \subset H$. Then $G$ is $C$-embedded in $M$, so $M$ is the Dieudonné completion of $G$, by Proposition 6.5.6. Hence, $G$ is a $PT$-group. □

In connection with Theorem 6.5.11 it is natural to introduce the following notion. A topological group $G$ is a *strong $PT$-group* if it is $C$-embedded in $\varrho_\omega G$. Since, by Proposition 6.5.2, $\varrho_\omega G$ is a Dieudonné complete topological group, it follows from Theorem 6.5.11 that every strong $PT$-group is a $PT$-group. The converse is not true, as we will show below in Example 6.5.30. The next assertion is obvious.

Proposition 6.5.12. *Every Raĭkov complete topological group is a strong $PT$-group.*

Proposition 6.5.2 implies the following statement:

Theorem 6.5.13. *Every Moscow topological group is a strong $PT$-group.*

The next result shows that the property of being (strong) $PT$-groups is hereditary with respect to $C$-embedded subgroups:

Proposition 6.5.14. *Let $H$ be a $C$-embedded subgroup of a (strong) $PT$-group $G$. Then $H$ is also a (strong) $PT$-group.*

Proof. Suppose that $G$ is a $PT$-group. Then, by Corollary 3.6.16, $\mu G$ is a topological subgroup of $\varrho G$ containing $G$, and $G$ is $C$-embedded in $\mu G$. It follows that $H$ is a $C$-embedded subgroup of the Dieudonné complete topological group $\mu G$, so Theorem 6.5.11 implies that $H$ is a $PT$-group.

If $G$ is a strong $PT$-group, then $G$ is $C$-embedded in $\varrho_\omega G$, so that $H$ is $C$-embedded in $\varrho_\omega G$. It follows from $H \subset G$ that $\varrho_\omega H \subset \varrho_\omega G$. We conclude that $H$ is $C$-embedded in $\varrho_\omega H$ and, therefore, $H$ is a strong $PT$-group. □

Here is a characterization of strong $PT$-groups which should be compared to Theorem 6.5.11.

THEOREM 6.5.15. *A topological group $H$ is a strong PT-group if and only if it is $C$-embedded in every topological group $H^*$ containing $H$ as a $G_\delta$-dense topological subgroup.*

PROOF. It suffices to observe that each group $H^*$ containing $H$ as a $G_\delta$-dense subgroup can be represented as a topological subgroup of the Raĭkov completion $\varrho H$ of $H$. Then, obviously, $H^*$ is contained in $\varrho_\omega H$, and $H$ is $C$-embedded in both $\varrho_\omega H$ and $H^*$. □

The next result follows immediately from Corollary 6.5.9 and the fact that $X$ is $G_\delta$-dense in $\mu X$, for each Tychonoff space $X$:

THEOREM 6.5.16. *A topological group $G$ is a PT-group iff $\mu G$ is a subgroup of $\varrho_\omega G$ containing $G$.*

It is worth mentioning that Theorem 6.5.16 does not hold for strong PT-group, since there exists a topological group $H$ such that $\mu H = H$, but $H$ is not $C$-embedded in $\varrho_\omega H$ (see Example 6.5.30 below).

Naturally, the following question arises: *Which groups $G$ satisfy the equality $\mu G = \varrho_\omega G$?* A topological group $G$ is called *completion friendly* if $\mu G = \varrho_\omega G$. The next statement follows from the definition of strong PT-groups:

PROPOSITION 6.5.17. *Every completion friendly group $G$ is a strong PT-group.*

It is clear from the definition that every Raĭkov complete group is completion friendly. It is not quite clear how to answer in $ZFC$ the next question: *Is every Moscow group completion friendly?*

To give a partial answer to this question, we first reformulate some of the results obtained above for the Hewitt–Nachbin completion $\upsilon G$ of a PT-group $G$. The following statement is well known, it follows from Proposition 6.5.6 and [**165**, 8.5.13 (h)]:

PROPOSITION 6.5.18. *Let $Y$ be a dense subspace of a Tychonoff space $X$ such that the cellularity $Y$ is Ulam non-measurable. Then the next three conditions are equivalent:*

1) $X = \upsilon Y$;
2) *$X$ is a Dieudonné complete space in which $Y$ is $C$-embedded;*
3) $X = \mu Y$.

*In particular, conditions 1)–3) are equivalent for every space $Y$ of Ulam non-measurable cardinality.*

According to Proposition 6.5.18, we can rephrase Theorem 6.5.1 as follows:

THEOREM 6.5.19. *Let $G$ be a Moscow group of Ulam non-measurable cellularity. Then the operations on $G$ can be extended to the Hewitt–Nachbin completion $\upsilon G$ of $G$ in such a way that $\upsilon G$ becomes a topological group containing $G$ as a topological subgroup.*

From Propositions 6.5.18, 6.5.2, and Theorem 6.5.19 we obtain:

PROPOSITION 6.5.20. *Every Moscow group $G$ of Ulam non-measurable cellularity satisfies $\mu G = \varrho_\omega G = \upsilon G$ and, hence, is completion friendly.*

According to item 5) of Corollary 6.4.11, every topological group of countable cellularity is a Moscow space. Therefore, Proposition 6.5.20 implies the next corollary:

COROLLARY 6.5.21. *Every topological group of countable cellularity is completion friendly.*

Now we establish a more general fact about topological groups of Ulam non-measurable cellularity.

THEOREM 6.5.22. *For a topological group $G$ of Ulam non-measurable cellularity, the following are equivalent:*

a) *$G$ is a strong $PT$-group;*
b) *$G$ is completion friendly;*
c) *$\mu G = \varrho_\omega G = \upsilon G$.*

PROOF. Evidently, c) $\Rightarrow$ b) $\Rightarrow$ a). To show that a) implies c), suppose that $G$ is a strong $PT$-group. Then $G$ is $C$-embedded in $\varrho_\omega G$ and, by Proposition 6.5.2, the group $\varrho_\omega G$ is Dieudonné complete. It remains to apply Proposition 6.5.18. □

In particular, a topological group $G$ of Ulam non-measurable cardinality is a strong $PT$-group if and only if $G$ is completion friendly. This fact complements Proposition 6.5.17. Here is a special case of Theorem 6.5.22.

COROLLARY 6.5.23. *An $\omega$-narrow topological group $G$ is a strong $PT$-group iff it satisfies $\mu G = \varrho_\omega G = \upsilon G$.*

PROOF. According to Theorem 5.4.10, the cellularity of every $\omega$-narrow topological group is not greater than $2^\omega$, so the conclusion follows from Theorem 6.5.22. □

In Section 6.4 it was established that many naturally defined classes of topological groups are contained in the class of Moscow groups. Therefore, we have the following corollaries to Theorem 6.5.13:

THEOREM 6.5.24. *Let $G$ be a topological group satisfying at least one of the following conditions:*

1) *$t(G) \leq \omega$;*
2) *$c(G) \leq \omega$;*
3) *$ot(G) \leq \omega$;*
4) *points in $G$ are $G_\delta$'s;*
5) *$G$ is perfectly $\kappa$-normal;*
6) *$G$ is extremally disconnected;*
7) *$G$ is a subgroup of a topological group $G^*$ such that $G^*$ is a $k$-space;*
8) *$G$ is precompact;*
9) *the $g$-tightness of $G$ is countable;*
10) *$G$ is separable.*

*Then $G$ is a strong $PT$-group.*

PROOF. It is enough to combine Theorems 6.4.2, 6.4.9, and Corollary 6.4.11 with Theorem 6.5.13. □

The above results can be used to identify conditions under which $G = \varrho_\omega G$. Clearly, this equality holds if and only if $G$ is $G_\delta$-closed in $\varrho G$, that is, if and only if $G$ is $G_\delta$-closed in every topological group in which it is dense.

COROLLARY 6.5.25. *Every Hewitt–Nachbin complete Moscow group $H$ is $G_\delta$-closed in $\varrho H$.*

PROOF. Indeed, $H$ is $C$-embedded in $\varrho_\omega H$, by Proposition 6.5.2. Since $H$ is Hewitt–Nachbin complete, it follows that $H = \varrho_\omega H$. Thus, $H$ is $G_\delta$-closed in $\varrho H$. □

The following simple lemma will be applied in the proof of Theorem 6.5.27.

LEMMA 6.5.26. *Every Lindelöf subspace $L$ of a Hausdorff space $X$ is $G_\delta$-closed in $X$.*

PROOF. Fix any point $x \in X \setminus L$. For every $y \in L$, take in $X$ disjoint open neighbourhoods $U_y$ and $V_y$ of $x$ and $y$, respectively. Since $L$ is Lindelöf, there exists a countable subset $C$ of $L$ such that $L \subset \bigcup_{y\in C} V_y$. Then $P = \bigcap_{y\in C} U_y$ is a $G_\delta$-set in $X$ that contains $x$ and is disjoint from $L$. □

THEOREM 6.5.27. *Let $G$ be a topological group such that $\mu G$ is a Lindelöf topological group (containing $G$ as a subgroup). Then $G$ is completion friendly and, therefore, a strong $PT$-group.*

PROOF. Indeed, $\mu G$ can be interpreted as a subgroup of $\varrho G$ such that $G \subset \mu G \subset \varrho_\omega G$. By Lemma 6.5.26, $\mu G$ is $G_\delta$-closed in $\varrho G$. It follows that $\mu G = \varrho_\omega G$. Thus, $G$ is completion friendly. □

The case when the Dieudonné completion $\mu G$ of a topological group $G$ is a Lindelöf space (not necessarily a group) will be considered in Theorem 6.6.12 and Corollary 6.6.13.

COROLLARY 6.5.28. *Every Lindelöf topological group $G$ is completion friendly and, therefore, a strong $PT$-group. In fact, the group $G$ satisfies the equalities $G = \mu G = \upsilon G = \varrho_\omega G$.*

We will now show how to construct a Lindelöf topological group that is not Moscow. Our construction makes use of the following simple result:

PROPOSITION 6.5.29. *Every regular $P$-space $X$ of weight $\leq \aleph_1$ is paracompact and strongly zero-dimensional.*

PROOF. Take any open covering $\gamma$ of $X$. Since the weight of $X$ does not exceed $\aleph_1$, we may assume that $|\gamma| \leq \aleph_1$ and $\gamma = \{U_\alpha : \alpha < \omega_1\}$. Since $X$ is a $P$-space, we may also assume that all sets $U_\alpha$ are closed. Now put $V_\alpha = U_\alpha \cap \bigcap_{\beta<\alpha}(X \setminus U_\beta)$, for each $\alpha < \omega_1$. Clearly, since $X$ is a $P$-space, $\{V_\alpha : \alpha < \omega_1\}$ is a disjoint open covering of $X$ refining $\gamma$. □

EXAMPLE 6.5.30. There exists a Lindelöf hereditarily paracompact topological group $G$ that fails to be a Moscow space. Further, $G$ contains a dense subgroup $H$ which is a $PT$-group, but not a strong $PT$-group.

In Example 4.4.11, take $K$ to be the discrete two-element group $\{0, 1\}$, and let $A = \omega_1$. Then the corresponding Lindelöf $P$-group $G = G_{\aleph_1}$ is Boolean and satisfies $w(G) = \aleph_1$. In particular, $G$ has a local base $\mathcal{B}$ of cardinality $\aleph_1$ at the neutral element $e$. Since $G$ is a regular $P$-space, we can assume that all elements of $\mathcal{B}$ are closed in $G$. In addition, the base $\mathcal{B} = \{U_\alpha : \alpha < \omega_1\}$ can be chosen to satisfy $U_\beta \subset U_\alpha$ whenever $\alpha < \beta < \omega_1$.

For every $\alpha < \omega_1$, let $V_\alpha = U_\alpha \setminus U_{\alpha+1}$. Since $G$ is non-discrete, we can find an uncountable subset $P$ of $\omega_1$ such that $V_\alpha \neq \varnothing$, for each $\alpha \in P$. Choose uncountable disjoint subsets $M_1$ and $M_2$ of $P$ and put $W_i = \bigcup_{\alpha\in M_i} V_\alpha$, where $i = 1, 2$. Then $W_1$ and $W_2$ are disjoint open subsets of the space $Y = G \setminus \{e\}$. It is clear that $e$ is the only accumulation point of the sets $W_1$ and $W_2$, so both sets are closed in $Y$. Hence, $\{e\} = \overline{W_1} \cap \overline{W_2}$, the closure is taken in $G$. Let us define a function $f$ on $Y$ by $f(y) = 1$ if $y \in W_1$, and $f(y) = 0$ if $y \in Y \setminus W_1$. In particular, $f(y) = 0$ for each $y \in W_2$. Since $W_1$ is open and closed in $Y$, this function is continuous on $Y$. On the other hand, $f$ cannot be extended to a continuous real-valued function on $G$, since $e$ is in the closure of both sets $W_1$ and $W_2$. Thus, $Y$ is not $C$-embedded in $G$. In fact, $Y$ is not even $C^*$-embedded in $G$.

Now, $Y$ is $G_\delta$-dense in $G$, since $G$ is a $P$-space and $Y$ is dense in $G$. It follows from Theorem 6.1.7 that $G$ is not a Moscow space and, therefore, the $g$-tightness of $G$ is uncountable, by Theorem 6.4.9.

Every proper dense subspace $Z$ of $G$ is not $C$-embedded in $G$ either. Indeed, by the homogeneity of $G$, we may assume that $e \in G \setminus Z$. Then $Z \subset G \setminus \{e\}$. Take the same function $f$ on $G \setminus \{e\}$ that cannot be continuously extended to $G$. Then, since $Z$ is dense in $G \setminus \{e\}$, it follows that the restriction of $f$ to $Z$ cannot be continuously extended over $G$.

Since the $P$-group $G$ is Lindelöf, it is Raĭkov complete, by Theorem 4.4.5. As in Example 4.4.11, we put $\mathrm{supp}(x) = \{\alpha < \omega_1 : x(\alpha) = 1\}$, for each $x \in G$. Evidently, $\mathrm{supp}(x)$ is a finite subset of $\omega_1$. Then

$$H = \{x \in G : |\,\mathrm{supp}(x)| \text{ is even}\}$$

is a proper subgroup of $G$. Since the index set $\omega_1$ is uncountable, $H$ is dense in $G$. Therefore, $G = \varrho H$ and $H$ is not $C$-embedded in $G$. It is also clear that $H$ is a $G_\delta$-dense subgroup of $G$, so we have that $G = \varrho_\omega H$. It follows that $H$ is not $C$-embedded in $\varrho_\omega H$. Hence, $H$ is not a strong $PT$-group. Now we can conclude that neither $G$ nor $H$ are Moscow groups, and that Theorem 6.1.7 and Corollary 6.1.10 cannot be extended from Moscow groups to the class of strong $PT$-groups. We also conclude that a Lindelöf topological group need not be a Moscow space.

Since $w(G) = \aleph_1$, Proposition 6.5.29 implies that every subspace of $G$ is paracompact. It follows that $H$ is paracompact and, hence, Dieudonné complete and Hewitt–Nachbin complete, by [**165**, 5.1.J (f) and 8.5.13 (h)]. So, $H$ is a $PT$-group. □

There is another way to demonstrate that the groups $G$ and $H$ in Example 6.5.30 are not Moscow. Since the cardinality of $G$ and $H$ is $\aleph_1$, and the groups $G$ and $H$ are not discrete, it is enough to refer to Corollary 6.2.8 saying that, for a space of Ulam non-measurable cardinality, to be both Moscow and a $P$-space is possible only if the space is discrete.

## Exercises

6.5.a. Is a closed subgroup of a (strong) $PT$-group a (strong) $PT$-group?

6.5.b. Suppose that a (strong) $PT$-group $H$ is a dense (or $G_\delta$-dense) subgroup of a topological group $G$. Is $G$ then a (strong) $PT$-group?

6.5.c. Use Corollary 6.5.21 (and item (a) of Problem 5.1.D) to show that a topological group $G$ of countable cellularity is topologically isomorphic to a closed subgroup of a product of second-countable topological groups if and only if the space $G$ is realcompact.

6.5.d. Give an example of an $\omega$-narrow $PT$-group that cannot be embedded as a closed subgroup into any product of metrizable topological groups.

## Problems

6.5.A. Let $p\colon G \to H$ be a continuous open homomorphism of topological groups with $p(G) = H$, and suppose that $G$ is a (strong) $PT$-group. Is $H$ a (strong) $PT$-group?

6.5.B. Give an example of a $G_\delta$-dense subgroup $H$ of a Raĭkov complete group such that $H$ fails to be a $PT$-group.
*Hint.* Such an example can be found in Section 6.7.

6.5.C. (S. Romaguera and M. Sanchis [**411**]) A subset $Y$ of a Tychonoff space $X$ is said to be *z-embedded* in $X$ if for every zero-set $C$ in $Y$, there exists a zero-set $F$ in $X$ such that $F \cap Y = C$. Prove that a topological group $G$ is a $PT$-group iff it is $z$-embedded as a dense subgroup in some Dieudonné complete topological group iff it is $z$-embedded as a subgroup in some Dieudonné complete topological subgroup. (This generalizes Theorem 6.5.11; see also Problem 6.5.4).

6.5.D. Apply item (a) of Problem 5.1.D to generalize the statement in Exercise 6.5.c by proving that for an $\omega$-narrow topological group $G$, the following conditions are equivalent:
a) $G$ is a realcompact strong $PT$-group;
b) $G = \varrho_\omega G$;
c) $G$ is topologically isomorphic to a closed subgroup of a product of second-countable topological groups.
(See also Problems 5.1.E, 6.5.E, 8.3.C, and Exercise 8.3.a.)

6.5.E. Prove that an $\omega$-balanced strong $PT$-group $G$ is topologically isomorphic to a closed subgroup of a product of metrizable topological groups provided that $G$ is realcompact (this complements Problem 6.5.D).

6.5.F. Let $G$ be a $PT$-group. Does the equality $\psi(\mu G) = \psi(G)$ hold?

## Open Problems

6.5.1. Is it true in $ZFC$ that every Moscow topological group is completion friendly?

6.5.2. Is the product of a strong $PT$-group and a compact group a strong $PT$-group? [Of course, the answer is "yes" for groups of Ulam non-measurable cardinality.]

6.5.3. Let $G$ be a topological group of countable tightness. Is $G \times G$ a strong $PT$-group? (See Problems 6.4.2 and 6.4.3.)

6.5.4. Suppose that $H$ is a $z$-embedded subgroup of a (strong) $PT$-group $G$. Is $H$ a (strong) $PT$-group? (See Proposition 6.5.14, Theorem 8.2.7, and Problem 6.5.C.)

6.5.5. Let $G$ be a paratopological group which is a completely regular Moscow space. Is $\mu G$ a paratopological group (containing $G$ as a dense subgroup)? What if, in addition, $G$ has countable pseudocharacter?

## 6.6. Pseudocompact groups and their completions

In this section we apply the techniques developed in the previous sections to pseudocompact topological groups and give transparent proofs of some classic results.

THEOREM 6.6.1. *Let $G$ be a precompact topological group, and $Y$ a dense subspace of $G$. Then $Y$ is $C$-embedded in the $G_\delta$-closure $Z$ of $Y$ in $G$.*

PROOF. By Theorem 6.4.2, $G$ is a Moscow space. It follows from $Y \subset Z \subset G$ that $Z$ is dense in $G$. Therefore, by Proposition 6.1.2, $Z$ is Moscow. Since $Y$ is $G_\delta$-dense in $Z$, Theorem 6.1.7 implies that $Y$ is $C$-embedded in $Z$. □

If $G$ is any compact space (not necessarily a topological group), and $Y$ is a dense subspace of $G$, then the $G_\delta$-closure $Z$ of $Y$ in $G$ is Hewitt–Nachbin complete [**165**, 3.12.25 (c)]. Therefore, from Theorem 6.6.1 we obtain:

THEOREM 6.6.2. *Let $G$ be a compact topological group, and $Y$ a dense subspace of $G$. Then the $G_\delta$-closure of $Y$ in $G$ is the Hewitt–Nachbin completion of the space $Y$.*

COROLLARY 6.6.3. [**W. W. Comfort and K. A. Ross**] *Let $G$ be a topological group, and $Y$ a dense subspace of $G$. Then the next three conditions are equivalent:*

a) *$Y$ is $G_\delta$-dense in $G$, and $G$ is pseudocompact;*
b) *$Y$ is $C$-embedded in $G$, and $G$ is pseudocompact;*
c) *$Y$ is pseudocompact.*

PROOF. By Theorem 6.6.1, a) implies b). Obviously, b) implies c). Finally, if $Y$ is pseudocompact, then $G$ is pseudocompact, since $Y$ is dense in $G$. Also, $Y$ is $G_\delta$-dense in $G$, by Proposition 3.7.20. Thus, c) implies a). □

THEOREM 6.6.4. *Let $G$ be a pseudocompact topological group. Then the Raĭkov completion $\varrho G$ of $G$ (and $\varrho_\omega G$) coincides, as a topological space, with the Čech–Stone compactification $\beta G$ of the space $G$.*

PROOF. The Raĭkov completion $\varrho G$ is a compact topological group containing $G$ as a topological subgroup, by Corollary 3.7.18. Since $G$ is pseudocompact, Corollary 3.7.21 implies that it is $G_\delta$-dense in $\varrho G$ and, therefore, $\varrho_\omega G = \varrho G$. Also, the space $G$ is Moscow, by Theorem 6.4.2. Since $\varrho G$ is homogeneous, it follows from Theorem 6.1.8 or Corollary 6.6.3 that $G$ is $C$-embedded in $\varrho G$. Hence, $\varrho G$ is the Čech–Stone compactification of $G$. □

Combining Theorem 6.6.4 and Corollary 6.6.3, we obtain the following strengthening of Corollary 3.7.21:

COROLLARY 6.6.5. *A precompact topological group $G$ is pseudocompact if and only if $G$ intersects every non-empty $G_\delta$-set in $\varrho G$.*

COROLLARY 6.6.6. *Every pseudocompact topological group $G$ is $G_\delta$-dense and $C$-embedded in its Raĭkov completion $\varrho G$.*

COROLLARY 6.6.7. *Every pseudocompact topological group $G$ is completion friendly, that is, satisfies the condition $\mu G = \varrho_\omega G$.*

PROOF. Indeed, the space $\mu G$ is pseudocompact, since it contains a dense pseudocompact subspace. Therefore, since $\mu G$ is Dieudonné complete, the space $\mu G$ is compact [**165**, 8.5.13 (c)]. It follows that $\mu G = \beta G$, as $\mu G$ is a dense subspace of $\beta G$. Now it follows from Theorem 6.6.4 that $\mu G$ coincides with $\varrho_\omega G = \varrho G$. □

Theorem 6.6.4 can be extended to precompact topological groups as follows:

COROLLARY 6.6.8. [**M. G. Tkachenko**] *Let $G$ be a precompact topological group. Then the Hewitt–Nachbin completion $\upsilon G$ of $G$ has a natural structure of a topological group, with continuous multiplication and inverse, extending the multiplication and inverse in $G$. Hence, $G$ is a $PT$-group.*

PROOF. Take the compact topological group $\varrho G$ containing $G$ as a dense subgroup. By Theorem 6.6.2, the $G_\delta$-closure $Z$ of $G$ in $\varrho G$ is the Hewitt–Nachbin completion of $G$, that is, $Z = \upsilon G$. It is also clear that $Z$ is a topological subgroup of $\varrho G$. □

COROLLARY 6.6.9. [**W. W. Comfort and K. A. Ross**] *Every continuous real-valued function on a pseudocompact topological group is uniformly continuous.*

PROOF. Let $f$ be a continuous real-valued function defined on a pseudocompact topological group $G$. By Theorem 6.6.4, $f$ can be extended to a continuous function $g$ on the compact group $\varrho G$ containing $G$ as a dense subgroup. Since every continuous function on a compact group is uniformly continuous by Proposition 1.8.11, the restriction $f = g \restriction G$ is uniformly continuous on $G$. □

COROLLARY 6.6.10. *The topological product of any family of pseudocompact groupy spaces is pseudocompact.*

PROOF. According to Corollary 2.4.2, every pseudocompact paratopological group is a topological group. Therefore, every pseudocompact groupy space is homeomorphic to a dense subspace of a compact topological group, by Corollary 3.7.18. It remains to refer to the fact that the product of any family of compact topological groups is a compact topological group, and to apply Corollary 6.6.3, since the $G_\delta$-characterization of pseudocompactness for dense subspaces of topological groups is obviously productive. □

Corollary 6.6.10 implies the next important statement known as the Comfort–Ross theorem:

COROLLARY 6.6.11. [**W. W. Comfort and K. A. Ross**] *The product of any family of pseudocompact topological groups is a pseudocompact topological group.*

THEOREM 6.6.12. *If $G$ is a topological group such that the space $\mu G$ is Lindelöf, then the topological group $\varrho_\omega G$ is also Lindelöf.*

PROOF. The group $\varrho G$ is a complete uniform space with respect to the two-sided group uniformity of $\varrho G$, and $\mu G$ is the completion of $G$ with respect to the maximal uniformity on $G$. Therefore, the identity mapping of $G$ onto itself can be extended to a continuous mapping $f$ of $\mu G$ to $\varrho G$. Put $X = f(\mu G)$. Then $X$ is a Lindelöf subspace of $\varrho G$, $G \subset X$, and $G$ is $G_\delta$-dense in $X$, since $G$ is $G_\delta$-dense in $\mu G$ and $f$ is continuous. Hence, $X \subset \varrho_\omega G$. However, since $X$ is Lindelöf, every point of $\varrho G \setminus X$ can be separated from $X$ by a $G_\delta$-set in $\varrho G$ (see [**165**, 3.12.24]). It follows that $\varrho_\omega G \subset X$, and therefore, $\varrho_\omega G = X$. Hence, $\varrho_\omega G$ is Lindelöf. □

COROLLARY 6.6.13. *If $G$ is a topological group such that the space $\mu G$ is Lindelöf, then $G$ is a dense topological subgroup of a Lindelöf topological group and therefore, the group $G$ is $\omega$-narrow.*

## Exercises

6.6.a. Apply Corollary 6.6.3, Problem 4.1.F, and Exercise 1.6.e to verify that the $G_\delta$-tightness of every pseudocompact topological group is countable.

6.6.b. Generalize Theorem 6.6.1 as follows: If $H$ is a locally precompact topological group, and $Y$ a dense subspace of $H$, then $Y$ is $C$-embedded in the $G_\delta$-closure $Z$ of $Y$ in $H$. (See Exercises 2.4.5 and 3.7.J.)

6.6.c. Suppose that $H$ is a locally precompact topological group. Prove that the following are equivalent:
a) $H$ is locally pseudocompact;
b) $H$ is $G_\delta$-dense in $\varrho H$;
c) $H$ is $C$-embedded in $\varrho H$.

6.6.d. Show that if a pseudocompact topological group $G$ is algebraically generated by a countable family of metrizable subgroups, then $G$ is compact and metrizable.

6.6.e. Show that the topological group $\mathbb{R}^{\omega_1}$ is not algebraically generated by a countable family of metrizable subgroups.

## Problems

6.6.A. Present an example of a pseudocompact topological group which is not countably compact (note that no $\Sigma$-product of compact topological groups has the required combination of properties).
*Hint.* One can construct such a group as a proper dense subgroup of $\Sigma D^{\mathfrak{c}}$, making use of Theorem 2.4.15.

6.6.B. (M. G. Tkachenko [**475**]) Prove that the product $G \times Y$ of a pseudocompact topological group with a pseudocompact space $Y$ is again pseudocompact. Extend the conclusion to the case when $G$ is a pseudocompact groupy space (see also Problem 6.10.E).

6.6.C. (M. G. Tkachenko [**475**]; for the special case of an invariant subgroup, W. W. Comfort and L. C. Robertson [**120**]) Let $H$ be a closed subgroup of a topological group $G$. Prove that if both spaces $H$ and $G/H$ are pseudocompact, then so is $G$ (see also Problem 6.10.B).

6.6.D. Give an example of a pseudocompact topological group $G$ such that $t_\delta(G) > \omega$. Notice that, in view of Exercise 6.6.a, such a group $G$ must satisfy $\omega = get(G) < t_\delta(G)$.

6.6.E. Prove that the space $Y$ in Theorem 6.6.2 satisfies $\mu Y = \upsilon Y$.

6.6.F. Apply Exercise 6.6.b to show that every locally precompact topological group is completion friendly.

## Open Problems

6.6.1. Does there exist a countably compact topological group $G$ with $t_\delta(G) > \omega$? (See also Problems 6.6.D and 4.1.F.)

6.6.2. Characterize the topological groups $G$ such that $\mu G$ is Lindelöf.
Here is a more concrete question in the same direction:

6.6.3. Suppose $G$ is a topological group such that $\mu G$ is Lindelöf. Is $\mu G$ a topological group containing $G$ as a subgroup or, equivalently, is $G$ a $PT$-group? (See Corollary 6.6.13.)

6.6.4. Characterize the topological groups $G$ such that $\mu G$ is a Lindelöf topological group containing $G$ as a subgroup.

6.6.5. Let a countably compact topological group $G$ be algebraically generated by two metrizable subspaces. Is $G$ metrizable? (See Exercises 6.6.d and 6.6.e.)

6.6.6. Is the topological group $\mathbb{R}^{\omega_1}$ algebraically generated by two (countably many) metrizable subspaces?

6.6.7. Do there exist in $ZFC$ two pseudocompact topological groups of countable tightness whose product has uncountable tightness? (See Problem 5.2.G and Open Problems 5.2.7 and 5.2.8.)

6.6.8. Do there exist pseudocompact quasitopological groups $G$ and $H$ such that both $G$ and $H$ are Moscow spaces, but the product $G \times H$ fails to be pseudocompact? (See Problem 2.4.G.)

## 6.7. Moscow groups and the formula $\upsilon(X \times Y) = \upsilon X \times \upsilon Y$

Under what restrictions on spaces $X$ and $Y$ does the formula $\upsilon(X \times Y) = \upsilon X \times \upsilon Y$ hold? This natural question was given considerable attention in a number of articles. In particular, the formula holds when $X$ is an arbitrary compact space of Ulam non-measurable cardinality or if the product space $X \times Y$ is pseudocompact [**165, 191**].

In this section we consider the case when the factors are topological groups. The key role belongs to Moscow groups and strong $PT$-groups.

PROPOSITION 6.7.1. *Let $G = \prod_{\alpha\in A} G_\alpha$ be the topological product of topological groups $G_\alpha$ such that $G$ is a strong $PT$-group, and the cellularity of $G$ is Ulam non-measurable. Then $\upsilon G = \prod_{\alpha\in A} \upsilon G_\alpha$.*

PROOF. Clearly, $G_\alpha$ is $G_\delta$-dense in $\upsilon G_\alpha$, for each $\alpha \in A$. Therefore, $G$ is $G_\delta$-dense in $G^* = \prod_{\alpha\in A} \upsilon G_\alpha$. Each $G_\alpha$ is a strong $PT$-group, since it is $C$-embedded in the product group $G$ (see Proposition 6.5.14). Obviously, the cellularity of each group $G_\alpha$ is Ulam non-measurable. Therefore, according to Proposition 6.5.18, each $\upsilon G_\alpha = \mu G_\alpha$ is a topological group, and $G^*$ is also a topological group.

Since $G$ is a strong $PT$-group and $G$ is $G_\delta$-dense in $G^*$, it follows from Theorem 6.5.15 that $G$ is $C$-embedded in $G^*$. Since the group $G^*$ is obviously Hewitt–Nachbin complete and the cellularity of $G$ is Ulam non-measurable, it follows that $\upsilon G = \prod_{\alpha\in A} \upsilon G_\alpha$. □

Proposition 6.7.1 and Theorem 6.5.13 imply the next statement:

THEOREM 6.7.2. *Let $G = \prod_{\alpha\in A} G_\alpha$ be a product of topological groups such that the space $G$ is Moscow and the cellularity of $G$ is Ulam non-measurable. Then $\upsilon G = \prod_{\alpha\in A} \upsilon G_\alpha$.*

Under the same restrictions as in Proposition 6.7.1 and Theorem 6.7.2, a similar formula holds for the Dieudonné completions. Indeed, by Proposition 6.5.18, the Dieudonné completion and the Hewitt–Nachbin completion coincide, for every Tychonoff space of Ulam non-measurable cardinality.

COROLLARY 6.7.3. *Let $\mathcal{F} = \{G_\alpha : \alpha \in A\}$ be a family topological groups $G_\alpha$ such that the cardinality of the product group $G = \prod_{\alpha\in A} G_\alpha$ is Ulam non-measurable. Then the formula*

$$\upsilon G = \prod_{\alpha\in A} \upsilon G_\alpha \tag{6.1}$$

*holds if at least one of the following conditions is satisfied:*

1) *every group in $\mathcal{F}$ is precompact;*
2) *every group in $\mathcal{F}$ is $k$-separable;*
3) *$\aleph_1$ is a precalibre of every space in $\mathcal{F}$;*
4) *the cellularity of the product space $G$ is countable;*
5) *the cellularity of every group in $\mathcal{F}$ is countable, and $MA + \neg CH$ holds;*
6) *the tightness of the product space $G$ is countable;*
7) *the $g$-tightness of the product group $G$ is countable;*
8) *the $\kappa$-tightness of the space $G$ is countable.*

PROOF. Notice that if the Souslin number of the product group $G$ is countable, then $G$ is Moscow, by item 5) of Corollary 6.4.11. According to Corollaries 4.1.8 and 5.3.22, this takes care of cases 1)–5). Similarly, in cases 6)–8) the group $G$ is also Moscow, by Corollary 6.4.11. Therefore, Theorem 6.7.2 applies. □

If the group $G$ in Corollary 6.7.3 satisfies $\mu G = \upsilon G$, then we can drop the assumption that the cardinality of $G$ is Ulam non-measurable. In particular, this is the case when the cellularity of every $G_\alpha \in \mathcal{F}$ is countable, as in items 1)–5) above (see Proposition 6.5.18).

It is also worth remarking that if the factors $G_\alpha$ in Corollary 6.7.3 are cosmic or have countable pseudocharacter, then each space $G_\alpha$ is realcompact, and the formula (6.1) holds trivially (in the case when $\psi(G_\alpha) \leq \omega$ for each $\alpha \in A$, one has to apply Corollary 6.10.10 below).

For the special case of the product of two groups more results are available. We need the following simple result.

PROPOSITION 6.7.4. *If $X_i$ is a minimal Dieudonné extension of $Y_i$ for $i = 1, \dots, k$, then $X = \prod_{i=1}^k X_i$ is a minimal Dieudonné extension of $Y = \prod_{i=1}^k Y_i$.*

PROOF. We may assume that $k = 2$, that is, $X = X_1 \times X_2$ and $Y = Y_1 \times Y_2$. Clearly, $X$ is Dieudonné complete and $Y$ is dense in $X$. Let $T$ be a Dieudonné complete space such that $Y \subset T \subset X$. First, we show that $Y_1 \times X_2 \subset T$. Assume the contrary. Then there exists $(a, b) \in Y_1 \times X_2$ such that $(a, b) \notin T$. Then

$$F = \{x \in X_2 : (a, x) \in T\} = (\{a\} \times X_2) \cap T$$

is a closed subspace of $T$ containing $Y_2$ and $F \neq X_2$, since $b \in X_2 \setminus F$. Clearly, $F$ is Dieudonné complete. This contradicts the minimality of $X_2$. It follows that $Y_1 \times X_2 \subset T$. Now it remains to repeat the above argument with $X_2$ in the role of $Y_1$ and $Y_1, X_1$ in the roles of $Y_2, X_2$, respectively. Hence, $T = X_1 \times X_2$. □

THEOREM 6.7.5. *Let $G_1$ and $G_2$ be two completion friendly groups. Then the next conditions are equivalent:*

1) *$G_1 \times G_2$ is a PT-group;*
2) *$\mu(G_1 \times G_2) = \mu G_1 \times \mu G_2$;*
3) *the group $G_1 \times G_2$ is completion friendly;*
4) *$G_1 \times G_2$ is a strong PT-group.*

PROOF. Clearly, 3) implies 4), and 4) implies 1). Let us show that 1) implies 2). We have $\varrho_\omega(G_1 \times G_2) = \varrho_\omega G_1 \times \varrho_\omega G_2$ and $\varrho_\omega G_i = \mu G_i$, for $i = 1, 2$. Therefore,

$$\varrho_\omega(G_1 \times G_2) = \mu G_1 \times \mu G_2.$$

Since $G_1 \times G_2$ is a $PT$-group, we have $G_1 \times G_2 \subset \mu(G_1 \times G_2) \subset \varrho_\omega(G_1 \times G_2)$. Therefore,

$$G_1 \times G_2 \subset \mu(G_1 \times G_2) \subset \mu G_1 \times \mu G_2.$$

However, by Proposition 6.7.4, $\mu G_1 \times \mu G_2$ is a minimal Dieudonné extension of $G_1 \times G_2$. It follows that $\mu(G_1 \times G_2) = \mu G_1 \times \mu G_2$.

To derive 3) from 2) is even easier. Indeed, we have:

$$\mu(G_1 \times G_2) = \mu G_1 \times \mu G_2 = \varrho_\omega G_1 \times \varrho_\omega G_2 = \varrho_\omega(G_1 \times G_2).$$

This completes the proof. □

Since, by virtue of Theorem 6.5.22, the Dieudonné completion coincides with the Hewitt–Nachbin completion for each space of Ulam non-measurable cellularity, and the notion of completion friendly group is equivalent to the notion of strong $PT$-group, the next statement is valid:

COROLLARY 6.7.6. *Let $G_1$ and $G_2$ be strong $PT$-groups of Ulam non-measurable cellularity. Then the formulae $\upsilon(G_1 \times G_2) = \upsilon G_1 \times \upsilon G_2$ and $\mu(G_1 \times G_2) = \mu G_1 \times \mu G_2$ hold if and only if $G_1 \times G_2$ is a $PT$-group. In this case, $G_1 \times G_2$ is automatically a strong $PT$-group.*

PROOF. It follows from [**262**, 4.6] that

$$c(G_1 \times G_2) \leq 2^{c(G_1)} \cdot 2^{c(G_2)}.$$

Since both cardinal numbers $c(G_1)$ and $c(G_2)$ are Ulam non-measurable, it follows from the above inequality that so is $c(G_1 \times G_2)$. According to Theorem 6.5.22, the groups $G_1$ and $G_2$ are completion friendly. Hence, it remains to apply Theorem 6.7.5. □

We will see in the end of the section that the product of two Moscow groups may be a strong $PT$-group which is not a Moscow group (Theorem 6.7.14). Now we need the following lemma:

LEMMA 6.7.7. *If $X$ is any Tychonoff space and $Y$ is a compact space, then $X \times Y$ is $C$-embedded in $\mu X \times Y$.*

PROOF. Take any continuous real-valued function $f$ on $X \times Y$. We have to extend $f$ continuously to $\mu X \times Y$. Let $C(X)$ and $C(Y)$ be the families of continuous real-valued functions on $X$ and $Y$, respectively.

With each $x \in X$ we associate a real-valued function $f_x \in C(Y)$ by the rule $f_x(y) = f(x, y)$. We endow $C(Y)$ with the metric $\varrho$ corresponding to the usual sup-norm, and with topology of uniform convergence, generated by $\varrho$. In other words, we put $\varrho(p, q) = \sup_{y \in Y} |p(y) - q(y)|$, for $p, q \in C(Y)$.

The mapping $g \colon X \to C(Y)$ given by the rule $g(x) = f_x$ is continuous. This follows from the compactness of $Y$ by an obvious standard argument. Since $C(X)$ is metrizable, it is Dieudonné complete. Therefore, there is a continuous extension $\tilde{g}$ of $g$ to $\mu X$, that is, there is a continuous mapping $\tilde{g} \colon \mu X \to C(Y)$ which coincides on $X$ with $g$.

Now we define a function $F$ on $\mu X \times Y$ by $F(z, y) = \tilde{g}(z)(y)$. Clearly, the restriction of $F$ to $X \times Y$ coincides with $f$. We claim that $F$ is continuous at each point $(a, b) \in \mu X \times Y$. Indeed, for any positive number $\varepsilon$, there exists an open neighbourhood $U$ of $a$ in $\mu X$ such that $\varrho(\tilde{g}(z), \tilde{g}(a)) < \varepsilon/2$ whenever $z \in U$. We also can choose an open neighbourhood $V$ of $b$ in $Y$ such that

$$|\tilde{g}(a)(y) - \tilde{g}(a)(b)| < \frac{\varepsilon}{2},$$

for each $y \in V$. Now it is clear that the neighbourhood $U \times V$ of $(a, b)$ in $\mu X \times Y$ has the property that

$$\begin{aligned} |F(z, y) - F(a, b)| &\leq |F(z, y) - F(a, y)| + |F(a, y) - F(a, b))| \\ &= |\tilde{g}(z)(y) - \tilde{g}(a)(y)| + |\tilde{g}(a)(y) - \tilde{g}(a)(b)| \\ &< \frac{\varepsilon}{2} + \frac{\varepsilon}{2} = \varepsilon \end{aligned}$$

whenever $(z, y) \in U \times V$. This completes the proof. □

THEOREM 6.7.8. [**W. W. Comfort and S. Negrepontis**] *If $X$ is a Tychonoff space and $Y$ is any compact space, then $\mu(X \times Y) = \mu X \times Y$.*

PROOF. It follows from Lemma 6.7.7 that $\mu X \times Y \subset \upsilon(X \times Y)$. It is also clear that the space $\mu X \times Y$ is Dieudonné complete and $\mu(X \times Y)$ is the smallest Dieudonné complete subspace of $\upsilon(X \times Y)$ containing $X \times Y$. Therefore, we have the inclusion $\mu(X \times Y) \subset \mu X \times Y$.

On the other hand, $\mu X \times Y$ is a minimal Dieudonné extension of $X \times Y$, by Proposition 6.7.4. It follows that $\mu(X \times Y) = \mu X \times Y$. □

THEOREM 6.7.9. *The product of a completion friendly group $G$ with a compact group $H$ is a completion friendly group, and therefore, a strong $PT$-group.*

PROOF. Indeed, $\mu(G \times H) = \mu G \times H = \mu G \times \mu H$, by Theorem 6.7.8. It remains to apply Theorem 6.7.5. □

We now give a criterion for the product of two topological groups to be a $PT$-group.

THEOREM 6.7.10. *The product $G \times H$ of topological groups $G$ and $H$ is a $PT$-group if and only if $G$ and $H$ are $PT$-groups and the formula $\mu(G \times H) = \mu G \times \mu H$ holds.*

PROOF. If $G$ and $H$ are $PT$-groups, then $\mu G$ and $\mu H$ are topological groups. Therefore, $\mu G \times \mu H$ is also a topological group. It follows from $\mu(G \times H) = \mu G \times \mu H$ that $\mu(G \times H)$ is a topological group. Hence, $G \times H$ is a $PT$-group.

Conversely, suppose that $G \times H$ is a $PT$-group. Clearly, the product group $G \times H$ contains $C$-embedded copies of the groups $G$ and $H$, so $G$ and $H$ are $PT$-groups, by Proposition 6.5.14. It remains to show that the formula $\mu(G \times H) = \mu G \times \mu H$ holds. Since $\mu(G \times H)$ is a topological group containing $G \times H$ as a dense subgroup, $\mu(G \times H)$ can be represented as a topological subgroup of the Raĭkov completion $\varrho(G \times H)$ of the group $G \times H$. Similarly, since $G \times H$ is a dense subgroup of the topological group $\mu G \times \mu H$, the group $\mu G \times \mu H$ can be also represented as a topological subgroup of $\varrho(G \times H)$. Both spaces $\mu(G \times H)$ and $\mu G \times \mu H$ are minimal Dieudonné extensions of the space $G \times H$, by Propositions 6.5.5 and 6.7.4. According to Proposition 6.5.3, the intersection of $\mu(G \times H)$ and $\mu G \times \mu H$, as subsets of $\varrho G \times H$, is again a Dieudonné complete extension of $G \times H$, so by the minimality of $\mu(G \times H)$ and $\mu G \times \mu H$, they coincide. Thus, $\mu(G \times H) = \mu G \times \mu H$. □

THEOREM 6.7.11. *The product of a $PT$-group $G$ and a compact group $H$ is a $PT$-group.*

PROOF. By Theorem 6.7.8, the formula $\mu(G \times H) = \mu G \times H = \mu G \times \mu H$ holds. Since $H$ is also a $PT$-group, the conclusion follows from Theorem 6.7.10. □

Theorem 6.7.10 can obviously be extended to finite products of $PT$-groups. However, for the product of an arbitrary family of topological groups, the corresponding criterion takes a slightly different form.

THEOREM 6.7.12. [**A. V. Arhangel'skii and M. Hušek**] *Suppose that $G = \prod_{\alpha \in A} G_\alpha$ is the product of topological groups $G_\alpha$'s. Then $G$ is a $PT$-group if and only if each $G_\alpha$ is a $PT$-group, and the formula*

$$\mu\Big(\prod_{\alpha \in A} G_\alpha\Big) \subset \prod_{\alpha \in A} \mu G_\alpha \tag{6.2}$$

*holds.*

PROOF. If (6.2) holds, and each $G_\alpha$ is a $PT$-group, then $\prod_{\alpha\in A}\mu G_\alpha$ is a Dieudonné complete topological group containing the Dieudonné completion $\mu G$ of the topological group $G$. Hence, $G$ is a $PT$-group, by Corollary 6.5.9.

Conversely, suppose $G$ is a $PT$-group. Then $G_\alpha$ is a $PT$-group, for each $\alpha \in A$. It remains to show that the formula (6.2) holds. Since $G$ is a $PT$-group, $\mu G$ can be represented as a topological subgroup of the Raĭkov completion $\varrho G$ of $G$. Similarly, the group $\prod_{\alpha\in A}\mu G_\alpha$ can be also represented as a topological subgroup of $\varrho G$. The space $\mu G$ is a minimal Dieudonné extension of the space $G$. Since the intersection of $\mu G$ and $\prod_{\alpha\in A}\mu G_\alpha$ (as subsets of $\varrho G$) is again a Dieudonné extension of $G$, it follows from the minimality of $\mu G$ that $\mu G$ is contained in $\prod_{\alpha\in A}\mu G_\alpha$. Thus, the inclusion (6.2) holds. □

We now present an example of two strong $PT$-groups whose product is not a $PT$-group. In addition, both factors are Moscow spaces, but the product space is not.

EXAMPLE 6.7.13. [**A. V. Arhangel'skii**] Let $X$ be a zero-dimensional pseudocompact non-compact topological group (see item b) of Example 1.6.39). Fix a covering $\eta$ of $X$ satisfying the next three conditions:

1) every element of $\eta$ is an open and closed subset of $X$;
2) no finite subfamily of $\eta$ covers $X$;
3) the union of any finite subfamily of $\eta$ belongs to $\eta$.

Consider the space $G = C_\eta(X)$ of all continuous functions on $X$ with values in the discrete two-element group $D = \{0, 1\}$ endowed with the topology of uniform convergence on elements of $\eta$. A basic open neighbourhood of the neutral element in $G$ has the form

$$O_P = \{f \in G : f(x) = 0 \text{ for each } x \in P\},$$

where $P$ is an arbitrary element of $\eta$. Clearly, $G$ is a topological group, and each $O_P$ is an open subgroup of $G$. Since all elements of $\eta$ are open in $X$, it is also clear that the group $G$ is Raĭkov complete.

By Theorem 3.7.2 and Corollary 4.1.8, the cellularity of $X$ is countable. Hence, there exists a countable subfamily $\gamma$ of $\eta$ such that $\bigcup\gamma$ is dense in $X$. For each $P \in \gamma$, the set $U_P$ of all $f \in G$ such that $f(x) = 0$ for every $x \in P$ is open in $G$ and contains the zero-function $\theta$ on $X$ which is the neutral element of $G$. It is obvious that $\theta$ is the only element in $\bigcap\{U_P : P \in \gamma\}$. Therefore, $\theta$ is a $G_\delta$-point in $G$. Since $G$ is a topological group, it follows that the pseudocharacter of $G$ is countable. Hence, $G$ is a Moscow group.

The group $X$ is also a Moscow space, by Theorem 6.4.2. Since $X$ is pseudocompact, Theorem 6.6.4 implies that $X$ is $C$-embedded in $\beta X$. It follows from Proposition 6.5.18 that $\mu X = \upsilon X = \beta X$, since $c(X) \leq \omega$.

Consider the natural evaluation mapping $\psi$ of the product space $X \times G$ to the discrete space $D = \{0, 1\}$ which on this occasion we treat as a subspace of $\mathbb{R}$. The mapping $\psi$ is defined by $\psi(x, f) = f(x)$, for all $x \in X$ and $f \in G$. Clearly, $\psi$ is continuous, since the elements of $\eta$ are open sets. Observe that $\beta X \times G$ is also a topological group and $X \times G$ is $G_\delta$-dense in $\beta X \times G$.

**Claim.** *The group $X \times G$ is not $C$-embedded in $\beta X \times G$.*

Let us check that $\psi$ cannot be extended to a continuous real-valued function on $\beta X \times G$. Here we will use property 2) of $\eta$. Since the closure in $\beta X$ of any element of $\eta$ is obviously open and $\beta X$ is compact, it follows from 2) that the closures of elements of $\eta$ in $\beta X$ do not cover $\beta X$. Therefore, we can choose $b \in \beta X \setminus X$ such that $b$ does not belong to the closure of any element of $\eta$. Consider the point $(b, \theta) \in \beta X \times G$ and the sets $B = \{(x, \theta) : x \in X\}$ and $C = \{(x, f_P) : P \in \eta,\ x \in X \setminus P\}$, where $f_P \in G$ is the characteristic function of $X \setminus P$, that is, $f_P(x) = 0$ for each $x \in P$ and $f_P(x) = 1$ for each $x \in X \setminus P$. Clearly, $\psi$ takes the value 1 at each element of $C$ and the value 0 at each element of $B$. Also, the point $(b, \theta)$ is in the closure of $B$. Therefore, if $\psi$ could be continuously extended to $(b, \theta)$, the value of this extension at $(b, \theta)$ should be 0. On the other hand, $b$ is not in the closure of any $P \in \eta$. Therefore, $(b, \theta)$ is in the closure of $C$ as well, and the extended function should be 1 at $(b, \theta)$, a contradiction.

Finally, let us show that the group $H = X \times G$ is not a $PT$-group, while both factors $X$ and $G$ are strong $PT$-groups. Indeed, assume that $H$ is a $PT$-group. Then $\mu H$ is a topological group. Therefore, $\mu H$ can be represented as a subgroup of the Raĭkov completion $\varrho H$ containing $H$. Since $\beta X = \varrho X$, by Theorem 6.6.4, we have the equalities $\varrho H = \varrho X \times \varrho G = \beta X \times G = \mu X \times \mu G$. Therefore, by Propositions 6.5.5 and 6.7.4, $\varrho H$ is a minimal Dieudonné extension of $H$. Since $H \subset \mu H \subset \varrho H$, it follows that $\mu H = \varrho H$. However, $H$ is $C$-embedded in $\mu H$. Hence, $H = X \times G$ is $C$-embedded in $\varrho H = \beta X \times G$, which contradicts the above Claim. This contradiction completes the proof that $H = X \times G$ is not a $PT$-group.

It also follows from Theorem 6.4.9 that the $g$-tightness of $\beta X \times G$ and of $H = X \times G$ is uncountable. □

Several other observations about the construction in Example 6.7.13 are in order. First, for the role of the group $X$ we may choose the $\Sigma$-product of $\omega_1$ copies of the discrete group $D = \{0, 1\}$. Then, according to Theorem 2.4.15 (or Theorem 6.6.4), $\beta X$ is the product $D^{\omega_1}$, the weight of $X$ and $\beta X$ is $\aleph_1$, and the cardinality of $G$ is $\aleph_1$. In this case $X$ is countably compact and Fréchet–Urysohn, by Corollary 1.6.35.

Also, the space $G$ is hereditarily Hewitt–Nachbin complete. Indeed, by virtue of Corollary 3.4.26, every Abelian topological group of countable pseudocharacter can be mapped by a continuous isomorphism onto a metrizable topological group $M$. Since the cardinalities of $G$ and $M$ are Ulam non-measurable, and $M$ is metrizable (hence, paracompact), it follows from [**165**, 5.5.10 (b)] that the space $M$ is hereditarily Hewitt–Nachbin complete. Therefore, $G$ is also hereditarily Hewitt–Nachbin complete, by [**165**, 3.11.B (a)]. Observe that $X$ and $G$ are strong $PT$-groups of Ulam non-measurable cardinality, so Theorem 6.7.10 and Proposition 6.5.20 imply that $\mu(X \times G) \neq \mu X \times \mu G$ and $\upsilon(X \times G) \neq \upsilon X \times \upsilon G$.

The group $\beta X \times G$ is not Moscow, since otherwise $X \times G$, as a dense subspace of $\beta X \times G$, would be a Moscow space and, therefore, a strong $PT$-group. On the other hand, $\beta X \times G$ is, obviously, Raĭkov complete and, hence, a strong $PT$-group. Thus, a strong $PT$-group need not be a Moscow group, and a $G_\delta$-dense subgroup of a strong $PT$-group need not be a $PT$-group.

We summarize the most important part of information collected while we discussed Example 6.7.13 in the next statement:

THEOREM 6.7.14. *There exist a countably compact topological group $X$ and a Raĭkov complete group $G$ of countable pseudocharacter with the following properties:*

1) *the product $X \times G$ is not a $PT$-group;*

2) *the product* $\beta X \times G$ *is not a Moscow group;*
3) *the product* $\beta X \times G$ *is a strong* $PT$*-group and even a completion friendly group;*
4) $\beta X \times G = \mu X \times G = \mu X \times \mu G = \upsilon X \times \upsilon G \neq \upsilon(X \times G) = \mu(X \times G)$;
5) *the groups* $X$, $\beta X$, *and* $G$ *are Moscow groups of countable g-tightness;*
6) *the g-tightness of* $\beta X \times G$ *(and of* $X \times G$*) is uncountable.*

The next corollary follows from Proposition 6.5.21 and Theorem 6.7.9:

COROLLARY 6.7.15. *The product of a topological group of countable cellularity and a compact group is a completion friendly group, and therefore, a strong* $PT$*-group.*

## Exercises

6.7.a. Verify that the group $G$ in Example 6.7.13 is not $\omega$-narrow.

6.7.b. Apply Corollary 6.6.10 to deduce the formula $\upsilon \prod_{i \in I} X_i = \prod_{i \in I} \upsilon X_i$ in the case when each $X_i$ is a pseudocompact groupy space.

6.7.c. Show that if $G = \prod_{i \in I} G_i$ is a product of topological groups and $G$ is a strong $PT$-group, then $\mu G = \prod_{i \in I} \mu G_i$. In particular, the equality holds if the group $G$ is Moscow (see Proposition 6.7.1).

6.7.d. (M. G. Tkachenko [**490**]) Prove that the product of a Moscow topological group with a metrizable group is a $PT$-group.
*Hint.* Verify that the product of a Moscow space with a first-countable space is Moscow.

## Problems

6.7.A. (M. G. Tkachenko [**490**]) Prove that if $X$ is a weakly Lindelöf completely regular space and $Y$ is a realcompact $P$-space, then $\upsilon(X \times Y) \cong \upsilon X \times Y$.

6.7.B. (M. G. Tkachenko [**490**]) Show that the product of a Lindelöf topological group $G$ with a precompact group $H$ is a $PT$-group. Extend the result to the case when $H$ is an arbitrary subgroup of a Lindelöf $\Sigma$-group.

6.7.C. (M. G. Tkachenko [**490**]) Extend the conclusion in Exercise 6.7.d to the product of a Moscow topological group with a feathered topological group.

6.7.D. (M. Sanchis [**416**]) Prove that the formula $\mu(\prod_{i \in I} G_i) \cong \prod_{i \in I} \mu G_i$ holds for an arbitrary product $\prod_{i \in I} G_i$ of locally pseudocompact topological groups.

6.7.E. (A. V. Arhangel'skii and M. Hušek [**54**]) Suppose that $S_1, \dots, S_n$ and $\mu(\prod_{i=1}^n S_i)$ are topological semigroups. Prove that the equality $\mu(\prod_{i=1}^n S_i) = \prod_{i=1}^n \mu S_i$ holds. Show that the similar equality remains valid for the Hewitt–Nachbin completion.

## Open Problems

6.7.1. Is every $\omega$-narrow topological group a $PT$-group?

6.7.2. Let $G = \prod_{n \in \omega} G_n$ be the product of completion friendly topological groups $G_n$. Assume also that $G$ is a $PT$-group. Is the formula $\mu G = \prod_{n \in \omega} \mu G_n$ valid?

6.7.3. Let $G$ be a topological group, and suppose that the formula $\mu(G \times H) \cong \mu G \times \mu H$ holds, for each (strong) $PT$-group $H$. Is $G$ locally compact?

6.7.4. Is every completely regular paratopological group of countable pseudocharacter Dieudonné complete? (See also Problems 5.7.2, 5.7.3, and Corollary 6.10.10.)

6.7.5. Let $G = \prod_{i \in I} G_i$ be a product of paratopological groups. Does the formula $\mu G = \prod_{i \in I} \mu G_i$ hold under the assumption that the $o$-tightness (or cellularity) of $G$ is countable?

## 6.8. Subgroups of Moscow groups

Since every topological group $G$ is a dense subgroup of its Raĭkov completion $\varrho G$, it follows from Example 6.7.13 that not every (dense) subgroup of a strong $PT$-group is a $PT$-group. However, we know that a $C$-embedded subgroup of a (strong) $PT$-group is a (strong) $PT$-group, by Proposition 6.5.14. In particular, if $G$ is a $PT$-group such that the space $G$ is normal, then every closed subgroup of $G$ is a $PT$-group. The next result follows from Proposition 6.5.2:

THEOREM 6.8.1. *If $G$ is a Moscow group, then every subgroup $H$ of $G$ which is $C$-embedded in $G$ is $C$-embedded in $\varrho_\omega H$ and, therefore, is a strong $PT$-group.*

COROLLARY 6.8.2. *If $G$ is a Moscow group, and the space $G$ is normal, then every closed subgroup $H$ of $G$ is $C$-embedded in $\varrho_\omega H$ and, therefore, is a strong $PT$-group.*

We will see in Example 6.8.8 below that not every closed subgroup of a Moscow group is a Moscow group, not even closed $C$-embedded subgroups of Moscow groups.

Now we present a general construction that shows that closed subgroups of topological groups need not be as nice as the groups themselves. Our construction is based on the following lemma:

LEMMA 6.8.3. *Every subgroup $H$ of a topological group $G$ can be represented as a closed invariant subgroup of a $G_\delta$-dense subgroup of the topological group $G^{\omega_1}$.*

PROOF. Let $M$ be the $\Sigma$-product of $\omega_1$ copies of the topological group $G$ (over the neutral element $e$ of $G^{\omega_1}$). Then, clearly, $M$ is a $G_\delta$-dense subgroup of $G^{\omega_1}$. Consider the mapping $i$ of $G$ to the diagonal $\Delta$ of $G^{\omega_1}$ defined by the rule $\pi_\alpha i(g) = g$ for all $g \in G$ and all $\alpha \in \omega_1$, where $\pi_\alpha \colon G^{\omega_1} \to G$ is the projection onto the $\alpha$th factor. It is clear that $i(G) = \Delta$ and $i$ is a topological isomorphism.

Put $H^* = i(H)$, and let $E$ be the smallest subgroup of $G^{\omega_1}$ containing both $M$ and $H^*$. It follows from the definition of $E$ that $\Delta \cap E = H^*$. Indeed, the inclusion $H^* \subset \Delta \cap E$ is evident. Conversely, suppose that $h = x_1 y_1 \cdots x_k y_k x_{k+1} \in \Delta \cap E$, where $x_1, \dots, x_{k+1} \in H^*$ and $y_1, \dots, y_k \in M$. Clearly, there exists $\alpha \in \omega_1$ such that $\pi_\alpha(y_i) = e$, for each $i = 1, \dots, k$. Hence, $\pi_\alpha(h) = \pi_\alpha(x_1) \cdots \pi_\alpha(x_{k+1}) \in H$. In addition, from $h \in \Delta$ it follows that $\pi_\beta(h) = \pi_\alpha(h)$ for each $\beta \in \omega_1$, so that $h \in H^*$. This implies that $\Delta \cap E \subset H^*$. Hence, $H^* = \Delta \cap E$.

Clearly, $\Delta$ is closed in $G^{\omega_1}$. It follows that $H^*$ is closed in $E$. Since $E$ contains $M$, the group $E$ is $G_\delta$-dense in $G^{\omega_1}$. The group $H^*$ is, obviously, topologically isomorphic to $H$. It remains to note that $\Delta$ is an invariant subgroup of $G^{\omega_1}$ because the coordinates of each element in $\Delta$ coincide. Therefore, $H^* = \Delta \cap E$ is an invariant subgroup of $E$. □

Here is an interesting corollary to Lemma 6.8.3:

THEOREM 6.8.4. *Every Abelian topological group $H$ can be represented as a closed subgroup of a $G_\delta$-dense subgroup of the product of some family of metrizable Abelian groups.*

PROOF. This follows from Lemma 6.8.3 and the fact that every Abelian topological group $H$ can be represented as a subgroup of the product of a family of metrizable Abelian groups (see Theorem 3.3.15). □

COROLLARY 6.8.5. *Every Abelian topological group $H$ can be represented as a closed subgroup of a Moscow Abelian group.*

PROOF. Indeed, according to Corollary 6.3.16, dense subspaces of products of metrizable spaces are Moscow. $\square$

By Example 6.7.13, there exists an Abelian topological group that is not Moscow. Hence, from Corollary 6.8.5 it follows that not every closed subgroup of a Moscow group is a $PT$-group. Let us show that even closed $C$-embedded subgroups of Moscow groups can fail to be Moscow. We start with two lemmas. The first of them will be extended to a larger class of groups in Theorem 8.1.6.

LEMMA 6.8.6. *Let $f$ be a continuous real-valued function defined on a Lindelöf $P$-group $H$. Then there exists an open invariant subgroup $N$ of $H$ such that $f$ is constant on every coset $xN$ of $N$ in $H$.*

PROOF. For every $r \in \mathbb{R}$, let $U_r = f^{-1}(r)$. Since the space $\mathbb{R}$ is first-countable, every $U_r$ is a $G_\delta$-set in $H$. As $H$ is a $P$-space, the sets $U_r$ are open in $H$. Hence, $\gamma = \{U_r : r \in \mathbb{R}\}$ is a disjoint open covering of $H$. Using the Lindelöf property of $H$, we can find a countable set $A \subset \mathbb{R}$ such that $H = \bigcup_{r\in A} U_r$.

Denote by $\mathcal{N}$ the family of all open invariant subgroups of $H$. Since the group $H$ is $\omega$-narrow, b) of Lemma 4.4.1 implies that $\mathcal{N}$ is a base at the identity of $H$. Fix an element $r \in A$. For every $x \in U_r$, there exists $N_x \in \mathcal{N}$ such that $xN_x \subset U_r$. Clearly, the set $U_r$ is Lindelöf as a closed subset of $H$, and $\{xN_x : x \in U_r\}$ is an open covering of $U_r$. Therefore, there exists a countable set $C_r \subset U_r$ such that $U_r = \bigcup_{x\in C_r} xU_x$. Then $N_r = \bigcap_{x\in C_r} N_x$ is an element of $\mathcal{N}$.

Clearly, $N = \bigcap_{r\in A} N_r$ belongs to $\mathcal{N}$. From our definition of $N$ it follows that the function $f$ is constant on the coset $xN$, for each $x \in H$. Indeed, let $y \in H$ be arbitrary. Then $y \in U_r$ for some $r \in A$, so we can find $x \in C_r$ such that $y \in xN_x$. Hence, $yN = xN \subset xN_x \subset U_r$, and we conclude that $f$ is constant on the coset $yN$. $\square$

LEMMA 6.8.7. *Let $H$ be a Lindelöf subgroup of an $\omega$-narrow topological group $G$. If $H$ is a $P$-space, then $H$ is $C$-embedded in $G$.*

PROOF. Suppose that $f\colon H \to \mathbb{R}$ is a continuous function. By Lemma 6.8.6, there exists an open invariant subgroup $N$ of $H$ such that $f$ is constant on the cosets of $N$ in $H$. Since $N$ is open in $H$, there exists an open neighbourhood $U$ of the identity $e$ in $G$ such that $U \cap H = N$. Choose a symmetric open neighbourhood $V$ of $e$ in $G$ such that $V^2 \subset U$.

**Claim 1.** *The function $f$ is constant on the set $xV \cap H$, for each $x \in G$.*

Indeed, suppose that $x \in G$ and $y_1, y_2 \in xV \cap H$. There exist $v_1, v_2 \in V$ such that $y_i = xv_i$ for $i = 1, 2$. Then $y_1^{-1}y_2 = v_1^{-1}v_2 \in V^2 \subset U$, whence it follows that $y_1^{-1}y_2 \in U \cap H = N$. So, $y_2 \in y_1N$ and, hence, $f(y_1) = f(y_2)$. This proves Claim 1.

Apply Corollary 3.4.19 to find a continuous homomorphism $\pi\colon G \to K$ onto a second-countable topological group $K$ such that $\pi^{-1}(W) \subset V$ for some open neighbourhood $W$ of the identity in $K$. Let $N_0$ be the kernel of $\pi$. Then $N_0 \subset V \subset U$, whence $N_0 \cap H \subset U \cap H = N$. It is clear that $xN_0 \cap H \subset xV \cap H$, so Claim 1 implies that $f$ is constant on the set $xN_0 \cap H$, for each $x \in G$. Therefore, there exists a function $h\colon \pi(H) \to \mathbb{R}$ satisfying the equality $f = h \circ \pi \restriction H$.

**Claim 2.** *The function $h$ is constant on the set $yW \cap \pi(H)$, for each $y \in K$.*

Indeed, let $y \in K$ and $z_1, z_2 \in yW \cap \pi(H)$ be arbitrary points. Take an element $x \in G$ with $\pi(x) = y$ and put $V_0 = \pi^{-1}(W)$. Then $V_0 \subset V$ and $xV_0 = \pi^{-1}\pi(xV_0)$. We thus have

$$yW \cap \pi(H) = \pi(xV_0 \cap H) \subset \pi(xV \cap H).$$

So, one can take elements $v_1, v_2 \in V$ such that $xv_i \in H$ and $\pi(xv_i) = z_i$ for $i = 1, 2$. Since $f = h \circ \pi \restriction H$, we have $f(xv_i) = h(z_i)$ for $i = 1, 2$. Since the elements $xv_1$ and $xv_2$ are in $xV \cap H$, it follows that $f(xv_1) = f(xv_2)$, by Claim 1. This implies immediately the equality $h(z_1) = h(z_2)$ and proves Claim 2.

It follows from Claim 2 that the function $h$ is continuous on the subgroup $H' = \pi(H)$ of $K$. Take an open neighbourhood $O$ of the identity in $K$ such that $OO^{-1} \subset W$. Claim 2 enables us to extend $h$ to a locally constant function on the open set $H'O \subset K$. To see this, take arbitrary points $z \in H'$, $y \in zO$ and put $h^*(y) = h(z)$. This definition is correct by Claim 2. Indeed, suppose that $y \in z_1O \cap z_2O$ for some $z_1, z_2 \in H'$. Then $z_1^{-1}z_2 \in OO^{-1} \subset W$, whence $z_2 \in z_1W$. Hence, Claim 2 implies that $h(z_1) = h(z_2)$, that is, the definition of $h^*(y)$ does not depend on the choice of an element $z \in H'$. It is also clear that $h^* \restriction H' = h$. In addition, our definition of the function $h^*$ implies that it is constant on the set $zO$ for each $z \in H'$. In particular, $h^*$ is continuous on $H'O$.

Denote by $F$ the closure of $H'$ in $K$. It is clear that $F \subset H'O$, so $h^* \restriction F$ is a continuous function on $F$. Since $F$ is a closed subset of the second-countable space $K$, $h^* \restriction F$ can be extended to a continuous function $\tilde{h}$ on $K$. Then $\tilde{f} = \tilde{h} \circ \pi$ is a continuous real-valued function on $G$ which satisfies $\tilde{f} \restriction H = f$. Thus, $H$ is $C$-embedded in $G$. □

Example 6.8.8. A closed $C$-embedded subgroup of a Moscow topological group may fail to be Moscow.

Let $H$ be a non-discrete Lindelöf Abelian $P$-group (see Example 4.4.11). Then $H$ is $\omega$-narrow, so it admits an embedding as a subgroup into a topological product $P$ of second-countable groups. Let $\Sigma \subset P^{\omega_1}$ be the $\Sigma$-product of $\omega_1$ copies of the group $P$ over the neutral element $e$. Consider the diagonal subgroup $H^*$ of $H$ in $P^{\omega_1}$ and denote by $G$ the subgroup of $P^{\omega_1}$ generated by $\Sigma \cup H^*$. It is easy to verify that $G$ is dense in $P^{\omega_1}$ and $H^*$ is closed in $G$ (see Lemma 6.8.3). In addition, $P^{\omega_1}$ is the product of second-countable groups, so the cellularity of $P^{\omega_1}$ is countable and, hence, this group is Moscow, by 3) of Corollary 6.4.11. Therefore, $G$ is also a Moscow group as a dense subgroup of $P^{\omega_1}$.

By Lemma 6.8.7, the group $H^*$ is $C$-embedded in $G$. It remains to note that the non-discrete Lindelöf $P$-group $H \cong H^*$ cannot be a Moscow space, by Corollary 6.2.9. □

The example above also shows that a closed non-discrete subgroup of a Moscow topological group can be a $P$-group (compare this with Corollary 6.2.8).

## Exercises

6.8.a. Suppose that $H_i$ is an open subgroup of a topological group $G_i$, for each $i \in I$. Show that if the product group $G = \prod_{i \in I} G_i$ is Moscow, then so is the group $H = \prod_{i \in I} H_i$.

6.8.b. Give an example of a topological group $G$ such that every subgroup of $G$ is a Moscow space, but $G$ has uncountable pseudocharacter.
*Hint.* Take $G$ to be the $\Sigma$-product of an uncountable family of non-trivial discrete groups and apply Theorem 1.6.24 and 3) of Corollary 6.4.11.

## Problems

6.8.A. Show that the condition of $\omega$-narrowness of the group $G$ in Lemma 6.8.7 can be omitted.

6.8.B. Is the Raĭkov completion of a Moscow topological group Moscow? (See also Problem 6.4.G).
*Hint.* The answer is "no". Indeed, as in Example 6.7.13, take $X$ to be a zero-dimensional pseudocompact non-compact topological group. In fact, one can additionally to choose $G$ to be separable (see Problem 6.6.A). Therefore, $G$ contains a countable dense subgroup $S$. Let also $G$ be the group from Example 6.7.13. Then $G$ is Raĭkov complete and has countable pseudocharacter. It follows that $S \times G$ is a dense subgroup of the Raĭkov complete group $\varrho X \times G$, so that $\varrho(S \times G) = \varrho X \times G$. It is also clear that the group $S \times G$ has countable pseudocharacter and, hence, is Moscow. However, the group $\varrho X \times G$ fails to be Moscow, as was shown in Example 6.7.13.

6.8.C. Suppose that every subgroup of a topological group $G$ is a Moscow space. Is the $o$-tightness of $G$ countable?
*Hint.* Construct a topological group of countable pseudocharacter which has uncountable $o$-tightness.

6.8.D. Prove that every topological group is a quotient of a Moscow topological group.
*Hint.* One can apply Theorem 7.6.18 of Chapter 7.

6.8.E. Let $H$ be a closed pseudocompact subgroup of a Moscow topological group $G$. Prove that the quotient space $G/H$ is Moscow.
*Hint.* Consider the Raĭkov completion $\varrho G$ of the group $G$, and denote by $G^*$ the subgroup of $\varrho G$ generated by $G \cup H^*$, where $H^*$ is the closure of $H$ in $\varrho G$. Show that $G$ is $G_\delta$-dense in $G^*$ and conclude, by Corollary 6.1.8, that $G^*$ is a Moscow group. Since $H^*$ is a compact subgroup of $G^*$, it remains to apply Theorem 6.3.1 (combined with Theorems 1.5.7 and 1.5.16).

## Open Problems

6.8.1. Can every topological group be embedded in a Moscow topological group?

6.8.2. Can every topological group be embedded as a closed subgroup in a Moscow topological group?

6.8.3. Are all subgroups of the product group $\mathbb{Z}^\tau$ Moscow spaces, for any cardinal $\tau$?

6.8.4. Let $G$ be a Moscow topological group and $H$ a closed precompact subgroup of $G$. Is the quotient space $G/H$ Moscow? (See also Problem 6.8.E.)

## 6.9. Pointwise pseudocompact and feathered groups

The theorem of W. W. Comfort and K. A. Ross, stating that every pseudocompact topological group is $C$-embedded in its Raĭkov completion, was discussed in Section 6.6 (see Corollary 6.6.6). Here it is extended to some new classes of topological groups.

A point $x \in X$ is called a *pseudocompactness point* of $X$ if there exists a sequence $\{U_n : n \in \omega\}$ of open neighbourhoods of $x$ in $X$ satisfying the condition:

(pp) *every sequence $\{V_n : n \in \omega\}$ of non-empty open sets in $X$ such that $V_n \subset U_n$ for each $n \in \omega$, has a point of accumulation in $X$.*

A space $X$ is said to be *pointwise pseudocompact* if each point of $X$ is a pseudocompactness point. It is easy to see that every pseudocompactness point of $X$ is a point of canonical weak pseudocompactness of $X$. Therefore, a pointwise pseudocompact space is pointwise canonically weakly pseudocompact.

Recall that a space $X$ is said to be of *countable type at a point* $x \in X$ if there exists a compact set $F \subset X$ with a countable neighbourhood base in $X$ such that $x \in F$. If $X$ is of countable type at every point $x \in X$, we say that $X$ is a space of *pointwise countable type*. Every Čech-complete space as well as any product of a first-countable space with a Čech-complete space is of pointwise countable type. Note that a topological group $G$ is feathered if and only if the space $G$ is of pointwise countable type (see Section 4.3).

A point $x$ of a space $X$ is said to be a *$q$-point* if there exists a sequence $\{U_n : n \in \omega\}$ of open neighbourhoods of $x$ in $X$ satisfying the following condition:

(qp) *For every sequence $\xi = \{x_n : n \in \omega\}$ of points in $X$ such that $x_n \in U_n$ for each $n \in \omega$, there exists a point of accumulation of $\xi$ in $X$.*

A space is called a *$q$-space* if all its points are $q$-points. Obviously, we have:

PROPOSITION 6.9.1. *Every $q$-space is pointwise pseudocompact.*

Since each space $X$ of pointwise countable type is, obviously, a $q$-space, all spaces of pointwise countable type are pointwise pseudocompact. The next two statements are obvious.

PROPOSITION 6.9.2. *If $Y$ is a dense subspace of a space $X$, and $y \in Y$ is a pseudocompactness point of $Y$, then $y$ is also a pseudocompactness point of $X$.*

PROPOSITION 6.9.3. *If a homogeneous space $X$ contains a dense pointwise pseudocompact subspace $Y$, then $X$ is also pointwise pseudocompact.*

COROLLARY 6.9.4. *If a topological group $G$ contains a dense pointwise pseudocompact subspace $Y$, then $G$ is pointwise pseudocompact.*

From Corollary 6.9.4 we obtain:

COROLLARY 6.9.5. *If $G$ is a pointwise pseudocompact topological group, then its Raĭkov completion $\varrho G$ is also a pointwise pseudocompact topological group.*

We can considerably improve Corollary 6.9.5. A subset $B$ of a space $X$ is said to be *bounded in $X$* (or simply *bounded*) if every continuous real-valued function on $X$ is bounded on $B$.

Bounded subsets of a Tychonoff space $X$ can be characterized by means of locally finite families of open sets in $X$ as follows.

LEMMA 6.9.6. *A subset $B$ of a Tychonoff space $X$ is bounded in $X$ iff for every locally finite family $\gamma$ of open sets in $X$, the set $B$ meets only finitely many elements of $\gamma$.*

PROOF. Suppose that $B$ intersects infinitely many elements of some locally finite family $\gamma$ of open sets in $X$. Then we can define sequences $\{x_n : n \in \omega\} \subset B$ and $\{U_n : n \in \omega\} \subset \gamma$ such that $x_n \in U_n$ for each $n \in \omega$ and $x_n \notin U_m$ whenever $n < m$. For every $n \in \omega$, take a continuous real-valued function $f_n$ on $X$ such that $f_n(x_n) = n$ and $f_n(X \setminus U_n) = 0$. Since the family $\gamma$ is locally finite, the function $f = \sum_{n\in\omega} f_n$ is continuous on $X$ and unbounded on $B$. Therefore, $B$ is not bounded in $X$.

Conversely, if $B$ is not bounded in $X$, there exists a continuous function $f\colon X \to \mathbb{R}$ such that $f(B)$ is unbounded. Then the family of open sets $\gamma = \{f^{-1}(n, n+2) : n \in \mathbb{Z}\}$ is locally finite in $X$ and $B$ intersects infinitely many elements of $\gamma$. □

We recall that if $(X, \mathcal{U})$ is a uniform space, then the topology of $X$ generated by $\mathcal{U}$ is Tychonoff [**165**, Coro. 8.1.13]. In particular, all Dieudonné complete spaces are automatically Tychonoff.

Proposition 6.9.7. *The closure of any bounded subset $B$ of a Dieudonné complete space $X$ is compact.*

Proof. We may assume that $B$ is closed in $X$, since the closure of a bounded subset is, obviously, bounded. Let us show that $B$ is also closed in the Čech–Stone compactification $\beta X$ of $X$. Assume the contrary. Then we can fix a point $z \in \beta X \setminus B$ such that $z$ is in the closure of $B$ in $\beta X$. Then, clearly, $x \in \beta X \setminus X$. Since $X$ is Dieudonné complete, there exists a locally finite open covering $\gamma$ of $X$ such that $z$ is not in the closure of $U$, for each $U \in \gamma$ [**165**, 8.5.13 (b)]. Since $B$ is bounded in $X$, Lemma 6.9.6 implies that the family $\eta$ of all $U \in \gamma$ such that $U \cap B \neq \varnothing$ is finite.

Since $B \subset \bigcup \eta$ and $\eta \subset \gamma$, from the choice of $\gamma$ it follows that $z$ is not in the closure of $\bigcup \eta$. Therefore, $z$ is not in the closure of $B$, a contradiction. □

Proposition 6.9.8. *Suppose that $x$ is a pseudocompactness point of a Dieudonné complete space $X$. Then $X$ is of countable type at $x$.*

Proof. Take a sequence $\xi = \{U_n : n \in \omega\}$ of open neighbourhoods of $x$ in $X$ witnessing that $x$ is a pseudocompactness point of $X$. Since $X$ is regular, we may assume that $\overline{U_{n+1}}$ is contained in $U_n$, for each $n \in \omega$. Then the set $P = \bigcap_{n=0}^{\infty} \overline{U_n} = \bigcap_{n=0}^{\infty} U_n$ is closed in $X$. From the choice of $\xi$ it follows that $P$ is bounded in $X$. Since $X$ is Dieudonné complete, Proposition 6.9.7 implies that $P$ is compact. Let us show that $\xi$ is a neighbourhood base of the set $P$ in $X$.

Assume the contrary. Then we can fix an open set $V$ in $X$ such that $P \subset V$ and $U_n \setminus \overline{V} \neq \varnothing$, for each $n \in \omega$. Put $W_n = U_n \setminus \overline{V}$, for each $n \in \omega$. By the choice of $\xi$, there exists a point of accumulation $y \in X$ for the family $\eta = \{W_n : n \in \omega\}$. Since $W_n \subset U_n$, it is clear that $y$ belongs to the closure of $U_n$, for each $n \in \omega$. It follows that $y \in P$. However, this is impossible, since $y$ is, obviously, in $X \setminus V$ and $P \subset V$. This contradiction shows that $\xi$ is a base of the set $P$ in $X$, so $X$ is of countable type at $x$. □

From Propositions 6.9.8 and 6.9.3 we obtain the next result:

Theorem 6.9.9. *If a Dieudonné complete homogeneous space $X$ contains a dense pointwise pseudocompact subspace $Y$, then $X$ is a space of pointwise countable type.*

Corollary 6.9.10. *Every pointwise pseudocompact topological group is Moscow and, therefore, a strong PT-group.*

Proof. This follows from Theorem 6.4.1, since every pointwise pseudocompact space is pointwise canonically weakly pseudocompact. □

The Raĭkov completion of a topological group may have better properties in comparison with the original group:

Theorem 6.9.11. *If a Dieudonné complete topological group $G^*$ contains a dense pointwise pseudocompact subgroup, then $G^*$ is a feathered topological group (therefore, a paracompact space).*

PROOF. Since $G^*$ is Dieudonné complete and homogeneous, from Theorem 6.9.9 it follows that $G^*$ is a space of point countable type. However, every topological group of pointwise countable type is paracompact, by Corollary 4.3.21. □

It turns out that the group $G^*$ in Theorem 6.9.11 can be replaced by the Dieudonné completion of a pointwise pseudocompact group:

THEOREM 6.9.12. *The Dieudonné completion $\mu G$ of every pointwise pseudocompact topological group $G$ is a feathered topological group (therefore, a paracompact space).*

PROOF. By Corollary 6.9.10, $G$ is Moscow. Therefore, according to Theorem 6.5.1, $\mu G$ is a topological group. Evidently, $\mu G$ is Dieudonné complete. To obtain the desired conclusion, it remains to apply Theorem 6.9.11. □

The next result can be considered as the converse to Theorem 6.9.12.

THEOREM 6.9.13. *Every $G_\delta$-dense subspace $X$ of a feathered topological group $G$ is pointwise pseudocompact.*

PROOF. Take any point $x \in X$. Since $G$ is a space of pointwise countable type, $x$ belongs to a compact subspace $F$ of $G$ such that there exists a countable base $\{V_n : n \in \omega\}$ of neighbourhoods of $F$ in $G$. Put $P = F \cap X$ and $U_n = V_n \cap X$, for each $n \in \omega$. Obviously, we have that $x \in P \subset U_n$, for each $n \in \omega$.

Now take any sequence $\xi = \{W_n : n \in \omega\}$ of non-empty open sets in $X$ such that $W_n \subset U_n$, for each $n \in \omega$. Let us show that there exists a point of accumulation of $\xi$ in $X$.

Since $X$ is dense in $G$, there exists an open subset $O_n$ of $G$ such that $O_n \cap X = W_n$ and $O_n \subset V_n$. Since $F$ is compact, and $\{V_n : n \in \omega\}$ is a base of open neighbourhoods of $F$ in $G$, some point $y$ of $F$ is an accumulation point of the sequence $\eta = \{O_n : n \in \omega\}$ in $G$. Put $E_k = \bigcup\{O_n : k \leq n\}$, for $k \in \omega$. Then, clearly, $y \in \overline{E_k}$, for each $k \in \omega$. Since every set $E_k$ is open in $G$, and the space $G$ is pointwise pseudocompact, it follows from Theorem 6.4.1 that the space $G$ is Moscow. Therefore, there exists a $G_\delta$-subset $B$ of $G$ such that $y \in B \subset F$ and every point $z$ of $B$ is an accumulation point of $\eta$ (note that $F$ is also a $G_\delta$-set in $G$). However, $X$ is $G_\delta$-dense in $G$. It follows that $B \cap X \neq \varnothing$. Obviously, every point of $B \cap X$ is an accumulation point of the sequence $\xi = \{W_n : n \in \omega\}$. Hence, $X$ is pointwise pseudocompact. □

From Theorems 6.9.12 and 6.9.13 we obtain immediately:

THEOREM 6.9.14. *A topological group $G$ is pointwise pseudocompact if and only if its Dieudonné completion $\mu G$ is a feathered topological group.*

Since every pointwise pseudocompact topological group is Moscow and, therefore, a strong $PT$-group, we can reformulate Theorem 6.9.14 as follows:

THEOREM 6.9.15. *A topological group $G$ is pointwise pseudocompact if and only if its $G_\delta$-closure $\varrho_\omega G$ in the Raĭkov completion of $G$ is a feathered topological group.*

PROOF. It suffices to apply Theorem 6.9.11 and the fact that the subgroup $\varrho_\omega G$ of $\varrho G$ is Dieudonné complete, by Proposition 6.5.2. □

The next result follows immediately from Corollaries 6.9.10 and 6.1.10.

THEOREM 6.9.16. *Let $G$ be a pointwise pseudocompact topological group, and $Y$ a dense subspace of $G$. Then $Y$ is $C$-embedded in the $G_\delta$-closure of $Y$ in $G$ and, therefore, the next two conditions are equivalent:*

a) *$Y$ is $G_\delta$-dense in $G$;*
b) *$Y$ is $C$-embedded in $G$.*

COROLLARY 6.9.17. *Let $G$ be a feathered topological group (in particular, Čech-complete or locally compact), and $Y$ a dense subspace of $G$. Then $Y$ is $C$-embedded in the $G_\delta$-closure of $Y$ in $G$. Therefore, under these assumptions, $Y$ is $C$-embedded in $G$ if and only if $Y$ is $G_\delta$-dense in $G$.*

PROOF. It suffices to recall that every feathered group is of point-countable type, and every space of point-countable type is pointwise pseudocompact. □

The above results allow to partially generalize the Comfort–Ross theorem on the preservation of pseudocompactness under products in the class of topological groups (Corollary 6.6.11). Indeed, we have:

THEOREM 6.9.18. *Let $G_i$ be a feathered topological group, and $Y_i$ a dense $C$-embedded subspace of $G_i$, for each $i \in \omega$. Then the product space $Y = \prod_{i\in\omega} Y_i$ is $C$-embedded in the product group $G = \prod_{i\in\omega} G_i$.*

PROOF. It follows from the assumptions about the factors that $Y$ is $G_\delta$-dense in $G$. Obviously, $G$ is a feathered topological group, since the product of any countable family of feathered topological groups is a feathered topological group, by Proposition 4.3.13. It remains to apply Corollary 6.9.17. □

THEOREM 6.9.19. *Let $G = \prod_{i\in\omega} G_i$ be the product of pointwise pseudocompact topological groups. Then the space $G$ is also pointwise pseudocompact.*

PROOF. Clearly, $\varrho_\omega G = \prod_{\in\omega} \varrho_\omega G_i$. By Theorem 6.9.15, each $\varrho_\omega G_i$ is a feathered topological group. Therefore, $\varrho_\omega G$ is also a feathered topological group. Applying Theorem 6.9.15 once again, we conclude that the topological group $G$ is also pointwise pseudocompact. □

While the cellularity of every pseudocompact topological group is countable, we cannot expect the same to be true for all pointwise pseudocompact topological groups, since every discrete group is in this class. However, we have the following curious result with an interesting corollary.

THEOREM 6.9.20. *If a topological group $G$ is pointwise pseudocompact, then the $o$-tightness of $G$ is countable.*

PROOF. By Theorem 6.9.12, $G$ is a dense subspace of the feathered topological group $\mu G$. Therefore, the $o$-tightness of $\mu G$ is countable, by Corollary 5.5.7. Since $G$ is dense in $\mu G$, it follows that the $o$-tightness of $G$ is countable. □

Theorem 6.9.20 is instrumental in establishing the following important fact:

THEOREM 6.9.21. *The product of any family of pointwise pseudocompact topological groups is a Moscow group.*

PROOF. Let $\xi$ be a family of pointwise pseudocompact topological groups. By Theorem 6.9.19 and Corollary 6.9.10, the product of any finite subfamily of the family $\xi$ is a Moscow group of countable $o$-tightness. Now it follows from Theorem 6.3.12 that the product of all spaces in $\xi$ is a Moscow topological group. □

Combining Theorem 6.9.21 with Theorem 6.7.2, we obtain:

COROLLARY 6.9.22. *For any family $\{G_\alpha : \alpha \in A\}$ of pointwise pseudocompact topological groups such that the cellularity of the product $G = \prod_{\alpha\in A} G_\alpha$ is Ulam non-measurable, we have:*

$$\upsilon G = \prod_{\alpha\in A} \upsilon G_\alpha.$$

A space $X$ is said to be *locally bounded* if it can be covered by open bounded subsets. We finish this section with considering locally bounded spaces and topological groups.

The next statement is obvious:

PROPOSITION 6.9.23. *Every locally bounded space is pointwise pseudocompact.*

PROPOSITION 6.9.24. *Let $G$ be a topological group, and $Y$ a $G_\delta$-dense subspace of $G$. Then the next conditions are equivalent:*

a) *$G$ is locally bounded;*
b) *$Y$ is locally bounded;*
c) *there exists a non-empty subset $V$ of $Y$ which is open in $Y$ and bounded in $G$.*

*Moreover, if at least one of the conditions a)–c) is satisfied, then $Y$ is $C$-embedded in $G$.*

PROOF. Clearly, b) implies c). Since $Y$ is dense in $G$, a) implies c). Let us show that c) implies a). Take any non-empty open subset $V$ of $Y$ such that $V$ is bounded in $G$. Then $\overline{V}$ is bounded in $G$, where the closure is taken in $G$. Since $Y$ is dense in $G$, $\overline{V}$ contains a non-empty open subset $U$ of $G$. Obviously, $U$ is bounded in $G$. Since $G$ is a topologically homogeneous, it follows that the space $G$ is locally bounded. Notice that we have not used $G_\delta$-denseness of $Y$ in $G$ so far. We are going to use it now to show that a) implies b).

Indeed, every locally bounded space is pointwise pseudocompact. Therefore, if a) holds, then $Y$ is $C$-embedded in $G$, by Theorem 6.9.16. Now take any $y \in Y$. Since $G$ is locally bounded, there exists an open neighbourhood $U$ of $y$ in $G$ such that $U$ is bounded in $G$. Then $V = U \cap Y$ is a non-empty open subset of $Y$ bounded in $G$. Since $Y$ is $C$-embedded in $G$, it follows that $V$ is bounded in $Y$.

The last assertion of the proposition follows from Theorem 6.9.16, as we just saw it in the last portion of the above argument. □

The next lemma generalizes Proposition 3.7.20.

LEMMA 6.9.25. *Suppose that $Y$ is a subspace of a Tychonoff space $X$. If $B$ is a bounded subset of $Y$, then the closure of $B$ in $X$ is contained in the $G_\delta$-closure of $Y$ in $X$.*

PROOF. Suppose that the conclusion of the lemma is false, and choose a point $b \in \overline{B}\setminus Z$, where $\overline{B}$ is the closure of $B$ in $X$ and $Z$ is the $G_\delta$-closure of $Y$ in $X$. It follows from the definition of $Z$ that there exists a sequence $\{U_n : n \in \omega\}$ of open neighbourhoods of $b$ in $X$ such that each $U_n$ contains $b$, and the set $P = \bigcap_{n\in\omega} U_n$ is disjoint from $Y$. Clearly, $b \in P$. We can assume that $U_{n+1} \subset U_n$, for each $n \in \omega$. There exist continuous real-valued functions $f_n$ on $X$ such that $0 \leq f_n \leq 1$, $f_n(b) = 0$, and $f_n(x) = 1$ for each $x \in X \setminus U_n$, where $n \in \omega$. Then the function $f = \sum_{n=0}^{\infty} 2^{-n} \cdot f_n$ is continuous on $X$ and $f(y) > 0$ for each $y \in Y$. Therefore, the function $g$ defined by $g(y) = 1/f(y)$ for each $y \in Y$, is continuous on $Y$. From $b \in \overline{B}$ and $f(b) = 0$ it follows that $g$ is unbounded on $B$. This contradiction shows that $\overline{B} \subset Z$. □

Proposition 6.9.24 is closely related to the next result:

PROPOSITION 6.9.26. *Let $G$ be a topological group and $Y$ a dense subgroup of $G$. Then the next assertions are equivalent:*

a) *$Y$ is locally bounded;*
b) *$G$ is locally bounded, and $Y$ is $G_\delta$-dense in $G$.*

PROOF. By Proposition 6.9.24, b) implies a). Assume that $Y$ is locally bounded. Then, as it was shown in the proof of Proposition 6.9.24, $G$ is locally bounded as well. Now take any non-empty open subset $V$ of the space $Y$ such that $V$ is bounded in $Y$. Let $Z$ be the $G_\delta$-closure of $Y$ in $G$. Since $Y$ is a subgroup of $G$, it follows that $Z$ is a subgroup of $G$. Since $V$ is bounded in $Y$, the closure $\overline{V}$ of $V$ in $G$ is contained in $Z$, by Lemma 6.9.25. But $\overline{V}$ contains a non-empty open subset of $G$, since $G$ is regular and $Y$ is dense in $G$. Therefore, $Z$ contains a non-empty open subset of $G$. It follows that $Z$ is an open subgroup of $G$. This implies that $Z$ is closed in $G$. Since $Z$ contains $Y$, $Z$ is dense in $G$. Hence, $Z = G$ or, equivalently, $Y$ is $G_\delta$-dense in $G$. □

Every Raĭkov complete locally bounded topological group is locally compact, since by Proposition 6.9.7, the closure of any bounded subset in a Dieudonné complete space is compact. This observation, combined with Propositions 6.9.24 and 6.9.26, brings us to the following conclusion:

THEOREM 6.9.27. *For every locally bounded topological group $G$, the Raĭkov completion $\varrho G$ of $G$ is a locally compact group which contains $G$ as a $C$-embedded and, therefore, $G_\delta$-dense subgroup.*

COROLLARY 6.9.28. *If $G$ is a locally bounded topological group, then $\mu G = \varrho G$. In particular, the group $G$ is completion friendly.*

PROOF. By Theorem 6.9.27, $G$ is a dense $C$-embedded subgroup of the locally compact group $\varrho G$. According to Theorem 3.6.24, every locally compact group is Raĭkov complete, so the space $\varrho G$ is Dieudonné complete. Therefore, Theorem 6.5.11 implies that $G$ is a $PT$-group, i.e., $\mu G$ is a topological subgroup of $\varrho G$. Clearly, $\mu G$ is dense in $\varrho G$ and is locally compact. Indeed, let $U$ be a bounded neighbourhood of the neutral element of $G$. Then the closure of $U$ in $\mu G$ is compact, by Proposition 6.9.7, so $\mu G$ is a locally compact group. However, the locally compact subgroup $\mu G$ of $\varrho G$ is closed in $\varrho G$, so that $\mu G = \varrho G$. Since $G$ is $G_\delta$-dense in $\mu G$, it follows that $\mu G = \varrho_\omega G = \varrho G$. Hence, $G$ is completion friendly. □

## Exercises

6.9.a. Give an example of a pointwise pseudocompact topological group which is not a $q$-space.
*Hint.* Let $G$ be a pseudocompact topological group which is not countably compact (see Problem 6.6.A). Then $G^{\omega_1}$ is as required.

6.9.b. Show that there exists a topological group $G$ which is a $q$-space, but fails to be feathered.
*Hint.* Take $G$ to be the $\Sigma$-product of an uncountable family of non-trivial finite discrete groups.

## Problems

6.9.A. Suppose that a topological group $G$ is algebraically generated by a pseudocompact subspace. Show that the group $\varrho_\omega G$ is $\sigma$-compact. (See also Problem 6.10.A.)

6.9.B. Give an example of a homogeneous locally compact space $X$ such that $\mu X$ is neither homogeneous nor locally compact.

6.9.C. (M. Sanchis [**416**]) Let $H$ be a closed bounded subgroup of a locally bounded topological group $G$. Then $\mu G = \varrho G$, by Corollary 6.9.28. Prove the following:
a) $\mu(G/H) \cong \mu G/\overline{H}$, where $\overline{H}$ is the closure of $H$ in $\mu G$;
b) $\mu(G/H \times X) \cong (\mu G/\overline{H}) \times \mu X$, where $X$ is an arbitrary locally pseudocompact space.
Here $Y \cong Z$ means that the spaces $Y$ and $Z$ are naturally homeomorphic.

## Open Problems

6.9.1. Is every first-countable completely regular paratopological group Dieudonné complete?

6.9.2. Let $G$ be a pointwise pseudocompact completely regular paratopological group. Is $\mu G$ a homogeneous space? What if $G$ is a $q$-space?

6.9.3. Is every pointwise pseudocompact paratopological group $G$ a Moscow space? What if $G$ is a $q$-space or has pointwise countable type?

6.9.4. Let $G$ and $H$ be topological groups and suppose that both groups are $q$-spaces. Is $G \times H$ a $q$-space?

## 6.10. Bounded and $C$-compact sets

In this section we consider three distinct classes of subsets of topological groups that arise when one weakens compactness to a special type of placement of a set in a space or topological group. One of these classes, formed by precompact sets in topological groups, was introduced and studied in Section 3.7. Bounded subsets appeared in Section 6.9; they form another class of sets which is especially interesting when related to topological groups (see Theorems 6.10.12 and 6.10.16). We introduce here the new class of $C$-compact sets and establish natural relations between these classes of sets in general and in the realm of topological groups.

We start with collecting the simplest properties of bounded sets in the next proposition.

PROPOSITION 6.10.1. *Let $B$ be a subset of a Tychonoff space $X$.*

a) *If $B \subset K \subset X$ and $K$ is compact, then $B$ is bounded in $X$.*
b) *If $B$ is bounded in $X$, then so is the closure of $B$ in $X$.*
c) *If $X$ is Dieudonné complete (in particular, metrizable) and $B$ is bounded in $X$, then the closure of $B$ in $X$ is compact.*

PROOF. Items a) and b) are obvious. Since every metrizable space is Dieudonné complete, c) follows from Proposition 6.9.7. □

We now establish a simple relation between precompact and bounded sets in topological groups.

PROPOSITION 6.10.2. *Every bounded subset of a topological group is precompact.*

PROOF. Let $B$ be a bounded subset of a topological group $G$. Suppose that there exists a neighbourhood $U$ of the identity $e$ in $G$ such that $B \setminus FU \neq \varnothing$, for each finite subset $F$ of $G$. By induction, define a sequence $\{x_n : n \in \omega\} \subset B$ such that $x_j \notin x_i U$ whenever $i < j$. Then choose an open symmetric neighbourhood $V$ of $e$ in $G$ such that $V^4 \subset U$. It is easy to see that the family $\gamma = \{x_n V : n \in \omega\}$ of open sets is discrete in $G$. In fact, for every $y \in G$, the set $yV$ intersects at most one element of $\gamma$.

The infinite family $\gamma$ is locally finite in $G$ and each element of $\gamma$ intersects $B$, which contradicts Lemma 6.9.6. Therefore, for every neighbourhood $U$ of $e$ in $G$, one can find a finite subset $F$ of $G$ such that $B \subset FU$, and a similar argument implies that the set $F$ can be chosen to satisfy $B \subset UF$. This finishes the proof. □

Since every pseudocompact subspace $B$ of a space $X$ is bounded in $X$, we obtain the following:

COROLLARY 6.10.3. *Pseudocompact subspaces of a topological group are precompact.*

It is easy to see that precompact subsets are not necessarily bounded. For example, consider the torsion subgroup $H$ of the circle group $\mathbb{T}$. Clearly, $H$ is a proper dense subgroup of the compact group $\mathbb{T}$ and, hence, $H$ is a precompact non-compact group. Since the groups $\mathbb{T}$ and $H$ are second-countable, $H$ is not pseudocompact. Therefore, the set $B = H$ is precompact in $H$, but $B$ is not bounded in $H$.

It was established in Theorem 6.4.2 that every precompact topological group is Moscow. Here is a more general fact.

PROPOSITION 6.10.4. *Suppose that a precompact subset $B$ of a topological group $G$ algebraically generates a dense subgroup of $G$. Then the group $G$ is Moscow and, furthermore, is completion friendly.*

PROOF. Denote by $H$ the subgroup of $G$ generated by $B$. By virtue of Corollary 3.7.12, $H$ is topologically isomorphic to a dense subgroup of a $\sigma$-compact topological group $K$, so Corollary 5.3.22 implies that the cellularity of the groups $K$ and $H$ is countable. Since $H$ is dense in $G$, the cellularity of $G$ is countable as well. To conclude that the group $G$ is Moscow, it suffices to apply item 5) of Corollary 6.4.11. Since, according to Theorem 6.5.13, every Moscow group is a strong $PT$-group, and the cellularity of $G$ is countable, it follows from the equivalence of items 1) and 2) of Theorem 6.5.22 that $G$ is completion friendly. □

The next fact is immediate from Propositions 6.10.2 and 6.10.4:

COROLLARY 6.10.5. *If a topological group $G$ contains a dense subgroup generated by a bounded set $B \subset G$, then $G$ is Moscow.*

In general, boundedness may fail to be productive — even the product of two pseudocompact spaces need not be pseudocompact [**165**, 3.10.19]. Nevertheless, boundedness becomes productive in topological groups, similarly to precompactness and pseudocompactness. To prove this fact, we need several auxiliary results.

The following lemma enables us to recognize bounded sets in arbitrary groups via quotient spaces of countable pseudocharacter. The notion of admissible subgroups introduced in Section 5.5 plays an important role here.

LEMMA 6.10.6. *Let $B$ be a subset of a topological group $G$ such that the set $\pi_H(B)$ is bounded in the quotient space $G/H$ for every admissible subgroup $H$ of $G$, where $\pi_H\colon G \to G/H$ is the quotient mapping. Then $B$ is bounded in $G$.*

PROOF. In what follows $G/H$ means the left coset space. Suppose to the contrary that $B$ is not bounded in $G$. Then, by Lemma 6.9.6, there exists a locally finite family $\gamma = \{V_n : n \in \omega\}$ of open sets in $G$ such that $B$ meets every $V_n$. For every $n \in \omega$, choose a point $x_n \in B \cap V_n$ and an open neighbourhood $W_n$ of the identity $e$ in $G$ such that $x_n W_n^2 \subset V_n$. According to b) and c) of Lemma 5.5.2, $W_n$ contains an admissible subgroup $H_n$ of $G$, and $H = \bigcap_{n\in\omega} H_n$ is also an admissible subgroup of $G$. Let $\pi\colon G \to G/H$ be the quotient mapping. Then

$$\pi^{-1}\pi(x_n W_n) = x_n W_n H \subset x_n W_n H_n \subset x_n W_n^2 \subset V_n, \tag{6.3}$$

for each $n \in \omega$. Since the mapping $\pi$ is open and $\gamma$ is locally finite in $G$, from (6.3) it follows that the family $\{\pi(x_n W_n) : n \in \omega\}$ of open sets is locally finite in $G/H$. Clearly, $\pi(B)$ meets every element of this family, so that $\pi(B)$ fails to be bounded in $G/H$, by Lemma 6.9.6. This contradiction completes the proof. □

The following lemma strengthens item a) of Lemma 5.5.2 and explains why quotient spaces $G/H$, with $H$ admissible in $G$, are important. We recall that a space $X$ is *submetrizable* if there exists a continuous one-to-one mapping of $X$ onto a metrizable space.

LEMMA 6.10.7. *If $H$ is an admissible subgroup of a topological group $G$, then the left coset space $G/H$ is submetrizable.*

PROOF. Let $\{U_n : n \in \omega\}$ be a sequence of symmetric open neighbourhoods of the identity $e$ in $G$ satisfying $U_{n+1}^3 \subset U_n$, for each $n \in \omega$, and such that $H = \bigcap_{n\in\omega} U_n$. By Lemma 3.3.10, there exists a prenorm $N$ on $G$ such that

$$\{x \in G : N(x) < 1/2^n\} \subset U_n \subset \{x \in G : N(x) \le 2/2^n\}, \tag{6.4}$$

for each $n \in \omega$. In particular, $N$ is continuous. From (6.4) it also follows that $N(x) = 0$ iff $x \in H$. We claim that

$$N(hx) = N(x) = N(xh), \text{ for all } x \in G \text{ and } h \in H. \tag{6.5}$$

Indeed, if $x \in G$ and $h \in H$, then

$$N(x) = N(h^{-1}hx) \le N(h^{-1}) + N(hx) = N(hx) \le N(h) + N(x) = N(x),$$

which gives the equality $N(hx) = N(x)$. A similar argument implies that $N(xh) = N(x)$.

Define a continuous pseudometric $d$ on $G$ by $d(x, y) = N(x^{-1}y)$, for all $x, y \in G$. A simple verification with the use of (6.5) implies that if $x, y \in G$ and $x' \in xH$, $y' \in yH$, then $d(x', y') = d(x, y)$. This enables us to define a pseudometric $\varrho$ on $G/H$ such that $d(x, y) = \varrho(\pi(x), \pi(y))$ for all $x, y \in G$, where $\pi\colon G \to G/H$ is the quotient mapping. Suppose that $\varrho(\pi(x), \pi(y)) = 0$ for some $x, y \in G$. Then $N(x^{-1}y) = d(x, y) = 0$, whence

$x^{-1}y \in H$ or, equivalently, $\pi(x) = \pi(y)$. Therefore, $\varrho$ is a metric. It remains to show that $\varrho$ is continuous when $G/H$ carries the quotient topology.

Let $x \in G$ and $n \in \omega$ be arbitrary. The set $V = \pi(xU_{n+1})$ is an open neighbourhood of $\pi(x)$ in $G/H$, and if $y \in xU_{n+1}$, then (2) implies that

$$\varrho(\pi(x), \pi(y)) = d(x, y) = N(x^{-1}y) \leq 1/2^n.$$

In other words, $\varrho(\bar{x}, \bar{y}) \leq 1/2^n$ for each $\bar{y} \in V$, where $\bar{x} = \pi(x)$. This means that the metric $\varrho$ is continuous on $G/H$. Therefore, the topology on $G/H$ induced by $\varrho$ is coarser than the quotient topology on $G/H$ and the space $G/H$ is submetrizable. □

We need to extend c) of Proposition 6.10.1 to bounded subsets of submetrizable spaces. This requires the following fact.

PROPOSITION 6.10.8. *Every completely regular submetrizable space $X$ is Dieudonné complete.*

PROOF. Our argument makes use of Proposition 1.6.1. Let $\mathcal{T}$ be the topology of $X$ and $\mathcal{T}_0 \subset \mathcal{T}$ a coarser metrizable topology on $X$. For every continuous real-valued function $f$ on $X$, consider the topology $\gamma_f$ on $X$ defined by

$$\gamma_f = \{f^{-1}(V) : V \text{ is open in } \mathbb{R}\}.$$

In other words, $\gamma_f$ is the coarsest topology on $X$ that makes $f$ continuous. Denote by $\mathcal{T}_f$ the upper bound of the topologies $\mathcal{T}_0$ and $\gamma_f$. It is clear that the topology $\mathcal{T}_f$ is submetrizable and $\mathcal{T}_f \subset \mathcal{T}$.

Let $C(X)$ be the family of all continuous real-valued functions on $X$. Since the space $X$ is completely regular, $\mathcal{T}$ is the upper bound of the topologies $\gamma_f$, with $f \in C(X)$. Therefore, since $\gamma_f \subset \mathcal{T}_f \subset \mathcal{T}$, for each $f \in C(X)$, it follows that $\mathcal{T}$ is the upper bound of the family $\{\mathcal{T}_f : f \in C(X)\}$. Clearly, each topology $\mathcal{T}_f$ contains the metrizable topology $\mathcal{T}_0$, so Proposition 1.6.1 implies that the space $X$ with the original topology $\mathcal{T}$ is homeomorphic to a closed subspace of the product $\prod_{f\in C(X)} X_f$ of metrizable spaces $X_f = (X, \mathcal{T}_f)$. Therefore, the space $X$ is Dieudonné complete. □

Combining c) of Proposition 6.10.1 and Proposition 6.10.8, we obtain the following important result:

COROLLARY 6.10.9. *If $B$ is a bounded subset of a Tychonoff submetrizable space $X$, then the closure of $B$ in $X$ is compact.*

If the neutral element $e$ of a topological group $G$ has countable pseudocharacter in $G$, then the trivial subgroup $\{e\}$ of $G$ is admissible, and Lemma 6.10.7 implies that $G$ itself is submetrizable. This together with Proposition 6.10.8 implies the following:

COROLLARY 6.10.10. *Any topological group of countable pseudocharacter is Dieudonné complete.*

The next auxiliary result involves admissible subgroups of products of topological groups.

LEMMA 6.10.11. *Let $H$ be an admissible subgroup of a product $G = \prod_{\alpha\in A} G_\alpha$ of topological groups and $\varphi\colon G \to G/H$ be the quotient mapping. Then one can find, for every $\alpha \in A$, an admissible subgroup $H_\alpha$ of $G_\alpha$ and a continuous mapping*

*$h\colon \prod_{\alpha\in A} G_\alpha/H_\alpha \to G/H$ such that $\varphi = h\circ\psi$, where $\psi\colon G \to \prod_{\alpha\in A} G_\alpha/H_\alpha$ is the product of the quotient mappings $\psi_\alpha\colon G_\alpha \to G_\alpha/H_\alpha$.*

PROOF. Since $H$ contains the identity $e$ of $G$ and is of type $G_\delta$ in $G$, we can find, for every $\alpha \in A$, a $G_\delta$-set $P_\alpha$ in $G_\alpha$ containing the identity of $G_\alpha$ such that $\prod_{\alpha\in A} P_\alpha \subset H$. By b) and c) of Lemma 5.5.2, $P_\alpha$ contains an admissible subgroup $H_\alpha$ of $G_\alpha$, where $\alpha \in A$. Clearly, $H^* = \prod_{\alpha\in A} H_\alpha \subset H$. Therefore, there exists a continuous mapping $p\colon G/H^* \to G/H$ such that $\varphi = p\circ\pi$, where $\pi\colon G \to G/H^*$ is the quotient mapping. Since $H^*$ is the product of the groups $H_\alpha$, there exists a natural homeomorphism $\lambda\colon G/H^* \to \prod_{\alpha\in A} G_\alpha/H_\alpha$ which assigns to every left coset $xH^* = \pi(x)$ the point $(\psi_\alpha(x_\alpha))_{\alpha\in A}$ in $\prod_{\alpha\in A} G_\alpha/H_\alpha$, where $x = (x_\alpha)_{\alpha\in A}$. It is easy to see that the composition $\lambda\circ\pi\colon G \to \prod_{\alpha\in A} G_\alpha/H_\alpha$ coincides with the mapping $\psi$, the product of the quotient mappings $\psi_\alpha\colon G_\alpha \to G_\alpha/H_\alpha$. Let $h = p\circ\lambda^{-1}$, $h\colon \prod_{\alpha\in A} G_\alpha/H_\alpha \to G/H$. Then $h\circ\psi = p\circ\lambda^{-1}\circ\psi = p\circ\pi = \varphi$, so the diagram below commutes.

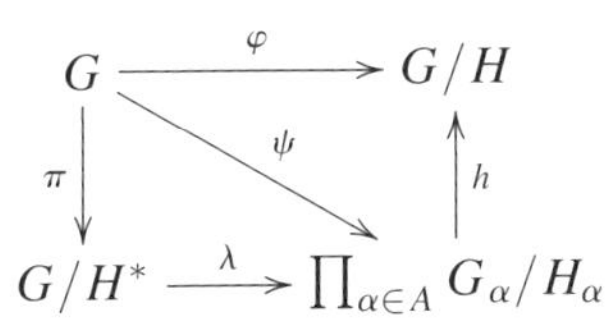


This finishes the proof. □

It is well known that pseudocompactness is not a productive property in topological spaces [**165**, 3.10.19]. The situation completely changes in the realm of topological groups — by Corollary 6.6.11, the topological product of an arbitrary family of pseudocompact groups is pseudocompact. We extend here this result to products of bounded sets in topological groups.

THEOREM 6.10.12. [**M. G. Tkachenko**] *Let $B_\alpha$ be a bounded subset of a topological group $G_\alpha$, for each $\alpha \in A$. Then the set $\prod_{\alpha\in A} B_\alpha$ is bounded in the product group $\prod_{\alpha\in A} G_\alpha$.*

PROOF. Suppose to the contrary that the set $B = \prod_{\alpha\in A} B_\alpha$ is not bounded in $G = \prod_{\alpha\in A} G_\alpha$. By Lemma 6.10.6, there exists an admissible subgroup $H$ of $G$ such that $\varphi(B)$ is not bounded in $G/H$, where $\varphi\colon G \to G/H$ is the quotient mapping. Apply Lemma 6.10.11 to find, for every $\alpha \in A$, an admissible subgroup $H_\alpha$ of $G_\alpha$ and a continuous mapping $h\colon \prod_{\alpha\in A} G_\alpha/H_\alpha \to G/H$ such that $\varphi = h\circ\psi$, where $\psi$ is the product of the quotient mappings $\psi_\alpha\colon G_\alpha \to G_\alpha/H_\alpha$ with $\alpha \in A$. Since $\psi_\alpha$ is continuous, the image $\psi_\alpha(B_\alpha)$ is bounded in $G_\alpha/H_\alpha$ for each $\alpha \in A$. By Lemma 6.10.7, the quotient space $G_\alpha/H_\alpha$ is submetrizable, so Corollary 6.10.9 implies that the closure of $\psi_\alpha(B_\alpha)$ in $G_\alpha/H_\alpha$, say, $K_\alpha$ is compact. Therefore, $\psi(B) = \prod_{\alpha\in A} \psi_\alpha(B_\alpha)$ is a subset of the compact set $K = \prod_{\alpha\in A} K_\alpha$ and, hence, $\varphi(B) = h(\psi(B))$ is contained in the compact set $h(K)$. However, this contradicts the fact that $\varphi(B)$ is not bounded in $G/H$. □

Note that if in the above theorem $B_\alpha = G_\alpha$ for all $\alpha \in A$ (i.e., if each group $G_\alpha$ is pseudocompact), we obtain, as a special case, Corollary 6.6.11 on the preservation of pseudocompactness under products of topological groups.

Theorem 6.10.12 has several applications. The first of them concerns group products of bounded subsets.

Corollary 6.10.13. *Let $B$ and $C$ be bounded subsets of a topological group $G$. Then the sets $B^{-1}$, $C^{-1}$, and $BC$ are bounded in $G$.*

Proof. Let $\theta\colon G\times G\to G$ be the group multiplication, $\theta(x, y) = x\cdot y$ for all $x, y\in G$. By Theorem 6.10.12, the set $B\times C$ is bounded in $G\times G$. Since $\theta$ is a continuous mapping, we conclude that the image $\theta(B\times C) = BC$ is bounded in $G$. Similarly, boundedness of the sets $B^{-1}$ and $C^{-1}$ in $G$ follows from the continuity of the inverse in the group $G$. □

A space $X$ is called *$\sigma$-bounded* if there exists a countable family $\gamma$ of bounded subsets of $X$ such that $X = \bigcup\gamma$.

Corollary 6.10.14. *If a bounded subset $B$ of a topological group $G$ algebraically generates $G$, then the group $G$ is $\sigma$-bounded.*

Proof. Let $C = B\cup B^{-1}\cup\{e\}$, where $e$ is the identity of $G$. By Corollary 6.10.13, the set $B^{-1}$ is bounded and, hence, so is $C$. The same corollary implies that the sets $C^2 = CC$, $C^3 = C^2C$, *etc.*, are bounded in $G$. It remains to note that $G = \bigcup_{n=1}^{\infty} C^n$, whence the conclusion follows. □

Modifying the proof of Proposition 1.8.11 slightly, one can show that every continuous real-valued function defined on a topological group $G$ is uniformly continuous on every compact subset of $G$. In fact, the same assertion remains valid for pseudocompact subsets of topological groups. Let us extend this result to bounded subsets. This requires a lemma.

Lemma 6.10.15. *Let $H = \bigcap_{n\in\omega} U_n$ be an admissible subgroup of a topological group $G$ and $C$ a bounded subset of the quotient space $G/H$. Then, for every neighbourhood $W$ of the diagonal $\Delta$ in $G/H\times G/H$, there exists $n\in\omega$ such that $C^2\cap\pi^2(W_n)\subset W$, where $W_n = \{(x, y)\in G\times G : x^{-1}y\in U_n\}$ and $\pi\colon G\to G/H$ is the quotient mapping.*

Proof. Denote by $K$ the closure of $C$ in $G/H$. By Lemma 6.10.7 and Corollary 6.10.9, $K$ is compact. It is easy to see that $\bigcap_{n\in\omega}\pi^2(W_n) = \Delta$. Therefore, the family

$$\gamma = \{K^2\cap\pi^2(W_n) : n\in\omega\}$$

forms a base of the diagonal $\Delta_K$ in $K\times K$. Indeed, it suffices to verify that the closure of $\pi^2(W_{n+1})$ in $G/H\times G/H$ is contained in $\pi^2(W_n)$, for each $n\in\omega$. Suppose that the point $(\pi(x), \pi(y))$ belongs to the closure of $\pi^2(W_{n+1})$ in $G/H\times G/H$, where $x, y\in G$. Since the mapping $\pi$ is open, we have

$$(\pi(xU_{n+2})\times\pi(yU_{n+2}))\cap\pi^2(W_{n+1})\neq\varnothing.$$

Therefore, there are $x_1, y_1\in G$ and $u, v\in U_{n+2}$ such that $x_1^{-1}y_1\in U_{n+1}$ and $\pi(xu) = \pi(x_1)$, $\pi(yv) = \pi(y_1)$. Then $x_1^{-1}xu\in H$ and $y_1^{-1}yv\in H$, whence $x\in x_1HU_{n+2}$ and $y\in y_1HU_{n+2}$ (we recall that the sets $U_k$ are symmetric). Consequently,

$$x^{-1}y\in U_{n+2}^{-1}Hx_1^{-1}\cdot y_1HU_{n+2}\subset U_{n+2}HU_{n+1}HU_{n+2}\subset U_{n+2}^2U_{n+1}U_{n+2}^2\subset U_{n+1}^3\subset U_n.$$

So, $(x, y)\in W_n$ and $(\pi(x), \pi(y))\in\pi^2(W_n)$, whence the required inclusion follows. □

Theorem 6.10.16. *Let $d$ be a continuous pseudometric on a topological group $G$ and $B$ be a bounded subset of $G$. Then $d$ is uniformly continuous on $B$ with respect to the left group uniformity $\mathscr{V}^l$ of $G$.*

PROOF. Suppose to the contrary that $d$ is not uniformly continuous with respect to $\mathcal{V}^l$. Then we can find $\varepsilon > 0$ such that, for every neighbourhood $U$ of the identity $e$ in $G$, there exist elements $x, y \in B$ satisfying $x^{-1}y \in U$ and $d(x, y) \geq \varepsilon$. We define by induction two sequences $\{U_n : n \in \omega\}$ and $\{(x_n, y_n) : n \in \omega\}$ satisfying the following conditions for all $z \in G$ and $n \in \omega$:

(i) $U_n$ is an open symmetric neighbourhood of $e$ in $G$;
(ii) $U_{n+1}^3 \subset U_n$;
(iii) $x_n, y_n \in B$ and $d(x_n, y_n) \geq \varepsilon$;
(iv) $x_n^{-1}z \in U_n \Rightarrow d(x_n, z) < \varepsilon/4$, and $y_n^{-1}z \in U_n \Rightarrow d(y_n, z) < \varepsilon/4$;
(v) $x_{n+1}^{-1}y_{n+1} \in U_n$.

It follows from (i) and (ii) that $H = \bigcap_{n\in\omega} U_n$ is an admissible subgroup of $G$. Let $\pi\colon G \to G/H$ be the quotient mapping onto the left coset space $G/H$. Put

$$W^* = \{(x, y) \in G \times G : d(x, y) < \varepsilon/2\}, \quad W = \pi^2(W^*)$$

and

$$W_n = \{(x, y) \in G \times G : x^{-1}y \in U_n\}.$$

Then $W$ is an open neighbourhood of the diagonal in $G/H \times G/H$, and we claim that

$$(C^2 \cap \pi^2(W_n)) \setminus W \neq \varnothing \text{ for each } n \in \omega, \tag{6.6}$$

where $C = \pi(B)$. Indeed, $(\pi(x_{n+1}), \pi(y_{n+1})) \in C^2 \cap \pi^2(W_n)$ in view of (iii) and (v). Let us verify that $(\pi(x_{n+1}), \pi(y_{n+1})) \notin W$. Otherwise there exists a pair $(x, y) \in W^*$ such that $\pi(x) = \pi(x_{n+1})$ and $\pi(y) = \pi(y_{n+1})$, i.e., $x_{n+1}^{-1}x \in H \subset U_{n+1}$ and $y_{n+1}^{-1}y \in H \subset U_{n+1}$. Therefore, $d(x, y) < \varepsilon/2$ and by (iv), $d(x_{n+1}, x) < \varepsilon/4$ and $d(y_{n+1}, y) < \varepsilon/4$. This implies that

$$d(x_{n+1}, y_{n+1}) \leq d(x_{n+1}, x) + d(x, y) + d(y, y_{n+1}) < \varepsilon,$$

thus contradicting (iii). This proves (6.6) which, however, contradicts Lemma 6.10.15. The proof is complete. □

COROLLARY 6.10.17. [**M. G. Tkachenko**] *Let $B$ be a bounded subset of a topological group $G$. Then every continuous real-valued function defined on $G$ is uniformly continuous on $B$ with respect to the left group uniformity of $G$.*

PROOF. Let $f\colon G \to \mathbb{R}$ be a continuous function. It suffices to apply Theorem 6.10.16 to the continuous metric $d$ on $G$ defined by $d(x, y) = |f(x) - f(y)|$ for all $x, y \in G$. □

Let $X$ be an arbitrary Tychonoff space. We recall that the *universal uniformity* on $X$ is the finest uniformity compatible with the topology of $X$. According to [**165**, Lemma 8.1.11], the universal uniformity of $X$ is generated by the family of all continuous pseudometrics on $X$.

COROLLARY 6.10.18. *Let $\mathcal{U}$, $\mathcal{V}$, $\mathcal{V}^l$ and $\mathcal{V}^r$ be the universal, two-sided, left and right uniformities of a topological group $G$, respectively. Then the restrictions of the four uniformities to every bounded subset of $G$ coincide.*

PROOF. It is clear that $\mathcal{V}^l \subset \mathcal{V}$, $\mathcal{V}^r \subset \mathcal{V}$, and $\mathcal{V} \subset \mathcal{U}$. Therefore, for a bounded subset $B$ of $G$, it suffices to verify that

$$\mathcal{V}^l \restriction B = \mathcal{U} \restriction B = \mathcal{V}^r \restriction B.$$

Since the uniformity $\mathcal{U}$ is generated by all continuous pseudometrics on $G$, Theorem 6.10.16 implies that $\mathcal{U}\restriction B = \mathcal{V}^l\restriction B$. To show that $\mathcal{U}\restriction B = \mathcal{V}^r\restriction B$, we need the following simple argument. Denote by $G'$ the group with the same underlying set $G$ whose group multiplication $\circ$ is defined by $x \circ y = y \cdot x$, for all $x, y \in G$. When $G'$ carries the same topology as $G$, it becomes a topological group which we denote by the same symbol $G'$. Let $\mathcal{U}_t$, $\mathcal{V}^l_t$, and $\mathcal{V}^r_t$ be the universal, left, and right uniformities of the group $G'$. Denote by $i$ the identity mapping of $G$ onto $G'$. It is easy to see that the mappings $i\colon (G, \mathcal{V}^r) \to (G', \mathcal{V}^l_t)$ and $i\colon (G, \mathcal{U}) \to (G', \mathcal{U}_t)$ are uniformly continuous. This implies that $\mathcal{V}^r = \mathcal{V}^l_t$ and $\mathcal{U} = \mathcal{U}_t$. Finally, from Theorem 6.10.16 it follows that $\mathcal{U}_t\restriction B = \mathcal{V}^l_t\restriction B$, so we conclude that $\mathcal{V}^r\restriction B = \mathcal{V}^l_t\restriction B = \mathcal{U}_t\restriction B = \mathcal{U}\restriction B$. □

There is one more kind of boundedness very close to pseudocompactness. A subset $B$ of a Tychonoff space $X$ is called *$C$-compact* in $X$ if $f(B)$ is a compact subspace of the real line, for every continuous real-valued function $f$ on $X$.

It is easy to see that every pseudocompact subspace of a space $X$ is $C$-compact in $X$ since pseudocompact subspaces of the real line are compact. However, $C$-compact subspaces need not be pseudocompact. In fact, we shall see in Example 6.10.26 that even $C$-compact subsets of topological groups can fail to be pseudocompact.

The next result follows immediately from the definition of $C$-compact sets:

PROPOSITION 6.10.19. *Every pseudocompact subspace of a space $X$ is $C$-compact, and every $C$-compact subset of $X$ is bounded in $X$.*

We now show that every $C$-compact subset of a Tychonoff space is $G_\delta$-dense in its closure.

LEMMA 6.10.20. *Let $B$ and $X$ be subsets of a Tychonoff space $Y$, and $B \subset X$. If $B$ is $C$-compact in $X$, then $B$ is $G_\delta$-dense in $cl_Y B$.*

PROOF. Let $y \in cl_Y B \setminus B$ be an arbitrary point. Suppose to the contrary that there exists a $G_\delta$-set $P$ in $Y$ such that $y \in P$ and $P \cap B = \varnothing$. Then we can find a continuous real-valued function $f$ on $Y$ such that $f \geq 0$, $f(y) = 0$ and $f^{-1}(0) \subset P$. Note that $0 = f(y) \in \overline{f(B)} \setminus f(B)$, so $f(B)$ is not compact and $B$ is not $C$-compact in $Y$. Since $f\restriction X$ is continuous on $X$, $B$ is not $C$-compact in $X$. □

Similarly to precompact subsets, $C$-compact sets in a topological group $G$ can be characterized by means of their closures in the Raĭkov completion of $G$.

LEMMA 6.10.21. *Let $\varrho G$ be the Raĭkov completion of a topological group $G$. Then the following conditions are equivalent for a bounded subset $B$ of $G$:*

a) *$B$ is $C$-compact in $G$;*
b) *$B$ is $G_\delta$-dense in $cl_{\varrho G} B$.*

PROOF. The implication a) $\Rightarrow$ b) follows from Lemma 6.10.20. Let us show that b) $\Rightarrow$ a). Suppose that $B$ is not $C$-compact in $G$. Then there exists a continuous real-valued function $f$ on $G$ such that $f(B)$ is not compact. Since $B$ is bounded in $G$, the set $f(B)$ is bounded in $\mathbb{R}$ and, hence, $f(B)$ is not closed in $\mathbb{R}$. We can assume without loss of generality that $0 \in \overline{f(B)} \setminus f(B)$.

By Corollaries 6.10.17 and 6.10.18, $f$ is uniformly continuous on $B$ with respect to the two-sided uniformity $\mathcal{V}$ of the group $G$. According to Proposition 1.8.4, the restriction of the two-sided uniformity of the group $\varrho G$ to $G$ is the two-sided uniformity of $G$. Therefore, it follows from [**165**, Th. 8.3.10] that the function $f \restriction B$ can be extended to a continuous function $f_K$ on $K = cl_{\varrho G} B$. By Proposition 6.10.2, $B$ is precompact in $G$, so Theorem 3.7.10 implies that $K$ is compact. Since compact sets are $C$-embedded, there exists a continuous real-valued function $g$ on $\varrho G$ such that $g \restriction K = f_K$. Therefore, $g \restriction B = f \restriction B$. Clearly, $g(K)$ is a compact subset of the real line which contains $f(B)$, whence it follows that $0 \in g(K)$. Let $P = g^{-1}(0)$. Then $P \cap K \neq \varnothing$, but $P \cap B = \varnothing$. This implies that $B$ is not $G_\delta$-dense in $K = cl_{\varrho G} B$. □

The following theorem shows that, similarly to boundedness, $C$-compactness is a productive property when considered in the class of topological groups.

THEOREM 6.10.22. *Let $\{G_i : i \in I\}$ be a family of topological groups and, for every $i \in I$, $B_i$ a $C$-compact subset of $G_i$. Then the set $B = \prod_{i \in I} B_i$ is $C$-compact in the product group $G = \prod_{i \in I} G_i$.*

PROOF. Since $C$-compact sets are bounded, the set $B$ is bounded in $G$, by Theorem 6.10.12. Lemma 6.10.21 implies that $B_i$ is $G_\delta$-dense in $K_i = cl_{\varrho G_i} B_i$, for each $i \in I$, where $\varrho G_i$ is the Raĭkov completion of the group $G_i$. Therefore, $B$ is $G_\delta$-dense in $K = \prod_{i \in I} K_i$. According to Corollary 3.6.23, we can identify the Raĭkov completion $\varrho G$ of the group $G$ with the topological product $\prod_{i \in I} \varrho G_i$. Under this identification, we have $cl_{\varrho G} B = K$. Since $B$ is bounded in $G$ and $G_\delta$-dense in $K$, Lemma 6.10.21 implies that $B$ is $C$-compact in $G$. □

According to Proposition 6.10.19, we have:

$$\text{pseudocompact} \Rightarrow C\text{-}compact \Rightarrow \text{bounded}.$$

Now we show that none of these implications can be inverted, not even in topological groups. First, we need a lemma.

LEMMA 6.10.23. *Every countable $C$-compact subset of a Tychonoff space $X$ is compact.*

PROOF. Let $B$ be a countable non-compact subset of $X$. Denote by $\beta X$ the Čech–Stone compactification of $X$. The set $K = cl_{\beta X} B$ is compact, so there exists a point $y \in K \setminus B$. Since $B$ is countable, we can find a closed $G_\delta$-set $P$ in $\beta X$ such that $y \in P$ and $P \cap B = \varnothing$. There exists a continuous real-valued function $f$ on $\beta X$ such that $f(y) = 0$ and $f^{-1}(0) \subset P$. Let $Q = f(B)$. Then $0 = f(y) \in \overline{Q} \setminus Q$, so that $Q$ is not compact. Therefore, $B$ is not $C$-compact in $X$. □

There are many pseudocompact spaces that fail to be countably compact (see [**165**, Example 3.10.29]). Since every Tychonoff space is homeomorphic to a closed subspace of a topological group, the next proposition gives us a lot of bounded subsets of topological groups that are not $C$-compact.

PROPOSITION 6.10.24. *Let $X$ be a pseudocompact, not countably compact subspace of a topological group $G$. Then $X$ contains a countably infinite closed discrete subset $B$ which is bounded in $G$ but is not $C$-compact in $G$.*

PROOF. Since $X$ is not countably compact, it contains a countably infinite closed discrete subset $B$. Clearly, every subset of the pseudocompact space $X$ is bounded in

$X$. Therefore, $B$ is bounded in $G$. The set $B$ is infinite and discrete, hence non-compact. By Lemma 6.10.23, $B$ is not $C$-compact in $G$. □

In Example 6.10.26 below we use the Franklin–Mrówka space $\Psi$ (see [**165**, 3.6.I (a)]). For our purpose, we need a detailed description of this space.

Denote by $\omega$ the discrete space of non-negative integers. A family $\gamma$ of infinite subsets of $\omega$ is called *almost disjoint* if the intersection $A \cap B$ is finite for all distinct $A, B \in \gamma$. If every almost disjoint family $\mu$ of infinite subsets of $\omega$ with $\gamma \subset \mu$ coincides with $\gamma$, then $\gamma$ is called *maximal almost disjoint* or, briefly, a *mad family*. Every almost disjoint family of infinite sets in $\omega$ is contained in a mad family. Let $\gamma_0$ be a countable, infinite, almost disjoint family of infinite subsets of $\omega$ (for example, take a partition of $\omega$ into countably many infinite subsets). Then there exists a mad family $\gamma$ containing $\gamma_0$. It is easy to see that $\gamma$ is uncountable since, otherwise, one can enumerate the family $\gamma$ as $\{A_n : n \in \omega\}$ and define, by induction, an infinite set $B = \{x_k : k \in \omega\} \subset \omega$ such that $x_k \in A_k \setminus \bigcup_{i<k}(A_k \cap A_i)$, for each $k \in \omega$. Then $B \notin \gamma$ and the family $\gamma \cup \{B\}$ is almost disjoint and contains $\gamma$, thus contradicting the fact that $\gamma$ is a mad family.

We topologize the set $\Psi = \omega \cup \gamma$ by declaring the points of $\omega$ isolated in $\Psi$ and taking the sets $\{A\} \cup (A \setminus F)$ as basic open neighbourhoods of a point $A \in \gamma$ in $X$, where $F$ is an arbitrary finite subset of $\omega$. Since the family $\gamma$ is almost disjoint, the space $\Psi$ is Hausdorff. Our definition of the topology on $\Psi$ implies that each of the basic open sets $\{A\} \cup (A \setminus F)$ is clopen in $\Psi$, so the space $\Psi$ is zero-dimensional and, hence, Tychonoff. Clearly, for each $A \in \gamma$, the set $A \subset \omega$ converges to the point $A \in \gamma \subset \Psi$. Note that $\omega$ is a dense open subset of $\Psi$ while $\gamma$ is a closed discrete subspace of the Franklin–Mrówka space $\Psi$.

LEMMA 6.10.25. *The space $\Psi$ is pseudocompact, locally compact, and non-compact. The subset $\gamma$ of $\Psi$ is $C$-compact in $\Psi$.*

PROOF. Let $B$ be an arbitrary infinite subset of $\omega$. Since $\gamma$ is a mad family, there exists $A \in \gamma$ such that $A \cap B$ is infinite. Hence $A$ is an accumulation point of the set $A \cap B$ in $\Psi$. We have thus proved that every infinite subset of $\omega$ has a cluster point in $\Psi$. Since $\omega$ is dense in $\Psi$ and consists of isolated points, we conclude that $\Psi$ is pseudocompact.

Let $f$ be a continuous real-valued function on $\Psi$. Since $\Psi$ is pseudocompact, the image $f(\Psi)$ is compact, and it suffices to verify that $C = f(\gamma)$ is closed in the real line. Suppose to the contrary that the set $C$ has a cluster point $t \notin C$. Since $f$ is continuous, we can define by induction sequences $\{A_n : n \in \omega\} \subset \gamma$ and $B = \{x_n : n \in \omega\} \subset \omega$ satisfying the following conditions for each $n \in \omega$:

(i) $|f(A_n) - t| < 1/2^n$;
(ii) $x_n \in A_n$ and $|f(x_n) - f(A_n)| < 1/2^n$;
(iii) $x_k \notin A_n$ and $x_k \neq x_n$ if $n < k$.

It follows from (i) and (ii) that the sequence $S = \{f(x_n) : n \in \omega\}$ converges to the point $t$, while (iii) implies that $B \subset \omega$ is infinite. Since $t \notin f(\gamma)$, the intersection of $B$ with each $A \in \gamma$ must be finite. However, this is impossible because $\gamma$ is a mad family. □

EXAMPLE 6.10.26. There exist a topological group $G$ and an infinite closed discrete subset $B$ of $G$ such that $B$ is $C$-compact in $G$. In particular, $B$ is not pseudocompact.

PROOF. The Mrówka–Franklin space $\Psi$ is the union of a dense open subset $\omega$ and a closed discrete uncountable subset $B = \gamma$. By Lemma 6.10.25, $B$ is $C$-compact in $\Psi$.

Clearly, $B$ is not pseudocompact. Apply Theorem 1.9.5 to find a topological group $G$ which contains $\Psi$ as a closed subspace. Then $B$ is a closed discrete subset of $G$. The set $B$ being $C$-compact in $\Psi$, remains $C$-compact in $G$. □

## Exercises

6.10.a. It is true that every bounded subset of a semitopological (paratopological) group is precompact?

6.10.b. (M. Bruguera and M. G. Tkachenko [**91**]) An infinite compact space with a single non-isolated point is called a *supersequence*. Suppose that a topological group $G$ contains one of the following sets:
   (a) Alexandroff duplicate of an uncountable compact space;
   (b) two arrows space;
   (c) a copy of $\omega_1$ with the order topology.
   Prove that $G$ contains an uncountable supersequence. Show that, in case (b), $G$ contains a supersequence of length $2^\omega$.

6.10.c. (M. Bruguera and M. G. Tkachenko [**91**]) Suppose that a paratopological group $G$ contains a copy of the ordinal space $\omega_1 + 1$ or an uncountable supersequence. Show that if a sequence $S = \{y_n : n \in \omega\} \subset G$ converges to an element $g \in G$, then $S$ is bounded in $G \setminus \{g\}$.

6.10.d. (M. Bruguera and M. G. Tkachenko [**91**]) Give an example of a topological group $G$ and a subgroup $H$ of $G$ satisfying the following conditions:
   (a) $G$ contains an uncountable supersequence;
   (b) $H$ contains a non-trivial sequence $S$ converging to an element of $G \setminus H$;
   (c) $S$ is not bounded in $H$.

6.10.e. (S. Hernández, M. Sanchis, and M. G. Tkachenko [**231**]) A subset $A$ of a space $X$ is called *relatively pseudocompact* or, for brevity, *r-pseudocompact* in $X$ if every infinite family $\gamma$ of open sets in $X$ such that $U \cap A \neq \varnothing$, for each $U \in \gamma$, has an accumulation point in $A$. Prove the following:
   (a) every $r$-pseudocompact subset of a Tychonoff space $X$ is $C$-compact in $X$;
   (b) every $C$-compact subset of a topological group $G$ is $r$-pseudocompact in $G$, so $r$-pseudocompactness is productive in topological groups (apply item (a) above and Theorem 6.10.22);
   (c) $C$-compact subsets of a Tychonoff space $X$ may fail to be $r$-pseudocompact in $X$.

6.10.f. Prove that if $G$ is a $\sigma$-bounded topological group, then so is the group $G^\bullet$ defined in Section 3.8.

6.10.g. Show that every $\sigma$-bounded topological group is $\omega$-stable.

## Problems

6.10.A. Suppose that a topological group $G$ is algebraically generated by a precompact set $B \subset G$. Prove that the group $\varrho_\omega G$ is $\sigma$-compact (see Problem 6.9.A and Proposition 6.10.4).

6.10.B. (M. G. Tkachenko [**475**]) Let $H$ be a closed subgroup of a topological group $G$. Suppose that $H$ is bounded in $G$ and that $X$ is a bounded subset of the quotient space $G/H$. Prove that the set $\pi^{-1}(X)$ is bounded in $G$, where $\pi\colon G \to G/H$ is the quotient mapping (this generalizes Problem 6.6.C).

6.10.C. (M. Bruguera and M. G. Tkachenko [**91**]) Let $K$ be a closed bounded subset of a topological group $H$, and let $A \subset K$. Prove that if $K$ does not contain non-empty $G_\delta$-subsets of the group $H$, then $A$ is bounded in $H \setminus (K \setminus A)$.

6.10.D. (M. Bruguera and M. G. Tkachenko [**91**]) Suppose that a topological group $H$ contains an uncountable supersequence. Apply Problem 6.10.C to show that if $K$ is a closed bounded subset of $H$ which does not contain uncountable supersequences and $A \subset K$ is an arbitrary set, then $A$ is bounded in $H \setminus (K \setminus A)$.

6.10.E. (M. G. Tkachenko [**475**]) Suppose that $A$ is a bounded subset of a topological group $G$ and $B$ is a bounded subset of a Tychonoff space $X$. Show that the set $A \times B$ is bounded in $G \times X$. Note that this implies the conclusion in Problem 6.6.B.

6.10.F. (S. Hernández and M. Sanchis [**230**]) Let $A$ be the union of a family of $G_\delta$-sets in a topological group $H$. Prove that if $A$ is bounded in $H$, then $cl_H A \times X$ is pseudocompact, for every pseudocompact space $X$. In particular, the space $cl_H A$ is pseudocompact.

6.10.G. (S. Hernández, M. Sanchis, and M. G. Tkachenko [**231**]) Let $A$ be a $C$-compact subset of a topological group $G$. Prove the following:
(a) the set $A$ is $C$-compact in the subgroup $\langle A \rangle$ of $G$ generated by $A$;
(b) in fact, $A$ is $C$-compact in the subspace $AA^{-1}A$ of $G$;
(c) every $C$-compact subgroup of $G$ is pseudocompact;
(d) there exist a topological group $G$ and a bounded subset $B$ of $G$ that fails to be bounded in $\langle B \rangle$.

6.10.H. (S. Hernández, M. Sanchis, and M. G. Tkachenko [**231**]) Show that if $A$ is an $r$-pseudocompact subset of a topological group $G$ and $B$ is $r$-pseudocompact in a Tychonoff space $Y$ (see Exercise 6.10.e), then $A \times B$ is $r$-pseudocompact in $G \times Y$.

6.10.I. (S. Hernández, M. Sanchis, and M. G. Tkachenko [**231**]) Prove that a Tychonoff space $X$ is weakly pseudocompact if and only if it can be embedded into a topological group as an $r$-pseudocompact subspace. Deduce that every uncountable discrete space is homeomorphic to an $r$-pseudocompact subspace of a topological group.

6.10.J. (S. Hernández, M. Sanchis, and M. G. Tkachenko [**231**]) Suppose that $A$ is a bounded subset of a topological group $G$ and $X$ is a Tychonoff space. Prove that $cl_{\beta(G\times X)}(A \times B) \cong cl_{\beta G} A \times cl_{\beta X} B$, for every bounded subset $B$ of $X$. (See also Problems 6.10.E and 6.10.K.)

6.10.K. (S. García-Ferreira, M. Sanchis, and S. Watson [**185**]) Let $A, B$ be bounded subsets of Tychonoff spaces $X$ and $Y$, respectively. Prove that $A \times B$ is bounded in $X \times Y$ provided that $cl_{\beta(X\times Y)}(A \times B) \cong cl_{\beta X} A \times cl_{\beta Y} B$.

6.10.L. (S. Hernández, M. Sanchis, and M. G. Tkachenko [**231**]) Let $A$ be a $C$-compact subset of a topological group $G$ and $B$ a $C$-compact subset of a Tychonoff space $Y$. Prove that $A \times B$ is $C$-compact in $G \times Y$.

6.10.M. (S. Hernández, M. Sanchis, and M. G. Tkachenko [**231**]) Let $G = \prod_{i\in I} G_i$ be the product of a family of topological groups. If $B_i$ is a bounded subset of $G_i$ and $C_i$ is the closure of $B_i$ in the Hewitt–Nachbin completion $\upsilon G_i$ of $G_i$, for each $i \in I$, then $cl_{\upsilon G}(\prod_{i\in I} B_i) \cong \prod_{i\in I} C_i$.

## Open Problems

6.10.1. Let $B_i$ be a bounded ($C$-compact) subset of a paratopological group $G_i$, for each $i \in I$. Is $\prod_{i\in I} B_i$ bounded ($C$-compact) in $\prod_{i\in I} G_i$? What if $|I| = 2$?

6.10.2. Suppose that $B$ is a bounded subset of a paratopological group $G$. Is the set $B^2$ bounded in $G$?

6.10.3. Can Theorem 6.10.16 and Corollary 6.10.17 be extended to bounded subsets of paratopological groups (taking the uniformity defined in Exercise 1.8.m)?

6.10.4. Is every $C$-compact subset of a paratopological group $G$ $r$-pseudocompact in $G$?

6.10.5. Which of items (a)–(c) of Problem 6.10.G remains valid for paratopological groups?

6.10.6. (S. García-Ferreira, M. Sanchis, and S. Watson [**185**]) Let $A, B$ be bounded subsets of Tychonoff spaces $X$ and $Y$, respectively. Does the equality $cl_{\beta(X\times Y)}(A \times B) \cong cl_{\beta X} A \times cl_{\beta Y} B$ hold whenever $A \times B$ is bounded in $X \times Y$?

## 6.11. Historical comments to Chapter 6

The problem posed at the beginning of the eighties by V. G. Pestov and M. G. Tkachenko in [**381**], whether it is always possible to continuously extend the operations in a topological group to its Dieudonné (or Hewitt–Nachbin) completion, became a source of the material for this chapter. The research in the direction of this basic problem was quite extensive, see, for example, [**517**], [**479**], [**481**], [**35**], [**37**], [**54**], [**43**]. Many results of Section 6.1 involve the notion of a Moscow space, introduced by A. V. Arhangel'skii in [**26**]. In particular, Propositions 6.1.1, 6.1.2, Theorems 6.1.8 and 6.1.11, Corollaries 6.1.9 and 6.1.10 are from [**37**]. One can find Corollaries 6.1.13 and 6.1.14 in [**39, 43**]. For Theorem 6.1.7 see [**517**] and [**468**]. Theorem 6.1.4 appeared in [**191**].

Many results in Section 6.2 are taken from [**36**]. In particular, Proposition 6.2.1 and Theorems 6.2.3, 6.2.4, 6.2.5 are from [**36**]. However, some prototypes of these results were obtained earlier by J. R. Isbell in [**254**]. For Theorem 6.2.2 see [**60**], [**263**], or [**30**]. Theorem 6.2.10 and Corollaries 6.2.6, 6.2.7, 6.2.8, 6.2.9, and 6.2.11 are also taken from [**36**]. Theorems 6.3.1, 6.3.3, Proposition 6.3.2, Corollary 6.3.4, and Example 6.3.5 are from [**39**]. Lemma 6.3.6 is also taken from [**39**]. Lemma 6.3.7, Proposition 6.3.8, Corollary 6.3.9 have their origins in [**468**]. Theorems 6.3.11, 6.3.12, and Corollaries 6.3.13, 6.3.14, 6.3.15, 6.3.16 are all from [**39**]. In connection with techniques involving products and used in this chapter, see also the article [**537**] by Y. Yajima.

The notion of a pointwise canonically weakly pseudocompact space was introduced in [**41**]. Theorems 6.4.1 and 6.4.2 are from there. The notion of $g$-tightness was introduced in [**37**]. Propositions 6.4.3, 6.4.4, 6.4.5, 6.4.8, Corollary 6.4.6, and Theorem 6.4.9 all originated in [**37**]. For Theorem 6.4.12, Corollaries 6.4.13, 6.4.14, and further references see [**40**].

The notion of a strong $PT$-group was introduced in [**37**]. For Theorem 6.4.20 see [**40**]. The key role in the proof of this result belongs to Corollary 5.3.22 stating that the cellularity of every $\sigma$-compact topological group is countable (see [**469**]).

For Theorem 6.5.1, Propositions 6.5.2, 6.5.3, 6.5.4, 6.5.5, 6.5.6, and further references see [**37, 40, 39**]. Corollaries 6.5.9, 6.5.10, Theorems 6.5.11, 6.5.13, and 6.5.15 are also either from [**37**] or [**40**]. Example 6.5.30 is from [**40**]; its prototype appeared in [**479**] for another purpose. Proposition 6.5.29 is from [**536**]. See also [**490**] for a discussion of problems related to $PT$-groups. In particular, the following problem from [**490**] still remains open: Is every $\omega$-narrow topological group a $PT$-group?

Theorems 6.6.1 and 6.6.2 come from [**35**]. Corollary 6.6.3, essentially, is a result of W. W. Comfort and K. A. Ross [**122**]. Corollary 6.6.8 appeared in [**479**]. Corollary 6.6.9 is again a result of Comfort and Ross, from the influential paper [**122**]. Corollary 6.6.11 is a famous theorem of Comfort and Ross; it shows that the influence of a compatible group structure on a pseudocompact space is so strong that pseudocompactness becomes totally productive. Theorem 6.6.12 and Corollary 6.6.13 are taken from [**40**].

Our approach in Section 6.7 follows, in general, [**37**]. In particular, Theorem 6.7.2 and Corollary 6.7.3 are taken from there. Some special cases of these results were known earlier, see [**37**] and [**40**] for further references. For Theorem 6.7.5 and Corollary 6.7.6 see [**40**] and [**37**]. Lemma 6.7.7 and Theorem 6.7.8 are due to W. W. Comfort and S. Negrepontis [**117**]. For Theorems 6.7.9, 6.7.10, 6.7.11, and further references see [**54**] and [**40**]. In connection with Theorem 6.7.12 see [**54**]. Example 6.7.13 is taken from [**37**]; the argument about the properties of objects involved in Example 6.7.13 provides a solution to the problem posed by V. G. Pestov and M. G. Tkachenko whether every topological group is a $PT$-group [**381**], a fundamental problem. Example 6.7.13 was constructed by M. Hušek in [**251**] to solve another problem. In connection with Corollary 6.7.15 and for further references see [**40**], [**54**], [**39**], and [**479**].

Many results in Section 6.8 originated in [**40**]. In particular, Theorems 6.8.1, 6.8.4 and Corollaries 6.8.2, 6.8.5 are from [**40**]. Lemmas 6.8.6, 6.8.7, and Example 6.8.8 are from [**490**].

The class of pointwise pseudocompact topological groups, introduced in [**44**], is much wider than the class of (locally) pseudocompact topological groups. On the other hand, it is still a subclass of the class of Moscow topological groups. Probably, any such a naturally defined class is of some interest, especially if it enjoys nice categorical properties, that is, is closed under certain operations. The importance of this particular class of groups is due to the fact that many of Comfort–Ross' results on pseudocompact groups can be extended to it.

Propositions 6.9.1, 6.9.2, 6.9.3, and Corollaries 6.9.4, 6.9.5 are from [**44**]. Proposition 6.9.7 is a part of the folklore. Proposition 6.9.8, Theorems 6.9.9, 6.9.11–6.9.16, 6.9.18–6.9.21, and Corollaries 6.9.17, 6.9.22 are taken from [**44**]. In connection with Proposition 6.9.24 and Theorem 6.9.27, and for further references see [**126**] and [**44**].

Most of the results in Section 6.10 come from [**231**]. The notion of relative pseudocompactness of a subset in a space was introduced by A. V. Arhangel'skii and K. Genedi in [**53**], under the name of strong relative pseudocompactness. See also [**45**] in connection with this concept. Propositions 6.10.1, 6.10.2, and Corollary 6.10.3 are a part of the folklore. For Proposition 6.10.4 see [**91**] and [**231**]. Proposition 6.10.8 and Corollary 6.10.9 also belong to the folklore (see [**165**] and [**32**]). For some further results and references see [**51**].

Theorem 6.10.12 appeared in [**475**]; it was a natural generalization of the fundamental result of W. W. Comfort and K. A. Ross in [**122**] about the preservation of pseudocompactness under taking topological products of topological groups. Lemma 6.10.15, Theorem 6.10.16, and Corollaries 6.10.17, 6.10.18 were also proved in [**475**].

## Chapter 7

# Free Topological Groups

In this chapter, we introduce the notion of a *free topological group* and familiarize the reader with basic properties of these groups that will be used in the rest of the book. Free topological groups were introduced in 1941 by A. A. Markov in [**305**] with the clear idea of extending the well-known construction of a free group from group theory to topological groups. It is easy to give a categorical definition of free topological groups as a kind of projective objects in the category of topological groups and continuous homomorphisms, but the existence proof of such objects is far from trivial, and this is just the first difficulty when studying free topological groups. After the complete (long and complicated) proof of the existence theorem had appeared in [**308**], S. Kakutani [**265**], T. Nakayama [**338**] and M. I. Graev [**201**] brought significant improvements into the original construction. Graev's approach, which involves an extension of continuous pseudometrics from the set of generators of a free topological group over the whole group, seems to be the most fruitful and constructive.

Free topological groups have become a powerful tool of investigation in the theory of topological groups that serve as a source of various examples and as an instrument for proving new theorems. We shall see below numerous facts confirming this statement.

### 7.1. Definition and basic properties

It is well known, and easy to prove (see [**409, 268**]), that for every non-empty set $X$, one can find a group $G$ and a mapping $\sigma\colon X \to G$ such that the image $\sigma(X)$ algebraically generates $G$ and the triple $(G, X, \sigma)$ satisfies the following condition:

(FG) For every mapping $f\colon X \to H$ of $X$ to an arbitrary group $H$, there exists a homomorphism $\tilde{f}\colon G \to H$ such that $\tilde{f} \circ \sigma = f$.

In other words, the homomorphism $\tilde{f}$ makes the diagram below commutative.

$$\begin{array}{ccc} X & \xrightarrow{\ \sigma\ } & G \\ {\scriptstyle f}\big\downarrow & \swarrow{\scriptstyle \tilde{f}} & \\ H & & \end{array}$$

It is not difficult to verify that the mapping $\sigma$ is injective. Such a triple $(G, X, \sigma)$ is called a *free group* on $X$ and is denoted by $F_a(X)$. It turns out that the group $F_a(X) = (G, X, \sigma)$ is unique in the following sense: If $(G', X, \sigma')$ is also a free group on $X$, then there exists an isomorphism $\varphi\colon G \to G'$ such that $\sigma' = \varphi \circ \sigma$. Therefore, identifying $X$ with its image $\sigma(X) \subset G$, we can say that the free group $F_a(X)$ on a set $X$ is an abstract group $G$

algebraically generated by its subset $X$ such that every mapping $f\colon X \to H$ to an arbitrary group $H$ admits an extension to a homomorphism $\tilde{f}\colon G \to H$. This group is then unique up to an isomorphism fixing the points of $X$. The set $X$ is called the *free basis* of $G$.

The change of the word "group" to "Abelian group" in (FG) gives the definition of the *free Abelian group* on $X$, which will be referred to as $A_a(X)$. The word "free" in the first definition appears due to the fact that there are no algebraic relations in $F_a(X)$ between elements of the set $X$ except for the trivial ones. Here are some details.

Since $X$ generates the free group $F_a(X)$, every element $g \in F_a(X)$ has the form $g = x_1^{\varepsilon_1} \dots x_n^{\varepsilon_n}$, where $x_1, \dots, x_n \in X$ and $\varepsilon_1, \dots, \varepsilon_n = \pm 1$. This expression or *word* for $g$ is called *reduced* if it contains no pair of consecutive symbols of the form $xx^{-1}$ or $x^{-1}x$. It turns out that if the word $g$ is reduced and non-empty, then it is different from the identity of $F_a(X)$. In particular, every element $g \in F_a(X)$ distinct from the identity can be uniquely written in the form $g = x_1^{r_1} x_2^{r_2} \dots x_n^{r_n}$, where $n \geq 1$, $r_i \in \mathbb{Z} \setminus \{0\}$, $x_i \in X$, and $x_i \neq x_{i+1}$ for each $i = 1, \dots, n-1$. Such an expression is called the *normal form* of $g$. Similar assertions (with the obvious changes for commutativity) are valid for $A_a(X)$. Note that the empty word is the identity of both $F_a(X)$ and $A_a(X)$.

There exists a profound analogy between abstract and topological groups. To see this, let us add the continuity requirement to (FG) in the case of topological groups, thus obtaining the definition of free topological groups. In what follows we say that a subset $Y$ of a topological group $G$ *topologically generates* $G$ if the subgroup $\langle Y \rangle$ of $G$ generated by $Y$ is dense in $G$.

Let $\sigma\colon X \to G$ be a continuous mapping of a space $X$ to a Hausdorff topological group $G$ that satisfies the following conditions:

1) the image $\sigma(X)$ topologically generates the group $G$;
2) for every continuous mapping $f\colon X \to H$ to a topological group $H$, there exists a continuous homomorphism $\tilde{f}\colon G \to H$ such that $\tilde{f} \circ \sigma = f$.

Then the triple $(G, X, \sigma)$ is denoted by $F(X)$ and is called the *free topological group* on $X$.

Again, if all the groups in the above definition are Abelian, the triple $(G, X, \sigma)$ is said to be the *free Abelian topological group* on $X$, and we designate it $A(X)$. In this section we shall usually prove the results only for the group $F(X)$, because the counterpart for $A(X)$ often requires just to add 'Abelian'before 'group'.

First we show that the free topological group $F(X)$ and the free Abelian topological group $A(X)$ on a space $X$ are, in a natural sense, unique (the existence theorem follows afterwards).

THEOREM 7.1.1. *The free topological group $F(X)$ and the free Abelian topological group $A(X)$ on a Tychonoff space $X$ are unique up to a topological isomorphism "fixing" the points of $X$. In other words, if $(G_1, X, \sigma_1)$ and $(G_2, X, \sigma_2)$ are free (Abelian) topological groups on $X$, then there exists a topological isomorphism $\varphi\colon G_1 \to G_2$ such that $\varphi \circ \sigma_1 = \sigma_2$.*

PROOF. By the definition of a free (Abelian) topological group, there exist continuous homomorphisms $\varphi_1\colon G_1 \to G_2$ and $\varphi_2\colon G_2 \to G_1$ such that $\varphi_1 \circ \sigma_1 = \sigma_2$ and $\varphi_2 \circ \sigma_2 = \sigma_1$. Consider the continuous homomorphism $\psi\colon G_1 \to G_1$ defined by $\psi = \varphi_2 \circ \varphi_1$. Then the restriction of $\psi$ to $Y_1 = \sigma_1(X)$ is the identity mapping, and since $\langle Y_1 \rangle$ is a dense subgroup of the Hausdorff group $G_1$, $\psi$ has to be the identity automorphism of $G_1$. A similar argument implies that the composition $\varphi_1 \circ \varphi_2$ is the identity automorphism of $G_2$. Therefore, both $\varphi_1$ and $\varphi_2$ are topological isomorphisms. Put $\varphi = \varphi_1$. Then the equality $\varphi \circ \sigma_1 = \sigma_2$ follows from the definition of $\varphi_1$. □

Clearly, if $X$ is a discrete space, then the free group $F_a(X)$ and the free Abelian group $A_a(X)$ on the set $X$, endowed with the discrete topology, become the free topological group and the free Abelian topological group on $X$, respectively. The existence and the structure of the free topological group $F(X)$ and of the free Abelian topological group $A(X)$ on a non-discrete Tychonoff space $X$ is far less obvious.

THEOREM 7.1.2. [**A. A. Markov**] *The free topological group $F(X) = (G, X, \sigma)$ on $X$ exists for every Tychonoff space $X$, and the mapping $\sigma\colon X \to G$ is a topological embedding. In addition, the image $\sigma(X)$ algebraically generates $G$. The same is true for the free Abelian topological group $A(X)$.*

PROOF. First we prove the existence of $F(X)$ for every Tychonoff space $X$. Put $\tau = \max\{\mathfrak{c}, |X| \cdot \aleph_0\}$, where $\mathfrak{c}$ is the power of the continuum, and consider the family $\mathscr{F}$ of all continuous mappings $f\colon X \to H_f$ of $X$ to topological groups $H_f$ satisfying $|H_f| \leq \tau$. Since the circle group $\mathbb{T}$ contains a copy of the unit interval $[0, 1]$, $\mathscr{F}$ separates points and closed sets in $X$, that is, $\mathscr{F}$ generates the topology of $X$. In fact, $\mathscr{F}$ is not a set but a proper class. To reduce the size of $\mathscr{F}$, we consider an equivalence relation on $\mathscr{F}$ defined by $f \sim g$ for $f, g \in \mathscr{F}$ if there exists a topological isomorphism $\psi\colon H_f \to H_g$ such that $g = \psi \circ f$. A standard cardinal estimate shows that the quotient set $\mathscr{F}/\!\sim$ of the equivalence classes has cardinality not greater than $\lambda = 2^{2^\tau}$. Therefore, we can choose a representative from each equivalence class, thus obtaining a family $\{f_i : i \in I\}$, where $|I| \leq \lambda$. Clearly, this family also generates the topology of $X$. Let $\sigma\colon X \to \prod_{i\in I} L_i$ be the diagonal product of the family $\{f_i : i \in I\}$, where $L_i = H_{f_i}$. The group $L = \prod_{i\in I} L_i$ with the Tychonoff product topology is a topological group, and $\sigma$ is a homeomorphism of $X$ onto $Y = \sigma(X)$. Denote by $G$ the subgroup of $L$ algebraically generated by $Y$.

It readily follows from the construction that the triple $(G, X, \sigma)$ satisfies condition (1) of the definition of free topological group. So, it suffices to verify (2). Let $f\colon X \to K$ be a continuous mapping to an arbitrary topological group $K$. Denote by $H_f$ the subgroup of $K$ generated by $f(X)$. Then $|H_f| \leq |X| \cdot \aleph_0 \leq \tau$, so $f$ (considered as a mapping to $H_f$) belongs to $\mathscr{F}$. Therefore, $f \sim f_i$, for some $i \in I$, and there exists a topological isomorphism $\psi\colon H_f \to L_i$ satisfying $f_i = \psi \circ f$. Denote by $p_i$ the projection of $L$ to the factor $L_i$. From the definition of $\sigma$ it follows that $f_i = p_i \circ \sigma$.

$$\begin{array}{ccc} X & \xrightarrow{\ f\ } & H_f \\ \Big\downarrow{\scriptstyle \sigma} & \searrow^{f_i} & \Big\downarrow{\scriptstyle \psi} \\ H & \xrightarrow{\ p_i\ } & L_i \end{array}$$

Put $\varphi = \psi^{-1} \circ p_i$. Then $\varphi\colon L \to H_f$ is a continuous homomorphism which satisfies $\varphi \circ \sigma = \psi^{-1} \circ p_i \circ \sigma = \psi^{-1} \circ f_i = f$. Therefore, the continuous homomorphism $\tilde{f} = \varphi \restriction G$ of $G$ to $H_f \subset K$ satisfies the similar equality $\tilde{f} \circ \sigma = f$, thus finishing the proof of the fact that $F(X) = (G, X, \sigma)$ is the free topological group on $X$, with $\sigma$ being a topological embedding.

The argument in the case of the free Abelian topological group $A(X)$ is completely analogous. □

If $F(X) = (G, X, \sigma)$ is the free topological group on a space $X$, then the mapping $\sigma\colon X \to G$ is a topological embedding, by Theorem 7.1.2. We can, therefore, identify $X$ with its image $\sigma(X) \subset G$. This permits us to call $G$ the *free topological group* on $X$, so that $G = F(X)$. The same applies to the free Abelian topological group $A(X)$. In the sequel we shall follow this agreement.

The study of free topological groups has algebraic and topological aspects. It turns out that the algebraic structure of free topological groups is very simple — all these groups are algebraically free. The proof of this result depends on two lemmas of a purely algebraic nature.

Let $p$ be a prime number. Denote by $\Gamma_2(p)$ the multiplicative group of all $2\times 2$ matrices

$$x = \begin{pmatrix} a & b \\ c & d \end{pmatrix}$$

with elements $a, b, c, d \in \mathbb{Z}$ such that $ad - bc = 1$ and $x \equiv e_2 \pmod p$, where $e_2$ is the identity $2 \times 2$ matrix. In other words, the integers $a, d$ are equal to 1 modulo $p$ while $b$ and $c$ are multiples of $p$.

LEMMA 7.1.3. *The group $\Gamma_2(p)$ contains a subgroup isomorphic to the free group $F_a(\mathbb{N})$ with infinitely many generators.*

PROOF. Consider the elements

$$u = \begin{pmatrix} 1 & p \\ 0 & 1 \end{pmatrix}, \quad v = \begin{pmatrix} 1 & 0 \\ p & 1 \end{pmatrix}$$

of $\Gamma_2(p)$. We claim that the subgroup $\langle u, v\rangle$ of $\Gamma_2(p)$ generated by $u$ and $v$ is isomorphic to the free group $F_a(u, v)$ with two generators $u, v$. Indeed, let $W$ be an alternating product of non-zero powers of $u$ and $v$ in $\Gamma_2(p)$. It suffices to show that $W \neq e_2$. If $v^k$ is the first factor in $W$, one can replace $W$ by the element $W' = v^{-k}Wv^k$. Clearly, $W = e_2$ iff $W' = e_2$, so we assume that the first factor in $W$ is a power of $u$. Let, therefore, $W = u^{r_1}v^{r_2}u^{r_3}\dots w^{r_n}$, where $w$ is either $u$ or $v$, and $r_i \neq 0$ for each $i \leq n$. Denote by $W_i$ the product of the first $i$ factors in $W$ (so that $W_1 = u^{r_1}$, $W_2 = u^{r_1}v^{r_2}, \dots, W_n = W$). Let $z_i$ be the upper row of the matrix $W_i$. If $z_{2k-1} = (y_{2k-1}, y_{2k})$, then

$$z_{2k} = z_{2k-1}v^{r_{2k}} = (y_{2k+1}, y_{2k}), \quad z_{2k+1} = z_{2k}u^{r_{2k+1}} = (y_{2k+1}, y_{2k+2}),$$

where $y_{2k+1} = y_{2k-1} + pr_{2k}y_{2k}$ and $y_{2k+2} = y_{2k} + pr_{2k+1}y_{2k+1}$. Here we use the fact that the second row of $v^{r_{2k}}$ is $(pr_{2k}, 1)$ and the first row of $u^{r_{2k+1}}$ is $(1, pr_{2k+1})$. Combining these formulas, we finally obtain

$$y_{i+2} = y_i + pr_{i+1}y_{i+1}, \quad i = 1, 2, \dots, n-1.$$

Let us show that the sequence $|y_1|, |y_2|, \dots$ is strictly increasing. Clearly, $|y_1| = 1$ and $|y_2| = p\cdot|r_1| \geq p \geq 2$. By induction, we have

$$|y_{i+2}| \geq p|r_{i+1}|\cdot|y_{i+1}| - |y_i| \geq 2|y_{i+1}| - |y_i| \geq |y_{i+1}| + 1.$$

Therefore, $y_i \neq 0$ for each $i$, whence $W = W_n \neq e_2$. This proves that the subgroup $\langle u, v\rangle$ of $\Gamma_2(p)$ is isomorphic with $F_a(u, v)$.

It remains to show that $F_a(u, v)$ contains a subgroup isomorphic to the free group $F_a(\mathbb{N})$. Indeed, for every $n \in \mathbb{N}$ put $x_n = v^n u v^n$ and let $X = \{x_n : n \in \mathbb{N}\}$. It is easy to verify that $X$ is a free basis of the subgroup $\langle X \rangle$ of $F_a(u, v)$, that is, $\langle X \rangle \cong F_a(\mathbb{N})$. The lemma is proved. □

LEMMA 7.1.4. *Let $F_a(X)$ be the abstract free group on a non-empty set $X$. Then the family of all homomorphism of $F_a(X)$ to finite groups separate elements of $F_a(X)$. In other words, for every element $g \in F_a(X)$ with $g \neq e$, there exists a homomorphism $h\colon F_a(X) \to K$ to a finite group $K$ such that $h(g) \neq e_K$.*

PROOF. Suppose that $g \in F_a(X)$, $g \neq e$, and $g = x_1^{\varepsilon_1} \dots x_n^{\varepsilon_n}$ is the reduced form of $g$, where $x_1, \dots, x_n \in X$ and $\varepsilon_1, \dots, \varepsilon_n = \pm 1$. Denote by $f$ any retraction of $X$ onto $X_0 = \{x_1, \dots, x_n\}$. Then there exists a homomorphism $\hat{f}\colon F_a(X) \to F_a(X_0)$ such that $f = \hat{f} \restriction X$. In particular, $id_{X_0} = \hat{f} \restriction X_0$. Since $X_0 \subset X$, we can identify $F_a(X_0)$ with the subgroup of $F_a(X)$ generated by $X_0$. This implies that $\hat{f}(g) = g \neq e$. Therefore, it remains to find a homomorphism $h_0$ of $F_a(X_0)$ to a finite group $K$ such that $h_0(g) \neq e_K$. Clearly, the composition $h = h_0 \circ \hat{f}\colon F_a(X) \to K$ will be the homomorphism we are looking for.

Let $p$ be a prime number. By Lemma 7.1.3, the group $\Gamma_2(p)$ contains a subgroup isomorphic to $F_a(\mathbb{N})$. Therefore, the group $F_a(X_0)$ can be identified with a subgroup of $\Gamma_2(p)$. For every $k \in \mathbb{N}$, put

$$\Gamma_2(p^k) = \{x \in \Gamma_2(p) : x \equiv e_2 \pmod{p^k}\}.$$

Then $\Gamma_2(p)$, $\Gamma_2(p^2), \dots$ is a decreasing sequence of invariant subgroups of $\Gamma_2(p)$, and the intersection of these subgroups contains only the identity $e_2$ of $\Gamma_2(p)$. So there exists $k \in \mathbb{N}$ such that $g \notin \Gamma_2(p^k)$. Let $\pi_k\colon \Gamma_2(p) \to \Gamma_2(p)/\Gamma_2(p^k)$ be the quotient homomorphism. It is clear that the group $K = \Gamma_2(p)/\Gamma_2(p^k)$ is finite (in fact, $|K| \leq (p^k)^4 = p^{4k}$). Therefore, $g$ does not belong to the kernel of the homomorphism $h_0 = \pi_k \restriction F_a(X_0)$. □

Let us go back to the study of the algebraic structure of free topological groups. The argument in the next theorem makes use the properties of unitary groups $U(n)$ (see Example 1.4.33).

THEOREM 7.1.5. *For every Tychonoff space $X$, the free topological group $F(X)$ is algebraically free, i.e., $X$ is a free algebraic basis for $F(X)$. In addition, continuous homomorphisms of $F(X)$ to finite-dimensional unitary groups $U(n)$ separate elements of $F(X)$.*

PROOF. Let $X$ be a non-empty Tychonoff space, and suppose that $w$ is a reduced element of the abstract free group $F_a(X)$ on $X$, $w \neq e$. Let also $i\colon F_a(X) \to F(X)$ be the homomorphism that extends the identity mapping $id_X\colon X \to X$. By Theorem 7.1.2, $X$ algebraically generates $F(X)$, so $i$ is an epimorphism. We shall find $n \in \mathbb{N}$ and a continuous homomorphism $\varphi$ of $F(X)$ to the unitary group $U(n)$ such that $\varphi(i(w)) \neq E_n$, where $E_n$ is the identity of $U(n)$. This will imply immediately that the kernel of $i$ is trivial or, equivalently, that the groups $F_a(X)$ and $F(X)$ are algebraically isomorphic.

By Lemma 7.1.4, there exists a homomorphism $h$ of the group $F_a(X)$ to a finite group $K$ such that $h(w) \neq e_K$. The group $K$ is isomorphic to a subgroup of the symmetric group $S_n$ of all permutations of the set $\{1, 2, \dots, n\}$, where $n = |K|$. In its turn, the group $S_n$ is isomorphic to a subgroup of $U(n)$. Indeed, for every element $\pi \in S_n$, consider the orthogonal operator $A_\pi\colon \mathbb{C}^n \to \mathbb{C}^n$ defined on the standard orthogonal basis $e_1, e_2, \dots, e_n$ of $\mathbb{C}^n$ by $A_\pi(e_i) = e_{\pi(i)}$, $i = 1, \dots, n$. Then the mapping $A\colon S_n \to U(n)$ defined by $A(\pi) = A_\pi$ for each $\pi \in S_n$ is a monomorphism of $S_n$ to $U(n)$. Therefore, we can identify $K$ with a subgroup of $U(n)$.

Let $w = x_1^{\varepsilon_1} \dots x_m^{\varepsilon_m}$, where $x_1, \dots, x_m \in X$ and $\varepsilon_1, \dots, \varepsilon_m = \pm 1$. It was established in Example 1.4.33 that the group $U(n)$ is path-connected, so there exists a continuous mapping $f\colon X \to U(n)$ such that $f(x_k) = h(x_k)$ for each $k = 1, \dots, m$. Indeed, for every $k \le m$, let $\varphi_k\colon [0,1] \to U(n)$ be a continuous mapping such that $\varphi_k(0) = E_n$ and $\varphi_k(1) = h(x_k) = a_k$. We can choose the mappings $\varphi_k$ in such a way that $a_i = a_j$ always implies $\varphi_i = \varphi_j$, $1 \le i, j \le m$. Then we choose open neighbourhoods $V_1, \dots, V_m$ of the points $x_1, \dots, x_m$ in $X$ such that $V_i \cap V_j = \varnothing$ if $x_i \ne x_j$ and $V_i = V_j$ if $x_i = x_j$. Since $X$ is a completely regular space, there exist continuous functions $\psi_1, \dots, \psi_m$ on $X$ with values in $[0,1]$ satisfying the following conditions for each $k = 1, \dots, m$:

(1) $\psi_k(x_k) = 1$;
(2) $\psi_k(X \setminus V_k) = 0$;
(3) $\psi_i = \psi_j$ if $x_i = x_j$, $1 \le i, j \le m$.

Put $F = X \setminus \bigcup_{k=1}^m V_k$. From (2) it follows that $\psi_k(x) = 0$ for all $x \in F$ and $k \le m$. Consider the mappings $\chi_k\colon X \to U(n)$, where $\chi_k = \varphi_k \circ \psi_k$ for $k = 1, \dots, m$. It is easy to see that $\chi_k(x_k) = a_k$ and $\chi_k(x) = E_n$ for all $x \in F$ and $k = 1, \dots, m$. Therefore, the mapping $f\colon X \to U(n)$ defined by

$$f(x) = \begin{cases} \chi_k(x) & \text{if } x \in V_k \\ E_n & \text{if } x \in F \end{cases}$$

is continuous on $X$ and satisfies $f(x_k) = a_k = h(x_k)$ for each $k \le m$.

Finally, extend $f$ to a continuous homomorphism $\tilde{f}\colon F(X) \to U(n)$. Then

$$\tilde{f}(i(w)) = f(x_1)^{\varepsilon_1} \cdot \dots \cdot f(x_m)^{\varepsilon_m} = a_1^{\varepsilon_1} \cdot \dots \cdot a_n^{\varepsilon_n} = h(w) \ne E_n,$$

whence $i(w) \ne e$. In other words, $i\colon F_a(X) \to F(X)$ is an algebraic isomorphism. □

Let us show that Theorem 7.1.5 has its counterpart in the Abelian case. First we need a simple auxiliary fact.

LEMMA 7.1.6. *The circle group $\mathbb{T}$ contains a subgroup algebraically isomorphic to the free Abelian group with $2^\omega$ generators.*

PROOF. Consider the additive group of the reals $\mathbb{R}$ as a vector space over the field $\mathbb{Q}$. Let $X = \{x_\alpha : \alpha < \mathfrak{c}\}$ be a Hamel basis for $\mathbb{R}$ such that $x_0 = 1$, where $\mathfrak{c} = 2^\omega$. Denote by $\pi$ the homomorphism of $\mathbb{R}$ onto $\mathbb{T}$ defined by $\pi(x) = e^{2\pi i x}$, for each $x \in \mathbb{R}$. Clearly, $\ker \pi = \mathbb{Z}$. For every $\alpha < \mathfrak{c}$, put $y_\alpha = \pi(x_\alpha)$. We claim that the set $Y = \{y_\alpha : 0 < \alpha < \mathfrak{c}\}$ is linearly independent in $\mathbb{T}$, that is, $Y$ generates the subgroup of $\mathbb{T}$ isomorphic to the free Abelian group of size $\mathfrak{c}$. Indeed, if $0 < \alpha_1 < \dots < \alpha_n < \mathfrak{c}$ and $k_1, \dots, k_n \in \mathbb{Z} \setminus \{0\}$, then $k_1 x_{\alpha_1} + \dots + k_n x_{\alpha_n} \notin \mathbb{Z}$ and, hence,

$$y_{\alpha_1}^{k_1} \cdot \dots \cdot y_{\alpha_n}^{k_n} = \pi(k_1 x_{\alpha_1} + \dots + k_n x_{\alpha_n}) \ne 1.$$

Therefore, the subgroup $\langle Y \rangle$ of $\mathbb{T}$ is isomorphic to the free Abelian group $A_a(Y)$. □

The above argument also shows that the abstract group $\mathbb{T}$ is isomorphic to the group $(\mathbb{Q}/\mathbb{Z}) \oplus A$, where $A$ is a free Abelian group of cardinality $2^\omega$.

THEOREM 7.1.7. *For a Tychonoff space $X$, the free Abelian topological group $A(X)$ is algebraically isomorphic to the group $A_a(X)$, with $X$ being the free algebraic basis for $A(X)$.*

PROOF. By Theorem 7.1.2, $X$ algebraically generates the group $A(X)$. It remains to verify that $X$ is the free basis of $A(X)$ in the sense that if $x_1, \ldots, x_n$ are pairwise distinct elements of $X$ and $k_1, \ldots, k_n \in \mathbb{Z} \setminus \{0\}$, then $x_1^{k_1} \cdots x_n^{k_n} \neq e$. By Lemma 7.1.6, we can find linearly independent elements $t_1, \ldots, t_n \in \mathbb{T}$. Clearly, there exists a continuous mapping $f\colon X \to \mathbb{T}$ such that $f(x_i) = t_i$, for $i = 1, \ldots, n$. Extend $f$ to a continuous homomorphism $\tilde{f}\colon A(X) \to \mathbb{T}$. If $k_1, \ldots, k_n \in \mathbb{Z} \setminus \{0\}$, then $\tilde{f}(x_1^{k_1} \cdots x_n^{k_n}) = t_1^{k_1} \cdot \ldots \cdot t_n^{k_n} \neq 1$, whence $x_1^{k_1} \cdots x_n^{k_n} \neq e$. □

Theorems 7.1.5 and 7.1.7 enable us to identify $F(X)$ with $F_a(X)$ and $A(X)$ with $A_a(X)$ from the algebraic point of view. We can now say that the free topological group $F(X)$ is merely the free group $F_a(X)$ endowed with the topological group topology that induces the original topology on the set $X$ of its generators (see Theorem 7.1.2) and such that every continuous mapping $f\colon X \to H$ to a topological group $H$ can be extended to a continuous homomorphism $\tilde{f}\colon F(X) \to H$. In fact, this assertion can be given a more elegant form.

COROLLARY 7.1.8. *The topology of the group $F(X)$ is the finest topological group topology on $F_a(X)$ that generates on $X$ its original topology. The same is valid for $A(X)$.*

PROOF. By Theorems 7.1.2 and 7.1.5, $X$ is a free algebraic basis for the group $F(X) \cong F_a(X)$, and the topology $\mathcal{T}$ of $F(X)$ generates on $X$ its original topology $\tau_X$. Let $\mathcal{T}'$ be a group topology on $F_a(X)$ such that $\mathcal{T}' \restriction X = \tau_X$. Extend the identity mapping $i\colon X \to X \subset H$ to a continuous homomorphism $\tilde{\imath}\colon F(X) \to H$, where $H$ is the group $F_a(X)$ endowed with the group topology $\mathcal{T}'$. Clearly, $\tilde{\imath}$ is an (algebraic) isomorphism, and the continuity of $\tilde{\imath}$ implies that $\mathcal{T}' \subset \mathcal{T}$. This argument also applies to the group $A(X)$. □

COROLLARY 7.1.9. *Let $f\colon X \to Y$ be a continuous mapping of Tychonoff spaces. Then $f$ admits an extension to continuous homomorphisms $F(f)\colon F(X) \to F(Y)$ and $A(f)\colon A(X) \to A(Y)$. In addition, if $f$ is quotient and $f(X) = Y$, then the homomorphisms $F(f)$ and $A(f)$ are open.*

PROOF. Since $Y$ is identified with the corresponding subspace of $F(Y)$, we can consider $f$ as a continuous mapping of $X$ to the free topological group $F(Y)$. Therefore, by the definition of $F(X)$, $f$ can be extended to a continuous homomorphism $F(f)\colon F(X) \to F(Y)$ which we shall denote by $\hat{f}$.

Now suppose that $f$ is quotient and onto. Denote by $\mathcal{T}_q$ the family of all images $\hat{f}(U)$, where $U$ is open in $F(X)$. It follows from the continuity of $\hat{f}$ that $\mathcal{T}_q$ is a group topology on the abstract group $F_a(Y)$ which is finer than the topology $\mathcal{T}$ of $F(Y)$. Let us show that $\mathcal{T}_q$ induces on $Y$ its original topology $\tau_Y$. Since $\mathcal{T} \restriction Y = \tau_Y$ (Corollary 7.1.8), it suffices to verify that $V = \hat{f}(U) \cap Y$ is open in $Y$ for every open subset $U$ of $F(X)$. Denote by $N$ the kernel of $\hat{f}$. It is easy to see that $f^{-1}(V) = X \cap NU$, so the set $f^{-1}(V)$ is open in $X$. Since the mapping $f$ is quotient and onto, we infer that $V$ is open in $Y$.

Thus, the group topology $\mathcal{T}_q$ on $F_a(Y)$ is finer than the topology $\mathcal{T}$ of $F(Y)$ and induces the original topology on $Y$. Hence Corollary 7.1.8 implies that $\mathcal{T}_q = \mathcal{T}$. In other words, $\hat{f}(U)$ is open in $F(Y)$ for every open set $U$ in $F(X)$. The argument in the case of the group $A(X)$ is exactly the same. □

The operations $X \mapsto F(X)$ and $f \mapsto F(f) = \hat{f}$ that assign to a Tychonoff space $X$ the free topological group $F(X)$ and to a continuous mapping $f\colon X \to Y$ its extension to the continuous homomorphism $\hat{f}\colon F(X) \to F(Y)$, define the *covariant functor* $F$ from the category of Tychonoff spaces and continuous mappings to the category of topological groups and continuous homomorphisms. Indeed, one easily verifies that if $g\colon Y \to Z$ is a continuous mapping (with $Y, Z$ being Tychonoff) and $h = g \circ f$, then $\hat{h} = \hat{g} \circ \hat{f}$, that is, $F(g \circ f) = F(g) \circ F(f)$. Similarly, one defines the covariant functor $A$ from the category of Tychonoff spaces to the category of Abelian topological groups.

The above corollary as well as the next one have many applications.

COROLLARY 7.1.10. *Let $f\colon X \to G$ be a quotient mapping of a Tychonoff space $X$ onto a topological group $G$. Then the continuous homomorphism $\tilde{f}\colon F(X) \to G$ extending $f$ is open. Similarly, if $G$ is Abelian, then the continuous homomorphism $\tilde{f}_A\colon A(X) \to G$ extending $f$ is also open.*

PROOF. Extend the identity mapping $i\colon G \to G$ to a continuous homomorphism $\tilde{\imath}\colon F(G) \to G$. Let also $\hat{f}\colon F(X) \to F(G)$ be the continuous open homomorphism extending $f$ (see Corollary 7.1.9). Then $\tilde{f} = \tilde{\imath} \circ \hat{f}$, so it suffices to show that the homomorphism $\tilde{\imath}$ is open.

Denote by $x \cdot y$ and $xy$ the product of the elements $x, y \in G$ in the groups $F(G)$ and $G$, respectively. Let $U$ be an open neighbourhood of the identity $e$ in $F(G)$. Choose an arbitrary element $x \in G$. Since $x^{-1} \cdot x = e$, there exists an open neighbourhood $V$ of $x$ in $G$ such that $x^{-1} \cdot V \subset U$. Then $\tilde{\imath}(U) \supseteq \tilde{\imath}(x^{-1} \cdot V) = x^{-1}V$ is an open neighbourhood of the identity in $G$. Therefore, the homomorphisms $\tilde{\imath}$ and $\tilde{f}$ are open. The same argument applies to the group $A(X)$ when $G$ is Abelian. □

Let us now establish a relation between the groups $F(X)$ and $A(X)$. We recall that the *derived* subgroup $G'$ of an abstract group $G$ is the subgroup of $G$ generated by all commutators $x^{-1}y^{-1}xy$, where $x, y \in G$. It is easy to verify that $G'$ is an invariant subgroup of $G$ and that the quotient group $G/G'$ is Abelian [**409**].

THEOREM 7.1.11. *The derived subgroup $K$ of the free topological group $F(X)$ is closed in $F(X)$ and $A(X) \cong F(X)/K$. In other words, $A(X)$ is a quotient of $F(X)$.*

PROOF. Let $i\colon X \to X$ be the identity mapping and $\varphi\colon F(X) \to A(X)$ be the extension of $i$ to a continuous homomorphism. It is easy to see that the kernel of $\varphi$ coincides with the derived subgroup $K$ of $F(X)$, so $K$ is closed in $F(X)$. To show that the homomorphism $\varphi$ is open, consider the family $\mathcal{T}_q$ of all images $\varphi(U)$, where $U$ is open in $F(X)$. The continuity of $\varphi$ implies that $\mathcal{T}_q$ is a group topology on the abstract group $A_a(X)$ which is finer than the topology $\mathcal{T}$ of the group $A(X)$. By Corollary 7.1.8, it suffices to verify that $\mathcal{T}_q$ induces on $X$ its original topology, since this will imply that $\mathcal{T}_q = \mathcal{T}$.

Let $U$ be an open subset of $F(X)$, and suppose that $x \in V = X \cap \varphi(U)$. Choose an element $z \in U$ such that $\varphi(z) = x$. Then $W = X \cap xz^{-1}U$ is an open neighbourhood of $x$ in $X$, and

$$W = \varphi(W) \subset X \cap \varphi(U) = V.$$

Therefore, $V$ is a neighbourhood of each point $x \in V$ and, hence, $V$ is open in $X$. □

There are several properties that free (Abelian) topological groups can never have. Precompactness and connectedness are among them. In what follows, we use $G(X)$ to denote the topological group $F(X)$ or $A(X)$. Thus, any statement about $G(X)$ applies to both $F(X)$ and $A(X)$.

PROPOSITION 7.1.12. *For a Tychonoff space $X$, the following are equivalent:*

a) *$G(X)$ is precompact;*
b) *$G(X)$ is connected;*
c) *$X = \varnothing$.*

PROOF. Let $X$ be a non-empty Tychonoff space. Define the mapping $f$ of $X$ to the discrete group $\mathbb{Z}$ by $f(x) = 1$ for each $x \in X$. Then $f$ is continuous and, hence, it admits an extension to a continuous homomorphism $\tilde{f}\colon G(X) \to \mathbb{Z}$. Clearly, $\tilde{f}(G(X)) = \mathbb{Z}$. Since the group $\mathbb{Z}$ is neither precompact nor connected, neither is $G(X)$. □

The existence proof given in Theorem 7.1.2 is the shortest one, but it is very non-constructive and does not provide effective tools for studying topological properties of free topological groups. Even the fact that $X$ is closed in the groups $F(X)$ and $A(X)$ (see a) of Theorem 7.1.13) is far from being obvious. We shall establish below several technical general results that shed more light on the topological structure of free (Abelian) topological groups. Let us introduce the necessary notation.

For every non-negative integer $n$, denote by $B_n(X)$ the subspace of the free (Abelian) topological group $G(X)$ that consists of all words of reduced length $\leq n$ with respect to the free basis $X$. Clearly, $B_0(X) = \{e\}$, where $e$ is the neutral element of $G(X)$. Put also $C_n(X) = B_n(X) \setminus B_{n-1}(X)$ for $n \geq 1$. It turns out that the topology of $C_n(X)$ inherited from $G(X)$ admits a clear description in terms of $X$. Denote by $\tilde{X}$ the free topological sum of $X$, its copy $X^{-1}$, and $\{e\}$: $\tilde{X} = X \oplus \{e\} \oplus X^{-1}$. For every $n \geq 1$, denote by $i_n$ the multiplication mapping of $\tilde{X}^n$ to $G(X)$, $i_n(y_1, \dots, y_n) = y_1 \cdot \ldots \cdot y_n$ for each point $(y_1, \dots, y_n) \in \tilde{X}^n$. From the continuity of the multiplication in $G(X)$ it follows that the mapping $i_n$ is continuous. It is easy to see that $i_n(\tilde{X}^n) = B_n(X)$. Finally, we denote by $C_n^*(X) = i_n^{\leftarrow}(C_n(X))$ the inverse image of $C_n(X)$ under the mapping $i_n$. We shall show in the next theorem that in the case of $F(X)$, the restriction of $i_n$ to $C_n^*(X)$ is a homeomorphism of $C_n^*(X)$ onto $C_n(X)$, for each $n \geq 1$. In other words, $C_n(X)$ is homeomorphic to a subspace of $\tilde{X}^n$.

THEOREM 7.1.13. [**A. V. Arhangel'skii**] *The following statements hold for every Tychonoff space $X$ and any integer $n \geq 1$:*

a) *The sets $B_n(X)$ and $i_n(X^n)$ are closed in $G(X)$. In particular, $X$ is closed in $G(X)$.*
b) *In the case of $F(X)$, $i_n$ homeomorphically maps $C_n^*(X)$ onto $C_n(X)$. In addition, $i_n(X^n)$ is a closed homeomorphic copy of $X^n$ in $F(X)$.*

PROOF. Consider the free (Abelian) topological group $G(\beta X)$ on the Čech–Stone compactification $\beta X$ of the space $X$. By the definition of $G(X)$, the canonical embedding $i\colon X \to \beta X$ can be extended to a continuous homomorphism $\hat{\imath}\colon G(X) \to G(\beta X)$. Since the restriction of $\hat{\imath}$ to $X$ is one-to-one and $G(\beta X)$ is algebraically free on $\beta X$ (see Theorems 7.1.5 and 7.1.7), $\hat{\imath}$ must be injective. In other words, $\hat{\imath}$ is a continuous isomorphism of $G(X)$ onto the subgroup of $G(\beta X)$ generated by $i(X)$.

a) Denote by $K$ the topological sum of $\beta X$, its copy $(\beta X)^{-1}$, and the identity $e$ of $G(\beta X)$, that is, $K = \beta X \oplus \{e\} \oplus (\beta X)^{-1}$. For every integer $n \geq 1$, consider the mapping

$i_n^*\colon K^n \to G(\beta X)$ defined by the formula $i_n^*(y_1, \ldots, y_n) = y_1 \cdot \ldots \cdot y_n$ for all $y_1, \ldots, y_n \in K$. Clearly, the mapping $i_n^*$ is continuous and the restriction of $i_n^*$ to $\tilde{X}^n$ coincides with $\hat{\imath} \circ i_n$ for each $n \geq 1$, where $\tilde{X} = X \oplus \{e\} \oplus X^{-1}$. Note that $B_n(\beta X) = i_n^*(K^n)$ is compact as a continuous image of the compact space $K^n$. Therefore, $B_n(\beta X)$ is closed in $G(\beta X)$. We have $\hat{\imath}(B_n(X)) = \hat{\imath}(G(X)) \cap B_n(\beta X)$, whence it follows that $\hat{\imath}(B_n(X))$ is closed in $\hat{\imath}(G(X))$. Since $\hat{\imath}$ is a continuous one-to-one mapping of $G(X)$ onto $\hat{\imath}(G(X))$, we conclude that $B_n(X)$ is closed in $G(X)$. Similarly, $\hat{\imath}(i_n(X^n)) = \hat{\imath}(G(X)) \cap i_n^*((\beta X)^n)$, so that $\hat{\imath}(i_n(X^n))$ is closed in $\hat{\imath}(G(X))$. Again, since $\hat{\imath}$ is continuous and one-to-one, $i_n(X^n)$ must be closed in $G(X)$. Therefore, $i_1(X) \cong X$ is closed in $G(X)$.

b) Since every element of $C_n(X)$ has length $n$, the restriction of the mapping $i_n\colon \tilde{X}^n \to F(X)$ to $C_n^*(X)$ is one-to-one. It suffices to show that this restriction is open as the mapping onto its image $C_n(X) = i_n(C_n^*(X))$. With notation as in a) and with $G(X) = F(X)$, note that the domain $K^n$ of $i_n^*$ is compact, so $i_n^*\colon K^n \to F(\beta X)$ is a continuous closed mapping. In addition, its restriction to $C_n^*(\beta X) = (i_n^*)^{-1}(C_n(\beta X))$ is one-to-one (again, every element of $C_n(\beta X)$ has length $n$). Therefore, the restriction of $i_n^*$ to $C_n^*(\beta X)$ is a homeomorphism of $C_n^*(\beta X)$ onto $C_n(\beta X)$. Let $U$ be an open subset of $C_n^*(X)$. Choose an open subset $V$ of $C_n^*(\beta X)$ such that $U = V \cap C_n^*(X)$. Since $i_n^* \restriction C_n^*(\beta X)$ is a homeomorphism and $\hat{\imath} \circ i_n = i_n^* \restriction \tilde{X}^n$, the set

$$\hat{\imath}(i_n(U)) = i_n^*(U) = i_n^*(V) \cap i_n^*(C_n^*(X)) = i_n^*(V) \cap \hat{\imath}(C_n(X))$$

is open in $\hat{\imath}(C_n(X))$. But $\hat{\imath}$ is a continuous monomorphism; hence, $i_n(U)$ is open in $C_n(X)$. Thus, the continuous bijection $i_n\colon C_n^*(X) \to C_n(X)$ is open, i.e., a homeomorphism. Since $X^n \subset C_n^*(X)$, it follows from a) that $i_n(X^n)$ is a closed homeomorphic copy of $X^n$ in $F(X)$. $\square$

The situation changes in the case of the group $A(X)$ — if $X$ contains at least two points, then the restriction of $i_n$ to $X^n$ is one-to-one iff $n = 1$. Indeed, if $n \geq 2$ and $x, y$ are distinct points of $X$, then

$$i_n(x, \ldots, x, y) = x^{n-1}y = yx^{n-1} = i_n(y, x, \ldots, x)$$

($x$ appears $n - 1$ times as argument). However, we show in the next proposition that the mapping $i_n\colon C_n^*(X) \to C_n(X)$ is perfect and open, for each $n \geq 1$ (we keep notation used in Theorem 7.1.13 and in its proof).

PROPOSITION 7.1.14. *In the case of $A(X)$, $i_n\colon C_n^*(X) \to C_n(X)$ and $i_n\colon X^n \to i_n(X^n)$ are perfect and open mappings, for every Tychonoff space $X$ and any integer $n \geq 1$.*

PROOF. Clearly, $i_n^*\colon K^n \to C_n(\beta X) \subset A(\beta X)$ is a perfect mapping, where $K = \beta X \oplus \{e\} \oplus (\beta X)^{-1}$. Put $P_n = \hat{\imath}(C_n(X))$. Then $P_n \subset C_n(\beta X)$ and $(i_n^*)^{\leftarrow}(P_n) = C_n^*(X)$. Therefore, $i_n^* \restriction C_n^*(X) = \hat{\imath} \circ i_n \restriction C_n^*(X)$ is also a perfect mapping. By Proposition 3.7.10 of [**165**], the mapping $i_n \restriction C_n^*(X)$ is perfect. Since $X^n$ is closed in $\tilde{X}^n$ and in $C_n^*(X)$, the restriction $i_n \restriction X^n$ is perfect.

It remains to verify that the mappings $i_n\colon C_n^*(X) \to C_n(X)$ and $i_n\colon X^n \to i_n(X^n)$ are open. Let $O$ be a non-empty open set in $C_n^*(X)$ and $x = (x_1^{\varepsilon_1}, \ldots, x_n^{\varepsilon_n}) \in O$ be an arbitrary point, where $x_1, \ldots, x_n \in X$ and $\varepsilon_1, \ldots, \varepsilon_n = \pm 1$. Since $C_n(X) = B_n(X) \setminus B_{n-1}(X)$ is open in $B_n(X)$ and $i_n$ is continuous, the set $C_n^*(X)$ is open in $\tilde{X}^n$. Therefore, we can find open sets $U_1, \ldots, U_n$ in $X$ satisfying the following conditions for all $i, j \leq n$:

1) $x_i \in U_i$;

2) if $x_i \neq x_j$, then $U_i \cap U_j = \varnothing$;
3) if $x_i = x_j$, then $U_i = U_j$;
4) $V = U_1^{\varepsilon_1} \times \cdots \times U_n^{\varepsilon_n} \subset O$.

Using 1)–4), we obtain that

$$i_n^{\leftarrow}(i_n(V)) = \bigcup\{U_{\pi(1)} \times \cdots \times U_{\pi(n)} : \pi \in S_n\},$$

where $S_n$ is the group of permutations of the set $\{1, \ldots, n\}$. Therefore, the set $i_n^{\leftarrow}(i_n(V))$ is open in $\tilde{X}^n$ and in $C_n^*(X)$. According to b) of Theorem 7.1.13, the mapping $i_n : C_n^*(X) \to C_n(X)$ is perfect (hence quotient), so we conclude that $i_n(V)$ is open in $C_n(X)$. It follows that $i_n(O)$ is the union of open sets in $C_n(X)$ and, hence, it is open in $C_n(X)$. So, the mapping $i_n : C_n^*(X) \to C_n(X)$ is open. Finally, the equality $X^n = i_n^{\leftarrow}(i_n(X^n))$ implies that the restriction of $i_n$ to $X^n$ is an open mapping of $X^n$ onto its image. □

Theorem 7.1.13 has several important corollaries. We know that every topological group is a Tychonoff space, by Theorem 3.3.11. It turns out that in this assertion, complete regularity cannot be strengthened to normality.

COROLLARY 7.1.15. *If $X$ is a Tychonoff non-normal space, then the groups $F(X)$ and $A(X)$ are not normal spaces either.*

PROOF. By a) of Theorem 7.1.13, $X$ is a closed subspace of the groups $F(X)$ and $A(X)$. If one of these groups were normal, its closed subspace $X$ would also be normal, a contradiction. □

According to b) of Theorem 7.1.13, $i_n(X^n)$ is a closed homeomorphic copy $X^n$ in $F(X)$. It turns out that $A(X)$ has a similar property, but the argument in this case is different.

COROLLARY 7.1.16. *The group $A(X)$ contains a closed homeomorphic copy of $X^n$, for each positive integer $n$.*

PROOF. For $n = 1$ the conclusion follows directly from Theorem 7.1.2 and a) of Theorem 7.1.13. Let $n \geq 2$ be an integer. In what follows we use the additive notation for the group multiplication in $A(X)$. Consider the mapping $f : X^n \to A(X)$ defined by

$$f(x_1, x_2, \ldots, x_n) = x_1 + 2x_2 + \cdots + 2^{n-1}x_n$$

for each $(x_1, x_2, \ldots, x_n) \in X^n$. From the continuity of the multiplication in $A(X)$ it follows that $f$ is continuous. Apply induction on $n$ along with the fact that $X$ is a free algebraic basis for $A(X)$ to show that $f$ is one-to-one.

Let $m = 2^n - 1$. Consider the embedding $g$ of $X^n$ to $X^m$ defined by the formula

$$g(x_1, x_2, \ldots, x_n) = (x_1, x_2, x_2, \ldots, x_n, \ldots, x_n),$$

where each $x_i$ appears in the right side of the equality $2^{i-1}$ times. Clearly, $g(X^n)$ is closed in $X^m$ and $f = i_m \circ g$, where $i_m : \tilde{X}^m \to A(X)$ is the natural multiplication mapping. In addition, the mapping $i_m : X^m \to i_m(X^m)$ is perfect by Proposition 7.1.14, so the composition $i_m \circ g$ is a closed mapping and, hence, is a homeomorphism. Finally, a) of Theorem 7.1.13 implies that $i_m(X^m)$ is closed in $A(X)$, so that the image $f(X^n) = i_m(g(X^n))$ is closed in $i_m(X^m)$ and in $A(X)$. □

COROLLARY 7.1.17. *For every space $X$, $nw(G(X)) = nw(X)$.*

PROOF. By Theorem 7.1.2, $X$ is a subspace of $G(X)$, so that $nw(X) \leq nw(G(X))$. Conversely, $nw(\tilde{X}) = nw(X)$ and $nw(\tilde{X}^n) = nw(X)$ for each $n \in \omega$, where $\tilde{X} = X \oplus \{e\} \oplus X^{-1}$. Since the multiplication mapping $i_n \colon \tilde{X}^n \to B_n(X)$ is continuous, we have $nw(B_n(X)) \leq nw(\tilde{X}^n) = nw(X)$ for each integer $n \geq 1$. Finally, from $G(X) = \bigcup_{n=1}^{\infty} B_n(X)$ it follows that $nw(G(X)) \leq nw(X) \cdot \omega = nw(X)$. This proves the equality $nw(G(X)) = nw(X)$. □

Now we can characterize the spaces $X$ for which the groups $F(X)$ and $A(X)$ are Lindelöf.

COROLLARY 7.1.18. *The free (Abelian) topological group $G(X)$ is Lindelöf iff the space $X^n$ is Lindelöf, for each $n \in \mathbb{N}$.*

PROOF. Suppose that the group $G(X)$ is Lindelöf. By a) of Theorem 7.1.13, the sets $i_n(X^n)$ are closed in $G(X)$; hence, they are Lindelöf. According to b) of Theorem 7.1.13 and Proposition 7.1.14, the mapping $i_n \colon X^n \to i_n(X)$ is perfect, for each $n \in \mathbb{N}$ (even is a homeomorphism in the case $G(X) = F(X)$), so that $X^n$ is Lindelöf, by [**165**, Theorem 3.8.9].

Suppose now that $X^n$ is Lindelöf for each $n \in \mathbb{N}$. Then $\tilde{X}^n$ is a finite union of closed copies of the spaces $X^k$ with $k \leq n$, hence Lindelöf. It remains to note that $G(X) = \bigcup_{n=1}^{\infty} B_n(X)$, where each $B_n(X) = i_n(\tilde{X}^n)$ is Lindelöf as a continuous image of the Lindelöf space $\tilde{X}^n$. □

Theorem 7.1.13 also enables us to describe a neighbourhood base at every element $g$ of the length $n$ in the subspace $B_n(X)$ of the groups $F(X)$ and $A(X)$.

COROLLARY 7.1.19. *Let $X$ be a Tychonoff space and $g = x_1^{\varepsilon_1} \ldots x_n^{\varepsilon_n}$ be a reduced element of $G(X)$. Suppose further that $U_i$ is an open neighbourhood of $x_i$ in $X$, and $U_i \cap U_j = \varnothing$ if $x_i \neq x_j$, $1 \leq i, j \leq n$. Then the set $W = U_1^{\varepsilon_1} \cdot \ldots \cdot U_n^{\varepsilon_n}$ is an open neighbourhood of $g$ in $B_n(X)$. Moreover, the family of such sets $W$ forms a local base of $B_n(X)$ at $g$.*

PROOF. By item a) of Theorem 7.1.13, $C_n(X) = B_n(X) \setminus B_{n-1}(X)$ is open in $B_n(X)$. Consider the point $x = (x_1^{\varepsilon_1}, \ldots, x_n^{\varepsilon_n}) \in \tilde{X}^n$. Then the set $U = U_1^{\varepsilon_1} \times \cdots \times U_n^{\varepsilon_n}$ is open in $\tilde{X}^n$, $x \in U$ and $i_n(U) = U_1^{\varepsilon_1} \cdot \ldots \cdot U_n^{\varepsilon_n} \subset C_n(X)$. Hence, $U \subset C_n^*$, and since the mapping $i_n \colon C_n^*(X) \to C_n(X)$ is open (see b) of Theorem 7.1.13 and Proposition 7.1.14), the set $W = i_n(U)$ is an open neighbourhood of $g = i_n(x)$ in $C_n(X)$ and in $B_n(X)$.

Clearly, the sets $U = U_1^{\varepsilon_1} \times \cdots \times U_n^{\varepsilon_n}$ form a base of $C_n^*(X)$ at $x$, so the sets $W = i_n(U)$ constitute a base of $C_n(X)$ at the point $g$. Since $C_n(X)$ is open in $B_n(X)$, the same conclusion remains valid for $B_n(X)$. □

Theorem 7.1.13 and Corollaries 7.1.17, 7.1.18, and 7.1.19 may suggest the idea that topological properties of the groups $F(X)$ and $A(X)$ are close to those of the space $X$. Let us show that this is not quite the case.

THEOREM 7.1.20. *If a Tychonoff space $X$ is non-discrete, then neither $F(X)$, nor $A(X)$ are first-countable.*

PROOF. Let $X$ be a non-discrete Tychonoff space with a non-isolated point $x^* \in X$. By Theorem 7.1.11, $A(X)$ is a quotient group of $F(X)$. Since open continuous mappings do not rise the character of topological space, it suffices to show that the character of $A(X)$ is uncountable. Assume the contrary, and take a countable decreasing base $\{U_n : n \in \mathbb{N}\}$ of the neutral element $e$ of $A(X)$. In what follows we use the additive notation for the group operation in $A(X)$.

There are open neighbourhoods $V_1, V_2, \dots, V_n, \dots$ of the point $x^*$ in $X$ such that $2^n(V_n - V_n) \subset U_n$ for each $n \geq 1$. We can choose points $x_n, y_n \in V_n$ such that $y_i \neq x_n \neq y_n \neq x_i$ for all $n, i \geq 1$. Then the sequence of elements $u_n = 2^n(x_n - y_n) \in U_n$ converges to $e$. Let us show that this leads to a contradiction.

By induction on $n$, we can easily define a sequence $\{f_n : n \in \mathbb{N}\}$ of continuous real-valued functions on $X$ satisfying the following conditions for every $n \geq 1$:

1) $|f_n| \leq 2^{-n}$;
2) $f_n(x^*) = f_n(x_k) = f_n(y_k) = f_n(y_n) = 0$ for each $k < n$;
3) $f_n(x_n) = \pm 2^{-n}$, where $f_n(x_n)$ and the sum $\sum_{k=1}^{n-1}(f_k(x_n) - f_k(y_n))$ have the same sign (that is, they are simultaneously positive or negative) if $n > 1$.

Consider the function $f = \sum_{n=1}^{\infty} f_n$ on $X$. From 1) it follows that $f$ is continuous; hence, it admits an extension to a continuous homomorphism $\tilde{f}\colon A(X) \to \mathbb{R}$. For every $n \geq 1$, we have

$$|\tilde{f}(u_n)| = 2^n \cdot |\sum_{k=1}^{\infty}(f_k(x_n) - f_k(y_n))|$$

$$= 2^n \cdot |f_n(x_n) + \sum_{k=1}^{n-1}(f_k(x_n) - f_k(y_n))| \geq 2^n \cdot |f_n(x_n)| = 1.$$

Therefore, the sequence $\{\tilde{f}(u_n) : n \in \mathbb{N}\}$ cannot converge to zero of the group $\mathbb{R}$. This contradicts the continuity of $\tilde{f}$. □

## Exercises

7.1.a. (M. I. Graev [**201**]) Let $X$ be a Tychonoff space with a fixed point $e \in X$. A topological group $G = F(X, e)$ is called the *free Graev topological group* on $(X, e)$ if $G$ satisfies the following conditions:

(G1) There exists a continuous mapping $\sigma\colon X \to G$ such that $\sigma(e) = e_G$ and $\sigma(X)$ algebraically generates $G$.

(G2) If a continuous mapping $f\colon X \to H$ to a topological group $H$ satisfies $f(e) = e_H$, then there exists a continuous homomorphism $\tilde{f}\colon G \to H$ such that $f = \tilde{f} \circ \sigma$.

Prove the following statements:

a) The group $F(X, e)$ exists for every Tychonoff space $X$ and is unique up to a natural topological isomorphism.

b) If $e_1, e_2 \in X$, then there exists a topological isomorphism $\varphi$ of $F(X, e_1)$ to $F(X, e_2)$ such that $\varphi \circ \sigma_1 = \sigma_2$, where $\sigma_i\colon (X, e_i) \to F(X, e_i)$ is a mapping satisfying condition (G1), for $i = 1, 2$.

c) The group $F(X, e)$ does not depend on the choice of $e \in X$, which permits us to denote the free Graev topological group on $X$ simply by $F^*(X)$ (apply a) and b)).

d) The mapping $\sigma\colon X \to F^*(X)$ is a topological embedding and $\sigma(X)$ is closed in $F^*(X)$.

e) There exists the Abelian analog $A^*(X)$ of the group $F^*(X)$, and $A^*(X)$ is a quotient group of $F^*(X)$.

f) $A(X) \cong A^*(X) \times \mathbb{Z}$, for every Tychonoff space $X$, where the group $\mathbb{Z}$ carries the discrete topology.

7.1.b. (M. I. Graev [**201**]) Let $X$ be a Tychonoff space, and $Y$ a space obtained by adding an isolated point to $X$. Verify that $F(X) \cong F^*(Y)$ and $A(X) \cong A^*(Y)$. Show that the corresponding isomorphisms $\varphi\colon F(X) \to F^*(Y)$ and $\psi\colon A(X) \to A^*(Y)$ can be chosen to satisfy $\varphi(X) \subset Y$ and $\psi(X) \subset Y$ (here we identify $X$ and $Y$ with the corresponding subspaces of $F(X)$ and $F^*(Y)$, respectively, and the same stands in the case of $A(X)$ and $A^*(Y)$).

7.1.c. (M. I. Graev [**201**]) Show that each of the groups $F^*(X)$, $A^*(X)$ is connected iff $X$ is connected.

7.1.d. (M. I. Graev [**201**]) Prove that if a Tychonoff space $X$ is disconnected, then there exists a subspace $Y$ of $F^*(X)$ such that $F^*(X) \cong F(Y)$ and $A^*(X) \cong A(Y)$.

7.1.e. Formulate and prove the analogs of Theorems 7.1.5, 7.1.7, 7.1.11, 7.1.13, 7.1.20, *etc.*, for the groups $F^*(X)$ and $A^*(X)$. (Note that the algebraic parts of Theorems 7.1.5 and 7.1.7 require small changes.)

7.1.f. Let $X$ be a Tychonoff space. The *free precompact group* $G = FP(X)$ on $X$ is a topological group defined by the following conditions:

(P1) There exists a continuous mapping $\sigma\colon X \to G$ such that $\sigma(X)$ topologically generates $G$;

(P2) the group $G$ is precompact;

(P3) for every continuous mapping $f\colon X \to H$ to a precompact group $H$, there exists a continuous homomorphism $\tilde{f}\colon G \to H$ such that $f = \tilde{f} \circ \sigma$.

Similarly, one defines the free Abelian precompact group $AP(X)$ on $X$. Prove the following:

a) The group $FP(X)$ exists, and is unique up to a topological isomorphism fixing points of $X$, for every Tychonoff space $X$.

b) The conclusions of Theorems 7.1.2, 7.1.5, and of Corollary 7.1.8 can be extended to $FP(X)$.

c) $X$ is closed in $FP(X)$, so that precompact topological groups need not be normal.

d) $AP(X)$ is a quotient group of $FP(X)$.

7.1.g. Let a non-empty class $\mathcal{V}$ of topological groups be an $\overline{S}C$*-variety*, i.e., let $\mathcal{V}$ be closed under taking closed subgroups and arbitrary topological products. For a space $X$, the $\mathcal{V}$*-free topological group* $G = F(X, \mathcal{V})$ on $X$ is a topological group which satisfies the following conditions:

(V1) There exists a continuous mapping $\sigma\colon X \to G$ such that $\sigma(X)$ topologically generates $G$;

(V2) $G \in \mathcal{V}$;

(V2) For every continuous mapping $f\colon X \to H$ to a group $H \in \mathcal{V}$, there exists a continuous homomorphism $\tilde{f}\colon G \to H$ such that $f = \tilde{f} \circ \sigma$.

Prove the following assertions:

a) The group $F(X, \mathcal{V})$ exists and is unique for every space $X$.

b) The equality $A(X) = F(X, \mathcal{A})$ holds for every space $X$, where $\mathcal{A}$ is the class of all Abelian topological groups. Show that $FP(X) = F(X, \mathcal{P})$, where $\mathcal{P}$ is the class of all precompact topological groups.

c) $\sigma\colon X \to F(X, \mathcal{V})$ is a topological embedding iff the continuous homomorphisms of $X$ to the groups in $\mathcal{V}$ separate points and closed sets in $X$.

d) If $\mathcal{V}$ is the $\overline{S}C$-variety of all zero-dimensional topological groups, then the group $F(I, \mathcal{V})$ is trivial, where $I = [0, 1]$.

e) The condition in c) does not guarantee that the image $\sigma(X)$ is closed in $F(X, \mathcal{V})$.

7.1.h. A non-empty class $\mathcal{V}$ of topological groups is called an $SC$*-variety* if it is closed under taking arbitrary subgroups and topological products.

a) Prove that if $\mathcal{V}$ is an $SC$-variety and $\sigma\colon X \to F(X, \mathcal{V})$ is a continuous mapping satisfying (V1) of Exercise 7.1.g, then $\sigma(X)$ algebraically generates the $\mathcal{V}$-free topological group $F(X, \mathcal{V})$.

b) Show that if $\mathcal{V}$ contains the circle group $\mathbb{T}$, then $\sigma$ is a closed embedding of $X$ into $F(X, \mathcal{V})$.

7.1.i. (O. Alas, *et al.* [**5**]) Show that if a Tychonoff space $X$ is connected and locally connected, then the groups $F(X)$ and $A(X)$ are locally connected. Find out whether the analogous result is valid for the free precompact group $FP(X)$ (see Exercise 7.1.f).

7.1.j. (O. G. Okunev [**358**]) A continuous onto mapping $f: X \to Y$ is called *R-quotient* if for every mapping $g: Y \to \mathbb{R}$, the composition $g \circ f$ is continuous iff $g$ is continuous. Clearly, every $R$-quotient mapping is continuous, and every quotient mapping is $R$-quotient. Generalize the second part of Corollary 7.1.9 by showing that if an onto mapping $f: X \to Y$ is $R$-quotient, then the homomorphism $F(f): F(X) \to F(Y)$ extending $f$ is open. Formulate and prove an analogous assertion in the Abelian case.

7.1.k. Prove that if $F(X)$ is Lindelöf, for some Tychonoff space $X$, then $F(X) \times F(X)$ is also Lindelöf. Extend the conclusion to the groups $A(X)$, $F^*(X)$, and $A^*(X)$.

## Problems

7.1.A. (B. V. S. Thomas [**464**]) Let $Z = X \oplus Y$ be the topological sum of Tychonoff spaces $X$ and $Y$. Prove that the free Abelian topological group $A(Z)$ is topologically isomorphic to the product group $A(X) \times A(Y)$.

7.1.B. In the class of Tychonoff spaces, define the notion of the free Tychonoff paratopological group of a Tychonoff space $X$. Show that it always exists, is unique, and has very similar properties to those of the free topological group of $X$. Do the same for the Abelian case.

7.1.C. In the class of Hausdorff spaces, follow the routine to define the notion of the free Hausdorff paratopological group of a Hausdorff space $X$. Does this object exist for every Hausdorff space? Is $X$ always naturally homeomorphic to a closed subspace of it?

7.1.D. Calculate the pseudocharacter of the free Abelian precompact topological group $AP(X)$ in terms of the space $X$.

7.1.E. Give an example of Lindelöf precompact topological groups $G$ and $H$ such that the product group $G \times H$ is not Lindelöf.

## Open Problems

7.1.1. Let $S$ be the Sorgenfrey line. Is there a Lindelöf topological group topology on the free Abelian group $A_a(S)$ such that $S$ is a closed subspace of $A_a(S)$ with this topology? Does there exist such a topological group topology on $F_a(S)$?

7.1.2. Let $X$ be an arbitrary normal space. Is there a normal topological group topology on the free Abelian group $A_a(X)$ such that $X$ is closed in $A_a(X)$ with this topology?

7.1.3. Suppose that $X$ is a space such that $X^n$ is normal, for every $n \in \omega$. Is the free topological group $F(X)$ normal? What about $A(X)$?

7.1.4. Suppose that $X$ is a space such that $X^n$ is paracompact, for every $n \in \omega$. Is the free topological group $F(X)$ paracompact?

7.1.5. Let $X$ be a sequential Tychonoff space. Is there a sequential group topology on the free Abelian group $A_a(X)$? Is there a sequential group topology on $A_a(X)$ such that $X$ is closed in $A_a(X)$ with this topology?

7.1.6. Characterize the Tychonoff spaces $X$ such that the quotient topological group $A(X)/2A(X)$ is extremally disconnected, where $2A(X) = \{2g : g \in A(X)\}$. Is every space $X$ with this property discrete?

7.1.7. Characterize Tychonoff spaces $X$ such that the free Abelian group $A_a(X)$ admits a pseudocompact group topology such that $X$ is a closed subspace of $A_a(X)$ with this topology. Consider the similar question for $F_a(X)$.

7.1.8. Suppose that $F(X)$ is (collectionwise) normal. Is $X$ countably paracompact?
7.1.9. Suppose that the space $F(X)$ has one of the following properties:
(a) sequentiality;
(b) countable tightness;
(c) countable cellularity;
(d) normality;
(e) paracompactness.
Does $F(X) \times F(X)$ have the same property?
7.1.10. When, in terms of the space $X$, does $F(X)$ admit a continuous homomorphism (a continuous mapping) onto $F(X) \times F(X)$?
7.1.11. When, in terms of $X$, is $F(X) \times F(X)$ a quotient group of $F(X)$?

## 7.2. Extending pseudometrics from $X$ to $F(X)$

A more detailed study of topological groups $F(X)$ and $A(X)$ requires understanding of how $X$ is placed in these groups. For example, we know that every continuous real-valued function on $X$ admits an extension to a continuous homomorphism of $A(X)$ to the additive group $\mathbb{R}$; thus, $X$ is $C$-embedded in $A(X)$ and $F(X)$. Theorem 7.2.2 below is a considerably stronger result: Every continuous pseudometric $\varrho$ on $X$ admits an extension to a continuous invariant pseudometric $\widehat{\varrho}$ on $F(X)$. An analogous assertion holds for $A(X)$. The invariance of $\widehat{\varrho}$ on $F(X)$ means that

$$\widehat{\varrho}(xg, xh) = \widehat{\varrho}(g, h) = \widehat{\varrho}(gx, hx)$$

for all $g, h, x \in F(X)$. The proof of Theorem 7.2.2 is based on a combinatorial work with the words that form the groups $F(X)$ and $A(X)$. This requires the concept of a *scheme* that plays a crucial role here.

Let $A$ be a subset of $\mathbb{N}$ such that $|A| = 2n$, for some $n \geq 1$. A *scheme on* $A$ is a bijection $\varphi\colon A \to A$ satisfying the following conditions:

(S1) $\varphi(i) \neq i$ and $\varphi(\varphi(i) = i$, for every $i \in A$;
(S2) there are no $i, j \in A$ such that $i < j < \varphi(i) < \varphi(j)$.

Thus, a scheme on $A$ is an idempotent permutation of $A$ without fixed points that satisfies (S2). Notice that (S2) is equivalent to the requirement that there are no $i, j \in A$ with $i < \varphi(j) < \varphi(i) < j$. We need the following technical statement:

PROPOSITION 7.2.1. *Suppose that $n$ is a positive integer and $\varphi$ is a scheme on $A_n = \{1, 2, \ldots, 2n\}$. If $i < \varphi(i)$ for some $i < 2n$, then there exists $j$ with $i \leq j < \varphi(i)$ such that $\varphi(j) = j + 1$. In particular, for every scheme $\varphi$ on $A_n$, there is $j < 2n$ such that $\varphi(j) = j + 1$.*

PROOF. Let $j$ be the maximal integer such that $i \leq j < \varphi(i)$ and $\varphi(j) < \varphi(i)$. We claim that $\varphi(j) = j + 1$. Indeed, assume the contrary. Then, clearly, $j + 1 < \varphi(j)$, and it follows from the definition of a scheme that $j + 1 < \varphi(j + 1) < \varphi(j)$, which contradicts the choice of $j$. □

Let $X$ be a non-empty set and $e$ be the identity element of the free group $F_a(X)$. A word $\mathfrak{X}$ in the alphabet $\tilde{X} = X \cup \{e\} \cup X^{-1}$ is said to be *almost reduced* if it satisfies the following two conditions:

(a) $\mathfrak{X}$ does not contains two adjacent letters of the form $x, x^{-1}$ or $x^{-1}, x$ for any $x \in X$ (but $\mathfrak{X}$ may contain several letters $e$);
(b) after deleting all $e$'s from the word $\mathfrak{X}$, one obtains an reduced word in the alphabet $X \cup X^{-1}$.

Since $e^{-1} = e$, an almost reduced word cannot contain two adjacent letters equal to $e$. Therefore, an almost reduced word of an even length $2n$ can contain at most $n$ letters $e$.

We are ready to prove one of the main results of this section, known as Graev's Extension Theorem.

THEOREM 7.2.2. [**M. I. Graev**] *Every pseudometric $\varrho$ on a non-empty set $X$ can be extended to invariant pseudometrics $\widehat{\varrho}$ and $\widehat{\varrho}_A$ on the groups $F_a(X)$ and $A_a(X)$, respectively. In addition, if $X$ is a Tychonoff space and $\varrho$ is continuous on $X$, then the pseudometrics $\widehat{\varrho}$ and $\widehat{\varrho}_A$ are continuous on $F(X)$ and $A(X)$, respectively.*

PROOF. We consider in detail the case of the groups $F_a(X)$ and $F(X)$, and then indicate the necessary changes for the groups $A_a(X)$ and $A(X)$.

The first step is to extend $\varrho$ to a pseudometric $\varrho^*$ on the subset $\tilde{X} = X \cup \{e\} \cup X^{-1}$ of $F_a(X)$, where $e$ is the neutral element of $F_a(X)$. Choose a point $x_0 \in X$ and for every $x \in X$, put

$$\varrho^*(e, x) = \varrho^*(e, x^{-1}) = 1 + \varrho(x_0, x).$$

Then for $x, y \in X$, define the distances $\varrho^*(x^{-1}, y^{-1})$, $\varrho^*(x^{-1}, y)$ and $\varrho^*(x, y^{-1})$ by

$$\varrho^*(x^{-1}, y^{-1}) = \varrho^*(x, y) = \varrho(x, y),$$
$$\varrho^*(x^{-1}, y) = \varrho^*(x, y^{-1}) = \varrho^*(x, e) + \varrho^*(e, y).$$

From our definition it follows immediately that $\varrho^* \restriction X = \varrho$ and $\varrho^*(z, t) = \varrho^*(t, z) \geq 0$, for all $z, t \in \tilde{X}$. Let us verify that $\varrho^*$ satisfies the triangle inequality

$$\varrho^*(u, w) \leq \varrho^*(u, v) + \varrho^*(v, w)$$

for all $u, v, w \in \tilde{X}$. This is clear if one of the points $u, v, w$ is equal to $e$. Therefore, we have to consider the following cases:

| | |
|---|---|
| (a) $u, v, w \in X$; | (a′) $u, v, w \in X^{-1}$; |
| (b) $u, v \in X,\ w \in X^{-1}$; | (b′) $u, v \in X^{-1},\ w \in X$; |
| (c) $u, w \in X,\ v \in X^{-1}$; | (c′) $u, w \in X^{-1},\ v \in X$; |
| (d) $v, w \in X,\ u \in X^{-1}$; | (d′) $v, w \in X^{-1},\ u \in X$. |

By the symmetry argument, it suffices to restrict our attention to the cases (b) and (c). In case (b), we have

$$\begin{aligned} \varrho^*(u, v) + \varrho^*(v, w) &= \varrho^*(u, v) + \varrho^*(v, e) + \varrho^*(e, w) \\ &\geq \varrho^*(u, e) + \varrho^*(e, w) = \varrho^*(u, w). \end{aligned}$$

Similarly, in case (c) we have

$$\begin{aligned} \varrho^*(u, v) + \varrho^*(v, w) &= \varrho^*(u, e) + \varrho^*(e, v) + \varrho^*(v, e) + \varrho^*(e, w) \\ &\geq \varrho^*(u, e) + \varrho^*(e, w) \geq \varrho^*(u, w). \end{aligned}$$

Now we have to extend the pseudometric $\varrho^*$ from $\tilde{X}$ to the whole group $F_a(X)$. Let $g$ be a reduced element of $F_a(X)$, and suppose that $\mathfrak{X} \equiv x_{i_1} x_{i_2} \dots x_{i_{2n}}$ is a word in the alphabet

$\tilde{X}$ of even length $l(\mathfrak{X}) = 2n$, where $i_1, \ldots, i_{2n}$ are pairwise distinct positive integers, such that all possible cancellations in $\mathfrak{X}$ (including deletions of the letter $e$ from $\mathfrak{X}$) transform $\mathfrak{X}$ to $g$ (we will write $[\mathfrak{X}] = g$ if all these conditions are satisfied). Denote by $\mathcal{S}_{\mathfrak{X}}$ the family of all schemes on $A = \{i_1, \ldots, i_{2n}\}$. We say that each $\varphi \in \mathcal{S}_{\mathfrak{X}}$ is a *scheme for* $\mathfrak{X}$. For every $\varphi \in \mathcal{S}_{\mathfrak{X}}$, put

$$\Gamma_\varrho(\mathfrak{X}, \varphi) = \frac{1}{2} \sum_{i \in A} \varrho^*(x_i^{-1}, x_{\varphi(i)}). \tag{7.1}$$

The factor $1/2$ in the above definition appears due to the fact that, for each $i \in A$, the sum in the right part of (7.1) contains two equal summands $\varrho^*(x_i^{-1}, x_{\varphi(i)})$ and $\varrho^*(x_{\varphi(i)}^{-1}, x_i)$. Then we define a number $N_\varrho(g)$ as follows:

$$N_\varrho(g) = \inf\{\Gamma_\varrho(\mathfrak{X}, \varphi) : l(\mathfrak{X}) = 2n \geq l(g),\ [\mathfrak{X}] = g,\ \varphi \in \mathcal{S}_{\mathfrak{X}},\ n \in \mathbb{N}\}.$$

It is clear that $N_\varrho(g) \geq 0$ for each $g \in F_a(X)$. We divide the rest of the proof into several steps.

**Claim 1.** *For every $g \in F_a(X)$ distinct from $e$, one can find an almost reduced word $\mathfrak{X}_g$ of even length $2n \geq 2$ in the alphabet $\tilde{X}$ and a scheme $\varphi_g$ for $\mathfrak{X}_g$ that satisfy the following conditions:*

(i) *$\mathfrak{X}_g$ contains only the letters of $g$ or the letter $e$;*
(ii) *$[\mathfrak{X}_g] = g$ and $l(\mathfrak{X}_g) \leq 2l(g)$;*
(iii) *$N_\varrho(g) = \Gamma_\varrho(\mathfrak{X}_g, \varphi_g)$.*

Indeed, let $\mathfrak{X}$ be a word of length $l(\mathfrak{X}) = 2n \geq l(g)$ with $[\mathfrak{X}] = g$ and $\varphi$ be a scheme for $\mathfrak{X}$. Suppose that $\mathfrak{X} = x_1 x_2 \ldots x_{2n}$, where $x_1, x_2, \ldots, x_{2n} \in \tilde{X}$. First, we show that there exist a non-empty almost reduced word $\mathfrak{X}_1$ of length $2m \leq 2n$ obtained after several simple transformations and cancellations in $\mathfrak{X}$ and a scheme $\varphi_1$ for $\mathfrak{X}_1$ that satisfy (i), (ii), and

(iv) $\Gamma_\varrho(\mathfrak{X}_1, \varphi_1) \leq \Gamma_\varrho(\mathfrak{X}, \varphi)$.

If $\mathfrak{X}$ is almost reduced, we simply put $\mathfrak{X}_1 = \mathfrak{X}$ and $\varphi_1 = \varphi$. Suppose, therefore, that $\mathfrak{X}$ contains either two adjacent symbols of the form $uu^{-1}$ or three adjacent symbols $ueu^{-1}$, for some $u \in \tilde{X}$ (in the latter case, deleting the letter $e$ from $\mathfrak{X}$ produces a new cancellation).

**Case I.** $\mathfrak{X}$ contains two adjacent symbols $uu^{-1}$ for some $u \in \tilde{X}$. Then $x_i = u$ and $x_{i+1} = u^{-1}$ for some $i < 2n$. Let us consider two subcases: $\varphi(i) = i+1$ and $\varphi(i) \neq i+1$. If $\varphi(i) = i+1$, we delete $uu^{-1}$ from $\mathfrak{X}$, thus obtaining the word $\mathfrak{X}'$, and define $\varphi'$ as the restriction of $\varphi$ to $\{1, \ldots, i-1, i+2, \ldots, 2n\}$. It is clear that $[\mathfrak{X}'] = g$ and $\Gamma_\varrho(\mathfrak{X}', \varphi') = \Gamma_\varrho(\mathfrak{X}, \varphi)$. If $\varphi(i) \neq i+1$, put $r = \varphi(i)$ and $s = \varphi(i+1)$. Then $\{r, s\} \cap \{i, i+1\} = \varnothing$. Again, we delete $uu^{-1}$ from $\mathfrak{X}$, thus obtaining the new word $\mathfrak{X}'$ with $[\mathfrak{X}'] = g$, and define a bijection $\varphi'$ of $A = \{1, \ldots, i-1, i+2, \ldots, 2n\}$ onto itself by $\varphi'(m) = \varphi(m)$ if $m \notin \{r, s\}$ and $\varphi'(r) = s$, $\varphi'(s) = r$. One easily verifies that $\varphi'$ is a scheme on $A$, and it follows from

$$\begin{aligned}\varrho^*(x_r^{-1}, x_s) &\leq \varrho^*(x_r^{-1}, x_i) + \varrho^*(x_i, x_s)\\ &= \varrho^*(x_r^{-1}, x_i) + \varrho^*(x_{i+1}^{-1}, x_s)\end{aligned}$$

that $\Gamma_\varrho(\mathfrak{X}', \varphi') \leq \Gamma_\varrho(\mathfrak{X}, \varphi)$. Evidently, each letter of $\mathfrak{X}'$ is also a letter of $\mathfrak{X}$.

**Case II.** The word $\mathfrak{X}$ contains three adjacent symbols $u^{-1}eu$ for some $u \in \tilde{X}$. Then there exists $i$ with $1 < i < 2n$ such that $u^{-1} = x_{i-1}$, $e = x_i$ and $u = x_{i+1}$. Let $r = \varphi(i-1)$, $s = \varphi(i)$ and $t = \varphi(i+1)$. As in Case I, two subcases are possible: $s \in \{i-1, i+1\}$ or

$s \notin \{i-1, i+1\}$. In the former subcase, we can assume without loss of generality that $s = i+1$ and $r < i-1$. Then $\mathfrak{X} \equiv Ax_r Bu^{-1}euC$, where the words $A$, $B$ and $C$ have lengths $r-1$, $i-r-2$ and $2n-i-1$, respectively. Put $\mathfrak{X}' \equiv Ax_r x_i BC$. Since $x_i = e$, we have $[\mathfrak{X}'] = [\mathfrak{X}] = g$, $l(\mathfrak{X}') = l(\mathfrak{X}) - 2$, and each letter of $\mathfrak{X}'$ is also a letter of $\mathfrak{X}$. Let $\varphi'$ be a bijection of the set $K = \{1, \ldots, i-1, i+2, \ldots, 2n\}$ which coincides with $\varphi$ on $K \setminus \{r, i-1\}$ and satisfies $\varphi'(r) = i$, $\varphi'(i) = r$. Then $\varphi'$ is a scheme on $K$, and an easy calculation shows that

$$\Gamma_\varrho(\mathfrak{X}, \varphi) - \Gamma_\varrho(\mathfrak{X}', \varphi') = \varrho(x_r, u) + \varrho(u, e) - \varrho(x_r, e) \geq 0.$$

Hence, we conclude that $\Gamma_\varrho(\mathfrak{X}', \varphi') \leq \Gamma_\varrho(\mathfrak{X}, \varphi)$.

Suppose now that $s \notin \{i-1, i+1\}$. Since $\varphi$ is a scheme, neither of the inequalities $r < t < i-1$, $i-1 < t < r$, $i+1 < t < s$, $s < t < i+1$ is possible. Suppose, for example, that $r < i-1 < i+1 < s < t$. Then $\mathfrak{X} \equiv Ax_r Bu^{-1}euCx_s^{-1}Dx_t E$, where $A, B, C, D$ and $E$ are words of the lengths $r-1$, $i-r-2$, $s-i-2$, $t-s-1$ and $2n-t$, respectively. Let $\mathfrak{X}' \equiv Ax_r BCx_s^{-1}Dx_i x_t E$ be the word obtained from $\mathfrak{X}$ by deleting $uu^{-1}$ and translating $e = x_i$ to the letter $x_t$. Clearly, $[\mathfrak{X}'] = g$. Let also $\varphi'$ be a bijection of the set $K = \{1, \ldots, i-2, i, i+2, \ldots, 2n\}$ that coincides with $\varphi$ on $K \setminus \{r, i, s, t\}$ and satisfies $\varphi'(r) = s$, $\varphi'(s) = r$, $\varphi(i) = t$ and $\varphi(t) = i$. Then $\varphi'$ is a scheme on $K$ (which connects $x_r$ with $x_s$ and $x_i = e$ with $x_t$). Our definition of $\varphi'$ implies that

$$\Gamma_\varrho(\mathfrak{X}, \varphi) - \Gamma_\varrho(\mathfrak{X}', \varphi') = \varrho(x_r, u) + \varrho(u, x_s) - \varrho(x_r, x_s) \geq 0.$$

Therefore, $\Gamma_\varrho(\mathfrak{X}', \varphi') \leq \Gamma_\varrho(\mathfrak{X}, \varphi)$. Notice that each letter of $\mathfrak{X}'$ is a letter of $\mathfrak{X}$. The other two cases, when $s < r < i-1$ or $i+1 < s < r$, are similar to the one considered and are left to the reader.

In each of Cases I and II, the length of $\mathfrak{X}'$ is strictly less than the length of $\mathfrak{X}$ and, in addition, $\mathfrak{X}'$ does not contain "new" letters. If the word $\mathfrak{X}'$ again fails to be almost reduced, we apply one of the operations described in Case I and Case II to $\mathfrak{X}'$, thus obtaining a word $\mathfrak{X}''$ and a scheme $\varphi''$ for $\mathfrak{X}''$ such that $[\mathfrak{X}''] = g$, $\Gamma_\varrho(\mathfrak{X}'', \varphi'') \leq \Gamma_\varrho(\mathfrak{X}', \varphi')$, and so on. Since $g = [\mathfrak{X}] = [\mathfrak{X}'] = [\mathfrak{X}''] = \ldots$ and $\Gamma_\varrho(\mathfrak{X}, \varphi) \geq \Gamma_\varrho(\mathfrak{X}', \varphi') \geq \Gamma_\varrho(\mathfrak{X}'', \varphi'') \geq \ldots$, we finally obtain an almost reduced word $\mathfrak{X}_1$ of even length and a scheme $\varphi_1$ for $\mathfrak{X}_1$ satisfying (i), (ii) and (iv). Notice that the inequality $l(\mathfrak{X}_1) \leq 2l(g)$ in (ii) is a consequence of the fact that the word $\mathfrak{X}_1$ is almost reduced.

To complete the proof of Claim 1, it is enough to observe that, given an element $g \in F_a(X)$, there exist only finitely many pairs $(\mathfrak{X}_1, \varphi_1)$ satisfying (i) and (ii). Therefore, one of these pairs, say, $(\mathfrak{X}_g, \varphi_g)$ satisfies (i)–(iii).

**Claim 2.** *The function $N_\varrho$ is an invariant prenorm on the group $F_a(X)$.*

Clearly, $N_\varrho(e) = 0$ and $N_\varrho(g) \geq 0$ for each $g \in F_a(X)$. Let us verify that $N_\varrho(g^{-1}) = N_\varrho(g)$ and $N_\varrho(gh) \leq N_\varrho(g) + N_\varrho(h)$ for all $g, h \in F_a(X)$.

For an element $g \in F_a(X)$, we can find, by Claim 1, a word $\mathfrak{X} = x_1 x_2 \ldots x_{2n}$ in the alphabet $\tilde{X}$ and a scheme $\varphi$ on $A_n = \{1, \ldots, 2n\}$ such that $[\mathfrak{X}] = g$ and $N_\varrho(g) = \Gamma_\varrho(\mathfrak{X}, \varphi)$. Consider the word $\mathfrak{Y} = y_1 \ldots y_{2n-1} y_{2n}$ with $y_i = x_{2n-i+1}^{-1}$, $1 \leq i \leq 2n$, and a scheme $\psi$ on $A_n$ defined by $\psi(i) = 2n - \varphi(2n-i+1) + 1$ for each $i \leq 2n$. Then $[\mathfrak{Y}] = g^{-1}$ and

$$N_\varrho(g^{-1}) \leq \Gamma_\varrho(\mathfrak{Y}, \psi) = \Gamma_\varrho(\mathfrak{X}, \varphi) = N_\varrho(g),$$

that is, $N_\varrho(g^{-1}) \le N_\varrho(g)$. With $g^{-1}$ instead of $g$, we have $N_\varrho(g) \le N_\varrho(g^{-1})$. Combining these inequalities, we finally obtain $N_\varrho(g^{-1}) = N_\varrho(g)$, for each $g \in F_a(X)$.

Let $g$ and $h$ be arbitrary elements of $F_a(X)$, and suppose that $\mathfrak{X} = x_1x_2\dots x_{2n}$ and $\mathfrak{Y} = y_1y_2\dots y_{2m}$ are words in the alphabet $\tilde{X}$ such that

$$[\mathfrak{X}] = g,\ [\mathfrak{Y}] = h,\ N_\varrho(g) = \Gamma_\varrho(\mathfrak{X}, \varphi) \text{ and } N_\varrho(h) = \Gamma_\varrho(\mathfrak{Y}, \psi),$$

where $\varphi \in \mathcal{S}_{A_n}$ and $\psi \in \mathcal{S}_{A_m}$. Put $\mathfrak{Z} = \mathfrak{X}\mathfrak{Y} = x_1\dots x_{2n}y_1\dots y_{2m}$ and rewrite $\mathfrak{Z}$ in the form $\mathfrak{Z} = z_1\dots z_{2n}z_{2n+1}\dots z_{2n+2m}$, where $z_i = x_i$ if $1 \le i \le 2n$ and $z_{2n+j} = y_j$ if $1 \le j \le 2m$. Define a scheme $\sigma \in \mathcal{S}_{A_{n+m}}$ by the formula

$$\sigma(k) = \begin{cases} \varphi(k) & \text{if } 1 \le k \le 2n; \\ 2n + \psi(k-2n) & \text{if } 2n < k \le 2n+2m. \end{cases}$$

It is clear that $[\mathfrak{Z}] = g \cdot h$ and

$$\Gamma_\varrho(\mathfrak{Z}, \sigma) = \Gamma_\varrho(\mathfrak{X}, \varphi) + \Gamma_\varrho(\mathfrak{Y}, \psi) = N_\varrho(g) + N_\varrho(h).$$

Therefore,

$$N_\varrho(g \cdot h) \le \Gamma_\varrho(\mathfrak{Z}, \sigma) = N_\varrho(g) + N_\varrho(h).$$

This proves that $N_\varrho$ is a prenorm on the group $F_a(X)$. It remains to verify that $N_\varrho$ is invariant, that is, $N_\varrho(h^{-1}gh) = N_\varrho(g)$ for all $g, h \in F_a(X)$. Evidently, it suffices to check the above equality in the case when $h$ has the length 1, say, $h = x \in X \cup X^{-1}$. Let $\mathfrak{X} = x_1\dots x_{2n}$ be a word in the alphabet $\tilde{X}$ such that $[\mathfrak{X}] = g$ and $\Gamma_\varrho(\mathfrak{X}, \varphi) = N_\varrho(g)$ for some $\varphi \in \mathcal{S}_{A_n}$. Consider the word $\mathfrak{Y} = y_1y_2\dots y_{2n+1}y_{2n+2}$, where $y_1 = x^{-1}$, $y_{2n+2} = x$ and $y_k = x_{k-1}$ if $2 \le k \le 2n+1$. Define a scheme $\psi \in \mathcal{S}_{A_{n+1}}$ by the formula

$$\psi(k) = \begin{cases} 2n+2 & \text{if } k = 1; \\ 1 & \text{if } k = 2n+2; \\ \varphi(k-1)+1 & \text{if } 2 \le k \le 2n+1. \end{cases}$$

Then $[\mathfrak{Y}] = x^{-1}gx$ and, hence,

$$N_\varrho(x^{-1}gx) \le \Gamma_\varrho(\mathfrak{Y}, \psi) = \Gamma_\varrho(\mathfrak{X}, \varphi) = N_\varrho(g).$$

Replace $x$ by $x^{-1}$ and $g$ by $x^{-1}gx$ in the above inequality to obtain $N_\varrho(g) \le N_\varrho(x^{-1}gx)$. The two inequalities imply that $N_\varrho(x^{-1}gx) = N_\varrho(g)$. This proves Claim 2.

**Claim 3.** $N_\varrho(x^{-1}y) = \varrho(x, y) = N_\varrho(xy^{-1})$ *for all* $x, y \in X$.

Fix two elements $x, y \in X$. Note that by Claim 2,

$$N_\varrho(xy^{-1}) = N_\varrho(x^{-1}xy^{-1}x) = N_\varrho(y^{-1}x) = N_\varrho(x^{-1}y),$$

so it suffices to show that $N_\varrho(x^{-1}y) = \varrho(x, y)$. If $x = y$, then, clearly,

$$\varrho(x, y) = 0 \text{ and } N_\varrho(x^{-1}y) = \Gamma_\varrho(\mathfrak{X}, \varphi) = 0,$$

where $\mathfrak{X} = x_1x_2$ is the word with $x_1 = x^{-1}$, $x_2 = y$ and $\varphi \in \mathcal{S}_{A_1}$ is the transposition of elements $1, 2$. Suppose, therefore, that $x \neq y$.

Put $g = x^{-1}y$. By Claim 1, one can find a word $\mathfrak{Y}$ of length $2m \le 3$ in the alphabet $\tilde{X}$ such that $[\mathfrak{Y}] = g$ and $\Gamma_\varrho(\mathfrak{Y}, \varphi) = N_\varrho(g)$ for some $\varphi \in \mathcal{S}_{A_m}$. Since the element $g \in F_a(X)$

is reduced, we must have $m = 1$ and $\mathfrak{Y} = x^{-1}y$. Again, if $\varphi$ is the transposition of 1 and 2, then

$$N_\varrho(x^{-1}y) = \Gamma_\varrho(\mathfrak{Y}, \varphi) = \varrho(x, y).$$

This proves Claim 3.

Define a pseudometric $\widehat{\varrho}$ on $F_a(X)$ by $\widehat{\varrho}(g, h) = N_\varrho(g^{-1}h)$ for all $g, h \in F_a(X)$.

**Claim 4.** *The pseudometric $\widehat{\varrho}$ is invariant on $F_a(X)$, and its restriction to $X$ coincides with $\varrho$.*

Indeed, by the invariance of $N_\varrho$ (Claim 2), we have

$$\widehat{\varrho}(gx, hx) = N_\varrho(x^{-1}g^{-1}hx) = N_\varrho(g^{-1}h) = \widehat{\varrho}(g, h)$$

for all $x, g, h \in F_a(X)$. From the definition of $N_\varrho$ it also follows that

$$\widehat{\varrho}(xg, xh) = N_\varrho((xg)^{-1}xh) = N_\varrho(g^{-1}h) = \widehat{\varrho}(g, h).$$

We conclude, therefore, that the pseudometric $\widehat{\varrho}$ is invariant. Claim 3 implies immediately that the restriction of $\widehat{\varrho}$ to $X$ coincides with $\varrho$, so Claim 4 is proved.

The next step is to establish the continuity of the extension $\widehat{\varrho}$ for every continuous pseudometric $\varrho$ on $X$. Denote by $\mathcal{P}$ the family of all continuous pseudometrics on $X$. For every $\varrho \in \mathcal{P}$, put

$$U_\varrho = \{g \in F_a(X) : N_\varrho(g) < 1\}.$$

First, we prove the following important result.

**Claim 5.** *The family $\mathcal{N} = \{U_\varrho : \varrho \in \mathcal{P}\}$ is a base at the identity $e$ for a Hausdorff group topology $\mathcal{T}_{inv}$ on $F_a(X)$. The restriction of $\mathcal{T}_{inv}$ to $X$ coincides with the original topology of the space $X$.*

We have to verify that the family $\mathcal{T}_{inv}$ satisfies the five conditions of the complete neighbourhood system at $e$ given in Theorem 1.3.12. Let us do it step by step.

1) $\{e\} = \bigcap\mathcal{N}$. Take any reduced element $g = x_1 \dots x_n \in F_a(X)$ distinct from the identity $e$. Since $X$ is completely regular, there exists a continuous pseudometric $\varrho$ on $X$ such that $\varrho^*(x_i^{-1}, x_j) \geq 1$ if $x_i^{-1} \neq x_j$. Indeed, $x_i = a_i^{\varepsilon_i}$ for some $a_i \in X$ and $\varepsilon_i = \pm 1$, $1 \leq i \leq n$. Take a continuous real-valued function $f$ on $X$ such that $|f(a_i) - f(a_j)| \geq 1$ if $a_i \neq a_j$ and define a continuous pseudometric $\varrho$ on $X$ by $\varrho(x, y) = |f(x) - f(y)|$ for all $x, y \in X$. We claim that $g \notin U_\varrho$. Indeed, by Claim 1, we can find a word $\mathfrak{X} = y_1 y_2 \dots y_{2m}$ of length $2m \leq l(g) + 1$ in the alphabet $\tilde{X}$ and a scheme $\varphi$ on $A_m = \{1, \dots, 2m\}$ such that $[\mathfrak{X}] = g$, $\{y_1, y_2 \dots, y_{2m}\} \subset \{e, x_1, x_2, \dots, x_n\}$ and $N_\varrho(g) = \Gamma_\varrho(\mathfrak{X}, \varphi)$. By Proposition 7.2.1, $\varphi(i) = i + 1$ for some $i < 2m$, whence

$$\varrho^*(y_i^{-1}, y_{i+1}) \leq \Gamma_\varrho(\mathfrak{X}, \varphi) = N_\varrho(g).$$

If both $y_i$, $y_{i+1}$ are distinct from $e$, then $y_i$ and $y_{i+1}$ are consecutive letters in $g$, for example, $y_i = x_j$ and $y_{i+1} = x_{j+1}$, for some $j < n$. Since the word $g$ is reduced, $x_j^{-1} \neq x_{j+1}$. Therefore,

$$\varrho^*(y_i^{-1}, y_{i+1}) = \varrho^*(x_j^{-1}, x_{j+1}) \geq 1.$$

If one of the elements $y_i$, $y_{i+1}$ is equal to $e$, say, $y_i = e$, then $y_{i+1} \neq e$ and, hence,

$$\varrho^*(y_i^{-1}, y_{i+1}) = \varrho^*(e, y_{i+1}) \geq 1$$

by the definition of $\varrho^*$. In either case, we have

$$1 \le \varrho^*(y_i^{-1}, y_{i+1}) \le N_\varrho(g),$$

so $g \notin U_\varrho$ by the definition of $U_\varrho$.

2) For every $U, V \in \mathcal{N}$, there exists $W \in \mathcal{N}$ such that $W \subset U \cap V$. Indeed, if $U = U_\varrho$ and $V = U_\sigma$ for some $\varrho, \sigma \in \mathcal{P}$, then put $d = \varrho + \sigma$. Clearly, $d \in \mathcal{P}$ and $W = U_d$ is as required.

3) For every $U \in \mathcal{N}$ there exists $V \in \mathcal{N}$ such that $VV^{-1} \subset U$. To find such a set $V$, suppose that $U = U_\varrho$ for some $\varrho \in \mathcal{P}$. Put $\sigma = 2\varrho$ and $V = U_\sigma$. If $g, h \in V$, then we have

$$N_\varrho(gh^{-1}) \le N_\varrho(g) + N_\varrho(h^{-1}) = N_\varrho(g) + N_\varrho(h) < 1/2 + 1/2 = 1.$$

This implies immediately that $VV^{-1} \subset U$.

4) For every $U \in \mathcal{N}$ and $g \in U$, there exists $V \in \mathcal{N}$ such that $gV \subset U$. Again, suppose that $U = U_\varrho$ for some $\varrho \in \mathcal{P}$, and let $g \in U_\varrho$ be arbitrary. Then $r = N_\varrho(g) < 1$, $s = 1 - r > 0$ and $\sigma = s^{-1} \cdot \varrho \in \mathcal{P}$. Put $V = U_\sigma$. Clearly, $V \in \mathcal{N}$ and if $h \in V$, then $N_\sigma(h) = s^{-1} \cdot N_\varrho(h) < 1$. So,

$$N_\varrho(gh) \le N_\varrho(g) + N_\varrho(h) < r + s = 1,$$

whence it follows that $gV \subset U$.

5) For every $U \in \mathcal{N}$ and $g \in F_a(X)$, there exists $V \in \mathcal{N}$ such that $gVg^{-1} \subset U$. Indeed, by Claim 2, the prenorm $N_\varrho$ is invariant for every $\varrho \in \mathcal{P}$, so one can take $V = U$.

We conclude that the family $\mathcal{N}$ satisfies 1)–5), so it forms a base at the identity for a Hausdorff group topology $\mathcal{T}_{inv}$ on $F_a(X)$. It remains to verify that $\mathcal{T}_{inv}$ induces on $X$ its original topology $\tau_X$.

Let $O$ be an open set in the group $(F_a(X), \mathcal{T}_{inv})$ such that $O \cap X \neq \varnothing$. Choose a point $x \in O \cap X$. Since $\mathcal{N}$ is a base for $\mathcal{T}_{inv}$ at the identity, we can find $\varrho \in \mathcal{P}$ such that $xU_\varrho \subset O$. The set

$$V = \{y \in X : \varrho(x, y) < 1\}$$

is an open neighbourhood of $x$ in $X$, and Claim 3 implies that $V \subset xU_\varrho \cap X \subset O$. This proves that $O \cap X$ is open in $X$ and, hence, $\tau_X$ is finer that $\mathcal{T}_{inv} \restriction X$.

Conversely, let $W$ be an open subset of $X$, and let $x_0 \in W$ be an arbitrary point. Since $X$ is completely regular, there exists a continuous real-valued function $f$ on $X$ such that $f(x_0) = 1$ and $f(x) = 0$ for each $x \in X \setminus W$. Then the pseudometric $\varrho$ on $X$ defined by $\varrho(x, y) = |f(x) - f(y)|$ for all $x, y \in X$ is continuous, so $\varrho \in \mathcal{P}$. Let us verify that $X \cap x_0U_\varrho \subset W$. If $x \in X \cap x_0U_\varrho$, then $x_0^{-1}x \in U_\varrho$. By Claim 3, we have

$$|f(x_0) - f(x)| = \varrho(x_0, x) = N_\varrho(x_0^{-1}x) < 1.$$

Since $f(x_0) = 1$, this implies that $f(x) \neq 0$, whence $x \in W$. This proves the inclusion $X \cap x_0U_\varrho \subset W$. We conclude, therefore, that $\mathcal{T}_{inv} \restriction X$ is finer than $\tau_X$. Summarizing, we have $\mathcal{T}_{inv} \restriction X = \tau_X$. Claim 5 is proved.

**Claim 6.** *If $\varrho$ is a continuous pseudometric on a Tychonoff space $X$, then the pseudometric $\widehat{\varrho}$ is continuous on the free topological group $F(X)$.*

Denote by $\mathcal{T}$ the topology of the free topological group $F(X)$. Then $\mathcal{T}$ is finer than $\mathcal{T}_{inv}$, by Claim 5 and Corollary 7.1.8. In particular, the set $U_{n\varrho}$ is open in $F(X)$, for each

$n \in \mathbb{N}$. This fact and invariance of $\widehat{\varrho}$ together imply the continuity of the extension $\widehat{\varrho}$. Claim 6 is proved.

Claims 4 and 6 prove the theorem for the groups $F_a(X)$ and $F(X)$. The argument for the Abelian groups $A_a(X)$ and $A(X)$ is similar to the one just given, but it requires one important change in the definition of a scheme. Given a finite subset $B$ of $\mathbb{N}$ with $|B| = 2n \geq 2$, we say that a bijection $\varphi\colon B \to B$ is an *Abelian scheme* on $B$ if $\varphi$ is an involution without fixed points, that is, $\varphi(i) = j$ always implies $j \neq i$ and $\varphi(j) = i$. Then one defines an invariant extension $\widehat{\varrho}_A$ of a given pseudometric $\varrho$ on $X$ to the group $A_a(X)$ using Abelian schemes. The rest of the proof goes the same way. □

Let us study Graev's extension of pseudometrics from $X$ to $A(X)$ in more detail. In what follows we use the additive notation for the group multiplication in $A(X)$. However, we will keep the symbol $e$ to stand for the neutral element of $A_a(X)$, to distinguish the last one from $0 \in \mathbb{Z}$. The next corollary follows immediately from the invariance of $\widehat{\varrho}_A$.

COROLLARY 7.2.3. *Let $\varrho$ be a continuous pseudometric on a space $X$. Graev's extension $\widehat{\varrho}_A$ of $\varrho$ over $A(X)$ satisfies $\widehat{\varrho}_A(g_1 + h, g_2 + h) = \widehat{\varrho}_A(g_1, g_2)$ and $\widehat{\varrho}_A(-h, e) = \widehat{\varrho}_A(h, e)$ for all $g_1, g_2, h \in A(X)$.*

Graev's extension $\widehat{d}_A$ of a pseudometric $d$ on $X$ to the free Abelian group $A_a(X)$ admits a clearer description given below. Note that every element $h \in A_a(X) \setminus \{e\}$ can be written as $h = k_1x_1 + \cdots + k_nx_n$, where $x_1, \ldots, x_n$ are distinct points of $X$ and $k_1, \ldots, k_n \in \mathbb{Z} \setminus \{0\}$. In the sequel we shall call such an expression the *normal form* of $h$.

COROLLARY 7.2.4. *Suppose that $d$ is a pseudometric on a set $X$ and $m_1x_1 + \cdots + m_nx_n$ is the normal form of an element $h \in A_a(X) \setminus \{e\}$ of the length $l = \sum_{i=1}^{n} |m_i|$. Then there exists a representation*

$$h = (u_1 - v_1) + \cdots + (u_k - v_k), \tag{7.2}$$

*where $2k = l$ if $l$ is even and $2k = l + 1$ if $l$ is odd, $u_1, v_1 \ldots, u_k, v_k \in \{\pm x_1, \ldots, \pm x_n\}$ (but $v_k = e$ if $l$ is odd), and such that*

$$\widehat{d}_A(h, e) = \sum_{i=1}^{k} d^*(u_i, v_i), \tag{7.3}$$

*where $d^*$ is the extension of $d$ over $\tilde{X} = X \cup \{e\} \cup (-X)$ constructed in Theorem 7.2.2. In addition, if $\widehat{d}_A(h, e) < 1$, then $l = 2k$, and one can choose $y_1, z_1, \ldots, y_k, z_k \in \{x_1, \ldots, x_n\}$ such that*

$$h = (y_1 - z_1) + \cdots + (y_k - z_k) \tag{7.4}$$

*and*

$$\widehat{d}_A(h, e) = \sum_{i=1}^{k} d(y_i, z_i). \tag{7.5}$$

PROOF. We have $h = t_1 + t_2 + \cdots + t_l$, where $t_1, t_2 \ldots, t_l \in \{\pm x_1, \ldots, \pm x_n\}$. Denote by $k$ the integer such that $2k - 1 \leq l \leq 2k$. If $l = 2k - 1$, then we additionally put $t_{2k} = e$. By

Claim 1 in the proof of Theorem 7.2.2, there exists an Abelian scheme $\varphi$ on $\{1, 2, \dots, 2k\}$ such that

$$\widehat{d}_A(h, e) = \frac{1}{2}\sum_{i=1}^{2k} d^*(-t_i, t_{\varphi(i)}).$$

Since the group $A_a(X)$ is Abelian, we can assume without loss of generality that $\varphi(2i-1) = 2i$ (hence, $\varphi(2i) = 2i - 1$) for each $i = 1, \dots, k$. Therefore, we have

$$h = (t_1 + t_2) + \cdots + (t_{2k-1} + t_{2k}) \tag{7.6}$$

and

$$\widehat{d}_A(h, e) = \sum_{i=1}^{k} d^*(-t_{2i-1}, t_{2i}) = \sum_{i=1}^{k} d^*(t_{2i-1}, -t_{2i}). \tag{7.7}$$

For every $i = 1, \dots, k$, we put $u_i = t_{2i-1}$ and $v_i = -t_{2i}$, thus obtaining (7.2) and (7.3) from (7.6) and (7.7), respectively.

Finally, suppose that $\widehat{d}_A(h, e) < 1$. Note that $d^*(x, e) \geq 1$ and $d^*(-x, y) = d^*(x, -y) \geq 1$ for all $x, y \in X$. From (7.7) it follows that $d^*(t_{2i-1}, -t_{2i}) \leq \widehat{d}_A(h, e) < 1$ for each $i = 1, \dots, k$ and, hence, one of the elements $t_{2i-1}, t_{2i}$ is in $X$ while the other is in $-X$. Therefore, for every $i \leq k$ we have $t_{2i-1} + t_{2i} = y_i - z_i$, where $y_i, z_i \in X$. It is clear that $y_i, z_i \in \{x_1, \dots, x_n\}$. Again, replacing $t_{2i-1}$ and $t_{2i}$ by the corresponding elements $\pm y_i$ and $\pm z_i$ in (7.6) and (7.7), we obtain (7.4) and (7.5), respectively. □

COROLLARY 7.2.5. *Let $m_1x_1 + \cdots + m_nx_n$ be the normal form of an element $g \in A_a(X) \setminus \{e\}$ and $d$ be a pseudometric on $X$.*

a) *If $\widehat{d}_A(g, e) < 1$, then $\sum_{i=1}^n m_i = 0$.*

b) *If $\sum_{i=1}^n m_i = 0$, then there exists an reduced representation of $g$ in the form*

$$g = (z_1 - t_1) + \cdots + (z_k - t_k)$$

*such that $2k = \sum_{i=1}^n |m_i|$, $z_j, t_j \in \{x_1, \dots, x_n\}$ for each $j \leq k$, and $\widehat{d}_A(g, e) = \sum_{j=1}^k d(z_j, t_j)$.*

PROOF. a) By Corollary 7.2.4, we can represent $g$ as the sum

$$g = (y_1 - z_1) + \cdots + (y_k - z_k)$$

such that $\{y_1, z_1, \dots, y_k, z_k\} \subset \{x_1, \dots, x_n\}$. Since the sum of the coefficients in the above representation of $g$ is zero, the same is valid for the representation $g = m_1x_1 + \cdots + m_nx_n$.

b) Since $\sum_{i=1}^n m_i = 0$, the number $l = \sum_{i=1}^n |m_i|$ has to be even, say, $l = 2k$ for some $k \in \mathbb{N}$. Again, we apply Corollary 7.2.4 to find an reduced representation $\varphi$ for $g$ of the form

$$g = (u_1 - v_1) + \cdots + (u_k - v_k) \tag{7.8}$$

such that

$$\widehat{d}_A(g, e) = \Gamma(\varphi) = \sum_{j=1}^{k} d^*(u_j, v_j), \tag{7.9}$$

where $u_j, v_j \in \{\pm x_1, \cdots, \pm x_n\}$ for each $j \leq k$. Each summand in (7.8) has the form $(a - b)$, $(-a + b) = (b - a)$, $(a + b)$, or $(-a - b)$, for some $a, b \in X$. Let us call the first and the second expressions *neutral* summands, while the third and the fourth ones will be called *positive* and *negative*, respectively.

Suppose that one of the summands in (7.8), for example, $(u_1 - v_1)$ is positive, so that $u_1 \in X$, $v_1 = -a$ for some $a \in X$ and, hence, $(u_1 - v_1) = (u_1 + a)$. Since $\sum_{i=1}^n m_i = 0$, the right part of (7.8) must contain a negative summand, say, $(u_2 - v_2)$. Then $v_2 \in X$ and $u_2 = -b$ for some $b \in X$. Therefore, from our definition of the pseudometric $d^*$ on $X \cup \{e\} \cup (-X)$ it follows that the sum $\Gamma(\varphi)$ in (7.9) contains the following part corresponding to $(u_1 - v_1)$ and $(u_2 - v_2)$:

$$\begin{aligned} d^*(u_1, v_1) + d^*(u_2, v_2) &= d^*(u_1, -a) + d^*(b, -v_2) \\ &= [d^*(u_1, e) + d^*(e, a)] + [d^*(b, e) + d^*(e, v_2)] \\ &= [d^*(u_1, e) + d^*(e, v_2)] + [d^*(b, e) + d^*(e, a)] \\ &\geq d^*(u_1, v_2) + d^*(a, b) = d(u_1, v_2) + d(a, b). \end{aligned}$$

Replace the sum $(u_1 - v_1) + (u_2 - v_2) = (u_1 + a) - (b + v_2)$ in (7.8) by $(u_1 - v_2) + (a - b)$. This gives us another reduced representation $\varphi'$ for $g$ with the corresponding sum $\Gamma(\varphi')$, and the above inequality implies immediately that $\Gamma(\varphi') \leq \Gamma(\varphi)$. Note that the representation $\varphi'$ for $g$ has less positive and negative summands than $\varphi$. If $\varphi'$ still has some positive or negative summands, then we apply the same procedure to $\varphi'$ and obtain one more reduced representation $\varphi''$ for $g$ with $\Gamma(\varphi'') \leq \Gamma(\varphi')$, which has less positive and negative summands than $\varphi'$, *etc.* Finally, we get an reduced representation $\psi$ for $g$ of the form $g = (z_1 - t_1) + \cdots + (z_k - t_k)$ with $z_j, t_j \in \{x_1, \ldots, x_n\}$, for each $j \leq k$, such that

$$\sum_{j=1}^{k} d(z_j, t_j) = \Gamma(\psi) \leq \Gamma(\varphi) = \widehat{d}_A(g, e).$$

However, the definition of $\widehat{d}_A$ implies that $\widehat{d}_A(g, e) \leq \Gamma(\psi)$, whence it follows that $\sum_{j=1}^k d(z_j, t_j) = \widehat{d}_A(g, e)$. □

Theorem 7.2.2 and Corollary 7.2.4 have many important applications in the theory of free topological groups. Let us describe, for example, a neighbourhood base at the neutral element of the free Abelian topological group $A(X)$ on an arbitrary Tychonoff space $X$. First, we need a simple auxiliary result.

LEMMA 7.2.6. *Let $V_1, V_2, \ldots, V_n, \ldots$ be a sequence of subsets of a group $G$ with identity $e$ such that $e \in V_i$ and $V_{i+1}^3 \subset V_i$ for each $i \geq 1$. If $k_1, \ldots, k_n$ are positive integers, $r \in \mathbb{N}$ and $\sum_{i=1}^n 2^{-k_i} \leq 2^{-r}$, then $V_{k_1} \cdots V_{k_n} \subset V_r$.*

PROOF. We apply induction on $n$. The lemma is trivially true for $n = 1$. Suppose that we have proved the lemma for $n = 1, \ldots, m$. Let us show that it remains valid for $n = m + 1$. Denote by $s$ the maximal positive integer such that $\sum_{i=1}^n 2^{-k_i} \leq 2^{-s}$. Clearly $r \leq s$, and since $n \geq 2$, we have $s + 1 \leq k_i$ for each $i \leq n$. Our definition of $s$ implies that $2^{-s-1} < \sum_{i=1}^n 2^{-k_i}$, so there exists the minimal positive integer $j \leq n$ such that $2^{-s-1} \leq \sum_{i=1}^j 2^{-k_i}$. Then

$$\sum_{i=1}^{j-1} 2^{-k_i} < 2^{-s-1} \quad \text{and} \quad \sum_{i=j+1}^{n} 2^{-k_i} \leq 2^{-s-1}.$$

Now the inductive hypothesis implies that

$$V_1 \cdots V_{k_{j-1}} \subset V_{s+1} \text{ and } V_{j+1} \cdots V_n \subset V_{s+1}.$$

Since $s + 1 \le k_j$ and $r \le s$ (so $V_{k_j} \subset V_{s+1}$ and $V_s \subset V_r$), we finally have

$$[V_{k_1} \cdots V_{k_{j-1}}]\, V_{k_j} [V_{k_{j+1}} \cdots V_{k_n}] \subset V_{s+1} V_{s+1} V_{s+1} \subset V_s \subset V_r.$$

□

THEOREM 7.2.7. *Let $X$ be a Tychonoff space and $\mathcal{P}_X$ the family of all continuous pseudometrics on $X$. Then the sets*

$$V_\varrho = \{g \in A(X) : \widehat{\varrho}_A(g, e) < 1\},$$

*with $\varrho \in \mathcal{P}_X$, form a local base at the neutral element $e$ of $A(X)$.*

PROOF. We use the additive notation for the group operation in $A(X)$. Let $V$ be an open neighbourhood of $e$ in $A(X)$. There exists a sequence $\{V_n : n \in \mathbb{N}\}$ of open neighbourhoods of $e$ in $A(X)$ such that $V_1 \subset V$, $-V_i = V_i$ and $V_{i+1} + V_{i+1} + V_{i+1} \subset V_i$ for each $i \ge 1$. For every $n \ge 1$, define an open entourage $U_n$ of the diagonal in $X^2$ by

$$U_n = \{(x, y) \in X^2 : x - y \in V_n\}.$$

Then $U_n$ is an element of the universal uniformity $\mathcal{U}_X$ on the space $X$ and $U_{n+1} \circ U_{i+1} \circ U_{n+1} \subset U_n$ for each $n \ge 1$, where $\circ$ denotes the "uniform multiplication" of entourages in the uniform space $(X, \mathcal{U}_X)$. Therefore, by [**165**, Th. 8.1.10], there exists a continuous pseudometric $d$ on $X$ such that

$$\{(x, y) \in X^2 : d(x, y) < 2^{-n}\} \subset U_n$$

for each $n \ge 1$. Put $\varrho = 4d$. We claim that $V_\varrho \subset V$.

Indeed, let $g \in V_\varrho$. By Corollary 7.2.4, the element $g$ can be written in the form

$$g = (x_1 - y_1) + \cdots + (x_p - y_p),$$

with $x_i, y_i \in X$ if $1 \le i \le p$, such that

$$\widehat{\varrho}_A(g, e) = \varrho(x_1, y_1) + \cdots + \varrho(x_p, y_p).$$

From $\varrho = 4d$ it follows that $\widehat{\varrho} = 4\widehat{d}$, whence

$$\widehat{d}_A(g, e) = d(x_1, y_1) + \cdots + d(x_p, y_p) < 1/4.$$

For every $i \le p$ such that $d(x_i, y_i) > 0$, choose a positive integer $k_i$ satisfying the inequality:

$$2^{-k_i - 1} \le d(x_i, y_i) < 2^{-k_i}.$$

To every $i \le p$ with $d(x_i, y_i) = 0$ we can assign a sufficiently large integer $k_i$ in such a way that $\sum_{i=1}^{p} 2^{k_i} < 1/2$. Since $x_i - y_i \in V_{k_i}$ for each $i \le p$, Lemma 7.2.6 implies that

$$g = (x_1 - y_1) + \cdots + (x_p - y_p) \in V_{k_1} + \cdots + V_{k_p} \subset V_1 \subset V.$$

This proves the inclusion $V_\varrho \subset V$. □

The above theorem can be reformulated as follows: The family of pseudometrics $\{\widehat{\varrho}_A : \varrho \in \mathcal{P}_X\}$, where $\mathcal{P}_X$ is the family of all continuous pseudometrics on $X$, generates the topology of the free Abelian topological group $A(X)$. It turns out that a similar assertion for the free topological group $F(X)$ is "almost always" wrong, except for a very special class of spaces $X$. In fact, the family $\{\widehat{\varrho} : \varrho \in \mathcal{P}_X\}$ generates the topology of the group $F(X)$ if and only if $F(X)$ is balanced or, equivalently, has an invariant basis (see Problem 7.2.B). There is, however, an analog of Theorem 7.2.7 for the non-Abelian case dealing with the *maximal* group topology $\mathcal{T}_{inv}$ with invariant basis on $F_a(X)$ that generates the original topology on $X$. Let us prove the existence of such a topology. We will also describe a base of this topology at the identity of $F_a(X)$. Our argument will be based on the following combinatorial lemma (we keep notation used in the proof of Theorem 7.2.2).

LEMMA 7.2.8. *Let $g = x_1 \dots x_{2n}$ be a reduced element of $F_a(X)$, where $x_1, \dots, x_{2n} \in X \cup X^{-1}$, and let $\varphi \in \mathcal{S}_n$ be a scheme. Then there are integers $1 \le i_1 < \dots < i_n \le 2n$ and elements $h_1, \dots, h_n \in F_a(X)$ satisfying the following two conditions:*

i) $\{i_1, \dots, i_n\} \cup \{i_{\varphi(1)}, \dots, i_{\varphi(n)}\} = \{1, 2, \dots, 2n\}$;
ii) $g = (h_1 x_{i_1} x_{\varphi(i_1)} h_1^{-1}) \cdot \dots \cdot (h_n x_{i_n} x_{\varphi(i_n)} h_n^{-1})$.

PROOF. We apply induction on $n$. The lemma is obviously true for $n = 0$ and $n = 1$. Suppose that we have proved it for each $m \le n$, for some $n \ge 1$. Let $g = x_1 \dots x_{2n+1} x_{2n+2}$ be an element of $F_a(X)$ where $x_i \in X \cup X^{-1}$, $1 \le i \le 2n + 2$. Consider a scheme $\varphi \in \mathcal{S}_{n+1}$. If $k = \varphi(1) \ne 2n + 2$, then $k$ has to be even, for example $k = 2m$, and we can apply the inductive hypothesis to the elements $g_1 = x_1 \dots x_{2m}$ and $g_2 = x_{2m+1} \dots x_{2n+2}$, thus obtaining two partitions

$$\begin{aligned} \{1, \dots, 2m\} &= \{i_1, \dots, i_m\} \cup \{\varphi(i_1), \dots, \varphi(i_m)\}, \\ \{2m + 1, \dots, 2n + 2\} &= \{j_1, \dots, j_{n-m}\} \cup \{\varphi(j_1), \dots, \varphi(j_{n-m})\} \end{aligned}$$

and two equalities

$$\begin{aligned} g_1 &= (p_1 x_{i_1} x_{\varphi(i_1)} p_1^{-1}) \cdot \dots \cdot (p_m x_{i_m} x_{\varphi(i_m)} p_m^{-1}), \\ g_2 &= (q_1 x_{j_1} x_{\varphi(j_1)} q_1^{-1}) \cdot \dots \cdot (q_{n-m} x_{j_{n-m}} x_{\varphi(j_{n-m})} q_{n-m}^{-1}). \end{aligned}$$

Combining in the obvious way these equalities, we finally get a required partition of the set $\{1, \dots, 2n + 2\}$ and a representation of $g = g_1 g_2$ satisfying i) and ii).

Suppose that $\varphi(1) = 2n + 2$. Then

$$g = (x_1 g_1 x_1^{-1}) \cdot (x_1 x_{2n+2}),$$

where $g_1 = x_2 \dots x_{2n+1}$. Since $l(g_1) = 2n$, by the inductive hypothesis there exist a partition

$$\{2, \dots, 2n + 1\} = \{i_1, \dots, i_n\} \cup \{\varphi(i_1), \dots, \varphi(i_n)\}$$

and a representation

$$g_1 = h_1 x_{i_1} x_{\varphi(1)} h_1^{-1} \cdot \dots \cdot h_n x_{i_n} x_{\varphi(n)} h_n^{-1}.$$

Finally, we have the partition

$$\{1, \dots, 2n + 2\} = \{1, i_1, \dots, i_n\} \cup \{2n + 2, \varphi(i_1), \dots, \varphi(i_n)\}$$

and the representation

$$g = (p_1 x_{i_1} x_{\varphi(i_1)} p_1^{-1}) \cdot \dots \cdot (p_n x_{i_n} x_{\varphi(i_n)} p_n^{-1}) \cdot (x_1 x_{2n+2}),$$

where $p_k = x_1 h_k$ for each $k \le n$. This proves the lemma. □

THEOREM 7.2.9. *For every Tychonoff space $X$, the abstract group $F_a(X)$ admits the maximal group topology $\mathcal{T}_{inv}$ with invariant basis in the sense that every continuous mapping $f\colon X \to H$ to a topological group $H$ with invariant basis can be extended to a continuous homomorphism $\tilde{f}\colon (F_a(X), \mathcal{T}_{inv}) \to H$. The family of all sets of the form*

$$U_\varrho = \{g \in F_a(X) : \widehat{\varrho}(g, e) < 1\},$$

*where $\varrho$ is an arbitrary continuous pseudometric on $X$, constitutes a base of the topology $\mathcal{T}_{inv}$ at the identity $e$ of $F_a(X)$.*

PROOF. Denote by $\mathcal{T}_{inv}$ the Hausdorff group topology on $F_a(X)$ with the family $\mathcal{N} = \{U_\varrho : \varrho \in \mathcal{P}\}$ as a base at the identity, where $\mathcal{P}$ is the family of all continuous pseudometrics on $X$ (see Claim 5 in the proof of Theorem 7.2.2). By Theorem 7.2.2, each pseudometric $\widehat{\varrho}$ is invariant on $F_a(X)$, so the group $F_{inv}(X) = (F(X), \mathcal{T}_{inv})$ has an invariant basis, and $\widehat{\varrho}$ is continuous on $F_{inv}(X)$.

Let $f\colon X \to H$ be a continuous mapping of $X$ to a topological group $H$ with invariant basis. Denote by $\tilde{f}$ the extension of $f$ to a homomorphism of $F_a(X)$ to $H$. We claim that $\tilde{f}\colon F_{inv}(X) \to H$ is also continuous. Indeed, if $V$ is an open neighbourhood of the identity in $H$, we can find an invariant prenorm $N$ on $H$ such that $W = \{h \in H : N(h) < 1\} \subset V$. Define a continuous pseudometric $\varrho$ on $X$ by $\varrho(x, y) = N(f(x)^{-1} \cdot f(y))$ for all $x, y \in X$. Let us show that $\tilde{f}(U_\varrho) \subset W$.

Take an arbitrary reduced element $g \in U_\varrho$ distinct from the identity $e$ of $F_a(X)$. Then $\widehat{\varrho}(g, e) < 1$. It is clear that $g$ has even length, say $g = x_1 \dots x_{2n}$, where $x_1, \dots, x_{2n} \in X \cup X^{-1}$. Since $\widehat{\varrho}(g, e) < 1$, there exists a scheme $\varphi \in \mathcal{S}_n$ such that $\widehat{\varrho}(g) = 2^{-1} \cdot \sum_{i=1}^{2n} \varrho^*(x_i^{-1}, x_{\varphi(i)}) < 1$. Apply Lemma 7.2.8 to find a partition $\{1, \dots, 2n\} = \{i_1, \dots, i_n\} \cup \{\varphi(i_1), \dots, \varphi(i_n)\}$ and a representation of $g$ as a product $g = g_1 \cdot \dots \cdot g_n$ such that every $g_k$ has the form $g_k = h_i x_{i_k} x_{\varphi(i_k)} h_i^{-1}$, where $h_i \in F_a(X)$. By the invariance of $N$, we have

$$\begin{aligned} N(\tilde{f}(g)) \le \sum_{k=1}^{n} N(\tilde{f}(g_k)) &= \sum_{k=1}^{n} N(\tilde{f}(x_k) \cdot \tilde{f}(x_{\varphi(k)})) \\ &= \varrho^*(x_1^{-1}, x_{\varphi(1)}) + \dots + \varrho^*(x_n^{-1}, x_{\varphi(n)}) < 1. \end{aligned}$$

This implies that $\tilde{f}(g) \in W$. Therefore, $\tilde{f}(U_\varrho) \subset W \subset V$, and the homomorphism $\tilde{f}$ is continuous. □

The "invariant" topology $\mathcal{T}_{inv}$ on the group $F_a(X)$ almost never coincides with the topology of $F(X)$. In spite of this difficulty, it is still possible to give a clear description of a neighbourhood base for the subspace $B_{n+2}(X)$ of $F(X)$ at any element $g \in C_n(X)$, $n \in \mathbb{N}$. For every $g = a_1 \dots a_n \in C_n(X)$ with $a_1, \dots, a_n \in X \cup X^{-1}$, denote by $\mathcal{P}_X(g)$ the subfamily of $\mathcal{P}_X$ consisting of all $\varrho$ such that $\widehat{\varrho}(a_i^{-1}, a_{i+1}) \ge 1$ for each $i < n$. Note that the latter inequality automatically holds if both letters $a_i$ and $a_{i+1}$ belong to $X$ or to $X^{-1}$. For a pseudometric $\varrho \in \mathcal{P}_X(g)$, put

$$U_\varrho(g) = \{x_1 \dots x_i\, y^\varepsilon z^{-\varepsilon} x_{i+1} \dots x_n : x_1, \dots, x_n \in X \cup X^{-1},\ y, z \in X,$$
$$\varepsilon = \pm 1,\ 0 \le i \le n, \text{ and } \varrho(y, z) + \sum_{k=1}^{n} \varrho(a_k, x_k) < 1\}.$$

First, we need the following simple fact about $U_\varrho(g)$.

PROPOSITION 7.2.10. *The reduced form of every element of $U_\varrho(g)$ has length equal to $n$ or $n+2$.*

PROOF. A word $w = x_1 \dots x_i\, y^\varepsilon z^{-\varepsilon} x_{i+1} \dots x_n \in U_\varrho(g)$ admits cancellations if either $y = z$ or $x_i\, y^\varepsilon = e$ or $z^{-\varepsilon} x_{i+1} = e$. If $y = z$, then $[w] = g$ is the reduced form of $w$ and $l(g) = n$. By the symmetry argument, it suffices to consider the third case. So, suppose that $x_{i+1} = z^\varepsilon$. We claim that the word $w' = x_1 \dots x_i\, y^\varepsilon x_{i+2} \dots x_n$ of the length $n$ is reduced. Indeed, assuming the contrary, there can be only two possible cancellations in $w'$: (a) $x_i\, y^\varepsilon = e$ or (b) $y^\varepsilon x_{i+2} = e$.

In case (a), we have $x_i^{-1} = y^\varepsilon$. As $w \in U_\varrho(g)$, we also have

$$\begin{aligned} \widehat{\varrho}(a_i^{-1}, a_{i+1}) &\le \widehat{\varrho}(a_i^{-1}, x_i^{-1}) + \widehat{\varrho}(x_i^{-1}, y^\varepsilon) \\ &+ \widehat{\varrho}(y^\varepsilon, z^\varepsilon) + \widehat{\varrho}(z^\varepsilon, x_{i+1}) + \widehat{\varrho}(x_{i+1}, a_{i+1}) \\ &= \widehat{\varrho}(a_i, x_i) + \widehat{\varrho}(y, z) + \widehat{\varrho}(x_{i+1}, a_{i+1}) < 1. \end{aligned}$$

This inequality contradicts the choice of the pseudometric $\varrho \in \mathcal{P}_X(g)$.

Similarly, in case (b), we have $x_{i+2} = y^{-\varepsilon}$. Again, from $w \in U_\varrho(g)$ it follows that

$$\begin{aligned} \widehat{\varrho}(a_{i+1}^{-1}, a_{i+2}) &\le \widehat{\varrho}(a_{i+1}^{-1}, x_{i+1}) + \widehat{\varrho}(x_{i+1}, z^\varepsilon) \\ &+ \widehat{\varrho}(z^\varepsilon, y^\varepsilon) + \widehat{\varrho}(y^\varepsilon, x_{i+2}) + \widehat{\varrho}(x_{i+2}, a_{i+2}) \\ &= \widehat{\varrho}(a_{i+1}^{-1}, x_{i+1}) + \widehat{\varrho}(z, y) + \varrho(x_{i+2}, a_{i+2}) < 1. \end{aligned}$$

This contradicts the choice of $\varrho \in \mathcal{P}_X(g)$. Thus, the word $w'$ is reduced. □

THEOREM 7.2.11. *The family $\{U_\varrho(g) : \varrho \in \mathcal{P}_X(g)\}$ is an open base for $B_{n+2}(X)$ at any point $g \in C_n(X)$.*

PROOF. Fix an element $g = a_1 \dots a_n \in C_n(X)$ and an open neighbourhood $U$ of $g$ in $F(X)$. For every $i = 1, \dots, n$, put $g_i = a_1 \dots a_i$. There exist open symmetric neighbourhoods $V_0$ and $V_1$ of $e$ in $F(X)$ such that $V_0^{3n+1} \cdot g \subset U$ and $g_i V_1 g_i^{-1} \subset V_0$ for each $i \le n$. Consider the set

$$O = \{(x, y) \in X^2 : x^{-1} y \in V_1 \text{ and } xy^{-1} \in V_1\}.$$

Then $O$ is an element of the universal uniformity $\mathcal{U}_X$ on the space $X$, so by [**165**, Lemma 8.1.11], there exists a continuous pseudometric $\varrho$ on $X$ such that

$$\{(x, y) \in X^2 : \varrho(x, y) < 1\} \subset O.$$

Without loss of generality, we can assume that $\varrho \in \mathcal{P}_X(g)$. Now we show that $U_\varrho(g) \subset U \cap B_{n+2}(X)$. The inclusion $U_\varrho(g) \subset B_{n+2}(X)$ is clear. Suppose that $h = x_1 \dots x_i\, y^\varepsilon z^{-\varepsilon} x_{i+1} \dots x_n$, where $x_1, \dots, x_n \in X \cup X^{-1}$, $y, z \in X$, $\varepsilon = \pm 1$ and $\varrho(y, z) + \sum_{k=1}^n \varrho(a_k, x_k) < 1$. Then $y^\varepsilon z^{-\varepsilon} \in V_1$ and, hence, $g_i y^\varepsilon z^{-\varepsilon} g_i^{-1} \in V_0$. Note that $\widehat{\varrho}(a_k, x_k) < 1$, so $x_k \in V_k a_k$ for each $k = 1, \dots, n$. Therefore,

$$\begin{aligned} x_1 x_2 \dots x_k \in V_1 a_1 V_1 a_2 \dots V_1 a_k = \\ V_1 \cdot (a_1 V_1 a_1^{-1}) \cdot \dots \cdot (a_1 a_2 \dots a_{k-1} V_1 a_{k-1}^{-1} \dots a_2^{-1} a_1^{-1}) \cdot a_1 a_2 \dots a_{k-1} a_k \subset \\ V_1 \underbrace{V_0 \cdots V_0}_{k-1 \text{ times}} \cdot g_k \subset V_0^k \cdot g_k \end{aligned}$$

In its turn, this implies that

$$x_1x_2\dots x_i\, y^\varepsilon z^{-\varepsilon} x_{i+1}\dots x_n = x_1\dots x_i\, y^\varepsilon z^{-\varepsilon} x_i^{-1}\dots x_1^{-1}x_1\dots x_n \in$$
$$V_0^i\cdot(g_i y^\varepsilon z^{-\varepsilon} g_i^{-1})\cdot V_0^i x_1\dots x_n \subset V_0^i\cdot V_0\cdot V_0^i\cdot V_0^n g\subset V_0^{3n+1}g\subset U,$$

for $i=0,1,\dots,n$. Hence, $U_\varrho(g)\subset U$.

It remains to verify that the sets of the form $U_\varrho(g)$ are open in the subspace $B_{n+2}(X)$ of $F(X)$. For a given pseudometric $\varrho\in\mathcal{P}_X(g)$, consider the set

$$W=\{h\in F(X):\widehat{\varrho}(h,e)<1\}.$$

We claim that

$$U_\varrho(g)=gW\cap B_{n+2}(X).$$

By Theorem 7.2.2, the set $W$ is open in $F(X)$, so the above equality means, in particular, that $U_\varrho$ is open in $B_{n+2}(X)$. The inclusion $U_\varrho(g)\subset gW\cap B_{n+2}(X)$ is almost evident. Indeed, every element of $U_\varrho(g)$ has length $\le n+2$. If $h=x_1\dots x_i\, y^\varepsilon z^{-\varepsilon}x_{i+1}\dots x_n\in U_\varrho(g)$, then a simple calculation shows that $\widehat{\varrho}(g^{-1}h,e)<1$. Indeed, let us write the element $g^{-1}h$ in the form $g^{-1}h=g_1\cdot h_1$, where

$$g_1=a_n^{-1}\dots a_1^{-1}x_1\dots x_n \text{ and } h_1=x_n^{-1}\dots x_{i+1}^{-1}y^\varepsilon z^{-\varepsilon}x_{i+1}\dots x_n.$$

Denote by $N_\varrho$ the prenorm on $F(X)$ associated with $\varrho$, that is, $N_\varrho(x)=\widehat{\varrho}(x,e)$ for each $x\in F(X)$. Then $N_\varrho(g_1)\le\varrho(a_1,x_1)+\dots+\varrho(a_n,x_n)$ and $N_\varrho(h_1)=\varrho(y,z)$, by invariance of $N_\varrho$. Therefore, from $h\in U_\varrho(g)$ it follows that

$$\widehat{\varrho}(g^{-1}h,e)=N_\varrho(g^{-1}h)\le N_\varrho(g_1)+N_\varrho(h_1)\le\varrho(y,z)+\widehat{\varrho}(a_1,x_1)+\dots+\widehat{\varrho}(a_n,x_n)<1.$$

Hence, $g^{-1}h\in W$ and $h\in gW$. This proves that $U_\varrho(g)\subset gW\cap B_{n+2}(X)$.

The proof of the inverse inclusion $gW\cap B_{n+2}(X)\subset U_\varrho(g)$ requires some work. First, we claim that $gW\cap B_{n-1}(X)=\varnothing$.

Indeed, let $w=v_1\dots v_{2k}$ be a reduced element of $W$, where the letters $v_1,\dots,v_{2k}$ belong to $X\cup X^{-1}$. The only possible cancellations in the word $\mathfrak{X}=gw=a_1\dots a_n v_1\dots v_{2k}$ can occur at the join of $g$ and $w$, so it suffices to show that there can be at most $k$ cancellations in the word $\mathfrak{X}$ or, in other words, the shortest possible form of $[\mathfrak{X}]$ is $\mathfrak{Y}=a_1\dots a_{n-k}v_{k+1}\dots v_{2k}$. Assume the contrary: that the word $\mathfrak{Y}$ is still reducible. This means that $a_n=v_1^{-1},\dots,a_{n-k+1}=v_k^{-1}$ and $a_{n-k}=v_{k+1}^{-1}$. Choose a scheme $\varphi\in\mathcal{S}_k$ such that $N_\varrho(w)=2^{-1}\cdot\sum_{i=1}^{2k}\widehat{\varrho}(v_i^{-1},v_{\varphi(i)})$. Clearly, there exists $i\le k$ such that $\varphi(i)\le k+1$ — otherwise $\varphi$ cannot be a bijection of $\{1,\dots,2k\}$ onto itself. Therefore, Proposition 7.2.1 implies that $\varphi(j)=j+1$ for some $j$ with $i\le j\le k$ and, hence,

$$\widehat{\varrho}(a_{n-j}^{-1},a_{n-j+1})=\widehat{\varrho}(v_{j+1},v_j^{-1})=\widehat{\varrho}(v_j^{-1},v_{j+1})\le N_\varrho(w)<1.$$

This contradicts the choice of $\varrho\in\mathcal{P}_X(g)$. Therefore, $gW\cap B_{n-1}(X)=\varnothing$. In addition, every reduced element $w\in W$ has even length, so that the length of any element $h\in gW\cap B_{n+2}(X)$ is either $n$ or $n+2$.

Let us show that every element $h\in gW\cap B_{n+2}(X)$ belongs to $U_\varrho(g)$, that is, $h$ has the form described before Proposition 7.2.10. Clearly, $h=g\cdot w$, where $w=v_1\dots v_{2k}\in W$ and $v_1,\dots,v_{2k}\in X\cup X^{-1}$. We have just established that either $l(h)=n+2$ or $l(h)=n$. Suppose that $l(h)=n+2$. Without loss of generality we can assume that the element $w$ is reduced. Consider the word $\mathfrak{X}=gw=a_1\dots a_n v_1\dots v_{2k}$. If $\mathfrak{X}$ is reduced, then either $w=e$ or $k=1$ and $w=v_1v_2$. In the second case, $1>\widehat{\varrho}(w,e)=\widehat{\varrho}(v_1^{-1},v_2)$, so one of the letters $v_1, v_2$ is in $X$ and the other is in $X^{-1}$. In other words, $v_1 = y^\varepsilon$ and $v_2 = z^{-\varepsilon}$ for some $y, z \in X$ and $\varepsilon = \pm 1$. In either case, $g \in U_\varrho(g)$. Suppose, therefore, that the word $\mathfrak{X}$ is reducible.

Again, all possible cancellations in the word $\mathfrak{X}$ can occur only at the join of $g$ and $w$, so that $[\mathfrak{X}] = a_1 \dots a_{n-l} v_{l+1} \dots v_{2k}$, where $1 \le l \le n$ and $a_n = v_1^{-1}, \dots, a_{n-l+1} = v_l^{-1}$. By the assumption, the length of the word $[\mathfrak{X}]$ is equal to $n+2$, so that $(n-l)+(2k-l) = n+2$ and $l = k-1$. Therefore, $[\mathfrak{X}] = a_1 \dots a_{n-k+1} v_k \dots v_{2k}$. Denote by $\varphi$ a scheme on $\{1, \dots, 2k\}$ such that $N_\varrho(w) = 2^{-1} \cdot \sum_{i=1}^{2k} \widehat{\varrho}(v_i^{-1}, v_{\varphi(i)})$. First, we claim that $\varphi(i) \ge k$ for each $i \le l$. This is clear if $l = 1$. If $l > 1$ and $\varphi(i) < k$ for some $i \le l$, we argue as above to find $j < l$ such that $\varphi(j) = j+1$. Then

$$\widehat{\varrho}(a_{n-j}^{-1}, a_{n-j+1}) = \widehat{\varrho}(v_{j+1}, v_j^{-1}) = \widehat{\varrho}(v_j^{-1}, v_{j+1}) \le N_\varrho(w) < 1,$$

and again this contradicts the fact that $\varrho \in \mathcal{P}_X(g)$.

Thus, $\varphi(1), \dots, \varphi(l)$ are distinct integers lying between $k$ and $2k$. Since $\varphi$ is a bijection of the set $\{1, \dots, 2k\}$ and $k = l+1$, there exists exactly one integer $i$ such that $k \le i < \varphi(i) \le 2k$. Put $j = \varphi(i)$. It is easy to see that $j = i+1$. Indeed, otherwise there exists an integer $q$ with $i < q < j$, and then $p < k \le i < q < j$, where $p = \varphi(q)$. Equivalently, we have $p < i < \varphi(p) < \varphi(i)$, which contradicts the fact that $\varphi$ is a scheme. So, $\varphi(i) = j = i+1$ and $\varphi(i+1) = i$.

Not only the integers $\varphi(1), \dots, \varphi(l)$ are distinct and lie between $k$ and $2k$, but none of them is equal to $i$ or $i+1$. The sets $\{1, \dots, l\}$ and $\{k, \dots, i-1, i+2, \dots, 2k\}$ have the same cardinality, so $\varphi$ maps the first of them onto the second one. Since $\varphi$ is a scheme, there is only one way to establish a bijection between these sets: $1 \le p < q \le l$ implies $k \le \varphi(q) < \varphi(p) \le 2k$. Therefore, we have

$$\varphi(1) = 2k, \quad \varphi(2) = 2k-1, \ \dots, \ \varphi(2k-i-1) = i+2,$$
$$\varphi(2k-i) = i-1, \quad \varphi(2k-i+1) = i-2, \ \dots, \ \varphi(k-1) = k.$$

Since $\widehat{\varrho}(v_i^{-1}, v_{i+1}) \le N_\varrho(w) < 1$, one of the letters $v_i, v_{i+1}$ belongs to $X$ and the other belongs to $X^{-1}$. Hence there exist $y, z \in X$ and $\varepsilon = \pm 1$ such that $v_i = y^\varepsilon$ and $v_{i+1} = z^{-\varepsilon}$.

Thus, we have

$$[\mathfrak{X}] = a_1 \dots a_{n-k+1} v_k \dots v_i v_{i+1} \dots v_{2k} = t_1 \dots t_j y^\varepsilon z^{-\varepsilon} t_{j+1} \dots t_n,$$

where $j = n - 2k + i + 1$ and $a_1 = t_1, \dots, v_{2k} = t_n$ (i.e., the two words in the above equality coincide letter by letter). The properties of the scheme $\varphi$ and the equalities $a_n = v_1^{-1}, \dots, a_{n-l+1} = v_l^{-1}$ together imply that

$$\varrho(y, z) + \sum_{r=1}^{n} \widehat{\varrho}(a_r, t_r) = \frac{1}{2} \sum_{s=1}^{2k} \widehat{\varrho}(v_s^{-1}, v_{\varphi(s)}) = N_\varrho(w) < 1.$$

This inequality means that the element $g \cdot w = [\mathfrak{X}]$ belongs to $U_\varrho(g)$.

Finally, if the length of $h = g \cdot w$ is equal to $n$ then we argue as above and show that the corresponding scheme $\varphi$ for $w$ satisfies $\varphi(i) = 2k - i + 1$ for each $i \le 2k$, so that

$$[\mathfrak{X}] = a_1 \dots a_{n-k} v_{k+1} \dots v_{2k} = t_1 \dots t_n$$

and

$$\sum_{r=1}^{n} \widehat{\varrho}(a_r, t_r) = \frac{1}{2} \sum_{s=1}^{2k} \widehat{\varrho}(v_s^{-1}, v_{\varphi(s)}) = N_\varrho(w) < 1.$$

Again, this implies immediately that $h = g \cdot w \in U_\varrho(g)$. The theorem is proved. □

For every continuous pseudometric $\varrho$ on $X$, put

$$U_\varrho(e) = \{x^\varepsilon y^{-\varepsilon} : x, y \in X,\ \varepsilon = \pm 1,\ \varrho(x, y) < 1\}.$$

In the case when $g$ is the identity $e$ of $F(X)$, Theorem 7.2.11 provides the following simple description of a neighbourhood base of $B_2(X)$ at $e$.

COROLLARY 7.2.12. *The family of the sets $\{U_\varrho(e) : \varrho \in \mathcal{P}_X\}$ is a base of the subspace $B_2(X) \subset F(X)$ at the identity e.*

## Exercises

7.2.a. (M. I. Graev [**201**]) Prove that for every continuous pseudometric $d$ on a Tychonoff space $X$, there exists the *maximal* invariant pseudometric $\hat{d}$ on the free Graev topological group $F^*(X)$ (see Exercise 7.1.a) extending $d$. Prove the continuity of $\hat{d}$. Formulate and prove a similar result for the group $A^*(X)$.

7.2.b. Give an alternative proof of Theorem 7.1.2 making use of Graev's extension of continuous pseudometrics and applying the fact that the upper bound of a family of group topologies on an abstract group $G$ is again a group topology on $G$.

7.2.c. Find out whether Corollary 7.1.19 and Theorem 7.1.20 remain valid for the groups $FP(X)$ and $AP(X)$ (see Exercise 7.1.f).

7.2.d. Prove that for every uniform space $(X, \mathcal{U})$, there exists a unique (up to a topological isomorphism) *free uniform group* $G = F(X, \mathcal{U})$ with the following properties:

(U1) There exists a uniformly continuous mapping $\sigma\colon (X, \mathcal{U}) \to (G, \mathcal{V}_G)$ such that $\sigma(X)$ generates a dense subgroup of $G$, where $\mathcal{V}_G$ is the two-sided group uniformity of $G$.

(U2) If $H$ is an arbitrary topological group with the two-sided uniformity $\mathcal{V}_H$, then for every uniformly continuous mapping $(X, \mathcal{U}) \to (H, \mathcal{V}_H)$, there exists a continuous homomorphism $\tilde{f}\colon G \to H$ such that $f = \tilde{f} \circ \sigma$.

Show that $\sigma$ is a uniform embedding of $(X, \mathcal{U})$ into $(F(X, \mathcal{U}), \mathcal{V}_G)$, the set $\sigma(X)$ is closed in $F(X, \mathcal{U})$ and algebraically generates $F(X, \mathcal{U})$. Define the *free uniform Abelian group* $A(X, \mathcal{U})$ and verify that similar assertions hold true for $A(X, \mathcal{U})$.

7.2.e. Let $\mathcal{C}$ be the family of all open coverings of a Tychonoff space $X$. For every sequence $s = (\gamma_0, \gamma_1, \ldots) \in \mathcal{C}^\omega$, put

$$V_s = \{\sum_{i=0}^{n}(x_i - y_i) : n \in \omega,\ x_i, y_i \in U_i \text{ for some } U_i \in \gamma_i,\ i = 0, \ldots, n\}.$$

Show that $V_s$ is an open neighbourhood of the identity $e$ in $A(X)$, for each $s \in \mathcal{C}^\omega$, and the family $\{V_s : s \in \mathcal{C}^\omega\}$ is a local base for $A(X)$ at $e$.

7.2.f. (Uspenskij [**520**]). Let $X$ be a Tychonoff space and $F_0(X)$ be the open subgroup of $F(X)$ which coincides with the kernel of the homomorphism $f\colon F(X) \to \mathbb{Z}$, $f(x) = 1$ for each $x \in X$.

(a) Verify that every element $v \in F_0(X)$ admits a representation

$$v = \prod_{i=1}^{n} g_i x_i^{\varepsilon_i} y_i^{-\varepsilon_i} g_i^{-1},\ \ x_i, y_i \in X,\ \varepsilon_i = \pm 1,\ g_i \in F(X). \qquad (*)$$

(b) Let $\mathcal{P}$ be the family of all continuous pseudometrics on $X$. For every $t = \{d_g : g \in F(X)\} \in \mathcal{P}^{F(X)}$ and every $v \in F_0(X)$, put

$$N_t(v) = \inf \sum_{i=1}^{n} d_{g_i}(x_i, y_i),$$

where the infimum is taken with respect to all representations $(*)$ of $v$. Prove that each $N_t$ is a continuous prenorm on $F_0(X)$ and the family $\{N_t : t \in \mathcal{P}^{F(X)}\}$ generates the topology of the group $F(X)$.

7.2.g. (K. Eda, H. Ohta, and K. Yamada [**156**]) Let $\omega_1 + 1$ be the space of all ordinals less than or equal to $\omega_1$, with the order topology.

(a) Suppose that both sets $S \subset \omega_1$ and $\omega_1 \setminus S$ are *stationary* in $\omega_1$ (i.e., each of them intersects every closed unbounded subset of $\omega_1$), and consider the subspace $X = S^2 \cup (\omega_1 \times \{\omega_1\}) \cup (\{\omega_1\} \times \omega_1)$ of $(\omega_1 + 1)^2$. Show that each of the groups $F(X)$ and $A(X)$ contains a topological copy of $\omega_1 + 1$, while $X$ does not.

(b) Apply Theorem 7.1.13 to prove that if $X$ is Tychonoff and $F(X)$ contains a copy of the space $\omega_1$, then $X$ also contains a topological copy of $\omega_1$.

*Remark.* The assertion in (b) remains valid for the group $A(X)$ under some extra axioms of the Set Theory [**156**].

## Problems

7.2.A. Let $\mathcal{T}_1$ and $\mathcal{T}_2$ be metrizable topologies on $X$, and let $\mathcal{T}_3$ be their supremum, that is, the smallest topology on $X$ containing both $\mathcal{T}_1$ and $\mathcal{T}_2$. Are the topologies of the free topological group on $X$ with these three topologies related in the same way?

7.2.B. (M. G. Tkachenko [**472**]) Let $X$ be a Tychonoff space and $\mathcal{P}_X$ be the family of all continuous pseudometrics on $X$. Prove that the family $\{\widehat{d} : d \in \mathcal{P}_X\}$ generates the topology on $F_a(X)$ coinciding with the topology of $F(X)$ iff the group $F(X)$ is balanced iff there exists a cardinal $\tau > \omega$ such that $X$ is pseudo-$\tau$-compact and the intersections of less than $\tau$ open sets in $X$ are open.

7.2.C. (M. G. Tkachenko [**482**]). Let $d$ be an arbitrary pseudometric on a set $X$ and $\widehat{d}_A$ the Graev extension of $d$ over the free Abelian group $A_a(X)$. Prove that $\widehat{d}_A(kg, kh) = |k| \cdot \widehat{d}(g, h)$ for all $g, h \in A(X)$ and all $k \in \mathbb{Z}$. (This generalizes Lemma 7.9.1.)

*Hint.* It suffices to prove that $\widehat{d}(kh, e) = k \cdot \widehat{d}(h, e)$ for all $h \in A_a(X)$ and all $k \in \mathbb{N}$, where $e$ is the neutral element of $A(X)$. Use Corollary 7.2.4 to write an element $h \in A_a(X)$ in an reduced form

$$h = \sum_{1 \le i < j \le n} k_{i,j}(x_i - \varepsilon_{i,j} x_j) + \sum_{1 \le i \le n} k_i x_i \tag{7.10}$$

with $x_i, x_j \in X$, $k_{i,j}, k_i \in \mathbb{Z}$ and $\varepsilon_{i,j} = \pm 1$, such that

$$\widehat{d}_A(h, e) = \sum_{i<j} |k_{i,j}| \cdot d^*(x_i, \varepsilon_{i,j} x_j) + \sum_{1 \le i \le n} |k_i| \cdot d^*(x_i, e) \tag{7.11}$$

and

$$l(h) = \sum_{1 \le i < j \le n} 2\,|k_{i,j}| + \sum_{1 \le i \le n} |k_i|. \tag{7.12}$$

In addition, $\sum_{i=1}^{n} |k_i| \le 1$. Call the representation (7.10) of $h$ satisfying (7.11) and (7.12) an *A-scheme* for $h$. To every such $A$-scheme $V$ for $h$, assign a non-oriented graph $G_V$ with vertices $\{x_1, \ldots, x_n\} \cup \{e\}$. The vertices $x_i$ and $x_j$ with $i < j$ are connected by an edge in the graph $G_V$ if $k_{i,j} \ne 0$ in (7.10). Similarly, the vertices $e$ and $x_i$ are connected by an edge in $G_V$ iff $k_i \neq 0$ in (7.10). Therefore, $e$ is connected with at most one vertex in the graph $G_V$. Assign to every edge $[x_i, x_j]$ (resp., $[x_i, e]$) in $G_V$ its *multiplicity* $|k_{i,j}|$ (resp., $|k_i|$). Verify the following:

(A) For every non-zero $h \in A(X)$, there exists an $A$-scheme $V$ for $h$ such that the graph $G_V$ does not contain circles.

Let $\Gamma$ be a finite graph, and each edge of $\Gamma$ is assigned a positive integer, its multiplicity. For every vertex $t \in \Gamma$, the sum $I_t$ of the multiplicities of all edges connecting $t$ with vertices in $\Gamma$ is called the *index* of $t$ in $\Gamma$. Show that the following fact is valid:

(B) Let $\Gamma$ be a finite graph without circles, and let the index $I_t$ of each vertex $t \in \Gamma$, except for at most one of them, denoted by $t^*$, be a multiple of an integer $k \geq 2$. Then the multiplicity of each edge in $\Gamma$, as well as the multiplicity index of $t^*$, are also multiples of $k$.

Apply (A) and (B) to deduce the necessary conclusion.

7.2.D. (P. Nickolas and M. G. Tkachenko [**348**]) Let $\omega^\omega$ be the family of function from $\omega$ to $\omega$. A family $\mathcal{D} \subset \omega^\omega$ is called *dominating* if for every $f \in \omega^\omega$, there exists $g \in \mathcal{D}$ such that $f(n) \leq g(n)$, for all $n \in \omega$. The minimal cardinality of a dominating family in $\omega^\omega$ is denoted by $\mathfrak{d}$. It is easy to verify that $\aleph_1 \leq \mathfrak{d} \leq \mathfrak{c}$, and each of the inequalities may be consistently strict (see [**263**, Chap. 17]). Prove the following for a Tychonoff space $X$:

1) If $X$ is not a $P$-space, then the character of the group $A(X)$ is at least $\mathfrak{d}$.
2) If $X$ is infinite, compact, and metrizable, then the weight of each of the groups $A(X)$, $F(X)$ is equal to $\mathfrak{d}$.
3) If $X$ is pseudo-$\aleph_1$-compact, then the weights of the groups $A(X)$ and $F(X)$ coincide.
4) There exists a Lindelöf space $Y$ such that the weight of both groups $A(Y)$ and $F(Y)$ is equal to $\aleph_1$.

*Hint.* To deduce 1), use Theorem 7.2.7. For 2), modify the description of a neighbourhood base at the neutral element of $F(X)$ given in Exercise 7.2.f (with an idea to have something resembling Exercise 7.2.e). Then show that the character of $F(X)$ does not exceed $\mathfrak{d}$, for any compact metrizable space $X$, and apply Corollary 5.2.4.

7.2.E. (P. Nickolas and M. G. Tkachenko [**349**]) Prove the following statements:

1) If $X$ is a completely metrizable separable space, then the character of the groups $A(X)$ and $F(X)$ is equal to $\mathfrak{d}$.
2) There exists a regular second-countable space $M$ such that both groups $A(M)$ and $F(M)$ have the character equal to $2^\omega$.
3) If $X$ and $Y$ are infinite compact spaces of the same weight, then the characters of the groups $A(X)$, $A(Y)$, $F(X)$, and $F(Y)$ coincide.

7.2.F. (S. García-Ferreira, M. Sakai, and M. Sanchis [**184**]) Let $\kappa$ be an uncountable regular cardinal. A base $\mathcal{B} = \{U_\alpha : \alpha < \kappa\}$ for a uniformity $\mathcal{U}$ on a set $X$ is called *telescopic* if it satisfies the following conditions:

(i) $U_\alpha = U_\alpha^{-1}$, for every $\alpha < \kappa$;
(ii) $U_\beta \circ U_\beta \subset U_\alpha$ whenever $\alpha < \beta < \kappa$;
(iii) $U_\beta = \bigcap_{\alpha<\beta} U_\alpha$ whenever $\beta < \kappa$ is a limit ordinal.

Prove the following statements:

a) $A(X)$ is topologically orderable and $\chi((A(X)) = \kappa$ iff the universal uniformity on $X$ has a telescopic base $\{U_\alpha : \alpha < \kappa\}$.
b) $F(X)$ is topologically orderable and $\chi((A(X)) = \kappa$ iff the universal uniformity on $X$ has a telescopic base $\{U_\alpha : \alpha < \kappa\}$ and every open covering of $X$ has a subcovering of size strictly less than $\kappa$.

## 7.3. Extension of metrizable groups by compact groups

In this section, we consider the following natural question. Suppose that $G$ is a topological group, and that $H$ is a closed invariant subgroup of $G$ such that both $H$ and the quotient group $G/H$ are feathered. Under the assumptions, must $G$ be feathered as well? Some optimism with regards to this question is nourished by Corollary 1.5.21: If a topological group $G$ is an extension of a first-countable group by a first-countable group, then $G$ is itself first-countable. Note also that if $H$ is compact and $G/H$ is metrizable, then, as we already know, the answer is "yes", since the natural quotient mapping of $G$ onto $G/H$ is perfect. So the next step is to consider the "dual" situation and to assume that $H$ is metrizable and $G/H$ is compact. Will $G$ be feathered in this case? We answer this question below, making use in a very essential way of the technique of free topological groups.

THEOREM 7.3.1. *There exists an Abelian topological group $P$ with a closed subgroup $H$ such that $H$ is metrizable, the quotient group $P/H$ is compact, and $P$ is not feathered.*

PROOF. Put $\mathfrak{c} = 2^\omega$, and let $G = D^\mathfrak{c}$ be the topological product of $\mathfrak{c}$ copies of the discrete Abelian group $D = \{0, 1\}$. Then $G$ is a compact Abelian group. Let $M$ be the group $G$ endowed with the discrete topology, and consider the topological product $M^\omega$ of $\omega$ copies of the discrete group $M$. Then $M^\omega$ is metrizable, non-discrete, and, clearly, $|M^\omega| = |G| = 2^\mathfrak{c}$. Let $L$ be the $\sigma$-product of $\omega$ copies of $M$ at the neutral element of $M^\omega$. Then $L$ is a dense subgroup of $M^\omega$, and $|M^\omega \setminus L| = |L| = |M| = 2^\mathfrak{c}$, since for any $z \in M^\omega \setminus L$, $z + L \subset M^\omega \setminus L$. Fix a bijection $g$ of $M^\omega \setminus L$ onto $G$, and define a mapping $f$ of the space $M^\omega$ onto the set $G$ by the following rule: $f(x) = e$, for each $x \in L$ (where $e$ is the neutral element of $G$), and $f(x) = g(x)$, for each $x \in M^\omega \setminus L$. Clearly, $f(M^\omega) = G$.

Consider the free Abelian group $A(M^\omega)$ of the set $M^\omega$. There exists a metrizable topology $\mathcal{T}_1$ on $A(M^\omega)$ which turns $A(M^\omega)$ into a topological group and induces on $M^\omega$ the product topology $\mathcal{T}$ of $M^\omega$. The mapping $f$, by the principal property of the group $A(M^\omega)$, can be extended to a homomorphism $f^*\colon A(M^\omega) \to G$. Put $H = (f^*)^{-1}(e)$. Since $L \subset H$ and $L$ is dense in $M^\omega$, it follows that $H$ is dense in $(A(M^\omega), \mathcal{T}_1)$. Of course, $f^*$ is not continuous with respect to $\mathcal{T}_1$. We introduce a new topology on $A(M^\omega)$ that makes $f^*$ continuous, as follows. Let $\mathcal{T}_G$ be the topology of $G$. Put $\mathcal{T}_2 = \{(f^*)^{-1}(V) : V \in \mathcal{T}_G\}$. Clearly, $\mathcal{T}_2$ is a non-Hausdorff topology on $A(M^\omega)$. However, it is a group topology on $A(M^\omega)$, since $f^*$ is a homomorphism and $G$ is a topological group.

The key idea is to blend the two topologies we defined on $A(M^\omega)$ into one. Let

$$\mathcal{B} = \{U \cap (f^*)^{-1}(V) : U \in \mathcal{T}_1,\ V \in \mathcal{T}_G\}.$$

Clearly, $\mathcal{T}_1 \subset \mathcal{B}$, and the family $\mathcal{B}$ is closed under intersections of finite subfamilies. Therefore, $\mathcal{B}$ is a base of a topology $\mathcal{T}_0$ on $A(M^\omega)$ that is stronger than $\mathcal{T}_1$. This topology $\mathcal{T}_0$ is the least upper bound of the group topologies $\mathcal{T}_1$ and $\mathcal{T}_2$ on $A(M^\omega)$, and therefore, $A(M^\omega)$, endowed with $\mathcal{T}_0$, is a topological group. Note that $(A(M^\omega), \mathcal{T}_0)$ is a Hausdorff space with a $G_\delta$-diagonal, since the topology $\mathcal{T}_0$ contains the metrizable topology $\mathcal{T}_1$. Therefore, all compact subspaces of $(A(M^\omega), \mathcal{T}_0)$ are metrizable.

Let us show that the mapping $f^*$ of $(A(M^\omega), \mathcal{T}_0)$ onto $G$ is open and continuous. The continuity of $f^*$ follows from the fact that $\mathcal{T}_2 = \{(f^*)^{-1}(V) : V \in \mathcal{T}_G\} \subset \mathcal{T}_0$. Hence, $H$ is closed in $(A(M^\omega), \mathcal{T}_0)$. Let us show that $f^*$ is open.

We will call a mapping $f: X \to Y$ of a space $X$ *superopen* if $f(U) = Y$, for each non-empty open subset $U$ of $X$.

**Claim 1.** *The mapping $f^*$ of $(A(M^\omega), \mathcal{T}_1)$ onto $G$ is superopen.*

Indeed, $(f^*)^{-1}(e)$ is dense in $(A(M^\omega), \mathcal{T}_1)$. Since $f^*(M^\omega) = G$, and $f^*$ is a homomorphism, it follows that $(f^*)^{-1}(a)$ is dense in $(A(M^\omega), \mathcal{T}_1)$, for each $a \in G$. Therefore, $f^*(U) = G$ for every non-empty $U \in \mathcal{T}_1$, and Claim 1 is verified.

Now consider an arbitrary element $W \in \mathcal{B}$. To show that $f^*$ is open, it is enough to check that $f^*(W)$ is open in $G$. We have $W = U \cap (f^*)^{-1}(V)$, for some $U \in \mathcal{T}_1$ and some $V \in \mathcal{T}_G$. Therefore, $f^*(W) = f^*(U) \cap V = G \cap V = V \in \mathcal{T}_G$. Thus, $f^*$ is an open continuous homomorphism of $(A(M^\omega), \mathcal{T}_0)$ onto $G$. Hence, the topological group $G$ can be identified with the quotient group of $(A(M^\omega), \mathcal{T}_0)$ with respect to the closed subgroup $H$.

**Claim 2.** *The subspace $H$ of $(A(M^\omega), \mathcal{T}_0)$ is metrizable.*

Indeed, the traces of elements of $\mathcal{B}$ on $H$ form the family

$$\{H \cap (U \cap (f^*)^{-1}(V)) : U \in \mathcal{T}_1,\ V \in \mathcal{T}_G\} = \{H \cap U : U \in \mathcal{T}_1\},$$

which is precisely the topology induced by $\mathcal{T}_1$ on $H$. This is so, since for each $V \in \mathcal{T}_G$ either $(f^*)^{-1}(V) \cap H = \varnothing$ or $H \subset (f^*)^{-1}(V)$. Therefore, the topology induced by $\mathcal{T}_0$ on $H$ coincides with the topology induced by $\mathcal{T}_1$ on $H$, and Claim 2 is proved.

Thus, we have proved that the topological group $P = (A(M^\omega), \mathcal{T}_0)$ is an extension of the metrizable group $H$ by the compact group $G$.

Let us, finally, verify that $(A(M^\omega), \mathcal{T}_0)$ is not feathered. Assume the contrary. Then $(A(M^\omega), \mathcal{T}_0)$ is a feathered space in which every point is a $G_\delta$; therefore, $(A(M^\omega), \mathcal{T}_0)$ is first-countable. Since the mapping $f^*$ is open and continuous, and $f^*(A(M^\omega)) = G$, it follows that $G$ is first-countable, a contradiction. □

Can the metrizable group $H$ in Theorem 7.3.1 be made separable? The next result provides us with a strong partial answer to this question.

THEOREM 7.3.2. *The following two statements are equivalent:*

1) $2^{\aleph_1} = 2^{\aleph_0}$;
2) *there exist an Abelian topological group $G$ with a closed subgroup $H$ such that the space $H$ has a countable base, the quotient group $G/H$ is compact, $|G| = 2^{\aleph_0}$, and $G$ is not feathered.*

PROOF. First, we show that 2) ⇒ 1). Assume to the contrary that $2^\omega = 2^{\omega_1}$. Since $|G/H| \leq |G| \leq 2^\omega$, and $G/H$ is compact and Hausdorff, it follows from Čech–Pospíšil's theorem [**165**, 3.12.11 (a)] that $G/H$ is first-countable at least at one point. Since $G/H$ is homogeneous, it follows that the space $G/H$ is first-countable. Then $G$ is first-countable, by Corollary 1.5.21, and Theorem 3.3.12 implies that $G$ is metrizable and, hence, is feathered, a contradiction.

Now we prove that 1) ⇒ 2). The argument is very similar to the proof of Theorem 7.3.1. Let $G = D^{\omega_1}$ be the topological product of $\omega_1$ copies of the discrete Abelian group $D = \{0, 1\}$. Then $G$ is a compact Abelian group, and $|G| = 2^{\aleph_1} = 2^{\aleph_0}$, by 1).

Let $\mathbb{R}$ be the Abelian group of real numbers, with the usual topology $\mathcal{T}$, $\mathbb{Q} \subset \mathbb{R}$ the set of rational numbers, and $e$ the neutral element of $G$. Since $|\mathbb{R} \setminus \mathbb{Q}| = 2^\omega = |G|$, we can fix a bijection $h$ of $\mathbb{R} \setminus \mathbb{Q}$ onto $G$. Define a mapping $f\colon \mathbb{R} \to G$ by $f(x) = e$, for each $x \in \mathbb{Q}$, and $f(x) = h(x)$, for each $x \in \mathbb{R} \setminus \mathbb{Q}$. Then $f(\mathbb{R}) = G$.

Consider the free Abelian group $A_a(\mathbb{R})$ of the set $\mathbb{R}$. Fix a separable metrizable topology $\mathcal{T}_1$ on $A_a(\mathbb{R})$ such that $\mathcal{T}_1$ induces the usual topology $\mathcal{T}$ on $\mathbb{R}$. The mapping $f$, by the principal property of $A_a(\mathbb{R})$, can be extended to a homomorphism $f^*\colon A_a(\mathbb{R}) \to G$. Note that $f^*$ is not continuous with respect to $\mathcal{T}_1$. We introduce a new topology on $A_a(\mathbb{R})$ that will make $f^*$ continuous, as follows. Let $\mathcal{T}_G$ be the topology of $G$. Put $\mathcal{T}_2 = \{(f^*)^{-1}(V) : V \in \mathcal{T}_G\}$. Clearly, $\mathcal{T}_2$ is a group topology on $A_a(\mathbb{R})$ since $f^*$ is a homomorphism and $G$ is a topological group.

Let us blend the topologies $\mathcal{T}_1$ and $\mathcal{T}_2$ on $A_a(\mathbb{R})$ into one. We put

$$\mathcal{B} = \{U \cap (f^*)^{-1}(V) : U \in \mathcal{T}_1,\ V \in \mathcal{T}_G\}.$$

Clearly, $\mathcal{T}_1 \subset \mathcal{B}$ and the family $\mathcal{B}$ is closed under intersections of finite subfamilies. Therefore, $\mathcal{B}$ is a base of a topology $\mathcal{T}_0$ on $A_a(\mathbb{R})$ which is stronger than $\mathcal{T}_1$. This topology $\mathcal{T}_0$ is the least upper bound of the group topologies $\mathcal{T}_1$ and $\mathcal{T}_2$ on $A_a(\mathbb{R})$ and, therefore, $A_a(\mathbb{R})$ endowed with $\mathcal{T}_0$ is a topological group. Note that $(A_a(\mathbb{R}), \mathcal{T}_0)$ is a Hausdorff space in which every point is a $G_\delta$, since the topology $\mathcal{T}_0$ contains the metrizable topology $\mathcal{T}_1$.

To show that the mapping $f^*$ of $(A_a(M^\omega), \mathcal{T}_0)$ onto $G$ is open and continuous we argue exactly as in the proof of Theorem 7.3.1, so we omit this part of the argument. Thus, $H$ is a closed subgroup of $(A_a(\mathbb{R}), \mathcal{T}_0)$, and $G$ can be interpreted as a quotient group of the topological group $(A_a(\mathbb{R}), \mathcal{T}_0)$ with respect to $H$.

Finally, $H$ is a separable metrizable subspace of $(A_a(\mathbb{R}), \mathcal{T}_0)$ since the topology induced by $\mathcal{T}_0$ on $H$ coincides with the topology induced by $\mathcal{T}_1$ on $H$ and $\mathcal{T}_1$ is separable metrizable.

Since each point in $(A_a(\mathbb{R}), \mathcal{T}_0)$ is a $G_\delta$, the space $(A_a(\mathbb{R}), \mathcal{T}_0)$ cannot be feathered. Indeed, otherwise $(A_a(\mathbb{R}), \mathcal{T}_0)$ and $G$ would have been first-countable, a contradiction. □

COROLLARY 7.3.3. *It is consistent with ZFC that there exists a topological group G which is not feathered and is an extension of a second-countable group by a compact group.*

## Exercises

7.3.a. Let $G$ be a topological group, and $H$ a closed metrizable subgroup of $G$ such that the quotient space $G/H$ is normal (paracompact). Is the space $G$ then normal (paracompact)?

7.3.b. (V. V. Uspenskij [**511**]) Let $G$ be an Abelian topological group and $H$ a closed subgroup of $G$ such that both $H$ and the quotient group $G/H$ are cosmic. Show that the space $G$ need not even be normal (compare with Problems 4.6.C and 4.6.D).

## Problems

7.3.A. Let $G$ be an Abelian topological group, and $H$ a closed subgroup of $G$ such that $H$ is metrizable and the quotient group $G/H$ is compact. Is $G$ paracompact? Is the topological group $(A(M^\omega), \mathcal{T}_0)$ constructed in Theorem 7.3.1, paracompact?

7.3.B. Let $G$ be an Abelian topological group and $H$ a closed subgroup of $G$ such that both $H$ and the quotient group $G/H$ are $\sigma$-compact. Is $G$ paracompact? (See also Problem 1.5.F.)

## Open Problems

7.3.1. Characterize the (Abelian) topological groups that can be obtained as an extension of a metrizable group by a compact group.

7.3.2. Characterize the class $\mathcal{CN}(2)$ of topological groups that can be obtained as an extension of a cosmic topological group by a cosmic topological group.

7.3.3. Give an example of a topological group $G$ with the following two properties:
   a) $G$ is not in $\mathcal{CN}(2)$;
   b) $G$ is an extension of a cosmic topological group $E$ by a topological group $H \in \mathcal{CN}(2)$.

7.3.4. Let $\mathcal{CM}$ be the smallest class of Abelian topological groups containing all compact Abelian groups, all metrizable Abelian groups, and closed under extensions, under taking quotients, and under taking closed subgroups. Give an example of an Abelian topological group that does not belong to this class.

7.3.5. Let $\mathcal{CSM}$ be the smallest class of Abelian topological groups containing all compact Abelian groups, all second-countable Abelian groups, and closed under extensions, under taking quotients, and under taking closed subgroups. Give an example of an $\omega$-narrow Abelian topological group that does not belong to this class. Is it true that every cosmic topological Abelian group $G$ belongs to $\mathcal{CSM}$?

7.3.6. Let $f$ be an open continuous homomorphism of a regular paratopological group $G$ onto a metrizable topological group $H$ such that the kernel of $f$ is metrizable. Is $G$ metrizable?

7.3.7. Does there exist a topological group $G$, as in Corollary 7.3.3, in $ZFC$ alone?

## 7.4. Direct limit property and completeness

The topology of sequential spaces is completely determined by convergent sequences in the sense that a subset $F$ of a sequential space $X$ is closed in $X$ iff the intersection of $F$ with every convergent sequence $C$ in $X$ (including its limit) is closed in $C$. The class of $k$-spaces is characterized by a similar property; one simply has to replace convergent sequences by the family of compact subsets of $X$. In general, we say that the topology of a space $X$ *is determined by a family* $\mathcal{C}$ of its subsets provided that a set $F \subset X$ is closed in $X$ iff $F \cap C$ is closed in $C$, for each $C \in \mathcal{C}$.

Denote by $G(X)$ the free topological group $F(X)$ or the free Abelian topological group $A(X)$ on a space $X$. This group is the union of the increasing sequence $\{B_n(X) : n \in \omega\}$ of its closed subsets $B_n(X)$, where $B_n(X)$ is the set of all elements in $G(X)$ which have reduced length $\leq n$ with respect to the basis $X$ (see Theorem 7.1.13). This suggests the following definition.

The free (Abelian) topological group $G(X)$ has the *direct limit property* if the topology of $G(X)$ is determined by the family $\{B_n(X) : n \in \omega\}$. If $G(X)$ has the direct limit property, we shall also say that $G(X)$ is the *direct limit* of its subspaces $B_n(X)$, $n \in \omega$.

Clearly, each $B_n(X)$ is compact if the space $X$ is compact. By analogy with $k$-spaces one may conjecture that $G(X)$ has the direct limit property for every compact space $X$. We will show in Theorem 7.4.1 that this is indeed the case for a wider class of spaces defined below.

Let a space $X$ be the union of an increasing sequence $\{X_n : n \in \omega\}$ of its compact subsets $X_n$. If the topology of $X$ is determined by the sequence $\{X_n : n \in \omega\}$, then $X$ is called a *$k_\omega$-space*, and $X = \bigcup_{n\in\omega} X_n$ is a *$k_\omega$-decomposition* of $X$.

Clearly, every $k_\omega$-space is $\sigma$-compact, but not vice versa (consider the space $\mathbb{Q}$ of rational numbers). However, every locally compact $\sigma$-compact space is a $k_\omega$-space. In particular, every open $F_\sigma$-subset of a compact space is a $k_\omega$-space.

For a subset $Y$ of a space $X$ and a positive integer $n$, we put $\langle Y\rangle_n = G(Y, X) \cap B_n(X)$, where $G(Y, X)$ is the subgroup of $G(X)$ generated by $Y$. This notation is used in the theorem below which is important in its own right and has a number of corollaries.

THEOREM 7.4.1. **[M.I. Graev, J. Mack, S.A. Morris, E.T. Ordman]** *The group $G(X)$ has the direct limit property for every $k_\omega$-space $X$. In addition, if $X = \bigcup_{n=1}^{\infty} X_n$ is a $k_\omega$-decomposition for $X$, then $G(X) = \bigcup_{n=1}^{\infty} \langle X_n\rangle_n$ is a $k_\omega$-decomposition for $G(X)$.*

PROOF. Let $X = \bigcup_{n\in\omega} X_n$ be a $k_\omega$-decomposition of $X$. Denote by $e$ be the identity of $G(X)$. For every $n \geq 1$, put $\tilde{X}_n = X_n \cup \{e\} \cup X_n^{-1}$ and $K_n = \tilde{X}_n \cdot \ldots \cdot \tilde{X}_n$ ($n$ times). Then $\langle X_n\rangle_n = K_n$ is a compact subset of $G(X)$, $K_n^{-1} = K_n$, $K_n \cdot K_n \subset K_{2n}$ for each $n \geq 1$, and $G(X) = \bigcup_{n=1}^{\infty} K_n$. Denote by $\mathcal{T}^*$ the new topology on $G_a(X)$ determined by the family $\{K_n : n \in \mathbb{N}\}$, where every $K_n$ carries the topology inherited from $G(X)$. In other words, a subset $O$ of $G(X)$ is in $\mathcal{T}^*$ iff $O \cap K_n$ is open in $K_n$ for each $n \geq 1$. Clearly, $\mathcal{T}^*$ is finer than the original topology $\mathcal{T}$ of the topological group $G(X)$, but the restrictions of $\mathcal{T}^*$ and $\mathcal{T}$ to $K_n$ coincide for each $n \in \mathbb{N}$. Our aim is to show that $\mathcal{T}^* = \mathcal{T}$. First, we establish the following fact.

**Claim.** *The family $\mathcal{T}^*$ is a group topology on $G_a(X)$ and the original topology of $X$ coincides with the one $X$ inherits from $(G_a(X), \mathcal{T}^*)$.*

Indeed, it is clear that $\mathcal{T}^*$ is a topology. Since $X \subset K_1$, the definition of $\mathcal{T}^*$ implies immediately the second part of our Claim. In addition, if $U \in \mathcal{T}^*$, then $U^{-1} \in \mathcal{T}^*$. Indeed, for every $n \geq 1$ there exists an open set $V_n$ in $F(X)$ such that $U \cap K_n = V_n \cap K_n$. Since $K_n$ is symmetric, the set

$$U^{-1} \cap K_n = U^{-1} \cap K_n^{-1} = (U \cap K_n)^{-1} = (V_n \cap K_n)^{-1} = V_n^{-1} \cap K_n$$

is open in $K_n$ and, hence, $U^{-1} \in \mathcal{T}^*$.

It suffices, therefore, to show that if $g, h \in G_a(X)$ and $g \cdot h \in U \in \mathcal{T}^*$, then there exist $V, W \in \mathcal{T}^*$ such that $g \in V$, $h \in W$ and $V \cdot W \subset U$. Choose $m \geq 1$ such that $g, h \in K_m$. We shall construct two sequences $\{V_n : n \geq m\}$ and $\{W_n : n \geq m\}$ satisfying the following conditions for each $n \geq m$:

(1) $V_n$ and $W_n$ are open in $K_n$;
(2) $A_n = cl_{K_n} V_n \subset V_{n+1}$ and $B_n = cl_{K_n} W_n \subset W_{n+1}$;
(3) $A_n \cdot B_n \subset U$.

By the continuity of the multiplication in $G(X)$, there exist open sets $V'_m$ and $W'_m$ in $K_m$ such that $g \in V'_m$, $h \in W'_m$ and $V'_m \cdot W'_m \subset U \cap K_{2m}$. Since $K_m$ is regular, one can find open sets $V_m$ and $W_m$ in $K_m$ such that $g \in V_m \subset cl_{K_m} V_m \subset V'_m$ and $h \in W_m \subset cl_{K_m} W_m \subset W'_m$.

Suppose that for some $n \geq m$, we have already defined sets $V_m, \ldots, V_n$ and $W_m, \ldots, W_n$ satisfying (1)–(3). By (3), the sets $A_n = cl_{K_n} V_n$ and $B_n = cl_{K_n} W_n$ satisfy $A_n \cdot B_n \subset U \cap K_{2n}$. Since the multiplication mapping $K_{n+1} \times K_{n+1} \to K_{2n+2}$ is continuous, there exist open set $V'_{n+1}$ and $W'_{n+1}$ in $K_{n+1}$ such that $A_n \subset V'_{n+1}$, $B_n \subset W'_{n+1}$ and $V'_{n+1} \cdot W'_{n+1} \subset U$. Using the normality of the compact space $K_{n+1}$, we can find open sets $V_{n+1}$ and $W_{n+1}$ in $K_{n+1}$ such that $A_n \subset V_{n+1} \subset cl_{K_{n+1}} V_{n+1} \subset V'_{n+1}$ and $B_n \subset W_{n+1} \subset cl_{K_{n+1}} W_{n+1} \subset W'_{n+1}$.

Continuing this process, we finally obtain the sequences

$$V_m \subset V_{m+1} \subset \cdots \subset V_n \subset \cdots \quad \text{and} \quad W_m \subset W_{m+1} \subset \cdots \subset W_n \subset \cdots$$

satisfying conditions (1)–(3). Put $V = \bigcup_{k=m}^{\infty} V_k$ and $W = \bigcup_{k=m}^{\infty} W_k$. Clearly, $g \in V$ and $h \in W$. From (1) and (2) it follows that the set $V \cap K_n = \bigcup_{k=n}^{\infty}(V_k \cap K_n)$ is open in $K_n$ for each $n \geq m$, so (1) implies that $V \in \mathcal{T}^*$. Similarly, $W \in \mathcal{T}^*$. In addition, (2) and (3) imply that

$$V \cdot W = \bigcup_{k=m}^{\infty} V_k \cdot W_k \subset \bigcup_{k=m}^{\infty} A_k \cdot B_k \subset U.$$

Thus, we have found the sets $V, W \in \mathcal{T}^*$ such that $g \in V$, $h \in W$ and $V \cdot W \subset U$. This proves our Claim.

Finally, by Corollary 7.1.8, the topology $\mathcal{T}$ of the group $G(X)$ is the finest group topology on $G_a(X)$ which induces on $X$ its original topology. Therefore, the inclusion $\mathcal{T} \subset \mathcal{T}^*$ and the above Claim together imply that $\mathcal{T}^* = \mathcal{T}$. This proves the theorem. □

COROLLARY 7.4.2. *If $X$ is compact, then the free (Abelian) topological group $G(X)$ has the direct limit property.*

Theorem 7.4.1 implies a number of interesting results about the topological properties of free topological groups. We start with two simple facts.

COROLLARY 7.4.3. *Let $X$ be an arbitrary space, and $C$ be any subset of $G(X)$. If $C \cap B_n(X)$ is finite for each $n \in \omega$, then $C$ is closed and discrete in $G(X)$.*

PROOF. Let $p\colon X \to K$ be a topological embedding of $X$ to a compact space $K$. Extend $p$ to a continuous monomorphism $\hat{p}\colon G(X) \to G(K)$ and consider the set $Q = \hat{p}(D)$, where $D$ is an arbitrary subset of $C$. Then the intersection $Q \cap B_n(K)$ is finite (hence closed) for each $n \in \omega$, so Theorem 7.4.1 implies that $Q$ is closed in $G(K)$. Since $\hat{p}$ is a continuous monomorphism, we conclude that $D$ is closed in $G(X)$. So, all subsets of $C$ are closed in $G(X)$ and, hence, $C$ is discrete. □

The corollary below will be generalized in Section 7.5 (see Theorem 7.5.3 and Corollary 7.5.4).

COROLLARY 7.4.4. *If $X$ is a space, and $K$ is a countably compact subspace of $G(X)$, then $K \subset B_n(X)$, for some $n \in \mathbb{N}$.*

PROOF. Suppose to the contrary that $K \setminus B_n(X) \neq \varnothing$ for each $n \in \mathbb{N}$. Then there exists an infinite subset $C$ of $K$ such that $C \cap B_n(X)$ is finite for each $n \in \mathbb{N}$. By Corollary 7.4.3, $C$ is closed and discrete in $G(X)$ and in $K$, which is a contradiction. □

The direct limit property of free (Abelian) topological groups on compact spaces implies the following general result.

THEOREM 7.4.5. *If $Y$ is a closed subspace of a Tychonoff space $X$, then the subgroup $G(Y, X)$ of $G(X)$ generated by $Y$ is closed in $G(X)$.*

PROOF. Denote by $bX$ a Hausdorff compactification of $X$, and extend the identity embedding of $X$ to $bX$ to a continuous monomorphism $\varphi\colon G(X) \to G(bX)$. Let $Y^*$ be the closure of $Y$ in $bX$ and $G(Y^*, bX)$ be the subgroup of $G(bX)$ generated by

$Y^*$. Since $Y^*$ is compact, the intersection $G(Y^*, bX) \cap B_n(bX)$ is compact for each $n \in \omega$. Apply Corollary 7.4.2 to deduce that $G(Y^*, bX)$ is closed in $G(bX)$. Therefore, $G(Y, bX) = G(Y^*, bX) \cap G(X, bX)$ is a closed subgroup of $G(X, bX)$. Since $\varphi$ is a continuous monomorphism and $\varphi(G(Y, X)) = G(Y, bX)$, we conclude that $G(Y, X)$ is closed in $G(X)$. □

COROLLARY 7.4.6. *If $K$ is a compact subset of a Tychonoff space $X$, then $G(K, X)$ is closed in $G(X)$ and $G(K, X) \cong G(K)$.*

PROOF. The fact that $G(K, X)$ is closed in $G(X)$ follows from Theorem 7.4.5. Let us show that $G(K, X) \cong G(K)$. Denote by $bX$ an arbitrary compactification of $X$ and let $f: K \hookrightarrow X$ and $g: X \hookrightarrow bX$ be the natural embeddings. Extend $f$ and $g$ to continuous monomorphisms $\hat{f}: G(K) \to G(X)$ and $\hat{g}: G(X) \to G(bX)$, respectively. By Theorem 7.4.1, $G(K)$ and $G(bX)$ are $k_\omega$-groups with $k_\omega$-decompositions $G(K) = \bigcup_{n=0}^{\infty} \langle K \rangle_n$ and $G(bX) = \bigcup_{n=0}^{\infty} \langle bX \rangle_n$.

Clearly, $\varphi = \hat{g} \circ \hat{f}$ is a continuous monomorphism of $G(K)$ to $G(bX)$. Let $C$ be an arbitrary closed set in $G(K)$. Then $C_n = C \cap \langle K \rangle_n$ is compact and, hence, $\varphi(C) \cap \langle bX \rangle_n = \varphi(C_n)$ is closed in $G(bX)$, for each $n \in \omega$. We conclude, therefore, that $\varphi(C)$ is closed in $G(bX)$. This proves that $\varphi$ is a closed mapping, whence it follows that $\varphi$ is a topological isomorphism between $G(K)$ and $G(K, bX)$. Finally, the equality $\varphi = \hat{g} \circ \hat{f}$ implies that $\hat{f}$ is a topological isomorphism between $G(K)$ and $G(K, X)$. □

Recall that $X$ is a *$P$-space* if every $G_\delta$-set in $X$ is open. It turns out that the groups $F(X)$ and $A(X)$ on an arbitrary $P$-space have the direct limit property. To show this, we present a useful auxiliary fact.

PROPOSITION 7.4.7. *The groups $F(X)$ and $A(X)$ are $P$-spaces iff $X$ is a $P$-space.*

PROOF. Since $X$ is a subspace of $F(X)$ and $A(X)$, the necessity is obvious. Suppose that $X$ is a $P$-space. Let $G(X)$ be either $F(X)$ or $A(X)$, and $\mathcal{T}$ be the topology of $G(X)$. Consider the topology $\mathcal{T}_\omega$ on $G(X)$ with the base consisting of all $G_\delta$-sets in $G(X)$. It is clear that $(G(X), \mathcal{T}_\omega)$ is a topological group. Since $X$ is a $P$-space, the restrictions of both $\mathcal{T}$ and $\mathcal{T}_\omega$ to $X$ coincide with the original topology of $X$. Therefore, Corollary 7.1.8 implies that $\mathcal{T}_\omega = \mathcal{T}$, i.e., $G(X)$ is a $P$-space. □

PROPOSITION 7.4.8. *If $X$ is a $P$-space, then the groups $F(X)$ and $A(X)$ have the direct limit property.*

PROOF. It suffices to prove the statement for $F(X)$, since the argument for $A(X)$ is similar. Suppose that $K$ is a subset of $F(X)$ such that $K \cap B_n(X)$ is closed in $B_n(X)$ for each $n \in \mathbb{N}$. By Theorem 7.1.13, the sets $B_n(X)$ are closed in $F(X)$, and so are the sets $K \cap B_n(X)$. Therefore, $K$ is an $F_\sigma$-set in $F(X)$. From Proposition 7.4.7 it follows that $F(X)$ is a $P$-space; therefore, $K$ is closed in $F(X)$. □

The direct limit property is very useful in arguments involving sequentiality or tightness of free topological groups.

COROLLARY 7.4.9. *Let $X$ be a $k_\omega$-space with one of the following properties:*

1) *sequentiality;*
2) *countable tightness.*

*Then the group $G(X)$ has the same property.*

PROOF. Let $X = \bigcup_{n\in\omega} X_n$ be a $k_\omega$-decomposition of $X$. Keeping notation of Theorem 7.4.1, we conclude that the sets $K_n = \tilde{X}_n \cdot \ldots \cdot \tilde{X}_n$ ($n$ times) determine the topology of the group $G(X)$, where $\tilde{X}_n = X_n \cup \{e\} \cup X_n^{-1}$ is a compact subspace of the space $\tilde{X} = X \oplus \{e\} \oplus X^{-1}$ and $e$ is the identity of $G(X)$. Note that if $X$ is sequential or has countable tightness, then so are $\tilde{X}_n$ and $(\tilde{X}_n)^k$ for all $n, k \in \omega$ (see Theorem 3.10.35 and Problem 3.12.8 (e) of [**165**]). In addition, continuous mappings preserve both properties in the class of compact spaces, according to Theorem 3.10.32 and Problem 3.12.8 (a) of [**165**]. Since the natural multiplication mapping $i_n\colon \tilde{X}^n \to G(X)$ is continuous, each $K_n = i_n((\tilde{X}_n)^n)$ is sequential (has countable tightness) if $X$ is sequential (has countable tightness).

1) Suppose that the space $X$ is sequential and let $P$ be a sequentially closed subset of $G(X)$. Then the intersection $P \cap K_n$ is closed in the sequential space $K_n$ for each $n \geq 1$. By Theorem 7.4.1, $G(X) = \bigcup_{n=1}^{\infty} K_n$ is a $k_\omega$-decomposition of $G(X)$, so $P$ is closed in $G(X)$. This proves the sequentiality of $G(X)$.

2) If $X$ has countable tightness, take an arbitrary subset $P$ of $G(X)$ which contains cluster points for all countable subsets of $P$. Arguing as in 1), we conclude that every intersection $P \cap K_n$ is closed in $K_n$ and, hence, $P$ is closed in $G(X)$. This means that $G(X)$ has countable tightness. □

The direct limit property of the free topological groups on compact spaces considerably improves the interaction between the topology and algebraic structure of these groups. In turns out, for example, that the groups in question are Raĭkov complete. To prove this fact, we need the following result.

LEMMA 7.4.10. *Let $\{K_n : n \in \omega\}$ be an increasing sequence of compact subspaces of a topological group $G$ such that $G = \bigcup_{n\in\omega} K_n$. If $G$ is the direct limit of the spaces $K_n$, then $G$ is Raĭkov complete.*

PROOF. Suppose to the contrary that the group $G$ fails to be Raĭkov complete. Then $G$ contains a Cauchy filter $\xi$ of closed subsets with empty intersection.

Denote by $e$ the identity of $G$. We can assume that the sets $K_n$ have the following properties:

(a) $e \in K_n$ and $K_n^{-1} = K_n$;
(b) $K_n \cdot K_n \subset K_{2n}$.

Indeed, it suffices to put $L_n = K_n \cup \{e\} \cup K_n^{-1}$ and $K_n^* = L_0 \cdot \ldots \cdot L_n$ for every $n \in \omega$. Since $K_n \subset K_n^*$, the compact sets $K_n^*$ generate the original topology of $G$. Thus, the sequence $\{K_n^* : n \in \omega\}$ is as required. In what follows we write $K_n$ instead of $K_n^*$.

Since the sets $K_n$ are compact, for every $n \in \omega$ there exists an element $C_n \in \xi$ such that $K_n \cap C_n = \varnothing$. Let us construct a sequence $\{V_i : i \in \mathbb{N}\}$ of sets in $G$ satisfying the following conditions for each $i \in \mathbb{N}$:

(1) $e \in V_i \subset K_i$ and $V_i$ is open in $K_i$;
(2) $V_i \subset V_{i+1}$;
(3) $(K_j \cdot \overline{V_i}) \cap C_{2j} = \varnothing$, for each $j \leq i$.

There exists an open neighbourhood $U_1$ of $e$ in $G$ such that $(K_1 \cdot \overline{U_1}) \cap C_2 = \varnothing$; we put $V_1 = U_1 \cap K_1$. Suppose that we have defined the sets $V_1, \ldots, V_n$ satisfying (1)–(3). By (3), we have $(K_i \cdot \overline{V_n}) \cap C_{2i} = \varnothing$, for each $i = 1, \ldots, n$. In addition, $(K_{n+1} \cdot \overline{V_n}) \subset K_{2n+2}$ by (1) and (b), so that the choice of the set $C_{2n+2}$ implies that $(K_{n+1} \cdot \overline{V_n}) \cap C_{2n+2} = \varnothing$. Since the product $K_{n+1} \cdot \overline{V_n}$ is compact, there exists an open neighbourhood $U$ of $e$ in $G$ such that

$$(K_i \cdot \overline{V_n} \cdot U^2) \cap C_{2i} = \varnothing$$

for each $i = 1, 2, \ldots, n+1$. Put $V_{n+1} = (V_n \cdot U) \cap K_{n+1}$. Clearly, then $\overline{V_{n+1}} \subset V_{n+1} \cdot U \subset V_n \cdot U^2$, whence it follows that the set $V_{n+1}$ satisfies (3). The verification of (1) and (2) for $V_{n+1}$ does not present any difficulty.

Consider the set $V = \bigcup_{n=1}^{\infty} V_n$. By (1) and (2), the intersection

$$V \cap K_n = \bigcup_{i=1}^{\infty} (V_i \cap K_n) = \bigcup_{i=n}^{\infty} (V_i \cap K_n)$$

is open in $K_n$ for each $n \in \mathbb{N}$. Since the sets $K_n$ determine the topology of $G$, we infer that $V$ is open in $G$. From (3) and our definition of $V$ it follows that

$$(K_j V) \cap C_{2j} = \varnothing, \quad j \in \mathbb{N}. \tag{7.13}$$

By the assumption, $\xi$ is a Cauchy filter in $G$, so one can find $C \in \xi$ and $x \in G$ such that $C \subset xV$. Then there is an integer $n \geq 1$ such that $x \in K_n$, and (7.13) implies that $xV \cap C_{2n} = \varnothing$. Since $C \subset xV$, we obtain $C \cap C_{2n} = \varnothing$, a contradiction with $C, C_{2n} \in \xi$. □

Combining Theorem 7.4.1 and Lemma 7.4.10, we obtain the first general result about the Raĭkov completeness of free topological groups.

THEOREM 7.4.11. *If $X$ is a $k_\omega$-space, then the groups $A(X)$ and $F(X)$ are Raĭkov complete.*

Here are two simple facts that follow from Theorem 7.4.11:

COROLLARY 7.4.12. *The groups $A(X)$ and $F(X)$ are Raĭkov complete, for every compact Hausdorff space $X$.*

COROLLARY 7.4.13. *Let $P$ be an open or closed subspace of the Euclidean space $\mathbb{R}^n$, for some integer $n \geq 1$. Then the groups $F(P)$ and $A(P)$ over the space $P$ are Raĭkov complete.*

PROOF. This follows from Theorem 7.4.11, since $P$ is locally compact and $\sigma$-compact, hence, is a $k_\omega$-space. □

A complete characterization of the spaces $X$ that generate a Raĭkov complete group $A(X)$ will be given in Section 7.9.

## Exercises

7.4.a. A subset $Y$ of $F(X)$ is said to be *regularly situated* in $F(X)$ if for every $n \in \omega$ there exists $m \in \omega$ such that $\langle Y \rangle \cap B_n(X) \subset \langle Y \rangle_m$, where $\langle Y \rangle_m$ is the set of all elements in $F(X)$ whose length is at most $m$ with respect to $Y$.

a) (M. I. Graev [**201**]) Let $X$ be a convergent sequence with its limit. Give an example of a compact subset $K$ of $F(X)$ such that $K$ is not regularly situated in $F(X)$, and the group $\langle K \rangle$ is closed in $F(X)$.

b) Prove that if $Y$ is a compact regularly situated subset of the free topological group $F(X)$ on a Tychonoff space $X$, then $\langle Y \rangle$ is closed in $F(X)$.

c) (S. A. Morris and B. V. S. Thompson [**331**]) Suppose that $Y$ is a compact regularly situated subset of $F(X)$. Show that if $Y$ is a free algebraic basis for $\langle Y \rangle$, then $\langle Y \rangle \cong F(Y)$.

7.4.b. (J. Mack, S. A. Morris, and E. T. Ordman [**297**]) Generalize c) of Exercise 7.4.a as follows. Let $X$ be a $k_\omega$-space with $k_\omega$-decomposition $X = \bigcup_{n\in\omega} X_n$ and $Y$ a closed subspace of $F(X)$ with $k_\omega$-decomposition $Y = \bigcup_{n\in\omega} Y_n$. Then the natural homomorphism of $F(Y)$ onto the subgroup $\langle Y \rangle$ of $F(X)$ is a topological isomorphism iff $Y$ is a free algebraic basis for $\langle Y \rangle$, and for every $n \in \omega$ there exists $m \in \omega$ such that $\langle Y \rangle \cap \langle X_n \rangle_n \subset \langle Y_m \rangle_m$. Formulate and prove a similar assertion for $A(X)$.

7.4.c. (J. Mack, S. A. Morris, and E. T. Ordman [**297**]) Apply Exercise 7.4.b to show that if $Y$ is a closed subset of a $k_\omega$-space $X$, then $F(Y, X) \cong F(Y)$ and $A(Y, X) \cong A(Y)$.

7.4.d. Let $X = X_0 \oplus X_1$, where $X_0$ is a $k_\omega$-space and $X_1$ is a discrete space. Prove that $A(X)$ is a $k$-space.

7.4.e. The following assertions complement Theorem 7.4.1.

a) Let $f\colon X \to Y$ be a quotient (more generally, an $R$-quotient, see Exercise 7.1.j) onto mapping. Show that if $G(X)$ has the direct limit property, then so does $G(Y)$. Apply this to the special case when $Y$ is a retract of $X$.

b) Verify that if $F(X)$ has the direct limit property, then $A(X)$ also has the direct limit property.

7.4.f. Modify the proof of Theorem 7.4.1 to show that if the subspace $B_n(X)$ of $G(X)$ is locally compact for each $n \in \omega$, then $G(X)$ has the direct limit property (see also Problem 7.9.G).

7.4.g. Give an example of a $P$-space $X$ and a closed subset $Y$ of $X$ such that the natural continuous isomorphism of $A(Y)$ onto the subgroup $A(Y, X)$ of $A(X)$ is not a homeomorphism (cf. 7.4.c and Theorem 7.7.4).

7.4.h. Let $Y$ be a subspace of $F(X)$ or $A(X)$. Prove that if $y \in Y$ and $\chi(y, Y) \le \omega$, then there exists $n \in \omega$ such that the interior of $Y \cap B_n(X)$ in $Y$ contains the point $y$.

## Problems

7.4.A. (T. H. Fay, E. T. Ordman, and B. V. S. Thomas [**167**]) Show that the free topological group of the space of rational numbers does not have the direct limit property.

7.4.B. Let $X$ be a zero-dimensional compact space. Show that the groups $A(X)$ and $F(X)$ are zero-dimensional.

7.4.C. (V. G. Pestov [**375**]) Give an example of a compact space $K$ such that the groups $A(K)$ and $F(K)$ are not homeomorphic.

*Hint.* Consider the topological sum $K = \beta\mathbb{N} \oplus I$, where $\beta\mathbb{N}$ is the Čech–Stone compactification of the discrete space $\mathbb{N}$ and $I = [0, 1]$ is the closed unit interval. Apply Problems 7.1.A and 7.4.B to show that the connected components of the neutral elements of the groups $A(K)$ and $F(K)$ have different cardinalities.

7.4.D. (S. Romaguera and M. Sanchis [**411**]) Let $X$ be a $k_\omega$-space. Prove that the left uniformity of the group $F(X)$ is cofinally complete (see Problem 1.8.E).

## Open Problems

7.4.1. Does there exist in $ZFC$ a $\sigma$-compact Tychonoff space $X$ such that $F(X)$ has the direct limit property, but $X$ fails to be a $k_\omega$-space?

7.4.2. Does there exist in $ZFC$ a cosmic Tychonoff space $X$ such that $F(X)$ has the direct limit property, but $X$ fails to be a $k_\omega$-space?

*Remark.* Assuming the existence of a free selective ultrafilter on $\omega$, O. V. Sipacheva constructed in [**451**] a countable subspace $Y$ of $\beta\omega$ such that $F(Y)$ has the direct limit property, but $Y$ is not a $k$-space (hence, not a $k_\omega$-space either).

7.4.3. Let $FPG(X)$ be the free Tychonoff paratopological group of a compact Hausdorff space $X$. Is $FPG(X)$ the direct limit of a countable family of compact spaces? Is $FPG(X)$ $\sigma$-compact?

7.4.4. Which results of this section can be generalized to $FPG(X)$?

## 7.5. Precompact and bounded sets in free groups

Every bounded subset of a topological group is precompact, but precompact sets need not be bounded (see Section 6.10). Here we show that in free (Abelian) topological groups, the two classes of sets coincide.

As in the preceding section, we use $G(X)$ to denote either $F(X)$ or $A(X)$. If $X$ is a subset of a space $Y$, then $G(X,Y)$ is the subgroup $\langle X\rangle$ of $G(Y)$ generated by $X$. The *support* of a reduced word $g = x_1^{\varepsilon_1}\cdot\dots\cdot x_n^{\varepsilon_n} \in G(Y)$ with $x_1,\dots,x_n \in Y$ is defined as follows:

$$\mathrm{supp}(g) = \{x_1,\dots,x_n\}.$$

Given a subset $K$ of $G(Y)$, we put

$$\mathrm{supp}(K) = \bigcup_{g\in K}\mathrm{supp}(g).$$

If $n\in\mathbb{N}$, the set $\{g\in G(Y) : |\,\mathrm{supp}(g)|\le n\}$ is denoted by $G_n(Y)$. For a subspace $X$ of $Y$ and $n\in\mathbb{N}$, we use the abbreviation $G_n(X,Y)$ for the subset $G_n(Y)\cap G(X,Y)$ of $F(Y)$.

We start with the following lemma which permits us to reduce the study of precompact subsets of the free topological group $G(X)$ on an arbitrary space $X$ to those of the free topological group on $\mathbb{R}$.

Lemma 7.5.1. *Let $S$ be a countable subset of a Tychonoff space $X$. If $S$ is not bounded in $X$, then there exists a continuous function $f\colon X\to\mathbb{R}$ such that $f(S)$ is unbounded in $\mathbb{R}$ and $f(x)\ne f(y)$, for any distinct $x,y\in S$.*

Proof. If $S$ is not bounded in $X$, there exists an infinite discrete family $\gamma$ of open sets in $X$ such that $U\cap S\ne\varnothing$ for each $U\in\gamma$. We can always find an infinite subfamily $\{U_n : n\in\omega\}$ of distinct elements of $\gamma$ and enumerate $S=\{y_n : n\in\omega\}$ in such a way that $y_{2n}\in U_n$ for each $n\in\omega$. Now we define by induction a sequence $\{f_n : n\in\omega\}$ of continuous real-valued functions on $X$ satisfying the following conditions for all $n\in\omega$, where $g_k=\sum_{i\le k} f_i$:

(a) $0\le f_{2n+1}\le 1/2^n$;
(b) $f_{2n}\ge 0$ and $f_{2n}(x)=0$ for each $x\in X\setminus U_n$;
(c) $f_n(y_k)=0$ if $k<n$;
(d) $|f_{2n}(y_{2n})|\ge n$;
(e) $f_{n+1}(y_{n+1})\ne g_k(y_k)-g_n(y_{n+1})$ for each $k\le n$.

The sum $\sum_{n\in\omega} f_{2n+1}$ is a continuous function on $X$ because of (a), and the sum $\sum_{n\in\omega} f_{2n}$ is continuous in view of (b) and of the choice of the family $\gamma$. Therefore, the function $f=\sum_{n\in\omega} f_n$ is also continuous. It follows from (d), (a), and (b) that $f(y_{2n})\ge f_{2n}(y_{2n})\ge n$, so that $f(S)$ is unbounded in $\mathbb{R}$. It remains to show that $f(y_k)\ne f(y_l)$, for any distinct $k,l\in\omega$. Suppose that $k<l$. Then $k\le n=l-1$. By (c), we have $f(y_k)=g_k(y_k)$ and

$f(y_l) = g_l(y_l)$. Since $g_{n+1} = g_n + f_{n+1}$, condition (e) implies that $g_{n+1}(y_{n+1}) \neq g_k(y_k)$ and, hence, $f(y_l) \neq f(y_k)$. □

LEMMA 7.5.2. *If $K$ is a precompact subset of $G(X)$, then $Y = \text{supp}(K)$ is bounded in $X$, and $K \subset G_n(Y, X)$ for some $n \in \mathbb{N}$.*

PROOF. Suppose that $Y = \text{supp}(K)$ is not bounded in $X$. Since boundedness of the set $Y$ is determined by its countable subsets, we can assume without loss of generality that both $K$ and $Y$ are countable. By Lemma 7.5.1, we can define a continuous function $f : X \to \mathbb{R}$ such that $Z = f(Y)$ is unbounded in $\mathbb{R}$ and $f(x) \neq f(y)$ for any distinct $x, y \in Y$. Let $\hat{f} : G(X) \to G(\mathbb{R})$ be the homomorphism extending the mapping $f$. Put $L = \hat{f}(K)$. Then $L$ is precompact in $G(\mathbb{R})$ and from the choice of $f$ it follows that $\text{supp}(L) = Z$. Since $\mathbb{R}$ is a $k_\omega$-space, the group $G(\mathbb{R})$ is Raĭkov complete by Theorem 7.4.11. Therefore, the closure $\overline{L}$ of $L$ in $G(\mathbb{R})$ is compact. For every $n \in \omega$, put $A_n = [-n, n] \subset \mathbb{R}$ and $C_n = G_n(A_n, \mathbb{R})$. Since $\mathbb{R} = \bigcup_{n\in\omega} A_n$ is a $k_\omega$-decomposition of $\mathbb{R}$, Theorem 7.4.1 implies that $G(\mathbb{R}) = \bigcup_{n\in\omega} C_n$ is a $k_\omega$-decomposition of the group $G(\mathbb{R})$. This implies immediately that $\overline{L} \subset C_n$ for some $n \in \omega$. Indeed, otherwise we can choose a sequence $S = \{g_n : n \in \omega\} \subset G(\mathbb{R})$ such that $g_n \in \overline{L} \setminus C_n$ for each $n \in \omega$. From Corollary 7.4.3 it follows that $S$ is a closed discrete subset of $\overline{L}$, thus contradicting the compactness of $\overline{L}$. The inclusion $\overline{L} \subset C_n$ implies that $Z = \text{supp}(L) \subset \text{supp}(\overline{L}) \subset A_n = [-n, n]$, which contradicts the unboundedness of $Z$ in $\mathbb{R}$. This proves that $Y$ is bounded in $X$. Finally, since $L \subset C_n \subset G_n(\mathbb{R})$ and the restriction of $f$ to $Y$ is one-to-one, we infer that $K \subset G_n(X) \cap G(Y, X) = G_n(Y, X)$. □

THEOREM 7.5.3. [**D. Dikranjan and M. G. Tkachenko**] *The following conditions are equivalent for a subset $K$ of the group $G(X)$:*

(1) *$K$ is bounded in $G(X)$;*
(2) *$K$ is precompact in $G(X)$;*
(3) *there exist an integer $n \in \omega$ and a bounded subset $Y$ of $X$ such that $K \subset G_n(Y, X)$.*

PROOF. The implications (1) ⇒ (2) and (2) ⇒ (3) follow from Proposition 6.10.2 and Lemma 7.5.2, respectively. Let us show that (3) ⇒ (1). If $Y$ is bounded in $X$ and $K \subset G_n(Y, X)$, then the set $\overline{Y} = Y \cup \{e\} \cup Y^{-1} = G_1(Y, X)$ is bounded in $G_1(X)$ and in $G(X)$. In its turn, $G_n(Y, X) = \overline{Y} \cdot \ldots \cdot \overline{Y}$ ($n$ times) is bounded in $G(X)$ by Corollary 6.10.13. Since $K \subset G_n(Y, X)$, the conclusion is immediate. □

Since every pseudocompact subspace of a space $Y$ is bounded in $Y$, we obtain the following generalization of Corollary 7.4.4:

COROLLARY 7.5.4. *If $K$ is a pseudocompact subspace of the group $G(X)$, then $K \subset B_n(X)$, for some $n \in \mathbb{N}$.*

COROLLARY 7.5.5. *Let $K$ be a precompact subset of the group $G(X)$. Then* $\text{supp}(K)$ *is precompact in $G(X)$.*

PROOF. By Lemma 7.5.2, $\text{supp}(K)$ is bounded in $X$ and in $G(X)$, so the conclusion follows from Proposition 6.10.2. □

COROLLARY 7.5.6. *If $K$ is a bounded subset of $G(X)$, then* $\text{supp}(K)$ *is bounded in $X$. In addition, if the space $X$ is Dieudonné complete, then the closure of* $\text{supp}(K)$ *in $X$ is compact.*

PROOF. The first claim follows directly from Theorem 7.5.3. The second one follows from the fact that the closure of a bounded subset of a Dieudonné complete space is compact. □

We recall that a space $X$ is said to be *$\sigma$-bounded* if it is the union of countably many bounded subsets (see Section 6.10).

COROLLARY 7.5.7. *The group $G(X)$ is $\sigma$-bounded iff the space $X$ is $\sigma$-bounded.*

PROOF. If $X$ is $\sigma$-bounded, then so is $G(X)$ by Corollary 6.10.14. Conversely, suppose that the group $G(X)$ is a union of a countable family of its bounded subsets, $G(X) = \bigcup_{n\in\omega} K_n$. By Corollary 7.5.6, $\mathrm{supp}(K_n)$ is bounded in $X$, for every $n$. Clearly, $X = \bigcup_{n\in\omega} \mathrm{supp}(K_n)$; hence, $X$ is $\sigma$-bounded. □

## Exercises

7.5.a. (K. Eda, H. Ohta, and K. Yamada [**156**]) Prove that the following conditions are equivalent for a Tychonoff space $X$:
1) $G(X)$ contains a non-trivial sequence converging to the identity $e$;
2) the subspace $B_2(X)$ of $G(X)$ contains a non-trivial sequence converging to $e$;
3) $X$ contains sequences $\{x_n : n \in \omega\}$ and $\{y_n : n \in \omega\}$ such that $x_n \neq y_n$ for each $n \in \omega$, and $d(x_n, y_n) \to 0$, for every continuous pseudometric $d$ on $X$.

7.5.b. (V. V. Tkachuk [**493**]) The *Alexandroff duplicate* of a space $X$ is the space $Y = X \times \{0, 1\}$ with the following topology: For each $x \in X$, a basic neighbourhood of $(x, 0)$ in $Y$ is the set of the form $(U \times \{0, 1\}) \setminus \{(x, 1)\}$, where $U$ is a neighbourhood of $x$ in $X$, and each point $(x, 1)$ is isolated in $Y$.
1) Prove that if $X$ is a compact space of cardinality $\kappa \geq \omega$, and $Y$ is the Alexandroff duplicate of $X$, then both $F(Y)$ and $A(Y)$ contain a copy of the one-point compactification of a discrete space of cardinality $\kappa$.
2) Deduce from 1) that there exists an infinite compact space $Z$ without non-trivial convergent sequences such that the groups $F(Z)$ and $A(Z)$ do contain non-trivial convergent sequences.

7.5.c. Apply Corollaries 6.10.9 and 7.5.7 to show that the groups $F(X)$ and $A(X)$ on a $\sigma$-bounded space $X$ are $\omega$-stable. In particular, $F(X)$ and $A(X)$ are $\omega$-stable, for every pseudocompact space $X$.

## Problems

7.5.A. Let $H$ be the free Tychonoff paratopological group on a pseudocompact space $X$.
1) Is $H$ $\sigma$-bounded?
2) Is $H$ $\omega$-stable?
3) Does $H$ have countable cellularity or countable $o$-tightness?
4) Is the space $H$ $G_\delta$-preserving or an Efimov space? (See Section 1.6.)
5) Is $H$ topologically isomorphic to a subgroup of the topological product of a family of $\sigma$-compact paratopological groups?

## Open Problems

7.5.1. Can Theorem 7.5.3 be generalized to the free Tychonoff paratopological group of a Tychonoff space $X$?

7.5.2. Is it true that the free Tychonoff paratopological group $FPG(X)$ on a Tychonoff space $X$ is $\sigma$-bounded if and only if $X$ is $\sigma$-bounded?

7.5.3. Let $X$ be a subset of a topological group $G$ such that the image $f(X)$ is countable, for every continuous real-valued function $f$ on $G$. Does the set $A^2 \subset G$ have the same property?

## 7.6. Free topological groups on metrizable spaces

Let $X$ be a metrizable space whose topology is generated by a metric $d$. By Theorem 7.2.2, $d$ can be extended to a continuous invariant metric $\widehat{d}$ on $F(X)$ and, hence, $F(X)$ admits a weaker metrizable group topology with invariant basis generated by the metric $\widehat{d}$. Thus, free topological groups of metrizable spaces are submetrizable in a strong way. Definitely, these groups constitute an important subject of study.

We know that for every Tychonoff space $X$ and any integer $n \in \mathbb{N}$, the subset $B_n(X)$ of $F(X)$ consisting of all words of reduced length $\leq n$ is closed (see a) of Theorem 7.1.13). It turns out that the sets $B_n(X)$ remain closed in $F(X)$ if $(X, d)$ is a metric space and $F(X)$ carries the weaker group topology generated by the metric $\widehat{d}$.

PROPOSITION 7.6.1. *Let $(X, d)$ be a metric space, and $\mathcal{T}_d$ be the topology on $F_a(X)$ generated by the Graev extension $\widehat{d}$ of $d$ to $F(X)$. Then $B_n(X)$ is closed in $(F_a(X), \mathcal{T}_d)$, for each integer $n \in \mathbb{N}$.*

PROOF. Suppose that $n \in \mathbb{N}$, and let $h = y_1 \dots y_k$ be the reduced form of an element of $F_a(X) \setminus B_n(X)$, where $y_1, \dots, y_k \in X \cup X^{-1} = Y$. It is clear that $n < k$. If $g$ is in the $\mathcal{T}_d$-closure of $B_n(X)$, we can find a sequence $\{g_i : i \in \omega\}$ of elements of $B_n(X)$ such that $\widehat{d}(g_i, h) < 1/i$ for each $i \in \omega$. Choosing a subsequence of the sequence $\{g_i : i \in \omega\}$, we can assume that all $g_i$ have the same length, say $m \leq n$, and all products $g_i^{-1}h$ also have equal length, say $p \leq m + k$. Let $q$ be the minimal positive integer with $2q \geq p$.

By Claim 1 in the proof of Theorem 7.2.2, for every $i$ there exist a word $\mathfrak{X}_i$ in the alphabet $Y \cup \{e\}$ and a scheme $\varphi_i$ on $\{1, 2, \dots, 2q\}$ such that $l(\mathfrak{X}_i) = 2q$, $[\mathfrak{X}_i] = g_i^{-1} \cdot h$ and $\widehat{d}(g_i, h) = \widehat{d}(e, g_i^{-1}h) = \Gamma_d(\mathfrak{X}_i, \varphi_i)$. Choosing a subsequence of $\{g_i : i \in \omega\}$ once again, we can assume that all schemes $\varphi_i$ are equal, say, to $\varphi$. Since $m \leq n < k$, Proposition 7.2.1 implies that there exists an integer $j$ such that $m < j \leq 2q$ and $\varphi(j) = j + 1$. Take an arbitrary positive integer $i$ with $1/i < \widehat{d}(y_j^{-1}, y_{j+1})$. By definition of $\Gamma_d$, we have

$$1/i < \widehat{d}(y_j^{-1}, y_{j+1}) \leq \Gamma_d(\mathfrak{X}_i, \varphi) < 1/i,$$

a contradiction. This proves that $B_n(X)$ is $\mathcal{T}_d$-closed in $F(X)$. □

The above result leads to an alternative proof of item a) of Theorem 7.1.13. Indeed, if $k > 0$ and $h = y_1^{\varepsilon_1} \dots y_k^{\varepsilon_k}$ (with $y_1, \dots, y_k \in X$ and $\varepsilon_1, \dots, \varepsilon_k = \pm 1$) is an reduced form of an element from $F_a(X)$, take a continuous function $f \colon X \to \mathbb{R}$ such that $f(y_i) \neq f(y_j)$ if $y_i \neq y_j$. Extend $f$ to a continuous homomorphism $\tilde{f} \colon F(X) \to F(\mathbb{R})$ and note that the length of $\tilde{f}(h)$ is equal to $k$. If $n < k$, then $\tilde{f}(h)$ does not belong to the closure of $B_n(\mathbb{R})$ in $F(\mathbb{R})$ by Proposition 7.6.1 and, hence, $B_n(X)$ is closed in $F(X)$.

To formulate the next result, we recall notation used in Theorem 7.1.13. Let $X$ be a space. Put $\tilde{X} = X \oplus \{e\} \oplus X^{-1}$. For $n \geq 1$, denote by $i_n$ the multiplication mapping of $\tilde{X}^n$ to $F(X)$, $i_n(y_1, \dots, y_n) = y_1 \cdot \dots \cdot y_n$ for each point $(y_1, \dots, y_n) \in \tilde{X}^n$. The mapping

$i_n$ is continuous, and $i_n(\tilde{X}^n) = B_n(X)$. Finally, we put $C_n(X) = B_n(X) \setminus B_{n-1}(X)$ and $C_n^*(X) = i_n^{\leftarrow}(C_n(X))$, for $n \geq 1$.

The following result complements item b) of Theorem 7.1.13.

THEOREM 7.6.2. *Let $(X, d)$ be a metric space, and $\mathcal{T}_d$ the topology on $F_a(X)$ generated by the Graev extension $\widehat{d}$ of $d$. Then the mapping $i_n$ is a homeomorphism of $C_n^*(X)$ onto the subspace $C_n(X)$ of $(F_a(X), \mathcal{T}_d)$.*

PROOF. By b) of Theorem 7.1.13, the restriction of $i_n$ to $C_n^*(X)$ is a homeomorphism of $C_n^*(X)$ onto the subspace $C_n(X)$ of $F(X)$. Therefore, it suffices to show that the topology on $C_n(X)$ inherited from $F(X)$ coincides with the restriction of $\mathcal{T}_d$ to $C_n(X)$.

Let $g = x_1^{\varepsilon_1} \dots x_n^{\varepsilon_n} \in C_n(X)$ be arbitrary. Suppose that $V$ is a neighbourhood of $g$ in $F(X)$. By Corollary 7.1.19, one can find open sets $U_1, \dots, U_n$ in $X$ such that $x_i \in U_i$ for each $i \leq n$, $U_i \cap U_j = \varnothing$ if $x_i \neq x_j$, and $U_1^{\varepsilon_1} \cdots U_n^{\varepsilon_n} \subset V$. It remains to find a real number $r > 0$ such that

$$C_n(X) \cap (g \cdot O_d(r)) \subset U_1^{\varepsilon_1} \cdots U_n^{\varepsilon_n},$$

where $O_d(r) = \{h \in F(X) : \widehat{d}(e, h) < r\}$. For every $i = 1, \dots, n$, there exists a positive number $r_i < 1$ such that $\{x \in X : d(x_i, x) < r_i\} \subset U_i$. We claim that $r = \min\{r_i : 1 \leq i \leq n\}$ works.

Indeed, suppose that $g \cdot h \in C_n(X)$ for some $h \in O_d(r)$. Since $r < 1$, the element $h$ has even length, $h = y_1^{\delta_1} \dots y_{2k}^{\delta_{2k}}$. By the assumption, the elements $g$ and $g \cdot h$ have the same length $n$, so we must have exactly $k$ reductions at the joint of $g$ and $h$ in the word $gh$, and $k \leq n$. In other words, $y_i = x_{n-i+1}$ and $\delta_i = -\varepsilon_{n-i+1}$, for each $i$ with $1 \leq i \leq k$. From the definition of $\widehat{d}$ (see Theorem 7.2.2) it follows that there exists a scheme $\varphi$ on $\{1, \dots, 2k\}$ such that

$$\frac{1}{2} \sum_{i=1}^{2k} d^*(y_i^{-\delta_i}, y_{\varphi(i)}^{\delta_{\varphi(i)}}) = \widehat{d}(e, h).$$

Therefore, $d^*(y_i^{-\delta_i}, y_{\varphi(i)}^{\delta_{\varphi(i)}}) \leq \widehat{d}(e, h) < r < 1$, whence $\delta_{\varphi(i)} = -\delta_i$ for each $i \leq 2k$. Let us verify that $\varphi(i) > k$ for each $i \leq k$. Suppose to the contrary that $\varphi(i) \leq k$ for some $i \leq k$. By Proposition 7.2.1, there exists an integer $j$ such that $i \leq j < \varphi(i)$ and $\varphi(j) = j + 1$. In particular, $j + 1 \leq \varphi(i) \leq k$ and $\delta_{j+1} = \delta_{\varphi(j)} = -\delta_j$. Since $j, j + 1 \leq k$, we have $y_j = x_{n-j+1}$, $y_{j+1} = x_{n-j}$, $\delta_j = \varepsilon_{n-j+1}$ and $\delta_{j+1} = \varepsilon_{n-j}$. In particular, $\varepsilon_{n-j+1} = -\varepsilon_{n-j}$ and, hence,

$$d(x_{n-j+1}, x_{n-j}) = d^*(x_{n-j+1}^{-\varepsilon_{n-j+1}}, x_{n-j}^{\varepsilon_{n-j}}) = d^*(y_j^{-\delta_j}, y_{j+1}^{\delta_{j+1}}) < r.$$

Since the word $g = x_1^{\varepsilon_1} \dots x_n^{\varepsilon_n}$ is reduced and $\varepsilon_{n-j+1} = -\varepsilon_{n-j}$, the points $x_{n-j}$ and $x_{n-j+1}$ must be distinct. By the choice of $r$, all points $x \in X$ with $d(x_{n-j}, x) < r$ are in $U_{n-j}$, so $x_{n-j+1} \in U_{n-j+1} \cap U_{n-j} \neq \varnothing$. This contradicts our choice of the sets $U_i$. Therefore, $\varphi(i) > k$ for each $i \leq k$. There exists, however, only one scheme $\varphi$ with this property, namely, $\varphi(i) = 2k - i + 1$ for each $i \in \{1, \dots, 2k\}$. We have, therefore, $\delta_{2k-i+1} = -\delta_i = \varepsilon_{n-i+1}$ for $i = 1, \dots, k$, that is, $\delta_{2k} = \varepsilon_n, \dots, \delta_{k+1} = \varepsilon_{n-k+1}$.

The final step is to see that the element

$$g \cdot h = x_1^{\varepsilon_1} \dots x_{n-k}^{\varepsilon_{n-k}} y_{k+1}^{\delta_{k+1}} \dots y_{2k}^{\delta_{2k}} = x_1^{\varepsilon_1} \dots x_{n-k}^{\varepsilon_{n-k}} y_{k+1}^{\varepsilon_{n-k+1}} \dots y_{2k}^{\varepsilon_n}$$

belongs to $U_1^{\varepsilon_1} \cdots U_n^{\varepsilon_n}$. Since $x_i \in U_i$ for $i = 1, \ldots, n-k$, it suffices to verify that $y_{k+1} \in U_{n-k+1}, \ldots, y_{2k} \in U_n$. Indeed, for every $i = 1, \ldots, k$, we have

$$d(x_{n-i+1}, y_{2k-i+1}) = d^*(x_{n-i+1}^{\varepsilon_{n-i+1}}, y_{2k-i+1}^{\varepsilon_{n-i+1}}) = d^*(y_i^{-\delta_i}, y_{2k-i+1}^{\delta_{2k-i+1}}) < r.$$

Therefore, our choice of $r$ implies that $y_{2k-i+1} \in U_{n-i+1}$ for every $i = 1, \ldots, k$. Hence $g \cdot h \in U_1 \cdots U_n$, and the proof is complete. □

Combining Theorem 7.6.2 and item b) of Theorem 7.1.13, we obtain the following result that will enable us to establish paracompactness of free topological groups on metrizable spaces.

COROLLARY 7.6.3. **[V. K. Bel'nov]** *Let $(X, d)$ be a metric space. Then the topology on $C_n(X)$ inherited from $F(X)$ coincides with the one induced by the Graev extension $\widehat{d}$ of the metric $d$.*

We noted at the beginning of this section that the free topological group $F(X)$ on a metrizable space $X$ admits a weaker metrizable group topology. It turns out that the relationship between these two topologies is much stronger than the simple inclusion. The above Corollary 7.6.3 gives the first idea about this special relation. To give the exact description of the situation we need a new concept and an auxiliary topological result.

Let $\mathcal{T}_1$ and $\mathcal{T}_2$ be two topologies on a set $X$ such that $\mathcal{T}_1$ is finer than $\mathcal{T}_2$. If the space $(X, \mathcal{T}_2)$ has a $\sigma$-discrete family of subsets which is a network for $(X, \mathcal{T}_1)$, then the topology $\mathcal{T}_2$ is called an *s-approximation* for $\mathcal{T}_1$.

The following properties of $s$-approximations are evident.

LEMMA 7.6.4. *Suppose that $\mathcal{T}_1$ and $\mathcal{T}_2$ are topologies on a set $X$ such that $\mathcal{T}_2$ is an s-approximation of the topology $\mathcal{T}_1$. Then the following hold:*

1) *for every subset $Y$ of $X$, $\mathcal{T}_2 \restriction Y$ is an s-approximation for $\mathcal{T}_1 \restriction Y$;*
2) *for every integer $n > 0$, the topology of the product $(X, \mathcal{T}_2)^n$ is an s-approximation for the topology of $(X, \mathcal{T}_1)^n$.*

Sometimes the existence of an $s$-approximation for a given topology $\mathcal{T}$ on a set $X$ implies that the space $(X, \mathcal{T})$ is paracompact.

LEMMA 7.6.5. *Let $\mathcal{T}_1$ and $\mathcal{T}_2$ be regular topologies on a set $X$ such that $\mathcal{T}_2$ is an s-approximation for $\mathcal{T}_1$ and the space $(X, \mathcal{T}_2)$ is collectionwise normal. Then the space $(X, \mathcal{T}_1)$ is paracompact.*

PROOF. By the assumption, there is a network $\mathcal{S} = \bigcup_{i\in\omega} \mathcal{S}_i$ for $(X, \mathcal{T}_1)$ such that every family $\mathcal{S}_i = \{P_{i,\alpha} : \alpha \in A_i\}$ is discrete in $(X, \mathcal{T}_2)$. Since $(X, \mathcal{T}_2)$ is collectionwise normal, for every $i \in \omega$ there exists a discrete family $\gamma_i = \{V_{i,\alpha} : \alpha \in A_i\}$ of open sets in $(X, \mathcal{T}_2)$ such that $P_{i,\alpha} \subset V_{i,\alpha}$ for each $\alpha \in A_i$.

Let $\xi \subset \mathcal{T}_1$ be a covering of $X$. For every $i \in \omega$, consider the set

$$B_i = \{\alpha \in A_i : \text{ there exists } W \in \xi \text{ such that } P_{i,\alpha} \subset W\}.$$

For every $i \in \omega$ and $\alpha \in B_i$, choose $W(i, \alpha) \in \xi$ such that $P_{i,\alpha} \subset W(i, \alpha)$. Since $\mathcal{S}$ is a network for $(X, \mathcal{T}_1)$, the family $\theta = \{P_{i,\alpha} : \alpha \in B_i, \ i \in \omega\}$ is a covering of $X$. For every $i \in \omega$, put $\mu_i = \{V_{i,\alpha} \cap W(i, \alpha) : \alpha \in B_i\}$. Since $\mathcal{T}_2 \subset \mathcal{T}_1$ and the family $\{V_{i,\alpha} : \alpha \in A_i\}$ is discrete in $(X, \mathcal{T}_2)$, we conclude that $\mu_i$ is a discrete family of open sets in $(X, \mathcal{T}_1)$. Clearly,

$\mu_i$ is a refinement of $\xi$. Note that $X = \bigcup \theta$ and $P_{i,\alpha} \subset V_{i,\alpha} \cap W(i, \alpha)$, for each $\alpha \in B_i$ and each $i \in \omega$, so $\mu = \bigcup_{i\in\omega} \mu_i$ is a covering of $X$. Therefore, $\mu$ is an open $\sigma$-discrete covering of $(X, \mathcal{T}_1)$ that refines $\xi$. Since the space $(X, \mathcal{T}_1)$ is regular, it has to be paracompact [**165**, Theorem 5.1.11]. □

Recall that a paracompact space with a $\sigma$-discrete network is said to be a *paracompact $\sigma$-space.* All metric spaces, as well as their arbitrary images under closed continuous mappings, are paracompact $\sigma$-spaces [**360**]. Paracompact $\sigma$-spaces can be characterized in terms of $s$-approximations.

THEOREM 7.6.6. *A regular $T_1$-space $(X, \mathcal{T})$ is a paracompact $\sigma$-space iff $\mathcal{T}$ admits a metrizable s-approximation.*

PROOF. Since every metrizable space is collectionwise normal, the sufficiency follows from Lemma 7.6.5. The proof of the necessity goes as follows. Let $\{\mathcal{S}_i : i \in \omega\}$ be a network of the space $(X, \mathcal{T})$, where every $\mathcal{S}_i = \{P_{i,\alpha} : \alpha \in A_i\}$ is a discrete family of subsets of $(X, \mathcal{T})$. We can assume without loss of generality that each $P_{i,\alpha}$ is $\mathcal{T}$-closed in $X$. Since $(X, \mathcal{T})$ is paracompact, for every $i \in \omega$ there exists a discrete family $\gamma_i = \{U_{i,\alpha} : \alpha \in A_i\}$ of open sets in $(X, \mathcal{T})$ such that $P_{i,\alpha} \subset U_{i,\alpha}$ for each $\alpha \in A_i$. Let

$$Z_i = \bigcup \{I_{i,\alpha} : \alpha \in A_i\}$$

be the metric hedgehog with "$A_i$" spines $I_{i,\alpha} = I \times \{\alpha\}$, where $I = [0, 1]$ is the usual unit interval, and all the points $(0, \alpha)$ of the spines are identified to a single point $\bar{0}$ [**165**, Example 4.1.5]. Every paracompact $\sigma$-space is perfectly normal. Therefore, for each $i \in \omega$ and each $\alpha \in A_i$, there exists a continuous function $f_{i,\alpha}\colon (X, \mathcal{T}) \to I$ such that $f_{i,\alpha}^{-1}(0) = X \setminus U_{i,\alpha}$ and $f_{i,\alpha}^{-1}(1) = P_{i,\alpha}$. We define a mapping $\varphi_i\colon X \to Z_i$ by $\varphi_i(x) = (f_{i,\alpha}, \alpha)$ if $x \in U_{i,\alpha}$, and $\varphi_i(x) = (0, \alpha) = \bar{0}$ if $x \in X \setminus U_{i,\alpha}$. The definition is correct since $(0, \alpha) = \bar{0} = (0, \alpha')$ for all $\alpha, \alpha' \in A_i$. It is clear that the mapping $\varphi_i\colon (X, \mathcal{T}) \to Z_i$ is continuous for each $i \in \omega$, and

$$\{\varphi_i(P_{i,\alpha}) : \alpha \in A_i\} = \{\{(1, \alpha)\} : \alpha \in A_i\}$$

is a discrete family of one-point sets in $Z_i$. Let $\psi\colon (X, \mathcal{T}) \to \prod_{i\in\omega} Z_i$ be the diagonal product of the mappings $\varphi_i$. One easily verifies that $\psi$ is a one-to-one continuous mapping of $(X, \mathcal{T})$ onto a subspace $Y$ of the metrizable space $\prod_{i\in\omega} Z_i$. In addition, the family $\{\psi(P_{i,\alpha}) : \alpha \in A_i\}$ is discrete in $Y$ for each $i \in \omega$, so the topology $\{\psi^{-1}(V) : V \text{ is open in } Y\}$ is the required metrizable $s$-approximation of the topology $\mathcal{T}$ on $X$. □

The next result shows that the class of paracompact $\sigma$-spaces is stable with respect to taking free topological groups.

THEOREM 7.6.7. *The free topological group $F(X)$ is a paracompact $\sigma$-space iff $X$ is a paracompact $\sigma$-space.*

PROOF. If $F(X)$ is a paracompact $\sigma$-space, then $X$ is closed in $F(X)$ by a) of Theorem 7.1.13 and, hence, $X$ is also a paracompact $\sigma$-space. Let us prove the sufficiency.

Suppose that the topology $\tau$ of $X$ admits a metrizable $s$-approximation $\tau_1$. Choose a metric $\varrho$ on $X$ which generates the topology $\tau_1$ and consider the topology $\mathcal{T}_d$ on $F_a(X)$ generated by the Graev extension $\widehat{\varrho}$ of $\varrho$ to $F_a(X)$ (see Theorem 7.2.2). Let $F_\varrho(X) = (F_a(X), \mathcal{T}_d)$. By Proposition 7.6.1, the set $B_n(X)$ of all elements of reduced length $\leq n$ with respect to the basis $X$ is closed in $F_\varrho(X)$ for each $n \in \omega$, so the set $C_n(X) = B_n(X) \setminus B_{n-1}(X)$ is open in the subspace $B_n(X)$ of $F_\varrho(X)$. Therefore, we can represent $C_n(X)$ as the union

$$C_n(X) = \bigcup_{k\in\omega} C_{n,k}$$

of closed subsets of $F_\varrho(X)$. For all $n, k \in \omega$, choose a $\sigma$-discrete network $\gamma_{n,k}$ for $(C_{n,k}, \widehat{\varrho})$. By Corollary 7.6.3, the topology on $C_n(X)$ inherited from $F(X)$ coincides with the one inherited from $F_\varrho(X)$, so the family $\gamma = \bigcup_{n,k\in\omega} \gamma_{n,k}$ is a network for $F(X)$. Since $\gamma$ is $\sigma$-discrete in $F_\varrho(X)$, we conclude that $\mathcal{T}_d$ is a metrizable $s$-approximation for the original topology $\mathcal{T}$ of $F(X)$. Therefore, Theorem 7.6.6 implies that $F(X)$ is a paracompact $\sigma$-space. □

We say that a Tychonoff space $X$ is *$\sigma$-closed-metrizable* if it can be represented as the union of countably many closed metrizable subspaces. The next result complements Theorem 7.6.7.

THEOREM 7.6.8. *The free topological group $F(X)$ is $\sigma$-closed-metrizable and paracompact iff the space $X$ is $\sigma$-closed-metrizable and paracompact.*

PROOF. The necessity is clear, so suppose that $X$ is a $\sigma$-closed-metrizable paracompact space. Let $X = \bigcup_{i\in\omega} X_i$, where each $X_i$ is a closed metrizable subspace of $X$. Every $X_i$ has a $\sigma$-discrete network $\gamma_i$, so $\gamma = \bigcup_{i\in\omega} \gamma_i$ is a $\sigma$-discrete network for $X$. Therefore, $F(X)$ is a paracompact $\sigma$-space, by Theorem 7.6.7. In particular, $F(X)$ is perfectly normal. Since $B_n(X)$ is closed in $F(X)$, and $C_n(X) = B_n(X) \setminus B_{n-1}(X)$ is open in $B_n(X)$, we can represent $C_n(X)$ as the union of a countable family of closed subsets of $B_n(X)$, say, $C_n(X) = \bigcup_{k\in\omega} C_{n,k}$. Applying Theorem 7.6.2 and Lemma 7.6.4, we conclude that every $C_{n,k}$ and hence, $C_n(X)$, is a union of a countable collection of closed metrizable subspaces of $F(X)$. Thus, $F(X) = \bigcup_{n=0}^{\infty} C_n$ is $\sigma$-closed-metrizable. □

COROLLARY 7.6.9. *The free topological group $F(X)$ on a metrizable space $X$ is $\sigma$-closed-metrizable and paracompact.*

Theorems 7.6.7 and 7.6.8 have several interesting applications. The first of them concerns dimension of free topological groups and needs a series of results regarding zero-dimensional spaces.

A continuous mapping $f\colon X \to Y$ will be called *gentle* if there is a network $\mathcal{S}$ in the space $X$ such that its image $\{f(P) : P \in \mathcal{S}\}$ is a $\sigma$-discrete family of sets in $Y$. We also recall that $\dim X$ denotes the *covering dimension* of a Tychonoff space $X$ defined in terms of finite cozero coverings of $X$, while $\operatorname{ind} X$ stands for the small inductive dimension of $X$ (see [**165**, Section 7.1]).

THEOREM 7.6.10. *Let $X$ be a paracompact $\sigma$-space such that* $\operatorname{ind} X = 0$. *Then* $\dim X = 0$ *if and only if there is a one-to-one gentle mapping of $X$ onto a metrizable space $Y$ such that* $\dim Y = 0$.

We naturally split this statement into two statements, one of which is slightly more general than the corresponding part of the theorem.

THEOREM 7.6.11. *Let $X$ be a paracompact $\sigma$-space such that* $\dim X = 0$. *Then there exists a one-to-one gentle mapping of $X$ onto a metrizable space $Y$ such that* $\dim Y = 0$.

PROOF. Fix a network $\mathcal{S} = \bigcup\{\mathcal{S}_i : i \in \omega\}$ of $X$ such that each $\mathcal{S}_i = \{P_{\alpha,i} : \alpha \in A_i\}$ is a discrete family of closed sets in $X$. Since $X$ is collectionwise normal, there is a discrete family $\{U_{\alpha,i} : \alpha \in A_i\}$ of open sets in $X$ such that $P_{\alpha,i} \subset U_{\alpha,i}$, for each $\alpha \in A_i$.

Clearly, $\dim X = 0$ is equivalent to saying that $X$ is strongly zero-dimensional [**165**, Section 7.1]. Therefore, according to [**165**, Theorem 6.2.4], we may assume that each $U_{\alpha,i}$ is open and closed in $X$. We may also assume that $0 \notin A_i$, for each $i \in \omega$. Put $D_i = A_i \cup \{0\}$ and consider $D_i$ as a discrete topological space. Let $W_i = \bigcup\{U_{\alpha,i} : i \in A_i\}$. The set $W_i$ is open and closed. We now define a mapping $g_i \colon X \to D_i$ as follows. Take any $x \in X$. If $x \notin W_i$, then $g_i(x) = 0$. If $x \in W_i$, then there is exactly one $\alpha \in A_i$ such that $x \in U_{\alpha,i}$, and we put $g_i(x) = \alpha$. Clearly, $g_i$ is a continuous mapping of $X$ to the discrete space $D_i$. It is also clear that the family $\{g_i(P_{\alpha,i}) : \alpha \in A_i\} = \{\{\alpha\} : \alpha \in A_i\}$ is discrete in $D_i$.

Let $B$ be the product of the discrete metrizable spaces $D_i$, where $i \in \omega$, with the product topology. Then $\operatorname{Ind} B = 0$, by [**165**, Example 7.3.14]. According to [**165**, Theorem 7.3.4], we have that $\operatorname{Ind} Y = 0$, for every non-empty subspace $Y$ of $B$. Consider the diagonal product $g$ of the mappings $g_i$, where $i \in \omega$. Then $g$ is a one-to-one continuous mapping of $X$ onto a metrizable space $Y = g(X) \subset B$ such that $\operatorname{Ind} Y = 0$. Notice that $Y$ is normal, so $\dim Y = \operatorname{Ind} Y = 0$, by virtue of [**165**, Theorem 7.1.10]). Clearly, $g(\mathcal{S}_i)$ is discrete in $B$, for every $i \in \omega$. Thus, the image under $g$ of the network $\mathcal{S}$ is a $\sigma$-discrete network of $Y$, and the mapping $g$ is gentle. □

LEMMA 7.6.12. *Every $\sigma$-locally finite covering of a space by closed and open sets admits a disjoint open refinement.*

PROOF. Let $\gamma = \bigcup \gamma_n$ be a covering of a space $X$ by closed and open sets, where each family $\gamma_n$ is locally finite in $X$. Suppose that $\gamma_n = \{V_\alpha : \alpha \in A_n\}$, where the index sets $A_i$ are pairwise disjoint. Given $n \in \omega$ and $\alpha \in A_n$, we put

$$U_\alpha = V_\alpha \setminus \bigcup_{k<n} \bigcup \gamma_k.$$

Since each family $\gamma_k$ is locally finite, the sets $U_\alpha$ are open and closed in $X$. Let $A = \bigcup_{n\in\omega} A_n$. Then the family $\mu = \{U_\alpha : \alpha \in A\}$ covers $X$, is a refinement of $\gamma$, and we claim that $\mu$ is locally finite in $X$.

Indeed, take an arbitrary point $x \in X$ and let $n$ be the smallest natural number such that $x \in \bigcup \gamma_n$. Choose $\alpha_0 \in A_n$ with $x \in V_{\alpha_0}$. It follows from the definition of the sets $U_\alpha$ that $V_{\alpha_0} \cap U_\alpha = \varnothing$ whenever $\alpha \in A_i$ and $i > n$. Since the families $\gamma_k$ are locally finite, we can choose, for each $i \le n$, an open neighbourhood $G_i$ of $x$ in $X$ that intersects only finitely many elements of $\gamma_i$. Then $G_0 \cap \cdots \cap G_n \cap V_{\alpha_0}$ is an open neighbourhood of $x$ that intersects only finitely many elements of $\mu$. Hence, the family $\mu$ is locally finite.

Let $\{W_\alpha : \alpha < \kappa\}$ be a well-ordering of $\mu$. Since the family $\mu$ is locally finite, the set $O_\alpha = W_\alpha \setminus \bigcup_{\beta<\alpha} W_\beta$ is open and closed in $X$, for each $\alpha < \kappa$. The family $\nu = \{O_\alpha : \alpha < \kappa\}$ covers $X$, refines $\mu$ and $\gamma$, consists of closed and open sets, and the elements of $\nu$ are pairwise disjoint. This finishes the proof. □

THEOREM 7.6.13. *Suppose that $g$ is a one-to-one gentle mapping of a space $X$ with* $\operatorname{ind} X = 0$ *onto a metrizable space $Y$ satisfying* $\dim Y = 0$. *Then* $\dim X = 0$ *and $X$ is paracompact.*

PROOF. Take any open covering $\eta$ of $X$. Since $\operatorname{ind} X = 0$, we may assume that the elements of $\eta$ are open and closed sets. Fix a network $\mathcal{S}$ in $X$ such that its image $\{f(P) : P \in \mathcal{S}\}$ is a $\sigma$-discrete family of sets in $Y$. We may assume that every element of $\mathcal{S}$ is contained in some element of $\eta$. Indeed, otherwise replace $\mathcal{S}$ with the family of all such elements of $\mathcal{S}$; since $\mathcal{S}$ is a network of $X$, enough elements will be left in $\mathcal{S}$.

We have $\{g(P) : P \in \mathcal{S}\} = \bigcup\{\xi_i : i \in \omega\}$, where each $\xi_i = \{F_{a,i} : a \in A_i\}$ is a discrete family of subsets of $Y$. Since $Y$ is collectionwise normal, there is a discrete family $\{V_{a,i} : a \in A_i\}$ of open sets in $Y$ such that $F_{a,i} \subset V_{a,i}$, for each $a \in A_i$. Since $\operatorname{Ind} Y = \dim Y = 0$, we may assume that each $V_{a,i}$ is open and closed in $Y$.

For each $i \in \omega$ and each $a \in A_i$, we can fix $P_{a,i} \in \mathcal{S}$ such that $g(P_{a,i}) = F_{a,i}$. By our assumption about $\mathcal{S}$, there is $W \in \eta$ such that $P_{a,i} \subset W$. By the assumption about $\eta$, the set $W$ is open and closed in $X$. The set $g^{-1}(V_{a,i})$ is also open and closed in $X$, since $g$ is continuous. It follows that the set $U_{a,i} = W \cap g^{-1}(V_{a,i})$ is open and closed in $X$ and $P_{\alpha,i} \subset U_{\alpha,i}$.

For $i \in \omega$, consider the family $\gamma_i = \{U_{a,i} : a \in A_i\}$. By the continuity of $g$, the family $\{g^{-1}(V_{a,i}) : a \in A_i\}$ is discrete in $X$, whence it follows that $\gamma_i$ is also discrete in $X$, and it is clear that $\gamma_i$ is a refinement of $\eta$. From the definition of $U_{a,i}$ it is immediate that $F_{a,i} = g(P_{a,i}) \subset g(U_{a,i})$. Since $g$ is one-to-one and $Y = \bigcup\{F_{a,i} : a \in A_i,\ i \in \omega\}$, it follows that $X = \bigcup\{U_{a,i} : a \in A_i,\ i \in \omega\}$.

Thus, $\gamma = \bigcup\{\gamma_i : i \in \omega\}$ is a $\sigma$-discrete covering of $X$ by open and closed sets, and $\gamma$ refines $\eta$. It follows from Lemma 7.6.12 that $\dim X \leq 0$ and that $X$ is paracompact. $\square$

It can be useful to note that the paracompactness of $X$ in the above theorem follows even without assuming that $\operatorname{ind} X = 0$ or $\dim X = 0$.

The proof of the following lemma is straightforward.

LEMMA 7.6.14. *The product of any countable family of (one-to-one) gentle mappings is a (one-to-one) gentle mapping.*

THEOREM 7.6.15. *Let $X_n$ be a paracompact $\sigma$-space such that* $\dim X_n = 0$, *for each $n \in \omega$. Then the product space $X = \prod_{n\in\omega} X_n$ satisfies* $\dim X = 0$, *and $X$ is paracompact.*

PROOF. By Theorem 7.6.11, we can fix, for each $n \in \omega$, a one-to-one gentle mapping $g_n$ of $X$ onto a metrizable space $Y_n$ such that $\dim Y_n = 0$. Then $\operatorname{Ind} Y_n = 0$, by [**165**, Theorem 7.1.10]. For the product space $Y = \prod_{n\in\omega} Y_n$, we have $\dim Y = \operatorname{Ind} Y = 0$, since each $Y_n$ is metrizable (see [**165**, Theorem 7.3.16]). Obviously, $\operatorname{ind} X = 0$. It remains to refer to Theorem 7.6.13 and Lemma 7.6.14. $\square$

THEOREM 7.6.16. *Let $X$ be a non-empty paracompact $\sigma$-space. Then* $\dim F(X) = 0$ *if and only if* $\dim X = 0$.

PROOF. Since every continuous function $f \colon X \to \mathbb{R}$ can be extended to a continuous homomorphism $\tilde{f} \colon F(X) \to \mathbb{R}$, the set $X$ is $C$-embedded in $F(X)$. Therefore, $\dim X \leq \dim F(X)$ by [**165**, Th. 7.1.8].

Now, suppose that $\dim X = 0$. Since $\widetilde{X} = X \oplus \{e\} \oplus X^{-1}$ is a paracompact $\sigma$-space, Theorem 7.6.2 and Lemma 7.6.4 imply that $C_n(X) = B_n(X) \setminus B_{n-1}(X)$ is a paracompact $\sigma$-space. As in the proof of Theorem 7.6.8, one can represent every $C_n(X)$ as the union $C_n(X) = \bigcup_{k\in\omega} C_{n,k}$, where each $C_{n,k}$ is closed in $B_n(X)$ (hence, is closed in $F(X)$). Then $C_{n,k}$ is homeomorphic to a closed subspace of $\widetilde{X}^n$. Hence, by Theorem 7.6.15,

$\dim C_{n,k} \leq \dim \widetilde{X}^n = \dim \widetilde{X} = 0$. Since $F(X) = \bigcup_{n,k\in\omega} C_{n,k}$, the countable sum theorem for the dimension dim (see [**165**, Th. 7.2.1]) implies that $\dim F(X) = 0$. □

The methods developed above enable us to show that every topological group $G$ is a quotient of a zero-dimensional topological group $H$ of countable pseudocharacter. In addition, the group $H$ can be always chosen in such a way that it admits a continuous isomorphism onto a metrizable topological group. This situation is in a sharp contrast with the case of locally compact groups, where quotient homomorphisms do not rise dimension (for the proof of this fact, the reader can consult [**543**] or [**456**]). Let us start with an auxiliary topological fact.

LEMMA 7.6.17. *Every $T_1$-space $X$ is an image under a quotient mapping of a topological sum $Z = \bigoplus_{\alpha\in A} Z_\alpha$, where each $Z_\alpha$ is a normal space of countable pseudocharacter with at most one non-isolated point.*

PROOF. First, suppose that the space $X = X_a$ has a single non-isolated point $a \in X$. Denote by $\mathcal{T}_a$ the topology of $X_a$, and put $Y_a = X_a \setminus \{a\}$ and $Z_a = \{a\} \cup (Y_a \times \mathbb{N})$. Let $\mathcal{T}_a^*$ be the topology on $Z_a$ such that all points of the set $Y_a \times \mathbb{N}$ are isolated, and the family

$$\gamma = \{V(k, W) : k \in \mathbb{N},\ a \in W \in \mathcal{T}_a\}$$

constitutes a base of $a$ in $(Z_a, \mathcal{T}_a^*)$, where

$$V(k, W) = \{a\} \cup ((W \setminus \{a\}) \times \{k, k+1, \ldots\})$$

It is clear that $\{a\} = \bigcap_{k\in\omega} V(k, X_a)$. Therefore, $(Z_a, \mathcal{T}_a^*)$ is a $T_1$-space with the single non-isolated point $a$, and this point has countable pseudocharacter in $(Z_a, \mathcal{T}_a^*)$. In particular, $Z_\alpha$ is normal.

Consider the natural mapping $g_a \colon (Z_a, \mathcal{T}_a^*) \to (X_a, \mathcal{T}_a)$ defined by $g_a(a) = a$ and $g_a(y, n) = y$ for all $y \in Y_a$ and all $n \in \mathbb{N}$. This mapping is quotient. Indeed, if $A \subset Y_a$, $a \in W \in \mathcal{T}_a$ and $W \cap A \neq \varnothing$, then

$$g_a^{-1}(A) \cap V(k, W) \supseteq (W \cap A) \times \{k\} \neq \varnothing.$$

Therefore, if $a$ belongs to the closure of $A$ in $X_a$, then $a = g_a^{-1}(a)$ also belongs to the closure of $g_a^{-1}(A)$ in $Z_a$. This proves that $g_a$ is quotient.

Let $(X, \mathcal{T})$ be an arbitrary $T_1$-space. For every point $a \in X$, define a topology $\mathcal{T}_a$ on $X_a = X \times \{a\}$ by declaring the point $(x, a) \in X_a$ isolated for each $x \in X \setminus \{a\}$ and taking the family $\{U \times \{a\} : a \in U \in \mathcal{T}\}$ as a base at the point $(a, a)$ in $(X_a, \mathcal{T}_a)$. Define a mapping $\pi_a \colon X_a \to X$ by $\pi_a(x, a) = x$ for each $x \in X$. The union of the mappings $\{\pi_a : a \in X\}$ gives a continuous mapping $\pi$ of the topological sum $X^* = \bigoplus_{a\in X} X_a$ onto $X$, $\pi(x, a) = \pi_a(x)$ for all $x, a \in X$. It is easy to verify that the mapping $\pi$ is quotient, and each $X_a$ is a $T_1$ space with a single non-isolated point. By the fact established in the first part of the proof, every $X_a$ is a quotient image of a normal space $Z_a$ of countable pseudocharacter and with at most one non-isolated point. Let $g_a \colon Z_a \to X_a$ be the corresponding quotient mapping. The union of the mappings $g_a$ is a quotient mapping $g$ of $Z^* = \bigoplus_{a\in X} Z_a$ onto $X^*$. Therefore, the composition $\varphi = \pi \circ g$ is the required quotient mapping of $Z^*$ onto $X$. □

In what follows we call a space $X$ *$\sigma$-closed-discrete* if $X$ is the union of countably many closed discrete subsets. The next result shows, in particular, that every topological group is a quotient group of a $\sigma$-closed-discrete group.

THEOREM 7.6.18. [**A. V. Arhangel'skii**] *Every topological group $G$ is a quotient of a topological group $H$ satisfying the following conditions:*

a) $\dim H = 0$ *and* $\dim Y = 0$, *for every* $Y \subset H$;
b) *$H$ is a paracompact $\sigma$-closed-discrete space, and every subspace $Y \subset H$ is a paracompact $\sigma$-space;*
c) *$H$ admits a continuous isomorphism onto a metrizable topological group.*

PROOF. Let $Z = \oplus_{a \in X} Z_a$ be a topological sum of normal spaces $Z_a$, each of which has countable pseudocharacter and contains at most one non-isolated point. Clearly, every $Z_a$ is $\sigma$-closed-discrete and hence, has a $\sigma$-discrete network. In fact, all one-point subsets of $Z_a$ form such a network. Therefore, the space $Z$ is also $\sigma$-closed-discrete. Every $Z_a$ is paracompact, as a space with a single non-isolated point and hence, so is $Z$. Thus, $Z$ is a paracompact $\sigma$-closed-discrete space and, in particular, $Z$ is a paracompact $\sigma$-space.

By Theorem 7.6.6, there is a metrizable topology $\tau^*$ on $Z$ weaker than the original topology $\tau$ of $Z$ that $s$-approximates $\tau$ (see the definition on page 458). Arguing as in the proof of Theorem 7.6.7, we take a metric $\varrho$ on $Z$ which generates the topology $\tau^*$ and extend it to a continuous left invariant metric $\widehat{\varrho}$ on $F(Z)$, thus obtaining the metric topological group $F_\varrho(Z) = (F_a(Z), \widehat{\varrho})$. The topology $\mathcal{T}_d$ of the group $F_\varrho(Z)$ is an $s$-approximation of the topology $\mathcal{T}$ of the group $F(Z)$, so $F(Z)$ is a paracompact $\sigma$-space, by Theorem 7.6.7. Lemma 7.6.4 and Theorem 7.6.6 imply that every subspace $Y$ of $F(Z)$ is also a paracompact $\sigma$-space.

The spaces $\tilde{Z} = Z \oplus \{e\} \oplus Z^{-1}$ and $\tilde{Z}^n$ are paracompact and $\sigma$-closed-discrete for each $n \in \mathbb{N}$. Therefore, every subspace of $\tilde{Z}^n$ is $\sigma$-closed-discrete, and from Theorem 7.6.2 it follows that so is the subspace $C_n(Z) = B_n(Z) \setminus B_{n-1}(Z)$ of $F(Z)$, $n \in \mathbb{N}$. The group $F(Z)$, being a paracompact $\sigma$-space, must be perfectly normal. Therefore, one can represent the open subset $C_n(Z)$ of $B_n(Z)$ as a union of countably many closed in $F(Z)$ subsets $C_{n,k}$, $k \in \omega$. Every $C_{n,k}$ is $\sigma$-closed-discrete, as a subspace of $C_n(Z)$. It follows that the group $F(Z) = \bigcup_{n \in \omega} C_n(Z)$ is also $\sigma$-closed-discrete. In other words, the paracompact space $F(Z)$ is the union of a countable family of closed discrete subsets. Now it follows from the countable sum theorem [**165**, Theorem 7.2.1] for the covering dimension that $\dim F(Z) = 0$. Since every subspace $Y$ of $F(Z)$ is paracompact (hence, is normal), the same argument gives the equality $\dim Y = 0$.

To complete the proof, consider an arbitrary topological group $G$ and apply Lemma 7.6.17 to find a strictly $\sigma$-discrete paracompact space $Z$ and a quotient mapping $\pi$ of $Z$ onto $G$. Extend $\pi$ to a continuous homomorphism $\tilde{\pi}\colon F(Z) \to G$. Then the homomorphism $\tilde{\pi}$ is open by Corollary 7.1.10. The above argument shows that the group $F(Z)$ satisfies a)–c). □

The theorem just proved admits a weaker, but still interesting, formulation:

COROLLARY 7.6.19. *Every topological group is a quotient of a strongly zero-dimensional topological group of countable pseudocharacter.*

A direct verification shows that Theorems 7.6.2, 7.6.7, 7.6.8, and 7.6.16 (as well as the auxiliary results they depend upon) remain valid for free Abelian topological groups.

According to Corollary 7.6.9, the group $F(X)$ on a metrizable space $X$ has good covering properties and can be covered by countably many closed metrizable subspaces. It turns out, nevertheless, that the groups $F(X)$ and $A(X)$ can have quite nasty convergence properties — for example, they fail to be $k$-spaces, even for the countable metric space $\mathbb{Q}$.

THEOREM 7.6.20. *If a space $X$ is metrizable and $A(X)$ is a $k$-space, then $X$ is locally compact.*

PROOF. Suppose that a point $x_0 \in X$ has no neighbourhood with compact closure in $X$. There exists a countable base $\{V_n : n \in \omega\}$ at the point $x_0$ in $X$ such that $\overline{V}_{n+1} \subset V_n$ and the set $F_n = \overline{V}_n \setminus V_{n+1}$ is not compact for each $n \in \omega$. Choose a closed discrete subset $\{x_{n,m} : m \in \omega\} \subset F_{2n}$ where $x_{n,m} \neq x_{n,m'}$ if $m \neq m'$. Then the set

$$M = \{x_{n,m} : m, n \in \omega\} \cup \{x_0\}$$

is closed in $X$ and all points of $M$, except for $x_0$, are isolated in $M$.

To each pair $k, l \in \omega$, assign the element

$$h_{k,l} = (x_{k,2l} - x_{k,2l+1}) + (x_{l,1} - x_{l,2}) + \cdots + (x_{l,2k-1} - x_{l,2k}) \in A(X)$$

and consider the sets $H_k = \{h_{k,l} : l > k\}$, $k \in \omega$ and $H = \bigcup_{k=0}^{\infty} H_k$. We claim that the intersection of $H$ with every compact set in $A(X)$ is finite (hence closed), but the identity $e$ of $A(X)$ is an accumulation point of $H$. Note that $e \notin H$.

Clearly, the length of $h_{k,l}$ equals $2k+2$ for each $l \in \omega$, so $H_k \subset A(X) \setminus B_{2k}(X)$. Let $K$ be a compact subset of $A(X)$. By Corollary 7.4.4, $K \subset B_n(X)$ for some $n \in \omega$. Hence $K$ intersects only finitely many sets $H_k$ and it suffices to verify that $K \cap H_k$ is finite for each $k \in \omega$.

The set $X_k = \{x_{k,m} : m \in \omega\}$ is closed and discrete in $X$. For $k, l \in \omega$ with $k < l$, set

$$\mathrm{supp}(h_{k,l}) = \{x_{k,2l}, x_{k,2l+1}, x_{l,1}, x_{l,2}, \ldots, x_{l,2k-1}, x_{l,2k}\}.$$

Clearly, $D_{k,l} = X_k \cap \mathrm{supp}(h_{k,l}) = \{x_{k,2l}, x_{k,2l+1}\}$, so the family $\{D_{k,l} : l > k\}$ is disjoint for each $k \in \omega$. Hence the intersection $X_k \cap \mathrm{supp}(P)$ is infinite for every infinite $P \subset H_k$, where $\mathrm{supp}(P) = \bigcup\{\mathrm{supp}(h_{k,l}) : h_{k,l} \in P\}$. Since $X_k$ is closed and discrete in the metrizable space $X$, the set $X_k \cap \mathrm{supp}(P)$ is not bounded in $X$. According to Corollary 7.5.6 this implies that the intersection $K \cap H_k$ is finite, for every compact subset $K$ of $A(X)$.

It remains to verify that $e$ is an accumulation point of $H$. Let $d$ be an arbitrary continuous metric on $X$. By Theorem 7.2.7, it suffices to show that $V_d \cap H \neq \varnothing$, where

$$V_d = \{g \in A(X) : \widehat{d}_A(e, g) < 1\}.$$

Since $\{V_n : n \in \omega\}$ is a base of $X$ at $x_0$, we can choose $k > 0$ such that $d(x_0, x) < 1/8$ for each $x \in V_{2k}$. Then $d(x_0, x_{k,i}) \leq 1/8$ and $d(x_{k,i}, x_{k,j}) \leq 1/4$ for all $i, j \in \omega$. Similarly, there exists $l > k$ such that $d(x_0, x) < 1/4k$ for each $x \in V_{2l}$. Clearly, $d(x_{l,i}, x_{l,j}) \leq 1/2k$ for all $i, j \in \omega$. Therefore,

$$\widehat{d}_A(e, h_{k,l}) \leq d(x_{k,2l}, x_{k,2l+1}) + \sum_{i=1}^{k} d(x_{l,2i-1}, x_{l,2i}) \leq \frac{1}{4} + \frac{k}{2k} < 1.$$

Hence, $h_{k,l} \in H \cap V_d$. The proof is complete. □

COROLLARY 7.6.21. *Neither $F(\mathbb{Q})$, nor $A(\mathbb{Q})$ are $k$-spaces.*

PROOF. The space of rationals $\mathbb{Q}$ with its usual metric topology is not locally compact, so the conclusion for $A(\mathbb{Q})$ follows directly from Theorem 7.6.20. In addition, Theorem 7.1.11 implies that $A(\mathbb{Q})$ is a quotient group of $F(\mathbb{Q})$. Since open continuous onto mappings preserve the property of being a $k$-space [**165**, Th. 3.3.23], we conclude that $F(\mathbb{Q})$ is not a $k$-space either. □

We shall see in Theorem 7.6.30 that even the local compactness of a metrizable space $X$ does not suffice to imply that $A(X)$ is a $k$-space. Another necessary condition on $X$ is that the set $X'$ of non-isolated points of $X$ has to be separable. To show this we need to establish several facts.

In the proof of the next auxiliary topological result we use the existence of so called *Hausdorff gaps* [**263**]. Let $V(\aleph_1) = \{b\} \cup \{a_{n,\alpha} : n \in \omega,\ \alpha < \omega_1\}$ be the Fréchet–Urysohn fan of cardinality $\aleph_1$, where the sequence $\{a_{n,\alpha} : n \in \omega\}$ converges to $b$ for each $\alpha < \omega_1$.

LEMMA 7.6.22. [**G. Gruenhage and Y. Tanaka**] *The tightness of the product $V(\aleph_1) \times V(\aleph_1)$ is uncountable.*

PROOF. By [**263**, Theorem 20.2], there exist families $\mathcal{A} = \{A_\alpha : \alpha < \omega_1\}$ and $\mathcal{B} = \{B_\alpha : \alpha < \omega_1\}$ of infinite subsets of $\omega$ such that

(a) $A_\alpha \cap B_\beta$ is finite for all $\alpha, \beta < \omega_1$;
(b) for no $A \subset \omega$, all sets $A_\alpha \setminus A$ and $B_\alpha \cap A$, $\alpha < \omega_1$ are finite.

For a pair $\mathcal{A}$, $\mathcal{B}$ of such families (known as a *Hausdorff gap*), put

$$X = \{(a_{n,\alpha}, a_{n,\beta}) \in V(\aleph_1) \times V(\aleph_1) : \alpha, \beta < \omega_1,\ n \in A_\alpha \cap B_\beta\}.$$

We claim that $(b, b)$ is an accumulation point for $X$, but not for any countable subset of $X$. It follows from the claim that the tightness of $V(\aleph_1) \times V(\aleph_1)$ is uncountable.

Given a function $\varphi\colon \omega_1 \to \omega$, put

$$O(\varphi) = \{b\} \cup \{a_{n,\alpha} : \alpha < \omega_1,\ n \geq \varphi(\alpha)\}.$$

From definition of $V(\aleph_1)$ is follows that the family $\{O(\varphi) : \varphi \in \omega^{\omega_1}\}$ is a base at the point $b$.

Take an arbitrary neighbourhood $U$ of the point $(b, b)$ in $V(\aleph_1) \times V(\aleph_1)$. One can assume that $U = O(\varphi) \times O(\varphi)$, for some function $\varphi\colon \omega_1 \to \omega$. For every $\alpha < \omega_1$, put

$$A'_\alpha = \{n \in A_\alpha : n \geq \varphi(\alpha)\},\ \ B'_\alpha = \{n \in B_\alpha : n \geq \varphi(\alpha)\}.$$

Then $A'_\alpha \cap B'_\beta \neq \varnothing$ for some $\alpha, \beta < \omega_1$, for otherwise

$$(\bigcup\{A'_\alpha : \alpha < \omega_1\}) \cap (\bigcup\{B'_\alpha : \alpha < \omega_1\}) = \varnothing$$

and, taking $A = \bigcup\{A'_\alpha : \alpha < \omega_1\}$, we get a contradiction with condition (b). So, choose $n \in A'_\alpha \cap B'_\beta$. Then

$$(a_{n,\alpha}, a_{n,\beta}) \in X \cap (O(\varphi) \times O(\varphi)) = X \cap U$$

and, hence, $X \cap U \neq \varnothing$. Thus $(b, b)$ is an accumulation point for $X$.

Let $K$ be an arbitrary countable subset of $X$. There exists an ordinal $\gamma < \omega_1$ such that

$$K \subset \{(a_{n,\alpha}, a_{n,\beta}) : \alpha, \beta < \gamma,\ n \in \omega\}.$$

We can assume without loss of generality that $\gamma = \omega$ (otherwise order $\gamma$ as $\omega$). Define a function $\psi\colon \omega_1 \to \omega$ by

$$\psi(m) = 1 + \max\{A_k \cap B_l : k, l \leq m\}$$

for each $m < \omega$ and $\psi(\alpha) = 0$ if $\omega \leq \alpha < \omega_1$. We claim that the neighbourhood $U = O(\psi) \times O(\psi)$ of $(b, b)$ is disjoint from $K$. Indeed, if $(a_{n,k}, a_{n,l}) \in K$, then $n \in A_k \cap B_l$. Put $m = \max\{k, l\}$. Clearly $\psi(m) > \max(A_k \cap B_l) \geq n$, whence it follows that $(a_{n,k}, a_{n,l}) \notin U$. This proves that $U \cap K = \varnothing$. Therefore, the tightness of $V(\aleph_1) \times V(\aleph_1)$ is uncountable. □

Since all compact subsets of $V(\aleph_1)$ are countable, Lemma 7.6.22 implies the following:

COROLLARY 7.6.23. *The product $V(\aleph_1) \times V(\aleph_1)$ is not a k-space.*

The next result shows that for $A(X)$ to have countable tightness, the space $X$ must be quite special. First, we need a purely topological fact.

LEMMA 7.6.24. *Let $f$ be an open mapping of a space $X$ onto a space $Y$. Then the tightness of $Y$ does not exceed the tightness of $X$.*

PROOF. Let $\tau = t(X)$. Take any $B \subset Y$ and any $y \in \overline{B}$, and put $A = f^{-1}(B)$, $P = f^{-1}(y)$. Then $P \cap \overline{A} \neq \varnothing$, since the mapping $f$ is open and continuous. Fix $x \in P \cap \overline{A}$. There is a subset $C$ of $A$ such that $x \in \overline{C}$ and $|C| \leq \tau$. Then $f(C) \subset B$, $|f(C)| \leq |C| \leq \tau$. Therefore, $t(Y) \leq \tau = t(X)$. □

PROPOSITION 7.6.25. *Let $X$ be a metrizable space. If the tightness of $A(X)$ is countable, then the set $X'$ of all non-isolated points in $X$ is separable.*

PROOF. If $X'$ is not separable, we can choose an uncountable discrete family $\{U_\alpha : \alpha < \omega_1\}$ of open sets in $X$ each of which contains a point $x_\alpha \in X'$. For every $\alpha < \omega_1$, choose a non-trivial sequence $C_\alpha \subset U_\alpha$ converging to $x_\alpha$ and put $D_\alpha = C_\alpha \cup \{x_\alpha\}$. Consider the sets

$$Y = \bigcup\{D_\alpha : \alpha < \omega_1\}, \quad Y_0 = \{x_\alpha : \alpha < \omega_1\}.$$

Then $Y$ and $Y_0$ are closed in $X$ and $Y$ is homeomorphic to the product $D \times T$, where $D$ is a convergent sequence (with its limit) and $T$ is a discrete space of cardinality $\aleph_1$. Clearly $Y_0$ is closed in $Y$. Let $p\colon X \to Z$ be the natural quotient mapping of $X$ onto the space $Z$ obtained from $X$ by identifying $Y_0$ to a point. Then $p$ is a closed mapping and so is the restriction of $p$ to $Y$. Therefore, the subspace $Z_1 = p(Y)$ of $Z$ is homeomorphic to the Fréchet–Urysohn fan $V(\aleph_1)$ of cardinality $\aleph_1$.

Assume to the contrary that the tightness of $A(X)$ is countable. The homomorphism $\hat{p}\colon A(X) \to A(Z)$ extending the quotient mapping $p$ is open by Corollary 7.1.9. Since, by Lemma 7.6.24, open mappings do not rise the tightness, $A(Z)$ has countable tightness as well. Corollary 7.1.16 implies, however, that $A(Z)$ contains a copy of $Z_1 \times Z_1$, which contradicts Lemma 7.6.22. This completes the proof. □

Theorem 7.6.27 below implies, in particular, that the implication in Lemma 7.6.25 can be reversed. First, we recall that a family $\mathcal{B}$ of open sets in a space $X$ is called an *external base* of a set $Y \subset X$ if for every point $y \in Y$ and every neighbourhood $U$ of $y$ in $X$ there exists $V \in \mathcal{B}$ such that $y \in V \subset U$. In other words, $\mathcal{B}$ is a base of $X$ at each point $y \in Y$.

For the sake of completeness, we supply the reader with the proof of the following simple fact.

LEMMA 7.6.26. *If $Y$ is a subset of a metrizable space $X$, then there exists an external base $\mathcal{B}$ for $Y$ in $X$ such that $|\mathcal{B}| = w(Y)$.*

PROOF. Denote by $\tau$ the weight of $Y$. Let $d$ be a metric on $X$ that generates the topology of $X$. For every $x \in X$ and an integer $n \geq 1$, denote by $O(y, 1/n)$ the open ball in $X$ with center $y$ and radius $1/n$ with respect to $d$. Then $\gamma_n = \{O(y, 1/n) : y \in Y\}$ is a covering of $Y$ by open sets in $X$. Since $w(Y) = \tau$, we can find a subfamily $\mu_n$ of $\gamma_n$ such that $Y \subset \bigcup \mu_n$ and $|\mu_n| \leq \tau$. Clearly, the family $\mathcal{B} = \bigcup_{n=1}^{\infty} \mu_n$ satisfies $|\mathcal{B}| \leq \tau$. Applying the triangle inequality, one can easily verify that $\mathcal{B}$ is an external base for $Y$ in $X$. □

THEOREM 7.6.27. [**A. V. Arhangel'skii, O. G. Okunev, V. G. Pestov**] *Let $X$ be a metrizable space, and $X'$ the set of all non-isolated points in $X$. Then the tightness of $A(X)$ does not exceed the weight of $X'$.*

PROOF. Denote by $\tau$ the weight of $X'$. It suffices to prove that if the identity $e$ of $A(X)$ is an accumulation point for a set $M \subset A(X)$, then $e \in \overline{M'}$ for some $M' \subset M$ such that $|M'| \leq \tau$. By Lemma 7.6.26, there exists an external base $\mathcal{B}$ for $X'$ in $X$ such that $|\mathcal{B}| = \tau$. Put

$$\mathcal{M} = \{\mathcal{A} \subset \mathcal{B} : X' \subset \bigcup \mathcal{A}\}.$$

Assign to every $\mathcal{A} \in \mathcal{M}$ an open covering of $X$ of the form

$$\gamma(\mathcal{A}) = \mathcal{A} \cup \{\{x\} : x \in X \setminus X'\}.$$

Clearly, for every open cover $\gamma$ of $X$ there exists $\mathcal{A} \in \mathcal{M}$ such that $\gamma(\mathcal{A})$ refines $\gamma$. Consider the family $\mathcal{M}^\omega$ of all sequences $(\mathcal{A}_0, \mathcal{A}_1, \ldots)$, where $\mathcal{A}_n \in \mathcal{M}$ for each $n \in \omega$. Assign to every sequence $s = (\mathcal{A}_0, \mathcal{A}_1, \ldots) \in \mathcal{M}^\omega$ the set $V_s$ of all $g \in A(X)$ that can be written in the form

$$g = x_0 - y_0 + x_1 - y_1 + \cdots + x_k - y_k,$$

where $k \in \omega$ and $x_i, y_i \in U_i$, for some $U_i \in \gamma(A_i)$, $i = 0, 1, \ldots, k$. Then 7.2.e implies that $\{V_s : s \in \mathcal{M}^\omega\}$ is a base at the identity $e$ of $A(X)$.

For every $U \in \mathcal{B}$, put

$$W(U) = \{\mathcal{A} \in \mathcal{M} : U \in \mathcal{A}\}.$$

Equip $\mathcal{M}$ with the topology with a subbase consisting of the sets $W(U)$, $U \in \mathcal{B}$. It is easy to see that $\mathcal{M}$ with this topology is naturally homeomorphic to the generalized Cantor cube $\{0, 1\}^{\mathcal{B}}$ (assign to each $\mathcal{A} \in \mathcal{M}$ its characteristic function $f_{\mathcal{A}} \in \{0, 1\}^{\mathcal{B}}$). Therefore, the space $\mathcal{M}$ is compact, Hausdorff, and $w(\mathcal{M}) \leq \tau$. Let us consider the space $S = \mathcal{M}^\omega$ with the product topology. Clearly, $w(S) \leq \tau$. The set $P_g = \{s \in S : g \notin V_s\}$ is closed in $S$ for each $g \in A(X)$. Indeed, suppose that $g \in A(X)$ and $g \in V_{s_0}$, for some $s_0 = (\mathcal{A}_0^0, \ldots, \mathcal{A}_n^0, \ldots) \in S$. We are going to find a neighbourhood $W$ of $s_0$ in $S$ such that $W \cap P_g = \varnothing$. By definition of $V_{s_0}$, $g$ can be written in the form $g = x_0 - y_0 + \cdots + x_k - y_k$, where $k \in \omega$ and for every $i \leq k$ there exists $U_i \in \gamma(\mathcal{A}_i^0)$ such that $x_i, y_i \in U_i$. We can assume that $x_i \neq y_i$ for each $i \leq k$. Then $U_i \in \mathcal{A}_i^0$ for $i = 0, \ldots, k$. Let

$$W = \{s = (\mathcal{A}_0, \ldots, \mathcal{A}_n, \ldots) \in S : U_i \in \mathcal{A}_i \text{ for each } i \leq k\}.$$

Clearly, $W$ is open in $S$, $s_0 \in W$, and $g \in W_s$, for all $s \in W$, so that $W \cap P_g = \varnothing$. This proves our claim.

For every subset $H$ of $A(X)$, put

$$P_H = \{s \in S : V_s \cap H = \varnothing\}.$$

Clearly, $P_H = \bigcap_{g \in H} P_g$, so $P_H$ is closed in $S$, for each $H \subset A(X)$. Note that if $H \subset K \subset A(X)$, then $P_K \subset P_H$.

Assume that the tightness of $A(X)$ is greater than $\tau$. Then there exists a set $M$ in $A(X)$ such that $e$ is in the closure of $M$ but $e \notin \overline{H}$ for each set $H \subset M$ with $|H| \leq \tau$. Then, for every subset $H$ of $M$ of cardinality $\leq \tau$ there exists a neighbourhood of $e$ disjoint with $H$. Therefore, $P_H \neq \varnothing$, and the family $\{P_H : H \subset M, \ |H| \leq \tau\}$ has the $\tau$-intersection property. All sets in this family are closed in $S$, and the weight of $S$ does not exceed $\tau$. Hence, the intersection $P_M = \bigcap\{P_H : H \subset M, \ |H| \leq \tau\}$ is not empty. Choose $s \in P_M$. Then $V_s$ is a neighbourhood of $e$ in $A(X)$ disjoint with $M$, which contradicts the assumption that $e$ is in the closure of $M$. The proof is complete. □

Combining Proposition 7.6.25 and Theorem 7.6.27, we obtain the following:

COROLLARY 7.6.28. *If $X$ is metrizable, then the tightness of $A(X)$ is countable iff the set $X'$ of all non-isolated points of $X$ is separable.*

Now we turn back to the study of the $k$-property in free Abelian topological groups. The following result is a special case of Theorem 7.6.30.

PROPOSITION 7.6.29. *If $X$ is a locally compact metrizable space, and the set $X'$ of all non-isolated points of $X$ is separable, then $A(X)$ is homeomorphic to the product of a $k_\omega$-space with a discrete space, and therefore, $A(X)$ is a $k$-space.*

PROOF. Let $\gamma$ be a countable covering of $X'$ by open (in $X$) sets with compact closures, and put $X_0 = X \setminus \bigcup\{\overline{U} : U \in \gamma\}$. Then $X_0$ is a $k_\omega$-space, and $X_1 = X \setminus X_0$ is a clopen discrete subset of $X$. Hence Exercise 7.4.d implies that $A(X)$ is topologically isomorphic to the product $A(X_0) \times A(X_1)$, where $A(X_0)$ is a $k_\omega$-space, by Theorem 7.4.1, and $A(X_1)$ is a discrete group. Therefore, $A(X)$ is a $k$-space. □

The following theorem completely characterizes metrizable spaces $X$ such that the group $A(X)$ is a $k$-space.

THEOREM 7.6.30. *If $X$ is metrizable, and $X'$ is the set of all non-isolated points of $X$, then the following conditions are equivalent:*

a) *$A(X)$ is a $k$-space;*
b) *$X$ is locally compact and $X'$ is separable.*

PROOF. The implication b) $\Rightarrow$ a) follows from Proposition 7.6.29. Let us show that a) $\Rightarrow$ b). Suppose that $A(X)$ is a $k$-space. Then the group $A(X)$ is sequential. Indeed, let $P$ be a non-closed subset of $A(X)$. There exists a compact set $K \subset A(X)$ such that $P \cap K$ is not closed in $K$. By Theorem 7.5.3, we can find an integer $n \in \omega$ and a bounded subset $Y$ of $X$ such that $K \subset A_n(Y, X)$. Let $C$ be the closure of $Y$ in $X$. Then $C$ is compact and second-countable. Clearly, $K \subset A_n(Y, X) \subset A_n(C, X)$ and the latter set is compact and second-countable. Therefore, the intersection $P \cap A_n(C, X)$ is not closed and there exists a non-trivial sequence lying in $P \cap A_n(C, X)$ and converging to a point outside of $P$.

This proves that the group $A(X)$ is sequential. Since every sequential space has countable tightness, Theorem 7.6.20 implies that the set $X'$ is separable. Finally, $X$ is locally compact by Theorem 7.6.20. □

Now we are going to characterize the metrizable spaces $X$ such that the free topological group $F(X)$ is a $k$-space. Surprisingly, the class of such spaces is narrower than the class of locally compact metrizable spaces with the separable set of non-isolated points (see Theorem 7.6.30). This is the first difference between the topological properties of the groups $F(X)$ and $A(X)$ we find here. Let us start with two preliminary results.

First, we present several conditions implying that a closed subset $Y$ of $F(X)$ generates a subgroup $\langle Y\rangle$ of $F(X)$ topologically isomorphic to $F(Y)$. Recall that $Y$ is a *$\mu$-space* if the closure of every bounded set in $Y$ is compact. According to Proposition 6.9.7, every Dieudonné complete space is a $\mu$-space.

PROPOSITION 7.6.31. *Let $Y$ be a subset of the group $F(X)$ on a Tychonoff space $X$. Suppose that $F(X)$ is a $k$-space, $Y$ is a $\mu$-space and $Y$ forms a free algebraic basis for the subgroup $\langle Y\rangle \subset F(X)$. Then the groups $\langle Y\rangle$ and $F(Y)$ are topologically isomorphic iff for every compact set $K \subset F(X)$, there exist $n \in \omega$ and a compact set $P \subset Y$ such that $\langle Y\rangle \cap K \subset \langle P\rangle_n$. If this is the case, then both $Y$ and $\langle Y\rangle$ are closed in $F(X)$.*

PROOF. Denote by $f$ the continuous isomorphism of $F(Y)$ onto $\langle Y\rangle$ whose restriction to $Y$ is the identity mapping.

Let us start with the necessity. Suppose that $f$ is a topological embedding. For every $g \in \langle Y\rangle$, let $\mathrm{supp}_Y(g)$ be the finite set of elements of $Y$ which appear in the reduced form of $g$ with respect to the basis $Y$. If $K$ is a compact subset of $F(X)$, then $\langle Y\rangle \cap K$ is a precompact subset of $\langle Y\rangle \cong F(Y)$. By Lemma 7.5.2, $S = \bigcup_{g\in K} \mathrm{supp}_Y(g)$ is bounded in $Y$ and $\langle Y\rangle \cap K \subset \langle S\rangle_n$, for some $n \in \omega$. Since $Y$ is a $\mu$-space, the closure $P$ of $S$ in $Y$ is compact. Clearly, $\langle Y\rangle \cap K \subset \langle P\rangle_n$, which proves the necessity.

To prove the sufficiency, suppose that for every compact set $K$ in $F(X)$ one can find a compact set $P \subset Y$ and $n \in \omega$ such that $\langle Y\rangle \cap K \subset \langle P\rangle_n$. First, we show that $F(Y)$ and $\langle Y\rangle$ have the same compact sets or, more strictly, $f^{-1}(C)$ is compact for each compact set $C \subset \langle Y\rangle$. Indeed, if $C$ is a compact subset of $\langle Y\rangle$, then by our assumption, there exist a compact set $P \subset Y$ and $n \in \omega$ such that $C \subset \langle P\rangle_n$. This implies that $f^{-1}(C) \subset F_n(P, Y)$. Since $F_n(P, Y)$ is compact and $f^{-1}(C)$ is closed in $F(Y)$, we conclude that $f^{-1}(C)$ is a compact subset of $F(Y)$.

Our next step is to verify that the subgroup $\langle Y\rangle$ is closed in $F(X)$. Let $C$ be a compact subset of $F(X)$. Then we can find a compact set $P \subset Y$ and $n \in \omega$ such that $\langle Y\rangle \cap C \subset \langle P\rangle_n$. Since $C$ and $\langle P\rangle_n$ are compact, we infer that the intersection $\langle Y\rangle \cap C = \langle P\rangle_n \cap C$ is closed in $C$. By the assumption, $F(X)$ is a $k$-space, so $\langle Y\rangle$ is a closed subgroup of $F(X)$. In particular, $\langle Y\rangle$ is a $k$-space.

Let $A$ be an arbitrary closed subset of $F(Y)$ and $B = f(A)$. If $C$ is a compact subset of $\langle Y\rangle$, then $f^{-1}(C)$ is a compact subset of $F(Y)$ and hence, $A_C = A \cap f^{-1}(C)$ is closed in $f^{-1}(C)$ and compact. So $B \cap C = f(A_C)$ is compact and closed in $\langle Y\rangle$. Thus, the intersection of $B$ with each compact set in $\langle Y\rangle$ is closed, so that $B = f(A)$ is closed in $\langle Y\rangle$. Hence, $f$ is a continuous closed bijection of $F(Y)$ onto $\langle Y\rangle$, that is, $f$ is a homeomorphism. In particular, $Y$ is closed in $\langle Y\rangle \cong F(Y)$. Since $\langle Y\rangle$ is closed in $F(X)$, so is $Y$. This finishes the proof. □

Corollary 7.6.32. *Suppose that a $\mu$-space $Y$ is a closed $C$-embedded subset of a space $X$. If $F(X)$ is a $k$-space, then $F(Y,X) \cong F(Y)$.*

Proof. By the assumption, $Y$ is closed in $X$, so Theorem 7.4.5 implies that $F(Y,X)$ is closed in $F(X)$. Clearly, $Y$ is a free algebraic basis for the subgroup $F(Y,X)$ of $F(X)$. In addition, if $K \subset F(X)$ is compact, then $P = K \cap F(Y,X)$ is a compact subset of $F(Y,X)$ and hence, $Z = \text{supp}(P) \subset Y$ is a bounded subset of $X$ and $P \subset \langle Z\rangle_n$, for some $n \in \omega$ (see Lemma 7.5.2).

We claim that $Z$ is bounded in $Y$. Indeed, otherwise there exists a continuous real-valued function $f$ on $Y$ such that $f$ is unbounded on $Z$. Since $Y$ is $C$-embedded in $X$, $f$ admits an extension to a continuous function $\tilde{f}$ on $X$. Hence, $Z$ is not bounded in $X$, which is a contradiction.

Denote by $C$ the closure of $Z$ in $X$. Since $Y$ is closed in $X$, we have $T \subset Y$. By the assumption, $Y$ is a $\mu$-space, so that $T$ is compact. Clearly,

$$K \cap F(Y,X) = P \subset \langle Z\rangle_n \subset \langle T\rangle_n.$$

Therefore, from Proposition 7.6.31 it follows that $F(Y,X) \cong F(Y)$. □

The following lemma is the main step towards the characterization of the $k$-property in free topological groups. Let us say that a covering $\mathscr{C}$ of a space $Z$ is *generating* if a subset $F$ of $Z$ is closed in $Z$ provided the intersection $F \cap C$ is closed in $C$ for each $C \in \mathscr{C}$. Clearly, $Z$ is a $k$-space iff the family of all compact subsets of $Z$ is generating.

Lemma 7.6.33. *Let $X = C \oplus D$, where $C = \{x_n : n \in \omega\}$ is the sequence converging to $x_0$ and $D$ is an uncountable discrete space. Then the tightness of $F(X)$ is uncountable, and $F(X)$ is not a $k$-space.*

Proof. Suppose that the tightness of $F(X)$ is countable. Then the covering $\lambda = \{F(C \oplus A, X) : A \subset D,\ |A| \leq \omega\}$ of $F(X)$ is generating. Indeed, for $g \in F(X)$ let $\text{supp}(g)$ be the set of all letters in the reduced form of $g$ with respect to the basis $X$. If $S$ is a non-closed subset of $F(X)$, then we can find a point $g \in F(X) \setminus S$ and a countable set $T \subset S$ such that $g \in \overline{T}$. Put

$$P = \text{supp}(g) \cup \bigcup_{h\in T} \text{supp}(h) \ \text{ and } \ A = P \setminus C.$$

Then $|A| \leq |P| \leq \omega$, $H = F(C \oplus A, X) \in \lambda$, and $T \cup \{g\} \subset H$; it follows that $S \cap H$ is not closed in $H$. This proves that the covering $\lambda$ is generating for $F(X)$. From Theorem 7.4.1 it follows that every element of $\lambda$ is a $k_\omega$-space and, hence, $F(X)$ is a $k$-space. So, if the tightness of $F(X)$ is countable then $F(X)$ is a $k$-space.

Assume that $F(X)$ is a $k$-space. Clearly, every bounded set in $X = C \oplus D$ is contained in a set of the form $C \oplus A$, where $A \subset D$ is finite. Therefore, the covering

$$\gamma = \{F(C \oplus A, X) : A \subset D,\ |A| < \omega\}$$

is generating for $F(X)$. Indeed, suppose that the intersection of a set $S \subset F(X)$ with every element of $\gamma$ is closed. Let $K$ be an arbitrary compact subset of $F(X)$. By Corollary 7.5.6, there exists a bounded subset $B$ of $X$ such that $K \subset F(B,X)$. In its turn, $B \subset C \oplus A$ for some finite set $A \subset D$, so $K \subset F(C \oplus A, X) \in \gamma$. Hence the intersection $S \cap K = (S \cap F(C \oplus A, X)) \cap K$ is closed in $K$. Since $F(X)$ is a $k$-space, we conclude that $S$ is closed in $F(X)$. In other words, $\gamma$ is generating for $F(X)$.

Put $C_a = \{x_0 a^{-1} x_0^{-1} x a : x \in C\}$, for each $a \in D$, and let $Y = \bigcup_{a \in D} C_a$. It is easy to see that each $C_a$ is homeomorphic to $C$ and $C_a \cap C_b = \{x_0\}$, for any distinct elements $a, b \in D$. For every finite set $A \subset D$, the set $Y_A = Y \cap F(C \oplus A, X) = \bigcup_{a \in A} C_a$ is compact and closed in $F(C \oplus A, X)$. Since the covering $\gamma$ is generating, $Y$ is closed in $F(X)$. Therefore, the covering $\gamma_Y = \{Y_A : A \subset D,\ |A| < \omega\}$ of $Y$ is generating. Note that each $C_a$ is a sequence converging to $x_0$, so $Y$ is homeomorphic to the Fréchet–Urysohn fan $V(\aleph_1)$ of cardinality $\aleph_1$.

Now we claim that $\langle Y \rangle$ is closed in $F(X)$ and $\langle Y \rangle \cong F(Y)$. Indeed, it is easy to see that $Y$ is a free algebraic basis for $\langle Y \rangle$. In addition, every compact subset $K$ of $F(X)$ is contained in $F_n(C \oplus A, X)$, for some finite set $A \subset D$ and some $n \in \omega$, and a direct verification shows that $\langle Y \rangle \cap K \subset \langle Y_A \rangle_n$, where $Y_A$ is a compact subset of $Y$. Since $X$ is metrizable, Corollary 7.6.9 implies that $F(X)$ is paracompact. In particular, the closed subset $Y$ of $F(X)$ is paracompact and hence, is a $\mu$-space. So we can apply Proposition 7.6.31 to deduce that $\langle Y \rangle \cong F(Y)$ is closed in $F(X)$. Therefore, $F(Y)$ is a $k$-space.

Finally, $F(Y)$ contains a closed copy of $Y^2$, by b) of Theorem 7.1.13, which contradicts Corollary 7.6.23. □

COROLLARY 7.6.34. *Suppose that $X$ is a non-discrete, non-separable metrizable space. Then $F(X)$ is not a $k$-space, and the tightness of $F(X)$ is uncountable.*

PROOF. Suppose to the contrary that $F(X)$ is a $k$-space. Clearly, $X$ contains a closed subspace $Y$ homeomorphic to $C \oplus D$, where $C$ is a convergent sequence with its limit, and $D$ is an uncountable discrete space. Then both $X$ and $Y$ are $\mu$-spaces, and $Y$ is $C$-embedded in $X$. Hence, $F(Y, X) \cong F(Y)$ by Corollary 7.6.32. Apply Lemma 7.6.33 to obtain a contradiction. □

In the next two theorems we characterize the metrizable spaces $X$ such that $F(X)$ has countable tightness or is a $k$-space. It is instructive to compare them with Corollary 7.6.28 and Theorem 7.6.20, respectively, that characterize similar properties of free Abelian topological groups. The non-commutativity of the group $F(X)$ requires $X$ to have better topological properties than those working for $A(X)$.

THEOREM 7.6.35. *If $X$ is metrizable, then the tightness of $F(X)$ is countable iff $X$ is separable or discrete.*

PROOF. Let $X$ be a metrizable space. If $X$ is discrete, then so is $F(X)$ and hence, the tightness of $F(X)$ is countable. If $X$ is separable, then $F(X)$ has a countable network by Corollary 7.1.17, and the tightness of $F(X)$ is countable by [**165**, 3.12.7 (e,f)].

Conversely, if $F(X)$ has countable tightness, then $X$ is either discrete or separable by Corollary 7.6.34. □

THEOREM 7.6.36. *The following conditions are equivalent for a metrizable space $X$:*

a) *$F(X)$ is a $k$-space;*
b) *$F(X)$ is a $k_\omega$-space or is discrete;*
c) *$X$ is locally compact separable or discrete.*

PROOF. Obviously, b) implies a). Every metrizable, locally compact, separable space is a $k_\omega$-space, so the implication c) $\Rightarrow$ b) follows from Theorem 7.4.1. It remains to show that a) $\Rightarrow$ c). Suppose that $F(X)$ is a $k$-space. We can assume without loss of generality that $X$ is not discrete. Then $X$ must be separable by Corollary 7.6.34. By Theorem 7.1.11, $A(X)$ is a quotient group of $F(X)$. It is easily seen that continuous open onto mappings preserve the property of being a $k$-space [**165**, Th. 3.3.23]. It follows that $A(X)$ is a $k$-space. Therefore, Theorem 7.6.20 implies that $X$ is locally compact. This finishes the proof. □

## Exercises

7.6.a. Let $X$ be a metrizable space, and $\tilde{X} = X \oplus \{e\} \oplus X^{-1}$ the topological sum of $X$, the identity $e$ of $F(X)$, and a copy $X^{-1}$ of $X$. Verify that the multiplication mapping $i_2 \colon \tilde{X}^2 \to F(X)$ defined by $i_2(x, y) = x \cdot y$ for $x, y \in \tilde{X}$, is quotient. Apply this fact to show that $i_2$ is a closed mapping. Present an example of a locally compact space $X$ such that the mapping $i_2 \colon \tilde{X}^2 \to F(X)$ is not quotient.

7.6.b. Let $S_1 = \{s_n : n \in \omega\}$ be a non-trivial sequence converging to its limit point $s_0$. The space $S_2$, called the *Arens space*, is obtained from $S_1$ by attaching to each isolated point $s_n$ of $S_1$ a sequence $s_{n,1}, s_{n,2}, \ldots$, converging to $s_n$. Thus, $S_2$ can be viewed as a quotient of a disjoint union of countably many convergent sequences, and we give it the quotient topology. Show that the group $F(S_1)$ contains a closed copy of the space $S_2$ and hence, $F(S_1)$ is not a Fréchet–Urysohn space. Verify that a similar result holds for the free Abelian topological group $A(X)$.

7.6.c. A space $X$ is called an $\aleph_0$-*space* if there exists a countable family $\mathcal{P}$ of subsets of $X$ such that for every compact set $B \subset X$ and every neighbourhood $V$ of $B$ in $X$, one can find $P \in \mathcal{P}$ with $B \subset P \subset V$. Show that the groups $F(X)$ and $A(X)$ are $\aleph_0$-spaces iff $X$ is an $\aleph_0$-space. Deduce that $F(X)$ is an $\aleph_0$-space, for every regular second-countable space $X$.

7.6.d. Let $X'$ be the set of all non-isolated points of a Tychonoff space $X$, and $\mathcal{C}$ be the family of all coverings of $X'$ by open sets in $X$. For a sequence $s = (\gamma_0, \gamma_1, \ldots) \in \mathcal{C}^\omega$, put

$$W_s = \left\{ \sum_{i=0}^{k} (x_i - y_i) : k \in \omega,\ x_i, y_i \in U_i \text{ for some } U_i \in \gamma_i,\ i = 0, \ldots, k \right\}.$$

Apply Exercise 7.2.e to deduce that $W_s$ is open in $A(X)$, and that the family of the sets $W_s$ with $s \in \gamma^\omega$ is a base at the identity of $A(X)$.

7.6.e. Verify that the sets $H_k$ defined in the proof of Theorem 7.6.20 are closed and discrete in $A(X)$.

7.6.f. Recall that a space $X$ is of *pointwise countable type* if $X$ can be covered by compact sets each of which has countable character in $X$. Apply Corollary 7.1.9 to generalize Theorem 7.6.20 as follows: If $X$ is a Tychonoff space of pointwise countable type and $A(X)$ is a $k$-space, then $X$ is locally pseudocompact.

7.6.g. Modify the proof of Theorem 7.6.20 to show that the closed subspace $B_8(X)$ of $A(X)$ is not a $k$-space.

7.6.h. Let $X$ be a Tychonoff space, and $G(X)$ be either $F(X)$ or $A(X)$.
   (a) Verify that if $G(X)$ is a $k$-space, then it has the direct limit property.
   (b) Show that if $B_n(X) \subset G(X)$ is a $k$-space, for each $n \in \omega$, and $G(X)$ has the direct limit property, then $G(X)$ is a $k$-space.

7.6.i. Let $Y = F(X)$ be the free topological group of a metrizable space $X$. Prove that the free topological group $F(Y)$ of $Y$ is paracompact and $\sigma$-closed metrizable.

7.6.j. Show that every closed bounded subspace of the free topological group of any metrizable space is metrizable and compact.

## Problems

7.6.A. (A. V. Arhangel'skii [**19**]) Prove that every second-countable topological group can be represented as a quotient group of a zero-dimensional second-countable topological group.

7.6.B. Let $X$ and $Y$ be metrizable spaces. Is $F(X) \times F(Y)$ paracompact?

7.6.C. Give an example of a Hausdorff space $X$ such that $X^n$ is strongly paracompact, for every $n \in \omega$, while the free Abelian topological group $A(X)$ is not paracompact.

7.6.D. Can an arbitrary paratopological group be represented as a quotient of a zero-dimensional paratopological group?

7.6.E. Is it true that the free topological group of a $\sigma$-closed-discrete space is $\sigma$-closed-discrete?

7.6.F. Give an example of a $\sigma$-closed-metrizable Tychonoff space $Y$ that cannot be topologically embedded into the free topological group of any metrizable space.

7.6.G. (A. V. Arhangel'skii, O. G. Okunev, and V. G. Pestov [**58**]). A space $X$ is called a *$k_R$-space* if a real-valued function $f$ on $X$ is continuous if and only if its restrictions to each compact subset of $X$ is continuous. Prove that for a regular second-countable space $X$, the group $F(X)$ is a $k_R$-space iff it is a $k$-space.

*Hint.* Apply Corollary 7.6.9, Exercise 7.6.c, and Michael's theorem [**320**]: If $Y$ is a $k_R$-space and $\aleph_0$-space representable as a countable union of its closed $k$-subspaces, then $Y$ is a $k$-space.

7.6.H. (O. V. Sipacheva and V. V. Uspenskij [**454**]; for the special case $|X| \leq \mathfrak{c}$, S. A. Morris and B. V. S. Thompson [**331**]). Prove that if $X$ is a submetrizable space, then $F(X)$ and $A(X)$ are *NSS*-groups, i.e., there exists a neighbourhood of the identity in each of these groups which does not contain non-trivial subgroups.

*Hint.* Since $A(X)$ is a quotient group of $F(X)$, it suffices to prove that $F(X)$ is an *NSS*-group. By assumption, there exists a continuous metric $d$ on $X$ which generates a topology weaker than the original topology of $X$. Extend $d$ to a continuous invariant metric $\widehat{d}$ on $F(X)$ and consider the open neighbourhood $V_d$ of the identity $e$ defined by

$$V_d = \{g \in F(X) : \widehat{d}(g, e) < 1\}.$$

Verify that if $g \in F(X)$ and $g \neq e$, then $\widehat{d}(kg, e) \to \infty$ for $k \to \infty$. Deduce that $V_d$ does not contain non-trivial subgroups.

7.6.I. (K. Yamada [**539**]) Show that the following statements hold:

(a) If $X$ is metrizable and the set $X'$ of all non-isolated points in $X$ is compact, then the subspace $B_n(X)$ of $A(X)$ is a $k$-space, for each $n \in \omega$.

(b) There exists a countable subspace $X$ of the real line $\mathbb{R}$ such that each $B_n(X) \subset A(X)$ is a $k$-space, but $A(X)$ is not.

*Hint.* For (a), suppose that there exist $n \in \omega$ and a subset $E$ of $B_n(X)$ such that $E \cap K$ is closed in $K$ for each compact subset $K$ of $B_n(X)$, but $e \in \overline{E} \setminus E$, where $e$ is the neutral element of $A(X)$. Let $d$ be a metric on $X$ generating the topology of $X$. For every $k \in \omega$, put $U_k = \{(x, y) \in X^2 : d(x, y) < 1/(k+1)\}$, $V_k = \{x \in X : d(x, X') < 1/(k+1)\}$ and consider the set

$$W_k = \{x_0 - y_0 + \cdots + x_r - y_r : r \in \omega,\ (x_i, y_i) \in U_k \cap (V_k \times V_k) \text{ for } i \leq r\}.$$

By Exercise 7.6.d, each $W_k$ is a neighbourhood of $e$ in $A(X)$, so there exists a point $a_k \in E \cap W_k$. Note that $\operatorname{supp}(a_k) \subset V_k$ for all $k \in \omega$, so the set $C = X' \cup \bigcup_{k=0}^{\infty} \operatorname{supp}(a_k) \subset X$ is compact. Therefore, $A_n(C, X) \cap E$ is closed in $A(X)$, and we have that $P = \{a_k : k \in \omega\} \subset A_n(C, X)$. To obtain a contradiction, verify that $e \in \overline{P}$.

To deduce (b), apply the above item (a) and Theorem 7.6.20.

7.6.J. (V. G. Pestov and K. Yamada [**382**]) Prove that the following assertions are valid for a metrizable space $X$:

(a) The group $A(X)$ has the direct limit property iff $X$ is locally compact and the set $X'$ of all non-isolated points of $X$ is separable;

(b) The group $F(X)$ has the direct limit property iff $X$ is either locally compact separable or discrete.

7.6.K. Prove that the following conditions are equivalent for a Tychonoff space $X$, where $G(X)$ stands for both $F(A)$ and $A(X)$:

(a) the group $G(X)$ is Čech-complete;

(b) the group $G(X)$ is feathered;

(c) $X$ is discrete.

7.6.L. Show that the groups $F(X)$ and $A(X)$ are weakly feathered (see Exercise 4.3.f) if and only if the space $X$ is submetrizable.

7.6.M. (M. G. Tkachenko [**480**]) Show that every topological group of weight $\tau$ can be represented as a quotient group of a zero-dimensional topological group of the same weight $\tau$. (See also Problem 7.6.A)

## Open Problems

7.6.1. Let $G$ be an arbitrary second-countable topological group. Does $G$ admit a finer second-countable zero-dimensional topological group topology? (Compare with Problem 7.6.A.)

7.6.2. Does there exist a metrizable space $X$ such that the $o$-tightness of the free topological group $F(X)$ is uncountable?

7.6.3. Let $X$ be a locally compact paracompact space. Is the free topological group $F(X)$ paracompact?

7.6.4. Let $X$ be a paracompact $p$-space. Is the free topological group $F(X)$ paracompact?

7.6.5. Suppose that $Z = X \times Y$, where $X$ is metrizable and $Y$ is compact Hausdorff. Is $F(X)$ paracompact?

7.6.6. Characterize the metrizable spaces $X$ such that the free topological group $F(X)$ is strongly paracompact.

7.6.7. Give an example of a Tychonoff space $X$ such that $F(X)$ is paracompact, while $A(X)$ is not.

7.6.8. Give an example of a Tychonoff space $X$ such that $A(X)$ is paracompact, while $F(X)$ is not.

7.6.9. Is the free topological group of a $\sigma$-closed-metrizable space $\sigma$-closed-metrizable?

7.6.10. Is the free topological group $F(X)$ of a first-countable Tychonoff space $X$ Moscow?

7.6.11. Is the free topological group $F(X)$ of a stratifiable space $X$ stratifiable?

*Comment.* The reader can consult the definition of a stratifiable space in [**450**], where O. V. Sipacheva showed that the free Abelian topological group $A(X)$ of any stratifiable space $X$ is stratifiable.

7.6.12. Suppose that $X$ is a $\sigma$-closed-discrete space without isolated points such that the tightness of $F(X)$ is countable. Is $X$ separable?

7.6.13. Suppose that $X$ is a normal $\sigma$-closed-discrete space without isolated points such that the tightness of $F(X)$ is countable. Is $X$ cosmic?

7.6.14. Suppose that $X$ is a Tychonoff space with a $\sigma$-discrete network. Does it follow that $F(X)$ has a $\sigma$-discrete network as well?

7.6.15. Characterize the Tychonoff spaces $X$ such that every closed subset of $F(X)$ is a $G_\delta$-set in $F(X)$.

7.6.16. Suppose that $X$ is a $\sigma$-discrete space, that is, the union of a countable family of discrete (not necessarily closed) subspaces. Is $F(X)$ $\sigma$-discrete as well?

7.6.17. When does the free topological group $F(X)$ (resp., $A(X)$) of a submetrizable space $X$ have the direct limit property?

7.6.18. Suppose that $X$ is a paracompact $p$-space such that the free Abelian topological group $A(X)$ is a $k$-space. Is $X$ locally compact?

7.6.19. When is the tightness of the free topological group $F(X)$ (or $A(X)$) over a paracompact $p$-space $X$ countable?

7.6.20. When does the free topological group $F(X)$ (or $A(X)$) over a paracompact $p$-space $X$ have the direct limit property?

## 7.7. Nummela–Pestov theorem

Suppose that $X$ is a subspace of a Tychonoff space $Y$. Then the embedding mapping $e_{X,Y}\colon X \to Y$ can be extended to a continuous monomorphism $\hat{e}_{X,Y}\colon G(X) \to G(Y)$, where $G(X)$ is either $A(X)$ or $F(X)$. It is very important to know when $\hat{e}_{X,Y}$ is a topological monomorphism, i.e., when $\hat{\imath}_{X,Y}$ is an embedding of $G(X)$ into $G(Y)$. We give a complete answer to this problem in the special case when $X$ is dense in $Y$. This requires the use of uniform structures on topological spaces (see [**165**, Ch. 8]).

Recall that the standard base of the *left uniformity* $\mathcal{L}_G$ on a topological group $G$ consists of the sets

$$W_U^l = \{(x, y) \in G \times G : x^{-1}y \in U\},$$

where $U$ is an arbitrary open neighbourhood of the identity in $G$. If $X$ is a subspace of $G$, then the base of the left induced uniformity $\mathcal{L}_X = \mathcal{L}_G \restriction X$ on $X$ consists of the sets

$$W_U^l \cap X^2 = \{(x, y) \in X \times X : x^{-1}y \in U\}.$$

We also recall that the *universal uniformity* of a space $X$ is the finest uniformity on $X$ that induces on $X$ its original topology [**165**, 8.1.C].

LEMMA 7.7.1. *The restriction $\mathcal{L}_X = \mathcal{L}_{G(X)} \restriction X$ of the left uniformity $\mathcal{L}_{G(X)}$ of the group $G(X)$ to the subspace $X \subset G(X)$ coincides with the universal uniformity $\mathcal{U}_X$ of $X$.*

PROOF. Let $G = G(X)$. Since the topology on $X$ generated by the left uniformity $\mathcal{L}_G$ of $G$ coincides with the original topology of the space $X$, we have the inclusion $\mathcal{L}_X \subset \mathcal{U}_X$. To prove the inverse inclusion, consider an arbitrary element $U \in \mathcal{U}_X$. By [**165**, Coro. 8.1.11], there exists a continuous pseudometric $d$ on $X$ such that

$$\{(x, y) \in X \times X : d(x, y) < 1\} \subset U.$$

Let $\hat{d}$ be the Graev extension of $d$ to a continuous invariant pseudometric on $G$ (see Theorem 7.2.2). Then by Theorem 7.2.7,

$$V = \{g \in G : \hat{d}(e, g) < 1\}$$

is an open neighbourhood of the identity $e$ in $G$. If $x, y \in X$ and $x^{-1}y \in V$, then

$$d(x, y) = \hat{d}(x, y) = \hat{d}(e, x^{-1}y) < 1,$$

from where it follows that the element

$$W_V^l = \{(g, h) \in G \times G : g^{-1}h \in V\}$$

of $\mathcal{L}_G$ satisfies $W_V^l \cap (X \times X) \subset U$. This proves that $\mathcal{U}_X \subset \mathcal{L}_X$ and, hence, the two uniformities on $X$ coincide. □

To characterize the pairs $(X, Y)$, where $X$ is a dense subspace of $Y$, such that the natural mapping $\hat{e}_{X,Y}: G(X) \to G(Y)$ is a topological monomorphism, we need two new concepts.

A subspace $X$ of a space $Y$ is called *$P^*$-embedded* in $Y$ if every bounded continuous pseudometric on $X$ admits a continuous extension over $Y$. If all continuous pseudometrics on $X$ admit continuous extensions over $Y$, then $X$ is said to be *$P$-embedded* in $Y$.

The following statement easily follows from [**165**, 8.5.6 (a)]. We present its proof for the sake of completeness. As in Chapter 6, $\mu X$ denotes the Dieudonné completion of a Tychonoff space $X$, while $\mathcal{U}_X$ stands for the universal uniformity of $X$.

LEMMA 7.7.2. *The following conditions are equivalent for a dense subset $X$ of a space $Y$:*

a) *$X$ is $P^*$-embedded in $Y$;*
b) *$X$ is $P$-embedded in $Y$;*
c) *$\mathcal{U}_Y \restriction X = \mathcal{U}_X$;*
d) *$X \subset Y \subset \mu X$.*

*In particular, if $X$ is a dense $P$-embedded subspace of a Dieudonné complete space $Z$, then $\mu X = Z$.*

PROOF. The implication b) $\Rightarrow$ a) is trivial. It suffices to show, therefore, that a) $\Rightarrow$ c) $\Rightarrow$ d) $\Rightarrow$ b).

a) $\Rightarrow$ c). Suppose that $X$ is $P^*$-embedded in $Y$. For every $U \in \mathcal{U}_X$, there exists a bounded continuous pseudometric $d_X$ on $X$ such that

$$W_X = \{(x, x') \in X \times X : d_X(x, x') < 1\} \subset U.$$

Denote by $d_Y$ an extension of $d_X$ to a continuous pseudometric over $Y$ and consider the set

$$W_Y = \{(y, y') \in Y \times Y : d_Y(y, y') < 1\}.$$

Clearly, $W_Y \in \mathcal{U}_Y$ and $W_Y \cap (X \times X) = W_X \subset U$. Hence, the uniformity $\mathcal{U}_Y \restriction X$ is finer than $\mathcal{U}_X$. The inverse inclusion $\mathcal{U}_Y \restriction X \subset \mathcal{U}_X$ is trivial. This proves that $\mathcal{U}_Y \restriction X = \mathcal{U}_X$.

c) $\Rightarrow$ d). Suppose that $\mathcal{U}_Y \restriction X = \mathcal{U}_X$. Denote by $(\widehat{Y}, \widehat{\mathcal{U}}_Y)$ the completion of the uniform space $(Y, \mathcal{U}_Y)$. Since $\widehat{\mathcal{U}}_Y \restriction Y = \mathcal{U}_Y$, we have $\widehat{\mathcal{U}}_Y \restriction X = \mathcal{U}_X$. In addition, $X$ is dense in both $Y$ and $\widehat{Y}$. Therefore, $(\widehat{Y}, \widehat{\mathcal{U}}_Y)$ is the completion of the uniform space $(X, \mathcal{U}_X)$. Hence, $X \subset Y \subset \widehat{Y} = \mu X$.

d) $\Rightarrow$ b). Suppose that $Y \subset \mu X$. Consider arbitrary continuous pseudometric $d$ on $X$, and denote by $(\bar{X}, \bar{d})$ the metric space obtained from $(X, d)$ by identifying the points of $X$ lying at zero distance one from another with respect to $d$. Let $\pi: X \to \bar{X}$ be the corresponding quotient mapping; then $d(x, y) = \bar{d}(\pi(x), \pi(y))$, for all $x, y \in X$. Let $\mathcal{U}_{\bar{X}}$ be the universal uniformity on $\bar{X}$. Since metrizable spaces are Dieudonné complete, the uniform space $(\bar{X}, \mathcal{U}_{\bar{X}})$ has to be complete. Clearly, the mapping $\pi: (X, \mathcal{U}_X) \to (\bar{X}, \mathcal{U}_{\bar{X}})$ is uniformly continuous, so $\pi$ admits a uniformly continuous extension $\bar{\pi}: (\mu X, \mathcal{U}_{\mu X}) \to (\bar{X}, \mathcal{U}_{\bar{X}})$ to the completion of $(X, \mathcal{U}_X)$. By the assumption, $Y \subset \mu X$; hence, we can define a continuous pseudometric $\varrho$ on $Y$ by $\varrho(x, y) = \bar{d}(\bar{\pi}(x), \bar{\pi}(y))$, for all $x, y \in Y$. It is easy to verify that the restriction of $\varrho$ to $X$ coincides with $d$. Thus, $X$ is $P$-embedded in $Y$.

Finally, suppose that $X$ is a dense $P$-embedded subspace of a Dieudonné complete space $Z$. Then $Z \subset \mu X$ by the first part of the lemma. However, a space $Y$ with $X \subset Y \subset \mu X$ is Dieudonné complete iff $Y = \mu X$. The proof is complete. □

THEOREM 7.7.3. [**E. C. Nummela, V. G. Pestov**] *Let $X$ be a dense subspace of a space $Y$. Then the natural mapping $\hat{e}_{X,Y}\colon F(X) \to F(Y)$ is a topological monomorphism iff $X$ is $P$-embedded in $Y$.*

PROOF. Suppose that the monomorphism $\hat{e}_{X,Y}\colon F(X) \to F(Y)$ extending the identity mapping $e_{X,Y}\colon X \to Y$ is a topological embedding. Then we can identify the group $F(X)$ with the subgroup $F(X, Y)$ of $F(Y)$ generated by the set $X$. Denote by $\mathcal{L}_Y$ and $\mathcal{L}_X$ the left uniformities of the groups $F(Y)$ and $F(X)$, respectively. Since $F(X)$ is a subgroup of $F(Y)$, we have $\mathcal{L}_Y \restriction F(X) = \mathcal{L}_X$. In addition, Lemma 7.7.1 implies that $\mathcal{L}_X \restriction X = \mathcal{U}_X$ and $\mathcal{L}_Y \restriction Y = \mathcal{U}_Y$. Therefore,

$$\mathcal{U}_Y \restriction X = \mathcal{L}_Y \restriction X = \mathcal{L}_X \restriction X = \mathcal{U}_X.$$

We conclude, by Lemma 7.7.2, that $X$ is $P$-embedded in $Y$.

Conversely, suppose that $X$ is $P$-embedded in $Y$. Then $X \subset Y \subset \mu X$, by Lemma 7.7.2. Let $H$ be the completion of the group $F(X)$, and $Z$ be the closure of $X$ in $H$. We claim that $X$ is $P$-embedded in $Z$.

Indeed, every continuous pseudometric $d$ on $X$ can be extended to a continuous invariant pseudometric $\hat{d}$ over $F(X)$, by Theorem 7.2.2. Note that $\hat{d}$ is uniformly continuous with respect to the two-sided uniformity $\mathcal{B}_X$ of the group $F(X)$. This fact follows immediately from the equivalence

$$\hat{d}(g, h) < \varepsilon \iff g^{-1}h \in V_d(\varepsilon) \iff gh^{-1} \in V_d(\varepsilon),$$

where $V_d(\varepsilon) = \{f \in F(X) : \hat{d}(f, e) < \varepsilon\}$ is an open neighbourhood of the identity $e$ in $F(X)$ (see Theorem 7.2.2). Since $F(X)$ is a dense subgroup of $H$, and the restriction of the two-sided uniformity $\mathcal{B}_H$ of $H$ to $F(X)$ coincides with $\mathcal{B}_X$, the pseudometric $\hat{d}$ admits an extension to a continuous pseudometric $\varrho$ on $H$ [**165**, Th. 8.3.10]. Therefore, $\varrho\restriction Z$ is a continuous extension of $d$ over $Z$. This proves that $X$ is $P$-embedded in $Z$.

Let us verify that $Z = \mu X$. Denote by $\mathcal{B}_Z$ the uniformity on $Z$ induced by $\mathcal{B}_H$. Since $Z$ is closed in $H$, the uniform space $(Z, \mathcal{B}_Z)$ is complete. In particular, the space $Z$ is Dieudonné complete. Finally, $X$ is a dense $P$-embedded subspace of the Dieudonné complete space $Z$, so that $Z = \mu X$ by Lemma 7.7.2. Applying Lemma 7.7.2 once again, we conclude that $Y \subset Z \subset H$. Let $f\colon F(Y) \to H$ be the continuous homomorphism which extends the natural embedding of $Y$ in $H$. Denote by $h$ the restriction of $f$ to the subgroup $G$ of $F(Y)$ generated by the set $X$, $h\colon G \to F(X) \subset H$. Then both $h$ and its inverse $\hat{e}_{X,Y}\colon F(X) \to G$ are continuous mappings. Therefore, $\hat{e}_{X,Y}$ is a topological embedding. □

In the Abelian case, we can considerably generalize Theorem 7.7.3 by omitting the assumption that $X$ is dense in $Y$. In fact, the next result completely characterizes the pairs $(X, Y)$ with $X \subset Y$ such that the mapping $\hat{e}_{X,Y}\colon A(X) \to A(Y)$ is a topological embedding.

THEOREM 7.7.4. [**M. G. Tkachenko**] *Let $X$ be an arbitrary subspace of a space $Y$. Then the natural mapping $\hat{e}_{X,Y}\colon A(X) \to A(Y)$ is a topological monomorphism iff $X$ is $P^*$-embedded in $Y$.*

PROOF. Denote by $e_{X,Y}$ the embedding of $X$ in $Y$. It is clear that the monomorphism $\hat{e}_{X,Y}$ is continuous. Let us verify the continuity of its inverse. Suppose that $U$ is a neighbourhood of the neutral element $e_X$ in $A(X)$. By Theorem 7.2.7, there exists a continuous pseudometric

$d$ on $X$ such that

$$V_d = \{g \in A(X) : \widehat{d}_A(g, e_X) < 1\} \subset U,$$

where $\widehat{d}_A$ is the Graev extension of $d$ over $A(X)$. We can assume without loss of generality that $d \leq 1$ — otherwise replace $d$ with $d' = \min\{d, 1\}$. Since $X$ is $P^*$-embedded in $Y$, $d$ can be extended to a continuous pseudometric $\varrho$ on $Y$. Let $\widehat{\varrho}_A$ be the Graev extension of $\varrho$ over $A(Y)$. Applying Theorem 7.2.7 once again, we conclude that

$$V_\varrho = \{h \in A(Y) : \widehat{\varrho}_A(h, e_Y) < 1\}$$

is an open neighbourhood of the neutral element $e_Y$ in $A(Y)$. Let us identify the abstract group $A_a(X)$ with the subgroup $\hat{e}_{X,Y}(A_a(X)) = A_a(X, Y)$ of $A_a(Y)$ generated by the subset $X$ of $A_a(Y)$. Since $\varrho \restriction X = d$, from Corollary 7.2.4 it follows that $\widehat{\varrho}_A(h, e_Y) = \widehat{d}_A(h, e_Y)$, for each $h \in A_a(X, Y)$. Therefore, $A_a(X, Y) \cap V_\varrho = V_d$ or, equivalently,

$$A(X, Y) \cap V_\varrho = \hat{e}_{X,Y}(V_d).$$

This implies immediately that the isomorphism $\hat{e}_{X,Y}^{-1} \colon A(X, Y) \to A(X)$ is continuous. □

For a space $X$, let $G(X)$ be either $F(X)$ or $A(X)$. Theorems 7.7.3 and 7.7.4 enable us to characterize the spaces $X$ with the property that $G(X)$ is topologically isomorphic to the subgroup $G(X, \beta X)$ of $G(\beta X)$ generated by $X$.

Corollary 7.7.5. *The natural mapping $\hat{e} \colon G(X) \to G(\beta X)$ is a topological monomorphism iff $X$ is pseudocompact.*

Proof. If $X$ pseudocompact, then $\mu X = \beta X$ [**255**], so Theorem 7.7.3 and Lemma 7.7.2 together imply that $\hat{e}$ is a topological embedding of $G(X)$ into $G(\beta X)$.

Conversely, if $\hat{e} \colon G(X) \to G(\beta X)$ is a topological embedding, then $X$ is $P$-embedded in $\beta X$ by Theorems 7.7.3 and 7.7.4 (note that $P$- and $P^*$-embedding properties coincide for dense subsets by Lemma 7.7.2). However, every $P$-embedded subset is $C$-embedded, so $X$ is $C$-embedded in the compact space $\beta X$. Therefore, $X$ is pseudocompact. □

The natural embedding of $G(X)$ into $G(\beta X)$ for a pseudocompact space $X$ has a very special property described in Theorem 7.7.7 below. It is based on the following simple lemma.

Lemma 7.7.6. *Suppose that a subspace $Y$ of a topological group $H$ algebraically generates $H$, and that $X$ is a $G_\delta$-dense subset of $Y$. Then the subgroup $\langle X \rangle$ of $H$ generated by $X$ is $G_\delta$-dense in $H$.*

Proof. Let $h \in H$ be an arbitrary element of $H$, and $P$ be a $G_\delta$-set in $H$ containing $h$. There exist a positive integer $n$, elements $y_1, \ldots, y_n \in Y$ and $\varepsilon_1, \ldots, \varepsilon_n = \pm 1$ such that $h = y_1^{\varepsilon_1} \cdot \ldots \cdot y_n^{\varepsilon_n}$. Consider the mapping $f \colon Y^n \to H$ defined by $f(x_1, \ldots, x_n) = x_1^{\varepsilon_1} \cdot \ldots \cdot x_n^{\varepsilon_n}$ for every $(x_1, \ldots, x_n) \in X^n$. Since $f$ is continuous and $X^n$ is $G_\delta$-dense in $Y^n$, one can find a point $(x_1, \ldots, x_n) \in X^n$ such that $f(x_1, \ldots, x_n) \in P$. Clearly, $f(x_1, \ldots, x_n) = x_1^{\varepsilon_1} \cdot \ldots \cdot x_n^{\varepsilon_n} \in \langle X \rangle \cap P$, so we conclude that $\langle X \rangle$ is $G_\delta$-dense in $H$. □

Theorem 7.7.7. *Let $X$ be a pseudocompact space. Then the Dieudonné completion of $G(X)$ is naturally homeomorphic to $G(\beta X)$. In particular, $G(X)$ is $P$-embedded in $G(\beta X)$.*

PROOF. By Corollary 7.4.12, the group $G(\beta X)$ is Raĭkov complete. In particular, $G(\beta X)$ is Dieudonné complete. Corollary 7.7.5 enables us to identify $G(X)$ with the subgroup $G(X, \beta X)$ of $G(\beta X)$ generated by $X$. Therefore, by d) of Lemma 7.7.2, it remains to verify that $G(X)$ is $P$-embedded in $G(\beta X)$. First, note that the pseudocompact space $X$ is $G_\delta$-dense in $\beta X$. Hence, Lemma 7.7.6 implies that $G(X)$ is $G_\delta$-dense in $G(\beta X)$. In its turn, this implies that $G(X)^2$ is $G_\delta$-dense in $G(\beta X)^2$. The groups $G(\beta X)$ and $G(\beta X)^2$ are $\sigma$-compact and hence, are perfectly $\kappa$-normal, by Corollary 5.3.29. Since perfectly $\kappa$-normal spaces are Moscow, from Theorem 6.1.7 it follows that $G(X)^2$ is $C$-embedded in $G(\beta X)^2$.

Finally, a continuous pseudometric $d$ on $G(X)$ can be considered as a continuous real-valued function on $G(X)^2$. By the above argument, $d$ can be extended to a continuous mapping $\tilde{d}\colon G(\beta X)^2 \to \mathbb{R}$. Since $G(X)$ is dense in $G(\beta X)$ and the restriction of $\tilde{d}$ to $G(X)^2$ coincides with $d$, we conclude that $\tilde{d}$ is a continuous pseudometric on $G(\beta X)$. This proves that $G(X)$ is $P$-embedded in $G(\beta X)$. □

We are now going to characterize the spaces $X$ such that the groups $A(X)$ and $F(X)$ are $\omega$-narrow or, more generally, $\tau$-narrow for some infinite cardinal $\tau$. Our characterization is based on Lemma 7.7.1 and Theorem 5.1.19.

Recall that a space $X$ satisfies $dc(X) \leq \tau$ if every discrete family of open sets in $X$ has cardinality strictly less than $\tau$ or, equivalently, if $X$ is pseudo-$\tau$-compact (see Section 5.1).

THEOREM 7.7.8. *The following conditions are equivalent for every space $X$:*

a) *the group $F(X)$ is $\tau$-narrow;*
b) *the group $A(X)$ is $\tau$-narrow;*
c) *$X$ is $\tau$-narrow in $F(X)$;*
d) *$X$ is $\tau$-narrow in $A(X)$;*
e) *$dc(X) \leq \tau^+$.*

PROOF. According to Theorem 7.1.11, $A(X)$ is a quotient group of $F(X)$, so a) implies b). By Theorem 5.1.19, c) implies a) and d) implies b). It also follows from Lemma 5.1.15 that e) implies both c) and d). Therefore, it suffices to prove that e) follows from b).

Assume that $X$ contains a discrete family $\gamma$ of non-empty open sets such that $|\gamma| > \tau$. Let $\gamma = \{U_\alpha : \alpha < \tau^+\}$. For every $\alpha < \tau^+$, choose a continuous real-valued function $f_\alpha$ on $X$ with values in $[0, 1]$ such that $f_\alpha(x_\alpha) = 1$, for some $x_\alpha \in U_\alpha$, and $f_\alpha(X \setminus U_\alpha) \subset \{0\}$. The function $f = \sum_{\alpha=0}^{\tau^+} f_\alpha$ is continuous, since the family $\gamma$ is discrete in $X$. Hence, the pseudometric $d$ on $X$ defined by $d(x, y) = |f(x) - f(y)|$ for $x, y \in X$, is also continuous. Put

$$V_d = \{g \in A(X) : \hat{d}_A(e, g) < 1\},$$

where $e$ is the neutral element of $A(X)$ and $\hat{d}_A$ is the Graev extension of $d$ over $A(X)$. Then $V_d$ is an open symmetric neighbourhood of $e$ in $A(X)$, by Theorem 7.2.7. We claim that $X \setminus FV_d \neq \varnothing$, for every $F \subset X$ such that $|F| \leq \tau$. Indeed, since $\gamma$ is discrete in $X$, there exists $\alpha < \tau^+$ such that $U_\alpha \cap F = \varnothing$. Hence, $d(x_\alpha, x) \geq 1$, for each $x \in F$. This implies that $x_\alpha \notin xV_d$, for each $x \in F$, which proves the claim.

The above claim implies, by Corollary 5.1.17, that $X$ fails to be $\tau$-narrow in $A(X)$, so that $A(X)$ is not $\tau$-narrow. Thus, b) implies e), as required. □

## Exercises

7.7.a. Let $Y$ be a retract of a Tychonoff space $X$. Show that the subgroup $G(Y, X)$ of $G(X)$ is topologically isomorphic to $G(Y)$.

7.7.b. Let $K$ be a compact subspace of a Tychonoff space $X$. Show that the subgroup $G(K, X)$ of $G(X)$ is topologically isomorphic to $G(K)$.

7.7.c. Suppose that $X$ is a $\sigma$-bounded space (see Section 6.10). Verify that the Dieudonné completion $\mu X$ of $X$ is $\sigma$-compact and $G(X) \cong G(X, \mu X)$.

7.7.d. Generalize Theorem 7.7.7 and show that the Dieudonné completion of the group $G(X)$ over a $\sigma$-bounded space $X$ is naturally homeomorphic to $G(\mu X)$.

7.7.e. (M. G. Tkachenko [**483**]) For a pseudocompact space $X$, we identify $G(X)$ with the subgroup $G(X, \beta X)$ of $G(\beta X)$. Apply Lemma 7.7.6 and Theorem 7.7.7 to show that if $K$ is a zero-set in $G(X)$, then the following assertions are valid:
(a) $K^* = cl_{G(\beta X)}(K)$ is a zero-set in $G(\beta X)$;
(b) $K^* \cap B_n(\beta X) = cl_{B_n(\beta X)}(K \cap B_n(X))$, for each $n \in \omega$.

7.7.f. (E. C. Nummela [**353**]) Let $(Y, \mathcal{V})$ be a uniform space, $X$ a dense subspace of $Y$, and $\mathcal{U} = \mathcal{V}\restriction X$. Modify the proof of Theorem 7.7.3 to show that the group $G(X, \mathcal{U})$ is topologically isomorphic to the dense subgroup of $G(Y, \mathcal{V})$ generated by $X$, where $G(X, \mathcal{U})$ denotes either $F(X, \mathcal{U})$ or $A(X, \mathcal{U})$ (see Exercise 7.2.d).

7.7.g. Let $X$ be a closed subset of a uniform space $(U, \mathcal{V})$, and $\mathcal{U} = \mathcal{V}\restriction X$. Generalize Theorem 7.7.4 as follows: The natural continuous isomorphism of $A(X, \mathcal{U})$ onto the subgroup of $A(Y, \mathcal{V})$ generated by $X$ is a homeomorphism iff every bounded uniformly continuous pseudometric on $(X, \mathcal{U})$ admits an extension to a uniformly continuous pseudometric on $(Y, \mathcal{V})$.

## Problems

7.7.A. (V. V. Uspenskij [**520**]) Let $Y$ be a closed subspace of a metrizable space $X$. Then the subgroup $F(Y, X)$ of $F(X)$ is topologically isomorphic to $F(Y)$.

7.7.B. (O. V. Sipacheva [**452**]) Let $Y$ be a subspace of a Tychonoff space $X$. Prove that the subgroup $G(Y, X)$ of $G(X)$ is topologically isomorphic to $G(Y)$ iff $Y$ is $P^*$-embedded in $X$.

7.7.C. Let $Y$ be a closed subspace of a $\sigma$-closed-metrizable space $X$. Is the subgroup $G(Y, X)$ of $G(X)$ topologically isomorphic to $G(Y)$?

7.7.D. (M. G. Tkachenko [**483**]) Let $X$ be a pseudocompact space and $n \geq 1$ be an integer. Prove that $\beta(B_n(X)) \cong B_n(\beta X)$ iff $X^n$ is pseudocompact.
*Hint.* By Corollary 7.7.5, one can identify $G(X)$ with the subgroup $G(X, \beta X)$ of $G(\beta X)$, so that $B_n(X)$ is a dense subspace of $B_n(\beta X)$. Suppose that $X^n$ is pseudocompact. Consider a continuous function $f\colon B_n(X) \to [0, 1]$ and put $g = f \circ j_n$, where $j_n\colon K^n \to B_n(\beta X)$ is the multiplication mapping and $K = \beta X \oplus \{e\} \oplus (\beta X)^{-1}$. Apply Glicksberg's theorem (see [**165**, 3.12.20 (d)]) to extend $g$ to a continuous function $\tilde{g}$ on $K^n$. Then verify that $\tilde{g}$ is constant on the fiber $j_n^{\leftarrow}(h)$, for each $h \in B_n(\beta X)$. For this purpose, apply the fact that if $a, b \in K^n$, $j_n(a) = j_n(b)$ and $U, V$ are neighbourhoods in $K^n$ of $a$ and $b$, respectively, then there exist $a' \in U \cap \bar{X}^n$ and $b' \in V \cap \bar{X}^n$ such that $j_n(a') = j_n(b')$. Conclude that there exists a continuous function $\tilde{f}\colon B_n(\beta X) \to [0, 1]$ such that $\tilde{g} = \tilde{f} \circ j_n$. Then $\tilde{f}\restriction B_n(X) = f$.
Conversely, suppose that $\beta(B_n(X)) \cong B_n(\beta X)$. Note that $j_n(X^n)$ is a clopen subset of $B_n(X)$ and apply Glicksberg's theorem to deduce that $X^n$ is pseudocompact.

7.7.E. (M. G. Tkachenko [**483**]) Suppose that $X^n$ is pseudocompact for some integer $n \geq 1$, and identify $G(X)$ with the subgroup $G(X, \beta X)$ of $G(\beta X)$. Prove that if $F$ is a zero-set in $B_n(X)$, then:
(a) $F^* = cl_{B_n(\beta X)}(F)$ is a zero-set in $B_n(\beta X)$;
(b) $F^* \cap B_m(\beta X) = cl_{B_m(\beta X)}(F \cap B_m(X))$, for each $m < n$.

*Hint.* Note that $B_n(X)$ is $G_\delta$-dense in $B_n(\beta X)$ and apply Problem 7.7.D.

7.7.F. (M. G. Tkachenko [**483**]) Suppose that all finite powers of a Tychonoff space $X$ are pseudocompact. Prove that a subset $F$ of $G(X)$ is a zero-set in $G(X)$ iff $F \cap B_n(X)$ is a zero-set in $B_n(X)$, for each $n \in \omega$.

*Hint.* Identify $G(X)$ with the dense subgroup $G(X, \beta X)$ of $G(\beta X)$. Let $F_n = F \cap B_n(X)$ be a zero-set in $B_n(X)$ for each $n \in \mathbb{N}$. Denote by $F_n^*$ the closure of $F_n$ in $B_n(\beta X)$, $n \in \mathbb{N}$. By induction on $n \in \omega$, define a sequence $\{g_n : n \in \omega\}$ satisfying the following conditions for all $n \in \omega$:

(i) $g_n : B_n(\beta X) \to \mathbb{R}$ is continuous;
(ii) $g_{n+1} \restriction B_n(\beta X) = g_n$;
(iii) $g_n^{-1}(0) \cap B_n(\beta X) = F_n^*$.

To construct the sequence $\{g_n : n \in \omega\}$ satisfying (i)–(iii), use 7.7.E along with the following lemma: If $Z$ is a closed subset of a normal space $Y$, $f$ is a continuous real-valued function on $Z$ and $F$ is a zero-set in $Y$ such that $Z \cap F = f^{-1}(0)$, then $f$ can be extended to a continuous function $\tilde{f}$ on $Y$ such that $F = \tilde{f}^{-1}(0)$.

Let $g$ be a function on $F(\beta X)$ such that $g \restriction B_n(\beta X) = g_n$ for each $n \in \omega$. Then $g$ is continuous by Corollary 7.4.2 and $g^{-1}(0) \cap G(X) = F$.

## 7.8. The direct limit property and countable compactness

Theorem 7.4.1 states that both groups $F(X)$ and $A(X)$ have the direct limit property for every $k_\omega$-space $X$. In particular, this is true for any compact Hausdorff space $X$. However, the class $\mathfrak{D}$ of spaces $X$ such that $F(X)$ and $A(X)$ have the direct limit property is considerably wider. For example, every Tychonoff $P$-space is in $\mathfrak{D}$, by Proposition 7.4.8. In this section we define a proper subclass of $\mathfrak{D}$ and study the properties of this new class of spaces.

Let us call $X$ an *NC-space* if $X^n$ is normal and countably compact, for each integer $n \geq 1$.

All finite powers of an *NC*-space are *NC*-spaces, and so are closed subsets of *NC*-spaces. Clearly, every compact space is an *NC*-space. However, the class of *NC*-spaces contains many non-compact spaces. For example, every ordinal space $W(\alpha)$ modeled on an ordinal $\alpha$ of uncountable cofinality and endowed with the order topology is an *NC*-space, and a closed subspace of a $\Sigma$-product of compact metrizable spaces is an *NC*-space (see Lemma 7.8.14 and Theorem 7.8.13, respectively).

First, we show that *NC*-spaces satisfy Wallace's theorem.

LEMMA 7.8.1. *Suppose that the space $X^n$ is normal and countably compact, for some $n \geq 1$. Let $F = F_1 \times \cdots \times F_n$ be the product of closed subsets of $X$ and $O$ be a neighbourhood of $F$ in $X^n$. Then there exist open sets $V_1, \ldots, V_n$ in $X$ such that $F \subset V_1 \times \cdots \times V_n \subset O$.*

PROOF. Since $X^n$ is countably compact, the spaces $\beta(X^n)$ and $(\beta X)^n$ are naturally homeomorphic by Glicksberg's theorem (see [**165**, 3.12.20 (d)]). Put $K = X^n \setminus O$. Then $F$ and $K$ are closed disjoint subsets of the normal space $X^n$ and, hence, their closures $\overline{F}$ and $\overline{K}$, respectively, in $(\beta X)^n$ are disjoint as well. Hence $\widetilde{O} = \beta X^n \setminus \overline{K}$ is an open neighbourhood of $\overline{F}$ in $(\beta X)^n$ which satisfies $\widetilde{O} \cap X^n = O$. It is clear that $\overline{F} = \overline{F_1} \times \cdots \times \overline{F_n}$, where $\overline{F_i}$ is the closure of $F_i$ in $(\beta X)^n$, $i = 1, \ldots, n$. Since the factors $\overline{F_1}, \ldots, \overline{F_n}$ are compact and $\overline{F} \subset \widetilde{O}$, we can apply Wallace's theorem (see [**165**, 3.2.10]) to find open sets $U_1 \ldots, U_n$ in $\beta X$ such that $\overline{F} \subset U_1 \times \cdots \times U_n \subset \widetilde{O}$. Then the sets $V_1 = U_1 \cap X, \ldots, V_n = U_n \cap X$ are open in $X$ and satisfy $F \subset V_1 \times \cdots \times V_1 \subset O$. $\square$

Our next step is to show that, for every integer $n \geq 1$, the natural multiplication mapping $i_n\colon \overline{X}^n \to G(X)$ is closed, where $G(X)$ is either $F(X)$ or $A(X)$, and $\overline{X} = X \oplus \{e\} \oplus X^{-1}$. Our proof of this fact makes use of the following useful concept.

Let $X$ be a subspace of a space $Y$, and let $f\colon X \to Z$ and $g\colon Y \to Z$ be continuous mappings, where $f = g{\restriction}X$. If $f^{-1}(z)$ is dense in $g^{-1}(z)$ for each $z \in f(X)$, then the mappings $f$ and $g$ are called *concordant*.

The following two results relate concordant and closed mappings.

LEMMA 7.8.2. *Let $f\colon X \to Z$ be a continuous closed onto mapping of Tychonoff spaces and $g\colon \beta X \to \beta Z$ be the continuous extension of $f$ over the Čech–Stone compactification $\beta X$ of $X$. Then the mappings $f$ and $g$ are concordant.*

PROOF. Let $z \in Z$ be arbitrary. Suppose to the contrary that there exist a point $x \in g^{-1}(z)$ and an open neighbourhood $U$ of $x$ in $\beta X$ such that $U \cap f^{-1}(z) = \varnothing$. Choose an open neighbourhood $V$ of $x$ in $\beta X$ such that $\overline{V} \subset U$. Clearly, $x \in \overline{V \cap X} = \overline{V} \subset U$. Then

$$z = g(x) \in g(\overline{V}) = \overline{g(V \cap X)} = \overline{f(V \cap X)} = f(\overline{V \cap X}) \subset f(U \cap X),$$

i.e., $z \in f(U \cap X)$. This contradicts the choice of the set $U$. □

LEMMA 7.8.3. *Let $f\colon X \to Z$ be a continuous onto mapping of Tychonoff spaces, and $g\colon \beta X \to \beta Z$ be a continuous extension of $f$ over the Čech–Stone compactification of $X$. If $X$ is normal and the mappings $f$ and $g$ are concordant, then $f$ is closed.*

PROOF. Suppose to the contrary that there exists a closed subset $F$ of $X$ and a point $z \in Z$ such that $z \in \overline{f(F)} \setminus f(F)$. Let $K$ be the closure of $F$ in $\beta X$. Since the mapping $g$ is closed, we have $K \cap g^{-1}(z) \neq \varnothing$. From $g^{-1}(z) = \overline{f^{-1}(z)}$ it follows that the closures in $\beta X$ of closed disjoint subsets $F$ and $f^{-1}(z)$ of $X$ intersect. This contradicts the normality of $X$. □

It is well known that closed continuous mappings preserve normality. Let us show that these mappings also preserve the property of being an $NC$-space.

PROPOSITION 7.8.4. *An image of an NC-space under a closed continuous mapping is an NC-space.*

PROOF. Let $f\colon X \to Y$ be a closed continuous mapping of an $NC$-space $X$ onto $Y$. For every $n \geq 1$, let $f^n\colon X^n \to Y^n$ be the $n$-th power of $f$. We claim that the mapping $f^n$ is closed. Indeed, denote by $g$ the continuous extension of $f$ over $\beta X$, $g\colon \beta X \to \beta Y$. Since $f$ is closed, from Lemma 7.8.2 it follows that $f$ and $g$ are concordant. It is easy to see that the mappings $f^n$ and $g^n$ are also concordant for each $n \in \mathbb{N}$. The spaces $X^n$ and $Y^n = f^n(X^n)$ are countably compact, so the Glicksberg theorem [**165**, 3.12.20 (d)] implies that $\beta(X^n) \cong (\beta X)^n$ and $\beta(Y^n) \cong (\beta Y)^n$ for each $n \in \mathbb{N}$. Hence, we can apply Lemma 7.8.3 to the mappings $f^n\colon X^n \to Y^n$ and $g^n\colon (\beta X)^n \to (\beta Y)^n$; it follows that the mapping $f^n$ is closed. Finally, closed onto mappings preserve normality (and countable compactness), so $Y^n$ is a normal countably compact space, for each $n \in \mathbb{N}$. Thus, $Y$ is an $NC$-space. □

Suppose that $X$ is a pseudocompact space. By Corollary 7.7.5, the natural monomorphism $\hat{\sigma}\colon G(X) \to G(\beta X)$ extending the embedding $\sigma\colon X \to \beta X$ is a topological isomorphism of $G(X)$ onto the subgroup $G(X, \beta X)$ of $G(\beta X)$ generated by $X$. Therefore, for every $n \in \mathbb{N}$, we can identify the subspace $B_n(X)$ of $G(X)$ with the subspace $\hat{\sigma}(B_n(X))$ of $B_n(\beta X) \subset G(\beta X)$. Let $i_n \colon \tilde{X}^n \to B_n(X)$ and $j_n \colon K^n \to B_n(\beta X)$ be the natural multiplication mappings, where $\tilde{X} = X \oplus \{e\} \oplus X^{-1}$, $K = \beta\tilde{X} = \beta X \oplus \{e\} \cup (\beta X)^{-1}$ and $e$ is the neutral element of $G(\beta X)$. Taking into account these identifications, we have the following.

LEMMA 7.8.5. *If $X$ is a pseudocompact space, then the mappings $i_n$ and $j_n$ are concordant, for each positive integer $n$.*

PROOF. Suppose that $g = j_n(\bar{x}) \in B_n(X)$, where $n \in \mathbb{N}$ and $\bar{x} = (x_1, \ldots, x_n) \in K^n$. If the length $l(g)$ of $g$ with respect to $X$ is equal to $n$, then $j_n^{\leftarrow}(g) = i_n^{\leftarrow}(g)$. Here $i_n^{\leftarrow}(g)$ and $j_n^{\leftarrow}(g)$ denote the preimages of the element $g$ under the mappings $i_n$ and $j_n$, respectively, in order to avoid ambiguity when using the inverse $^{-1}$ in the group $G(X)$ or in $G(\beta X)$. Suppose therefore, that $l(g) < n$. Then the word $g$ is obtained after several cancellations in the word $w = x_1 \ldots x_n$, and the letters in the reduced form of $w$ belong to $X$. Let $O$ be an arbitrary neighbourhood of $\bar{x}$ in $K^n$. We can find open sets $U_1, \ldots, U_n$ in $K$ satisfying the following conditions for all $k, l \leq n$:

(a) $x_k \in U_k$ and $U_1 \times \cdots \times U_n \subset O$;
(b) if $x_k = e$, then $U_k = \{e\}$;
(c) if $x_l = x_k^{-1}$, then $U_l = U_k^{-1}$.

Since $X$ is dense in $\beta X$, we can choose, applying (c), a point $y_k \in \tilde{X} \cap U_k$ for $k = 1, \ldots, n$ in such a way that $y_l = y_k^{-1}$ whenever $x_l = x_k^{-1}$. Then from (b) it follows that $y_k = e$ iff $x_k = e$ for each $k \leq n$. Therefore, $\bar{y} = (y_1, \ldots, y_n) \in (U_1 \times \cdots \times U_n) \cap \tilde{X}^n \subset O$ and $i_n(\bar{y}) = j_n(\bar{x}) = g$. This proves that $i_n^{\leftarrow}(g)$ is dense in $j_n^{\leftarrow}(g)$. □

THEOREM 7.8.6. *If $X$ is an NC-space, then the natural multiplication mapping $i_n \colon \tilde{X}^n \to G(X)$ is closed and $B_n(X) = i_n(\tilde{X}^n)$ is an NC-space, for each integer $n \geq 1$.*

PROOF. Every *NC*-space is pseudocompact, so we can identify $G(X)$ with the subgroup $G(X, \beta X)$ of $G(\beta X)$ by Corollary 7.7.5. Using Lemma 7.8.5 (and notation in its proof), we conclude that the mappings $i_n \colon \tilde{X}^n \to B_n(X) \subset B_n(\beta X)$ and $j_n \colon K^n \to B_n(\beta X)$ are concordant for each $n \in \mathbb{N}$. In addition, since $\tilde{X}^n$ is pseudocompact, Glicksberg's theorem implies that $\beta(\tilde{X}^n) \cong K^n$ and, hence, we can apply Lemma 7.8.3 to conclude that the mapping $i_n \colon \tilde{X}^n \to B_n(X)$ is closed. Therefore, $B_n(X)$ is an *NC*-space by Proposition 7.8.4. It remains to note that $B_n(X)$ is a closed subset of $G(X)$ by a) of Theorem 7.1.13, so $i_n \colon \tilde{X}^n \to G(X)$ is a closed mapping. □

LEMMA 7.8.7. *Let $X$ be an NC-space. Suppose that $F_1$ and $F_2$ are closed subsets of $B_n(X)$, for some $n \in \mathbb{N}$, and that $F_1 \cdot F_2 \subset O$, where $O$ is open in $B_{2n}(X)$. Then there exist open sets $V_1$ and $V_2$ in $B_n(X)$ such that $F_1 \subset V_1$, $F_2 \subset V_2$, and $V_1 \cdot V_2 \subset O$.*

PROOF. Let $\varphi \colon G(X) \times G(X) \to G(X)$ be the multiplication mapping, $\varphi(g, h) = g \cdot h$ for all $g, h \in G(X)$. For $n \in \mathbb{N}$, denote by $\varphi_n$ the restriction of $\varphi$ to $B_n(X) \times B_n(X)$, so that $\varphi_n$ is a continuous mapping of $B_n(X) \times B_n(X)$ to $B_{2n}(X)$. Clearly, the set $W = \varphi_n^{-1}(O)$ is open in $B_n(X) \times B_n(X)$ and $F_1 \times F_2 \subset W$. Then $B_n(X)$ is an *NC*-space, by Theorem 7.8.6, so we can apply Lemma 7.8.1 to find open sets $V_1$ and $V_2$ in $B_n(X)$ such that $F_1 \times F_2 \subset V_1 \times V_2 \subset W$. Hence $V_1 \cdot V_2 = \varphi_n(V_1 \times V_2) \subset \varphi_n(W) = O$. This finishes the proof. □

The following result extends Corollary 7.4.2 to free topological groups on *NC*-spaces.

THEOREM 7.8.8. [**M. G. Tkachenko**] *If $X$ is an NC-space, then the groups $F(X)$ and $A(X)$ have the direct limit property.*

PROOF. Let $G(X)$ be either $F(X)$ or $A(X)$ and $e$ be the identity of $G(X)$. For every $n \in \mathbb{N}$, $K_n = B_n(X)$ is a closed symmetric subset of $G(X)$, $K_n \cdot K_n \subset K_{2n}$, and $G(X) = \bigcup_{n=0}^{\infty} K_n$. As in the proof of Theorem 7.4.1, denote by $\mathcal{T}^*$ the new topology on the abstract group $G_a(X)$ which consists of all sets $O \subset G_a(X)$ such that $O \cap K_n$ is open in the subspace $K_n$ of $G(X)$ for each $n \in \mathbb{N}$. Clearly, $\mathcal{T}^*$ is finer than the original topology $\mathcal{T}$ of the topological group $G(X)$, but the restrictions of $\mathcal{T}^*$ and $\mathcal{T}$ to $K_n$ coincide for each $n \in \mathbb{N}$. In particular, $\mathcal{T}^* \restriction X = \mathcal{T} \restriction X$. Therefore, it remains to show that $\mathcal{T}^* = \mathcal{T}$. Since $\mathcal{T}$ is the finest group topology on $G_a(X)$ which induces the original topology on $X$, all we need is to verify that $\mathcal{T}^*$ is a group topology on $G_a(X)$.

Clearly, if $U \in \mathcal{T}^*$, then $U^{-1} \in \mathcal{T}^*$. So, it suffices to show that if $g, h \in G_a(X)$ and $g \cdot h \in U \in \mathcal{T}^*$, then there exist $V, W \in \mathcal{T}^*$ such that $g \in V$, $h \in W$ and $V \cdot W \subset U$. Choose $m \in \mathbb{N}$ such that $g, h \in K_m$. As in the proof of Theorem 7.4.1, we shall construct two sequences $\{V_n : n \geq m\}$ and $\{W_n : n \geq m\}$ satisfying the following conditions for each $n \geq m$:

(1) $g \in V_m$ and $h \in W_m$;
(2) $V_n$ and $W_n$ are open in $K_n$;
(3) $A_n = cl_{K_n} V_n \subset V_{n+1}$ and $B_n = cl_{K_n} W_n \subset W_{n+1}$;
(4) $A_n \cdot B_n \subset U \cap K_{2n}$.

By the continuity of the multiplication in $G(X)$, there exist open sets $V'_m$ and $W'_m$ in $K_m$ such that $g \in V'_m$, $h \in W'_m$ and $V'_m \cdot W'_m \subset U \cap K_{2m}$. Since $K_m$ is regular, one can find open sets $V_m$ and $W_m$ in $K_m$ such that $g \in V_m \subset cl_{K_m} V_m \subset V'_m$ and $h \in W_m \subset cl_{K_m} W_m \subset W'_m$.

Suppose that for some $n \geq m$, we have defined the sets $V_m, \ldots, V_n$ and $W_m, \ldots, W_n$ satisfying (1)–(4). By (4), the sets $A_n = cl_{K_n} V_n \subset K_{n+1}$ and $B_n = cl_{K_n} W_n \subset K_{n+1}$ satisfy $A_n \cdot B_n \subset U \cap K_{2n} \subset U \cap K_{2n+2}$. Therefore, by Lemma 7.8.7, there exist open sets $V'_{n+1}$ and $W'_{n+1}$ in $K_{n+1}$ such that $A_n \subset V'_{n+1}$, $B_n \subset W'_{n+1}$ and $V'_{n+1} \cdot W'_{n+1} \subset U$. Since the space $\tilde{X}^{n+1}$ is normal and the surjective mapping $i_{n+1} \colon \tilde{X}^{n+1} \to K_{n+1}$ is closed by Theorem 7.8.6, we conclude that $K_{n+1}$ is also normal. So, we can find open sets $V_{n+1}$ and $W_{n+1}$ in $K_{n+1}$ such that $A_n \subset V_{n+1} \subset cl_{K_{n+1}} V_{n+1} \subset V'_{n+1}$ and $B_n \subset W_{n+1} \subset cl_{K_{n+1}} W_{n+1} \subset W'_{n+1}$.

Continuing this process, we obtain sequences

$$V_m \subset V_{m+1} \subset \cdots \subset V_n \subset \cdots \quad \text{and} \quad W_m \subset W_{m+1} \subset \cdots \subset W_n \subset \cdots$$

that satisfy conditions (1)–(4). Put $V = \bigcup_{k=m}^{\infty} V_k$ and $W = \bigcup_{k=m}^{\infty} W_k$. Then $g \in V$ and $h \in W$ by (1). It follows from (2) and (3) that the set $V \cap K_n = \bigcup_{k=n}^{\infty} (V_k \cap K_n)$ is open in $K_n$ for each $n \geq m$, so $V \in \mathcal{T}^*$. Similarly, $W \in \mathcal{T}^*$. In addition, (3) and (4) imply that

$$V \cdot W = \bigcup_{k=m}^{\infty} V_k \cdot W_k \subset \bigcup_{k=m}^{\infty} A_k \cdot B_k \subset U.$$

Thus, we have defined the sets $V, W \in \mathcal{T}^*$ such that $g \in V$, $h \in W$, and $V \cdot W \subset U$. Therefore, the multiplication in $(G_a(X), \mathcal{T}^*)$ is continuous. This proves that $\mathcal{T}^*$ is a group topology and hence, $\mathcal{T}^* = \mathcal{T}$. □

To present a wide subclass of *NC*-spaces, we need to generalize the construction of $\sigma$- and $\Sigma$-products given in Section 1.6 and establish some auxiliary facts.

Let $\xi = \{X_I : i \in I\}$ be a family of spaces, $\Pi = \prod_{i \in I} X_i$ be the product of $\xi$, and $b \in \Pi$ an arbitrary point. For every $x \in \Pi$, we define the *support* of $x$ (with respect to $b$) by

$$supp_b(x) = \{i \in I : x_i \neq b_i\}.$$

Given an infinite cardinal $\tau$, one can define the $\Sigma_{<\tau}$*-product* and $\Sigma_\tau$*-product* of the family $\xi$ with center at $b$ as follows:

$$\Sigma_{<\tau}\Pi\xi = \{x \in \Pi : |supp_b(x)| < \tau\} \text{ and } \Sigma_\tau\Pi\xi = \{x \in \Pi : |supp_b(x)| \leq \tau\}.$$

It is clear that $\Sigma_{<\omega}$-products are the usual $\sigma$-products, while $\Sigma_{<\omega_1}$-products are the $\Sigma$-products. Note also that any $\Sigma_{<\tau^+}$-product and the corresponding $\Sigma_\tau$-product coincide, for each cardinal $\tau$. Many results of Section 1.6 can be extended to $\Sigma_{<\tau}$-products; here we formulate without proof a complement to Corollary 1.6.34.

COROLLARY 7.8.9. *If $\tau$ is an uncountable regular cardinal, then the $\Sigma_{<\tau}$-product $P$ of any family of compact spaces is $<\tau$-bounded in the sense that the closure of every subset $B$ of $P$ with $|B| < \tau$ is compact.*

The following fact about $\Sigma_{<\tau}$-product of compact metrizable spaces is quite useful.

THEOREM 7.8.10. *Let $\xi = \{X_i : i \in I\}$ be a family of compact metrizable spaces and $\tau \geq \omega$ a regular cardinal. Then the $\Sigma_{<\tau}$-product of $\xi$ is a normal space.*

PROOF. If $\tau = \omega$, then the $\Sigma_{<\tau}$-product of $\xi$ is the usual $\sigma$-product which is $\sigma$-compact, by Proposition 1.6.41. We can assume, therefore, that $\tau > \omega$.

Let $\Pi = \prod_{i \in I} X_i$ be the product space and $b \in \Pi$ the center of $\Sigma_{<\tau}(b) = \Sigma_{<\tau}\Pi\xi$. It follows from the definition of $\Sigma_{<\tau}(b)$ that $x \in \Sigma_{<\tau}(b)$ iff the support $supp_b(x)$ of $x$ has cardinality less than $\tau$. For every non-empty set $J \subset I$, let $\pi_J$ be the projection of $\Pi$ onto the subproduct $\Pi_J = \prod_{i \in J} X_i$. The restriction of $\pi_J$ to $\Sigma_{<\tau}(b)$ will be denoted by $p_J$. It follows from Corollary 7.8.9 that $\Sigma_{<\tau}(b)$ is a $<\tau$-bounded space. Hence, for every set $J \subset I$ with $|J| < \tau$, the projection $p_J$ of $\Sigma_{<\tau}(b)$ onto the space $\Pi_J$ of weight $< \tau$ is a closed mapping.

Suppose that $F_1$ and $F_2$ are closed disjoint subsets of $\Sigma_{<\tau}(b)$. We claim that there exists a subset $J$ of the index set $I$ such that $|J| < \tau$ and the projections $p_J(F_1)$ and $p_J(F_2)$ of the sets $F_1$ and $F_2$ are disjoint. Since the projection $p_J$ is closed and the compact space $\Pi_J$ is normal, this will imply the conclusion of the lemma. To this end, we are going to construct a sequence $\{J_n : n \in \omega\}$ of subsets of $I$ and a sequence $\{S_{i,n} : n \in \omega\}$ of subsets of $F_i$, for $i = 1, 2$, satisfying the following conditions for all $n \in \omega$ and $i = 1, 2$:

(i) $|J_n| < \tau$;
(ii) $J_n \subset J_{n+1} \subset I$;
(iii) $S_{i,n} \subset F_i$ and $|S_{i,n}| < \tau$;
(iv) $supp_b(x) \subset J_{n+1}$ for each $x \in S_{i,n}$;
(v) $p_{J_n}(S_{i,n})$ is dense in $p_{J_n}(F_i)$.

To start with, we take an arbitrary element $i_0 \in I$ and put $J_0 = \{i_0\}$. Since the space $X_{i_0}$ has a countable base, we can choose, for $i = 1, 2$, a countable set $S_{i,0} \subset F_i$ such that $p_{J_0}(S_{i,0})$ is dense in $p_{J_0}(F_i)$. Suppose that for some $n \in \omega$, we have defined sequences $\{J_k : k \leq n\}$ and $\{S_{i,k} : k \leq n\}$, where $i = 1, 2$, that satisfy (i)–(v). Put

$$J_{n+1} = J_n \cup \bigcup\{supp_b(x) : x \in S_{1,n} \cup S_{2,n}\}.$$

Since (i) and (iii) hold at the stage $n$, and $\tau > \omega$ is a regular cardinal, the set $J_{n+1} \subset I$ has cardinality less than $\tau$. Hence, the space $\Pi_{J_{n+1}}$ has weight $< \tau$, and we can choose, for $i = 1, 2$, a set $S_{i,n+1} \subset F_i$ such that $|S_{i,n+1}| < \tau$ and $p_{J_{n+1}}(S_{i,n+1})$ is dense in $p_{J_{n+1}}(F_i)$. It is easy to see that conditions (i)–(v) hold at the stage $n+1$. This finishes our construction.

Put $J = \bigcup_{n=0}^{\infty} J_n$ and $S_i = \bigcup_{n=0}^{\infty} S_{i,n}$ for $i = 1, 2$. It follows from (i) that the set $J \subset I$ satisfies $|J| < \tau$, while (ii) and (iv) imply that $supp_b(x) \subset J$ for each $x \in S_1 \cup S_2$. It also follows from (v) that $p_J(S_i)$ is dense in $p_J(F_i)$ for $i = 1, 2$. Denote by $C_i$ the closure of $S_i$ in $\Sigma_{<\tau}(b)$ for $i = 1, 2$. Then $C_i \subset F_i$ and $supp_b(x) \subset J$ for each $x \in C_1 \cup C_2$. Since the projection $p_J \colon \Sigma_{<\tau}(b) \to \Pi_J$ is closed, we also have that $p_J(C_i) = p_J(F_i)$ for $i = 1, 2$.

Suppose by the way of contradiction that there exists a point $y \in p_J(F_1) \cap p_J(F_2)$. For $i = 1, 2$, choose a point $x_i \in C_i$ such that $p_J(x_i) = y$. It follows from our choice of $x_1$ and $x_2$ that $p_{I\setminus J}(x_1) = p_{I\setminus J}(b) = p_{I\setminus J}(x_2)$, which in its turn implies that $x_1 = x_2$. Therefore, $F_1 \cap F_2 \neq \varnothing$, a contradiction. This proves that the space $\Sigma_{<\tau}(b)$ is normal. □

COROLLARY 7.8.11. *The $\Sigma_\tau$-product of a family of compact metrizable spaces is normal, for each cardinal $\tau$.*

PROOF. Let $\xi = \{X_i : i \in I\}$ be a family of compact metrizable spaces. Put $\Pi = \prod_{i \in I} X_i$, and choose a point $b \in \Pi$. If the cardinal $\tau$ is finite, then $\Sigma_\tau \Pi \xi$ with center at $b$ is a closed subset of $\Pi$. Hence, $\Sigma_\tau \Pi \xi$ is compact and normal. If $\tau \geq \omega$, then $\tau^+$ is a regular uncountable cardinal, and the conclusion follows from Theorem 7.8.10. □

We will also need the next simple lemma whose proof is left to the reader.

LEMMA 7.8.12. *Suppose that $\tau$ is an infinite regular cardinal and $X$ is a $\Sigma_{<\tau}$-product of a family of compact metrizable spaces. Then, for every cardinal $\kappa < \tau$, the space $X^\kappa$ is naturally homeomorphic to a closed subspace of another $\Sigma_{<\tau}$-product of compact metrizable spaces.*

Now we can show that the class of $NC$-spaces contains all closed subsets of $\Sigma_{<\tau}$-products of compact metrizable spaces.

THEOREM 7.8.13. *Every closed subset of a $\Sigma_{<\tau}$-product of compact metrizable spaces is an $NC$-space, where $\tau > \omega$ is a regular cardinal. In particular, the same is valid for closed subsets of $\Sigma_\tau$-products of compact metrizable spaces, for each cardinal $\tau \geq \omega$.*

PROOF. Since a closed subset of an $NC$-space is again an $NC$-space, it suffices to verify that any $\Sigma_{<\tau}$-product $X$ of compact metrizable spaces is an $NC$-space. It follows from Corollary 7.8.9 that the closure of every countable subset of $X$ is compact, that is, $X$ is $\omega$-bounded. Clearly, this property is productive and implies countable compactness. Therefore, all finite powers of $X$ are countably compact. Finally, Theorem 7.8.10 and Lemma 7.8.12 together imply that $X^n$ is normal, for each integer $n \geq 1$. This means that $X$ is an $NC$-space. □

For an ordinal $\alpha$, we denote by $W(\alpha)$ the set $\alpha$ endowed with the order topology. It is clear that the space $W(\alpha)$ is compact iff the ordinal $\alpha$ is successor. In the next lemma we characterize the ordinals $\alpha \geq \omega$ such that $W(\alpha)$ is an $NC$-space.

LEMMA 7.8.14. *Let $\alpha$ be an infinite ordinal. Then $W(\alpha)$ is an $NC$-space iff $\alpha$ is either successor or a regular uncountable cardinal.*

PROOF. If $\alpha$ is successor, then the space $W(\alpha)$ is compact and, consequently, an $NC$-space. Suppose that $\alpha$ is a regular uncountable cardinal. To every ordinal $\beta < \alpha$ we assign a point $x(\beta) \in D^\alpha$, where $D = \{0, 1\}$, defined by $x(\beta)_\gamma = 1$ if $\gamma < \beta$ and $x(\beta)_\gamma = 0$ otherwise. It is easy to see that the correspondence $\beta \mapsto x(\beta)$ is a homeomorphism of $W(\alpha)$ onto the subspace $X = \{x(\beta) : \beta < \alpha\}$ of $D^\alpha$. Denote by $\bar{0}$ the point of $D^\alpha$ with zero coordinates. Clearly, $X$ is a closed subset of $\Sigma_{<\alpha}(\bar{0})$, the $\Sigma_{<\alpha}$-product of $\alpha$ copies of the discrete space $D$ with the basic point $\bar{0}$, considered as a subspace of $D^\alpha$. Therefore, Theorem 7.8.10 implies that both $X$ and $W(\alpha)$ are $NC$-spaces.

Finally, suppose $\alpha$ is a limit ordinal and $\beta = cf(\alpha) < \alpha$. If $\beta = \omega$, then $W(\alpha)$ contains an infinite closed discrete subset (a cofinal set in $\alpha$), so $W(\alpha)$ is not countably compact. If $\beta > \omega$, then $W(\alpha)$ contains a closed subspace $C$ cofinal in $\alpha$ which is homeomorphic to $W(\beta)$ (take any strictly increasing function $f\colon \beta \to \alpha$ whose image $B$ is cofinal in $\alpha$ and let $C$ be the closure of $B$ in $W(\alpha)$). Consider the closed subspace $Z = W(\beta+1) \times C$ of $W(\alpha) \times W(\alpha)$ which is homeomorphic to $W(\beta + 1) \times W(\beta)$. We claim that $Z$ is not normal. Indeed, since $\beta > \omega$, the argument of [**165**, Example 3.1.27] implies that every continuous real-valued function on $W(\beta)$ is eventually constant and, hence, the Čech–Stone compactification of the space $W(\beta)$ is homeomorphic to $W(\beta + 1)$. Since $W(\beta)$ is not paracompact according to [**165**, 5.5.22 (f)], we apply Tamano's theorem [**165**, Theorem 5.1.38] to conclude that $Z$ fails to be normal. Since $Z$ is a closed subset of $W(\alpha) \times W(\alpha)$, the latter space cannot be normal either. This proves that $W(\alpha)$ is not an $NC$-space whenever $\omega \le cf(\alpha) < \alpha$. □

COROLLARY 7.8.15. *Let $\alpha > 0$ be an ordinal. The groups $F(W(\alpha))$ and $A(W(\alpha))$ have the direct limit property in each of the following cases:*

a) *$\alpha$ is a successor ordinal;*
b) *$\alpha$ is limit and $cf(\alpha) = \omega$;*
c) *$\alpha$ is an uncountable regular cardinal.*

PROOF. If $\alpha$ is a successor ordinal, then the space $W(\alpha)$ is compact. If $cf(\alpha) = \omega$, then $W(\alpha)$ is a $k_\omega$-space. In either case, the free (Abelian) topological group $G(X)$ has the direct limit property, by Theorem 7.4.1. Therefore, we can assume that $\alpha = cf(\alpha) > \omega$. Then Lemma 7.8.14 implies that $W(\alpha)$ is an $NC$-space, so it remains to apply Theorem 7.8.8. □

## Exercises

7.8.a. Is the topological sum of two $NC$-spaces an $NC$-space? Is the product of two $NC$-spaces an $NC$-space?

7.8.b. Show that the topological sum of an $NC$-space and a compact sequential space is an $NC$-space.

7.8.c. Let $X$ be an $NC$-space and $K$, $L$ closed subsets of $G(X)$.
   (a) Verify that if $K, L \subset B_n(X)$ for some $n \in \omega$, then $KL$ is closed in $G(X)$.
   (b) Apply (a) and Theorem 7.8.8 to show that if $K \subset B_n(X)$ for some $n \in \omega$, then $KL$ is closed in $G(X)$.

7.8.d. Let us say that $X = \bigcup_{n\in\omega} X_n$ is an *$NC_\omega$-decomposition* of a space $X$ if $X_n \subset X_{n+1}$, each $X_n$ is a closed $NC$-subspace of $X$, and the family $\{X_n : n \in \omega\}$ determines the topology of $X$. A space with an $NC_\omega$-decomposition is called an *$NC_\omega$-space.*
   (a) Generalize Theorem 7.8.8 and show that if $X = \bigcup_{n\in\omega} X_n$ is an $NC_\omega$-decomposition of $X$, then $G(X) = \bigcup_{n\in\omega} \langle X_n \rangle_n$ is an $NC_\omega$-decomposition of $G(X)$.

(b) Verify that if $X = \bigcup_{n\in\omega} X_n$ is an $NC_\omega$-decomposition of $X$, then the natural monomorphism $\varphi_n : G(X_n) \to G(X_n, X)$ is a topological embedding, for each $n \in \omega$.

7.8.e. Apply (b) of Exercise 7.8.d to show that if $X$ is an $NC_\omega$-space, then the multiplication mapping $i_n : \tilde{X}^n \to G(X)$ is quotient, for each $n \in \omega$, where $\tilde{X} = X \oplus \{e\} \oplus X^{-1}$. Give an example of an $NC_\omega$-space $X$ for which the mappings $i_n$ fail to be closed (cf. Theorem 7.8.6).

7.8.f. Prove that if $X$ is a Lindelöf $P$-space, then the multiplication mapping $i_n : \tilde{X}^n \to G(X)$ is closed, for each $n \in \omega$.

7.8.g. A continuous mapping $f : Y \to Y$ is called *z-closed* if $f(P)$ is closed in $Y$, for each zero-set $P$ in $X$.

(a) Let $X$ be a dense $C^*$-embedded subset of a space $\tilde{X}$ and let $f : X \to Y$, $\tilde{f} : \tilde{X} \to Y$ be concordant continuous mappings. Show that if $\tilde{f}$ is $z$-closed and the fiber $f^{-1}(y)$ is pseudocompact, for each $y \in Y$, then $f$ is also $z$-closed.

(b) Apply the above item (a), Lemma 7.8.5, and Glicksberg's theorem [**165**, 3.12.20 (d)] to prove that if $X^n$ (resp., $X^{2n}$) is pseudocompact for some $n \in \mathbb{N}$, then the multiplication mapping $i_n : \tilde{X}^n \to G(X)$ (resp., $i_n \times i_n$) is $z$-closed.

(c) Use (b) to verify that if $X^{2n}$ is pseudocompact and $K, L$ are zero-sets in $B_n(X)$, then the set $KL$ is closed in $G(X)$.

## Problems

7.8.A. Suppose that an $NC$-space $X$ is a closed subspace of a Tychonoff space $Y$. Prove that the free topological group $G(X)$ is naturally topologically isomorphic to the topological subgroup $G(X, Y)$ of $G(Y)$.

7.8.B. (M. G. Tkachenko [**483**]) Suppose that $X$ is a pseudocompact space such that the free topological group $F(X)$ has the direct limit property. Then $X$ is an $NC$-space.

7.8.C. Suppose that $X$ is a Tychonoff space such that whenever $X$ is represented as a closed subspace of a Tychonoff space $Y$, the free topological group $A(X)$ (or $F(X)$) is naturally topologically isomorphic to the topological group $A(X, Y)$ (or to $F(X, Y)$). Is $X$ an $NC$-space?

7.8.D. (O. V. Sipacheva [**451**]) Let $X = \bigoplus_{\alpha<\omega_1} X_\alpha$ be the topological sum of some family of zero-dimensional Tychonoff spaces. Show that if $A(X)$ has the direct limit property, then all $X_\alpha$'s, except, perhaps, for countably many of them, are $P$-spaces.

*Hint.* Suppose that $A(X)$ has the direct limit property, while there are uncountably many $X_\alpha$'s that are not $P$-spaces. By Exercise 7.4.e, one can assume that no $X_\alpha$ is a $P$-space. For every $\alpha < \omega_1$, choose a point $x_\alpha \in X_\alpha$ and a decreasing family $\{U_{\alpha,n} : n \in \omega\}$ of clopen neighbourhoods of $x_\alpha$ in $X_\alpha$ such that $x_\alpha \notin \mathrm{Int}(\bigcap_{n\in\omega} U_{\alpha,n})$, and put $C_{\alpha,n} = U_{\alpha,n} \setminus U_{\alpha,n+1}$. For every $\alpha \in \omega_1 \setminus \omega$, enumerate the set $\alpha$ as $\alpha = \{\beta_{\alpha,i} : i \in \omega\}$. Consider the sets

$$F_{\alpha,n} = \{(x - x_\alpha) + n(y - x_{\beta_{\alpha,i}}) : x \in C_{\alpha,n},\ y \in C_{\beta_{\alpha,i},m},\ n < m < i\},$$

where $\alpha \in \omega_1 \setminus \omega$ and $n \in \omega$. Put also $F_n = \bigcup_{\omega\le\alpha<\omega_1} F_{\alpha,n}$. Note that $F_n \subset B_{2n+2}(X) \setminus B_{2n+1}(X)$ for each $n \in \omega$, so $B_k(X) \cap F_n = \varnothing$ if $n > 2k + 1$. Therefore, to obtain a contradiction it suffices to verify that each $F_n$ is closed in $A(X)$, but the neutral element $e$ of $A(X)$ is in the closure of $F = \bigcup_{n\in\omega} F_n$.

Note that if $\gamma$ is a disjoint open covering of $X$, then

$$H(\gamma) = \{\sum_{i=0}^{k}(x_i - y_i) : k \in \omega \text{ and } (\forall\, i \le k)(\exists O_i \in \gamma)(x_i, y_i \in O_i)\}$$

is an open subgroup of $A(X)$. For given $\alpha < \omega_1$ and $n \in \omega$, choose a disjoint open covering $\gamma_{\alpha,n}$ of $X$ such that $F_{\alpha,n} \cap (g + H(\gamma_{\alpha,n}))$ is closed in $g + H(\gamma_{\alpha,n})$, for each $g \in A(X)$. Conclude that $F_{\alpha,n}$ is closed in $A(X)$. Then apply a similar argument to verify that $F_n$ is closed in $A(X)$, for each $n \in \omega$.

To show that $e \in \overline{F}$, consider a continuous pseudometric $d$ on $X$ and for every $\alpha < \omega_1 \setminus \omega$ choose $n_\alpha \in \omega$ such that $B_d(x_\alpha, 1/2)$ intersects $C_{\alpha,n_\alpha}$, where $B_d(x, \varepsilon) = \{y \in X : d(x, y) < \varepsilon\}$, for $x \in X$ and $\varepsilon > 0$. Find $n \in \omega$ such that the set $A = \{\alpha < \omega_1 : n_\alpha = n\}$ is uncountable. Then find an uncountable set $B \subset A$ and $m > n$ such that $B_d(x_\alpha, 1/2n) \cap C_{\alpha,m} \neq \varnothing$ for each $\alpha \in B$. Finally, take $\alpha \in B$ such that $\alpha \cap B$ is infinite and choose $i > m$ with $\beta_{\alpha,i} \in B$. Then pick $x \in B_d(x_\alpha, 1/2)$ and $y \in B_d(x_{\beta_{\alpha,i}}, 1/2n) \cap C_{\beta_{\alpha,i},m}$. Show that $(x - x_\alpha) + n(y - x_{\beta_{\alpha,i}})$ belongs to $V_d \cap F_n$, where $V_d = \{g \in A(X) : \widehat{d}(e, g) < 1\}$. Apply Theorem 7.2.7 to conclude that $e \in \overline{F}$.

## Open Problems

7.8.1. Suppose that $X$ is a pseudocompact space such that the topological group $A(X)$ has the direct limit property. Is $X$ necessarily an $NC$-space? (Compare with Problem 7.8.B.)

## 7.9. Completeness of free Abelian topological groups

By Theorem 7.4.11, the groups $A(X)$ and $F(X)$ are Raĭkov complete if $X$ is compact or, more generally, a $k_\omega$-space. However, this result does not characterize those spaces $X$ for which the groups $A(X)$ and $F(X)$ are Raĭkov complete. Our aim in this section is to show that the free Abelian topological group $A(X)$ is Raĭkov complete if and only if $X$ is Dieudonné complete. This requires several preliminary steps. In what follows we shall use the additive notation for the group operation in $A(X)$. However, the neutral element of $A(X)$ will be denoted by $e$. We start with a result that sheds a new light on the properties of the Graev extensions of pseudometrics from $X$ over the abstract free Abelian group $A_a(X)$.

LEMMA 7.9.1. *Let $d$ be arbitrary pseudometric on $X$, and $\widehat{d}_A$ be the Graev extension of $d$ on $A_a(X)$. Then $\widehat{d}_A(kx, ky) = |k| \cdot d(x, y)$, for all $x, y \in X$ and all $k \in \mathbb{Z}$.*

PROOF. The lemma is trivially true if $x = y$. If $k = 0$, then $kx = ky = e$ and there is nothing to prove. In addition, Corollary 7.2.3 implies that

$$\widehat{d}_A(-kx, -ky) = \widehat{d}_A(ky - kx, e) = \widehat{d}_A(kx - ky, e) = \widehat{d}_A(kx, ky)$$

for an arbitrary $k \in \mathbb{Z}$. Therefore, we can assume that $x \neq y$ and $k \geq 1$. Consider the element $g = kx - ky \in A_a(X)$ of the even length $2k$. By Corollary 7.2.5, there exists an reduced representation for $g$,

$$g = (z_1 - t_1) + \cdots + (z_k - t_k) \tag{7.14}$$

such that

$$\widehat{d}_A(g, e) = \sum_{i=1}^{k} d(z_i, t_i), \tag{7.15}$$

where $z_i, t_i \in \{x, y\}$ for each $i = 1, \ldots, k$. Since the above representation of $g$ is reduced and $k > 0$, each summand $(z_i - t_i)$ in (7.14) is equal to $(x - y)$, so that the equality $\widehat{d}_A(kx, ky) = \widehat{d}_A(g, e) = k \cdot d(x, y)$ follows from (7.15). □

The above lemma implies that the multiplication by a non-zero integer $k$ in the free Abelian topological group $A(X)$ is a uniform isomorphism of $X$ onto $kX$. More precisely, we have the following result that generalizes Lemma 7.7.1 in the case of free Abelian topological groups.

LEMMA 7.9.2. *For every non-zero integer $k$, the mapping $\varphi_k\colon X \to A(X)$ defined by $\phi_k(x) = kx$ for each $x \in X$, is a uniform isomorphism of $(X, \mathcal{U})$ onto the subspace $kX$ of $(A(X), \mathcal{V})$, where $\mathcal{U}$ is the universal uniformity of $X$, and $\mathcal{V}$ is the group uniformity of $A(X)$.*

PROOF. By Theorem 7.2.7, the sets

$$V_d = \{g \in A(X) : \widehat{d}_A(g, e) < 1\}$$

form a base of open neighbourhoods at the neutral element $e$ of $A(X)$, where $d$ runs through the family $\mathcal{P}_X$ of all continuous pseudometrics on $X$. Therefore, the corresponding entourages of the diagonal in $A(X) \times A(X)$, defined by

$$W_d = \{(g, h) \in A(X) \times A(X) : \widehat{d}_A(g, h) < 1\},$$

constitute a base of the group uniformity $\mathcal{V}$ on $A(X)$, where $d \in \mathcal{P}_X$.

Let $d \in \mathcal{P}_X$ and $k \in \mathbb{Z} \setminus \{0\}$ be arbitrary. Lemma 7.9.1 implies that

$$\widehat{d}_A(kx, ky) < 1 \Longleftrightarrow d(x, y) < 1/|k|$$

for all $x, y \in X$. Therefore, $(kx, ky) \in W_d$ iff $d(x, y) < 1/|k|$. This implies immediately that both $\varphi_k$ and its inverse $\varphi_k^{-1}$ are uniformly continuous with respect to the uniformities $\mathcal{U}$ and $\mathcal{V}$ on $X$ and $A(X)$, respectively. □

The next simple fact plays a crucial role in the proofs of Lemma 7.9.4 and Theorem 7.9.6.

LEMMA 7.9.3. *Let $K$ be a subset of an Abelian topological group $G$ with the group uniformity $\mathcal{V}$. If the uniform space $(K, \mathcal{V}\restriction K)$ is complete, then every Cauchy filter $\xi$ in $G$ satisfies one of the following two conditions:*

a) *$\xi$ converges to some point $x \in K$;*
b) *there exist $F \in \xi$ and a neighbourhood $V$ of the neutral element $e$ in $G$ such that $K \cap (F + V) = \varnothing$.*

PROOF. Suppose that a) does not hold and consider the family

$$\gamma = \{F + V : F \in \xi,\ V \in \mathcal{N}(e)\},$$

where $\mathcal{N}(e)$ is the family of all open neighbourhoods of $e$ in $G$. It is easy to see that $\gamma$ is a base of some Cauchy filter $\tilde{\xi}$ in $(G, \mathcal{V})$. Indeed, let $\tilde{\xi}$ be the family of all subsets of $G$ containing at least one set from $\gamma$, and take arbitrary $U \in \mathcal{N}(e)$. Choose a symmetric neighbourhood $V \in \mathcal{N}(e)$ such that $V + V + V \subset U$. There exists $F \in \xi$ such that $F - F \subset V$. Then $F + V \in \tilde{\xi}$, and

$$(F + V) - (F + V) = (F - F) + (V - V) \subset V + V + V \subset U.$$

This implies that $\tilde{\xi}$ is a Cauchy filter in $(G, \mathcal{V})$.

If the intersection $K \cap P$ is not empty for each $P \in \tilde{\xi}$, then the family $\{K \cap P : P \in \tilde{\xi}\}$ forms a Cauchy filter $\mu$ in the complete uniform space $(K, \mathcal{V}\restriction K)$ and, hence, $\mu$ converges to a point $x \in K$. However, this implies immediately that both filters $\tilde{\xi}$ and $\xi$ converge to $x$, which contradicts our assumption about $\xi$. □

Let us establish the completeness of "small" subspaces of the group $A(X)$ on a Dieudonné complete space $X$. If $n \in \mathbb{N}$, and $v = (k_1, \ldots, k_n)$ is an element of $\mathbb{Z}^n$, consider the mapping $\varphi_v \colon X^n \to A(X)$ defined by

$$\varphi_v(x_1, \ldots, x_n) = k_1x_1 + \cdots + k_nx_n$$

for each $(x_1, \ldots, x_n) \in X^n$. Let $\mathcal{V}$ be the group uniformity of $A(X)$.

LEMMA 7.9.4. *Let $X$ be a Dieudonné complete space. Then the subspace $\varphi_v(X^n)$ of $A(X)$ with the uniformity inherited from $(A(X), \mathcal{V})$ is complete, for all $n \in \mathbb{N}$ and $v \in \mathbb{Z}^n$.*

PROOF. We prove the lemma by induction on $n$. Let $n \in \mathbb{N}$ and $v = (k_1, \ldots, k_n) \in \mathbb{Z}^n$ be arbitrary. Put $|v| = |k_1| + \cdots + |k_n|$. If $n = 1$, we have $\varphi_v(X) = k_1X$ and, hence, the completeness of $\varphi_v(X)$ follows directly from Lemma 7.9.2. Suppose that, for some $n \geq 2$, we already know that $\varphi_w(X^m)$ is complete, for all $m < n$ and all $w \in \mathbb{Z}^m$. Clearly, it suffices to consider the case when $k_1 \cdot \ldots \cdot k_n \neq 0$. Let $\xi$ be a Cauchy filter in $\varphi_v(X^n)$. By Lemma 7.9.3, we can assume that there exist an element $F_0 \in \xi$ and a neighbourhood $V_0$ of the neutral element $e$ in $A(X)$ such that $\varphi_w(X^m) \cap (F_0 + V_0) = \varnothing$ for all $m < n$ and all $w \in \mathbb{Z}^m$ satisfying $|w| \leq |v|$.

According to Theorem 7.2.7, there exists a continuous pseudometric $d$ on $X$ such that

$$V_d = \{g \in A(X) : \widehat{d}_A(g, e) < 1\} \subset V_0.$$

This implies immediately that if $m < n$ and $w \in \mathbb{Z}^m$, $|w| \leq |v|$, then

$$\widehat{d}_A(a, b) \geq 1 \text{ for all } a \in F_0 \text{ and } b \in \varphi_w(X^m). \tag{7.16}$$

Suppose that $a = k_1x_1 + \cdots + k_nx_n \in F_0$, where $(x_1, \ldots, x_n) \in X^n$. For $i, j$ with $1 \leq i < j \leq n$, let $a_{i,j} = a + k_j(x_i - x_j)$. Then $a_{i,j} \in \varphi_w(X^{n-1})$, where

$$w = (k_1, \ldots, k_{i-1}, k_i + k_j, k_{i+1}, \ldots, k_{j-1}, k_{j+1}, \ldots, k_n) \in \mathbb{Z}^{n-1}.$$

Clearly, $|w| \leq |v|$. Therefore, it follows from (7.16), the invariance of $\widehat{d}_A$, and Lemma 7.9.1 that

$$1 \leq \widehat{d}_A(a, a_{i,j}) = \widehat{d}_A(k_jx_i, k_jx_j) = |k_j| \cdot d(x_i, x_j).$$

In other words, we have proved that if $k_1x_1 + \cdots + k_nx_n \in F_0$ for some $(x_1, \ldots, x_n) \in X^n$, then

$$d(x_i, x_j) \geq 1/|k_j| \geq 1/|v| \text{ whenever } 1 \leq i < j \leq n. \tag{7.17}$$

In particular, $x_i \neq x_j$ if $i \neq j$.

Recall that $S_n$ is the group of permutations of the set $\{1, 2, \ldots, n\}$. Let us show that the following holds.

**Claim A.** *Let $\varrho$ be a continuous pseudometric on $X$ such that $\varrho \geq |v| \cdot d$, and let $F_1 \in \xi$ satisfy $F_1 \subset F_0$ and $F_1 - F_1 \subset V_{2\varrho}$. Suppose that $g, h \in F_1 \cap \varphi_v(X^n)$ have normal forms $g = k_1x_1 + \cdots + k_nx_n$ and $h = k_1y_1 + \cdots + k_ny_n$. Then there exists a permutation $\pi \in S_n$ such that*

$$k_{\pi(i)} = k_i \textit{ for each } i = 1, \ldots, n \tag{7.18}$$

*and*

$$\widehat{\varrho}_A(g, h) = \sum_{i=1}^{n} |k_i| \cdot \varrho(x_i, y_{\pi(i)}). \tag{7.19}$$

Indeed, from $F_1 - F_1 \subset V_{2\varrho}$ it follows that $\widehat{\varrho}_A(g, h) < 1/2$. Therefore, Corollary 7.2.4 implies that $g - h$ can be written in an reduced form as follows:

$$g - h = (z_1 - z_2) + \cdots + (z_{2p-1} - z_{2p}), \tag{7.20}$$

where $p \leq |v|$, $z_1, \ldots, z_{2p} \in \{x_1, \ldots, x_n\} \cup \{y_1, \ldots, y_n\}$, and

$$\sum_{i=1}^{p} \varrho(z_{2i-1}, z_{2i}) = \widehat{\varrho}_A(g, h) < 1/2. \tag{7.21}$$

In particular, $\varrho(z_{2i-1}, z_{2i}) < 1/2$ for each $i = 1, \ldots, p$. Since $\varrho \geq |v| \cdot d$, from (7.17) (which applies to both $g$ and $h$) it follows that for every $i \leq p$, one of the elements $z_{2i-1}, z_{2i}$ is in $\{x_1, \ldots, x_n\}$ while the other is in $\{y_1, \ldots, y_n\}$. Suppose that there are two summands $(z_{2i-1} - z_{2i})$ and $(z_{2j-1} - z_{2j})$ with $i \neq j$ in the right part of (7.20) such that each of the pairs $(z_{2i-1}, z_{2i})$ and $(z_{2j-1}, z_{2j})$ contains $x_1$, say, $z_{2i-1} = x_1 = z_{2j-1}$, but the second components are distinct, say, $z_{2i} = y_r$ and $z_{2j} = y_s$, where $r \neq s$ (note that the cases $z_{2i-1} = x_1 = z_{2j}$ and $z_{2i} = x_1 = z_{2j-1}$ are impossible, since the representation (7.20) of $g - h$ is reduced). This implies that

$$|v| \cdot d(y_r, y_s) \leq \varrho(y_r, y_s) \leq \varrho(y_r, x_1) + \varrho(x_1, y_s) < 1/2 + 1/2 = 1.$$

However, $y_r$ and $y_s$ are distinct letters in the normal form of the element $h \in F_1 \cap \varphi_v(X^n) \subset F_0$, which contradicts (7.17). Similarly, if $i \leq n$ and (7.20) contains two summands $(x_r - y_i)$ and $(x_s - y_i)$ (or the summands $(y_i - x_r)$ and $(y_i - x_s)$), then $r = s$. In addition, the representation (7.20) of $g - h$ is reduced, so it contains both summands $(x_1 - y_i)$ and $(y_j - x_1)$ for no $i, j \leq n$. This shows that the right part of (7.20) contains $|k_1|$ equal summands of the form $(x_1 - y_i)$ or $(y_i - x_1)$ for some $i \leq n$, and the rest of the summands contain neither $x_1$ nor $y_i$. So, we conclude that $k_1 = k_i$.

Consider $g_1 = k_2x_2 + \cdots + k_nx_n$ and $h_1 = k_1y_1 + \cdots + k_{i-1}y_{i-1} + k_{i+1}y_{i+1} + \cdots + k_ny_n$. From (7.21) it follows that $\widehat{\varrho}_A(g_1, h_1) \leq \widehat{\varrho}_A(g, h) < 1/2$. Therefore, the above argument applied to $g_1$ and $h_1$ implies the existence of an index $j \leq n$ with $j \neq i$ such that the right part of (7.20) contains $|k_2|$ equal summands of the form $(x_2 - y_j)$ or $(y_j - x_2)$. Consequently, $k_2 = k_j$, and the rest of the summands in (7.20) do not contain either $x_2$ or $y_j$. Continuing in this way, we finally obtain a permutation $\pi \in S_n$ such that $k_i = k_{\pi(i)}$ and the right part of (7.20) contains $|k_i|$ equal summands of the form $(x_i - y_{\pi(i)})$ or $(y_{\pi(i)} - x_i)$ for each $i = 1, \ldots, n$. This implies (7.18), and then (7.19) follows from (7.21). Claim A is proved.

Let $\varrho = |v| \cdot d$. Since $\xi$ is a Cauchy filter in the subspace $\varphi_v(X^n)$ of $(A(X), \mathscr{V})$, we can find $F_1 \in \xi$ such that $F_1 \subset F_0$ and $F_1 - F_1 \subset V_{3\varrho}$. Fix $g^* = k_1x_1^* + \cdots + k_nx_n^* \in F_1$ where $x_1^*, \ldots, x_n^* \in X$, and for every $i \leq n$ put

$$P_i = \{x \in X : \varrho(x_i^*, x) \leq 1/3\}.$$

Then $P = P_1 \times \cdots \times P_n$ is a closed subset of $X^n$.

**Claim B.** *The set $\varphi_v(P)$ contains $F_1$. In addition, if $a, b \in P \cap \varphi_v^{-1}(F_1)$ and $a = (x_1, \ldots, x_n)$, $b = (y_1, \ldots, y_n)$, then $\varrho(x_i, y_j) \geq 1/3$, for any distinct $i, j \leq n$.*

Indeed, from $F_1 - F_1 \subset V_{3\varrho}$ it follows that $\widehat{\varrho}_A(g^*, h) < 1/3$, for each $h = k_1y_1 + \cdots + k_ny_n \in F_1$, where $y_1, \ldots, y_n \in X$. By Claim A, there exists a permutation $\pi \in S_n$ such that $\widehat{\varrho}_A(g^*, h) = \sum_{i=1}^{n} |k_i|\varrho(x_i^*, y_{\pi(i)})$ and $k_i = k_{\pi(i)}$ for each $i = 1, \ldots, n$.

Since $k_i \neq 0$, we deduce that $\varrho(x_i^*, y_{\pi(i)}) \leq \widehat{\varrho}_A(g^*, h) < 1/3$ and, hence, $y_{\pi(i)} \in P_i$ for each $i \leq n$. This implies that $c = (y_{\pi(1)}, \dots, y_{\pi(n)}) \in P$ and

$$\varphi_v(c) = \sum_{i=1}^n k_i y_{\pi(i)} = \sum_{i=1}^n k_{\pi(i)} y_{\pi(i)} = \sum_{i=1}^n k_i y_i = h.$$

So, $F_1 \subset \varphi_v(P)$.

Suppose that $a, b \in P \cap \varphi_v^{-1}(F_1)$, where $a = (x_1, \dots, x_n)$ and $b = (y_1, \dots, y_n)$. From $x_j, y_j \in P_j$ it follows that

$$\varrho(x_j, y_j) \leq \varrho(x_j, x_j^*) + \varrho(x_j^*, y_j) \leq 1/3 + 1/3 = 2/3 \tag{7.22}$$

for $j = 1, \dots, n$. Suppose that $\varrho(x_i, y_j) < 1/3$, for distinct $i, j \leq n$. Then (7.22) implies that

$$\varrho(x_i, x_j) \leq \varrho(x_i, y_j) + \varrho(y_j, x_j) < 1/3 + 2/3 = 1.$$

Since $\varrho = |v| \cdot d$, this contradicts (7.17). The proof of Claim B is complete.

Our next step is to refine Claim A.

**Claim C.** *Let $m$ be a continuous pseudometric on $X$ with $m \geq \varrho$, and suppose that $F \in \xi$ satisfies $F \subset F_1$ and $F - F \subset V_{3m}$. If $a, b \in P \cap \varphi_v^{-1}(F)$ and $a = (x_1, \dots, x_n)$, $b = (y_1, \dots, y_n)$, then*

$$\widehat{m}_A(\varphi_v(a), \varphi_v(b)) = \sum_{i=1}^n |k_i| \cdot m(x_i, y_i) < 1/3. \tag{7.23}$$

Indeed, suppose that $a, b \in P \cap \varphi_v^{-1}(F)$, where $a = (x_1, \dots, x_n)$ and $b = (y_1, \dots, y_n)$. Let $g = \varphi_v(a)$ and $h = \varphi_v(b)$. By Claim A, there exists a permutation $\sigma \in S_n$ such that

$$\widehat{m}_A(g, h) = \sum_{i=1}^n |k_i| \cdot m(x_i, y_{\sigma(i)}). \tag{7.24}$$

From $g, h \in F$ and $F - F \subset V_{3m}$ it follows that $\widehat{m}_A(g, h) < 1/3$. In particular,

$$\varrho(x_i, y_{\sigma(i)}) \leq m(x_i, y_{\sigma(i)}) \leq \widehat{m}_A(g, h) < 1/3,$$

for each $i \leq n$. On the other hand, Claim B implies that $\varrho(x_i, y_j) \geq 1/3$ whenever $i \neq j$. Therefore, $\sigma(i) = i$ for each $i \leq n$ and, hence, (7.23) follows from (7.24). Claim C is proved.

**Claim D.** *The family $\lambda = \{P \cap \varphi_v^{-1}(F) : F \in \xi\}$ is a base of a Cauchy filter $\eta$ in the uniform space $(X^n, \mathcal{U}^n)$, where $\mathcal{U}^n$ is the $n$-fold product of the universal uniformity $\mathcal{U}$ on $X$.*

Indeed, from Claim B it follows that $F_1 \subset \varphi_v(P)$, so that $\lambda$ is a base of a filter $\eta$. Let $U$ be an element of $\mathcal{U}^n$. There exists a pseudometric $m \in \mathcal{P}_X$ such that $m \geq \varrho$ and

$$\{(x_1, \dots, x_n, y_1, \dots, y_n) \in X^{2n} : m(x_i, y_i) < 1 \text{ for each } i \leq n\} \subset U.$$

Choose an element $F \in \xi$ such that $F \subset F_1$ and $F - F \subset V_{3m}$. Then $F' = P \cap \varphi_v^{-1}(F) \in \eta$. Suppose that $a = (x_1, \dots, x_n) \in F'$ and $b = (y_1, \dots, y_n) \in F'$. By Claim C,

$$m(x_i, y_i) \leq |k_i| \cdot m(x_i, y_i) \leq \widehat{m}_A(g, h) < 1/3 < 1$$

for each $i \leq n$. Consequently, our choice of $m$ implies that $(a, b) \in U$; therefore, $F' \times F' \subset U$. So, $\eta$ is a Cauchy filter in $(X^n, \mathcal{U}^n)$. This proves Claim D.

Finally, the uniform space $(X, \mathcal{U}_X)$ is complete, since $X$ is Dieudonné complete. Hence the $n$-th power $(X^n, \mathcal{U}^n)$ of the space $(X, \mathcal{U})$ is also complete. Since $P$ is closed in $X^n$, we can apply Claim D to conclude that the filter $\eta$ with the base $\lambda = \{P \cap \varphi_v^{-1}(F) : F \in \xi\}$ converges to a point $x^* = (x_1^*, \dots, x_n^*) \in P$. By Claim B, the filter $\xi$ contains the family $\varphi_v(\lambda) = \{\varphi_v(P) \cap F : F \in \xi\}$ and hence, $\xi$ converges to the element $\varphi_v(x^*) = k_1 x_1^* + \dots + k_n x_n^* \in \varphi_v(X^n)$. Thus, the subspace $\varphi_v(X^n)$ of $(A(X), \mathcal{V})$ is complete. □

LEMMA 7.9.5. *Let $\mathcal{V}$ be the group uniformity on $A(X)$, where $X$ is a Dieudonné complete space. Then the uniformity $\mathcal{V}_n = \mathcal{V} \restriction B_n(X)$ is complete, for each $n \in \mathbb{N}$.*

PROOF. Let $\xi$ be a Cauchy filter in the uniform space $(B_n(X), \mathcal{V}_n)$. Clearly,

$$B_n(X) = \bigcup \{\varphi_v(X^m) : m \leq n,\ v \in \mathbb{Z}^m,\ |v| \leq n\}. \tag{7.25}$$

Since the number of summands on the right side of (7.25) is finite, one can find $m \leq n$ and $v \in \mathbb{Z}^m$ with $|v| \leq n$ such that $F \cap \varphi_v(X^m) \neq \varnothing$, for each $F \in \xi$. By Lemma 7.9.4, $\varphi_v(X^m)$ is a complete subspace of the space $(B_n(X), \mathcal{V}_n)$. Therefore, $\xi$ converges to a point $g \in \varphi_v(X^m) \subset B_n(X)$. □

THEOREM 7.9.6. [**M. G. Tkachenko**] *The free Abelian topological group $A(X)$ is Raĭkov complete iff the space $X$ is Dieudonné complete.*

PROOF. By Lemma 7.7.1, the group uniformity $\mathcal{V}$ of $A(X)$ induces on $X$ its universal uniformity $\mathcal{U}$. In addition, $X$ is closed in $A(X)$, by item a) of Theorem 7.1.13. Therefore, if the group $A(X)$ is Raĭkov complete, then the uniform subspace $(X, \mathcal{U})$ of $(A(X), \mathcal{V})$ is also complete. In other words, the space $X$ is Dieudonné complete.

Suppose that a Cauchy filter $\xi$ in $(A(X), \mathcal{V})$ does not converge, where $X$ is Dieudonné complete space. For every $n \in \mathbb{N}$, denote by $\mathcal{V}_n$ the uniformity on

$$B_n = \{g \in A(X) : l(g) \leq n\}$$

inherited from $(A(X), \mathcal{V})$. By Lemma 7.9.5, the uniform space $(B_n, \mathcal{V}_n)$ is complete. Hence one can apply Lemma 7.9.3 to find an $F_1 \in \xi$ and a neighbourhood $V_1$ of the neutral element $e$ in $A(X)$ such that $(F_1 + V_1 + V_1) \cap B_1 = \varnothing$.

Suppose that for some $n \in \mathbb{N}$ we have defined elements $F_1, \dots, F_n$ of $\xi$ and symmetric neighbourhoods $V_1, \dots, V_n$ of $e$ in $A(X)$ satisfying the following conditions:

(i) $F_{i+1} \subset F_i$ if $i < n$;
(ii) $V_{i+1}^2 \subset V_i$ if $i < n$;
(iii) $(F_i + V_i + V_i) \cap B_{i^2} = \varnothing$, for each $i \leq n$.

Then we choose an element $F_{n+1} \in \xi$ and a symmetric neighbourhood $V_{n+1}$ of $e$ in $A(X)$ such that $F_{n+1} \subset F_n$, $V_{n+1}^2 \subset V_n$ and $(F_{n+1} + V_{n+1} + V_{n+1}) \cap B_{(n+1)^2} = \varnothing$. This is possible because the space $(B_{n+1}, \mathcal{V}_{n+1})$ is complete by Lemma 7.9.3.

Suppose that the sequences $\{F_n : n \in \mathbb{N}\}$ and $\{V_n : n \in \mathbb{N}\}$ satisfying (i)–(iii) have been defined. For every $n \in \mathbb{N}$, put $W_n = V_n \cap B_{2n}$ and consider the set

$$V = W_1 + W_2 + \dots + W_n + \cdots.$$

We claim that the following conditions are fulfilled:

(a) $(F_n + V) \cap B_n = \varnothing$ for each $n \in \mathbb{N}$;
(b) $V$ is a symmetric neighbourhood of $e$ in $A(X)$.

Let us verify (a). First, from (ii) it easily follows that

$$V_{i+1} + V_{i+2} + \cdots + V_{i+k} \subset V_i$$

for all $i, k \in \mathbb{N}$. Hence,

$$V_i + V_{i+1} + \cdots + V_{i+k} \subset V_i + V_i.$$

Since $W_j \subset V_j$ for each $j \in \mathbb{N}$, the above inclusion implies that

$$W_i + W_{i+1} + \cdots + W_{i+k} \subset V_i + V_i \tag{7.26}$$

whenever $i, k \in \mathbb{N}$. In particular, $V \subset V_1 + V_1$, so from (iii) (with $i = 1$) it follows that $(F_1 + V) \cap B_1 = \varnothing$. Assume that (a) does not hold. Then $(F_{n+1} + V) \cap B_{n+1} \neq \varnothing$, for some integer $n \geq 1$. The definition of $V$ implies that

$$(F_{n+1} + W_1 + \cdots + W_m) \cap B_{n+1} \neq \varnothing \tag{7.27}$$

for some $m \in \mathbb{N}$. We can assume that $m > n$. Note that $W_i \subset B_{2i}$, for each $i = 1, \ldots, n$ and $\sum_{i=1}^{n} 2i = n(n+1)$. Therefore,

$$W_1 + W_2 + \cdots + W_n \subset B_{n(n+1)}. \tag{7.28}$$

From (7.26) and (7.28) it follows that

$$F_{n+1} + W_1 + \cdots + W_n + W_{n+1} + \cdots + W_m \subset F_{n+1} + B_{n(n+1)} + V_{n+1} + V_{n+1}.$$

This inclusion and (7.27) together imply that

$$((F_{n+1} + V_{n+1} + V_{n+1}) + B_{n(n+1)}) \cap B_{n+1} \neq \varnothing.$$

Hence, $(F_{n+1} + V_{n+1} + V_{n+1}) \cap B_{(n+1)^2} \neq \varnothing$, which contradicts (iii). Thus, (a) holds.

Let us verify (b). Note that $W_n^{-1} = (V_n \cap B_{2n})^{-1} = V_n^{-1} \cap B_{2n}^{-1} = V_n \cap B_{2n}$, for each $n \in \mathbb{N}$, so that $V$ is symmetric as the sum of symmetric sets $W_n$. For every $n \in \mathbb{N}$, set $p(n) = 2^n + n$ and

$$U_n = \{(x, y) \in X \times X : x - y \in V_{p(n)}\}. \tag{7.29}$$

By [**165**, Theorem 8.1.10], there exists a continuous pseudometric $\varrho$ on $X$ such that

$$\{(x, y) \in X \times X : \varrho(x, y) < 2^{-n}\} \subset U_n, \tag{7.30}$$

for each $n \in \mathbb{N}$. Clearly, (b) will follow if we show that $V_{2\varrho} \subset V$.

Suppose that $g$ is an arbitrary element of $V_{2\varrho}$. Then $\widehat{\varrho}_A(g, e) < 1/2$ and Corollary 7.2.4 implies that there exists an reduced representation

$$g = (x_1 - y_1) + \cdots + (x_m - y_m) \tag{7.31}$$

with $x_i, y_i \in X$, such that

$$\widehat{\varrho}_A(g, e) = \sum_{i=1}^{m} \varrho(x_i, y_i). \tag{7.32}$$

For every $i \leq m$ with $\varrho(x_i, y_i) \neq 0$, let $k_i \geq 1$ be an integer such that $2^{-k_i - 1} \leq \varrho(x_i, y_i) < 2^{-k_i}$. Apply (7.32) to find sufficiently large numbers $k_i \in \mathbb{N}$ for the rest of indices $i \leq m$ such that $\sum_{i=1}^{m} 2^{-k_i} < 2 \cdot \widehat{\varrho}_A(g, e) < 1$. Then, for every $j \in \mathbb{N}$, the sum $\sum_{i=1}^{m} 2^{-k_i}$ contains less than $2^j$ summands equal to $2^{-j}$. Let $k_{i_1}, \ldots, k_{i_r}$ be the list of all $k_i$ satisfying

$k_i = j$; then $r < 2^j$. From (7.30), (7.29), and the choice of the numbers $k_i$ it follows that $x_{i_1} - y_{i_1} \in V_{p(j)}, \ldots, x_{i_r} - y_{i_r} \in V_{p(j)}$. In addition, a simple calculation making use of (ii) shows that the sum $V_{p(j)} + \cdots + V_{p(j)}$ ($2^j$ times) is contained in $V_{2^j}$. Therefore,

$$(x_{i_1} - y_{i_1}) + \cdots + (x_{i_r} - y_{i_r}) \in B_{2^j} \cap (\underbrace{V_{p(j)} + \cdots + V_{p(j)}}_{r \text{ times}}) \subset B_{2^j} \cap V_{2^j} = W_{2^j}.$$

This inclusion is valid for each $j \leq k = \max\{k_1, \ldots, k_m\}$. So, we have

$$g = (x_1 - y_1) + \cdots + (x_m - y_m) \in W_2 + W_4 + \cdots + W_{2^k} \subset V.$$

Hence, $V_{2_\varrho} \subset V$ and (b) holds.

Finally, since $\xi$ is a Cauchy filter in $(A(X), \mathcal{V})$, there is $F \in \xi$ such that $F - F \subset V$. Pick $g \in F$. Then $g \in B_n$, for some $n \in \mathbb{N}$; therefore, $F \subset g + V \subset B_n + V$. However, $(F_n + V) \cap B_n = \varnothing$ by (a) or, equivalently, $F_n \cap (B_n + V) = \varnothing$, by the symmetry of the set $V$. Consequently, $F \cap F_n = \varnothing$. This contradicts the fact that both $F$ and $F_n$ belong to the filter $\xi$. □

Corollary 7.9.7. *If $X$ is metrizable or paracompact, then the free Abelian topological group $A(X)$ is Raĭkov complete.*

Proof. Each metrizable space is paracompact and all paracompact spaces are Dieudonné complete [**255**]. Therefore, the conclusion follows from Theorem 7.9.6. □

It turns out that the completion of the group $A(X)$ for an arbitrary space $X$ has a similar structure.

Theorem 7.9.8. *The Raĭkov completion of $A(X)$ is topologically isomorphic to $A(\mu X)$, for every space $X$, where $\mu X$ is the Dieudonné completion of $X$.*

Proof. Clearly, $X$ is $P$-embedded in $\mu X$. Therefore, Theorem 7.7.4 implies that the natural monomorphism $\hat{e}\colon A(X) \to A(\mu X)$ is a topological embedding, where $e\colon X \to \mu X$ is the identity embedding. As $X$ is dense in $\mu X$, the subgroup $A(X, \mu X) = \hat{e}(A(X))$ of $A(\mu X)$ is dense in $A(\mu X)$. In addition, the group $A(\mu X)$ is Raĭkov complete, by Theorem 7.9.6. It follows that the Raĭkov completion of $A(X)$ is topologically isomorphic to $A(\mu X)$. □

## Exercises

7.9.a. Extend Theorem 7.9.6 to free uniform Abelian groups as follows: If a uniform space $(X, \mathcal{U})$ is complete, then the group $A(X, \mathcal{U})$ is Raĭkov complete.

7.9.b. Apply Exercise 7.7.c to generalize Theorem 7.9.8 as follows. If $(X, \mathcal{U})$ is a uniform space, then the Raĭkov completion of the free uniform Abelian group $A(X, \mathcal{U})$ is topologically isomorphic to $A(\widehat{X}, \widehat{\mathcal{U}})$, where $(\widehat{X}, \widehat{\mathcal{U}})$ is the completion of $(X, \mathcal{U})$.

## Problems

7.9.A. Let $X$ be a non-discrete Dieudonné complete first-countable space such that the group $F(X)$ has the direct limit property. Is $X$ then $\sigma$-compact?

7.9.B. Show that every Abelian topological group is a quotient group of a Raĭkov complete Abelian topological group.

7.9.C. Let $X$ be a Tychonoff space. Is $A(X)$ $C$-embedded in $A(\mu(X))$?

7.9.D. Is the free Abelian topological group $A(X)$ of the Alexandroff one-point compactification of a discrete space of cardinality $\aleph_1$ hereditarily realcompact?

7.9.E. (V. V. Uspenskij [**520**]) Suppose that $X$ is the topological product of a family of metrizable spaces. Show that $F(X)$ is complete with respect to the left uniformity of $F(X)$ (that is, $F(X)$ is Weil complete).

7.9.F. (V. V. Uspenskij [**520**]) Suppose that $X$ is a Dieudonné complete pseudo-$\aleph_1$-compact Tychonoff space. Show that $F(X)$ is Weil complete.

7.9.G. Let $X$ be a Tychonoff space such that the subspace $B_n(X)$ of $F(X)$ is locally compact, for each $n \in \omega$.

(a) (V. G. Pestov and K. Yamada [**382**]). Prove that if $X$ is not discrete, then it is pseudocompact.

(b) (P. Nickolas and M. G. Tkachenko [**347**]). Apply (a) to show that $X$ is either discrete or compact.

*Hint.* (a) Suppose that $X$ is neither discrete nor pseudocompact. Choose a discrete family $\{U_n : n \in \omega\}$ of non-empty open sets in $X$ and a set $\{x_n : n \in \omega\}$ such that $x_n \in U_n$ for each $n \in \omega$. Since $X$ is non-discrete and locally compact, there exists an infinite compact set $C \subset X$ with a non-isolated point $a \in C$. It is not restrictive to assume that $C \cap U_n = \varnothing$ for each $n \in \omega$. Then $Z = C \cup \{x_n : n \in \omega\}$ is a closed $C$-embedded subset of $X$. Note that $Z$ is a $\mu$-space and apply Corollary 7.6.32 to conclude that $F(Z) \cong F(Z, X)$. Deduce that $F(Z)$ is a $k$-space.

Put $C_n = x_n^{-1}a^{-1}Cx_n$ for each $n \in \omega$ and let $Y = \bigcup_{n=0}^{\infty} C_n$. Then $Y \subset B_4(Z)$ and $C_n \cap C_m = \{e\}$ for distinct $n, m \in \omega$, where $e$ is the identity of $F(Z)$. Verify that for every compact subset $K$ of $F(Z)$ there is $n \in \omega$ such that $K \subset F(C \cup E_n, Z)$, where $E_n = \{x_k : k \le n\}$. Conclude that

$$K \cap Y \subset F(C \cup E_n, Z) \cap Y = \bigcup_{k \le n} C_k$$

is a compact subset of $Y$, so that $Y$ is closed in $F(Z)$ and, hence, is a $k$-space. Therefore, $Y$ is the inductive limit of the compact sets $C_n$, $n \in \omega$. Apply this fact to show that $Y$ fails to be locally compact at the point $e$, thus contradicting the fact that $Y$ is a closed subspace of $B_4(Z)$.

(b) Suppose that $X$ is not discrete and apply (a) to conclude that it is pseudocompact. Denote by $\beta X$ the Čech–Stone compactification of $X$. By Theorem 7.7.3, $F(X)$ can be identified with the subgroup $F(X, \beta X)$ of $F(\beta X)$. Then $B_2(X)$ is a dense locally compact subspace of $B_2(\beta X)$ and, hence, it is open in $B_2(\beta X)$.

Note that the mapping $j_2 \colon (\beta X)^2 \to B_2(\beta X)$ defined by $j_2(x, y) = x \cdot y^{-1}$ for $x, y \in \beta X$ is perfect. Therefore, the preimage $j_2^{\leftarrow}(B_2(X)) = X^2 \cup \Delta$ is open in $(\beta X)^2$, where $\Delta = \{(y, y) : y \in \beta X\}$ is the diagonal in $(\beta X)^2$. Show that this implies the equality $X = \beta X$.

7.9.H. Is every topological group $G$ a topological quotient of a Raĭkov complete topological group? *Hint.* See [**452**].

7.9.I. Give an example of a Tychonoff space $X$ such that $F(X)$ is not $C$-embedded in $F(\mu(X))$. *Hint.* See [**42**, Theorem 7.26].

## Open Problems

7.9.1. Is the free topological group $F(X)$ of a pseudo-$\aleph_1$-compact space a $PT$-group?

7.9.2. Is the free (Abelian) topological group of each first-countable Tychonoff space Moscow?

7.9.3. Is the free (Abelian) topological group of a first-countable Tychonoff space a $PT$-group?

7.9.4. When is $F(X)$ $C$-embedded in $F(\mu(X))$? What if $X$ is first-countable? (See Problem 7.9.I.)

7.9.5. Characterize the Tychonoff spaces $X$ such that the free (Abelian) topological group over $X$ is hereditarily Dieudonné complete.

7.9.6. Let $FPG(X)$ be the free Tychonoff paratopological group over a Dieudonné complete space $X$. Suppose further that $FPG(X)$ is topologically isomorphic to a subgroup $H$ of a (Tychonoff or Hausdorff) paratopological group $G$. Is $H$ closed in $G$?

## 7.10. $M$-equivalent spaces

In this section we assume all spaces considered to be Tychonoff if nothing to the contrary is specified. Suppose that the free topological groups $F(X)$ and $F(Y)$ are topologically isomorphic. Are then the spaces $X$ and $Y$ homeomorphic? It turns out that there is a wealth of non-homeomorphic spaces $X$ and $Y$ with topologically isomorphic groups $F(X)$ and $F(Y)$ (see Examples 7.10.5 and 7.10.18). This makes it natural considering the relations of *$M$-equivalence* and *$A$-equivalence* in Tychonoff spaces defined as follows.

Two spaces $X$ and $Y$ are called *$M$-equivalent* if the groups $F(X)$ and $F(Y)$ are topologically isomorphic. Similarly, $X$ and $Y$ are said to be *$A$-equivalent* if the free Abelian topological groups $A(X)$ and $A(Y)$ are topologically isomorphic. A topological property $\mathcal{P}$ is called *$M$-invariant* (*$A$-invariant*) if every space $Y$ $M$-equivalent (respectively, $A$-equivalent) to a space $X$ with $\mathcal{P}$ also has the property $\mathcal{P}$.

It is clear that the relations of $M$- and $A$-equivalence are reflexive, symmetric and transitive, so they are equivalence relations. First, we show that $M$-equivalent spaces are always $A$-equivalent.

Proposition 7.10.1. *$M$-equivalent spaces $X$ and $Y$ are $A$-equivalent.*

Proof. Suppose that there exists a topological isomorphism $\varphi\colon F(X) \to F(Y)$. Denote by $K_X$ and $K_Y$ the derived subgroups of $F(X)$ and $F(Y)$, respectively. Clearly, $\varphi(K_X) = K_Y$. By Theorem 7.1.11, $A(X) \cong F(X)/K_X$ and $A(Y) \cong F(Y)/K_Y$. Therefore, $A(X) \cong A(Y)$ and hence, $X$ and $Y$ are $A$-equivalent. □

Let us show that connectedness is both $M$- and $A$-invariant. By Proposition 7.10.1, it suffices to verify this only for $A$-equivalence. Given a space $X$, we put

$$A_0(X) = \{x_1 + \cdots + x_n - y_1 - \cdots - y_n \in A(X) : x_1, y_1, \ldots, x_n, y_n \in X, n \in \mathbb{N}\}.$$

Let us establish several useful properties of the subgroup $A_0(X)$ and relate $A_0(X)$ to the connected component $C_X$ of the neutral element in $A(X)$. To this end, we recall one useful concept from the theory of groups.

Suppose that $A$ is a non-empty subset of an Abelian group $G$ with neutral element $e$, and that each element of $A$ has infinite order. The set $A$ is called *linearly independent* or simply *independent* if for all distinct elements $a_1, \ldots, a_n$ of $A$ and integers $m_1, \ldots, m_n$, the equality $m_1a_1 + \cdots + m_na_n = e$ implies that $m_i = 0$, for each $i \leq n$.

An application of Zorn's lemma shows that every independent subset of $G$ is contained in a maximal independent subset. The *torsion-free rank* $r_0(G)$ of $G$ is the cardinality of a maximal independent subset of $G$. An argument very close to that in [**409**, 4.2.1] shows that the cardinalities of every two maximal independent subsets of $G$ coincide, so our definition of $r_0(G)$ is correct.

Lemma 7.10.2. *Let $X$ be a space. Then:*

a) *the subgroup $A_0(X)$ of $A(X)$ is open and the quotient group $A(X)/A_0(X)$ is isomorphic to the discrete group $\mathbb{Z}$;*
b) *if $X$ is disconnected, then the torsion-free rank of the group $A(X)/C_X$ is greater than or equal to 2;*
c) *the group $A_0(X)$ is connected iff $X$ is connected;*
d) *if $X$ is connected, then $A_0(X)$ coincides with the connected component $C_X$ of the neutral element in $A(X)$.*

PROOF. a) Let $f\colon X \to \mathbb{Z}$ be the constant mapping, $f(x) = 1$ for each $x \in X$. Extend $f$ to a continuous homomorphism $\varphi\colon A(X) \to \mathbb{Z}$. Suppose that $n, m \in \mathbb{N}$ and $g \in A(X)$,

$$g = x_1 + \cdots + x_n - y_1 - \cdots - y_m,$$

where $x_i, y_j \in X$ for each $i \le m$. Then $\varphi(g) = n - m$, so the kernel of $\varphi$ coincides with the subgroup $A_0(X)$ of $A(X)$. Since $\mathbb{Z}$ is discrete, $A_0(X) = \varphi^{-1}(0)$ is an open invariant subgroup of $A(X)$. In addition, $\varphi(A(X)) = \mathbb{Z}$, so that $A(X)/A_0(X) \cong \mathbb{Z}$.

b) Suppose that the space $X$ is disconnected, $X = A \cup B$, where $A$ and $B$ are disjoint non-empty closed subsets of $X$. Let $D = \{a, b\}$ be a discrete two-point space. Define a mapping $f\colon X \to D$ by $f(x) = a$ if $x \in A$ and $f(x) = b$ if $x \in B$. Then $f$ is continuous, so it can be extended to a continuous homomorphism $\hat{f}\colon A(X) \to A(D)$. Since $A(D)$ is discrete, the kernel $N$ of $\hat{f}$ is an open and closed subgroup of $A(X)$. In particular, the component $C_X$ of the neutral element $e$ in $A(X)$ is contained in $N$. Therefore, there exists a homomorphism $h\colon A(X)/C_X \to A(D)$ such that $h \circ \pi = \hat{f}$, where $\pi\colon A(X) \to A(X)/C_X$ is the quotient homomorphism. Clearly, $\hat{f}$ and $h$ are epimorphisms, so the torsion-free rank of $A(X)/C_X$ is not less than the torsion-free rank of $A(D)$, while the latter is equal to 2.

c) Suppose that $X$ is connected. For $n \in \mathbb{N}$, consider the mapping $j_n\colon X^{2n} \to A(X)$, where

$$j_n(x_1, \ldots, x_n, y_1, \ldots, y_n) = x_1 + \cdots + x_n - y_1 - \cdots - y_n,$$

for each point $(x_1, \ldots, x_n, y_1, \ldots, y_n) \in X^{2n}$. Clearly, $j_n$ is continuous and $j_n(X^{2n})$ is a connected subspace of $A_0(X)$ containing the neutral element $e$ of $A(X)$. Since $A_0(X)$ is the union of the sets $j_n(X^{2n})$, where $n \in \mathbb{N}$, the group $A_0(X)$ is connected.

Conversely, suppose that $X$ is disconnected. Then there are non-empty closed disjoint subsets $A$ and $B$ of $X$ such that $X = A \cup B$. As in (b), consider the mapping $f\colon X \to D = \{a, b\}$ defined by $f(x) = a$ if $x \in A$ and $f(x) = b$ if $x \in B$. Then $f$ admits an extension to a continuous homomorphism $\hat{f}\colon A(X) \to A(D)$. Since $A(D)$ is discrete, the kernel $N$ of $\hat{f}$ is a clopen subgroup of $A(X)$. Choose points $x \in A$, $y \in B$ and consider the element $g = 2x - 2y \in A_0(X)$. The image $\hat{f}(g) = 2a - 2b$ is distinct from the neutral element of $A(D)$ and hence, $g \notin N$. Therefore, $A_0(X) \cap N$ is a proper clopen subgroup of $A_0(X)$. Thus, $A_0(X)$ is disconnected.

d) Let $X$ be connected. It follows from a) and c) that $A_0(X)$ is an open connected subgroup of $A(X)$, so that $A_0(X)$ is the connected component of $e$ in $A(X)$. □

THEOREM 7.10.3. *The relations of $M$- and $A$-equivalence preserve connectedness.*

PROOF. Let $X$ and $Y$ be $A$-equivalent spaces, and suppose that $X$ is connected. Then there exists a topological isomorphism $\varphi\colon A(X) \to A(Y)$. Denote by $C_X$ and $C_Y$ the connected components of zero in the groups $A(X)$ and $A(Y)$, respectively. Clearly, $\varphi(C_X) = C_Y$, and Lemma 7.10.2 implies that the quotient group $A(X)/C_X$ is isomorphic to

$\mathbb{Z}$. Therefore, $A(Y)/C_Y \cong A(X)/C_X \cong \mathbb{Z}$. In particular, the torsion-free rank of the quotient group $A(Y)/C_Y$ is equal to one, so b) of Lemma 7.10.2 implies that $Y$ is connected. □

Clearly, two homeomorphic spaces are $M$-equivalent, but the converse is false as we shall see in Example 7.10.5. Below we present the first general method of obtaining $M$-equivalent spaces.

LEMMA 7.10.4. *Let $X = X_1 \oplus X_2$ be the topological sum of non-empty spaces $X_1$ and $X_2$. Then the subspaces $Y_1 = b_1a^{-1}X_1 \cup X_2$ and $Y_2 = b_2a^{-1}X_1 \cup X_2$ of $F(X)$ are $M$-equivalent, for all $a \in X_1$ and $b_1, b_2 \in X_2$.*

PROOF. Clearly, $a^{-1}X_1 \cup X_2 \subset \langle Y_i \rangle$ and $Y_i \subset \langle a^{-1}X_1 \cup X_2 \rangle$ for $i = 1, 2$, so $\langle Y_1 \rangle = \langle Y_2 \rangle$. Therefore, it remains to show that $\langle Y_1 \rangle \cong F(Y_1)$ and $\langle Y_2 \rangle \cong F(Y_2)$. By the symmetry argument, it suffices to verify the former relation.

The identity mapping of $Y_1$ onto $Y_1$ can be extended to a continuous monomorphism $h\colon F(Y_1) \to F(X)$. Suppose that $f\colon Y_1 \to G$ is a continuous mapping of $Y_1$ to an arbitrary topological group $G$. We claim that $f$ admits an extension to a continuous homomorphism of $F(X)$ to $G$. Indeed, let $g\colon X \to G$ be the mapping defined by $g(x) = f(x)$ for each $x \in X_2$ and $g(x) = f(b_1a^{-1}x)$, for each $x \in X_1$. Since $X = X_1 \oplus X_2$, the mapping $g$ is continuous. From our definition of $g$ it follows that $g(b_1) = f(b_1)$. In addition, if $x = a \in X_1$, then $g(a) = f(b_1a^{-1}a) = f(b_1)$. In particular, $g(a) = g(b_1)$. Let $\tilde{g}\colon F(X) \to G$ be a continuous homomorphism extending $g$. Then the restriction of $\tilde{g}$ to $Y_1$ coincides with $f$. Indeed, the equalities $\tilde{g}(x) = g(x) = f(x)$ are evident for each $x \in X_2$. If $x \in X_1$, then from $g(a) = g(b_1)$ it follows that

$$\tilde{g}(b_1a^{-1}x) = g(b_1) \cdot g(a)^{-1} \cdot g(x) = f(b_1a^{-1}x)$$

and, hence, $\tilde{g}\restriction Y_1 = f$. This proves our claim.

We can now apply the claim to the natural embedding mapping of $Y_1$ to $G = F(Y_1)$ and extend it to a continuous homomorphism $\tilde{g}\colon F(X) \to F(Y_1)$. Then the restriction $g^*$ of $\tilde{g}$ to $\langle Y_1 \rangle$ satisfies $g^* \circ h = id_{F(X)}$ and $h \circ g^* = id_{\langle Y_1 \rangle}$. Since the homomorphisms $h$ and $g^*$ are continuous, $h$ must be a topological isomorphism of $F(Y_1)$ onto the subgroup $\langle Y_1 \rangle$ of $F(X)$. Similarly, the subgroup $\langle Y_2 \rangle$ of $F(X)$ is topologically isomorphic to $F(Y_2)$. Since $\langle Y_1 \rangle = \langle Y_2 \rangle$, we finally conclude that the spaces $Y_1$ and $Y_2$ are $M$-equivalent. □

The above lemma enables us to present two non-homeomorphic $M$-equivalent compact subspaces of the Euclidean plane $\mathbb{R}^2$.

EXAMPLE 7.10.5. The closed unit interval and the letter $T$ considered as a subspace of the plane $\mathbb{R}^2$ are $M$-equivalent but not homeomorphic.

Indeed, let $X = X_1 \oplus X_2$, where $X_1 = [0, 1]$ and $X_2 = [2, 3]$ are subspaces of $\mathbb{R}$. Let also $b_1$ and $b_2$ be an end point and an interior point of the interval $X_2$, respectively. By Lemma 7.10.4, the subspaces $Y_1 = b_1a^{-1}X_1 \cup X_2$ and $Y_2 = b_2a^{-1}X_1 \cup X_2$ of $F(X)$ are $M$-equivalent for each point $a \in X_1$. For $i = 1, 2$, consider the mapping $f_i\colon X \to Y_i$ defined by $f_i(x) = x$ if $x \in X_2$ and $f_i(x) = b_ia^{-1}x$ for $x \in X_1$. Clearly, $f_i$ is continuous and $f_i(X) = Y_i$. In addition, the only non-trivial fiber of the mapping $f_i$ is $f_i^{-1}(f_i(a)) = \{a, b_i\}$, $i = 1, 2$. Since $X$ is compact and $Y_i$ is Hausdorff, this implies that the space $Y_i$ is obtained from the space $X$ by identifying the points $a$ and $b_i$. Therefore, $Y_1$ is homeomorphic with a closed interval of $\mathbb{R}$ while $Y_2$ is homeomorphic to the letter $T$ considered as a subspace of $\mathbb{R}^2$. Evidently, $Y_1$ and $Y_2$ are not homeomorphic. □

A different problem is to establish whether two given spaces are not $M$-equivalent. The notion of a *homotopy class* can help here. Let $X$ and $Y$ be spaces and $C(X, Y)$ be the family of all continuous mappings of $X$ to $Y$. Then two mappings $f, g \in C(X, Y)$ are called *homotopic* if there exists a continuous mapping $\varphi\colon X \times [0, 1] \to Y$ such that $\varphi(x, 0) = f(x)$ and $\varphi(x, 1) = g(x)$, for each $x \in X$. The mapping $\varphi$ is said to be a *homotopy* between $f$ and $g$. Denote by $[f]$ the class of all $h \in C(X, Y)$ homotopic to $f$. It is easy to see that either $[f] = [g]$ or $[f] \cap [g] = \varnothing$, for any $f, g \in C(X, Y)$. In other words, we have the equivalence relation $f \sim g$ if $f$ and $g$ are homotopic. The set of all *homotopy classes* in $C(X, Y)$ is denoted by $[X, Y]$:

$$[X, Y] = \{[f] : f \in C(X, Y)\}.$$

Suppose that $Y = G$ is a topological group. Then we can define a group structure in $[X, G]$ as follows:

$$[f] * [g] = [f \cdot g], \quad f, g \in C(X, G).$$

This definition is correct — if $f_1 \sim f_2$ and $g_1 \sim g_2$, then clearly $f_1 \cdot g_1 \sim f_2 \cdot g_1 \sim f_2 \cdot g_2$, whence $f_1 \cdot g_1 \sim f_2 \cdot g_2$. A standard argument (see [**459**]) shows that $[X, G]$ with the binary operation $*$ is a group.

Let us show that if two spaces $X$ and $Y$ are $M$-equivalent, then the groups $[X, G]$ and $[Y, G]$ are isomorphic, for every topological group $G$.

LEMMA 7.10.6. *Let $X$ be a compact space and $\varphi\colon X \times [0, 1] \to G$ a continuous mapping to a topological group $G$. For every $t \in [0, 1]$, denote by $\varphi_t$ the restriction of $\varphi$ to $X \times \{t\}$, and let $\Phi_t\colon F(X) \to G$ be the continuous homomorphism extending $\varphi_t$. Then the mapping $\Phi\colon F(X) \times [0, 1] \to G$ defined by $\Phi(g, t) = \Phi_t(g)$, for all $g \in F(X)$ and $t \in [0, 1]$, is continuous.*

PROOF. By Corollary 7.4.2, the group $F(X)$ has the direct limit property. So, $F(X) \times [0, 1]$ is a $k_\omega$-space with the $k_\omega$-decomposition $F(X) \times [0, 1] = \bigcup_{n\in\omega} B_n(X) \times [0, 1]$. It suffices, therefore, to verify the continuity of the mapping $\Phi$ on every subspace $B_n(X) \times [0, 1]$.

Let $h \in B_n(X)$ be arbitrary, and let $O$ be a neighbourhood of $g_0 = \Phi(h, t)$ in $G$, where $t \in [0, 1]$. Suppose that a point $y = (y_1, \ldots, y_n) \in \overline{X}^n$ satisfies $i_n(y) = h$, where $\overline{X} = X \oplus \{e\} \oplus X^{-1}$, $e$ is the identity of $F(X)$ and $i_n\colon \overline{X}^n \to B_n(X)$ is the multiplication mapping. Then $g_0 = \Phi_t(y_1) \cdots \Phi_t(y_n)$, so we can find, for every $i \leq n$, an open neighbourhood $O_i$ of the point $\Phi_t(y_i)$ in $G$ in such a way that $O_1 \cdots O_n \subset O$. By the continuity of $\varphi$, there are open neighbourhoods $V_i$ and $U_i$ of $y_i$ and $t$, respectively, in $\overline{X}$ and $[0, 1]$ such that $\Phi(V_i \times U_i) \subset O_i$, $i = 1, \ldots, n$. Put $V_y = V_1 \times \cdots \times V_n$ and $U_y = \bigcap_{i=1}^n U_i$. Then $y \in V_y$, $t \in U_y$, and

$$\Phi(i_n(V_y) \times U_y) = \Phi(V_1 \cdots V_n \times U_y) \subset O_1 \cdots O_n \subset O.$$

By the Wallace theorem, we can find open neighbourhoods $W$ and $W_t$ of the set $i_n^{\leftarrow}(h)$ and of the point $t$, respectively, in $\overline{X}^n$ and $[0, 1]$ satisfying

$$W \times W_t \subset \bigcup\{V_y \times U_y : y \in i_n^{\leftarrow}(h)\}.$$

Since the mapping $i_n\colon \overline{X}^n \to B_n(X)$ is perfect, there exists an open neighbourhood $W_h$ of $h$ in $B_n(X)$ such that $i_n^{\leftarrow}(W_h) \subset W$. The sets $W_h$ and $W_t$ satisfy $\Phi(W_h \times W_t) \subset O$. Hence, $\Phi$ is continuous on $B_n(X) \times [0, 1]$, and the lemma is proved. □

THEOREM 7.10.7. *If $X$ and $Y$ are $M$-equivalent compact spaces, then the groups $[X, G]$ and $[Y, G]$ of homotopy equivalent classes of mappings from $X$ and $Y$, respectively, to $G$ are isomorphic.*

PROOF. Since $X$ and $Y$ are $M$-equivalent, there exists a topological isomorphism $\varphi\colon F(X) \to F(Y)$. Therefore, we can identify the groups $F(X)$ and $F(Y)$; then each of the spaces $X$ and $Y$ becomes a topological basis of this topological group, that will be denoted by $H$. Every continuous mapping $f\colon X \to G$ defines a continuous mapping $\varphi(f)\colon Y \to G$ by the rule $\varphi(f) = \tilde{f}\restriction Y$, where $\tilde{f}$ is the extension of $f$ to a continuous homomorphism of $H$ to $G$. Similarly, every continuous mapping $g\colon Y \to G$ defines a continuous mapping $\psi(g)\colon X \to G$ by the rule $\psi(g) = \tilde{g}\restriction X$, where $\tilde{g}$ is the extension of $g$ to a continuous homomorphism of $H$ to $G$. It is clear that $\psi(\varphi(f)) = f$ and $\varphi(\psi(g)) = g$ for all $f \in C(X, G)$ and all $g \in C(Y, G)$. Moreover, from the definition of $\varphi$ it follows that $\varphi(f_1 \cdot f_2) = \varphi(f_1)\cdot\varphi(f_2)$ for all $f_1, f_2 \in C(X, G)$. Since the mapping $\varphi\colon C(X, G) \to C(Y, G)$ is surjective, it must be a homeomorphism. Similarly, $\psi\colon C(Y, G) \to C(X, G)$ is a homeomorphism. Since $\psi \circ \varphi$ and $\varphi \circ \psi$ are the identity automorphisms of $C(X, G)$ and $C(Y, G)$, respectively, we conclude that $\varphi$ and $\psi$ are isomorphisms.

Finally, Lemma 7.10.6 implies that if $f_1 \sim f_2$ in $C(X, G)$ and $g_1 \sim g_2$ in $C(Y, G)$, then $\varphi(f_1) \sim \varphi(f_2)$ in $C(Y, G)$ and $\psi(g_1) \sim \psi(g_2)$ in $C(X, G)$. Therefore, the mapping $\varphi^*\colon [X, G] \to [Y, G]$ defined by $\varphi^*([f]) = [\varphi(f)]$, is an isomorphism between the groups $[X, G]$ and $[Y, G]$. □

The above theorem permits to show that many non-homeomorphic compact spaces are not $M$-equivalent either.

EXAMPLE 7.10.8. The closed unit interval of the real line and the unit circle in the plane are not $M$-equivalent.

Indeed, denote by $X$ and $Y$ the closed unit interval and the circle $S^1$, respectively, and let $G = S^1$. Then the group $[X, G]$ is trivial since every continuous mapping of $X$ to $S^1$ is homotopic to a constant mapping (see [**459**]). However, the group $[Y, G]$ is isomorphic to $\mathbb{Z}$ [**459**]. Therefore, Theorem 7.10.7 implies that $X$ and $Y$ are not $M$-equivalent. □

A subspace $Y$ of the free (Abelian) topological group $G(X)$ is called a *topological basis* of $G(X)$ if $Y$ is a free algebraic basis of $G(X)$ and the maximal topological group topology on the abstract group $G_a(X)$ which induces on $Y$ its original topology coincides with the topology of the group $G(X)$.

Clearly, if $Y$ is a topological basis of $F(X)$, then the groups $F(X)$ and $F(Y)$ are topologically isomorphic, that is, $X$ and $Y$ are $M$-equivalent. Conversely, if $\varphi\colon F(X) \to F(Y)$ is a topological isomorphism, then $\varphi^{-1}(Y)$ is a topological basis for $F(X)$. Similar assertions are valid in the Abelian case. Therefore, to find non-trivial examples of $M$- and $A$-equivalent spaces it suffices to know how to construct topological bases of free (Abelian) topological groups. The following fact about topological bases is almost immediate.

LEMMA 7.10.9. *Every topological basis $Y$ of the group $G(X)$ is closed in $G(X)$.*

PROOF. Let $\varphi\colon G(Y) \to G(X)$ be a topological isomorphisms extending the identity embedding of $Y$ into $G(X)$. By a) of Theorem 7.1.13, $Y$ is closed in $G(Y)$, so the image $\varphi(Y) = Y$ is closed in $G(X)$. □

The notion of a topological basis is especially useful when one tries to establish that certain topological properties are invariant under $M$- or $A$-equivalence relations.

THEOREM 7.10.10. *Pseudocompactness is an $A$-invariant property.*

PROOF. Let $X$ and $Y$ be some $A$-equivalent spaces, and suppose that $Y$ is pseudocompact. Then we can assume that $Y$ is a topological basis of the group $A(X)$. Assume that $X$ is not pseudocompact. Then $X$ contains a discrete family $\gamma = \{U_i : i \in \omega\}$ of non-empty open sets. For every $i \in \omega$, choose a point $x_i \in U_i$.

For $g \in A(X)$ and $x \in X$, denote by $d(x, g)$ the coefficient $k$ that stands at $x$ in the normal form of $g$ with respect to the basis $X$. In other words, if $g = kx + k_1x_1 + \cdots + k_nx_n$ where $x, x_1, \ldots, x_n$ are pairwise distinct elements of $X$ and $k, k_1, \ldots, k_n \in \mathbb{Z}$, then $d(x, g) = k$. In particular, $d(x, g) = 0$ iff $x$ does not appear in the normal form of $g$.

Since $Y$ is an algebraic basis of $A(X)$, for every $i \in \omega$ there exists $y_i \in Y$ such that $d(x_i, y_i) \neq 0$. Let $t_{i,1}, \ldots, t_{i,n_i}$ be all letters distinct from $x_i$ which appear in the normal form of $y_i$ (possibly, $n_i = 0$). Choosing a subsequence of $\{y_i : i \in \omega\}$, we can additionally assume that $d(x_j, y_i) = 0$ whenever $i < j$.

By induction on $n \in \omega$, we define continuous real-valued functions $f_n$ on $X$ as follows. First, we put $f_0 \equiv 0$. Suppose that for some $n \geq 1$ we have defined the functions $f_0, \ldots, f_{n-1}$. Put $g_n = \sum_{i=0}^{n-1} f_i$. Then there exists a continuous real-valued function $f_n$ on $X$ that takes the value 0 on $X \setminus U_n$ and at each point $t_{i,j}$ with $i \leq n$ that belongs to $U_n$, and also satisfies

$$f_n(x_n) = n + \sum_j |b_{n,j} g_n(t_{n,j})|, \tag{7.33}$$

where $b_{n,j} = d(t_{n,j}, y_n)$ and the sum in (7.33) is taken over all $j$ such that $t_{n,j} \in U_0 \cup \cdots \cup U_{n-1}$. This completes our construction.

Since the family $\gamma$ is discrete, the function $f = \sum_{n \in \omega} f_n$ is continuous on $X$ and the functions $f$, $f_n$, $g_{n+1}$ coincide on $U_n$, for each $n \in \omega$. In addition, the definition of $f$ implies that for all $n \geq 1$ and all $j$,

$$f(t_{n,j}) = 0 \text{ whenever } t_{n,j} \notin U_0 \cup \cdots \cup U_{n-1}. \tag{7.34}$$

Extend $f$ to a continuous homomorphism $\tilde{f} \colon A(X) \to \mathbb{R}$. From (7.33) and (7.34) it follows that $|\tilde{f}(y_n)| \geq n$, for each $n \in \omega$. As $\{y_n : n \in \omega\} \subset Y$, we conclude that $Y$ is not pseudocompact. This contradiction completes the proof. □

Let us show that a similar statement holds for compact spaces.

THEOREM 7.10.11. [**M. I. Graev**] *Let $X$ and $Y$ be any $A$-equivalent spaces.*

a) *If $X$ is compact, then so is $Y$.*
b) *If $X$ is compact and metrizable, then so is $Y$.*

*In other words, compactness is $A$-invariant and, in the presence of compactness, metrizability is $A$-invariant as well.*

PROOF. a) We can assume that $Y$ is a topological basis in the group $A(X)$. By Lemma 7.10.9, $Y$ is closed in $A(X)$. Theorem 7.10.10 implies that $Y$ is pseudocompact. Hence, $Y \subset B_n(X)$ for some $n \in \mathbb{N}$, by Corollary 7.5.4. Therefore, $Y$ is a closed subset of the compact space $B_n(X)$ and hence, $Y$ is compact.

b) Suppose that $X$ is a compact metrizable space. As in a), $Y$ is a closed subset of $B_n(X)$, for some $n \in \mathbb{N}$. The spaces $\overline{X} = X \oplus \{e\} \oplus X^{-1}$ and $\overline{X}^n$ are also compact and metrizable. Therefore, $B_n(X)$ is compact and metrizable, as the image of $\overline{X}^n$ under the continuous mapping $i_n$. The same conclusion for $Y$ is immediate. □

COROLLARY 7.10.12. *The class of NC-spaces is A-invariant.*

PROOF. Let $X$ be an $NC$-space, and suppose that $Y$ is a topological basis of the group $A(X)$. Since $X$ is pseudocompact, Theorem 7.10.10 implies that so is $Y$. Hence, by Corollary 7.5.4, $Y \subset B_n(X)$, for some $n \in \mathbb{N}$. By Lemma 7.10.9, $Y$ is closed in $A(X)$. Therefore, $Y$ is a closed subset of $B_n(X)$. Similarly to b) of Theorem 7.10.11, consider the multiplication mapping $i_n\colon \overline{X}^n \to B_n(X)$. Clearly, $\overline{X}^n$ is an $NC$-space. Theorem 7.8.6 implies that $i_n$ is a closed continuous mapping of $\overline{X}^n$ onto $B_n(X)$ and hence, $B_n(X)$ is an $NC$-space, by Proposition 7.8.4. The same conclusion is valid for the closed subspace $Y$ of $B_n(X)$. □

Similar methods enable us to establish the $A$-invariance of a number of topological properties.

PROPOSITION 7.10.13. *Let $\tau$ be an infinite cardinal. The following topological properties or classes of spaces are A-invariant:*

a) *$\sigma$-compactness;*
b) *$\sigma$-boundedness;*
c) *Dieudonné completeness;*
d) *the class of paracompact $\sigma$-closed-metrizable spaces;*
e) *the class of paracompact $\sigma$-spaces;*
f) *the class of spaces $X$ with $nw(X) \le \tau$;*
g) *the class of spaces $X$ satisfying $d(X) \le \tau$;*
h) *the class of spaces $X$ with $dc(X) \le \tau^+$.*

PROOF. Let $Y$ be a topological basis of the group $A(X)$.
a) If a space $X$ is $\sigma$-compact, then so is the group $A(X)$. The topological basis $Y$ for $A(X)$ is closed in $A(X)$ and, hence, is $\sigma$-compact.

b) Let $Y = \bigcup_{n\in\omega} K_n$ be a $\sigma$-bounded space, where each $K_n$ is a bounded subset of $Y$. For every $g \in A(X)$, denote by supp$(g)$ the finite set of letters that appear in the normal form of $g$ with respect to the basis $X$. For every $n \in \omega$, put

$$B_n = \bigcup\{\mathrm{supp}(g) : g \in K_n\}.$$

By Corollary 7.5.6, every $B_n$ is bounded in $X$. Since $X$ and $Y$ are algebraic bases of $A(X)$, we have that $X = \bigcup_{n\in\omega} B_n$. Therefore, the space $X$ is $\sigma$-bounded.

c) Suppose that $X$ is a Dieudonné complete space. By Theorem 7.9.6, the group $A(X)$ is Raĭkov complete. Since $Y$ is a closed subset of $A(X)$, the group uniformity $\mathscr{V}$ of $A(X)$ induces on $X$ a complete uniformity $\mathscr{V}_Y$. The universal uniformity $\mathscr{U}_Y$ of the space $Y$ is finer than $\mathscr{V}_Y$, so the uniform space $(Y, \mathscr{U}_Y)$ is complete. Hence, $Y$ is Dieudonné complete.

d) If $X$ is a paracompact $\sigma$-closed-metrizable space, then so is the group $A(X)$, by Theorem 7.6.8. Since $Y$ is a closed subspace of $A(X)$, it follows that $Y$ is a paracompact $\sigma$-closed-metrizable space as well.

e) Suppose that $X$ is a paracompact $\sigma$-space. Then $A(X)$ is also a paracompact $\sigma$-space, according to Theorem 7.6.7, and the conclusion about $Y$ follows as in d).

f) Follows directly from Corollary 7.1.17.

g) Let $\mathcal{D}(\tau)$ be the class of spaces $X$ satisfying $d(X) \le \tau$. It is clear that this class is closed under countable sums, finite products and taking continuous images. It follows that $d(A(X)) \le \tau$, for each $X \in \mathcal{D}(\tau)$.

Conversely, suppose that $A(X) \in \mathcal{D}(\tau)$, for some space $X$. Let $S$ be a dense subspace of $A(X)$ with $|S| \le \tau$. For every $g \in S$, take a finite set $D(g) \subset X$ such that $g \in \langle D(g) \rangle$, and let $D = \bigcup_{g \in S} D(g)$. We claim that $D$ is dense in $X$ and, hence, $d(X) \le \tau$. Indeed, it follows from the definition of $D$ that $S \subset \langle D \rangle$. Denote by $F$ the closure of $D$ in $X$. By Theorem 7.4.5, the subgroup $A(F, X)$ of $A(X)$ generated by $F$ is closed in $A(X)$ and, evidently, $S \subset \langle D \rangle \subset A(F, X)$. Since $S$ is dense in $A(X)$, it follows that $A(F, X) = A(X)$, whence $F = X$. We conclude that $d(X) \le \tau$.

Combining the two inequalities, we obtain that $d(A(X)) = d(X)$. This finishes the proof of g).

h) By Theorem 7.7.8, $dc(X) \le \tau^+$ is equivalent to the $\tau$-narrowness of the group $A(X) \cong A(Y)$, so $dc(Y) \le \tau^+$, by the same theorem. □

Intuitively, $M$- or $A$-equivalent spaces have to have many properties in common. The following result provides some precise general information in this direction.

THEOREM 7.10.14. *If $X$ and $Y$ are $M$-equivalent spaces, then $Y$ can be represented as the union of countably many subspaces each of which is homeomorphic to a subspace of $X$.*

PROOF. Let $X$ and $Y$ be some topological bases of the topological group $G \cong F(X) \cong F(Y)$. For $g \in G$, denote by $l_X(g)$ and $l_Y(g)$ the reduced lengths of $g$ with respect to the bases $X$ and $Y$, respectively. For $n \in \mathbb{N}$, put

$$C_n(Y) = \{g \in G : l_Y(g) = n\}.$$

In addition, for $n \in \mathbb{N}$ and $\nu = (m_1, \dots, m_n) \in \mathbb{N}^n$, we define a subset $Y_\nu$ of $Y$ as follows:

$$Y_\nu = Y \cap \prod_{i=1}^{n} (X \cap C_{m_i}(Y)),$$

where the product is taken in the group $G$. Clearly, $Y = \bigcup\{Y_\nu : \nu \in \Sigma\}$, where $\Sigma = \bigcup\{\mathbb{N}^n : n \in \mathbb{N}\}$.

Let $n \in \mathbb{N}$ and $\nu = (m_1, \dots, m_n) \in \mathbb{N}^n$ be arbitrary. For every $i = 1, \dots, n$, denote by $\varphi_i$ the mapping from $Y_\nu$ to $X$ assigning to $g \in Y_\nu$ the point of $X$ that appears at the $i$th place in the reduced form of $g$ with respect to the basis $X$. Similarly, for every $j = 1, \dots, m_i$, denote by $\psi_{i,j}$ the mapping from $C_{m_i}(Y)$ to $Y$ assigning to an element $h \in C_{m_i}(Y)$ the letter of $Y$ that appears at the $j$-th place in the reduced form of $h$ with respect to the basis $Y$. By b) of Theorem 7.1.13, the mappings $\varphi_i$ and $\psi_{i,j}$ are continuous.

Suppose that $y \in Y_\nu$, where $\nu = (m_1, \dots, m_n) \in \mathbb{N}^n$. Then we can write

$$y = x_1^{\varepsilon_1} \cdots x_n^{\varepsilon_n}, \quad \text{where } x_i \in X, \ \varepsilon_i = \pm 1$$

and for every $i = 1, \dots, n$,

$$x_i = y_{i,1}^{\delta_{i,1}} \cdots y_{i,m_i}^{\delta_{i,m_i}}, \quad \text{where } y_{i,j} \in Y, \ \delta_{i,j} = \pm 1.$$

Therefore,

$$y = \prod_{i=1}^{n}(\prod_{j=1}^{m_i} y_{i,j}^{\delta_{i,j}})^{\varepsilon_i}.$$

Since $Y$ is a free basis in $G$, we must have $y = y_{i,j}$ for some $i$, $j$. Hence $y$ belongs to the set $C_{i,j}^{v}$ of fixed points of the mapping $\psi_{i,j} \circ \varphi_i$. In addition, $\varphi_i\colon C_{i,j}^{v} \to X$ is a homeomorphic embedding: its inverse is $\psi_{i,j}$. It remains to note that $Y$ is the union of the subspaces $C_{i,j}^{v}$, where $v = (m_1, \ldots, m_n) \in \mathbb{N}^n$, $n \in \mathbb{N}$, $i = 1, \ldots, n$, and $j = 1, \ldots, m_i$. The proof is complete. □

Recall that a topological property $\mathcal{P}$ is *hereditary* if every subspace of a space with $\mathcal{P}$ also has $\mathcal{P}$. The property $\mathcal{P}$ is said to be *countably additive* if every space $X$, that can be represented as the union of a countable family of its subspaces $X_n$ with the property $\mathcal{P}$, also has $\mathcal{P}$. Theorem 7.10.14 implies the following fact immediately.

COROLLARY 7.10.15. *Let $\mathcal{P}$ be a hereditary countably additive topological property. Then $\mathcal{P}$ is $M$-invariant.*

Since the classes of Lindelöf perfectly normal spaces and of $\sigma$-discrete spaces are hereditary and countably additive, we have two more results.

COROLLARY 7.10.16. *If Lindelöf spaces $X$ and $Y$ are $M$-equivalent and $X$ is perfectly normal, so is $Y$.*

COROLLARY 7.10.17. *The class of $\sigma$-discrete spaces is $M$-invariant.*

We have seen in Theorems 7.10.3, 7.10.11, and 7.10.11 that $A$-equivalence preserves connectedness, compactness and pseudocompactness. Many other topological properties, such as Dieudonné completeness, $\sigma$-compactness, $\sigma$-boundedness, $\sigma$-discreteness, as well as the classes of paracompact $\sigma$-spaces and of Lindelöf perfectly normal spaces, are also stable with respect to the $A$-equivalence, according to Proposition 7.10.13 and Corollaries 7.10.16 and 7.10.17.

However, many important topological properties fail to be $M$- or $A$-invariant. In particular, metrizability, local compactness, the first and second countability conditions are among them. Let us show this.

Recall that the *countable fan* is a space obtained as follows. Let $\{C_i : n \in \omega\}$ be a countable family of pairwise disjoint convergent sequences $C_i = \{x_i\} \cup \{x_{i,j} : j \in \omega\}$, where $x_i$ is the limit of the sequence $\{x_{i,j} : j \in \omega\}$, $i \in \omega$. Denote by $X$ the topological sum of these sequences, $X = \oplus_{i\in\omega} C_i$. Let $Z$ be the quotient space of $X$ obtained by gluing the closed subset $F = \{x_i : i \in \omega\}$ of $X$ to a point. Then $Z$ is called a *countable fan*. It is well known (and easy to verify) that $Z$ is a $k_\omega$-space, but it is neither first-countable nor locally compact. The next example shows that the spaces $X$ and $Z$ are $M$-equivalent.

EXAMPLE 7.10.18. [**M. I. Graev**] The topological sum $X$ of countably many convergent sequences and the countable fan $Z$ are $M$-equivalent spaces. Therefore, $M$-equivalence does not preserve metrizability, local compactness, the first and second countability.

Indeed, let $X = \oplus_{i\in\omega} C_i$ be the topological sum of pairwise disjoint sequences $C_i = \{x_i\} \cup \{x_{i,j} : j \in \omega\}$, $i \in \omega$. Then $X$ is evidently second-countable, metrizable, locally compact space. Clearly, $X$ is a $k_\omega$-space with $k_\omega$-decomposition $X = \bigcup_{n=0}^{\infty} D_n$, where

$D_n = \bigcup_{i=0}^n C_i$ for each $n \in \omega$. Therefore, Theorem 7.4.1 implies that $F(X)$ is a $k_\omega$-space with $k_\omega$-decomposition $F(X) = \bigcup_{n=0}^\infty \langle D_n \rangle_n$. Since $\langle D_n \rangle_n \subset F(D_n, X)$ and each subgroup $F(D_n, X)$ is closed in $F(X)$ by Theorem 7.4.5, we conclude that $\{F(D_n, X) : n \in \omega\}$ is a generating family for $F(X)$.

Consider the subspace

$$Y = \{x_i : i \in \omega\} \cup \bigcup_{i \in \omega} x_0 x_i^{-1} C_i$$

of the group $F(X)$. It is easy to verify that $Y$ is a free algebraic basis of $F(X)$. We claim that $Y$ is a topological basis of $F(X)$. Indeed, suppose that $\mathcal{T}'$ is the maximal group topology on $F_a(X)$ which induces on $Y$ its original topology, i.e., the subspace topology it inherits from $F(X)$. Clearly, $\mathcal{T}'$ is finer than the topology $\mathcal{T}$ of $F(X)$. Since translations in every topological group are homeomorphisms, $\mathcal{T}'$ induces the original topology on each subspace $C_i$ of $F(X)$. In addition, $X$ is the topological sum of its subspaces $C_i$'s, so from $\mathcal{T} \subset \mathcal{T}'$ it follows that $\mathcal{T}' \restriction X = \mathcal{T} \restriction X$. However, $\mathcal{T}$ is the maximal group topology on $X$ which induces on $X$ its original topology, so that $\mathcal{T}' = \mathcal{T}$. This proves our claim. Therefore, $X$ and $Y$ are $M$-equivalent. By Lemma 7.10.9, $Y$ is closed in $F(X)$.

It remains to verify that $Y$ is homeomorphic to the countable fan $Z$. Put $Y_0 = \bigcup_{i=0}^\infty x_0 x_i^{-1} C_i$ and consider the onto mapping $f \colon X \to Y_0$ defined by $f(x) = x_0 x_i^{-1} x$ for each $x \in C_i$, $i \in \omega$. Then $f$ is continuous since its restriction to every $C_i$ is continuous. Put $P = \{x_i : i \in \omega\}$. Note that $f^{-1} f(x) = \{x\}$ for each $x \in X \setminus P$ and $P = f^{-1} f(x_i)$ for each $i \in \omega$. In other words, $Y_0$ is obtained from $X$ by collapsing the set $P$ to a point. In particular, for every $i \in \omega$, $f(C_i)$ is a non-trivial sequence in $Y_0$ converging to the point $x_0 \in Y$, and $f(C_i) \cap f(C_j) = \{x_0\}$, for all distinct $i, j \in \omega$. Note that $Y_0 \cap F(D_n, X) = f(D_n)$ is compact and closed in $F(D_n, X)$ for each $n \in \omega$, so $Y_0$ is closed in $F(X)$ and the family $\{f(D_n) : n \in \omega\}$ is generating for $Y_0$. Hence, the mapping $f \colon X \to Y_0$ is quotient, and $Y_0$ is homeomorphic to $Z$.

Finally, $P = X \cap Y$ is closed in $Y$, $Y_0$ is closed in $F(X)$ and in $Y$, and $P \cap Y_0 = \{x_0\}$. Since the set $P$ is discrete, we conclude that $Y$ is homeomorphic to the topological sum of $Y_0$ with the countable discrete space $P' = P \setminus \{x_0\}$. Hence, $Y$ is homeomorphic to the countable fan $Z$. This completes the proof. □

## Exercises

7.10.a. Let us call Tychonoff spaces $X$ and $Y$ *$H$-equivalent* if the spaces $F(X)$ and $F(Y)$ are homeomorphic. Prove the following statements for $H$-equivalent spaces $X$ and $Y$:

(a) If $X$ is a cosmic space, then $Y$ is also cosmic;
(b) If $X$ is separable, then $Y$ is separable;
(c) If $X$ and $Y$ are compact spaces, then the weight of $X$ is equal to the weight of $Y$;
(d) If $X$ is $\sigma$-compact, then so is $Y$;
(e) If $X$ is a $k_\omega$-space, then $Y$ is also $k_\omega$-space;
(f) If $X$ is a Lindelöf $P$-space, then so is $Y$;
(g) If $X$ is submetrizable, then $Y$ is also submetrizable;
(h) If $X$ and $Y$ are compact spaces and the tightness of $X$ is countable, then the tightness of $Y$ is countable as well;
(i) The class of paracompact $\sigma$-closed-metrizable spaces is $H$-invariant.

## Problems

7.10.A. (S. A. Morris and H. B. Thompson [**331**]) Let $S = \{x_n : n \in \omega\} \subset F(X)$ be a non-trivial sequence converging to the identity $e$ of $F(X)$. Show that $\langle S \rangle$ contains a closed subgroup topologically isomorphic to $F(S')$, where $S' = S \cup \{e\}$.
*Hint.* Construct an infinite subset $T$ of $S$ satisfying the following conditions:
(i) $T$ is a free algebraic basis for $\langle T \rangle$;
(ii) $T$ is regularly situated in $F(X)$.
Then apply 3) of Exercise 7.5.a.

7.10.B. (S. A. Morris and H. B. Thompson [**331**]) Let $X$ be an arbitrary Tychonoff space. Prove that every metrizable subgroup of $F(X)$ is discrete.
*Hint.* Use Problem 7.10.A and Exercise 7.4.b.

7.10.C. Prove that every Fréchet-Urysohn subgroup of $F(X)$ is discrete, for any Tychonoff space $X$.
*Hint.* Use the facts established in Problem 7.10.A and Exercise 7.6.b.

7.10.D. (P. Nickolas [**346**]) Show that if $X$ is a $k_\omega$-space, then $F(X)$ contains a closed subgroup topologically isomorphic to $F(X \times X)$.
*Hint.* Apply Exercise 7.4.b to verify that the subgroup of $F(X)$ generated by the set $\{xyx : x, y \in X\}$ is as required.

7.10.E. (E. Katz, S. A. Morris, and P. Nickolas [**274**]) A subspace $Y$ of the free topological group $F(X)$ is *enjoyably embedded* in $F(X)$ if for some $a \in X$, the subspace $Z = \{y^{-1}ay : y \in Y\}$ of $F(X)$ is homeomorphic to $Y$, and $\langle Z \rangle \cong F(Z)$. Prove that if $X$ is a $k_\omega$-space, and a closed subset $Y$ of $F(X)$ satisfies $\langle Y \rangle \subset \langle X \setminus \{a\} \rangle$, then $Y$ is enjoyably embedded in $F(X)$.
*Hint.* Consider the subspace $Z = \{y^{-1}ay : y \in Y\}$ of $F(X)$. Note that $Z$ is closed in $F(X)$ and forms a free algebraic basis of the subgroup $\langle Z \rangle$ of $F(X)$. Then apply Exercise 7.4.b to deduce that $\langle Z \rangle \cong F(Z)$. Finally, to show that $Y$ and $Z$ are homeomorphic, verify that if $X = \bigcup_{n\in\omega} X_n$ is a $k_\omega$-decomposition of $X$, then $Z = \bigcup_{n\in\omega} Z_n$ is a $k_\omega$-decomposition of $Z$, where $Z_n = Z \cap B_n(X_n)$ for each $n \in \omega$.

7.10.F. (E. Katz, S. A. Morris, and P. Nickolas [**274**]) Let $X$ be a $k_\omega$-space with at least two points. Prove that if $Y$ is a closed subspace of $F(X)$, then $F(X)$ contains a closed subgroup topologically isomorphic to $F(Y)$.
*Hint.* Apply Problem 7.10.D to find a closed subgroup of $F(X)$ topologically isomorphic to $F(X \times X)$. For a fixed point $b \in X$, the closed subspace $X \times \{b\}$ of $X \times X$ generates a closed subgroup of $F(X \times X)$ topologically isomorphic to $F(X)$. Therefore, if $a \in X$ and $a \neq b$, then the subgroup $\langle X \times X \setminus \{(a, b)\} \rangle$ of $F(X \times X)$ contains a closed copy of $Y$. Apply Problem 7.10.E to conclude that this copy of $Y$ is enjoyably embedded in $F(X \times X)$.

7.10.G. (E. Katz, S. A. Morris, and P. Nickolas [**274**]) Show that if the free topological group $F(X)$ on a Tychonoff space $X$ contains a non-trivial convergent sequence, then $F(X)$ contains a closed subgroup topologically isomorphic to $F(Z)$, where $Z$ is a countable fan.
*Hint.* By Problem 7.10.A, one can assume that $X = \{x_0\} \cup \{x_n : n \in \mathbb{N}\}$ is a convergent sequence with its limit $x_0$, where $x_i \neq x_j$ if $i \neq j$. For every natural $n$, let $y_n = x_0^{-1}x_n$ and $Z_1 = \{y_n : n \in \omega\}$. Then $Z_1 \cong X$ is a compact subset of $F(X)$ which contains the identity of $F(X)$. Apply Exercise 7.4.b and Problem 7.10.F to verify that $Z = \bigcup_{m=1}^{\infty} Z_m$ is as required, where $Z_m = \{y_n^m : n \in \omega\}$ for each $m \geq 1$. Use the fact that $Z = \bigcup_{m=1}^{\infty} Z_m^*$ is a $k_\omega$-decomposition of $Z$, where $Z_m^* = \bigcup_{k=1}^{m} Z_k$ for each $m \geq 1$.

7.10.H. (E. Katz, S. A. Morris, and P. Nickolas [**273**]) Prove that $A(J)$ is topologically isomorphic to a closed subgroup of $A(I)$, where $I = [0, 1]$ and $J = (0, 1)$.
*Hint.* First, construct a closed subgroup of $A(I)$ topologically isomorphic to $A(J')$, where $J' = [0, 1)$. For every $n \in \omega$, define a mapping $f_n$ of $I_n = [n/(n+1), (n+1)/(n+2)]$ to $J_n = [0, (n+1)/(n+2)]$ by $f_n(x) = (n+1)^2 x - n(n+1)$. Then consider a mapping

$\varphi\colon J' \to A(I)$ defined by

$$\varphi(x) = (n+1)x + f_n(x) \text{ for } x \in I_n,$$

where "+" between $(n+1)x$ and $f_n(x)$ denotes addition in $A(I)$. Verify that $\varphi$ is continuous and one-to-one. For every $n \in \omega$, put $Y_n = \varphi(J_n)$. Then $\varphi\colon J_n \to Y_n$ is a homeomorphism. Note that $B_{n+2}(I) \cap \varphi(J') = Y_n$ for each $n \in \omega$, and deduce that $Y = \varphi(J')$ is a $k_\omega$-space with a $k_\omega$-decomposition $Y = \bigcup_{n\in\omega} Y_n$. Therefore, $\varphi$ is a closed mapping and, hence, $\varphi$ is a homeomorphism of $J'$ onto $Y$. Then show that $Y$ is a free algebraic basis for the subgroup $\langle Y\rangle$ of $F(I)$ and apply Exercise 7.5.b to conclude that $\langle Y\rangle \cong A(Y)$.

Modify the above argument to show that $A(J')$ contains a closed subgroup topologically isomorphic to $A(J)$.

7.10.I. Is $A(J)$ a topological quotient of $A(I)$, where $I = [0,1]$ and $J = (0,1)$?

7.10.J. Is $A(I)$ a topological quotient of $A(J)$?

7.10.K. (K. Eda, H. Ohta, and K. Yamada [**156**]). Let $\kappa$ be an infinite cardinal and $Y$ be one of the spaces $\mathbb{R}$, $\mathbb{Q}$, $\mathbb{R}\setminus\mathbb{Q}$, $\beta\omega$, $\beta\omega\setminus\omega$, $2^\kappa$, where each of them carries its usual topology (so that the last three spaces are compact). Prove that if $X$ is an arbitrary Tychonoff space and $F(X)$ or $A(X)$ contains a topological copy of $Y$, then so does $X$.

*Hint.* Note that every non-empty open subset of $Y$ contains a copy of $Y$, i.e., $Y$ is *densely self-embeddable.* Apply Exercise 7.4.h to show that if $Y$ is first-countable and $Y \subset F(X)$ (or $Y \subset A(X)$), then the interior of $Y \cap B_n(X)$ in $Y$ is not empty for some $n \in \omega$. The Baire category theorem implies the same conclusion if $Y$ is compact. Since $Y$ is densely self-embeddable, there is the minimal $k \in \omega$ such that $B_{k+1}(X)\setminus B_k(X)$ contains a copy of $Y$. Then apply b) of Theorem 7.1.13 and Proposition 7.1.14 along with the fact that the space $Y$ is *prime* in the sense that if $Y \subset P \times Q$, then either $P$ or $Q$ contains a copy of $Y$.

7.10.L. (P. Gartside, E. A. Reznichenko, and O. V. Sipacheva [**187**]) Show that there exists a Lindelöf topological group $G$ satisfying $c(G) = 2^\omega$.

7.10.M. Let $\tau$ be an infinite cardinal and $X$ be an arbitrary pseudo-$\tau^+$-compact space. Prove that the groups $F(X)$ and $A(X)$ are $\tau$-steady if and only if the space $X$ is $\tau$-stable.

7.10.N. Show that the unit circle $S^1$ and the closed unit interval $I$ are not $H$-equivalent.

7.10.O. Is the closed unit interval $I = [0,1]$ $H$-equivalent to the square $I \times I$?

7.10.P. Give an example of $M$-equivalent Tychonoff spaces $X$ and $Y$ such that $X$ is a $k$-space, but $Y$ is not.

7.10.Q. Give an example of $M$-equivalent Tychonoff spaces $X$ and $Y$ such that the tightness of $X$ is countable and the tightness of $Y$ is uncountable.

7.10.R. (O. G. Okunev [**359**]) Show that the $M$-equivalence relation does not preserve the Fréchet–Urysohn property in the class of compact spaces.

*Hint.* Let $X$ and $Y$ be compact scattered spaces constructed by P. Simon in [**448**]. Then $X$ and $Y$ are Fréchet–Urysohn, but the product space $X \times Y$ fails to be Fréchet–Urysohn. Apply Okunev's method from [**358**] of constructing $M$-equivalent spaces to take a quotient space $Z$ of $X \times Y$ which is Fréchet–Urysohn and is $M$-equivalent to $X \times Y$.

7.10.S. Prove that if $X$ and $Y$ are $M$-equivalent, then the spaces $C_p(X)$ and $C_p(Y)$ of continuous real-valued functions on $X$ and $Y$, respectively, with the topology of pointwise convergence, are linearly homeomorphic, i.e., there exists a linear homeomorphism of $C_p(X)$ onto $C_p(Y)$.

7.10.T. Show that the class of regular Lindelöf spaces is $M$-invariant.

*Hint.* See [**524**].

7.10.U. Let $I = [0,1]$ be the closed unit interval. Prove that $A(X)$ admits an embedding into $A(I)$ as a topological subgroup if and only if $X$ is homeomorphic to a closed subspace of $A(I)$.

*Hint.* See [**290**].

7.10.V. Prove that for any finite-dimensional compact metrizable space $X$, the free topological group $F(X)$ is topologically isomorphic to a closed subgroup of $F(I)$, where $I$ is the closed unit interval.
*Hint.* See [**290**].

7.10.W. Does $H$-equivalence preserve connectedness?

## Open Problems

7.10.1. Let $X$ and $Y$ be $M$-equivalent completely regular spaces, and suppose that $X$ is countably compact. Is $Y$ countably compact?

7.10.2. Give an example of a Tychonoff space $X$ such that $F(X)$ does not contain a subgroup topologically isomorphic to $F(X) \times F(X)$.

7.10.3. Are the Sorgenfrey line and its square $H$-equivalent?

7.10.4. Is there an uncountable Tychonoff space $X$ such that the topological group $A(X)$ can be represented as a continuous image of the Sorgenfrey line?

7.10.5. Is the class of regular hereditarily Lindelöf spaces $H$-invariant?

7.10.6. Is the class of regular Lindelöf spaces $H$-invariant?

## 7.11. Historical comments to Chapter 7

Free topological groups were introduced in 1941 by A. A. Markov in the short note [**305**]. The complete construction, about 50 pages long, appeared four years latter, in [**308**], where several basic properties of free topological groups over Tychonoff spaces were established. In particular, Theorems 7.1.1, 7.1.2, 7.1.5, 7.1.7, and the last part of item a) of Theorem 7.1.13 can be found in [**308**]. The latter result, i.e., the fact that every Tychonoff space $X$ is (homeomorphic to) a closed subspace of $F(X)$, enabled Markov to answer in the negative Kolmogorov's question as to whether every topological group is a normal space (see Corollary 7.1.15).

The proof of the existence of free topological groups given in Theorem 7.1.2 is due to S. Kakutani [**265**] which seems to be the shortest one. The importance of free topological groups was immediately recognized by specialists and, in general, by the mathematical community. Two more existence proofs appeared quite soon after the original Markov's article [**305**] had been published. One of them, by T. Nakayama [**338**], generalizes Markov's construction to free topological groups over uniform spaces (see Exercise 7.2.d), while the other, due to M. I. Graev [**201**], is based on extending continuous pseudometrics from a space $X$ to invariant continuous pseudometrics on $F(X)$ (or $A(X)$). Such an extension is described in Theorem 7.2.2 and is widely used in Chapter 7. Graev's extension is maximal among all invariant extensions of a given pseudometric from $X$ to $F(X)$, but there are quite a few natural constructions of invariant extensions of pseudometrics — the article [**328**] by S. A. Morris and P. Nickolas contains descriptions of the most important of them.

Theorem 7.1.5 is usually formulated by saying that the free topological group $F(X)$ on a Tychonoff space $X$ has a complete system of continuous unitary representations. This result appeared in [**338**]; our proof of it follows the scheme outlined in [**202**]. The major part of the results of Section 7.1 can be found in [**308**] or in [**201, 202**], with just a few exceptions. For example, items a) and b) of Theorem 7.1.13 appeared for the first time in [**18**] and in [**19**], and several years later, in [**216**]. It is worth noting that a closed embedding of $X^n$ into $A(X)$, for a Tychonoff space $X$ and an arbitrary integer $n \geq 1$, was constructed by B. V. S. Thomas in [**464**]. Corollary 7.1.19 is due to Arhangel'skii's [**19**]; this result was rediscovered by C. Joiner in [**261**].

Theorem 7.2.2 on extension of pseudometrics is, as we have just mentioned, due to M. I. Graev [**201**]. Apparently, the description of a local base at the identity of the free Abelian topological group $A(X)$ in terms of continuous pseudometric on $X$ in Theorem 7.2.7 was known to Graev, but appeared explicitly in [**345**] and [**309**] (it can also be found in [**470**]). Theorem 7.2.9 was proved by Graev in [**202**] (see also [**329**]). Theorem 7.2.11 is a part of the folklore; as far as we know, its proof appears here for the first time. Corollary 7.2.12, a special case of Theorem 7.2.11, is due to V. G. Pestov [**373**].

Theorem 7.3.1 on an extension of a compact group by a metrizable group appeared in [**410**]; our exposition follows V. G. Pestov's article [**374**]. Theorem 7.3.2 and Corollary 7.3.3 are apparently new.

Direct limits of topological spaces appeared at the very beginning of the development of general topology. The concept of $k_\omega$-space widely used in Section 7.4 plays an important role both in General Topology and Topological Algebra. The reader can find useful information on this subject in [**174**]. Theorem 7.4.1 on free topological groups over $k_\omega$-spaces was proved by J. Mack, S. A. Morris, and E. T. Ordman in [**297**], thus generalizing an earlier Graev's theorem in [**201**] about free topological groups on compact spaces (see Corollary 7.4.2). Our proof of Theorem 7.4.1 mimics the argument from [**19**]. Theorem 7.4.5 is due to M. I. Graev's [**201**]. Proposition 7.4.7 is a part of the folklore; for example, the fact that the free (Abelian) topological group on a $P$-space is a $P$-group was used in [**22**] and [**469**]. Proposition 7.4.8 appeared in [**473**]. Lemma 7.4.10 and Corollary 7.4.12 were proved by M. Graev in [**201**], while Theorem 7.4.11 follows immediately from Theorem 7.4.1 and Lemma 7.4.10 (see [**247**]).

Lemma 7.5.1 on a continuous function separating points of a countable set in a Tychonoff space appeared in [**142**]; it follows from a slightly more general fact established by A. V. Arhangel'skii, O. G. Okunev, and V. G. Pestov in [**58**]. Lemma 7.5.2 and Theorem 7.5.3 were proved in [**142**]. Corollary 7.5.6 on bounded subsets of free topological groups appeared in [**102**], the corresponding argument was corrected in [**58**]. Corollary 7.5.6 also follows from the results obtained in [**453**]. Corollary 7.5.7 characterizing $\sigma$-boundedness of free topological groups was proved in [**476**].

Metrizable subspaces of free topological groups is a subject of many research articles. Decomposing a topological group into the union of a countable family of metrizable subspaces frequently helps to clarify the topological properties of the group. V. K. Bel'nov in [**71**] established Proposition 7.6.1, Theorem 7.6.2, and Corollary 7.6.3 (it was pointed out in [**24**] that the original arguments by Bel'nov from [**71**] had some small and easily corrected inaccuracy). Theorems 7.6.6, 7.6.7, and 7.6.8 appeared in [**24**], as well as Corollary 7.6.9. In fact, the latter result had originally been proved in [**23**], and then the argument was refined in [**24**].

Theorem 7.6.15 and its proof based on Theorems 7.6.10, 7.6.11, and 7.6.13 are apparently new. A subtle combination of these facts together with arguments from [**24**] resulted in the proof of Theorem 7.6.16 which appears here for the first time. Theorem 7.6.18 and Corollary 7.6.19 are from [**24**].

The study of the question of when the free (Abelian) topological group on a given Tychonoff space $X$ is sequential or a $k$-space goes back to T. H. Fay, E. T. Ordman, and

B. V. S. Thomas. They showed in [**167**] that the free topological group over the space of rationals $\mathbb{Q}$ is neither sequential nor a $k$-space (which is Corollary 7.6.21). C. Borges observed in [**79**] that the free topological group over a locally compact metrizable space need not be a $k$-space either. A. V. Arhangel'skii, O. G. Okunev, and V. G. Pestov gave in [**58**] a complete characterization of metrizable spaces $X$ for which the free topological group $F(X)$ or the free Abelian topological group $A(X)$ is a $k$-space. Theorems 7.6.20, 7.6.30, 7.6.36, and Proposition 7.6.29 are all from [**58**]. It is worth noting that the corresponding characterizations in the Abelian and non-Abelian cases are quite different, as Theorems 7.6.30 and 7.6.36 show. To some extent, Proposition 7.6.31 and Corollary 7.6.32 are new. Their proofs are built on ideas from [**201, 58, 142**].

Additional information on the $k$-property in subspaces of free topological groups is given in the problem sections of Chapter 7 and in the articles [**538, 539, 541, 542**] by K. Yamada.

An interesting open problem is to characterize the class of Tychonoff spaces $X$ such that the group $A(X)$ (or $F(X)$) has countable tightness. According to Corollary 7.4.9, every $k_\omega$-space of countable tightness is in this class. The case of metrizable spaces is especially important; the results of Section 7.6 resolve it completely. Every non-separable metrizable space admits a quotient mapping onto a space containing a copy of a sequential fan $V(\aleph_1)$ with uncountably many spines. Hence, the first step toward the solution of the problem is Lemma 7.6.22, on the tightness of $V(\aleph_1)^2$, which is due to G. Gruenhage and Y. Tanaka [**205**]. Proposition 7.6.24, Theorem 7.6.27, Corollary 7.6.28, as well as Lemma 7.6.33 and Theorem 7.6.35 are all from [**58**]. Further calculations of the tightness of free Abelian topological groups on metrizable space are given in [**540**].

When working with free topological groups, it is very important to know under which conditions on a subspace $X$ of a Tychonoff space $Y$, the subgroup $F(X, Y)$ of $F(Y)$ generated by $X$ is topologically isomorphic to the group $F(X)$, under the natural isomorphism extending the identity embedding of $X$ to $Y$. Some instances of this situation were presented in Corollaries 7.4.6 and 7.6.32. One of the main results of Section 7.7, Theorem 7.7.3, which deals with the case when $X$ is dense in $Y$, appeared (in an equivalent form) in the article [**369**] by V. G. Pestov, and in a more general form, for free uniform topological groups, in [**354**] by E. C. Nummela. A slightly weaker form of Lemma 7.7.1, as a step towards the proof of Theorem 7.7.3, was given in [**369**]. In the Abelian case, the assumption on the density of $X$ in $Y$ can be omitted. This is shown in Theorem 7.7.4 which was proved in [**470**]. Corollary 7.7.5 is from [**369**], while Theorem 7.7.7 appeared in [**474**]. Theorem 7.7.8 characterizing $\tau$-narrowness of free topological groups is essentially due to A. V. Arhangel'skii and I. I. Guran (for quite a while, no published version of the proof existed, see in this respect [**471**, p. 550] and [**142**, Lemma 3.2]).

The class of $NC$-spaces as well as concordant mappings introduced in Section 7.8 were considered in [**473, 483**] (no name was given to $NC$-spaces there). Lemma 7.8.1, Proposition 7.8.4, Lemma 7.8.5, Theorem 7.8.6, Lemma 7.8.7, and Theorem 7.8.8 are all from [**473**]. Lemmas 7.8.2 and 7.8.3 can be found, with similar formulations, in [**60**]. Theorem 7.8.10, in a more general form, appeared in [**280**]. Theorem 7.8.13 is new, while Corollary 7.8.15 is essentially from [**473**]. Some results of Section 7.8 were extended to universal free topological algebras in [**103**].

The results of Section 7.9 are almost entirely from [**470**]. Lemma 7.9.1 is a weak form of a result from [**482**], it fills in a small gap in the proof of Theorem 7.9.6 presented in [**470**]. O. V. Sipacheva extended Theorem 7.9.6 to free topological groups (see [**452**]).

The definition of $M$-equivalence relation go back to A. A. Markov's article [**308**]. M. I. Graev was the first to undertake a thorough study of $M$- and $A$-equivalent spaces in [**201**]. He proved Proposition 7.10.1, Theorem 7.10.3, and Lemma 7.10.4. Examples 7.10.5, 7.10.8, and 7.10.18 are also essentially from [**201**]. Lemma 7.10.6 and Theorem 7.10.7 were proved by B. A. Pasynkov and V. Vylov in [**367**]. Lemma 7.10.9 and Theorem 7.10.10 on the preservation of pseudocompactness under $A$-equivalence are due to Graev's [**201**] (the latter fact was not formulated by Graev in the form given here, but our argument in the proof of Theorem 7.10.10 is an adaptation of that in [**201**, Lemma 9.1]). Theorem 7.10.11 is again from [**201**]. Corollary 7.10.12 is new. Proposition 7.10.13 is a collection of several results obtained by distinct authors. For example, items a), f), and g) of Proposition 7.10.13 go back to [**22**] (the corresponding results were formulated there for $M$-equivalence, but the same arguments work in the Abelian case as well). Item b) appeared (implicitly) in [**476**], item c) originated in [**470**] (see also [**29**]). Items d) and e) are due to A. V. Arhangel'skii (see [**24**]), and item h) is a part of the folklore (based on Guran's results from [**208**]).

A prototype of Theorem 7.10.14 was proved by M. I. Graev in [**201**] and, in the present form, by V. G. Pestov in [**372**]. Corollaries 7.10.15, 7.10.16, and 7.10.17 are also from [**372**].

# Chapter 8

# $\mathbb{R}$-Factorizable Topological Groups

In this chapter the reader is introduced to the theory of $\mathbb{R}$-factorizable topological groups. Roughly, these are the groups whose algebraic and topological structure is most closely related to continuous real-valued functions over them. When working with complicated mathematical objects, a common but fruitful idea is to reduce their study to the study of some kind of "countable reflections" of these objects and then to go back and apply new knowledge to clarify the structure and properties of the original objects. Thousands of implementations of this idea in Set Theory, General Topology, and Algebra have been given. We present one of these here; it starts with the definition of $\mathbb{R}$-factorizable groups.

Let $G$ be a compact topological group and $f$ a continuous real-valued function on $G$. By Theorem 8.1.1 below, $G$ contains a closed invariant subgroup $N$ such that the quotient group $G/N$ is metrizable and $f$ is constant on each coset $xN$ of $N$ in $G$. In other words, one can define a real-valued function $h$ on $G/N$ such that $f = h \circ \pi$, where $\pi\colon G \to G/N$ is the quotient homomorphism. Since $\pi$ is open, the function $h$ is continuous. Clearly, the group $G/N$ has a countable base.

We isolate this property of compact groups and study the resulting class of $\mathbb{R}$-factorizable groups. We shall see in Sections 8.1 and 8.2 that this class contains, apart from compact groups, arbitrary subgroups of $\sigma$-compact groups, all Lindelöf groups, and dense subgroups of topological products of second-countable groups.

All spaces in this chapter are assumed to be Tychonoff unless a different separation axiom is specified.

## 8.1. Basic properties

We start this section with a theorem on factorization of continuous real-valued functions on compact topological groups that justifies the introduction of the class of $\mathbb{R}$-factorizable groups. The theorem will be considerably generalized in several directions in this section, but it is a good point to start with.

THEOREM 8.1.1. **[L. S. Pontryagin]** *Let $f$ be a continuous real-valued function on a compact topological group $G$. Then there exists a closed invariant subgroup $N$ of $G$ such that the quotient group $G/N$ is metrizable, and $f$ is constant on every coset of $N$ in $G$.*

PROOF. We define a binary relation $\sim$ for elements $a, b \in G$ by the rule $a \sim b$ if $f(xay) = f(xby)$, for all $x, y \in G$. It is immediate from the definition that this relation is reflexive, symmetric, and transitive. Let $N$ be the equivalence class containing the identity $e$ of $G$. Let us show that $N$ is a closed invariant subgroup of $G$ and the equivalence classes are cosets of $N$ in $G$.

Clearly,

$$N = \{a \in G : f(xay) = f(xey) = f(xy) \text{ for all } x, y \in G\}.$$

For $x, y \in G$, define a function $\varphi_{x,y}\colon G \to \mathbb{R}$ by $\varphi_{x,y}(z) = f(xzy)$, for each $z \in G$. Since the functions $\varphi_{x,y}$ are continuous, the set $N = \bigcap_{x,y\in G} \varphi_{x,y}^{-1}(\varphi_{x,y}(e))$ is closed in $G$.

To show that $N$ is a subgroup of $G$, take an arbitrary element $a \in G$. Then $f(xay) = f(xy)$ for all $x, y \in G$. Replacing $x$ by $xa^{-1}$ in the latter equality, we obtain $f(xy) = f(xa^{-1}y)$; hence, $a^{-1} \in N$. Similarly, if $b$ is another element of $N$, then $f(xaby) = f(xa(by)) = f(xby) = f(xy)$ for all $x, y \in G$. It follows that $ab \in N$ and, hence, $N$ is a subgroup of $G$. In addition, if we take an arbitrary $z \in G$ and replace $x$ by $xz^{-1}$ and $y$ by $zy$ in the equality $f(xay) = f(xy)$, we obtain immediately that $f(xz^{-1}azy) = f(xy)$. Therefore, $z^{-1}az \in N$, for each $z \in G$, and $N$ is an invariant subgroup of $G$.

Finally, if $c, d \in G$ and $c \sim d$, then $f(xcy) = f(xdy)$, for all $x, y \in G$. Replacing $y$ by $d^{-1}y$ in this equality, we obtain $f(xcd^{-1}y) = f(xy)$, that is, $cd^{-1} \in N$. Conversely, if $cd^{-1} \in N$, then $f(xcd^{-1}y) = f(xy)$, and replacing $y$ by $dy$ we deduce that $f(xcy) = f(xdy)$, for all $x, y \in G$. This proves that the equivalence classes of the relation $\sim$ coincide with the cosets of $N$ in $G$.

Therefore, for any fixed $x, y \in G$, the function $f(xay)$ with argument $a$ is constant on the coset $A = aN$. So, if $B = bN$ for some $b \in G$, we can define

$$\varrho(A, B) = \sup_{x,y\in G} |f(xay) - f(xby)|.$$

The function $\varrho$ is correctly defined and we claim that $\varrho$ is a metric on $G/N$.

Clearly, $\varrho(A, B) \geq 0$. Suppose that $\varrho(A, B) = 0$, where $A = aN$ and $B = bN$. Then $f(xay) = f(xby)$ for all $x, y \in G$, whence $a \sim b$ and $A = B$. It is also clear that $f(B, A) = f(A, B)$. To verify the triangle inequality, we take arbitrary $A = aN$, $B = bN$, and $C = cN$. Then

$$\begin{aligned}\varrho(A, C) &= \sup_{x,y\in G} |f(xay) - f(xcy)| \\ &\leq \sup_{x,y\in G} \big(|f(xay) - f(xby)| + |f(xby) - f(xcy)|\big) \\ &\leq \sup_{x,y\in G} |f(xay) - f(xby)| + \sup_{x,y\in G} |f(xby) - f(xcy)| \\ &= f(A, B) + f(B, C).\end{aligned}$$

The next step is to verify the continuity of $\varrho$ with respect to the quotient topology on $G/N$. Let $U = \{B \in G/N : \varrho(A, B) < \varepsilon\}$ be a ball with center $A = aN \in G/N$ and radius $\varepsilon > 0$. The function $\varphi\colon G^3 \to \mathbb{R}$ defined by $\varphi(x, b, y) = |f(xay) - f(xby)|$ is continuous, and $\varphi(x, a, y) = 0$ for all $x, y \in G$. Therefore, one can find open neighbourhoods $U(x, y)$, $V(x, y)$ and $W(x, y)$ of $x$, $a$ and $y$, respectively, in $G$ such that

$$\varphi(U(x, y) \times V(x, y) \times W(x, y)) \subset [0, \varepsilon).$$

Since the family $\{U(x, y) \times W(x, y) : x, y \in G\}$ is an open covering of the compact space $G \times G$, there exist points $(x_1, y_1), \dots, (x_n, y_n)$ in $G \times G$ such that

$$G \times G = \bigcup_{i=1}^{n} (U(x_i, y_i) \times W(x_i, y_i)).$$

Hence, $V^* = \bigcap_{i=1}^n V(x_i, y_i)$ is an open neighbourhood of $a$ in $G$, and if $b \in V^*$, then $\varphi(x, b, y) < \varepsilon$ for all $x, y \in G$. So, $\varrho(A, B) < \varepsilon$, where $B = bN$. In particular, the open set $V = \pi(V^*)$ in $G/N$, where $\pi\colon G \to G/N$ is the quotient homomorphism, satisfies $A \in V \subset U$. Hence, the metric $\varrho$ is continuous on the quotient group $G/N$, which means that the topology generated by $\varrho$ on the set $G/N$ is contained in the quotient topology of $G/N$. Since the quotient group $G/N$ is compact and Hausdorff, it follows that the quotient topology of $G/N$ and the topology generated by the metric $\varrho$ coincide. □

It is clear that the quotient group $G/N$ in the above theorem is compact and metrizable. Therefore, $G/N$ has a countable base. This is the basic feature of the above construction that we want to preserve in the general definition of $\mathbb{R}$-*factorizable groups* that follows.

A topological group $G$ is called $\mathbb{R}$-*factorizable* if, for every continuous real-valued function $f$ on $G$, there exist a continuous homomorphism $\pi\colon G \to K$ onto a second-countable topological group $K$ and a continuous function $h$ on $K$ such that $f = h \circ \pi$. The homomorphism $\pi$ and the functions $f, h$ in this definition make the following triangular diagram commutative.

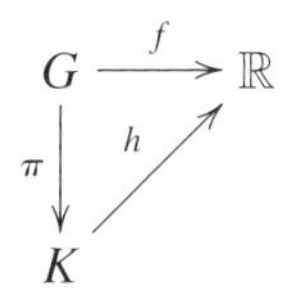


We also say that the homomorphism $\pi$ *factorizes* $f$ or, equivalently, we write $\pi \prec f$ (see also Section 1.7).

It is immediate from the above definition that every second-countable topological group is $\mathbb{R}$-factorizable. By Theorem 8.1.1, compact groups are $\mathbb{R}$-factorizable as well. It also follows from Lemma 1.7.1 and Corollary 1.7.5 that an arbitrary topological product of second-countable topological groups is $\mathbb{R}$-factorizable. To extend these results to wider classes of topological groups, we have to establish several simple but useful facts.

The next result shows, in particular, that the real line $\mathbb{R}$ in the definition of $\mathbb{R}$-factorizability can be replaced by the space $\mathbb{R}^\omega$.

LEMMA 8.1.2. *Suppose that $f\colon G \to X$ is a continuous mapping of an $\mathbb{R}$-factorizable group $G$ to a Tychonoff space $X$ with $w(X) \leq \tau \geq \omega$. Then one can find a continuous homomorphism $\pi\colon G \to K$ onto a topological group $K$ with $w(K) \leq \tau$ such that $\pi \prec f$.*

PROOF. According to [**165**, Theorem 2.3.23], we can identify $X$ with a subspace of $\mathbb{R}^\tau$. For every $\alpha < \tau$, denote by $p_\alpha$ the projection of $\mathbb{R}^\tau$ to the $\alpha$th factor. Then $p_\alpha \circ f\colon G \to \mathbb{R}$ is a continuous real-valued function, so we can find a continuous homomorphism $\pi_\alpha\colon G \to K_\alpha$ onto a second-countable topological group $K_\alpha$ and a continuous real-valued function $g_\alpha$ on $K_\alpha$ such that $p_\alpha \circ f = g_\alpha \circ \pi_\alpha$. Denote by $\pi$ the diagonal product of the homomorphisms $\pi_\alpha$, with $\alpha < \tau$. Then $\pi\colon G \to \prod_{\alpha<\tau} K_\alpha$ is a continuous homomorphism and the image $K = \pi(G)$ is a subgroup of the group $\Pi = \prod_{\alpha<\tau} K_\alpha$ satisfying $w(\Pi) \leq \tau$. Therefore, $w(K) \leq \tau$. For every $\alpha < \tau$, let $q_\alpha\colon \Pi \to K_\alpha$ be the projection. Then $\pi_\alpha = q_\alpha \circ \pi$, for each $\alpha < \tau$. Finally, denote by $g$ the Cartesian product of the family $\{g_\alpha : \alpha < \tau\}$. Then the mapping $g\colon \Pi \to \mathbb{R}^\tau$ is continuous. In addition, the diagram below commutes.

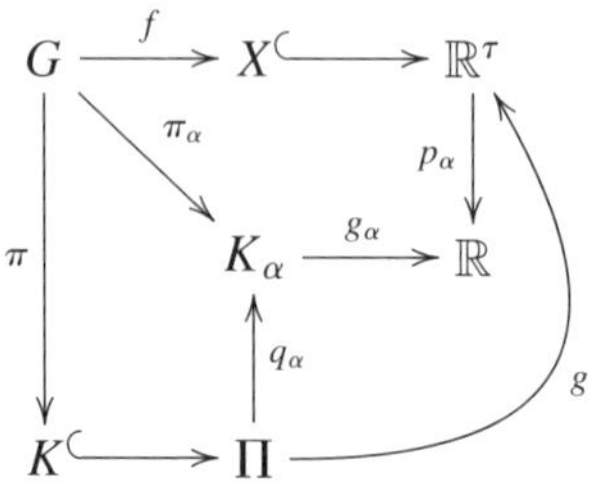


Indeed, it suffices to verify that $f = g \circ \pi$ or, equivalently, that $p_\alpha \circ f = p_\alpha \circ g \circ \pi$ for each $\alpha < \tau$. This, however, follows from our definition of $g$ and $\pi$:

$$p_\alpha \circ g \circ \pi = g_\alpha \circ q_\alpha \circ \pi = g_\alpha \circ \pi_\alpha = p_\alpha \circ f.$$

Therefore, the homomorphism $\pi \colon G \to K$ and the mapping $h = g \restriction K$ satisfy the equality $f = h \circ \pi$, i.e., $\pi \prec f$. □

On one hand, every compact topological group is $\mathbb{R}$-factorizable, by Theorem 8.1.1. On the other hand, uncountable discrete groups seem to be very far from being $\mathbb{R}$-factorizable. Theorem 8.1.9 shows that this is indeed so. It requires several preliminary steps. The first of them relates $\mathbb{R}$-factorizable and $\omega$-narrow groups.

PROPOSITION 8.1.3. *Every $\mathbb{R}$-factorizable group is $\omega$-narrow.*

PROOF. Let $\{f_i : i \in I\}$ be the family of continuous real-valued functions on an $\mathbb{R}$-factorizable topological group $G$. For every $i \in I$, define a continuous homomorphism $\pi_i \colon G \to K_i$ to a second-countable topological group $K_i$ and a continuous function $h_i$ on $K_i$ such that $f_i = h_i \circ \pi_i$. Since the family $\{f_i : i \in I\}$ separates points and closed sets in $G$, so does $\{\pi_i : i \in I\}$. Therefore, the diagonal product $\pi$ of the family $\{\pi_i : i \in I\}$ is a topological isomorphism of $G$ onto the subgroup $H = \pi(G)$ of $\prod_{i \in I} K_i$. Every subgroup of a topological product of second-countable groups is $\omega$-narrow, by Proposition 3.4.3 and Theorem 3.4.4. It follows that $G$ is $\omega$-narrow. □

We shall see in Example 8.2.1 that $\omega$-narrowness does not imply $\mathbb{R}$-factorizability. However, the stronger condition of being Lindelöf does imply this. To prove this fact, we need the following general lemma on factorization of continuous functions (no separation axiom on the factors is assumed):

LEMMA 8.1.4. *Let $S$ be a subspace of the product space $X = \prod_{i \in I} X_i$ that satisfies $l(S) \leq \tau$. Then, for every continuous real-valued function $f$ on $S$, one can find a set $J \subset I$ with $|J| \leq \tau$ and a continuous function $h \colon \pi_J(S) \to \mathbb{R}$ such that $f = h \circ \pi_J \restriction S$, where $\pi_J \colon X \to X_J = \prod_{i \in J} X_i$ is the projection.*

PROOF. For every $n \in \mathbb{N}$, let $\gamma_n$ be a covering of $\mathbb{R}$ by open intervals of length less than $1/n$. Since $f$ is continuous, for every $x \in S$ and every $n \in \mathbb{N}$ there exists a canonical open set $U_n(x)$ in $X$ such that $x \in U_n(x)$ and $f(U_n(x) \cap S) \subset V$, for some $V \in \gamma_n$ containing $f(x)$. It follows from $l(S) \leq \tau$ that, for every $n \in \mathbb{N}$, the covering $\{U_n(x) : x \in S\}$ of $S$ contains a subcovering $\mu_n$ with $|\mu_n| \leq \tau$. Each element $U \in \mu_n$ depends on a finite coordinate set $J(U) \subset I$, so we define $J_n = \bigcup\{J(U) : U \in \mu_n\}$ and $J = \bigcup\{J_n : n \in \mathbb{N}\}$. It is clear that $|J| \leq \tau$ and $U = \pi_J^{-1}\pi_J(U)$, for each $U \in \mu = \bigcup_{n=1}^{\infty} \mu_n$.

We claim that $f(x) = f(y)$ for all $x, y \in S$ satisfying $\pi_J(x) = \pi_J(y)$. Indeed, assume the contrary, and choose $x, y \in S$ and $n \in \mathbb{N}$ such that $\pi_J(x) = \pi_J(y)$ and $1/n < |f(x) - f(y)|$. By the definition of $\mu_n$, there are $U \in \mu_n$ and $V \in \gamma_n$ such that $x \in U$ and $f(U \cap S) \subset V$. From $U = \pi_J^{-1}\pi_J(U)$ and $\pi_J(x) = \pi_J(y)$ it follows that $y \in U$; hence $f(x), f(y) \in V$. Therefore, $|f(x) - f(y)| < 1/n$, contradicting the choice of $n$.

By the above claim, there is a function $h\colon \pi_J(S) \to \mathbb{R}$ such that $f = h \circ \pi_J \restriction S$. It remains to show that $h$ is continuous. Let $y$ be an arbitrary point of $Y = \pi_J(S)$, $z = h(y)$, and $\varepsilon > 0$ a real number. We can find $n \in \mathbb{N}$ such that $1/n < \varepsilon$. Choose a point $x \in S$ with $\pi_J(x) = y$ and an element $U \in \mu_n$ containing $x$. Clearly, $f(x) = h(\pi_J(x)) = h(y) = z$, so there exists $V \in \gamma_n$ such that $f(U \cap S) \subset V \subset O_\varepsilon(z)$, where $O_\varepsilon(z) = \{t \in \mathbb{R} : |t - z| < \varepsilon\}$. Then $W = \pi_J(U) \cap Y$ is an open neighbourhood of $y$ in $Y$, and we have

$$h(W) = h(\pi_J(U) \cap Y) = f(U \cap S) \subset O_\varepsilon(z).$$

Thus, $h$ is continuous. □

PROPOSITION 8.1.5. *Let $S$ be a Lindelöf subspace of an $\omega$-narrow topological group $H$ and $f$ a continuous real-valued function on $S$. Then there exists a continuous homomorphism $p\colon H \to K$ onto a second-countable group $K$ such that $p \restriction S \prec f$.*

PROOF. Since the group $H$ is $\omega$-narrow, Theorem 3.4.23 implies that $H$ can be embedded as a topological subgroup into a product $\Pi = \prod_{i \in I} H_i$ of second-countable topological groups. Let us identify $H$ with its image under this embedding. By Lemma 8.1.4, for every continuous function $f\colon S \to \mathbb{R}$, we can find a countable set $J \subset I$ and a continuous function $h\colon \pi_J(S) \to \mathbb{R}$ such that $f = h \circ \pi_J \restriction S$, where $\pi_J\colon \Pi \to \prod_{i \in J} H_i$ is the projection. Note that both $\pi_J$ and its restriction $p = \pi_J \restriction H$ are continuous homomorphisms, and the image $K = \pi_J(H)$ is a second-countable topological group as a subgroup of the second-countable group $\prod_{i \in J} H_i$. This completes the proof. □

The next theorem follows immediately from Proposition 8.1.5 in the special case when $S = H$. It supplies us with many non-trivial examples of $\mathbb{R}$-factorizable groups.

THEOREM 8.1.6. [**M. G. Tkachenko**] *Every Lindelöf topological group is $\mathbb{R}$-factorizable.*

Since every space with a countable network is Lindelöf, the following fact follows directly from Theorem 8.1.6:

COROLLARY 8.1.7. *Every cosmic topological group is $\mathbb{R}$-factorizable.*

The notion of $\mathbb{R}$-factorizability can naturally be extended to the classes of semitopological, quasitopological, and paratopological groups. However, unlike the case of topological groups, we have to specify an appropriate axiom of separation. For example, we say that a regular paratopological group $G$ is *$\mathbb{R}_3$-factorizable* if, for every continuous real-valued function $f$ on $G$, one can find a continuous homomorphism $p\colon G \to H$ onto a regular paratopological group $H$ with a countable base and a continuous real-valued function $h$ on $H$ such that $f = h \circ p$. Similarly, one defines *$\mathbb{R}_2$-factorizable* and *$\mathbb{R}_1$-factorizable* paratopological groups, by restricting attention to Hausdorff and $T_1$ paratopological groups, respectively. It turns out that not all results established in this section can be extended to paratopological groups, as the following example shows.

EXAMPLE 8.1.8. The Sorgenfrey line $Z$, with the usual sum operation, is a Lindelöf paratopological group which is not $\mathbb{R}_1$-factorizable.

Indeed, let $f\colon Z \to \mathbb{R}$ be the function of $Z$ defined by $f(x) = 1$ if $0 \le x < 1$ and $f(x) = 0$ otherwise. Since the set $J = [0, 1)$ is clopen in $Z$, $f$ is continuous. Suppose that there exist a continuous homomorphism $\pi\colon Z \to G$ of $Z$ onto a $T_1$ paratopological group $G$ and a continuous function $g$ on $G$ such that $f = g \circ \pi$. We claim that $\pi$ is a homeomorphism.

If $U$ is open in $\mathbb{R}$, then $\pi(f^{-1}(U)) = g^{-1}(U)$ is open in $G$ and, clearly, $f^{-1}(U) = \pi^{-1}\pi(f^{-1}(U))$. Since $J = f^{-1}(0, 2)$, it follows that $V = \pi(J)$ is open in $G$ and $J = \pi^{-1}(V)$. Taking into account that $\pi$ is a homomorphism, we have that $x + J = \pi^{-1}\pi(x + J) = \pi^{-1}(y + V)$, where $x \in Z$ and $y = \pi(x)$. The sets $J \cap (x + J)$, with $-1 < x < 0$, form a local base at zero in $Z$, and the above equalities imply that

$$[0, 1 - x) = J \cap (x + J) = \pi^{-1}(V \cap (y + V)),$$

where $y = \pi(x)$. Each of the sets $V \cap (y + V)$ is an open neighbourhood of the neutral element $0_G$ in $G$, so the preimages under $\pi$ of elements of a neighbourhood base at $0_G$ in $G$ constitute a local base at zero in $Z$. This means that $\pi$ is an isomorphism and $\pi^{-1}$ is continuous, that is, $\pi$ is a homeomorphism. Since $Z$ is not second-countable, it is not $\mathbb{R}_1$-factorizable either. □

The following theorem characterizes $\mathbb{R}$-factorizability in the class of locally compact groups.

THEOREM 8.1.9. *A locally compact topological group is $\mathbb{R}$-factorizable iff it is $\sigma$-compact.*

PROOF. Every $\sigma$-compact topological group is Lindelöf and hence, $\mathbb{R}$-factorizable, by Theorem 8.1.6. Conversely, let $G$ be an $\mathbb{R}$-factorizable locally compact group. Then Proposition 8.1.3 implies that $G$ is $\omega$-narrow. Choose an open neighbourhood $U$ of the identity in $G$ with compact closure. There exists a countable set $F \subset G$ such that the translates $xU$, with $x \in F$, cover $G$. Since the closures of the sets $xU$ are compact, we conclude that $G$ is $\sigma$-compact. □

COROLLARY 8.1.10. *A discrete topological group is $\mathbb{R}$-factorizable iff it is countable.*

Theorem 8.1.6 extends the frontier of the class of $\mathbb{R}$-factorizable groups by weakening compactness to the Lindelöf property. However, we can move to another direction and consider countably compact or pseudocompact topological groups. In what follows we present more general results which imply, in particular, that *all subgroups* of compact topological groups are $\mathbb{R}$-factorizable. This will require, however, some work. We start with the following useful lemma that will enable us to shorten arguments.

LEMMA 8.1.11. *Let $G$ be a topological group with the property that for every continuous function $f\colon G \to \mathbb{R}$, there exists a continuous homomorphism $\pi\colon G \to H$ onto an $\mathbb{R}$-factorizable group $H$ such that $\pi \prec f$. Then the group $G$ is $\mathbb{R}$-factorizable.*

PROOF. Let $f\colon G \to \mathbb{R}$ be a continuous function. By the assumptions, we can find a continuous homomorphism $\pi\colon G \to H$ onto an $\mathbb{R}$-factorizable group $H$ and a continuous function $g\colon H \to \mathbb{R}$ such that $f = g \circ \pi$. By the $\mathbb{R}$-factorizability of $H$, there are a continuous homomorphism $p: H \to K$ onto a second-countable topological group $K$ and a continuous real-valued function $h$ on $K$ such that $g = h \circ p$.

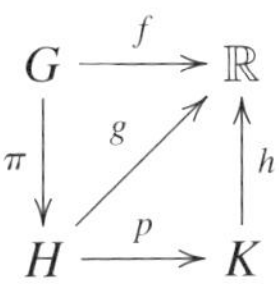


Clearly, the continuous homomorphism $\varphi = p \circ \pi$ of $G$ onto $K$ satisfies $f = h \circ \varphi$. □

The next result refines Corollary 5.2.14 slightly.

Lemma 8.1.12. *Let $G$ be a cosmic topological group and $\gamma$ a countable family of open sets in $G$. Then there exists a continuous isomorphism $p: G \to H$ onto a second-countable topological group $H$ such that $p(U)$ is open in $H$, for each $U \in \gamma$.*

Proof. By the assumptions, we have $ib(G) \leq l(G) \leq nw(G) \leq \omega$ and $\psi(G) \leq nw(G) \leq \omega$, so Proposition 5.2.11 implies that there exists a continuous isomorphism $\psi_0: G \to H_0$ of $G$ onto a second-countable topological group $H_0$.

By Theorem 3.4.23, there is a base $\mathcal{B}$ for the topology of $G$ which consists of the sets of the form $\psi^{-1}(V)$, where $\psi: G \to K$ is a continuous homomorphism onto a second-countable group $K$ and $V$ is open in $K$. Let $U$ be an arbitrary non-empty open set in $G$. Since $nw(G) \leq \omega$, the group $G$ is hereditarily Lindelöf, so we can find a countable family $\{U_n : n \in \omega\} \subset \mathcal{B}$ such that $U = \bigcup_{n\in\omega} U_n$. Let $\psi_n: G \to K_n$ be a continuous homomorphism onto a second-countable group $K_n$ corresponding to $U_n$, $n \in \omega$. Then the diagonal product $\psi_U$ of the family $\{\psi_n : n \in \omega\}$ is a continuous homomorphism of $G$ to the second-countable group $\prod_{n\in\omega} K_n$, so the subgroup $H_U = \psi_U(G)$ of $\prod_{n\in\omega} K_n$ is also second-countable. By the construction, we have $U_n = \psi_U^{-1}\psi_U(U_n)$, where $\psi_U(U_n)$ is open in $H_U$ for each $n \in \omega$.

Finally, let $p$ be the diagonal product of the family $\{\psi_U : U \in \gamma\} \cup \{\psi_0\}$, and take $H = p(G)$. Then $H$ is second-countable as a subgroup of the second-countable group $H_0 \times \prod_{U\in\gamma} H_U$. Hence, $p$ and $H$ are as required. □

One of the main results of this section is a factorization theorem for dense subgroups of topological products of Lindelöf $\Sigma$-groups (see Theorem 8.1.14 below). First we present a version of this result for arbitrary subgroups of a single Lindelöf $\Sigma$-group.

Proposition 8.1.13. *Every subgroup of a Lindelöf $\Sigma$-group is $\mathbb{R}$-factorizable.*

Proof. Let $H$ be a subgroup of a Lindelöf $\Sigma$-group $G$ and $H^*$ be the closure of $H$ in $G$. Then $H^*$ is also a Lindelöf $\Sigma$-group, so taking $H^*$ in place of $G$ we can assume that $H$ is dense in $G$.

By the continuity of $f$, for every $x \in H$ we can find a closed $G_\delta$-set $F$ in $G$ such that $x \in F$, $f$ admits a continuous extension over $H \cup F$, and this extension is constant on $F$. Denote by $\gamma$ the family of these sets $F$. Since $H$ is dense in $G$, we can apply [**165**, 3.2.A(b)] to conclude that there exists a continuous extension of $f$ over $S = \bigcup \gamma$, and this extension (denoted by the same letter $f$) is constant on each $F \in \gamma$. By a) of Theorem 5.3.30, we can find a countable subfamily $\gamma_0$ of $\gamma$ such that $\bigcup \gamma_0$ is dense in $S$. Let $\gamma_0 = \{F_n : n \in \omega\}$. According to c) of Theorem 5.3.30, for every $n \in \omega$ there exists a continuous homomorphism $\varphi_n\colon G \to K_n$ onto a topological group $K_n$ with a countable network such that $F_n = \varphi_n^{-1}\varphi_n(F_n)$. Denote by $\varphi$ the diagonal product of the family $\{\varphi_n : n \in \omega\}$. Then $\varphi$ is a continuous homomorphism of $G$ to the product group $K = \prod_{n\in\omega} K_n$ which has a countable network. Denote by $N$ the kernel of $\varphi$ and consider the quotient homomorphism $p\colon G \to G/N$. Since $p \prec \varphi$, we have that $F_n = p^{-1}p(F_n)$ for each $n \in \omega$. Clearly, the pseudocharacter of the identity in $K$ is countable, so $N$ is a $G_\delta$-set in $G$. Hence, $nw(G/N) \leq \aleph_0$, by a) of Lemma 5.3.24.

By our choice, the union $Q = \bigcup_{n\in\omega} F_n$ is dense in $S \supseteq H$, so we can apply Lemma 1.7.6 (with $X = G$, $S = H$, $Y = \mathbb{R}$, $g = p$ and $T = p(Q)$) to define a continuous function $h\colon p(H) \to \mathbb{R}$ such that $f = h \circ p{\restriction}H$.

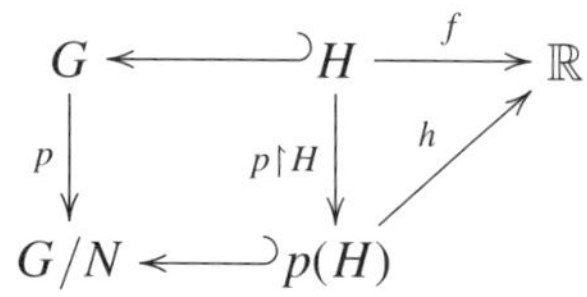


Clearly, the group $p(H)$ is cosmic as a subgroup of $G/N$. By Corollary 8.1.7, the group $p(H)$ is $\mathbb{R}$-factorizable. It remains to apply Lemma 8.1.11 to conclude that the group $H$ is $\mathbb{R}$-factorizable as well. □

We shall see in Example 8.2.1 that Proposition 8.1.13 cannot be extended to subgroups of Lindelöf groups.

THEOREM 8.1.14. [**M. G. Tkachenko**] *Dense subgroups of topological products of Lindelöf $\Sigma$-groups are $\mathbb{R}$-factorizable.*

PROOF. Let $G = \prod_{i\in I} G_i$ be the topological product of Lindelöf $\Sigma$-groups $G_i$, $H$ a dense subgroup of $G$, and $f\colon H \to \mathbb{R}$ a continuous function. It follows from Corollary 5.3.31 that the space $G$ is $\omega$-cellular. By Theorem 1.7.7, we can find a countable set $K \subset I$ and a continuous function $h\colon p_K(H) \to \mathbb{R}$ such that $f = h \circ p_K{\restriction}H$, where $p_K$ is the projection of $G$ onto $G_K = \prod_{i\in K} G_i$. It follows from $|K| \leq \omega$ that $Nag(G_K) \leq \omega$ (we apply Proposition 5.3.9 here). Since $H_K = p_K(H)$ is a subgroup of the Lindelöf $\Sigma$-group $G_K$, the group $H_K$ is $\mathbb{R}$-factorizable, by Proposition 8.1.13.

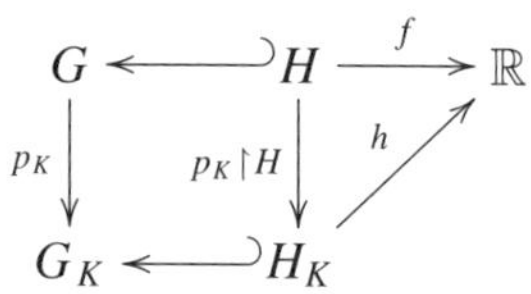


Hence Lemma 8.1.11 implies that $H$ is also $\mathbb{R}$-factorizable. □

Several special cases of Theorem 8.1.14 deserve mentioning here. The first of them corresponds to the fact that every second-countable space is a Lindelöf $\Sigma$-space.

COROLLARY 8.1.15. *A dense subgroup of a topological product of second-countable topological groups is $\mathbb{R}$-factorizable.*

Since the class of Lindelöf $\Sigma$-groups contains all $\sigma$-compact groups, we have the following:

Corollary 8.1.16. *Any subgroup of a $\sigma$-compact topological group is $\mathbb{R}$-factorizable.*

By Corollary 3.7.17, precompact groups are subgroups of compact groups, so Corollary 8.1.16 implies the following fact which generalizes Theorem 8.1.1.

Corollary 8.1.17. *Every precompact topological group is $\mathbb{R}$-factorizable.*

Let us show that weakly Lindelöf topological groups possess a property somewhat weaker than $\mathbb{R}$-factorizability. We present a more general result for topological groups $G$ satisfying $wl(G) \leq \tau$.

Theorem 8.1.18. **[E. V. Schepin]** *For every continuous real-valued function $f$ on a topological group $G$ with $wl(G) \leq \tau$, there exists an open continuous homomorphism $\pi\colon G \to K$ onto a topological group $K$ satisfying $\psi(K) \leq \tau$ such that $\pi \prec f$.*

Proof. Let $\varepsilon > 0$ be arbitrary. For every $x \in G$, choose a neighbourhood $U_x$ of $x$ in $G$ such that $|f(y) - f(x)| < \varepsilon$ for each $y \in U_x$. We can also find an open neighbourhood $V_x$ of the identity $e$ in $G$ such that $V_x^2 x \subset U_x$. Since $wL(G) \leq \tau$, the family $\{V_x x : x \in G\}$ contains a subfamily $\mu$ with $|\mu| \leq \tau$ whose union is dense in $G$. Let $\mu = \{V_x x : x \in C\}$, where $C \subset G$ and $|C| \leq \tau$. Then $P(\varepsilon) = \bigcap\{V_x : x \in C\}$ is a $G_\tau$-set in $G$, $e \in P(\varepsilon)$, and we claim that

$$|f(gx) - f(x)| \leq 2\varepsilon \text{ for all } x \in G \text{ and } g \in P(\varepsilon). \tag{8.1}$$

Assume to the contrary that $|f(gz) - f(z)| > 2\varepsilon$ for some $z \in G$ and $g \in P(\varepsilon)$. Then there exists a neighbourhood $O$ of $z$ such that the inequality $|f(gy) - f(y)| > 2\varepsilon$ holds for all $y \in O$. Since the set $\cup\mu$ is dense in $G$, we can find $x \in C$ such that $O \cap (V_x x) \neq \varnothing$. Pick a point $y \in O \cap (V_x x)$. Then our choice of $V_x$ implies that $|f(y) - f(x)| < \varepsilon$. In addition, from $y \in V_x x$ and $g \in P(\varepsilon) \subset V_x$ it follows that $gy \in V_x^2 x \subset U_x$. Therefore, $|f(gy) - f(x)| < \varepsilon$, and we have

$$|f(gy) - f(y)| \leq |f(gy) - f(x)| + |f(x) - f(y)| < 2\varepsilon.$$

This contradiction completes the proof of our claim.

Put $P = \bigcap_{n=1}^{\infty} P(1/n)$. Then $P$ is a $G_\tau$-set in $G$ and $e \in P$. From (8.1) it follows immediately that $f(gx) = f(x)$ for all $x \in G$ and all $g \in P$. Let $\{W_i : i \in I\}$ be a family of neighbourhoods of $e$ in $G$ such that $P = \bigcap_{i\in I} W_i$, where $|I| \leq \tau$. By Proposition 5.2.8, the group $G$ is $\tau$-narrow. It follows from Corollary 5.1.7 that, for every $i \in I$, there exists a continuous homomorphism $\varphi_i\colon G \to H_i$ onto a group $H_i$ with $w(H_i) \leq \tau$ such that $N_i = \ker\varphi_i \subset W_i$. Denote by $\varphi$ the diagonal product of the homomorphisms $\varphi_i$, $i \in I$. Then the image $H = \varphi(G)$ is a subgroup of the product $\prod_{i\in I} H_i$ and hence, $w(H) \leq \tau$. Let $N$ be the kernel of $\varphi$. Clearly, $N = \bigcap_{i\in I} N_i$, so that $N \subset P$. Consider the quotient homomorphism $\pi\colon G \to G/N$. Evidently, there exists an algebraic isomorphism $i\colon G/N \to H$ such that $\varphi = i \circ \pi$. Since $\pi$ is open, the isomorphism $i$ is continuous. Therefore, the quotient group $K = G/N$ satisfies $\psi(K) \leq \tau$. Finally, the inclusion $N \subset P$ implies that $f$ is constant on all cosets $Nx$ in $G$. This fact enables us to define a function $h\colon G/N \to \mathbb{R}$ such that $f = h \circ \pi$. Again, $h$ is continuous because $\pi$ is open. So, $\pi \prec f$. □

Corollary 8.1.19. *For every continuous real-valued function $f$ on a topological group $H$ of countable cellularity, there exists a closed invariant subgroup $N$ of type $G_\delta$ in $H$ such that $f$ is constant on every coset of $N$ in $H$.*

Theorem 8.1.18 shows that weakly Lindelöf topological groups are close to being $\mathbb{R}$-factorizable. It will be shown in Example 8.6.16, however, that $\mathbb{R}$-factorizable groups need not be weakly Lindelöf. On the other hand, weakly Lindelöf $\omega$-steady groups are $\mathbb{R}$-factorizable:

PROPOSITION 8.1.20. *Every weakly Lindelöf $\omega$-steady group is $\mathbb{R}$-factorizable.*

PROOF. Let $f\colon G \to \mathbb{R}$ be a continuous function defined on a weakly Lindelöf $\omega$-steady group $G$. By Theorem 8.1.18, we can find a quotient homomorphism $\pi\colon G \to H$ onto a topological group $H$ of countable pseudocharacter and a continuous function $g\colon H \to \mathbb{R}$ such that $f = g \circ \pi$. Evidently, the group $H$ is $\omega$-steady as a quotient of the $\omega$-steady group $G$. Since the group $H$ is $\omega$-steady and $\psi(H) \leq \omega$, it must have a countable network. Then $H$ is $\mathbb{R}$-factorizable, by Corollary 8.1.7. The required conclusion now follows from Lemma 8.1.11. □

The results of Section 5.6 about the relationship between $\omega$-steady and $\omega$-stable topological groups can be complemented as follows.

THEOREM 8.1.21. *Every $\mathbb{R}$-factorizable $\omega$-steady topological group is $\omega$-stable.*

PROOF. Let $G$ be an $\mathbb{R}$-factorizable $\omega$-steady topological group. Suppose that $f\colon G \to X$ and $g\colon X \to Y$ are continuous onto mappings, where $g$ is one-to-one and $w(Y) \leq \omega$. The mapping $h = g \circ f$ of $G$ to $Y$ is continuous, so we can apply Lemma 8.1.2 (with $\tau = \omega$) to find a continuous homomorphism $p\colon G \to H$ onto a second-countable topological group $H$ and a continuous mapping $r\colon H \to Y$ such that $h = r \circ p$. Since $g$ is one-to-one, there exists a (not necessarily continuous) mapping $s\colon H \to X$ satisfying $r = g \circ s$. In fact, $s = g^{-1} \circ r$.

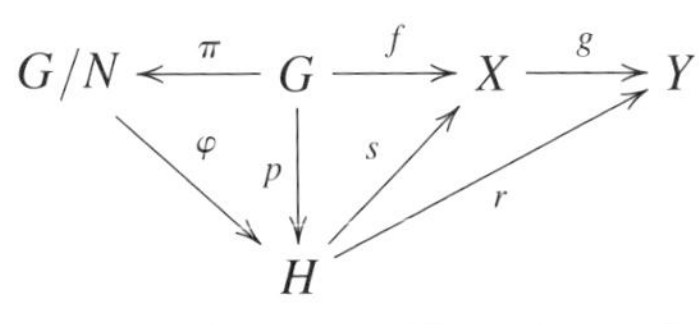


Denote by $N$ the kernel of $p$. Then $N$ is a closed invariant subgroup of type $G_\delta$ in $G$. Since $G$ is $\omega$-steady, the quotient group $G/N$ has a countable network. Let $\pi\colon G \to G/N$ be the quotient homomorphism. Denote by $\varphi$ the natural isomorphism of $G/N$ onto $H$. Then, clearly, $p = \varphi \circ \pi$. The isomorphism $\varphi$ is continuous since $\pi$ is open.

Let $\tilde{s} = s \circ \varphi$. We claim that the mapping $\tilde{s}\colon G/N \to X$ is continuous. Indeed, we have

$$f = s \circ p = s \circ \varphi \circ \pi = \tilde{s} \circ \pi,$$

where $f$ is continuous and $\pi$ is open. Therefore, the mapping $\tilde{s}$ is also continuous. Since $nw(G/N) \leq \omega$, the image $X = \tilde{s}(G/N)$ satisfies $nw(X) \leq \omega$ as well. The latter inequality implies that $G$ is $\omega$-stable. □

Combining Theorem 8.1.21 and Proposition 8.1.20, we deduce the next corollary:

COROLLARY 8.1.22. *A weakly Lindelöf $\omega$-steady topological group is $\omega$-stable.*

It turns out that a stronger form of countable cellularity in topological groups does imply $\mathbb{R}$-factorizability, but only in a special model of $ZFC$.

PROPOSITION 8.1.23. *It is consistent with ZFC that every topological group G with $cel_\omega(G) \leq \omega$ is $\mathbb{R}$-factorizable.*

PROOF. Suppose that a topological group $G$ satisfies $cel_\omega(G) \leq \omega$, and let $f: G \to \mathbb{R}$ be a continuous function. Since $c(G) \leq cel_\omega(G) \leq \omega$, we can apply Theorem 8.1.18 to find a continuous homomorphism $\pi: G \to K$ onto a group $K$ of countable pseudocharacter and a continuous function $h: K \to \mathbb{R}$ such that $f = h \circ \pi$. Then $cel_\omega(K) \leq cel_\omega(G) \leq \omega$. We conclude, therefore, that the group $K$ is hereditarily separable. By Theorem 9.3 of [**494**], there is a model of $ZFC$ in which every hereditarily separable regular space is hereditarily Lindelöf. In this model, the group $K$ is Lindelöf and hence, $\mathbb{R}$-factorizable, by Theorem 8.1.6. So the group $G$ is $\mathbb{R}$-factorizable, by Lemma 8.1.11. □

## Exercises

8.1.a. Complement Exercise 8.1.8 as follows: Every continuous automorphism $f$ of the paratopological group $Z$ (the Sorgenfrey line) has the form $f(x) = ax$, for each $x \in Z$, where $a = f(1) > 0$.

8.1.b. Suppose that a real-valued function $f$ on an $\omega$-narrow topological group $G$ is uniformly continuous with respect to the two-sided uniformity of $G$. Prove that there exists a continuous homomorphism $\pi: G \to K$ onto a second-countable group $K$ such that $\pi \prec f$. Deduce that if every continuous real-valued function on an $\omega$-narrow topological group $G$ is uniformly continuous with respect to the two-sided uniformity of $G$, then $G$ is $\mathbb{R}$-factorizable.

8.1.c. Prove that the free topological group $F(X)$ and the free Abelian topological group $A(X)$ are $\mathbb{R}$-factorizable, for every pseudocompact space $X$.

8.1.d. Let $X$ be a Tychonoff space. Prove the following:
 a) If the free topological group $F(X)$ is $\mathbb{R}$-factorizable, then so is the free Abelian topological group $A(X)$.
 b) If the free Abelian topological group $A(X)$ is $\mathbb{R}$-factorizable, then $X$ is pseudo-$\aleph_1$-compact.

8.1.e. Show that the topological group $C_p(X)$ defined in Section 1.9 is $\mathbb{R}$-factorizable, for every Tychonoff space $X$.

8.1.f. A Mal'tsev space $X$ is called $\mathbb{R}$-*factorizable* if, for every continuous real-valued function $f$ on $X$, there exists a continuous homomorphism $\pi: X \to Y$ onto a second-countable Mal'tsev space $Y$ such that $\pi \prec f$. Prove that every $\sigma$-compact Mal'tsev space is $\mathbb{R}$-factorizable.

8.1.g. Show that a precompact Tychonoff paratopological group need not be $\mathbb{R}_1$-factorizable. *Hint.* Consider the circle group $\mathbb{T}$ with the Sorgenfrey topology, and modify the argument in Example 8.1.8.

## Problems

8.1.A. Let $G$ be a topological group algebraically generated by a pseudocompact subspace. Show that all subgroups of $G$ are $\mathbb{R}$-factorizable.

8.1.B. Can Theorem 8.1.1 be extended to all compact left topological groups?

8.1.C. If every continuous mapping of a topological group $G$ to a metric space $Y$ can be factorized by a continuous homomorphism of $G$ onto a metrizable topological group, we will say that $G$ is $\mathcal{M}$-*factorizable.* Show that a paracompact topological group need not be $\mathcal{M}$-factorizable.

8.1.D. The notion of uniform continuity of mappings is naturally defined in the classes of paratopological and semitopological groups.

Suppose that $f$ is a uniformly continuous real-valued function on a regular (commutative) Lindelöf paratopological group. Can $f$ be factorized through a continuous homomorphism to a regular second-countable paratopological group?

8.1.E. (M. Sanchis and M. G. Tkachenko [**418**]) Let $G$ be a paratopological group which is a Lindelöf $\Sigma$-space. Prove that $G$ is $\mathbb{R}_3$-factorizable.

## Open Problems

8.1.1. Is it true that every topological group $G$ with countable Souslin number is $\mathbb{R}$-factorizable? What if $G$ is separable?

8.1.2. Must every paracompact $\mathbb{R}$-factorizable topological group be Lindelöf?

8.1.3. Is it true in $ZFC$ that every topological group $G$ with $cel_\omega(G) \leq \omega$ is $\mathbb{R}$-factorizable? (See Proposition 8.1.23.)

8.1.4. Let $G$ be a topological group such that $\aleph_1$ is a calibre of the space $G$ (that is, every uncountable family of non-empty open sets in $X$ has an uncountable subfamily with non-empty intersection). Is $G$ necessarily $\mathbb{R}$-factorizable?

8.1.5. Does there exist in $ZFC$ a (normal) $\mathbb{R}$-factorizable topological group of countable extent (see Section 5.2) which fails to be Lindelöf?

*Remark.* Under $CH$, A. Hajnal and I. Juhász constructed in [**215**] a normal, countably compact, hereditarily separable, non-compact (hence, non-Lindelöf) topological group $G$. Since countably compact groups are precompact, the group $G$ is $\mathbb{R}$-factorizable, by Corollary 8.1.17.

8.1.6. Is every (normal) topological group of countable extent $\mathbb{R}$-factorizable?

8.1.7. Does the $\mathbb{R}$-factorizability of $A(X)$ imply that of $F(X)$?

8.1.8. Characterize the spaces $X$ such that the free topological group $F(X)$ or the free Abelian topological group $A(X)$ is $\mathbb{R}$-factorizable.

8.1.9. Let $X$ be a Tychonoff space such that the cellularity of $F(X)$ is countable. Is $F(X)$ then $\mathbb{R}$-factorizable? What if $X$ is separable?

## 8.2. Subgroups of $\mathbb{R}$-factorizable groups. Embeddings

In view of Corollary 8.1.16 and Propositions 8.1.3 and 8.1.13, one can ask whether subgroups of Lindelöf groups or, more generally, of $\omega$-narrow groups are $\mathbb{R}$-factorizable. In both cases the answer is "no", as the following example shows.

EXAMPLE 8.2.1. There exist an Abelian $P$-group $G$ and a dense subgroup $H$ of $G$ such that $G$ is Lindelöf, but $H$ is not $\mathbb{R}$-factorizable. In particular, $H$ is an $\omega$-narrow $P$-group which is not Raĭkov complete.

Indeed, let $G = G_{\omega_1}$ be the Lindelöf $P$-group in Example 4.4.11 (with $A = \omega_1$ and $K = \mathbb{Z}(2) = \{0, 1\}$), constructed as a subgroup of the product group $\mathbb{Z}(2)^{\omega_1}$. More precisely,

$$G = \{x \in \mathbb{Z}(2)^{\omega_1} : |\operatorname{supp}(x)| < \omega\},$$

where $\operatorname{supp}(x) = \{\alpha < \omega_1 : x(\alpha) = 1\}$. Below we use the additive notation for the group operation in $G$. The group $G$ carries the $\omega$-box topology with the standard base

$$\mathcal{B} = \{x + U_\alpha : x \in G,\ \alpha < \omega_1\},$$

where each $U_\alpha = \{x \in G : \operatorname{supp}(x) \cap \alpha = \varnothing\}$ is an open subgroup of $G$. Note that $U_\alpha = \bigcap_{\beta<\alpha} U_\beta$ for every limit ordinal $\alpha < \omega_1$. Now we put

$$H = \{x \in G : |\operatorname{supp}(x)| \text{ is even}\}.$$

It is easy to see that $H$ is a proper dense subgroup of $G$. In particular, $H$ is an $\omega$-narrow $P$-group that fails to be Raĭkov complete. Let us show that the group $H$ is not $\mathbb{R}$-factorizable.

Clearly, $\{U_\alpha : \alpha < \omega_1\}$ is a decreasing sequence of open subgroups of $G$ which forms a base for $G$ at the identity. In particular, $G$ is zero-dimensional. For every $\alpha < \omega_1$, there exist non-empty disjoint clopen sets $B_\alpha$ and $C_\alpha$ in $G$ such that $U_\alpha \setminus U_{\alpha+1} = B_\alpha \cup C_\alpha$. Consider disjoint open sets

$$V = (G \setminus U_0) \cup \bigcup_{\alpha<\omega_1} B_\alpha \text{ and } W = \bigcup_{\alpha<\omega_1} C_\alpha.$$

Clearly, $V \cup W = G \setminus \{e\}$, where $e$ is the neutral element of $G$. Pick an arbitrary element $g \in G \setminus H$ and put

$$V' = (g + V) \cap H, \;\; W' = (g + W) \cap H.$$

Then $V'$ and $W'$ are non-empty disjoint open subsets of $H$ and $H = V' \cup W'$. Let a function $f\colon H \to [0, 1]$ be defined by $f(x) = 0$ if $x \in V'$ and $f(x) = 1$ if $x \in W'$. Clearly, $f$ is continuous. Suppose that $\pi\colon H \to K$ is a continuous homomorphism onto a second-countable group $K$. Then the kernel $N$ of $\pi$ is of type $G_\delta$ in $H$ and, hence, $N$ is open in $H$. Therefore, there exists $\alpha < \omega_1$ such that $U_\alpha \cap H \subset N$. Choose elements $x \in H \cap (g + B_\alpha)$ and $y \in H \cap (g + C_\alpha)$. Then $x - y \in H \cap U_\alpha \subset N$; it follows that $\pi(x) = \pi(y)$. However, $f(x) = 0$ and $f(y) = 1$, so $\pi$ does not factorize $f$. We conclude that $H$ is not $\mathbb{R}$-factorizable. □

Since Lindelöf topological groups are $\mathbb{R}$-factorizable, by Theorem 8.1.6, it follows from Example 8.2.1 that subgroups of $\mathbb{R}$-factorizable groups need not be $\mathbb{R}$-factorizable. One can ask, however, whether closed subgroups of $\mathbb{R}$-factorizable groups inherit this property. Again, the answer is "no".

THEOREM 8.2.2. *Every $\omega$-narrow topological group can be embedded into an $\mathbb{R}$-factorizable group as a closed invariant subgroup.*

PROOF. Let $G$ be an $\omega$-narrow group. By Theorem 3.4.23, $G$ can be embedded as a subgroup into a topological product of second-countable groups, say $K$. Let $\Pi = \prod_{n\in\omega} K_n$, where $K_n = K$ for each $n \in \omega$. Denote by $\sigma$ the $\sigma$-product of the groups $K_n$'s, i.e. the subgroup of $\Pi$ consisting of all points that do not coincide with the neutral element of $\Pi$ on at most finitely many coordinates. Obviously, $\sigma$ is dense in $\Pi$ when the latter is endowed with the usual product topology. Consider the embedding $i$ of $G$ into the diagonal $\Delta$ of $\Pi$ defined by the rule $\pi_n i(g) = g$ for all $g \in G$ and all $n \in \omega$, where $\pi_n\colon \Pi \to K_n$ is the projection. It is clear that $i$ is a topological isomorphism of $G$ onto $i(G)$. Put $G^* = \langle i(G) \cup \sigma\rangle$. Then $G^*$ is a dense subgroup of $\Pi$. The group $K$ is a product of second-countable groups, and so is $\Pi$. Since $G^*$ is dense in $\Pi$, Theorem 8.1.14 implies that the group $G^*$ is $\mathbb{R}$-factorizable. Now one can apply the argument in the proof of Lemma 6.8.3 to show that $i(G) = G^* \cap \Delta$, whence it follows that $G \cong i(G)$ is a closed invariant subgroup of $G^*$. □

Since every Lindelöf topological group is $\mathbb{R}$-factorizable, it is natural to ask, after Theorem 8.2.2, whether every $\mathbb{R}$-factorizable group is topologically isomorphic to a subgroup of a Lindelöf group. Again, the answer is in the negative. Indeed, the group $G = \mathbb{Z}^{\omega_1}$ with the Tychonoff product topology is $\mathbb{R}$-factorizable, by Corollary 8.1.15, but it does not admit an isomorphic embedding into any Lindelöf topological group. The reason for that is the Raĭkov completeness of $G$ — if $G$ were a subgroup of a Lindelöf group $G^*$, it would be closed in $G^*$ and hence, $G$ would be Lindelöf.

Example 8.2.1 shows that there are $\omega$-narrow Abelian groups that are not $\mathbb{R}$-factorizable. Hence, by Theorem 8.2.2, closed subgroups of $\mathbb{R}$-factorizable groups can fail to be $\mathbb{R}$-factorizable. On the other hand, it is easy to verify that $C$-embedded subgroups of $\mathbb{R}$-factorizable groups are $\mathbb{R}$-factorizable:

PROPOSITION 8.2.3. *A $C$-embedded subgroup of an $\mathbb{R}$-factorizable group is $\mathbb{R}$-factorizable.*

PROOF. Let $H$ be a $C$-embedded subgroup of an $\mathbb{R}$-factorizable group $G$. Then every continuous function $f\colon H \to \mathbb{R}$ admits an extension to a continuous function $g\colon G \to \mathbb{R}$. Since $G$ is $\mathbb{R}$-factorizable, we can find a continuous homomorphism $\pi\colon G \to K$ onto a second-countable group $K$ such that $\pi \prec g$. Then the homomorphism $\pi_0 = \pi\restriction H$ of $H$ onto the subgroup $\pi(H)$ of $K$ satisfies $\pi_0 \prec f$. □

Since every retract of a space $X$ is $C$-embedded in $X$, Proposition 8.2.3 implies the following.

COROLLARY 8.2.4. *If a subgroup $H$ of an $\mathbb{R}$-factorizable group is a retract of $G$, then $H$ is $\mathbb{R}$-factorizable.*

In general, an $\mathbb{R}$-factorizable subgroup of a topological group $G$ need not be $C$-embedded nor $C^*$-embedded in $G$. For example, let $G = \mathbb{T}$ be the circle group and $H$ an arbitrary countable dense subgroup of $G$. Then $H$ is second-countable and, hence, is $\mathbb{R}$-factorizable. However, $H$ is not $C^*$-embedded in $G$.

This makes it desirable to find a topological property responsible for the preservation of $\mathbb{R}$-factorizability when taking subgroups with that property. To present such a property, we recall that the complement of a zero-set is a *cozero-set.* A subspace $Y$ of $X$ is said to be *z-embedded* in $X$ if for every zero-set $F$ in $Y$, there exists a zero-set $\Phi$ in $X$ such that $F = Y \cap \Phi$. Clearly, $C$- and $C^*$-embedded subsets are $z$-embedded. It turns out that the property we are looking for is exactly the property of being $z$-embedded. We start the proof of this fact with the following general result.

THEOREM 8.2.5. *An $\mathbb{R}$-factorizable subgroup $H$ of arbitrary topological group $G$ is $z$-embedded in $G$.*

PROOF. Let $F$ be a zero-set in $H$, and consider a continuous real-valued function $f$ on $H$ such that $F = f^{-1}(0)$. Since $H$ is $\mathbb{R}$-factorizable, we can find a continuous homomorphism $\pi\colon H \to P$ onto a second-countable group $P$ and a continuous function $g\colon P \to \mathbb{R}$ such that $f = g \circ \pi$. Denote by $L$ the kernel of $\pi$. Let $\{O_n : n \in \omega\}$ be a countable base at the identity of $P$. We can define by induction a sequence $\{U_n : n \in \omega\}$ of open symmetric neighbourhoods of the identity $e$ in $G$ satisfying the following conditions for each $n \in \omega$:

(i) $U_{n+1}^2 \subset U_n$;
(ii) $U_n \cap H \subset \pi^{-1}(O_n)$.

It is clear that $K = \bigcap_{n\in\omega} U_n$ is a closed subgroup of $G$ and $K \cap H \subset L$. Let $\phi$ be the canonical mapping of $G$ onto the left coset space $G/K$. Since $K \cap H$ is a subgroup of $L$, there exists a function $\psi\colon \phi(H) \to P$ satisfying $\psi \circ \phi\restriction H = \pi$ (see the diagram below).

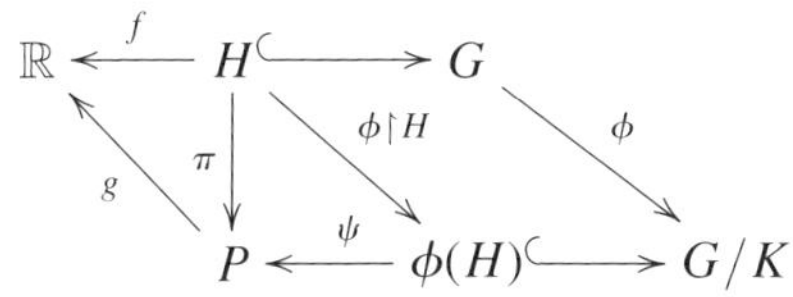

Since the open sets $U_n$ satisfy $e \in U_n = U_n^{-1}$ and $U_{n+1}^2 \subset U_n$ for each $n \in \omega$, Lemma 3.3.10 implies that there exists a continuous prenorm $N$ on $G$ such that

$$\{x \in G : N(x) < 1/2^n\} \subset U_n \subset \{x \in G : N(x) \leq 2/2^n\} \tag{8.2}$$

for all $n \in \omega$. Define a continuous left-invariant pseudometric $d$ on $G$ by $d(x, y) = N(x^{-1}y)$ for $x, y \in G$. One easily verifies that $d(x, y) = 0$ iff $x^{-1}y \in K$. The latter enables us to define a metric $\varrho$ on $G/K$ such that $d(x, y) = \varrho(\phi(x), \phi(y))$ for all $x, y \in G$.

For every $x \in G$, $y \in G/K$, and $\varepsilon > 0$, we put $B_\varepsilon(x) = \{x' \in G : d(x', x) < \varepsilon\}$ and $C_\varepsilon(y) = \{y' \in G/K : \varrho(y', y) < \varepsilon\}$. By the definition of $\varrho$, we have that

(a) $\phi(B_\varepsilon(x)) = C_\varepsilon(\phi(x))$, for each $x \in G$.

In other words, the images under $\phi$ of open balls in $G$ are open in the metric space $(G/K, \varrho)$. One can easily verify that the balls $B_\varepsilon(x)$ satisfy the condition:

(b) $B_\varepsilon(x) = \phi^{-1}\phi(B_\varepsilon(x))$ for all $x \in G$ and $\varepsilon > 0$.

Let $t_\varrho$ be the topology on $G/K$ generated by $\varrho$. Note that $t_\varrho$ is weaker than the quotient topology on $G/K$. We claim that the homomorphism $\psi$ of $\phi(H)$ to $P$ remains continuous if $\phi(H)$ is considered as a subspace of $(G/K, t_\varrho)$, and this is the key point of the proof. Indeed, let a point $y \in \phi(H)$ and an open set $O \subset P$ with $z = \psi(y) \in O$ be arbitrary. There exists $n \in \omega$ such that $zO_n \subset O$. Choose $x \in H$ with $\phi(x) = y$; then $\pi(x) = z$. The set $U = B_{1/2^n}(e)$ is contained in $U_n$ by (8.2), and the image $\phi(xU)$ is an open neighbourhood of $y$ in $(G/K, t_\varrho)$ by (a), so (b), (ii), and the equality $H \cap xU = x(H \cap U)$ together imply that

$$\begin{aligned}\psi(\phi(xU) \cap \phi(H)) &= \psi(\phi(H \cap xU)) = \pi(H \cap xU)\\ &= y\pi(H \cap U) \subset y\pi(H \cap U_n) \subset yO_n \subset O.\end{aligned}$$

This proves the continuity of $\psi$ on the subspace $\phi(H)$ of $(G/K, t_\varrho)$.

It remains to find a zero-set $F_0$ in $G$ such that $F_0 \cap H = F$. Let $\Phi = g^{-1}(0) \subset P$. From $f = g \circ \pi$ it follows that $\pi^{-1}(\Phi) = F$. It is clear that $\Phi^* = \psi^{-1}(\Phi)$ is a zero-set in $\phi(H)$. Being a subspace of the metric space $(G/K, \varrho)$, $\phi(H)$ is $z$-embedded in $G/K$. Therefore, there exists a zero-set $F^*$ in $(G/K, \varrho)$ such that $F^* \cap \phi(H) = \Phi^*$. Let us verify that $F = F_0 \cap H$, where $F_0 = \phi^{-1}(F^*)$ is a zero-set in $G$. Since $\psi \circ \phi\restriction H = \pi$, we have $\phi^{-1}(\psi^{-1}(\Phi)) \cap H = \pi^{-1}(\Phi)$, that is, $\phi^{-1}(\Phi^*) \cap H = F$. Consequently,

$$F_0 \cap H = \phi^{-1}(F^*) \cap H = \phi^{-1}(F^* \cap \phi(H)) \cap H = \phi^{-1}(\Phi^*) \cap H = F.$$

This finishes the proof of the theorem. □

We now present a counterpart of Theorem 8.2.5 for $z$-embedded subgroups of ℝ-factorizable groups.

THEOREM 8.2.6. *Let $X$ be a $z$-embedded subset of an $\mathbb{R}$-factorizable topological group $G$ and $f\colon X \to \mathbb{R}$ a continuous function. Then there exists a continuous homomorphism $\pi\colon G \to K$ onto a second-countable topological group $K$ such that $\pi\restriction X \prec f$. Hence, every $z$-embedded subgroup of an $\mathbb{R}$-factorizable group is $\mathbb{R}$-factorizable. The analogous result remains valid for paratopological, quasitopological, and semitopological groups.*

PROOF. We consider only the case when $G$ is a topological group — the argument for paratopological, quasitopological, and semitopological groups is exactly the same.

Let $\gamma = \{U_n : n \in \omega\}$ be the family of all open intervals in $\mathbb{R}$ with rational end points. For every $n \in \omega$, choose a cozero-set $V_n$ in $G$ such that $V_n \cap X = f^{-1}(U_n)$ and define a continuous function $g_n\colon G \to \mathbb{R}$ such that $g_n^{-1}(U_n) = V_n$. The diagonal product $g$ of the family $\{g_n : n \in \omega\}$ is a continuous mapping of $G$ to $\mathbb{R}^\omega$, so we can apply Lemma 8.1.2 to find a continuous homomorphism $\pi\colon G \to K$ onto a second-countable topological group $K$ and a continuous function $g^*\colon K \to \mathbb{R}^\omega$ such that $g = g^* \circ \pi$.

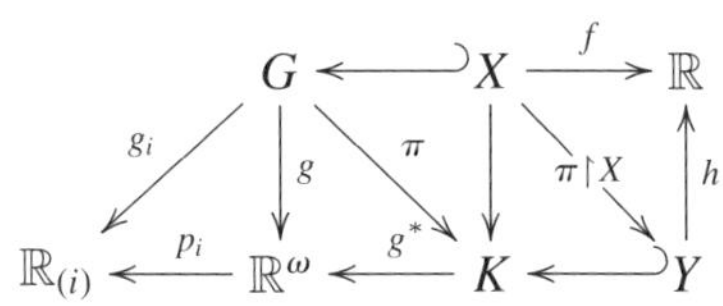


We claim that for all $x_0, x_1 \in X$, $\pi(x_0) = \pi(x_1)$ implies $f(x_0) = f(x_1)$. Assume to the contrary that $f(x_0) \neq f(x_1)$, for some $x_0, x_1 \in X$ with $\pi(x_0) = \pi(x_1)$. Suppose that $f(x_0) < f(x_1)$. If $r_0, r_1, r_2$ are rationals with $r_0 < f(x_0) < r_1 < f(x_1) < r_2$, consider the intervals $U_k = (r_0, r_1) \in \gamma$ and $U_l = (r_1, r_2) \in \gamma$. For $i \in \omega$, let $p_i\colon \mathbb{R}^\omega \to \mathbb{R}_{(i)}$ be the projection. Then $g_i = p_i \circ g$. The sets $g_k^{-1}(U_k) \cap X = f^{-1}(U_k)$ and $g_l^{-1}(U_l) \cap X = f^{-1}(U_l)$ are obviously disjoint. Equivalently, the sets $g^{-1}(O_k) \cap X$ and $g^{-1}(O_l) \cap X$ are disjoint, where $O_k = p_k^{-1}(U_k) \ni g(x_0)$ and $O_l = p_l^{-1}(U_l) \ni g(x_1)$. In particular, $g(x_0) \neq g(x_1)$. We have, however, that $g(x_0) = g^*\pi(x_0) = g^*\pi(x_1) = g(x_1)$, which is a contradiction.

Put $Y = \pi(X)$. The statement just proved implies that there is a function $h\colon Y \to \mathbb{R}$ such that $f = h \circ \pi\restriction X$. It remains to verify that $h$ is continuous. Let $U_n \in \gamma$ be arbitrary. Then the set

$$h^{-1}(U_n) = \pi\big(f^{-1}(U_n)\big) = \pi\big(g_n^{-1}(U_n) \cap X\big) = (g^*)^{-1}\big(p_n^{-1}(U_n)\big) \cap Y$$

is open in $Y$. Since $\gamma = \{U_n : n \in \omega\}$ is a base for $\mathbb{R}$, it follows that $h$ is continuous. So, $\pi\restriction X \prec f$. □

Combining Theorems 8.2.5 and 8.2.6, we conclude that a subgroup $H$ of an $\mathbb{R}$-factorizable topological group $G$ is $\mathbb{R}$-factorizable iff $H$ is $z$-embedded in $G$. In fact, we have obtained the following characterization of $\mathbb{R}$-factorizability in topological terms.

THEOREM 8.2.7. *An $\omega$-narrow topological group $G$ is $\mathbb{R}$-factorizable iff $G$ is $z$-embedded in every topological group that contains $G$ as a topological subgroup.*

PROOF. The necessity follows directly from Theorem 8.2.5. Let us prove the sufficiency. Suppose that $G$ is $z$-embedded in every topological group that contains $G$ as a subgroup. Since $G$ is $\omega$-narrow, we can embed $G$ as a subgroup into a topological product $\Pi$ of second-countable topological groups (see Theorem 3.4.23). The group $\Pi$ is $\mathbb{R}$-factorizable by Corollary 8.1.15, so the conclusion follows from Theorem 8.2.6. □

In the following example we show that the condition of $\omega$-narrowness of $G$ in Theorem 8.2.7 cannot be dropped.

EXAMPLE 8.2.8. Let $G$ be an uncountable discrete group. By Corollary 8.1.10, $G$ is not $\mathbb{R}$-factorizable. Suppose that $G^*$ is a topological group that contains $G$ as a subgroup. Choose open symmetric neighbourhoods $U$ and $V$ of the identity in $G^*$ such that $U \cap G = \{e\}$ and $V^4 \subset U$. It follows from Lemma 1.4.22 that $\{xV : x \in G\}$ is a discrete family of open sets in $G^*$, so $G$ is $C$-embedded (hence, $z$-embedded) in $G^*$. □

## Exercises

8.2.a. Prove that every $\omega$-narrow $P$-group $G$ with $w(G) = \aleph_1$ contains a proper dense subgroup which is not $\mathbb{R}$-factorizable.

8.2.b. Show that an open subgroup of an $\mathbb{R}$-factorizable group is $\mathbb{R}$-factorizable.

8.2.c. Generalize Proposition 8.2.3 (without recourse to Theorem 8.2.6) as follows: Every $C^*$-embedded subgroup of an $\mathbb{R}$-factorizable group is $\mathbb{R}$-factorizable.

8.2.d. Use Corollary 5.3.27 and Theorem 8.2.6 to give an alternative proof of Proposition 8.1.13.

8.2.e. Show that for every $\omega$-narrow group $G$, there exists a "test group" $G^*$ which contains $G$ as a subgroup and has the property that if $G$ is $z$-embedded in $G^*$, then $G$ is $z$-embedded in every topological group that contains $G$ as a subgroup.

8.2.f. Let $G$ be a topological group, and $F(G)$ the free topological group of the space $G$. Prove that if $F(G)$ is $\mathbb{R}$-factorizable, so is $G$.

8.2.g. Prove that the product $\Pi$ of an arbitrary family of regular second-countable paratopological groups is $\mathbb{R}_3$-factorizable, and that every dense subgroup of $\Pi$ is again $\mathbb{R}_3$-factorizable.

## Problems

8.2.A. (C. Hernández and M. G. Tkachenko [**227**]) Present an example of a closed invariant subgroup $H$ of an $\mathbb{R}$-factorizable group $G$ such that $H$ has type $G_\delta$ in $G$ but is not $\mathbb{R}$-factorizable.

8.2.B. Show that every subgroup of a $\sigma$-product of Lindelöf $\Sigma$-groups is $\mathbb{R}$-factorizable.

8.2.C. Prove that every Tychonoff $\mathbb{R}_3$-factorizable paratopological group is topologically isomorphic to a subgroup of a product of regular second-countable paratopological groups.

8.2.D. Prove that if a Tychonoff paratopological group $G$ is $\mathbb{R}_3$-factorizable and first-countable, then $G$ is metrizable.

8.2.E. Use item c) of Problem 5.2.A to show that every cosmic Hausdorff paratopological group is $\mathbb{R}_3$-factorizable. Apply this fact to generalize Exercise 8.2.g to topological products of cosmic Hausdorff paratopological groups.

8.2.F. (M. Sanchis and M. G. Tkachenko [**418**]) Suppose that a paratopological group $G$ is a Lindelöf $\Sigma$-space. Prove that every subgroup of $G$ is $\mathbb{R}_3$-factorizable.
*Hint.* Apply Theorem 8.2.6 and the fact established in Problem 5.7.Q.

8.2.G. Apply Corollary 3.4.29 and Problem 8.2.C to deduce that the Sorgenfrey line is not topologically isomorphic to any subgroup of a Tychonoff $\mathbb{R}_3$-factorizable paratopological group.

## Open Problems

8.2.1. Suppose that $G$ is a topological group such that every subgroup of $G$ is $\mathbb{R}$-factorizable.
(a) Is the cellularity of $G$ countable?
(b) Is $G$ pseudo-$\aleph_1$-compact?
(c) Is $G$ Moscow?

(One can use Problem 8.2.B and some results from [**61**] to deduce that such a group $G$ need not be a subgroup of a Lindelöf $\Sigma$-group.)

8.2.2. Let $\tau$ be an arbitrary uncountable cardinal.
   (a) Is every subgroup of $\mathbb{Z}^\tau$ $\mathbb{R}$-factorizable?
   (b) Is every subgroup of $\mathbb{Z}^\tau$ pseudo-$\aleph_1$-compact?

8.2.3. Let $X$ be a compact Hausdorff space. Is every subgroup of $C_p(X)$ $\mathbb{R}$-factorizable?

8.2.4. Is there a first-countable Hausdorff paratopological group $G$ such that every continuous real-valued function on $G$ is constant?

## 8.3. Dieudonné completion of $\mathbb{R}$-factorizable groups

As we have already seen in Chapter 6, it is a highly non-trivial question whether the Dieudonné completion $\mu G$ of a given topological group $G$ is again a topological group (containing $G$ as a dense topological subgroup). In general, $\mu G$ can fail to be a topological group (see Example 6.7.13). However, we show in Corollary 8.3.7 below that the Dieudonné completion of every $\mathbb{R}$-factorizable group is again a topological group, and this group is $\mathbb{R}$-factorizable.

First, we prove that the Dieudonné completion $\mu X$ and the Hewitt–Nachbin completion $\upsilon X$ of a Tychonoff space $X$ coincide, unless the cardinality of $X$ is very large. Recall that an infinite cardinal $\tau$ is said to be *Ulam measurable* if there exists a countably closed free ultrafilter on a set of cardinality $\tau$ (see Section 6.2). The next lemma generalizes Proposition 6.5.18 slightly.

LEMMA 8.3.1. *Let $X$ be a space such that the cardinality of every discrete family of open sets in $X$ is Ulam non-measurable. Then $\upsilon X = \mu X$.*

PROOF. Let $d$ be a continuous pseudometric on $X$. It suffices to show that $d$ can be extended to a continuous pseudometric on $\upsilon X$. Denote by $X^* = (X/d, d^*)$ the metric space associated with $(X, d)$, and let $p\colon X \to X^*$ be the projection which assigns to a point $x \in X$ the equivalence class of all $y \in X$ satisfying $d(x, y) = 0$. The mapping $p$ is continuous, so the cardinality of every discrete family of open sets in $X^*$ is Ulam non-measurable. Let $\tau = c(X^*) = w(X^*)$. Since $X^*$ has a $\sigma$-discrete base, we infer that either $X^*$ contains a discrete family $\gamma$ of open sets with $|\gamma| = \tau$, or $cf\,\tau = \omega$, say, $\tau = \sup_{n\in\omega}\tau_n$, and for every $n \in \omega$, $X^*$ contains a discrete family $\gamma_n$ of open sets with $|\gamma_n| = \tau_n$. In both cases, b) of Theorem 6.2.2 implies that $\tau$ is Ulam non-measurable. Since the space $X^*$ is metric, we have $|X^*| \leq w(X^*)^\omega = \tau^\omega$, so that the cardinality of $X^*$ is Ulam non-measurable as well.

We conclude that $X^*$ is realcompact, as every metric space whose cardinality is Ulam non-measurable [**165**, 8.5.13 (h)]. Hence, one can extend $p$ to a continuous mapping $\tilde{p}\colon \upsilon X \to X^*$. It remains to define a continuous pseudometric $\varrho$ on $\upsilon X$ by $\varrho(x, y) = d^*(\tilde{p}(x), \tilde{p}(y))$ for all $x, y \in \upsilon X$. If $x, y \in X$, then

$$\varrho(x, y) = d^*(\tilde{p}(x), \tilde{p}(y)) = d^*(p(x), p(y)) = d(x, y).$$

Therefore, the restriction of $\varrho$ to $X$ coincides with $d$. Since $\upsilon X$ is Dieudonné complete, this proves that $\mu X = \upsilon X$. □

COROLLARY 8.3.2. *Let $H$ be a topological group such that the cardinal $ib(H)$ is Ulam non-measurable. Then $\mu H = \upsilon H$.*

Proof. By Theorem 5.4.7, we have $c(H) \le 2^{ib(H)}$, so Theorem 6.2.2 implies that the cardinal $c(H)$ is Ulam non-measurable. Since the cardinality of every discrete family of open sets in $H$ does not exceed $c(H)$, we can apply Lemma 8.3.1 to conclude that $\mu H = \upsilon H$. □

Corollary 8.3.3. *If a topological group $H$ is $\omega$-narrow or, in particular, has countable cellularity, then $\mu H = \upsilon H$.*

Sometimes the existence of a dense $\mathbb{R}$-factorizable subgroup of a topological group implies the $\mathbb{R}$-factorizability of the whole group:

Proposition 8.3.4. *Let $H$ be a $G_\delta$-dense $\mathbb{R}$-factorizable subgroup of a topological group $G$. Then $H$ is $C$-embedded in $G$ and $G$ is $\mathbb{R}$-factorizable.*

Proof. Let $f$ be a continuous real-valued function on $H$. We can find a continuous homomorphism $p\colon H \to L$ onto a second-countable topological group $L$ and a continuous function $g\colon L \to \mathbb{R}$ such that $f = g \circ p$. Denote by $\varrho G$ the Raĭkov completion of the group $G$. Since $H$ is dense in $\varrho G$, one can extend $p$ to a continuous homomorphism $\tilde{p}\colon \varrho G \to \varrho L$, where $\varrho L$ is the Raĭkov completion of $L$. It is clear that the group $\varrho L$ is second-countable, and since $H$ is $G_\delta$-dense in $G$, we conclude that $\tilde{p}(G) = p(H) = L$. So, $\tilde{f} = g \circ \tilde{p}\restriction G$ is a continuous extension of $f$ over $G$. This proves that $H$ is $C$-embedded in $G$.

A similar argument shows that the group $G$ is $\mathbb{R}$-factorizable. Indeed, let $f\colon G \to \mathbb{R}$ be a continuous function. Then we can find a continuous homomorphism $\pi\colon H \to K$ onto a second-countable group $K$ and a continuous function $h\colon K \to \mathbb{R}$ such that $f\restriction H = h \circ \pi$. As above, $\pi$ admits an extension to a continuous homomorphism $\tilde{\pi}\colon G \to K$. The function $g = h \circ \tilde{\pi}$ is continuous on $G$ and $g\restriction H = f\restriction H$, so the density of $H$ in $G$ implies that $g = f$. This proves the $\mathbb{R}$-factorizability of $G$. □

In Theorem 8.3.6 below we show that the Dieudonné completion $\mu K$ of an $\mathbb{R}$-factorizable group $K$ is a well-defined subgroup of its Raĭkov completion $\varrho K$. To present this description of $\mu K$ in terms of $K$ and $\varrho K$, we need a lemma about realcompact (equivalently, Hewitt–Nachbin complete) spaces which follows immediately from Corollaries 3.11.7 and 3.11.8 of [**165**].

Lemma 8.3.5. *Let $X$ be an arbitrary Tychonoff space.*

a) *If $\gamma$ is a family of realcompact subspaces of $X$, then $\bigcap \gamma$ is realcompact;*
b) *If $X$ is realcompact, then every cozero-set in $X$ is realcompact.*

*In particular, the intersection of an arbitrary family of cozero sets in a realcompact space is realcompact.*

Denote by $\varrho_\omega K$ the $G_\delta$-closure of $K$ in $\varrho K$, that is, the set of all $x \in \varrho K$ such that every $G_\delta$-set in $\varrho K$ containing $x$ meets $K$. It is easy to verify that $\varrho_\omega K$ is a subgroup of $\varrho K$ and $K \subset \varrho_\omega K$ (see Section 6.4).

Theorem 8.3.6. [**S. Hernández, M. Sanchis, and M. G. Tkachenko**] *Every $\mathbb{R}$-factorizable group $K$ satisfies $\upsilon K = \mu K = \varrho_\omega K$. In addition, the group $\varrho_\omega K$ is $\mathbb{R}$-factorizable.*

Proof. Clearly, the group $K$ is $G_\delta$-dense in $L = \varrho_\omega K$, so Proposition 8.3.4 implies that $K$ is $C$-embedded in $L$, and $L$ is $\mathbb{R}$-factorizable. The group $\varrho K$ is Raĭkov complete, hence, it is Dieudonné complete. In addition, Proposition 8.1.3 implies that the group $K$ is

$\omega$-narrow, and since $K$ is dense in $\varrho K$, the group $\varrho K$ is also $\omega$-narrow, by Theorem 3.4.9. Therefore, by Corollary 8.3.2, $\varrho K$ is realcompact.

The complement $\varrho K \setminus L$ is the union of $G_\delta$-sets in $\varrho K$; equivalently, $L$ is the intersection of cozero sets in $\varrho K$. By Lemma 8.3.5, $L$ is realcompact. Since $K$ is $C$-embedded in $L$, we have $\upsilon K = L$. Finally, Corollary 8.3.2 implies that $\mu K = \upsilon K = L$. □

COROLLARY 8.3.7. *If $G$ is an $\mathbb{R}$-factorizable group, then the group operations in $G$ can be continuously extended to the Dieudonné completion $\mu G$ of $G$, thus making $\mu G$ into a topological group.*

COROLLARY 8.3.8. *The following conditions are equivalent for a $G_\delta$-dense subgroup $H$ of an $\mathbb{R}$-factorizable group $G$:*

1) *$H$ is $\mathbb{R}$-factorizable;*
2) *$H$ is $C$-embedded in $G$.*

PROOF. Since the group $G$ is $\mathbb{R}$-factorizable, the implication 2) ⇒ 1) follows directly from Proposition 8.2.3. To see that 1) ⇒ 2), apply Proposition 8.3.4. □

It is worth noting at this point that $G_\delta$-dense subgroups of $\mathbb{R}$-factorizable groups are not necessarily $\mathbb{R}$-factorizable (see Example 8.2.1).

The last result of this section follows from Theorem 8.3.6 and the definition of completion friendly groups given in Section 6.5:

COROLLARY 8.3.9. *Every $\mathbb{R}$-factorizable group is completion friendly and, therefore, is a strong $PT$-group.*

## Exercises

8.3.a. Apply Corollary 8.3.9 and item (a) of Problem 5.1.D to show that an $\mathbb{R}$-factorizable group $G$ is topologically isomorphic to a closed subgroup of the product of some family of second-countable topological groups iff the space $G$ is realcompact. Show that this equivalence is no longer valid for the wider class of $\omega$-narrow topological groups (see item (f) of Problem 5.1.D).
*Hint.* Apply Theorem 8.1.18 and item (c) of Problem 5.1.D.

8.3.b. Show that a dense $\mathbb{R}$-factorizable subgroup $H$ of a topological group $G$ need not be either $C$-embedded or $C^*$-embedded in $G$.

## Problems

8.3.A. Let $H$ be a closed invariant subgroup of a topological group $G$ and suppose that the spaces $H$ and $G/H$ are realcompact. Is the space $G$ realcompact? What if the group $H$ is compact?

8.3.B. (M. G. Tkachenko [**490**]) Apply Theorem 8.1.18 along with Corollaries 5.2.9 and 8.3.3 to prove that every weakly Lindelöf topological group is completion friendly.

8.3.C. Show that a weakly Lindelöf topological group $G$ is topologically isomorphic to a closed subgroup of the product of some family of second-countable topological groups iff the space $G$ is realcompact. (See Problem 5.1.D and Exercises 8.3.a and 6.5.c.)
*Hint.* Use Problem 8.3.C.

8.3.D. Let $H$ be a closed invariant subgroup of a topological group $G$, and suppose that the spaces $H$ and $G/H$ are $G_\delta$-closed in $\varrho H$ and $\varrho(G/H)$, respectively. Is the group $G$ $G_\delta$-closed in $\varrho G$?

8.3.E. Prove that a completely regular $\mathbb{R}_3$-factorizable paratopological group $G$ is topologically isomorphic to a closed subgroup of the product of some family of regular second-countable paratopological groups if and only if the space $G$ is realcompact. (Compare with Problems 5.1.D and 8.2.C.)

## Open Problems

8.3.1. According to Corollary 8.3.8, the following conditions are equivalent for an arbitrary $G_\delta$-dense subgroup $H$ of an $\mathbb{R}$-factorizable topological group $G$:
1) the group $H$ is $\mathbb{R}$-factorizable;
2) the subgroup $H$ is $C$-embedded in $G$.
Are conditions 1) and 2) equivalent for any $G_\delta$-dense subgroup $H$ of a Tychonoff $\mathbb{R}_3$-factorizable paratopological group $G$?

8.3.2. Let $G$ be a realcompact topological group of countable extent. Is $G$ topologically isomorphic to a closed subgroup of the product of some family of second-countable topological groups?

## 8.4. Homomorphic images of $\mathbb{R}$-factorizable groups

The results of Sections 8.1 and 8.2 give us a good idea how $\mathbb{R}$-factorizable groups look like. Nevertheless, the answers to many questions about the behaviour of this class of groups remain unknown. We do not know, for instance, whether continuous homomorphic images of $\mathbb{R}$-factorizable groups are $\mathbb{R}$-factorizable. In many special cases, however, $\mathbb{R}$-factorizability turns out to be stable with respect to continuous homomorphisms.

Let us show that continuous open homomorphisms preserve $\mathbb{R}$-factorizability. The proof of this fact requires a simple lemma.

LEMMA 8.4.1. *Let $\{U_n : n \in \omega\}$ be a family of neighbourhoods of the identity $e$ in an $\omega$-narrow group $G$. Then one can find a continuous homomorphism $p\colon G \to H$ onto a second-countable topological group $H$ and a family $\{V_n : n \in \omega\}$ of open neighbourhoods of the identity $e_H$ in $H$ such that $p^{-1}(V_n) \subset U_n$, for each $n \in \omega$.*

PROOF. For every $n \in \omega$, apply Corollary 3.4.19 to find a continuous homomorphism $\pi_n$ of $G$ onto a second-countable topological group $H_n$ and an open neighbourhood $W_n$ of the neutral element in $H_n$ such that $\pi_n^{-1}(W_n) \subset U_n$. Let $\pi$ be the diagonal product of the family $\{\pi_n : n \in \omega\}$. Then $H = \pi(G)$ is a subgroup of the product $P = \prod_{n\in\omega} H_n$, so the group $H$ is second-countable. Denote by $p_n$ the projection of $P$ onto the factor $H_n$. Obviously, $\pi_n = p_n \circ \pi$, for each $n \in \omega$. The open neighbourhoods $V_n = H \cap p_n^{-1}(W_n)$ of the identity in $H$ satisfy the condition $\pi^{-1}(V_n) = \pi_n^{-1}(W_n) \subset U_n$, as required. □

THEOREM 8.4.2. [**M. G. Tkachenko**] *A quotient group of an $\mathbb{R}$-factorizable group is $\mathbb{R}$-factorizable.*

PROOF. Let $\pi\colon G \to H$ be an open continuous homomorphism of an $\mathbb{R}$-factorizable group $G$ onto a topological group $H$, and $f\colon H \to \mathbb{R}$ be a continuous function. Then $f \circ \pi$ is a continuous real-valued function on $G$, and since $G$ is $\mathbb{R}$-factorizable, we can find a continuous homomorphism $\varphi\colon G \to K$ onto a second-countable topological group $K$ and a continuous function $g\colon K \to \mathbb{R}$ such that $f \circ \pi = g \circ \varphi$. Choose a countable local base $\{U_n : n \in \omega\}$ at the identity of $K$ and put $V_n = \pi(\varphi^{-1}(U_n))$, for each $n \in \omega$. The open neighbourhoods $V_n$ of the identity in $H$ have the following property:

($*$) *for every* $h \in H$ *and every* $\varepsilon > 0$, *there exists* $n \in \omega$ *such that* $f(hV_n) \subset (f(h) - \varepsilon, f(h) + \varepsilon)$.

Indeed, let $h \in H$ and $\varepsilon > 0$ be arbitrary. Choose $g \in G$ with $\pi(g) = h$ and put $x = \varphi(g)$. Since $g$ is continuous, there exists $n \in \omega$ such that $g(xU_n) \subset (g(x) - \varepsilon, g(x) + \varepsilon)$. Then $f(h) = g(x)$ and one easily verifies that $f(hV_n) \subset (f(h) - \varepsilon, f(h) + \varepsilon)$.

The group $H$, being a quotient of $G$, is $\omega$-narrow by Proposition 3.4.2. Therefore, we can apply Lemma 8.4.1 to find a continuous homomorphism $p\colon H \to L$ onto a second-countable topological group $L$ with a local base $\{W_n : n \in \omega\}$ at the identity such that $p^{-1}(W_n) \subset V_n$ for each $n \in \omega$. Let $N$ be the kernel of $p$. Then $N \subset \bigcap_{n\in\omega} V_n$, and from ($*$) it follows that the function $f$ is constant on every coset $hN$ in $H$. This enables us to define a function $h\colon L \to \mathbb{R}$ such that $h \circ p = f$.

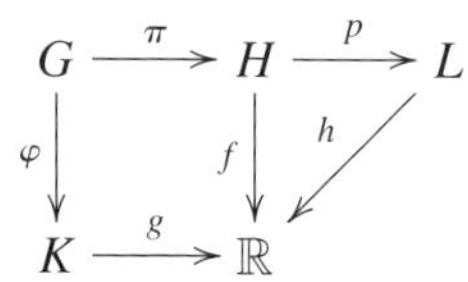

Finally, the choice of the local base $\{W_n : n \in \omega\}$ and ($*$) imply immediately that for every $y \in L$ and $\varepsilon > 0$, there exists $n \in \omega$ such that $h(yW_n) \subset (h(y) - \varepsilon, h(y) + \varepsilon)$. This gives the continuity of $h$. □

Clearly, continuous homomorphic images of Lindelöf topological groups are Lindelöf and, hence, $\mathbb{R}$-factorizable (see Theorem 8.1.6). By Propositions 8.1.13 and 5.3.11, this holds true for arbitrary subgroups of Lindelöf $\Sigma$-groups:

PROPOSITION 8.4.3. *Continuous homomorphic images of subgroups of Lindelöf $\Sigma$-groups are $\mathbb{R}$-factorizable.*

Corollary 8.1.15 suggests considering a wider class of $\mathbb{R}$-factorizable groups, namely, dense subgroups of topological products of Lindelöf $\Sigma$-groups. It turns out that continuous homomorphic images of the groups from this class remain $\mathbb{R}$-factorizable (see Corollary 8.4.7).

To present a result about the preservation of $\mathbb{R}$-factorizability in a general form, we introduce the minimal class $\mathcal{V}(\omega)$ of topological groups which contains all Lindelöf $\Sigma$-groups and is closed under taking topological products, continuous homomorphic images, and dense subgroups. Our aim is to show that all groups in $\mathcal{V}(\omega)$ are $\mathbb{R}$-factorizable, thus generalizing Theorem 8.1.14.

Since every continuous homomorphism $p\colon G \to H$ of topological groups admits an extension to a continuous homomorphism $\tilde{p}\colon \varrho G \to \varrho H$, the following lemma is immediate.

LEMMA 8.4.4. *A topological group $H$ belongs to $\mathcal{V}(\omega)$ if and only if $H$ is a dense subgroup of a continuous homomorphic image of a topological product of Lindelöf $\Sigma$-groups.*

We recall that a space $X$ is called *perfectly $\kappa$-normal* if the closure of every open set is a zero-set in $X$ (see Section 6.1). Evidently, this class of spaces is hereditary with respect to taking open subsets; a similar fact for dense subsets follows from Lemma 8.4.5 given below.

LEMMA 8.4.5. **[R. L. Blair]** *Every dense subspace of a perfectly $\kappa$-normal space $X$ is $z$-embedded in $X$.*

PROOF. Let $S$ be a dense subspace of $X$ and $F$ be a zero-set in $S$. We can find a continuous function $f\colon S \to \mathbb{R}$ such that $F = f^{-1}(0)$. For every $n \in \mathbb{N}$, put $U_n = f^{-1}(-1/n, 1/n)$ and choose an open set $V_n$ in $X$ such that $U_n = V_n \cap S$. Then

$$F = \bigcap_{n=1}^{\infty} cl_S U_n \subseteq \bigcap_{n=1}^{\infty} cl_X V_n.$$

Since $X$ is perfectly $\kappa$-normal, each set $cl_X V_n$ is a zero-set in $X$ and, hence, $P = \bigcap_{n=1}^{\infty} cl_X V_n$ is also a zero-set in $X$. Clearly, $F = P \cap S$, which proves the lemma. □

THEOREM 8.4.6. *Every group in $\mathcal{V}(\omega)$ is $\mathbb{R}$-factorizable and perfectly $\kappa$-normal.*

PROOF. Let $G = \prod_{i \in I} G_i$ be a product of Lindelöf $\Sigma$-groups, and $\varphi\colon G \to H$ be a continuous homomorphism of $G$ onto a topological group $H$. By Corollary 5.6.17, the group $G$ is stable, and so is the continuous image $H$ of $G$. In addition, Theorem 5.3.30 implies that $cel_\omega(G) \leq \omega$, so that $c(G) \leq \omega$ and $c(H) \leq \omega$. According to Corollary 5.6.19, every $\omega$-stable topological group of countable cellularity is perfectly $\kappa$-normal; in particular, this is true for $H$. It follows from $c(H) \leq \omega$ that $H$ is weakly Lindelöf, and we can apply Proposition 8.1.20 to conclude that the group $H$ is $\mathbb{R}$-factorizable.

Finally, consider an arbitrary dense subgroup $K$ of $H$. By Lemma 8.4.4, it suffices to show that $K$ is perfectly $\kappa$-normal and $\mathbb{R}$-factorizable. The first property of $K$ is immediate since $K$ is a dense subgroup of the perfectly $\kappa$-normal group $H$. In addition, Lemma 8.4.5 implies that $K$ is $z$-embedded in $H$. Now the required conclusion about $K$ follows from Theorem 8.2.6. □

COROLLARY 8.4.7. *Continuous homomorphic images of dense subgroups of topological products of Lindelöf $\Sigma$-groups are $\mathbb{R}$-factorizable.*

COROLLARY 8.4.8. *Let $G$ be a dense subgroup of a product of cosmic topological groups. Then every continuous homomorphic image of $G$ is $\mathbb{R}$-factorizable.*

Homomorphic images of $\mathbb{R}$-factorizable groups will be also considered in Section 8.5 (see Proposition 8.5.7 and Corollary 8.5.10).

## Exercises

8.4.a. Show that every continuous homomorphic image of a weakly Lindelöf $\omega$-steady group is $\mathbb{R}$-factorizable (compare with Proposition 8.1.20).

8.4.b. Verify that every dense subgroup of a perfectly $\kappa$-normal $\mathbb{R}$-factorizable group is $\mathbb{R}$-factorizable.

8.4.c. Show that the topological group $\mathbb{Z}^{\mathfrak{c}}$ has countable cellularity but it cannot be represented as a continuous homomorphic image of any subgroup of a Lindelöf $\Sigma$-group.

8.4.d. Prove that if $X$ is a dense $\omega$-stable subspace of a product of second-countable spaces, then the free topological group $F(X)$ and the free Abelian topological group $A(X)$ are $\mathbb{R}$-factorizable. Show that the assertion remains valid for every $\omega$-stable space $X$ that contains a dense $\sigma$-compact subspace.

8.4.e. Let $\mu X$ be the Dieudonné completion of a Tychonoff space $X$. Verify the following:

a) If $F(X)$ is $\mathbb{R}$-factorizable, then so is $F(\mu X)$.
b) If $X$ contains a dense $\sigma$-compact subset and $F(\mu X)$ is $\mathbb{R}$-factorizable, then $F(X)$ is also $\mathbb{R}$-factorizable.

## Problems

8.4.A. (M. G. Tkachenko [**488**]) Let $G$ be a topological group and $f$ a continuous real-valued function on $G$. For every $a \in G$, let $f_a$ be the function on $G$ defined by $f_a(x) = f(a^{-1}x)$, for each $x \in G$. Consider the set $Gf = \{f_a : a \in G\}$ as a subset of $C_p(G)$, with the subspace topology (see Section 1.9). Show that if the group $G$ is $\mathbb{R}$-factorizable, then the space $Gf$ is cosmic, for each $f \in C_p(G)$.

8.4.B. (M. G. Tkachenko [**488**]) Apply Problem 8.4.A to prove that if $G$ is a $P$-group satisfying $\chi(G) = \aleph_1$ and $G$ is not Raĭkov complete, then $G$ cannot be a continuous homomorphic image of any $\mathbb{R}$-factorizable group. Deduce that not every $\omega$-narrow topological group is a continuous homomorphic image of an $\mathbb{R}$-factorizable group.

8.4.C. Give an example of an uncountable abstract group $G$ such that every Hausdorff topological group topology on $G$ is discrete.
*Remark.* A group $G$ with this combination of properties was constructed under the additional assumption of $CH$ by S. Shelah in [**442**]. Afterwards the assumption of $CH$ was eliminated by G. Hesse in [**233**].

## Open Problems

8.4.1. Let $G$ be an $\mathbb{R}$-factorizable topological group and $f\colon G \to H$ a continuous homomorphism of $G$ onto a topological group $H$. Is $H$ $\mathbb{R}$-factorizable? What if every subgroup of $G$ is $\mathbb{R}$-factorizable?

8.4.2. Does there exist a continuous mapping of a subgroup of a Lindelöf $\Sigma$-group onto the space $\mathbb{Z}^{\mathfrak{c}}$? (See Exercise 8.4.c.)

8.4.3. Suppose that a topological group $H$ is an image of an $\mathbb{R}$-factorizable group $G$ under a continuous homomorphism. Is $H$ a *PT*-group?

8.4.4. Is any quotient group of an $\mathcal{M}$-factorizable topological group $\mathcal{M}$-factorizable? ($\mathcal{M}$-factorizable groups were defined in Problem 8.1.C.)

8.4.5. Suppose that $H$ is a quotient paratopological group of an $\mathbb{R}_i$-factorizable paratopological group, where $i \in \{1, 2, 3\}$. Is $H$ then $\mathbb{R}_i$-factorizable provided it satisfies the $T_i$ separation axiom?

8.4.6. Suppose that a paratopological group $H$ satisfies the $T_i$ separation axiom and is an image of an $\mathbb{R}_i$-factorizable topological group $G$ under a continuous homomorphism. Is $H$ $\mathbb{R}_i$-factorizable?

## 8.5. Products with a compact factor and *m*-factorizability

Recall that a space $X$ is said to be *pseudo-$\aleph_1$-compact* if every discrete family of open sets in $X$ is countable. All theorems about "countable" factorization of continuous functions defined on a subspace of a Cartesian product have the inherent assumption that the subspace has to be at least pseudo-$\aleph_1$-compact (or has to satisfy a stronger condition such as being Lindelöf, of countable cellularity, *etc.*). Since every $\mathbb{R}$-factorizable group $G$ is $\omega$-narrow and, hence, is topologically isomorphic to a subgroup of a product $\Pi$ of second-countable topological groups (Theorem 3.4.23), the subspace $G$ of $\Pi$ has to satisfy a kind of a factorization theorem. It is natural, therefore, to ask whether every $\mathbb{R}$-factorizable group is pseudo-$\aleph_1$-compact, weakly Lindelöf, or has countable cellularity.

First, we note that the Lindelöf $P$-group $G_\tau$ in Example 4.4.11 is $\mathbb{R}$-factorizable by Theorem 8.1.6. However, $G_\tau$ has uncountable cellularity if $\tau > \omega$. Indeed, in this case $G_\tau$ is non-discrete, so we can define by recursion a strictly decreasing sequence $\{U_\alpha : \alpha < \omega_1\}$ of clopen neighbourhoods of the identity in $G_\tau$. Then the disjoint family $\{U_\alpha \setminus U_{\alpha+1} : \alpha < \omega_1\}$ consists of non-empty open sets and, hence, $c(G_\tau) > \omega$. In fact, we show in Example 8.6.16 that $\mathbb{R}$-factorizable groups can even fail to be weakly Lindelöf.

A property stronger than $\mathbb{R}$-factorizability arises if we require that every continuous mapping of a topological group $G$ to a metrizable space admits a factorization via a continuous homomorphism onto a second-countable topological group. More precisely, we say that a topological group $G$ is *$m$-factorizable* if for every continuous mapping $f\colon G \to M$ to a metrizable space $M$, there exists a continuous homomorphism $\pi\colon G \to K$ onto a second-countable group $K$ such that $\pi \prec f$.

Clearly, every $m$-factorizable group is $\mathbb{R}$-factorizable. The next result characterizes $m$-factorizability in terms of continuous pseudometrics.

PROPOSITION 8.5.1. *A topological group $G$ is $m$-factorizable iff for every continuous pseudometric $d$ on $G$, one can find a continuous homomorphism $\pi\colon G \to K$ onto a second-countable topological group $K$ and a continuous pseudometric $\varrho$ on $K$ such that $d(x, y) = \varrho(\pi(x), \pi(y))$, for all $x, y \in G$.*

PROOF. Suppose that $G$ is $m$-factorizable, and let $d$ be a continuous pseudometric on $G$. Consider the metric space $M = G/d$ with the associated metric $d^*$, obtained from $G$ by identifying points at zero distance with respect to $d$. Let also $p\colon G \to G/d$ be the projection assigning to a point $x \in G$ the equivalence class $\bar{x}$ consisting of all $z \in G$ with $d(x, z) = 0$. Then $d(x, y) = d^*(p(x), p(y))$, for all $x, y \in G$. Since $G$ is $m$-factorizable, we can find a continuous homomorphism $\pi\colon G \to K$ onto a second-countable group $K$ and a continuous mapping $h\colon K \to M$ such that $p = h \circ \pi$. Define a continuous pseudometric $\varrho$ on $K$ by $\varrho(s, t) = d^*(h(s), h(t))$ for all $s, t \in K$. It is easy to verify that $d(x, y) = \varrho(\pi(x), \pi(y))$, for $x, y \in G$.

Conversely, suppose that $G$ has the above property of factorization of continuous pseudometrics, and consider a continuous mapping $f\colon G \to M$ to a metric space $M$ with a metric $\kappa$. Define a continuous pseudometric $d$ on $G$ by $d(x, y) = \kappa(f(x), f(y))$ for all $x, y \in G$. By the assumption, we can find a continuous homomorphism $\pi\colon G \to K$ onto a second-countable group $K$ and a continuous pseudometric $\varrho$ on $K$ such that $d(x, y) = \varrho(\pi(x), \pi(y))$, for all $x, y \in G$. We conclude that

$$\varrho(\pi(x), \pi(y)) = \kappa(f(x), f(y)), \text{ for all } x, y \in G. \tag{8.3}$$

Therefore, the equality $\pi(x) = \pi(y)$ always implies $f(x) = f(y)$. So, there exists a mapping $h\colon K \to M$ such that $f = h \circ \pi$. Then (8.3) implies that $\kappa(h(z), h(t)) < \varepsilon$ whenever points $z, t \in K$ satisfy $\varrho(z, t) < \varepsilon$. Hence the mapping $h$ is continuous. □

Let us show that pseudo-$\aleph_1$-compactness is exactly what we need to add to $\mathbb{R}$-factorizability in order to obtain $m$-factorizability.

THEOREM 8.5.2. *A topological group $G$ is $m$-factorizable iff $G$ is $\mathbb{R}$-factorizable and pseudo-$\aleph_1$-compact.*

PROOF. Suppose that a topological group $G$ is $m$-factorizable. Then $G$ is $\mathbb{R}$-factorizable, so we have to show that $G$ is pseudo-$\aleph_1$-compact. Let $\gamma = \{U_i : i \in I\}$ be a discrete family of non-empty open sets in $G$. For every $i \in I$, choose a point $x_i \in U_i$ and a continuous function $f_i\colon G \to \mathbb{R}$ such that $f_i(x_i) = 1$ and $f_i(x) = 0$ for each $x \in X \setminus U_i$. Then $d_i$ defined by the rule $d_i(x, y) = |f_i(x) - f_i(y)|$, for all $x, y \in G$, is a continuous pseudometric on $G$. Since the family $\gamma$ is discrete, the pseudometric $d = \sum_{i\in I} d_i$ is also continuous. By Proposition 8.5.1, we can find a continuous homomorphism $\pi\colon G \to K$ onto a second-countable topological group $K$ and a continuous pseudometric $\varrho$ on $K$ such that $d(x, y) = \varrho(\pi(x), \pi(y))$ for all $x, y \in G$. For every $i \in I$, define $y_i = \pi(x_i)$ and $V_i = \{y \in K : \varrho(y_i, y) < 1\}$. Then $V_i$ is open in $K$ and $\pi^{-1}(V_i) = U_i$ for each $i \in I$. Therefore, the family $\{V_i : i \in I\}$ is disjoint and, hence, countable. It follows that the family $\gamma$ is also countable, so $G$ is pseudo-$\aleph_1$-compact.

Suppose now that the group $G$ is $\mathbb{R}$-factorizable and pseudo-$\aleph_1$-compact. Let $f\colon G \to M$ be a continuous mapping of $G$ onto a metrizable space $M$. Then $M$ is pseudo-$\aleph_1$-compact and, hence, $w(M) \leq \omega$. Since $G$ is $\mathbb{R}$-factorizable, we can apply Lemma 8.1.2 to find a continuous homomorphism $\pi\colon G \to K$ onto a second-countable topological group $K$ such that $\pi \prec f$. This proves that the group $G$ is $m$-factorizable. □

Since continuous mappings preserve pseudo-$\aleph_1$-compactness, Theorems 8.5.2 and 8.4.2 imply that quotients of $m$-factorizable groups are $m$-factorizable:

COROLLARY 8.5.3. *Let $\pi\colon G \to H$ be an open continuous homomorphism of a topological group $G$ onto $H$. If $G$ is $m$-factorizable, so is $H$.*

It is a general problem, yet unsolved, whether the product of two $m$- or $\mathbb{R}$-factorizable groups is $m$- or $\mathbb{R}$-factorizable. We solve it here for $m$-factorizability in the special case when one of the factors is compact.

First, we prove that continuous homomorphisms defined on subgroups of arbitrary topological products with values in first-countable groups depend on at most countably many coordinates.

LEMMA 8.5.4. *Let $G$ be a subgroup of the topological product $\Pi = \prod_{i\in I} G_i$ of left (right) topological groups and $\pi\colon G \to H$ a continuous homomorphism to a left (right) topological group $H$ satisfying $\chi(H) \leq \kappa$. Then one can find a set $J \subset I$ with $|J| \leq \kappa$ and a continuous homomorphism $\varphi\colon p_J(G) \to H$ such that $\pi = \varphi \circ p_J{\restriction}G$, where $p_J$ is the projection of $\Pi$ to $\Pi_J = \prod_{i\in J} G_i$. If, in particular, the space $H$ is first-countable, then the set $J$ can be chosen to be countable.*

PROOF. It suffices to consider the case of left topological groups. Let $\{V_\alpha : \alpha < \kappa\}$ be a local base at the identity $e_H$ of the group $H$. Since $\pi$ is continuous, for every $\alpha < \kappa$ there exists a canonical open neighbourhood $U_\alpha$ of the identity in $\Pi$ such that $\pi(U_\alpha \cap G) \subset V_\alpha$. Let $J_\alpha$ be a finite subset of $I$ such that $U_\alpha = p_{J_\alpha}^{-1} p_{J_\alpha}(U_\alpha)$. Put $J = \bigcup_{\alpha<\kappa} J_\alpha$. Evidently, $|J| \leq \kappa$. We claim that if $x, y \in G$ and $p_J(x) = p_J(y)$, then $\pi(x) = \pi(y)$. Indeed, suppose that $p_J(x) = p_J(y)$. Then $x^{-1}y \in U_\alpha$ and $\pi(x^{-1}y) \in V_\alpha$, for each $\alpha < \omega$. Since the intersection of the sets $V_\alpha$ contains only $e_H$, we conclude that $\pi(x^{-1}y) = e_H$, i.e., $\pi(x) = \pi(y)$.

Therefore, there exists a homomorphism $\varphi\colon p_J(G) \to H$ such that $\pi = \varphi \circ p_J{\restriction}G$. It remains to verify that $\varphi$ is continuous. Let $\alpha < \kappa$ be arbitrary. Then $W_\alpha = p_J(U_\alpha)$ is an open neighbourhood of the identity in the group $\Pi_J$. Since the canonical set $U_\alpha$ depends only on the indices in $J_\alpha \subset J$, we have $U_\alpha = p_J^{-1}(W_\alpha)$. It follows from $\pi = \varphi \circ p_J \restriction G$ that $\varphi(W_\alpha \cap p_J(G)) = \pi(U_\alpha \cap G) \subset V_\alpha$. This implies that the homomorphism $\varphi$ is continuous at the identity of the left topological group $p_J(G)$. By Proposition 1.3.4, it is continuous on $p_J(G)$. $\square$

THEOREM 8.5.5. *Let G be a topological group.*

a) *If G is m-factorizable and K is any compact group, then the group $G \times K$ is m-factorizable.*
b) *Conversely, if $G \times \mathbb{Z}(2)^{\omega_1}$ is $\mathbb{R}$-factorizable, then G is m-factorizable.*

PROOF. a) Suppose that $G$ is $m$-factorizable, and let $f: G \times K \to \mathbb{R}$ be a continuous function. Denote by $C(K)$ the space of all continuous real-valued functions on $K$ with the sup-norm topology, and consider the mapping $\Psi: G \to C(K)$ defined by $\Psi(x)(y) = f(x, y)$, for all $x \in G$ and $y \in K$. Since $K$ is compact, $\Psi$ is continuous. By Theorem 8.5.2, $G$ is $\mathbb{R}$-factorizable and pseudo-$\aleph_1$-compact, so the subspace $\Psi(G)$ of the metric space $C(K)$ is pseudo-$\aleph_1$-compact and, hence, second-countable. By the $\mathbb{R}$-factorizability of $G$ and Lemma 8.1.2, there are a continuous homomorphism $\pi: G \to H$ onto a second-countable topological group $H$ and a continuous mapping $\psi: H \to C(K)$ such that $\Psi = \psi \circ \pi$.

We claim that if $x_1, x_2 \in G$ and $\pi(x_1) = \pi(x_2)$, then $f(x_1, y) = f(x_2, y)$ for each $y \in K$. Indeed, if $f(x_1, y) \neq f(x_2, y)$ for some $x_1, x_2 \in G$ and $y \in K$, then $\Psi(x_1)(y) \neq \Psi(x_2)(y)$, i.e., $\Psi(x_1) \neq \Psi(x_2)$. Therefore, the equality $\Psi = \psi \circ \pi$ implies that $\pi(x_1) \neq \pi(x_2)$.

The fact just proved enables us to define a mapping $h: H \times K \to \mathbb{R}$ such that $h \circ (\pi \times id_K) = f$, where $id_K$ is the identity mapping of $K$ onto itself.

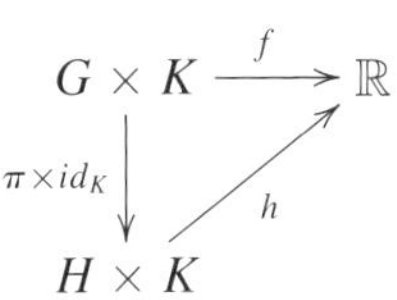


Let us verify that $h$ is continuous. Choose an arbitrary point $(s, y) \in H \times K$ and a number $\varepsilon > 0$. Let also $x^* \in G$, $\pi(x^*) = s$. Since $\psi$ is continuous, there exists an open neighbourhood $U$ of $s$ in $H$ such that $||\psi(t) - \psi(s)|| < \varepsilon/2$ for each $t \in U$, i.e., $|f(x, z) - f(x^*, z)| < \varepsilon/2$ whenever $\pi(x) \in U$ and $z \in K$. There exists a neighbourhood $V$ of $y$ in $K$ such that $|f(x^*, z) - f(x^*, y)| < \varepsilon/2$ for each $z \in V$. Let $(t, z) \in U \times V$ be arbitrary, and choose $x \in G$ with $\pi(x) = t$. Then we have:

$$\begin{aligned} |h(t, z) - h(s, y)| &= |f(x, z) - f(x^*, y)| \\ &\leq |f(x, z) - f(x^*, z)| + |f(x^*, z) - f(x^*, y)| < \varepsilon/2 + \varepsilon/2 = \varepsilon. \end{aligned}$$

This implies the continuity of $h$.

The group $H \times K$ is Lindelöf as the product of the second-countable group $H$ and the compact group $K$. By Theorem 8.1.6, $H \times K$ is $\mathbb{R}$-factorizable. Since $\pi \times id_K \prec f$, Lemma 8.1.11 implies that the group $G \times K$ is $\mathbb{R}$-factorizable. Finally, the product of a pseudo-$\aleph_1$-compact space with a compact space is pseudo-$\aleph_1$-compact, so the group $G \times K$ is $m$-factorizable, by Theorem 8.5.2. This proves a) of the theorem.

b) Suppose that the group $G \times K$ is $\mathbb{R}$-factorizable, where $K = \mathbb{Z}(2)^{\omega_1}$ and $\mathbb{Z}(2) = \{0, 1\}$ is the discrete group. Let $e_K$ be the identity of the group $K$. It is clear that $G \cong G \times \{e_K\}$

is a retract of $G \times K$, so Corollary 8.2.4 implies that the group $G$ is $\mathbb{R}$-factorizable. Let us show that $G$ is pseudo-$\aleph_1$-compact.

Assume to the contrary that $G$ contains a discrete family $\{U_\alpha : \alpha < \omega_1\}$ of non-empty open sets. For every $\alpha < \omega_1$, choose a point $x_\alpha \in U_\alpha$ and define a continuous function $f_\alpha \colon G \to [0, 1]$ such that $f_\alpha(x_\alpha) = 1$ and $f_\alpha(x) = 0$ if $x \in G \setminus U_\alpha$. We also consider the function $h \colon \mathbb{Z}(2) \to [0, 1]$ such that $h(0) = 0$ and $h(1) = 1$. Let $\pi \colon G \times K \to G$ be the projection. If $y \in \mathbb{Z}(2)^{\omega_1}$ and $\alpha < \omega_1$, denote by $y_\alpha$ the $\alpha$th coordinate of $y$. For every $\alpha < \omega_1$, we define a function $g_\alpha \colon G \times K \to [0, 1]$ by $g_\alpha(x, y) = f_\alpha(x) \cdot h(y_\alpha)$, for every $(x, y) \in G \times K$. Clearly, $g_\alpha$ is continuous. Since the family $\{U_\alpha : \alpha < \omega_1\}$ is discrete, the function $g = \sum_{\alpha<\omega_1} g_\alpha$ is also continuous. By the $\mathbb{R}$-factorizability of $G \times K$, we can find a continuous homomorphism $\varphi \colon G \times K \to L$ to a second-countable group $L$ and a continuous function $\tilde{g} \colon L \to \mathbb{R}$ such that $g = \tilde{g} \circ \varphi$. From Lemma 8.5.4 it follows that there exists a countable subset $J$ of $\omega_1$ and a continuous homomorphism $\psi \colon G \times \mathbb{Z}(2)^J \to L$ such that $\varphi = \psi \circ (\mathrm{id}_G \times p_J)$, where $id_G$ is the identity automorphism of $G$ and $p_J \colon \mathbb{Z}(2)^{\omega_1} \to \mathbb{Z}(2)^J$ is the projection.

$$\begin{array}{ccc} G \times \mathbb{Z}(2)^{\omega_1} & \xrightarrow{\;g\;} & \mathbb{R} \\ {\scriptstyle id_G \times p_J}\downarrow & \searrow^{\varphi} & \uparrow{\scriptstyle \tilde{g}} \\ G \times \mathbb{Z}(2)^J & \xrightarrow{\;\psi\;} & L \end{array}$$

Since the above diagram commutes, we conclude that if $x \in G$ and $y, y' \in \mathbb{Z}(2)^{\omega_1}$ satisfy $p_J(y) = p_J(y')$, then $g(x, y) = g(x, y')$. Choose an ordinal $\alpha \in \omega_1 \setminus J$. Now we define two points $y, y' \in \mathbb{Z}(2)^{\omega_1}$ by $y_\beta = y'_\beta = 0$ if $\beta \neq \alpha$ and $y_\alpha = 0$, $y'_\alpha = 1$. Evidently, $p_J(y) = p_J(y')$. A simple calculation shows that $g(x_\alpha, y) = h(y_\alpha) = 0$ and $g(x_\alpha, y') = h(y'_\alpha) = 1$; it follows that $g(x_\alpha, y) \neq g(x_\alpha, y')$. This contradiction shows that $G$ is pseudo-$\aleph_1$-compact. Finally, Theorem 8.5.2 implies that $G$ is $m$-factorizable. □

COROLLARY 8.5.6. *The product of an $\mathbb{R}$-factorizable group $G$ and a compact metrizable group $K$ is $\mathbb{R}$-factorizable.*

PROOF. Let $f \colon G \times K \to \mathbb{R}$ be a continuous function. As in the proof of item a) of Theorem 8.5.5, consider the mapping $\Psi \colon G \to C(K)$, where $C(K)$ is taken with the sup-norm topology, defined by $\Psi(x)(y) = f(x, y)$ for all $x \in G$ and all $y \in K$. Since $K$ is compact, the mapping $\Psi$ is continuous. In addition, $w(C(K)) = w(K) \le \omega$. By the $\mathbb{R}$-factorizability of $G$ and Lemma 8.1.2, there are a continuous homomorphism $\pi \colon G \to H$ onto a second-countable topological group $H$ and a continuous mapping $\psi \colon H \to C(K)$ such that $\Psi = \psi \circ \pi$. The equality $\pi(x_1) = \pi(x_2)$ implies that $f(x_1, y) = f(x_2, y)$, for all $x_1, x_2 \in G$ and all $y \in K$. Therefore, there is a continuous mapping $h \colon H \times K \to \mathbb{R}$ such that $h \circ (\pi \times id_K) = f$, where $id_K$ is the identity automorphism of $K$ onto itself. Since the group $H \times K$ is second-countable, we conclude that the product $G \times K$ is $\mathbb{R}$-factorizable. □

The following result shows that the property of being pseudo-$\aleph_1$-compact for an $\mathbb{R}$-factorizable group is determined by its quotient groups of countable pseudocharacter. We recall that the preservation of $\mathbb{R}$-factorizability under taking continuous homomorphic images is an open problem (see Problem 8.4.1).

PROPOSITION 8.5.7. *An $\mathbb{R}$-factorizable group $G$ is pseudo-$\aleph_1$-compact iff so is every quotient group $G/N$ of countable pseudocharacter.*

PROOF. Continuous mappings preserve pseudo-$\aleph_1$-compactness, so we prove only the sufficiency. Suppose that every quotient $G/N$ of countable pseudocharacter is pseudo-$\aleph_1$-compact. By Theorems 8.5.5 and 8.5.2, it suffices to show that the product $G \times K$ is $\mathbb{R}$-factorizable for $K = \mathbb{Z}(2)^{\omega_1}$. According to [**165**, Theorem 2.3.15], $K$ contains a countable dense subset $S$. Consider an arbitrary continuous function $f\colon G \times K \to \mathbb{R}$. For every $y \in K$, denote by $i_y$ the natural embedding of $G$ into $G \times K$ defined by $i_y(x) = (x, y)$ for each $x \in G$. Since the group $G$ is $\mathbb{R}$-factorizable and $S$ is countable, we can apply Lemma 8.1.2 to find a continuous homomorphism $\pi\colon G \to G_0$ onto a second-countable group $G_0$ and a family $\{f_y : y \in S\}$ of continuous real-valued functions on $G_0$ such that $f \circ i_y = f_y \circ \pi$, for each $y \in S$. Let us verify that for all $x_1, x_2 \in G$ and all $y \in K$, $\pi(x_1) = \pi(x_2)$ implies $f(x_1, y) = f(x_2, y)$.

Assume to the contrary that there exist $x_1, x_2 \in G$ with $\pi(x_1) = \pi(x_2)$ and $y \in K$ such that $f(x_1, y) \neq f(x_2, y)$. We can find disjoint open sets $U_1, U_2$ in $G$ and an open set $V$ in $K$ such that $x_i \in U_i$, for $i = 1, 2$, $y \in V$, and $f(U_1 \times V) \cap f(U_2 \times V) = \varnothing$. Since $S$ is dense in $K$, there is a point $z \in V \cap S$. Clearly, $f(x_1, z) \neq f(x_2, z)$. However,

$$f(x_1, z) = f_z(\pi(x_1)) = f_z(\pi(x_2)) = f(x_2, z),$$

which is a contradiction.

Therefore, there exists a function $g\colon G_0 \times K \to \mathbb{R}$ such that $f = g \circ (\pi \times id_K)$, where $id_K$ is the identity automorphism of $K$. The function $g$ is not necessarily continuous, so we proceed as follows. Denote by $N$ the kernel of $\pi$ and consider the quotient homomorphism $\pi^*\colon G \to G/N$. Clearly, there exists a continuous isomorphism $j\colon G/N \to G_0$ such that $\pi = j \circ \pi^*$. The pseudocharacter of the space $G^* = G/N$ is countable, and the function $g^* = g \circ (j \times id_K)$ satisfies the equality $f = g^* \circ (\pi^* \times id_K)$.

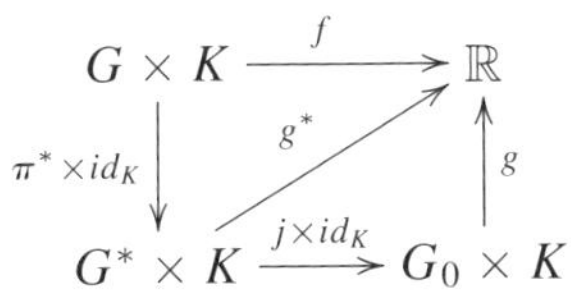


Since $\pi^*$ and $\pi^* \times id_K$ are open homomorphisms, the function $g^*$ is continuous. By the assumptions, the quotient group $G^*$ is pseudo-$\aleph_1$-compact. In addition, $G^*$ is $\mathbb{R}$-factorizable, by Theorem 8.4.2. Therefore, Theorem 8.5.2 implies that the group $G^*$ is $m$-factorizable. According to a) of Theorem 8.5.5, the product group $G^* \times K$ is $m$-factorizable and, therefore, $\mathbb{R}$-factorizable. Since $\pi^* \times id_K \prec f$, Lemma 8.1.11 implies that the group $G \times K$ is also $\mathbb{R}$-factorizable. Finally, apply b) of Theorem 8.5.5 to conclude that the group $G$ is $m$-factorizable and, by Theorem 8.5.2, pseudo-$\aleph_1$-compact. □

We are going to show, under the additional assumption that $\mathfrak{c} < 2^{\aleph_1}$, that any "small" $\mathbb{R}$-factorizable group $G$ is pseudo-$\aleph_1$-compact, so the product $G \times K$ of such a group $G$ with any compact group $K$ remains $\mathbb{R}$-factorizable. As usual, $C(X)$ denotes the set of continuous real-valued functions on a space $X$.

THEOREM 8.5.8. *If an $\mathbb{R}$-factorizable group $G$ satisfies $w(G) \leq \tau \geq \aleph_0$, then $|C(G)| \leq \tau^\omega$. In particular, every $\mathbb{R}$-factorizable group $G$ with $w(G)^\omega < 2^{\aleph_1}$ is $m$-factorizable.*

PROOF. The $\mathbb{R}$-factorizable group $G$ is $\omega$-narrow by Proposition 8.1.3, so we can apply Theorem 3.4.23 to embed $G$ as a subgroup to the product $\Pi = \prod_{i\in I} G_i$ of second-countable groups $G_i$. Since every $\omega$-narrow group is range-$\mathcal{P}$ for the class $\mathcal{P}$ of separable metrizable groups, Theorem 3.4.21 enables us to choose the index set $I$ satisfying $|I| \leq \chi(G) \leq w(G) \leq \tau$.

Since $G$ is $\mathbb{R}$-factorizable, for every $f \in C(G)$ there exist a continuous homomorphism $\pi\colon G \to K$ onto a second-countable group $K$ and a continuous function $h\colon K \to \mathbb{R}$ such that $f = h \circ \pi$. By Lemma 8.5.4, we can find a countable set $J \subset I$ and a continuous homomorphism $\varphi\colon p_J(G) \to K$ such that $\pi = \varphi \circ p_J\restriction G$, where $p_J\colon \Pi \to \Pi_J = \prod_{i\in J} G_i$ is the projection.

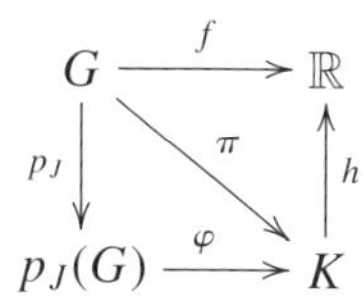


Then the function $g = h \circ \varphi$ is continuous and satisfies $f = g \circ p_J$.

For every countable subset $J$ of $I$, put $C_J = \{g \circ p_J : g \in C(p_J(G))\}$. We have just proved that

$$C(G) = \bigcup\{C_J : J \subset I,\ |J| \leq \omega\}.$$

Since the weight of the group $p_J(G)$ is countable, for every countable set $J \subset I$, we have that $|C(p_J(G))| \leq 2^\omega$. Therefore, $|C(G)| \leq |I|^\omega \cdot 2^\omega \leq \tau^\omega$. This proves the first part of the theorem.

Assume that $w(G)^\omega < 2^{\aleph_1}$ and that $G$ has an uncountable discrete family $\{U_\alpha : \alpha < \omega_1\}$ of non-empty open sets. For every $\alpha < \omega_1$, choose a point $x_\alpha \in U_\alpha$. Then the set $X = \{x_\alpha : \alpha < \omega_1\}$ is $C$-embedded in $G$ and, hence, $C(G)$ contains a subset that can be identified with $\mathbb{R}^X$. Hence $|C(G)| \geq 2^{|X|} = 2^{\aleph_1}$, which contradicts the inequalities $|C(G)| \leq w(G)^\omega < 2^{\aleph_1}$. Thus, $G$ is pseudo-$\aleph_1$-compact. Finally, every $\mathbb{R}$-factorizable pseudo-$\aleph_1$-compact group is $m$-factorizable, by Theorem 8.5.2. □

COROLLARY 8.5.9. *Let $G$ be an $\mathbb{R}$-factorizable group that satisfies $w(G)^\omega < 2^{\aleph_1}$. Then the product $G \times K$ is $\mathbb{R}$-factorizable, for every compact group $K$. In particular, under $\mathfrak{c} < 2^{\aleph_1}$, this conclusion holds for every $\mathbb{R}$-factorizable group $G$ with $w(G) \leq \mathfrak{c}$.*

PROOF. By Theorem 8.5.8, the group $G$ is $m$-factorizable. It remains to apply a) of Theorem 8.5.5. □

The above corollary applies to small $\mathbb{R}$-factorizable groups only. One can, however, extend it to a much wider class of groups whose continuous homomorphic images of small weight are pseudo-$\aleph_1$-compact.

COROLLARY 8.5.10. *Suppose that every continuous homomorphic image $H$ of an $\mathbb{R}$-factorizable group $G$ satisfying $w(H) \leq \aleph_1$ is pseudo-$\aleph_1$-compact. Then $G$ is also pseudo-$\aleph_1$-compact, and the product $G \times K$ is $\mathbb{R}$-factorizable, for each compact group $K$.*

PROOF. By Theorems 8.5.2 and 8.5.5 a), it suffices to verify that $G$ is pseudo-$\aleph_1$-compact. Let $K = \mathbb{Z}(2)^{\omega_1}$, and suppose that $f\colon G \times K \to \mathbb{R}$ is a continuous function. As in the proof of Theorem 8.5.5, consider the mapping $\Psi\colon G \to C(K)$ defined by $\Psi(x)(y) = f(x, y)$, for every $x \in G$ and every $y \in K$, where $C(K)$ is the Banach space of all continuous real-valued functions on $K$ with the sup-norm. The compactness of $K$ implies that $w(C(K)) = w(K) = \aleph_1$. Since $G$ is $\mathbb{R}$-factorizable, we can apply Lemma 8.1.2 to find a continuous homomorphism $\pi\colon G \to H$ onto a topological group $H$ with $w(H) \le \aleph_1$ and a continuous mapping $\psi\colon H \to X$ such that $\Psi = \psi \circ \pi$. By the assumptions, the group $H$ is pseudo-$\aleph_1$-compact.

Let $id_K$ be the identity mapping of $K$ onto itself. As in the proof of Theorem 8.5.5, there exists a continuous mapping $h\colon H \times K \to \mathbb{R}$ such that $h \circ (\pi \times id_K) = f$. Define the mapping $\Phi\colon H \to C(K)$ by $\Phi(x)(y) = h(x, y)$, for all $x \in H$ and all $y \in K$. The group $H$ is pseudo-$\aleph_1$-compact and the mapping $\Phi$ is continuous, so the image $\Phi(H) = \Psi(G)$ is second-countable. The latter fact and Lemma 8.1.2 permit us to find a continuous homomorphism $\varphi\colon G \to P$ onto a second-countable group $P$ and a continuous mapping $p\colon P \to C(K)$ such that $\Psi = p \circ \varphi$. In its turn, we use the homomorphism $\varphi$ to define a continuous mapping $h^*\colon P \times K \to \mathbb{R}$ such that $f = h^* \circ (\varphi \times id_K)$. Since the group $P \times K$ is Lindelöf, we conclude, as in Theorem 8.5.5 a), that the product $G \times K$ is $\mathbb{R}$-factorizable. Therefore, $G$ is pseudo-$\aleph_1$-compact, by b) of Theorem 8.5.5. □

The next result demonstrates one more time the importance of pseudo-$\aleph_1$-compactness when studying products of $\mathbb{R}$-factorizable groups. In fact, it completely characterizes $\mathbb{R}$-factorizability of the product $G \times K$ in the case when the second factor is compact.

THEOREM 8.5.11. *The product $G \times K$ of an $\mathbb{R}$-factorizable group $G$ and a compact group $K$ is $\mathbb{R}$-factorizable iff either $G$ is pseudo-$\aleph_1$-compact or $K$ is metrizable.*

PROOF. If the group $G$ is pseudo-$\aleph_1$-compact, then the product $G \times K$ is $\mathbb{R}$-factorizable, by Theorem 8.5.5. In addition, if $K$ is metrizable, then the product $G \times K$ is $\mathbb{R}$-factorizable for an arbitrary $\mathbb{R}$-factorizable group $G$, by Corollary 8.5.6.

Conversely, suppose that $G$ contains a discrete family $\{U_\alpha : \alpha < \omega_1\}$ of non-empty open sets, while $K$ is not metrizable. We can assume without loss of generality that each $U_\alpha$ is a cozero set. For every $\alpha < \omega_1$, let $f_\alpha$ be a continuous real-valued function on $G$ such that $U_\alpha = G \setminus f_\alpha^{-1}(0)$. Since $K$ is a compact non-metrizable group, there exists a continuous mapping $\varphi$ of $K$ onto $I^{\omega_1}$, where $I$ is the closed unit interval (see Corollary 4.2.5). For every $\alpha < \omega_1$, denote by $\pi_\alpha$ the projection of $I^{\omega_1}$ to the $\alpha$-th factor, and let $g_\alpha = \pi_\alpha \circ \varphi$. Clearly, each $g_\alpha$ is a continuous real-valued function on $K$. Since the family $\{U_\alpha : \alpha < \omega_1\}$ is discrete, the function

$$f(x, y) = \sum_{\alpha<\omega_1} f_\alpha(x) \cdot g_\alpha(y),$$

with $x \in G$ and $y \in K$, is continuous on the product $G \times K$.

Assume that $f$ is factorizable, that is, there exist a continuous homomorphism $p\colon G \times K \to H$ onto a second-countable topological group $H$ and a continuous function $h\colon H \to \mathbb{R}$ such that $f = h \circ p$. Let $e_G$ and $e_K$ be the identities of $G$ and $K$, respectively. Since $\ker p$ is a $G_\delta$-set in $G \times K$, we can find $G_\delta$-sets $P_G$ and $P_K$ in $G$ and $K$, respectively, such that $e_G \in P_K$, $e_K \in P_K$, and $P_G \times P_K \subset \ker p$. The group $K$ is compact, so there exists a closed invariant $G_\delta$-subgroup $N$ in $K$ such that $N \subset P_K$. Let $\pi\colon K \to L$ be the quotient homomorphism, where $L = K/N$. Then $\pi$ is open; hence, from $\{e_G\} \times N \subset P_G \times P_K \subset \ker p$ it follows that $i \times \pi \prec p$, where $i$ is the identity automorphism of $G$. Therefore, we can find a continuous homomorphism $q\colon G \times L \to H$ such that $p = q \circ (i \times \pi)$.

$$\begin{array}{ccc} G \times K & \xrightarrow{\ f\ } & \mathbb{R} \\ {\scriptstyle i\times\pi}\downarrow & \searrow^{p} & \uparrow{\scriptstyle h} \\ G \times L & \xrightarrow{\ q\ } & H \end{array}$$

The commutativity of the above diagram implies that if $\pi(y_1) = \pi(y_2)$ for some $y_1, y_2 \in K$, then $g_\alpha(y_1) = g_\alpha(y_2)$ for each $\alpha < \omega_1$. Indeed, suppose that $\pi(y_1) = z = \pi(y_2)$ for some $y_1, y_2 \in K$. Pick arbitrary point $x \in U_\alpha$. Then

$$(i \times \pi)(x, y_1) = (x, z) = (i \times \pi)(x, y_2).$$

Therefore, $f(x, y_i) = h \circ q \circ (i \times \pi)(x, y_i) = h(q(x, z))$ for $i = 1, 2$; hence, $f(x, y_1) = f(x, y_2)$. The definition of $f$ implies that $f(x, y_i) = f_\alpha(x) \cdot g_\alpha(y_i)$, for $i = 1, 2$. Since $f_\alpha(x) \neq 0$, we have $g_\alpha(y_1) = g_\alpha(y_2)$.

Since the homomorphism $\pi\colon K \to L$ is open, the claim just proved implies that $\pi \prec g_\alpha$ for each $\alpha < \omega_1$. The mapping $\varphi\colon K \to I^{\omega_1}$ is the diagonal product of the functions $g_\alpha$, so $\pi \prec \varphi$ as well. Hence, there exists a continuous mapping $\psi\colon L \to I^{\omega_1}$ such that $\varphi = \psi \circ \pi$. Clearly, $\psi(L) = I^{\omega_1}$. Finally, the kernel $N$ of the homomorphism $\pi$ is of type $G_\delta$ in the compact group $K$, so that the group $L = K/N$ is first-countable and, hence, metrizable. Thus, the Tychonoff cube $I^{\omega_1}$ is metrizable, as a continuous image of the compact metrizable space $L$, a contradiction. □

In Theorem 8.5.13 below we show that the product of a weakly Lindelöf ℝ-factorizable group with a pseudocompact group is ℝ-factorizable. This requires a topological lemma in which no separation axiom on the factors $X$ and $Y$ is assumed.

LEMMA 8.5.12. *Let $f\colon X \times Y \to \mathbb{R}$ be a continuous function. If the space $Y$ is weakly Lindelöf, then for every $x \in X$ there exists a $G_\delta$-set $P_x$ in $X$ containing $x$ such that $f(x, y) = f(x', y)$, for all $x' \in P_x$ and $y \in Y$.*

PROOF. Take an arbitrary point $x_0 \in X$. For every $n \in \mathbb{N}$ and every $y \in Y$, we can find open sets $U(y, n)$ and $V(y, n)$ in $X$ and $Y$, respectively, such that $x_0 \in U(y, n)$, $y \in V(y, n)$ and $|f(x', y') - f(x_0, y)| < 1/n$ for all $x' \in U(y, n)$ and $y' \in V(y, n)$. Then the family $\{V(y, n) : y \in Y\}$ covers $Y$ and since $Y$ is weakly Lindelöf, there exists a countable set $A_n \subset Y$ such that the union $V(n) = \bigcup\{V(y, n) : y \in A_n\}$ is dense in $Y$. Then the set $P(n) = \bigcap\{U(y, n) : y \in A_n\}$ is a $G_\delta$-set in $X$ and contains $x_0$. Let us verify that

$$|f(x, y) - f(x_0, y)| \le 1/n \text{ for all } x \in P(n) \text{ and } y \in Y. \tag{8.4}$$

Assume to the contrary that $d = |f(x, y) - f(x_0, y)| > 1/n$, for some $x \in P(n)$ and some $y \in Y$. Let $\varepsilon = d - 1/n$. Then there exist open sets $U_1, U_2$ in $X$ and an open set $V$ in $Y$ such that $(x, y) \in U_1 \times V$, $(x_0, y) \in U_2 \times V$, $U_1 \cup U_2 \subset U(n)$, and

$$|f(x', y') - f(x, y)| < \varepsilon/2, \quad |f(x'', y'') - f(x_0, y)| < \varepsilon/2 \tag{8.5}$$

whenever $(x', y') \in U_1 \times V$, $(x'', y'') \in U_2 \times V$. Since $V(n)$ is dense in $Y$, we can find $z \in A_n$ such that $V \cap V(z, n) \neq \varnothing$. Pick a point $t \in V \cap V(z, n)$. Then (8.5) and our choice of $x \in P(n) \subset U(z, n)$ and $t \in Y$ together imply that

$$\begin{aligned} d = |f(x, y) - f(x_0, y)| &\le |f(x, y) - f(x, t)| + |f(x, t) - f(x_0, t)| + |f(x_0, t) - f(x_0, y)| \\ &< \varepsilon/2 + 1/n + \varepsilon/2 = d, \end{aligned}$$

which is a contradiction. This proves (8.4).

Clearly, $P = \bigcap_{n=1}^{\infty} P(n)$ is a $G_\delta$-set in $X$ that contains $x_0$. From (8.4) it follows immediately that $f(x, y) = f(x_0, y)$, for all $x \in P$ and $y \in Y$. The lemma is proved. □

THEOREM 8.5.13. *The product $G \times H$ of a weakly Lindelöf $\mathbb{R}$-factorizable group $G$ and a pseudocompact group $H$ is $\mathbb{R}$-factorizable.*

PROOF. Let $f: G \times H \to \mathbb{R}$ be a continuous function. Applying Lemma 8.5.12, we find, for every point $y \in H$, a $G_\delta$-set $K_y$ in $H$ such that $y \in K_y$ and $f(x, y) = f(x, y')$ whenever $x \in G$ and $y' \in K_y$. Then the completion $\varrho H$ of $H$ is a compact group, and $H$ is a dense subgroup of $\varrho H$. For every $y \in H$, choose a closed $G_\delta$-set $F_y$ in $\varrho H$ such that $y \in F_y \cap H \subset K_y$. By Corollary 5.3.19, there exists a countable set $M \subset H$ such that $\bigcup_{x \in M} F_y$ is dense in $\bigcup_{y \in H} F_y$. Since $H$ intersects every non-empty $G_\delta$-set in $\varrho H$ (see Corollary 6.6.6), the set $R = \bigcup_{y \in M} K_y$ is dense in $H$. The family $\{F_y : y \in M\}$ consists of zero-sets in the compact group $\varrho H$ and, since $\varrho H$ is $\mathbb{R}$-factorizable, we can find a continuous homomorphism $\pi: \varrho H \to H_0$ onto a second-countable topological group $H_0$ such that $F_y = \pi^{-1}\pi(F_y)$, for each $y \in M$. Note that the homomorphism $\pi$ is open because the group $\varrho H$ is compact. Let $p = \pi \restriction H$ be the restriction of $\pi$ to $H$. The pseudocompactness of $H$ implies that $p(H) = \pi(H) = H_0$. In addition, since $H$ intersects every non-empty $G_\delta$-set in $\varrho H$, we have $p(U \cap H) = \pi(U)$, for each open set $U$ in $\varrho H$, i.e., the homomorphism $p$ is open. Note that the homomorphisms $\widehat{\varphi} = id_G \times \pi$ and $\varphi = id_G \times p$ are open as products of open homomorphisms (see the diagram below).

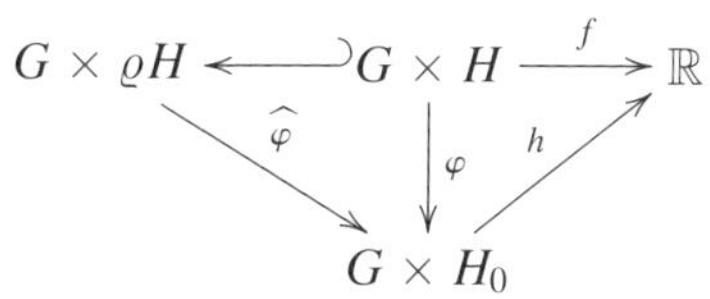


By Lemma 1.7.6 (with $X = S = G \times H$, $Y = \mathbb{R}$, $g = \varphi$, $Z = G \times H_0$ and $T = G \times p(R)$), there exists a continuous mapping $h: G \times H_0 \to \mathbb{R}$ such that $f = h \circ \varphi$. Since $G$ is $\mathbb{R}$-factorizable and $H_0$ is compact and second-countable, the group $G \times H_0$ is $\mathbb{R}$-factorizable, by Corollary 8.5.6. Clearly, $\varphi \prec f$, so Lemma 8.1.11 implies that the product group $G \times H$ is $\mathbb{R}$-factorizable. □

COROLLARY 8.5.14. *The product $G \times H$ of a Lindelöf group $G$ and a pseudocompact group $H$ is $m$-factorizable.*

PROOF. Since every Lindelöf group is weakly Lindelöf and $\mathbb{R}$-factorizable (see Theorem 8.1.6), from Theorem 8.5.13 it follows that the group $G \times H$ is $\mathbb{R}$-factorizable.

Let $K$ be an arbitrary compact group. Then the group $G \times H \times K \cong (G \times K) \times H$ is $\mathbb{R}$-factorizable because the first factor $G \times K$ is Lindelöf and the second factor $H$ is pseudocompact. Hence, $G \times H$ is $m$-factorizable, by b) of Theorem 8.5.5. □

COROLLARY 8.5.15. *The product $G \times H$ of a Lindelöf topological group $G$ and a pseudocompact group $H$ is $C$-embedded in $G \times \varrho H$.*

PROOF. Evidently, $G \times H$ is $G_\delta$-dense in $G \times \varrho H$. Since every $m$-factorizable group is $\mathbb{R}$-factorizable, it suffices to apply Proposition 8.3.4 and Corollary 8.5.14. □

We finish this section with a result close to Theorem 8.5.13. Its proof requires the following auxiliary fact.

LEMMA 8.5.16. *The product $X \times Y$ of a weakly Lindelöf space $X$ and a dense subspace $Y$ of a space $Y^*$ with $cel_\omega(Y^*) \leq \omega$ is weakly Lindelöf.*

PROOF. Let $\gamma$ be a covering of $X \times Y$ by open sets in $X \times Y^*$. We can assume without loss of generality that the elements of $\gamma$ are rectangular, that is, each element of $\gamma$ has the form $U \times V$. Since $X$ is weakly Lindelöf, for every $y \in Y$ there exists a countable subfamily $\gamma(y)$ of $\gamma$ such that $p_Y^{-1}(y) \cap \bigcup \gamma(y)$ is dense in $p_Y^{-1}(y)$, where $p_Y$ is the projection of $X \times Y^*$ to $Y^*$. Clearly, we can assume that all elements of $\gamma(y)$ intersect the fiber $p_Y^{-1}(y)$. Denote by $K_y$ the intersection of the family $\{p_Y(E) : E \in \gamma(y)\}$. Then $K_y$ is a $G_\delta$-set in $Y^*$ and $y \in K_y$, by our choice of $\gamma(y)$. Since $cel_\omega(Y^*) \leq \omega$, there exists a countable set $M \subset Y$ such that $\bigcup_{y \in M} K_y$ is dense in $\bigcup_{y \in Y} K_y$. It follows from $Y \subset \bigcup_{y \in Y} K_y$ that $\bigcup_{y \in M} K_y$ is dense in $Y^*$. Let $\gamma^* = \bigcup_{y \in M} \gamma(y)$. Then $\gamma^*$ is a countable subfamily of $\gamma$, and we claim that $(X \times Y) \cap \bigcup \gamma^*$ is dense in $X \times Y$.

Indeed, take an arbitrary non-empty open set $O \times W$ in $X \times Y$. Let $W^*$ be an open subset of $Y^*$ such that $W = W^* \cap Y$. Since $\bigcup_{y \in M} K_y$ is dense in $Y^*$, the intersection $K_y \cap W^*$ is not empty, for some $y \in M$. It follows from our choice of the family $\gamma(y)$ that $O \cap U \neq \varnothing$, for some element $U \times V \in \gamma(y)$. Since $K_y \subset V$, we have that $W^* \cap V \supset W^* \cap K_y \neq \varnothing$. However, the sets $W^*$ and $V$ are open in $Y^*$ and $Y$ is dense in $Y^*$, whence it follows that the set $W \cap V = (W^* \cap Y) \cap V = (W^* \cap V) \cap Y$ is not empty. We conclude, therefore, that $(O \times W) \cap (U \times V) \neq \varnothing$. This proves that the set $(X \times Y) \cap \bigcup \gamma^*$ is dense in $X \times Y$. Since the subfamily $\gamma^*$ of $\gamma$ is countable, the product space $X \times Y$ is weakly Lindelöf. □

THEOREM 8.5.17. *Let $G$ be a weakly Lindelöf $\omega$-steady topological group and $H$ an arbitrary subgroup of a Lindelöf $\Sigma$-group. Then the product group $G \times H$ is $\mathbb{R}$-factorizable.*

PROOF. According to Corollaries 5.2.9 and 8.1.22, the group $G$ is $\omega$-narrow and $\omega$-stable. By our assumption, $H$ is a topological subgroup of a Lindelöf $\Sigma$-group, say, $\widetilde{H}$. Taking the closure of $H$ in $\widetilde{H}$, we can assume without loss of generality that $H$ is dense in $\widetilde{H}$. Suppose that $f$ is a continuous real-valued function on the group $G \times H$. It follows from Corollary 5.3.19 that the group $\widetilde{H}$ satisfies $cel_\omega(\widetilde{H}) \leq \omega$, so Lemma 8.5.16 implies that the product group $G \times H$ is weakly Lindelöf. Hence, by Theorem 8.1.18 (with $\tau = \omega$), the group $G \times H$ contains a closed invariant subgroup $N$ of type $G_\delta$ such that $f$ is constant on each coset of $N$ in $G \times H$. Let $e_H$ be the neutral element of $H$. Since the group $G$ is $\omega$-narrow and $N \cap (G \times \{e_H\})$ is a $G_\delta$-set in $G \times \{e_H\} \cong G$, Corollary 5.1.8 implies that there exists a quotient homomorphism $p \colon G \to G_0$ onto a topological group $G_0$ of countable pseudocharacter such that $\ker p \times \{e_H\} \subset N$.

Let $\pi$ be the product of $p$ and the identity automorphism $id_H$ of the group $H$. Then the kernel $P$ of $\pi$ is contained in $N$, so that $f$ is constant on each coset of $P$ in $G \times H$. In other words, there exists a real-valued function $g$ on the group $G_0 \times H$ such that $f = g \circ \pi$. Since the homomorphism $\pi$ is open, the function $g$ is continuous. By the assumption, the group $G$ is $\omega$-steady, while the continuous homomorphic image $G_0$ of $G$ has countable pseudocharacter. Hence, the group $G_0$ has a countable network and, by Proposition 5.3.3 , is a Lindelöf $\Sigma$-space. It follows that $G_0 \times H$ is a (dense) subgroup of the Lindelöf $\Sigma$-group $G_0 \times \widetilde{H}$, so Proposition 8.1.13 implies that the group $G_0 \times H$ is $\mathbb{R}$-factorizable. Since $\pi \prec f$, the group $G \times H$ is $\mathbb{R}$-factorizable according to Lemma 8.1.11. □

Since every Lindelöf $P$-group is $\omega$-stable and, hence, $\omega$-steady (see Corollary 5.6.10 and Proposition 5.6.13), the following corollary is immediate from Theorem 8.5.17:

COROLLARY 8.5.18. *The product of a Lindelöf $P$-group and a precompact group is $\mathbb{R}$-factorizable.*

## Exercises

8.5.a. Let $G$ and $H$ be topological group. Show that if the product group $G \times H$ is $\mathbb{R}$-factorizable, then either $G$ or $H$ is pseudo-$\aleph_1$-compact. Extend this result to completely regular paratopological groups.
*Hint.* Suppose not, and choose discrete families $\{U_\alpha : \alpha < \omega_1\}$ and $\{V_\alpha : \alpha < \omega_1\}$ of non-empty open sets in $G$ and $H$, respectively. Then define a continuous real-valued function $f$ on $G \times H$ such that $f(U_\alpha \times V_\beta) \subset (1, \infty)$ if $\alpha < \beta$ and $f(U_\alpha \times V_\beta) \subset (-\infty, 0)$ if $\alpha \geq \beta$. Consider an arbitrary continuous homomorphism $p \colon G \times H \to K$ to a second-countable topological group $K$ and find elements $x, y \in K$ such that every neighbourhood of $x$ in $K$ intersects uncountably many elements of the family $\{p(U_\alpha \times \{e_H\}) : \alpha < \omega_1\}$, while every neighbourhood of $y$ intersects uncountably many elements of the family $\{p(\{e_G\} \times V_\alpha) : \alpha < \omega_1\}$. To finish the argument, show that every real-valued function $h$ on $K$ satisfying $f = h \circ p$ is discontinuous at the point $xy \in K$. Notice that the same argument works for Tychonoff paratopological groups $G$ and $H$.

8.5.b. Let $G$ be any collectionwise normal topological group such that the group $G^2$ is $\mathbb{R}$-factorizable. Show that the extent of $G$ is countable.

8.5.c. Prove that if an $\mathbb{R}$-factorizable group is $\omega$-stable, then it is $m$-factorizable.

8.5.d. Let $G$ be an $m$-factorizable group and $H$ a pseudocompact group.
 (a) Show that the product group $G \times H$ is $m$-factorizable iff $G \times H$ is $C$-embedded in $G \times \varrho H$.
 (b) Apply (a) to prove that if $ot(G) \leq \omega$, then $ot(G \times H) \leq \omega$ and the group $G \times H$ is $\mathbb{R}$-factorizable.
 (c) Verify that if $G$ is a $k$-group, then $ot(G \times H) \leq \omega$ and the group $G \times H$ is $m$-factorizable.

8.5.e. Let $G$ be a Lindelöf topological group and $X$ a pseudocompact space. Modify the proof of Theorem 8.5.13 to show that the product groups $G \times F(X)$ and $G \times A(X)$ are $\mathbb{R}$-factorizable.

8.5.f. Verify that the product $G \times C_p(X)$ is $\mathbb{R}$-factorizable, for every precompact topological group $G$ and every Tychonoff space $X$.

8.5.g. Prove that the product $G \times H$ of a Lindelöf $\omega$-stable group $G$ with a $\sigma$-bounded topological group $H$ is $\mathbb{R}$-factorizable.

8.5.h. Modify the proof of Theorem 8.5.5 to show that if the product group $G \times C_p(X)$ is $\mathbb{R}$-factorizable, for an uncountable Tychonoff space $X$, then $G$ is $m$-factorizable.

8.5.i. Show that every subgroup of a Lindelöf $\Sigma$-group is $m$-factorizable.

## Problems

8.5.A. Give an example of a connected, locally connected $\omega$-narrow topological group that is not $\mathbb{R}$-factorizable.
*Hint.* For a topological group $G$, let $G^\bullet$ be the connected, locally connected group containing $G$ as a subgroup that was constructed in Section 3.8. It follows from Corollary 3.8.5 that $G$ is $C^*$-embedded in $G^\bullet$. Apply Example 8.2.1 and Theorem 8.2.6 to construct the required topological group.

8.5.B. (M. G. Tkachenko [**479**]) Prove that every locally connected $\mathbb{R}$-factorizable group is pseudo-$\aleph_1$-compact.

*Hint.* First, establish the following fact. If $f: G \to X$ is a continuous mapping of a locally connected $\mathbb{R}$-factorizable group to a second-countable space $X$, then there exists a continuous homomorphism $\pi: G \to K$ onto a topological group $K$ with a countable base $\mathcal{B}$ such that $\pi \prec f$ and $\pi^{-1}(V)$ is connected for each $V \in \mathcal{B}$. Then modify the argument in the proof of Theorem 8.5.2.

8.5.C. Let $G$ be a topological group such that $G^\omega$ is $\mathbb{R}$-factorizable. Prove that the product group $G^\tau$ and the $\Sigma$-product $\Sigma G^\tau$ are $\mathbb{R}$-factorizable, for each cardinal $\tau > \omega$.
*Hint.* Apply Exercise 8.5.a and Proposition 1.6.22 to show that the groups $G^\omega$, $G^\tau$ and $\Sigma G^\tau$ are pseudo-$\aleph_1$-compact. Then extend Theorem 1.7.2 to $\Sigma$-products.

8.5.D. (C. Hernández and M. G. Tkachenko [**227**]) Let $N$ be a closed subgroup of a topological group $G$. Suppose that a topology $\mathcal{T}$ on the left coset space $G/N$ is weaker than the quotient topology of $G/N$. We say that $\mathcal{T}$ is *left-invariant* if the left translations $\phi_a: G/N \to G/N$ defined by $\phi_a(xN) = axN$ for all $a, x \in G$ are continuous with respect to $\mathcal{T}$. The group $G$ is called *semi-$\mathbb{R}$-factorizable* if, for every continuous function $f: G \to \mathbb{R}$, one can find a closed subgroup $N$ of $G$ and a second-countable left-invariant $T_1$ topology $\mathcal{T}$ on $G/N$ such that the natural projection $p: G \to (G/N, \mathcal{T})$ satisfies $p \prec f$. Prove that every semi-$\mathbb{R}$-factorizable group is $\mathbb{R}$-factorizable.
*Hint.* Show that every semi-$\mathbb{R}$-factorizable group is $\omega$-narrow.

## Open Problems

8.5.1. Let $G$ be any $\mathbb{R}$-factorizable topological group. Is $G$ is pseudo-$\aleph_1$-compact?

8.5.2. Is the product of an $\mathbb{R}$-factorizable group with a second-countable group $\mathbb{R}$-factorizable?

8.5.3. Is the product of two $\mathbb{R}$-factorizable topological groups of countable cellularity $\mathbb{R}$-factorizable?

8.5.4. Suppose that $G$ and $H$ are weakly Lindelöf $\mathbb{R}$-factorizable topological groups. Is the group $G \times H$ $\mathbb{R}$-factorizable?

8.5.5. Is the product of an $\mathbb{R}$-factorizable group with (a subgroup of) a $\sigma$-compact group $\mathbb{R}$-factorizable?

8.5.6. Is the product of an $\mathbb{R}$-factorizable group with (a subgroup of) a Lindelöf $\Sigma$-group $\mathbb{R}$-factorizable?

8.5.7. Is the product of two $\mathbb{R}$-factorizable topological groups a $PT$-group?

8.5.8. Let $i \in \{1, 2, 3\}$. Is the product of an $\mathbb{R}_i$-factorizable paratopological group with a compact topological group $\mathbb{R}_i$-factorizable?

8.5.9. Is the product of two separable $\mathbb{R}_i$-factorizable paratopological groups an $\mathbb{R}_i$-factorizable paratopological group, where $i \in \{1, 2, 3\}$?

8.5.10. Suppose that $G$ is a topological group and $G = \bigcup_{i=0}^\infty G_i$, where each $G_i$ is an $\mathbb{R}$-factorizable subgroup of $G$. Must $G$ be $\mathbb{R}$-factorizable? Is $G$ a $PT$-group?

## 8.6. $\mathbb{R}$-factorizability of $P$-groups

Unlike the general case, the $\mathbb{R}$-factorizability of $P$-groups admits a complete description — by Theorem 8.6.12, a $P$-group is $\mathbb{R}$-factorizable if and only if it is pseudo-$\aleph_1$-compact. The class of $P$-groups serves as a source of numerous examples that help to understand relationship between various classes of topological groups. It was not accidental, for instance, that the $\omega$-narrow non-$\mathbb{R}$-factorizable group $H$ in Example 8.2.1 was constructed as a dense subgroup of a Lindelöf $P$-group. We develop further the technique used in the previous sections and construct an $\mathbb{R}$-factorizable $P$-group that is not weakly Lindelöf (see Example 8.6.16).

The following lemma is a part of Theorem 8.6.12 that characterizes $\mathbb{R}$-factorizability of $P$-groups.

LEMMA 8.6.1. *Every $\mathbb{R}$-factorizable $P$-group $G$ is pseudo-$\aleph_1$-compact.*

PROOF. Assume the contrary. Then $G$ contains an uncountable discrete family $\{U_\alpha : \alpha < \omega_1\}$ of non-empty open sets. Let $R = \{r_\alpha : \alpha < \omega_1\}$ be a set of pairwise distinct real numbers. For every $\alpha < \omega_1$, define a continuous real-valued function $f_\alpha$ on $G$ such that $f_\alpha(x_\alpha) = r_\alpha$, for some $x_\alpha \in U_\alpha$, and $f_\alpha(x) = 0$ if $x \in G \setminus U_\alpha$. Then the function $f = \sum_{\alpha<\omega_1} f_\alpha$ is continuous on $G$. Clearly, $R \subset f(G)$, so that $f(G)$ is uncountable. Since the group $G$ is $\mathbb{R}$-factorizable, we can find a continuous homomorphism $\pi\colon G \to K$ onto a second-countable group $K$ and a continuous function $h\colon K \to \mathbb{R}$ such that $f = h \circ \pi$. By Lemma 4.4.2, the group $K$ is countable, whence $|f(G)| \leq |K| \leq \omega$, a contradiction. □

The notion of a *$\tau$-complete family* of mappings plays an important role in the arguments to follow. Let $X$ be a space and $\mathscr{F}$ be a family of continuous mappings from $X$ elsewhere. Given a subfamily $\gamma$ of $\mathscr{F}$, we denote by $h_\gamma = \otimes\gamma$ the diagonal product of the mappings from $\gamma$ considered as a mapping of $X$ onto its image $h_\gamma(X)$. Clearly, $h_\gamma$ is continuous for every $\gamma \subset \mathscr{F}$. If $\tau$ is an infinite cardinal, we say that $\mathscr{F}$ is *$\tau$-complete* if for every sequence $\{f_\alpha : \alpha < \tau\} \subset \mathscr{F}$ satisfying $f_\beta \prec f_\alpha$ when $\alpha < \beta < \tau$, the mapping $\otimes_{\alpha<\tau} f_\alpha$ belongs to $\mathscr{F}$.

LEMMA 8.6.2. *Let $G$ be a Lindelöf topological group, and $\mathscr{L}$ be the family of all continuous open homomorphisms $\varphi\colon G \to K$ onto topological groups $K$ satisfying $w(K) \leq \aleph_1$. Then the family $\mathscr{L}$ is $\aleph_1$-complete.*

PROOF. Suppose that $\{f_\alpha : \alpha < \omega_1\} \subset \mathscr{L}$ satisfies $f_\beta \prec f_\alpha$, whenever $\alpha < \beta < \omega_1$. Let $f = \otimes_{\alpha<\omega_1} f_\alpha$ and $H = f(G)$. We also put $H_\alpha = f_\alpha(G)$, for every $\alpha < \omega_1$. Since $H$ is a subgroup of the product $\prod_{\alpha<\omega_1} H_\alpha$, and $w(H_\alpha) \leq \aleph_1$ for each $\alpha < \omega_1$, we have $w(H) \leq \aleph_1$. So, it suffices to show that the homomorphism $f$ is open.

Let $U$ and $V$ be neighbourhoods of the identity $e$ in $G$ such that $V^2 \subset U$. For every $\alpha < \omega_1$, denote by $N_\alpha$ the kernel of the homomorphism $f_\alpha$. Then $N_\beta \subset N_\alpha$ if $\alpha < \beta < \omega_1$, and $N = \bigcap_{\alpha<\omega_1} N_\alpha$ is the kernel of $f$. Since the group $G$ is Lindelöf, there exists an ordinal $\alpha < \omega_1$ such that $N_\alpha \subset VN$. Then

$$f_\alpha^{-1} f_\alpha(V) = VN_\alpha \subset VVN \subset UN = f^{-1} f(U). \tag{8.6}$$

If $\alpha < \omega_1$, then $f \prec f_\alpha$, so there exists a continuous homomorphism $g_\alpha\colon H \to H_\alpha$ such that $f_\alpha = g_\alpha \circ f$. Therefore, from (8.6) it follows that $g_\alpha^{-1}(f_\alpha(V)) \subset f(U)$. Since the homomorphism $f_\alpha$ is open, $g_\alpha^{-1}(f_\alpha(V))$ is an open neighbourhood of the identity in $H$. Thus, the interior of $f(U)$ is non-empty and contains the identity of $H$. So, the homomorphism $f$ is open and, hence, $f \in \mathscr{L}$. □

Neither pseudo-$\aleph_1$-compact $P$-spaces, nor $P$-groups have to be Lindelöf (see Example 8.2.1). However, we have the following:

LEMMA 8.6.3. *Any regular pseudo-$\aleph_1$-compact $P$-space $X$ satisfying $l(X) \leq \aleph_1$ is Lindelöf.*

PROOF. Assume the contrary. Then there exists an open covering $\gamma$ of $X$ such that $X \setminus \bigcup \mu \neq \varnothing$, for each countable subfamily $\mu$ of $\gamma$. Since $l(X) \leq \aleph_1$, we can assume that $|\gamma| = \aleph_1$. Clearly, the space $X$ has a base of clopen sets, so we can also assume that all elements of $\gamma$ are clopen in $X$. Let $\gamma = \{U_\alpha : \alpha < \omega_1\}$. For every $\alpha < \omega_1$, the set $V_\alpha = U_\alpha \setminus \bigcup_{\beta<\alpha} U_\beta$ is clopen in $X$. Consider the disjoint family $\lambda = \{V_\alpha : \alpha < \omega_1\}$. It is easy to see that $\bigcup \lambda = \bigcup \gamma = X$. Since $\lambda$ refines $\gamma$, we have $|\lambda| = |\gamma| = \aleph_1$. So, $\lambda$ is an uncountable disjoint open covering of $X$ and hence, the space $X$ is not pseudo-$\aleph_1$-compact. □

Now we apply Lemma 8.6.1 and 8.6.3 to deduce the following.

COROLLARY 8.6.4. *An* $\mathbb{R}$*-factorizable* $P$*-group* $G$ *with* $l(G) \leq \aleph_1$ *is Lindelöf. In particular, every* $\mathbb{R}$*-factorizable* $P$*-group of weight* $\leq \aleph_1$ *is Lindelöf.*

Similarly to Lindelöf $P$-groups, every non-discrete $\mathbb{R}$-factorizable $P$-group has many open homomorphisms onto topological groups of weight $\aleph_1$.

LEMMA 8.6.5. *Let* $\mathcal{H}$ *be the family of all continuous homomorphisms of an* $\omega$*-narrow* $P$*-group* $G$ *onto* $P$*-groups of weight* $\leq \aleph_1$. *If all images* $f(G)$ *with* $f \in \mathcal{H}$ *are Lindelöf, then every* $f \in \mathcal{H}$ *is open.*

PROOF. Consider a homomorphism $f \in \mathcal{H}$, $f: G \to H$. Let $U$ be a neighbourhood of the identity $e$ in $G$. Since the group $G$ is $\omega$-narrow, (b) of Lemma 4.4.1 implies that there exists an open invariant subgroup $P$ of $G$ with $P \subset U$. Let $\varphi: G \to G/P$ be the quotient homomorphism. The group $G/P$ is discrete and, hence, countable by Lemma 4.4.2. Let $\psi = f \triangle \varphi$ be the diagonal product of $f$ and $\varphi$. Then $K = \psi(G)$ is a subgroup of $H \times G/P$, whence it follows that $w(K) \leq \aleph_1$. Since $H$ is a $P$-group and $G/P$ is discrete, the product $H \times G/P$ and its subgroup $K$ are $P$-groups. By the assumptions, $K$ is Lindelöf. Denote by $g$ the restriction to $K$ of the projection of the product $H \times G/P$ to the first factor. Then $f = g \circ \psi$, so $g(K) = H$. Since $K$ and $H$ are Lindelöf $P$-groups, Lemma 4.4.6 implies that the homomorphism $g$ is open.

Let $p$ be the projection of the product $H \times G/P$ to the second factor and $h = p{\restriction}K$ be the restriction of $p$ to $K$. Then from $P = \varphi^{-1}(e^*)$ and $\varphi = h \circ \psi$ it follows that $\psi(P) = h^{-1}(e^*)$ is open in $K$, where $e^*$ is the identity of the discrete group $G/P$. Therefore, $f(P) = g(\psi(P))$ is open in $H$. We conclude that $f(U)$ contains the open neighbourhood $f(P)$ of the identity in $H$, which implies that $f$ is open. □

The following result plays the key role in the proof of Theorem 8.6.12. In a sense, it enables us to replace a given pseudo-$\aleph_1$-compact $P$-group by its quotient of weight less than or equal to $\aleph_1$, and this quotient group is Lindelöf, by Lemma 8.6.3.

LEMMA 8.6.6. *Let* $G$ *be a pseudo-*$\aleph_1$*-compact* $P$*-group and* $\mathcal{L}$ *be the family of all continuous open homomorphisms* $f: G \to K$ *onto topological groups* $K$ *satisfying* $w(K) \leq \aleph_1$. *Then the family* $\mathcal{L}$ *is* $\aleph_1$*-complete.*

PROOF. It is easy to see that $G$ satisfies the conditions of Lemma 8.6.5, i.e., if a continuous homomorphic image $H = f(G)$ of weight $\leq \aleph_1$ is a $P$-group, then $H$ is Lindelöf. Indeed, the group $H$ is pseudo-$\aleph_1$-compact being a continuous image of $G$, so $H$ is Lindelöf by Lemma 8.6.3. Therefore, all such homomorphisms $f: G \to H$ are open by Lemma 8.6.5.

Suppose that the family $\{f_\alpha : \alpha < \omega_1\} \subset \mathcal{L}$ satisfies $f_\beta \prec f_\alpha$ whenever $\alpha < \beta < \omega_1$. Let $f$ be the diagonal product of the mappings $f_\alpha$'s and $H = f(G)$. Put $H_\alpha = f_\alpha(G)$ for every $\alpha < \omega_1$. Since $H$ is a subgroup of the product $\prod_{\alpha<\omega_1} H_\alpha$, and $w(H_\alpha) \leq \aleph_1$ for each

$\alpha < \omega_1$, we have $w(H) \leq \aleph_1$. Therefore, by Lemma 8.6.5, it suffices to show that $H$ is a $P$-group.

If $\alpha < \omega_1$, then $f \prec f_\alpha$, so there exists a continuous homomorphism $g_\alpha \colon H \to H_\alpha$ such that $f_\alpha = g_\alpha \circ f$. Clearly, the family

$$\mathcal{B} = \{g_\alpha^{-1}(V) : \alpha < \omega_1,\ V \text{ is open in } H_\alpha\}$$

constitutes a base of $H$. In addition, if $\alpha < \beta < \omega_1$, then $g_\beta \prec g_\alpha$, so there exists a continuous homomorphism $p_{\beta,\alpha} \colon H_\beta \to H_\alpha$ such that $g_\alpha = p_{\beta,\alpha} \circ g_\beta$. Therefore, for every countable subfamily $\gamma = \{g_{\alpha_n}^{-1}(V_n) : n \in \omega\}$ of $\mathcal{B}$, we can find an ordinal $\beta < \omega_1$ satisfying $\alpha_n < \beta$ for each $n \in \omega$ and a countable family $\{W_n : n \in \omega\}$ of open sets in $H_\beta$ such that $\gamma = \{g_\beta^{-1}(W_n) : n \in \omega\}$. Since $H_\beta$ is a $P$-group by (c) of Lemma 4.4.1, the set $W = \bigcap_{n=0}^{\infty} W_n$ is open in $H_\beta$ and hence, $\bigcap \gamma = g_\beta^{-1}(W)$ is open in $H$. Therefore, $H$ is a $P$-group. □

Corollary 8.6.7. *Let $H$ be a pseudo-$\aleph_1$-compact $P$-group satisfying $\psi(H) \leq \aleph_1$. Then $w(H) \leq \aleph_1$.*

Proof. By the assumption, there exists a family $\{U_\alpha : \alpha < \omega_1\}$ of open neighbourhoods of the identity $e$ in $H$ such that $\{e\} = \bigcap_{\alpha<\omega_1} U_\alpha$. Using (b) of Lemma 4.4.1, it is easy to define by recursion a family $\{N_\alpha : \alpha < \omega_1\}$ of open invariant subgroups of $H$ such that $N_\beta \subset N_\alpha \cap U_\alpha$ whenever $\alpha < \beta < \omega_1$. For every $\alpha < \omega_1$, let $p_\alpha \colon H \to H/N_\alpha$ be the quotient homomorphism. Clearly, $p_\beta \prec p_\alpha$ if $\alpha < \beta$. Denote by $p$ the diagonal product of the family $\{p_\alpha : \alpha < \omega_1\}$. Then $K = p(H)$ is a subgroup of the product $\prod_{\alpha<\omega_1} H/N_\alpha$ of countable discrete groups and hence, $w(K) \leq \aleph_1$. From Lemma 8.6.6 it follows that the homomorphism $p \colon H \to K$ is open. Since $N_\beta$ is the kernel of the homomorphism $p_\beta$ for each $\beta < \omega_1$, we have

$$\ker p = \bigcap_{\beta<\omega_1} N_\beta \subset \bigcap_{\alpha<\omega_1} U_\alpha = \{e\}.$$

So, $\ker p = \{e\}$ and hence, $p$ is a topological isomorphism between $H$ and $K$. Therefore, $w(H) = w(K) \leq \aleph_1$. □

Every pseudo-$\aleph_1$-compact group $G$ is $\omega$-narrow, by Proposition 3.4.31, so Theorem 5.4.10 implies that $c(G) \leq 2^\omega$. We improve this conclusion in the next lemma.

Lemma 8.6.8. *Every pseudo-$\aleph_1$-compact $P$-group $G$ satisfies $c(G) \leq \aleph_1$.*

Proof. Denote by $\mathcal{N}$ the family of all open invariant subgroups of $G$. By b) of Lemma 4.4.1, $\mathcal{N}$ is a base at the identity of $G$. Since $G$ is $\omega$-narrow by Proposition 3.4.31, the quotient group $G/N$ is countable for each $N \in \mathcal{N}$. In addition, the family $\mathcal{N}$ is closed under countable intersections because $G$ is a $P$-group.

Let $\gamma$ be an arbitrary family of open sets in $G$. We have to show that $\gamma$ contains a subfamily $\lambda$ with $|\lambda| \leq \aleph_1$ such that $\bigcup \lambda$ is dense in $\bigcup \gamma$. Without loss of generality, we can assume that every $U \in \gamma$ has the form $U = xN$, for some $x \in G$ and some $N \in \mathcal{N}$. We shall define by recursion two sequences $\{N_\alpha : \alpha < \omega_1\}$ and $\{\lambda_\alpha : \alpha < \omega_1\}$ satisfying the following conditions for all $\alpha, \beta < \omega_1$:

(1) $N_\alpha \in \mathcal{N}$ and $N_\alpha \subset N_\beta$ if $\beta < \alpha$;
(2) $\lambda_\alpha \subset \gamma$;
(3) $|\lambda_\alpha| \leq \omega$;

(4) $\pi_\alpha(\bigcup \lambda_\alpha) = \pi_\alpha(\bigcup \gamma)$;
(5) $U = \pi_{\alpha+1}^{-1}\pi_{\alpha+1}(U)$ for each $U \in \lambda_\alpha$.

In (4) and (5), we use $\pi_\beta$ to denote the quotient homomorphism of $G$ onto $G/N_\beta$, where $\beta$ is $\alpha$ or $\alpha+1$, respectively.

Let $N_0 \in \mathcal{N}$ be arbitrary. Since the group $G/N_0$ is countable, there exists a countable subfamily $\lambda_0 \subset \gamma$ such that $\pi_0(\bigcup \lambda_0) = \pi_0(\bigcup \gamma)$, where $\pi_0 \colon G \to G/N_0$ is the quotient homomorphism. Suppose that, for some $\alpha < \omega_1$, we have defined the sequences $\{N_\beta : \beta < \alpha\}$ and $\{\lambda_\beta : \beta < \alpha\}$ satisfying (1)–(5). If $\alpha$ is limit, put $N_\alpha = \bigcap_{\beta<\alpha} N_\beta$. Since the quotient group $H_\alpha = G/N_\alpha$ is countable, there exists a countable family $\lambda_\alpha \subset \gamma$ such that $\pi_\alpha(\bigcup \lambda_\alpha) = \pi_\alpha(\bigcup \gamma)$. Suppose now that $\alpha = \beta + 1$. Since $\lambda_\beta \subset \gamma$ is countable, we can find $N_{\beta+1} \in \mathcal{N}$ such that $N_{\beta+1} \subset N_\beta$ and $U = \pi_{\beta+1}^{-1}\pi_{\beta+1}(U)$, for each $U \in \lambda_\beta$. Again, the quotient group $G/N_{\beta+1}$ is countable, so there is a countable family $\lambda_{\beta+1} \subset \gamma$ such that $\pi_{\beta+1}(\bigcup \lambda_{\beta+1}) = \pi_{\beta+1}(\bigcup \gamma)$. In either case, the sequences $\{N_\nu : \nu \le \alpha\}$ and $\{\lambda_\nu : \nu \le \alpha\}$ satisfy (1)–(5). This completes the recursive construction.

Put $\lambda = \bigcup_{\alpha<\omega_1} \lambda_\alpha$. It follows from (2) and (3) that $\lambda \subset \gamma$ and $|\lambda| \le \aleph_1$. Denote by $\pi$ the diagonal product of the quotient homomorphisms $\pi_\alpha \colon G \to G/N_\alpha$ with $\alpha < \omega_1$, and let $H = \pi(G)$. Then (5) implies that $U = \pi^{-1}\pi(U)$ for each $U \in \lambda$. Therefore, the open set $O = \bigcup \lambda$ satisfies $O = \pi^{-1}\pi(O)$. In addition, from (4) it follows that $\pi_\alpha(O) = \pi_\alpha(\bigcup \gamma)$ for each $\alpha < \omega_1$, so $\pi(O)$ is dense in $\pi(\bigcup \gamma)$. By Lemma 8.6.6, the homomorphism $\pi \colon G \to H$ is open. Therefore, we have

$$\bigcup \gamma \subset \pi^{-1}\pi(\bigcup \gamma) \subset \pi^{-1}\overline{\pi(O)} = \overline{\pi^{-1}\pi(O)} = \overline{O} = \overline{\bigcup \lambda}.$$

This proves that $\bigcup \lambda$ is dense in $\bigcup \gamma$, so $c(G) \le \aleph_1$. □

Since every Lindelöf space is pseudo-$\aleph_1$-compact, we have the following:

Corollary 8.6.9. *Every Lindelöf P-group G satisfies $c(G) \le \aleph_1$.*

Now we apply Lemma 8.6.8 to deduce that the "location complexity" of regular closed sets in a pseudo-$\aleph_1$-compact $P$-group is at most $\aleph_1$.

Lemma 8.6.10. *For every open subset $U$ of a pseudo-$\aleph_1$-compact P-group $G$, there exists a continuous open homomorphism $\pi \colon G \to H$ onto a topological group $H$ with $w(H) \le \aleph_1$ such that $\overline{U} = \pi^{-1}\pi(\overline{U})$.*

Proof. By (b) of Lemma 4.4.1, the family $\mathcal{N}$ of all open invariant subgroups of $G$ forms a base at the identity of $G$. Therefore, for every $x \in U$, there exists $N_x \in \mathcal{N}$ such that $xN_x \subset U$. Since $c(G) \le \aleph_1$ by Lemma 8.6.8, the family $\gamma = \{xN_x : x \in U\}$ contains a subfamily $\lambda = \{xN_x : x \in C\}$ such that $|C| \le \aleph_1$ and $O = \bigcup \lambda$ is dense in $U = \bigcup \gamma$.

If $|C| \le \aleph_0$, then $\bigcap_{x\in C} N_x = N \in \mathcal{N}$ and $xN_x = \pi^{-1}\pi(xN_x)$ for each $x \in C$, where $\pi \colon G \to G/N$ is the quotient homomorphism. In particular, $O = \pi^{-1}\pi(O)$. If $|C| = \aleph_1$, then we can enumerate the family $\gamma_C = \{N_x : x \in C\}$ in order type $\omega_1$, say, $\gamma_C = \{N_\alpha : \alpha < \omega_1\}$. For every $\alpha < \omega_1$, put $P_\alpha = \bigcap_{\beta\le\alpha} N_\beta$. Then $P_\alpha \in \mathcal{N}$ and, hence, the quotient group $G/P_\alpha$ is countable and discrete, for each $\alpha < \omega_1$. Let $\pi_\alpha \colon G \to G/P_\alpha$ be the quotient homomorphism. Denote by $\pi$ the diagonal product of the family $\{\pi_\alpha : \alpha < \omega_1\}$, and let $H = \pi(G)$. Then $H$ is a subgroup of the product $\prod_{\alpha<\omega_1} G/P_\alpha$ of countable discrete groups $G/P_\alpha$ and hence, $w(H) \le \aleph_1$. From the definition of $\pi$ it follows that

$xN_x = \pi^{-1}\pi(xN_x)$, for each $x \in C$, and again we have $O = \pi^{-1}\pi(O)$. By Lemma 8.6.6, the homomorphism $\pi$ is open.

In either case, we use the density of $O$ in $U$ to conclude that

$$\pi^{-1}\pi(\overline{U}) = \pi^{-1}\pi(\overline{O}) \subset \pi^{-1}\overline{\pi(O)} = \overline{\pi^{-1}\pi(O)} = \overline{O} = \overline{U}.$$

This shows that $\overline{U} = \pi^{-1}\pi(\overline{U})$. □

Dense subgroups of Lindelöf $P$-groups need not be $\mathbb{R}$-factorizable (see Example 8.2.1). The following theorem shows that such subgroups have a weaker property that could be called $\aleph_1$-*factorizability*.

THEOREM 8.6.11. *Let $S$ be a dense subspace of a pseudo-$\aleph_1$-compact $P$-group $G$. Then for every continuous function $f: S \to \mathbb{R}$, there exists a continuous open homomorphism $\pi: G \to H$ onto a group $H$ with $w(H) \leq \aleph_1$ such that $\pi \restriction S \prec f$.*

PROOF. Consider the family $\gamma = \{f^{-1}(r) : r \in \mathbb{R}\}$. Since $S$ is a $P$-space, $\gamma$ is a disjoint open covering of $S$. From Lemma 8.6.8 it follows that $c(G) \leq \aleph_1$ and hence, since $S$ is dense in $G$, we have $c(S) = c(G) \leq \aleph_1$. So, $|\gamma| \leq \aleph_1$. For every non-empty $U \in \gamma$, choose an open set $V_U$ in $G$ such that $V_U \cap S = U$. Applying Lemma 8.6.10, we find a continuous open homomorphism $\pi_U: G \to H_U$ onto a group $H_U$ with $w(H_U) \leq \aleph_1$ such that $\overline{V}_U = \pi_U^{-1}\pi_U(\overline{V}_U)$. Denote by $\pi_0$ the diagonal product of the family $\{\pi_U : U \in \gamma\}$. Then $H_0 = \pi_0(G)$ is a subgroup of the product $\Pi = \prod_{U\in\gamma} H_U$, so $w(H_0) \leq \aleph_1$. For every $U \in \gamma$, let $p_U: \Pi \to H_U$ be the projection. Then $\pi_U = p_U \circ \pi_0$. The definition of $\pi_0$ implies that $\overline{V}_U = \pi_0^{-1}\pi_0(\overline{V}_U)$, for each $U \in \gamma$.

Let $N$ be the kernel of $\pi_0$ and $\pi: G \to H$ be the quotient homomorphism, where $H = G/N$. Then there exists an algebraic isomorphism $i: H \to H_0$ such that $\pi_0 = i \circ \pi$. Since $\pi$ is open, the isomorphism $i$ is continuous. Hence, $\psi(H) \leq \chi(H_0) \leq w(H_0) \leq \aleph_1$, and Corollary 8.6.7 implies that $w(H) \leq \aleph_1$. Since $\pi \prec \pi_0$, we have $\overline{V}_U = \pi^{-1}\pi(\overline{V}_U)$, for each $U \in \gamma$. It remains to verify that $\pi \restriction S \prec f$.

First, suppose that $x, y \in S$ satisfy $\pi(x) = \pi(y)$. Then $\pi_U(x) = \pi_U(y)$ for each $U \in \gamma$. Suppose that $x \in U$ for some $U \in \gamma$. Since $U = S \cap V_U = S \cap \overline{V}_U$, the equality $\overline{V}_U = \pi_U^{-1}\pi_U(\overline{V}_U)$ implies that $y \in S \cap \overline{V}_U = U$, i.e., both $x$ and $y$ lie in $U$. This implies immediately that $f(x) = f(y)$. Therefore, there exists a function $h: \pi(S) \to \mathbb{R}$ such that $f = h \circ \pi \restriction S$. We claim that $f$ is constant on elements of $\gamma$, so that $h$ is constant on the open subset $\pi(V_U) \cap \pi(S)$ of $\pi(S)$, for each $U \in \gamma$. Indeed, if $s, t \in \pi(V_U) \cap \pi(S)$, choose $x, y \in S$ such that $\pi(x) = s$ and $\pi(y) = t$. From $\overline{V}_U = \pi^{-1}\pi(\overline{V}_U)$ it follows that $x, y \in S \cap \overline{V}_U = S \cap V_U = U$; hence,

$$h(s) = h(\pi(x)) = f(x) = f(y) = h(\pi(y)) = h(t).$$

Since the sets $\pi(V_U) \cap \pi(S)$ with $U \in \gamma$ cover $\pi(S)$, the function $h$ is continuous. Thus, $\pi \restriction S \prec f$. □

The following theorem characterizes the $\mathbb{R}$-factorizability of $P$-groups in purely topological terms. It is not surprising, after the above series of results, that the characteristic property is precisely pseudo-$\aleph_1$-compactness.

THEOREM 8.6.12. *For an arbitrary $P$-group $G$, the following conditions are equivalent:*

1) *$G$ is $\mathbb{R}$-factorizable;*

2) *$G$ is $\omega$-narrow, and every continuous homomorphic image $H$ of $G$ with $w(H) \leq \aleph_1$ is Lindelöf;*
3) *$G$ is pseudo-$\aleph_1$-compact;*
4) *$G$ is $\omega$-stable.*

PROOF. The implication 1) $\Rightarrow$ 3) is exactly Lemma 8.6.1. The equivalence of 3) and 4) follows from Propositions 5.6.8 and 5.6.9. Therefore, it suffices to show that 3) $\Rightarrow$ 1) and 2) $\Leftrightarrow$ 3).

3) $\Rightarrow$ 1). Let $G$ be a pseudo-$\aleph_1$-compact $P$-group, and suppose that $f\colon G \to \mathbb{R}$ is a continuous function. Denote by $\mathcal{L}$ the family of continuous open homomorphisms $\pi\colon G \to H$ onto topological groups $H$ with $w(H) \leq \aleph_1$ and apply Theorem 8.6.11 to find $\pi \in \mathcal{L}$, $\pi\colon G \to H$, and a continuous function $g\colon H \to \mathbb{R}$ such that $f = g \circ \pi$. By c) of Lemma 4.4.1, $H$ is a pseudo-$\aleph_1$-compact $P$-group. Therefore, by Lemma 8.6.3, $H$ is Lindelöf. Since Lindelöf groups are $\mathbb{R}$-factorizable, by Theorem 8.1.6, and $\pi \prec f$, Lemma 8.1.11 implies that the group $G$ is $\mathbb{R}$-factorizable.

3) $\Rightarrow$ 2). Suppose that the group $G$ is pseudo-$\aleph_1$-compact. Proposition 3.4.31 implies that $G$ is $\omega$-narrow. Let $f\colon G \to H$ be a continuous homomorphism of $G$ onto a group $H$ with $w(H) \leq \aleph_1$. Denote by $N$ the kernel of $f$. Then the quotient homomorphism $\pi\colon G \to G/N$ is open, so $G/N$ is a $P$-group, by c) of Lemma 4.4.1. Clearly, there is a continuous isomorphism $p\colon G/N \to H$ such that $f = p \circ \pi$. Since $w(H) \leq \aleph_1$, the pseudocharacter of the quotient group $G/N$ does not exceed $\aleph_1$. Note that $G/N$ is pseudo-$\aleph_1$-compact, as a continuous image of $G$, so Corollary 8.6.7 implies that $w(G/N) \leq \aleph_1$. Therefore, $G/N$ is Lindelöf by Lemma 8.6.3, and so is $H = p(G/N)$.

2) $\Rightarrow$ 3). Suppose that every continuous homomorphic image of $G$ of weight $\leq \aleph_1$ is Lindelöf. If $G$ contains a discrete family $\{U_\alpha : \alpha < \omega_1\}$ of non-empty open sets, then we apply b) of Lemma 4.4.1 to choose, for every $\alpha < \omega_1$, a point $x_\alpha \in U_\alpha$ and an open invariant subgroup $P_\alpha$ of $G$ such that $V_\alpha = x_\alpha P_\alpha \subset U_\alpha$. Clearly, the family $\gamma = \{V_\alpha : \alpha < \omega_1\}$ is discrete in $G$. Put $P = \bigcap_{\alpha<\omega_1} P_\alpha$ and consider the quotient homomorphism $\pi\colon G \to G/P$. Then $\psi(G/P) \leq \aleph_1$, so Corollary 8.6.7 implies that $w(G/P) \leq \aleph_1$ and hence, the group $G/P$ is Lindelöf, by the assumption. Since the homomorphism $\pi$ is open and $V_\alpha = \pi^{-1}\pi(V_\alpha)$, for each $\alpha < \omega_1$, the family $\{\pi(V_\alpha) : \alpha < \omega_1\}$ is discrete in $G/P$. This contradicts the Lindelöf property of $G/P$. Therefore, $G$ is pseudo-$\aleph_1$-compact. □

COROLLARY 8.6.13. *Let $H$ be a pseudo-$\aleph_1$-compact $P$-group. Then the Raĭkov completion $\varrho H$ of $H$ is $\mathbb{R}$-factorizable and $\upsilon H = \mu H = \varrho H$. In particular, Dieudonné complete pseudo-$\aleph_1$-compact $P$-groups are Raĭkov complete.*

PROOF. By Theorem 8.6.12, the group $H$ is $\mathbb{R}$-factorizable. In addition, d) of Lemma 4.4.1 implies that $\varrho H$ is also a $P$-group, so $H$ is $G_\delta$-dense in $\varrho H$. By Proposition 8.3.4, $H$ is $C$-embedded in $\varrho H$, and the group $\varrho H$ is $\mathbb{R}$-factorizable. Therefore, $\varrho H \subset \upsilon G$. Since $H$ is $\mathbb{R}$-factorizable, from Theorem 8.3.6 it follows that $\upsilon H = \mu H$ is a topological group that contains $H$ as a dense subgroup. Consequently, $\mu H \subset \varrho H$. Hence, $\upsilon H = \mu H = \varrho H$. If $H$ is Dieudonné complete, then $H = \mu H$ and the above equalities imply that $H = \mu H = \varrho H$. So, $H$ is complete. □

COROLLARY 8.6.14. *The following conditions are equivalent for any $P$-group $H$:*

a) *$H$ is $\mathbb{R}$-factorizable;*

b) *$\varrho H$ is $\mathbb{R}$-factorizable, and $H$ intersects every non-empty $G_{\omega_1}$-set in $\varrho H$.*

PROOF. Suppose that $H$ is $\mathbb{R}$-factorizable. Then $\varrho H$ is also $\mathbb{R}$-factorizable, by Corollary 8.6.13. In addition, $H$ is pseudo-$\aleph_1$-compact, by Theorem 8.6.12, while d) of Lemma 4.4.1 implies that $\varrho H$ is a $P$-group. In particular, $\varrho H$ is zero-dimensional. Consider a non-empty $G_{\aleph_1}$-set $Q$ in $\varrho H$. If $Q \cap H = \varnothing$, pick a point $x \in Q$ and define a strictly decreasing sequence $\{U_\alpha : \alpha < \omega_1\}$ of clopen neighbourhoods of $x$ in $\varrho H$ such that $\bigcap_{\alpha<\omega_1} U_\alpha \subset Q$. For every $\alpha < \omega_1$, put $V_\alpha = (U_\alpha \setminus U_{\alpha+1}) \cap H$. Then $\{V_\alpha : \alpha < \omega_1\}$ is a discrete family of non-empty open sets in $H$, which contradicts the pseudo-$\aleph_1$-compactness of $H$. Therefore, $H$ intersects every non-empty $G_{\aleph_1}$-set in $\varrho H$.

Conversely, suppose that $H$ and its completion $\varrho H$ satisfy b). Let $f\colon H \to \mathbb{R}$ be a continuous function. By Theorem 8.6.11, there exists a continuous open homomorphism $\pi\colon \varrho H \to K$ onto a group $K$ satisfying $w(K) \leq \aleph_1$, and a continuous function $g\colon K \to \mathbb{R}$ such that $f = g \circ \pi \restriction H$. Since $H$ intersects every non-empty $G_{\aleph_1}$-set in $\varrho H$, we conclude that $\pi(H) = \pi(\varrho H) = K$. So, $\widetilde{f} = g \circ \pi$ is a continuous extension of $f$ over $\varrho H$. Hence, $H$ is $C$-embedded in $\varrho H$ and $H$ is $\mathbb{R}$-factorizable, by Proposition 8.2.3. □

By Lemma 8.6.1, an $\mathbb{R}$-factorizable $P$-group $G$ is pseudo-$\aleph_1$-compact. In addition, if $G$ satisfies $w(G) \leq \aleph_1$, then Corollary 8.6.4 implies that $G$ is Lindelöf. Now we present an example showing that the restriction on the weight of $G$ is essential. In fact, we shall construct an $\mathbb{R}$-factorizable $P$-group that is not weakly Lindelöf. Let us start with a simple fact.

LEMMA 8.6.15. *A regular weakly Lindelöf $P$-space is Lindelöf.*

PROOF. Let $\gamma$ be an open covering of a regular weakly Lindelöf $P$-space $X$. Clearly, $X$ has a base of clopen sets, so we can assume that all elements of $\gamma$ are clopen. By the assumption, $\gamma$ contains a countable subfamily $\lambda$ such that the union $\bigcup \lambda$ is dense in $X$. Since $X$ is a $P$-space, this union is closed in $X$ and hence, $X = \bigcup \lambda$. □

EXAMPLE 8.6.16. For every cardinal $\tau > \aleph_1$, there exists an $\mathbb{R}$-factorizable $P$-group $H$ of cardinality $\tau$ which is not weakly Lindelöf.

Let $\mathbb{Z}(2)$ be the two-element group $\{0, 1\}$ with the discrete topology, and $\tau > \aleph_1$ be a cardinal. Consider the Lindelöf $P$-group $G = G_\tau$ defined in Example 4.4.11 as the $\sigma$-product in the product $\mathbb{Z}(2)^\tau$. The group $G$ carries the $\omega$-box topology inherited from $\mathbb{Z}(2)^\tau$. In other words, a base at the neutral element of $G$ consists of subgroups

$$\{G \cap \pi_A^{-1}(0_A) : A \subset \tau,\ |A| \leq \omega\},$$

where $\pi_A\colon \mathbb{Z}(2)^\tau \to \mathbb{Z}(2)^A$ is the projection, and $0_A$ is the neutral element of $\mathbb{Z}(2)^A$. In what follows we use the fact that the restriction of $\pi_B$ to $G$ is an open homomorphism of $G$ onto $\pi_B(G)$, for each $B \subset \tau$. The group $G$ is $\mathbb{R}$-factorizable, by Theorem 8.1.6. Put

$$H = \{x \in G : |\operatorname{supp}(x)| \text{ is even}\}.$$

Clearly, $H$ is a $P$-group being a subgroup of $G$. We claim that $H$ is an $\mathbb{R}$-factorizable group which fails to be weakly Lindelöf. We split the proof of this fact in two steps.

Clearly, $H$ is dense in $G$, because basic open sets in $G$ depend on countably many coordinates. In fact, $H$ has the following stronger property:

$$\pi_B(H) = \pi_B(G) \text{ for each } B \subset \tau \text{ with } |B| \leq \aleph_1. \tag{8.7}$$

Indeed, given $B \subset \tau$ with $|B| \le \aleph_1$ and a point $y \in \pi_B(G)$, one can always find $x \in \mathbb{Z}(2)^\tau$ with a finite support of even cardinality such that $x$ and $y$ coincide at each coordinate in $B$. Then $x \in H$ and $\pi_B(x) = y$. This proves (8.7).

It follows immediately from (8.7) that $H$ intersects every non-empty $G_{\omega_1}$-set in $G$. The group $G$ is Raĭkov complete, by Proposition 4.4.5. Since $H$ is dense in $G$, it follows that $\varrho H = G$. Since the group $G$ is $\mathbb{R}$-factorizable, Corollary 8.6.14 implies that its subgroup $H$ is also $\mathbb{R}$-factorizable.

The group $H$ is not Lindelöf as a proper dense subgroup of the $P$-group $G$. Therefore, $H$ is not weakly Lindelöf, by Lemma 8.6.15. □

We conclude this section with two more results showing that the class of $\mathbb{R}$-factorizable $P$-groups is stable under taking topological products and continuous homomorphic images if we restrict ourselves to considering $P$-groups.

The next lemma is an essential step in the proof of Theorem 8.6.18; it generalizes Corollary 8.6.7.

LEMMA 8.6.17. *Let $G = \prod_{n\in\omega} G_n$ be the product of a countable family of $\mathbb{R}$-factorizable $P$-groups. If $\varphi\colon G \to K$ is a continuous homomorphism onto a group $K$ with $\psi(K) \le \aleph_1$, then $K$ is Lindelöf and $nw(K) \le \aleph_1$.*

PROOF. By Lemma 5.6.2 we can fix, for every $n \in \omega$, a quotient homomorphism $p_n$ of $G_n$ onto a group $H_n$ with $\psi(H_n) \le \aleph_1$ and a continuous homomorphism $h\colon \prod_{n\in\omega} H_n \to K$ such that $\varphi = h \circ p$, where $p = \prod_{n\in\omega} p_n\colon G \to \prod_{n\in\omega} H_n$ is the product homomorphism. Then, by (c) of Lemma 4.4.1, $H_n$ is a $P$-group, for each $n \in \omega$. From Lemma 8.6.1 it follows that the groups $G_n$'s are pseudo-$\aleph_1$-compact, and so are the groups $H_n = p_n(G_n)$. Also $w(H_n) \le \aleph_1$ by Corollary 8.6.7. Therefore, Corollary 8.6.4 implies that each group $H_n$ is Lindelöf. Since a countable product of Lindelöf $P$-spaces is Lindelöf by Theorem 4.4.10, we conclude that the group $H = \prod_{n\in\omega} H_n$ is Lindelöf and $w(H) \le \aleph_1$. So, the group $K = h(H)$ is also Lindelöf and $nw(K) \le \aleph_1$. □

THEOREM 8.6.18. *An arbitrary topological product $G = \prod_{\alpha\in A} G_\alpha$ of $\mathbb{R}$-factorizable $P$-groups is $\mathbb{R}$-factorizable. Moreover, for every continuous real-valued function $f$ on $G$, there exists a quotient homomorphism $\pi\colon G \to K$ onto a second-countable group $K$ such that $\pi \prec f$.*

PROOF. For a non-empty $B \subset A$, let $\pi_B$ be the projection of $G$ onto $G_B = \prod_{\alpha\in B} G_\alpha$. Since $\mathbb{R}$-factorizable groups are $\omega_0$-narrow and the class of $\aleph_0$-narrow groups is productive, $G_B$ is an $\omega_0$-narrow $P$-group for each finite $B \subset A$. Hence, Lemma 8.6.17 and the equivalence 3) $\Leftrightarrow$ 1) $\Leftrightarrow$ 2) of Theorem 8.6.12 together imply that the group $G_B$ is $\mathbb{R}$-factorizable and pseudo-$\aleph_1$-compact, for each finite $B \subset A$. So, the product group $G$ is pseudo-$\aleph_1$-compact, by Proposition 1.6.22. Therefore, every continuous function $f\colon G \to \mathbb{R}$ depends on, at most, countably many coordinates, by Theorem 1.7.2. In other words, there exist a countable set $C \subset A$ and a function $g\colon G_C \to \mathbb{R}$ such that $f = g \circ \pi_C$. The continuity of $g$ follows from the fact that the projection $\pi_C$ is open. Thus, we may assume without loss of generality that the index set $A$ is countable.

Our next step is to find an upper bound for the cellularity of $G$. If $B \subset A$ is finite, then the $P$-group $G_B$ is pseudo-$\aleph_1$-compact and Lemma 8.6.8 implies that $c(G_B) \le \aleph_1$. Therefore, $c(G) \le \aleph_1$, by Theorem 1.6.21.

Let $f\colon G \to \mathbb{R}$ be a continuous function. Note that $wl(G) \leq c(G) \leq \aleph_1$; applying Theorem 8.1.18, we find a continuous homomorphism $\varphi\colon G \to H$ onto a group $H$ with $\psi(H) \leq \aleph_1$ and a continuous function $g\colon H \to \mathbb{R}$ such that $f = g \circ \varphi$. By Lemma 5.6.2, for every $\alpha \in A$ there is a quotient homomorphism $p_\alpha\colon G_\alpha \to H_\alpha$ onto a group $H_\alpha$ with $\psi(H_\alpha) \leq \aleph_1$ such that $p \prec \varphi$, where $p\colon G \to \prod_{\alpha\in A} H_\alpha$ is the product of the homomorphisms $p_\alpha$'s. Since $\psi(\prod_{\alpha\in A} H_\alpha) \leq \aleph_1$, Lemma 8.6.17 implies that the product group $\prod_{\alpha\in A} H_\alpha$ is Lindelöf. To simplify notation, we can assume that $\varphi = p$ and $H = \prod_{\alpha\in A} H_\alpha$. The homomorphism $p$ is open as the product of open homomorphisms $p_\alpha$'s.

The Lindelöf group $H$ is $\mathbb{R}$-factorizable, by Theorem 8.1.6. So, there exist a continuous homomorphism $\psi\colon H \to K$ onto a second-countable topological group $K$ and a continuous function $h\colon K \to \mathbb{R}$ such that $g = h \circ \psi$.

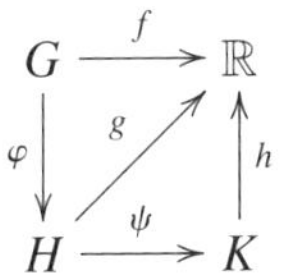


Applying again Lemma 5.6.2, we find a family $\{q_\alpha : \alpha \in A\}$ of quotient homomorphisms $q_\alpha\colon H_\alpha \to K_\alpha$ onto groups $K_\alpha$ of countable pseudocharacter such that the product $q$ of the homomorphisms $q_\alpha$'s satisfies $q \prec \psi$. Clearly, the homomorphism $q$ is open. Since each $H_\alpha$ is an $\omega_0$-narrow $P$-group, from Lemma 4.4.2 it follows that $K_\alpha$ is a countable discrete group. The group $\prod_{\alpha\in A} K_\alpha$ is second-countable, so we can assume that $q = \psi$ and $K = \prod_{\alpha\in A} K_\alpha$. Therefore, the continuous homomorphism $\pi = \psi \circ \varphi$ of $G$ to the second-countable group $K$ is open and $f = h \circ \pi$. □

Now we extend Theorem 8.6.18 to continuous homomorphic images of topological products of $\mathbb{R}$-factorizable $P$-groups.

THEOREM 8.6.19. *Continuous homomorphic images of arbitrary topological products of $\mathbb{R}$-factorizable $P$-groups are $\mathbb{R}$-factorizable.*

PROOF. Let $G = \prod_{i\in I} G_i$ be a product of $\mathbb{R}$-factorizable $P$-groups and $\varphi\colon G \to H$ be a continuous homomorphism of $G$ onto a topological group $H$. Consider a continuous function $f\colon H \to \mathbb{R}$. By Theorem 8.6.18, we can find a quotient homomorphism $\pi\colon G \to K$ onto a second-countable group $K$ and a continuous function $g\colon K \to \mathbb{R}$ such that $f \circ \varphi = g \circ \pi$. Put $N = \ker \pi$ and $M = \overline{\varphi(N)}$. Clearly, $N$ and $M$ are closed normal subgroups of $G$ and $H$, respectively. The equality $f \circ \varphi = g \circ \pi$ implies that $f$ is constant on every coset of $\varphi(N)$ in $H$ and, by the continuity of $f$, the same holds for cosets of $M$ in $H$. Let $p\colon H \to H/M$ be the quotient homomorphism. Then there exists a function $h\colon H/M \to \mathbb{R}$ satisfying $f = h \circ p$. Since $p$ is open, $h$ is continuous. Let $\psi = p \circ \varphi$.

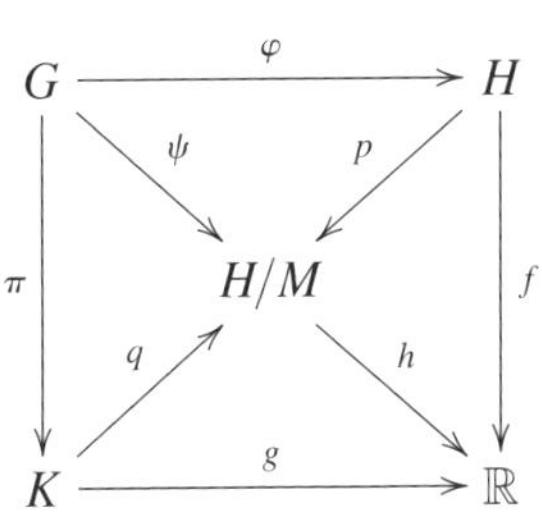

Note that $\ker \pi = N \subset \ker \psi$, so there exists a homomorphism $q\colon K \to H/M$ such that $\psi = q \circ \pi$. Again, $\pi$ is open, so $q$ is continuous. Therefore, the group $H/M = q(K)$ is Lindelöf and hence, $\mathbb{R}$-factorizable. We conclude that the homomorphism $p\colon H \to H/M$ to the $\mathbb{R}$-factorizable group $H/M$ satisfies $p \prec f$, so, by Lemma 8.1.11, $H$ is also $\mathbb{R}$-factorizable. □

COROLLARY 8.6.20. *Continuous homomorphic images of $\mathbb{R}$-factorizable $P$-groups are $\mathbb{R}$-factorizable.*

## Exercises

8.6.a. Verify that every $\mathbb{R}$-factorizable $P$-group is $m$-factorizable.

8.6.b. Suppose that $X$ is a non-empty Tychonoff space. Show that the Abelian topological group $(C_p(X))_\omega$ is not $\mathbb{R}$-factorizable, where $(C_p(X))_\omega$ is the $G_\delta$-modification of the group $C_p(X)$ (see Problem 3.6.H).

8.6.c. Let $X$ be a compact space. Prove that the groups $(F(X))_\omega$ and $(A(X))_\omega$ are $\mathbb{R}$-factorizable if and only if $X$ is scattered, that is, every non-empty subspace of $X$ has an isolated point.

8.6.d. Let $G$ and $H$ be topological groups. Verify that if $G$ is a $P$-group, then (a) $\Rightarrow$ (b), (a) $\Rightarrow$ (c) and (b) & (c) $\Rightarrow$ (a), where (a), (b), and (c) are the following statements:

(a) $G \times H$ is $\mathbb{R}$-factorizable;
(b) $\varrho G \times H$ is $\mathbb{R}$-factorizable;
(c) $G \times H$ is $C$-embedded in $\varrho G \times H$.

8.6.e. Prove that the product $G \times H$ of an $\mathbb{R}$-factorizable $P$-group $G$ and a precompact topological group $H$ is $\mathbb{R}$-factorizable (see also Problem 8.5.g).

8.6.f. Show that if $G$ is an $\mathbb{R}$-factorizable $P$-group and $X$ is an arbitrary space, then the product group $G \times C_p(X)$ is $\mathbb{R}$-factorizable (cf. Exercise 8.5.h).

8.6.g. Prove that if all subgroups of a $P$-group $G$ are $\mathbb{R}$-factorizable, then $G$ is countable.

8.6.h. Let $G$ be a Lindelöf paratopological group. Prove that if $G$ is a Hausdorff $P$-space, then $G$ is a topological group.
*Hint.* Use Lemma 5.7.10, Proposition 4.4.9, and then apply Lemma 4.4.3.

## Problems

8.6.A. Show that closed subgroups of $\mathbb{R}$-factorizable $P$-groups need not be $\mathbb{R}$-factorizable, not even in the Abelian case.

8.6.B. (M. G. Tkachenko, [**488**]) If a $P$-group $G$ is a continuous homomorphic image of an $\mathbb{R}$-factorizable group, then $G$ is also $\mathbb{R}$-factorizable.
*Hint.* First, refine Theorem 8.6.12 and show that an $\omega$-narrow $P$-group $G$ is $\mathbb{R}$-factorizable iff every continuous homomorphic image $H$ of weight $\leq \aleph_1$ is Lindelöf provided $H$ is a $P$-group. Then, for a non-Lindelöf $\omega$-narrow $P$-group $H$ with $w(H) = \aleph_1$, construct by recursion two clopen complementary subsets $W_0$ and $W_1$ of $H$ such that the family $\{xW_i : x \in H,\ i = 0, 1\}$ is a subbase for the topology of $H$. Let $f$ be a real-valued function on $H$ defined by $f(x) = i$ if $x \in W_i$, where $i = 0, 1$. Note that $f$ is continuous and show that the orbit $Hf$ of $f$ in $C_p(H)$ has uncountable network weight. Apply Problem 8.4.A to deduce that $H$ cannot be a continuous homomorphic image of an $\mathbb{R}$-factorizable group. This, along with the refined version of Theorem 8.6.12, implies the required conclusion.

8.6.C. Show that an $\omega$-narrow topological Abelian group of countable pseudocharacter need not be $\mathbb{R}$-factorizable.
*Hint.* Strengthen the topology of the group $H$ constructed in Example 8.2.1 and apply Problem 8.6.B.

8.6.D. Let $G$ and $K$ be $\mathbb{R}$-factorizable groups. Prove that if $G$ is a $P$-group and $K$ is weakly Lindelöf, then the product group $G \times K$ is $\mathbb{R}$-factorizable.
*Hint.* Let $f: G \times K \to \mathbb{R}$ be a continuous function. Apply Lemma 8.5.12 to find, for every $x \in G$, an open neighbourhood $U_x$ of $x$ in $G$ such that $f(x, y) = f(x', y)$ for all $x' \in U(x)$ and $y \in K$. For $a, b \in G$, put $a \sim b$ if there exists a finite sequence $x_0 = a, x_1, \ldots, x_n, x_{n+1} = b$ in $G$ such that $U(x_i) \cap U(x_{i+1}) \neq \varnothing$ for each $i = 0, 1, \ldots, n$. Verify that $\sim$ is an equivalence relation on $G$ and denote by $O(x)$ the equivalence class containing $x \in G$. Verify that $f(a, y) = f(b, y)$ whenever $a \sim b$ and $y \in K$. Use Lemma 8.6.1 to find a countable set $\{x_n : n \in \omega\} \subset G$ such that $G = \bigcup_{n\in\omega} O(x_n)$. For every $n \in \omega$, define the function $f_n$ on $K$ by $f_n(y) = f(x_n, y)$ for each $y \in K$. Then continue as in Exercise 8.5.g and find a continuous homomorphism $p: K \to L$ onto a second-countable group $L$ such that $p \prec f_n$ for each $n \in \omega$. Let $r: G \to \mathbb{R}$ be a mapping defined by $r(x) = n$ iff $x \in O(x_n)$. Then $r$ is continuous, so there exists a continuous homomorphism $\pi: G \to H$ onto a countable discrete group $H$ such that $\pi \prec r$. Show that the homomorphism $\varphi = \pi \times p$ of $G$ to $H \times L$ factorizes $f$, i.e., $\varphi \prec f$. Finally, note that the product group $H \times L$ is second-countable.

8.6.E. Let $G$ be a regular $\omega$-narrow paratopological group, and suppose that $G$ is a $P$-space. Must $G$ be a topological group? (Compare with Exercise 8.6.h.)

8.6.F. (A. Dow, personal communication.) As Theorem 8.6.18 shows, topological products of pseudo-$\aleph_1$-compact $P$-groups are pseudo-$\aleph_1$-compact. Prove that under the assumption $2^{\aleph_0} = 2^{\aleph_1}$, the product of two regular pseudo-$\aleph_1$-compact $P$-spaces can fail to be pseudo-$\aleph_1$-compact.

## Open Problems

8.6.1. Do there exist in $ZFC$ two regular pseudo-$\aleph_1$-compact $P$-spaces whose product is not pseudo-$\aleph_1$-compact? (See Problem 8.6.F.)

8.6.2. Is every regular $P$-space homeomorphic to a closed subspace of an $\mathbb{R}$-factorizable $P$-group?

8.6.3. Suppose that $G$ is a pseudo-$\aleph_1$-compact topological group such that every countable subset of $G$ is closed. Is $G$ $\mathbb{R}$-factorizable?

8.6.4. Suppose that $G$ is an $\mathbb{R}$-factorizable topological group such that every countable subset of $G$ is closed. Is $G$ then $m$-factorizable?

8.6.5. Suppose that $G$ is an $\mathbb{R}$-factorizable topological group such that every countable subset of $G$ is closed. Suppose further that a topological group $H$ is an image of $G$ under a continuous homomorphism. Is $H$ $\mathbb{R}$-factorizable?

8.6.6. Suppose that $G$ is the topological product of a family $\{G_i : i \in I\}$ of $\mathbb{R}$-factorizable topological groups such that every countable subset of $G_i$ is closed, for each $i \in I$. Is $G$ $\mathbb{R}$-factorizable?

8.6.7. Suppose that $f: G \to H$ is a continuous mapping (not necessarily a homomorphism) of an $\mathbb{R}$-factorizable group $G$ onto a $P$-group $H$. Is $H$ then $\mathbb{R}$-factorizable?

8.6.8. Suppose that $G$ is the product of a family of pseudo-$\aleph_1$-compact paratopological groups that are $P$-spaces. Is $G$ pseudo-$\aleph_1$-compact?

8.6.9. Let $G$ be a regular pseudo-$\aleph_1$-compact paratopological group which is a $P$-space. Is $G$ topologically isomorphic to a subgroup of the product of a family of second-countable paratopological groups? Must $G$ be $\mathbb{R}_3$-factorizable?

## 8.7. Factorizable groups and projectively Moscow groups

In this section we show that the idea of $\mathbb{R}$-factorizability can be fruitfully applied to the study of $PT$-groups (see Section 6.5 for the related definition).

Let $\mathcal{P}$ be a class of topological groups. A topological group $G$ is said to be *factorizable over* $\mathcal{P}$ or simply $\mathcal{P}$*-factorizable* if, for every continuous real-valued function $f$ on $G$, there exists a continuous homomorphism $g$ of $G$ to a topological group $H \in \mathcal{P}$ and a continuous real-valued function $h$ on $H$ such that $f = h \circ g$. Clearly, a topological group $G$ is $\mathbb{R}$-factorizable iff it is factorizable over the class of second-countable groups.

Neither the class of Moscow groups contains the class of $\mathbb{R}$-factorizable groups, nor the class of $\mathbb{R}$-factorizable groups contains the class of Moscow groups. Indeed, every Lindelöf group is $\mathbb{R}$-factorizable by Theorem 8.1.6, while not every Lindelöf group is a Moscow space, as it is shown in Example 6.5.30. In addition, any discrete group is, obviously, a Moscow space, while a discrete $\mathbb{R}$-factorizable group is countable, by Corollary 8.1.10.

THEOREM 8.7.1. *If a topological group $G$ is factorizable over the class $\mathcal{PT}$ of all $PT$-groups, then $G$ is a $PT$-group.*

To prove this theorem, we need two lemmas.

LEMMA 8.7.2. *If a topological group $G$ is factorizable over the class $\mathcal{PT}$ of $PT$-groups, then for each continuous real-valued function $f$ on $G$, there exists a Dieudonné complete topological group $G_f$ containing $G$ as a topological subgroup such that $f$ can be continuously extended to $G_f$.*

PROOF. By the assumption, there exist a continuous homomorphism $g$ of $G$ to a $PT$-group $H$, and a continuous real-valued function $h$ on $H$ such that $f = h \circ g$. Let $g_\varrho \colon \varrho G \to \varrho H$ be the continuous extension of $g$ to the Raĭkov completions of $G$ and $H$. Since $H$ is a $PT$-group, there exists a Dieudonné complete topological group $Z$ such that $H \subset Z \subset \varrho H$ and $h$ admits an extension to a continuous real-valued function $h^*$ on $Z$.

Put $Y = g_\varrho^{-1}(Z)$ and $f^* = h^* \circ g_\varrho$. Then $Y$ is a subgroup of $\varrho G$, $G$ is a subgroup of $Y$, and $f^*$ is a continuous real-valued function on $Y$, the restriction of which to $G$ coincides with $f$. We claim that the space $Y$ is Dieudonné complete.

Let $F = \{(x, g_\varrho(x)) : x \in \varrho G\}$ be the graph of $g_\varrho$. Then $F$ is closed in $\varrho G \times \varrho H$, and the canonical mapping $\phi_\varrho$ of $\varrho G$ to $\varrho G \times \varrho H$, given by the formula $\phi_\varrho(x) = (x, g_\varrho(x))$, for each $x \in \varrho G$, is a homeomorphism of $\varrho G$ onto the space $F$. Clearly, $\phi_\varrho(Y) = F \cap (\varrho G \times Z)$. Hence, $Y$ is homeomorphic to the closed subspace $F \cap (\varrho G \times Z)$ of the product space $\varrho G \times Z$. Since $\varrho G$ and $Z$ are Dieudonné complete, it follows that $\varrho G \times Z$ and $Y$ are also Dieudonné complete. □

LEMMA 8.7.3. *A topological group $G$ is a $PT$-group if and only if, for each continuous real-valued function $f$ on $G$, there exists a Dieudonné complete topological group $G_f$ containing $G$ as a topological subgroup such that $f$ can be continuously extended to $G_f$.*

PROOF. The necessity is obvious, since each space is $C$-embedded in its Dieudonné completion. Let us prove the sufficiency.

By Lemma 8.7.2, for each $f \in C(G)$ we can fix a Dieudonné complete topological group $G_f$ such that $G$ is a topological subgroup of $G_f$ and $f$ admits a continuous extension over the space $G_f$. Obviously, we may also assume that $G$ is dense in $G_f$. Then $G_f$ can be represented as a topological subgroup of the Raĭkov completion $\varrho G$ such that $G \subset G_f$. Put $G^* = \bigcap\{G_f : f \in C(G)\}$. Then $G^*$ is a topological group containing $G$ as a subgroup, each $f \in C(G)$ can be continuously extended to $G^*$, and $G^*$ is Dieudonné complete, since it is canonically homeomorphic to a closed subspace of the product space $\prod\{G_f : f \in C(G)\}$ (to the "diagonal" of this product). Therefore, $G^*$ is a *PT*-group. Now it follows from Proposition 6.5.14 that $G$ is a *PT*-group. □

Theorem 8.7.1 follows immediately from Lemma 8.7.2 and 8.7.3.

THEOREM 8.7.4. *If a topological group $H$ is factorizable over the class of strong PT-groups, then $H$ is also a strong PT-group.*

PROOF. In our proof we rely upon Theorem 6.5.15. Take any topological group $G$ such that $H$ is a $G_\delta$-dense subgroup of $G$. Let us show that $H$ is $C$-embedded in $G$.

Let $f$ be any continuous real-valued function on $H$. By the assumption, there exist a strong *PT*-group $M$, a continuous homomorphism $\phi$ of $H$ to $M$, and a function $g \in C(M)$ such that $f = g \circ \phi$. Let $M^* = \varrho_\omega M$. Since $M$ is $C$-embedded in $M^*$, $g$ can be extended to a continuous real-valued function $g^*$ on $M^*$.

Further, since $H$ is $G_\delta$-dense in the group $G$, we may assume that $G$ is a subgroup of $\varrho_\omega H$. The homomorphism $\phi$ can be extended to a continuous homomorphism $\phi_1$ of $\varrho H$ to $\varrho M$. Clearly, by the continuity of $\phi_1$ we have that $\phi_1(H^*) \subset M^*$.

Now put $f^* = g^*\phi_1$. Then $f^*$ is a continuous function on $H^*$ the restriction of which to $H$ coincides with $f$. Since $G \subset H^*$, the function $f$ is continuously extended to $G$ as well. Thus, $H$ is $C$-embedded in $G$. Therefore, according to Theorem 6.5.15, $H$ is a strong *PT*-group. □

COROLLARY 8.7.5. *If a topological group $G$ is factorizable over the class of Moscow groups, then it is a strong PT-group.*

After Theorem 8.7.4 and Corollary 8.7.5, it is natural to introduce the following definition. Let us call a topological group $G$ *projectively Moscow* if it is factorizable over the class of Moscow groups. Theorem 8.7.4 and Example 6.5.30 show that the class of projectively Moscow groups is strictly smaller than the class of *PT*-groups. Indeed, the topological group $H$ in Example 6.5.30 is not factorizable over the class of Moscow topological groups, that is, $H$ is not projectively Moscow, since $H$ is not a strong *PT*-group. On the other hand, the group $G$ in Example 6.5.30 is $\mathbb{R}$-factorizable by Theorem 8.1.6, since it is Lindelöf. Therefore, $\mathbb{R}$-factorizability is not inherited by $G_\delta$-dense subgroups.

The fact that the "bad" (meaning non-Moscow) topological group we constructed in Example 6.5.30 turned out to be a $P$-space is rather suggestive. Indeed, we already saw in Section 6.2 that topological groups that are $P$-spaces are almost never Moscow spaces, and therefore, we might pay a special attention to the class of groups that are $P$-spaces when we are looking for non-*PT*-groups.

It is not clear if every $\omega$-narrow topological group is a *PT*-group (see Problem 6.7.1). It is not clear either whether every topological group of countable cellularity is $\mathbb{R}$-factorizable (we refer to Problem 8.1.1). However, every topological group of countable cellularity is Moscow, by Corollary 6.4.11. We should also mention that the non-*PT*-group in Example 6.7.13 is not $\omega$-narrow.

We show in the next theorem that closed subgroups of $\omega$-narrow Moscow groups can be as "bad" as possible.

THEOREM 8.7.6. *Every $\omega$-narrow topological group $G$ can be represented as a closed invariant subgroup of an $\mathbb{R}$-factorizable (hence, $\omega$-narrow) Moscow topological group.*

PROOF. By Theorem 8.2.2, every $\omega$-narrow group $G$ can be represented as a closed invariant subgroup of a dense subgroup of an $\mathbb{R}$-factorizable group $G^*$. In addition, the group $G^*$ in Theorem 8.2.2 is constructed as a dense subgroup of the product of some family of separable metrizable groups. Then, since the Souslin number of the product is countable, the product group is Moscow and therefore, by Proposition 6.1.2, the group $G^*$ is Moscow. □

COROLLARY 8.7.7. *A closed subgroup of an $\mathbb{R}$-factorizable Moscow group need not be a strong PT-group; therefore, it need not be a Moscow group.*

PROOF. The group $H$ in Example 6.5.30 is $\omega$-narrow, since it is a subgroup of a Lindelöf group (see Theorem3.4.4). Therefore, by Theorem 8.7.6, $H$ can be represented as a closed subgroup of an $\mathbb{R}$-factorizable Moscow group. □

The notion of $\mathbb{R}$-factorizability and the technique related to it allow us to obtain an additional information on the validity of the formula $\upsilon X \times \upsilon Y = \upsilon(X \times Y)$.

THEOREM 8.7.8. *Let $G$ be a Lindelöf group and $H$ be a pseudocompact group. Then:*

1) *$G \times H$ is completion friendly and, therefore, a strong PT-group;*
2) *$\upsilon G \times \upsilon H = \upsilon(G \times H)$.*

PROOF. Under the restrictions on $G$ and $H$ in the theorem, the group $G \times H$ is $\mathbb{R}$-factorizable, by Theorem 8.5.13. Clearly, both factors $G$ and $H$ are $\mathbb{R}$-factorizable as well. It follows from Corollary 8.3.9 that the groups $G$, $H$, and $G \times H$ are completion friendly. Since these groups are $\omega$-narrow, it remains to apply Theorem 6.7.5 and Corollary 8.3.3. □

THEOREM 8.7.9. *If $G$ is a Lindelöf P-group and $H$ is a precompact group, then the product $G \times H$ is a strong PT-group and $\upsilon G \times \upsilon H = \upsilon(G \times H)$.*

PROOF. The proof is the same as in the case of Theorem 8.7.8, the only difference is that we have to refer to Corollary 8.5.18 stating that under the assumptions about $G$ and $H$, the group $G \times H$ is $\mathbb{R}$-factorizable. □

THEOREM 8.7.10. *The product of a precompact group $H$ with a topological group $G$ of countable o-tightness and of Ulam non-measurable cardinality is a completion friendly group, and $\upsilon G \times \upsilon H = \upsilon(G \times H)$.*

PROOF. It follows from item 4) of Corollary 6.4.11 and Proposition 6.5.20 that $G$ is completion friendly. By Corollary 4.1.8 and Proposition 6.5.21, $H$ is also completion friendly. According to Corollary 5.5.10, the product group $G \times H$ has countable $o$-tightness and, therefore, is Moscow. By Theorem 6.5.13, $G \times H$ is a (strong) $PT$-group. It remains to refer to Theorem 6.7.5. □

## Exercises

8.7.a. Show that if a topological group $G$ is factorizable over the class of $\mathbb{R}$-factorizable groups, then $G$ is $\mathbb{R}$-factorizable.

8.7.b. Show that the product of a Moscow topological group and an $\mathbb{R}$-factorizable group need not be a $PT$-group.

8.7.c. Let $G$ be a topological group factorizable over the class of topological groups of countable cellularity. Show that $c(G) \leq 2^{\omega}$.

8.7.d. For every infinite cardinal $\tau$, give an example of a topological group $G$ factorizable over the class of metrizable topological groups such that $ot(G) \geq \tau$.

### Problems

8.7.A. Show that every Abelian group admits a Hausdorff topology turning it into a Moscow $\mathbb{R}$-factorizable topological group.

8.7.B. (L. de Leo and M. G. Tkachenko [**130**]) Prove that every uncountable Abelian group admits a non-$\mathbb{R}$-factorizable $\omega$-narrow topological group topology.

### Open Problems

8.7.1. (A. V. Arhangel'skii [**40**]) Is every strong $PT$-group projectively Moscow?
Since every Raĭkov complete group is a strong $PT$-group, the next question is just a version of Problem 8.7.1.

8.7.2. (A. V. Arhangel'skii [**40**]) Is every Raĭkov complete group projectively Moscow?

8.7.3. Is every Raĭkov complete Abelian group projectively Moscow?

8.7.4. Suppose that a group $G$ admits an $\omega$-narrow Hausdorff topological group topology. Does $G$ then admit a topology turning $G$ into an $\mathbb{R}$-factorizable topological group? (The answer is "Yes" for Abelian groups.)

8.7.5. Is every $\omega$-narrow topological group factorizable over the class of topological groups of weight $\leq 2^{\omega}$?

## 8.8. Zero-dimensionality of $\mathbb{R}$-factorizable groups

Recall that a cozero set is the complement of a zero-set, and a clopen set is a set that is open and closed. A space $X$ with a base of clopen sets is called *zero-dimensional* or, in symbols, $\operatorname{ind} X = 0$. If every finite covering of $X$ by cozero sets admits a disjoint refinement by clopen sets, then $X$ is called *strongly zero-dimensional.* This fact is usually abbreviated to $\dim X = 0$. Every strongly zero-dimensional space is zero-dimensional [**165**, Th. 6.2.6], but the converse is false [**165**, Example 6.2.20]. By a cozero covering we mean below a covering consisting of cozero sets.

In this section we show that the above implication can be reversed for $\mathbb{R}$-factorizable groups and for arbitrary subgroups of locally compact groups. Our arguments require some preliminary facts from the dimension theory of topological spaces that are usually not included in standard topological courses. All spaces in this section are assumed to be Tychonoff. The three general results that follow will be applied only to zero-dimensional spaces and topological groups, so the reader does not really have to be familiar with the dimension theory in the more general situation.

The first fact we need is the combination of Theorems 7.1.1 and 7.3.3 of [**165**].

THEOREM 8.8.1. *Every regular second-countable space $Y$ satisfies the equality* $\operatorname{ind} Y = \dim Y$*, and if $M$ is a subspace of $Y$, then* $\dim M \leq \dim Y$.

The next result is known as the *Mardešić factorization theorem* (see [**165**, 7.4.14]):

THEOREM 8.8.2. [**S. Mardešić**] *Let $f\colon X \to Z$ be a continuous mapping of a compact space $X$ with* $\dim X \le n$ *onto a space $Z$ with $w(Z) \le \tau$. Then there exist a compact space $Y$ and continuous onto mappings $g\colon X \to Y$ and $h\colon Y \to Z$ such that $f = h \circ g$,* $\dim Y \le n$*, and $w(Y) \le \tau$.*

In the proof of the following lemma, we shall apply Theorem 8.8.2 in the special case when $\tau = \omega$.

LEMMA 8.8.3. *The following conditions are equivalent for an arbitrary space $X$ and any integer $n \ge 0$:*

i) $\dim X \le n$;
ii) *for every continuous mapping $f\colon X \to Z$ onto a space $Z$ of countable weight, there exist a space $Y$ of countable weight satisfying* $\dim Y \le n$ *and continuous onto mappings $g\colon X \to Y$ and $h\colon Y \to Z$ such that $f = h \circ g$.*

PROOF. Suppose that $\dim X \le n$ and consider a continuous mapping $f\colon X \to Z$ onto a space $Z$ of countable weight. Let $i\colon Z \to I^\omega$ be an embedding of $Z$ into the Hilbert cube $I^\omega$. Then the composition $i \circ f\colon X \to I^\omega$ is continuous, so it admits an extension to a continuous mapping $\tilde{f}\colon \beta X \to I^\omega$, where $\beta X$ is the Čech–Stone compactification of $X$. Since $\dim \beta X = \dim X = n$, by virtue of [**165**, 7.1.17], we can apply Theorem 8.8.2 to find a compact space $Y_0$ and continuous mappings $g_0\colon \beta X \to Y_0$ and $h_0\colon Y_0 \to I^\omega$ such that $h_0 \circ g_0 = \tilde{f}$, $\dim Y_0 \le \dim \beta X = n$ and $w(Y_0) \le \omega$. Let us put $Y = g_0(X)$, $g = g_0{\restriction}X$ and $h = i^{-1}{\restriction}i(Z) \circ h_0{\restriction}Y$. Then $w(Y) \le \omega$ and $f = h \circ g$. In addition, Theorem 8.8.1 implies that $\dim Y \le \dim Y_0 \le n$. Thus, i) $\Rightarrow$ ii).

To prove ii) $\Rightarrow$ i), take any finite cozero covering $\{U_1, \ldots, U_m\}$ of $X$ by cozero sets. For every $i \le m$, there exists a continuous function $f_i\colon X \to \mathbb{R}$ such that $U_i = f_i^{-1}(V_i)$ for some open set $V_i \subset \mathbb{R}$. Denote by $W_i$ the product $\prod_{j=1}^m W_{i,j} \subset \mathbb{R}^m$, where $1 \le i \le m$ and $W_{i,j} = V_i$ if $j = i$, and $W_{i,j} = \mathbb{R}$ otherwise. Clearly, $W_1, \ldots, W_m$ are open sets in $\mathbb{R}^m$. Let $f$ be the diagonal product of the functions $f_1, \ldots, f_m$, and $Z = f(X) \subset \mathbb{R}^m$. Notice that $U_i = f^{-1}(W_i)$ for each $i \le m$ and hence, $Z \subset \bigcup_{i=1}^m W_i$. Applying (ii), we choose a space $Y$ of countable weight with $\dim Y \le n$ and continuous onto mappings $g\colon X \to Y$ and $h\colon Y \to Z$ such that $f = h \circ g$. Since

$$\gamma = \{h^{-1}(W_i \cap Z) : 1 \le i \le m\}$$

is a covering of $Y$ by cozero sets, it follows from $\dim Y \le n$ that $\gamma$ has a refinement $\{O_1, \ldots, O_k\}$ by cozero sets of order $\le n+1$. Clearly, $\{g^{-1}(O_1), \ldots, g^{-1}(O_k)\}$ is a cozero refinement of the covering $\{U_1, \ldots, U_n\}$ of order $\le n+1$. Hence, $\dim X \le n$. □

The next result is a version of the Mardešić factorization theorem for zero-dimensional $\mathbb{R}$-factorizable groups.

THEOREM 8.8.4. *Let $\pi\colon G \to K$ be a continuous homomorphism of topological groups, where the group $G$ is $\mathbb{R}$-factorizable and satisfies* $\operatorname{ind} G = 0$*. Then there exist a topological group $H$ and continuous onto homomorphisms $g\colon G \to H$ and $h\colon H \to K$ such that $\pi = h \circ g$,* $\operatorname{ind} H = 0$*, and $w(H) \le w(K)$.*

PROOF. Let $\mathfrak{m} = w(K)$. We will define by induction topological groups $G_n$ and continuous homomorphisms $g_n\colon G \to G_n$ and $h_{n+1}\colon G_{n+1} \to G_n$ satisfying the following conditions for each $n \in \omega$:

(i) $G_0 = K$ and $g_0 = \pi$;
(ii) $w(G_n) \le \mathfrak{m}$;
(iii) $g_n = h_{n+1} \circ g_{n+1}$;
(iv) for every open neighbourhood $U$ of the neutral element $e_n$ of $G_n$, there exists a clopen neighbourhood $W$ of the neutral element $e_{n+1}$ of $G_{n+1}$ such that $W \subset h_{n+1}^{-1}(U)$.

Choose $G_0$ and $g_0$ as in (i). Suppose that, for some $n \in \omega$, we have defined groups $G_k$ and continuous homomorphisms $g_k, h_k$ satisfying (i)–(iv) for each $k \le n$. Since $w(G_n) \le \mathfrak{m}$ by (ii), we can take a base $\{U_{n,\alpha} : \alpha < \mathfrak{m}\}$ of open neighbourhoods of $e_n$ in $G_n$. Let $\alpha < \mathfrak{m}$ be arbitrary. Since $\operatorname{ind} G = 0$, there exists a clopen neighbourhood $V_{n,\alpha}$ of the neutral element $e$ of $G$ such that $V_{n,\alpha} \subset g_n^{-1}(U_{n,\alpha})$. Define a continuous function $f_\alpha : G \to \mathbb{R}$ by $f_\alpha(x) = 1$ if $x \in V_{n,\alpha}$ and $f_\alpha(x) = 0$ otherwise. Since the group $G$ is $\mathbb{R}$-factorizable, there exist a second-countable topological group $H_\alpha$, a continuous homomorphism $p_\alpha : G \to H_\alpha$ and a continuous function $\varphi_\alpha : H_\alpha \to \mathbb{R}$ such that $f_\alpha = \varphi_\alpha \circ p_\alpha$. Denote by $g_{n+1}$ the diagonal product of $g_n$ and the family $\{p_\alpha : \alpha < \mathfrak{m}\}$, and put $G_{n+1} = g_{n+1}(G)$. Since $w(G_n) \le \mathfrak{m}$ and $w(H_\alpha) \le \omega$ for each $\alpha < \mathfrak{m}$, we have $w(G_{n+1}) \le \mathfrak{m}$. Let also $h_{n+1} : G_{n+1} \to G_n$ and $h_{n,\alpha} : G_{n+1} \to H_\alpha$, with $\alpha < \mathfrak{m}$, be the natural projections. It follows from the definition of $g_{n+1}$ and $h_{n,\alpha}$ that $g_n = h_{n+1} \circ g_{n+1}$ and $h_\alpha = h_{n,\alpha} \circ g_{n+1}$, for each $\alpha < \mathfrak{m}$. Hence (ii) and (iii) hold at the stage $n+1$. In addition, for every $\alpha < \mathfrak{m}$, the set $W_{n,\alpha} = (\varphi_\alpha \circ h_{n,\alpha})^{-1}(0)$ is a clopen neighbourhood of $e_{n+1}$ in $G_{n+1}$, and we have

$$\begin{aligned} g_{n+1}^{-1}(W_{n,\alpha}) &= g_{n+1}^{-1}((\varphi_\alpha \circ h_{n,\alpha})^{-1}(0)) = (\varphi_\alpha \circ h_{n,\alpha} \circ g_{n+1})^{-1}(0) \\ &= (\varphi_\alpha \circ p_\alpha)^{-1}(0) = f_\alpha^{-1}(0) = V_{n,\alpha} \subset g_n^{-1}(U_{n,\alpha}). \end{aligned}$$

Hence, $W_{n,\alpha} \subset g_{n+1}(g_n^{-1}(U_{n,\alpha})) = h_{n+1}^{-1}(U_{n,\alpha})$, thus implying (iv). The inductive step of the construction is complete.

Finally, we put $g = \Delta\{g_n : n \in \omega\}$ and $H = g(G)$. Since $G_0 = K$ and $g_0 : G \to K$, the natural projection $h : H \to K$ is defined and satisfies the equality $\pi = h \circ g$. It also follows from (ii) that $w(H) \le \mathfrak{m}$. By (iii), the group $H$ is topologically isomorphic to a subgroup of the limit $H^*$ of the inverse sequence $\{G_n,\ p_m^n : m.n \in \omega,\ m \le n\}$, where $p_m^n = h_{m+1} \circ h_{m+2} \circ \cdots \circ h_n$ if $m < n$, and $p_n^n = id_{G_n}$. For every $n \in \omega$, let $\pi_n : H^* \to G_n$ be the limit projection; clearly, $\pi_n = h_{n+1} \circ \pi_{n+1}$.

We claim that $\operatorname{ind} H^* = 0$. Take an arbitrary neighbourhood $V$ of $e$ in $H^*$. Then we can find $n \in \omega$ and an open neighbourhood $U$ of $e_n$ in $G_n$ such that $\pi_n^{-1}(U) \subset V$. By (iv), there exists a clopen neighbourhood $W$ of $e_{n+1}$ in $G_{n+1}$ such that $W \subset h_{n+1}^{-1}(U)$. Therefore, $O = \pi_{n+1}^{-1}(W)$ is an clopen neighbourhood of the neutral element of $H^*$ which satisfies

$$O = \pi_{n+1}^{-1}(W) \subset \pi_{n+1}^{-1}(h_{n+1}^{-1}(U)) = \pi_n^{-1}(U) \subset V.$$

This proves the claim. Since $\operatorname{ind}(H) \le \operatorname{ind} H^* = 0$, the group $H$ and the homomorphisms $g$ and $h$ are as required. □

THEOREM 8.8.5. [**D. B. Shakhmatov**] *Every zero-dimensional $\mathbb{R}$-factorizable topological group $G$ is strongly zero-dimensional. In other words,* $\operatorname{ind} G = 0$ *and* $\dim G = 0$ *are equivalent for each $\mathbb{R}$-factorizable group $G$.*

PROOF. Clearly, $\dim G = 0$ implies $\operatorname{ind} G = 0$. Let us prove the reverse implication. Suppose that $G$ is an $\mathbb{R}$-factorizable group satisfying $\operatorname{ind} G = 0$, and let $\{U_1, \dots, U_n\}$ be any finite cozero covering of $G$. For every $i \le n$, take a continuous real-valued function $f_i$ on $G$ such that $U_i = G \setminus f_i^{-1}(0)$, and denote by $f$ the diagonal product of the family $\{f_i : 1 \le i \le n\}$. Since $G$ is $\mathbb{R}$-factorizable, we can find, by Lemma 8.1.2, a continuous homomorphism $\pi\colon G \to K$ onto a second-countable topological group $K$ and continuous functions $g_1, \ldots, g_n$ on $K$ such that $f_i = g_i \circ \pi$ for each $i \le n$. Hence, $\{\pi(U_i) : 1 \le i \le n\}$ is a covering of $K$ and each $\pi(U_i) = K \setminus g_i^{-1}(0)$ is a cozero set in $K$. By Theorem 8.8.4, we can find a second-countable topological group $H$ and continuous surjective homomorphisms $g\colon G \to H$ and $h\colon H \to K$ such that $\pi = h \circ g$ and $\dim H = 0$. Since $\gamma = \{h^{-1}(\pi(U_i)) : 1 \le i \le n\}$ is a cozero covering of $H$, there exists a disjoint open covering $\{V_k : 1 \le k \le m\}$ of $H$ that refines $\gamma$. Hence $\{g^{-1}(V_k) : 1 \le k \le m\}$ is a disjoint open covering of $G$ refining $\{U_i : 1 \le i \le n\}$. Thus, $\dim G = 0$. □

Since every subgroup of a $\sigma$-compact topological group is $\mathbb{R}$-factorizable by Corollary 8.1.16, the next curious result follows immediately from Theorem 8.8.5.

COROLLARY 8.8.6. *The equalities* $\dim H = 0$ *and* $\operatorname{ind} H = 0$ *are equivalent for an arbitrary subgroup $H$ of a $\sigma$-compact topological group.*

Locally compact topological groups need not be $\mathbb{R}$-factorizable; in fact, a locally compact group is $\mathbb{R}$-factorizable if and only if it is $\sigma$-compact (see Theorem 8.1.9). Nevertheless, Corollary 8.8.6 remains valid for arbitrary subgroups of locally compact groups:

THEOREM 8.8.7. *Every zero-dimensional subgroup of a locally compact topological group is strongly zero-dimensional.*

PROOF. Let $H$ be a zero-dimensional subgroup of a locally compact topological group $G$. Take an open neighbourhood $U$ of the neutral element $e$ in $G$ with compact closure and denote by $G^*$ the subgroup of $G$ generated by the compact set $\overline{U}$. Then the group $G^*$ is $\sigma$-compact, so its subgroup $H^* = H \cap G^*$ is $\mathbb{R}$-factorizable by Corollary 8.1.16. Clearly, the group $H^*$ is zero-dimensional as a subgroup of the zero-dimensional group $H$. Hence, by virtue of Theorem 8.8.5, $H^*$ is strongly zero-dimensional. Notice that $G^*$ is an open subgroup of $G$, so $H^*$ is an open subgroup of $H$. Since $H$ is the disjoint union of left cosets of $H^*$, and each of the cosets is strongly zero-dimensional, the group $H$ is also strongly zero-dimensional. □

## Exercises

8.8.a. Give an example of a zero-dimensional Tychonoff space that cannot be embedded as a closed subspace into any $\mathbb{R}$-factorizable zero-dimensional topological group.

8.8.b. Prove that the space $\mathbb{Z}^\tau$ is strongly zero-dimensional, for each cardinal $\tau$.

8.8.c. Show that every zero-dimensional locally precompact topological group is strongly zero-dimensional.

## Problems

8.8.A. (M. G. Tkachenko [**485**]) Show that there exists a Raĭkov complete $\omega$-narrow group which is not $\mathbb{R}$-factorizable.

*Hint.* Let $K$ be the free Abelian group on the set of rationals with the discrete topology. Consider the group $G = K^{\omega_1}$ with the $\omega$-box topology and define a closed non-Lindelöf subgroup $H$ of $G$ such that $|\pi_B(H)| \le \omega$, for every countable set $B \subset \omega_1$, where

$\pi_B \colon K^{\omega_1} \to K^B$ is the projection. Show that such a group $H$ is Raĭkov complete and $\omega$-narrow. Finally, apply Corollary 8.6.4 to conclude that $H$ is not $\mathbb{R}$-factorizable.

8.8.B. Let $G$ be an $\omega$-steady $\mathbb{R}$-factorizable group. Prove that the group $G^\bullet$ (see Section 3.8) is $\omega$-steady and $\mathbb{R}$-factorizable as well.

8.8.C. Prove that every realcompact space $X$ admits a closed embedding into a realcompact $\mathbb{R}$-factorizable topological group.

*Hint.* By Proposition 1.9.12, $X$ is naturally homeomorphic to a closed subspace of the group $G = C_p(C_p(X))$. There exists a tightness type topological property $\mathcal{P}$ such that $X$ is realcompact if and only if $C_p(X)$ has the property $\mathcal{P}$ [**510**]. It also happens to be true for this property $\mathcal{P}$ that a Tychonoff space $X$ has $\mathcal{P}$ if and only if $C_p(X)$ is realcompact (see [**32**]). It follows that $X$ is realcompact if and only if $G = C_p(C_p(X))$ is realcompact. Since, for each Tychonoff space $Y$, $C_p(Y)$ is a dense subgroup of $\mathbb{R}^Y$, it follows from Corollary 8.1.15 that $G$ is $\mathbb{R}$-factorizable.

## Open Problems

8.8.1. Let $G$ be an $\mathbb{R}$-factorizable group. Is the group $G^\bullet$ (see Section 3.8) $\mathbb{R}$-factorizable?

8.8.2. (D. B. Shakhmatov [**431**]) Is it true that the three classical dimensions dim, ind, and Ind coincide for $\sigma$-compact topological groups?

8.8.3. Is an arbitrary zero-dimensional $\mathbb{R}$-factorizable paratopological group strongly zero-dimensional?

8.8.4. Let $G$ be a zero-dimensional topological group of countable cellularity. Is $G$ strongly zero-dimensional?

8.8.5. Is there an $\mathbb{R}$-factorizable topological group $G$ of the weight $\aleph_1$ such that every $\mathbb{R}$-factorizable topological group $H$ of the weight $\aleph_1$ is topologically isomorphic to a subgroup of $G$?

8.8.6. Is every (strongly) $\sigma$-discrete topological group strongly zero-dimensional?

8.8.7. Let a topological group $G$ be the union of a countable family of locally compact subgroups, and suppose that $H$ is a zero-dimensional subgroup of a $G$. Is $H$ strongly zero-dimensional?

## 8.9. Historical comments to Chapter 8

Theorem 8.1.1 is a starting point of the study of $\mathbb{R}$-factorizable groups. It was obtained by L. S. Pontryagin in slightly different terms (see [**387**, Example 37]). Proposition 8.1.3 appeared in [**481**, p. 26] as a simple observation and, with a proof, in [**484**]. Lemma 8.1.4 originated in [**465**], while Theorem 8.1.6 was proved in [**476**]. Example 8.1.8, which shows that Theorem 8.1.6 cannot be extended to paratopological groups, is new. Theorem 8.1.9 and Corollary 8.1.10 are from [**484**]. Lemma 8.1.12 is very close to the results from [**21**] (we refer especially to Theorem 5.2.13 and Corollary 5.2.14 that were originally proved in [**21**]). Proposition 8.1.13, Theorem 8.1.14, and Corollaries 8.1.15 and 8.1.16 are from [**481**]. Corollary 8.1.17 appeared in [**475**]. Theorem 8.1.18 was proved, in a more general form, by E. V. Schepin in [**419**]. Proposition 8.1.20, Theorem 8.1.21, and Corollary 8.1.22 are new. Proposition 8.1.23 is from [**484**].

The study of subgroups of $\mathbb{R}$-factorizable groups originated in [**479**] where Example 8.2.1 appeared. Theorem 8.2.2 and Proposition 8.2.3 are also from [**479**]. Theorems 8.2.5 and 8.2.6 characterizing the $\mathbb{R}$-factorizability of topological groups in terms of $z$-embeddings were proved in [**231**] and [**227**], respectively, while Theorem 8.2.7 and Example 8.2.8 are from [**484**].

A thorough study of the Dieudonné completion of topological groups was undertaken by A. V. Arhangel'skii in [**37, 39, 40**]. Section 8.3 contains several results regarding the Dieudonné completion of $\mathbb{R}$-factorizable groups. A prototype of Lemma 8.3.1 was proved by T. Shirota in [**446**], and, in the present form, it appeared in [**487**]. Corollary 8.3.2, Proposition 8.3.4, Theorem 8.3.6, and Corollary 8.3.8 are also from [**487**]. Corollary 8.3.7 was originally proved in [**231**].

Theorem 8.4.2 appeared in [**479**], while Proposition 8.4.3 follows from the results obtained in [**481**]. Lemma 8.4.5 is due to R. L. Blair, see [**76**]. Theorem 8.4.6 and Corollaries 8.4.7 and 8.4.8 are from [**487**].

The notion of $m$-factorizability was introduced in [**484**] as a natural strengthening of $\mathbb{R}$-factorizability. Proposition 8.5.1 and Theorem 8.5.2 were proved in [**484**], as well as item a) of Theorem 8.5.5. Item b) of Theorem 8.5.5 is new. Lemma 8.5.4 appeared in [**487**] for the special case when the factors were topological groups and the subgroup $G$ was the whole product group. Corollary 8.5.6 and Proposition 8.5.7 were mentioned (without proof) in [**484**]. Theorem 8.5.8 and Corollaries 8.5.9 and 8.5.10 are also from [**484**]. Theorem 8.5.11 and Lemma 8.5.12 are new. Theorem 8.5.13 is also new. In the special case when the first factor $G$ is Lindelöf, it appeared in [**484**] (see Corollary 8.5.14 here). Lemma 8.5.16 and Theorem 8.5.17 appear here for the first time. The latter has a prototype in [**484**].

Almost all the results of Section 8.6, with a few exceptions, are from [**487**]. Lemma 8.6.2 and Corollary 8.6.9 were proved by V. V. Uspenskij in [**512**]. The second part of Corollary 8.6.4 appeared in [**485**].

Section 8.7 is based on the article [**40**] by A. V. Arhangel'skii.

The dimension theory of separable metrizable spaces was created by efforts of several outstanding mathematicians (see historical comments in the book [**166**] by R. Engelking). The equality $\dim X = \operatorname{ind} X$ in Theorem 8.8.1, for separable metrizable spaces $X$, goes back to Hurewicz's article [**248**], while the monotonicity of the covering dimension dim in the same class of spaces follows from the above equality and the simple fact that the small inductive dimension ind is monotone in regular spaces. Theorem 8.8.2 was proved by S. Mardešić in [**303**]. Lemma 8.8.3 is a part of the folklore; it appeared, for example, in [**371**], but its prototype can be found on page 368 of [**8**]. Theorems 8.8.4, 8.8.5, and Corollary 8.8.6 were proved by D. B. Shakhmatov in [**430**]. In fact, Theorem 8.8.4 is analogous to a result proved in [**480**] for the covering dimension dim in place of ind. Theorem 8.8.7 appeared in [**433**].

# Chapter 9

# Compactness and its Generalizations in Topological Groups

This chapter introduces the reader to the Pontryagin–van Kampen duality theory, the Bohr topology on Abelian groups, and the study of algebraic and topological structure of compact, countably compact, and pseudocompact Abelian groups. A special emphasis is given to the study of Abelian groups in each of the directions just mentioned. The main objective in Sections 9.1–9.4 is to prepare all necessary material for the proof of an important corollary to Peter–Weyl's theorem on irreducible representations of compact topological groups. We refer to Theorem 9.4.11 saying that the family of continuous homomorphisms of a compact Abelian group to the circle group separates points of the group.

In the initial four sections we will have to use some basic notions and techniques from various parts of mathematics. For the sake of completeness we introduce all that material in the course of the chapter. Section 9.5 contains the duality theorem in the compact-discrete case, while Section 9.6 familiarizes the reader with several basic facts regarding the structure theory of compact Abelian groups and dual algebraic characterizations of various topological properties of compact groups (such as connectedness, total disconnectedness, weight, dimension). In Section 9.7 we extend Theorem 9.4.11 to locally compact Abelian groups.

Section 9.8 contains the proof of a special case of the celebrated Varopoulos theorem — we establish that every sequentially continuous homomorphism of compact Abelian groups is continuous if the cardinalities of the groups are Ulam non-measurable. We also prove Arhangel'skii's theorem stating that a strongly sequentially continuous isomorphism of a countably compact topological Abelian group $G$ onto a compact topological group is a homeomorphism provided that $G$ has Ulam non-measurable cardinality.

A detailed study of the Bohr topology (equivalently, the maximal precompact topological group topology) on Abelian groups is presented in Section 9.9. Among other facts we prove van Douwen's theorem on $C$-embedded subsets of the groups $G^\#$ and Trigos-Arrieta's theorem saying that the group $G^\#$ is a normal space iff $G$ is countable.

The rest of the chapter is devoted to the thorough study and comparison of the algebraic and topological structure of pseudocompact and countably compact topological Abelian groups.

## 9.1. Krein–Milman Theorem

In this section we establish a fundamental property of compact convex subsets of locally convex topological vector spaces. Recall that a *real topological vector space* $L$ is defined as an Abelian topological group $L$ on which a scalar multiplication is given — with every $x \in X$ and every real number $\lambda$ an element $\lambda x$ of $L$ is associated in such a way that the corresponding mapping $\mathbb{R} \times L \to L$ is continuous and that the set $L$ taken with this multiplication and with the addition given in $L$ is a vector space in the algebraic sense. Two real topological vector spaces $L_1$ and $L_2$ are called *topologically isomorphic* is there exists a homeomorphism $f$ of $L_1$ onto $L_2$ that preserves the linear and additive structures of the spaces, that is, $f(x+y) = f(x) + f(y)$ and $f(\lambda x) = \lambda f(x)$, for all $x, y \in L_1$ and $\lambda \in \mathbb{R}$.

A subset $F$ of a real vector space is said to be *convex* if, whenever $x$ and $y$ belong to $F$ and $0 \leq \lambda \leq 1$, then $\lambda x + (1 - \lambda)y \in F$. If a real topological vector space $L$ has a base consisting of convex open sets, then we say that $L$ is *locally convex*. The proof of the next simple fact is left to the reader.

PROPOSITION 9.1.1. *Let $L$ be a real topological vector space, and $F$ a convex subset of $L$. Then the closure of $F$ in $L$ is also convex.*

If a vector space $L$ has a finite basis, then the number of elements in this basis is called the *dimension* of $L$ and denoted by $\dim L$ (see [**236**]). If $\dim L = n$, where $n$ is a natural number, we say that the vector space $L$ is $n$-dimensional. A standard example of an $n$-dimensional real topological vector space is $\mathbb{R}^n$, with the natural operations.

The importance of the notion of dimension can be already seen from the following simple statement a straightforward proof of which is left to the reader. We will also see in Theorem 9.2.2 that one can drop "locally convex" in this statement.

PROPOSITION 9.1.2. *Every one-dimensional locally convex real topological vector space is topologically isomorphic to $\mathbb{R}$.*

Sometimes we use the abbreviation *r.t.v.s.* to stand for "real topological vector space".

A subset $M$ of a r.t.v.s. $L$ is called a *vector subspace* of $L$ if $M$ is a subgroup of the additive group $L$, and $\lambda x \in M$ for all $x \in M$ and $\lambda \in \mathbb{R}$. It is clear that every vector subspace $M$ of an r.t.v.s. $L$ is again an r.t.v.s. with the operations from $L$ restricted to $M$. We will also need the next statement the proof of which is obvious and is omitted.

PROPOSITION 9.1.3. *If $M$ is a closed vector subspace of a r.t.v.s. $L$, then the quotient group $L/M$, with naturally defined scalar multiplication, is also a r.t.v.s. (called the quotient topological vector space). Moreover, the natural quotient mapping $\pi\colon L \to L/M$ preserves the scalar multiplication, that is, $\pi(\lambda x) = \lambda \pi(x)$, for all $x \in L$ and $\lambda \in \mathbb{R}$. Therefore, if $F$ is a convex subset of $L$, then $\pi(F)$ is a convex subset of $L/M$, and if $L$ is locally convex, then $L/M$ is also locally convex.*

To establish some basic facts about separation in locally convex spaces, we need the next technical lemma:

LEMMA 9.1.4. *Let $E$ be a real topological vector space and $U$ a convex open subset of $E$ which does not contain zero vector $\theta$ of $E$. Suppose also that $E$ is neither zero-dimensional, nor one-dimensional. Then there exists a one-dimensional vector subspace $H$ of $E$ such that $H \cap U = \varnothing$.*

PROOF. Since $\dim E \geq 2$, the subspace $Z$ of $E$ consisting of all non-zero vectors is connected. Indeed, any non-collinear vectors $x$ and $y$ in $E$ can be obviously joined by a topological copy of the closed interval $[0, 1]$ that misses $\theta$. If $x, y \in Z$ are collinear, then, by our assumption, there exists a vector $z \in Z$ that does belong to the set $\{\lambda x : \lambda \in \mathbb{R}\}$. Hence, $x$ and $z$ are not collinear, and neither are $y$ and $z$. Again, it follows that $x$ and $y$ can be joined by a copy of the interval $[0, 1]$ in $Z$. It follows that $Z$ is (pathwise) connected.

Consider the set

$$A = \{\lambda x : x \in U, \ \lambda \in \mathbb{R}, \ \lambda > 0\}.$$

Clearly, $A$ is an open convex subset of $E$ and $\theta \notin A$. Hence, the set $-A = \{-x : x \in A\}$ is also open in $E$, since $E$ is a topological group, and $\theta \notin -A$. It is easy to see that $A \cap (-A) = \varnothing$. Indeed, assume the contrary, and fix $y \in A \cap (-A)$. Then $y \in A$ and $-y \in A$. Since $A$ is convex, it follows that $\theta = y/2 + (-y)/2 \in A$, a contradiction. Hence, $A \cap (-A) = \varnothing$.

Since the sets $A$ and $-A$ are open and disjoint, the connected subspace $Z$ of $E$ cannot be the union of $A$ and $-A$. Therefore, there exists a non-zero element $b \in E$ such that $b \notin A$ and $b \notin -A$. Then $b \notin A$ and $-b \notin A$ which obviously implies that $\lambda b \notin U$, for any $\lambda \in \mathbb{R}$. Hence, the one-dimensional subspace $H = \{\lambda b : \lambda \in \mathbb{R}\}$ of $E$ is disjoint from $U$. This completes the argument. □

We will now formulate an important result on separation of convex sets in locally convex topological vector spaces. It is a geometric version of the *Hahn–Banach theorem.*

THEOREM 9.1.5. [**H. Hahn**, **S. Banach**] *Let $L$ be a real topological vector space and $V$ a non-empty open convex subset of $L$ such that $\theta \notin V$, where $\theta$ is the zero vector of $L$. Then there is a closed topological vector subspace $P$ of $L$ such that the quotient topological vector space $L/P$ is topologically isomorphic to $\mathbb{R}$ and $P \cap V = \varnothing$.*

PROOF. Consider the family $\mathscr{E}$ of all vector subspaces of $L$ disjoint from $V$ and ordered by inclusion. Clearly, the union of any linearly ordered subfamily of $\mathscr{E}$ belongs to $\mathscr{E}$. Therefore, by Zorn's lemma, there is a maximal element $P$ in $\mathscr{E}$. Clearly, the closure $\overline{P}$ of $P$ in $L$ is a vector subspace of $L$ as well. Since $V$ is open in $L$, $\overline{P}$ is disjoint from $V$. Hence, $\overline{P} \in \mathscr{E}$. Since $P$ is a maximal element of $\mathscr{E}$ and $P \subset \overline{P}$, it follows that $P = \overline{P}$. Thus, $P$ is closed in $L$.

Since $P \neq L$, the quotient topological vector space $L/P$ (see Proposition 9.1.3) is not trivial. Let us show that the space $L/P$ is one-dimensional. Assume the contrary. Then $\dim L/P \geq 2$. Denote by $W$ the image of $V$ under the quotient mapping $\pi \colon L \to L/P$. By Proposition 9.1.3, $W$ is a convex open subset of $L/P$. Observe that the zero-vector of $L/P$ does not belong to $W$, since $V \cap P = \varnothing$. Hence, by Lemma 9.1.4, the quotient space $L/P$ contains a one-dimensional vector subspace $H$ disjoint from $W$. Then $\pi^{-1}(H)$ is an element of $\mathscr{E}$ strictly larger than $P$, thus contradicting the maximality of $P$ in $\mathscr{E}$. This contradiction completes the argument. □

Let $L$ be a r.t.v.s. A real-valued function $f$ on $L$ is called *linear functional* if $f(x + y) = f(x) + f(y)$ and $f(\lambda x) = \lambda f(x)$, for all $x, y \in L$ and $\lambda \in \mathbb{R}$. Here is an important corollary from Theorem 9.1.5 on separation of compact convex sets and points not belonging to them by continuous linear functionals.

COROLLARY 9.1.6. *Suppose that $K$ is a compact convex subset of a real topological vector space $L$ and $a$ is an element of $L \setminus K$. Then there exists a continuous linear functional $f \colon L \to \mathbb{R}$ such that $f(a) \notin f(K)$.*

PROOF. It suffices to consider the case when $a$ is the zero-vector $\theta$ of $L$, by means of a suitable translation of $L$. There is a symmetric convex open neighbourhood $W$ of $\theta$ such that $W \cap K = \varnothing$ (note that $K$ is closed in $L$ since it is compact). Then the set $V = KW$ is an open convex subset of $L$, $K \subset V$, and $\theta \notin V$. From Proposition 9.1.2 and Theorem 9.1.5 it follows immediately that there exists a continuous linear functional $f\colon L \to \mathbb{R}$ such that $0 \notin f(V)$. Since $f(a) = f(\theta) = 0$ and $f(K) \subset f(V)$, it follows that $f(a) \notin f(K)$. □

Now we have to introduce a concept playing a crucial role in the study of compact convex sets. Let $K$ be a convex subset of a real topological vector space $L$. A subset $B$ of $K$ is called an *extreme subset* of $K$ if whenever $x \in K$, $y \in K$, and the midpoint $(x+y)/2$ of the segment joining $x$ and $y$ in $L$ belongs to $B$, it follows that $x \in B$ and $y \in B$. Clearly, $K$ is an extreme subset of itself, and the empty set is an extreme subset of any convex set.

A point $b$ of a convex set $K$ is said to be an *extreme point* of $K$ if the singleton $\{b\}$ is an extreme subset of $K$. The next statement follows immediately from the definition of extreme set.

PROPOSITION 9.1.7. *Let $K$ be a convex subset of a real topological vector space $L$, and $\mathcal{F}$ be a family of extreme subsets of $K$. Then the intersection $\bigcap \mathcal{F}$ of this family is again an extreme subset of $K$.*

The existence of non-trivial extreme subsets of compact convex sets is guaranteed by the next statement.

PROPOSITION 9.1.8. *Let $K$ be a non-empty compact convex subset of a real topological vector space $L$ and $f$ a continuous linear functional from $L$ to $\mathbb{R}$. Put $H = \{x \in L : f(x) = c\}$, where $c$ is the maximum of $f$ on $K$ ($c$ is well-defined since $K$ is compact and $f$ is continuous). Then $H \cap K$ is a non-empty extreme subset of $K$.*

PROOF. Take any $v \in K$ with $f(v) = c$. Clearly, $v \in H \cap K \neq \varnothing$. Suppose that $x \in K$ and $y \in K$ are such that the point $z = (x+y)/2$ is in $H \cap K$. Then $f(x) \leq c$ and $f(y) \leq c$. Therefore, by the linearity of $f$ and the definition of $H$, $c = f(z) = [f(x) + f(y)]/2 \leq (c+c)/2 = c$. It follows that $f(x) = c$ and $f(y) = c$, which implies that $x \in H \cap K$ and $y \in H \cap K$. Thus, $H \cap K$ is an extreme subset of $K$. □

Now we can show that the family of extreme subsets of an arbitrary compact convex set is quite rich.

PROPOSITION 9.1.9. *Let $K$ be a compact convex subset of a real topological vector space $L$ and $F$ be a compact convex subset of $K$ such that $F \neq K$. Then there exists a closed non-empty convex extreme subset $B$ of $K$ such that $B \cap F = \varnothing$.*

PROOF. Since $F$ is a proper subset of $K$, we can fix $a \in K \setminus F$. By Corollary 9.1.6, there is a continuous linear functional $f\colon L \to \mathbb{R}$ such that $f(a) \notin f(F)$.

The set $F$ is connected, since it is convex. Therefore, $f(F)$ is a connected subset of $\mathbb{R}$. It follows now from $f(a) \notin f(F)$ that either $f(y) < f(a)$ for each $y \in F$, or $f(a) < f(y)$ for every $y \in F$. We may assume without loss of generality that the first alternative holds.

Since $K$ is compact and $f$ is continuous, it follows that $f(K)$ is a closed bounded subset of $\mathbb{R}$. Therefore, there exist $b \in K$ and $c \in \mathbb{R}$ such that $f(b) = c$ and $c$ is the supremum of $f$ on $K$. Put $H = \{x \in L : f(x) = c\}$. Then $f(y) < f(a) \leq c$ for each $y \in F$, which implies that $F \cap H = \varnothing$. On the other hand, $H \cap K$ is an extreme subset of $K$, by Proposition 9.1.8. Observe that $b \in H \cap K$, by the definition of $H$. Therefore, the set $B = H \cap K$ is an extreme subset of $K$ we are looking for. □

PROPOSITION 9.1.10. *Let $K$ be a compact convex subset of a real topological vector space $L$. Then every non-empty closed convex extreme subset $B$ of $K$ contains an extreme point of $K$.*

PROOF. Let $\mathscr{E}$ be the family of all non-empty closed convex extreme subsets of $K$ contained in a non-empty closed convex subset $B$ of $K$. We partially order $\mathscr{E}$ by the inclusion. Let $\mathscr{C}$ be any non-empty chain in $\mathscr{E}$. Then $\bigcap \mathscr{C}$ is a non-empty closed convex subset of $K$, since $K$ is compact. By Proposition 9.1.7, $\bigcap \mathscr{C}$ is an extreme subset of $K$. Clearly, $\bigcap \mathscr{C} \subset B$. Therefore, $\bigcap \mathscr{C} \in \mathscr{E}$, by the definition of $\mathscr{E}$. Now it follows from Zorn's Lemma that there is a minimal element $F$ in the partially ordered set $\mathscr{E}$.

Clearly, $F$ is a non-empty closed convex extreme subset of $K$ and $F \subset B$. Let us show that $|F| = 1$. Assume the contrary, and fix two distinct points $a$ and $b$ of $F$. Then $A = \{a\}$ is a proper compact convex subset of $F$. Therefore, by Proposition 9.1.9, there exists a non-empty closed convex extreme subset $P$ of $F$ such that $P \cap A = \varnothing$.

Since $P$ is an extreme subset of $F$, and $F$ is an extreme subset of $K$, it obviously follows that $P$ is an extreme subset of $K$. Clearly, $P \subset B$. These statements imply that $P \in \mathscr{E}$. However, $P$ is a proper subset of $F$ since $a$ is not in $P$. This contradicts the minimality of $F$ in $\mathscr{E}$. Hence, $|F| = 1$, and by the definition, the only point of $F$ is an extreme point of $K$ belonging to $B$. □

COROLLARY 9.1.11. *Let $K$ be a compact convex subset of a real topological vector space $L$ and $F$ be a proper compact convex subset of $K$. Then the set $K \setminus F$ contains at least one extreme point of $K$.*

PROOF. This follows immediately from Propositions 9.1.9 and 9.1.10. □

The above results easily imply the following version of the Krein–Milman theorem:

THEOREM 9.1.12. [**M. G. Krein and D. P. Milman**] *Let $K$ be a compact convex subset of a real topological vector space $L$, and $E$ be the set of all extreme points of $K$. Then the closure of the set $S$ of all vectors in $L$ that can be represented as finite linear combinations of vectors in $E$ contains $K$.*

PROOF. Let $H$ be the closure of $S$ in $L$. Clearly, $H$ is a closed vector subspace of $L$. Therefore, $F = K \cap H$ is a closed convex subset of $L$. Assume that $K$ is not a subset of $H$. Then, clearly, $F \neq K$, and it follows from Corollary 9.1.11 that some extreme point $b$ of $K$ belongs to $K \setminus F$. However, by the definition of $F$, $b \in E \subset S \subset H$ and $b \in K$. Hence, $b \in F$, a contradiction. □

## Exercises

9.1.a. Prove Proposition 9.1.1.

9.1.b. Show that the set of all extreme points of a compact convex subset of a locally convex real vector space is not necessarily compact.

9.1.c. Suppose that the set of all extreme points of a compact convex subset $K$ of a locally convex real vector space is finite. Prove that $K$ is second-countable.

9.1.d. Give an example of a non-metrizable convex compact subspace $K$ of a locally convex real vector space $L$ such that the set of extreme points of $K$ is metrizable.

## Problems

9.1.A. Does there exist an infinite-dimensional locally convex vector space? (See also Exercise 5.1.h.)

9.1.B. Suppose that $K$ is a convex compact subset of a locally convex real vector space $L$ such that the set of extreme points of $K$ is cosmic. Show that $K$ is second-countable.

9.1.C. Suppose that $K$ is a convex compact subset of a locally convex real vector space $L$ such that the network weight of the space consisting of all extreme points of $K$ does not exceed an infinite cardinal $\tau$. Show that the weight of $K$ is not greater than $\tau$.

9.1.D. Take away a point from an arbitrary locally convex topological vector space $L$. Show that the space $X$ so obtained is homogeneous. (See Problems 1.4.B, 1.4.C, and 1.4.1.)

*Hint.* Use the Hahn–Banach theorem to show that every finite-dimensional vector subspace has a direct complement in $L$. Now reduce the problem to the finite-dimensional case, assuming that the point taken away is the neutral element.

9.1.E. Take away a point from an arbitrary locally convex topological vector space $L$. Is the space $X$ so obtained homeomorphic to a topological group? What if $L = C_p(X)$?

## 9.2. Gel'fand–Mazur Theorem

We will need the concepts of a Banach space (that appeared once in Section 5.1) and of a Banach algebra. But first we have to generalize the concept of a real vector space discussed in Section 9.1, and to generalize accordingly some results obtained there.

Let $F$ be a field with multiplicative identity 1 and $E$ an Abelian group, in additive notation. Suppose that for each $\alpha \in F$ and each $x \in E$, an element $\alpha x$ of $E$ is defined in a such a way that the following conditions are satisfied for all $\alpha, \beta \in F$ and all $x, y \in E$:

(V1) $\alpha(x + y) = \alpha x + \alpha y$;
(V2) $(\alpha + \beta)x = \alpha x + \beta x$;
(V3) $\alpha(\beta x) = \alpha\beta x$;
(V4) $1x = x$.

Then $E$ is called a *linear* or a *vector* space over the field $F$. The zero element of $F$ is denoted by 0, the zero element of $E$ is denoted by $\theta$. If $F$ is the field $\mathbb{R}$ of real numbers, then $E$ is a real vector space (see Section 9.1). If $F = \mathbb{C}$, where $\mathbb{C}$ is the field of complex numbers, we call $E$ a *complex vector space.*

In the next lemma we establish almost obvious properties of operations in a vector space.

LEMMA 9.2.1. *Let $E$ be a vector space over a field $F$. Then the following hold for all $x \in E$ and $\alpha \in F$:*

1) $0x = \theta$;
2) $(-1)x = -x$;
3) $\alpha\theta = \theta$;
4) *if $\alpha x = \theta$, then either $\alpha = 0$ or $x = \theta$.*

PROOF. It follows from (V2) that if $x \in E$, then $0x = (0 + 0)x = 0x + 0x$, whence $0x = \theta$. This gives 1). The equalities $\theta = 0x = (1 - 1)x = 1x + (-1)x$ and (V4) imply that $(-1)x = -x$. This proves 2). Further, let $y \in E$ be arbitrary. Then, according to (V1), (V3), and 2), we have that $\alpha\theta = \alpha(y - y) = \alpha y - \alpha y = \theta$, whence 3) follows.

Finally, if $\alpha x = \theta$ and $\alpha \neq 0$, then it follows from (V3) and 3) that $\alpha^{-1}\alpha x = \alpha^{-1}\theta = \theta$ or, equivalently, $1x = \theta$, so that $x = \theta$. □

A *topological vector space* over a topological field $F$ is a vector space $E$ with a topology $\mathcal{T}$ on $E$ which turns $E$ into an Abelian topological group and satisfies the condition:

(M) The natural mapping $m\colon F \times E \to E$ given by the rule $m(\alpha, x) = \alpha x$, is continuous.

Since we consider only $T_1$-topologies, every topological vector space is Hausdorff. A topological vector space $E$ over a field $F$ is said to be *one-dimensional* if it is one-dimensional as a vector space, that is, $E = \{\alpha a : \alpha \in F\}$, for some non-zero $a \in E$. Clearly, every topological field $F$ can be considered as a one-dimensional topological vector space over itself.

The next statement is not as trivial as it appears on the surface. It demonstrates how strong are, in fact, the axioms of a linear topological space — it turns out that if $E$ is a one-dimensional vector space over $\mathbb{R}$ or $\mathbb{C}$, then there is only one topology on $E$ which makes it into a topological vector space.

THEOREM 9.2.2. [**N. Bourbaki**] *Suppose that $E$ is a one-dimensional topological vector space over the topological field $\mathbb{R}$ (or $\mathbb{C}$). Then $E$ is topologically isomorphic to the topological vector space $\mathbb{R}$ (respectively, $\mathbb{C}$).*

PROOF. Fix a non-zero $a \in E$, and put $g(\alpha) = \alpha a$, for every $\alpha \in \mathbb{R}$. Then $g$ is a one-to-one linear continuous mapping of $\mathbb{R}$ onto $E$. It remains to verify that the inverse mapping $g^{-1}$ is also continuous at $\theta$. Assume the contrary. Then we can find a sequence $\xi = \{\alpha_n : n \in \omega\}$ of non-zero real numbers $\alpha_n$ such that the sequence $\{\alpha_n a : n \in \omega\}$ converges to $\theta$ and 0 is not a limit point for $\xi$. Observe also that no non-zero $\beta \in \mathbb{R}$ can be a limit point for $\xi$. Otherwise, by the continuity of $g$, we would have $\beta a = \theta$, a contradiction with 4) of Lemma 9.2.1. Hence, the sequence $\xi$ does not have limit points in $\mathbb{R}$ at all. It follows that the sequence $\eta = \{(\alpha_n)^{-1} : n \in \omega\}$ converges in $\mathbb{R}$ to 0. Since $a \neq \theta$, we can select an open neighbourhood $W$ of $\theta$ in $E$ such that $a \notin W$. By axiom (M) of topological vector spaces, there exist an open neighbourhood $V$ of 0 in $\mathbb{R}$ and an open neighbourhood $U$ of $\theta$ in $E$ such that $VU \subset W$. Since the sequence $\eta$ converges to 0, and the sequence $\{\alpha_n a : n \in \omega\}$ converges to $\theta$, there exists $k \in \omega$ such that $(\alpha_k)^{-1} \in V$ and $\alpha_k a \in U$. Then $a = (\alpha_k)^{-1}\alpha_k a \in VU \subset W$, that is, $a \in W$, a contradiction. In the case of the field $\mathbb{C}$ the argument is exactly the same. □

Let $E$ be a vector space over a field $F$, where $F$ is either $\mathbb{R}$ or $\mathbb{C}$. A *norm* on $E$ is a real-valued function $|\cdot|$ on $E$ such that:

1) $|x| > 0$, for each $x \neq \theta$;
2) $|\alpha x| = |\alpha| \cdot |x|$, for each $\alpha \in F$ and each $x \in E$;
3) $|x + y| \leq |x| + |y|$, for all $x, y \in E$.

Notice that the absolute value of $\alpha$, when $\alpha \in \mathbb{R}$ or $\alpha \in \mathbb{C}$, is also denoted by $|\alpha|$. A vector space $E$ over $\mathbb{R}$ or $\mathbb{C}$, with a norm on it, is called a *normed vector space*, and the set $\{x \in E : |x| \leq 1\}$ is called the *unit ball* in $E$.

Let $E$ be a normed vector space. Put $d(x, y) = |x - y|$, for all $x, y \in E$. Clearly, $d$ is a metric on $E$. It is easy to verify that the operations in $E$ are continuous with respect to the topology on $E$ generated by this metric. This means that $E$ is a *topological vector space* with regard to the topology generated by $d$, also called the *norm topology*. If the metric space $(E, d)$ is complete, the normed space $E$ is said to be a *Banach space*, real or complex, depending on whether the field $F$ under consideration is $\mathbb{R}$ or $\mathbb{C}$.

Every vector space $E$ over the field of complex numbers $\mathbb{C}$ can be also considered as a real vector space, that is, as a vector space over the field $\mathbb{R}$. Of course, formally these two vector spaces are not the same.

Suppose that $E$ is a real or complex vector space. Then the concepts of a segment in $E$ and of a convex subset of $E$ are defined exactly as in Section 9.1. The next fact is obvious.

PROPOSITION 9.2.3. *For every normed vector space $E$, convex open sets form a base of the topology of $E$.*

Topological vector spaces, real or complex, in which the above statement holds, are called *locally convex* (see also Section 9.1). If $E_1$ is a subgroup of the additive group of a vector space $E$ over a field $F$, and $\alpha x \in E_1$ for all $\alpha \in F$ and $x \in E_1$, we will call $E_1$ a *vector* or *linear* subspace of $E$. It is clear that if $E_1$ is a vector subspace of a locally convex space $E$, then $E_1$ is locally convex as well.

Suppose that $H$ is a closed vector subspace of a topological vector space $E$ over a field $K$, where $K$ is $\mathbb{R}$ or $\mathbb{C}$. Then $H$ is a closed subgroup of the additive topological group $E$, and we can consider the quotient group $E/H$, with the usual quotient group topology. Let $\pi\colon E \to E/H$ be the canonical quotient mapping. We define a multiplication in $E/H$ by scalars from $K$ as follows: $\alpha\pi(x) = \pi(\alpha x)$, for all $\alpha \in K$ and $x \in E$. A simple verification shows that this definition is correct, and the corresponding multiplication mapping of $K \times E/H$ to $E/H$ is continuous. This makes $E/H$ into a topological vector space over the same field $K$. Furthermore, the quotient mapping $\pi\colon E \to E/H$ is a *homomorphism* (or a *linear mapping*) of topological vector spaces in the sense that it satisfies the following conditions for all $x, y \in E$ and $\alpha \in K$:

(H1) $\pi(x + y) = \pi(x) + \pi(y)$;
(H2) $\pi(\alpha x) = \alpha\pi(x)$.

Recall that a continuous linear functional on a topological vector space $E$ over a topological field $F$ is a continuous linear mapping of $E$ to $F$. In particular, every linear functional is a homomorphism of the corresponding topological vector spaces. Here is an important theorem on the existence of non-trivial linear functionals.

PROPOSITION 9.2.4. *Suppose that $E$ is a real topological vector space, $M$ a vector subspace of $E$, and $U$ a non-empty open convex subset of $E$ disjoint from $M$. Then there exists a continuous linear functional $f$ on $E$ such that $f(x) = 0$, for each $x \in M$, and $f(x) \neq 0$, for each $x \in U$.*

PROOF. Let $\mathscr{E}$ be the family of all vector subspaces of $E$ disjoint from $U$ and containing $M$. Clearly, the closure of any $H \in \mathscr{E}$ belongs to $\mathscr{E}$, and the union of any chain of elements of $\mathscr{E}$ is again an element of $\mathscr{E}$. Therefore, by Zorn's Lemma, there is a maximal element $H_M$ in $\mathscr{E}$, and $H_M$ is closed in $E$.

Consider the quotient space $K = E/H_M$, with the natural quotient topology. It is easily verified that $K$ is a real topological vector space. The natural quotient mapping $p\colon E \to K$ is open and continuous, since $K$ is the quotient group of the topological group $E$. Hence, $p(U)$ is a non-empty open convex subset of $K$, and it follows from $H_M \cap U = \varnothing$ that $p(U)$ does not contain zero vector of the vector space $K$. In particular, we see that the vector space

$K$ is not zero-dimensional. Let us show that the vector space $K$ is one-dimensional. Assume the contrary. Then, by Lemma 9.1.4, there exists a one-dimensional vector subspace $B$ of $K$ such that $B \cap p(U) = \varnothing$. Put $L = p^{-1}(B)$. Then $L$ is a vector subspace of $E$, and $M \subset H_M \subset L$. Also $U \cap L = \varnothing$. It follows that $L \in \mathcal{E}$. Since, obviously, $H_M$ is a proper subset of $L$, we conclude that $H_M$ is not a maximal element of $E$, a contradiction.

Since $K$ is one-dimensional, Theorem 9.2.2 implies that the topological vector space $K$ is topologically isomorphic to $\mathbb{R}$, and $p$ can be interpreted as a continuous linear functional on $E$ we are looking for. □

Finally, we can formulate and prove a version of the Hahn–Banach theorem which we will need below.

THEOREM 9.2.5. *Suppose that $E$ is a locally convex, real or complex, topological vector space. Then, for every non-zero $a \in E$, there exists a continuous linear functional $g$ on $E$ such that $g(a) \neq 0$.*

PROOF. If $E$ is a real vector space, this follows from Proposition 9.2.4, since $a$ can be separated from zero vector $\theta$ by a convex open neighbourhood $U$. It remains to consider the case when $E$ is a complex vector space.

Put $b = ia$. Let us denote by $E_{\mathbb{R}}$ the vector space $E$ treated as a vector space over $\mathbb{R}$. Note that the vectors $a$ and $b$ are not collinear in $E_{\mathbb{R}}$, though they are, clearly, collinear in $E$. Let $M = \{\alpha b : \alpha \in \mathbb{R}\}$. Then $M$ is a one-dimensional vector subspace of $R_{\mathbb{R}}$, and so $M \cong \mathbb{R}$, by Theorem 9.2.2. Since the additive group $\mathbb{R}$ is locally compact, it follows from Proposition 1.4.19 that $M$ is closed in $E_{\mathbb{R}}$. Clearly, $a$ is not in $M$. The assumption that $E$ is locally convex means exactly that the vector space $E_{\mathbb{R}}$ is locally convex. Therefore, there exists an open convex neighbourhood $U$ of $a$ in $E_{\mathbb{R}}$ such that $U \cap M = \varnothing$.

It follows from Proposition 9.2.4 that there exists a continuous linear functional $f\colon E_{\mathbb{R}} \to \mathbb{R}$ on $E_{\mathbb{R}}$ such that $f(x) = 0$, for each $x \in M$, and $f(x) \neq 0$, for each $x \in U$. In particular, $f(a) \neq 0$, and $f(b) = 0$.

Let us now define a complex-valued function $g$ on $E$ by the rule $g(x) = f(x) - i f(ix)$, for each $x \in E$. Using the fact that $f$ is continuous and linear, it is easily verified that $g$ is a continuous linear functional on $E$. We also have that $g(a) = f(a) - i f(ia) = f(a) - i f(b) = f(a) \neq 0$. Thus, $g$ is what we are looking for. □

Suppose that $A$ is a complex vector space. Suppose further that a multiplication is defined in $A$ making $A$ into a semigroup satisfying the following conditions for all $x, y, z \in A$ and all $\alpha \in \mathbb{C}$:

a) $x(y + z) = xy + xz$;
b) $(x + y)z = xz + yz$;
c) $(\alpha x)y = x(\alpha y) = \alpha(xy)$.

Then $A$ is called a *complex algebra*. If $A \setminus \{\theta\}$ has a unit element with respect to multiplication, we say that $A$ is an *algebra with the unit element*. The unit element is also called the *identity* in $A$. If the multiplication in $A$ is commutative, we call $A$ a *commutative algebra*.

An algebra $A$ over the field $\mathbb{C}$ is called a *Banach* or *normed algebra* if $A$ is a Banach space and $|xy| \leq |x||y|$, for all $x, y \in A$. If there is a unit element $e$ in $A$, we also require that $|e| = 1$. Two Banach algebras are called *isomorphic* if there is a bijection between them preserving the operations and the norm. It is clear that $\mathbb{C}$ itself, with the operations and the norm given in $\mathbb{C}$, is a Banach algebra and that the norm on $\mathbb{C}$ is unique, in the natural sense (this easily follows from the axioms of a normed space).

In the proof of the main result of this section we need the following general fact from the theory of Banach algebras:

THEOREM 9.2.6. *Suppose $B$ is any Banach algebra with identity $e$. Then every element of the set $U = \{x \in B : |x - e| < 1\}$ has the inverse in $B$, and the inverse operation is continuous at every $a \in B$ such that $a^{-1}$ is defined.*

PROOF. By a standard argument, it is enough to check the continuity of the inverse operation at $a = e$. Clearly, $U = \{e + y : y \in B,\ |y| < 1\}$ and $U$ is an open neighbourhood of $e$. Fix $y \in B$ such that $|y| < 1$, and consider the series

$$e - y + y^2 - y^3 + \cdots .$$

This series is obviously norm-dominated by the convergent series

$$1 + |y| + |y|^2 + |y|^3 + \cdots .$$

Therefore, since the Banach space $B$ is complete, the first series converges in $B$, that is, there exists $z \in B$ such that

$$z = e - y + y^2 - y^3 + \cdots .$$

Then, clearly,

$$z = e - y(e - y + y^2 - y^3 + \cdots) = e - yz,$$

which implies that $z + yz = e$ and $(e + y)z = e$. Similarly, $z(e + y) = e$. Hence, $z$ is the inverse element to $e + y$. We have shown that every element of $U$ has an inverse.

It follows from the definition of $z$ that

$$|z| \le 1 + |y| + |y|^2 + \cdots = 1/(1 - |y|).$$

On the other hand, since $z - e = -yz$, we have: $|z - e| = |yz| \le |y||z|$. Therefore,

$$|z - e| \le |y||z| \le |y|/(1 - |y|),$$

and if $|y| < \varepsilon < 1$, we have $|z - e| \le \varepsilon/(1 - \varepsilon)$, which implies the continuity of the inverse at $e$. □

THEOREM 9.2.7. [**I. M. Gel'fand**, **S. Mazur**] *Suppose that $B$ is a Banach algebra over $\mathbb{C}$ with identity $e$ such that every non-zero element of $B$ has an inverse. Then $B$ is isomorphic to the field $\mathbb{C}$ of complex numbers.*

PROOF. Take any non-zero element $a \in B$. It suffices to show that there exists $\lambda \in \mathbb{C}$ such that $a = \lambda e$. Assume the contrary. Then $a_\lambda = (a - \lambda e)^{-1}$ exists for every $\lambda \in \mathbb{C}$. Let us define a mapping $\phi$ of $\mathbb{C}$ to $B$ by $\phi(\lambda) = a_\lambda$, for each $\lambda \in \mathbb{C}$. Since, by Theorem 9.2.6, the inverse operation is continuous on $B \setminus \{0\}$, the mapping $\phi$ is continuous as well.

Fix $\mu \in \mathbb{C}$, and take any $\lambda \in \mathbb{C}$. Clearly, $(a - \mu e) - (a - \lambda e) = (\lambda - \mu)e$. Multiplying both sides first by $a_\mu$ and then by $a_\lambda$, we obviously obtain:

$$a_\lambda - a_\mu = (\lambda - \mu)a_\lambda a_\mu .$$

It follows that $(a_\lambda - a_\mu)/(\lambda - \mu) = a_\lambda a_\mu$, whenever $\lambda \ne \mu$. Since $\phi$ is continuous, the limit of $(a_\lambda - a_\mu)/(\lambda - \mu)$, when $\lambda$ converges to $\mu$ is $a_\mu a_\mu$, for each $\mu \in \mathbb{C}$. Since $a_\mu a_\mu$ is distinct from the zero element of $B$, the mapping $\phi$ of $\mathbb{C}$ to $B$ is not constant. According to the above formula, the function $\phi$ is differentiable, in the natural sense.

Take any $\lambda \in \mathbb{C}$ and put $a = a_\lambda = \varphi(\lambda)$. Then $a \neq 0 \neq a^2$ and, by Theorem 9.2.5, there exists a continuous linear functional $f$ on $B$ such that $f(a^2) \neq 0$. Since $f$ is linear and continuous, $f$ is differentiable. Therefore, the function $g\colon \mathbb{C} \to \mathbb{C}$ is differentiable, at every $\mu \in \mathbb{C}$. It follows that the composition $g = f \circ \phi$ is not constant, because the derivative of $g$ at $\lambda$ is equal to $f(a^2) \neq 0$.

Claim. *The function $g$ converges to* 0 *when* $\lambda \to \infty$.

Since $f$ is continuous and $f(\theta) = 0$, it suffices to establish that $\phi(\lambda)$ converges to $\theta$ when $\lambda \to \infty$. For any $\lambda \neq 0$ we have: $a_\lambda = (a - \lambda e)^{-1} = (\lambda(\lambda^{-1}a - e))^{-1} = \lambda^{-1}(\lambda^{-1}a - e)^{-1}$. It follows that

$$|(a - \lambda e)^{-1}| = |\lambda^{-1}(\lambda^{-1}a - e)^{-1}| \leq |\lambda^{-1}||(\lambda^{-1}a - e)^{-1}|.$$

Under $\lambda \to \infty$ we have that $\lim(\lambda^{-1}a - e) = e$ and $\lim|\lambda^{-1}| = 0$, which implies that $\lim|(\lambda^{-1}a - e)^{-1}| = 1$ and $\lim|\phi(\lambda)| = |(a - \lambda e)^{-1}| = 0$. Therefore, $\phi(\lambda)$ converges to $\theta$ when $\lambda \to \infty$. The claim is proved.

From the above claim, using the Cauchy formula, we have, for any $z \in \mathbb{C}$:

$$g(z) = (1/2\pi i)\int_S g(w)/(w - z)dw,$$

where $S$ is a circumference of some radius $r > 0$ with center at $z$. Let $M_r$ be the maximum of $|g(w)|$ when $w \in S$. Then

$$|g(z)| = \left|\int_S g(w)/(w - z)dw\right| \leq \frac{M_r}{2\pi r} \cdot 2\pi r = M_r.$$

Now, since $g(w)$ converges to 0 when $\lambda$ tends to infinity, $M_r$ becomes as small as we wish if we choose a sufficiently large $r$. Hence $|g(z)| < \varepsilon$, for each $\varepsilon > 0$, that is, $g(z) = 0$, for every $z \in \mathbb{C}$. However, $g$ is not constant, which is a contradiction. □

### Exercises

9.2.a. Verify that the quotient space $E/H$ defined on page 578 is indeed a topological vector space. Show that if $E$ is locally convex, so is $E/H$.

9.2.b. Show that the complex-valued function $g$ defined in the proof of Theorem 9.2.5 is a continuous linear functional on $E$.

## 9.3. Invariant integral on a compact group

Let $G$ be a compact topological group. Throughout the section, we denote by $\mathcal{B}_e$ the family of open neighbourhoods of the neutral element $e$ in $G$. By a function on $G$ we understand in this section a real-valued function on $G$, unless the contrary is clearly specified.

We recall that a function $f$ on $G$ is called *right uniformly continuous* if for every $\varepsilon > 0$, there exists $U \in \mathcal{B}_e$ such that $|f(x) - f(y)| < \varepsilon$ whenever $x, y \in G$ satisfy $xy^{-1} \in U$. It was established in Proposition 1.8.11 (along with Lemma 1.8.10) that every continuous function on a compact topological group is (right) uniformly continuous.

Let $\xi$ be a family of continuous real-valued functions on a topological group $G$. Then $\xi$ is called *right uniformly continuous* if for every $\varepsilon > 0$, there exists $V \in \mathcal{B}_e$ such that $|f(x) - f(y)| < \varepsilon$ for all $f \in \xi$, whenever $x, y \in G$ satisfy the condition $xy^{-1} \in V$. In the sequel we omit the word "right" and call such a family $\xi$ *uniformly continuous* (this agreement also applies to individual functions on a topological group).

Suppose that $f \in C(G)$ is a continuous function on a topological group $G$, and $a \in G$. Then $f_a$ is a function on $G$ given by the rule $f_a(x) = f(xa)$, for each $x \in G$. Accordingly, ${}_af$ is a function on $G$ given by ${}_af(x) = f(ax)$, for each $x \in G$.

PROPOSITION 9.3.1. *Let $f: G \to \mathbb{R}$ be a uniformly continuous function on a topological group $G$. Then the family $\{f_a : a \in G\}$ is uniformly continuous.*

PROOF. Take any $\varepsilon > 0$. Choose $V \in \mathcal{B}_e$ such that $|f(x) - f(y)| < \varepsilon$, whenever $xy^{-1} \in V$. If $xy^{-1} \in V$, then $(xa)(ya)^{-1} = xaa^{-1}y^{-1} \in V$. Hence, $|f_a(x) - f_a(y)| = |f(xa) - f(ya)| < \varepsilon$. □

Suppose that for each $\alpha \in \Lambda$, a function $f_\alpha : G \to \mathbb{R}$ is given. The family $\{f_\alpha : \alpha \in \Lambda\}$ is *uniformly bounded* if there exists a positive number $M \in \mathbb{R}$ such that $|f_\alpha(x)| \leq M$, for all $x \in G$ and $\alpha \in \Lambda$.

Now we are going to introduce some notation which will be used throughout this chapter.

Let $f \in C(G)$ be a continuous function on a topological group $G$. Suppose that $A = \{a_1, a_2, \ldots, a_n\}$ is a finite subset of $G$. Then

$$\langle A, f\rangle(x) = \frac{1}{n}\sum_{i=1}^{n} f(xa_i) = \frac{1}{n}\sum_{i=1}^{n} f_{a_i}(x)$$

and

$$[A, f](x) = \frac{1}{n}\sum_{i=1}^{n} f(a_i x) = \frac{1}{n}\sum_{i=1}^{n} {}_{a_i}f(x).$$

Let also put

$$\mathcal{E}_f = \{\langle A, f\rangle : A \subset G,\ |A| < \omega\} \text{ and } \mathcal{F}_f = \{[A, f] : A \subset G,\ |A| < \omega\}.$$

PROPOSITION 9.3.2. *If $G$ is a compact topological group, then for every $f \in C(G)$, the families $\mathcal{E}_f$ and $\mathcal{F}_f$ are uniformly continuous.*

PROOF. The statement will be only proved for $\mathcal{E}_f$, since the proof for $\mathcal{F}_f$ is similar. Take any $\varepsilon > 0$. Since $G$ is compact and $f$ is continuous, the family $\{f_a : a \in G\}$ is uniformly continuous, by Proposition 1.8.11 and 9.3.1. So there exists $V \in \mathcal{B}_e$ such that $|f_a(x) - f_a(y)| < \varepsilon$, whenever $xy^{-1} \in V$ and $a \in G$. Now let $A = \{a_1, \ldots, a_n\} \subset G$, and take $x, y \in G$ such that $xy^{-1} \in V$. Then

$$\begin{aligned} |\langle A, f\rangle(x) - \langle A, f\rangle(y)| &= \Big|\frac{1}{n}\sum_{i=1}^{n} f_{a_i}(x) - \frac{1}{n}\sum_{i=1}^{n} f_{a_i}(y)\Big| \\ &\leq \frac{1}{n}\sum_{i=1}^{n} |f_{a_i}(x) - f_{a_i}(y)| < \frac{1}{n}\sum_{i=1}^{n} \varepsilon = \varepsilon. \end{aligned}$$

This implies the required conclusion. □

The following statement is clearly true.

PROPOSITION 9.3.3. *Let $G$ be a compact topological group, and $A$ be a finite subset of $G$. Then* $\min(f) \leq \min(\langle A, f\rangle) \leq \max(\langle A, f\rangle) \leq \max(f)$, *and* $\min(f) \leq \min([A, f]) \leq \max([A, f]) \leq \max(f)$, *for each continuous function $f$ on $G$.*

Let $G$ be a compact topological group. We define a metric $\rho$ on the set $C(G)$ of all continuous functions on $G$ in the usual way — if $g_1, g_2 \in C(G)$, then $\rho(g_1, g_2) = \max\{|g_1(x) - g_2(x)| : x \in G\}$. This definition turns $C(G)$ into a metric space.

For a finite subset $A$ of $G$ and for any $f \in C(G)$, put $\psi_A(f) = \langle A, f\rangle$. In this way we have defined a mapping $\psi_A$ of $C(G)$ to itself.

PROPOSITION 9.3.4. *Let $A$ be a finite subset of $G$. Then:*

1) *the mapping $\psi_A$ of $C(G)$ to $C(G)$ is linear;*
2) $\rho(\langle A, g_1\rangle, \langle A, g_2\rangle) \leq \rho(g_1, g_2)$ *and* $\rho([A, g_1], [A, g_2]) \leq \rho(g_1, g_2)$, *for any* $g_1, g_2 \in C(G)$;
3) *the mapping $\psi_A$ of $C(G)$ to $C(G)$ is continuous.*

PROOF. From the definition it is obvious that $\langle A, f + g\rangle = \langle A, f\rangle + \langle A, g\rangle$ and $\langle A, \lambda f\rangle = \lambda\langle A, f\rangle$, for each $\lambda \in \mathbb{R}$. This takes care of 1).

Now we prove 2). We have $\rho(\langle A, g_1\rangle, \langle A, g_2\rangle) = \max|\langle A, g_1\rangle - \langle A, g_2\rangle| = \max|\langle A, g_1 - g_2\rangle| \leq \max|g_1 - g_2| = \rho(g_1, g_2)$, by 1) and Proposition 9.3.3. The argument for $[A, f]$ is the same. Thus, 2) is proved.

We can rewrite 2) as follows: $\rho(\psi_A(f), \psi_A(g)) \leq \rho(f, g)$, for any $f, g \in C(G)$, that is, the mapping $\psi_A$ does not increase distances. This implies the continuity of $\psi_A$. □

PROPOSITION 9.3.5. *Let $G$ be a compact topological group and $f \in C(G)$. If $A$, $B$ are finite subsets of $G$, then the following conditions are satisfied:*

1) $\langle A, \langle B, f\rangle\rangle = \langle AB, f\rangle$;
2) $[A, [B, f]] = [AB, f]$;
3) $\langle A, [B, f]\rangle = [B, \langle A, f\rangle]$.

PROOF. All three statements are verified by routine direct calculations. □

PROPOSITION 9.3.6. *For each compact topological group $G$, each finite subset $A$ of $G$ and each $f \in C(G)$, the families $\mathscr{E}_f$ and $\mathscr{P}_f$ are invariant under the mapping $\psi_A$, that is,* $\psi_A(\mathscr{E}_f) \subset \mathscr{E}_f$ *and* $\psi_A(\mathscr{P}_f) \subset \mathscr{P}_f$.

PROOF. The invariance of $\mathscr{E}_f$ follows from 1) and 2) of Proposition 9.3.5. Using now the continuity of $\psi_A$ (see 3) of Proposition 9.3.4), we conclude that $\mathscr{P}_f$ is also invariant under $\psi_A$. □

Let $G$ be a compact topological group. For each $f \in C(G)$, let $\mathscr{P}_f$ be the closure of $\mathscr{E}_f$ in the space $(C(G), \rho)$, and let $\mathscr{Q}_f$ be the closure of $\mathscr{F}_f$ in the same space.

PROPOSITION 9.3.7. *For each compact topological group $G$ and each $f \in C(G)$, the families $\mathscr{P}_f$ and $\mathscr{Q}_f$ are uniformly continuous.*

PROOF. We will prove that the family $\mathscr{P}_f$ is uniformly continuous. The proof for $\mathscr{Q}_f$ is similar. Let $\varepsilon > 0$ be arbitrary. Since, by Proposition 9.3.2, the family $\mathscr{E}_f$ is uniformly continuous, there exists $V \in \mathscr{B}_e$ such that whenever $h \in \mathscr{E}_f$ and $xy^{-1} \in V$, we have that

$|h(x) - h(y)| < \varepsilon$. Take any $g \in \mathcal{P}_f$, and choose $h \in \mathcal{E}_f$ with $\rho(g, h) < \varepsilon$. Then for every $x, y \in G$ such that $xy^{-1} \in V$, we have:

$$\begin{aligned} |g(x) - g(y)| &= |g(x) - h(x) + h(x) - h(y) + h(y) - g(y)| \\ &\leq |g(x) - h(x)| + |h(x) - h(y)| + |h(y) - g(y)| < 3\varepsilon. \end{aligned}$$

Therefore, $\mathcal{P}_f$ is uniformly continuous. □

Recall that a metric space $(X, \rho)$ is said to be *totally bounded* if, for each $\varepsilon > 0$, there exists a finite $\varepsilon$-net $A$ in $X$, that is, a finite subset $A$ of $X$ such that the family of $\varepsilon$-balls around points of $A$ cover $X$ or, equivalently, $\rho(x, A) < \varepsilon$, for each $x \in X$.

For each $g \in C(G)$, put $M(g) = \max_{x\in G} g(x)$ and $m(g) = \min_{x\in G} g(x)$ (we recall that $G$ is a compact group).

LEMMA 9.3.8. *Let $G$ be a compact topological group and let $f \in C(G)$. Then the metric space $\mathcal{P}_f$ is totally bounded.*

PROOF. Take any $\varepsilon > 0$. By Proposition 9.3.3, since $\mathcal{P}_f = \overline{\mathcal{E}_f}$, we have $m(f) \leq m(g) \leq M(g) \leq M(f)$, for all $g \in \mathcal{P}_f$. Let $S \subset \mathbb{R}$ be a finite $\varepsilon$-net in the interval $[m(f), M(f)]$. Apply Proposition 9.3.7 to choose $V \in \mathcal{B}_e$ such that $|g(x) - g(y)| < \varepsilon$ whenever $g \in \mathcal{P}_f$ and $xy^{-1} \in V$. Since $G$ is compact, there exists a finite set $A \subset G$ such that $G = VA$. Let $\mathcal{F}$ be the set of all functions from $A$ to $S$ and note that $\mathcal{F}$ is finite. For every $\varphi \in \mathcal{F}$, choose (if possible) a function $f_\varphi \in \mathcal{P}_f$ such that

$$|f_\varphi(a) - \varphi(a)| < \varepsilon, \text{ for each } x \in A. \tag{9.1}$$

If such a function $f_\varphi$ does not exist, denote by $f_\varphi$ the zero function on $G$. Then $\mathcal{N} = \{f_\varphi : \varphi \in \mathcal{F}\}$ is a finite subset of $\mathcal{P}_f$, and we claim that $\mathcal{N}$ is a $4\varepsilon$-net in $\mathcal{P}_f$.

Indeed, let $g \in \mathcal{P}_f$ be arbitrary. It follows from our choice of $S$ that, for every $a \in A$, there exists $r_a \in S$ such that $|g(a) - r_a| < \varepsilon$. We define a function $\varphi_g \in \mathcal{F}$ by the rule $\varphi_g(a) = r_a$, for each $a \in A$. Then the corresponding function $f_{\varphi_g} \in \mathcal{N}$ satisfies (9.1), where $\varphi = \varphi_g$ and $f_\varphi = f_{\varphi_g}$. To finish the proof, it remains to show that $\rho(g, f_{\varphi_g}) < 4\varepsilon$. Let $x \in G$ be arbitrary. There exists $a \in A$ such that $x \in Va$, whence $xa^{-1} \in V$. Since $g$ and $f_{\varphi_g} = f^*$ are elements of $\mathcal{P}_f$, it follows that

$$|g(x) - g(a)| < \varepsilon \text{ and } |f^*(a) - f^*(x)| < \varepsilon. \tag{9.2}$$

It also clear from the definition of $\varphi_g$ and $f^*$ that

$$|g(a) - f^*(a)| \leq |g(a) - \varphi_g(a)| + |\varphi_g(a) - f^*(a)| < \varepsilon + \varepsilon = 2\varepsilon. \tag{9.3}$$

Therefore, applying (9.2) and (9.3), we obtain:

$$|g(x) - f^*(x)| \leq |g(x) - g(a)| + |g(a) - f^*(a)| + |f^*(a) - f^*(x)| < 4\varepsilon.$$

Since the above inequality holds for each $x \in G$, the distance $\rho(g, f^*) = \rho(g, f_{\varphi_g})$ is less than $4\varepsilon$ (again, we use the compactness of $G$). This proves that $\mathcal{N}$ is a $4\varepsilon$-net in $\mathcal{P}_f$. □

Since the metric space $C(G)$ is complete, its closed subspace $\mathcal{P}_f$ is also complete. It is well known (see [**165**, Theorem 4.3.29]) that every totally bounded complete metric space is compact. This fact and Lemma 9.3.8 yield the following theorem:

THEOREM 9.3.9. *Let $G$ be a compact topological group. Then $\mathcal{P}_f$ is compact, for any $f \in C(G)$.*

PROPOSITION 9.3.10. *Let $G$ be a compact topological group, and let $g \in C(G)$. If $g$ is not constant, then there exists a finite set $A \subset G$ such that $M(\langle A, g\rangle) < M(g)$ and, similarly, there exists a finite set $B \subset G$ such that $M([B, g]) < M(g)$.*

PROOF. Let $M = M(g)$. Since $g$ is not constant, there exist $x_0 \in G$ and $h \in \mathbb{R}$ such that $g(x_0) < h < M$. Let $U$ be an open neighbourhood of $x_0$ such that $g(x) \le h$ for all $x \in U$. Since $G$ is compact, there exists a finite set $A \subset G$ such that $G = UA$. Put $k = |A|$, and consider the function $\langle A^{-1}, g\rangle$. For each $x \in G$, there exist elements $a \in A$ and $u \in U$ such that $x = ua$. Then $g_{a^{-1}}(x) = g(uaa^{-1}) = g(u) \le h$. Hence, $M(\langle A^{-1}, g\rangle) \le 1/k \cdot [M(k-1)+h] = M - 1/k \cdot (M-h) < M$. The argument in the case of $[B, g]$ is similar. □

It is easy to see that the functions $M$ and $m$ on $C(G)$ defined before Lemma 9.3.8 are continuous. Since $\mathcal{P}_f$ is compact, there exists $p_f \in \mathcal{P}_f$ such that $\min_{g\in\mathcal{P}_f} M(g) = M(p_f)$. Similarly, one can find a function $q_f \in \mathcal{Q}_f$ with $\min_{g\in\mathcal{Q}_f} M(g) = M(q_f)$.

THEOREM 9.3.11. *Let $G$ be a compact topological group, and let $f \in C(G)$. Then:*

a) *the functions $p_f$ and $q_f$ are constant;*
b) *there is only one constant function in $\mathcal{P}_f$;*
c) $p_f = q_f$.

PROOF. Item a) obviously follows from Propositions 9.3.10 and 9.3.6.

To prove b), suppose that $p \in \mathcal{P}_f$ is a constant function. In view of a), it suffices to show that $p = q_f$, which will also prove c). Take any $\varepsilon > 0$. Choose finite sets $A, B \subset G$ such that $\rho(\langle A, f\rangle, p) < \varepsilon$ and $\rho([B, f], q_f) < \varepsilon$. Then, by Propositions 9.3.5 and 9.3.4,

$$\rho(\langle A, [B, f]\rangle, p) = \rho([B, \langle A, f\rangle], [B, p]) \le \rho(\langle A, f\rangle, p) < \varepsilon$$

and

$$\rho(\langle A, [B, f]\rangle, q_f) = \rho(\langle A, [B, f]\rangle, \langle A, q_f\rangle) \le \rho([B, f], q_f) < \varepsilon.$$

Hence, $\rho(p, q_f) < 2\varepsilon$. Therefore, since $\varepsilon > 0$ is arbitrary, we have that $p_f = p = q_f$. □

COROLLARY 9.3.12. *Let $G$ be a compact topological group, $f, g \in C(G)$, $\alpha \in \mathbb{R}$, and let $A \subset G$ be a finite subset of $G$. Then:*

a) $p_{\alpha f} = \alpha p_f$;
b) $p_{\langle A, f\rangle} = p_f$ *and, in general,* $p_g = p_f$ *for each* $g \in \mathcal{P}_f$;
c) $p_{f+g} = p_f + p_g$;
d) $m(f) \le p_f(x) \le M(f)$, *for each* $x \in G$.

PROOF. Indeed, $\alpha p_f \in \mathcal{P}_{\alpha f}$, by the linearity of the mappings $\psi_B$ (see Proposition 9.3.4). Since, by a) of Theorem 9.3.11, $\alpha p_f$ is a constant function, it follows from b) of the same theorem that $p_{\alpha f} = \alpha p_f$.

Clearly, $\mathcal{P}_{\langle A, f\rangle} \subset \mathcal{P}_f$. Therefore, $p_{\langle A, f\rangle} \in \mathcal{P}_f$. Now b) of Theorem 9.3.11 implies that $p_{\langle A, f\rangle} = p_f$. Since $g \in \mathcal{P}_f$ implies that $\mathcal{P}_g \subset \mathcal{P}_f$, we have that $p_g \in \mathcal{P}_f$, and, by b) of Theorem 9.3.11, $p_g = p_f$.

To prove c), we argue as follows. For a given function $g \in C(G)$ and an arbitrary $\varepsilon > 0$, there exists a finite set $B \subset G$ such that $|\langle B, g\rangle - p_g| < \varepsilon$. Put $h = \langle B, g\rangle$. It follows that $|\langle A', h\rangle - q| < \varepsilon$, for any finite set $A' \subset G$. By 1) of Proposition 9.3.5, we have $\langle A', h\rangle = \langle A', \langle B, g\rangle\rangle = \langle A'B, g\rangle$, whence

$$|\langle A'B, g\rangle - p_g| < \varepsilon. \tag{9.4}$$

It follows from item b) that $p_{\langle B,f\rangle} = p_f$, so there exists a finite set $A \subset G$ such that $|\langle A, \langle B, f\rangle\rangle - p_f| < \varepsilon$ or, equivalently,

$$|\langle AB, f\rangle - p_f| < \varepsilon. \tag{9.5}$$

Combining (9.4) and (9.5) and taking $A' = A$, we obtain (making use of the linearity of $\psi_{AB}$):

$$|\langle AB, f + g\rangle - (p_f + p_g)| \leq |\langle AB, f\rangle - p_f| + |\langle AB, g\rangle - p_g| < 2\varepsilon.$$

Since the above inequality holds for all $\varepsilon > 0$, this proves that $p_{f+g} = p_f + p_g$.

Finally, the inequality in d) is evident. □

Let $G$ be a compact topological group, and let $f \in C(G)$. Then $p_f(e)$ is called the *von Neumann integral* or simply the *invariant integral* of $f$ and is denoted by $\int f(x)\,dx$. Thus, $\int f(x)\,dx = p_f(e)$.

THEOREM 9.3.13. **[von Neumann]** *Let $G$ be a compact topological group, and let $f, g \in C(G)$. Then the following conditions are satisfied:*

1) *for each $\alpha \in \mathbb{R}$, $\displaystyle\int \alpha f(x)\,dx = \alpha \int f(x)\,dx$;*

2) $\displaystyle\int (f(x) + g(x))dx = \int f(x)\,dx + \int g(x)\,dx$;

3) *if $f(x) \geq 0$ for all $x \in G$, then $\displaystyle\int f(x)\,dx \geq 0$;*

4) *if $f(x) = 1$ for all $x \in G$, then $\displaystyle\int f(x)\,dx = 1$;*

5) $\displaystyle\int f(x)\,dx = \int f(xa)\,dx$ *for all $a \in G$;*

6) $\displaystyle\int f(x)\,dx = \int f(ax)\,dx$ *for all $a \in G$;*

7) $\displaystyle\int f(x^{-1})\,dx = \int f(x)\,dx$;

8) *if $f(x) \geq 0$ for all $x \in G$, and $f(y) > 0$ for some $y \in G$, then $\displaystyle\int f(x)\,dx > 0$.*

PROOF. Properties 1) and 2) follow from Corollary 9.3.12. Property 3) follows from the obvious observation that if $f \in C(G)$ is a non-negative function, then every function $g \in \mathcal{P}_f$ is non-negative. To prove 4), it is enough to note that if $f \in C(G)$ is a constant function, then $p_f = f$. Properties 5) and 6) follow from the obvious equalities $p_f = p_{\langle\{a\},f\rangle}$ and $p_f = p_{[\{a\},f]}$, according to b) of Corollary 9.3.12.

Let us deduce property 7). Define $g\colon G \to \mathbb{R}$ by $g(x) = f(x^{-1})$. Clearly, $g$ is continuous, and $\mathcal{P}_g = \mathcal{Q}_f$. Therefore, $p_g = q_f$. However, $q_f = p_f$, by Theorem 9.3.11. Hence, $p_g(e) = p_f(e)$, that is, $\int f(x^{-1})\,dx = \int f(x)\,dx$.

Finally, let us prove 8). Take any function $f \in C(G)$ such as in 8).

CLAIM. *There exists a function $g \in \mathcal{P}_f$ such that $g(x) > 0$, for each $x \in G$.*

Indeed, there exists $y \in G$ such that $f(y) > 0$. Let $U$ be an open neighbourhood of $y$ such that $f(x) > 0$, for each $x \in U$. Take a finite set $A \subset G$ such that $UA = G$. Let $z \in G$ be arbitrary, and consider $\langle A^{-1}, f\rangle(z)$. Since $UA = G$, there exist $a \in A$, and $u \in U$ such that $z = ua$. Then $f(za^{-1}) = f(uaa^{-1}) = f(u) > 0$. Since $f(x) \geq 0$ for all $x \in G$, we have that $\langle A^{-1}, f\rangle(z) > 0$. Clearly, the function $g = \langle A^{-1}, f\rangle$ belongs to $\mathcal{P}_f$, which implies the claim.

Take the function $g \in \mathcal{P}_f$ such as in the above claim. Then $m(g) > 0$, by the compactness of $G$. We also have $p_f = p_g$ and $0 < m(g) \leq p_g(e)$, by b) and c) of Corollary 9.3.12. It follows that $\int f(x)\,dx = p_f(e) = p_g(e) \geq m(g) > 0$, and the theorem is proved. □

We will now formulate and prove an important theorem on the uniqueness of the invariant integral.

THEOREM 9.3.14. *Let $G$ be a compact topological group. If a function $\phi\colon C(G) \to \mathbb{R}$ satisfies conditions 1)–5) of Theorem 9.3.13, then $\phi(f) = \int f(x)\,dx$, for all $f \in C(G)$.*

PROOF. It follows from 2) and 3) of Theorem 9.3.13 that if $f(x) \leq g(x)$, for all $x \in G$, then $\phi(f) \leq \phi(g)$, which in turn implies that $\phi(|f|) \geq |\phi(f)|$. Also, by 1) and 4), $\phi(\mathbf{c}) = c$, for all $c \in \mathbb{R}$, where $\mathbf{c}$ is the constant function on $G$ with value $c$.

We also note that $\phi(\langle A, f\rangle) = \phi(f)$, for every $f \in C(G)$ and every finite set $A \subset G$. Indeed, this follows immediately from the properties 1), 2), and 5) of $\phi$ and from the definition of $\langle A, f\rangle$. Now it is easy to show that $\phi(f) = p_f$. Take any $\varepsilon > 0$, and let $A$ be a finite subset of $G$ such that $\rho(\langle A, f\rangle, p_f) < \varepsilon$. Then

$$\begin{aligned} |\phi(f) - p_f(e)| &= |\phi(\langle A, f\rangle) - p_f(e)| = |\phi(\langle A, f\rangle) - \phi(p_f)| \\ &= |\phi(\langle A, f\rangle - p_f)| \leq \phi(|\langle A, f\rangle - p_f|) \leq \phi(\boldsymbol{\varepsilon}) = \varepsilon. \end{aligned}$$

Hence, $\phi(f) = p_f(e) = \int f(x)\,dx$. The theorem is proved. □

Suppose that $G$ is a compact topological group with neutral element $e$ and that $K\colon G \times G \to \mathbb{R}$ is a continuous function of two variables $x$ and $y$. Put $F(y) = \int K(x, y)\,dx$, for each $y \in G$. Since $G \times G$ is a compact group, the function $K(x, y)$ is uniformly continuous. Therefore, for each $\varepsilon > 0$ there exists an open neighbourhood $V$ of $e$ in $G$ such that $|K(x, y) - K(x, z)| < \varepsilon$ whenever $yz^{-1} \in V$. Then, if $y, z \in G$ and $yz^{-1} \in V$, we have:

$$\begin{aligned} |F(y) - F(z)| &= \left| \int K(x, y)\,dx - \int K(x, z)\,dx \right| \\ &= \left| \int [K(x, y) - K(x, z)]\,dx \right| \leq \int |K(x, y) - K(x, z)|\,dx \leq \varepsilon, \end{aligned}$$

by 1), 2), 3), and 4) of Theorem 9.3.13. It follows that the function $F$ is continuous, that is, $F \in C(G)$. Therefore, the integral $\int F(y)\,dy$ is defined. We then put

$$\iint K(x, y)\,dx\,dy = \int F(y)\,dy = \int \left( \int K(x, y)\,dx \right) dy$$

and

$$\iint K(x, y)\,dy\,dx = \int \left( \int K(x, y)\,dy \right) dx.$$

THEOREM 9.3.15. *For every compact topological group $G$ and every continuous real-valued function $K(x, y)$ on the topological product $G \times G$, we have:*

$$\int\int K(x, y)\,dxdy = \int\int K(x, y)\,dy\,dx.$$

PROOF. Indeed, each of the functions $\varphi_1(K) = \int\int K(x, y)\,dx\,dy$ and $\varphi_2(K) = \int\int K(x, y)\,dy\,dx$ on $C(G \times G)$ has the properties 1)–5) formulated in Theorem 9.3.13. Applying Theorem 9.3.13 once again, we conclude that $\varphi_1(K) = \varphi_2(K)$, for each $K \in C(G \times G)$. □

So far in this section we have considered only real-valued functions on $G$. For many applications, however, it is important to have similar results for complex-valued functions. We show below how to derive these more general statements from Theorems 9.3.13 and 9.3.14.

By $C^*(G)$ we denote the space of all continuous complex-valued functions on a compact topological group $G$, with usual operations and sup-norm, turning it into a Banach space.

Every function $f \in C^*(G)$ can be represented in a unique way as $f = f_1 + i f_2$, where $f_1$ and $f_2$ are continuous real-valued functions, that is, $f \in C(G)$ and $g \in C(G)$. Now we define the integral on $C^*(G)$ by linearity: $\int f(x)\,dx = \int f_1(x)\,dx + i \int f_2(x)\,dx$. Clearly, in this way we extended the definition of the invariant integral to all complex-valued continuous functions on $G$.

Some important properties of the invariant integral of complex-valued functions are collected in the next theorem.

THEOREM 9.3.16. *Let $G$ be any compact topological group. Then for every $f, g \in C^*(G)$ and for every $\kappa \in \mathbb{C}$ we have:*

1) $\int \kappa f(x)\,dx = \kappa \int f(x)\,dx$;
2) $\int (f(x) + g(x))\,dx = \int f(x)\,dx + \int g(x)\,dx$;
3) *if $f$ is real-valued, $f \neq 0$, and $f(x) \geq 0$ for all $x \in G$, then $\int f(x)\,dx$ is a positive real number;*
4) *if $f(x) = 1$ for all $x \in G$, then $\int f(x)\,dx = 1$;*
5) $\int f(x)\,dx = \int f(xa)\,dx$ *for all $a \in G$;*
6) $\int f(x)\,dx = \int f(ax)\,dx$ *for all $a \in G$;*
7) $\int f(x^{-1})\,dx = \int f(x)\,dx$;
8) $\overline{\int f(x)\,dx} = \int \overline{f(x)}\,dx$, *where the bar denotes the complex conjugation;*
9) $Re\left[\int f(x)\,dx\right] = \int Re[f(x)]\,dx$, *where $Re[z]$ is the real part of $z \in \mathbb{C}$;*
10) $\left|\int f(x)\,dx\right| \leq \int |f(x)|\,dx$.

PROOF. Properties 2) and 8) follow directly from the definition of invariant integral on $C^*(G)$. Property 3) follows from the definition and 3) of Theorem 9.3.13. Property 4) follows from 4) of Theorem 9.3.13. Similarly, properties 5), 6), and 7) follow from the definition and corresponding statements in Theorem 9.3.13.

To prove 1), let $\kappa = \alpha + i\beta$ and $f = f_1 + if_2$, where $\alpha, \beta \in \mathbb{R}$ and $f_1, f_2 \in C(G)$. Then $\kappa f = (\alpha+i\beta)(f_1+if_2) = (\alpha f_1 - \beta f_2) + i(\beta f_1 + \alpha f_2)$. It follows from the definition of the invariant integral that

$$\begin{aligned}\int \kappa f(x)\,dx &= \int(\alpha f_1(x) - \beta f_2(x))\,dx + i\int(\beta f_1(x) + \alpha f_2(x))\,dx \\ &= \alpha\int f_1(x)\,dx - \beta\int f_2(x)\,dx + i\left(\beta\int f_1(x)\,dx + \alpha\int f_2(x)\,dx\right) \\ &= (\alpha + i\beta)\left(\int f_1(x)\,dx + i\int f_2(x)\,dx\right) \\ &= \kappa\int(f_1(x) + if_2(x))\,dx = \kappa\int f(x)\,dx.\end{aligned}$$

This implies 1).

Clearly, 9) follows from the definition of the integral of a complex-valued function. To finish the proof, it suffices to deduce 10). If $\int f(x)\,dx = 0$, then the conclusion follows from 3) of Theorem 9.3.13. Suppose therefore that $re^{i\theta} = \int f(x)\,dx$, where $r$ is a positive real number and $0 \le \theta < 2\pi$. Hence, by 1), $r = \int e^{-i\theta} f(x)\,dx$. Applying 9), we obtain the equality $r = \int Re[e^{-i\theta} f(x)]\,dx$. We also have, for each $x \in G$:

$$Re[e^{-i\theta} f(x)] \le |e^{-i\theta} f(x)| = |f(x)|,$$

Therefore, from 2) and 3) of Theorem 9.3.13 it follows that

$$r = \int Re[e^{-i\theta} f(x)]\,dx \le \int |f(x)|\,dx.$$

Since $r = |\int f(x)\,dx|$, we conclude that $|\int f(x)\,dx| \le \int |f(x)|\,dx$. □

Theorem 9.3.16 is complemented by the following uniqueness result, similar to Theorem 9.3.14.

THEOREM 9.3.17. *If a complex-valued function $\mu$ on a compact topological group $G$ satisfies conditions 1)–5) of Theorem 9.3.16, then $\mu$ coincides with the invariant integral $\int$, that is, $\mu(f) = \int f(x)\,dx$, for each $f \in C^*(G)$.*

PROOF. Every real-valued continuous function can be represented as the difference of two non-negative continuous real-valued functions. Therefore, it follows from conditions 1), 2), and 3) of Theorem 9.3.16, imposed on $\mu$, that $\mu(f)$ is a real number, for each real-valued continuous function $f$ on $G$. Hence, the restriction $\phi = \mu \restriction C(G)$ of $\mu$ to $C(G)$ is a real-valued function on $C(G)$ satisfying conditions 1)–5) of Theorem 9.3.13. Again, by Theorem 9.3.13, $\mu(f) = \phi(f) = \int f(x)\,dx$, for each $f \in C(G)$. Now, by the linearity of $\mu$ and $\int$, we conclude that $\mu(f) = \int f(x)\,dx$, for each $f \in C^*(G)$. □

Theorem 9.3.17 implies that Theorem 9.3.15 remains true for functions in $C^*(G \times G)$. This result (as well as Theorem 9.3.15) is called the *Fubini theorem.* It plays an important role in Functional Analysis.

The invariant integral allows us to make a further important step — to introduce a scalar product in $C(G)$ and in $C^*(G)$, as follows.

For any $f, h \in C^*(G)$ we put

$$(f, h) = \int f(x)\overline{h(x)}\,dx,$$

and call $(f, h)$ the *scalar product* of $f$ and $h$. From the properties of the invariant integral it is immediate that the following usual properties of the scalar product are satisfied for any $f, h, h_1, h_2 \in C^*(G)$ and any complex number $\lambda$:

1) $(h, f) = \overline{(f, h)}$;
2) $(\lambda f, h) = \lambda(f, h)$;
3) $(f, h_1 + h_2) = (f, h_1) + (f, h_2)$;
4) $(f, f) > 0$ if $f$ is a non-zero function.

Note that $(f, f)$ is always a non-negative real number. This follows directly from the definition of the scalar product, or from the properties 1), 2), and 4). We has to mention that the properties 1)–4) of the scalar product lead to the *Cauchy–Bunyakovski inequality:* $(f, h)^2 \leq (f, f)(h, h)$ which in this case can be also written as follows:

$$\left|\int f(x)\overline{h(x)}\,dx\right| \leq \sqrt{\int |f(x)|^2\,dx}\,\sqrt{\int |h(x)|^2\,dx}.$$

A standard proof of the Cauchy–Bunyakovski inequality can be found in almost any book on Analysis or Linear Algebra.

Using the scalar product, we can define the new length, or norm, of a vector $f$ in $C^*(G)$ and the new distance between elements $f$, $h$ of $C^*(G)$ by $|f|_I = \sqrt{(f, f)}$, and $d_I(f, h) = |f - h|_I$. It follows by a standard argument from the Cauchy–Bunyakovski inequality that $C^*(G)$, endowed with the distance $d_I$, becomes a metric space. Applying the properties of the invariant integral, one can easily see that

$$d_I(f, g) \leq \max_{x\in G} |f(x) - g(x)| = \varrho(f, g),$$

for all $f, g \in C^*(G)$. Therefore, the topology of $C^*(G)$ generated by $d_I$ is weaker than the topology generated by the uniform convergence metric $\varrho$. If the group $G$ is infinite, then the metric space $(C^*(G), d_I)$ is not complete, in a sharp contrast with the case of the uniform convergence metric.

The notion of invariant integral on a compact group $G$ allows us to define the concept of *Haar measure* on $G$. Below we just make a few first steps in this direction.

If $F$ is a closed subset of $G$, we put

$$m(F) = \inf\left\{\int f(x)\,dx : f \in \mathscr{E}_F\right\},$$

where $\mathscr{E}_F$ is the family of all non-negative continuous real-valued functions $f$ on $G$ such that $f(x) \geq 1$, for each $x \in F$. Clearly, $0 \leq m(F) \leq 1$, for every closed subset $F$ of $G$. The number $m(F)$ is called the *Haar measure* of $F$. In the rest of the section $G$ is a compact group. The next property of the Haar measure is obvious.

**Property 1.** *If $P$ and $F$ are closed subsets of $G$ such that $P \subset F$, then $m(P) \leq m(F)$.*

For any subset $A$ of $G$, we define a number $m(A) \geq 0$ by the rule

$$m(A) = \sup\{m(F) : F \subset A,\ F \text{ is closed in } G\}.$$

If $A \subset G$ in the above definition is closed, then the two definitions are easily seen to be equivalent. If $A$ is open, we call $m(A)$ the *Haar measure of $A$*. The next property of the function $m(A)$ is also obvious.

**Property 2.** *The number $m(A)$ is invariant under translations, that is, $m(xA) = m(Ax) = m(A)$, for any $x \in G$ and any $A \subset G$.*

Let $\chi_U$ be the *characteristic function* of a set $U \subset G$, that is, $\chi_U(y) = 1$ if $y \in U$, and $\chi_U(y) = 0$ if $y \in G \setminus U$. We put $\int \chi_U(x)\,dx = m(U)$ and say that $m(U)$ is the invariant integral of the characteristic function $\chi_U$.

We list below several properties of $m(A)$; all of them are natural, but not all of them are obvious. Some of these properties we need below and in the next section.

**Property 3.** *For any closed subsets $F$ and $P$ of $G$, we have $m(F \cup P) \leq m(F) + m(P)$.*

PROOF. Here and in some other arguments below we use notation from the definition of $m(F)$. Fix a positive number $\varepsilon$. There are functions $f \in \mathscr{E}_F$ and $g \in \mathscr{E}_P$ such that $\int f(x)\,dx \leq m(F) + \varepsilon$ and $\int g(x)\,dx \leq m(P) + \varepsilon$. Clearly, the function $h = f + g$ is continuous, non-negative, and $h(x) \geq 1$, for every $x \in F \cup P$. Therefore, $h \in \mathscr{E}_{F \cup P}$. By a standard property of the invariant integral, we have that $\int h(x)\,dx = \int f(x)\,dx + \int g(x)\,dx$. Hence,

$$\begin{aligned} m(P \cup F) \leq \int h(x)\,dx &= \int f(x)\,dx + \int g(x)\,dx \leq m(F) + \varepsilon + m(P) + \varepsilon \\ &= m(F) + m(P) + 2\varepsilon. \end{aligned}$$

Since $\varepsilon$ is any positive number, the required conclusion follows. □

**Property 4.** *For every closed subset $F$ of $G$ and every positive number $\varepsilon$, there is an open subset $U$ of $G$ such that $F \subset U$ and $m(\overline{U}) \leq m(F) + \varepsilon$.*

PROOF. We can find a function $f \in \mathscr{E}_F$ such that $m(F) \geq \int f(x)\,dx - \delta$, where $\delta = \varepsilon/2$. We have $f(x) \geq 1$, for each $x \in F$. Therefore, by the continuity of $f$, there is an open neighbourhood $U$ of $F$ such that $f(x) \geq 1 - \delta$, for every $x \in \overline{U}$. Put $h(x) = f(x) + \delta$, for $x \in G$. Clearly, $\int h(x)\,dx = \int f(x)\,dx + \delta$, and $h \in \mathscr{E}_{\overline{U}}$. Therefore,

$$m(\overline{U}) \leq \int h(x)\,dx \leq \int f(x)\,dx + \delta.$$

We also have the inequality $\int f(x)\,dx - \delta \leq m(F) \leq m(\overline{U})$, since $F \subset \overline{U}$. It follows that $m(\overline{U}) - m(F) \leq 2\delta = \varepsilon$. □

The last statement is conveniently complemented by the following one:

**Property 5.** *For any disjoint subsets $A_1$ and $A_2$ of $G$, we have that $m(A_1 \cup A_2) \geq m(A_1) + m(A_2)$.*

PROOF. Let $A = A_1 \cup A_2$ and take an arbitrary $\varepsilon > 0$. It suffices to show that $m(A) \geq m(A_1) + m(A_2) - 2\varepsilon$. We can find closed sets $F_1 \subset A_1$ and $F_2 \subset A_2$ such that $m(F_1) \geq m(A_1) - \varepsilon$ and $m(F_2) \geq m(A_2) - \varepsilon$. The sets $F_1$ and $F_2$ are disjoint and the set

$F = F_1 \cup F_2$ is closed. Obviously, it remains to show that $m(F) \geq m(F_1) + m(F_2)$. Let $\delta$ be any positive number. There is a function $f \in \mathscr{E}_F$ such that $m(F) \geq \int f(x)\,dx - \delta$. Since the space $G$ is Tychonoff and the sets $F_1$, $F_2$ are compact and disjoint, we can find non-negative continuous real-valued functions $h_1$ and $h_2$ on $G$ such that the following conditions are satisfied for every $i = 1, 2$:

1) $h_i(x) \leq f(x)$, for each $x \in G$;
2) $h_i(x) = 1$, for each $x \in F_i$;
3) for every $x \in G$, either $h_1(x) = 0$ or $h_2(x) = 0$.

Put $h = h_1 + h_2$. From conditions 1) and 3) it follows that $h(x) \leq f(x)$, for each $x \in G$. Therefore, $\int f(x)\,dx \geq \int h(x)\,dx$ and $m(F) \geq \int f(x)\,dx - \delta \geq \int h(x)\,dx - \delta$. However, $\int h(x)\,dx = \int h_1(x)\,dx + \int h_2(x)\,dx$ and $\int h_i(x)\,dx \geq m(F_i)$ for $i = 1, 2$, since, clearly, $h_i \in \mathscr{E}_{F_i}$. Hence,

$$m(F) \geq \int h_1(x)\,dx + \int h_2(x)\,dx - \delta \geq m(F_1) + m(F_2) - \delta.$$

Since this inequality holds for every positive number $\delta$, we conclude that $m(F) \geq m(F_1) + m(F_2)$. □

Notice that Properties 3 and 5 imply the following one:

**Property 6.** *For any disjoint closed sets $F_1$ and $F_2$ in $G$, we have $m(F_1 \cup F_2) = m(F_1) + m(F_2)$.*

**Property 7.** *Let $F$ and $H$ be arbitrary closed subsets of $G$ such that $H \subset F$. Then $m(F) = m(F \setminus H) + m(H)$.*

PROOF. Take an arbitrary $\varepsilon > 0$. There exists an open neighbourhood $U$ of $H$ such that $m(P) \leq m(H) + \varepsilon$, where $P = \overline{U}$. Put $F_1 = F \setminus U$. Clearly, $F_1$ is closed and $F_1 \subset F \setminus H$. Therefore, $m(F_1) \leq m(F \setminus H)$. Since $F \subset F_1 \cup P$, we apply Property 4 to deduce that

$$m(F) \leq m(F_1) + m(P) \leq m(F \setminus H) + m(P) \leq m(F \setminus H) + m(H) + \varepsilon.$$

Since this is true for every $\varepsilon \geq 0$, it follows that $m(F) \leq m(F \setminus H) + m(H)$. On the other hand, by Property 5, $m(F) \geq m(F \setminus H) + m(H)$. Hence, $m(F) = m(F \setminus H) + m(H)$. □

A similar property holds for open sets.

**Property 8.** *Let $V$ be any open subset of $G$, and $A$ an arbitrary subset of $G$. Then $m(A) = m(A \setminus V) + m(V \cap A)$.*

PROOF. By Property 5, we have $m(A) \geq m(A \setminus V) + m(V \cap A)$. Let us prove the reverse inequality. Fix an arbitrary positive number $\varepsilon$ and pick a closed subset $F$ of $A$ such that $m(A) \leq m(F) + \varepsilon$. It follows from Property 7 that $m(F \setminus V) + m(F \cap V) = m(F) \geq m(A) - \varepsilon$. Therefore, $m(A \setminus V) + m(A \cap V) \geq m(A) - \varepsilon$. Since this is true for every $\varepsilon > 0$, we have $m(A \setminus V) + m(A \cap V) \geq m(A)$. □

The general concept playing a central role in the theory of Haar measure can be now introduced as follows. A subset $A$ of a compact group $G$ is said to be *Haar measurable* if $m(A) + m(G \setminus A) = 1$. We always have $m(A) + m(G \setminus A) \leq 1$, by Property 5. It follows from Properties 7 and 8 that open sets and closed sets are Haar measurable. Some properties of Haar measurable sets are discussed in problems to this section.

## Exercises

9.3.a. Let $G$ be a compact topological group. For every $f, g \in C^*(G)$, let $d_I(f, g) = \sqrt{\int |f(x) - g(x)|^2 \, dx}$. Use the Cauchy–Bunyakovski inequality to prove that $d_I$ is a metric on $C^*(G)$.

9.3.b. Verify that if $G$ is an infinite compact group, then the metric space $(C^*(G), d_I)$ (see the previous exercise) is not complete.

9.3.c. Prove, using the properties of the invariant integral, that the cellularity of every compact topological group $G$ is countable.
*Hint.* Assume the contrary, and fix an uncountable family $\xi = \{U_\alpha : \alpha \in A\}$ of non-empty open sets. For each $\alpha \in A$, choose a continuous real-valued function $f_\alpha$ on $G$ such that
1) $0 \le f_\alpha(x) \le 1$, for every $x \in G$;
2) $f_\alpha(x) = 0$, for each $x \in G \setminus U_\alpha$;
3) $f_\alpha(x) > 0$, for some $x \in U_\alpha$.
Put $r_\alpha = \int f_\alpha(x) \, dx$, for each $\alpha \in A$. Then, by 3) of Theorem 9.3.13, each $r_\alpha$ is positive. It follows that there exist a positive number $\varepsilon$ and an uncountable subset $B$ of $A$ such that $r_\alpha > \varepsilon$, for each $\alpha \in B$. Take a natural number $N$ such that $N \cdot \varepsilon > 1$, and a subset $K$ of $B$ with $|K| = N$. Define a function $h_K \colon G \to \mathbb{R}$ as the sum of the functions $f_\alpha$, where $\alpha$ runs over $K$. Clearly, $0 \le |h_K(x)| \le 1$, for each $x \in G$. Therefore, $\int h_K(x) \, dx \le 1$. On the other hand, it is easy to see that $\int h_K(x) \, dx \ge N \cdot \varepsilon > 1$, which is a contradiction.

## Problems

9.3.A. Let $U$ be an open subset of a compact group $G$, $B$ an arbitrary subset of $G$, and $q$ a function on $G$ defined by $q(x) = m(xU \cap B)$, for each $x \in G$. Then $q$ is continuous.
*Solution.* Fix $x \in G$, and take any number $\varepsilon > 0$. We have to find an open neighbourhood $W$ of the neutral element $e$ in $G$ such that if $xy^{-1} \in W$, then $|q(y) - q(x)| \le \varepsilon$. Clearly, there is a compact subset $F$ of $xU$ such that $m(F) \ge m(xU) - \varepsilon$. Now we can find a symmetric open neighbourhood $W$ of the neutral element $e$ of $G$ such that $F \subset yU$ for each $y \in G$ such that $xy^{-1} \in W$. We assume below that $y$ satisfies this condition. Then $m(xU \cap yU) \ge m(F) \ge m(xU) - \varepsilon$. It follows from Property 8 that $m(xU \setminus (xU \cap yU)) \le \varepsilon$. Since $m(xU) = m(yU)$, we also have that $m(yU \setminus (xU \cap yU)) \le \varepsilon$. It follows that $m(xU \cap yU \cap B) \le q(x) = m(xU \cap B) \le m(xU \cap yU \cap B) + \varepsilon$ and that $m(xU \cap yU \cap B) \le q(y) = m(yU \cap B) \le m(xU \cap yU \cap B) + \varepsilon$. Hence, $|q(x) - q(y)| \le \varepsilon$ whenever $xy^{-1} \in W$. Thus, the function $q$ is continuous.

9.3.B. Show that the intersection of any two Haar measurable subsets of a compact group $G$ is again a Haar measurable subset of $G$.

9.3.C. Show that the union of two Haar measurable subsets of a compact group $G$ is again a Haar measurable subset of $G$.

9.3.D. Prove that if $A$ and $B$ are Haar measurable subsets of a compact group $G$, then $m(A \cup B) = m(A) + m(B) - m(A \cap B)$.

9.3.E. If $A$ and $B$ are any Haar measurable subsets of a compact group $G$, then the set $A \setminus B$ is also Haar measurable.

9.3.F. The union of a countable family of Haar measurable subsets of a compact group $G$ is again a Haar measurable subset of $G$.

## 9.4. Existence of non-trivial continuous characters on compact Abelian groups

The main result of this section is Theorem 9.4.11 which implies that continuous homomorphisms of a compact Abelian topological group $G$ to the circle group $\mathbb{T}$ separate points of $G$.

Throughout this section $G$ is an Abelian compact topological group and $e$ is the neutral element of $G$. A complex-valued function $\phi$ on $G$ is said to be *positive definite* if

$$\sum_{i,j=1}^{n} \lambda_i \overline{\lambda_j} \phi(g_i g_j^{-1})$$

is a non-negative real number for any integer $n \geq 1$, any system of complex numbers $\lambda_1, \ldots, \lambda_n$, and any elements $g_1, \ldots, g_n \in G$.

The next statement is perfectly obvious:

PROPOSITION 9.4.1. *Let $f$, $g$ be positive definite functions on $G$, and $\mu$ a non-negative real number. Then:*

a) *$f + g$ is a positive definite function;*
b) *$\mu f$ is also a positive definite function.*

In the next statement several useful properties of positive definite functions are collected.

PROPOSITION 9.4.2. *Let $\phi$ be a positive definite function on $G$. Then:*

1) *$\phi(e) \geq 0$, that is, $\phi(e)$ is a non-negative real number;*
2) *$\phi(g^{-1}) = \overline{\phi(g)}$, for any $g \in G$;*
3) *$|\phi(g)| \leq \phi(e)$, for any $g \in G$.*

PROOF. To prove 1), it suffices to put $n = 1$, $g_1 = e$, and $\lambda_1 = 1$ in the definition of a positive definite function.

Let us prove 2). Put $n = 2$, $g_1 = g$, $g_2 = e$, $\lambda_1 = 1$, and $\lambda_2 = \lambda$. It follows that

$$0 \leq \phi(e) + \phi(g)\overline{\lambda} + \phi(g^{-1})\lambda + \phi(e)|\lambda|^2 \geq 0 \in \mathbb{R}, \tag{9.6}$$

for any complex number $\lambda$. Taking first $\lambda = 1$ and then $\lambda = i$ in (9.6), we conclude, by 1), that $\phi(g^{-1}) + \phi(g)$ and $i[\phi(g^{-1}) - \phi(g)]$ are real numbers. It follows that $\phi(g^{-1}) = \overline{\phi(g)}$. Indeed, let $\phi(g^{-1}) = a + ib$ and $\phi(g) = c + id$, where $a, b, c, d$ are real numbers. Then $\phi(g^{-1})+\phi(g) = (a+c)+i(b+d)$. Since $\phi(g^{-1})+\phi(g)$ is real, it follows that $b+d = 0$, that is, $b = -d$. Similarly, we can show that $a = c$. Hence, $\phi(g^{-1}) = a + ib = c - id = \overline{\phi(g)}$.

It remains to prove 3). If $\phi(e) = 0$, then we put $\lambda = -\phi(g)$ in (9.6). By 2), we obtain $-2|\phi(g)|^2 \geq 0$, whence $\phi(g) = 0$, and 3) is satisfied. Assume now that $\phi(e) > 0$, and put $\lambda = -\phi(g)\phi(e)^{-1}$ in (9.6). It follows that 3) is again satisfied. □

The next statement provides us with a wide variety of positive definite functions.

THEOREM 9.4.3. *For any continuous real-valued function $f$ on a compact Abelian group $G$, the function $\phi_f$ defined by the rule $\phi_f(x) = \int f(xy)f(y)\,dy$, for each $x \in G$, is positive definite.*

PROOF. Take any system of $n$ complex numbers $\lambda_1, \ldots, \lambda_j$, and any $n$ elements $g_1, \ldots, g_n$ in $G$, and consider the number $b = \Sigma_{i,j=1}^{n} \lambda_i \overline{\lambda_j} \phi_f(g_i g_j^{-1})$. We have to show that $b$ is a non-negative real number. Fix any pair $i, j \in \{1, \ldots, n\}$. We have $\phi_f(g_i g_j^{-1}) =$

$\int f(g_i g_j^{-1} y) f(y)\, dy = \int f(g_i y) f(g_j y)\, dy$, since $G$ is Abelian and since the translation by $g_j^{-1}$ preserves the integral (see 6) of Theorem 9.3.13). Since $f$ is real-valued, it follows that $\phi_f(g_i g_j^{-1}) = ({}_{g_i}f, {}_{g_j}f)$, where on the right side stands the scalar product of the functions ${}_{g_i}f$ and ${}_{g_j}f$ in the space $C^*(G)$ (see Section 9.3) and $({}_g f)(y) = f(gy)$, for all $g, y \in G$. Therefore, by the properties of the scalar product, we have

$$b = \sum_{i,j=1}^{n} \lambda_i \overline{\lambda_j} ({}_{g_i}f, {}_{g_j}f) = (h, h),$$

where $h = \sum_{i=1}^{n} \lambda_i \cdot ({}_{g_i}f) \in C^*(G)$. Since $(h, h) \geq 0$, it follows that $b \geq 0$. □

The requirement of the continuity of $f$ in the above statement can be weakened considerably, but we do not need this fact for our purposes.

Some general ideas in the proof of the existence of non-trivial continuous characters on every compact Abelian group can be expressed (in a rather vague way) as follows. First, characters can be represented as fixed points of some natural transformations on the set of complex functions on $G$. Second, characters can be interpreted as extreme points of some suitably defined compact convex sets of functions. Of course, how to define such convex compact subset is far from obvious. It is also not clear immediately how to guarantee the continuity of the character we construct in this way, since it turns out that the functions we have to consider are not all continuous. These are the difficulties to overcome.

We begin with defining a certain rich enough family of transformations of functions. For any $s \in G$, $\theta \in [0, 2\pi]$, any non-negative real number $a$, and any function $f \in C^*(G)$, define a function $T(a, s, \theta)f$ on $G$ by the rule

$$[T(a, s, \theta)f](x) = a^2[2f(x) + e^{2i\theta} f(sx) + e^{-2i\theta} f(s^{-1}x)],$$

where $e^{i\alpha} = \cos\alpha + i\sin\alpha$, for each $\alpha \in \mathbb{R}$. It is clear from the definition that $T(a, s, \theta)f = a^2 T(1, s, \theta)f$, for each function $f$ on $G$. In other words, the transformation $T(a, s, \theta)$ satisfies $T(a, s, \theta) = a^2 T(1, s, \theta)$, for all $s \in G$ and $\theta \in [0, 2\pi]$. In the rest of the section $\theta, \theta_1, \theta_2$ always denote elements of $[0, 2\pi]$, and we call $\theta_1, \theta_2$ *strictly distinct* if $\theta_2 - \theta_1$ is not equal to $k\pi/2$, for any integer $k$.

A few technical lemmas below establish some connection between the transformations so defined and the existence of characters.

LEMMA 9.4.4. *Let $f$ be a function on $G$ such that $f(e) = 1$, and let $s \in G$. Suppose further that $a_1$ and $a_2$ are non-negative real numbers, and that $\theta_1, \theta_2$ are strictly distinct elements of $[0, 2\pi]$ such that*

$$f = T(a_1, s, \theta_1)f = T(a_2, s, \theta_2)f.$$

*Then*

$$f(xs) = f(x)f(s) \text{ and } f(xs^{-1}) = f(x)f(s^{-1}),$$

*for every $x \in G$.*

PROOF. Suppose that $f$ satisfies the condition $T(a, s, \theta)f = f$. From $f(e) = 1$ it follows that $a \neq 0$ and $2f(x) + e^{2i\theta} f(sx) + e^{-2i\theta} f(s^{-1}x) = a^{-2} f(x)$. Substituting $x = e$, we obtain: $a^{-2} = 2 + e^{2i\theta} f(s) + e^{-2i\theta} f(s^{-1})$. It follows that $2f(x) + e^{2i\theta} f(sx) + e^{-2i\theta} f(s^{-1}x) = (2 + e^{2i\theta} f(s) + e^{-2i\theta} f(s^{-1}))f(x)$. Simplifying, we arrive at the equality:

$$[f(sx) - f(s)f(x)]e^{2i\theta} + [f(s^{-1}x) - f(s^{-1})f(x)]e^{-2i\theta} = 0.$$

It is given that the last equality holds for the two distinct values $\theta_1$ and $\theta_2$ of $\theta$. Thus, we have a homogeneous system of two linear equations

$$\begin{cases} c_{11}(f(sx) - f(s)f(x)) + c_{12}(f(s^{-1}x) - f(s^{-1})f(x)) = 0 \\ c_{21}(f(sx) - f(s)f(x)) + c_{22}(f(s^{-1}x) - f(s^{-1})f(x)) = 0, \end{cases}$$

where $c_{11} = e^{2i\theta_1}$, $c_{12} = e^{-2i\theta_1}$, $c_{21} = e^{2i\theta_2}$, and $c_{22} = e^{-2i\theta_2}$. Since $\theta_1$ and $\theta_2$ are strictly distinct, for the determinant of this system of equations we have:

$$c_{11}c_{22} - c_{21}c_{12} = 2i \sin 2(\theta_1 - \theta_2) \neq 0.$$

Hence, the system has only trivial solutions, that is, $f(sx) - f(s)f(x) = 0$ and $f(s^{-1}x) - f(s^{-1})f(x) = 0$. This finishes the proof. □

LEMMA 9.4.5. *For any transformation $T(a, s, \theta)$ with $a \neq 0$, there is a transformation $T^* = T^*(a, s, \theta)$ and there are positive real numbers $b, c, \lambda, \mu$ such that*

$$\lambda + \mu = 1,\ b^2 + c^2 = \mu^{-1}(1 - 2\lambda a^2),\ bc = \mu^{-1}\lambda a^2, \tag{9.7}$$

$$[T^* f](x) = (b^2 + c^2)f(x) + bc\,[e^{i(2\theta+\pi)} f(sx) + e^{-i(2\theta+\pi)} f(s^{-1}x)] \tag{9.8}$$

*and*

$$\lambda T(a, s, \theta)f + \mu T^* f = f, \tag{9.9}$$

*for every complex-valued function $f$ on $G$. Furthermore, $\lambda$ can be chosen to be any real number satisfying the inequalities $0 < \lambda < 1$ and $\lambda \leq a^{-2}/4$.*

PROOF. Clearly, the equality (9.9) will be satisfied if $2\lambda a^2 + \mu(b^2 + c^2) = 1$ and $\lambda a^2 - \mu bc = 0$, that is, if the last two equalities in (9.7) are valid. To guarantee the two conditions it suffices to make sure that

$$(b + c)^2 = \mu^{-1} \text{ and } (b - c)^2 = \mu^{-1}(1 - 4\lambda a^2).$$

Indeed, suppose that the above equalities hold. Then

$$\mu(b^2 + 2bc + c^2) = 1 \text{ and } \mu(b^2 - 2bc + c^2) = 1 - 4\lambda a^2.$$

The sum of the left sides equals to the sum of the right sides, that is, $2\mu(b^2 + c^2) = 2 - 4\lambda a^2$, which implies that $2\lambda a^2 + \mu(b^2 + c^2) = 1$. On the other hand, using the subtraction, we obtain $4\mu bc = 4\lambda a^2$. Hence, $\lambda a^2 - \mu bc = 0$.

Now we proceed as follows. We fix any positive real number $\lambda$ such that $\lambda \leq a^{-2}/4$ and $\lambda < 1$, and put $\mu = 1 - \lambda$. Then we find real numbers $b$ and $c$ from the following system of linear equations:

$$\begin{cases} b + c = \sqrt{\mu^{-1}} \\ b - c = \sqrt{\mu^{-1}(1 - 4\lambda a^2)}. \end{cases}$$

It follows that $1 - 4\lambda a^2 \geq 0$ and $\mu^{-1} > 0$, due to the choice of $\lambda$ and $\mu$. Since the determinant of the system of equations is non-zero and all the coefficients in it are real numbers, we can indeed find real values of $b$ and $c$ satisfying the system. An easy verification shows that both $b$ and $c$ are positive. □

For each transformation $T(a, s, \theta)$, there may exist many transformations $T^*$ such as in the above lemma, depending on the choice of a positive parameter $\lambda$ in (9.7). Each of them will be called an *adjoint transformation* of $T(a, s, \theta)$ and will be denoted by $T^*(a, s, \theta)$ or simply $T^*$.

PROPOSITION 9.4.6. *Let $f$ be any positive definite function on $G$. Then, for each transformation $T(a, s, \theta)$, the function $T(a, s, \theta) f$ is positive definite and, for each adjoint transformation $T^* = T^*(a, s, \theta)$, the function $T^* f$ is also positive definite.*

PROOF. The case $a = 0$ is trivial, so we can assume that $a \neq 0$. Put $z = e^{2i\theta}$, for brevity. Consider the function $\phi$ on $G$ defined by

$$\phi(x) = 2f(x) + zf(sx) + \bar{z} f(s^{-1}x),$$

for each $x \in G$. Let us verify the condition of positive definiteness of $\phi$ for the arrays $g_1, \dots, g_n \in G$ and $\lambda_1, \dots, \lambda_n \in \mathbb{C}$. Since $z\bar{z} = |z|^2 = 1$, and $f$ is positive definite at the arrays $g_1, \dots, g_n, sg_1, \dots, sg_n$ and $\lambda_1, \dots, \lambda_n, z\lambda_1, \dots, z\lambda_n$ of the length $2n$, we obtain, using the commutativity of $G$:

$$\sum_{i,j=1}^{n} \lambda_i \overline{\lambda_j} \phi(g_i g_j^{-1}) = \left[ \sum_{i,j=1}^{n} 2\lambda_i \overline{\lambda_j} f(g_i g_j^{-1}) + \sum_{i,j=1}^{n} z\lambda_i \overline{\lambda_j} f(sg_i g_j^{-1}) + \sum_{i,j=1}^{n} \lambda_i \overline{z\lambda_j} f(s^{-1} g_i g_j^{-1}) \right] \geq 0.$$

Therefore, the function $T(a, s, \theta) f = a^2 \phi$ is positive definite as well.

To prove the second part of the statement, we observe that, for each real number $\gamma$ with $0 \leq \gamma \leq 1$, every $s \in G$, and every $z \in \mathbb{C}$ with $|z| = 1$, the complex function

$$\psi(x) = 2f(x) + \gamma(zf(sx) + \bar{z} f(s^{-1}x))$$

is positive definite. This follows from Proposition 9.4.1 and the obvious equality

$$\psi(x) = \gamma[2f(x) + zf(sx) + \bar{z} f(s^{-1}s)] + (2 - 2\gamma) f(x),$$

where each of two summands on the right side of the equality is a positive definite function.

Finally, according to (9.8), $T^* f$ can be represented in the form

$$[T^* f](x) = d^2[2f(x) + \gamma(z_1 f(sx) + \overline{z_1} f(s^{-1}x))],$$

where $d^2 = (b^2 + c^2)/2$, $\gamma = 2bc/(b^2 + c^2)$ for some real numbers $b, c > 0$, and $z_1 \in \mathbb{C}$, $|z_1| = 1$. Since $d^2 > 0$ and $0 \leq \gamma \leq 1$, the function $T^* f$ is also positive definite. □

Now we are going to define a concept of a $T$-invariant set of functions, and a more narrow version of it, the concept of a strongly $T$-invariant set. A set $A$ of complex functions on $G$ is said to be *$T$-invariant* if for each $f \in A$ and all $s \in G$, $\theta \in [0, 2\pi]$ such that $[T(1, s, \theta) f](e) \neq 0$, there is a real number $a \neq 0$ such that $T(a, s, \theta) f \in A$ and for at least one adjoint $T^*$ of $T(a, s, \theta)$, we have that $T^* f \in A$.

Let us say that a set $A$ of complex-valued functions on $G$ is *strongly $T$-invariant* if there are positive real numbers $m$ and $M$ such that for each $f \in A$, each $s \in G$, and for every $\theta \in [0, 2\pi]$ such that $[T(1, s, \theta) f](e) \neq 0$, one can find a real number $a > 0$ satisfying the following conditions:

1) $0 < m \leq a^2 \leq M \cdot |[T(1, s, \theta) f](e)|^{-1}$;
2) $T(a, s, \theta) f \in A$;
3) $T^* f \in A$, for every choice of the adjoint transformation $T^*$ of $T(a, s, \theta)$.

It is clear from the definitions that every strongly $T$-invariant set of functions is $T$-invariant. The next statement and its proof reveal some technical reasons for introducing the two notions of $T$-invariance.

We recall that a partially ordered set $(B, <)$ is called *directed* if for every $x, y \in B$, there exists $z \in B$ such that $x < z$ and $y < z$. It is a common practice to "transfer" the partial order from the directed set $B$ to a faithfully ordered set $\{p_x : x \in B\}$.

PROPOSITION 9.4.7. *If $A$ is a strongly $T$-invariant set of functions on $G$, then the closure $\overline{A}$ of $A$ in the space $C_p^*(G)$ with the pointwise convergence topology is $T$-invariant.*

PROOF. Let $f \in \overline{A}$ and suppose that $|[T(1, s, \theta) f](e)| > 0$. Then there is a directed set $\{f_\beta : \beta \in B\}$ in $A$ which converges to $f$. Then $\{f_\beta(x) : \beta \in B\}$ converges to $f(x)$, for each $x \in G$, and we may assume that there is a positive number $\varepsilon$ such that $|[T(1, s, \theta) f_\beta](e)| \geq \varepsilon$, for all $\beta \in B$.

Since $A$ is strongly $T$-invariant, for each $\beta \in B$ we can find a real number $a_\beta > 0$ such that $0 < m \leq a_\beta^2 \leq M \cdot \varepsilon^{-1}$, $T(a_\beta, s, \theta) f_\beta \in A$, and $T^*(a_\beta, s, \theta) f_\beta \in A$, for every choice of the adjoint $T^*(a_\beta, s, \theta)$ of the transformation $T(a_\beta, s, \theta)$. From Lemma 9.4.5 it follows that we can take positive real numbers $\lambda^*$ and $\mu^*$ such that $\lambda^* + \mu^* = 1$ and

$$f_\beta = \lambda^* T(a_\beta, s, \theta) f_\beta + \mu^* T^*(a_\beta, s, \theta) f_\beta,$$

for each $\beta \in B$. The directed set $\{a_\beta : \beta \in B\}$ is bounded in $\mathbb{R}$ and, hence, it has a limit point $a \in \mathbb{R}$. Clearly, $0 < m \leq a^2 \leq M\varepsilon^{-1}$. From the definition of $T(a, s, \theta)$ it follows immediately that the function $T(a, s, \theta) f$ belongs to the closure of the set $\{T(a_\beta, s, \theta) f_\beta : \beta \in B\}$ in the topology of pointwise convergence. Let $T^*(a, s, \theta)$ be an adjoint transformation of $T(a, s, \theta)$ corresponding to $\lambda^*$ and $\mu^*$ and defined in Lemma 9.4.5. Then, by the choice of $\lambda^*$ and $\mu^*$, $T^*(a, s, \theta) f$ belongs to the closure of the set $\{T^*(a_\beta, s, \theta) f_\beta : \beta \in B\}$ in the space $C_p^*(G)$. Hence, $\overline{A}$ is $T$-invariant. □

It is convenient to introduce the following notation. Let $f$ be a complex-valued function on $G$ such that $f(e) = 1$. Denote by $\mathscr{E}[f]$ the smallest family of functions on $G$ such that $f \in \mathscr{E}[f]$ and $\mathscr{E}[f]$ is invariant under every transformation of the form $T(a, s, \theta)$ and every adjoint transformation $T^*(a, s, \theta)$ of $T(a, s, \theta)$. Then we define $\Delta[f]$ to be the set of all functions on $G$ of the form $\sum_{i=1}^n \lambda_i h_i$, where $n \in \mathbb{N}$, $h_i \in \mathscr{E}[f]$, $h_i(e) = 1$, and $\lambda_i > 0$, for $i = 1, \dots, n$, and $\sum_{i=1}^n \lambda_i = 1$. It easily follows from the definition that $h(e) = 1$, for each $h \in \Delta[f]$.

PROPOSITION 9.4.8. *Suppose that $\phi$ is a positive definite function on $G$. Then the family $\Delta[\phi]$ satisfies the following conditions:*

1) *each function $f \in \Delta[\phi]$ is positive definite;*
2) *$|f(x)| \leq 1$, for each $f \in \Delta[\phi]$ and each $x \in G$;*
3) *$[T(1, s, \theta) f](e) \geq 0$, for all $s \in G$, $\theta \in [0, 2\pi]$, and any $f \in \Delta[\phi]$;*
4) *the set $\Delta[\phi]$ is strongly $T$-invariant with $m = 1/4$ and $M = 1$;*
5) *the closure of $\Delta[\phi]$ in the topology of pointwise convergence on $C^*(G)$ is $T$-invariant.*

PROOF. Take any $f \in \Delta[\phi]$. By Propositions 9.4.1 and 9.4.6, $f$ is positive definite. It follows from the definition of $\Delta[\phi]$ that $f(e) = 1$. Therefore, by Proposition 9.4.2, $|f(x)| \leq f(e) = 1$, for each $x \in G$. By Proposition 9.4.6, the function $T(1, s, \theta) f$ is also positive definite. Hence, by Proposition 9.4.2, $[T(1, s, \theta) f](e) \geq 0$. It remains to prove (4), since (5) follows from (4), by Proposition 9.4.7.

Again, let $f \in \Delta[\phi]$. If $[T(1, s, \theta) f](e) = c \neq 0$, then $c > 0$, by (3). Hence, for $a^2 = c^{-1}$ we have $[T(a, s, \theta) f](e) = 1$, which implies that $T(a, s, \theta) f \in \Delta[\phi]$. Indeed, since $f \in \Delta[\phi]$, there exist functions $h_1, \ldots, h_n \in \mathscr{E}(\phi)$ and positive reals $\lambda_1, \ldots, \lambda_n$ such that $f = \sum_{i=1}^n \lambda_i h_i$ and $\sum_{i=1}^n \lambda_i = 1$, where $h_i(e) = 1$ for each $i \leq n$. It follows from the definition of $\mathscr{E}(\phi)$ that $g_i = T(a, s, \theta) h_i \in \mathscr{E}(\phi)$, for $i = 1, \ldots, n$. We can assume without loss of generality that the numbers $g_i(e)$ are distinct from zero. Indeed, if $g_i(e) = 0$, for some $i \leq n$, then item 3) of Proposition 9.4.2 implies that $g_i(x) = 0$, for all $x \in G$ (note that by 1), each $g_i$ is positive definite). Hence, we can assume that the real numbers $r_i = g_i(e)$ are positive. From $[T(a, s, \theta) f](e) = 1$ and the linearity of the transformation $T(a, s, \theta)$ it follows that $\sum_{i=1}^n \lambda_i r_i = 1$. Put $\mu_i = \lambda_i r_i$ and $g_i^* = r_i^{-1} g_i$, for each $i \leq n$. Then $f = \sum_{i=1}^n \mu_i g_i^*$ and $\sum_{i=1}^n \mu_i = 1$, where $\mu_i > 0$, $g_i^* \in \mathscr{E}(\phi)$, and $g_i^*(e) = 1$, for each $i \leq n$. This means that $f \in \Delta[\phi]$.

From $f(e) = [T(a, s, \theta) f](e) = 1$ and the equality

$$f(e) = \lambda [T(a, s, \theta) f](e) + \mu [T^*(a, s, \theta) f](e)$$

(see (9.9)) it follows that $[T^*(a, s, \theta) f](e) = 1$, since $\lambda + \mu = 1$. Consequently, $T^*(a, s, \theta) f \in \Delta[\phi]$. Finally, since $|f(x)| \leq 1$ for every $x \in G$, we have $a^{-2} = [T(1, s, \theta) f](e) \leq 4$. It follows that $\Delta[\phi]$ is strongly $T$-invariant with $m = 1/4$ and $M = 1$. □

A real-valued function $k$ on the group $G$ is called *symmetric* if $k(x^{-1}) = k(x)$, for each $x \in G$. Clearly, the function $f(x) + f(x^{-1})$ is symmetric and continuous, for any function $f \in C(G)$.

PROPOSITION 9.4.9. *For every $s \in G$ with $s \neq e$, there exist a positive definite continuous real-valued function $\phi$ on $G$ with $\phi(e) = 1$ and a continuous real-valued non-negative symmetric function $k$ on $G$ such that the function $f$ on $G$ defined by*

$$f(x) = \int k(xy)\phi(y^{-1})\, dy$$

*satisfies the condition $f(s) \neq f(e)$.*

PROOF. Take a symmetric open set $O$ in $G$ such that $e \in O$ and $s \notin O^3$. This is obviously possible. By Urysohn's lemma, there is a continuous real-valued function $h$ on $G$ such that $h(e) = 1$, $0 \leq h(x) \leq 1$ for every $x \in G$, and $h(x) = 0$, for every $x \in G \setminus O$. We can also assume that $h(y) = h(y^{-1})$ for every $y \in G$, since the set $O$ is symmetric. Indeed, otherwise replace $h$ with the function $h^*$ defined by the rule $h^*(y) = [h(y) + h(y^{-1})]/2$, for each $y \in G$.

For each $x \in G$, put

$$\phi_1(x) = \int h(xy)h(y)\, dy.$$

Clearly, the function $\phi_1$ is real-valued and, by Theorem 9.4.3, $\phi_1$ is positive definite. Furthermore, since $h(xy)h(y) \geq 0$ for all $x, y \in G$, it follows from 3) of Theorem 9.3.13 that the function $\phi_1(x)$ is non-negative.

**Claim 1.** *$\phi_1(e) > 0$ and $\phi_1(x) = 0$, for each $x \in G \setminus O^2$.*

Clearly, the function $h(y)h(y)$ is non-negative, real-valued, and continuous. We also have $h(e)h(e) = 1 > 0$. It follows from 8) of Theorem 9.3.13 that $\int h(y)h(y)\,dy > 0$, that is, $\phi_1(e) > 0$. Suppose now that $x \in G \setminus O^2$, and consider an arbitrary $y \in G$. If $y \notin O$, then $h(y) = 0$ which implies that $h(xy)h(y) = 0$. If $y \in O$, then $xy \notin O$, since otherwise $x$ would belong to $O^2$. Therefore, $h(xy) = 0$ and hence, $h(xy)h(y) = 0$. It follows that $h(xy)h(y) = 0$ for every $y \in G$ if $x \in G \setminus O^2$. This means precisely that $\phi_1(x) = 0$, for every $x \in G \setminus O^2$. Claim 1 is verified.

By Urysohn's lemma, we can also fix a continuous real-valued function $k$ on $G$ such that $k(x) = 0$, for each $x \in G \setminus O$, $k(e) = 1$, and $0 \le k(x) \le 1$, for every $x \in G$. As in the case of the function $h$, we can also assume that $k(y) = k(y^{-1})$, for each $y \in G$. Put

$$f(x) = \int k(xy)\phi_1(y^{-1})\,dy.$$

We claim that $f(s) \neq f(e)$. In fact, we will show that $f(s) = 0$ and $f(e) \neq 0$. Suppose that $x \in G \setminus O^3$, and consider an arbitrary $y \in G$. If $y \notin O^2$, then $y^{-1} \notin O^2$ and $\phi_1(y^{-1}) = 0$, which in turn implies that $k(xy)\phi_1(y^{-1}) = 0$. If $y \in O^2$, then $xy \notin O$, since otherwise $x$ would belong to $O^3$. Therefore, $k(xy) = 0$ and $k(xy)\phi_1(y^{-1}) = 0$. It follows that $k(xy)\phi_1(y^{-1}) = 0$ for every $y \in G$, if $x \in G \setminus O^3$. This means that $f(x) = 0$, for every $x \in G \setminus O^3$. In particular, $f(s) = 0$, as $s \notin O^3$.

Since the function $k(y)\phi_1(y^{-1})$ is real-valued, continuous and non-negative, and its value at $e$ is the positive number $k(e)\phi_1(e) = \phi_1(e)$, it follows from 8) of Theorem 9.3.13 that

$$f(e) = \int k(y)\phi_1(y^{-1})\,dy > 0.$$

Hence, $f(s) \neq f(e)$. To complete the proof, we have to replace $\phi_1$ with a function $\phi$ that satisfies $\phi(e) = 1$. Clearly, it is enough to put $\phi = \lambda\phi_1$, where $\lambda = 1/\phi_1(e))$ is a positive real number, by Claim 1. According to Proposition 9.4.1, $\phi$ is positive definite. □

Recall that the *envelope* of a set $A$ of functions is the set of all finite linear combinations of functions in $A$. We also say that a continuous homomorphism of $G$ to the topological group $\mathbb{T}$ is a *character* on a topological group $G$. The latter concept is of extreme importance for the Pontryagin–van Kampen duality theory developed to some extent in Section 9.5.

THEOREM 9.4.10. *Let $\phi$ be a continuous positive definite function on $G$ such that $\phi(e) = 1$, and let $k$ be a non-negative continuous symmetric real-valued function on $G$. Then the function $f$ on $G$ defined by the formula $f(x) = \int k(xy)\phi(y^{-1})\,dy$ belongs to the closure of the envelope of the set of continuous characters on $G$ in the topology of pointwise convergence.*

PROOF. Consider the set $\Delta[\phi]$, and for each $h \in \Delta[\phi]$ put

$$(k * h)(x) = \int k(xy)h(y^{-1})\,dy.$$

Put also $K = \{k * h : h \in \Delta[\phi]\}$. Clearly, $f = k * \phi$, so that $f \in K$. Let $h \in \Delta[\phi]$ be arbitrary. It follows from Proposition 9.4.6 and 1) of Proposition 9.4.8 that the functions $h$ and $T(a, s, \theta)h$ are positive definite. Since $h(e) = 1$, we have, by 3) of Proposition 9.4.2, that $|h(x)| \le 1$ and $|[T(a, s, \theta)h](x)| \le [T(a, s, \theta)h](e)$, for any $x \in G$.

Since $|h(x)| \leq 1$ for each $x \in G$ and the function $k$ is non-negative, it follows from 9), 2), 3), and 6) of Theorem 9.3.16 that

$$|(k * h)(x)| = \left| \int k(xy)h(y^{-1})\,dy \right| \leq \int |k(xy)h(y^{-1})|\,dy \leq \int k(xy)\,dy = \int k(y)\,dy.$$

Therefore, every function in $K$ is bounded by the number $C = \int k(y)\,dy$.

Clearly, the operation $*$ satisfies the condition

$$k * (\lambda_1 h_1 + \lambda_2 h_2) = k * h_1 + k * h_2, \tag{9.10}$$

for any continuous functions $h_1$ and $h_2$ on $G$ and for any complex numbers $\lambda_1$ and $\lambda_2$. Therefore, we have

$$T(a, s, \theta)(k * h) = k * T(a, s, \theta)h,$$

and a similar formula holds for every adjoint transformation $T^*$ of $T(a, s, \theta)$. This implies that the set $K$ is $T$-invariant, but we need a stronger conclusion.

Using again Theorem 9.3.16 and the inequality $|[T(a, s, \theta)h](x)| \leq [T(a, s, \theta)h](e)$ valid for each $x \in G$, we obtain that

$$|T(a, s, \theta)(k * h)(e)| = |k * [T(a, s, \theta)h](e)| = \left| \int k(xy)[T(a, s, \theta)h](y^{-1})\,dy \right| \leq \int k(xy)[T(a, s, \theta)h](e)\,dy \leq C \cdot [T(a, s, \theta)h](e).$$

Thus, we have shown that each $h \in \Delta[\phi]$ satisfies

$$|T(a, s, \theta)(k * h)(e)| \leq C \cdot [T(a, s, \theta)h](e). \tag{9.11}$$

Take an arbitrary $h \in \Delta[\phi]$, and let $f = k * h \in K$. If $[T(1, s, \theta)f](e) \neq 0$, then $[T(1, s, \theta)h](e) \neq 0$ — otherwise Propositions 9.4.2 and 9.4.6 imply that $T(1, s, \theta)f = T(1, s, \theta)(k * h) = k * [T(1, s, \theta)h] \equiv 0$, which contradicts our assumption about $f$. Hence, By 4) of Proposition 9.4.8, there is a real number $a > 0$ with $1/4 \leq a^2 \leq 1/[T(1, s, \theta)h](e)$ such that $T(a, s, \theta)h \in \Delta[\phi]$ and $T^*(a, s, \theta)h \in \Delta[\phi]$, for each adjoint $T^*(a, s, \theta)$ of $T(a, s, \theta)$. It follows from (9.11) that $1/[T(1, s, \theta)h](e) \leq C/|T(1, s, \theta)(k*h)(e)|$. We conclude that the set $K$ is strongly $T$-invariant with $m = 1/4$ and $M = C$. Therefore, by Proposition 9.4.7, the closure $F$ of the set $K$ in the topology of pointwise convergence on $\mathbb{C}^G$ is $T$-invariant.

We also see that all functions in $F$ are bounded by the number $C$, since the same is true for all functions in $K$. From the Tychonoff compactness theorem it follows that $F$ is a compact subspace of the space of complex-valued functions on $G$ endowed with the topology of pointwise convergence. Since $\Delta[\phi]$ is convex and, according to (9.10), the operation $k * h$ is linear with respect to $h$, it follows that the set $K$ is also convex. Therefore, according to Proposition 9.1.1, $F$ is convex as well, since $C^*(G)$ can be also considered as a real topological vector space. Thus, $F$ is a compact convex set of functions.

By Theorem 9.1.12, the envelope of the set $E$ of extreme points of $F$ is dense in $F$. Since $f \in K \subset F$, it suffices to establish that every function in $F$ is continuous, and that every element of $E$ is a character on $G$ multiplied by a constant. We do this below in several steps.

**Claim 1.** *The family of functions $F$ is uniformly equicontinuous. Hence, every function in $F$ is continuous.*

To prove this, take any $h \in \Delta[\phi]$ and put $p = k * h \in K$. Since $|h(x)| \leq 1$ for each $x \in G$, we have that

$$|p(tx) - p(x)| = \left| \int (k(txy) - k(xy))h(y^{-1})\, dy \right|$$
$$\leq \int |k(txy) - k(xy)||h(y^{-1})|\, dy \leq \int |k(tz) - k(z)|\, dz,$$

for all $x, t \in G$. Clearly, the right side depends neither on $x$ nor $h$ and can be made smaller than any given positive $\varepsilon$ by requiring that $t$ be in a sufficiently small neighbourhood of the neutral element $e$ of $G$ — it suffices to note that the function $k$ is uniformly continuous.

Hence, functions in $K$ are uniformly equicontinuous with respect to the natural uniform structure on $G$. Obviously, it follows that the functions in $F$ are also uniformly equicontinuous.

**Claim 2.** *Suppose that $f_0$ is an extreme point of $F$ and that $s \in G$ and $\theta \in [0, 2\pi]$ satisfy $T(1, s, \theta) f_0(e) \neq 0$. Then there exists a positive real number $a$ such that $T(a, s, \theta) f_0 = f_0$.*

Indeed, since $F$ is $T$-invariant, it follows from the assumption about $f_0, s, \theta$ that there is a positive real number $a$ such that $T(a, s, \theta) f_0 \in F$ and that, for at least one choice of the adjoint transformation $T^*(a, s, \theta)$, we have $T^*(a, s, \theta) f_0 \in F$. By the definition of $T^*(a, s, \theta)$,

$$f_0 = \lambda T(a, s, \theta) f_0 + \mu T^*(a, s, \theta) f_0,$$

for some positive real numbers $\lambda$ and $\mu$ such that $\lambda + \mu = 1$. Since $f_0$ is an extreme point of $F$ and both functions $T(a, s, \theta) f_0$ and $T^*(a, s, \theta) f_0$ are in $F$, this implies that $T(a, s, \theta) f_0 = f_0$. Claim 2 is proved.

From now on we suppose that $f_0$ is an arbitrary extreme point of $F$.

**Claim 3.** *Let $s$ be any element of $G$ such that at least one of the numbers $f_0(e)$, $f_0(s)$, $f_0(s^{-1})$ is non-zero. Then there is $\theta \in [0, 2\pi]$ such that $[T(1, s, \theta) f_0](e) \neq 0$.*

We have

$$[T(1, s, \theta) f_0](e) = a + bz + c\bar{z},$$

where $a = 2 f_0(e)$, $b = f_0(s)$, $c = f_0(s^{-1})$, and the complex number $z = e^{2i\theta}$ corresponding to $\theta$ satisfies $|z| = 1$. It remains to show that there is a complex number $z$ such that $|z| = 1$ and $a + bz + c\bar{z} \neq 0$. Indeed, suppose that $|z| = 1$, and multiply both sides of the equation $a + bz + c\bar{z} = 0$ by $z$. We obtain $az + bz^2 + c = 0$, since $\bar{z}z = |z|^2 = 1$. By the assumption, at least one coefficient in the quadratic equation is non-zero; therefore, it can have two solutions at most. Since there are infinitely many complex numbers $z$ such that $|z| = 1$, Claim 3 is established.

**Claim 4.** *If $f_0$ is not identically zero, then for each $s \in G$ at least one of the numbers $f_0(e)$, $f_0(s)$, and $f_0(s^{-1})$ is not zero. In fact, $f_0(e) \neq 0$.*

Fix $s_0 \in G$ such that $f_0(s_0) \neq 0$. By Claim 3 applied to $s_0$, there is $\theta \in [0, 2\pi]$ such that $[T(1, s_0, \theta) f_0](e) \neq 0$. It follows from Claim 2 that

$$f_0(e) = [T(a, s_0, \theta) f_0](e) = a^2 [T(1, s_0, \theta) f_0](e) \neq 0,$$

for some positive real number $a$. Thus, $f_0(e) \neq 0$. This proves Claim 4.

We are ready to formulate and to prove the crucial fact:

**Claim 5.** *Each extreme point $f_0$ of the convex compact set $F$ is of the form $f_0 = f_0(e)\chi$, for some continuous character $\chi$ on $G$.*

If $f_0(s) = 0$ for each $s \in G$, then $f_0 = f_0(e)\chi_T$ where $\chi_T$ is the trivial character on $G$. Thus, Claim 5 trivially holds in this case. Hence, we may assume that $f_0$ is not identically zero on $G$. Then $f_0(e) \neq 0$, by Claim 4.

Take any $s \in G$. By Claim 3, we can find $\theta \in [0, 2\pi]$ such that $[T(1, s, \theta) f_0](e) \neq 0$. Since $[T(1, s, \theta) f_0](e)$ is a continuous function of $\theta$, there are two strictly distinct values $\theta_1$ and $\theta_2$ of $\theta$ such that $[T(1, s, \theta_1) f_0](e) \neq 0$ and $[T(1, s, \theta_2) f_0](e) \neq 0$.

By Claim 2, there are positive real numbers $a_1$ and $a_2$ such that $T(a_1, s, \theta_1) f_0 = f_0$ and $T(a_2, s, \theta_2) f_0 = f_0$. Put $c = f_0(e)$. By Claim 4, $c \neq 0$. Let $f_1 = c^{-1} f_0$. Clearly, $T(a_1, s, \theta_1) f_1 = f_1$ and $T(a_2, s, \theta_2) f_1 = f_1$. We also have $f_1(e) = 1$, and the function $f_1$ is continuous, since $f_0$ is continuous. Now it follows from Lemma 9.4.4 that $f_1(xs) = f_1(x) f_1(s)$ and $f_1(xs^{-1}) = f_1(x) f_1(s^{-1})$, for every $x \in G$. Since $s$ is also an arbitrary element of $G$, we conclude that $f_1$ is a continuous character $\chi$ on $G$. Now we have that $f_0(x) = c f_1(x) = f_0(e)\chi(x)$, for each $x \in G$. Claim 5 is proved.

This finishes the proof of Theorem 9.4.10. □

The main result of this section is now easy to prove.

THEOREM 9.4.11. **[F. Peter and H. Weyl]** *For every compact Abelian group $G$ and every $a \in G$ distinct from the neutral element $e$, there exists a continuous character $\chi$ on $G$ such that $\chi(a) \neq 1$.*

PROOF. By Proposition 9.4.9, there exist a positive definite continuous real-valued function $\phi$ on $G$ with $\phi(e) = 1$ and a continuous real-valued symmetric non-negative function $k$ on $G$ such that the function $f$ on $G$ defined by

$$f(x) = \int k(xy)\phi(y^{-1})\, dy$$

satisfies the condition $f(a) \neq f(e)$. By Theorem 9.4.10, the function $f$ belongs to the closure of the envelope of the set of all continuous characters on $G$ in the topology of pointwise convergence. Since $f(a) \neq f(e)$, it follows that at least one of these continuous characters must take distinct values at $a$ and $e$. □

## Exercises

9.4.a. Show that on every infinite precompact Abelian group $G$ there is a non-trivial continuous character.

9.4.b. Prove that every compact Abelian group $G$ is topologically isomorphic to a (closed) subgroup of some power $\mathbb{T}^\tau$ of the topological group $\mathbb{T}$.

9.4.c. Show that every positive constant real-valued function on a compact Abelian topological group is positive definite, but not every positive real-valued function on a compact Abelian topological group is positive definite.

9.4.d. Show that translations do not preserve, in general, positive definiteness of functions on a compact Abelian topological group.

### Problems

9.4.A. Is it true that every locally compact, $\sigma$-compact topological Abelian group can be embedded as a topological subgroup into the group $\mathbb{R}^\tau \times \mathbb{T}^\tau$, for some cardinal $\tau$?
*Hint.* The answer is "no". One can take the free Abelian group $G$ over an infinite countable set and endow $G$ with the discrete topology.

9.4.B. Show that not every infinite Abelian topological group admits a non-trivial continuous character.

9.4.C. A subset $U$ of an Abelian topological group $G$ is said to be *big* if a finite number of translations of $U$ cover $G$, that is, if there exists a finite subset $K$ of $G$ such that $KU = G$. Let $G$ be an Abelian topological group, $U$ a big open neighbourhood of the neutral element $e$ of $G$, and let $s \in G \setminus U^6$. Then there exists a continuous character $\chi$ on $G$ such that $\chi(s) \neq 1$.
*Hint.* See [**127**].

## 9.5. Pontryagin–van Kampen duality theory for discrete and for compact groups

The duality theory brings into natural correspondence compact Abelian groups and discrete Abelian groups. The definitions, constructions, and the proofs of the two main duality theorems presented in this section are especially elegant in this case.

Recall that a continuous character on a topological group $G$ is a continuous homomorphism of $G$ to the topological group $\mathbb{T}$.

We need some elementary facts concerning the topological group $\mathbb{T}$. Geometrically, $\mathbb{T}$ is the unit circumference in the complex plane, with center at 0, and the multiplication in $\mathbb{T}$ by an arbitrary $\alpha \in \mathbb{T}$ can be interpreted as the rotation of $\mathbb{T}$ by an angle represented by $\alpha$. On the other hand, $\mathbb{T}$ is topologically isomorphic to the quotient group of the topological group $\mathbb{R}$ of real numbers with respect to the discrete subgroup $\mathbb{Z}$ of all integers in $\mathbb{R}$, that is, $\mathbb{T} = \mathbb{R}/\mathbb{Z}$. We denote by $p$ the natural quotient mapping of $\mathbb{R}$ onto $\mathbb{T}$. Keeping in mind these two interpretations of $\mathbb{T}$, we can easily establish some basic properties of subgroups of $\mathbb{T}$.

PROPOSITION 9.5.1. *Every infinite closed subgroup $H$ of $\mathbb{T}$ coincides with $\mathbb{T}$.*

PROOF. Clearly, $H$ is compact. Since $H$ is infinite and compact, $H$ cannot be discrete. Therefore, for each $n \in \mathbb{N}$, we can find $h_n \in H \setminus \{1\}$ such that $|h_n - 1| \leq 1/n$, where $|z|$ denotes the modulus of $z \in \mathbb{C}$, that is, the distance between the origin and $z$ in the complex plain $\mathbb{C}$. Put $H_n = \{(h_n)^k : k \in \mathbb{Z}\}$. Then $H_n \subset H$, and it is immediate from the geometric description of $\mathbb{T}$ that, for every $z \in \mathbb{T}$, there exists $h \in H_n$ such that $|z - h| \leq 1/n$. Hence, the set $A = \bigcup\{H_n : n \in \mathbb{N}\}$ is dense in $\mathbb{T}$. Since $A \subset H$ and $H$ is closed in $\mathbb{T}$, it follows that $H = \mathbb{T}$. □

PROPOSITION 9.5.2. *Every proper closed subgroup $H$ of $\mathbb{T}$ is a finite cyclic group.*

PROOF. By Proposition 9.5.1, $H$ is finite. Let $h$ be the element of $H \setminus \{1\}$ closest to 1 with respect to the usual metric of the complex plane $\mathbb{C}$, and put $M = \{h^n : n \in \mathbb{Z}\}$. It is clear from the geometric description of $\mathbb{T}$ that for every $z \in \mathbb{T}$, there exists $y \in M$ such that $|z - y| < |h - 1|$. On the other hand, $M \subset H$, and, since $H$ is a subgroup of $\mathbb{T}$, the distance between any two distinct elements of $H$ is not less than $|h - 1|$. It follows that $H = M$. Hence, $H$ is a finite cyclic subgroup of $\mathbb{T}$, with the generating element $h$. □

Obviously, from the geometric description of $\mathbb{T}$ it follows that every finite cyclic group can be realized as a subgroup of $\mathbb{T}$.

The above statements are complemented by the next one, the proof of which is clear after the proof of Proposition 9.5.2.

PROPOSITION 9.5.3. *For each positive integer $n$, there exists exactly one subgroup of $\mathbb{T}$ consisting of exactly $n$ elements.*

COROLLARY 9.5.4. *For every finite group $K$, the group of characters on $K$ is finite.*

PROOF. This follows from Proposition 9.5.3. □

If $G$ is an Abelian compact topological group, then we define $G^*$ to be the group of continuous characters on $G$, with the discrete topology. If $G$ is an Abelian discrete group, then $G^*$ is defined to be the group of all characters on $G$, with the topology of pointwise convergence.

In both cases, the group $G^*$ is called the *Pontryagin dual* or simply the *dual group* to $G$. Clearly, finite Abelian groups are covered by both cases. However, the group of characters is finite in this case, and the two definitions give the same result.

PROPOSITION 9.5.5. *For any Abelian discrete group $G$, the dual topological group $G^*$ is compact and Hausdorff.*

PROOF. Since $G$ is discrete, $C_p(G, \mathbb{T})$ coincides with $\mathbb{T}^G$, the topological product of $|G|$ copies of $\mathbb{T}$. Since $\mathbb{T}$ is compact and Hausdorff, it follows that so is the group $C_p(G, \mathbb{T})$. On the other hand, the topological group $G^*$ coincides with $Hom_p(G, \mathbb{T})$, which is closed in $C_p(G, \mathbb{T})$ by Proposition 1.9.13. It follows that $G^*$ is compact and Hausdorff. □

PROPOSITION 9.5.6. *For any Abelian compact topological group $G$, the evaluation mapping $\Psi: G \to (G^*)^*$ is a continuous homomorphism.*

PROOF. This follows directly from Theorem 1.9.15. □

The statement in Proposition 9.5.6 plays a technical role in constructions and arguments below. One of the two main results of this section will considerably strengthen Proposition 9.5.6 — we will show that $\Psi$ is, actually, a topological isomorphism. This is the Pontryagin–van Kampen duality theorem for compact Abelian groups.

In the next two examples, we leave it to the reader to write down the argument formally.

EXAMPLE 9.5.7. Pontryagin dual group $\mathbb{Z}^*$ to the discrete group $\mathbb{Z}$ of integers is the compact group $\mathbb{T}$. Indeed, every character $h: \mathbb{Z} \to \mathbb{T}$ is completely determined by its value $h(1)$ at $1 \in \mathbb{Z}$, and $h(1)$ can be any element $z$ of $\mathbb{T}$, since 1 is a free generator of $\mathbb{Z}$. A basic open neighbourhood of the character $h$ in the topology of pointwise convergence on $\mathbb{Z}^*$ corresponds to an arbitrary open neighbourhood of $z = h(1)$ in the space $\mathbb{T}$. Thus, $\mathbb{Z}^*$ is naturally isomorphic and homeomorphic to the topological group $\mathbb{T}$. □

EXAMPLE 9.5.8. Pontryagin's dual $\mathbb{T}^*$ to the group $\mathbb{T}$ is the discrete group $\mathbb{Z}$ of the integers. Indeed, every continuous homomorphism $h: \mathbb{T} \to \mathbb{T}$ is determined by its kernel up to the natural reflection $z \to \bar{z}$ of $\mathbb{T}$. If the homomorphism $h$ is non-trivial, then its kernel is a finite subgroup of $\mathbb{T}$. However, we know after Proposition 9.5.3 that for each $n \in \mathbb{N}$, there exists exactly one subgroup of $\mathbb{T}$ consisting of $n$ elements. Hence, every character

$f: \mathbb{T} \to \mathbb{T}$ has the form $f(z) = z^n$, for some fixed $n \in \mathbb{Z}$. It follows that there exists a natural correspondence between the continuous characters on $\mathbb{T}$ and the elements of $\mathbb{Z}$. This correspondence, obviously, preserves operations. Thus, $\mathbb{T}^* = \mathbb{Z}$. □

Now it is easy to establish the following fact.

PROPOSITION 9.5.9. *The evaluation mapping of $\mathbb{T}$ to $(\mathbb{T}^*)^*$ is a topological isomorphism.*

PROOF. The evaluation mapping $\Psi$ of $\mathbb{T}$ to $(\mathbb{T}^*)^*$ is one-to-one, since the identity character separates the points of $\mathbb{T}$. Since $\mathbb{T}$ is compact, it follows that $\Psi(\mathbb{T})$ is an infinite closed subgroup of $(\mathbb{T}^*)^*$. Besides, Examples 9.5.7 and 9.5.8 show that $(\mathbb{T}^*)^*$ is topologically isomorphic to $\mathbb{T}$. Now it follows from Proposition 9.5.1 that $\Psi(\mathbb{T}) = (\mathbb{T}^*)^*$. Since $\Psi$ is continuous and $\mathbb{T}$ is compact, we conclude that the evaluation mapping $\Psi$ is a topological isomorphism. □

The above proposition means that Pontryagin–van Kampen's duality theorem holds for $\mathbb{T}$.

In Examples 9.5.7 and 9.5.8 we have established several natural isomorphisms. These isomorphisms are instrumental in uncovering and proving certain important facts. However, sometimes too much invoking natural isomorphisms, or various interpretations of the same group, might make the argument too cumbersome, introducing into it too many unnecessary details. To avoid this, we can use an approach to duality described below.

EXAMPLE 9.5.10. Put $[z, n] = z^n \in \mathbb{T}$, for each $z \in \mathbb{T}$ and each $n \in \mathbb{Z}$. Then $[\cdot, \cdot]$ is a mapping of $\mathbb{T} \times \mathbb{Z}$ to $\mathbb{T}$ satisfying the following conditions:

a) for each $z \in \mathbb{T}$, the correspondence $n \to [z, n]$, where $n$ runs over $\mathbb{Z}$, is a character on the group $\mathbb{Z}$;
b) for each character $f: \mathbb{Z} \to \mathbb{T}$ on $\mathbb{Z}$, there exists $z \in \mathbb{T}$ such that $f(n) = [z, n]$, for each $n \in \mathbb{Z}$;
c) for each $n \in \mathbb{Z}$, the correspondence $z \to [z, n]$, where $z$ runs over $\mathbb{T}$, is a continuous character on the compact topological group $\mathbb{T}$;
d) for each continuous character $f: \mathbb{T} \to \mathbb{T}$, there exists $n \in \mathbb{Z}$ such that $f(z) = [z, n]$, for each $z \in \mathbb{T}$.

The statements a)–d) are just reformulations of what we have already established in Examples 9.5.7, 9.5.8, and Proposition 9.5.9. □

From Example 9.5.10 and the definition of evaluation mapping we obtain immediately:

PROPOSITION 9.5.11. *The evaluation mapping of $\mathbb{Z}$ to $(\mathbb{Z}^*)^*$ is an isomorphism.*

EXAMPLE 9.5.12. For every finite cyclic group $K$, the dual group $K^*$ is isomorphic to the group $K$ and, clearly, $K^*$ is discrete. Note that $K$ can be interpreted as a subgroup of $\mathbb{T}$. Then every element of $K$ can be identified with a character on $K$. We leave to the reader to fill the remaining details in this argument. □

PROPOSITION 9.5.13. *The evaluation mapping of $K$ to the discrete group $(K^*)^*$ is a (topological) isomorphism, for any finite cyclic group $K$.*

PROOF. Let $a$ be the generating element of $K$. Put $n = |K|$, and fix $b \in \mathbb{T}$ such that $b^n = 1$ and $b^i \neq 1$, for each $i = 1, \dots, n-1$. From the geometric description of $\mathbb{T}$ it is clear that $F = \{b^i : 1 = 1, \ldots, n\} = \{z \in \mathbb{T} : z^n = 1\}$. Now take any character $\chi$ on $K$. Since $K$ is cyclic, $\chi$ is determined by its value $\chi(a)$ on $a$. Besides, we have $(\chi(a))^n = 1$, since $a^n$ is the neutral element of $K$. It follows that $\chi(a) = b^k$, for some $k \in \{1, \ldots, n\}$. Put $\chi_1(a^i) = b^i$, for each $i = 1, \ldots, n$. Clearly, $\chi_1$ is a character on $K$ and it follows from the above remarks that $\chi = (\chi_1)^k$. Hence $K^*$ is a cyclic group with $n$ elements. Note that the character $\chi_1$ separates distinct elements of $K$. Therefore, the evaluation mapping is one-to-one in this case. It must be also onto, since the number of elements in $K^*$ and in $(K^*)^*$ is the same as in $K$. It remains to apply Proposition 9.5.6. □

Suppose that $G$ is a compact or a discrete topological group. We will say that $G$ *satisfies Pontryagin's duality* if the evaluation mapping of $G$ to $(G^*)^*$ is a topological isomorphism. Thus, the circle group $\mathbb{T}$, each cyclic group $K$, and the discrete group $\mathbb{Z}$ satisfy Pontryagin's duality.

For the sake of brevity, the topological group $\mathbb{T}$ and all finite cyclic groups, that is, all closed subgroups of $\mathbb{T}$, will be called *elementary compact groups.*

According to a basic theorem in the theory of Abelian groups, every finitely generated Abelian group $G$ (considered without topology) is the product of a finite number of cyclic groups, finite or infinite (see [**409**, 4.2.10]). Recall that a group is said to be *finitely generated* if there exists a finite subset $A$ of $G$ such that the smallest subgroup $H$ of $G$ containing $A$ coincides with $G$.

The next statement allows us to find the group of characters for every finitely generated Abelian group. (We will need below only the part d) of this theorem.)

THEOREM 9.5.14. *Suppose that $G$ is the product of a finite collection of Abelian discrete groups $G_i$, $i = 1, \ldots, m$, where each $G_i$ satisfies Pontryagin's duality, $X_i$ is the group of characters of $G_i$, $X = \prod_{i=1}^m X_i$, and $[x, a] = x_1(a_1) \cdot \ldots \cdot x_m(a_m)$, for each $x = (x_1, \ldots, x_m) \in X$ and each $a = (a_1, \ldots, a_m) \in G$. Then $[\cdot, \cdot]$ is a Pontryagin duality between the groups $G$ and $X$, that is, the following conditions hold:*

a) *for each $x \in X$, the correspondence $a \to [x, a]$, where $a$ runs over $G$, is a character on the group $G$;*
b) *for each character $f : G \to \mathbb{T}$ on $G$, there exists $x \in X$ such that $f(a) = [x, a]$, for all $a \in G$;*
c) *for each $a \in G$, the correspondence $x \to [x, a]$, where $x$ runs over $X$, is a continuous character on the compact topological group $X$;*
d) *for each continuous character $\varphi : X \to \mathbb{T}$, there exists $a \in G$ such that $\varphi(x) = [x, a]$, for all $x \in X$.*

PROOF. Statements a) and c) are verified directly without any difficulty. For example, we have:

$$[x, a + b] = \prod_{k=1}^{m} x_k(a_k + b_k) = \prod_{k=1}^{m} x_k(a_k) \cdot \prod_{k=1}^{m} x_k(b_k) = [x, a] \cdot [x, b],$$

where the group operation in $G$ is written additively. To prove b), notice that $G_i$ can be canonically identified with the subgroup of $G$ consisting of all $a \in G$ such that each coordinate $a_j$ of $a$, except for $a_i$, is the zero-element of $G_j$. Let $x_i = f \restriction G_i$, for each $i = 1, \ldots, m$. Then $x_i$ is a character on $G_i$ and we put $x = (x_1, \ldots, x_m) \in X$. Take any

$a = (a_1, \ldots, a_m) \in G$. Under our interpretation of $G_i$, $G_i \subset G$ and $a = a_1 + \cdots + a_m$. Now we have that $[x, a] = \prod_{k=1}^{m} x_k(a_k) = \prod_{k=1}^{m} f(a_k) = f(a)$. Thus, b) is proved.

Let us prove d). Each $X_i$ can be canonically interpreted as a topological subgroup of $X$. Then $\varphi_i = \varphi \restriction X_i$ is a continuous character on $X_i$, for each $i = 1, \ldots, m$. Since $G_i$ satisfies Pontryagin's duality, there exists $a_i \in G_i$ such that $\varphi_i(x_i) = x_i(a_i)$, for each $x_i \in X_i$. Let $a = (a_1, \ldots, a_m) \in G$. Then, clearly, $\varphi(x) = [x, a]$, for each $x \in X$. □

COROLLARY 9.5.15. *Suppose that $G$ is the product of finitely many discrete Abelian groups $G_i$, $i = 1, \ldots, m$, where each $G_i$ satisfies Pontryagin's duality. Then the character group $G^*$ of $G$ is the product of the character groups $G_i{}^*$, $i = 1, \ldots, m$, and the groups $G$ and $G^*$ also satisfy Pontryagin's duality.*

In particular, from Corollary 9.5.15 and Examples 9.5.8 and 9.5.12 we obtain:

COROLLARY 9.5.16. *Suppose that $G$ is the product of finitely many elementary compact groups $G_i$, $i = 1, \ldots, m$. Then $G$ satisfies Pontryagin's duality.*

PROPOSITION 9.5.17. *Suppose that $F$ is a closed subgroup of the topological group $\mathbb{T}^n$, for some $n \in \mathbb{N}$. Then $F$ is topologically isomorphic to the product of a finite number of elementary compact groups.*

PROOF. We prove it by induction on $n$. Consider the natural projection $p$ of $\mathbb{T}^n$ onto $\mathbb{T}^{n-1}$, where $n > 1$. We can assume that $p(F)$ satisfies the conclusion of the theorem. By Proposition 9.5.1, either $K = \ker p \restriction F$ is finite or $K = \ker p = \mathbb{T}$, since $K$ is (topologically isomorphic to) a closed subgroup of $\mathbb{T}$. In both cases, it is clear that $F$ is topologically isomorphic to the product $K \times p(F)$. □

We will now make a major step towards the proof of the duality theorem in the compact versus discrete case.

PROPOSITION 9.5.18. *Suppose that $G$ is an Abelian compact group, and $f$ is a character on $G^*$, that is, a homomorphism of $G^*$ to $\mathbb{T}$. Then, for each finite set $h_1, \ldots, h_m$ of elements of $G^*$, there exists $a \in G$ such that $f(h_i) = h_i(a)$, for each $i = 1, \ldots, m$.*

PROOF. Put $F = \bigcap_{i=1}^{m} \ker h_i$. Then $F$ is a closed subgroup of $G$, and the quotient group $G/F$ is topologically isomorphic to a closed subgroup of the group $\mathbb{T}^m$ (the diagonal product of the homomorphisms $h_1, \ldots, h_m$ serves to establish such an isomorphism). Hence, $G/F$ is topologically isomorphic to the product of a finite number of elementary compact groups, by Proposition 9.5.17. Now Corollary 9.5.16 implies that $G/F$ satisfies Pontryagin's duality. Let $p\colon G \to G/F$ be the quotient homomorphism. The mapping $p^*\colon (G/F)^* \to G^*$ where $p^*(q) = q \circ p$, for each $q \in (G/F)^*$, is an isomorphism of $(G/F)^*$ onto a subgroup $M$ of $G^*$ such that $h_i \in M$, for each $i = 1, \ldots, m$. Pick $q_1, \ldots, q_m$ in $(G/F)^*$ such that $p^*(q_i) = h_i$, for $i = 1, \ldots, m$.

Clearly, $\phi = f \circ p^*$ is a character on $(G/F)^*$. Since $(G/F)^*$ satisfies Pontryagin's duality, there exists $c \in G/F$ such that $\phi(q_i) = q_i(c)$, for $i = 1, \ldots, m$. We have that $\phi(q_i) = fp^*(q_i) = f(h_i)$, for $i = 1, \ldots, m$. Pick $a \in G$ such that $p(a) = c$. Since $h_i = q_i \circ p$, it follows that $h_i(a) = q_i(p(a)) = q_i(c)$. Hence, $f(h_i) = h_i(a)$, for each $i = 1, \ldots, m$. □

We need one more elementary property of the group $\mathbb{T}$ which is obvious from the geometric interpretation of $\mathbb{T}$.

PROPOSITION 9.5.19. *There exists an open neighbourhood $V$ of the neutral element* 1 *of the group $\mathbb{T}$ such that the only subgroup of $\mathbb{T}$ contained in $V$ is $\{1\}$. In fact, one can take $V = \{e^{\pi i x} : -1/2 < x < 1/2\}$.*

Now we are ready to prove the main result of this section.

THEOREM 9.5.20. [**L. S. Pontryagin**, **E. van Kampen**] *Suppose that $G$ is a compact Abelian topological group. Then the evaluation mapping $\Psi$ of $G$ to $(G^*)^*$ is a topological isomorphism of $G$ onto $(G^*)^*$.*

PROOF. The evaluation mapping $\Psi$ of $G$ to $(G^*)^*$ is a continuous homomorphism, by Proposition 9.5.6. According to Theorem 9.4.11, there are enough of continuous characters on $G$ to separate points of $G$, so the mapping $\Psi$ is one-to-one. Since $G$ is compact, it follows that $\Psi$ is a topological isomorphism of $G$ onto a closed subgroup $B = \Psi(G)$ of $(G^*)^*$. It remains to establish that $B = (G^*)^*$.

Assume the contrary. Put $M = (G^*)^*$. Then $M/B$ is a non-trivial compact Abelian group. Therefore, applying Theorem 9.4.11 once again, we can find a non-trivial continuous character on $M/B$. It follows that there is a non-trivial continuous character $\xi$ on $M$ such that $\xi(b) = 1$, for each $b \in B$. We fix such a $\xi$. Fix also an open neighbourhood $V$ of 1 in $\mathbb{T}$ which does not contain non-trivial subgroups (see Proposition 9.5.19).

Since $\xi$ is continuous, there exists an open neighbourhood $W$ of the neutral element $e$ of $M$ such that $\xi(W) \subset V$. By the definition of the topology of pointwise convergence, there exists a finite collection $h_1, \ldots, h_m$ of elements of $G^*$ and $\varepsilon > 0$ such that the following condition is satisfied:

($\delta$) if $f \in M$ and $|f(h_i) - 1| < \varepsilon$ for each $i = 1, \ldots, m$, then $f \in W$.

Put $L = \{f \in M : f(h_i) = 1 \text{ for all } i = 1, \ldots, m\}$. Clearly, $L$ is a subgroup of $M$, and $L \subset W$. Hence, $\xi(L) \subset V$. Since $\xi$ is a homomorphism, $\xi(L)$ is a subgroup of $\mathbb{T}$. Thus, $\xi(L)$ is a subgroup of $\mathbb{T}$ contained in $V$. It follows from the choice of $V$ that $\xi(L) = \{1\}$.

Take now any $f \in M$. We are going to show that $\xi(f) = 1$. By Proposition 9.5.18, there exists $a \in G$ such that $f(h_i) = h_i(a)$, for each $i = 1, \ldots, m$. By the definition of the evaluation mapping $\Psi$, for $g = \Psi(a)$ we also have $g(h_i) = h_i(a)$, for each $i = 1, \ldots, m$. Therefore, $(fg^{-1})(h_i) = 1$, for $i = 1, \ldots, m$, that is, $fg^{-1} \in L$. Hence, $\xi(fg^{-1}) = 1$ and $\xi(f) = \xi(g)$. Since $g = \Psi(a) \in \Psi(G) = B$, we have that $\xi(g) = 1$, by the choice of $\xi$. It follows that $\xi(f) = 1$. This is a contradiction, since the character $\xi$ is non-trivial, and $f$ is arbitrary element of $M$. □

Finally, let us consider the case of discrete Abelian groups from the point of view of Pontryagin's duality. We will rely upon Corollary 1.1.8: For every discrete Abelian group $G$ and every element $a \in G$ distinct from the neutral element, there exists a character $f : G \to \mathbb{T}$ such that $f(a) \neq 1$. In fact, we also need a slightly different statement:

COROLLARY 9.5.21. *For every discrete Abelian group $G$, and every proper subgroup $H$ of $G$, there exists a non-trivial character $g$ on $G$ such that $g(h) = 1$, for every $h \in H$.*

PROOF. By Corollary 1.1.8, there exists a non-trivial character $f$ on the quotient group $G/H$. Put $g = f \circ p$, where $p$ is the quotient homomorphism of $G$ onto $G/H$. Clearly, $g$ is a character on $G$ we are looking for. □

THEOREM 9.5.22. *For any discrete Abelian group $G$, the evaluation mapping $\Psi\colon G \to (G^*)^*$ is an isomorphism.*

PROOF. From Corollary 1.1.8 it follows that $\Psi$ is one-to-one. We know that $\Psi$ is a homomorphism. It remains to show that $\Psi$ is onto. Assume the contrary, and put $H = \Psi(G)$. Then $H$ is a proper subgroup of the discrete Abelian group $(G^*)^*$ and, by Corollary 9.5.21, there exists a non-trivial character $f$ on $(G^*)^*$ such that $f(h) = 1$, for every $h \in H$.

By Proposition 9.5.5, the Abelian group $G^*$ is compact. Therefore, by Theorem 9.5.20, the evaluation mapping of $G^*$ to its second dual group is onto. Hence, there exists $\chi \in G^*$ such that $f(y) = y(\chi)$, for each $y \in (G^*)^*$. Since the character $f$ is non-trivial, it follows that $\chi$ is a non-trivial character on $G$. Therefore, there exists $a \in G$ such that $\chi(a) \neq 1$. Then, for $h_0 = \Psi(a)$ we have that $h_0(\chi) = \chi(a) \neq 1$. On the other hand, $f(h_0) = 1$, since $h_0 \in H = \Psi(G)$. Since $h_0(\chi) = f(h_0)$, it follows that $h_0(\chi) = 1$, a contradiction. □

We can reformulate the last result as follows:

COROLLARY 9.5.23. *Every discrete Abelian group $G$ satisfies Pontryagin's duality.*

## Exercises

9.5.a. Using the "algebraic" definition of the topological group $\mathbb{T}$ as the quotient group $\mathbb{T} = \mathbb{R}/\mathbb{Z}$ of the group $\mathbb{R}$, with the usual topology, prove the main properties of $\mathbb{T}$ and of its subgroups, established in this section on the basis of the "geometric" definition of $\mathbb{T}$. In particular, prove in this way Propositions 9.5.1, 9.5.2, 9.5.3, and 9.5.19.

9.5.b. Complete the arguments given in Examples 9.5.7, 9.5.8, and 9.5.12.

9.5.c. Suppose that $G$ is a compact Abelian group. Let $H$ be the group of continuous characters on $G$ taken with the topology of pointwise convergence. Is $H$ discrete?

9.5.d. Suppose that $G$ is a compact Abelian group and $H$ is the group of continuous characters on $G$ taken with the topology of pointwise convergence. Denote by $H'$ the group of continuous characters on $H$, with the topology of pointwise convergence. Is $H'$ topologically isomorphic to $G$?

## Problems

9.5.A. For a prime number $p$, let $\mathbb{Z}(p^\infty)$ be the *quasicyclic subgroup* of $\mathbb{T}$ of all elements $z$ satisfying $z^{p^n} = 1$, for some integer $n \geq 0$.

(a) Prove that Pontryagin's dual group $(\mathbb{Z}_p)^*$ is topologically isomorphic to the discrete group $\mathbb{Z}(p^\infty)$, where $\mathbb{Z}_p$ is the compact group of $p$-adic integers (see Examples 1.1.10 and 1.3.16).

(b) Calculate the dual group $(\mathbb{Z}_r)^*$, for an arbitrary integer $r > 1$.

## 9.6. Some applications of Pontryagin–van Kampen's duality

Here we study the topological and algebraic properties of compact Abelian topological groups by methods of Pontryagin–van Kampen's duality theory. The next technical result is sometimes very helpful.

COROLLARY 9.6.1. *Suppose that $G$ is an Abelian group, compact or discrete, and $H$ a closed subgroup of $G^*$ separating elements of $G$. Then $H = G^*$.*

PROOF. Assume the contrary. Since the quotient group $G^*/H$ is either compact or discrete, there exists a non-trivial character $f$ on $G^*$ such that $f(h) = 1$, for each $h \in H$. By Theorems 9.5.20 and 9.5.22, there exists $a \in G$ such that $f(y) = y(a)$, for every $y \in G^*$. Note that $a$ is not the neutral element of $G$, since $f$ is non-trivial. Since $H$ separates elements of $G$, there exists $h \in H$ such that $h(a) \neq 1$. On the other hand, by the choice of $f$ and of $a$ we have $h(a) = f(h) = 1$, since $h \in H$, a contradiction. □

PROPOSITION 9.6.2. *Let $G$ be an Abelian topological group, compact or discrete, and $H$ a closed subgroup of $G$. Then the natural restriction mapping $\phi\colon G^* \to H^*$ given by the rule $\phi(f) = f\restriction H$, for each $f \in G^*$, is a continuous open homomorphism of the topological group $G^*$ onto the topological group $H^*$.*

PROOF. Clearly, $\phi$ is a continuous homomorphism. Since $G^*$ and $H^*$ are both either discrete or compact, the mapping $\phi$ must be closed, and therefore, quotient. Hence, $\phi$ is open. It remains to show that $\phi$ is onto. The subgroup $\phi(G^*)$ separates elements of $H$, since $G^*$ separates elements of $G$, and $\phi(G^*)$ is closed in $H^*$, since the mapping $\phi$ is closed. It follows from Corollary 9.6.1 that $\phi(G^*) = H^*$. □

Sometimes it is important to know whether a character of a subgroup can be extended to a continuous character on the whole group. The next statement is, obviously, a corollary to Proposition 9.6.2.

THEOREM 9.6.3. *Suppose that $G$ is an Abelian group, compact or discrete, $H$ is a closed subgroup of $G$, and $f$ is a continuous character on $H$. Then there exists a continuous character $g$ on $G$ such that $g\restriction H = f$.*

The essence of Pontryagin–van Kampen's duality theory is to establish the relationships between the structure and properties of a (compact or discrete) topological group $G$ and the structure and properties of the group of characters $G^*$. The theory is rich and contains many deep results. Here we present only a sample of more elementary statements of this kind, just to provide a glimpse of the theory.

THEOREM 9.6.4. *For any finite Abelian group $G$, the group $G^*$ is isomorphic to $G$.*

PROOF. Every finite Abelian group is the product of a finite number of cyclic groups. According to Proposition 9.5.13, the group of characters of a finite cyclic group $K$ is isomorphic to $K$ (see also Example 9.5.12). Now it follows from Corollary 9.5.15 that $K^*$ is isomorphic to $K$. □

In a more general case of finitely generated Abelian groups, we have:

THEOREM 9.6.5. *For every finitely generated discrete Abelian group $G$, the dual group $G^*$ is the product of a finite family of groups each of which is either a finite cyclic group or the circle group $\mathbb{T}$.*

PROOF. Every finitely generated discrete Abelian group is the product of a finite collection of groups, each of which is either a finite cyclic group or the discrete group $\mathbb{Z}$. Applying Proposition 9.5.13 and Corollary 9.5.15, and taking into account that $\mathbb{Z}^*$ is topologically isomorphic to $\mathbb{T}$, we arrive at the desired conclusion. □

Our next result is of a slightly different nature. Instead of comparing the exact structure of $G$ with that of $G^*$, we show that, under Pontryagin's duality, to a certain basic property of $G$ corresponds a very different basic property of the dual group $G^*$.

THEOREM 9.6.6. *The weight of an infinite compact Abelian group $G$ coincides with the cardinality of $G^*$.*

PROOF. Let $e$ be the neutral element of $G$, $X = G \setminus \{e\}$, and $F_f = X \cap \ker f$, for each $f \in G^*$. Then $\xi = \{F_f : f \in G^*\}$ is a family of closed subsets of $X$ such that $\bigcap \xi = \varnothing$, since continuous characters on $G$ separate elements of $G$. Let $\tau$ be the weight of $G$. Then there exists a subfamily $\eta$ of $\xi$ such that $\bigcap \eta = \varnothing$ and $|\eta| \leq \tau$. In other words, there exists a subset $H$ of $G^*$ such that $|H| \leq \tau$ and $\bigcap\{\ker f : f \in H\} = \{e\}$. Clearly, we may assume that $H$ is a subgroup of the group $G^*$. Since $H$ separates elements of $G$, it follows from Corollary 9.6.1 that $H = G^*$. Hence, $|G^*| \leq \tau$.

Conversely, if $|G^*| = \tau$ for some $\tau \geq \omega$, then the space $(G^*)^*$ is homeomorphic to a subspace of $\mathbb{T}^{G^*} \cong \mathbb{T}^\tau$. Since $G$ is homeomorphic to $(G^*)^*$, it follows that $w(G) \leq \tau$. □

COROLLARY 9.6.7. *A compact Abelian group $G$ is metrizable if and only if its dual group $G^*$ is countable.*

Applying duality theorems for compact and discrete Abelian groups, we derive the following result immediately from Corollary 9.6.7:

COROLLARY 9.6.8. *A discrete Abelian group $G$ is countable if and only if its dual $G^*$ is a second-countable group.*

We need one more elementary property of the group $\mathbb{T}$:

PROPOSITION 9.6.9. *The only closed connected subgroups of the topological group $\mathbb{T}$ are $\{1\}$ and $\mathbb{T}$ itself.*

PROOF. This follows immediately from Proposition 9.5.1. □

LEMMA 9.6.10. *Suppose that $f$ is a non-trivial continuous character on a compact Abelian group $G$. Then $f$ is of infinite order in $G^*$ if and only if $f(G)$ is a connected subspace of $\mathbb{T}$.*

PROOF. Suppose that $f(G)$ is connected. Since $f(G)$ is a closed subgroup of $\mathbb{T}$ (recall that $G$ is compact), it follows from Proposition 9.6.9 that $f(G) = \mathbb{T}$. Therefore, $f^n(G) = \mathbb{T}$, for each positive integer $n$, since $\{z^n : z \in \mathbb{T}\} = \mathbb{T}$ for every such $n$. It follows that the homomorphism $f^n$ is non-trivial. Thus, $f$ is of infinite order.

Suppose now that $f \in G^*$ and $n \in \mathbb{N}$ satisfy the condition that $f^n(x) = 1$, for each $x \in G$, and that $f$ is distinct from the neutral element of $G^*$. Let $K_n = \{z \in \mathbb{T} : z^n = 1\}$ and $H = f(G)$. Clearly, $H \subset K_n$ and $H$ contains 1 and at least one more element of $\mathbb{T}$. Since $K_n$ contains exactly $n$ elements, we conclude that $H$ is a finite space containing more than one element. It follows that $H = f(G)$ is disconnected. □

THEOREM 9.6.11. [**L. S. Pontryagin**] *A compact Abelian group $G$ is connected if and only if the dual group $G^*$ is torsion-free.*

PROOF. Suppose that $G$ is connected, and take any non-trivial continuous character $f$ on $G$. Then $f(G)$ is connected and, therefore, $f$ is of infinite order in $G^*$, by Lemma 9.6.10.

Suppose now that $G$ is not connected. Then there exists a proper open and closed subset $U$ of $G$ containing the neutral element $e$ of $G$. By Proposition 3.1.8, there exists an open and closed subgroup $H$ of $G$ such that $H \subset U$. The quotient group $G/H$ is discrete, compact, and contains more than one element. Therefore, there exists a non-trivial continuous character $\phi$ on $G/H$. Put $f = \phi \circ p$, where $p\colon G \to G/H$ is the natural quotient homomorphism. Then $f$ is a non-trivial continuous character on $G$, and $f(G) = \phi(G/H)$ is a finite subset of $\mathbb{T}$, containing more than one element. Therefore, $f(G)$ is disconnected. It follows from Lemma 9.6.10 that $f$ is of finite order in $G^*$. □

Recall that a topological space $X$ is said to be *totally disconnected* if every connected subspace of $X$ is trivial (that is, contains at most than one point). According to Proposition 3.1.7, a compact Hausdorff space $X$ is totally disconnected if and only if $X$ is zero-dimensional, that is, has a base consisting of open and closed sets.

THEOREM 9.6.12. *A compact Abelian group $G$ is totally disconnected if and only if $G^*$ is a torsion group.*

PROOF. Suppose that $G$ is totally disconnected. Then $G$ is zero-dimensional. Take any non-trivial continuous character $f$ on $G$, and put $F = f(G)$ and $H = \ker f$. The mapping $f$ is closed, since $G$ is compact. Therefore, the subgroup $F$ of $\mathbb{T}$ is topologically isomorphic to the quotient group $G/H$, and the mapping $f$ of $G$ onto the subspace $F$ of $\mathbb{T}$ is also open. Since, by Theorem 3.1.14, the quotient group of any totally disconnected compact group is always zero-dimensional, it follows that $F$ is zero-dimensional and, hence, $F \neq \mathbb{T}$. Since $f$ is non-trivial, it follows that $|F| > 1$. Therefore, $F$ is disconnected and, by Lemma 9.6.10, $f$ is of finite order.

Suppose now that $G$ is not totally disconnected. Then there exists a connected subset $A$ of $G$ such that $|A| > 1$. Clearly, we can assume that the neutral element $e$ of $G$ is in $A$. Fix $a \in A$ distinct from $e$, and take a continuous character $f$ on $G$ such that $f(a) \neq 1$. Then $B = f(A)$ is a connected subset of $\mathbb{T}$ containing more than one element. Also, $f(G)$ is a closed subgroup of $\mathbb{T}$ and $B \subset f(G)$. Hence, $f(G)$ is an infinite closed subgroup of $\mathbb{T}$ and, by Proposition 9.6.9, $f(G) = \mathbb{T}$, that is, $f(G)$ is connected. Therefore, by Lemma 9.6.10, $f$ is of infinite order in $G^*$. □

The basic duality theorems for compact and discrete Abelian groups allow to turn around Theorems 9.6.11 and 9.6.12. In this way we obtain the following statements:

COROLLARY 9.6.13. *A discrete Abelian group $G$ has no elements of finite order distinct from the identity if and only if the dual group $G^*$ is connected.*

COROLLARY 9.6.14. *A discrete Abelian group $G$ is torsion if and only if the dual group $G^*$ is totally disconnected (equivalently, zero-dimensional).*

Theorem 9.6.11 can be used to characterize connectedness of compact Abelian groups in terms of their purely algebraic property, namely, divisibility.

THEOREM 9.6.15. *A compact Abelian group $G$ is divisible if and only if $G$ is connected.*

PROOF. Suppose that $G$ is divisible, and take any non-trivial continuous character $f$ on $G$. By Theorem 9.6.11, the connectedness of $G$ will follow if we establish that $f$ is of infinite order in $G^*$.

Fix $a \in G$ such that $f(a) \neq 1$. Now take an arbitrary integer $n > 0$. Since $G$ is divisible, there exists $b \in G$ such that $b^n = a$. Then $f^n(b) = (f(b))^n = f(b^n) = f(a) \neq 1$. Hence, $f^n(b) \neq 1$; it follows that $f^n$ is not the neutral element of $G^*$. Since this is true for every positive integer $n$, we conclude that $f$ is of infinite order. Therefore, $G$ is connected.

Suppose now that $G$ is not divisible, and fix $a \in G$ and $n \in \mathbb{N}$ such that $y^n \neq a$, for each $y \in G$. Put $F = \{y^n : y \in G\}$. Clearly, $F$ is a subgroup of $G$, and $a \notin F$. Also, $F$ is a compact subspace of $G$, since it is a continuous image of a closed subspace (the diagonal) of the compact product space $G^n$. Therefore, $F$ is closed in $G$. Hence, there exists a continuous character $f$ on $G$ such that $f(a) \neq 1$ and $f(y) = 1$, for every $y \in F$. By Theorem 9.6.11, to show that $G$ is disconnected, it is enough to establish that $f^n$ is the neutral element of $G^*$. Take any $x \in G$. Then $x^n \in F$. Therefore, $f^n(x) = (f(x))^n = f(x^n) = 1$. Since this is true for each $x \in G$, we conclude that $f^n$ is the neutral element of $G^*$. □

Below we prove, on the basis of duality theory, an important result on quotients of compact Abelian groups. In the next statement we introduce some terminology and notation convenient to use in the argument.

PROPOSITION 9.6.16. *Let $G$ be a compact or discrete Abelian group, and $H$ a closed subgroup of $G$. Put $A(G^*, H) = \{f \in G^* : f(H) = \{1\}\}$. Then $A(G^*, H)$ is a closed subgroup of $G^*$ called the annihilator of $H$ in $G^*$, and $H^*$ is topologically isomorphic to the quotient group $G^*/A(G^*, H)$.*

PROOF. Consider the restriction mapping $r\colon G^* \to H^*$ given by the rule $r(f) = f\restriction H$, for each $f \in G^*$. Then $r$ is a continuous homomorphism of $G^*$ onto $H^*$. Clearly, $\ker r = A(G^*, H)$, so the required conclusion is just a reformulation, in the new terms, of Proposition 9.6.2. □

We also need the next easy to prove statement.

PROPOSITION 9.6.17. *Let $G$ be a discrete Abelian group, and $U$ an open neighbourhood of the neutral element in the dual group $G^*$. Then there exists a finitely generated subgroup $H$ of $G$ such that $A(G^*, H) \subset U$.*

PROOF. Obviously, we can assume that $U$ is a standard open neighbourhood of the neutral element in $G^*$, in the topology of pointwise convergence, of the form

$$U = \{f \in G^* : |f(x_i) - 1| < \varepsilon \text{ for each } i = 1, \dots, k\},$$

for some $x_1, \dots, x_k \in G$ and some $\varepsilon > 0$. Take the subgroup $H$ of $G$ generated by $x_1, \dots, x_k \in G$, and let $f \in A(G^*, H)$. Then, clearly, $f(x_i) = 1$ for each $i \leq k$. Hence, $f \in U$ and $A(G^*, H) \subset U$. □

THEOREM 9.6.18. *Let $F$ be a compact Abelian group and $U$ an open neighbourhood of the neutral element $e$ in $F$. Then there exists a closed subgroup $E$ of $F$ such that $E \subset U$ and the quotient group $F/E$ is topologically isomorphic to $\mathbb{T}^n \times K$, for some non-negative integer $n$ and some finite group $K$.*

PROOF. Due to the duality, we can assume that $F = G^*$, for some discrete Abelian group $G$. Then $F^*$ can be identified with $G$ by means of the evaluation mapping. By Proposition 9.6.17, there exists a finitely generated subgroup $H$ of $G$ such that $A(G^*, H) = A(F, H) \subset U$. According to Proposition 9.6.16, $G^*/A(G^*, H) = F/A(F, H) = H^*$. Since $H$ is a finitely generated discrete group, from Theorem 9.6.5 it follows that $H^*$ is topologically isomorphic to $\mathbb{T}^n \times K$, for some non-negative integer $n$ and some finite group $K$. Now for $E = A(F, H)$, we have that $E \subset U$, $E$ is a closed subgroup of $F$, and $F/E \cong \mathbb{T}^n \times K$. □

We recall that the *torsion-free rank* of an Abelian group $G$ is *finite* and is equal to $m \in \omega$, if $m$ is the smallest element of $\omega$ such that the number of elements in every independent subset of $G$ does not exceed $m$ (see Section 7.10).

There is an interesting connection between the torsion-free rank of a discrete Abelian group and the dimension of its dual group. We describe below the simplest case, leaving more complete results for exercises.

LEMMA 9.6.19. *Suppose that $G$ is a discrete Abelian group of finite torsion-free rank, and $F = \{x_1, \dots, x_n\}$ is a maximal independent subset of $G$. Put $V = \{z \in \mathbb{T} : |z - 1| \leq \sqrt{2}\}$. Then there exists a homeomorphism $\phi$ of $V^F$ onto a subspace of $G^*$ such that $h = \phi(h)\restriction F$, for each $h \in V^F$ (we identify elements of $V^F$ with functions from $F$ to $V$, and consider the topology of pointwise convergence on $V^F$). This homeomorphism can be selected in such a way that if $h(F) = \{1\}$, then $\phi(h)$ is the neutral element of $G^*$.*

PROOF. With each function $h\colon F \to V$ we associate $f_h\colon G \to \mathbb{T}$ as follows. Take any $x \in G$. Then, by the maximality of $F$, the family $\{x_1, \dots, x_n, x\}$ is not independent. Therefore, since $F$ is independent, we have $mx = m_1x_1 + \cdots + m_nx_n$, for some integers $m, m_1, \dots, m_n$, where $m \neq 0$, and where $m_1, \dots, m_n$ are uniquely determined by $m$ (though, in general, $m$ cannot be chosen to be any non-zero integer). Put

$$f_h(x) = (h(x_1))^{m_1/m} \cdot \ldots \cdot (h(x_n))^{m_n/m},$$

where the exponent $z^{k/m}$, for each $z = e^{2\pi i x}$ with $|x| \leq 1/2$, is defined as $z^{k/m} = e^{2\pi k i x/m}$. It is straightforward to verify that this definition is correct and $f_h$ is a homomorphism of $G$ to $\mathbb{T}$. Thus, $f_h \in G^*$, and the rule $\phi(h) = f_h$ describes a mapping $\phi$ of the $n$-dimensional cube $V^F$ to $G^*$. Since both $G^*$ and $V^F$ have the topology of pointwise convergence (on $G$ and $F$, respectively), it is easy to see that $\phi$ is a continuous mapping of $V^F$ onto a subspace of $G^*$. Clearly, $(\phi(h))\restriction F = h$, for each $h \in V^F$. Since the restriction mapping $r\colon G^* \to V^F$ (which associates with a character on $G$ its restriction to $F$) is continuous and $r \circ \phi = id_{V^F}$, we conclude that $\phi$ is a homeomorphism of $V^F$ onto a subspace of $G^*$. □

Below we will use the following general statement on the group of characters of a quotient group.

PROPOSITION 9.6.20. *Suppose that $G$ is a discrete Abelian group, and $H$ is a subgroup of $G$. Then $(G/H)^*$ is topologically isomorphic to the subgroup $A(G^*, H)$ of $G^*$.*

PROOF. Let $p$ be the quotient homomorphism of $G$ onto $G/H$. With an arbitrary $f \in (G/H)^*$ we associate a homomorphism $p \circ f$ of $G$ to $\mathbb{T}$. Clearly, $(p \circ f)\restriction H \in A(G^*, H)$. Put $\phi(f) = p \circ f$. Then $\phi$ is a monomorphism of $(G/H)^*$ to the subgroup $A(G^*, H)$ of $G^*$. Since both $G^*$ and $(G/H)^*$ are taken with the topology of pointwise convergence, $\phi$

is a homeomorphism of $(G/H)^*$ onto the subspace $\phi((G/H)^*)$ of $G^*$. It remains to show that $A(G^*, H) \subset \phi((G/H)^*)$. Take any $q \in A(G^*, H)$. Then $q(H) = \{1\}$ and $q$ is a homomorphism of $G$ to $\mathbb{T}$. It follows that $q$ can be factored through $p$, that is, $q = f' \circ p$, for some homomorphism $f' \colon G/H \to \mathbb{T}$. Then $q = \phi(f') \in \phi((G/H)^*)$. □

We also need the following lemma.

LEMMA 9.6.21. *Suppose that $G$ is a discrete Abelian group and $x_1, \ldots, x_m$ is a finite independent subset of $G$. Then there exists a subgroup $H$ of $G$ such that $p_H(x_1), \ldots, p_H(x_m)$ is a maximal linearly independent system of elements of the group $G/H$, where $p_H$ is the natural quotient mapping of $G$ onto $G/H$.*

PROOF. Denote by $\mathcal{H}$ the family of all subgroups $H$ of $G$ such that the system $p_H(x_1), \ldots, p_H(x_m)$ is linearly independent in the group $G/H$. We partially order $\mathcal{H}$ by inclusion and observe that the union of any chain in $\mathcal{H}$ is, obviously, an element of $\mathcal{H}$. Therefore, Zorn's lemma is applicable, and there exists a maximal element $H$ in $\mathcal{H}$. Then, by a standard argument, $p_H(x_1), \ldots, p_H(x_m)$ is a maximal linearly independent system of elements of $G/H$. □

THEOREM 9.6.22. *Suppose that $G$ is a discrete Abelian group such that the dual compact group $G^*$ is finite-dimensional. Then the torsion-free rank of $G$ is finite and does not exceed $m = \dim(G^*)$.*

PROOF. Take any finite linearly independent system $F = \{x_1, \ldots, x_n\}$ of elements of $G$. We have to show that $n \leq m$. By Lemma 9.6.21, there exists a subgroup $H$ of $G$ such that $p_H(x_1), \ldots, p_H(x_n)$ is a maximal linearly independent system of elements of the group $G/H$, where $p_H$ is the natural quotient mapping of $G$ onto $G/H$.

By Proposition 9.6.20, $(G/H)^*$ is topologically isomorphic to the closed subgroup $A(G^*, H)$ of $G^*$. Hence, since $G^*$ is compact (and normal), we have that $\dim(G/H)^* \leq \dim(G^*) = m$ [**165**, Theorem 7.1.8], and it remains to show that $n \leq \dim(G/H)^*$. Thus, we have reduced the argument about the system $F$ to the case when $F$ is a maximal linearly independent system of elements of the group $G$. So, to simplify the notation, we now make this assumption.

Put $V = \{z \in \mathbb{T} : |z - 1| \leq \sqrt{2}\}$. Clearly, $V$ is homeomorphic to the closed interval $[0, 1]$, and the dimension of $V^F$ is $n$, by [**165**, Coro. 7.3.20]. It follows from Lemma 9.6.19 that there exists a homeomorphism $\phi$ of $V^F$ onto a closed subspace of $G^*$. Hence, $\dim(G^*) \geq \dim(V^F) = |F| = n$, that is, $m \geq n$. □

THEOREM 9.6.23. *Let $G$ be a discrete Abelian group of finite torsion-free rank, and let the dual group $G^*$ be locally connected. Then the group $G$ is finitely generated.*

PROOF. Fix a maximal linearly independent system of elements of $G$ consisting of $m$ elements, say $F = \{x_1, \ldots, m\}$. Let $H$ be the subgroup generated by $F$, and $\Phi = A(G^*, H)$. We claim that the quotient group $G/H$ is finite.

Indeed, assume that $G/H$ is infinite. Clearly, every element of $G/H$ is of finite order. Therefore, the dual group $(G/H)^*$ is totally disconnected, by Corollary 9.6.14. However, $(G/H)^* = A(G^*, H) = \Phi$, by Proposition 9.6.20. Therefore, $\Phi$ is totally disconnected. Put $V = \{z \in \mathbb{T} : |z - 1| \leq \sqrt{2}\}$. By Lemma 9.6.19, there exists a homeomorphism $\phi$ of $V^F$ onto a subspace of $G^*$ such that $h = \phi(h)\restriction F$, for each $h \in V^F$. Put $W = \{f \in G^* : f(F) \subset V\}$. Then $W$ is a neighbourhood of the neutral element in $G^*$.

Let us define a mapping $r$ of $W$ to the subspace $A(G^*, H)$ of $G^*$ as follows. For any $f \in W$, put $r(f) = f \cdot (\phi(f\restriction F))^{-1}$. Clearly, $r$ is continuous. From $f \in W$ it follows that $f\restriction F \in V^F$ and, therefore, $\phi(f\restriction F)$ is defined. Since $h = \phi(h)\restriction F$, for each $h \in V^F$, we have that $(r(f))(x_i) = 1$, for $i = 1, \dots, m$. Hence, $r(f) \in A(G^*, H) = \Phi$, for each $f \in W$. Clearly, $\Phi \subset W$. Take any $g \in \Phi$. Then $r(g) = g \cdot (\phi(g\restriction F))^{-1} = g$, since $\phi(g\restriction F)$ is the neutral element of $G^*$ (see Lemma 9.6.19). Thus, the restriction of $r$ to $\Phi$ is the identity mapping of $\Phi$ onto itself. (In fact, we have shown that $r$ is a continuous retraction of $W$ onto its subspace $\Phi$). Since $G^*$ is locally connected, we can take an open connected neighbourhood $U$ of the neutral element $e$ of $G^*$ such that $e \in U \subset W$. Corollary 9.6.7 implies that $\Phi$ is infinite, since $G/H$ was assumed to be infinite. Since $\Phi$ is also compact, it follows that $\Phi$ is not discrete. Therefore, $e$ is non-isolated in $\Phi$. Hence $|U \cap \Phi| \geq 2$. Since $r\restriction\Phi$ is one-to-one, we have $|r(U)| \geq 2$. However, $r(U)$ is connected and $r(U) \subset \Phi$. Thus, $r(U)$ is a non-trivial connected subset of the totally disconnected space $\Phi$, a contradiction. Hence, the group $G/H$ is finite. Since $H$ is finitely generated, it follows that so is $G$. □

Now we can prove an interesting result on the structure of compact locally connected groups.

THEOREM 9.6.24. [**L. S. Pontryagin**] *Every compact finite-dimensional locally connected Abelian group $M$ is topologically isomorphic to $\mathbb{T}^n \times K$, where $n = \dim(M)$ and $K$ is a finite group.*

PROOF. By Theorem 9.5.20, we can represent $M$ as the dual group $G^*$ to some discrete Abelian group $G$. Since $M = G^*$ is finite-dimensional, it follows from Theorem 9.6.22 that $G$ is of finite torsion-free rank. Since $G^*$ is locally connected, Theorem 9.6.23 implies that the group $G$ is finitely generated. Hence, $G$ is the product of a finite number of cyclic groups, which implies, by Theorem 9.6.5, that $M = G^*$ is topologically isomorphic to $\mathbb{T}^n \times K$, where $n$ is a natural number and $K$ is a finite group. Clearly, $n$ must be equal to $\dim(M)$. □

As another application of the duality theory, we present below a somewhat unexpected result, Theorem 9.6.26, which requires the following generalization of Theorem 9.5.14:

PROPOSITION 9.6.25. *Let $G = \oplus_{i\in I} G_i$ be a direct sum of Abelian groups, and suppose that $G$ carries the discrete topology. Then the dual group $G^*$ is topologically isomorphic to the topological product $X = \prod_{i\in I} G_i^*$ of dual groups.*

PROOF. For every $\chi = (\chi_i)_{i\in I}$ in $X$ and every $g \in G$, let

$$[\chi, g] = \prod_{i\in I} \chi_i(g_i), \tag{9.12}$$

where $g = \sum_{i\in I} g_i$ and $g_i \in G_i$, for each $i \in I$. Since the sum $g = \sum_{i\in I} g_i$ contains finitely many elements, for each $g \in G$, so does the product on the right hand side in (9.12). It is clear that the function $[\chi, \cdot]$ is a character on $G$, for each $\chi \in X$. Given any $j \in I$, denote by $H_j$ the subgroup of $G$ whose elements $h = \sum_{i\in I} g_i$ satisfy $g_i = 0_{G_i}$ for each $i \neq j$. Clearly, $H_i$ is just a copy of $G_i$. For every character $\chi$ on $G$, let $\chi_j$ be the restriction of $\chi$ to $H_j$, $j \in I$. Then $\chi_j$ is a character of $H_j$ and $\varphi(\chi) = (\chi_j)_{j\in I}$ is an element of $X$. An easy verification shows that $\chi(g) = [\varphi(\chi), g]$ for each $g \in G$, that is, $\chi = [\varphi(\chi), \cdot]$. Evidently, the mapping $Is\colon (\chi_i)_{i\in I} \mapsto [\varphi(\chi), \cdot]$ is an algebraic isomorphism of $X = \prod_{i\in I}(H_i)^*$ onto $G^*$, where each group $H_i \cong G_i$ carries the discrete topology. It also follows from the definition of the product topology of $X$ and the topology of the dual group $G^*$ that the isomorphism $Is$ is continuous. Since the group $X$ is compact, we conclude that $Is$ is a topological isomorphism. □

A torsion Abelian group $G$ is called *bounded torsion* if there exists a positive integer $m$ such that $mx = 0_G$, for each $x \in G$. The minimal $m$ with this property is called the *exponent* of $G$. The following two facts clarify the topological and algebraic structure of compact Abelian groups of a prime exponent.

THEOREM 9.6.26. *Let $G$ be a compact Abelian topological group of a prime exponent $p$. Then $G$ is topologically isomorphic to the product group $\mathbb{Z}(p)^\kappa$, for some cardinal $\kappa \geq 0$.*

PROOF. The dual group $G^*$ is discrete and every element of $G^*$ distinct from zero has order $p$. Considering $G^*$ as a linear vector space over the field $\mathbb{Z}(p)$ and taking a Hamel basis of $G^*$, we conclude that $G^*$ is the direct sum of $\kappa$ copies of the group $\mathbb{Z}(p)$ for some cardinal $\kappa$, say, $G^* \cong \oplus_{\alpha<\kappa}\mathbb{Z}(p)_\alpha$. Since, by Proposition 9.6.25, the dual group to a direct sum of groups is the topological product of dual groups to summands, it follows that $G$ is topologically isomorphic to the product of $\kappa$ copies of the dual group $(\mathbb{Z}(p))^* = \mathbb{Z}(p)$. □

COROLLARY 9.6.27. *If $G$ is a compact Abelian group, then for every prime $p$, the subgroup $G[p] = \{x \in G : px = 0\}$ of $G$ is topologically isomorphic to the group $\mathbb{Z}(p)^\kappa$, for some cardinal $\kappa$.*

PROOF. For a prime $p$, consider the homomorphism $\varphi_p$ of $G$ to $G$ defined by $\varphi(x) = px$ for each $x \in G$. It is clear that $\varphi_p$ is continuous and the kernel of $\varphi_p$ coincides with $G[p]$. Therefore, $G[p]$ is a closed subgroup of $G$. We conclude that $G[p]$ is a compact Abelian group of the prime exponent $p$, so one applies Theorem 9.6.26 to finish the argument. □

Let us characterize the structure of compact Abelian torsion groups. First we recall the Prüfer–Baer theorem on bounded torsion groups (see [**409**, Theorem 4.3.5]):

THEOREM 9.6.28. *Every bounded torsion Abelian group is isomorphic to a direct sum of cyclic groups with bounded finite orders.*

Clearly, the result below generalizes Theorem 9.6.26.

THEOREM 9.6.29. *A compact Abelian torsion group $G$ is topologically isomorphic to the finite product $\mathbb{Z}(n_1)^{\kappa_1} \times \cdots \times \mathbb{Z}(n_r)^{\kappa_r}$, where $n_1, \ldots, n_r$ are pairwise distinct positive integers and $\kappa_1, \ldots, \kappa_r$ are arbitrary cardinal numbers.*

PROOF. For every integer $n \in \mathbb{N}$, let $G[n] = \{x \in G : nx = 0_G\}$. Since $G$ is a torsion group, we have that $G = \bigcup_{n=1}^\infty G[n]$. It is also clear that each $G[n]$ is a closed subgroup of $G$. Every compact space has the Baire property, so $G[m]$ has a non-empty interior, for some $m \in \mathbb{N}$. Hence, $G[m]$ is an open subgroup of $G$ and the quotient group $G/G[m]$ is finite. Let $s$ be the size of $G/G[m]$. Since $G$ is Abelian, it follows that $msx = 0_G$ for each $x \in G$, that is, $G$ is a bounded torsion group. Then the dual group $G^*$ is also bounded torsion, so

Theorem 9.6.28 implies that

$$G^* \cong \bigoplus_{i=1}^{r} \mathbb{Z}(n_i)^{(\kappa_i)},$$

where $n_1, \ldots, n_r$ are distinct positive integers and $\kappa_1, \ldots, \kappa_r$ are cardinal numbers. It remains to apply Proposition 9.6.25 along with Example 9.5.12 to deduce the required conclusion. □

COROLLARY 9.6.30. *Every compact Abelian torsion group is zero-dimensional.*

There are many interesting concrete duality theorems establishing connections between the properties of $G$ and $G^*$. For example, one can characterize the dimension of a compact Abelian group $G$ by a property of $G^*$. It is also possible to characterize the local connectedness of a compact Abelian group $G$ by a property of $G^*$. The last characterization can be effectively used to give a complete description of the topological and algebraic structures of locally connected metrizable compact Abelian groups. However, we do not go deeper into this subject, since this area is enormous and already well developed, and though the arguments are based on the principal results of this and preceding section, the technical details we omit are mostly algebraic in nature and require a more extensive knowledge of the abstract theory of groups than we presume. So we refer the interested reader to [**387**], [**236**], and [**337**]. In particular, Theorem 9.6.18 is Theorem 9.5 in [**236**], where it plays a key role in establishing the structure of locally compact compactly generated Abelian groups.

## Exercises

9.6.a. Prove that every compact metrizable Abelian group is topologically isomorphic to a closed subgroup of the group $\mathbb{T}^\omega$.

9.6.b. Suppose that $G = \prod_{i\in I} G_i$ is the topological product of compact Abelian groups. Then the discrete dual group $G^*$ is isomorphic to the direct sum $\bigoplus_{i\in I}(G_i)^*$ of the dual groups.

9.6.c. Suppose that $G$ is a discrete Abelian group of a finite torsion-free rank $n$. Show that the dimension of the dual group $G^*$ does not exceed $n$ (and, therefore, $\dim(G^*) = n$).

9.6.d. Let $G$ be a compact Abelian group with a closed subgroup $H$. Prove that for every continuous character $h$ on $H$ and for every $a \in G \setminus H$, there exists a continuous character $f$ on $G$ such that $f \restriction H = h$ and $f(a) \neq 1$.

9.6.e. Let $G$ be a compact Abelian torsion group. Prove that the closure of every countable subset of $G$ has a countable base.

9.6.f. Let us call a topological group $G$ *monothetic* if $G$ contains a dense cyclic subgroup. Verify that a compact Abelian group $G$ is monothetic iff the dual group $G^*$ is isomorphic to a subgroup of the group $\mathbb{T}$ endowed with the discrete topology.

9.6.g. Suppose that $G$ is a pseudocompact Abelian topological group. Show that for every element $a \in G$ distinct from $e$, there exists a continuous character $f$ on $G$ such that $f(a) \neq 1$.

9.6.h. Suppose that $G$ is a pseudocompact connected Abelian group. Show that the dual group $G^*$ is torsion-free.

9.6.i. Apply Theorem 9.6.29 to show that for every prime $p$ and every cardinal $\kappa$, the group $\mathbb{Z}(p)^{(\omega)} \oplus \mathbb{Z}(p^2)^{(\kappa)}$ does not admit a compact Hausdorff group topology, where $G^{(\alpha)}$ denotes the direct sum of $\alpha$ copies of the group $G$.

## Problems

9.6.A. Let $G$ be a compact Abelian group with a closed subgroup $H$. Prove that $(G/H)^*$ is topologically isomorphic to $G^*/A(G^*, H)$.

9.6.B. Prove that if $G$ is a non-metrizable compact Abelian group, then there exists a continuous homomorphism of $G$ onto a product group $P = \prod_{i\in I} P_i$, where each $P_i$ is a non-trivial compact metrizable Abelian group and $|I| = w(G)$.

*Hint.* Let $\tau = |G^*|$. Then $\tau = w(G) > \aleph_0$, by Theorem 9.6.6, so we can find an independent subset $X$ of $G^*$ satisfying $|X| = \tau$ (see Proposition 9.9.20 below). Then the subgroup $S$ of $G^*$ generated by the set $X$ is a direct sum of non-trivial cyclic groups, say, $S = \bigoplus_{\alpha<\tau} C_\alpha$. Since the group $G^*$ is discrete, we can apply Proposition 9.6.2 to conclude that there exists a continuous homomorphism of $(G^*)^* \cong G$ onto $S^* \cong \prod_{\alpha<\tau} C_\alpha^*$. Clearly, each $C_\alpha^*$ is a non-trivial compact metrizable Abelian group.

9.6.C. Let $H$ be a compact zero-dimensional monothetic group (see Exercise 9.6.f). Then $H$ is topologically isomorphic to a product group $\prod_{p\in\mathbb{P}} A_p$, where for each prime $p$, $A_p$ is either trivial, or the finite cyclic group $\mathbb{Z}(p^{n_p})$ with $n_p \in \mathbb{N}$, or the group of $p$-adic integers.

*Hint.* Apply Exercise 9.6.f along with Theorem 9.6.12 to conclude that the dual group $H^*$ is algebraically isomorphic to a torsion subgroup of $\mathbb{T}$. It follows from [**409**, Theorem 4.1.1] (see also Theorem 9.9.14 below) that $H^*$ is the direct sum of groups $C_p$, with $p \in \mathbb{P}$, where each $C_p$ is the *p-primary component* of $H^*$, that is, the subgroup of $H^*$ consisting of elements of $p$-power order. Therefore, for each prime $p$, $C_p$ is either the $p$-torsion subgroup $\mathbb{Z}(p^\infty)$ of $\mathbb{T}$, or the finite cyclic group $\mathbb{Z}(p^n)$, for some integer $n \geq 0$. According to the duality theorem and Proposition 9.6.25, the group $H$ is topologically isomorphic to the product $\prod_{p\in\mathbb{P}} A_p$ of the dual groups $A_p = (C_p)^*$. Since $(\mathbb{Z}(p^n))^* = \mathbb{Z}(p^n)$ and $(\mathbb{Z}(p^\infty))^* = \mathbb{Z}_p$ (see Problem 9.5.A), the required conclusion follows.

9.6.D. Show that the additive group $\mathbb{R}$ of reals admits a compact Hausdorff topological group topology.

*Hint.* Apply item e) of Problem 1.1.G and Exercise 1.3.i.

9.6.E. Prove that a non-trivial free Abelian group does not admit a compact Hausdorff group topology.

*Hint.* Suppose to the contrary that there exists a compact Hausdorff group topology $\mathcal{T}$ on an infinite free Abelian group $G$. Then the compact group $K = (G, \mathcal{T})$ is zero-dimensional. Indeed, let $C$ be the connected component of the neutral element of $K$. Then $C$ is a compact connected topological group, so Theorem 9.6.15 implies that $C$ is divisible. Since, algebraically, $C$ is a subgroup of the free Abelian group $G$, it is itself a free Abelian group, by Nielsen's theorem (see [**409**, Theorem 6.1.1]). Hence, the group $C$ is trivial and $K$ is zero-dimensional. Let $H$ be the closure in $K$ of an arbitrary infinite cyclic subgroup of $K$. Then $H$ is a compact zero-dimensional monothetic group, and it follows from Problem 9.6.C that $H$ is topologically isomorphic to the product $\prod_{p\in\mathbb{P}} A_p$, where for each prime $p$, $A_p$ is either a finite cyclic group, or the group $\mathbb{Z}_p$ of $p$-adic integers. The first possibility is excluded because the subgroup $H \subset G$ is torsion-free. The second possibility is excluded because the element $(1, 0, 0, \ldots)$ of $\mathbb{Z}_p$ is divisible in $\mathbb{Z}_p$ by $q^n$, for every prime $q$ distinct from $p$ and every integer $n \geq 1$ (see Problem 1.1.F).

9.6.F. Suppose that $G$ is a pseudocompact non-compact Abelian group, and let $G_p^*$ be the group of continuous characters on $G$, with the topology of pointwise convergence. Let $(G_p^*)_p^*$ be the group of continuous characters on $G_p^*$, also with the topology of pointwise convergence. Consider the evaluation mapping $\Psi$ of $G$ to $(G_p^*)_p^*$. Then $\Psi$ is a continuous homomorphism.

(a) Verify that $\Psi$ is one-to-one.

(b) Must $\Psi$ be a homeomorphism of $G$ onto the subspace $\Psi(G)$ of $(G_p^*)_p^*$?

(c) Must $\Psi(G)$ be closed in $(G_p^*)_p^*$?

9.6.G. Let $G$ be a compact torsion topological group, not necessarily Abelian. Then every finitely generated subgroup of $G$ is finite.
*Comment.* This deep result was proved by E. I. Zel'manov in [**549**].

## Open Problems

9.6.1. (E. Hewitt and K. A. Ross [**237**]) Let $G$ be a compact torsion topological group, not necessarily Abelian. Is $G$ bounded torsion?

## 9.7. Non-trivial characters on locally compact Abelian groups

In this section we extend the principal result on the existence of non-trivial continuous characters on compact Abelian groups, Theorem 9.4.11, to locally compact Abelian groups. To reach this goal, we establish several results on the structure of certain locally compact Abelian groups.

A topological group $G$ is called *monothetic* if it contains a dense cyclic subgroup. Evidently, every monothetic group is Abelian.

THEOREM 9.7.1. *Every monothetic locally compact non-compact group $G$ is discrete and isomorphic to the group $\mathbb{Z}$ of integers.*

PROOF. Assume that $G$ is not discrete. Clearly, $G$ contains a dense cyclic subgroup algebraically isomorphic to $\mathbb{Z}$, since $\mathbb{Z}$ is the only infinite cyclic group. Hence, we may assume that $\mathbb{Z}$ is a dense subgroup of $G$. We put $N = \{n \in \mathbb{Z} : n < 0\}$ and $P = \{n \in \mathbb{Z} : n > 0\}$. The neutral element of $G$ is also denoted by 0.

We claim that each of the sets $N$ and $P$ is dense in $G$. Since $\mathbb{Z}$ is dense in $G$, it is enough to show that both $N$ and $P$ are dense in the subspace $\mathbb{Z}$ of $G$. Since under taking the inverse the space $\mathbb{Z}$ is mapped homeomorphically onto $\mathbb{Z}$ and $P$ is mapped onto $N$, it suffices to show that $P$ is dense in the subspace $\mathbb{Z}$. Since $G$ has a base at 0 consisting of symmetric open sets, and $\mathbb{Z}$ is dense in $G$, we have that $0 \in \overline{P}$. Therefore, by the continuity of translations, $n \in \overline{n + P}$, for each $n \in \mathbb{Z}$. Since $n + P \subset P$, for each non-negative $n \in \mathbb{Z}$, it follows that $n \in \overline{n + P} \subset \overline{P}$, for every non-negative $n \in \mathbb{Z}$. If $n < 0$, we put $P_n = \{k \in P : k > |n|\}$. The set $n + (P \setminus P_n)$ is finite and does not contain $n$. Hence, $n \in \overline{n + P_n}$. Since $n + P_n \subset P$, it follows that $n \in \overline{P}$. Thus, the sets $N$ and $P$ are both dense in $G$. It follows that, for any open neighbourhood $U$ of 0, the family $\{n + U : n \in N\}$ is an open covering of $G$. Indeed, we may assume that $U$ is symmetric. Take any $g \in G$. Since $g + U$ is an open neighbourhood of $g$ and $N$ is dense in $G$, there exists $j \in N$ such that $j \in g + U$. Then $g \in j + U$, since $U$ is symmetric. Similarly, the family $\{n + U : n \in P\}$ is also an open covering of $G$.

Now we fix an open symmetric neighbourhood $U$ of 0 in $G$ such that $\overline{U}$ is compact. Put $F = \overline{U}$. Since $\{n+U : n \in N\}$ is an open covering of $G$, there is a finite subset $K$ of $N$ such that $F \subset \bigcup\{n+U : n \in K\}$. Put $m = \max\{|n| : n \in K\}$ and $Y = \bigcup\{n+U : 1 \leq n \leq m\}$. Observe that $\overline{Y} = \bigcup\{n + F : 1 \leq n \leq m\}$ and therefore, $\overline{Y}$ is compact.

Let us show that $Y = X$. Take any $g \in G$, and let $n_g$ be the first element of $P$ such that $g \in n_g + U$. Then $g = n_g + a$, for some $a \in U$. Since $U \subset F$, there exists $j \in K$ such that $a \in j + U$. Then $g \in n_g + j + U$. Since $j$ is negative, from the definition of $n_g$ it follows that $n_g + j$ does not belong to $P$. Hence, $n_g + j \leq 0$. Therefore, $n_g \leq |j| \leq m$, which implies that $g \in Y$. Since the closure of $Y$ is compact, it follows that $G$ is compact, a contradiction. Thus, $G$ is discrete. □

Recall that a topological group $G$ is said to be *compactly generated* if there exists a compact subset $F$ of $G$ such that $G$ is the smallest subgroup of $G$ containing $F$. We will say that the *degree of non-compactness* of a topological group $G$ is finite if there exist a compact set $F \subset G$ and a finite subset $K \subset G$ such that $F + \langle K \rangle = G$, where $\langle K \rangle$ is the smallest subgroup of $G$ containing $K$. In fact, we can define the *degree of non-compactness* $nc(G)$ for any topological group $G$ as the smallest cardinal number $\kappa$ (which can be also finite) such that $G = F + H$, where $F$ is a compact subset of $G$ and $H$ is a subgroup of $G$ algebraically generated by a set $K \subset G$ of the cardinality $\kappa$. Clearly, $nc(G) = 0$ if and only if $G$ is compact.

THEOREM 9.7.2. *For any compactly generated locally compact Abelian topological group $G$, the degree of non-compactness of $G$ is finite.*

PROOF. Fix a compact subset $F$ of $G$ which algebraically generates $G$. Since $G$ is locally compact, we can assume that $F$ is the closure of a symmetric open neighbourhood $U$ of 0 such that $U$ algebraically generates $G$. The subspace $F + F$ of $G$ is also compact, since it is a continuous image of the compact space $F \times F$. By the compactness of $F + F$, there exists a finite subset $K$ of $F + F$ such that $F + F \subset K + U$. Then $F + F \subset K + F$. Put $F_0 = F$ and $F_{n+1} = F_n + F$, for each $n \in \omega$. Take the smallest subgroup $H$ of $G$ containing $K$. Clearly, $F_1 = F + F \subset F + K \subset F + H$, and $G = \bigcup_{n=1}^{\infty} F_n$. By induction it follows that $F_n \subset F + H$, for each $n \in \omega$. Therefore, $G = F + H$, which implies that the degree of non-compactness of $G$ is finite. □

PROPOSITION 9.7.3. *For any locally compact Abelian topological group $G$ of finite non-compactness degree and any non-zero $b \in G$, there exists a locally compact Abelian topological group $G_b$ of smaller or zero non-compactness degree and a continuous open homomorphism $p\colon G \to G_b$ such that $p(b) \neq 0$.*

PROOF. Let $m = nc(G)$. There exist a finite subset $K \subset G$ and a compact set $F \subset G$ such that $|K| = m$ and $G = F + H_K$, where $H_K$ is the smallest subgroup of $G$ containing $K$. Clearly, we may assume that $m > 0$. Then $K \neq \varnothing$.

Take any $a \in K$ and put $M = K \setminus \{a\}$. Let $H_a$ be the closure of in $G$ of the cyclic group $\langle a \rangle$. Then $H_a$ is not compact, since otherwise $\Phi = F + H_a$ is a compact subset of $G$ such that $\Phi + H_M = G$, where $H_M$ is the subgroup of $G$ algebraically generated by $M$. Then $nc(G) \leq |M| < |K| = m$, a contradiction. Therefore, by Theorem 9.7.1, $H_a = \langle a \rangle$ is an infinite closed discrete subgroup of $G$. Take any infinite subgroup $H$ of $H_a$ such that $b \notin H_a$, and let $p\colon G \to G/H$ be the natural quotient homomorphism. Clearly, $G/H$ is a locally compact Abelian group, $p(b) \neq 0$, since $b \notin H$, and $p(H_a) \cong H_a/H$ is a finite subset of $G/H$. Therefore, the set $B = p(F) + p(H_a)$ is compact. We also have:

$$
\begin{aligned}
G/H = p(G) = p(F + H_K) &= p(F + H_a + H_M) \\
&= p(F) + p(H_a) + p(H_M) = B + p(H_M).
\end{aligned}
$$

Since $p(H_M)$ is a subgroup of $G/H$ generated by the finite set $p(M)$, it follows that the non-compactness degree of $G/H$ does not exceed the cardinality of $p(M)$. Thus, $nc(G/H) \leq |p(M)| \leq |M| < m = nc(G)$. □

We are ready to prove a theorem for the sake of which we introduced the notion of non-compactness degree. This result provides a bridge from arbitrary locally compact Abelian groups to compact Abelian groups.

THEOREM 9.7.4. *For every locally compact Abelian group $G$ of finite non-compactness degree and each non-zero $b \in G$, there exists a compact Abelian group $G_b$ and an open continuous homomorphism $p\colon G \to G_b$ such that $p(b) \neq 0$.*

PROOF. We apply Proposition 9.7.3 $nc(G)$ times, or less, in an obvious way. Taking the composition of the quotient mappings that arise in the process, we obtain the desired quotient homomorphism of $G$ onto a locally compact Abelian topological group $G_b$ such that the non-compactness degree of $G_b$ is zero. However, the last condition implies that $G_b$ is compact. □

Finally, we are in a position to establish a result on the existence of non-trivial continuous characters on which the study of the structure of locally compact Abelian topological groups rests.

THEOREM 9.7.5. *For any locally compact Abelian group $G$ and each non-zero element $b \in G$, there exists a continuous character $f\colon G \to \mathbb{T}$ on $G$ such that $f(b) \neq 1$.*

PROOF. Take an open and closed subgroup $G_1$ of $G$ such that $G_1$ is compactly generated.
*Case 1.* Suppose that $b$ is not in $G_1$. Consider the quotient group $G/G_1$ (which is discrete) and the quotient mapping $\pi\colon G \to G/G_1$. Then $\pi(b) \neq 0$. Since $G/G_1$ is a discrete Abelian group, there exists a homomorphism $h\colon G/G_1 \to \mathbb{T}$ such that $h(\pi(b)) \neq 1$, by Corollary 1.1.8. Then $h$ is continuous, and the homomorphism $f = h \circ \pi$ is the character on $G$ we are looking for.
*Case 2.* Suppose that $b \in G_1$. In view of Theorem 9.7.2, the non-compactness degree of $G_1$ is finite. Hence, by Theorem 9.7.4, there exist a compact Abelian group $G_b$ and a continuous homomorphism $p_1\colon G_1 \to G_b$ such that $p_1(b) \neq 0$.

Since the group $\mathbb{T}$ is divisible, $p_1$ can be extended to a homomorphism $p$ of $G$ to $\mathbb{T}$, by Theorem 1.1.6. Since $G_1$ is an open and closed subgroup of $G$, and $p_1$ is continuous, the homomorphism $p$ must be continuous as well. We also have $p(b) = p_1(b) \neq 1$. □

## Exercises

9.7.a. Let $G$ be a locally compact Abelian group. We denote by $G^*$ the group of continuous characters on $G$ endowed with the compact-open topology. Prove that $G^*$ is a locally compact Abelian topological group.

9.7.b. Let $G$ be a locally compact Abelian group. Define $G^*$ and $(G^*)^*$ as in Exercise 9.7.a. Consider the evaluation mapping $\Psi\colon G \to (G^*)^*$ and prove that $\Psi$ is one-to-one and continuous.

## Problems

9.7.A. (V. G. Pestov [**375**]) Prove that every $\sigma$-compact topological group $G$ is a closed subgroup of some compactly generated topological group $H$. Furthermore, the compactly generated group $H$ can be chosen in such a way that $G$ will be a retract of $H$ under a continuous open homomorphism.

## 9.8. Varopoulos' theorem: the Abelian case

A mapping $f\colon X \to Y$ of topological spaces is called *sequentially continuous* if $f(x) = \lim_{n\to\infty} f(x_n)$ for every sequence $\{x_n : n \in \omega\}$ in $X$ converging to $x \in X$. It is immediate from the definition that every continuous mapping is sequentially continuous, but not vice versa. Indeed, every mapping of a non-discrete space $X$ without non-trivial convergent sequences to elsewhere is sequentially continuous, so there are lots of sequentially continuous mappings of $\beta\omega$ to a discrete two-point space that fail to be continuous. On the other hand, every sequentially continuous mappings defined on a sequential space is evidently continuous. In particular, a sequentially continuous mapping of a first-countable space to elsewhere is continuous.

The aim of this section is to show that in "almost all" cases, every sequentially continuous homomorphism defined on a compact Abelian group $G$ is continuous. This happens if the cardinality of the group $G$ is not Ulam measurable (see Theorem 9.8.5). The statement remains true for an arbitrary compact group, but the proof is more involved (see [**522**]). The proof of this deep theorem requires several preliminary steps, even in the Abelian case. The first of them is interesting in itself; it enables us to reduce the general case of a compact Abelian group $G$ to the case when $G$ is a topological product of compact metrizable groups. In fact, the following result refines the conclusion of Theorem 4.1.7 for the special case of compact Abelian groups.

THEOREM 9.8.1. *Let $G$ be an arbitrary compact Abelian group. Then there exists a continuous onto homomorphism $h\colon \prod_{i\in J} K_i \to G$, where $|J| \le w(G)$ and each $K_i$ is a compact metrizable Abelian group.*

PROOF. According to [**409**, 4.1.6], there exists a discrete divisible Abelian group $H$ that contains the discrete dual group $G^*$ to $G$ as a subgroup. It also follows from [**409**, 4.1.5] that the group $H$ is a direct sum of countable divisible groups, say, $H = \oplus_{i\in I} H_i$. Hence, by Proposition 9.6.25, the dual compact group $H^*$ is topologically isomorphic to the product $P = \prod_{i\in I} H_i^*$, where each group $H_i^*$ is metrizable, by Corollary 9.6.7. Therefore, the compact group $G$ is topologically isomorphic to the quotient group $P/N$, where $N = A(P, G^*)$ is the annihilator of the subgroup $G^* \subset H$ in $P$ (see Proposition 9.6.16). Let $\pi\colon P \to P/N$ be the quotient homomorphism. By Lemma 8.5.4, one can find a set $J \subset I$ with $|J| \le w(G)$ and a continuous homomorphism $h\colon P_J \to P/N$ such that $\pi = h \circ p_J$, where $p_J$ is the projection of $P$ onto $P_J = \prod_{i\in J} H_i^*$. It remains to put $K_i = H_i^*$, for each $i \in J$, and note that the homomorphism $h\colon \prod_{i\in J} K_i \to P/N \cong G$ is as required. □

Let us consider a topological product $P = \prod_{i\in I} G_i$ of compact metrizable groups and a sequentially continuous homomorphism $f\colon P \to H$ to a first-countable topological group $H$. For every non-empty set $J \subset I$, we denote by $\pi_J$ the natural projection of $P$ onto the subproduct $P_J = \prod_{i\in J} G_i$. We will also identify $P_J$ with the subgroup of $P$ obtained by complementing the elements of $P_J$ by the neutral elements of the factors $G_i$ on all coordinates $i \in I \setminus J$. Therefore, each projection $\pi_J$ is an open homomorphic retraction of $P$ onto the closed subgroup $P_J$ of $P$. This notation will be used in the three lemmas that follow (we do not assume that the groups are Abelian).

LEMMA 9.8.2. *Let $e$ be the neutral element of the group $P$ and $\Sigma P$ the $\Sigma$-product of the groups $G_i$ with center at $e$ considered as a subgroup of $P$. Then the restriction $g = f{\restriction}\Sigma P$*

*of $f$ to $\Sigma P$ is continuous and there exists a continuous homomorphism $f^*$ of $P$ to $H$ which depends on at most countably many coordinates and whose restriction to $\Sigma P$ coincides with $g$.*

PROOF. By Corollary 1.6.33, the space $\Sigma P$ is Fréchet–Urysohn (hence, sequential), so the restriction $g = f\restriction\Sigma P$ is continuous on $\Sigma P$. Since the group $H$ is first-countable, Lemma 8.5.4 implies that $g$ depends on at most countably many coordinates or, more precisely, one can find a countable set $J \subset I$ and a continuous homomorphism $h\colon P_J \to H$ such that $g = h \circ \pi_J\restriction\Sigma P$. Then the homomorphism $f^* = h \circ \pi_J$ is as required. □

Since $H$ is a group, there exists a mapping $\varphi\colon P \to H$ satisfying $f(x) = f^*(x)\cdot\varphi(x)$, for all $x \in P$. It is clear that $\varphi$ is sequentially continuous and is identically equal to $e_H$ on the subgroup $\Sigma P$ of $P$. In the next technical lemma we establish a special property of the mapping $\varphi$; it is crucial for the proof of Theorem 9.8.5.

LEMMA 9.8.3. *Suppose that $H$ is a topological group with no small subgroups. Then there exists a finite partition $I = I_1 \cup I_2 \cup \ldots \cup I_k$ of the index set $I$ such that for every $i \le k$ and every two disjoint sets $A$ and $B$ with $I_i = A \cup B$, the mapping $\varphi$ is identically equal to $e_H$ on one of the groups $P_A$, $P_B$ (or possibly both).*

PROOF. If the required partition does not exist, one easily defines a sequence $\{I_n : n \in \omega\}$ of non-empty disjoint subsets of $I$ and a sequence $\{x_n : n \in \omega\}$ such that $x_n \in P_{I_n}$ and $\varphi(x_n) \ne e_H$ for each $n \in \omega$. For a given $n \in \omega$, denote by $K_n$ the product group $\prod_{i\in I_n} K_{n,i}$, where $K_{n,i}$ is the closure in $G_i$ of the cyclic group generated by the element $x_n(i) = \pi_i(x_n)$. The group $K_n$ is evidently Abelian, $x_n \in K_n$, and the intersection $K_n \cap \Sigma P$ is dense in $K_n$. Then the closure $L_n$ of $f(K_n)$ in $H$ is an Abelian subgroup of $H$ and $f(x_n) \in L_n$. In addition, since the homomorphism $f^*\colon P \to H$ depends only on a countable set $J \subset I$ and $f^*\restriction\Sigma P = f\restriction\Sigma P$, we conclude that the element $f^*(x_n) = f^*(\pi_J(x_n)) = f(\pi_J(x_n))$ also belongs to the group $L_n$ (we recall that $P_J \subset \Sigma P \subset P$ and, therefore, $\pi_J(x_n) \in K_n \cap \Sigma P$). Since $K_n$ is Abelian, the latter fact and the definition of $\varphi$ imply that the restriction of $\varphi$ to $K_n$ is a homomorphism. Thus, $L_n$ contains the elements $f(x_n^p)$ and $f^*(x_n^q)$, for all integers $p$ and $q$. Finally, since $H$ has no small subgroups, there exists an open neighbourhood $U$ of $e_H$ in $H$ such that $\varphi(x_n^{p_n}) \notin U$ for some non-zero $p_n \in \mathbb{Z}$, where $n \in \omega$. The sets $I_n$ are mutually disjoint, so the sequence $\{x_n^{p_n} : n \in \omega\}$ converges to $e$ in $P$. This, however, contradicts the sequential continuity of $\varphi$. □

LEMMA 9.8.4. *Let $\varphi$ and $I_1, \ldots, I_k$ be as in Lemma 9.8.3. If $\varphi$ is not constant on $P_{I_m}$ for some $m \le k$, then the family $\mathcal{F}$ of the subsets $A$ of $A_m$ such that $\varphi$ is not constant on $P_A$ is a countably closed free ultrafilter on the set $A_m$.*

PROOF. First we claim that if $A = B \cup C \subset I$ and $\varphi$ is constant on both groups $P_B$ and $P_C$, then $\varphi$ is also constant on $P_A$. Indeed, since $e \in P_B \cap P_C$ and $\varphi(e) = e_H$, it follows from the definition of $\varphi$ that $f(x) = f^*(x)$, for each $x \in P_B \cup P_C$. Since $f$ and $f^*$ are homomorphisms and the set $P_B \cup P_C$ algebraically generates the group $P_A$, we conclude that both $f$ and $f^*$ are identically equal to $e_H$ on $P_A$. This implies immediately that $\varphi(x) = e_H$, for each $x \in P_A$.

Suppose that $\varphi$ is not constant on $A_m$, for some $m \le k$. Since each $G_i$ is a subgroup of $\Sigma P$, the mapping $\varphi$ is constant on $G_i$ and $\{i\} \notin \mathcal{F}$, for each $i \in A_m$. It is also clear that if

$A \in \mathcal{F}$ and $A \subset B \subset A_m$, then $B \in \mathcal{F}$. If $A_m = A \cup B$ and $A \cap B = \varnothing$, then Lemma 9.8.3 and the above claim together imply that exactly one of the sets $A$, $B$ belongs to $\mathcal{F}$.

It remains to show that $\mathcal{F}$ contains intersections of arbitrary countable subfamilies. Take an arbitrary sequence $\{B_n : n \in \omega\} \subset \mathcal{F}$. Then $C_n = I_m \setminus B_n \notin \mathcal{F}$, for each $n \in \omega$. For every $n \in \omega$, let $D_n = C_0 \cup \cdots \cup C_n$. Then $D_0 \subset D_1 \subset \cdots \subset D_n \subset \cdots$ is an increasing sequence of subsets of $I_m$ and it follows from the above claim that $D_n \notin \mathcal{F}$, for each $n \in \omega$. Let $D = \bigcup_{n\in\omega} D_n$ and take an arbitrary element $x \in P_D$. For every $n \in \omega$, denote by $x_n$ the restriction of $x$ to $D_n$. Then $x_n \in P_{D_n}$ and the sequence $\{x_n : n \in \omega\}$ converges to $x$ in $P$. Since $\varphi(x_n) = e_H$ for each $n \in \omega$, we must have $\varphi(x) = e_H$, so that $\varphi$ is constant on $P_D$ and, consequently, $D \notin \mathcal{F}$. We have thus proved that $\bigcap_{n\in\omega} B_n = I_m \setminus D \in \mathcal{F}$. Therefore, $\mathcal{F}$ is a countably complete free ultrafilter on $I_m$. □

Here is the main theorem of this section, the Abelian case of *Varopoulos' theorem*, which states that sequential continuity implies continuity for homomorphisms of locally compact groups unless the groups have Ulam measurable cardinalities. We assume, in addition, that the groups in question are compact.

THEOREM 9.8.5. [**N. Varopoulos**] *Let $f: G \to H$ be a sequentially continuous homomorphism of compact Abelian topological groups. If the cardinality of the group $G$ is Ulam non-measurable, then $f$ is continuous.*

PROOF. By Theorem 9.8.1, there exists a continuous onto homomorphism $h: K \to G$, where $K = \prod_{i\in I} K_i$ is a product of compact metrizable topological groups and $|I| \leq w(G)$. Since $w(G) \leq |G|$, the former inequality implies that the cardinality of the index set $I$ is Ulam non-measurable. Clearly, the composition $g = f \circ h$ is a sequentially continuous homomorphism of the product group $K$ to $H$. It suffices to verify that $g$ is continuous, since the fact that $h$ is open will then imply the continuity of $f$.

First, we prove the theorem in the special case when $H = \mathbb{T}$. Evidently, the group $\mathbb{T}$ has no small subgroups. As in Lemma 9.8.2, we define a continuous homomorphism $g^*$ of $K$ to $\mathbb{T}$ which coincides with $g$ on the corresponding $\Sigma$-product $\Sigma K \subset K$. Then $g^* = h^* \circ \pi_J$, for some countable set $J \subset I$, where $\pi_J$ is the projection of $K$ to $K_J = \prod_{i\in J} K_i$ and $h^*: K_J \to \mathbb{T}$ is a continuous homomorphism. Let a mapping $\varphi: K \to \mathbb{T}$ be defined by $g(x) = g^*(x) \cdot \varphi(x)$, for each $x \in K$. Take the partition $I_1, \ldots, I_k$ of the set $I$ as in Lemma 9.8.3 (defined for $K$ in place of $P$). Since $|I_l| \leq |I|$ for each $l \leq k$, no free ultrafilter on $I_l$ is countably complete. Hence, Lemma 9.8.4 implies that $\varphi \equiv 1$ on $K_{I_l}$, for each $l \leq k$. Since the subgroups $K_{I_l}$ with $l \leq k$ algebraically generate the group $K$, the mapping $\varphi$ has to be identically equal to 1 on $K$. This and the definition of $\varphi$ together imply that $g = g^*$ is continuous.

Finally, let $H$ be an arbitrary compact Abelian group and $\chi$ a continuous character of $H$. Then the composition $\chi \circ g$ is a sequentially continuous homomorphism of $K$ to $\mathbb{T}$ which is continuous as we have just shown above. Since the topology of $H$ is generated by the family of all continuous characters, we conclude that the homomorphism $g$ is continuous. Hence, the homomorphism $f: G \to H$ is also continuous. □

The next example shows that the Ulam non-measurability restriction on the cardinality of the compact group $G$ in Theorem 9.8.5 is essential.

EXAMPLE 9.8.6. Suppose that $\mathcal{F}$ is a countably closed free ultrafilter on a set $I$ and $K = \prod_{i\in I} C_i$, where $C_i = C$ is a compact Abelian group for each $i \in I$, and $|C| \geq 2$. Then there exists a non-trivial sequentially continuous character $\chi$ on $K$ which is constant on the dense subgroup $\Sigma K$ of $K$ and, hence, is discontinuous.

Since $|C| \geq 2$, one can take a non-trivial character $\mu$ on $C$. Given a point $x = (x_i)_{i\in I}$ of $K$, we define the value $\chi(x) \in \mathbb{T}$ by the rule

$$\chi(x) = \lim_{\mathcal{F}} \mu(x_i) \tag{9.13}$$

or, equivalently, $\chi(x)$ is the unique common point of the sets $\overline{\{\mu(x_i) : i \in A\}}$, where $A \in \mathcal{F}$ and the closure is taken in $\mathbb{T}$. Since $\mathcal{F}$ is an ultrafilter and $\mathbb{T}$ is compact, the limit in (9.13) exists and is unique. Let us verify the following:

CLAIM. *For every sequence $\{x^{(n)} : n \in \omega\}$ in $K$, there exists $A \in \mathcal{F}$ such that $\chi(x^{(n)}) = \mu(x_i^{(n)})$, for all $n \in \omega$ and $i \in A$.*

Indeed, take an arbitrary $n \in \omega$ and an open neighbourhood $U$ of $y_n = \chi(x^{(n)})$ in $\mathbb{T}$. Since $y_n = \lim_{\mathcal{F}} \mu(x_i^{(n)})$, there exists $A(U, n) \in \mathcal{F}$ such that $\{\mu(x_i^{(n)}) : i \in A(U, n)\} \subset U$. Let $\mathcal{B}_n$ be a countable base of $\mathbb{T}$ at $y_n$. Since the ultrafilter $\mathcal{F}$ is countably complete, we conclude that the set $A = \bigcap\{A(U, n) : U \in \mathcal{B}_n,\ n \in \omega\}$ is as required.

As $\mu$ is a character of $C$, it follows immediately from our Claim that $\chi$ is a character on $K$, and that both $\mu$ and $\chi$ are non-trivial. Since $\mu$ is continuous, the above claim also implies that the character $\chi$ is sequentially continuous. Finally, $\{i\} \notin \mathcal{F}$ for each $i \in I$, so $B \notin \mathcal{F}$ for each countable set $B \subset I$. This fact and Claim together imply that $\chi \equiv 1$ on the dense subgroup $\Sigma K$ of $K$. Hence, the continuity of the character $\chi$ would imply that $\chi$ is constant on $K$, which is not the case. □

We will now apply Varopoulos' theorem to prove, under a mild restriction on the cardinality of a compact Abelian group $G$, that there is no strictly stronger countably compact group topology on $G$. In fact, the same is true for compact non-Abelian groups as well, and the proof is virtually the same, we just need the general, non-Abelian version of Varopoulos' theorem. However, below we prove the result only for the Abelian case.

Every topological space considered below, until the end of the section, is assumed to be Hausdorff.

Let us call a sequence $\{x_n : n \in \omega\}$ in a space $X$ *accumulating* if there exists an accumulation point for this sequence in $X$. A mapping $f : X \to Y$ will be called *strongly sequentially continuous* if, for every accumulating sequence $\xi = \{x_n : n \in \omega\}$ in $X$, there exists an accumulation point $x \in X$ for $\xi$ such that $f(x)$ is an accumulation point for $\{f(x_n) : n \in \omega\}$ in $Y$. A mapping $f : X \to Y$ is said to be $\aleph_0$*-continuous* if, for every countable subset $A$ of $X$ and for each point $x \in \overline{A}$, $f(x)$ belongs to the closure of $f(A)$. Clearly, every $\aleph_0$-continuous mapping is strongly sequentially continuous.

PROPOSITION 9.8.7. *Every strongly sequentially continuous mapping is sequentially continuous.*

PROOF. Every convergent sequence has a unique accumulation point, so the conclusion follows from the definition of strong sequential continuity. □

PROPOSITION 9.8.8. *Let $f$ be a strongly sequentially continuous mapping of a countably compact space $X$ onto a space $Y$. Then $Y$ is also countably compact.*

PROOF. Take an arbitrary sequence $\eta = \{y_n : n \in \omega\}$ in $Y$ and, for each $n \in \omega$, fix a point $x_n \in X$ such that $f(x_n) = y_n$. Since $X$ is countably compact, the sequence $\xi = \{x_n : n \in \omega\}$ is accumulating in $X$. Therefore, by the strong sequential continuity of $f$, there exists an accumulation point $x$ for $\xi$ in $X$ such that $f(x)$ is an accumulation point for $\eta$ in $Y$. Thus, $Y$ is countably compact. □

PROPOSITION 9.8.9. *Let $f$ be a one-to-one strongly sequentially continuous mapping of a space $X$ onto a countably compact space $Y$. Then the inverse mapping $f^{-1}$ is strongly sequentially continuous if and only if $X$ is countably compact.*

PROOF. Suppose that $X$ is countably compact. Let $\eta = \{y_n : n \in \omega\}$ be an accumulating sequence in $Y$. For every $n \in \omega$, choose a point $x_n \in X$ such that $f(x_n) = y_n$. Since $X$ is countably compact, the sequence $\xi = \{x_n : n \in \omega\}$ is accumulating in $X$. By the strong sequential continuity of $f$, there exists an accumulation point $x \in X$ for $\xi$ such that $y = f(x)$ is an accumulation point for $\eta$. Let $g = f^{-1}$. Since $g(y) = x$ and $g(y_n) = x_n$, for each $n \in \omega$, the mapping $g$ is strongly sequentially continuous. Observe that in this part of the argument we have not used the assumption that $Y$ is countably compact, though $Y$ has this property automatically, by Proposition 9.8.8.

Conversely, suppose that $g = f^{-1}$ is strongly sequentially continuous. Then $X$ is countably compact, by Proposition 9.8.8. □

We are ready to present one of the main results of the section.

THEOREM 9.8.10. *Let $f$ be a strongly sequentially continuous isomorphism of a countably compact Abelian topological group $H$ onto a compact topological group $G$. Assume also that the cardinality of $G$ is Ulam non-measurable. Then $H$ is compact, and $f$ is a topological isomorphism.*

PROOF. Clearly, the group $G$ is Abelian as a homomorphic image of the Abelian group $H$. The mapping $f^{-1}$ is strongly sequentially continuous, by Proposition 9.8.9. Therefore, it is sequentially continuous. Since $H$ is countably compact, it is topologically isomorphic to a topological subgroup of a compact topological group $F$; therefore, $f^{-1}$ can be considered as a homomorphism of the compact Abelian group $G$ to the compact group $F$. Since $f^{-1}$ is sequentially continuous, it remains to apply Theorem 9.8.5. □

THEOREM 9.8.11. *Every strongly sequentially continuous homomorphism $f$ of a compact Abelian group $G$, whose cardinality is Ulam non-measurable, onto an arbitrary topological group $H$ is continuous, and $H$ is compact.*

PROOF. By Proposition 9.8.8, $H$ is countably compact. Therefore, $H$ can be treated as a subgroup of a compact group $F$, and $f$ can be considered as a homomorphism of the compact group $G$ to a compact group $F$. By Proposition 9.8.7, $f$ is sequentially continuous. By Varopoulos' theorem, $f$ is continuous. Hence, $H$ is compact. □

EXAMPLE 9.8.12. Let $f$ be a one-to-one mapping of the Čech–Stone compactification $\beta\mathbb{N}$ of the discrete space of natural numbers onto a discrete space $Y$. Then $f$ is sequentially continuous, since all convergent sequences in $\beta\mathbb{N}$ are trivial. The space $Y$ is not countably compact, while the space $\beta\mathbb{N}$ is compact. Thus, in Proposition 9.8.8, it is not enough to assume $f$ to be sequentially continuous. The mapping $f$ is not strongly sequentially continuous. This example also shows that Proposition 9.8.9 cannot be generalized to sequentially continuous mappings. □

The conclusion of Theorem 9.8.10 won't hold if we replace the condition that $G$ is compact by the condition that $G$ is countably compact.

EXAMPLE 9.8.13. Let $D = \{0, 1\}$ be the two-element additive group, and let $A$ be a set of Ulam measurable cardinality. The last condition means that there exists a free ultrafilter $\xi$ on $A$ closed under countable intersections. For $x = (x_a)_{a\in A} \in D^A$, we put $B(x) = \{a \in A : x_a = 0\}$. Let $g(x) = 0$ if $B(x) \in \xi$, and $g(x) = 1$ if $B(x) \notin \xi$. The product group $D^A$, taken with the Tychonoff product topology $\mathcal{T}$, is a compact topological group, and the mapping $g$ defined above is a homomorphism of $D^A$ to $D$. Clearly, if only countably many coordinates of a point $x \in D^A$ are equal to zero, then $g(x) = 0$. The points of this kind form a dense subset of $D^A$. On the other hand, $g(a) = 1$ for the point $a \in X$ with all coordinates equal to 1. It follows that the mapping $g\colon D^A \to D$ is not continuous. Let us show that $g$ is $\aleph_0$-continuous. Let $M$ be a countable subset of $D^A$ and $x \in \overline{M}$. Evidently, we only need consider the case when all coordinates of $x$ are equal to zero. Then $g(x) = 0$. Assume to the contrary that $g(y) = 1$ for each $y \in M$. Then, for each $y \in M$, the set $B(y)$ is not in $\xi$. On the other hand, from $x \in \overline{M}$ it follows that $A = \bigcup\{B(y) : y \in M\}$. Then $A \setminus B(y) \in \xi$ for each $y \in M$ and $\bigcap\{A \setminus B(y) : y \in M\} = \varnothing$, contradicting the Ulam measurability of the ultrafilter $\xi$. Thus, $g$ is $\aleph_0$-continuous (and strongly sequentially continuous), but not continuous homomorphism of the compact group $D^A$ to the group $D$.

Let us now take the smallest topological group topology $\mathcal{T}_1$ on the group $D^A$ containing the topology $\mathcal{T}$ and such that $g$ is $\mathcal{T}_1$-continuous. Since $g$ is $\aleph_0$-continuous with respect to $\mathcal{T}$, we have that $\mathcal{T}\restriction M = \mathcal{T}_1\restriction M$, for every countable subset $M$ of $D^A$. It follows that the identity mapping of the space $D^A$ with the original topology $\mathcal{T}$ onto the set $D^A$ provided with the topology $\mathcal{T}_1$, is $\aleph_0$-continuous. Therefore, this mapping is also strongly sequentially continuous. By Proposition 9.8.8, the space $(D^A, \mathcal{T}_1)$ is countably compact. Thus, the restriction on the cardinality of $G$ in Theorems 9.8.10 and 9.8.11 cannot be omitted. □

EXAMPLE 9.8.14. In the compact subspace $X = \beta\mathbb{N} \setminus \mathbb{N}$ of the space $\beta\mathbb{N}$ we fix a countable infinite discrete subset $A$ satisfying the condition that, for each $x \in A$, there exists a countable set $B_x \subset X \setminus A$ such that $x \in \overline{B_x}$. Let us now provide the set of points of the space $X$ with a new topology, defined as the smallest topology containing the topology of $X$ and all subsets of $A$. Clearly, $X$ with the new topology is countably compact. Hence, by Proposition 9.8.9, the identity mapping of the compact space $X$ onto this new space is strongly sequentially continuous, but evidently, it is not $\aleph_0$-continuous. □

EXAMPLE 9.8.15. In the product group $X = D^{\omega_1}$, where $D = \{0, 1\}$ is the discrete two-element additive group, we put $C(x) = \{\alpha \in \omega_1 : x_\alpha = 1\}$ and take subsets $Y$ and $Z$ defined by

$$Y = \{x \in X : |C(x)| \le \omega\} \text{ and } Z = \{x \in X : |\omega_1 \setminus C(x)| \le \omega\}.$$

Then $Y$ and $H = Y \cup Z$ are countably compact topological subgroups of the compact group $X$ (by Corollary 1.6.34, the closure in $X$ of every countable subset of $Y$ or $Z$ is a compact and metrizable subset of $H$). Adding to the topology of $H$ induced from $X$ two new open sets $Y$ and $Z$, we introduce a new finer topological group topology $\mathcal{T}_1$ on $H$, and obtain a countably compact topological group $G = (H, \mathcal{T}_1)$. Clearly, the space $G$ is the disjoint topological union of the spaces $Y$ and $Z$, and the identity mapping $f$ of $H$ onto $G$ is $\aleph_0$-continuous (hence, strongly sequentially continuous), but not continuous. It follows that the conclusion of Theorem 9.8.10 does not hold if we replace the condition that $G$ is compact by the condition that $G$ is countably compact.

Now we come to a curious feature of the mapping $f$ — it cannot be extended to a sequentially continuous mapping of $\beta H$ to $\beta G$. Indeed, $\beta H$ is homeomorphic to the product space $D^{\omega_1}$, and every sequentially continuous mapping of $D^{\omega_1}$ to an arbitrary Tychonoff space is continuous [**313**]. □

THEOREM 9.8.16. *Let $g$ be a continuous homomorphism of a topological group $G$ onto a compact Abelian topological group $H$ of Ulam non-measurable cardinality, and let $B$ be a countably compact subspace of $G$ such that $g(B) = H$. Then the homomorphism $g$ is open.*

PROOF. We can represent $g$ as the composition of a quotient homomorphism $f\colon G \to H_1$ and a continuous monomorphism $h\colon H_1 \to H$, so that $g = h \circ f$. Clearly, $H_1$ is Abelian and $f(B) = f(G) = H_1$. By the continuity of $f$ it follows that $H_1$ is countably compact. Therefore, by Theorem 9.8.10, $h$ is a topological isomorphism. Hence, $g$ is a quotient homomorphism, which implies that $g$ is open. □

COROLLARY 9.8.17. *Every continuous homomorphism $g$ of a countably compact topological group onto a compact Abelian group of Ulam non-measurable cardinality is quotient, that is, the mapping $g$ is open.*

THEOREM 9.8.18. *Let $g$ be a continuous homomorphism of a topological group $G$ onto a compact Abelian topological group $F$ such that the cardinality of $F$ is Ulam non-measurable and the kernel $g^{-1}(e)$ is compact, and there exists a countably compact subspace $B$ of $G$ such that $g(B) = F$. Then $G$ is compact, and the mapping $g$ is open and closed.*

PROOF. By Theorem 9.8.16, the mapping $g$ is open. Since the kernel of $g$ is compact, the quotient homomorphism $g$ is also a closed mapping. It follows that $g$ is perfect, which implies that $G = g^{-1}(F)$ is compact (see [**165**, Theorem 3.7.2]. □

COROLLARY 9.8.19. *Let $g$ be a continuous homomorphism of a countably compact topological group $G$ onto a compact Abelian topological group $F$ such that the kernel of $g$ is compact and the cardinality of $F$ is Ulam non-measurable. Then $G$ is compact and the mapping $g$ is open and closed.*

## Exercises

9.8.a. (M. Hušek [**252**]) A topological group $G$ is called an *$s$-group* if every sequentially continuous homomorphism of $G$ to any topological group $H$ is continuous. Verify the following assertions:

(a) The product of two $s$-groups is an $s$-group;

(b) Apply (a) to show that the topological product of countably many $s$-groups is again an $s$-group.

9.8.b. (M. Hušek [**252**]) A prenorm $N$ on a topological group $G$ is called *sequentially continuous* if for every sequence $\{x_n : n \in \omega\}$ in $G$ converging to an element $x \in G$, the values $N(x_n)$ converge to $N(x)$. Let $D = \{0, 1\}$ be the discrete two-element group.

(a) Prove that for every cardinal $\kappa \geq \omega$, a prenorm $N$ on the product group $D^\kappa$ is sequentially continuous iff for every sequence $\{x_n : n \in \omega\}$ in $G$ converging to the neutral element of $D^\kappa$, the values $N(x_n)$ converge to zero.

(b) Let $\kappa$ be an infinite cardinal. For $x, y \in D^\kappa$, we write $x \leq y$ if $x(\alpha) \leq y(\alpha)$, for each $\alpha \in \kappa$. Suppose that $f$ is a non-negative, sequentially continuous, bounded real-valued function on the group $D^\kappa$. Is the function $N$, defined by $N(x) = \sup_{y \leq x} f(y)$, sequentially continuous on $D^\kappa$? Is $N$ a prenorm?

## Problems

9.8.A. Generalize Theorem 9.8.5 by showing that every sequentially continuous homomorphism $f: G \to H$ of a compact Abelian group $G$ to a locally compact group $H$ is continuous unless the cardinality of $G$ is Ulam measurable. Deduce from this fact that the result remains valid for every group $H$ topologically isomorphic to a closed subgroup of a topological product of locally compact groups.

9.8.B. (N. Noble [**351**]) In connection with Theorem 9.8.5 prove that if $f: X \to F$ is a sequentially continuous mapping of a topological product $X = \prod_{i \in I} X_i$ of first-countable spaces to a finite discrete space $F$ and the cardinal $|I|$ is Ulam non-measurable, then $f$ is continuous.

9.8.C. Let $\varphi: G \to H$ be a sequentially continuous homomorphism of a compact Abelian group $G$ onto a topological group $H$. Find out which of the following general statements are valid:
(a) $H$ is compact;
(b) $H$ is countably compact;
(c) $H$ is pseudocompact;
(d) $H$ is precompact.
What happens if, additionally, the group $H$ satisfies $w(G) \leq \mathfrak{c} = 2^\omega$ or $w(G) \leq 2^{\mathfrak{c}}$?
*Hint.* See the next problem.

9.8.D. Let $f$ be a sequentially continuous homomorphism of a compact Abelian group $G$ onto a topological group $H$. Show that, consistently, $H$ need not be precompact, even if $G = D^{\mathfrak{c}}$, where $\mathfrak{c} = 2^\omega$.
*Remark.* The articles [**55**] and [**56**] contain several counterexamples regarding Problems 9.8.C and 9.8.D.

9.8.E. Under the constructibility axiom $V = L$, every continuous mapping $f$ of a countably compact topological Abelian group $G$ onto a compact space $F$ is $R$-quotient.
*Hint.* The Čech–Stone compactification $\beta G$ of the space $G$ is homeomorphic to a compact topological group. Hence, $\beta G$ is a dyadic compactum. The mapping $f$ can be extended to a continuous mapping of $\beta G$ onto $F$. Therefore, $F$ is a dyadic compactum as well. See further details in [**34**].

9.8.F. It is consistent with $ZFC$ that every one-to-one continuous mapping of a countably compact Abelian group onto a compact Hausdorff space is a homeomorphism.
*Hint.* Every one-to-one continuous $R$-quotient mapping of a Tychonoff space onto another Tychonoff space is a homeomorphism. It remains to refer to the preceding problem. Thus, under $V = L$, if $G$ is a countably compact Abelian group which is not compact, then there is no one-to-one continuous mapping of the space $G$ onto a compact Hausdorff space. Example 9.8.13 shows that this assertion cannot be proved in $ZFC$.

9.8.G. Let $f$ be a continuous mapping of an $\omega$-bounded Abelian group $G$ onto a compact space $X$, where the cardinality of $G$ is Ulam non-measurable. Then $f$ is $R$-quotient.
*Hint.* There are a Tychonoff space $Y$ and an $R$-quotient mapping $h$ of $G$ onto $Y$ such that $f = g \circ h$, where $g$ is a one-to-one continuous mapping of $Y$ onto $X$. The space $Y$ is $\omega$-bounded, since $G$ is $\omega$-bounded. One can show that $X$ is a dyadic compactum. The cardinality

of $Y$ is Ulam non-measurable. Now it is possible to show that $g$ is a homeomorphism. Since $h$ is an $R$-quotient mapping, it follows that the mapping $g$ is also $R$-quotient.

9.8.H. Let $G$ be an $\omega$-bounded Abelian group of Ulam non-measurable cardinality, and let $f$ be a one-to-one continuous mapping of $G$ onto a compact space $X$. Then $f$ is a homeomorphism and $G$ is compact.

*Hint.* Obviously, every one-to-one $R$-quotient continuous mapping of one Tychonoff space onto another Tychonoff space is a homeomorphism. It remains to apply Problem 9.8.G.

9.8.I. Let $G$ be a countably compact $\aleph_0$-monolithic Abelian group (see Exercise 4.1.b), and let $f$ be a continuous mapping of $G$ onto a compact space $X$. Then $X$ is metrizable.

*Hint.* It follows from Theorem 6.6.4 that $f$ can be extended to a continuous mapping $g\colon \varrho G \to X$. Hence, $X$ is a dyadic compactum. The space $X$ is also $\aleph_0$-monolithic. It remains to observe that every $\aleph_0$-monolithic dyadic compactum is metrizable, since $D^{\omega_1}$ is not $\aleph_0$-monolithic (another way to finish the proof is to apply Proposition 10.4.16).

9.8.J. Show that every totally disconnected pseudocompact topological group $G$ admits a continuous isomorphism of $G$ onto a zero-dimensional topological group.

*Hint.* The Stone–Čech compactification $\beta G$ of $G$ is a compact group, and $G$ is a topological subgroup of $\beta G$. The connected component $C$ of the neutral element of the group $\beta G$ is a closed invariant subgroup of $\beta G$. The space $G$ is totally disconnected, that is, the neutral element $e$ of $G$ can be represented as the intersection of a family of clopen subsets of $G$. Therefore, $G \cap C = \{e\}$. The quotient group $\beta G/C$ is a zero-dimensional compact group, and the restriction to $G$ of the quotient homomorphism of $\beta G$ onto $\beta G/C$ is a one-to-one continuous homomorphism of $G$ onto a zero-dimensional subgroup of $\beta G/C$.

9.8.K. Prove that every non-metrizable compact Abelian group admits a strictly finer pseudocompact topological group topology. Extend the result to non-metrizable pseudocompact Abelian groups.

*Hint.* For the first part, see the article [**120**] by W. W. Comfort and L. C. Robertson. The stronger second part is a recent theorem proved by W. W. Comfort and J. van Mill in [**116**].

9.8.L. If a compact space $Y$ is a continuous image of an $\omega$-bounded topological group $G$ of countable tightness, then $Y$ is metrizable.

*Hint.* Clearly, the group $G$ is pseudocompact. Let $f$ be a continuous mapping of $G$ onto a compact space $Y$. First, the space $Y$ is dyadic. Indeed, the Čech–Stone compactification $\beta G$ of the space $G$ is the Raĭkov completion $\varrho G$ of the group $G$ (Theorem 6.6.4), the compact group $\varrho G$ is dyadic (Theorem 4.1.7), and $f$ admits an extension to a continuous mapping of $\beta G$ onto $Y$. Second, the tightness of each separable subspace of $Y$ is countable. To show this, take any countable subset $B$ of $Y$, and let $P$ be the closure of $B$ in $Y$. It suffices to show that the tightness of $P$ is countable. Clearly, there exists a countable subset $A$ of $G$ such that $f(A) = B$. Then the closure $H$ of $A$ in $G$ is a compact space of countable tightness and $f(H) = P$. It follows that the tightness of $P$ is countable [**165**, 3.12.8 (a)]. If $Y$ is not metrizable, then $Y$ contains a topological copy of some separable non-metrizable dyadic compactum $Z$. But this is impossible, since the tightness of $Z$ is countable and every dyadic compactum of countable tightness is metrizable.

9.8.M. (A. V. Arhangel'skii [**34**]) Let $G$ be a topological group of countable tightness such that for each $n \in \mathbb{N}$, the space $G^n$ is countably compact and normal, and let $f$ be a continuous mapping of $G$ onto a compact space $X$. Then $X$ is metrizable.

9.8.N. (A. V. Arhangel'skii [**34**]) If $f$ is a one-to-one continuous mapping of a countably compact topological group $G$ of countable tightness onto a compact space $X$, then $X$ is metrizable and $f$ is a homeomorphism.

9.8.O. Suppose that $G$ is an initially $\aleph_1$-compact (that is, every open covering of $G$ of cardinality $\leq \aleph_1$ contains a finite subcovering) topological group of countable tightness. Show that $G$ is compact and metrizable.

*Hint.* Prove, using free sequences (see [**20**] or [**262**]), that the tightness of $\beta G$ is countable. Apply the fact that $\beta G$ is a topological group, since $G$ is pseudocompact. Observe that every compact group of countable tightness is metrizable. Use also the fact that every metrizable countably compact space is compact.

9.8.P. (A. V. Arhangel'skii [**34**]) Let $f$ be a continuous homomorphism of a topological group $G$ onto a feathered Abelian group $H$, and assume that for each compact subset $P$ of $H$ there exists a countably compact subspace $B$ of $G$ such that $f(B) = P$. Then the mapping $f$ is quotient.

9.8.Q. (A. V. Arhangel'skii [**34**]) Let $X$ be a countably compact topological group such that the space $C_p(X)$ is normal. Show that every compact space $Y$ that can be represented as a continuous image of $X$ is metrizable.

## Open Problems

9.8.1. Is it true that every compact topological group contains a dense countably compact sequential subgroup?

9.8.2. Is it true that every compact topological group contains a dense sequential subspace? A dense countably compact subgroup of countable tightness?

9.8.3. Let $f$ be a continuous mapping of a countably compact sequential topological group $G$ onto a compact Hausdorff space $Y$. Is $Y$ metrizable?

9.8.4. Let $f$ be a continuous mapping of a countably compact topological group of countable tightness onto a compact space $X$. Is $X$ then metrizable? What if one weakens 'countably compact' to 'pseudocompact'?

9.8.5. Let $X$ be a countably compact topological group such that the space $C_p(X)$ is normal (or even Lindelöf). Is $X$ then $\aleph_0$-monolithic?

9.8.6. Let $G$ be a separable countably compact topological group such that the space $C_p(G)$ is normal (or Lindelöf). Is $G$ then metrizable?

9.8.7. Let $G$ be a countably compact topological group of countable tightness. Is $C_p(G)$ normal or Lindelöf?

9.8.8. Let $G$ be a hereditarily separable countably compact topological group. Is $C_p(G)$ normal or Lindelöf?

9.8.9. Let $G$ be an $\omega$-bounded topological group of countable tightness. Is $C_p(G)$ normal or Lindelöf?

## 9.9. Bohr topology on discrete Abelian groups

The *Bohr topology* of a topological Abelian group $G$ is the topology on $G$ induced by the family $G^*$ of all continuous homomorphisms of $G$ to the circle group $\mathbb{T}$. In other words, the Bohr topology of $G$ is the coarsest group topology $\tau_b(G)$ on $G$ which satisfies $(G^+)^* = G^*$, where $G^+$ is the underlying group $G$ endowed with the topology $\tau_b(G)$. It is clear from the definition of the Bohr topology that $\tau_b(G)$ is coarser than the original topology of the group $G$ and that the topological group $G^+$ is precompact. According to Example 9.9.61, the topology $\tau_b(G)$ can fail to be Hausdorff, but this never happens if the group $G$ is locally compact (see Proposition 9.9.1).

Denote by $r_G$ the diagonal product of the family $G^*$. Then $r_G$ is a continuous homomorphism of $G$ to the product group $\mathbb{T}^{G^*}$. Clearly, $r_G$ is a monomorphism if and only if the family $G^*$ *separates elements of* $G$, that is, if for any two distinct elements $x, y \in G$ there exists $h \in G^*$ such that $h(x) \neq h(y)$. The closure of $r_G(G)$ in $\mathbb{T}^{G^*}$, denoted by $bG$, is a compact Hausdorff topological group. The group $bG$ is called the *Bohr compactification* of $G$. Note that $r_G\colon G \to bG$ is a continuous homomorphism of $G$ onto a dense subgroup of $bG$. In general, $r_G$ need not be injective. However, from Theorem 9.7.5 and the definition of the group $bG$ we obtain:

PROPOSITION 9.9.1. *The mapping $r_G\colon G \to bG$ is a continuous monomorphism, for every locally compact topological Abelian group $G$. Hence the Bohr topology $\tau_b(G)$ of $G$ is Hausdorff and precompact.*

The next result follows from the definition of the group $G^+$.

PROPOSITION 9.9.2. *Let $G$ be a topological Abelian group. Then every continuous homomorphism $f\colon G \to K$ to a compact Hausdorff group $K$ remains continuous when considered as a homomorphism of $G^+$ to $K$.*

PROOF. The closure in $K$ of the image $f(G)$ is a compact Abelian group, so we can assume without loss of generality that $K$ is Abelian. For every continuous character $\chi$ on $K$, the composition $\chi \circ f$ is a continuous character on $G$, so $\chi \circ f$ remains continuous on the group $G^+$. Since, by Theorem 9.4.11, the topology of the compact group $K$ is generated by the family of all continuous characters of $K$, the homomorphism $f\colon G^+ \to K$ is also continuous. □

In the next result, we give an internal characterization of the Bohr topology.

PROPOSITION 9.9.3. *The Bohr topology $\tau_b(G)$ of a topological Abelian group $G$ is the maximal precompact group topology on $G$ coarser than the original topology of $G$.*

PROOF. It follows from our definition that the topology $\tau_b(G)$ is precompact and coarser than the original topology of $G$. Conversely, let $t$ be an arbitrary precompact group topology on $G$ coarser than the original topology $\tau$ of $G$ (we do not require that $t$ be Hausdorff). Denote by $N$ the closure of the neutral element $e_G$ in $G_t = (G, t)$ and consider the quotient homomorphism $\pi\colon G_t \to G_t/N$. It is immediate from our definition of $N$ that every closed subset $F$ of $G_t$ satisfies $F = \pi^{-1}\pi(F)$. The quotient group $H = G_t/N$ is Hausdorff and precompact, so the Raĭkov completion $\varrho H$ of $H$ is a compact Hausdorff topological group.

Let $f = \pi \circ i$, where $i\colon G \to G_t$ is the identity isomorphism. Then $i$ is continuous by the choice of $t$, so $f$ is a continuous homomorphism of $G$ onto the subgroup $H$ of $\varrho H$. Since the group $\varrho H$ is compact and Hausdorff, it follows from Proposition 9.9.2 that $f\colon G^+ \to \varrho H$ is also continuous. We have thus proved that $f\colon G^+ \to H$ is continuous as well.

Take an arbitrary closed subset $F$ of the group $G_t$. Since the quotient homomorphism $\pi\colon G_t \to H$ is open, it follows from the equality $F = \pi^{-1}\pi(F)$ that $\pi(F)$ is closed in $H$. Hence, by the continuity of the homomorphism $f\colon G^+ \to H$, the set $i^{-1}(F) = f^{-1}(\pi(F))$ is closed in $G$. This implies the continuity of the isomorphism $i\colon G^+ \to G_t$ and shows that $t \subset \tau_b(G)$. Therefore, $\tau_b(G)$ is the maximal precompact group topology on $G$ coarser than $\tau$. □

COROLLARY 9.9.4. *If $G$ is a precompact Abelian topological group, then $G^+ = G$.*

Let $f\colon G \to H$ be a homomorphism of Abelian topological groups. Denote by $f^+$ the same homomorphism $f$ considered as a mapping of $G^+$ to $H^+$. It turns out that the continuity is preserved when passing from $f$ to $f^+$.

Proposition 9.9.5. *If a homomorphism $f\colon G \to H$ of topological Abelian groups is continuous, then so is $f^+\colon G^+ \to H^+$.*

Proof. Let $\tau_b(G)$ and $\tau_b(H)$ be the Bohr topologies of the groups $G$ and $H$, respectively. If $h\colon H \to \mathbb{T}$ is a continuous homomorphism, then the composition $h \circ f$ is a continuous homomorphism of $G$ to $\mathbb{T}$. Since the topologies $\tau_b(G)$ and $\tau_b(H)$ are initial with respect to the families $G^*$ and $H^*$ of continuous characters of $G$ and $H$, respectively, we conclude that the homomorphism $f^+$ is continuous. □

In the special case when an Abelian group $G$ carries the discrete topology, we denote the group $G$ with the Bohr topology $\tau_b(G)$ by $G^\#$ in place of $G^+$. An equivalent description of $G^\#$ is as follows: $G^\#$ is the underlying group $G$ equipped with the initial topology with respect to the family of *all homomorphisms* of $G$ to $\mathbb{T}$.

The next corollary follows immediately from Proposition 9.9.1 and the definition of the Bohr compactification of a topological Abelian group:

Corollary 9.9.6. *$G^\#$ is a precompact Hausdorff topological group, for every discrete Abelian group $G$. In addition, $r_G\colon G^\# \to bG$ is an isomorphic topological embedding, that is, $G^\#$ is topologically isomorphic to a dense subgroup of the group $bG$, the Bohr compactification of $G$.*

In what follows, we identify the group $G^\#$ with the dense subgroup $r_G(G^\#)$ of $bG$.

Corollary 9.9.7. *If $G$ is a discrete Abelian group, then every homomorphism $f\colon G^\# \to H$ to a precompact topological group $H$ is continuous.*

Proof. Let $\varrho H$ be the Raĭkov completion of $H$. Since the group $\varrho H$ is compact and $G^\# = G^+$, Proposition 9.9.2 implies that $f\colon G^\# \to \varrho H$ is continuous. The continuity of $f\colon G^\# \to H$ is now obvious. □

Given a homomorphism $f\colon G \to H$ of discrete Abelian groups, we denote by $f^\#$ the same mapping $f$ considered as a homomorphism of $G^\#$ to $H^\#$.

Corollary 9.9.8. *For every homomorphism $f\colon G \to H$ of discrete Abelian groups, the homomorphism $f^\#\colon G^\# \to H^\#$ is continuous. If $f(G) = H$, then $f^\#$ is open.*

Proof. Since the group $H^\#$ is precompact, it follows from Corollary 9.9.7 that $f^\#$ is continuous. Suppose now that $f(G) = H$. Let $K$ be the kernel of $f$ and $\pi\colon G^\# \to G^\#/K^\#$ be the quotient homomorphism. Then there exists an isomorphism $i\colon G^\#/K^\# \to H^\#$ satisfying $f^\# = i \circ \pi$. Clearly, $i$ is continuous since $\pi$ is open. Since the quotient group $G^\#/K^\#$ is precompact, the inverse isomorphism $i^{-1}\colon H^\# \to G^\#/K^\#$ is also continuous by Corollary 9.9.7. Hence $i$ is a topological isomorphism and it follows from the equality $f^\# = i \circ \pi$ that $f^\#$ is open. □

A topological Abelian group $G$ is called *maximally almost periodic* if the corresponding group $G^+$ is Hausdorff or, equivalently, if the continuous characters of $G$ separate points in $G$. According to Theorem 9.7.5, every locally compact Abelian group is maximally almost periodic. Therefore, every discrete Abelian group is maximally almost periodic.

If $f\colon G \to H$ and $g\colon H \to K$ are continuous homomorphisms of topological Abelian groups, then $(g \circ f)^+ = g^+ \circ f^+$ is a continuous homomorphism of $G^+$ to $K^+$. In particular, if the groups $G$, $H$, and $K$ are discrete, then $(g \circ f)^\# = g^\# \circ f^\#$ is a continuous homomorphism of $G^{\#}$ to $K^{\#}$. Therefore, $^{+}$ and $^{\#}$ are covariant functors from the categories of maximally almost periodic groups and discrete Abelian groups, respectively, to the category of precompact Abelian topological groups.

We present below three useful properties of Abelian groups with the Bohr topology. Conditions b) and c) of the next proposition mean that the functor $^{\#}$ "respects" subgroups and quotient groups.

PROPOSITION 9.9.9. *Let $H$ be a subgroup of a discrete Abelian group $G$. Then:*

a) *$H$ is closed in $G^{\#}$;*
b) *$H^{\#}$ is a topological subgroup of $G^{\#}$;*
c) *$(G/H)^{\#} = G^{\#}/H^{\#}$.*

PROOF. Let $\pi\colon G \to G/H$ be the canonical homomorphism.

a) The group $(G/H)^{\#}$ is Hausdorff by Corollary 9.9.6, while the homomorphism $\pi^{\#}\colon G^{\#} \to (G/H)^{\#}$ is continuous by Corollary 9.9.8. Hence, the group $H$ is closed in $G^{\#}$ as the kernel of $\pi^{\#}$.

b) The restriction to $H$ of a character of $G$ is a character of $H$, so the topology $\tau_b(H)$ of the group $H^{\#}$ is always finer than the topology of $H$ inherited from $G^{\#}$. On the other hand, since the circle group $\mathbb{T}$ is divisible, Theorem 1.1.6 implies that every character of $H$ can be extended to a character of $G$. This means that $\tau_b(H)$ is coarser than the topology of $H$ inherited from $G^{\#}$. Therefore, the two topologies on $H$ coincide.

c) The groups $(G/H)^{\#}$ and $G^{\#}/H^{\#}$ are algebraically isomorphic and, since $H$ is closed in $G^{\#}$ by a), the quotient group $G^{\#}/H^{\#}$ is Hausdorff and precompact. Hence the identity isomorphism $i\colon (G/H)^{\#} \to G^{\#}/H^{\#}$ is continuous by Corollary 9.9.7. Let $p\colon G^{\#} \to G^{\#}/H^{\#}$ be the quotient homomorphism. It is clear that $p = i \circ \pi^{\#}$.q

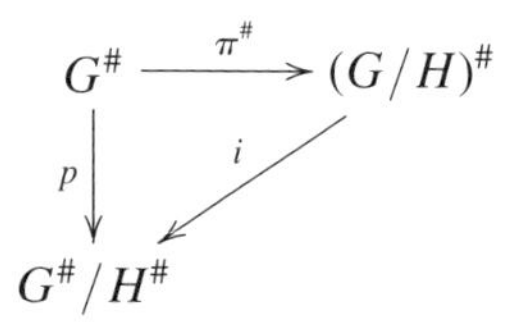


If $U$ is an open set in $(G/H)^{\#}$, then $i(U) = p((\pi^{\#})^{-1}(U))$ is open in $G^{\#}/H^{\#}$, so $i$ is a topological isomorphism. □

COROLLARY 9.9.10. *If $H$ is a subgroup of a discrete Abelian group $G$, then the Bohr compactification $bH$ of $H$ is topologically isomorphic to the closure of $H$ in the Bohr compactification $bG$ of the group $G$.*

PROOF. It follows from Corollary 9.9.6 that $G^{\#}$ is a topological subgroup of the compact group $bG$. Hence, b) of Proposition 9.9.9 implies that $H^{\#}$ is a dense subgroup of the compact group $K = cl_{bG} H$. Then, again by Corollary 9.9.6, $H^{\#}$ is a dense subgroup of the compact topological group $bH$. Therefore, the compact groups $bH$ and $K$ are topologically isomorphic, by Theorem 3.6.14. □

Let us show that the functor $^{\#}$ preserves finite products.

PROPOSITION 9.9.11. *Let $G = H_1 \times \cdots \times H_n$ be a product of discrete Abelian groups. Then the identity mapping $i$ of $G^{\#}$ onto $H_1^{\#} \times \cdots \times H_n^{\#}$ is a topological isomorphism.*

PROOF. Since the group $H_1^\# \times \cdots \times H_n^\#$ is precompact, it follows from Corollary 9.9.7 that $i$ is continuous. Conversely, for every $k \leq n$, let $j_k$ be the canonical embedding of $H_k$ to the product group $G = H_1 \times \cdots \times H_n$. If $\varphi$ is a character on $G^\#$, denote by $\varphi_k$ the composition $\varphi \circ j_k$ for $k = 1, \ldots, n$. Then $\varphi_k$ is a continuous character on $H_k^\#$, and we can define a continuous character $\psi$ on $H_1^\# \times \cdots \times H_n^\#$ by letting $\psi(x) = \sum_{k=1}^n \varphi_k(x_k)$ for each $x = (x_1, \ldots, x_n) \in G$. It is easy to verify that $\varphi = \psi$; thus, every character on $G^\#$ remains continuous as a character on $H_1^\# \times \cdots \times H_n^\#$. Hence, $i$ is a topological isomorphism. □

Proposition 9.9.11 is not valid either for infinite direct sums or for infinite topological products of discrete Abelian groups (see Exercise 9.9.j).

COROLLARY 9.9.12. *Let $G = G_1 \times G_2$ be the product of discrete Abelian groups. Then the groups $bG$ and $bG_1 \times bG_2$ are topologically isomorphic.*

PROOF. It follows from Corollary 9.9.6 and Proposition 9.9.11 that $G^\#$ is topologically isomorphic to a dense subgroup of the compact topological groups $bG$ and $bG_1 \times bG_2$. Since compact topological groups are Raĭkov complete, Theorem 3.6.14 implies that the groups $bG$ and $bG_1 \times bG_2$ are topologically isomorphic. □

For the further study of Abelian groups with the Bohr topology, we need several concepts and facts from the theory of abstract Abelian groups (see [**409**, Section 4.2]).

Given an Abelian group $G$, we denote by $tor(G)$ the *torsion subgroup* of $G$ which consists of all elements of $G$ of finite order (see Section 1.1). For example, the torsion subgroup of the circle group $\mathbb{T}$ is the group

$$tor(\mathbb{T}) = \{e^{2\pi qi} : q \in \mathbb{Q}\}.$$

Note that the quotient group $G/tor(G)$ is *torsion-free*, that is, all non-zero elements of $G/tor(G)$ have infinite order.

Here we extend the concept of linearly independent subsets of Abelian groups defined in Section 7.10 to subsets which contain elements of finite order, and define the $p$-rank of $G$, for each prime $p$.

Suppose that $A$ is a non-empty subset of an Abelian group $G$ with neutral element $e$. If, for any distinct elements $a_1, \ldots, a_n \in A$ and any integers $m_1, \ldots, m_n$, the equality $m_1 a_1 + \cdots + m_n a_n = e$ implies that $m_i x_i = e$, for each $i = 1, \ldots, n$, we say that $A$ is an *independent subset* of $G$.

Zorn's lemma implies that every independent subset of $G$ is contained in a maximal independent subset. One can choose independent subsets consisting either of elements of infinite order or of elements having prime power orders, for a given prime $p$. In the first case, this gives rise to the *torsion-free rank* $r_0(G)$ of $G$ considered in Section 7.10, while in the second case we obtain the definition of the *p-rank* of $G$. More precisely, the $p$-rank $r_p(G)$ of $G$ is the cardinality of a maximal independent subset of elements of $p$-power orders in $G$. Similarly to the torsion-free rank $r_0(G)$, the definition of $r_p(G)$ does not depend on the choice of a corresponding maximal independent subset (see [**409**, 4.2.1]).

LEMMA 9.9.13. *Let $f : G \to H$ be a homomorphism of Abelian groups. If $H = f(G)$, then $r_0(H) \leq r_0(G)$.*

PROOF. Let $B$ be a maximal independent subset of $H$ of elements of infinite order. Choose a subset $A$ of $G$ such that $f(A) = B$ and the restriction of $f$ to $A$ is one-to-one.

Then $|A| = |B|$, the set $A$ is independent in $G$ and each element of $A$ has infinite order. This proves that $r_0(H) \leq r_0(G)$. □

Given an Abelian group $G$ and a prime number $p$, we denote by $G_p$ the set of all elements in $G$ of a $p$-power order. It is easy to see that $G_p$ is a subgroup of $G$ which is called the *p-primary component* of $G$, or a *primary component* of $G$ if we do not need to specify $p$. It is clear from the definitions that $r_p(G) = r_p(G_p)$. If $G = G_p$ for some prime $p$, then $G$ is called a *torsion p-group*.

A torsion Abelian group $G$ is said to be *bounded torsion* if there exists an integer $n \in \mathbb{N}$ such that $nx = 0_G$, for each $x \in G$.

THEOREM 9.9.14. *Every torsion Abelian group $G$ is the direct sum of the primary components of $G$.*

PROOF. First, we have to verify that $G = \sum_{p \in \mathbb{P}} G_p$. Indeed, let $x$ be an arbitrary non-zero element of $G$ and $n$ be the order of $x$. Then $n = p_1^{k_1} \dots p_m^{k_m}$, where $p_1, \dots, p_m$ are distinct primes and $k_1, \dots, k_m$ are positive integers. For every $i \leq m$, let $N_i = n/p_i^{k_i}$. Then the integers $N_1, \dots, N_m$ are relatively prime, so we can find integers $s_1, \dots, s_m$ such that $s_1 N_1 + \dots + s_m N_m = 1$. Put $y_i = s_i N_i x$ for each $i = 1, \dots, m$. Then $p_i^{m_i} y_i = s_i n\, x = 0_G$; it follows that $y_i \in G_{p_i}$ for all $i \leq m$. In addition, the choice of $s_1, \dots, s_m$ implies that $x = y_1 + \dots + y_m$ and hence, $x \in G_{p_1} + \dots + G_{p_m}$.

We claim that $G_p \cap (G_{q_1} + \dots + G_{q_k}) = \{0_G\}$, for any pairwise distinct primes $p, q_1, \dots, q_k$. Indeed, let $x = y_1 + \dots + y_k$, where $x \in G_p$ and $y_i \in G_{q_i}$ for each $i \leq k$. Then the order of $x$ is a power of $p$, while the order of $y_1 + \dots + y_k$ is $q_1^{n_1} \cdots q_k^{n_k}$, for some $n_1, \dots, n_k \in \mathbb{N}$. Therefore, $x = 0_G$. Thus, $G$ is isomorphic to the direct sum $\oplus_{p \in \mathbb{P}} G_p$. □

For an Abelian group $G$ and an integer $n > 0$, let

$$G[n] = \{x \in G : nx = 0_G\}.$$

Clearly, $G[n]$ is a subgroup of $G$ and the order of every non-zero element $x \in G[n]$ divides $n$.

LEMMA 9.9.15. *Let $G$ be an uncountable Abelian torsion $p$-group. Then $|G[p]| = |G|$.*

PROOF. Denote by $\varphi$ the homomorphism of $G$ to $G$ defined by $\varphi(g) = pg$ for each $g \in G$. It is clear that $\ker \varphi = G[p]$. Note that if $g, h \in G$ and $\varphi(h) = g$, then $\varphi^{-1}(g) = h + G[p]$. Therefore,

$$|\varphi^{-1}(g)| \leq |G[p]| \text{ for each } g \in G. \tag{9.14}$$

Consider an increasing sequence $H_1 \subset H_2 \subset \cdots$ of subgroups of $G$ defined by $H_1 = G[p]$ and $H_{n+1} = \varphi^{-1}(H_n)$, for each integer $n \geq 1$. Let $\tau = |H_1|$ and suppose that we have proved the inequality $|H_n| \leq \tau \cdot \omega$ for some integer $n \geq 1$. By (9.14) and the inductive hypothesis, we have $|H_{n+1}| \leq \tau \cdot \omega$. Note that $H_n = G[p^n]$ for each $n \geq 1$, so $G = \bigcup_{n=1}^{\infty} H_n$. Since $|G| > \omega$, we conclude that $|G| = \tau = |G[p]|$. □

COROLLARY 9.9.16. *Every uncountable torsion $p$-group $G$ satisfies $r_p(G) = |G|$.*

PROOF. By Lemma 9.9.15, the cardinalities of $G$ and $G[p]$ coincide. Let $A$ be a maximal independent subset of the group $G[p]$. It is easy to verify that $A$ is also a maximal independent subset of $G$, so it suffices to show that $|A| = |G[p]|$. By the choice of

$A$, for every $x \in G[p] \setminus A$ the set $A \cup \{x\}$ is dependent, so that $nx \in \langle A \rangle$, for some $n \notin p\mathbb{Z}$. Therefore, there are integers $a, b \in \mathbb{Z}$ such that $an + bp = 1$; it follows that $x = (an + bp)x = anx \in \langle A \rangle$. Hence, $G[p] = \langle A \rangle$, so $A$ is uncountable and $|A| = |\langle A \rangle| = |G[p]| = |G|$. □

Another useful characteristic of an Abelian group $G$ is the *complete rank* which we will often simply call the *rank* of $G$:

$$r(G) = r_0(G) + \sum_{p \in \mathbb{P}} r_p(G),$$

where $\mathbb{P}$ is the set of all prime integers. It is clear that $r(G) = r_0(G)$ if the group $G$ is torsion-free, and $r(G) = r_p(G)$ if $G$ is a torsion $p$-group, for some $p \in \mathbb{P}$.

Let us establish several elementary facts concerning the rank of Abelian groups. The first auxiliary fact is almost immediate.

LEMMA 9.9.17. *Let $A_0$ be a maximal independent subset of an Abelian group $G$ consisting of elements of infinite order. For every prime number $p$, let $A_p$ be a maximal independent subset of the subgroup $G_p$ of $G$. Then $A = A_0 \cup \bigcup_{p \in \mathbb{P}} A_p$ is a maximal independent subset of $G$.*

PROOF. A simple argument about the orders of elements shows that the set $A$ is independent. Let us verify the maximality of $A$. Take an arbitrary $x \in G \setminus A$.

**Case 1.** $x \in tor(G)$. By Theorem 9.9.14, there are distinct primes $p_1, \ldots, p_n$ such that $x \in G_{p_1} + \cdots + G_{p_n}$. Then $x = y_1 + \cdots + y_n$, where $y_i \in G_{p_i}$ for each $i \leq n$. Take any $i \leq n$ and let $p_i^{s_i}$ be the order of $y_i$. Then either $y_i \in A_{p_i}$ or the set $\{y_i\} \cup A_{p_i}$ is dependent. In either case, there exists an integer $m_i$ with $0 < m_i < p_i^{s_i}$ such that $m_i y_i \in \langle A_{p_i} \rangle$. Put $M = m_1 \cdots m_n$. Then $My_i \in \langle A_{p_i} \rangle$ for each $i \leq n$, and it follows that

$$Mx = My_1 + \cdots + My_n \in \langle A_{p_1} \cup \ldots \cup A_{p_n} \rangle \subset \langle A \rangle.$$

In addition, $M = m_1 \cdots m_n < p_1^{s_1} \cdots p_n^{s_n} = o(x)$, so that $Mx \neq 0_G$. Hence, the set $A \cup \{x\}$ is dependent.

**Case 2.** $x \notin tor(G)$. Then the set $A_0 \cup \{x\}$ is dependent, and so is the set $A \cup \{x\}$.

Thus, $A$ is maximal. □

The next result relates the rank and the cardinality of maximal independent sets.

PROPOSITION 9.9.18. *Let $B$ be a maximal independent subset of an Abelian group $G$. Then $|B| = r(G)$.*

PROOF. Let the sets $A_0$, $A_p$ and $A$ be as in Lemma 9.9.17. Then $|A_0| = r_0(G)$ and $|A_p| = r_p(G)$ for each $p \in \mathbb{P}$. Since $A_0 \cap A_p = \varnothing$ and $A_p \cap A_q = \varnothing$ for distinct $p, q \in \mathbb{P}$, it follows from the definition of $r(G)$ that $r(G) = |A|$. Since any two maximal independent subsets of $G$ have the same cardinality, we have $|B| = |A| = r(G)$. □

COROLLARY 9.9.19. *If $H$ is a subgroup of an Abelian group $G$, then $r(H) \leq r(G)$.*

PROOF. Take a maximal independent subset $B$ of the group $H$. By Zorn's Lemma, there exists a maximal independent subset $A$ of $G$ which contains $B$. Now it follows from Proposition 9.9.18 that $r(H) = |B| \leq |A| = r(G)$. □

Notice that the group $\mathbb{Q}$ is infinite, while $r(\mathbb{Q}) = r_0(\mathbb{Q}) = 1$. It turns out that for uncountable groups, the rank and the cardinality always coincide.

PROPOSITION 9.9.20. *Every uncountable Abelian group $G$ contains an independent subset $A$ satisfying $|A| = |G|$. Hence, $r(G) = |G|$ for such a group $G$.*

PROOF. Again, let $A_0$ and $A_p$ for $p \in \mathbb{P}$ be as in Lemma 9.9.17. Then $A = A_0 \cup \bigcup_{p\in\mathbb{P}} A_p$ is a maximal independent subset of $G$. Suppose to the contrary that $|A| < |G|$. By the definition of $A$, for every non-zero element $x \in G$ there exists an integer $n$ such that $0_G \neq nx \in \langle A\rangle$. Since $|\langle A\rangle| \leq |A| \cdot \omega < |G|$, we can find an element $g \in \langle A\rangle \setminus \{0_G\}$, an integer $m \neq 0$ and a set $D \subset G$ with $|D| > |A| \cdot \omega$ such that $g = mx$ for each $x \in D$. Let $x_0 \in D$ be arbitrary. Then $m(x - x_0) = 0_G$ for each $x \in D$; therefore, $|G[m]| \geq |D|$. By Theorem 9.9.14, the group $H = G[m]$ is the direct sum of the $p$-primary components $H_p$, where $p$ runs through the prime divisors of $m$. Hence $|H_p| = |H|$ for some $p$. It follows from Corollary 9.9.16 that $r_p(H_p) = |H_p| = |H| \geq |D|$, so $|A_p| = r_p(H_p) \geq |D| > |A| \geq |A_p|$. This contradiction proves the equality $|A| = |G|$. Proposition 9.9.18 implies that $r(G) = |G|$. □

Let us go back to Bohr topologies. The next simple result has many applications.

PROPOSITION 9.9.21. *Let $G$ be an Abelian group. Then every independent subset of $G$ is closed and discrete in $G^\#$.*

PROOF. Suppose that a set $A \subset G$ is independent and take an arbitrary $x \in A$. Since, by a) of Proposition 9.9.9, the subgroup $H_x = \langle A \setminus \{x\}\rangle$ is closed in $G^\#$ and $A \cap (G \setminus H_x) = \{x\}$, we conclude that the point $x$ is isolated in $A$. Hence the set $A$ is discrete as a subspace of $G^\#$. To show that $A$ is closed in $G^\#$, take a point $y \in \overline{A}$. Then $y \in \overline{A} \subset \overline{\langle A\rangle} = \langle A\rangle$, so there exists a finite set $B \subset A$ such that $y \in \langle B\rangle$. Hence, $U = G \setminus \langle A \setminus B\rangle$ is an open neighbourhood of $y$ in $G^\#$ which satisfies $U \cap A \subset B$, that is, $|U \cap A| < \omega$. Therefore, $A$ is closed in $G^\#$. □

COROLLARY 9.9.22. [**K. P. Hart and J. van Mill**] *Let $G$ be a discrete Abelian group. Then $G^\#$ contains a closed discrete subset of size $|G|$.*

PROOF. If $G$ is uncountable, the conclusion follows from Propositions 9.9.20 and 9.9.21. The case of a finite group $G$ is trivial, so we can assume that $|G| = \omega$. Since every compact topological group is either finite or uncountable, by b) of Corollary 5.2.7, $G^\#$ cannot be compact. In addition, since $G^\#$ is countable (hence, Lindelöf), it is not countably compact. Hence, $G^\#$ contains an infinite closed discrete subset. □

It will be shown below in Theorem 9.9.30 that, for every infinite Abelian group $G$, the topological group $G^\#$ is very far from being compact. It turn out, however, that the group $G^\#$ is realcompact (that is, Hewitt–Nachbin complete), for each $G$ of a not extremely large cardinality. A proof of this fact, presented below, is based on Theorem 9.8.5. We start with two lemmas. The first of them is evident, so we omit its proof.

LEMMA 9.9.23. *Let $H$ be an abstract group and $F \subset \mathbb{T}^H$ a set. If every $f \in F$ satisfies $f(xy) = f(x)f(y)$ for all $x, y \in H$, then so does every $p \in cl_{\mathbb{T}^H}(F)$.*

Recall that, for a discrete Abelian group $G$, the image $G^\# = r_G(G)$ and its closure $bG$ in $\mathbb{T}^{G^*}$ are topological subgroups of the compact group $\mathbb{T}^{G^*}$, so every element of $G^\#$ can be considered as a character of the dual group $G^*$. Hence, by Lemma 9.9.23, every element of $bG$ is a (perhaps, discontinuous) character of the compact dual group $G^*$.

LEMMA 9.9.24. *Let $G$ be a discrete Abelian group and $p \in bG$. Then $p$ belongs to the $G_\delta$-closure of $G^\#$ in $bG$ iff $p$ is continuous on every countable subset of $G^*$.*

PROOF. Let $V$ be a $G_\delta$-set in $bG$ containing $p$. There exists a countable subset $C$ of the compact group $G^*$ such that $V_C(p) = \{q \in bG : q \restriction C = p \restriction C\} \subset V$. Then $V_C(p)$ is also a $G_\delta$-set in $bG$. Since $q \restriction C$ is continuous on $C$ for each $q \in r_G(G)$, we have that $V_C(p) \cap r_G(G) \neq \varnothing$ if and only if $p$ is continuous on $C$. □

THEOREM 9.9.25. [**W. W. Comfort**, **S. Hernández**, **and F. J. Trigos- Arrieta**] *Let $G$ be a discrete Abelian group. Then $G^\#$ is realcompact if and only if the cardinality of $G$ is Ulam non-measurable.*

PROOF. Suppose that the cardinality of $G$ is Ulam non-measurable. From the inequality $|G^*| \leq 2^{|G|}$ it follows that the cardinality of the dual group $G^*$ is also Ulam non-measurable. Take an arbitrary element $p \in bG$ that belongs to the $G_\delta$-closure of the subgroup $G^\#$. By Lemma 9.9.24, the character $p$ is continuous on every countable subset of the compact dual group $G^*$, so $p$ is continuous on $G^*$ by Theorem 9.8.5. By the Pontryagin–van Kampen duality theorem, the latter means that $p = r_G(g)$ for some $g \in G$, that is, $p \in G^\#$. Thus, the $G_\delta$-closure of $G^\#$ in $bG$ coincides with $G^\#$ or, equivalently, the complement $bG \setminus G^\#$ is a union of closed $G_\delta$-sets in $bG$. Hence, $G^\#$ is the intersection of cozero sets in $bG$ and Lemma 8.3.5 implies that the space $G^\#$ is realcompact.

Conversely, suppose that the space $G^\#$ is realcompact. If $|G| \leq \omega$, there is nothing to prove. We can assume, therefore, that $|G| > \omega$. By Corollary 9.9.22, the group $G^\#$ contains a closed discrete subspace $D$ with $|D| = |G|$. Then $D$ is realcompact as a closed subspace of the realcompact space $G^\#$, so the cardinality of $D$ cannot be Ulam measurable according to [**165**, 8.5.13 (h)]. □

All compact subsets of a discrete Abelian group $G$ are finite. When one takes the Bohr topology $\tau_b(G)$ of $G$, there might appear infinite compact sets in the group $G^\#$. However, we will show in Theorem 9.9.30 that this is not the case. Our proof of this important result is based on the study of the subsets of an Abelian group that can be sent to a dense subset of $\mathbb{T}$ by means of a character of the group. We start with subsets of the group $\mathbb{Z}$.

LEMMA 9.9.26. *For every infinite set $A \subset \mathbb{Z}$, there exists a homomorphism $\varphi\colon \mathbb{Z} \to \mathbb{T}$ such that the image $\varphi(A)$ is dense in $\mathbb{T}$.*

PROOF. Our aim is to show that there exists an element $t \in \mathbb{T}$ such that the set $A_t = \{t^n : n \in A\}$ is dense in $\mathbb{T}$. Let $\{s_k : k \in \omega\}$ be a dense subset of $\mathbb{T}$. It suffices to choose an element $t \in \mathbb{T}$ satisfying the following condition:

$$\text{For every } k \in \omega, \text{ there exists } n \in A \text{ such that } ||t^n - s_k|| \leq 2^{-k}, \tag{9.15}$$

where $||\cdot||$ denotes the usual norm in the complex plane $\mathbb{C}$. Denote by $I_0$ a non-trivial closed connected arc in $\mathbb{T}$ of length $l(I_0) \leq 1$ which does not contain 1. There exists $n_0 \in A$ such that $|n_0| \cdot l_0 \geq 2\pi$ and, hence, $[I_0^{n_0}] = \{x^{n_0} : x \in I_0\} = \mathbb{T}$. Suppose that we have defined a decreasing sequence $I_0 \supseteq \cdots \supseteq I_k$ of non-trivial connected arcs in $\mathbb{T}$ such that the length $l_i$ of $I_i$ is less than or equal to $2^{-i}$ for each $i \leq k$. There exists $n_k \in A$ such that $|n_k| \cdot l_k \geq 2\pi$, so that $[I_k^{n_k}] = \mathbb{T}$. Pick a point $x_k$ in $I_k$ such that $x_k^{n_k} = s_k$ and choose a non-trivial closed connected arc $I_{k+1}$ containing $x_k$ such that $I_{k+1} \subset I_k$, $l(I_{k+1}) \leq 2^{-k-1}$ and $||x^{n_k} - s_k|| \leq 2^{-k}$ for each $x \in I_{k+1}$. This finishes our construction.

Since $l(I_k) \leq 2^{-k}$ for each $k \in \omega$, the intersection $\bigcap_{k=0}^{\infty} I_k$ consists of a single point, say $t$. Our construction guarantees that $t$ satisfies (9.15), so that $\overline{A_t} = \mathbb{T}$. It remains to define $\varphi(n) = t^n$ for each $n \in \mathbb{Z}$. Then the image $\varphi(A) = A_t$ is dense in $\mathbb{T}$. □

Lemma 9.9.26 is not valid for torsion groups, since the elements of an infinite set $A$ can have bounded orders. It turns out that this is the only obstacle for the existence of a character sending $A$ to a dense subset of $\mathbb{T}$.

LEMMA 9.9.27. *Let $A$ be an infinite subset of a torsion Abelian group $G$. If the orders of the elements of $A$ are unbounded, then there exists a homomorphism $\varphi\colon G \to \mathbb{T}$ such that the image $\varphi(A)$ is dense in $\mathbb{T}$.*

PROOF. Since $G$ is a torsion group, every finite subset of $G$ generates a finite subgroup. Choose a sequence $\{x_n : n \in \omega\}$ of elements of $A$ such that for each $n \in \omega$, the order $k_{n+1}$ of $x_{n+1}$ satisfies $k_{n+1} \geq 2^{n+1} \cdot |H_n|$, where $H_n$ is the subgroup of $G$ generated by $x_0, \ldots, x_n$. As in Lemma 9.9.26, we fix a countable dense subset $\{s_n : n \in \omega\}$ of $\mathbb{T}$. Our aim is to define a homomorphism $\varphi\colon G \to \mathbb{T}$ satisfying

$$||\varphi(x_n) - s_n|| \leq 2\pi/2^n \text{ for each } n \in \omega. \tag{9.16}$$

It is clear that (9.16) will imply the density of the image $\varphi(A)$ in $\mathbb{T}$.

Choose an element $t_0 \in \mathbb{T}$ of order $k_0 = \mathrm{ord}(x_0)$ and put $\varphi_0(x_0) = t_0$. This defines the homomorphism $\varphi_0$ on the subgroup $H_0$ of $G$ generated by the element $x_0$. The condition $||t_0 - s_0|| \leq 2\pi$ holds trivially. Suppose that for some $n \in \omega$, we have defined a homomorphism $\varphi_n$ on the subgroup $H_n$ of $G$ generated by $x_0, \ldots, x_n$. Denote by $m$ the minimal positive integer $l$ such that $l\,x_{n+1} \in H_n$. Since $k_{n+1} \geq 2^{n+1} \cdot |H_n|$, we conclude that $2^{n+1} \leq m \leq k_{n+1}$. Let $h = mx_{n+1}$ and $b = \varphi_n(h)$. Since the solutions of the equation $y^m = b$ in $\mathbb{T}$ form a right polygon with $m$ vertices on the unit circle, we can choose one of them $t_{n+1}$ so that $||t_{n+1} - s_{n+1}|| \leq 2\pi/m \leq 2\pi/2^{n+1}$. As in the proof of Theorem 1.1.6, we extend $\varphi_n$ to a homomorphism $\varphi_{n+1}$ of $H_{n+1} = \langle x_0, \ldots, x_n, x_{n+1}\rangle$ to $\mathbb{T}$ such that $\varphi_{n+1}(x_{n+1}) = t_{n+1}$.

Continuing in this way, we obtain a homomorphism $\varphi_\infty$ defined on the subgroup $H = \bigcup_{n\in\omega} H_n$ of $G$ which extends $\varphi_n$, for each integer $n \geq 0$. It follows from the construction that $\varphi_\infty$ satisfies (9.16); hence, $\varphi_\infty(A)$ is dense in $\mathbb{T}$. Finally, we apply Theorem 1.1.6 to extend $\varphi_\infty$ over the whole group $G$. □

LEMMA 9.9.28. *Suppose that $G$ is a bounded torsion Abelian group. Then all compact sets in $G^\#$ are finite.*

PROOF. First, we consider the case when the exponent of $G$ is a prime number $p$. Obviously, every infinite set $C \subset G$ contains an infinite independent subset. Indeed, if $A$ is a finite independent subset of $C$, then $H = \langle A\rangle$ is a finite subgroup of $G$, so we can take an arbitrary element $x \in C \setminus H$, thus obtaining the independent set $A \cup \{x\} \subset C$. This implies immediately that every maximal independent subset of $C$ is infinite. Since independent subsets of $G$ are closed and discrete in $G^\#$ by Proposition 9.9.21, no infinite subset of $G^\#$ can be compact.

Let $m$ be the exponent of $G$, and suppose that we have proved the lemma for all torsion Abelian groups of any exponent less than $m$. By the way of contradiction, assume that $C$

is an infinite compact subset of the group $G^\#$. It follows from Proposition 9.9.21 that $C$ does not contain infinite independent subsets. Let $A$ be a maximal independent subset of $C$. Denote by $H$ the subgroup of $G$ generated by $A$. Clearly, $A$ and $H$ are finite. Since the set $A \cup \{x\}$ is dependent for every $x \in C \setminus H$, there exists a proper divisor $k$ of $m$ such that $0_G \neq kx \in H$. Hence we can find a non-zero element $h_0 \in H$, a proper divisor $k_0$ of $m$ and an infinite subset $S$ of $C$ such that $k_0x = h_0$ for each $x \in S$. Since the multiplication by $k_0$ is continuous in $G^\#$, the set $P = \{x \in G : k_0x = h_0\}$ is closed in $G^\#$. This implies that $D = C \cap P$ is compact in $G^\#$. Clearly, $S \subset D$, so $D$ is infinite. Choose an element $x_0 \in D$. Then $k_0(x - x_0) = 0_G$ for each $x \in D$; it follows that $E = D - x_0$ is an infinite compact subset of the group $G[k_0]^\#$. Since the exponent of $G[k_0]$ is not greater than $k_0$ and $k_0 < m$, this contradicts the inductive hypothesis. □

LEMMA 9.9.29. *If $G$ is a finitely generated Abelian group, then all compact sets in $G^\#$ are finite.*

PROOF. It follows from Lemma 9.9.26 that all compact sets in the group $\mathbb{Z}^\#$ are finite. Indeed, if $A$ is an infinite subset of $\mathbb{Z}$, then there exists a homomorphism $\varphi\colon \mathbb{Z} \to \mathbb{T}$ such that $\varphi(A)$ is dense in $\mathbb{T}$. Clearly, the character $\varphi\colon \mathbb{Z}^\# \to \mathbb{T}$ is continuous, so if $A$ were compact in $\mathbb{Z}^\#$, the image $\varphi(A)$ would be the whole group $\mathbb{T}$, which is impossible, since $G$ is countable and $\mathbb{T}$ is uncountable.

Finally, let $G$ be an arbitrary finitely generated Abelian group. Then $G$ is algebraically isomorphic to the product $H_1 \times \cdots \times H_n$ of cyclic groups $H_1, \ldots, H_n$, by [**409**, 4.2.10]. In other words, $G$ is isomorphic to the group $\mathbb{Z}^m \times K$, where $K$ is a finite Abelian group and $m$ is the number of infinite groups among $H_1, \ldots, H_n$. Hence, by Proposition 9.9.11, the groups $G^\#$ and $(\mathbb{Z}^\#)^m \times K_d$ are topologically isomorphic, where $K_d$ is the group $K$ with the discrete topology. Let $C$ be a compact set in the group $(\mathbb{Z}^\#)^m \times K_d$. Taking projections of $C$ onto the factors and applying the fact that all compact sets in $\mathbb{Z}^\#$ are finite, we conclude that $C$ is also finite. □

We are now in a position to prove the first main result of this section.

THEOREM 9.9.30. [**H. Leptin**] *For every Abelian group $G$, all compact subsets of $G^\#$ are finite.*

PROOF. Suppose to the contrary that $C$ is an infinite compact subset of $G^\#$. We can assume without loss of generality that the group $G$ is countable. Indeed, choose a countable infinite subset $S$ of $C$ and consider the subgroup $L = \langle S \rangle$ of $G^\#$. By a) and b) of Proposition 9.9.9, $L^\#$ is a closed topological subgroup of $G^\#$. It is clear that $S \subset C \cap L$, so the intersection $D = C \cap L$ is an infinite compact subset of the countable group $L^\#$.

Let $A$ be a maximal independent subset of $C$. Since $C$ is compact and $A$ is closed and discrete in $G^\#$ by Proposition 9.9.21, it follows that $A$ is finite. Let $H = \langle A \rangle$ and consider the canonical homomorphism $\pi\colon G \to G/H$. Corollary 9.9.8 implies that the homomorphism $\pi^\#\colon G^\# \to (G/H)^\#$ is continuous. We claim that the compact subset $D = \pi(C)$ of the group $(G/H)^\#$ is infinite and the subgroup $K$ of $G/H$ generated by $D$ is a torsion group. Indeed, if $D$ is finite, take a finite set $F \subset G$ such that $\pi(F) = D$. Then $C \subset H + F \subset \langle A \cup F \rangle$, which contradicts Lemma 9.9.29. To show that $K$ is a torsion group, take an arbitrary element $y \in K$ and choose $x \in \langle C \rangle$ with $\pi(x) = y$. Since $A$ is a maximal independent subset of

$C$, there exists an integer $n > 0$ such that $nx \in \langle A \rangle = H$. Then $ny = \pi(nx)$ is the neutral element of $G/H$. This proves the claim.

If the orders of elements of $D$ are bounded, then $K$ is a bounded torsion group. Since $D$ is an infinite compact subset of $K^\#$, this contradicts Lemma 9.9.28. If the orders of the elements of $D$ are unbounded, then, by Lemma 9.9.27, there exists a character $\varphi\colon K \to \mathbb{T}$ such that $\varphi(D)$ is dense in $\mathbb{T}$. Since $\varphi$ is continuous as a homomorphism of $K^\#$ to $\mathbb{T}$ and $D \subset K^\#$ is compact, we must have $\varphi(D) = \mathbb{T}$. This contradicts the fact that $D$ is countable. Thus, the set $C$ is finite. □

In the next theorem we establish another important property of Bohr topologies. To avoid confusion, we use notation

$$h^{\leftarrow}(U) = \{x \in G : h(x) \in U\}$$

for a character $h \in G^*$ and a set $U \subset \mathbb{T}$.

THEOREM 9.9.31. [**E. van Douwen**] *The group $G^\#$ is zero-dimensional, for every discrete Abelian group $G$.*

PROOF. Let us show that $G^\#$ has a base of clopen sets at the neutral element $0_G$. Let $\mathcal{N}(1)$ be the family of open neighbourhoods of the identity 1 in the circle group $\mathbb{T}$. Since the family

$$\{h^{\leftarrow}(U) : h \in G^*,\ U \in \mathcal{N}(1)\}$$

is a subbase for the topology of $G^\#$ at $0_G$, it suffices to find, for each subbasic neighbourhood $h^{\leftarrow}(U)$ of $0_G$ in $G^\#$, a clopen set $V$ in $G^\#$ with $0_G \in V \subset h^{\leftarrow}(U)$.

If $G$ is a torsion group, then the image $h(G)$ is a subgroup of the countable group $tor(\mathbb{T})$ for each $h \in G^*$. Hence $G^\#$ is topologically isomorphic to a subgroup of $tor(\mathbb{T})^\kappa$, for some cardinal $\kappa$. Since the subgroup $tor(\mathbb{T})$ of $\mathbb{T}$ is zero-dimensional, so are the groups $tor(\mathbb{T})^\kappa$ and $G^\#$.

In the general case, notice that $\mathbb{T}$ contains two proper subgroups $A$ and $B$ such that $A \cap B = \{1\}$ and $\mathbb{T} = A \cdot B$. Indeed, take $A = t\mathbb{T}$. The group $A$ is divisible, so there exists a homomorphism $\varphi\colon \mathbb{T} \to A$ which extends the identity automorphism $id_A$ of $A$. Then the groups $A$ and $B = \ker \varphi$ are as required. Clearly, the groups $A$ and $B$ are zero-dimensional, since they are proper subgroups of $\mathbb{T}$.

The function $m\colon A \times B \to \mathbb{T}$ defined by $m(a, b) = a \cdot b$ for all $a \in A$ and $b \in B$ is continuous since it is the restriction of the multiplication in $\mathbb{T}$. It follows from our choice of $A$ and $B$ that $m$ is an isomorphism of $A \times B$ onto $\mathbb{T}$.

We claim that, for a given $h \in G^*$, the composition $f = m^{-1} \circ h$ is a continuous homomorphism of $G^\#$ to $A \times B$ (even if the function $m^{-1}$ is discontinuous as a mapping of the connected space $\mathbb{T}$ onto the zero-dimensional space $A \times B$). Indeed, $A \times B$ as a subgroup of the compact group $\mathbb{T} \times \mathbb{T}$, so $m^{-1} \circ h\colon G^\# \to A \times B$ is continuous by Corollary 9.9.7.

Let $U \in \mathcal{N}(1)$ be arbitrary. Since $m$ is continuous and $A \times B$ is zero-dimensional, we can find a clopen neighbourhood $V$ of $(1, 1)$ in $A \times B$ such that $m(V) \subset U$. Then, by the continuity of $f$, the set $W = f^{-1}(V)$ is a clopen neighbourhood of the neutral element in $G^\#$, and $W = f^{-1}(V) = h^{\leftarrow}(m(V)) \subset h^{\leftarrow}(U)$. □

Since every group of the form $G^\#$ is precompact (hence, is a subgroup of a compact topological group), one can invoke Corollary 8.8.6 to strengthen the conclusion of Theorem 9.9.31 as follows:

THEOREM 9.9.32. *Every discrete Abelian group endowed with the Bohr topology is strongly zero-dimensional.*

One can show that closed subsets of an uncountable Abelian group $G^{\#}$ need not be $C^*$-embedded in $G^{\#}$ (see Exercise 9.9.f). It turns out, however, that all subgroups of $G^{\#}$ are $C$-embedded in $G^{\#}$. This will be established in Theorem 9.9.40.

Let $S$ be a space. A subset $A$ of a space $X$ is called $S$-*embedded* in $X$ if every continuous mapping of $A$ to $S$ can be extended to a continuous mapping of $X$ to $S$. It follows from the definition that the following implications hold:

$$\mathbb{R}\text{-embedded} \Leftrightarrow C\text{-embedded}, \quad I\text{-embedded} \Leftrightarrow C^*\text{-embedded},$$

where $I = [0, 1]$ is the closed unit interval. For the further study of Bohr topologies we need several facts about $S$-embedded subsets, where $S$ is one of the spaces $\mathbb{R}$, $I$, $\mathbb{N}$ or $2 = \{0, 1\}$. The following implications are trivial:

$$\mathbb{R}\text{-embedded} \Rightarrow I\text{-embedded}, \quad \mathbb{N}\text{-embedded} \Rightarrow 2\text{-embedded}.$$

Below we present several implications which are less obvious.

LEMMA 9.9.33. *Let $A$ be a strongly zero-dimensional subspace of a space $X$. Then:*

i) *if $A$ is 2-embedded in $X$, then it is $I$-embedded in $X$;*
ii) *if $A$ is $\mathbb{N}$-embedded in $X$, then it is $\mathbb{R}$-embedded in $X$;*

*If, in addition, $X$ is strongly zero-dimensional, then every $\mathbb{R}$-embedded subset of $X$ is $\mathbb{N}$-embedded in $X$.*

PROOF. i) This follows from Theorem 6.1.5.

ii) Suppose that $A$ is $\mathbb{N}$-embedded in $X$ and consider a continuous function $f\colon A \to \mathbb{R}$. Since $A$ is strongly zero-dimensional, we can choose, for every $n \in \mathbb{N}$, a clopen set $U_n$ in $A$ such that $f^{-1}([n, n+1]) \subset U_n \subset f^{-1}(n-1, n+2)$. For every $x \in A$, set

$$h(x) = \min\{n \in \mathbb{Z} : x \in U_n\}.$$

Then $h\colon A \to \mathbb{Z}$ is a continuous function and since $A$ is $\mathbb{N}$-embedded in $X$, $h$ admits a continuous extension $\tilde{h}\colon X \to \mathbb{Z}$. The family $\gamma = \{\tilde{h}^{-1}(n) : n \in \mathbb{Z}\}$ is a disjoint covering of $X$ by clopen sets such that $f\restriction E$ is bounded for each $E \in \gamma$. As $A$ is $\mathbb{N}$-embedded in $X$, it is 2-embedded in $X$. Hence $A \cap E$ is 2-embedded in $E$ for each $E \in \gamma$. Since $A \cap E$ is strongly zero-dimensional (as a clopen subset of $A$), it follows from i) that $A \cap E$ is $C^*$-embedded in $E$. So we can find a continuous function $\varphi_E\colon E \to \mathbb{R}$ which extends $f\restriction A \cap E$. Then $\varphi = \bigcup_{E\in\gamma} \varphi_E$ is a continuous real-valued function on $X$ that extends $f$.

Finally, let $B$ be an $\mathbb{R}$-embedded subset of $X$, and take a continuous function $f\colon B \to \mathbb{N}$. There exists a continuous function $g\colon X \to \mathbb{R}$ such that $g\restriction B = f$. If $T = \{n + \frac{1}{2} : n \in \mathbb{Z}\}$, then $g^{-1}(\mathbb{Z})$ and $g^{-1}(T)$ are disjoint zero-sets in the strongly zero-dimensional space $X$, so we can find a clopen set $U$ in $X$ such that $g^{-1}(\mathbb{Z}) \subset U \subset X \setminus g^{-1}(T)$. For every $x \in U$, choose an integer $n_x$ satisfying $n_x - \frac{1}{2} < g(x) < n_x + \frac{1}{2}$. Then the function $h\colon X \to \mathbb{N}$ given by $h(x) = n_x$ for $x \in U$ and $h(x) = 0$ for $x \in X \setminus U$ is continuous and its restriction to $B$ coincides with $f$. So $B$ is $\mathbb{N}$-embedded in $X$. □

Since every countable regular space is normal and strongly zero-dimensional, by [**165**, Coro. 6.2.8], the next corollary follows immediately from the last claim of Lemma 9.9.33:

COROLLARY 9.9.34. *Every closed subspace of a countable Tychonoff space $X$ is $\mathbb{N}$-embedded in $X$.*

Let us discuss more closely relations between $z$-embeddings, $C^*$-embeddings and $C$-embeddings. The results of this discussion with be applied to deduce the fact that $H^\#$ is $C$-embedded in $G^\#$, for every subgroup $H$ of a discrete Abelian group $G$ (see Theorem 9.9.40).

First, we present a necessary and sufficient condition for a $C^*$-embedded subset to be $C$-embedded.

LEMMA 9.9.35. *A $C^*$-embedded subset $Y$ of a space $X$ is $C$-embedded in $X$ if and only if $Y$ is completely separated from every zero-set in $X$ disjoint from it.*

PROOF. Suppose that $Y$ is $C$-embedded in $X$. Let $F$ be a zero-set in $X$ disjoint from $Y$ and $h$ be a continuous function on $X$ with values in the closed unit interval $[0, 1]$ such that $F = h^{-1}(0)$. Define a function $f$ on $Y$ by $f(y) = 1/h(y)$ for each $y \in Y$. Then $f$ is continuous on $Y$ and since $Y$ is $C$-embedded in $X$, there exists a continuous function $g\colon X \to \mathbb{R}$ extending $f$. Then the function $gh$ is continuous on $X$ and is equal to 0 on $F$ and to 1 on $Y$. This proves the necessity.

Conversely, let $f$ be a continuous real-valued function on $Y$. Take a homeomorphism $\varphi$ of $\mathbb{R}$ onto the open interval $(0, 1)$. Then $\varphi \circ f$ is a continuous bounded function on $Y$, so it admits an extension to a continuous function $g$ on $X$. Clearly,

$$Z = \{x \in X : |g(x)| \geq 1\}$$

is a zero-set in $X$ disjoint from $Y$, so by our hypothesis, there exists a continuous function $h$ on $X$ with values in $[0, 1]$ equal to 0 on $Z$ and to 1 on $Y$. Then the restriction of $g \cdot h$ to $Y$ coincides with $\varphi \circ f$ and satisfies $|(g \cdot h)(x)| \leq 1$ for each $x \in X$. Therefore, $\varphi^{-1} \circ (g \cdot h)$ is a continuous extension of $f$ over $X$. □

The following theorem helps to recognize when a $z$-embedded subset is $C$-embedded. Note that the theorem below strengthens Lemma 9.9.35.

THEOREM 9.9.36. *Let $Y$ be a $z$-embedded subset of a space $X$. Then:*

a) *$Y$ is $C$-embedded in $X$ if and only if $Y$ is completely separated in $X$ from every zero-set disjoint from it;*
b) *if $Y$ is a zero-set in $X$, then $Y$ is $C$-embedded in $X$.*

PROOF. a) The condition is necessary, by Lemma 9.9.35. For the sufficiency, suppose that $Y$ is completely separated in $X$ from every zero-set disjoint from it. Let us show that every two completely separated sets $F_1$ and $F_2$ in $Y$ are completely separated in $X$ and, hence, $Y$ is $C^*$-embedded in $X$ by Theorem 6.1.5. Clearly, we can assume that $F_1$ and $F_2$ are zero-sets in $Y$. Since $Y$ is $z$-embedded in $X$, there exist zero-sets $T_1$ and $T_2$ in $X$ such that $Y \cap T_i = F_i$ for $i = 1, 2$. Then $T_1 \cap T_2$ is a zero-set in $X$ and $Y \cap T_1 \cap T_2 = \varnothing$. By our assumption about $Y$, one can find a zero-set $C$ in $X$ such that $Y \subset C$ and $C \cap T_1 \cap T_2 = \varnothing$. Then the zero-sets $Z_1 = T_1 \cap C$ and $Z_2 = T_2 \cap C$ in $X$ are disjoint. Therefore, $Z_1$ and $Z_2$ are completely separated in $X$ by [**165**, Theorem 1.5.13]. Since $F_i \subset Z_i$ for $i = 1, 2$, the sets $F_1$ and $F_2$ are also completely separated in $X$, and hence, $Y$ is $C^*$-embedded in $X$. Finally, from Lemma 9.9.35 it follows that $Y$ is $C$-embedded in $X$.

b) Suppose that $Y$ is a zero-set in $X$, and let $F$ be any zero-set in $X$ disjoint from $Y$. According to [**165**, Theorem 1.5.13], the sets $Y$ and $F$ are completely separated in $X$, so condition a) of the theorem implies that $Y$ is $C$-embedded in $X$. □

If a topological group $G$ admits a continuous isomorphism onto a second-countable topological group $H$, then $|G| = |H| \leq \mathfrak{c}$. It turns out that for an Abelian topological group of the form $G^{\#}$, the restriction $|G| \leq \mathfrak{c}$ is also sufficient to guarantee the existence of such an isomorphism.

PROPOSITION 9.9.37. *If a discrete Abelian group $G$ satisfies $|G| \leq 2^{\kappa}$ for some $\kappa \geq \omega$, then there exists a continuous isomorphism of $G^{\#}$ onto a topological group $H$ with $w(H) \leq \kappa$. In particular, if $|G| \leq \mathfrak{c}$, then $G^{\#}$ admits a continuous isomorphism onto a second-countable topological group.*

PROOF. The dual group $G^*$ is compact by Proposition 9.5.5 and $w(G^*) \leq |G| \leq 2^{\kappa}$ by Theorem 9.6.6. It follows from c) of Corollary 5.2.7 that $d(G^*) \leq \kappa$, so we can choose a dense subset $S$ of $G^*$ with $|S| \leq \kappa$. Then $K = \langle S \rangle$ is a dense subgroup of $G^*$ and $|K| \leq \kappa$. Let $i \colon G^{\#} \to \mathbb{T}^K$ be the mapping given by $i(x)(h) = h(x)$ for all $x \in G$ and $h \in K$. Obviously, $i$ is a continuous homomorphism and $i$ is a monomorphism, since $K$ is dense in $G^*$. The group $H = i(G^{\#}) \subset \mathbb{T}^K$ satisfies $w(H) \leq |K| \leq \kappa$, so the isomorphism $i \colon G^{\#} \to H$ is as required. □

In the next theorem we present more topological consequences of the inequality $|G| \leq \mathfrak{c}$.

THEOREM 9.9.38. [**W. W. Comfort, S. Hernández, and F. J. Trigos-Arrieta**] *For a discrete Abelian group $G$, the following are equivalent:*

a) *$G^{\#}$ is hereditarily realcompact;*
b) *$\psi(G^{\#}) \leq \omega$;*
c) *$|G| \leq \mathfrak{c}$;*
d) *$G^{\#}$ admits a continuous isomorphism onto a second-countable topological group.*

PROOF. a) ⇒ b). Since $G^{\#}$ is a subgroup of the compact group $bG$ by Corollary 9.9.6, it follows from Corollary 5.3.29 that the group $G^{\#}$ is perfectly $\kappa$-normal and, hence, is a Moscow space. If the neutral element $e$ of $G^{\#}$ has uncountable pseudocharacter in $G^{\#}$, then $G^{\#} \setminus \{e\}$ is $G_\delta$-dense in $G^{\#}$. Therefore, Theorem 6.1.7 implies that $G^{\#} \setminus \{e\}$ is $C$-embedded in $G^{\#}$. In its turn, this implies that $G^{\#} \subset \upsilon(G^{\#} \setminus \{e\})$ and, hence, $G^{\#} \setminus \{e\} \neq \upsilon(G^{\#} \setminus \{e\})$. In other words, the subspace $G^{\#} \setminus \{e\}$ of $G^{\#}$ is not realcompact.

b) ⇒ c). If $\psi(G^{\#}) \leq \omega$, then $|G| \leq 2^{\omega} = \mathfrak{c}$ by Corollary 5.2.16.

c) ⇒ d). This is Proposition 9.9.37.

d) ⇒ a). Let $i \colon G^{\#} \to H$ be a continuous isomorphism of $G^{\#}$ onto a second-countable topological group $H$. Since $H$ is hereditarily realcompact, so is $G^{\#}$ by [**165**, 3.11.B]. □

The following fact plays an auxiliary role; it will be given a final form in Theorem 9.9.40.

COROLLARY 9.9.39. *Let $H$ be an arbitrary subgroup of a discrete Abelian group $G$, where $|G| \leq \mathfrak{c}$. Then $H^{\#}$ is $C$-embedded in $G^{\#}$.*

PROOF. It follows from Corollary 8.1.17 that the group $H^{\#}$ is $\mathbb{R}$-factorizable, so Theorem 8.2.5 implies that $H^{\#}$ is $z$-embedded in $G^{\#}$. The group $H^{\#}$ is closed in $G^{\#}$ by Proposition 9.9.9, so the quotient group $G^{\#}/H^{\#}$ is Hausdorff and satisfies $|G^{\#}/H^{\#}| \leq$

$|G^{\#}| \leq \mathfrak{c}$. Hence, $\psi(G^{\#}/H^{\#}) \leq \omega$ by Theorem 9.9.38. Since $H^{\#}$ is the kernel of the natural quotient homomorphism $\pi\colon G^{\#} \to G^{\#}/H^{\#}$, we conclude that $H^{\#}$ is a zero-set in $G^{\#}$ that is clearly $z$-embedded in $G^{\#}$. Hence, by b) of Theorem 9.9.36, $H^{\#}$ is $C$-embedded in $G^{\#}$. □

In general, subgroups of a topological group $G$ need not be $C$-embedded or $C^*$-embedded in $G$ — take the subgroup $\mathbb{Q}$ of $\mathbb{R}$ or the quasicyclic subgroup $\mathbb{Q}_{p^\infty}$ of $\mathbb{T}$, for a prime $p$. In fact, even closed subgroups can fail to be $z$-embedded in a larger Abelian topological group (see Example 9.9.e). Let us show that this never happens in the groups endowed with the Bohr topology.

THEOREM 9.9.40. *Let $H$ be a subgroup of a discrete Abelian group $G$. Then $H^{\#}$ is $C$-embedded and $\mathbb{N}$-embedded in $G^{\#}$.*

PROOF. Let $f\colon H^{\#} \to \mathbb{R}$ be an arbitrary continuous function. The groups $H^{\#}$, $G^{\#}$ and $bG$ are $\mathbb{R}$-factorizable by Corollary 8.1.17, so $H^{\#}$ is $z$-embedded in the groups $G^{\#}$ and $bG$ by Theorem 8.2.5. We now apply Theorem 8.2.6 to find a continuous homomorphism $\pi\colon bG \to K$ onto a second-countable topological group $K$ and a continuous function $h\colon \pi(H^{\#}) \to \mathbb{R}$ such that $f = h \circ \pi\restriction H^{\#}$. Clearly, we have

$$|\pi(G^{\#})| \leq |K| \leq 2^\omega = \mathfrak{c}.$$

It now follows from Corollary 9.9.39 (applied to the abstract groups $\pi(H)$ and $\pi(G)$ in place of $H$ and $G$) that $\pi(H)^{\#}$ is $C$-embedded in $\pi(G)^{\#}$. Note that the topology of the group $\pi(H)^{\#}$ is finer than that of $\pi(H^{\#}) \subset K$, so $h$ remains continuous when considered as a function on $\pi(H)^{\#}$. Hence $h$ can be extended to a continuous function $h^*\colon \pi(G)^{\#} \to \mathbb{R}$. Since the homomorphism $\pi^{\#}\colon G^{\#} \to \pi(G)^{\#}$ is continuous by Corollary 9.9.8, $f^* = h^* \circ \pi\restriction G^{\#}$ is a continuous extension of $f$ over $G^{\#}$. Thus $H^{\#}$ is $C$-embedded in $G^{\#}$.

Finally, since the group $G^{\#}$ is strongly zero-dimensional, by Theorem 9.9.32, it remains to apply the last claim of Lemma 9.9.33 to conclude that $H^{\#}$ is $\mathbb{N}$-embedded in $G^{\#}$. □

The above result has several interesting applications. For example, it enables us to strengthen the conclusion of Theorem 9.9.30 and show that all bounded subsets of the groups with the Bohr topology are finite. We need a simple topological fact.

LEMMA 9.9.41. *Let $X$ be a countable regular space such that all compact subsets of $X$ are finite. Then every infinite set $A \subset X$ contains an infinite subset $B$ which is closed and discrete in $X$.*

PROOF. Suppose that $A \subset X$ is infinite. The infinite set $F = cl_X A$ is not compact. Since $F$ is countable, it follows that $F$ is not pseudocompact. Take any unbounded continuous function $f\colon F \to \mathbb{R}$. Since $A$ is dense in $F$, the function $g$ is unbounded on $A$. For every integer $n$, choose a point $a_n \in A$ such that $|g(a_n)| \geq n$. Then the infinite set $B = \{a_n : n \in \omega\} \subset A$ is closed and discrete in $X$. □

THEOREM 9.9.42. [**F. J. Trigos – Arrieta**] *Let $G$ be a discrete Abelian group. Then all bounded subsets of $G^{\#}$ are finite.*

PROOF. Let $A$ be an infinite subset of $G$. We have to find a continuous function $f\colon G^{\#} \to \mathbb{R}$ with unbounded image $f(A) \subset \mathbb{R}$. Choose a countable infinite subset $B$ of $A$ and denote by $H$ the subgroup of $G$ generated by $B$. Then $H^{\#}$ is $\mathbb{N}$-embedded in $G^{\#}$ by Theorem 9.9.40. Consider $B$ as a subspace of $H^{\#}$ and apply Lemma 9.9.41 to choose an infinite set $C = \{x_n : n \in \omega\} \subset B$ which is closed and discrete in the countable group $H^\#$. Then $C$ is $\mathbb{N}$-embedded in $H^\#$ according to Corollary 9.9.34, which in turns implies that $C$ is $\mathbb{N}$-embedded in $G^\#$. Define a function $g\colon C \to \mathbb{N}$ by $g(x_n) = n$ for each $n \in \omega$. Clearly, $g$ is continuous and, hence, admits a continuous extension $f$ over $G^\#$. Then $f(A) \supseteq f(C) = g(C) = \mathbb{N}$ is unbounded in $\mathbb{R}$, as required. □

For a group $G$ of Ulam non-measurable cardinality, the conclusion of the above theorem can be deduced from Theorem 9.9.30 and 9.9.25, since the closure of a bounded subset $B$ of a realcompact space is compact.

Two important facts regarding independent subsets of Abelian groups were established in Propositions 9.9.20 and 9.9.21. Below we present some further delicate topological properties of independent sets.

THEOREM 9.9.43. *Let $G$ be a discrete Abelian group. Then every independent subset of $G$ is $\mathbb{N}$-embedded and $C$-embedded in $G^\#$.*

PROOF. Let $A$ be an independent subset of $G$ and $H$ be the subgroup of $G$ generated by $A$. According to Theorem 9.9.40, the subgroup $H^\#$ of $G^\#$ is $\mathbb{N}$-embedded in $G^\#$. Since the relation of being $\mathbb{N}$-embedded is transitive, the first claim of the theorem will follow if we show that $A$ is $\mathbb{N}$-embedded in $H^\#$.

Suppose that $f\colon A \to \mathbb{N}$ is an arbitrary mapping to the discrete space $\mathbb{N}$. Note that the set $A$ is closed and discrete in $G^\#$ by Proposition 9.9.21, so $f$ is continuous. For every integer $m \geq 2$, denote by $A_m$ the set of all elements $a \in A$ of order $m$. For all $m \geq 2$ and $n \in f(A_m)$, choose an element $a_{m,n} \in A_m$ such that $f(a_{m,n}) = n$. Then

$$B = \{a_{m,n} : m \in \mathbb{N},\ m \geq 2,\ n \in f(A_m)\}$$

is a countable subset of $A$. Let us define a mapping $g\colon A \to A$ by the rule:

$$g(a) = a_{m,n} \text{ whenever } a \in A_m \text{ and } f(a) = n.$$

Since the family $\{A_m \cap f^{-1}(n) : n \in f(A_m)\}$ is a partition of $A_m$ for each $m \geq 2$, the mapping $g$ is correctly defined. It follows directly from the definition of $g$ that $f(g(a)) = f(a)$ and, furthermore, the orders of $a$ and $g(a)$ coincide for each $a \in A$.

Let $L$ be the subgroup of $G$ generated by $B$. Clearly, $L$ is countable and $L \subset H$. Since the set $A$ is independent and $g$ does not change orders of the elements of $A$, one can extend $g$ to a homomorphism $h\colon H \to L$. By Corollary 9.9.8, the corresponding homomorphism $h^\#\colon H^\# \to L^\#$ is continuous. Notice that $B = A \cap L$ is a closed discrete subset of the countable group $L^\#$, so the restriction of $f$ to $B$ admits an extension to a continuous mapping $\tilde{f}\colon L^\# \to \mathbb{N}$, by Corollary 9.9.34. Then the composition $\tilde{f} \circ h^\#$ is a continuous extension of $f$ over $H^\#$, thus implying that $A$ is $\mathbb{N}$-embedded in $H^\#$. Thus, $A$ is $\mathbb{N}$-embedded in $G^\#$.

Finally, since $A$ is a discrete subspace of $G^\#$, by Proposition 9.9.21, it follows from ii) of Lemma 9.9.33 that $A$ is $\mathbb{R}$-embedded (that is, $C$-embedded) in $G^\#$. □

The next step is to show that independent subsets of a discrete Abelian group $G$ are $C^*$-embedded in the compact group $bG$, the Bohr compactification of $G$. A simple lemma below is a part of the argument in the proof of Theorem 9.9.45.

LEMMA 9.9.44. *Let $G$ be a discrete Abelian group and $H_1$, $H_2$ be subgroups of $G$ with trivial intersection. Then the intersection of the closures of $H_1$ and $H_2$ in the Bohr compactification $bG$ of $G$ is also trivial.*

PROOF. Consider the subgroup $H = H_1 + H_2$ of $G$. It follows from Corollary 9.9.10 that the Bohr compactification $bH$ of $H$ is topologically isomorphic to the closure of $H$ in $bG$. Since the intersection $H_1 \cap H_2$ is trivial, the group $H$ is algebraically isomorphic to the product group $H_1 \times H_2$. Hence $bH$ is topologically isomorphic to the group $bH_1 \times bH_2$ by Corollary 9.9.12. Let us identify the closure of $H$ in $bG$ with the product group $bH_1 \times bH_2$. It is clear that $H_1 \subset bH_1 \times \{0_2\}$ and $H_2 \subset \{0_1\} \times bH_2$, where $0_1$ and $0_2$ are the neutral elements of the groups $bH_1$ and $bH_2$, respectively. So the closures of the groups $H_1$ and $H_2$ in $bH$ (and in $bG$) have the unique common element, the neutral element of $bH$. □

THEOREM 9.9.45. *Let $G$ be a discrete Abelian group. Then every independent subset $A$ of $G$ is $C^*$-embedded in $bG$. Hence, the closure of $A$ in $bG$ is naturally homeomorphic to the Čech–Stone compactification $\beta A$ of $A$.*

PROOF. First we prove that $A$ is $C^*$-embedded in $bG$. By Theorem 6.1.5, it suffices to verify that, for every subset $B$ of $A$, the sets $cl_{bG} B$ and $cl_{bG}(A \setminus B)$ are disjoint. Denote by $H_1$ and $H_2$ the subgroups of $G$ generated by $A \setminus B$ and $B$, respectively. Since the set $A$ is independent, the intersection of the groups $H_1$ and $H_2$ is trivial. By Lemma 9.9.44, this implies that the intersection of the closures of $H_1$ and $H_2$ in $bG$ is also trivial. According to Proposition 9.9.21, the sets $A \setminus B$ and $B$ are closed in $G^\#$ and, clearly, none of them contains the neutral element $0_G$ of the group $G^\# \subset bG$. Hence their closures in $bG$ do not contain $0_G$ either. Since $cl_{bG}(A \setminus B) \subset cl_{bG} H_1$ and $cl_{bG} B \subset cl_{bG} H_2$, we conclude that the sets $cl_{bG}(A \setminus B)$ and $cl_{bG} B$ are disjoint. This proves the first claim of the theorem.

Since the set $A$ is $C^*$-embedded in $bG$, it is also $C^*$-embedded in its closure $K = cl_{bG} A$. The space $K$ being compact, there exists a homeomorphism $h\colon K \to \beta A$ such that $h(x) = x$ for each $x \in A$ (see [**165**, Coro. 3.6.3]). □

Combining Theorems 9.9.43 and 9.9.45, we conclude that every independent subset of a discrete Abelian group $G$ is $\mathbb{N}$-embedded in $G^\#$ and $C^*$-embedded in $bG$. In fact, the subsets of $G$ with these properties are abundant — we will show in Theorem 9.9.51 that every uncountable set $A$ in an Abelian group $G$ contains a subset $B$ of the same cardinality as $A$ which is $\mathbb{N}$-embedded in $G^\#$ and $C^*$-embedded in $bG$. This fact shows that the relationship of the subgroup $G^\#$ to the group $bG$ is very similar to the relationship of an infinite discrete space $D$ to its Čech–Stone compactification $\beta D$ (even though the group $G^\#$ is very far from being discrete when $G$ is infinite). The argument that follows requires a series of auxiliary results. The first of them is purely geometric in nature, but it introduces one of the main ideas in the proof of Theorem 9.9.51.

LEMMA 9.9.46. *The torus $\mathbb{T}^2$ contains two closed disjoint subsets $K$ and $L$ such that, for every integer $m \geq 2$ and every $b \in \mathbb{T}^2$, the equation $x^m = b$ has solutions in both $K$ and $L$.*

PROOF. Let us represent $\mathbb{T}^2$ as the square $OXTY$ with side $OX$ of length 1 (see the diagram below). We glue the sides $OX$ and $YT$ (identifying the points of $OX$ and $YT$ with the same $x$-projections) as well as the sides $OY$ and $XT$ (identifying the points of $OY$ and $XT$ with the same $y$-projections). We introduce the coordinate system in $OXTY$ by taking

$O$ as the origin, $OX$ as the $x$-axis and $OY$ as the $y$-axis, respectively. Therefore, a point $P(x, y)$ in the square corresponds to the point $(e^{2\pi xi}, e^{2\pi yi})$ in $\mathbb{T}^2$. In particular, each of the points $O, X, Y, T$ represents the same point $(1, 1)$ of $\mathbb{T}^2$.

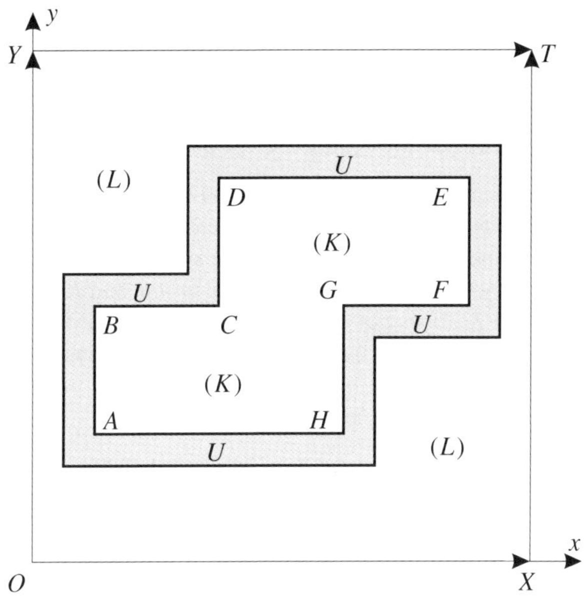


Let $K$ be the union of the two congruent white closed rectangles $ABGH$ and $CDEF$, where $|AB| = |CD| = 1/4$ and $|AH| = |CF| = 1/2$. The vertices of the rectangles are $A(1/8, 1/4)$, $B(1/8, 1/2)$, $C(3/8, 1/2)$, $D(3/8, 3/4)$, $E(7/8, 3/4)$, $F(7/8, 1/2)$, $G(5/8, 1/2)$, and $H(5/8, 1/4)$. A narrow open shadowed "channel" $U$ around $K$ has constant width $1/16$, and $L$ is the set $\mathbb{T}^2 \setminus (K \cup U)$. Clearly, the union $K \cup U$ is open, so $L$ is closed in $\mathbb{T}^2$.

We claim that the sets $K$ and $L$ are as required. For every integer $m$, let $f_m : \mathbb{T}^2 \to \mathbb{T}^2$ be a mapping defined by $f_m(e^{2\pi xi}, e^{2\pi yi}) = (e^{2\pi mxi}, e^{2\pi myi})$ for all $x, y \in \mathbb{R}$. It suffices to show that for every $m \geq 2$, the image of each of the sets $K$ and $L$ under the mapping $f_m$ covers the torus $\mathbb{T}^2$. Notice that in the "real" plane $Oxy$, the mapping $f_m$ is the radial magnification (modulo 1 in both coordinates) with center at $O$ and coefficient $m$. Geometrically, one can represent the images $f_m(K)$ and $f_m(L)$ as the sets $K$ and $L$ but magnified $m$ times in the square of size $m \times m$, where each of the $m^2$ small unit squares is canonically identified with the unit square $OXTY$. Let us verify that $f_2(K) = \mathbb{T}^2 = f_2(L)$, leaving the rest to the reader (the verification of the equalities $f_m(K) = \mathbb{T}^2 = f_m(L)$ for $m \geq 3$ is even easier than that for $m = 2$, since the radial magnification with coefficient $m$ magnifies small areas $m^2$ times).

Clearly, the restriction of $f_2$ to the interior of $K$ is one-to-one, so the area of the image $f_2(K)$ is four times greater than that of $K$. Therefore, the area of $f_2(K)$ is equal to 1 which means that $f_2(K)$ is dense in the unit square $OXYT$. Since $K$ is compact and $f_2$ is continuous, we have the equality $f_2(K) = \mathbb{T}^2$. Notice also that for each $b \in \mathbb{T}^2$, the four solutions of the equation $x^2 = b$, when represented in the square $OXTY$, have the form $(z_1, z_2)$, $(z_1 + 1/2, z_2)$, $(z_1, z_2 + 1/2)$ and $(z_1 + 1/2, z_2 + 1/2)$ for some $z_1, z_2$ (modulo 1).

Whenever $b \in \mathbb{T}^2$, the area $K \cup U$ doesn't contain at least one of the four points. Hence, $f_2(L) = \mathbb{T}^2$. □

The result below easily follows from Lemma 9.9.46. It plays an important role in the proof of Lemma 9.9.50.

LEMMA 9.9.47. *There exist a countable (multiplicatively written) divisible subgroup $\Sigma$ of $\mathbb{T}^\omega$ and a countable infinite discrete family $\mathcal{K}$ of closed subsets of $\Sigma$ such that for every integer $m \geq 2$ and every $b \in \Sigma$, the equation $x^m = b$ has solutions in each $K \in \mathcal{K}$.*

PROOF. Let $H$ be the torsion subgroup of $\mathbb{T}$, and let $G$ the square $H^2$. In what follows we write $G$ multiplicatively, so $e = (1, 1)$ is the identity of $G$. We put $K^* = K \cap G$ and $L^* = L \cap G$, where $K$ and $L$ are the closed disjoint subsets of $\mathbb{T}^2$ constructed in the proof of Lemma 9.9.46. Clearly, $G$ is a countable divisible group, and $K^*$, $L^*$ are closed disjoint subsets of $G$; notice that $K^*$ does not contain $e$. It follows from the choice of the sets $K^*$ and $L^*$ that for every integer $m \geq 2$ and every $b \in G$, the equation $x^m = b$ has solutions in both $K^*$ and $L^*$.

Let $\Sigma$ be the direct sum of $\omega$ copies of the group $G$, that is,

$$\Sigma = \{x \in G^\omega : x_n = e \text{ for all but finitely many } n \in \omega\}.$$

The group $\Sigma$ is also countable and divisible. For every $n \in \omega$, we define a subset $K_n$ of $\Sigma$ by

$$K_n = \{x \in \Sigma : x_k \in K^* \text{ for each } k < n \text{ and } x_n \in L^*\}.$$

It follows from the above definition that the sets $K_n$ are closed in $\Sigma$ and disjoint. Let us show that the group $\Sigma$ and the family $\mathcal{K} = \{K_n : n \in \omega\}$ are as required. If $x \in G^\omega$ is an accumulation point of the family $\mathcal{K}$, then every neighbourhood of $x$ meets infinitely many $K_n$'s and, hence, $x_n \in K^*$ for each $n \in \omega$. This implies that $x \notin \Sigma$ because $e \notin K^*$ or, in other words, the family $\mathcal{K}$ has no accumulation points in $\Sigma$. Since any two distinct elements of $\mathcal{K}$ are disjoint, $\mathcal{K}$ is a discrete family in $\Sigma$.

Finally, let $m, n \in \omega$ be integers, where $m \geq 2$, and let $b \in \Sigma$ be an arbitrary element of $\Sigma$. Choose $N \in \omega$ such that $N > n$ and $b_k = e$, for each $k > N$. Then there exists $x \in G^\omega$ satisfying the following conditions:

(1) $x_k \in K^*$ if $k < n$;
(2) $x_n \in L^*$;
(3) $x_k = e$ if $k > N$;
(4) $(x_k)^m = b_k$ whenever $k \leq N$.

It follows from (3) that $x \in \Sigma$, while (1) and (2) imply that $x \in K_n$. Clearly, $x^m = b$ by (4). □

The following lemma enables us, in a sense, to replace an Abelian group $G$ by a homomorphic image of $G$ when we are looking for $C^*$-embedded subsets of $bG$ lying in $G$.

LEMMA 9.9.48. *Let $\pi\colon G \to H$ be a homomorphism of discrete Abelian groups, and suppose that $A \subset G$ and $B \subset H$ are infinite sets such that the restriction $\pi\restriction A$ is a bijection of $A$ onto $B$. If $B$ is discrete and $C^*$-embedded in $bH$, then $A$ is $C^*$-embedded in $bG$.*

PROOF. Take an arbitrary function $f\colon A \to I$, where $I$ is the closed unit interval. Denote by $i$ the inverse of $\pi\restriction A$; then $i\colon B \to A$ is a bijection and $i \circ \pi\restriction A = id_A$. Since $B$

is discrete in $bH$ and in $H^\#$, the function $g = f \circ i$ is continuous on the subspace $B \subset H^\#$. By the assumptions of the lemma, $g$ can be extended to a continuous function $\tilde{g}\colon bH \to I$. By Corollary 9.9.6, $G^\#$ and $H^\#$ are dense subgroups of the compact groups $bG$ and $bH$, respectively. Hence, the continuous homomorphism $\pi^\#\colon G^\# \to H^\#$ admits an extension to a continuous homomorphism $\tilde{\pi}\colon bG \to bH$. We claim that $\tilde{f} = \tilde{g} \circ \tilde{\pi}$ is a continuous extension of $f$ over the whole $bG$. Indeed, since $\pi^\#(A) = B$, we have that

$$\tilde{g} \circ \tilde{\pi}\restriction A = \tilde{g} \circ \pi^\#\restriction A = g \circ \pi^\#\restriction A = f \circ i \circ \pi\restriction A = f.$$

This proves that $A$ is $C^*$-embedded in $bG$. □

The next algebraic result is almost obvious, since every subgroup of the group $\mathbb{Z}$ is cyclic.

LEMMA 9.9.49. *Let $H_0 \subset H_1 \subset \cdots \subset H_n \subset \cdots$ be an increasing sequence of subgroups of the group $\mathbb{Z}$. Then the sequence stabilizes, that is, there exists $k \in \omega$ such that $H_n = H_k$ for each $n \geq k$.*

PROOF. The union $H = \bigcup_{n=0}^{\infty} H_n$ is a subgroup of $\mathbb{Z}$, so there exists $x \in H$ such that $H = \langle x \rangle$. Then $x \in H_k$ for some $k \in \omega$, whence it follows that $H = \langle x \rangle \subset H_k \subset H$. Therefore, $H_k = H$ and $H_n = H_k$ for each $n \geq k$. □

The lemma below gives a key to a proof of Theorem 9.9.51.

LEMMA 9.9.50. *Let $\kappa$ be an infinite cardinal and $X = \{x_\alpha : \alpha < \kappa\}$ a subset of an Abelian group $G$ such that for every $\alpha < \kappa$, the element $x_\alpha$ does not belong to the subgroup of $G$ generated by the set $\{x_\nu : \nu < \alpha\}$. Then $X$ is closed and discrete in $G^\#$, is $\mathbb{N}$-embedded in $G^\#$, and $C^*$-embedded in $bG$.*

PROOF. We start with the last assertion of the lemma.

**Claim 1.** *$X$ is $C^*$-embedded in $bG$.*

Let us show that for every partition $X = X_1 \cup X_2$ of the set $X$, there exists a homomorphism $f\colon G \to H$ to the compact group $\mathbb{T}^2$ such that the closures of the sets $f(X_1)$ and $f(X_2)$ in $\mathbb{T}^2$ are disjoint.

By Lemma 9.9.46, the group $\mathbb{T}^2$ contains two disjoint closed sets $K$ and $L$ such that the equation $y^m = a$ has solutions in both $K$ and $L$ for every integer $m \geq 2$ and every $a \in \mathbb{T}^2$. We use the sets $K$ and $L$ to define by recursion a homomorphism $f\colon H \to \mathbb{T}^2$ such that $f(X_1) \subset K$ and $f(X_2) \subset L$, where $H$ is the subgroup of $G$ generated by $X$. Suppose that for some $\alpha < \kappa$, we have defined a family $\{f_\beta : \beta < \alpha\}$, where each $f_\beta$ is a homomorphism of $H_\beta$ to $\mathbb{T}^2$, and suppose that this family satisfies the following conditions for all $\gamma < \beta < \alpha$:

(1) if $\gamma < \beta < \alpha$, then $f_\beta\restriction H_\gamma = f_\gamma$;
(2) $f_\beta(x_\gamma) \in K$ if $x_\gamma \in X_1$ and $f_\beta(x_\gamma) \in L$ if $x_\gamma \in X_2$.

If $\alpha$ is a limit ordinal, we simply put $f_\alpha = \bigcup_{\beta<\alpha} f_\beta$. Let $\alpha = \beta + 1$. We can assume without loss of generality that $x_\beta \in X_1$. Since $x_\beta \notin H_\beta$, the number $m = \min\{n \in \mathbb{N} : nx_\beta \in H_\beta\}$ is either an integer $\geq 2$ or $m$ is not defined. In the first case, choose an element $y \in K$ satisfying $y^m = f_\beta(mx_\beta)$. Then, by Lemma 1.1.5, extend $f_\beta$ to a homomorphism $f_\alpha\colon H_\alpha \to \mathbb{T}^2$ satisfying $f_\alpha(x_\beta) = y$. If $m$ is not defined, then take an arbitrary element $y \in K$ and, by

Lemma 1.1.5, extend $f_\beta$ to a homomorphism $f_\alpha\colon H_\alpha \to \mathbb{T}^2$ such that $f_\alpha(x_\beta) = y$. Then the family $\{f_\gamma : \gamma \leq \alpha\}$ satisfies conditions (1) and (2). This completes the construction.

Let $f^* = \bigcup_{\alpha<\kappa} f_\alpha$. Then, by (1), $f^*$ is a homomorphism of $H$ to $\mathbb{T}^2$. Since the group $\mathbb{T}^2$ is divisible, $f^*$ admits an extension to a homomorphism $f\colon G \to \mathbb{T}^2$. Since $f\restriction H_\alpha = f_\alpha$ for each $\alpha < \kappa$, it follows from (2) that $f(X_1) \subset K$ and $f(X_2) \subset L$. Since every character of $G$ can be extended to a continuous character of $bG$, we conclude that the closures of $X_1$ and $X_2$ in $bG$ are disjoint. Hence, Theorem 6.1.5 implies that $X$ is $C^*$-embedded in $bG$. This proves Claim 1.

**Claim 2.** *$X$ is closed and discrete in $G^\#$.*

The discreteness of $X$ is almost immediate. Indeed, take an arbitrary element $y \in X$ and put $X_1 = \{y\}$ and $X_2 = X \setminus \{y\}$. Then the closures of $X_1$ and $X_2$ in $bG$ are disjoint by Claim 1, so that the point $y$ is isolated in the subspace $X$ of $G^\# \subset bG$. Let $x \in G \setminus X$ be an arbitrary point. We have to find an open neighbourhood of $x$ in $G^\#$ disjoint from $X$. Let $F = \langle x\rangle \cap X$. Then $\langle F \cup \{x\}\rangle \subset \langle x\rangle$, so we have that $\langle F \cup \{x\}\rangle \cap X = F$. Let $Y = \{x\} \cup (X \setminus F)$. It is clear that the set $F$ is countable, so $|Y| = \kappa$. It follows from the choice of the sets $X$ and $Y$ that there exists an enumeration $Y = \{y_\nu : \nu < \kappa\}$ such that $y_0 = x$ and $y_\mu \notin \langle y_\nu : \nu < \mu\rangle$ for each $\mu < \kappa$. As $Y \subset X$ and $X$ is a discrete subspace of the group $G^\#$, there exists an open neighbourhood $U_1$ of $x = y_0$ in $G^\#$ disjoint from $Y \setminus \{x\} = X \setminus F$. It remains to find an open neighbourhood of $x$ in $G^\#$ which does not meet $F$. This is trivial if the cyclic group $\langle x\rangle$ is finite since then the set $F \subset \langle x\rangle$ is also finite. Suppose, therefore, that $x$ has infinite order. We claim that the set $F$ is again finite. Assume to the contrary that $F$ is infinite. Since $F \subset X$, there exists a sequence $\{z_n : n \in \omega\} \subset F$ such that $z_{n+1} \notin \langle z_0, \ldots, z_n\rangle = H_n$ for each $n \in \omega$. Hence $\{H_n : n \in \omega\}$ is a strictly increasing sequence of subgroups of the cyclic group $\langle x\rangle \cong \mathbb{Z}$, which contradicts Lemma 9.9.49. Thus, in either case, $F$ is finite and $U_2 = G \setminus F$ is a neighbourhood of $x$ in $G^\#$ disjoint from $F$. So $x \in U_1 \cap U_2 \subset G^\# \setminus X$, as required.

**Claim 3.** *The set $X$ is $\mathbb{N}$-embedded in $G^\#$.*

Consider an arbitrary function $f\colon X \to \mathbb{N}$. Let $\Sigma$ and $\mathcal{K} = \{K_n : n \in \omega\}$ be as in Lemma 9.9.47. Since $\Sigma$ is a countable regular space, the closed subset $F = \bigcup \mathcal{K}$ is $\mathbb{N}$-embedded in $\Sigma$ by Corollary 9.9.34. Hence there exists a continuous function $r\colon \Sigma \to \mathbb{N}$ such that $r(x) = n$ for each $x \in K_n$, where $n \in \omega$. As in the proof of Lemma 9.9.50, one can define a homomorphism $h\colon G \to \Sigma$ such that $h(x) \in K_{r(x)}$ for each $x \in X$. Obviously, the function $r \circ h$ extends $f$. Since the group $\Sigma$ is precompact, the homomorphism $h\colon G^\# \to \Sigma$ is continuous by Corollary 9.9.7. Hence, $r \circ h$ is a continuous extension of $f$ over $G^\#$. □

THEOREM 9.9.51. *Every uncountable subset $A$ of a discrete Abelian group $G$ contains a subset $B$ with $|B| = |A|$ such that $B$ is closed and discrete in $G^\#$, is $\mathbb{N}$-embedded in $G^\#$, and $C^*$-embedded in $bG$.*

PROOF. Let $A$ be a subset of $G$ such that $|A| = \kappa > \omega$. By transfinite recursion of the length $\kappa$, we can construct a subset $B = \{x_\alpha : \alpha < \kappa\}$ of $A$ such that, for every $\alpha < \kappa$, the element $x_\alpha$ does not belong to the subgroup $H_\alpha$ of $G$ generated by the set $\{x_\nu : \nu < \alpha\}$. It is clear that $|B| = \kappa$. Then, by Lemma 9.9.50, the set $B$ is as required. □

The existence of infinite independent sets or sets as in Lemma 9.9.50 cannot be guaranteed for very simple infinite Abelian groups, such as the group $\mathbb{Z}$. Indeed, the group $\mathbb{Z}$ is infinite, but every independent subset of $\mathbb{Z}$ has cardinality at most 1, that is, $r_0(\mathbb{Z}) = r(\mathbb{Z}) = 1$. Furthermore, Lemma 9.9.49 prohibits any construction of a sequence $\{x_n : n \in \omega\}$ in $\mathbb{Z}$ such that $x_{n+1} \notin \langle x_0, \dots, x_n \rangle$ for each $n \in \omega$. This means that the strategy for a construction of $C^*$-embedded subsets in the case of countable Abelian groups has to be quite different.

We will find infinite $C^*$-embedded sets in $bG$ among closed discrete subsets of the group $G^{\#}$. Notice that if $A \subset G^{\#}$ is closed and discrete, then $A$ is $C^*$-embedded in $bG$ if and only if every two disjoint subsets of $A$ have disjoint closures in $bG$. To guarantee the latter property of $A$, we have to separate disjoint subsets of $A$ by means of a character of the group $G$. The next lemma explains this idea in a greater detail.

LEMMA 9.9.52. *Let $A = \{a_n : n \in \omega\}$ be a subset of $\mathbb{Z}$ such that $|a_0| \geq 2$ and $|a_{n+1}| \geq 4|a_n|$, for each $n \in \omega$. Then for every partition $A = A_1 \cup A_2$, there exists a character $\chi$ on $\mathbb{Z}$ such that the closures of the sets $\chi(A_1)$ and $\chi(A_2)$ in $\mathbb{T}$ are disjoint. Therefore, the set $A$ is discrete in $\mathbb{Z}^{\#}$ and $C^*$-embedded in $b\mathbb{Z}$.*

PROOF. For every $n \in \omega$, denote by $i_n$ an element of $\{1, 2\}$ such that $a_n \in A_{i_n}$. Our aim is to define a decreasing sequence $\{C_n : n \in \omega\}$ of closed arcs in the circle group $\mathbb{T}$ whose lengths tend to zero and then take a number $z_0 \in \bigcap_{n\in\omega} C_n$ as the value of $\chi$ at $1 \in \mathbb{Z}$.

Let $J = \{e^{i\varphi} : \pi/2 \leq \varphi \leq 3\pi/2\}$. Clearly, $J$ is a closed arc of the length $\pi$ in $\mathbb{T}$. and $J$ does not contain 1. Put also

$$D_1 = \{e^{i\varphi} : -\pi/2 \leq \varphi \leq 0\} \text{ and } D_2 = \{e^{i\varphi} : \pi/2 \leq \varphi \leq \pi\}.$$

Then $D_1$, $D_2$ are disjoint closed arcs in $\mathbb{T}$, each of the length $\pi/2$. Since $|a_0| \geq 2$, we have that $[J^{a_0}] = \{x^{a_0} : x \in J\} = \mathbb{T}$. Choose a closed arc $C_0 \subset J$ of the length $\pi/(2|a_0|)$ such that $[C_0^{a_0}] = D_{i_0}$. Suppose that we have defined closed arcs $C_0 \supseteq \dots \supseteq C_n$ in $\mathbb{T}$ satisfying the following two conditions for each $k \leq n$:

(1) $[C_k^{a_k}] = D_{i_k}$;
(2) the length of $C_k$ is equal to $\pi/(2|a_k|)$.

It follows from $|a_{n+1}| \geq 4|a_n|$ and (1) (with $k = n$) that $[C_n^{a_{n+1}}] = \mathbb{T}$. Hence we can choose a closed arc $C_{n+1} \subset C_n$ of the length $\pi/(2|a_{n+1}|)$ satisfying $[C_{n+1}^{a_{n+1}}] = D_{i_{n+1}}$. Our construction of the sequence $\{C_k : k \in \omega\}$ satisfying (1) and (2) for all $k \in \omega$ is complete.

Since $\mathbb{T}$ is compact, it follows from (2) that there exists a unique element $z_0 \in \bigcap_{k\in\omega} C_k$. We define a homomorphism $\chi\colon \mathbb{Z} \to \mathbb{T}$ by $\chi(n) = z_0^n$ for each $n \in \omega$. Then (1) and the definition of the numbers $i_n$ imply that $\chi(A_i) \subset D_i$, for each $i = 1, 2$. Hence, the closures of the sets $\chi(A_1)$ and $\chi(A_2)$ in $\mathbb{T}$ are disjoint.

The character $\chi$ is continuous on the group $\mathbb{Z}^{\#}$, so it can be extended to a continuous character $\bar{\chi}$ of the compact group $bG$. Then $\bar{\chi}(A_1) \subset D_1$ and $\bar{\chi}(A_2) \subset D_2$; it follows that the sets $cl_{bG} A_1$ and $cl_{bG} A_2$ are disjoint as well. If $a \in A$, we apply this property to the sets $A_1 = \{a\}$ and $A_2 = A \setminus \{a\}$ and conclude that the point $a$ is isolated in the subspace $A$ of $\mathbb{Z}^{\#}$. Hence $A$ is discrete in $\mathbb{Z}^{\#}$ and by [**165**, Coro. 3.6.4], the closure $K = cl_{bG} A$ is homeomorphic to the Čech–Stone compactification $\beta A$ of the discrete space $A$. It follows that $A$ is $C^*$-embedded in $K$ and in $bG$. □

Our next step is to prove Theorem 9.9.54 in the special case of Abelian torsion groups. In this case, we can directly apply Lemma 9.9.50.

LEMMA 9.9.53. *Let $G$ be a discrete Abelian torsion group and $A \subset G$ be an infinite set. Then $A$ contains an infinite subset $B$ such that $B$ is closed and discrete in $G^{\#}$, $\mathbb{N}$-embedded in $G^{\#}$, and $C^*$-embedded in $bG$.*

PROOF. Since $G$ is a torsion group, every finite subset of $G$ generates a finite subgroup of $G$. Therefore, one can easily define a set $B = \{b_n : n \in \omega\} \subset A$ such that $b_{n+1} \notin \langle b_0, \ldots, b_n \rangle$ for each $n \in \omega$. Then Lemma 9.9.50 implies that the set $B$ is as required. □

THEOREM 9.9.54. *Let $A$ be an infinite subset of a discrete Abelian group $G$. Then $A$ contains an infinite subset $B$ which is closed and discrete in $G^{\#}$, $\mathbb{N}$-embedded in $G^{\#}$, and $C^*$-embedded in $bG$.*

PROOF. We can assume that $A$ is countable and generates the group $G$. Theorem 9.9.30 implies that all compact subsets of $G^{\#}$ are finite, so by Lemma 9.9.41, we can additionally assume that $A$ is closed and discrete in $G^{\#}$.

Let $D$ be a maximal independent subset of $A$. It follows from Theorem 9.9.45 that $D$ is $C^*$-embedded in $bG$. If $D$ is infinite, we are done. Suppose, therefore, that $D$ is finite. Denote by $K$ the subgroup of $G$ generated by $D$. The set $D \cup \{x\}$ is dependent for every element $x \in G \setminus K$, so the intersection $\langle x \rangle \cap K$ is non-trivial. Hence every element of the quotient group $G/K$ has finite order, that is, $H = G/K$ is a torsion group. Let $\pi \colon G \to H$ be the canonical homomorphism. We consider the following two cases.

*Case 1.* The set $\pi(A)$ is infinite. Lemma 9.9.53 permits to choose an infinite discrete subset $C$ of $\pi(A)$ such that $C$ is $\mathbb{N}$-embedded in $H^{\#}$ and $C^*$-embedded in $bH$. Now we can take a set $B \subset A$ such that the restriction of $\pi$ to $B$ is a bijection of $B$ onto $C$. The set $B$ is $C^*$-embedded in $bG$, by Lemma 9.9.48.

*Case 2.* The set $\pi(A)$ is finite. Then $A \cap (x + K)$ is infinite, for some $x \in G$ or, equivalently, the set $A_x = K \cap (A - x)$ is infinite. Since the group $K$ is finitely generated, it follows from [**409**, 4.2.10] that $K$ can be represented as a direct sum $K = K_1 \oplus \cdots \oplus K_n$, where $K_1, \ldots, K_n$ are cyclic subgroups of $K$. For every $i = 1, \ldots, n$, let $\varphi_i$ be the canonical projection of $K$ onto the subgroup $K_i$. Since the set $A_x$ is infinite, one can find $i \leq n$ such that $\varphi_i(A_x)$ is also infinite. This means that $K_i \cong \mathbb{Z}$. Choose an infinite subset $Y = \{y_k : k \in \omega\}$ of $\varphi_i(A_x)$ such that $|y_0| \geq 2$ and $|y_{k+1}| \geq 4|y_k|$ for each $k \in \omega$. Then $Y$ is $C^*$-embedded in $b\mathbb{Z} = bK_i$ by Lemma 9.9.52. As in Case 1, take an infinite set $B_x \subset A_x$ such that the restriction of $\varphi_i$ to $B_x$ is a bijection of $B_x$ onto $Y$. Then $B_x$ is $C^*$-embedded in $bG$, by Lemma 9.9.48. Therefore, the infinite set $B = x + B_x$ satisfies $B \subset A$ and is $C^*$-embedded in $bG$.

To complete the proof, observe that in each of the above cases $B$ is closed and discrete in $G^{\#}$ as a subset of the closed and discrete set $A$. Hence, $B$ is $\mathbb{N}$-embedded in $G^{\#}$ by Corollary 9.9.34. □

Combining Theorems 9.9.51, 9.9.54, and ii) of Lemma 9.9.33, we obtain the following.

THEOREM 9.9.55. [**E. van Douwen**] *Let $A$ be an infinite subset of a discrete Abelian group $G$. Then $A$ contains a subset $B$ with $|B| = |A|$ such that $B$ is closed and discrete in $G^{\#}$, is $\mathbb{N}$-embedded in $G^{\#}$ and $C^*$-embedded in $bG$. Also, $B$ is $C$-embedded in $G^{\#}$.*

We conclude this section with a deep result on normality of certain Abelian topological groups. Corollary 7.1.15 provides many of examples of non-normal Abelian topological groups — the free Abelian topological group $A(X)$ on arbitrary Tychonoff non-normal space $X$ cannot be normal, since it contains as a closed subspace $X$. From the algebraic point of view, all free Abelian topological groups look very much alike — all these groups are algebraically direct sums of infinite cyclic groups. This gives rise to the question of when an uncountable Abelian group admits a non-normal Hausdorff topological group topology. Clearly, every Hausdorff topological group topology on a countable group is normal. The following result shows that the countability of an Abelian group $G$ is the only obstacle for the existence of a non-normal Hausdorff group topology on $G$. It is not surprising, after Theorems 9.9.30 and 9.9.54, that the Bohr topology works.

THEOREM 9.9.56. **[F. J. Trigos-Arrieta]** *If $G$ is a discrete Abelian group, then $G^\#$ is normal iff $G$ is countable.*

PROOF. If $G$ is countable, then $G^\#$ is a regular Lindelöf space and, hence, the space $G^\#$ is normal.

Conversely, suppose that the group $G$ is uncountable. By Proposition 9.9.20, $G$ contains an uncountable independent subset $D$. Choose two disjoint subsets $D_1$ and $D_2$ of $D$ such that $|D_1| = |D| = |D_2|$. Every independent subset of $G$ is $C^*$-embedded in the Bohr compactification $bG$ of the group $G$, by Theorem 9.9.45. Since the set $D \subset G^\#$ is discrete, by Proposition 9.9.21, and every subset of $D$ is independent, it follows that the compact space $K_i = cl_{bG} D_i$ is homeomorphic to the Čech–Stone compactification $\beta D_i$ of the discrete set $D_i$, for each $i = 1, 2$.

Denote by $m$ the mapping of $bG \times bG$ to $bG$ defined by $m(x, y) = x + y$, for all $x, y \in bG$. Since $D_1$ and $D_2$ are disjoint subsets of the independent set $D$, the groups $H_1 = \langle D_1 \rangle$ and $H_2 = \langle D_2 \rangle$ have the trivial intersection. It follows from Lemma 9.9.44 and Corollary 9.9.10 that the same holds for the subgroups $bH_1 \cong cl_{bG} H_1$ and $bH_2 \cong cl_{bG} H_2$ of the group $bG$. Therefore, the restriction of $m$ to the product $bH_1 \times bH_2$ is a topological embedding into $bG$. Since $K_i \subset bH_i$ for $i = 1, 2$, we can identify the product $K_1 \times K_2$ with its image $K_1 + K_2 = cl_{bG}(D_1 + D_2)$ in $bG$. In addition, since the discrete sets $D_1$ and $D_2$ have the same cardinality, we can also identify $D_1$ with $D_2$ (both denoted by $S$) and $K_1 \cong \beta D_1$ with $K_2 \cong \beta D_2$ (both denoted by $K$). This gives us the natural embeddings $S^2 \hookrightarrow K^2 \hookrightarrow bG$, where $K^2 \cap G^\# = S^2$ is a closed discrete subset of $G^\#$.

Let $\Delta_K = \{(x, x) : x \in K\}$ be the diagonal in $K^2$ and $\Delta_S = S^2 \cap \Delta_K$ be the diagonal in $S^2$. Let also $P = S^2 \setminus \Delta_S$. We claim that the closed subsets $\Delta_S$ and $P$ of $G^\#$ cannot be separated by open neighbourhoods. Indeed, take an arbitrary open neighbourhood $U$ of $P$ in $G^\#$. There exists an open set $V$ in $bG$ such that $U = V \cap bG$. Then, by Corollary 5.3.29, $cl_{bG} V$ is a zero-set in $bG$. Clearly, the complement $W = K^2 \setminus cl_{bG} V$ is a cozero-set in $K^2$, so the space $W$ is Lindelöf as an $F_\sigma$-set in the compact space $K^2$. Note that $\Delta_S$ is dense in $\Delta_K$ and $S^2$ is dense in $K^2$. Since $S^2 = \Delta_S \cup P$ and $P \subset U \subset V$, we have $K^2 \subset \Delta_K \cup cl_{bG} V$. This inclusion and the definition of $W$ imply that $W \subset \Delta_K$. Since no point of $\Delta_K \setminus \Delta_S$ has a neighbourhood in $K^2$ lying in $\Delta_K$, we have $W \subset \Delta_S$. Hence, the discrete Lindelöf space $W$ is countable. Finally, since $\Delta_S$ is uncountable, the set $cl_{bG} V$ intersects $\Delta_S$. Since $S^2$ is dense in $K^2$, it follows that every open neighbourhood of $\Delta_S$ in $G^\#$ intersects the sets $V$ and $U$. This proves the claim and shows that the space $G^\#$ is not normal. □

We will now present several results about cardinal characteristics of Abelian groups with the Bohr topology. The first simple observation is that the equalities $c(G^\#) = c(bG) = \omega$

hold for every infinite Abelian group $G$. This directly follows from Corollary 4.1.8. Let us calculate the character and the weight of $G^{\#}$ and $bG$ in terms of $G$.

THEOREM 9.9.57. $\chi(G^{\#}) = w(G^{\#}) = w(bG) = \chi(bG) = 2^{|G|}$, *for every infinite discrete Abelian group* $G$.

PROOF. Since $G^{\#}$ is a dense subgroup of $bG$, the equality $\chi(G^{\#}) = \chi(bG)$ follows from Lemma 1.4.15, while $\chi(bG) = w(bG)$ and $\chi(G^{\#}) = w(G^{\#})$ hold by Corollary 5.2.4. The inequality $\chi(G^{\#}) \leq 2^{|G|}$ is evident. It remains to show that $w(bG) \geq 2^{|G|}$. We divide the proof in two parts.

1) The group $G$ is uncountable. By Proposition 9.9.20, $G$ contains an independent subset $A$ satisfying $|A| = |G|$. It follows from Proposition 9.9.21 that $A$ is discrete in $G^{\#}$, while Theorem 9.9.45 implies that the closure of $A$ in $bG$ is homeomorphic with the Čech–Stone compactification $\beta A$ of $A$. Since $A$ is infinite, we have that $w(\beta A) = 2^{|A|} = 2^{|G|}$, by [**165**, Theorem 3.6.11]. Since $bG$ contains a topological copy of $\beta A$, we also have $w(bG) \geq w(\beta A) = 2^{|G|}$.

2) The group $G$ is countable. Then, by Theorem 9.9.54, the group $G^{\#}$ contains an infinite, closed, discrete subset $A$ which is $C^*$-embedded in $bG$. Hence the closure of $A$ in $bG$ is homeomorphic to the Čech–Stone compactification $\beta A$, and again we have $w(bG) \geq w(\beta A) = 2^{\omega} = 2^{|G|}$. □

COROLLARY 9.9.58. *If* $G$ *is an infinite discrete Abelian group, then* $|bG| = 2^{2^{|G|}}$.

PROOF. Since the group $bG$ is compact, it follows from b) of Corollary 5.2.7 that $|bG| = 2^{w(bG)}$. It remains to note that $w(bG) = 2^{|G|}$, by Theorem 9.9.57. □

The exact values of the density and pseudocharacter of $G^{\#}$ are given in the next theorem.

THEOREM 9.9.59. *Let* $G$ *be an infinite discrete Abelian group. Then:*

a) $d(Y) = |Y|$, *for every subspace* $Y$ *of* $G^{\#}$;
b) $\psi(G^{\#}) = \mathrm{Ln}\,|G|$.

PROOF. a) Take any dense subset $A$ of $Y$. We may assume that $Y$ is infinite, since otherwise there is nothing to prove. Then $A$ is infinite as well, and the subgroup $H = \langle A \rangle$ of $G^{\#}$ has the same cardinality as $A$. Hence, $|H| = |A| \leq |Y|$, and $H$ is closed in the space $G^{\#}$, as a subgroup of $G^{\#}$. Since $A \subset H$ and $A$ is dense in $Y$, we have $Y \subset \overline{A} \subset H$. Therefore, $|Y| \leq |H| = |A| \leq |Y|$. It follows that $|Y| = |A|$. Notice that the equality $d(G^{\#}) = |G|$ follows from a) of Proposition 9.9.9.

b) To prove the equality $\psi(G^{\#}) = \mathrm{Ln}\,|G|$, suppose that $G$ is uncountable (otherwise the equality is trivially satisfied). Note that $\psi(G^{\#})$ coincides with the minimum number $|I|$ of homomorphisms $f_i\colon G \to \mathbb{T}$, with $i \in I$, that separate points of $G$. For every point-separating family $\{f_i : i \in I\}$, the diagonal product $\triangle_{i \in I} f_i$ is an injective homomorphism of $G$ to $\mathbb{T}^I$; hence, $|G| \leq |\mathbb{T}^I| = 2^{|I|}$ and, consequently, $\mathrm{Ln}\,|G| \leq |I|$.

Conversely, Proposition 9.9.37 implies that the group $G^{\#}$ admits a continuous isomorphism onto a topological group $H$ such that $w(H) \leq \mathrm{Ln}\,|G|$, so $\psi(G^{\#}) \leq w(H) \leq \mathrm{Ln}\,|G|$. Therefore, $\psi(G^{\#}) = \mathrm{Ln}\,|G|$. □

Recall that the *extent* $e(X)$ of a space $X$ is the smallest cardinal number $\tau$ such that the cardinality of every closed discrete subspace $Y$ of $X$ does not exceed $\tau$. Clearly, Theorem 9.9.55 implies immediately the following statement:

THEOREM 9.9.60. [**K. P. Hart and J. van Mill**] *Let $G$ be a discrete Abelian group. Then $l(G^\#) = e(G^\#) = |G|$. Moreover, the same equalities hold for every subspace $A$ of the space $G^\#$.*

We conclude the section with some observations on the Bohr topologies of non-discrete topological groups.

By Proposition 9.9.1, the Bohr topology of every locally compact Abelian group is Hausdorff. The next example shows that the local compactness requirement is essential. Furthermore, it also shows that there can exist continuous characters on a compact subgroup of an Abelian topological group $G$ which do not admit an extension to a continuous character over the group $G$ (compare this with Theorem 9.6.3).

EXAMPLE 9.9.61. [**R. C. Hooper**] There exists a second-countable topological Abelian group $G$ such that the continuous characters of $G$ do not separate points of $G$, so the group $G^+$ is not Hausdorff. In addition, $G$ contains a subgroup $C$ isomorphic to the discrete group $\mathbb{Z}(2)$ such that the unique non-trivial character on $C$ cannot be extended to a continuous character on $G$.

Indeed, denote by $\sigma$ the linear space of all sequences $x = (x_n)_{n\in\omega}$ where $x_n \in \mathbb{R}$ for each $n \in \omega$, and $x_n = 0$ for all but finitely many values of $n$. The sum and multiplication by a constant in $\sigma$ are defined coordinatewise. For every $x \in \sigma$, let $\|x\| = \max\{|x_n| : n \in \omega\}$. Then $\|\cdot\|$ is a norm on $\sigma$, and we define an invariant metric $d$ on $G$ by $d(x, y) = \|x - y\|$ for all $x, y \in \sigma$. Take the topology on $\sigma$ generated by the metric $d$. With this topology, $\sigma$ becomes a separable metric Abelian topological group. Note that $\sigma$ is connected as a linear space over $\mathbb{R}$. Put

$$H = \{(x_n)_{n\in\omega} \in \sigma : x_n \in \mathbb{Z} \text{ for each } n \in \omega\}$$

and

$$K = \{(x_n)_{n\in\omega} \in H : \sum_{n=0}^{\infty} x_n \text{ is even}\}.$$

The summation in the definition of $K$ is finite. Clearly, $H$ and $K$ are subgroups of $\sigma$ and $K \subset H$. In addition, the subgroup $H$ is discrete. Indeed, let $U = \{x \in \sigma : \|x\| < 1\}$. Then $U$ is an open neighbourhood of the neutral element $e$ in $\sigma$, and $H \cap U = \{e\}$. This implies that the subgroups $H$ and $K$ are closed in $\sigma$.

Let $G = \sigma/K$ and $\varphi$ be the quotient homomorphism of $\sigma$ onto $G$. Observe that $G$ is a connected, metrizable, Abelian topological group. We claim that the family of continuous characters of $G$ does not separate points of $G$. To show this, for every $n \in \omega$ define an element $b(n) \in H$ by $b(n)_k = 1$ if $k = n$ and $b(n)_k = 0$ if $k \neq n$. Then $b(n) \in H \setminus K$ and $b(n) + K = H \setminus K$ for each $n \in \omega$. Hence, $\varphi(b(n)) = g \in G$ for all $n \in \omega$, where $g$ is distinct from the neutral element of $G$. Let us verify that $\chi(g) = 1$ for all $\chi \in G^*$, where $G^*$ is the family of all continuous characters of $G$.

Suppose that $\chi(g) \neq 1$, for some $\chi \in G^*$. Since $b(n) + b(n) \in K$ and $\chi$ is a homomorphism, we have $\chi(g) = -1$. Given any $n, k \in \omega$ with $k \neq 0$, the number $t = (\chi \circ \varphi)((1/k) \cdot b(n))$ satisfies $t^k = -1$, so there exists $m \in \mathbb{N}$ such that $t = \exp((2m + 1)\pi i/k)$. We conclude, therefore, that the continuous function $f$ from the closed interval $[0, 1/k]$ to $\mathbb{T}$ defined by $f(r) = \chi \circ \varphi(r \cdot b(n))$ sends this interval to an arc with the end points 1 and $\exp((2m + 1)\pi i/k)$. Hence, by the connectivity argument, there exists a number $r_{n,k} \in [0, 1/k]$ such that $(\chi \circ \varphi)(r_{n,k} \cdot b(n)) = \exp(\pi i/k)$. For every integer $k > 0$, define an element $c(k) \in \sigma$ by $c(k) = \sum_{n<k} r_{n,k} \cdot b(n)$. Then, on one hand,

$$(\chi \circ \varphi)(c(k)) = \prod_{n<k}(\chi \circ \varphi)(r_{n,k} \cdot b(n))) = (\exp(\pi i/k))^k = -1.$$

On the other hand, from $\|c(k)\| \leq 1/k$ we have that $c(k)$ tends to the neutral element $0 \in \sigma$, so that $(\chi \circ \varphi)(c(k)) \to 1$ in $\mathbb{T}$, which is a contradiction. This proves that the continuous characters do not separate points of $G$.

Since $H = \langle b(0) \rangle + K$, the subgroup $C = H/K$ of $G$ is isomorphic to the finite discrete group $\mathbb{Z}(2)$. It follows from $b(0) \in H \setminus K$ that $g = \varphi(b(0)) \in C$ and $C = \langle g \rangle$. Hence, the character on $C$ sending $g$ to $-1$ cannot be extended to a continuous character on $G$. □

## Exercises

9.9.a. Let $G$ be a discrete Abelian group. Show that the topological group $G^{\#}$ is discrete if and only if $G$ is finite.

9.9.b. Let $G$ be a discrete Abelian group. Show that the space $G^{\#}$ is a quotient of a locally compact Hausdorff space if and only if $G$ is finite.

9.9.c. Let $G$ be a discrete Abelian group. Show that the space $G^{\#}$ is a quotient of a metrizable space if and only if $G$ is finite.

9.9.d. Let $G$ be a discrete Abelian group. Show that the group $G^{\#}$ is extremally disconnected iff $G$ is finite.

9.9.e. Give an example of an $\omega$-narrow Abelian group $G$ and a closed subgroup $H$ of $G$ such that $H$ is not $z$-embedded in $G$.

9.9.f. Verify that closed subsets of the groups $G^{\#}$ can fail to be $C^*$-embedded in $G^{\#}$. Are they $z$-embedded in $G^{\#}$?

9.9.g. (G. Reid [**402**], P. Flor [**172**]) Prove that no sequences in $G$ converges to a point of $bG \setminus G$.

9.9.h. Is the set $\{n^2 : n \in \mathbb{N}\}$ dense in $\mathbb{Z}^{\#}$?

9.9.i. Show that, consistently, there is an Abelian group $G$ such that $d(bG) < d(G^{\#})$.

9.9.j. Given a family $\{G_i : i \in I\}$ of topological groups, we denote by $\sigma\Pi_{i\in I}G_i$ the $\sigma$-product of this family considered as a topological subgroup of the product group $\prod_{i\in I} G_i$ (see Section 1.6). Prove that if $G = \oplus_{i\in I}G_i$ is a direct sum of non-trivial discrete Abelian groups, then $G^{\#}$ is topologically isomorphic to the group $\sigma\Pi_{i\in I}G_i{}^{\#}$ if and only if the index set $I$ is finite. Extend this result to products of Abelian groups.

9.9.k. (K. P. Hart and J. van Mill [**219**]) Let $G$ be a Boolean group endowed with the linear topological group topology $\mathcal{L}_G$ whose base at zero consists of all subgroups $H$ of $G$ with $|G : H| < \omega$ (see also Exercise 3.7.h). Prove that $\mathcal{L}_G$ is the Bohr topology of $G$.

9.9.l. Show that there is a second-countable Čech-complete topological Abelian group $G$ such that the Bohr topology of $G$ is not Hausdorff.

## Problems

9.9.A. Let $G$ be a discrete Abelian group. Show that the space $G^{\#}$ is Čech-complete if and only if $G$ is finite.

9.9.B. Let $B$ be a countable, infinite, Boolean group and $G$ the group of continuous automorphisms of $B^{\#}$ onto itself, with the topology of pointwise convergence. Describe the group $G$ in algebraic and topological terms.

9.9.C. Let $G$ be an infinite discrete Abelian group. Prove that there exist disjoint closed subsets $A$ and $B$ of the group $G^{\#}$ such that $cl_{bG} A \cap cl_{bG} B \neq \varnothing$.

9.9.D. Suppose that $G$ is an uncountable precompact topological group such that all subgroups of $G$ are normal spaces. Is $G$ metrizable? What if $G$ is hereditarily normal?

*Remark.* Under $CH$, the answer to both questions above is "No" (see the remark to Problem 8.1.5). Therefore, we ask for counterexamples in $ZFC$.

9.9.E. Describe in topological and algebraic terms the Bohr compactifications $b\mathbb{R}$ and $b\mathbb{Z}$ of the discrete additive groups $\mathbb{R}$ and $\mathbb{Z}$, respectively.

9.9.F. Let $\kappa$ be an infinite cardinal. Prove that every Abelian group of cardinality less than or equal to $2^\kappa$ is isomorphic to a subgroup of the (abstract) group $\mathbb{T}^\kappa$.

*Hint.* Apply Theorem 9.9.59 and Proposition 5.2.11.

9.9.G. Let $G$ be a discrete Abelian group such that $C_p(G^\#)$ is Lindelöf. Prove that $G$ is countable.

9.9.H. Prove that the tightness of the space $G^\#$ is countable, for every discrete Abelian group $G$.

*Hint.* From the definition of the Bohr topology it follows that $G^\#$ is topologically isomorphic (under the natural evaluation mapping) to a topological subgroup of the topological group $C_p(G^*, \mathbb{T})$, where $G^*$ is the Pontryagin dual to $G$ with the usual topology, which is the topology of pointwise convergence, since the group $G$ is discrete. The space $G^*$ is compact; therefore, the tightness of the space $C_p(G^*, \mathbb{T})$ is countable [**32**]. It follows that the tightness of $G^*$ is countable as well, since the tightness is monotonous with respect to subspaces.

9.9.I. Prove that the group $G^\#$ is monolithic, for every discrete Abelian group $G$, that is, the density of $Y$ and the network weight of $Y$ coincide, for every subspace $Y$ of the space $G^\#$.

9.9.J. A *neighbourhood assignment* on a topological space $X$ is a function $f$ from $X$ to the set of all subsets of $X$ such that $f(x)$ is an open neighbourhood of $x$, for each $x \in X$. A space $X$ is said to be a *D-space* (after E. van Douwen) if, for every neighbourhood assignment $f$ on $X$, there exists a closed discrete subset $A$ of $X$ such that $\bigcup\{f(x) : x \in X\} = X$.

Let $G$ be a discrete Abelian group. Prove that $G^\#$ is a $D$-space hereditarily, that is, every subspace $Y$ of $D^\#$ is a $D$-space.

*Hint.* Since $G^*$ is compact, this follows from remarkable results of R. Z. Buzyakova in [**95**]. They imply immediately that $C_p(G^*, \mathbb{T})$ is a $D$-space hereditarily. Since $G^\#$ is homeomorphic to a subspace of $C_p(G^*, \mathbb{T})$, the required conclusion follows.

9.9.K. (K. P. Hart and J. van Mill [**219**]) There exists a countable Abelian group $G$ such that some discrete subspace $M$ of $G^\#$ is not closed in $G^\#$.

*Hint.* Apply Theorem 3.7.27. Another way is to take an infinite Boolean group $G$, and pick a maximal independent subset $H$ of $G$. Show that the set $(H + H) \setminus \{0\}$ is discrete, and that the neutral element 0 of $G$ belongs to the closure of this set. Observe that this example shows that the group $G^\#$ need not be maximal or submaximal, since in every submaximal space all nowhere dense subsets are closed.

9.9.L. (W. W. Comfort and J. van Mill [**115**]) Prove that every abstract infinite Abelian group $G$ with $r_2(G) < \infty$ is strongly resolvable.

9.9.M. (E. van Douwen [**150**]) Let $G$ be an infinite discrete Abelian group. Show that the space $G^\#$ does not have the Baire property.

*Hint.* There exists a homomorphism $f$ of $G$ onto a countable infinite group $F$. The kernel $H$ of the homomorphism $f$ is a subgroup of $G$. Hence, $H$ is closed in $G^\#$. The group $F$ is precompact and infinite. Therefore, $F$ is not discrete. Hence, $H$ is not open in $G^\#$, which implies that the interior of $H$ in $G^\#$ is empty. Thus, $H$ is nowhere dense in $G^\#$, and $\{f^{-1}(y) : y \in F\}$ is a countable covering of the space $G^\#$ by nowhere dense subsets. Hence, $G^\#$ is not Baire.

9.9.N. Prove that in the category of locally compact Abelian groups, the covariant functor $^+$ defined on page 633 preserves topological subgroups, quotients and topological products.

*Hint.* Let $H$ be a closed subgroup of a locally compact Abelian group $G$. We have to prove that the subgroup $H^+$ is a topological subgroup of $G^+$. The continuity of the inclusion

$\iota\colon H^+ \to G^+$ follows from the functoriality of $^+$. To see that $\iota$ is an embedding recall that by [**236**, 24.12], every continuous character $H \to \mathbb{T}$ can be extended to a continuous character of $G$.

Now we show that $(G/H)^+ = G^+/H^+$, where the group $G^+/H^+$ carries the quotient topology. The continuity of the quotient homomorphism $f\colon G^+ \to (G/H)^+$ and of the identity isomorphism $(G/H)^+ \to G^+/H^+$ follows from the functoriality of $^+$. The continuity of the identity $G^+/H^+ \to (G/H)^+$ follows from the continuity of the quotient homomorphism $f$ and the universal property of the quotient topology. This proves the preservation of quotients under $^+$.

Let $G$ and $H$ be locally compact Abelian groups. The continuity of the identity map $(G \times H)^+ \to G^+ \times H^+$ follows easily from the functoriality of $^+$. Again, by functoriality of $^+$, both inclusions $i\colon G^+ \hookrightarrow (G \times H)^+$ and $j\colon H^+ \hookrightarrow (G \times H)^+$ are continuous. Since the product topology of $G^+ \times H^+$ has the universal property with respect to the monomorphisms $i$ and $j$ (see [**139**, Exercise 2.10.4(i)]), it follows that the identity $G^+ \times H^+ \to (G \times H)^+$ is continuous. This proves the preservation of finite products under $^+$. The case of arbitrary products may lead out of the category of locally compact groups. Nevertheless, the preservation is still available since the Bohr compactification commutes with arbitrary products (see [**253**]).

9.9.O. For every locally compact Abelian group $G$, the groups $G^+$ and $G$ have the same compact sets.

*Hint.* See Glicksberg's article [**198**].

9.9.P. (F. J. Trigos-Arrieta [**501**]) The functor $^+$ "respects" compactness-like properties, such as pseudocompactness, realcompactness, the Lindelöf property, *etc.*

9.9.Q. (F. J. Trigos-Arrieta [**501**]) Prove that for a locally compact Abelian group $G$, the group $G^+$ is normal iff $G$ is $\sigma$-compact.

9.9.R. (V. G. Pestov [**379**]) A subset $S$ of an Abelian group $G$ is said to be *big* if $F + S = G$, for some finite $F \subset G$. Let $S$ be a big subset of the group $\mathbb{Z}$ of integers. Prove that the set $S - S + S$ contains an open neighbourhood of 0 in the space $\mathbb{Z}^\#$.

9.9.S. (K. Kunen [**286**]) There are countable infinite Abelian groups $G$ and $H$ such that the spaces $G^\#$ and $H^\#$ are not homeomorphic.

*Hint.* The groups in question are $G = \mathbb{Z}(2)^{(\omega)}$ and $H = \mathbb{Z}(3)^{(\omega)}$, the countable direct sums of the groups $\mathbb{Z}(2)$ and $\mathbb{Z}(3)$, respectively. More information on this subject can be found in the article [**138**] by D. Dikranjan and L. de Leo.

## Open Problems

9.9.1. Does every infinite Abelian group $G$ admit a Hausdorff topological group topology in which every continuous character on $G$ is trivial?

*Remark.* It was shown in [**1**] that every infinite Abelian group $G$ admits a Hausdorff topological group topology in which continuous characters do not separates points of $G$.

9.9.2. Let $G$ be a discrete Abelian group such that the group $G^\#$ is subparacompact. Is $G$ countable?

9.9.3. Let $G$ be a discrete Abelian group and $f$ a continuous mapping of the space $G^\#$ onto a compact Hausdorff space $F$. Is $F$ dyadic?

9.9.4. Let $G$ be an infinite discrete Abelian group. Is $G^\#$ a left-separated space? (See Problems 4.2.C and 4.2.D.)

9.9.5. (V. G. Pestov [**379**]) Let $S$ be a big subset of the group $\mathbb{Z}$ of integers (see Problem 9.9.R). Is it true that the set $S - S = \{a - b : a, b \in S\}$ contains an open neighbourhood of 0 in the space $\mathbb{Z}^\#$?

9.9.6. Let $G$ be a precompact Abelian group such that every countable subgroup of $G$ is a retract of $G$ under a continuous homomorphism. Is $G$ zero-dimensional?

9.9.7. Let $G$ be a precompact torsion Abelian group such that every finite subgroup of $G$ is a retract of $G$ under a continuous homomorphism. Is $G$ zero-dimensional?

## 9.10. Bounded sets in extensions of groups

Let $H$ be a closed invariant subgroup of a topological group $G$ and suppose that all bounded subsets of the groups $H$ and $G/H$ are finite. Are the bounded subsets of $G$ then finite? Here we answer this question in the negative by presenting an example which destroys a number of tempting conjectures about bounded sets in extensions of topological groups.

Let us call a space $X$ *B-closed* if every bounded subset of $X$ is closed. It is evident that no infinite compact space is $B$-closed. The next result describes $B$-closed topological groups.

LEMMA 9.10.1. *A topological group $G$ is $B$-closed if and only if all bounded subsets of $G$ are finite.*

PROOF. Suppose that $G$ contains an infinite bounded subset $X$. We claim that the closure of the set

$$P = \{x^{-1}y : x, y \in X, x \neq y\}$$

contains the identity $e$ of $G$. If not, choose an open symmetric neighbourhood $U$ of $e$ in $G$ such that $U^4 \cap P = \varnothing$. An easy verification shows that the family of open sets $\{xU : x \in X\}$ is discrete in $G$. Clearly, each element of this family intersects $X$. However, by Lemma 6.9.6, only finitely many elements of a discrete family of open sets in $G$ can meet a bounded set, which gives a contradiction. Hence, $e$ is in the closure of $P$ and $P$ is dense in $X^{-1}X = P \cup \{e\}$.

The sets $X^{-1}$ and $X^{-1}X$ are bounded in $G$, by Corollary 6.10.13. Since $e \notin P \subset X^{-1}X$, we conclude that $P$ is a non-closed bounded subset of $G$. This proves the necessity of the condition. The sufficiency is evident. □

To show that the class of $B$-closed topological groups is not stable under taking extensions, we need an auxiliary fact given below. As usual, $\mathfrak{c} = 2^{\omega}$ is the power of the continuum. We denote by $2^{\mathfrak{c}}$ the product of $\mathfrak{c}$ copies of the discrete group $2 = \{0, 1\}$.

LEMMA 9.10.2. *There exists a dense subgroup $H$ of $2^{\mathfrak{c}}$ satisfying the following conditions:*

a) $|H| = \mathfrak{c}$;
b) *all bounded subsets of $H$ are finite;*
c) $|\ker f| = \mathfrak{c}$, *for every continuous homomorphism $f : H \to P$ to a first-countable topological group $P$.*

PROOF. Let $X$ be a set of cardinality $\mathfrak{c}$. Denote by $H_X$ the family of all finite subsets of $X$ with the binary operation $A \Delta B = (A \setminus B) \cup (B \setminus A)$, the symmetric difference of $A$ and $B$, where $A, B \in H_X$. Then $H_X$ is a Boolean group of cardinality $\mathfrak{c}$ whose identity is the empty set. Given a homomorphism $\varphi : H_X \to 2$ and a subset $Y$ of $X$, we say that $\varphi$ *depends only on* $Y$ if $\varphi(\{x\}) = 0$ for each $x \in X \setminus Y$. Denote by $\mathcal{F}$ the family of all homomorphisms of $H_X$ to 2 that depend only on a countable subset of $X$. It is easy to see that $|\mathcal{F}| = \mathfrak{c}$. Let $\tau$ be the coarsest group topology on $H_X$ which makes all homomorphisms of the family $\mathcal{F}$ continuous. Then $H_X = (H_X, \tau)$ is a precompact Hausdorff topological group and $w(H_X) \leq |\mathcal{F}| = \mathfrak{c}$. Hence, by Theorem 9.6.26, the Raĭkov completion $\varrho H_X$ of $H_X$ is topologically isomorphic to the group $2^\lambda$, for some cardinal $\lambda$ satisfying $\omega \leq \lambda \leq \mathfrak{c}$. Since every element of $\mathcal{F}$ depends only on countably many indices, the pseudocharacter of the neutral element $\bar{0}$ of $H_X$ is equal to $\mathfrak{c}$. Therefore, $\lambda = \mathfrak{c}$, that is, $\varrho H_X$ is topologically isomorphic to $2^\mathfrak{c}$. This enables us considering $H = H_X$ as a dense subgroup of $2^\mathfrak{c}$.

Clearly, $|H_X| = \mathfrak{c}$, thus implying a) of the lemma. To deduce b), we argue as follows. First, for every non-empty subset $Y$ of $X$, let $p_Y\colon H_X \to H_Y$ be defined by $p_Y(A) = A \cap Y$, for each $A \in H_X$, where $H_Y$ is the Boolean group of all finite subsets of $Y$. Clearly, $p_Y$ is a homomorphic retraction of $H_X$ onto its subgroup $H_Y$. Our definition of the topology $\tau$ on $H_X$ implies that $p_Y$ is continuous (when $H_Y$ is taken with the subspace topology) and, in particular, $H_Y$ is closed in $H_X$. Note also that, for countable $Y \subset X$, the group $H_Y$ is countable and carries the Bohr topology. Hence, all bounded subsets of $H_Y$ are finite, by Theorem 9.9.30. Let $K$ be an infinite subset of $H_X$. We can assume without loss of generality that $K$ is countable. Then there exists a countable subset $Y$ of $X$ such that $A \subset Y$, for each $A \in K$. Since $K$ is infinite, there exists a continuous real-valued function $f$ on $H_Y$ such that $f(K)$ is an unbounded subset of the reals. Then $g = f \circ p_Y$ is a continuous function on $H_X$ and $g(K) = f(K)$ is unbounded in $\mathbb{R}$. This implies b).

Finally, we verify c). Let $f\colon H_X \to P$ be a continuous homomorphism to a first-countable topological group $P$. Then the kernel of $f$ is of type $G_\delta$ in $H_X$ and, since the family $\mathcal{F}$ generates the topology $H_X$, we can find a countable family $\{f_n : n \in \omega\} \subset \mathcal{F}$ such that $\bigcap_{n\in\omega} \ker f_n \subset \ker f$. Each $f_n$ depends only on a countable subset of $X$, so there exists a countable set $C \subset X$ such that $H_{X\setminus C} \subset \ker f$. Evidently, $|\ker f| \geq |H_{X\setminus C}| = |X \setminus C| = \mathfrak{c}$, whence c) follows. □

It is worth comparing the next example with Theorem 3.3.24 or with Exercise 3.3.h saying that if all compact subsets of a closed subgroup $N$ of a topological group $G$ as well as those of the quotient space $G/N$ are finite, then the same holds valid for the group $G$.

EXAMPLE 9.10.3. There exist a precompact Boolean topological group $G$ and an infinite, closed, bounded subgroup $N$ of $G$ such that both groups $N$ and $G/N \cong N$ are $B$-closed. So, $G$ fails to be $B$-closed, and $B$-closedness is not a three space property.

We will define $G$ to be a dense subgroup of the product group $K \times K$, where $K = 2^\mathfrak{c}$. Let $H$ be a dense subgroup of $K$ as in Lemma 9.10.2. Put $N = \{\mathbf{0}\} \times H$, where $\mathbf{0}$ is the neutral element of $K$. Our aim is to define an algebraic homomorphism $\varphi\colon H \to K$ and to take $G = N + P$, where $P = \{(y, \varphi(y)) : y \in H\}$ is the graph of $\varphi$. For every $A \subset \mathfrak{c}$, let $\pi_A\colon 2^\mathfrak{c} \to 2^A$ be the projection and $\mathbf{0}_A$ the neutral element of the group $2^A$. To guarantee the boundedness of $N$ in $G$, the homomorphism $\varphi$ has to satisfy the following condition:

(a) For every non-empty countable set $A \subset \mathfrak{c}$ and every $x \in 2^A$, there exists $y \in H$ such that $\pi_A(y) = \mathbf{0}_A$ and $\pi_A(\varphi(y)) = x$.

Consider the family

$$\gamma = \{(A, x) : \varnothing \neq A \subset \mathfrak{c},\ |A| \leq \omega,\ x \in 2^A\}.$$

It easy to see that $|\gamma| = \mathfrak{c}$, so we can write $\gamma = \{(A_\alpha, x_\alpha) :\ \alpha < \mathfrak{c}\}$. One can define by recursion two sets $Y = \{y_\alpha :\ \alpha < \mathfrak{c}\} \subset H$ and $Z = \{z_\alpha :\ \alpha < \mathfrak{c}\} \subset K$ satisfying the following conditions for each $\alpha < \mathfrak{c}$:

(i) $y_\alpha \notin \langle Y_\alpha \rangle$, where $Y_\alpha = \{y_\nu : \nu < \alpha\}$;
(ii) $\pi_{A_\alpha}(y_\alpha) = \mathbf{0}_{A_\alpha}$;
(iii) $\pi_{A_\alpha}(z_\alpha) = x_\alpha$.

Such a construction is possible since $|\pi_A^{-1}(\mathbf{0}_A) \cap H| = \mathfrak{c}$, for each countable set $A \subset \mathfrak{c}$ (see c) of Lemma 9.10.2). It follows from (i) that the set $Y$ is independent in $K$. Hence, there exists a homomorphism $p\colon \langle Y \rangle \to K$ such that $p(y_\alpha) = z_\alpha$ for each $\alpha < \mathfrak{c}$. Since the group $H$ is Boolean, $p$ admits an extension to a homomorphism $\varphi\colon H \to K$. Note that (a) follows immediately from (ii), (iii) and the definition of $\varphi$.

Let $P = \{(y, \varphi(y)) : y \in H\}$ be the graph of $\varphi$. It is easy to verify that the subgroup $G = N + P$ of $K \times K$ satisfies $G \cap (\{\mathbf{0}\} \times K) = N = \{\mathbf{0}\} \times H$, where $H$ is dense in $K$. Hence, by virtue of Lemma 1.5.16, the quotient group $G/N$ is topologically isomorphic to the projection of $G$ onto the first factor of the product $K \times K$, that is, $G/N \cong H \cong N$. In particular, all bounded subsets of $G/N$ are finite, so $G/N$ is $B$-closed. Note that the density of $H$ in $K$ implies that $G = N + P$ is dense in $K \times K$.

It remains to show that $N$ is bounded in $G$. It follows from (a) that

$$(\pi_A \times \pi_A)(G) \supseteq (\pi_A \times \pi_A)(P) \supseteq \{\mathbf{0}_A\} \times 2^A,$$

for each countable set $A \subset \mathfrak{c}$. It is also clear that

$$(\pi_A \times \pi_A)(N) = \{\mathbf{0}_A\} \times \pi_A(H) \subset \{\mathbf{0}_A\} \times 2^A.$$

In other words, for every countable $A \subset \mathfrak{c}$, the set $(\pi_A \times \pi_A)(N)$ is contained in a compact subset of $(\pi_A \times \pi_A)(G)$. Let $f$ be a continuous real-valued function on $G$. Since $G$ is dense in $K \times K = 2^{\mathfrak{c}} \times 2^{\mathfrak{c}}$, it follows from Corollary 1.7.8 that $f$ depends on at most countably many coordinates or, equivalently, one can find a non-empty countable set $A \subset \mathfrak{c}$ and a continuous real-valued function $g$ on $(\pi_A \times \pi_A)(G)$ such that $f = g \circ (\pi_A \times \pi_A)$. Since $(\pi_A \times \pi_A)(N)$ is contained in a compact subset of $(\pi_A \times \pi_A)(G)$, the image $f(N) = g((\pi_A \times \pi_A)(N))$ is a bounded subset of the reals. So, $N$ is bounded in $G$.

Finally, since $N$ is infinite, it follows from Lemma 9.10.1 that $G$ fails to be $B$-closed. In fact, one can avoid the use of Lemma 9.10.1 by noting that $N \setminus \{(\mathbf{0}, \mathbf{0})\}$ is a non-closed bounded subset of $G$. □

## Exercises

9.10.a. (M. Bruguera and M. G. Tkachenko [**90**]) Let $\mathcal{P}$ be a topological property. We say that $\mathcal{P}$ is an *inverse fiber property* if, given an arbitrary continuous onto mapping $f\colon X \to Y$ such that the space $Y$ has $\mathcal{P}$ and the fibers $f^{-1}(y)$ have $\mathcal{P}$, it follows that $X$ also has $\mathcal{P}$. Verify that every inverse fiber property is a three space property. Prove the following:

a) all pseudocompact subsets are finite (see also Exercise 3.3.h);
b) all convergent sequences are trivial;
c) all supersequences (see Exercise 6.10.b) have length strictly less than a given infinite cardinal $\tau$;
d) there is no subspace homeomorphic to the ordinal space $\omega_1$ (or $\omega_1 + 1$);
e) there is no subspace homeomorphic to $\beta\omega$;
f) all compact (countably compact) subsets are scattered (see Problem 4.2.B);
g) all compact (countably compact) subsets are left-separated (see Problem 4.2.C);
h) all compact (countably compact) subsets are first-countable;
i) all compact (countably compact) subsets are $C$-closed (see Problem 4.2.F);

j) all countably compact subsets are compact (this property of a space is called *C-compactness*);
k) all compact sets are zero-dimensional;
l) all compact subsets have countable tightness;
m) all compact subsets are sequential, under $2^{\aleph_0} < 2^{\aleph_1}$.

Therefore, each of the properties described in a)–m) is a three space property.

## Problems

9.10.A. (M. Bruguera and M. G. Tkachenko [**90**]) Let $K$ be a one-point compactification of the Franklin–Mrówka space $X$ (see [**165**, 3.6.I]). Show that the free Abelian topological group $A(K)$ over $K$ contains a closed subgroup $N$ such that all compact subsets of the groups $N$ and $A(K)/N$ are $\aleph_0$-monolithic, while this is false for the group $A(K)$. Therefore, "all compact subsets are $\aleph_0$-monolithic" is not a three space property.

9.10.B. (M. Bruguera and M. G. Tkachenko [**90**]) Construct a pseudocompact non-compact topological Abelian group $G$ and a closed countable subgroup $N$ of $G$ such that the quotient group $G/N$ is compact and all bounded subsets of $N$ are finite (equivalently, $N$ is $B$-closed, see Lemma 9.10.1). Therefore, none of the following is a three space property:
a) realcompactness;
b) Dieudonné completeness;
c) "every bounded subset has compact closure";
d) "every pseudocompact subset has compact closure".

9.10.C. (M. Bruguera and M. G. Tkachenko [**90**]) Let $G$ be the free Abelian topological group over the Franklin–Mrówka space $X$. Show that $G$ contains a closed subgroup $K$ such that all pseudocompact subsets of the groups $K$ and $G/K$ are compact, while $G$ contains a pseudocompact non-compact subspace $X$. Therefore, "all pseudocompact subsets are compact" is not a three space property.

9.10.D. (M. Bruguera and M. G. Tkachenko [**90**]) Give an example of a topological Abelian group $G$ and a closed subgroup $N$ of $G$ such that all bounded subsets of the groups $N$ and $G/N$ have compact closures, but $G$ contains a closed copy of the ordinal space $\omega_1$.

## Open Problems

9.10.1. In what follows $N$ denotes a closed invariant subgroup of a topological group $G$. Suppose that all compact subsets of the groups $N$ and $G/N$ are sequentially compact. Does $G$ have the same property?

9.10.2. Suppose that all compact subsets of $N$ and $G/N$ are separable. Are the compact subsets of $G$ separable (or have countable cellularity)?

9.10.3. Let all compact subsets of the groups $N$ and $G/N$ be Fréchet–Urysohn. Does the same hold for compact subsets of $G$?

9.10.4. Does there exist in $ZFC$ a topological group $G$ and a closed invariant subgroup $N$ of $G$ such that all pseudocompact subspaces of $N$ and $G/N$ are metrizable, but $G$ contains a non-metrizable (and/or non-closed) pseudocompact subspace?

## 9.11. Pseudocompact group topologies on Abelian groups

We have seen in the preceding chapters that the existence of the structure of a topological group on a topological space improves considerably topological properties of the space and makes easier their investigation. For example, by the Birkhoff–Kakutani theorem, (see Theorem 3.3.12), every first-countable topological group is metrizable, while Corollary 6.6.11, due to Comfort and Ross, implies that pseudocompactness becomes invariant under taking arbitrary topological products of topological groups.

Conversely, the topological properties of a topological group have a notable influence on the algebraic structure of the group, which is especially strong in the case of compact, countably compact and pseudocompact topological groups. (This phenomenon is, however, almost invisible in the case of Lindelöf groups or Lindelöf $P$-groups, see Problems 4.4.1 and 4.4.2.) This section and several sections to follow will elucidate this influence, especially in the case of Abelian groups of cardinality $\mathfrak{c} = 2^{\omega}$.

First, we show that the existence of a pseudocompact Hausdorff topological group topology on an Abelian group implies several restrictions on the cardinality and on algebraic properties of the group.

Recall that every Čech-complete space $X$ has the Baire property (see [**165**, Theorem 3.9.3]). We extend this result to all weakly pseudocompact spaces, i.e., $G_\delta$-dense subspaces of compact spaces.

LEMMA 9.11.1. *Every weakly pseudocompact space $X$ has the Baire property.*

PROOF. Let $\mathcal{F} = \{F_n : n \in \omega\}$ be a family of nowhere dense subsets of the space $X$. It suffices to show that $V \setminus \bigcup \mathcal{F} \neq \varnothing$, for every non-empty open subset $V$ of $X$. Let $bX$ be a compactification of $X$ such that $X$ is $G_\delta$-dense in $bX$. Choose an open set $U$ in $bX$ such that $U \cap X = V$.

Since $F_0$ is nowhere dense in $X$ and $bX$, there is a non-empty open set $U_0$ in $bX$ such that $U_0 \subset U$ and $U_0 \cap F_0 = \varnothing$. Suppose we have defined non-empty open sets $U_0, \dots, U_n$ in $bX$ such that $U_i \cap F_i = \varnothing$ and $\overline{U_i} \subset U_{i-1}$ for each $i \leq n$ (we assume that $U_{-1} = bX$). Since $bX$ is regular and $F_{n+1}$ is nowhere dense in $bX$, there exists a non-empty open set $U_{n+1}$ in $bX$ such that $U_{n+1} \cap F_{n+1} = \varnothing$ and $\overline{U_{n+1}} \subset U_n$.

Consider the sequence $\xi = \{U_n : n \in \omega\}$ and put $K = \bigcap_{n \in \omega} \overline{U_n}$. Since the space $bX$ is compact and $\overline{U_{n+1}} \subset U_n$ for each $n \in \omega$, it follows that $K \neq \varnothing$. In addition, our construction of the sequence $\xi$ implies that $\overline{U_{n+1}} \cap F_n = \varnothing$ for each $n \in \omega$, so that $K$ does not meet any element of $\mathcal{F}$. Since $X$ is $G_\delta$-dense in $bX$, the intersection $K \cap X$ is non-empty. It remains to note that $K \cap X \subset U \cap X = V$, whence it follows that $V \setminus \bigcup \mathcal{F} \neq \varnothing$. □

According to b) of Corollary 5.2.7, the cardinality of an infinite compact topological group $G$ satisfies $|G| = 2^{w(G)}$. This is no longer valid neither for pseudocompact nor for countably compact topological groups — it suffices to take the $\Sigma$-product $H$ of $\mathfrak{c}$ copies of the circle group $\mathbb{T}$. By Corollary 1.6.34, $H$ with the topology inherited from $\mathbb{T}^{\mathfrak{c}}$ is countably compact and $|H| = w(H) = \mathfrak{c}$. In fact, the cardinality of a countably compact topological group can even be less than the weight of the group (see Exercise 5.2.c). However, there are certain restrictions on the cardinality of a pseudocompact topological group. Some of them are given in Theorem 9.11.2 below.

Let us call a cardinal $\lambda$ *strong limit* if $2^{\tau} < \lambda$ for each $\tau < \lambda$. Clearly, $\aleph_0 = \omega$ is the first strong limit cardinal. For every infinite cardinal $\tau$, one can find a strong limit cardinal $\lambda > \tau$ as the limit of the sequence $\{\tau_n : n \in \omega\}$, where $\tau_0 = \tau$ and $\tau_{n+1} = 2^{\tau_n}$ for each $n \in \omega$.

THEOREM 9.11.2. *Let $G$ be an infinite pseudocompact topological group such that $|G| = \kappa$. Then $\kappa \geq \mathfrak{c}$, and if $cf(\kappa) = \omega$, then the cardinal $\kappa$ cannot be strong limit.*

PROOF. Since the group $G$ is infinite and pseudocompact, it has no isolated points. The Raĭkov completion $\varrho G$ of $G$ is a non-discrete compact topological group, by Corollary 3.7.18. Therefore, by Proposition 4.2.3, there exists a continuous mapping $f$ of the space $\varrho G$ onto the unit segment [0, 1]. It follows from Corollary 3.7.21 that $G$ intersects every non-empty $G_\delta$-set in $\varrho G$. In particular, $G$ intersects each fiber of the mapping $f$ and, therefore, $f(G) = [0, 1]$. This gives immediately the inequality $\kappa = |G| \geq \mathfrak{c}$.

Assume that the cardinal $\kappa = |G|$ is strong limit and countably cofinal, that is, $\kappa = \sup_{n\in\omega} \kappa_n$, where $\kappa_n < \kappa$ for each $n \in \omega$. We already know that $\kappa \geq \mathfrak{c}$, so one can assume that all cardinals $\kappa_n$ are infinite. Represent $G$ as the union $G = \bigcup_{n\in\omega} X_n$, where $|X_n| = \kappa_n$, for each $n \in \omega$. Let $Y_n$ be the closure of $X_n$ in $G$. Since $\kappa$ is strong limit, we have $w(Y_n) \leq 2^{\kappa_n} < \kappa$ and $|Y_n| \leq 2^{w(Y_n)} < \kappa$, according to Theorems 1.5.6 and 1.5.1 of [**165**], respectively. By Lemma 9.11.1, the space $G$ has the Baire property. Hence, there exists $m \in \omega$ such that $Y_m$ contains a non-empty open subset $U$ of $G$. Theorem 3.7.2 implies that the group $G$ is precompact, so $G$ can be covered by finitely many translations of the set $U$. Hence, the cardinalities of $U$ and $G$ coincide. This contradicts the fact that $U$ is a subset of $Y_m$, where $|Y_m| < \kappa$. □

COROLLARY 9.11.3. *Let $G$ be a pseudocompact topological Abelian group. Then, for every integer $d \in \mathbb{N}$, the subgroup $dG = \{dx : x \in G\}$ of the group $G$ is either finite or has cardinality at least $\mathfrak{c}$.*

PROOF. The homomorphism $\varphi_d$ of $G$ to $G$ defined by $\varphi_d(x) = dx$ for each $x \in G$, is continuous. It follows the homomorphic image $\varphi(G) = dG$ of $G$ is again a pseudocompact group. Therefore, if $dG$ is infinite, Theorem 9.11.2 implies that $|dG| \geq \mathfrak{c}$. □

Every infinite Abelian group admits a non-discrete precompact Hausdorff topological group topology — it suffices to endow the group with the Bohr topology (this also follows from the proof of Theorem 1.4.25, since only the homomorphisms to the circle group $\mathbb{T}$ were used there). However, for every torsion Abelian group $G$, the product group $G \times \mathbb{Z}$ does not admit a pseudocompact topological group topology. This follows from Theorem 9.11.5 below whose proof depends on the next special case of it.

LEMMA 9.11.4. *If $G$ is a compact Abelian topological group with an element of infinite order, then $r_0(G) \geq \mathfrak{c}$.*

PROOF. Clearly, the group $G$ is infinite and, hence, non-discrete. For every $n \in \mathbb{N}$, let

$$Z_n = \{(k_0, \ldots, k_n) \in \mathbb{Z}^{n+1} \colon \sum_{i=0}^{n} |k_i| \leq n(n+1) \text{ and } k_j \neq 0 \text{ for some } j \leq n\}.$$

First we establish the following auxiliary fact.

CLAIM. *For every sequence $U_0, \ldots, U_n$ of non-empty open sets in $G$, there exists a sequence $V_0, \ldots, V_n$ of non-empty open subsets of $G$ such that $V_i \subset U_i$ for each $i \leq n$, and if $(k_0, \ldots, k_n) \in Z_n$ and $x_i \in V_i$ for each $i \leq n$, then $k_0x_0 + \cdots + k_nx_n \neq 0_G$.*

To prove Claim, fix an arbitrary element $(k_0, \ldots, k_n)$ of $Z_n$. The mapping $\pi \colon G^{n+1} \to G$ defined by $\pi(x_0, \ldots, x_n) = k_0x_0 + \cdots + k_nx_n$ is a continuous homomorphism. Put $H = \pi(G)$. Since $G$ is compact, the homomorphism $\pi \colon G \to H$ is open. Hence, $U = \pi(U_0 \times \cdots \times U_n)$ is a non-empty open subset of $H$. It follows from $(k_0, \ldots, k_n) \in Z_n$

that $k_j \neq 0$ for some $j \leq n$. By the assumption, $G$ contains a non-torsion element $z$; therefore, $k_j z$ is a non-torsion element of $H$. So the compact group $H$ and its open subset $U$ are infinite. Therefore, we can choose an element $y_i \in U_i$ for every $i \leq n$, such that $k_0 y_0 + \cdots + k_n y_n \neq 0_G$. Since $\pi$ is continuous, there exist open sets $W_0, \ldots, W_n$ in $G$ such that $y_i \in W_i \subset U_i$ for each $i \leq n$ and $0_G \notin \pi(W_0 \times \cdots \times W_n)$. Since the set $Z_n$ is finite, we can repeat this construction finitely many times (for every element of $Z_n$), thus obtaining the required open sets $V_0, \ldots, V_n$. Our Claim is proved.

Let us go back to the main argument. As in the proof of Proposition 4.2.3, let $2^n$ be the family of all mappings from $n = \{0, \ldots, n-1\}$ to $\{0, 1\} = 2$, where $n \in \mathbb{N}$. Let also $\mathscr{C} = \bigcup_{n=1}^{\infty} 2^n$. The above Claim permits us to define a family $\{V_f : f \in \mathscr{C}\}$ of non-empty open sets in $G$ satisfying the following conditions for each $n \in \mathbb{N}$:

(1) $V_f \cap V_g = \varnothing$ for all distinct $f, g \in 2^n$;
(2) if $f, g \in \mathscr{C}$ and $g$ is a proper extension of $f$, then $\overline{V_g} \subset V_f$;
(3) if $F \subset 2^n$ and $k_f \in \mathbb{Z}$ and $x_f \in G$ satisfy $|k_f| \leq n$ and $x_f \in V_f$ for each $f \in F$, then the equality $\sum_{f \in F} k_f x_f = 0_G$ holds only if $k_f = 0$ for all $f \in F$.

By (2) and compactness of $G$, the set $K_f = \bigcap_{n=1}^{\infty} \overline{V_{f \restriction n}}$ is non-empty for each $f \in 2^\omega$. Choose $x_f \in K_f$, $f \in 2^\omega$. It is clear from (3) that all elements of $X = \{x_f : f \in 2^\omega\}$ are of infinite order, and that $|X| = \mathfrak{c}$. Let us verify that the set $X$ is independent. For any pairwise distinct elements $y_0, \ldots, y_m$ of $X$ and $(k_0, \ldots, k_m) \in \mathbb{Z}^{m+1}$ with $k_j \neq 0$ for some $j \leq m$, we can find $n \in \mathbb{N}$ with $m + 1 \leq n$ and a set $F \subset 2^n$ of size $m + 1$ such that $|k_i| \leq n$, for each $i = 0, \ldots, m$, and $\{y_0, \ldots, y_m\} = \{x_f : f \in F\}$. By condition (3), $k_0 y_0 + \cdots + k_m y_m \neq 0_G$. Hence, $X$ is an independent subset of $X$. □

The following theorem gives a restraint on the structure of pseudocompact Abelian groups very different from those established in Theorem 9.11.2.

THEOREM 9.11.5. *Let $G$ be a pseudocompact Abelian topological group. Then either $r_0(G) \geq \mathfrak{c}$ or $G$ is a bounded torsion group.*

PROOF. Suppose first that $G$ is not a torsion group. Then there exists an element $a \in G$ of infinite order, so the cyclic subgroup $H = \langle a \rangle$ of $G$ generated by $a$ is infinite. The Raĭkov completion $K = \varrho G$ of $G$ is a compact Abelian group, by Corollary 3.7.18. According to Theorem 9.4.11, the continuous characters on $K$ separate elements of $K$. Therefore, for any distinct $x, y \in H$, we can choose a continuous character $h_{x,y} \in K^*$ such that $h_{x,y}(x - y) \neq 1$ or, equivalently, $h_{x,y}(x) \neq h_{x,y}(y)$. Denote by $f$ the diagonal product of the family $\mathscr{H} = \{h_{x,y} : x, y \in H,\ x \neq y\}$. It is clear that $|\mathscr{H}| = |H| = \omega$, so $f(K)$ is a subgroup of the product group $\mathbb{T}^\omega$. Since the homomorphism $f$ is continuous, the group $L = f(K)$ is compact and metrizable. It also follows from the choice of the characters $h_{x,y}$ and the definition of the family $\mathscr{H}$ that $f(x) \neq f(y)$ for all distinct $x, y \in H$. Hence, the group $f(H) = \langle f(a) \rangle$ is infinite and the element $f(a)$ has infinite order in $L$. We conclude that the group $L$ is not torsion. Now it follows from Lemma 9.11.4 that $r_0(L) \geq \mathfrak{c}$. Let $\{b_\alpha : \alpha < \mathfrak{c}\}$ be an independent family of elements of infinite order in $L$.

Since $G$ intersects all non-empty $G_\delta$-sets in $K = \varrho G$, the image of $G$ under the homomorphism $f$ coincides with the group $L$, that is, $f(G) = f(K) = L$. For every $\alpha < \mathfrak{c}$, choose an element $a_\alpha \in G$ with $f(a_\alpha) = b_\alpha$. Then $\{a_\alpha : \alpha < \mathfrak{c}\}$ is an independent family of elements of infinite order in $G$, so that $r_0(G) \geq \mathfrak{c}$.

Finally, suppose that $G$ is a torsion group. For every positive integer $n$, $G[n] = \{x \in G : nx = 0\}$ is a closed subgroup of $G$ and $G = \bigcup_{n=1}^{\infty} G[n]$. The space $G$ is Baire by Lemma 9.11.1, so the group $G[k]$ has a non-empty interior for some $k \in \mathbb{N}$. Clearly, $G[k]$ is open in $G$ and, since $G$ is precompact by Theorem 3.7.2, it can be covered by finitely many translations of the subgroup $G[k]$. Choose $x_1, \ldots, x_m \in G$ such that $G = \bigcup_{i=1}^{m}(x_i + G[k])$. The group $G$ being torsion, there exists an integer $N > 0$ such that $Nx_i = 0$, for each $i \leq m$. Then $Nkx = 0_G$, for each $x \in G$. Indeed, choose $i \leq m$ such that $x \in x_i + G[k]$. Then $x = x_i + y$ for some $y \in G[k]$; hence, $Nkx = Nkx_i + Nky = 0_G$. Therefore, $G$ is a bounded torsion group. □

In the theorem below we present a sufficient condition for an Abelian group to admit a pseudocompact Hausdorff topological group topology. Such a topology will be additionally chosen to be connected and locally connected.

THEOREM 9.11.6. [**D. Dikranjan and D. B. Shakhmatov**] *Let $G$ be an infinite Abelian group satisfying $|G|^{\omega} = |G|$ and suppose that either $r_0(G) = |G|$ or $G$ is torsion-free. Then $G$ admits a pseudocompact, connected, locally connected Hausdorff topological group topology.*

PROOF. In either case, we are going to identify $G$ with a dense pseudocompact subgroup of the group $\mathbb{T}^{\kappa}$, where $\kappa = |G|$. Since $\kappa^{\omega} = \kappa \geq \omega$, it follows that $\kappa \geq \mathfrak{c}$. Therefore, by Proposition 9.9.20, there exists an independent set $X \subset G$ of elements of infinite order such that $|X| = \kappa$.

Denote by $[\kappa]^{\leq\omega}$ the family of all non-empty countable subsets of $\kappa$, and put

$$\mathcal{B} = \bigcup\{\mathbb{T}^{A} : A \in [\kappa]^{\leq\omega}\}.$$

It follows from $\kappa^{\omega} = \kappa$ that $|\mathcal{B}| = \kappa$, so $\mathcal{B} = \{b_\beta : \beta < \kappa\}$. For every $\beta < \kappa$, denote by $A_\beta$ the unique element of $[\kappa]^{\leq\omega}$ such that $b_\beta \in \mathbb{T}^{A_\beta}$. Let also $\{y_\beta : \beta < \kappa\}$ be a one-to-one enumeration of the set $G \setminus \{0_G\}$.

Our aim is to construct a family $\mathcal{H} = \{h_\alpha : \alpha < \kappa\}$ of homomorphisms of $G$ to $\mathbb{T}$ that separates elements of $G$ and such that the image $h(G)$ is $G_\delta$-dense in the product group $\mathbb{T}^{\kappa}$, where $h$ is the diagonal product of the family $\mathcal{H}$. To this end, we have to define, for each $\beta < \kappa$, a set $E_\beta \subset \kappa$, a subgroup $G_\beta$ of $G$, and a family $\mathcal{H}_\beta = \{h_{\alpha,\beta} : \alpha \in E_\beta\}$ of homomorphisms of $G_\beta$ to $\mathbb{T}$ satisfying the following conditions:

(i) $|E_\beta| \leq |\beta| + \omega$ and $|G_\beta| \leq |\beta| + \omega$;
(ii) $E_\gamma \subset E_\beta$ and $G_\gamma \subset G_\beta$ for each $\gamma < \beta$;
(iii) $h_{\alpha,\beta} \restriction G_\gamma = h_{\alpha,\gamma}$, whenever $\gamma < \beta$ and $\alpha \in E_\gamma$;
(iv) $\beta \subset E_\beta$ and $A_\beta \subset E_\beta$;
(v) $y_\beta \in G_\beta$;
(vi) $h_{\alpha,\beta}(y_\beta) \neq 1$ for some $\alpha \in E_\beta$;
(vii) there exists $x_\beta \in G_\beta \cap X$ such that $h_{\alpha,\beta}(x_\beta) = b_\beta(\alpha)$, for each $\alpha \in A_\beta$.

Choose $\alpha_0 \in \kappa \setminus A_0$ and $x_0 \in X$. Then both the set $E_0 = A_0 \cup \{\alpha_0\}$ and the subgroup $G_0 = \langle x_0, y_0 \rangle$ of $G$ are countable. By Lemma 1.1.5, there exists a homomorphism $h_{\alpha_0,0}$ of $G_0$ to $\mathbb{T}$ such that $h_{\alpha_0,0}(y_0) \neq 1$. Similarly, for every $\alpha \in A_0$, Lemma 1.1.5 permits to take a homomorphism $h_{\alpha,0}\colon G_0 \to \mathbb{T}$ such that $h_{\alpha,0}(x_0) = b_0(\alpha)$. Therefore, $E_0$, $G_0$ and $\mathcal{H}_0 = \{h_{\alpha,0} : \alpha \in E_0\}$ satisfy (i)–(vii).

Suppose that $E_\nu$, $G_\nu$ and $\mathcal{H}_\nu$ satisfying (i)–(vii) have been defined for each $\nu < \beta$, where $0 < \beta < \kappa$. Put $D_\beta = \bigcup_{\nu<\beta} E_\nu$. It follows from (i) that $|D_\beta| \leq |\beta| + \omega < \kappa$, so we can pick $\alpha_\beta \in \kappa \setminus D_\beta$. Put $E_\beta = \beta \cup A_\beta \cup D_\beta \cup \{\alpha_\beta\}$ and $H_\beta = \bigcup_{\nu<\beta} G_\nu$. Then $H_\beta$ is a subgroup of $G$, by (ii). It follows from (i) that $|H_\beta| \leq |\beta| \cdot \omega < \kappa$ and, since $|X| = \kappa$, we can take an $x_\beta \in X$ such that the intersection $\langle x_\beta \rangle \cap H_\beta$ is trivial. Define $G_\beta$ to be the subgroup of $G$ generated by the set $H_\beta \cup \{x_\beta, y_\beta\}$. Clearly, $E_\beta$ and $G_\beta$ satisfy (i), (ii), (iv) and (v). Now choose a homomorphism $h_{\alpha_\beta,\beta}$ of $G_\beta$ to $\mathbb{T}$ such that $h_{\alpha_\beta,\beta}(y_\beta) \neq 1$. According to (iii), for every $\alpha \in D_\beta$ there exists a homomorphism $g_{\alpha,\beta}$ of $H_\beta$ to $\mathbb{T}$ such that $g_{\alpha,\beta} \restriction G_\gamma = h_{\alpha,\gamma}$ whenever $\gamma < \beta$ and $\alpha \in E_\gamma$. In other words, $g_{\alpha,\beta}$ extends every homomorphism $h_{\alpha,\gamma}$ with $\gamma < \beta$ and $\alpha \in E_\gamma$. It follows from the choice of $x_\beta \in X$ and Lemma 1.1.5 that we can find, for every $\alpha \in A_\beta$, a homomorphism $h_{\alpha,\beta}$ of $G_\beta$ to $\mathbb{T}$ such that $h_{\alpha,\beta}(x_\beta) = b_\beta(\alpha)$ and, in addition, $h_{\alpha,\beta} \restriction H_\beta = g_{\alpha,\beta}$ if $\alpha \in D_\beta$. In the latter case, we have $h_{\alpha,\beta} \restriction G_\gamma = h_{\alpha,\gamma}$ whenever $\gamma < \beta$ and $\alpha \in E_\gamma$. Hence $E_\beta$, $G_\beta$ and $\mathcal{H}_\beta = \{h_{\alpha,\beta} : \alpha \in E_\beta\}$ satisfy (i)–(vii). The recursive construction is complete.

It follows from (iv) that $\kappa = \bigcup_{\beta<\kappa} E_\beta$, while (ii) and (v) imply that $G = \bigcup_{\beta<\kappa} G_\beta$. Therefore, by (ii) and (iii), for every $\alpha < \kappa$ there exists a homomorphism $h_\alpha \colon G \to \mathbb{T}$ such that $h_\alpha \restriction G_\gamma = h_{\alpha,\gamma}$ whenever $\alpha \in E_\gamma$. It remains to verify that the family $\mathcal{H} = \{h_\alpha : \alpha < \kappa\}$ has the required properties.

First, the family $\mathcal{H}$ separates points of $G$ since the kernel of the homomorphism $h = \Delta_{\alpha<\kappa} h_\alpha$ of $G$ to $\mathbb{T}^\kappa$ is trivial. Indeed, if $\beta < \kappa$, we use (vi) to find $\alpha \in E_\beta$ such that $h_{\alpha,\beta}(y_\beta) \neq 1$. Hence, $h_\alpha(y_\beta) = h_{\alpha,\beta}(y_\beta) \neq 1$. Thus, $h$ is an isomorphism of $G$ onto the subgroup $h(G)$ of $\mathbb{T}^\kappa$. To show that $h(G)$ is $G_\delta$-dense in $\mathbb{T}^\kappa$, it suffices to verify that $\pi_A(h(G)) = \mathbb{T}^A$, for every countable subset of the index set $\kappa$, where $\pi_A \colon \mathbb{T}^\kappa \to \mathbb{T}^A$ is the projection. Indeed, suppose that $b \in \mathbb{T}^A$ for some $A \in [\kappa]^{\leq\omega}$. Then $b = b_\beta$, for some $\beta < \kappa$, whence $A = A_\beta$. Since $h_\alpha \restriction G_\beta = h_{\alpha,\beta}$ for each $\alpha \in A_\beta \subset E_\beta$, it follows from (vii) that $\pi_A(h(x_\beta)) = b_\beta = b$. Therefore, projections of $h(G)$ fill in all countable subproducts of $\mathbb{T}^\kappa$, and $h(G)$ is $G_\delta$-dense in $\mathbb{T}^\kappa$. Now it follows from Theorem 2.4.15 that $h(G)$ is a dense pseudocompact subgroup of $\mathbb{T}^\kappa$. In addition, the group $h(G)$ is connected and locally connected, by the same theorem.

Finally, identifying $G$ with its image $h(G)$ under the isomorphism $h$, we obtain a pseudocompact Hausdorff topological group topology on $G$ with the required properties. □

As a combination of Theorems 9.11.5 and 9.11.6, we obtain:

Corollary 9.11.7. *Let $G$ be a non-torsion Abelian group such that $|G| = \mathfrak{c}$. Then $G$ admits a pseudocompact Hausdorff topological group topology iff $r_0(G) = \mathfrak{c}$.*

Corollary 9.11.8. *A free Abelian group $G$ of cardinality $\kappa$ satisfying $\kappa^\omega = \kappa$ admits a pseudocompact Hausdorff topological group topology.*

Let us briefly come across pseudocompact torsion groups. The next result generalizes Corollary 9.6.30 and is based on it.

Proposition 9.11.9. *Every pseudocompact torsion Abelian group $G$ is zero-dimensional.*

Proof. By Theorem 9.11.5, $G$ is bounded torsion, that is, $nG = \{0\}$, for some integer $n > 1$. Let $K$ be the Raĭkov completion of $G$. Then $K$ is a compact Abelian group satisfying the same equality $nK = \{0\}$. By Corollary 9.6.30, the space $K$ is zero-dimensional. Hence, the subspace $G$ of $K$ is also zero-dimensional. □

The theorem below reduces the problem of existence of a pseudocompact topologization of a given bounded torsion group to a similar problem for primary components of the group.

THEOREM 9.11.10. *For a bounded torsion Abelian group $G$, the following conditions are equivalent:*

a) *$G$ admits a Hausdorff pseudocompact topological group topology;*
b) *each $p$-primary component $G_p$ of $G$ admits a Hausdorff pseudocompact topological group topology;*
c) *for every integer $m \geq 1$, the group $mG = G/G[m]$ admits a pseudocompact Hausdorff topological group topology.*

PROOF. First we show that a) implies c). Suppose that $\tau$ is a Hausdorff topological group topology on $G$, and let $G^* = (G, \tau)$. Given an integer $m \geq 1$, consider the mapping $\varphi_m \colon G^* \to G^*$ defined by $\varphi_m(x) = mx$, for each $x \in G^*$. Clearly, $\varphi_m$ is a continuous homomorphism and the kernel of $\varphi_m$ is the closed subgroup $G[m] = \{x \in G^* : mx = 0\}$ of $G^*$. Therefore, if the group $G^*$ is pseudocompact, so is the continuous homomorphic image $mG^* = \varphi_m(G^*)$ of $G^*$, that is, the group $mG$ admits a pseudocompact Hausdorff topological group topology (inherited from $G^*$).

To see that c) implies b), we assume that c) holds. In particular, the group $G$ admits a Hausdorff pseudocompact topological group topology $\tau$ — it suffices to take $m = 1$ in c). By our assumption, the period $n$ of $G$ is finite. Let $n = p_1^{k_1} \cdots p_r^{k_r}$, where $r \in \mathbb{N}$ and $p_1, \ldots, p_r$ are pairwise distinct prime numbers. For an integer $i \leq r$, let $m_i = n/p_i^{k_i}$. Then $m_i G = G_{p_i}$, where

$$G_{p_i} = \{x \in G : p_i^{k_i} x = 0\}$$

is the $p_i$-primary component of $G$. Hence, by our assumption, $G_{p_i}$ admits a Hausdorff pseudocompact topological group topology.

It remains to verify that b) implies a). By virtue of Theorem 9.9.14, the group $G$ is the direct sum of its primary components $G_p$. Since $G$ is bounded torsion, at most finitely many of the primary components are non-trivial, say, $G_{p_1}, \ldots, G_{p_r}$. Suppose that for every $i \leq r$, the group $G_{p_i}$ admits a pseudocompact Hausdorff topological group topology $\tau_i$. Let $G_i^* = (G_{p_i}, \tau_i)$, for each $i \leq r$. Then, by Corollary 6.6.11, the Hausdorff topological group $G^* = \prod_{i=1}^r G_i^*$ is pseudocompact. This finishes the proof. □

To complete the algebraic characterization of bounded torsion Abelian groups admitting a pseudocompact Hausdorff topological group topology, it suffices, after Theorem 9.11.10, to consider bounded torsion $p$-groups. This part will be done in the problem section below (see Exercise 9.11.g and Problem 9.11.E). We need, however, a short preliminary discussion prior to this job.

According to Theorem 9.11.2, there exist cardinals $\kappa > \mathfrak{c}$ such that no group of cardinality $\kappa$ admits a Hausdorff pseudocompact topological group topology. For example, let $k_0 = \aleph_0$ and $\kappa_{n+1} = 2^{\kappa_n}$, for each $n \in \omega$. Then $\kappa = \sup_{n \in \omega} \kappa_n$ is such a cardinal.

Let us call an infinite cardinal $\tau$ *admissible* if there exists a pseudocompact Hausdorff topological group $G$ such that $|G| = \kappa$. By Theorem 9.11.2, no cardinal $\tau$ less than $\mathfrak{c}$ is admissible, while Theorem 9.11.6 implies that every infinite cardinal $\tau$ with $\tau = \tau^\omega$ is admissible. A characterization of admissible cardinals is not a simple task at all. Several results in this direction require additional axioms of $ZFC$. Suppose, for example, that the Singular Cardinals Hypothesis ($SCH$, for brevity) is valid (see [**260**, Chapter 8]):

$$\text{if } \tau \geq \mathfrak{c} \text{ is a cardinal and } cf(\tau) \neq \omega, \text{ then } \tau^\omega = \tau.$$

Since every cardinal $\tau \geq \mathfrak{c}$ satisfying $\tau^\omega = \tau$ is admissible, we conclude that under $SCH$, all cardinals $\tau \geq \mathfrak{c}$ with $cf(\tau) \neq \omega$ are admissible. However, it was shown by A. H. Tomita in [**498**] that the existence of a countably compact (hence, pseudocompact) topological group of any size $\kappa \geq \mathfrak{c}$ is consistent with $ZFC$. Under $GCH$, however, every cardinal $\kappa$ of countable cofinality is strong limit, so no pseudocompact topological group has cardinality $\kappa$, by Theorem 9.11.2. Summing up, the existence of a pseudocompact topological group of cardinality $\kappa > \mathfrak{c}$ with $cf(\kappa) = \omega$ is independent of $ZFC$. A couple of complementary results on admissible cardinals is given in Exercise 9.11.g.

## Exercises

9.11.a. Give an example of an infinite pseudocompact Abelian group that does not have a proper dense pseudocompact subgroup.

9.11.b. (D. B. Shakhmatov [**426**]) Show that every pseudocompact topological field is finite.

9.11.c. Does the Abelian group $\mathbb{Z}(2)^{(\mathfrak{c})} \oplus \mathbb{Z}(4)^{(\omega)}$ admit a pseudocompact Hausdorff topological group topology?

9.11.d. (D. Dikranjan and D. B. Shakhmatov [**140**]) Modify the proof of Theorem 9.11.6 to show that for every infinite cardinal $\tau$ with $\tau^\omega = \tau$, the free group of cardinality $\tau$ admits a pseudocompact Hausdorff topological group topology. In other words, Corollary 9.11.8 admits an extension to the non-Abelian case.

9.11.e. (D. Dikranjan and D. B. Shakhmatov [**140**]) Show that Proposition 9.11.9 can be extended to pseudocompact Abelian groups $G$ satisfying $r_0(G) < \mathfrak{c}$.

9.11.f. Let $H$ be a dense divisible subgroup of a compact topological group $G$. Show that $G$ is divisible.

9.11.g. (W. W. Comfort and L. C. Robertson [**119**]) Let $Ps(\tau, \sigma)$, where $\tau$ and $\sigma$ are infinite cardinals, abbreviate the statement that $\{0, 1\}^\sigma$ contains a $G_\delta$-dense subset of cardinality $\tau$. Prove the following:

a) If $P(\tau, \sigma)$ holds, then $\mathfrak{c} \leq \tau \leq 2^\sigma$.

b) A cardinal $\tau \geq \omega$ is admissible if and only if $Ps(\tau, \sigma)$ holds, for some $\sigma \geq \omega$.

*Hint.* Item a) is trivial, and it suffices to verify the direct implication in b). Suppose that there exists a pseudocompact topological group $H$ of cardinality $\tau$, i.e., $\tau$ is admissible. The Raĭkov completion $\varrho H$ of $H$ is a compact group. By virtue of Theorem 4.2.4, there exists a continuous mapping $f$ of $\varrho H$ onto the Tychonoff cube $I^\sigma$, where $\sigma = w(\varrho G) = w(G)$. Since $H$ is $G_\delta$-dense in $\varrho H$, the image $X = f(G)$ is $G_\delta$-dense in $I^\sigma$ and, clearly, $|X| \leq |G| = \tau$. Let $h$ be an arbitrary mapping of $I$ onto $\{0, 1\}$. Then the product mapping $h^\sigma$ sends $X$ to a $G_\delta$-dense subset $Y = h^\sigma(X)$ of $\{0, 1\}^\sigma$, and $|Y| \leq |X| \leq \tau$. By a), $|\{0, 1\}^\sigma| = 2^\sigma \geq \tau$, so we can enlarge $Y$ to a $G_\delta$-dense subset $Z$ of $\{0, 1\}^\sigma$ satisfying $|Z| = \tau$.

## Problems

9.11.A. (H. Wilcox [**535**]) Show that every pseudocompact divisible group is connected (this extend the "if" part of Theorem 9.6.15 to arbitrary pseudocompact groups).

*Hint.* Suppose to the contrary that a pseudocompact divisible group $H$ is disconnected. Then $H = U \cup V$, where $U$ and $V$ are disjoint open non-empty subsets of $H$. We can assume that $U$ contains the neutral element $e$ of $H$. Since $H$ intersects every non-empty

$G_\delta$-set in the Raĭkov completion $\varrho H$ of $H$ and the compact group $\varrho H$ is a Moscow space (see item 1) of Theorem 6.4.2), the sets $cl_{\varrho H} U$ and $cl_{\varrho H} V$ are disjoint. Hence, $cl_{\varrho H} U$ is a compact open neighbourhood of $e$ in $\varrho H$, and Proposition 3.1.10 implies that $cl_{\varrho H} U$ contains an open invariant subgroup $K$ of $\varrho H$. Clearly, $\varrho H/K$ is a finite group. The group $\varrho H$ is divisible, by 9.11.f. It follows that the quotient group $\varrho H/K$ is also divisible. However, a finite divisible group is trivial, a contradiction.

9.11.B. Give an example of an Abelian topological group $G$ such that all closed subgroups of $G$ are pseudocompact, but $G$ is not countably compact.

*Hint.* Let $\kappa > \aleph_0$ be a cardinal. We say that an element $x$ of the product group $\mathbb{T}^\kappa$ is *metrizable* if the closure in $\mathbb{T}^\kappa$ of the cyclic group $\langle x \rangle$ is metrizable. Verify that the set $G$ of all metrizable elements of $\mathbb{T}^\kappa$ is a subgroup of $\mathbb{T}^\kappa$ with the required properties.

9.11.C. Show that every Tychonoff space can be embedded as a closed subspace into a pseudocompact topological Abelian group.

9.11.D. (D. Dikranjan and D. B. Shakhmatov [**140**]) Prove that every finitely generated subgroup of a pseudocompact torsion (not necessarily Abelian) topological group is finite.

*Hint.* Apply Problem 9.6.G.

9.11.E. (D. Dikranjan and D. B. Shakhmatov [**140**]) Let $G$ be a bounded torsion $p$-group, that is, $G = G_p$, for a prime $p$. Prove that the group $G$ admits a pseudocompact Hausdorff topological group topology iff each cardinal $\beta_k = |p^k G|$ is either finite or admissible, $k \in \omega$.

*Hint.* It follows from our assumption about $G$ and Theorem 9.6.28 that $G = \bigoplus_{k=1}^{r} \mathbb{Z}(p^k)^{(\alpha_k)}$, where $\alpha_1, \ldots, \alpha_r$ are cardinals, finite or infinite. Apply Exercise 9.11.g to show that if $\alpha_k \le \alpha_r$, for each $k \le r$, and the cardinal $\alpha_r$ is either finite or admissible, then $G$ admits a pseudocompact Hausdorff topological group topology. Then extend the same conclusion to the case when $\gamma_k = \max\{\alpha_{k+1}, \ldots, \alpha_r\}$ is either finite or admissible, for each $k = 0, 1, \ldots, r-1$. Finally, in the general case, use the assumption about cardinals $\beta_k$ to show that the sequence $(\alpha_1, \ldots, \alpha_r)$ satisfies the above condition involving cardinals $\gamma_k$.

9.11.F. (W. W. Comfort and J. van Mill [**116**]) Prove that every non-metrizable pseudocompact Abelian group contains a proper dense pseudocompact subgroup.

9.11.G. (M. G. Tkachenko [**489**])) Let $G$ be an infinite Abelian group satisfying $r_0(G) = |G| = |G|^\omega$. Prove that $G$ admits a pseudocompact Hausdorff topological group topology $\mathcal{T}$ such that the space $(G, \mathcal{T})$ is Fréchet–Urysohn. Deduce that every torsion-free Abelian group $G$ with $|G| = |G|^\omega$ admits a pseudocompact Hausdorff topological group topology making it into a Fréchet–Urysohn space.

*Hint.* Embed $G$ into an appropriate $\Sigma$-product of circle groups as a $G_\delta$-dense subgroup.

9.11.H. (M. G. Tkachenko [**489**])) Let $G = H \oplus \mathbb{Z}(n)^{(\kappa)}$, where $n \in \mathbb{N}$, $\kappa$ is an infinite cardinal satisfying $\kappa^\omega = \kappa$, and $H$ is a torsion Abelian group of a finite period $m$ dividing $n$, $|H| \le \kappa$. Prove that $G$ admits a pseudocompact Hausdorff topological group topology making it into a Fréchet–Urysohn space.

*Hint.* Embed $G$ into the group $\Sigma\mathbb{Z}(n)^\kappa$ as a $G_\delta$-dense subgroup.

## Open Problems

9.11.1. Let $G$ be an infinite torsion-free Abelian group satisfying $|G|^\omega = |G|$. Does $G$ admit a pseudocompact Hausdorff group topology of countable tightness? (See Problems 9.11.G and 9.11.H.)

9.11.2. Is it possible to embed an arbitrary sequential Tychonoff space into a pseudocompact sequential topological group?

9.11.3. Is the product of two sequential pseudocompact groups a sequential space?

9.11.4. Let $G$ be a non-metrizable pseudocompact quasitopological Abelian group. Does $G$ contain a proper dense pseudocompact subgroup? Does it admit a strictly finer pseudocompact quasitopological group topology? (See Problems 9.8.K and 9.11.F.)

## 9.12. Countably compact topologies on Abelian groups

In this section we consider the question of when an abstract group admits a countably compact Hausdorff topological group topology. Henceforth all groups are assumed to be Abelian (except for in Example 9.12.19), so the additive notation is used.

We saw in Section 9.6 that the algebraic structure of a compact topological Abelian group is subject to various strong restrictions. When the assumption of compactness of an Abelian group is weakened to countable compactness or to pseudocompactness, the resulting constraints on the algebraic structure of the group are becoming considerably softer. For example, it follows from Corollary 9.12.11 that the existence of a countably compact Hausdorff topological group topology on the group $H = \mathbb{Z}(2)^{(\omega)} \oplus \mathbb{Z}(4)^{(\mathfrak{c})}$ is consistent with $ZFC$, where $G^{(\alpha)}$ denotes the direct sum of $\alpha$ copies of the group $G$. On the other hand, no Hausdorff topological group topology on $H$ is compact, by Exercise 9.6.i.

Under Martin's Axiom, we completely characterize the algebraic structure of Abelian groups of cardinality $\mathfrak{c}$ that admit a countably compact Hausdorff topological group topology. It turns out that, in the torsion case, these are exactly the groups that admit a pseudocompact Hausdorff topological group topology (see Theorem 9.12.9). The algebraic constraints for the existence of a countably compact Hausdorff topological group topology on an Abelian group $G$ of size $\mathfrak{c}$ are relatively simple: For every integer $n > 1$, the subgroup $G[n]$ of $G$ has to be finite or to satisfy $|G[n]| = \mathfrak{c}$, and the same has to be true for the subgroup $dG[n]$, where $d$ is any divisor of $n$.

We start with the following general result, where no upper bound on the cardinality of a group is imposed. Let us recall that a space $X$ is called *$\omega$-bounded* if the closure of every countable set in $X$ is compact. It is clear that every $\omega$-bounded space is countably compact.

PROPOSITION 9.12.1. *Let $G$ be an Abelian group of a prime exponent $p$ such that $|G|^\omega = |G|$. Then $G$ admits an $\omega$-bounded Hausdorff topological group topology. In particular, $G$ admits a countably compact topological group topology.*

PROOF. We can assume that $G$ is infinite. Let $\kappa = |G|$. Denote by $H$ the $\Sigma$-product of $\kappa$ copies of the discrete group $\mathbb{Z}(p)$. Hence, $H \subset \mathbb{Z}(p)^\kappa$ and $H$ is a dense $\omega$-bounded subgroup of $\mathbb{Z}(p)^\kappa$, by Corollary 1.6.34. Clearly, $|H| = \kappa^\omega = \kappa$, so that $|G| = |H|$. Consider both $G$ and $H$ as vector spaces over the field $\mathbb{Z}(p)$ and take Hamel bases $A = \{g_\alpha : \alpha < \kappa\}$ and $B = \{h_\alpha : \alpha < \kappa\}$ in $G$ and $H$, respectively. Then the mapping $i\colon A \to B$, where $i(g_\alpha) = h_\alpha$ for each $\alpha < \kappa$, can be extended to an isomorphism $f\colon G \to H$ of the Abelian groups $G$ and $H$. Identifying $G$ with $H$ by means of $f$, we obtain an $\omega$-bounded Hausdorff topological group topology on $G$. □

For the further study of countably compact topological groups we introduce special subsets of infinite products and establish their properties.

Let $\lambda$ be an infinite cardinal. An infinite subset $Y$ of a topological product $X = \prod_{\alpha<\lambda} X_\alpha$ is called *finally dense* in $X$ if there exists $\beta < \lambda$ such that $\pi_{\lambda\setminus\beta}(Y)$ is dense in $\prod_{\beta\le\alpha<\lambda} X_\alpha$, where $\pi_{\lambda\setminus\beta}$ is the projection of $X$ onto $\prod_{\beta\leq\alpha<\lambda} X_\alpha$. If every infinite subset of $Y$ is finally dense in $X$, then $Y$ is called *hereditarily finally dense* (briefly, HFD) in $X$.

The following purely topological result will help us to guarantee countable compactness of the topological groups we are going to construct.

PROPOSITION 9.12.2. *Let $\lambda$ be an uncountable regular cardinal and $X = \prod_{\alpha<\lambda} X_\alpha$ be the product of compact metrizable spaces $X_\alpha$ each of which contains at least two points. Suppose that $Y$ is a subset of $X$ such that $\pi_\beta(Y) = \prod_{\alpha<\beta} X_\alpha$, for each $\beta < \lambda$. If $S \subset Y$ is an HFD set in $X$, then $S$ has a cluster point in $Y$, but no sequence in $S$ converges in $X$. In particular, if $Y$ is HFD in $X$, then $Y$ is a countably compact dense subspace of $X$ which does not contain non-trivial convergent sequences.*

PROOF. Let us show that $S$ has a cluster point in $Y$. Since $S$ is an HFD set in $X$, for every infinite subset $T$ of $S$ there exists an ordinal $\beta = \beta_T < \lambda$ such that $\pi_{\lambda\setminus\beta}(T)$ is dense in $X_{\lambda\setminus\beta} = \prod_{\beta\leq\alpha<\lambda} X_\alpha$. For $\beta < \lambda$, define $Z_\beta = \prod_{\alpha<\beta} X_\alpha$ and let $\pi_\beta \colon X \to Z_\beta$ be the canonical projection. Define an increasing sequence $\beta_S = \beta_0 < \beta_1 < \cdots < \beta_n < \cdots < \lambda$ as follows. For an ordinal $\beta_n < \lambda$, let $\mathcal{B}_n$ be a base of $Z_{\beta_n}$ satisfying $|\mathcal{B}_n| \leq |\beta_n| \cdot \omega < \lambda$. Consider the family

$$\gamma_n = \{S \cap \pi_{\beta_n}^{-1}(U) : U \in \mathcal{B}_n\}$$

and choose an ordinal $\beta_{n+1} < \lambda$ such that $\beta_n < \beta_{n+1}$ and $\beta_T < \beta_{n+1}$ for each infinite $T \in \gamma_n$.

Once the sequence $\{\beta_n : n \in \omega\}$ has been defined, put $\beta^* = \sup_{n\in\omega} \beta_n$. Then $\beta^* < \lambda$. Since $Z_{\beta^*} = \pi_{\beta^*}(Y)$ is compact, the set $\pi_{\beta^*}(S)$ has a cluster point $x^* \in Z_{\beta^*}$. Choose a point $x \in Y$ such that $\pi_{\beta^*}(x) = x^*$. We claim that $x$ is a cluster point of $S$. Indeed, let $O$ be a neighbourhood of $x$ in $X$. There exist open sets $U \subset Z_{\beta^*}$ and $V \subset X_{\lambda\setminus\beta^*}$ such that $x \in U \times V \subset O$. We can assume without loss of generality that the set $U$ is canonical, i.e., depends on finitely many coordinates $\alpha_1 < \alpha_2 < \cdots < \alpha_k$ with $\alpha_k < \beta^*$. Therefore, we can find $n \in \omega$ such that $\alpha_k < \beta_n$. Since $\mathcal{B}_n$ is a base for $Z_{\beta_n}$, there exists $U_0 \in \mathcal{B}_n$ such that $\pi_{\beta_n}(x) \in U_0 \subset \pi_{\beta_n}(U \times V)$. Then $T = S \cap \pi_{\beta_n}^{-1}(U_0)$ is an element of the family $\gamma_n$ and $\beta_T \leq \beta_{n+1} < \beta^*$; therefore, $\pi_{\lambda\setminus\beta^*}(T)$ is dense in $X_{\lambda\setminus\beta^*}$. So $V \cap \pi_{\lambda\setminus\beta^*}(T) \neq \varnothing$ and hence, $S \cap (U \times V) \neq \varnothing$ and $S \cap O \neq \varnothing$, thus the claim holds.

Suppose that $S$ contains a non-trivial sequence $T = \{x_n : n \in \omega\}$ converging to a point $x \in X$. Then $T^* = T \cup \{x\}$ is a countable infinite compact space. Since $S$ is HFD in $X$, $\pi_{\lambda\setminus\alpha}(T) \subset \pi_{\lambda\setminus\alpha}(T^*)$ is dense in $X_{\lambda\setminus\alpha}$ for some $\alpha < \lambda$. Therefore, from compactness of $T^*$ it follows that $\pi_{\lambda\setminus\alpha}(T^*) = X_{\lambda\setminus\alpha}$, contradicting $|X_{\lambda\setminus\alpha}| \geq 2^{|\lambda\setminus\alpha|} = 2^\lambda > \lambda > \omega$.

Obviously, $Y$ is dense in $X$. If $Y$ is HFD in $X$, then every infinite subset $S$ of $Y$ has a cluster point in $Y$ by the above argument and hence, $Y$ is countably compact and does not contain non-trivial convergent sequences. □

For an Abelian group $G$ and an integer $d \in \mathbb{N}$, denote by $\varphi_d$ the homomorphism of $G$ to $G$ associated with the multiplication by $d$, i.e., $\varphi_d(x) = dx$ for each $x \in G$. The following technical concepts appear in the proofs of all main results of this section. They will enable us to treat both torsion and non-torsion Abelian groups in Theorems 9.12.9 and 9.12.16.

Let $G$ be an Abelian group, and $n$ be an integer greater than 1. A countable infinite subset $S$ of $G$ is called:

a) *$n$-round* if $S \subset G[n]$ and the restriction of the homomorphism $\varphi_d$ to $S$ is finite-to-one, for any proper divisor $d$ of $n$;

b) *sharp* if the restriction of the homomorphism $\varphi_d$ to $S$ is finite-to-one, for each $d \in \mathbb{N}$.

Clearly, a countable infinite subset $S$ of $G$ is $n$-round iff $S \subset G[n]$ and, for every proper divisor $d$ of $n$, $S$ meets every coset of the subgroup $G[d]$ in a finite set. Similarly, $S$ is sharp iff $S$ has finite intersections with all cosets of the form $x + G[d]$, for each $d \in \mathbb{N}$. In particular, an sharp set can never be $n$-round.

The proof of the following simple lemma is left to the reader.

LEMMA 9.12.3. *Let $G$ be an Abelian group and $n, k \in \mathbb{N}$, $n > 1$. Then:*

a) *for each $p \in \mathbb{P}$, every countable infinite subset of $G[p]$ is $p$-round;*
b) *infinite subsets of $n$-round sets are $n$-round;*
c) *if the integers $n$ and $k$ are coprime, then for every $nk$-round set $S$, the set $kS$ is $n$-round.*

We need more facts about $n$-round and sharp subsets of Abelian groups. The next one is almost evident.

LEMMA 9.12.4. *Every infinite set in an Abelian group $G$ of exponent $n$ contains a subset of the form $T + z$, where $z \in G$ and $T$ is a $d$-round subset of $G$, for some divisor $d > 1$ of $n$.*

PROOF. Let $S$ be a countable infinite subset of $G$. If $S$ is $n$-round, there is nothing to prove. Suppose that there exists a proper divisor $d$ of $n$ such that the restriction of $\varphi_d$ to $S$ is not finite-to-one. We can assume that $d$ is the minimal divisor of $n$ with this property. Therefore, there are an element $g \in G$ and an infinite subset $S'$ of $S$ such that $dx = g$ for each $x \in S'$. Choose an element $z \in S'$ and put $T = S' - z$. Then $T \subset G[d]$ and from the minimality of $d$ it follows that $T$ is $d$-round in $G$. Clearly, $T + z \subset S$. □

The definition of $n$-good and sharp sets and Lemma 9.12.4 imply the following result immediately:

LEMMA 9.12.5. *Every infinite set in an Abelian group $G$ contains either an sharp subset or a set of the form $T + z$, where $z \in G$ and $T$ is a $d$-round subset of $G$ for some $d \in \mathbb{N}$.*

For a discrete Abelian group $G$, let $G^*$ be the group of all homomorphisms $f\colon G \to \mathbb{T}$ endowed with the topology of pointwise convergence, i.e., Pontryagin's dual of $G$. Recall that the standard base of $G^*$ consists of the sets

$$W(\varphi, x_1, \ldots, x_n, \varepsilon) = \{f \in G^* : |f(x_i) - \varphi(x_i)| < \varepsilon \text{ for each } i = 1, \ldots, n\},$$

where $\varphi \in G^*$, $x_1, \ldots, x_n \in G$ and $\varepsilon > 0$. Here and in the sequel, $|x - y|$ denotes the minimal length of the arcs connecting the points $x, y$ in $\mathbb{T}$. The topological group $G^*$ is compact and Hausdorff, by Proposition 9.5.5. The group $Hom(G, \mathbb{T}^\omega)$ of homomorphisms of $G$ to $\mathbb{T}^\omega$ is naturally identified with $(G^*)^\omega$. We will also identify $\mathbb{Z}(n)$ with the subgroup $\mathbb{T}[n]$ of $\mathbb{T}$. Therefore, if $G$ has exponent $n$, then $G^* = Hom(G, \mathbb{T}) = Hom(G, \mathbb{Z}(n))$.

The following result prepares ground for applications of Martin's Axiom and is used in the proof of Lemma 9.12.7.

LEMMA 9.12.6. *Let $S$ be an $n$-round subset of a discrete Abelian group $G$. Then the set*

$$H_S = \{f \in (G^*)^\omega : f(S) \text{ is dense in } \mathbb{Z}(n)^\omega\}$$

*is the intersection of countably many open dense sets in $(G^*)^\omega$ and, hence, is dense in $(G^*)^\omega$.*

PROOF. For every $r \in \mathbb{N}$ and $g \in \mathbb{Z}(n)^{r+1}$, consider the set

$$U_g = \{f \in (G^*)^\omega \colon \exists x \in S \text{ such that } \pi_i(f(x)) = g(i) \text{ for each } i = 0, \ldots, r\},$$

where $\pi_i \colon \mathbb{T}^\omega \to \mathbb{T}_i$ is the projection to the $i$th factor. Since $nx = 0$ for each $x \in S$, the sets $U_g$ are open in $(G^*)^\omega$. To see this, it suffices to replace $\pi_i(f(x)) = g(i)$ by the equivalent condition $|\pi_i(f(x)) - g(i)| < 1/n$ in the definition of the set $U_g$. The equality

$$H_S = \bigcap\{U_g : g \in \mathbb{Z}(p)^{r+1} \text{ for some } r \in \mathbb{N}\}$$

is also evident. It remains to show that $U_g$ is dense in $(G^*)^\omega$ for each $g \in \mathbb{Z}(n)^{r+1}$, $r \in \mathbb{N}$. Equivalently, we have to show that the set

$$V_g = \{(f_0, \ldots, f_r) \in (G^*)^{r+1} : \exists x \in S \text{ such that } f_i(x) = g(i),\ i = 0, \ldots, r\}$$

is dense in $(G^*)^{r+1}$. Let $r \in \mathbb{N}$ and $g \in \mathbb{Z}(p)^{r+1}$ be arbitrary. Consider a basic open set $W = W_0 \times \cdots \times W_r$ in $(G^*)^{r+1}$, where

$$\begin{aligned} W_i &= W(\psi_i, x_{i,1}, \ldots, x_{i,m_i}, \varepsilon_i) \\ &= \{f \in G^* : |f(x_{i,k}) - \psi_i(x_{i,k})| < \varepsilon_i,\ k = 1, \ldots, m_i\}. \end{aligned}$$

Here $\varepsilon_i > 0$, $\psi_i \in G^*$, and $x_{i,k} \in G$ whenever $0 \le i \le r$ and $1 \le k \le m_i$. Denote by $N$ the subgroup of $G$ generated by the elements $x_{i,k}$ with $0 \le i \le r$ and $1 \le k \le m_i$. Since $N$ is a finitely generated Abelian group, its torsion part $tor(N)$ is finite. We claim that there exists a non-zero element $x \in S$ of order $n$ such that $\langle x\rangle \cap N = \{0\}$. Indeed, the intersection $S \cap G[d]$ is finite for each proper divisor $d$ of $n$, so almost all elements of $S$ have order $n$. We can assume, therefore, that every element of $S$ is of order $n$. Assume that $\langle x\rangle \cap N \neq \{0\}$ for each $x \in S$, and find a divisor $d_x$ of $n$ with $d_x \neq n$ such that $d_x x \in N$. Since $d_x x \in tor(N)$, there exist a proper divisor $d$ of $n$, an infinite subset $S'$ of $S$ and an element $a \in N$ such that $dx = a$ for all $x \in S'$. This contradicts our assumption that $S$ is $n$-round. So, we can pick an element $x \in S$ of order $n$ with $\langle x\rangle \cap N = \{0\}$. For every $i \le r$, define a homomorphism $h_i \colon G \to \mathbb{T}$ satisfying $h_i \restriction N = \psi_i \restriction N$ and $h_i(x) = g(i)$. Clearly, $(h_0, \ldots, h_r) \in W \cap V_g$. □

Martin's Axiom is equivalent to the following purely topological assertion: *If $X$ is a compact Hausdorff space of countable cellularity, then the intersection of less than $\mathfrak{c}$ open dense sets in $X$ is dense in $X$* (see **[262, 285]**). We shall use this topological form of $MA$ in the proofs of all main results of this section. It is known that, under $MA$, every infinite cardinal $\kappa < \mathfrak{c}$ satisfies $2^\kappa = \mathfrak{c}$ **[262, 285]**. In particular, $MA$ implies that $\mathfrak{c}$ is a regular cardinal.

The next lemma is the key to the proof of Theorem 9.12.9.

LEMMA 9.12.7. *Assume that $MA$ holds. Let $G$ be a discrete Abelian group, $x^*$ a non-zero element of $G$, and $\alpha < \mathfrak{c}$ an ordinal. For every $\gamma < \alpha$, let $f_\gamma \colon G \to G_\gamma$ be a homomorphism of $G$ to a topological group $G_\gamma$. Suppose that:*

a) *for every $\gamma < \alpha$, $S_\gamma \subset G$ is an $n_\gamma$-round in $G$ subset of $G$, for some $n_\gamma \in \mathbb{N}$;*
b) *$f_\gamma(S_\gamma)$ is dense in a non-discrete subgroup $K_\gamma$ of $G_\gamma$ with $w(K_\gamma) < \mathfrak{c}$.*

*Then there exists a homomorphism $f \colon G \to \mathbb{T}^\omega$ such that $f(x^*) \neq 0$ and, for every $\gamma < \alpha$, the image $(f_\gamma \Delta f)(S_\gamma)$ is dense in $K_\gamma \times \mathbb{Z}(n_\gamma)^\omega$, where $f_\gamma \Delta f$ is the diagonal product of the homomorphisms $f_\gamma$ and $f$.*

PROOF. For every $\gamma < \alpha$, denote by $\mathcal{B}_\gamma$ a base for $K_\gamma$ such that $|\mathcal{B}_\gamma| < \mathfrak{c}$. Let also $\mathcal{B}$ be a countable base for $L = \mathbb{T}^\omega$. For $\gamma < \alpha$ and $U \in \mathcal{B}_\gamma$, put $S_\gamma(U) = S_\gamma \cap f_\gamma^{-1}(U \cap f_\gamma(S_\gamma))$. Since $f_\gamma(S_\gamma)$ is dense in the non-discrete group $K_\gamma$, the family

$$\mathcal{S}_\gamma = \{S_\gamma(U) : U \in \mathcal{B}_\gamma\}$$

consists of infinite subsets of $G$, and $|\mathcal{S}_\gamma| \leq |\mathcal{B}_\gamma| < \mathfrak{c}$. Note that $S_\gamma(U)$ is $n_\gamma$-round for every $U \in \mathcal{B}_\gamma$ as an infinite subset of the $n_\gamma$-round set $S_\gamma$. Denote by $H_\gamma$ the set of all homomorphisms $h\colon G \to K$ such that $h(S)$ is dense in $\mathbb{Z}(n_\gamma)^\omega$ for each $S \in \mathcal{S}_\gamma$. By Lemma 9.12.6, $H_\gamma$ is the intersection of at most $|\mathcal{S}_\gamma| \cdot \omega < \mathfrak{c}$ open dense subsets of the compact group $(G^*)^\omega$. Since $\alpha < \mathfrak{c}$, the set $H = \bigcap_{\gamma<\alpha} H_\gamma$ is the intersection of less than $\mathfrak{c}$ open dense subsets of $(G^*)^\omega$ (here we use the fact that $\mathfrak{c}$ is a regular cardinal under $MA$). The dual group $G^*$ of the discrete Abelian group $G$ is compact, hence dyadic, by Theorem 4.1.7. So the Souslin number of $(G^*)^\omega$ is countable. Therefore, $MA$ implies that $H$ is dense in $(G^*)^\omega$.

Consider the non-empty open subset

$$W = \{h \in (G^*)^\omega : h(x^*) \neq 0_K\}$$

of $(G^*)^\omega$. Then $W \cap H \neq \varnothing$, so there exists $f \in W \cap H$. Clearly, $f(x^*) \neq 0_K$. For every $\gamma < \alpha$, denote by $h_\gamma$ the diagonal product of $f_\gamma$ and $f$. We claim that $h_\gamma(S_\gamma)$ is dense in $K_\gamma \times \mathbb{Z}(n_\gamma)^\omega$ for each $\gamma < \alpha$. Indeed, let $U \times V$ be a non-empty open subset of $K_\gamma \times L$, where $U \in \mathcal{B}_\gamma$, $V \in \mathcal{B}$ and $V \cap \mathbb{Z}(n_\gamma)^\omega \neq \varnothing$. Since $S = S_\gamma(U) \in \mathcal{S}_\gamma$ and $f \in H$, we infer that $f(S)$ is dense in $\mathbb{Z}(n_\gamma)^\omega$. Choose a point $y \in S$ such that $f(y) \in V$. Then $h_\gamma(y) = (f_\gamma(y), f(y)) \in U \times V$, whence $h_\gamma(S) \cap (U \times V) \neq \varnothing$. So $h_\gamma(S_\gamma)$ is dense in $K_\gamma \times \mathbb{Z}(n_\gamma)^\omega$. □

Given two integers $d, n \in \mathbb{N}$, we write $d|n$ if $d$ divides $n$. If $\alpha$ is an ordinal and $H$ is a group, then $H^{(\alpha)}$ denotes the direct sum of $\alpha$ copies of the group $H$. This notation is used in the next corollary to Theorem 9.6.28:

LEMMA 9.12.8. *Let $G$ be a bounded torsion Abelian group of exponent $n$ with $|G| = \kappa > \omega$. Suppose that for every divisor $d$ of $n$, the group $dG$ either is finite or is of cardinality $\kappa$. Then there exist a divisor $m > 1$ of $n$ and a finite subgroup $F$ of $G$ such that $G$ is isomorphic to the direct sum $F \oplus \bigoplus_{d|m} \mathbb{Z}(d)^{(\alpha_d)}$, where $\alpha_m = \kappa$ and $\alpha_d = 0$ or $\omega \leq \alpha_d \leq \kappa$, for each proper divisor $d$ of $m$.*

PROOF. According to Theorem 9.6.28, $G$ is isomorphic to a direct sum $\oplus_{i\in I} H_i$, where each $H_i$ is a finite cyclic group whose order $d_i$ divides $n$. It is clear that $|I| = \kappa$. For every divisor $d > 1$ of $n$, let $\alpha_d$ be the number of summands in the direct sum $G \cong \oplus_{i\in I} H_i$ isomorphic to the group $\mathbb{Z}(d)$. Denote by $m$ the biggest divisor of $n$ with $\alpha_m \geq \omega$. Now we can represent $G$ as follows: $G \cong F \oplus \bigoplus_{d|m} \mathbb{Z}(d)^{(\alpha_d)}$, where $F$ is the direct sum of the groups $\mathbb{Z}(d)^{(\alpha_d)}$ with finite $\alpha_d$. Evidently, $\alpha_d \leq \kappa$ for each divisor $d$ of $m$. Therefore, it remains to verify that $\alpha_m = \kappa$.

If $m$ is prime, it follows immediately from $|I| = \kappa$ that $\alpha_m = \kappa$ as well. In the other case, let $p$ be a prime divisor of $m$. Then $m = pr$, for some integer $r > 1$. Obviously,

$$rG \cong rF \oplus \bigoplus_{d|m} r\mathbb{Z}(d)^{(\alpha_d)} \cong rF \oplus \mathbb{Z}(p)^{(\alpha_m)},$$

and since $\alpha_m$ is infinite, the assumptions of the lemma imply that $|rG| = \kappa$. The summand $rF$ is finite, so we must have $\alpha_m = \kappa$. □

In the theorem below, we present necessary and sufficient conditions for the existence of a countably compact topological group topology on "small" Abelian torsion groups.

THEOREM 9.12.9. [**D. Dikranjan and M. G. Tkachenko**] *Under Martin's Axiom, the following conditions are equivalent for every torsion Abelian group $G$ of cardinality at most $\mathfrak{c}$:*

a) *$G$ admits a pseudocompact Hausdorff topological group topology;*
b) *$G$ admits a countably compact Hausdorff topological group topology;*
c) *$G$ admits a countably compact Hausdorff topological group topology without non-trivial convergent sequences;*
d) *$G$ has finite exponent $n$ and $dG$ is either finite or has cardinality $\mathfrak{c}$, for every proper divisor $d$ of $n$.*

PROOF. Implications c) ⇒ b) ⇒ a) are obvious. The implication a) ⇒ d) is Corollary 9.11.3. So we shall only prove the implication d) ⇒ c).

Suppose that $G$ satisfies d), and let $n$ be the exponent of $G$. If $G$ is finite there is nothing to prove, so assume from now on that $G$ is infinite. Then d) (with $d = 1$) implies that $|G| = \mathfrak{c}$.

By virtue of Lemma 9.12.8, one can find a finite subgroup $F$ of $G$ and a divisor $m > 1$ of $n$ such that $G \cong F \oplus \bigoplus_{d|m} \mathbb{Z}(d)^{(\alpha_d)}$, where $\alpha_m = \mathfrak{c}$ and for each proper divisor $d$ of $m$, either $\alpha_d = 0$ or $\omega \le \alpha_d \le \mathfrak{c}$. Since the product of a finite group and a countably compact group is countably compact, we can assume that $F = \{0\}$. Therefore, we may assume that $m = n$.

Put $K = \mathbb{Z}(n)^\omega$. Our aim is to construct an injective homomorphism $h\colon G \to K^\mathfrak{c}$ such that the subgroup $h(G)$ of $K^\mathfrak{c}$ will fill all $\kappa$-faces of $K^\mathfrak{c}$ for every $\kappa < \mathfrak{c}$. We shall carry out the construction of $h$ by transfinite recursion of length $\mathfrak{c}$ with the help of Lemma 9.12.7.

Given a subset $A$ of $\mathfrak{c}$, $\pi_A$ will stand for the projection of $K^\mathfrak{c}$ onto $K^A$. In particular, $\pi_\alpha$ is the projection of $K^\mathfrak{c}$ onto $K^\alpha$. The construction of $h$ will guarantee the following two properties of the subgroup $h(G)$ of $K^\mathfrak{c}$.

(A) if $d|n$ and $S$ is a $d$-round subset of $G$, then $h(S)$ is finally dense in $(K[d])^\mathfrak{c}$;
(B) for every $\alpha < \mathfrak{c}$ and a divisor $d$ of $n$, we have $\pi_\alpha(h(G[d])) = (K[d])^\alpha$.

It follows from (B) that $h(G)$ is dense in $K^\mathfrak{c}$.

**Claim 1.** *$h(G)$ is countably compact and does not contain non-trivial convergent sequences.*

Let us deduce from (A) that $h(G)$ does not contain non-trivial convergent sequences. Take any non-trivial sequence $\{h(x_n) : n \in \omega\}$ in $h(G)$. Put $S = \{x_n : n \in \omega\}$. By Lemma 9.12.4, we can find $z \in G$, a divisor $d$ of $n$ with $d > 1$, and a $d$-round subset $T$ of $G$ such that $T + z \subset S$. All infinite subsets of $T$ are also $d$-round, so (A) implies that $h(T)$ is HFD in $K[d]^\mathfrak{c}$. By Proposition 9.12.2, neither $h(T)$ nor $h(S) \supseteq h(T) + h(z)$ converges in $h(G) \subset K^\mathfrak{c}$.

Let us prove that $h(G)$ is countably compact. First, (A) and (B) imply that $h(G[p])$ is countably compact, for every prime divisor $p$ of $n$. Indeed, by (A) and a) of Lemma 9.12.3, $h(G[p])$ is an HFD subgroup of $(K[p])^\mathfrak{c}$ and, by (B), the projections of $h(G[p])$ fill all "initial" facets of $(K[p])^\mathfrak{c}$. Therefore, by Proposition 9.12.2, $h(G[p])$ is countably compact.

Suppose that we have proved the countable compactness of $h(G[d])$ for every proper divisor $d$ of $n$, and consider a countable infinite subset $S$ of $G$. If $S$ is $n$-round, then every infinite subset of $S$ is also $n$-round, so (A) implies that $h(S)$ is an HFD subset of $K^{\mathfrak{c}}$. Since $h(G)$ fills all "initial" facets of $K^{\mathfrak{c}}$, Proposition 9.12.2 implies again that $h(S)$ has a cluster point in $h(G)$. If $S$ is not $n$-round, we apply Lemma 9.12.4 to find a proper divisor $d$ of $n$, an element $z \in G$, and a $d$-round subset $T$ of $S$ such that $T+z \subset S$. Since $h(G[d])$ is countably compact by the inductive hypothesis, $h(T)$ has a cluster point $h(y)$ in $h(G[d])$. Therefore, $h(z+y)$ is a cluster point of $h(S)$. This proves that $h(G)$ is countably compact.

Let us return to the main proof. To obtain the required group topology on $G$ as in c), one has to identify $G$ with its image $h(G) \subset K^{\mathfrak{c}}$.

To construct $h$, some preliminary work is still needed. The group $G$ can be represented as the union $G = \bigcup_{\nu<\mathfrak{c}} N_\nu$ of the increasing chain of its subgroups $N_\nu = \bigoplus_{d|n} \mathbb{Z}(d)^{(\alpha(\nu,d))}$, where $\alpha(\nu, n) = \omega + \nu$ (so $\alpha(0, n) = \omega$) and $\alpha(\nu, d) = \min\{\nu, \alpha_d\}$ for every $d|n$ with $d \neq n$. Note that $N_0 = \mathbb{Z}(n)^{(\omega)}$ and $N_\nu = \bigcup_{\mu<\nu} N_\mu$ for each limit ordinal $\nu \geq \omega$. For every $\alpha < \mathfrak{c}$, we shall define a homomorphism $h_\alpha \colon G \to K$. Then we will take $h = \Delta_{\alpha<\mathfrak{c}} h_\alpha$, the diagonal product of the homomorphisms $h_\alpha$. In fact, at a step $\alpha < \mathfrak{c}$ of the construction, we shall define a family $\{h_{\gamma,\nu} : \gamma, \nu \leq \alpha\}$, where $h_{\gamma,\nu}$ is a homomorphism of $N_\nu$ to $K$ for each $\gamma \leq \alpha$. In addition, the homomorphism $h_{\gamma,\nu}$ will extend $h_{\gamma,\mu}$ whenever $\mu < \nu$, so the restriction of $h_\gamma$ to $N_\nu$ will coincide with $h_{\gamma,\nu}$ for all $\gamma, \nu < \mathfrak{c}$.

For every $x \in G$, denote by $\xi(x)$ the minimal ordinal $\xi < \mathfrak{c}$ such that $x \in N_\xi$. Note that either $\xi(x)$ is non-limit or $\xi(x) = 0$.

Denote by $\mathcal{S}$ the family of all countable infinite subsets of $G \setminus \{0\}$ which are $d$-round for some divisor $d$ of $n$. Since $|\mathcal{S}| = \mathfrak{c}$, there exists an enumeration $\mathcal{S} = \{S_\mu : \mu < \mathfrak{c}\}$ such that $S_\mu \subset N_\mu$ for each $\mu < \mathfrak{c}$. In addition, for every $\mu < \mathfrak{c}$, let $d_\mu$ be the divisor of $n$ such that $S_\mu$ is $d_\mu$-round. These preliminary steps are needed to make sure that (A) is satisfied.

Let $\Sigma$ be the subgroup of $K^{\mathfrak{c}}$ consisting of all $x \in K^{\mathfrak{c}}$ such that $|\{\alpha < \mathfrak{c} : x(\alpha) \neq 0_K\}| < \mathfrak{c}$. Since $MA$ implies that $2^\kappa = \mathfrak{c}$ for every infinite cardinal $\kappa < \mathfrak{c}$, we can enumerate $\Sigma$ as follows: $\Sigma = \{b_\nu : \nu < \mathfrak{c}\}$ (this is needed satisfy (B)).

Suppose now that we have already defined a family $\{h_{\alpha,\nu} : \alpha, \nu < \mathfrak{c}\}$, satisfying the following conditions for all $\alpha, \nu < \mathfrak{c}$:

(1) $h_{\alpha,\nu} \colon N_\nu \to K$ is a homomorphism and $h_{\alpha,\nu}$ extends $h_{\alpha,\mu}$ if $\mu < \nu$;
(2) the image $(\Delta_{\mu\leq\gamma<\alpha} h_{\gamma,\mu})(S_\mu)$ is dense in $(K[d_\mu])^{\alpha\setminus\mu}$ whenever $\mu < \alpha$;
(3) if $b_\nu \neq 0$ and $d$ is the order of $b_\nu$, then there exists $x \in (N_{\nu+1} \setminus N_\nu) \cap G[d]$ such that $h_{\gamma,\nu+1}(x) = b_\nu(\gamma)$ for each $\gamma < \nu$;
(4) if $\alpha > 0$ and $\xi = \min\{\xi(x) : x \in S_\alpha\}$, then there exists $z_\alpha \in S_\alpha \cap N_\xi$ such that $h_{\alpha,\alpha}(z_\alpha) \neq 0_K$.

(See Claim 2 below for the explicit construction). By (1), for every $\alpha < \mathfrak{c}$, there exists a homomorphism $h_\alpha \colon G \to K$ whose restriction to $N_\nu$ coincides with $h_{\alpha,\nu}$ for each $\nu < \mathfrak{c}$. Denote by $h$ the diagonal product of the homomorphisms $h_\alpha$, where $\alpha < \mathfrak{c}$.

We claim that $h$ is injective. Indeed, let $z \in G$ be an arbitrary non-zero element. Since $|N_\nu| < \mathfrak{c}$ for each $\nu < \mathfrak{c}$, there exists an $n$-round subset $S$ of $G$ such that $z \in S$ and $\xi(z) < \xi(x)$, for each $x \in S \setminus \{z\}$. Find $\alpha < \mathfrak{c}$ such that $S = S_\alpha$. Because $\xi(z)$ is minimal on $S$, we must have $z = z_\alpha$ by (4), and so $h_{\alpha,\alpha}(z) \neq 0$. Hence, $z \notin \ker(h)$.

It is easy to see that $h(G)$ satisfies (A). Indeed, if $S$ is a $d$-round subset of $G$, then $S = S_\mu$ for some $\mu < \mathfrak{c}$. For every $\alpha$ satisfying $\mu < \alpha < \mathfrak{c}$, let $f_{\mu,\alpha} = \Delta_{\mu\le\gamma<\alpha} h_{\gamma,\alpha}$ be the diagonal product of the family $\{h_{\gamma,\alpha} : \mu \le \gamma < \alpha\}$. From the equality $\pi_{\alpha\setminus\mu} \circ h = f_{\mu,\alpha}$ and (1), (2) it follows that $\pi_{\alpha\setminus\mu}(h(S_\mu))$ is dense in $(K[d])^{\alpha\setminus\mu}$ for each $\alpha > \mu$, and hence, $h(S_\mu)$ is dense in $(K[d])^{\mathfrak{c}\setminus\mu}$.

Let us show that $h(G)$ satisfies (B). Suppose that $\alpha < \mathfrak{c}$, $d|n$ and that $z \in (K[d])^\alpha$. Since the cardinality of the set $\{\nu < \mathfrak{c} : ord(b_\nu) = d \text{ and } \pi_\alpha(b_\nu) = z\}$ is $\mathfrak{c}$, there exists $\nu < \mathfrak{c}$ such that $\alpha \le \nu$, $ord(b_\nu) = d$ and $b_\nu(\gamma) = z(\gamma)$ for each $\gamma < \alpha$. By (3), there exists a point $x \in N_{\nu+1} \cap G[d]$ such that $h_{\gamma,\nu+1}(x) = b_\nu(\gamma)$ for each $\gamma < \nu$. This and the definition of $h$ together imply that $\pi_\alpha(h(x)) = z$. Since $z$ is an arbitrary point of $(K[d])^\alpha$, we have proved that $\pi_\alpha(h(G[d])) = (K[d])^\alpha$.

**Claim 2.** *The required family $\mathcal{H}$ exists under MA.*

By Lemma 9.12.6, one can find a homomorphism $h_{0,0}$ of $N_0$ to $K$ such that $h_{0,0}(S_0)$ is dense in $K$. Let $0 < \alpha < \mathfrak{c}$, and suppose that for every $\beta < \alpha$, we have defined a family $\{h_{\gamma,\nu} : \gamma, \nu \le \beta\}$ satisfying (1)–(4). If $\alpha$ is a limit ordinal, then the equality $N_\alpha = \bigcup_{\nu<\alpha} N_\nu$ and (1) together imply that for every $\gamma < \alpha$, there exists a homomorphism $h_{\gamma,\alpha} \colon N_\alpha \to K$ whose restriction to $N_\nu$ coincides with $h_{\gamma,\nu}$ for each $\nu < \alpha$. In this case, it remains to define a homomorphism $h_{\alpha,\alpha} \colon N_\alpha \to K$ satisfying (2) and (4). Then one can put $h_{\alpha,\nu} = h_{\alpha,\alpha} \restriction N_\nu$ for each $\nu < \alpha$. The existence of such a homomorphism $h_{\alpha,\alpha}$ can be established by the argument given below in the case of a non-limit ordinal $\alpha$, so we will now consider this case.

Suppose that $\alpha = \beta + 1$ and that the family $\{h_{\gamma,\nu} : \gamma, \nu \le \beta\}$ has been defined in such a way that conditions (1)–(4) hold. We have to extend the homomorphisms $h_{\gamma,\beta}$ (with $\gamma \le \beta$) over $N_\alpha$, thus obtaining the homomorphisms $h_{\gamma,\alpha}$, and construct a homomorphism $h_{\alpha,\alpha} \colon N_\alpha \to K$.

By the definition of $N_\alpha$, there exist subgroups $P, Q$ of $N_\alpha$ such that $P \cong \mathbb{Z}(n)$ and $N_\alpha = N_\beta \oplus P \oplus Q$. Denote by $d$ the order of $b_\beta \neq 0$ and choose an element $x \in P$ of order $d$. For every $\gamma \le \beta$, define a homomorphism $h_{\gamma,\alpha} \colon N_\alpha \to K$ extending $h_{\gamma,\beta}$ and satisfying $h_{\gamma,\alpha}(x) = b_\beta(\gamma)$. (If $b_\beta = 0$, then the extension $h_{\gamma,\alpha}$ of $h_{\gamma,\beta}$ can be chosen arbitrarily).

To define a homomorphism $h_{\alpha,\alpha}$ satisfying (2) and (4), choose a point $z \in S_\alpha$ such that $\xi(z) = \min\{\xi(y) : y \in S_\alpha\}$. For every $\mu \le \beta$, denote by $f_{\mu,\alpha}$ the diagonal product of the homomorphisms $h_{\gamma,\alpha}$ where $\mu \le \gamma \le \beta$. Then $f_{\mu,\alpha}$ is a homomorphism of $N_\alpha$ to $K^{\alpha\setminus\mu}$. Applying Lemma 9.12.7 to the family $\{f_{\mu,\alpha} : \mu < \alpha\}$ and to the element $z \in G \setminus \{0\}$, we find a homomorphism $f \colon N_\alpha \to K$ such that $f(z) \neq 0_K$ and, for every $\mu \le \beta$, the image $(h_{\mu,\alpha}\Delta f)(S_\mu)$ is dense in $(K[n_\mu])^{\alpha\setminus\mu} \times \mathbb{Z}(n_\mu)^\omega$. It remains to put $h_{\alpha,\alpha} = f$ and $h_{\alpha,\nu} = f \restriction N_\nu$ for each $\nu < \alpha$. The recursive construction is complete.

It is straightforward to verify that the family $\{h_{\gamma,\nu} : \gamma, \nu \le \alpha\}$ satisfies conditions (1)–(4). This completes the proof of Claim 2 and of the theorem. □

We will call a subset $H$ of an abstract group $G$ *unconditionally closed* in $G$ if $H$ is closed in $(G, \tau)$, for each topological group topology $\tau$ on $G$. Every subgroup $G[n]$ of $G$ is unconditionally closed in $G$. Indeed, $G[n]$ is the kernel of the homomorphism $\varphi_n \colon G \to G$ defined by $\varphi_n(x) = nx$, for all $x \in G$, and the homomorphism $\varphi_n$ is continuous for any topological group topology $\tau$ on $G$, i.e., $\varphi_n$ is *unconditionally continuous.*

Corollary 9.12.10. *Under MA, a torsion Abelian group G admits a countably compact Hausdorff topological group topology if and only if G is of finite exponent and every p-primary component $G_p$ of G admits a countably compact Hausdorff topological group topology.*

Proof. If $G$ admits a countably compact group topology, then $G$ is of finite exponent by Theorem 9.11.5. If $p^n$ is the exponent of $G_p$, then the subgroup $G_p = G[p^n]$ of $G$ is unconditionally closed, so $G_p$ also admits a countably compact group topology. Conversely, if $G$ is of finite exponent and $G_p$ admits a countably compact group topology for every prime $p$, then $G = \prod_{p\in\mathbb{P}} G_p$, by Theorem 9.9.14 (as $G_p \neq \{0\}$ only for finitely many prime $p$). Therefore, $G$ is a torsion group that admits a pseudocompact group topology, since pseudocompactness is invariant under topological products by Corollary 6.6.11. By the equivalence of a) and b) of Theorem 9.12.9, $G$ admits a countably compact group topology. □

Corollary 9.12.11. *Suppose that MA holds. Then for every prime p and every integer $n > 1$, the group $G = \mathbb{Z}(p)^{(\omega)} \oplus \mathbb{Z}(p^n)^{(\mathfrak{c})}$ admits a countably compact topological group topology.*

Proof. Notice that $p^n$ is the exponent of $G$ and every divisor of $p^n$ has the form $p^m$, for some $m \le n$. Since the group $p^m G \cong \mathbb{Z}(p^{n-m})^{(\mathfrak{c})}$ is of cardinality $\mathfrak{c}$ for each $m < n$, the conclusion follows from Theorem 9.12.9. □

The next step is to consider the algebraic structure of countably compact Abelian groups of size $\mathfrak{c}$ with a non-trivial torsion part. It turns out that there is a wealth of Abelian groups $G$ of size $\mathfrak{c}$ that admit a pseudocompact Hausdorff topological group topology, while no Hausdorff topological group topology on $G$ is countably compact (see Example 9.12.18).

By technical and notational reasons, we identify the circle group $\mathbb{T}$ with the quotient group $\mathbb{R}/\mathbb{Z}$. In particular, $\mathbb{T}$ is considered as the additive group of the reals with addition modulo one.

For $p \in \mathbb{N}$, every continuous homomorphism $g\colon \mathbb{T}^p \to \mathbb{T}$ has the form $g(t) = \sum_{i=1}^p k_i t_i$, where $t = (t_1, \dots, t_p) \in \mathbb{T}^p$ and $k_i \in \mathbb{Z}$ for each $i \le p$. Set $\|g\| = \sum_{i=1}^p |k_i|$. This notation is used in the following lemma on the relative density of the solutions of linear equations on the tori.

Lemma 9.12.12. *Let $g\colon \mathbb{T}^p \to \mathbb{T}$ be a non-trivial continuous homomorphism such that $2\delta \cdot \|g\| > 1$, where $p \in \mathbb{N}$ and $\delta > 0$ is a positive number. Then, for every $a \in \mathbb{T}$ and $c \in \mathbb{T}^p$, there exists $t = (t_1, \dots, t_p) \in \mathbb{T}^p$ such that $g(t) = a$ and $|t_i - c_i| < \delta$ for each $i = 1, \dots, p$.*

Proof. Consider the linear function $f\colon \mathbb{R}^p \to \mathbb{R}$ defined by the formula $f(x_1, \dots, x_p) = k_1 x_1 + \dots + k_p x_p$ for every $(x_1, \dots, x_p) \in \mathbb{R}^p$. Let $\pi\colon \mathbb{R} \to \mathbb{T}$ be the natural quotient map, $\pi(x) = \mathbb{Z} + x$. Choose a point $b = (b_1, \dots, b_p) \in \mathbb{R}^p$ such that $\pi(b_i) = c_i$ for each $i = 1, \dots, p$. Take $\mu > 0$ satisfying $1/\|g\| < 2\mu < 2\delta$ and denote by $\Pi_\mu$ the $p$-dimensional cube with the center at $b$, $\Pi_\mu = \prod_{i=1}^p [b_i - \mu, b_i + \mu]$. One easily verifies that $f(\Pi_\mu) = [f(b) - M, f(b) + M]$, where $M = \mu \cdot \|g\|$. It is clear that $2M > 1$, so $\pi(f(\Pi_\mu)) = \mathbb{T}$. Therefore, we can find a point $x = (x_1, \dots, x_p) \in \Pi_\mu$ such that $\pi(f(x)) = a$. Then the point $t = (t_1, \dots, t_p) \in \mathbb{T}^p$ with $t_i = \pi(x_i)$, $i = 1, \dots, p$, is as required. □

The next result serves as a substitute for Lemma 9.12.6 in the case of non-torsion groups.

LEMMA 9.12.13. *Let $S$ be an sharp subset of an Abelian group $G$. Then the set*

$$H_S = \{h \in G^* : h(S) \text{ is dense in } \mathbb{T}\}$$

*is the intersection of countably many open dense subsets of $G^*$ and, hence, $H_S$ is dense in $G^*$.*

PROOF. Let $\{s_n : n \in \mathbb{N}\}$ be a countable dense subset of $\mathbb{T}$. For every $n \in \mathbb{N}$, consider the set

$$U_n = \{h \in G^* : \exists x \in S \text{ such that } |h(x) - s_n| < 1/n\}.$$

Clearly, the sets $U_n$ are open in $G^*$. Let us show that $H_S = \bigcap_{n=1}^{\infty} U_n$ and each $U_n$ is dense in $G^*$.

We start with a proof of the inclusion $P = \bigcap_{n=1}^{\infty} U_n \subset H_S$. It suffices to verify that $h(S)$ is dense in $\mathbb{T}$ for each $h \in P$. If $h \in P$ and $O$ is a non-empty open set in $\mathbb{T}$, choose $n \in \mathbb{N}$ such that $s_n \in O$ and the open interval with the center at $s_n$ and of radius $1/n$ is contained in $O$. From $h \in U_n$ it follows that $|h(x) - s_n| < 1/n$, for some $x \in S$; hence $h(S) \cap O \neq \varnothing$. This proves the density of $h(S)$ in $\mathbb{T}$, so $P \subset H_S$. The inverse inclusion $H_S \subset P$ is immediate.

It remains to show that every $U_n$ is dense in $G^*$. Let $n \in \mathbb{N}$ be arbitrary. Take a basic open set $W_0 = W(f, z_1, \ldots, z_r, \varepsilon)$ in $G^*$, where $f \in G^*$, $z_1, \ldots, z_r \in G$ and $\varepsilon > 0$. We claim that $W_0 \cap U_n \neq \varnothing$.

Denote by $H$ the subgroup of $G$ generated by $z_1, \ldots, z_r$. Since $H$ is a finitely generated Abelian group, there are $y_1, \ldots, y_m \in H$ such that $H = \bigoplus_{i=1}^{m} \langle y_i \rangle$ [**409**, 4.2.10]. Note that $m \leq r$. Since $z_1, \ldots, z_p$ are linear combinations of $y_1, \ldots, y_m$, we can find $\delta > 0$ such that $W_1 = W(f, y_1, \ldots, y_m, \delta) \subset W(f, z_1, \ldots, z_p, \varepsilon)$. Therefore, it suffices to show that $W_1 \cap U_n \neq \varnothing$.

For every $x \in S$, set $||x|| = \min\{k \in \mathbb{N} : kx \in H\}$ (if $kx \notin H$ for every $k \in \mathbb{N}$, we put $||x|| = \infty$). Let us consider the following cases.

1) There exists $x \in S$ such that $||x|| > n$.

1a) If $k = ||x|| > n$ is finite, then $1/k < 1/n$, so we can find an element $t_0 \in \mathbb{T}$ such that $kt_0 = f(kx)$ $(mod\ 1)$ and $|t_0 - s_n| < 1/n$. Define a mapping $h\colon H \cup \{x\} \to \mathbb{T}$ by $h{\restriction}H = f{\restriction}H$ and $h(x) = t_0$. We have $kh(x) = kt_0 = f(kx) = h(kx)$ $(mod\ 1)$, so, by Lemma 1.1.5, $h$ can be extended to a homomorphism of $G$ to $\mathbb{T}$ (denotes below also by $h$). Clearly, $h \in W_1 \cap U_n \neq \varnothing$.

1b) If $||x|| = \infty$, we simply put $h(y_i) = f(y_i)$ for $i = 1, \ldots, m$, and $h(x) = s_n$. Since $kx \notin H$ for each $k \in \mathbb{N}$, we can apply Lemma 1.1.5 to extend $h$ to a homomorphism $G \to \mathbb{T}$, so $h \in W_1 \cap U_n$.

2) $||x|| \leq n$ for all $x \in S$.

Since $S$ is infinite, we can assume that all $||x||$ are equal to some $d \in \mathbb{N}$. We claim that at least one of $y_1, \ldots, y_m$ has infinite order. Assume that the order $d_i$ of $y_i$ is finite, for each $i = 1, \ldots, m$. Put $D = d_1 \cdot \cdots \cdot d_m$. Then, for every $x \in S$, we have $dx \in H$. Hence, $Ddx = 0_G$. This implies immediately that $S \subset G[Dd]$, a contradiction.

Suppose that $y_1, \ldots, y_p$ are all of infinite order, and that $y_{p+1}, \ldots, y_m$ have finite orders $d_{p+1}, \ldots, d_m$, respectively, where $1 \leq p \leq m$. Let $S = \{x_j : j \in \omega\}$. For every

$j \in \omega$, there exist integers $k_{j,1}, \dots, k_{j,m}$ such that $dx_j = \sum_{i=1}^m k_{j,i} y_i$ and $0 \le k_{j,i} < d_i$ for each $i = p+1, \dots, m$. Again, we can find an infinite subset $J$ of $\omega$ and non-negative integers $k_{p+1}, \dots, k_m$ such that $k_{j,i} = k_i$ whenever $j \in J$ and $p+1 \le i \le m$. For every $j \in J$, put $r(j) = \sum_{i=1}^p |k_{j,i}|$. Since $J$ is infinite, $r(j) > p/2 \cdot \delta$, for some $j \in J$. If not, then $\sum_{i=1}^m k_{j,i} y_i = dx_j$ takes only finitely many values. So $\varphi_d(S)$ is finite, which contradicts the admissibility of $S$. Taking such $j$ and applying Lemma 9.12.12 to the homomorphism $g : \mathbb{T}^p \to \mathbb{T}$, defined by $g(u_1, \dots, u_p) = \sum_{i=1}^p k_{j,i} u_i$, and to the element $a = ds_n - \sum_{i=p+1}^m k_i f(y_i) \in \mathbb{T}$, we get $t_1, \dots, t_p \in \mathbb{T}$ satisfying

(i) $|t_i - f(y_i)| < \delta$ for $i = 1, \dots, p$;
(ii) $\sum_{i=1}^p k_{j,i} t_i = ds_n - \sum_{i=p+1}^m k_i f(y_i)$.

It remains to put $h(y_i) = t_i$, for $i = 1, \dots, p$, $h(y_i) = f(y_i)$, for $i = p+1, \dots, m$ (this defines $h$ on $H$), and $h(x_j) = s_n$. Then, by (ii), we have

$$h(dx_j) = \sum_{i=1}^{p} k_{j,i} t_i + \sum_{i=p+1}^{m} k_i f(y_i) = ds_n = dh(x_j)$$

and, hence, $h$ admits an extension to a homomorphism of $G$ to $\mathbb{T}$. Since $x_j \in S$, from (i) and the definition of $h$ it follows that $h \in W_1 \cap U_n \neq \varnothing$. □

The proof of Lemma 9.12.14 below is completely analogous to that of Lemma 9.12.7, so we omit it. This auxiliary lemma is exactly what we need to establish the criterion for the existence of countably compact topological group topologies on non-torsion Abelian groups, such as in Theorem 9.12.16.

LEMMA 9.12.14. *Assume that Martin's Axiom holds. Let $G$ be a discrete Abelian group, $\alpha < \mathfrak{c}$ be an ordinal, and $x^*$ a non-zero element of $G$. For every $\gamma < \alpha$, let $f_\gamma \colon G \to G_\gamma$ be a homomorphism of $G$ to a topological group $G_\gamma$. Suppose that*

a) *for every $\gamma < \alpha$, $S_\gamma$ is a subset of $G$ that is either $n_\gamma$-round or sharp, and also satisfies*
b) *$f_\gamma(S_\gamma)$ is dense in a non-discrete subgroup $K_\gamma$ of $G_\gamma$ with $w(K_\gamma) < \mathfrak{c}$.*

*Then there exists a homomorphism $f \colon G \to \mathbb{T}^\omega$ such that $f(x^*) \neq 0$ and, for every $\gamma < \alpha$, the image $(f_\gamma \Delta f)(S_\gamma)$ is dense in $K_\gamma \times \mathbb{Z}(n_\gamma)^\omega$ if $S_\gamma$ is $n_\gamma$-round, and dense in $K_\gamma \times \mathbb{T}^\omega$ otherwise.*

We write $K \le G$ if $K$ is a subgroup of $G$. We need the following algebraic fact:

LEMMA 9.12.15. *Let $G$ be an Abelian group such that for all $d, n \in \mathbb{N}$ with $d|n$, the group $dG[n] \cong G[n]/G[d]$ is either finite or has cardinality $\mathfrak{c}$. If $G$ contains an $n$-round subset for some $n \in \mathbb{N}$, then $G$ also contains an isomorphic copy of the group $\mathbb{Z}(n)^{(\mathfrak{c})}$.*

PROOF. Indeed, $S \subset G[n]$; hence $\omega = |S| \le |G[n]|$. By the assumptions in the lemma, this implies that $|G[n]| = \mathfrak{c}$. For every proper divisor $d$ of $n$, $dS$ is an infinite subset of $dG[n]$, so again $|dG[n]| = \mathfrak{c}$. Let us show that $\mathbb{Z}(n)^{(\mathfrak{c})} \le G[n] \le G$ by induction on the number of divisors of $n$.

If $n = p$ is prime, then $G[p]$ is the direct sum of $\mathfrak{c}$ copies of the group $\mathbb{Z}(p)$, i.e., $\mathbb{Z}(p)^{(\mathfrak{c})} \cong G[p] \le G$. Now suppose that $n = mp$, where $m > 1$ and $p \in \mathbb{P}$. Then $pS$ is an $m$-round subset of $G$ by c) of Lemma 9.12.3, and the inductive hypothesis implies that $\mathbb{Z}(m)^{(\mathfrak{c})} \le G[m]$. We have $m = m_0 p^k$, where $m_0$ and $p$ are coprime. Then $G[n] = G[m_0 p^{k+1}] \cong G[m_0] \oplus G[p^{k+1}]$. By the Prüfer–Baer theorem (see

Theorem 9.6.28), the group $G[p^{k+1}]$ is isomorphic to the direct sum $\bigoplus_{i=1}^{k+1} \mathbb{Z}(p^i)^{(\alpha_i)}$, where $\alpha_1, \ldots, \alpha_{k+1}$ are cardinals $\leq \mathfrak{c}$. Since the group

$$\begin{aligned} mG[n] = mG[m_0 p^{k+1}] &\cong m_0 p^k (G[m_0] \oplus G[p^{k+1}]) \\ &\cong p^k G[p^{k+1}] \cong (p^k \mathbb{Z}(p^{k+1}))^{(\alpha_{k+1})} \end{aligned}$$

has cardinality $\mathfrak{c}$, this implies that $\alpha_{k+1} = \mathfrak{c}$. Therefore, $\mathbb{Z}(p^{k+1})^{(\mathfrak{c})} \leq G[p^{k+1}] \leq G$. Finally, if $n \neq p^{k+1}$ (i.e., if $m_0 \neq 1$), then $m_0 S$ is a $p^{k+1}$-round subset of $G$, and the inductive hypothesis implies that $\mathbb{Z}(m_0)^{(\mathfrak{c})} \leq G[m_0]$. We now have

$$\mathbb{Z}(n)^{(\mathfrak{c})} \cong \mathbb{Z}(m_0)^{(\mathfrak{c})} \oplus \mathbb{Z}(p^{k+1})^{(\mathfrak{c})} \leq G[m_0] \oplus G[p^{k+1}] \cong G[n],$$

and the proof is complete. □

THEOREM 9.12.16. *Under MA, the following conditions are equivalent for any Abelian non-torsion group $G$ of cardinality at most $\mathfrak{c}$:*

1) *$G$ admits a countably compact Hausdorff topological group topology;*
2) *$G$ admits a countably compact Hausdorff topological group topology without non-trivial convergent sequences;*
3) *$r_0(G) = |G/tor(G)| = \mathfrak{c}$ and, for all $d, n \in \mathbb{N}$ with $d|n$, the group $dG[n] \cong G[n]/G[d]$ is either finite or has cardinality $\mathfrak{c}$.*

PROOF. Clearly, 2) implies 1). Further, 1) implies 3) according to Proposition 9.11.5. Let us show that 3) implies 2). This is the only implication we need $MA$ for.

The argument below follows the line of the proof of Theorem 9.12.9, but it requires several important modifications. Suppose that 3) holds for the group $G$. Denote by $D$ the set of all integers $n > 1$ such that $G$ contains an isomorphic copy of the group $\mathbb{Z}(n)^{(\mathfrak{c})}$. Clearly, if $n \in D$, $d|n$, and $d > 1$, then $d \in D$. Notice that $D$ can be empty.

Put $K = \mathbb{T}^\omega$. We will construct an injective homomorphism $h\colon G \to K^{\mathfrak{c}}$ satisfying the following conditions:

($A_1$) if $n \in \mathbb{N}$ and $S$ is an $n$-round subset of $G$, then $h(S)$ is finally dense in $(K[n])^{\mathfrak{c}}$;
($A_2$) $h(S)$ is finally dense in $K^{\mathfrak{c}}$ for every sharp subset of $G$;
($B_1$) if $n \in D$, then $\pi_\alpha(h(G[n])) = (K[n])^\alpha$ for every $\alpha < \mathfrak{c}$;
($B_2$) $\pi_\alpha(h(G)) = K^\alpha$, for each $\alpha < \mathfrak{c}$.

Recall that in ($B_1$), $\pi_\alpha$ stands for the projection of $K^{\mathfrak{c}}$ onto $K^\alpha$.

**Claim 3.** *The subgroup $h(G)$ of $K^{\mathfrak{c}}$ satisfying ($A_1$), ($A_2$), ($B_1$), and ($B_2$) is countably compact and has no convergent sequences other than the trivial ones.*

Indeed, consider any countable infinite subset $S$ of $G$. By Lemma 9.12.5, $S$ contains either a set of the form $T + z$, where $T$ is a $d$-round subset of $G$ for an integer $d > 1$, or an sharp subset $S'$. In the first case, the image $h(T)$ is finally dense in $(K[d])^{\mathfrak{c}}$ by ($A_1$) and, in the second case, ($A_2$) implies that $h(S')$ is finally dense in $K^{\mathfrak{c}}$. In either case, $h(S)$ cannot converge, since compact countable subsets of $K^{\mathfrak{c}}$ or $(K[d])^{\mathfrak{c}}$ are not finally dense.

The countable compactness of $h(G)$ can be established in a similar way. First, we note that $h(G[n])$ is countably compact for each $n \in D$. This follows from ($A_1$), ($B_1$), Lemma 9.12.15, and Proposition 9.12.2; the argument here is the same as in the corresponding part of the proof of Theorem 9.12.9. Suppose now that $S$ is a countable infinite subset of $G$. If $S$ contains a subset of the form $T + z$, where $T$ is $n$-round in $G$, then $n \in D$ by Lemma 9.12.15, $T \subset G[d]$ and, hence, $h(T)$ has a cluster point $h(x)$ in

$h(G[n])$. Therefore, $h(x+z)$ is a cluster point of $h(S)$. In the remaining case, $S$ contains an sharp subset $S'$, by Lemma 9.12.5, so ($A_2$) implies that $h(S')$ is finally dense in $K^{\mathfrak{c}}$. In fact, every infinite subset of $S'$ is sharp, so $h(S')$ is an HFD subset of $K^{\mathfrak{c}}$. From ($B_2$) and Proposition 9.12.2 it follows that $h(S') \subset h(S)$ has a cluster point in $h(G)$, so that $h(G)$ is countably compact. Thus, Claim 3 is established.

By Claim 3, we obtain the required countably compact topological group topology on $G$ identifying the group $G$ with its image $h(G)$.

We shall construct the monomorphism $h\colon G \to K^{\mathfrak{c}}$ by recursion of length $\mathfrak{c}$. By the definition of $D$ and the assumption that $r_0(G) = \mathfrak{c}$, one can easily define recursively a family $\{N_\xi : \xi < \mathfrak{c}\}$ of subgroups of $S$ such that $G = \bigcup_{\xi<\mathfrak{c}} N_\xi$ and the subgroups $N_\xi$ satisfy the following conditions for all $\xi < \mathfrak{c}$:

(a) $N_0 \cong \mathbb{Z}$, and $N_\zeta \subset N_\xi$ if $\zeta < \xi$;
(b) $|N_\xi| = |\xi| \cdot \omega$;
(c) $N_\xi = \bigcup_{\zeta<\xi} N_\zeta$ if $\xi \geq \omega$ is limit;
(d) $N_{\xi+1}$ contains a copy of the group $\mathbb{Z}$ which trivially intersects $N_\xi$;
(e) for every $n \in D$, $N_{\xi+1}$ contains a copy of the group $\mathbb{Z}(n)$ which trivially intersects $N_\xi$.

As in the proof of Theorem 9.12.9, we shall define, for every $\alpha < \mathfrak{c}$, a homomorphism $h_\alpha\colon G \to K$. We will use the decomposition $G = \bigcup_{\xi<\mathfrak{c}} N_\xi$ in order to define homomorphisms $h_{\alpha,\xi} = h_\alpha \restriction N_\xi$ of $N_\xi$ to $K$. Then $h_\alpha = \bigcup_{\xi<\mathfrak{c}} h_{\alpha,\xi}$ for each $\alpha < \mathfrak{c}$. The homomorphism $h\colon G \to K^{\mathfrak{c}}$ will be the diagonal product of the family $\{h_\alpha : \alpha < \mathfrak{c}\}$.

Denote by $\mathcal{S}$ the family of all countable infinite subsets of $G \setminus \{0\}$ such that every $S \in \mathcal{S}$ is either sharp or $n$-round for some $n \in \mathbb{N}$. Clearly, $|\mathcal{S}| = \mathfrak{c}$, so there exists an enumeration $\mathcal{S} = \{S_\mu : \mu < \mathfrak{c}\}$ such that $S_\mu \subset N_\mu$, for each $\mu < \mathfrak{c}$. If $S_\mu$ is $n$-round for some $n \in \mathbb{N}$, we put $d_\mu = n$; otherwise, write $d_\mu = \infty$ (this means that $S_\mu$ is sharp).

Let $\Sigma$ be the subgroup of $K^{\mathfrak{c}}$ consisting of all $x \in K^{\mathfrak{c}}$ that satisfy $|\{\alpha < \mathfrak{c} : x(\alpha) \neq 0_K\}| < \mathfrak{c}$. We use $MA$ to enumerate $\Sigma = \{b_\nu : \nu < \mathfrak{c}\}$ in such a way that every $x \in \Sigma$ appears in this enumeration $\mathfrak{c}$ times. Finally, for every $x \in G$, denote by $\xi(x)$ the minimal ordinal $\xi < \mathfrak{c}$ such that $x \in N_\xi$. Then $\xi(x)$ is either zero or non-limit.

Our aim now is to define a family $\mathcal{H} = \{h_{\alpha,\nu} : \alpha, \nu < \mathfrak{c}\}$ satisfying the following conditions for all $\alpha, \nu < \mathfrak{c}$:

(1) $h_{\alpha,\nu}\colon N_\nu \to K$ is a homomorphism and $h_{\alpha,\nu}$ extends $h_{\alpha,\mu}$ if $\mu < \nu$;
(2) for every $\mu < \alpha$, the image $\Delta_{\mu\leq\gamma<\alpha} h_{\gamma,\mu}(S_\mu)$ is dense in $(K[d_\mu])^{\alpha\setminus\mu}$ if $d_\mu < \infty$; otherwise this image is dense in $K^{\alpha\setminus\mu}$;
(3) if $b_\nu \neq 0$, then there exists a point $x \in N_{\nu+1} \setminus N_\nu$ such that $h_{\gamma,\nu+1}(x) = b_\nu(\gamma)$ for each $\gamma < \nu$; in addition, if the order $n$ of $b_\nu$ is finite and $n \in D$, then such a point $x$ belongs to $(N_{\nu+1} \setminus N_\nu) \cap G[n]$;
(4) if $\alpha > 0$ and $\xi = \min\{\xi(x) : x \in S_\alpha\}$, then there exists a point $z \in S_\alpha \cap N_\xi$ such that $h_{\alpha,\alpha}(z) \neq 0_K$.

Conditions (1)–(4) almost coincide with those in the proof of Theorem 9.12.9. The more complicated form of the new conditions (2) and (3) is due to the necessity of considering sharp sets.

Arguing as in the proof of Theorem 9.12.9 (taking in account sharp sets), we see that the homomorphism $h = \Delta_{\alpha<\mathfrak{c}} h_\alpha\colon G \to K^{\mathfrak{c}}$ is injective. Conditions (1)–(3) imply the validity of ($A_1$), ($A_2$), ($B_1$) and ($B_2$); the verification of this fact is similar to the corresponding part of the proof of Theorem 9.12.9, so we omit it. Therefore, it remains to prove the following:

**Claim 4.** *The family $\mathcal{H}$ satisfying (1)–(4) exists under MA.*

Indeed, $S_0 \subset N_0 = \mathbb{Z}$ by (a), so $S_0$ is sharp, and we apply Lemma 9.12.13 $\omega$ times (or a special case of Lemma 9.12.14) to choose a homomorphism $h_{0,0}$ which sends $S_0$ to a dense subset of $K$. Let $\alpha$ be an ordinal where $0 < \alpha < \mathfrak{c}$, and assume that for every $\beta < \alpha$, we have defined a family $\{h_{\gamma,\nu} : \gamma, \nu \leq \beta\}$ satisfying (1)–(4). As in the proof of Theorem 9.12.9, we omit considering the case when $\alpha$ is limit. Suppose, therefore, that for $\alpha = \beta + 1$, the family $\{h_{\gamma,\nu} : \gamma, \nu \leq \beta\}$ has been defined and satisfies conditions (1)–(4). We have to extend the homomorphisms $h_{\gamma,\beta}$ (with $\gamma \leq \beta$) over $N_\alpha$, thus obtaining the homomorphisms $h_{\gamma,\alpha}$, and to define a homomorphism $h_{\alpha,\alpha}\colon N_\alpha \to K$.

Suppose that the element $b_\beta \neq 0$ of $\Sigma$ has a finite order $d$. If $d \in D$, then, by (e), there exists an element $x \in N_\alpha$ of order $d$ such that $\langle x \rangle \cap N_\beta = \{0\}$. For every $\gamma \leq \beta$, define a homomorphism $h_{\gamma,\alpha}\colon N_\alpha \to K$ extending $h_{\gamma,\beta}$ and satisfying $h_{\gamma,\alpha}(x) = b_\beta(\gamma)$. If $d \notin D$ or $b_\beta$ has infinite order, then by (d), we can choose an element $x \in N_\alpha$ of infinite order such that $\langle x \rangle \cap N_\beta = \{0\}$. Again, for every $\gamma \leq \beta$, define a homomorphic extension $h_{\gamma,\alpha}\colon N_\alpha \to K$ of $h_{\gamma,\beta}$ such that $h_{\gamma,\alpha}(x) = b_\beta(\gamma)$.

Finally, we define a homomorphism $h_{\alpha,\alpha}\colon N_\alpha \to K$. First, choose $z \in S_\alpha$ such that $\xi(z) = \min\{\xi(y) : y \in S_\alpha\}$. For every $\mu \leq \beta$, denote by $f_{\mu,\alpha}$ the diagonal product of the homomorphisms $h_{\gamma,\beta}$ where $\mu \leq \gamma \leq \beta$. Then $f_{\mu,\alpha}\colon N_\alpha \to K^{\alpha\setminus\mu}$. Applying Lemma 9.12.14, we find a homomorphism $f\colon N_\alpha \to K$ such that $f(z) \neq 0_K$ and, for every $\mu \leq \beta$, the image $(f_{\mu,\alpha}\Delta f)(S_\mu)$ is dense in $(K[d_\mu])^{\alpha\setminus\mu} \times \mathbb{Z}(d_\mu)^\omega$ if $d_\mu < \infty$, or is dense in $K^{\alpha\setminus\mu} \times K$ if $S_\mu$ is sharp. It remains to put $h_{\alpha,\alpha} = f$ and $h_{\alpha,\nu} = f\restriction N_\nu$, for each $\nu < \alpha$. The recursive construction is complete.

It is easy to verify that the family $\mathcal{H} = \{h_{\alpha,\nu} : \alpha, \nu < \mathfrak{c}\}$ satisfies conditions (1)–(4). This proves Claim 4 and the theorem. □

The free Abelian group of cardinality $\mathfrak{c}$ satisfies all the conditions in Theorem 9.12.16. Therefore, we have:

COROLLARY 9.12.17. *Under MA, the free Abelian group of cardinality $\mathfrak{c}$ admits a countably compact Hausdorff topological group topology.*

The next example shows that the existence of a countably compact Hausdorff topological group topology on an Abelian non-torsion group $G$ of size $\mathfrak{c}$ is a considerably stronger condition than the existence of a pseudocompact Hausdorff topological group topology.

EXAMPLE 9.12.18. For every prime $p$, the group $G = \mathbb{R} \oplus \mathbb{Z}(p)^{(\omega)}$ admits a pseudocompact Hausdorff topological group topology, while no Hausdorff topological group topology on $G$ is countably compact, by Theorem 9.12.16.

Indeed, it follows from Corollary 9.11.7 that the group $G$ admits a pseudocompact Hausdorff topological group topology since $|G/tor(G)| = \mathfrak{c}$, or equivalently, $r_0(G) = \mathfrak{c}$. In fact, this conclusion does not require $MA$ because the countable subgroup $H = \{0\} \times \mathbb{Z}(p)^{(\omega)}$ of $G$ is unconditionally closed in $G$ (as the kernel of the unconditionally continuous homomorphism $\varphi_p\colon G \to G$, where $\varphi_p(x) = px$ for each $x \in G$). Therefore, if $G$ had a countably compact group topology $\tau$, the subgroup $H$ would be closed in $(G, \tau)$ and therefore, $H$ would be countably compact. It remains to note that every infinite countably compact topological group is uncountable, by Theorem 9.11.2, while $H$ is countable. □

It is known that for every infinite cardinal $\tau$ with $\tau^\omega = \tau$, the free group of cardinality $\tau$ admits a pseudocompact Hausdorff topological group topology (see Exercise 9.11.d). We show in the next example that this is no longer valid for countably compact Hausdorff topological group topologies. This explains why in the above statements very often the groups considered were assumed to be Abelian.

EXAMPLE 9.12.19. No abstract free group $F_a(X)$ consisting of more than one element admits a countably compact Hausdorff topological group topology. Furthermore, every countably compact subgroup of a free group, provided with a Hausdorff topological group topology, is trivial.

Indeed, suppose that $\mathcal{T}$ is a Hausdorff topological group topology on $F = F(X)$ and that $G$ is a non-trivial countably compact subgroup of the group $(F, \mathcal{T})$. Take an arbitrary element $a \in G$ distinct from the neutral element $e$. Clearly, the cyclic subgroup $\langle a \rangle$ of $G$ is Abelian, and so is the closure $H = \overline{\langle a \rangle}$ of $\langle a \rangle$ in $G$. So $H$ is a countably compact Abelian subgroup of $G$. However, by Nielsen's theorem (see [**409**, Theorem 6.1.1]), a subgroup of a free group is free, so $H$ must be an infinite cyclic group. This contradicts the fact that every infinite countably compact group has cardinality at least $\mathfrak{c}$. Thus, the difference between pseudocompact and countably compact Hausdorff topological group topologizations of groups becomes even more drastic in the non-Abelian case. □

Denote by $\mathfrak{C}$ the class of Abelian groups that admit a countably compact Hausdorff topological group topology. The methods of this section permit us to establish, under $MA$, that many Abelian non-torsion groups $G$ of cardinality $\mathfrak{c}$ are in $\mathfrak{C}$ by means of an appropriate embedding of $G$ into $\mathbb{T}^\mathfrak{c}$. Since, according to Problem 9.9.F, every Abelian group of size $\leq 2^\mathfrak{c}$ can be embedded as a subgroup in $\mathbb{T}^\mathfrak{c}$, the techniques developed above allow us to find countably compact Hausdorff topological group topologies on many groups of the cardinality greater than $\mathfrak{c}$. For example, if $K$ is an Abelian group that admits a compact Hausdorff topological group topology, then $G = K \times H \in \mathfrak{C}$ whenever $H \in \mathfrak{C}$. This follows immediately from the fact that the product of a countably compact group with a compact group is countably compact (see also Problem 9.12.D).

Many natural problems remain, however, open. It is not known, for example, whether the free Abelian group of size $\mathfrak{c}$ belongs to $\mathfrak{C}$ in $ZFC$ only (see Corollary 9.12.17 and Problem 9.12.4). In other words, even very simple Abelian groups present serious difficulties when searching for non-trivial countably compact topological group topologizations. In this respect, Theorem 9.11.6 and Corollary 9.11.8 show that pseudocompact topologizations are much easier to achieve due to the description of pseudocompact topological groups as $G_\delta$-dense subgroups of compact groups (see Corollary 6.6.5).

## Exercises

9.12.a. Show that the free Abelian group $A(X)$ of a set $X$ with some topology $\mathcal{T}$ on $A(X)$ is a countably compact left topological group if and only if $A(X)$ with the same topology $\mathcal{T}$ is a topological group.

9.12.b. Show that in item 1) of Theorem 9.12.16, one can add "connected and locally connected" to the properties of the group $G$.

9.12.c. Let $G$ be a countably compact Abelian topological group, and $H$ be the set of all *metrizable elements* of $G$, that is, $H$ consists of all $x \in G$ such that the closure in $G$ of the subgroup generated by $x$ is metrizable. Prove that $H$ is a subgroup of $G$. (See also Problem 9.11.B.)

9.12.d. A topological group $G$ is called *sequentially complete* if no sequence in $G$ converges to a point of $\varrho G \setminus G$. Show that if a group $G$ does not contain non-trivial convergent sequences, then it is sequentially complete.

9.12.e. Show that every infinite Abelian group admits a non-discrete $\omega$-narrow Hausdorff topological group topology of countable tightness and countable cellularity.

## Problems

9.12.A. A group $G$ is called *reduced* if $G$ has no divisible subgroups beyond $\{0\}$. Show that if an Abelian group $G$ admits a totally disconnected countably compact Hausdorff topological group topology, then $G$ is reduced.

*Hint.* A totally disconnected countably compact group $G$ is zero-dimensional [**135**]. The completion $\varrho G$ of $G$ is a compact zero-dimensional group [**468**]. Hence, $\varrho G$ is a reduced group and consequently, $G$ itself is reduced.

9.12.B. Prove that in c) of Theorem 9.12.9 and in 3) of Theorem 9.12.16, one can add "hereditarily separable" to the properties of the group $G$, under the stronger assumption of $CH$..

9.12.C. (D. Dikranjan and M. G. Tkachenko [**144**]) Let $G$ be a divisible Abelian group of cardinality at most $\mathfrak{c}$. Then the following are equivalent:

(a) $G$ admits a countably compact Hausdorff topological group topology;

(b) $G$ admits a compact Hausdorff topological group topology;

(c) $r_0(G) = \mathfrak{c}$ and, for every prime $p$, either $r_p(G) < \omega$ or $r_p(G) = \mathfrak{c}$.

9.12.D. (D. Dikranjan and M.G. Tkachenko [**144**]) Let $\{G_i : i \in I\}$ be a family of Abelian groups, where each $G_i$ admits a countably compact Hausdorff topological group topology and $|G_i| \leq \mathfrak{c}$, for each $i \in I$. Prove, assuming $MA$, that if $|I| < \mathfrak{c}$, then $G = \prod_{i \in I} G_i$ also admits a countably compact Hausdorff topological group topology.

9.12.E. (A. H. Tomita [**497**]) A space $X$ is called *initially* $\aleph_1$*-compact* if every covering of $X$ by at most $\aleph_1$ open sets contains a finite subcovering. Prove that the existence of an initially $\aleph_1$-compact Hausdorff topological group topology on some infinite free Abelian group is independent of $ZFC$.

9.12.F. (A. H. Tomita [**497**]) For no Hausdorff topological group topology on an infinite free Abelian group $F$, the power $F^\omega$ is countably compact. Therefore, a non-trivial free Abelian group $F$ does not admit a sequentially compact Hausdorff topological group topology.

9.12.G. (E. van Douwen [**149**]) Prove under $GCH$ that for every infinite countably compact topological group $G$, the equality $|G|^\omega = |G|$ holds.

*Remark.* According to Proposition 9.12.1, there are no obstacles for an Abelian group $G$ of a prime exponent to admit a countably compact Hausdorff topological group topology other than the equality $|G|^\omega = |G|$ (this statement does not require any extra set-theoretic axioms).

9.12.H. (A. H. Tomita [**498**]) The existence of a countably compact topological group $G$ satisfying $|G| = \aleph_\omega$ is consistent with $ZFC$.

9.12.I. (D. Dikranjan and D. B. Shakhmatov [**141**]) It is consistent with $ZFC$ that the free Abelian group of cardinality $2^{\mathfrak{c}}$ admits a countably compact Hausdorff topological group topology.

## Open Problems

9.12.1. It is consistent with $ZFC$ that every precompact (Abelian) topological group without non-trivial convergent sequences is topologically isomorphic to a closed subgroup of the product of a family of countably compact (Abelian) topological groups?

9.12.2. Is it consistent with $ZFC$ that a precompact Abelian topological group $G$ is topologically isomorphic to a closed subgroup of a product of countably compact Abelian topological groups if and only if $G$ is sequentially complete? (See Exercise 9Ex12.4.)

9.12.3. Can every countably compact Tychonoff space be embedded as a closed subspace into a countably compact topological group?

9.12.4. (M. G. Tkachenko [**478**], W. W. Comfort [**110**]) Does there exist in $ZFC$ a countably compact Hausdorff topological group topology on the free Abelian group of cardinality $\mathfrak{c}$?

9.12.5. (D. Dikranjan and M. G. Tkachenko [**144**]) Characterize the algebraic structure of Abelian groups of cardinality $\mathfrak{c}$ that admit an $\omega$-bounded Hausdorff topological group topology.

9.12.6. (D. Dikranjan and M. G. Tkachenko [**144**]) Does the group $\mathbb{Z}(2)^{(\omega)} \oplus \mathbb{Z}(4)^{(\mathfrak{c})}$ admit an $\omega$-bounded Hausdorff topological group topology under the assumption of $MA$?

9.12.7. (D. Dikranjan and M. G. Tkachenko [**144**]) Characterize the algebraic structure of the non-torsion Abelian groups of size $\mathfrak{c}$ that admit a totally disconnected countably compact Hausdorff topological group topology.

9.12.8. (D. Dikranjan and M. G. Tkachenko [**144**]) Can the equivalence of a), b), d), and e) of Theorem 9.12.9 be proved in $ZFC$? Equivalently, is the implication e) $\Rightarrow$ b) a theorem of $ZFC$?

9.12.9. (D. Dikranjan and M. G. Tkachenko [**144**]) Does the equivalence of conditions 1) and 3) in Theorem 9.12.16 remain valid in $ZFC$?

9.12.10. (E. van Douwen [**148**]) Under which restrictions on an infinite countably compact group $G$, the equality $|G|^{\omega} = |G|$ holds in $ZFC$?

9.12.11. (E. van Douwen [**148**]) Under which restrictions on an infinite countably compact group $G$, we have $\mathrm{cf}\,|G| \neq \omega$ in $ZFC$?

9.12.12. Does there exist in $ZFC$ a countably compact Hausdorff topological group topology of countable tightness on the free Abelian group of cardinality $\mathfrak{c}$?

9.12.13. Does every infinite Abelian group admits a precompact sequential Hausdorff topological group topology? (Compare with Exercise 9.12.e.)

9.12.14. Can every Abelian group be represented as a subgroup of a countably compact topological group of countable tightness?

9.12.15. Does there exist in $ZFC$ an $\omega$-bounded topological group $G$ such that $\mathfrak{c} = |G| < w(G)$? (Compare this with Exercise 5.2.c.)

9.12.16. Is there a $ZFC$ example of a countably compact topological group $G$ such that $G \times G$ is not countably compact?

*Hint.* See [**300**] and [**218**] where consistent examples of such groups are constructed. See also [**149**].

9.12.17. Suppose that $G$ is an Abelian group admitting a countably compact Hausdorff topological group topology. Does the group $G \times G$ also admit a countably compact Hausdorff topological group topology?

9.12.18. Suppose that $G$ and $H$ are Abelian groups admitting a countably compact Hausdorff topological group topology. Does the group $G \times H$ admit a countably compact Hausdorff topological group topology?

9.12.19. Is every countably compact sequential topological group a Fréchet–Urysohn space?

9.12.20. Is every separable countably compact sequential (Fréchet–Urysohn) topological group metrizable?

9.12.21. (R. Z. Buzyakova [**94**]) Give an example of a compact Abelian group $G$ and a countably compact normal topological Abelian group $H$ such that there is no one-to-one continuous mapping of the group $G \times H$ onto a normal space.

*Solution.* Put $G = D^{\omega_1}$, where $D = \{0, 1\}$. Let $H$ be the subgroup of $G$ consisting of all points of $G$ with at most countably many non-zero coordinates. The space $X = G \times H$ is pseudocompact, and it cannot be mapped onto a normal space $Z$ by a one-to-one continuous mapping. Assume the contrary, and fix a mapping $f\colon X \to Z$ with these properties. The Čech–Stone compactification of $X$ is the space $G \times G$ (see [**165**, 3.12.20 (c)]), so $f$ admits an extension to a continuous mapping $f^*$ of $G \times G$ to $\beta Z$. Take any $p \in G \setminus H$.

*Claim.* $f^*((p, p)) \in Z$. Indeed, since the space $G$ has a base of cardinality $\omega_1$ at $p$, and $H$ is dense in $G$, and the closure in $G$ of each countable subset of $H$ is compact and is contained in $H$, it is easy to construct by a transfinite recursion a subset $A$ of $H$ such that $\overline{A} = A \cup \{p\}$ (take the free sequence in $G$ converging to $p$ and contained in $H$, see [**20**]).

Put $B = \{p\} \times A$ and $C = \{(y, y) : y \in A\}$. Clearly, $B \subset X$, $C \subset X$, and the sets $B^* = B \cup \{(p, p)\}$ and $C^* = C \cup \{(p, p)\}$ are compact. Assume now that $f^*((p, p))$ is not in $Z$. Then $f(B) = Z \cap f^*(B^*)$ and $f(C) = Z \cap f^*(C^*)$. Since $f$ is one-to-one and $B^*$ and $C^*$ are compact, it follows that $f(B)$ and $f(C)$ are disjoint closed subsets of $Z$. Since $Z$ is normal, there exists a continuous real-valued function $g$ on $Z$ separating $f(B)$ and $f(C)$. Then $gf$ is a continuous function on $X$ separating $B$ and $C$. However, this is impossible, since the closures of $B$ and $C$ in $\beta X = G \times G$ are not disjoint. Our Claim is proved.

Put $\Delta_H = \{(y, y) : y \in H\}$ and $\Delta_G = \{(y, y) : y \in G\}$. The subspace $\Delta_G$ of $G \times G$ is compact and separable, since it is naturally homeomorphic to $D^{\omega_1}$. Put $E = f^*(\Delta_G)$. Then $E$ is also compact and separable. From the above Claim it follows that $E$ is a subspace of $Z$. Now we can fix a countable subset $M$ of $X$ such that $f(M)$ is dense in $E$. It is easy to check that the closure of $M$ in $G \times G$ is a compact subspace $F$ of $X$. Obviously, $f(F) = E$. Since $H$ is not compact, the image of $F$ under the natural projection mapping of $G \times H$ onto the second factor $H$ is not compact. Therefore, there exists $u \in H$ such that $(u, u) \notin F$. Clearly, $f((u, u)) \in E$. On the other hand, from $f(F) = E$ it follows that $f(w) = f((u, u))$, for some $w \in F$. Since $(u, u) \notin F$ and $w \in F$, we conclude that $w \neq (u, u)$, a contradiction with the assumption that $f$ is one-to-one.

## 9.13. Historical comments to Chapter 9

Our approach to the proof of Krein–Milman's theorem follows the one adopted in [**81**]. In particular, one finds there some versions of Propositions 9.1.2, 9.1.3, 9.1.5, 9.1.9, and 9.1.10. The first proof of Krein–Milman's theorem (see Theorem 9.1.12) was given in [**283**]. This important result has many applications; see [**157**] for those and for further references.

One can find Theorem 9.2.2, some versions of Lemma 9.1.4, and Proposition 9.2.4 in [**81**]. A combination of these results leads to a proof of Theorem 9.2.5 called the *Hahn–Banach theorem*. This theorem was first established in [**212**] and [**66**]. Hahn–Banach's theorem is one of the main principles in the theory of topological vector spaces. In particular, it provides a basis for the study of convexity problems, and demonstrates that the set of continuous linear functionals in most spaces is immensely rich.

Theorem 9.2.7 is called the *Gel'fand–Mazur theorem*; it was proved, for a special case, in [**188, 189**], and in [**312**]. Our proof of it is taken from [**337**].

The existence of the invariant integral on a compact Lie group was established by F. Peter and H. Weyl in [**383**]. Haar measure and invariant integral serve as a basis for Harmonic Analysis. A. Haar made an important step in 1933; he proved the existence of a left-invariant Haar measure on every locally compact topological group with a countable base [**211**]. For compact groups, the existence and uniqueness of the Haar integral, its two-sided invariance, and its invariance under the inverse, was proved by J. von Neumann [**344**].

Our treatment of the subject in this chapter follows Pontryagin's book [**387**]. In particular, Propositions 9.3.1, 9.3.2, 9.3.3, and Theorem 9.3.9 are taken from there. Theorem 9.3.13, as we already mentioned, is due to J. von Neumann [**344**].

Characters of finite Abelian groups were introduced by G. Frobenius. Pontryagin [**385**] molded them into topological Abelian groups and developed the theory of characters. An outstanding role in this theory belongs to a corollary to Peter–Weyl's theorem [**383**]. The theorem tells that there are sufficiently many irreducible representations of compact groups, and the especially important corollary from this result is the theorem that there are sufficiently many continuous characters on every compact Abelian group. This is precisely Theorem 9.4.11. Our proof of this basic fact is different from the one given in [**387**]. In broad lines we follow the approach of M. Cotlar and R. Ricabarra developed in [**127**]. Though we have introduced several modifications in the argument to make it even more elementary (and slightly less general), almost every statement in Section 9.4 has its prototype in [**127**]. In fact, it was established in [**127**] that an element $g$ of an Abelian topological group $G$ is separated from the neutral element $e$ by a continuous character if and only if there exists a symmetric open neighbourhood $U$ of $e$ such that $g \notin U^6$ and $KU = G$, for some finite subset $K$ of $G$. Later, this result was improved in [**163**] (the number 6 was replaced by 4). In treating Pontryagin's duality between compact and discrete Abelian groups, we follow the original treatment of this matter by Pontryagin in [**387**], as well as the small book [**327**] by S. A. Morris. For example, one can find Propositions 9.5.1, 9.5.2, 9.5.3, and some other similar statements in [**327**]. On the other hand, Theorem 9.5.14, Corollaries 9.5.15, 9.5.16, and Proposition 9.5.18 have their prototypes in [**387**]. The main result, Theorem 9.5.20, was proved by Pontryagin for compact metrizable Abelian groups in [**385**]. Soon afterwards it was extended to locally compact not necessarily metrizable Abelian groups by E. van Kampen in [**266**].

An alternative treatment of the subject can be found in [**236**] and in [**337**], where the theory of Banach algebras is used. Our goal was to make the proof of the main theorem as elementary as possible. More general results with applications and references can be found in [**236**] and in [**337**]. See more on the history of the subject and references in [**236**].

In Section 9.6 we mostly follow Pontryagin's book [**387**]. In particular, Corollary 9.6.1, Proposition 9.6.2, Theorems 9.6.3, 9.6.4, 9.6.5, 9.6.6, Corollaries 9.6.7, 9.6.8, Theorems 9.6.11, 9.6.12, 9.6.15, 9.6.22, 9.6.23, and many other results of the section come from Pontryagin's [**385, 387**]. See also [**236**] for a somewhat different treatment of the Pontryagin–van Kampen duality and for further references and comments.

E. van Kampen performed the task of extending Pontryagin's duality theory to the general case of locally compact Abelian groups. The proof of Theorem 9.7.1 follows [**236**]. We used some further material from [**236**] to extend the theorem on the existence of non-trivial characters to all locally compact Abelian groups (Theorem 9.7.5). Usually, this theorem is proved on the basis of Peter–Weyl's theorem on irreducible representations, but we gave a more elementary proof. A detailed study of the structure of locally compact Abelian groups, as well as the duality theory for this class of groups, on which the structure theory is based, can be found in many books, in particular, in [**236**]. We restricted ourselves to the compact–discrete case with the sole goal of introducing the reader to certain important ideas in topological algebra that are already well developed in the existing literature. There are also many interesting and profound results in the duality theory beyond the class of locally compact groups. In particular, we recommend the reader to look into [**232**], [**377**], and [**310**].

Theorem 9.8.1 is an important result based on the Pontryagin–van Kampen duality theory; it appeared, for example, in [**522**]. Lemmas 9.8.2, 9.8.3, 9.8.4, and Theorem 9.8.5 (known as Varopoulos' theorem) were also proved in [**522**] as was Example 9.8.6. Propositions 9.8.7, 9.8.8, 9.8.9, Theorems 9.8.10 and 9.8.11, as well as Example 9.8.12 were given by A. V. Arhangel'skii in [**34**]. Example 9.8.13, in a slightly different form, was described by N. T. Varopoulos in [**522**]. Examples 9.8.14 and 9.8.15 are taken from [**34**]. Theorems 9.8.16, 9.8.18, and Corollaries 9.8.17, 9.8.19 are also from [**34**].

When the duality theory for locally compact Abelian groups had been created, the importance of the group $Hom(G, \mathbb{T})$ of continuous characters of a given topological Abelian group $G$, even if $G$ is not locally compact, became clear. The weaker topological group topology on $G$ generated by all continuous homomorphisms $f\colon G \to \mathbb{T}$ was called the *Bohr topology* of $G$, and the group $G$ with the Bohr topology was denoted by $G^+$ in the fundamental article [**150**] by E. van Douwen. However, the study of the Bohr topology and Bohr compactification originated much earlier (see [**9, 11, 121, 123, 172, 198, 238, 244, 326**]). Elementary Propositions 9.9.1, 9.9.2, and 9.9.3 are a part of the folklore. Items a) and b) of Proposition 9.9.9 as well as Proposition 9.9.11 appeared in [**123**]. Historical comments on the structure theory of abstract Abelian groups concerning the series of results starting Lemma 9.9.13 to Proposition 9.9.20 can be found in [**409**]. Proposition 9.9.21 and Corollary 9.9.22 were proved by K. P. Hart and J. van Mill in [**219**]. Lemma 9.9.24 and Theorem 9.9.25 are due to W. W. Comfort, S. Hernández, and F. J. Trigos-Arrieta [**112**]. In fact, the observation that the group $G^\#$ is not realcompact if the cardinality of $G$ is not Ulam-measurable was attributed to A. Dow in [**150**, Fact 4.15].

Lemma 9.9.27 appeared (in a slightly more general form) in [**492**]. Theorem 9.9.30 was proved by H. Leptin in [**292**]. It is a special case of a more general result proved by I. Glicksberg in [**198**]: If $G$ is a locally compact Abelian group, then every compact subset of $G^+$ is compact in $G$ (i.e., the compact subsets of $G$ and $G^+$ coincide). The original argument in [**198**] used some methods of functional analysis, while W. W. Comfort and F. J. Trigos-Arrieta presented in [**125**] an alternative proof of Theorem 9.9.30 based exclusively on topological group type arguments. Theorem 9.9.31 appeared in the article [**150**] by E. van Douwen, while Theorem 9.9.32 that refines it, leans on a general result about precompact groups (see Corollary 8.8.6) obtained by D. B. Shakhmatov in [**433**].

Items i) and ii) of Lemma 9.9.33 appeared in [**150**], while the last claim of the lemma appeared in [**112**]. Lemma 9.9.35 came from [**191**], while Theorem 9.9.36 was proved by R. Blair in [**76**]. Proposition 9.9.37 (implicitly) and Theorems 9.9.38 and 9.9.40 are results from [**112**]. Theorem 9.9.42 is a special case of a more general result proved by F. J. Trigos-Arrieta in [**502, 501**] (see Problem 9.9.P).

Theorem 9.9.43 generalizes a couple of results proved by J. Galindo and S. Hernández in [**182**] and, apparently, appears in print for the first time (even if the result itself was known to specialists, see the proof of the main theorem in [**503**]). The proof of Theorem 9.9.43 given here is a slight refinement of the argument in [**182**, Lemma 1]. Theorem 9.9.45 can be considered as a natural combination of some results in [**150**] and [**219**].

Theorems 9.9.51, 9.9.54, and 9.9.55 are essentially due to E. van Douwen (see [**150**]). Some important technical improvements to the original arguments from [**150**] were made by K. P. Hart and J. van Mill in [**219**], and by J. Galindo and S. Hernández in [**182**]. Our proofs of Theorems 9.9.51, 9.9.54, and 9.9.55 contain some new ingredients that clarify main ideas and the whole strategy developed in [**150**].

Theorem 9.9.56 is an important contribution to the study of separation properties of topological groups. It was proved by F. J. Trigos-Arrieta in [**503**]. Theorems 9.9.57 and 9.9.59 on cardinal invariants of the groups $G^{\#}$ are essentially from [**145**]. An assertion equivalent to Theorem 9.9.60 was proved by K. P. Hart and J. van Mill in [**219**]. Example 9.9.61 appeared in [**245**]. An example of topological Abelian group $K$ with the stronger property that every continuous character on $K$ is trivial was presented by I. Prodanov in [**388**] (earlier, an example with the same property appeared in [**236**], but its construction was based on some ideas from functional analysis). It is worth mentioning that in [**1**], M. Ajtai, I. Havas, and J. Komlós proved that every infinite Abelian group admits a Hausdorff topological group topology in which the continuous characters on the group do not separate points.

Many results of Section 9.9 can be extended to maximally almost periodic groups. In this respect, see [**111**].

Lemmas 9.10.1, 9.10.2, and Example 9.10.3 are taken from [**91**].

The study of cardinal restrictions on pseudocompact topological groups and of the algebraic structure of these groups, which is the subject of Section 9.11, was initiated by E. van Douwen in [**148**]. Auxiliary Lemma 9.11.1 can be found in [**311**]. Proposition 4.2.3 is difficult to attribute to someone; it appeared, for example, in [**424**]. Theorem 9.11.2 was proved in [**148**]. Lemma 9.11.4 goes back to W. W. Comfort and J. van Mill's [**114**, Remark 2.17]; the first printed proof of Lemma 9.11.4 is due to D. Dikranjan and D. B. Shakhmatov, see [**140**]. Theorem 9.11.5 appeared, in the present form, in [**140**]. However, its new ingredient compared to Lemma 9.11.4 — the fact that every pseudocompact torsion Abelian group is of bounded torsion — was proved in [**120**]. Theorem 9.11.6 as well as Corollaries 9.11.7 and 9.11.8 are from [**140**]. The same article [**140**] contains a complete characterization of abstract groups admitting a Hausdorff pseudocompact topological group topology for the following special classes of groups: a) free Abelian groups; b) torsion-free Abelian groups; c) torsion Abelian groups; d) divisible Abelian groups. Proposition 9.11.9 was proved by D. Dikranjan in [**135**]. Theorem 9.11.10 is a weaker form of a result from [**140**].

The existence of countably compact Hausdorff topological group topologies on Abelian groups is quite a subtle matter. Again, it was E. van Douwen who established several fundamental results for this class of groups in [**148, 149**]. Proposition 9.12.1 is one of them. Proposition 9.12.2, that serves as a base for further constructions in Section 9.12, appeared in [**213, 214**]. Lemmas 9.12.6, 9.12.7, 9.12.8, as well as Theorem 9.12.9 and Corollaries 9.12.10 and 9.12.11 are from [**144**]. Lemmas 9.12.13, 9.12.14, 9.12.15, Theorem 9.12.16, and Corollary 9.12.17 are also from [**144**]. Under the weaker assumption of $CH$, Corollary 9.12.17 was proved in [**478**]. Example 9.12.19 is a weaker form of Theorem 5.13 from [**140**].

The problem of characterizing the algebraic structure of the countably compact topological groups of cardinality higher than $2^{\mathfrak{c}}$ seems to be far from a reasonable solution. In [**150**], E. van Douwen asked whether every infinite countably compact group $G$ satisfies $|G|^{\omega} = |G|$ or at least $\mathrm{cf}(|G|) \neq \omega$. In the same article [**149**], van Douwen proved that this hypothesis is valid under $GCH$. According to Proposition 9.12.1, for Abelian groups of a prime exponent, there are no other restrictions (this does not require any extra set-theoretic axioms). However, A. H. Tomita managed to prove in [**498**] that the existence of a countably compact topological group $G$ satisfying $|G| = \aleph_\omega$ is consistent with $ZFC$. Finally, it is consistent with $ZFC$ that the free Abelian group of cardinality $2^{\mathfrak{c}}$ is in $\mathfrak{C}$, according to a theorem of D. Dikranjan and D. B. Shakhmatov in [**141**].

# Chapter 10

# Actions of Topological Groups on Topological Spaces

In this short chapter we introduce an important topic of continuous actions of topological groups on topological spaces. No attempt is made at a systematic treatment of the subject; this would require a separate book. Some such books already exist (see, in particular, [**530, 86**]). Our goal is much more modest — to give the reader just the flavour of the topic by establishing several important results on dyadicity or similar properties of compact spaces in this context. One of these results concerns compact $G_\delta$-sets in quotient spaces of $\omega$-balanced topological groups. Even the corollary dealing with the case of the quotient space itself is extremely interesting and highly non-trivial. Another theorem provides a deep insight into the topological structure of compact spaces admitting a continuous transitive action of an $\omega$-narrow topological group. In fact, all compact spaces just mentioned have the following strong property — they are *Dugundji spaces*. Our arguments require several topological facts which usually do not form a part of standard courses on general topology, so the first section of the chapter familiarizes the reader with the concepts of Dugundji spaces, 0-soft mappings, and nearly open mappings. We also develop further the techniques involving inverse spectra (in Chapter 4 we have already made the first steps in this direction). We also introduce some basic concepts and elementary results on actions of topological groups on topological spaces.

## 10.1. Dugundji spaces and 0-soft mappings

A compact space $X$ is called *Dugundji* if for every zero-dimensional compact space $Z$ and every continuous mapping $f\colon A \to X$, where $A$ is a closed subset of $Z$, there exists a continuous mapping $g\colon Z \to X$ extending $f$.

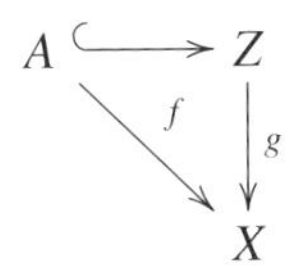


In the proposition below we establish two basic properties of Dugundji spaces.

Proposition 10.1.1. *The class of Dugundji spaces has the following properties:*

a) *every compact metrizable space is Dugundji;*
b) *the product of an arbitrary family of Dugundji spaces is Dugundji.*

PROOF. a) Let $f: A \to X$ be a continuous mapping, where $X$ is a compact metrizable space and $A$ is a closed subset of a zero-dimensional compact space $Z$. Consider a mapping $F: Z \to Exp(X)$ of $Z$ to the family $Exp(X)$ of closed subsets of $X$ defined by $F(z) = \{f(z)\}$ for each $z \in A$ and $F(z) = X$ for $z \in Z \setminus A$. Then the mapping $F$ is lower semicontinuous, so Theorem 4.1.1 implies that there exists a continuous selection $g: Z \to X$ for $F$. It follows from the definition of $F$ that $g(z) = f(z)$, for each $z \in A$, so $g$ is a continuous extension of $f$ over $Z$. Hence, $X$ is Dugundji.

b) Suppose that $X = \prod_{i \in I} X_i$ is the product of a family $\{X_i : i \in I\}$ of Dugundji spaces. Since each space $X_i$ is compact, the product space $X$ is also compact. Let $f: A \to X$ be a continuous mapping, where $A$ is a closed subset of a zero-dimensional compact space $Z$. For every $i \in I$, consider the mapping $f_i = p_i \circ f$, where $p_i: X \to X_i$ is the natural projection. Since $f_i: A \to X_i$ is a continuous mapping to the Dugundji space $X_i$, it admits a continuous extension $g_i: Z \to X_i$. Let $g$ be the diagonal product of the family $\{g_i : i \in I\}$. Then the mapping $g: Z \to X$ is continuous and $g \restriction A = f$; thus, $X$ is Dugundji. □

Combining items a) and b) of the above proposition, we deduce the following:

COROLLARY 10.1.2. *The product of an arbitrary family of second-countable compact spaces is Dugundji.*

Let us now establish a simple but very important fact:

THEOREM 10.1.3. [**R. Haydon**] *Every Dugundji space is dyadic.*

PROOF. Let $X$ be an arbitrary Dugundji space of weight $\tau \geq \omega$. By Theorem 4.1.5, we can find a closed subset $A$ of the space $D^\tau$, where $D = \{0, 1\}$ is the discrete two-point space, and a continuous onto mapping $f: A \to X$. Since $X$ is Dugundji and $D^\tau$ is compact and zero-dimensional, $f$ can be extended to a continuous mapping $g: D^\tau \to X$. Clearly, $g(D^\tau) = g(A) = f(A) = X$, so $X$ is dyadic. □

A continuous mapping $f: X \to Y$ is called 0-*soft* if for every zero-dimensional compact space $Z$, every continuous mapping $g: Z \to Y$ and a continuous mapping $h: A \to X$ of a closed subset $A$ of $Z$ satisfying $g \restriction A = f \circ h$, there exists a continuous mapping $\varphi: Z \to X$ extending $h$ which makes the following diagram commutative.

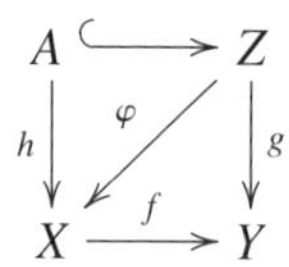


It follows immediately from the above definition that a compact space $X$ is Dugundji if and only if a mapping of $X$ to a singleton (i.e., a constant mapping) is 0-soft. *A priori*, 0-soft mappings are not assumed to be onto. However, every 0-soft mapping of compact spaces is surjective.

PROPOSITION 10.1.4. *If $f: X \to Y$ is a* 0-*soft mapping of compact spaces $X$ and $Y$, then $f(X) = Y$.*

PROOF. By Theorem 4.1.5, there exists a zero-dimensional compact space $Z$ and a continuous onto mapping $g: Z \to Y$. Let $y \in f(X)$ be an arbitrary point. Choose $x \in X$ and $z \in Z$ such that $f(x) = y = g(z)$, and put $A = \{z\}$. Define a mapping $h: A \to X$

by $h(z) = x$. Since $f$ is 0-soft, $h$ admits an extension to a continuous mapping $\varphi\colon Z \to X$ satisfying $g = f \circ \varphi$. Hence, $Y = g(Z) = f(\varphi(Z))$; it follows that $Y = f(X)$. □

One more important property of 0-soft mappings is given below.

PROPOSITION 10.1.5. *The composition of* 0*-soft mappings of compact spaces is* 0*-soft.*

PROOF. Let $f_1\colon X_1 \to X_2$ and $f_2\colon X_2 \to X_3$ be 0-soft mappings of compact spaces $X_1, X_2, X_3$, and $f = f_2 \circ f_1$. Suppose that $g\colon Z \to X_3$ and $h\colon A \to X_1$ are a continuous mappings such that $f \circ h = g \circ i$, where $A$ is a closed subset of a zero-dimensional compact space $Z$, and $i\colon A \to Z$ is the natural embedding. Put $h_1 = f_1 \circ h$. Since $f_2$ is 0-soft, $h_1$ can be extended to a continuous mapping $\varphi_1\colon Z \to X_2$ such that $f_2 \circ \varphi_1 = g$.

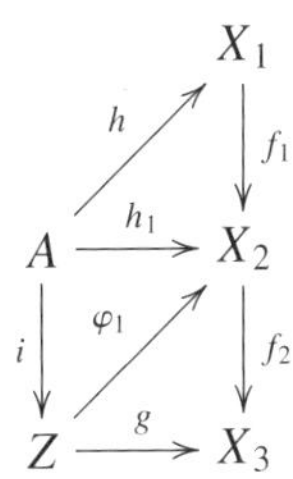


Again, since $f_1$ is 0-soft, we can find a continuous mapping $\varphi\colon Z \to X_1$ extending $h$ such that the diagram below commutes.

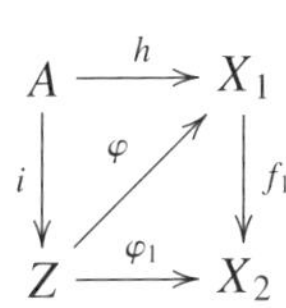


Clearly, the mapping $\varphi$ satisfies the equality $f \circ \varphi = f_2 \circ \varphi_1 = g$, so the mapping $f$ is 0-soft. □

The further study of Dugundji spaces essentially involves inverse spectra. Therefore, we will present the facts about inverse spectra we need and will define some new concepts.

In Section 4.1, we introduced the concept of an inverse spectrum as an object assigned to a given spectral representation $\mathcal{F} = \{f_\alpha : \alpha \in A\}$ of a space $X$, where each $f_\alpha$ was a quotient mapping of $X$ to some space $X_\alpha$. It turns out, however, that one can start with an inverse spectrum $S$ and then define a limit space $X$ of $S$ and a spectral representation $\mathcal{P} = \{p_\alpha : \alpha \in A\}$ of $X$ in such a way that the inverse spectrum assigned to $\mathcal{P}$ will be $S$. Here are the details of the corresponding construction. The reader can find such a construction in a more general setting, for inverse spectra on arbitrary directed index sets, in [**165**, Section 2.5].

In what follows, $\kappa$ stands for a limit ordinal and we use ordinals $\alpha < \kappa$ to enumerate the spaces of a given inverse spectrum. Let $S = \{X_\alpha, p^\beta_\alpha : \alpha < \beta < \kappa\}$ be a family, where each $X_\alpha$ is a space and each $p^\beta_\alpha\colon X_\beta \to X_\alpha$ is a continuous mapping. We say that $S$ is a well-ordered inverse spectrum if the mappings $p^\beta_\alpha$ satisfy the following commutativity condition:

(Com) $p^\beta_\alpha \circ p^\gamma_\beta = p^\gamma_\alpha$, whenever $\alpha < \beta < \gamma < \kappa$.

The mappings $p^\beta_\alpha$ are called *connecting*, while each mapping $p^{\alpha+1}_\alpha$ with $\alpha < \kappa$ is called *bonding*. It is a common practice to denote by $p^\alpha_\alpha$ the identity mapping of $X_\alpha$ onto itself, which permits to use the alternative notation $\{X_\alpha, p^\beta_\alpha : \alpha \leq \beta < \kappa\}$ for the spectrum $S$.

To define the limit space of $S$, consider the product space $\Pi = \prod_{\alpha<\kappa} X_\alpha$ and denote by $\pi_\alpha$ the projection of $\Pi$ to the factor $X_\alpha$, where $\alpha < \kappa$. Let $X$ be the subspace of $\Pi$ defined by

$$X = \{(x_\alpha)_{\alpha<\kappa} \in \Pi : p^\beta_\alpha(x_\beta) = x_\alpha \text{ whenever } \alpha < \beta < \kappa\}.$$

Then $X$ is the *limit space* of the spectrum $S$, and every $x \in X$ is called a *thread* of $S$. Clearly, the restriction $p_\alpha = \pi_\alpha \restriction X$ is a continuous mapping of $X$ to $X_\alpha$, for each $\alpha < \kappa$; $p_\alpha$ is called a *limit projection* of $S$.

Two simple properties of the limit projections of an inverse spectrum are given in the next proposition.

PROPOSITION 10.1.6. *Let $\{p_\alpha : \alpha < \kappa\}$ be the family of the limit projections of a well-ordered inverse spectrum $S = \{X_\alpha, p^\beta_\alpha : \alpha < \beta < \kappa\}$. Then:*

a) *the equalities $p_\alpha = p^\beta_\alpha \circ p_\beta = p^\beta_\alpha \circ \pi_\beta \restriction X$ are valid for all $\alpha, \beta$ with $\alpha < \beta < \kappa$;*
b) *the family $\{p_\alpha : \alpha < \kappa\}$ is a spectral representation of the space $X$.*

PROOF. Condition a) follows immediately from the definition of the limit space $X$ of $S$. To deduce b), we have to verify conditions (S1) and (S2) of Section 4.1. Suppose that $x, y \in X$, $\alpha < \beta < \kappa$, and $p_\beta(x) = p_\beta(y)$. By a), we have $p_\alpha(x) = p^\beta_\alpha(p_\beta(x))$ and $p_\alpha(x) = p^\beta_\alpha(p_\beta(x))$, so that $p_\alpha(x) = p_\alpha(y)$. This implies (S1).

Suppose that $x, y \in X$ and $x \neq y$. Since $X \subset \Pi$, there exists $\alpha < \kappa$ such that $\pi_\alpha(x) \neq \pi_\alpha(y)$. But $p_\alpha(x) = \pi_\alpha(x)$ and $p_\alpha(y) = \pi_\alpha(y)$, so $p_\alpha(x) \neq p_\alpha(y)$, and (S2) follows. Therefore, $\{p_\alpha : \alpha < \kappa\}$ is a spectral representation of the space $X$. □

By a) of the above proposition, one can reconstruct the connecting mappings of the inverse spectrum $S$ using the limit projections of $S$ — it suffices to put $p^\beta_\alpha = p_\alpha \circ (p_\beta)^{-1}$, for all $\alpha, \beta$ satisfying $\alpha < \beta < \kappa$. Hence, the spectrum $S$ is completely defined by the limit space $X$ and by the family $\{p_\alpha : \alpha < \kappa\}$ of limit projections under the additional condition that each $p_\alpha$ maps $X$ onto $X_\alpha$ (the space $X$ can be empty, even if all connecting mappings $p^\beta_\alpha$ of $S$ are surjective; see [**165**, 2.5.A]).

The next result provides more details on the topology of the limit space of an inverse spectrum.

PROPOSITION 10.1.7. *Let $X$ be the limit space of a well-ordered inverse spectrum $S = \{X_\alpha, p^\beta_\alpha : \alpha < \beta < \kappa\}$ with Hausdorff spaces $X_\alpha$. Then $X$ is a closed subspace of the product space $\Pi = \prod_{\alpha<\kappa} X_\alpha$. In addition, the sets of the form $(p_\alpha)^{-1}(W)$, where $\alpha < \kappa$, $p_\alpha \colon X \to X_\alpha$ is the limit projection, and $W$ is open in $X_\alpha$, constitute a base for the topology of $X$.*

PROOF. Take an arbitrary $x = (x_\alpha)_{\alpha<\kappa}$ in $\Pi \setminus X$. By the definition of $X$, there exist $\alpha, \beta$ with $\alpha < \beta < \kappa$ such that $p^\beta_\alpha(x_\beta) \neq x_\alpha$. Since $X_\alpha$ is a Hausdorff space and the connecting mapping $p^\beta_\alpha$ is continuous, we can find open neighbourhoods $U$ and $V$ of $x_\beta$ and $x_\alpha$ in $X_\beta$ and $X_\alpha$, respectively, such that $p^\beta_\alpha(U) \cap V = \varnothing$. It is clear that the set $O = (\pi_\beta)^{-1}(U) \cap (\pi_\alpha)^{-1}(V)$ is open in $\Pi$ and $O \cap X = \varnothing$, where $\pi_\alpha$ and $\pi_\beta$ are the projections of $\Pi$ to the factors $X_\alpha$ and $X_\beta$, respectively. Therefore, the complement $\Pi \setminus X$ is open in $\Pi$.

To prove the second part of the proposition, take any $x \in X$ and any neighbourhood $U$ of $x$ in $X$. Since $X$ is a subspace of $\Pi$, there exists a canonical open set $V$ in $\Pi$ such that $x \in V \cap X \subset U$. Choose ordinals $\alpha_1 < \ldots < \alpha_n < \kappa$ and open sets $V_1, \ldots, V_n$ in $X_{\alpha_1}, \ldots, X_{\alpha_n}$, respectively, such that $V = (\pi_{\alpha_1})^{-1}(V_1) \cap \cdots \cap (\pi_{\alpha_n})^{-1}(V_n)$. Since each limit projection $p_\alpha \colon X \to X_\alpha$ is the restriction of $\pi_\alpha$ to $X$, the open neighbourhood $O = (p_{\alpha_1})^{-1}(V_1) \cap \cdots \cap (\pi_{\alpha_n})^{-1}(V_n)$ of $x$ in $X$ is contained in $U$. Clearly, the set $W = (p^{\alpha_n}_{\alpha_1})^{-1}(V_1) \cap \cdots \cap (p^{\alpha_n}_{\alpha_{n-1}})^{-1}(V_{n-1}) \cap V_n$ is open in $X_{\alpha_n}$ and it follows from the definition of $O$ and from a) of Proposition 10.1.6 that $x \in O = (p_{\alpha_n})^{-1}(W) \subset U$. □

THEOREM 10.1.8. *Let $S = \{X_\alpha, p^\beta_\alpha : \alpha < \beta < \kappa\}$ be an inverse spectrum of compact spaces $X_\alpha$ and continuous connecting mappings $p^\beta_\alpha$. Then the limit space $X$ of $S$ is compact. In addition, if each $p^\beta_\alpha$ is surjective, then so are the limit projections $p_\alpha \colon X \to X_\alpha$.*

PROOF. The product space $\Pi = \prod_{\alpha<\kappa} X_\alpha$ is compact and, by Proposition 10.1.7, $X$ is a closed subspace of $\Pi$. Hence $X$ is compact.

Suppose that each connecting mapping $p^\beta_\alpha$ is surjective. For every $\beta < \kappa$, let

$$\Pi_\beta = \{x \in \Pi : p^\beta_\alpha(\pi_\beta(x)) = \pi_\alpha(x) \text{ for each } \alpha < \beta\},$$

where $\pi_\alpha \colon \Pi \to X_\alpha$ is the natural projection of $\Pi$ onto $X_\alpha$. Clearly, $\Pi_\beta$ is compact, as a closed subset of $\Pi$ and, in addition, $\pi_\beta(\Pi_\beta) = X_\beta$, for each $\beta < \kappa$. Indeed, let $y \in X_\beta$, and take $x = (x_\alpha)_{\alpha<\kappa} \in \Pi$ such that $x_\beta = y$, $x_\alpha = p^\beta_\alpha(y)$, for each $\alpha < \beta$, where the coordinates $x_\gamma$ with $\beta < \gamma < \kappa$ are chosen arbitrary. Then $x \in \Pi_\beta$ and $\pi_\beta(x) = y$.

It follows from the definition of the limit space $X$ of the spectrum $S$ that $X = \bigcap_{\beta<\kappa} \Pi_\beta$. Let $\alpha < \kappa$ and $y \in X_\alpha$ be arbitrary. Then $y \in \pi_\alpha(\Pi_\beta)$ or, equivalently, $(\pi_\alpha)^{-1}(y) \cap \Pi_\beta \neq \varnothing$, for each $\beta$ where $\alpha \leq \beta < \kappa$. Since $\{\Pi_\beta : \alpha \leq \beta < \kappa\}$ is a decreasing sequence of compact subsets of $\Pi$, and the set $(\pi_\alpha)^{-1}(y)$ is closed in $\Pi$, we conclude that the intersection $(\pi_\alpha)^{-1}(y) \cap \bigcap_{\alpha\leq\beta<\kappa} \Pi_\beta = (\pi_\alpha)^{-1}(y) \cap X$ is not empty. Hence, $y \in \pi_\alpha(X) = p^\alpha(X)$ and, therefore, $p_\alpha(X) = X_\alpha$. □

Let $S = \{X_\alpha, p^\beta_\alpha : \alpha < \beta < \kappa\}$ be a well-ordered inverse spectrum with continuous projections $p^\beta_\alpha$, and $X$ be the limit space of $S$. For every $\alpha < \kappa$, denote by $p_\alpha$ the limit projection of $X$ to $X_\alpha$. If $Y \subset X$ is a subspace of $X$, we put $Y_\alpha = p_\alpha(Y)$ for each $\alpha < \kappa$ and define $q^\beta_\alpha$ as the restriction of $p^\beta_\alpha$ to $Y_\beta$, where $\alpha < \beta < \kappa$. Clearly, $S_Y = \{Y_\alpha, q^\beta_\alpha : \alpha < \beta < \kappa\}$ is a well-ordered inverse spectrum; this spectrum is called the *subspectrum* of $S$ generated by $Y$.

PROPOSITION 10.1.9. *Suppose that $X$ is a limit space of a well-ordered inverse spectrum $S = \{X_\alpha, p^\beta_\alpha : \alpha < \beta < \kappa\}$, and that $Y$ is a closed subspace of $X$. Then the limit space of the inverse spectrum $S_Y$ is naturally homeomorphic to $Y$.*

PROOF. It follows from the definition of the subspectrum $S_Y = \{Y_\alpha, q^\beta_\alpha : \alpha < \beta < \kappa\}$ of $S$ that $Y$ can be identified with a subspace of the limit space $Y^*$ of $S_Y$ which is, in its turn, a subspace of $X$. Therefore, it suffices to show that $Y^* \subset Y$. Suppose to the contrary that there exists a point $x \in Y^* \setminus Y$. By Proposition 10.1.7, $Y$ is closed in $X$, so we can find an ordinal $\alpha < \kappa$ and an open set $U \subset X_\alpha$ such that $x \in (p_\alpha)^{-1}(U) \subset X \setminus Y$. Hence $(p_\alpha)^{-1}(U) \cap Y = \varnothing$ and $U \cap p_\alpha(Y) = \varnothing$, that is, $U$ does not intersect the set $Y_\alpha$, a contradiction with $p_\alpha(x) \in U \cap Y_\alpha$. □

Given an inverse spectrum $S = \{X_\alpha,\ p^\beta_\alpha : \alpha < \beta < \kappa\}$ and an ordinal $\gamma < \kappa$, let $S_\gamma = \{X_\alpha,\ p^\beta_\alpha : \alpha < \beta < \gamma\}$ be the initial part of $S$ of the length $\gamma$. The spectrum $S$ is called *continuous* if the following conditions are satisfied:

(C1) $p^{\alpha+1}_\alpha(X_{\alpha+1}) = X_\alpha$, for each $\alpha < \kappa$;

(C2) if $\beta > 0$ is a limit ordinal, then the diagonal product of the family $\{p^\beta_\alpha : \alpha < \beta\}$ is a homeomorphism of $X_\beta$ onto the limit space of the spectrum $S_\beta$.

The simplest examples of continuous spectra are constructed on the basis of Tychonoff products of spaces and of compact subspaces of products:

EXAMPLE 10.1.10. Let $\kappa$ be an infinite cardinal and $\Pi = \prod_{\alpha<\kappa} Y_\alpha$ be a product space. For every $\alpha < \kappa$, put $\Pi_\alpha = \prod_{\nu<\alpha} Y_\nu$ and denote by $\pi_\alpha$ the projection of $\Pi$ onto $\Pi_\alpha$. If $\alpha < \beta < \kappa$, let also $\pi^\beta_\alpha \colon \Pi_\beta \to \Pi_\alpha$ be the corresponding projection. Clearly, the spectrum $S = \{\Pi_\alpha,\ \pi^\beta_\alpha : \alpha < \beta < \kappa\}$ is continuous, and the limit space of $S$ is naturally homeomorphic to $\Pi$, since the equality $\pi_\alpha = \pi^\beta_\alpha \circ \pi_\beta$ holds for all $\alpha, \beta$, where $\alpha < \beta < \kappa$.

Further, take an arbitrary compact subspace $X$ of $\Pi$ and consider the subspectrum $S_X$ of $S$ generated by $X$. Since $X$ is closed in $\Pi$, Proposition 10.1.9 implies that the limit space of $S_X$ is naturally homeomorphic to $X$. Similarly, if $\beta < \kappa$ is a limit ordinal, then the limit space of the spectrum $S_{X,\beta}$, the initial part of $S_X$ of the length $\beta$, is naturally homeomorphic to $X_\beta = \pi_\beta(X)$. We have thus shown that the spectrum $S_X$ is also continuous. □

The next example gives a general way of constructing continuous spectra of compact spaces.

EXAMPLE 10.1.11. Let $Y$ be a compact space, $\kappa$ an infinite cardinal, and $\{f_\alpha : \alpha < \kappa\}$ a family of continuous mappings of $Y$ to some spaces. For every $\alpha < \kappa$, denote by $h_\alpha$ the diagonal product of the family $\{f_\nu : \nu < \alpha\}$ and put $X_\alpha = h_\alpha(Y)$. As usual, we assume that $h_0$ is a mapping of $Y$ to a one-point space $X_0$. It is clear that $h_\beta \prec h_\alpha$ if $\alpha < \beta < \kappa$, so there exists a continuous mapping $h^\beta_\alpha \colon X_\beta \to X_\alpha$ satisfying $h_\alpha = h^\beta_\alpha \circ h_\beta$. Therefore, we have constructed a well-ordered spectrum $S = \{X_\alpha,\ h^\beta_\alpha : \alpha < \beta < \kappa\}$ with compact spaces $X_\alpha$ and continuous onto mappings $h^\beta_\alpha$. We claim that the spectrum $S$ is continuous and the limit space of $S$ is naturally homeomorphic to the image $h(Y)$, where $h$ is the diagonal product of the family $\{f_\alpha : \alpha < \kappa\}$.

Indeed, by the definition of $h$, the image $X = h(Y)$ is a compact subspace of the product space $\Pi = \prod_{\alpha<\kappa} Y_\alpha$, where $Y_\alpha = f_\alpha(Y)$. For every $\alpha < \kappa$, let $\pi_\alpha \colon \Pi \to \Pi_\alpha$ be the natural projection, where $\Pi_\alpha = \prod_{\nu<\alpha} Y_\nu$. Obviously, $h_\alpha = \pi_\alpha \circ h$, so $X_\alpha = \pi_\alpha(X)$ is a subspace of $\Pi_\alpha$, and the restriction to $X_\beta$ of the projection $\pi^\beta_\alpha \colon \Pi_\beta \to \Pi_\alpha$ coincides with $h^\beta_\alpha$, where $\alpha < \beta < \kappa$. Therefore, the inverse spectrum $S = \{X_\alpha,\ h^\beta_\alpha : \alpha < \beta < \kappa\}$ is continuous and the limit space of $S$ is naturally homeomorphic to $X$, as is shown in Example 10.1.10. □

PROPOSITION 10.1.12. *If a spectrum $S = \{X_\alpha,\ p^\beta_\alpha : \alpha < \beta < \kappa\}$ is continuous, then the connecting mappings $p^\beta_\alpha$ and the limit projections $p_\alpha \colon X \to X_\alpha$ are surjective, where $X$ is the limit space of $S$.*

PROOF. Since $p_\alpha = p^\beta_\alpha \circ p_\beta$, for all $\alpha, \beta$ where $\alpha < \beta < \kappa$, it suffices to verify that each $p_\alpha$ is an onto mapping. Let $\alpha < \kappa$ and $x_\alpha \in X_\alpha$ be arbitrary. We will define by recursion a transfinite sequence $\{x_\beta : \alpha \le \beta < \kappa\}$ satisfying the following two conditions:

(i) $x_\beta \in X_\beta$ if $\alpha \le \beta < \kappa$;

(ii) $p^\gamma_\beta(x_\gamma) = x_\beta$, whenever $\alpha \le \beta < \gamma < \kappa$.

Suppose that we have defined a family $\{x_\beta : \alpha \le \beta < \delta\}$ satisfying (i) and (ii), where $\alpha < \delta < \kappa$. If $\delta$ is limit, we use (C2) to choose a point $x_\delta$ in $X_\delta$ such that $p^\delta_\beta(x_\delta) = x_\beta$ for each $\beta$ with $\alpha \le \beta < \delta$. Clearly, the family $\{x_\beta : \alpha \le \beta \le \delta\}$ satisfies (i) and (ii). If $\delta = \gamma + 1$, condition (C1) permits to choose a point $x_{\gamma+1} \in X_{\gamma+1}$ such that $p^{\gamma+1}_\gamma(x_{\gamma+1}) = x_\gamma$. Again, $\{x_\beta : \alpha \le \beta \le \delta\}$ satisfies (i) and (ii).

Once the transfinite sequence $\{x_\beta : \alpha \le \beta < \kappa\}$ satisfying (i) and (ii) is defined, it follows from the definition of the limit space $X$ that there exists $x \in X$ such that $p_\beta(x) = x_\beta$ for each $\beta$ with $\alpha \le \beta < \kappa$. In particular, $p_\alpha(x) = x_\alpha$. Hence, the projection $p_\alpha$ is surjective, for each $\alpha < \kappa$. □

In the next proposition we establish an important relation between 0-soft mappings and continuous spectra of compact spaces. This relation plays a crucial role in the arguments to follow.

PROPOSITION 10.1.13. *Let $\gamma$ be an ordinal and $S = \{X_\alpha, \pi^\beta_\alpha : \alpha < \beta < \gamma\}$ be a continuous inverse spectrum of compact spaces $X_\alpha$ and of* 0*-soft bonding mappings $\pi^{\alpha+1}_\alpha$, $\alpha + 1 < \gamma$. Then the limit projection $\pi_\alpha\colon X \to X_\alpha$ is* 0*-soft, for each $\alpha < \gamma$, where $X$ is the limit space of $S$.*

PROOF. We apply induction on the length $\gamma$ of $S$. If $\gamma$ is a successor, that is, $\gamma = \beta + 1$, then $X = X_\beta$. First, suppose that $\beta$ is a limit ordinal. By the inductive assumption, the limit projections $\pi^\beta_\alpha$ of the spectrum $S_\beta = \{X_\alpha, \pi^\delta_\alpha : \alpha < \delta < \beta\}$ are 0-soft (and coincide with connecting mappings of $S$). If $\beta$ is a successor, that is, $\beta = \alpha + 1$, then the limit projection $\pi_\nu\colon X \to X_\nu$ coincides with the mapping $\pi^\beta_\nu = \pi^\alpha_\nu \circ \pi^{\alpha+1}_\alpha$, for each $\nu \le \alpha$, where both mappings $\pi^{\alpha+1}_\alpha$ and $\pi^\alpha_\nu$ are 0-soft (here we use the inductive hypothesis again). Hence, $\pi^\beta_\nu$ is 0-soft, by Proposition 10.1.5.

Finally, suppose that $\gamma$ is a limit ordinal. By the inductive hypothesis, the mappings $\pi^\beta_\alpha$ are 0-soft for all $\alpha, \beta$ satisfying $\alpha < \beta < \gamma$. Fix an ordinal $\alpha < \gamma$, and suppose that $g\colon Z \to X_\alpha$ and $h\colon A \to X$ are continuous mappings such that $\pi_\alpha \circ h = g{\restriction}A$, where $A$ is a closed subset of a zero-dimensional compact space $Z$. Let $i$ be the natural embedding of $A$ to $Z$.

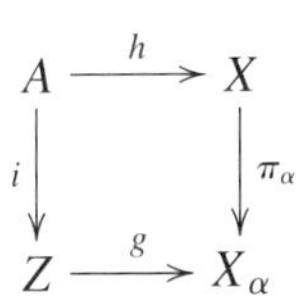


For every $\beta$ with $\alpha \le \beta < \gamma$, put $h_\beta = \pi_\beta \circ h$. Since the spectrum $S$ is continuous, one can define, by a transfinite recursion, a spectral mapping $\Phi = \{\varphi_\beta : \alpha \le \beta < \gamma\}$ of $Z$ to the spectrum $S' = \{X_\beta, \pi^\delta_\beta : \alpha \le \beta < \delta < \gamma\}$ such that $\varphi_\alpha = g$ and the following diagram commutes, for all $\beta, \delta$ where $\alpha < \beta < \delta < \gamma$.

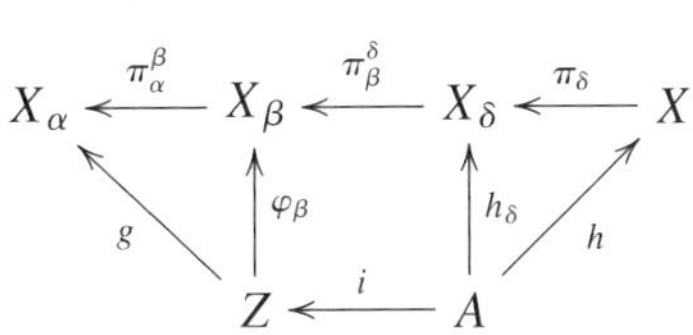


By Theorem 4.1.6, there exists a continuous mapping $\varphi\colon Z \to X$ satisfying $\varphi_\beta = \pi_\beta \circ \varphi$ whenever $\alpha \le \beta < \gamma$. The commutativity of the above diagrams implies that the next diagram commutes as well.

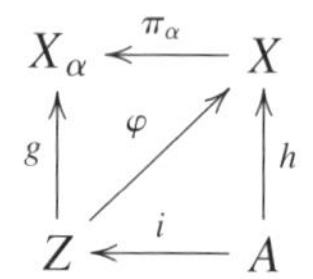

Therefore, $\varphi$ is the required continuous extension of $h$ over $Z$, and $\pi_\alpha$ is 0-soft. □

Our next step is to introduce a special class of mappings that will frequently appear in the proofs of the main results of this section.

A continuous mapping $f\colon X \to Y$ has *metrizable kernel* if there exists a metrizable compact space $C$ and a topological embedding $i\colon X \to Y \times C$ such that $f = p \circ i$, where $p\colon Y \times C \to Y$ is the projection.

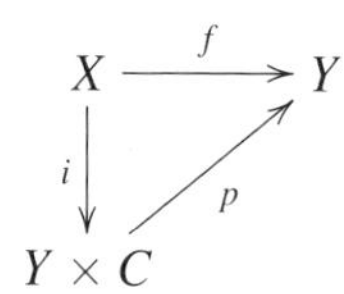

A standard way of constructing mappings with metrizable kernel is presented in the example below.

EXAMPLE 10.1.14. Let $f_0\colon X_0 \to Y$ and $r_0\colon X_0 \to C$ be continuous mappings, where $C$ is a second-countable compact space. Denote by $\varphi$ the diagonal product of $f_0$ and $r_0$, $\varphi\colon X_0 \to Y \times C$, and put $X = \varphi(X_0)$. Let $i$ be the identity embedding of $X$ to $Y \times C$. Clearly, there are continuous mappings $f\colon X \to Y$ and $r\colon X \to C$ that make the following diagram to commute, where $p_Y\colon Y \times C \to Y$ and $p_C\colon Y \times C \to C$ are projections.

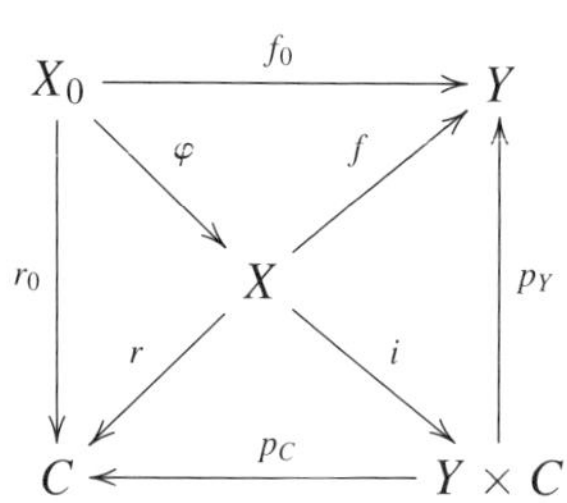

Therefore, the mapping $f\colon X \to Y$ has metrizable kernel. □

It turns out that open continuous mappings with metrizable kernel are 0-soft:

THEOREM 10.1.15. *Let $f\colon X \to Y$ be an open continuous onto mapping of compact spaces. If $f$ has metrizable kernel, then $f$ is 0-soft.*

PROOF. Suppose that $A$ is a closed subset of a zero-dimensional compact space $Z$ and that $h\colon A \to X$ and $g\colon Z \to Y$ are continuous mappings such that $f \circ h = g \circ j$, where $j\colon A \to Z$ is the identity embedding. Since $f$ has metrizable kernel, we can find a second-countable compact space $C$ and a topological embedding $i\colon X \to Y \times C$ that make the diagram below to commute, where $p\colon Y \times C \to Y$ is the projection.

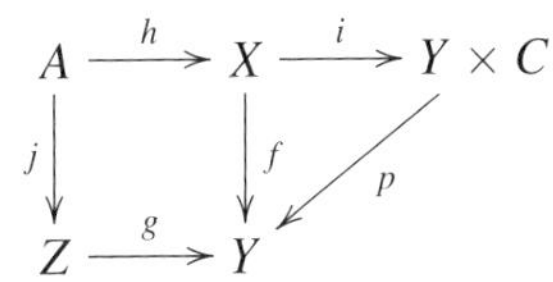


We now define a mapping $\varphi\colon Z \to Exp(C)$, where $Exp(C)$ is the family of closed subsets of $C$, as follows. For every $z \in A$, let $\varphi(z) = \pi i h(z)$, where $\pi\colon Y \times C \to C$ is the projection. If $z \in Z \setminus A$, put $\varphi(z) = \pi i f^{-1}(g(z))$. We claim that the mapping $\varphi$ is lower semicontinuous. Indeed, take an arbitrary open set $U \subset C$. It follows from the definition of $\varphi$ that

$$\begin{aligned}\varphi^{-1}(U) &= \{z \in Z : U \cap \varphi(z) \neq \varnothing\}\\ &= \{z \in Z \setminus A : f^{-1}(g(z)) \cap i^{-1}\pi^{-1}(U) \neq \varnothing\} \cup \{z \in A : h(z) \in i^{-1}\pi^{-1}(U)\}.\end{aligned}$$

The set $V = i^{-1}\pi^{-1}(U)$ is open in $X$ and $\varphi^{-1}(U) = h^{-1}(V) \cup (g^{-1} f(V) \setminus A)$. The set $g^{-1} f(V) \setminus A$ is open in $Z$ since $f$ is an open mapping and $A$ is closed in $Z$. Choose an open set $W$ in $Z$ such that $W \cap A = h^{-1}(V)$. Then the inclusion $h^{-1}(V) \subset g^{-1} f(V)$ implies that the set

$$\varphi^{-1}(U) = (W \cap \pi^{-1} f(V)) \cup (g^{-1} f(V) \setminus A)$$

is open in $X$. This proves the claim about $\varphi$.

By Theorem 4.1.1, there exists a continuous selection $s\colon Z \to C$ for the mapping $\varphi$. Let $\lambda_1\colon Z \to Y \times C$ be the diagonal product of the mappings $g$ and $s$. It follows from the equality $f \circ h = g{\restriction}A$ that $s(z) \in \pi i f^{-1}(g(z))$ for each $z \in Z$, and we claim that $\{g(z)\} \times \pi i f^{-1}(g(z)) \subset i(X)$. Indeed, put $y = g(z)$ and take an arbitrary point $c \in \pi i f^{-1}(y)$. There exists $x \in X$ such that $f(x) = y$ and $c = \pi i(x)$. Notice that $i$ is the diagonal product of the mappings $p \circ i = f$ and $\pi \circ i$, so we have:

$$(g(z), c) = (y, c) = (f(x), \pi i(x)) = (pi(x), \pi i(x)) = i(x) \in i(X).$$

Hence, $\lambda_1(Z) \subset i(X)$ and the mapping $\lambda = i^{-1} \circ \lambda_1\colon Z \to X$ is correctly defined. The definition of $\lambda$ implies that the diagram below commutes.

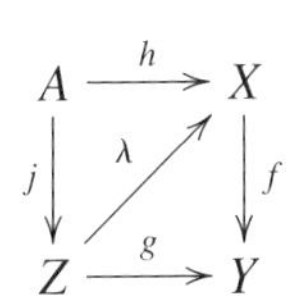


Clearly, $\lambda$ is continuous, so the mapping $f$ is 0-soft. □

The following theorem on a special spectral representation of certain compact spaces will be applied to topological groups in the next section.

THEOREM 10.1.16. [**R. Haydon**] *Let $S = \{X_\alpha, \pi^\beta_\alpha : \alpha < \beta < \kappa\}$ be a continuous inverse spectrum, where every space $X_\alpha$ is compact and every bonding mapping $\pi^{\alpha+1}_\alpha$ is continuous, open, and has metrizable kernel. If $X_0$ is Dugundji, then so is the limit space $X$ of the spectrum $S$.*

PROOF. The limit space of an inverse spectrum with compact spaces is compact, so all we need to verify is that $X$ is Dugundji. By Theorem 10.1.15, all bonding mapping $\pi_\alpha^{\alpha+1}$ of the spectrum $S$ are 0-soft, so Proposition 10.1.13 implies that the limit projections $\pi_\alpha \colon X \to X_\alpha$ of $S$ are also 0-soft. In particular, the projection $\pi_0 \colon X \to X_0$ is 0-soft. Let $p_0$ be a mapping of $X_0$ to a one-point space and $p = p_0 \circ \pi_0$. Since $X_0$ is Dugundji, the mapping $p_0$ is 0-soft. Therefore, the constant mapping $p$ is 0-soft by Proposition 10.1.5; it follows that $X$ is Dugundji. □

An inverse spectrum, such as in Theorem 10.1.16 in which $X_0$ is a compact metrizable space, is called a *Haydon spectrum.*

Now we present some basic facts about nearly open mappings that will be applied to the study of continuous actions of topological groups on compact spaces.

We recall that a continuous mapping $f \colon X \to Y$ is called *nearly open* if, for every open set $U \subset X$, there exists an open set $V \subset Y$ such that $f(U)$ is a dense subset of $V$ (see Section 4.3). Evidently, $f$ is nearly open if and only if $f(U) \subset \mathrm{Int}_Y\, cl_Y(f(U))$, for each open set $U \subset X$. Every open mapping is nearly open, but not vice versa. Usually, nearly open mappings appear as restrictions of open mappings to dense subspaces of the domains:

PROPOSITION 10.1.17. *Suppose that $f \colon X \to Y$ is a continuous open mapping and $S$ is a dense subspace of $X$. Then the mapping $f \restriction S \colon S \to Y$ is nearly open.*

PROOF. Let $g = f \restriction S$. If $O$ is an open subset of $S$, take an open set $U \subset X$ such that $O = U \cap S$. Clearly, $O$ is dense in $U$ and $g(O)$ is dense in the open set $V = f(U) \subset Y$. □

COROLLARY 10.1.18. *Let $S$ be a dense subspace of a product space $X \times Y$. Then the restriction to $S$ of the projection $p \colon X \times Y \to X$ is a nearly open mapping of $S$ to $Y$.*

It is well known that a continuous mapping $f \colon X \to Y$ is open if and only if the equality $\overline{f^{-1}(B)} = f^{-1}(\overline{B})$ holds for each set $B \subset Y$ (see [**165**, 1.4.C]). A similar characterization of nearly open mappings is given below.

LEMMA 10.1.19. *The following conditions are equivalent for a continuous mapping $f \colon X \to Y$:*

a) *$f$ is nearly open;*
b) *the equality $\overline{f^{-1}(O)} = f^{-1}(\overline{O})$ holds for each open set $O \subset Y$.*

PROOF. a) $\Rightarrow$ b). Let $O$ be an open subset of $Y$. Since $\overline{f^{-1}(O)} \subset f^{-1}(\overline{O})$, it suffices to verify the inverse inclusion. Suppose that $x \in X$ satisfies $f(x) \in \overline{O}$. If $x \notin \overline{f^{-1}(O)}$, then $U = X \setminus \overline{f^{-1}(O)}$ is an open neighbourhood of $x$ in $X$, so $f(U)$ is a dense subset of an open set $V \subset Y$ and, hence, $\overline{f(U)} = \overline{V}$. It follows from $f(U) \cap O = \varnothing$ that $\overline{V} \cap O = \overline{f(U)} \cap O = \varnothing$, a contradiction with $f(x) \in V \cap \overline{O}$.

b) $\Rightarrow$ a). Let $U$ be an open subset of $X$. It suffices to show that $f(U)$ is contained in the interior of the closed set $F = \overline{f(U)}$. Since $O = Y \setminus F$ is open in $Y$, we have, by the assumption, $f^{-1}(\overline{O}) = \overline{f^{-1}(O)}$. It follows from the definition of $O$ that the open sets $U$ and $f^{-1}(O)$ are disjoint, so $U \cap f^{-1}(\overline{O}) = U \cap \overline{f^{-1}(O)} = \varnothing$. Hence, $f(U) \cap \overline{O} = \varnothing$ and, therefore, $f(U)$ is contained in the interior of $F$. □

Under some additional conditions, nearly open mappings turn out to be open. Let us say that a mapping $f \colon X \to Y$ is *locally closed at a point* $x \in X$, if for every open set $U$ in $X$

containing $x$, there exists a neighbourhood $N$ of $x$ in $X$ such that $N \subset U$ and $f(N)$ is closed in $Y$. If $f$ is locally closed at every point of $X$, we say that $f$ is *locally closed*. If the space $X$ is regular and $f$ is continuous, then the set $N$ in the above definition can be chosen to be closed. Indeed, take an open neighbourhood $V$ of $x$ with $\overline{V} \subset U$ and find a neighbourhood $N$ of $x$ such that $N \subset V$ and $f(N)$ is closed in $Y$. Then $f(N) \subset f(\overline{N}) \subset \overline{f(N)} = f(N)$, so the image of the closed neighbourhood $\overline{N}$ of $x$ is closed in $Y$. In addition, $\overline{N} \subset \overline{V} \subset U$.

PROPOSITION 10.1.20. *Every nearly open locally closed mapping $f\colon X \to Y$ is open.*

PROOF. Let $U$ be an open subset of $X$. For every point $x \in U$, choose a neighbourhood $N_x$ of $x$ such that $N_x \subset U$ and $f(N_x)$ is closed in $Y$. Then the interior $U_x$ of $N_x$ contains $x$. Since $f$ is nearly open, there exists an open set $V_x \subset Y$ containing $f(U_x)$ as a dense subset. Since the image $f(N_x)$ is closed in $Y$ and $U_x \subset N_x \subset U$, we have:

$$f(x) \in V_x \subset \overline{f(U_x)} \subset \overline{f(N_x)} = f(N_x) \subset f(U).$$

This implies that the set $f(U) = \bigcup_{x \in U} V_x$ is open in $Y$. □

COROLLARY 10.1.21. *Every nearly open mapping of a locally pseudocompact Tychonoff space $X$ to a regular space $Y$ of countable pseudocharacter is open.*

PROOF. By Proposition 10.1.20, it suffices to verify that every continuous mapping $f\colon X \to Y$ is locally closed. Let $U$ be a non-empty open set in $X$ and $x \in U$ an arbitrary point. There exists an open neighbourhood $O$ of $x$ in $X$ such that $\overline{O}$ is a pseudocompact subset of $U$. We claim that $Z = f(\overline{O})$ is closed in $Y$. If not, choose a point $y \in \overline{Z} \setminus Z$ and take a countable family $\{V_n : n \in \omega\}$ of open neighbourhoods of $y$ in $Y$ such that $\overline{V_{n+1}} \subset V_n$ for each $n \in \omega$ and $\{y\} = \bigcap_{n=0}^{\infty} V_n$. Then $\{O \cap f^{-1}(V_n) : n \in \omega\}$ is an infinite locally finite family of non-empty open sets in $\overline{O}$, a contradiction with pseudocompactness of $\overline{O}$. Therefore, $Z = f(\overline{O})$ is closed in $Y$ and the mapping $f$ is locally closed. □

COROLLARY 10.1.22. *Let $X$ and $Y$ be Hausdorff spaces. If $X$ is locally compact, then every nearly open mapping of $X$ to $Y$ is open.*

PROOF. Every continuous mapping of a locally compact space to a Hausdorff space is locally closed. It remains to apply Proposition 10.1.20. □

COROLLARY 10.1.23. *Every closed nearly open mapping is open. In particular, nearly open mappings of compact Hausdorff spaces are open.*

## Exercises

10.1.a. Give an example of a continuous nearly open mapping that is not open.

10.1.b. Can a continuous homomorphism of a metrizable topological group $G$ onto a metrizable group $H$ be nearly open but not open?

10.1.c. Show that the character cannot increase under a continuous nearly open mapping, in the class of regular spaces.

### Problems

10.1.A. Give an example of an open perfect mapping $f$ of a zero-dimensional compact Hausdorff space $X$ onto the Cantor set $Y$ such that every fiber $f^{-1}(y)$ is metrizable but $f$ is not 0-soft.

10.1.B. Give an example of a dyadic compactum that is not a Dugundji compactum.

10.1.C. (E. V. Schepin [**420**]) Prove that a compact Hausdorff space $X$ is homeomorphic to $D^\tau$, for some $\tau \geq \omega$, if and only if the character of $X$ at every point is exactly $\tau$, $X$ is zero-dimensional and Dugundji.

## 10.2. Continuous action of topological groups on spaces

The fundamental concept of action of a group on a topological space appears occasionally in Chapters 2 and 3. Here we define this concept in full generality and establish some general facts which are applied in the next section to the study of the structure of compact Hausdorff spaces admitting a continuous transitive action of an $\omega$-narrow topological group.

Let $X$ be a non-empty set, and $G$ be an abstract group with identity $e$. Suppose that a mapping $\theta\colon G \times X \to X$ satisfies the following conditions for all $g, h \in G$ and $x \in X$:

(A1) $\theta(e, x) = x$;

(A2) $\theta(h, \theta(g, x)) = \theta(hg, x)$.

Then we say that $G$ *acts* on $X$ and that $\theta$ is a (left) *action* of $G$ on $X$. It is a usual practice to write $gx$ (or $g * x$) in place of $\theta(g, x)$. Using the short form of notation, we can rewrite (A1) and (A2) as $ex = x$ and $h(gx) = (hg)x$, respectively.

Let $\theta$ be an action of a group $G$ on a set $X$. Every element $g \in G$ determines a *translation* $\theta_g$ of $X$ defined by $\theta_g(x) = \theta(g, x)$ for each $x \in X$ or, equivalently, $\theta_g(x) = gx$. It follows from (A1) and (A2) that $\theta_g$ is a bijection of $X$, for each $g \in G$. Indeed, $\theta_e$ is the identity mapping of $X$ onto itself, and $\theta_{hg} = \theta_h \circ \theta_g$ for all $g, h \in G$. Hence, $(\theta_g)^{-1} = \theta_{g^{-1}}$, that is, the inverse mapping of $\theta_g$ is $\theta_{g^{-1}}$, thus implying that $\theta_g$ is a bijection of $X$.

Suppose that $H \subset G$ and $A \subset X$ are non-empty sets, and that $x \in X$. We put

$$HA = \{ha : h \in H,\ a \in A\} \text{ and } Hx = \{hx : h \in H\}.$$

We say that $A$ is $H$-*invariant* if $HA \subset A$. For brevity, $G$-invariant subsets of $X$ are called *invariant*. The *orbit* of a point $x \in X$ is the set $Gx$. It follows from (A1) and (A2) that $Gx$ is an invariant subset of $X$ containing $x$. Clearly, every invariant subset $A$ of $X$ is the union of the orbits of elements of $A$.

Given $x, y \in X$, we write $x \sim y$ if $y$ belongs to the orbit of $x$ under the action of the group $G$. Let us show that the binary relation $\sim$ on $X$ determines the partition of $X$ into the orbits.

PROPOSITION 10.2.1. *The relation $\sim$ is an equivalence relation on $X$. The equivalence class $[x]$ of an arbitrary element $x \in X$ with respect to $\sim$ is the orbit $Gx$.*

PROOF. It follows from $ex = x$ that $x \sim x$ for each $x \in X$. Suppose that $x \sim y$, that is, $y \in Gx$. Then there is an element $g \in G$ with $y = gx$, so (A2) and (A1) imply that $g^{-1}y = g^{-1}gx = ex = x$. Hence $y \sim x$, that is, the relation $\sim$ is symmetric. To verify the transitivity of $\sim$, take any points $x, y, z \in X$ such that $x \sim y$ and $y \sim z$. Choose $g, h \in G$ satisfying $y = gx$ and $z = hy$. Then (A2) implies that $(hg)x = hy = z$ and, therefore, $x \sim z$. This proves the first part of the proposition. The second part follows immediately from definition of the relation $\sim$ on $X$. □

In view of Proposition 10.2.1, we can define the quotient set

$$X/G = \{[x] : x \in X\}$$

called the *set of orbits*. The natural quotient mapping $\pi\colon X \to X/G$ defined by $\pi(x) = [x]$ for each $x \in X$, is called the *orbital projection*. It is easy to verify that a set $A \subset X$ is invariant if and only if $A = \pi^{-1}\pi(A)$.

For every $x \in X$, the set

$$G_x = \{g \in G : gx = x\}$$

is called the *stabilizer* of $x$. Two important properties of stabilizers are given in the next proposition.

PROPOSITION 10.2.2. *The stabilizer $G_x$ of any point $x \in X$ is a subgroup of $G$. In addition, $G_{gx} = gG_xg^{-1}$ for each $g \in G$.*

PROOF. Take an arbitrary point $x \in X$. Clearly, $ex = x$, so that $e \in G_x$. Suppose that $g, h \in G_x$. Then $gx = x = hx$ and $h^{-1}x = h^{-1}hx = ex = x$. Consequently, $(h^{-1}g)x = h^{-1}(gx) = h^{-1}x = x$ and hence, $h^{-1}g \in G_x$. Thus, $G_x$ is a subgroup of $G$.

To prove that $G_{gx} = gG_xg^{-1}$, take any $h \in G_x$. Then $ghg^{-1}(gx) = ghx = gx$; hence, $ghg^{-1} \in G_{gx}$ and $gG_xg^{-1} \subset G_{gx}$. Substituting in the above inclusion $g^{-1}$ in place of $g$ and $gx$ in place of $x$, we obtain $g^{-1}G_{gx}g \subset G_{g^{-1}gx} = G_x$. Consequently, $G_{gx} \subset gG_xg^{-1}$. □

COROLLARY 10.2.3. *The stabilizer $G_x$ is an invariant subgroup of $G$ if and only if $G_x = G_y$, for each $y \in Gx$.*

An action $\theta\colon G \times X \to X$ is called *free* if $G_x = \{e\}$, for each $x \in X$, or equivalently, if the equality $gx = x$ holds only for $g = e$. In other words, the translation $\theta_g$ of $X$ has no fixed points if $g \neq e$.

The action of $G$ on $X$ is *effective* if $\bigcap_{x\in X} G_x = \{e\}$. The latter is equivalent to saying that the translation $\theta_g$ of $X$, with $g \in G$, coincides with the identity mapping of $X$ only if $g = e$. Obviously, every free action is effective.

The action of $G$ is called *transitive* if $Gx = X$ for each (equivalently, for some) $x \in X$. In this case, the only orbit of $X$ coincides with $X$. If the action of $G$ is transitive, then for any pair of points $x, y \in X$, there exists $g \in G$ such that $gx = y$. The converse is true as well.

Three examples below clarify the concept of action of a group.

EXAMPLE 10.2.4. Let $G$ be a group with multiplication $\theta\colon G \times G \to G$, $\theta(x, y) = xy$. Then $\theta$ is an action of $G$ on the set $X = G$ by left translations, that is, $\theta_g = \lambda_g$, for each $g \in G$. This action is free, since if $gx = x$ for $g, x \in G$, then $g = e$. In addition, $\theta$ is transitive.

Given a subgroup $H$ of $G$, one defines an action $\theta_H$ of $H$ on $G$ as the restriction of $\theta$ to the product $H \times G$. Again, $H$ acts on $G$ by left translations. The orbits of $G$ under the action $\theta_H$ are the right cosets $Hx$, with $x \in G$. Therefore, the orbital projection coincides with the quotient mapping of $G$ onto the set $G/H$ of right cosets. Clearly, the action $\theta_H$ is free, and it is transitive iff $H = G$. □

EXAMPLE 10.2.5. Let $G$ be a group, and $H$ a subgroup of $G$. We define an action $\varphi$ of $G$ on the quotient set $G/H$ of left cosets by the rule $\varphi(g, xH) = gxH$, for all $g, x \in G$. Clearly,

$\varphi$ satisfies conditions (A1) and (A2). The action $\varphi$ is effective iff $\bigcap_{x\in G} x^{-1}Hx = \{e\}$ and $\varphi$ is always transitive. □

EXAMPLE 10.2.6. Let $X$ be a non-empty set, and $\mathcal{S}_X$ be the permutation group on $X$. The natural action $\sigma$ of $\mathcal{S}_X$ on $X$ is defined by $\sigma(f, x) = f(x)$, for all $f \in \mathcal{S}_X$ and all $x \in X$. For any subgroup $G$ of $\mathcal{S}_X$, one defines the corresponding action $\sigma_G$ of $G$ on $X$ as the restriction of $\sigma$ to the product $G \times X$. The action $\sigma_G$ is always effective, but it may or may not be transitive, depending on the choice of $G$. For example, if $X = \{1, \ldots, n\}$ for some $n \in \mathbb{N}$ and $G = A_n$ is the alternating subgroup of the permutation group $S_n = \mathcal{S}_X$ ($A_n$ consists of all even permutations), then the action of $A_n$ on $X$ is transitive iff $n \neq 2$. □

Let now $G$ be a topological group and $X$ be a topological space. An action $\theta$ of $G$ on $X$ is called *continuous* if $\theta$ is continuous as a mapping of $G \times X$ to $X$. The space $X$ (with a given continuous action of $G$ on $X$) is called a $G$-*space*. The first simple but important observation about $G$-spaces is that the left translations $\theta_g$ with $g \in G$ are homeomorphisms, that is, $G$ acts on the $G$-space $X$ by homeomorphisms. Indeed, each translation $\theta_g$ is a bijection of $X$ onto itself and the continuity of both mappings $\theta_g$ and $(\theta_g)^{-1} = \theta_{g^{-1}}$ is immediate, by the continuity of $\theta$. This fact has an immediate application:

PROPOSITION 10.2.7. *Every continuous action* $\theta\colon G \times X \to X$ *of a topological group* $G$ *on a space* $X$ *is an open mapping.*

PROOF. It suffices to verify that the images under $\theta$ of the elements of some base for $G \times X$ are open in $X$. Let $O = U \times V \subset G \times X$, where $U$ and $V$ are open sets in $G$ and $X$, respectively. Then $\theta(O) = \bigcup_{g\in U} \theta_g(V)$ is open in $X$ since every $\theta_g$ is a homeomorphism of $X$ onto itself. Since the open sets $U \times V$ form a base for $G \times X$, the mapping $\theta$ is open. □

Similarly to the case of homomorphisms of topological groups, the continuity of an action $\theta$ of a topological group $G$ can be deduced from its continuity at the identity of $G$.

PROPOSITION 10.2.8. *The continuity of an action* $\theta\colon G\times X \to X$ *of a topological group* $G$ *with identity* $e$ *on a space* $X$ *is equivalent to the continuity of* $\theta$ *at the points of the set* $\{e\} \times X \subset G \times X$.

PROOF. It suffices to verify the sufficiency of the condition. Let $g \in G$ and $x \in X$ be arbitrary, and $U$ be a neighbourhood of $gx$ in $X$. Since $\theta_h$ is a homeomorphism of $X$ for each $h \in G$, the set $V = \theta_{g^{-1}}(U)$ is a neighbourhood of $x$ in $X$. By the continuity of $\theta$ in $(e, x)$, we can find a neighbourhood $O$ of $e$ in $G$ and a neighbourhood $W$ of $x$ in $X$ such that $hy \in V$ for all $h \in O$ and $y \in W$. Clearly, if $h \in O$ and $y \in W$, then $(gh)(y) = g(hy) \in gV = \theta_g(V) = U$. Thus, $ky \in U$, for all $k \in gO$ and all $y \in W$, where $O' = gO$ is a neighbourhood of $g$ in $G$. Hence, the action $\theta$ is continuous. □

Here are several examples of continuous actions of topological groups.

EXAMPLE 10.2.9. Any topological group $G$ acts on itself by left translations, that is, $\theta(x, y) = xy$ for all $x, y \in G$ (see Example 10.2.4). The continuity of this action follows from the continuity of the multiplication in $G$. □

EXAMPLE 10.2.10. Let $G$ be a topological group, $H$ a closed subgroup of $G$, and let $G/H$ be the corresponding left coset space. The action $\varphi$ of $G$ on $G/H$, defined in Example 10.2.5, is continuous. Indeed, take any $y_0 \in G/H$, and fix an open neighbourhood

$O$ of $y_0$ in $G/H$. Choose $x_0 \in G$ such that $\pi(x_0) = y_0$, where $\pi\colon G \to G/H$ is the quotient mapping. There exist open neighbourhoods $U$ and $V$ of the identity $e$ in $G$ such that $\pi(Ux_0) \subset O$ and $V^2 \subset U$. Clearly, $W = \pi(Vx_0)$ is open in $G/H$ and $y_0 \in W$. By the choice of $U$ and $V$, if $g \in V$ and $y \in W$, then $\varphi(g, y) \in O$. Indeed, take $x_1 \in Vx_0$ with $\pi(x_1) = y$. Then $y = x_1 H$ and $\varphi(g, y) = gx_1 H \in VVx_0 H \subset \pi(Ux_0) \subset O$. It follows that $\varphi$ is continuous at $(e, y_0) \in G \times G/H$; hence, $\varphi$ is continuous, by Proposition 10.2.8. □

EXAMPLE 10.2.11. Let $G = G(n, \mathbb{R})$ be the general linear group with the topology inherited from $\mathbb{R}^{n^2}$, where $n \in \mathbb{N}$ (see e) of Example 1.2.5). The group $G$ acts on the $n$-dimensional Euclidean space $\mathbb{R}^n$ by multiplication. Each $g \in G$ is an invertible $n \times n$ matrix with real entries, and the action $\theta(g, x)$ of $g$ on $x \in \mathbb{R}^n$ is the result of the usual matrix multiplication $y = gx \in \mathbb{R}^n$. The continuity of $\theta$ follows immediately from the formula $y_i = \sum_{k=1}^{n} g_{i,k} x_k$ for $i = 1, \ldots, n$, expressing the coordinates of $y = (y_1, \ldots, y_n)$ in terms of the coordinates of $x = (x_1, \ldots, x_n)$ and of the entries of the matrix $g = (g_{i,k})$. The orbit of the zero vector $\bar{0} \in \mathbb{R}^n$ under this action is the one-point set $\{\bar{0}\}$, while the orbit $Gx$ of an arbitrary point $x \in \mathbb{R}^n$ distinct from $\bar{0}$ is the set $\mathbb{R}^n \setminus \{\bar{0}\}$. Therefore, $\mathbb{R}^n \setminus \{\bar{0}\}$ is an invariant subset of $\mathbb{R}^n$ under this action. The action $\theta$ is effective but not transitive. □

Suppose that a topological group $G$ acts continuously on a space $X$ and that $Y = X/G$ is the corresponding orbit set. Let $Y$ carry the quotient topology generated by the orbital projection $\pi\colon X \to X/G$ — a set $U \subset Y$ is open in $Y$ if and only if the preimage $\pi^{-1}(U)$ is open in $X$. The topological space $X/G$ so obtained is called the *orbital space* or the *orbit space* of the $G$-space $X$. The orbital projection is always an open mapping:

PROPOSITION 10.2.12. *If $\theta\colon G \times X \to X$ is a continuous action of a topological group $G$ on a space $X$, then the orbital projection $\pi\colon X \to X/G$ is open.*

PROOF. Take an arbitrary open set $U \subset X$, and consider the set $\pi^{-1}\pi(U) = GU$. Every left translation $\theta_g$ is a homeomorphism of $X$ onto itself, so the set $GU = \bigcup_{g \in G} \theta_g(U)$ is open in $X$. Since $\pi$ is a quotient mapping, $\pi(U)$ is open in $Y$. □

When $G$ is a compact group, the conclusion in Proposition 10.2.12 can be considerably strengthened.

THEOREM 10.2.13. *If a compact topological group $H$ acts continuously on a Hausdorff space $X$, then the orbital projection $\pi\colon X \to X/H$ is an open and perfect mapping.*

PROOF. Let $Y = X/H$. If $y \in Y$, choose $x \in X$ such that $\pi(x) = y$ and note that $\pi^{-1}(y) = Hx$ is the orbit of $x$ in $X$. Since the mapping of $H$ onto $Hx$ assigning to every $g \in H$ the point $gx \in X$ is continuous, the image $Hx$ of the compact group $H$ is also compact. Hence, all fibers of $\pi$ are compact.

To verify that the mapping $\pi$ is closed, let $y \in Y$ and $x \in X$ be as above, and let $O$ be an open set in $X$ containing $\pi^{-1}(y) = Hx$. Since the action of $H$ on $X$ is continuous, we can find, for every $g \in H$, open neighbourhoods $U_g \ni g$ and $V_g \ni x$ in $H$ and $X$, respectively, such that $U_g V_g \subset O$. By the compactness of $H$ and of the orbit $Hx$, there exists a finite set $F \subset H$ such that $H = \bigcup_{g \in F} U_g$ and $Hx \subset \bigcup_{g \in F} gV_g$. Then $V = \bigcap_{g \in F} V_g$ is an open neighbourhood of $x$ in $X$, and we claim that $HV \subset O$. Indeed, if $h \in H$ and $z \in V$, then $h \in U_g$, for some $g \in F$, so that $hz \in U_g V \subset U_g V_g \subset O$. Thus, $W = \pi(V)$ is an open neighbourhood of $y$ in $Y$, and we have $\pi^{-1}\pi(V) = HV \subset O$. Hence, the mapping $\pi$ is closed by [**165**, Theorem 1.4.13]. Finally, $\pi$ is open, by Proposition 10.2.12. □

Theorem 1.5.7 follows from the above result in the special case when the compact group $H$ is a subgroup of a bigger topological group $G$ and $H$ acts on $G$ by left translations (see Examples 10.2.4 and 10.2.9).

Let $X$ be a $G$-space, with action $\theta$ of $G$ on $X$. For every $x \in X$, denote by $\theta^x$ the mapping of $G$ to $X$ defined by the rule $\theta^x(g) = gx$, for each $g \in G$. Then $\theta^x$ is continuous, as the restriction of $\theta$ to the subspace $G \times \{x\}$ of the product $G \times X$. Clearly, the image $\theta^x(G)$ is the orbit $Gx$ of $x$. Therefore, each orbit in $X$ under the action of $G$ is a continuous image of $G$.

In general, the mappings $\theta^x$ need not be either open or quotient, even if the action $\theta$ is transitive. Indeed, consider an arbitrary non-discrete topological group $G$, and let $G_d$ be the same group $G$ endowed with the discrete topology. Then $G_d$ acts continuously and transitively on $G$ by left translations and every mapping $\theta^x$ is a continuous bijection of $G_d$ onto $G$. Therefore, $\theta^x$ fails to be quotient, for each $x \in G$.

However, under some addition restrictions on $G$ and $X$, the mappings $\theta^x$ might become quite nice; we will see this at the beginning of the next section.

Let $X$ and $Y$ be $G$-spaces with continuous actions $\theta_X : G \times X \to X$ and $\theta_Y : G \times Y \to Y$. A continuous mapping $f : X \to Y$ is called *$G$-equivariant* (or *equivariant*) if $\theta_Y(g, f(x)) = f(\theta_X(g, x))$, that is, $gf(x) = f(gx)$, for all $g \in G$ and all $x \in X$. Clearly, $f$ is equivariant if and only if the diagram below commutes,

$$\begin{array}{ccc} G \times X & \xrightarrow{\theta_X} & X \\ {\scriptstyle F}\downarrow & & \downarrow{\scriptstyle f} \\ G \times Y & \xrightarrow{\theta_Y} & Y \end{array}$$

where $F = id_G \times f$ is the product of the identity mapping $id_G$ of $G$ and the mapping $f$.

EXAMPLE 10.2.14. Let $H$ be a closed subgroup of a topological group $G$, and $Y = G/H$ be the left coset space. Denote by $\theta_G$ the action of $G$ on itself by left translations defined in Example 10.2.9, and by $\theta_Y$ the natural continuous action of $G$ on $Y$ (see Examples 10.2.5 and 10.2.10). Then the quotient mapping $\pi : G \to G/H$ defined by $\pi(x) = xH$ for each $x \in G$, is equivariant. Indeed, the equality $g(\pi(x)) = gxH = \pi(gx)$ holds for all $g, x \in G$. Equivalently, the diagram

$$\begin{array}{ccc} G \times G & \xrightarrow{\theta_G} & G \\ {\scriptstyle \Pi}\downarrow & & \downarrow{\scriptstyle \pi} \\ G \times Y & \xrightarrow{\theta_Y} & Y \end{array}$$

is commutative, where $\Pi = id_G \times \pi$. □

Let $\eta = \{X_i : i \in I\}$ be a family of $G$-spaces. Then the product space $X = \prod_{i \in I} X_i$ has a natural structure of a $G$-space. To define the action of $G$ on $X$, take any $g \in G$ and any $x = (x_i)_{i \in I} \in X$, and put $gx = (gx_i)_{i \in I}$. Thus, $G$ acts on $X$ coordinatewise. The following result guarantees the continuity of this action.

PROPOSITION 10.2.15. *The coordinatewise action of $G$ on the product $X = \prod_{i \in I} X_i$ of $G$-spaces is continuous, that is, $X$ is a $G$-space.*

PROOF. By Proposition 10.2.8, it suffices to verify the continuity of the action of $G$ on $X$ at the neutral element $e \in G$. Let $x = (x_i)_{i\in I} \in X$ be an arbitrary point and $O \subset X$ a neighbourhood of $gx$ in $X$. Since canonical open sets form a base of $X$, we can assume that $O = \prod_{i\in I} O_i$, where each $O_i$ is an open neighbourhood of $x_i$ in $X_i$ and the set $F = \{i \in I : O_i \neq X_i\}$ is finite. Since all factors are $G$-spaces, we can choose, for every $i \in F$, open neighbourhoods $U_i \ni e$ and $V_i \ni x_i$ in $G$ and $X_i$, respectively, such that $U_i V_i \subset O_i$. Put $U = \bigcap_{i\in F} U_i$ and $W = \prod_{i\in I} W_i$, where $W_i = V_i$ if $i \in F$ and $V_i = X_i$ otherwise. It follows immediately from the definition of the sets $U$ and $W$ that $UW \subset O$. Therefore, the action of $G$ on $X$ is continuous. □

The $G$-space $X$ defined above is called the *Cartesian product* of the family $\eta$ of $G$-spaces.

The definition of the coordinatewise action of a group $G$ on a product of $G$-spaces can be further extended to limit spaces of inverse spectra. Let $S = \{X_\alpha, p^\beta_\alpha : \alpha < \beta < \kappa\}$ be a well-ordered inverse spectrum, where each $X_\alpha$ is a $G$-space and the connecting mappings $p^\beta_\alpha$ are $G$-equivariant. Then the limit space of the spectrum $S$ is the subspace $X$ of the product space $\Pi = \prod_{\alpha<\kappa} X_\alpha$ consisting of all elements $(x_\alpha)_{\alpha<\kappa}$ of $\Pi$ satisfying $p^\beta_\alpha(x_\beta) = x_\alpha$ whenever $\alpha < \beta < \kappa$. By Proposition 10.2.15, $\Pi$ is a $G$-space. We use this fact and this notation to define a continuous action of $G$ on $X$ as follows:

PROPOSITION 10.2.16. *The subspace $X$ of $\Pi$ is $G$-invariant; therefore, the restriction to $X \subset \Pi$ of the coordinatewise action of $G$ on $\Pi$ is a continuous action of $G$ on $X$. In addition, all limit projections $p_\alpha\colon X \to X_\alpha$ of the spectrum $S$ are $G$-equivariant.*

PROOF. Take arbitrary $x \in X$ and $g \in G$. Then $x = (x_\alpha)_{\alpha<\kappa}$ is a point of $\Pi$ and $p^\beta_\alpha(x_\beta) = x_\alpha$, whenever $\alpha < \beta < \kappa$. It follows from the definition of the action of $G$ on $\Pi$ that $gx = (gx_\alpha)_{\alpha<\kappa}$, so every projection $\pi_\alpha\colon \Pi \to X_\alpha$ is $G$-equivariant. In addition, if $\alpha < \beta < \kappa$, then $p^\beta_\alpha(gx_\beta) = gp^\beta_\alpha(x_\beta) = gx_\alpha$ by the $G$-equivariance of $p^\beta_\alpha$. Hence, $gx$ is an element of $X$, that is, $X$ is $G$-invariant in $\Pi$, and the action of $G$ on $X$ is can be defined as the one inherited from $\Pi$. Clearly, this induced action is continuous, since $\Pi$ is a $G$-space.

Finally, for every $\alpha < \kappa$, the limit projection $p_\alpha = \pi_\alpha \restriction X$ of $X$ to $X_\alpha$ is $G$-equivariant as the restriction of the $G$-invariant mapping $\pi_\alpha$ to the $G$-invariant subset $X$ of $\Pi$. □

Thus, the limit space of an inverse spectrum of $G$-spaces with $G$-equivariant projections is a $G$-space as well. Speaking about the structure of a $G$-space in the case of limits of inverse spectra, we will always mean the one defined in Proposition 10.2.16.

## Exercises

10.2.a. Fill details in the proof of Proposition 10.2.1.

10.2.b. Suppose that $X$ is a $G$-space, for some topological group $G$.
   a) Does $X$ remain a $G$-space, when we introduce the discrete topology on $G$?
   b) Does $X$ remain a $G$-space, when we introduce a stronger group topology on $G$?

10.2.c. Suppose that $X$ is a $G$-space, for some topological group $G$. Prove the following statements:
   a) The stabilizer $G_x$ is a closed subgroup of $G$, for each $x \in X$.
   b) If $G$ is either countably compact, pseudocompact, $\sigma$-compact, Lindelöf, or connected, then so are the orbits $Gx$ in $X$.
   c) If the group $G$ is compact and the space $X$ is Hausdorff, then the orbit $Gx$ is homeomorphic to the quotient space $G/G_x$, for every $x \in X$.

10.2.d. Let $X$ be a compact $G$-space, where $G$ is a second-countable topological group. Are the orbits $Gx$ first-countable?

10.2.e. Let $G$ be a topological group and $\kappa$ a limit ordinal. Show that if $S = \{X_\alpha, p_\alpha^\beta : \alpha < \beta < \kappa\}$ is a continuous inverse spectrum, where each $X_\alpha$ is a $G$-space and the bonding mappings $p_\alpha^{\alpha+1}$ are $G$-equivariant, then the limit projections $p_\alpha$ of $S$ are $G$-equivariant as well.

10.2.f. Let $\mathbb{R}$ be the additive group of real numbers, with the usual topology, and $\mathbb{Z}$ the group of integers. Fix an irrational number $a \in \mathbb{R}$, and put $H = \{an : n \in \mathbb{Z}\}$. Both $\mathbb{Z}$ and $H$ are closed subgroups of the Abelian group $\mathbb{R}$. Hence, $\mathbb{R}$ acts continuously on both quotient groups $X = \mathbb{R}/\mathbb{Z}$ and $Y = \mathbb{R}/H$. Then the diagonal product of these actions is a continuous action $\theta$ of $\mathbb{R}$ on $X \times Y$. Since each of the groups $X$ and $Y$ is topologically isomorphic to the circle group $\mathbb{T}$, $\theta$ is a continuous action of the topological group $\mathbb{R}$ on the torus $\mathbb{T}^2$. Prove the following:

a) each orbit under $\theta$ is a proper dense subset of $\mathbb{T}^2$;
b) the orbital space of the action $\theta$ is antidiscrete, that is, the orbital space has only two open sets, both trivial;
c) the natural mapping of $\mathbb{R}$ onto an arbitrary orbit is one-to-one, continuous, but not a homeomorphism.

10.2.g. Let $X$ be an infinite discrete space, and $S_X$ the group of permutations of $X$, in the topology of pointwise convergence. Is the natural action of $S_X$ on $X$ continuous?

10.2.h. Let $X$ be an infinite discrete space, and $S_X$ be the group of permutations of $X$, with the topology of pointwise convergence. Then $S_X$ acts in a natural way, by translations, on the space $\mathbb{R}^X$. Is this action continuous?

10.2.i. Suppose that a $G$-space $Y$ is embedded as a subspace into a space $X$. This embedding is called *$G$-equivariant* or simply *equivariant* if there exists a continuous action of the group $G$ on $X$ which extends the action of $G$ on $Y$. Show that the natural embedding of the unit interval $I$ into the real line is $G$-equivariant, for any continuous action of an arbitrary finite discrete group $G$ on $I$.

## Problems

10.2.A. Let $G$ be a feathered topological group acting continuously and effectively on a first-countable separable space $X$. Prove that $G$ is metrizable. Verify that the conclusion remains valid if 'feathered' is replaced by 'pseudocompact'.
*Hint.* Take a countable dense set $S \subset X$ and show that $\{e_G\} = \bigcap_{x \in S} G_x$. Deduce that the identity $e_G$ of $G$ is the intersection of countably many open subsets of $G$.

10.2.B. Suppose that $G$ is a finite group acting continuously on the real line considered with the usual topology. Prove that $|G| = 2$.

10.2.C. Prove that the circle group $\mathbb{T}$ does not admit a non-trivial continuous action on the space $\mathbb{R}$.

10.2.D. (V. G. Pestov [**379**]) Let $G = H_+(\mathbb{R})$ be the group of all orientation-preserving homeomorphisms of the space $\mathbb{R}$ onto itself, with the topology of pointwise convergence. Prove that if the group $G$ acts continuously on a non-empty compact Hausdorff space $X$, then there exists $a \in X$ such that $ga = a$, for each $g \in G$ (any such a point $a \in X$ is called a *fixed point* of the action).

10.2.E. Let $G$ be a discrete group, and $\beta G$ the Čech–Stone compactification of $G$. Then, by the principal property of Čech–Stone compactifications, the natural action of $G$ on itself can be extended to an action of $G$ on $\beta G$. Is this action of $G$ on the space $\beta G$ continuous?

10.2.F. Suppose that a topological group $G$ acts continuously on a compact space $X$. A non-empty subspace $M$ of $X$ is said to be *minimal* if $M$ is closed in $X$, $GM \subset M$ (that is, $M$ is $G$-invariant), and no proper closed non-empty subset of $M$ is $G$-invariant. Apply Zorn's lemma to prove that for any continuous action of a topological group $G$ on a compact space $X$, there exists a minimal subspace of $X$.

10.2.G. Suppose that a topological group $G$ acts continuously on a space $X$. Prove that a closed non-empty subspace $M$ of $X$ is minimal if and only if the orbit $Gx$ is dense in $M$, for every $x \in M$.

10.2.H. Let $G$ be a discrete group, and $M_g$ a minimal subspace of $\beta G$ defined for the action of $G$ on the space $\beta G$ as in Problem 10.2.E. Prove that $M_G$ is a retract of $\beta G$ and that $M_G$ is extremally disconnected.

10.2.I. (J. de Vries [**529**]) A topological group $G$ is called a $V$-*group* if every Tychonoff $G$-space can be equivariantly embedded into a compact Hausdorff $G$-space (see Exercise 10.2.i). Prove that every locally compact group is a $V$-group.

10.2.J. (M. G. Megrelishvili and T. Scarr [**315**]) Prove that each $\omega$-narrow $V$-group is locally precompact.

10.2.K. (S. A. Antonyan and M. Sanchis [**10**]) Let $G$ be a locally pseudocompact group. Prove that every $G$-space $X$ admits an equivariant embedding into a compact Hausdorff $G$-space in each of the following cases:

a) $X$ is first countable;
b) $X$ is a $k_R$-space;
c) $X$ is locally pseudocompact.

10.2.L. (S. A. Antonyan and M. Sanchis [**10**]) Let $G$ be a locally pseudocompact (respectively, pseudocompact) group and $X$ a first-countable Dieudonné complete space. Then for each point $x \in X$, the stabilizer $G_x = \{g \in G : gx = x\}$ is a closed locally pseudocompact (respectively, pseudocompact) subgroup of $G$. Furthermore, $\mu(G_x) = (\mu G)_x$.

10.2.M. (F. González and M. Sanchis [**199**]) Every pseudocompact space $X$ can be equivariantly embedded into a compact Hausdorff $G$-space provided that one of the following holds:

a) $G$ is first-countable;
b) $G$ is a $k_R$-space;
c) $G$ is locally pseudocompact.

10.2.N. (S. A. Antonyan and M. Sanchis [**10**]) Let $G$ be a Moscow topological group. Then every continuous action $\alpha\colon G \times X \to X$ on a first-countable Dieudonné complete space $X$ admits an extension to a continuous action $\bar{\alpha}\colon \mu G \times X \to X$.

## Open Problems

10.2.1. (S. Glasner [**194**]) Let $G$ be an Abelian topological group without non-trivial continuous characters. Is it true that for every compact Hausdorff space $X$ and every continuous action of $G$ on $X$, there exists a point $a \in X$ such that $Ga = \{a\}$ (that is, a fixed point of the action)?

10.2.2. (S. A. Antonyan and M. Sanchis [**10**]) Is every locally pseudocompact group a $V$-group? (See Problems 10.2.I and 10.2.J.)

10.2.3. (S. A. Antonyan and M. Sanchis [**10**]) Let $G$ be a Moscow topological group. Does every continuous action of $G$ on a Dieudonné complete $k_R$-space (or locally pseudocompact space) admit a continuous extension to an action of $\mu G$ on $X$?

10.2.4. Is the group of rational numbers with the usual interval topology a $V$-group?

## 10.3. Uspenskij's theorem on continuous transitive actions of $\omega$-narrow groups on compacta

In this section, making use of techniques developed in the preceding sections and in Chapter 2, we prove an amazing theorem: Every compact Hausdorff space admitting a continuous transitive action of an $\omega$-narrow topological group turns out to be a Dugundji compactum. Another general theorem on continuous actions of $\omega$-balanced groups is also established. Several auxiliary results on which the proofs are based seem to be interesting in themselves.

PROPOSITION 10.3.1. *Suppose that an $\omega$-narrow topological group $G$ acts continuously and transitively on a space $Y$ with the Baire property. Then, for any $y \in Y$, the mapping $\theta^y: G \to Y$ defined by $\theta^y(g) = gy$, for each $g \in G$, is nearly open.*

PROOF. Let $U$ be an arbitrary open neighbourhood of the identity $e$ in $G$. Take a symmetric open neighbourhood $V$ of $e$ in $G$ such that $V^2 \subset U$. Since $G$ is $\omega$-narrow, there exists a countable set $A \subset G$ such that $AV = G$. Hence, the sets $aVy$, with $a \in A$, cover the space $Y$. By the Baire property of $Y$, there exists $a \in A$ such that the set $\overline{aVy}$ has a non-empty interior. Since the translation $\theta_a$ is a homeomorphism of $Y$, the set $\overline{Vy}$ also has a non-empty interior. Pick $v \in V$ such that $vy \in \operatorname{Int} \overline{Vy}$. It follows that

$$y \in v^{-1} \operatorname{Int} \overline{Vy} = \operatorname{Int} \overline{v^{-1}Vy} \subset \operatorname{Int} \overline{Uy}.$$

Thus, for every open neighbourhood $U$ of $e$ in $G$ and every $y \in Y$, the interior of the closure of $Uy$ contains $y$.

Now take any $z \in Uy$. Choose $g \in U$ such that $z = gy$. There is an open neighbourhood $W$ of $e$ in $G$ such that $Wg \subset U$. Then $Wz = Wgy \subset Uy$, so it follows from the above that $z \in \operatorname{Int} \overline{Wz} \subset \operatorname{Int} \overline{Uy}$. Thus, $Uy \subset \operatorname{Int} \overline{Uy}$ and the mapping $\theta^y$ is nearly open at the identity of $G$.

Finally, let $U$ be an arbitrary non-empty open set in $G$ and $y \in Y$. Choose $g \in U$ and put $V = g^{-1}U$. Since $V$ is an open neighbourhood of $e$ in $G$, we have $Vy \subset \operatorname{Int} \overline{Vy}$. Applying the homeomorphism $\theta_g$, we obtain:

$$Uy = gVy \subset g \operatorname{Int} \overline{Vy} = \operatorname{Int} \overline{gVy} = \operatorname{Int} \overline{Uy}$$

or, equivalently, $\theta^g(U) \subset \operatorname{Int} \overline{\theta^g(U)}$. □

The lemma below complements the material about groups of isometries in Section 3.5.

LEMMA 10.3.2. *Suppose that an $\omega$-narrow topological group $G$ acts continuously on a metric space $(X, d)$ by isometries. Then the orbit $Gx$ is separable, for each $x \in X$.*

PROOF. Let $x \in X$ be arbitrary. For every integer $n \geq 1$, choose an open neighbourhood $U_n$ of the neutral element $e$ in $G$ such that $d(gx, x) < 1/n$ for each $g \in U_n$. Since the group $G$ is $\omega$-narrow, there exists a countable set $C_n \subset G$ such that $G = C_nU_n$. Then the set $C = \bigcup_{n=1}^{\infty} C_n$ is also countable, and the set $Cx$ is dense in $Gx$. Indeed, for every $g \in G$ and $n \in \mathbb{N}$, we can find elements $h_n \in C_n$ and $g_n \in U_n$ such that $g = h_ng_n$. Then $d(gx, h_nx) = d(g_nx, x) < 1/n$, so every neighbourhood of $gx$ in $X$ contains elements of the countable set $Cx$. □

The following important example is, in fact, an essential part of the proof of the central Theorem 10.3.5 below. However, the construction presented in Example 10.3.3 is fairly general and interesting in itself.

EXAMPLE 10.3.3. Let $\theta\colon G \times X \to X$ be a continuous action of a topological group $G$ on a compact space $X$. Denote by $C(X)$ the space of all continuous real-valued functions on $X$ endowed with the topology generated by the sup-norm: $||f|| = \sup_{x\in X} |f(x)|$, for each $f \in C(X)$. The action $\theta$ admits a natural "extension" to the action $\Theta$ of $G$ on $C(X)$ defined by the rule $\Theta(g, f)(x) = f(\theta(g^{-1}, x))$, for all $f \in C(X)$ and $g, x \in G$. One can rewrite the above definition in the form $(gf)(x) = f(g^{-1}x)$. Then $\Theta_g$ is an isometry of $C(X)$, for each $g \in G$, that is, $\Theta$ acts on $C(X)$ by isometries. Indeed, if $f \in C(X)$ and $g \in G$, then

$$||gf|| = \sup_{x\in X} |(gf)(x)| = \sup_{x\in X} |f(g^{-1}x)| = \sup_{y\in X} |f(y)| = ||f||.$$

This readily implies that $||gf - gf'|| = ||f - f'||$, for all $f, f' \in C(X)$.

We claim that the action $\Theta$ is continuous. Indeed, take an arbitrary $f_0 \in C(X)$ and fix a real number $\varepsilon > 0$. Consider the open neighbourhood $U = \{f \in C(X) : ||f - f_0|| < \varepsilon\}$ of $f_0$ in $C(X)$. For every $x \in X$, choose an open set $U_x$ in $X$ such that $x \in U_x$ and $|f_0(y) - f_0(x)| < \varepsilon/3$ for each $y \in U_x$. Since the action of $G$ on $X$ is continuous and $ex = x$, we can choose an open neighbourhood $V_x$ of $x$ in $X$ and an open symmetric neighbourhood $O_x$ of $e$ in $G$ such that $O_x V_x \subset U_x$. Since $e \in O_x$, we have $V_x \subset U_x$. By the compactness of $X$, there exists a finite subset $F$ of $X$ such that the family $\{V_x : x \in F\}$ covers $X$. Clearly, $O = \bigcap_{x\in F} O_x$ is an open symmetric neighbourhood of $e$ in $G$. If $g \in O$ and $y \in X$, then $y \in V_x$ for some $x \in F$; it follows that $g^{-1}y \in OV_x \subset O_x V_x \subset U_x$. Therefore, if $f \in C(X)$ and $||f - f_0|| < \varepsilon/3$, then

$$\begin{aligned} |(gf)(y) - f_0(y)| &= |f(g^{-1}y) - f_0(y)| \le \\ |f(g^{-1}y) - f_0(g^{-1}y)| + |f_0(g^{-1}y) &- f_0(x)| + |f_0(x) - f_0(y)| < \\ &||f - f_0|| + \varepsilon/3 + \varepsilon/3 < \varepsilon. \end{aligned}$$

This implies that $||gf - f_0|| < \varepsilon$, so $gf \in U$ and hence, the action $\Theta$ is continuous at $(e, f_0)$. Finally, by Proposition 10.2.8, $\Theta$ is continuous. □

We also need the following property of the spaces $C(X)$, for compact $X$:

LEMMA 10.3.4. *Let $X$ be a compact space, and $C(X)$ be the space of all continuous real-valued functions on $X$ with the sup-norm metric. If $F$ is a separable subspace of $C(X)$ and $\varphi$ is the diagonal product of the functions in $F$, then the image $\varphi(X)$ is compact and metrizable.*

PROOF. Take a countable dense subset $S$ of $F$ and denote by $\psi$ the diagonal product of the functions of $S$, $\psi\colon X \to \mathbb{R}^S$. Then $Z = \psi(X)$ is a compact subspace of the metrizable space $\mathbb{R}^S$, so $Z$ is metrizable. Denote by $p$ the restriction to $Y = \varphi(X)$ of the projection $\mathbb{R}^F \to \mathbb{R}^S$. Clearly, $p$ is continuous, and the equality $\psi = p \circ \varphi$ implies that $p(Y) = Z$. Let us verify that $p$ is a bijection.

Suppose to the contrary that there exist two distinct points $y_1, y_2 \in Y$ such that $p(y_1) = p(y_2)$. Take $x_1, x_2 \in X$ satisfying $\varphi(x_i) = y_i$ for $i = 1, 2$. Since $y_1 \ne y_2$, we can find $f \in F$ such that $f(x_1) \ne f(x_2)$. Put $\varepsilon = |f(x_1) - f(x_2)|$. Since $S$ is dense in $F$, there exists $g \in S$ such that $||f - g|| < \varepsilon/2$. Then $|f(x_i) - g(x_i)| < \varepsilon/2$ for

$i = 1, 2$, whence it follows that $g(x_1) \neq g(x_2)$. Since $g \in S$, we must have $\psi(x_1) \neq \psi(x_2)$, a contradiction with $\psi(x_1) = p(y_1) = p(y_2) = \psi(x_2)$. Thus, $p\colon Y \to Z$ is a homeomorphism, as a continuous bijection of compact Hausdorff spaces, and $w(Y) = w(Z) \leq \omega$. □

The next theorem is one of the main results of the section. It can be considered as a generous generalization of Theorem 4.1.7 about dyadicity of compact topological groups.

THEOREM 10.3.5. [**V. V. Uspenskij**] *If a compact space $X$ admits a continuous transitive action of an $\omega$-narrow topological group $G$, then $X$ is Dugundji. In particular, $X$ is dyadic.*

PROOF. We are going to construct a Haydon spectrum whose limit space will be $X$. Denote by $C(X)$ the space of all continuous real-valued functions on $X$, with the sup-norm metric. According to Example 10.3.3, $C(X)$ has the natural structure of a $G$-space: the action $\Theta$ of $G$ on $C(X)$ is defined by $(gf)(x) = f(g^{-1}x)$, for all $g \in G$, $f \in C(X)$ and $x \in X$. In addition, every translation $\Theta_g$ is an isometry of $C(X)$. Let $\kappa$ be the cardinality of $C(X)$. Then $C(X) = \{f_\nu : \nu < \kappa\}$. Denote by $C_0$ the set of all constant functions on $X$, $C_0 \subset C(X)$. For every ordinal $\alpha$, where $0 < \alpha < \kappa$, let

$$C_\alpha = C_0 \cup \{gf_\nu : g \in G,\ \nu < \alpha\}.$$

Clearly, each $C_\alpha$ is a $G$-invariant subset of $C(X)$. Let $\pi_\alpha$ be the diagonal product of the functions in $C_\alpha$, $\pi_\alpha\colon X \to \mathbb{R}^{C_\alpha}$. The mapping $\pi_\alpha$ is continuous, so the image $X_\alpha = \pi_\alpha(X)$ is a compact subspace of $\mathbb{R}^{C_\alpha}$. If $\alpha < \beta < \kappa$, let $\pi^\beta_\alpha$ be the restriction to $X_\beta$ of the natural projection of $\mathbb{R}^{C_\beta}$ to $\mathbb{R}^{C_\alpha}$. It follows from the definition of the mappings $\pi_\alpha$, $\pi_\beta$, and $\pi^\beta_\alpha$ that $\pi_\alpha = \pi_\beta \circ \pi^\beta_\alpha$. In particular, we have $X_\alpha = \pi^\beta_\alpha(X_\beta)$. Clearly, $C_\alpha = \bigcup_{\nu<\alpha} C_\nu$, for each limit ordinal $\alpha < \kappa$. This equality and the compactness of $X$ imply that the inverse spectrum $S = \{X_\alpha, \pi^\beta_\alpha : \alpha < \beta < \kappa\}$ is continuous, and that the limit of $S$ is naturally homeomorphic to $X$ (see Example 10.1.11). We claim that $S$ is a Haydon spectrum, that is, the space $X_0$ is compact metrizable, and all bonding mappings $\pi^{\alpha+1}_\alpha$ are open and have metrizable kernel.

First, we verify that each $\pi^{\alpha+1}_\alpha$ has metrizable kernel. If $\alpha < \kappa$, we have $C_{\alpha+1} = C_\alpha \cup Gf_\alpha$. Denote by $\varphi_\alpha$ the diagonal product of the functions in $Gf_\alpha$. It follows from the definition that $\pi_{\alpha+1} = \pi_\alpha \Delta \varphi_\alpha$. Since $Gf_\alpha$ is a separable subspace of $C(X)$, by Lemma 10.3.2, the image $Y_\alpha = \varphi_\alpha(X)$ is compact and metrizable, by Lemma 10.3.4. Clearly,

$$X_{\alpha+1} = \pi_{\alpha+1}(X) \subset X_\alpha \times Y_\alpha \subset \mathbb{R}^{C_\alpha} \times \mathbb{R}^{Gf_\alpha},$$

so the bonding mapping $\pi^{\alpha+1}_\alpha\colon X_{\alpha+1} \to X_\alpha$ is the restriction to $X_{\alpha+1}$ of the projection of the product space $X_\alpha \times Y_\alpha$ to the first factor. The claim follows, since $Y_\alpha$ is metrizable. Also $|X_0| = 1$, since all functions in $C_0$ are constant.

It remains to prove that all bonding mappings of the spectrum $S$ are open. We can define an action of $G$ on the space $X_\alpha$ by $g\pi_\alpha(x) = \pi_\alpha(gx)$, for all $g \in G$ and all $x \in X$. Clearly, the mapping $\pi_\alpha\colon X \to X_\alpha$ is equivariant, for every $\alpha < \kappa$. This action of $G$ on $X_\alpha$ is continuous. Indeed, since the mapping $\pi_\alpha$ is equivariant, the diagram below commutes.

$$\begin{array}{ccc} G \times X & \xrightarrow{\theta} & X \\ {\scriptstyle \Pi_\alpha}\downarrow & & \downarrow{\scriptstyle \pi_\alpha} \\ G \times X_\alpha & \xrightarrow{\theta_\alpha} & X_\alpha \end{array}$$

Here $\theta$ and $\theta_\alpha$ are the actions of $G$ on $X$ and $X_\alpha$, respectively, and $\Pi_\alpha$ is the product of the identity mapping $id_G$ and $\pi_\alpha$. Clearly, $\theta$ is continuous, and $\pi_\alpha$ is perfect, so the mapping $\Pi_\alpha$ is perfect, as a product of two perfect mappings [**165**, Theorem 3.7.7]. In particular, $\Pi_\alpha$ is quotient and, hence, $\theta_\alpha$ is continuous. Thus, each $X_\alpha$ is a $G$-space.

All projections $\pi^\beta_\alpha$ in the spectrum $S$ are equivariant. Indeed, suppose that $\alpha < \beta < \kappa$, and let $g \in G$ and $x_\beta \in X_\beta$ be arbitrary. Choose $x \in X$ where $\pi_\beta(x) = x_\beta$. Since $\pi_\beta$ and $\pi_\alpha$ are equivariant, we have

$$\pi^\beta_\alpha(gx_\beta) = \pi^\beta_\alpha \pi_\beta(gx) = \pi_\alpha(gx) = g\pi_\alpha(x) = g\pi^\beta_\alpha \pi_\beta(x) = g\pi^\beta_\alpha(x_\beta).$$

Let us show that the mapping $\pi^\beta_\alpha$ is open. Pick $x_\beta \in X_\beta$ and $x_\alpha \in X_\alpha$ such that $\pi^\beta_\alpha(x_\beta) = x_\alpha$. Consider the mappings $\varphi_\beta \colon G \to X_\beta$ and $\varphi_\alpha \colon G \to X_\alpha$ defined by $\varphi_\beta(g) = gx_\beta$ and $\varphi_\alpha(g) = gx_\alpha$, for each $g \in G$. Clearly, both $\varphi_\beta$ and $\varphi_\alpha$ are continuous, and Proposition 10.3.1 (applied to $X_\beta$ and $X_\alpha$, respectively) implies that these mappings are nearly open. Since $\varphi_\alpha$ is an onto mapping and the composition $\varphi_\alpha = \pi^\beta_\alpha \circ \varphi_\beta$ is nearly open, we infer that $\pi^\beta_\alpha$ is also nearly open. It follows from Corollary 10.1.22 that $\pi^\beta_\alpha$ is open. Thus, $S$ is a Haydon spectrum whose limit space is $X$, so the space $X$ is a Dugundji compactum, by Theorem 10.1.16. □

Theorem 10.3.8 below states that every compact $G_\delta$-set in a quotient space of an $\omega$-balanced topological group is Dugundji. Again, this important result generalizes Theorem 4.1.7 on dyadicity of compact topological groups.

Recall that a Tychonoff space $X$ admitting a continuous bijection $f \colon X \to M$ onto a metrizable space $M$ is said to be *submetrizable*, and that $f$ is said to be a *condensation* of $X$ onto $M$. Recall also that every $\omega$-narrow topological group of countable pseudocharacter is submetrizable, by Proposition 5.2.11 (in this case, the condensation can be chosen to be an isomorphism of topological groups).

In what follows, we fix a topological group $G$ and its closed subgroup $H$. Let $\pi \colon G \to G/H$ be the quotient mapping of $G$ onto the space $G/H$ of the left cosets of $H$ in $G$. Denote by $\mathcal{H}$ the family of all closed subgroups $K$ of $G$ such that $H \subset K$ and the quotient space $G/K$ is submetrizable. If $K, L \in \mathcal{H}$ and $K \subset L$, let $\pi_K \colon G \to G/K$, $\pi_L \colon G \to G/L$ and $\pi^K_L \colon G/K \to G/L$ be the natural mappings, where $\pi^K_L(xK) = xL$ for each $x \in G$. Since $\pi_L = \pi^K_L \circ \pi_K$ and both mappings $\pi_K$ and $\pi_L$ are open, so is $\pi^K_L$.

LEMMA 10.3.6. *If $\gamma$ is a countable subfamily of $\mathcal{H}$, then $\bigcap \gamma \in \mathcal{H}$.*

PROOF. Put $N = \bigcap \gamma$. Clearly, $H \subset N$, so it suffices to verify that the quotient space $G/N$ is submetrizable. For every $K \in \gamma$, the submetrizable space $G/K$ is an image of $G/N$ under the continuous the mapping $\pi^N_K$, so the diagonal product $\varphi$ of the family $\{\pi^N_K : K \in \gamma\}$ is a continuous injective mapping of $G/N$ to the product space $\Pi = \prod_{K \in \gamma} G/K$. The space $\Pi$ is submetrizable, as a countable product of submetrizable spaces. Since $\varphi$ is a continuous injective mapping and any subspace of a submetrizable space is submetrizable, the space $G/N$ is also submetrizable, that is, $N \in \mathcal{H}$. □

LEMMA 10.3.7. *If the group $G$ is $\omega$-balanced, then the family $\{\pi_K^H : K \in \mathcal{K}\}$ generates the topology of the quotient space $G/H$.*

PROOF. It suffices to verify that for every neighbourhood $U$ of the neutral element $e$ in $G$, one can find a neighbourhood $V$ of $e$ in $G$ and $K \in \mathcal{K}$ such that $VK \subset UH$. This will imply that $(\pi_K)^{-1}\pi_K(V) \subset \pi^{-1}\pi(U)$ and hence, $(\pi_K^H)^{-1}(\pi_K(V)) \subset \pi(U)$.

So, let $U$ be a neighbourhood of $e$ in $G$. By Theorem 3.4.18, one can find a continuous homomorphism $\varphi\colon G \to M$ onto a metrizable topological group $M$, and an open neighbourhood $W$ of the neutral element in $M$ such that $\varphi^{-1}(W) \subset U$. Denote by $N$ the closure of $\varphi(H)$ in $M$. Then $K = \varphi^{-1}(N)$ is a closed subgroup of $G$, and $H \subset K$. Further, let $i\colon G/K \to M/N$ be a mapping defined by the rule $i(xK) = \varphi(x)N$, for each $x \in G$. Clearly, the mapping $i$ is defined correctly. Since the mapping $\pi_K$ is open, and $i \circ \pi_K = p_N \circ \varphi$, where $p_N\colon M \to M/N$ is the quotient homomorphism, the mapping $i$ is continuous.

$$\begin{array}{ccc} G & \xrightarrow{\pi_K} & G/K \\ {\scriptstyle\varphi}\downarrow & & \downarrow{\scriptstyle i} \\ M & \xrightarrow{p_N} & M/N \end{array}$$

We claim that $i$ is a bijection of $G/K$ onto $M/N$. Indeed, suppose that $x, y \in G$ and $i(xK) = i(yK)$. Then $\varphi(x)N = \varphi(y)N$ and, consequently, $\varphi(x^{-1}y) \in N$. Hence, $x^{-1}y \in \varphi^{-1}(N) = K$ and $\pi_K(x) = \pi_K(y)$, that is, $xK = yK$.

Thus, $i$ is a condensation of $G/K$ onto the quotient space $M/N$ which is metrizable, by Proposition 3.3.19. Therefore, $K \in \mathcal{K}$.

Finally, $V = \varphi^{-1}(W)$ is an open neighbourhood of $e$ in $G$ and $V \subset U$. Since $W$ is open in $M$ and $\varphi(H)$ is dense in $N$, we have

$$VK = \varphi^{-1}(W)\varphi^{-1}(N) = \varphi^{-1}(WN) = \varphi^{-1}(W\varphi(H)) = \varphi^{-1}(W)H \subset UH.$$

As we mentioned at the beginning of the proof, this means that the family $\{\pi_K^H : K \in \mathcal{K}\}$ generates the topology at the point $\pi(e)$ of the quotient space $G/H$. The conclusion of the lemma follows, since the natural action of $G$ on $G/H$ is continuous and transitive (see Example 10.2.10). □

THEOREM 10.3.8. [**V. V. Uspenskij**] *Let $H$ be a closed subgroup of an $\omega$-balanced topological group $G$. Then every compact $G_\delta$-set in the quotient space $G/H$ is Dugundji.*

PROOF. Let $X$ be a compact $G_\delta$-set in $G/H$. There exists a sequence $\{U_n : n \in \omega\}$ of open sets in $G/H$ such that $X = \bigcap_{n\in\omega} U_n$. Let $n \in \omega$ be arbitrary. Since $X$ is compact, we can find, by Lemma 10.3.7, an element $L_n \in \mathcal{K}$ and an open set $V_n$ in $G/L_n$ such that $X \subset (\pi_{L_n}^H)^{-1}(V_n) \subset U_n$. Let $L = \bigcap_{n\in\omega} L_n$. Then $L \in \mathcal{K}$ by Lemma 10.3.6 and it follows from the choice of the sets $L_n$ and $V_n$ that the following holds:

$$X \subset (\pi_L^H)^{-1}\pi_L^H(X) \subset \bigcap_{n=0}^{\infty}(\pi_{L_n}^H)^{-1}\pi_{L_n}^H(X) \subset \bigcap_{n=0}^{\infty}(\pi_{L_n}^H)^{-1}(V_n) \subset \bigcap_{n=0}^{\infty} U_n = X.$$

Thus, $X = (\pi_L^H)^{-1}\pi_L^H(X)$. Therefore, since the mapping $\pi_L^H$ is open, the restriction of $\pi_L^H$ to $X$ is also an open mapping of $X$ onto $\pi_L^H(X)$.

Again, we will construct a Haydon spectrum $S$ with the limit space $X$. Consider the family $\mathcal{K}$ of closed subgroups of $G$ defined before Lemma 10.3.6. We can enumerate the family $\mathcal{K}$ as $\mathcal{K} = \{K_\alpha : \alpha < \kappa\}$, where $\kappa$ is the cardinality of $\mathcal{K}$. Let $N_0 = L$, and $N_\alpha = L \cap \bigcap_{\nu<\alpha} K_\nu$, if $0 < \alpha < \kappa$. Given ordinals $\alpha, \beta$ with $\alpha < \beta < \kappa$, we shorten $\pi_{N_\alpha}^{N_\beta}$ to $\pi_\alpha^\beta$ and $\pi_{N_\alpha}^H$ to $\pi_\alpha$. Put also $X_\alpha = \pi_\alpha(X)$, and $p_\alpha^\beta = \pi_\alpha^\beta \restriction X_\beta$. Let us show that $S = \{X_\alpha, p_\alpha^\beta : \alpha < \beta < \kappa\}$ is a Haydon spectrum, with the limit space $X$.

Since the family $\{p_\alpha^\beta : \alpha < \beta < \kappa\}$ generates the topology of the compact space $X$ (we apply Lemma 10.3.7 here), $X$ is homeomorphic to the limit space of $S$. The spectrum $S$ is continuous because $N_\alpha = \bigcap_{\nu<\alpha} N_\nu$, for each limit ordinal $\alpha < \kappa$. It follows from $N_0 = L$ and from the choice of $L$ that $X = (\pi_0)^{-1}\pi_0(X) = \pi_0^{-1}(X_0)$, so that $X_\beta = (\pi_\alpha^\beta)^{-1}(X_\alpha)$, whenever $\alpha < \beta < \kappa$. Since the mappings $\pi_\alpha^\beta$ are open, the projections $p_\alpha^\beta$ of the spectrum $S$ are open as well.

It remains to verify that each bonding mapping $p_\alpha^{\alpha+1}$ has a metrizable kernel. It follows from $N_{\alpha+1} = N_\alpha \cap K_\alpha$ that the diagonal product of the mappings $\pi_\alpha^{\alpha+1} : G/N_{\alpha+1} \to G/N_\alpha$ and $\pi_{K_\alpha}^{N_{\alpha+1}} : G/N_{\alpha+1} \to G/K_\alpha$ is injective. Let $f_\alpha$ be the restriction of $\pi_{K_\alpha}^{N_\alpha+1}$ to $X_{\alpha+1}$. The compact space $Y_\alpha = f_\alpha(X_{\alpha+1})$ is metrizable, as a subspace of the submetrizable space $G/K_\alpha$. Since the diagonal product $h = p_\alpha^{\alpha+1} \Delta f_\alpha$ is a condensation of $X_{\alpha+1}$ to the product space $X_\alpha \times Y_\alpha$, the mapping $h$ is a homeomorphic embedding of $X_{\alpha+1}$ and, hence, $p_\alpha^{\alpha+1}$ has metrizable kernel. □

The following fact is an almost immediate corollary of Theorem 10.3.8:

COROLLARY 10.3.9. [**M. M. Choban**] *Every compact $G_\delta$-set in a topological group is Dugundji.*

PROOF. Let $X$ be a compact $G_\delta$-set in a topological group $K$ with identity $e$. Denote by $G$ the subgroup of $K$ generated by $X$. Then the group $G$ is $\sigma$-compact and, hence, $\omega$-narrow and $\omega$-balanced. It remains to apply Theorem 10.3.8 in the special case when $H = \{e\}$. □

The following theorem provides us with a variety of natural examples of topological groups which are not $\omega$-balanced.

THEOREM 10.3.10. *Suppose that $X$ is a zero-dimensional homogeneous compact space such that the group $Homeo(X)$ of all homeomorphisms of $X$ onto itself, with the compact-open topology, is $\omega$-balanced. Then $X$ is Dugundji.*

PROOF. Every zero-dimensional homogeneous compactum is homeomorphic to a quotient space of the topological group $G = Homeo(X)$ of homeomorphisms of $X$ onto itself, with the compact-open topology (see Section 3.5). Since $G$ is $\omega$-balanced, it remains to apply Theorem 10.3.8. □

COROLLARY 10.3.11. *Suppose that $X$ is a zero-dimensional homogeneous compact space of countable tightness such that the group $Homeo(X)$ of all homeomorphisms of $X$ onto itself, with the compact-open topology, is $\omega$-balanced. Then $X$ is metrizable.*

PROOF. By Theorem 10.3.10, $X$ is Dugundji and therefore, $X$ is dyadic. According to [**165**, 3.12.12 (h)], every dyadic compactum of countable tightness is metrizable, whence the conclusion follows. □

If $X$ is a metrizable compact space, then, by Theorem 3.5.5, the group $Homeo(X)$ of all homeomorphisms of $X$ onto itself, with the compact-open topology, has a countable base and therefore, is $\omega$-narrow and $\omega$-balanced. So the converse to Corollary 10.3.11 holds.

EXAMPLE 10.3.12. The group $G$ of all homeomorphisms of the two arrows space onto itself, with the compact-open topology, is not $\omega$-balanced. Indeed, the two arrows space is a zero-dimensional homogeneous first-countable compact Hausdorff space. If $G$ were $\omega$-balanced, this compactum would be metrizable, by Corollary 10.3.11. □

## Exercises

10.3.a. Let $G$ be a topological group acting continuously on a Tychonoff space $X$. Take the associated action of $G$ on the space $C^0(X)$ of all bounded continuous real-valued functions on $X$, with the sup-norm topology. Is this action continuous?

10.3.b. Show that for any homogeneous compact Hausdorff space $X$, there exists an $\omega$-balanced topological group acting on $X$ continuously and transitively.

10.3.c. Suppose that $G$ is a separable topological group acting continuously on a compact Hausdorff space $X$ and that the orbital space $X/G$ is metrizable. Must the space $X$ be metrizable?

## Problems

10.3.A. Suppose that $G$ is a second-countable topological group acting continuously on a Tychonoff space $X$. If the orbital space $X/G$ is regular and second-countable, must $X$ be metrizable?

10.3.B. Let $G$ be a topological group acting continuously on a Tychonoff space $X$, and suppose that each of the spaces $G$ and $X/G$ is cosmic. Does it follow that the space $X$ is cosmic?

10.3.C. Suppose that $G$ is a second-countable topological group acting continuously and transitively on a Tychonoff space $X$. Does it follow that the space $X$ is second-countable?

10.3.D. Let a Lindelöf topological group $G$ act continuously and transitively on a compact Hausdorff space $X$. Prove that $X$ is metrizable.

## Open Problems

10.3.1. Let $G$ be the group of homeomorphisms of a Tychonoff cube $I^\tau$, where $\tau > \omega$, onto itself. Take $G$ with the compact-open topology, and suppose that $G$ acts continuously and transitively on a compact Hausdorff space $X$. Is $X$ dyadic?

10.3.2. Suppose that $G$ is a second-countable topological group acting continuously on a compact Hausdorff space $X$. Suppose further that $X/G$ is second-countable. Does it follow that the space $X$ has a countable base?

10.3.3. Suppose that $G$ is an $\omega$-narrow group acting continuously on a compact Hausdorff space $X$. Suppose further that the orbital space $X/G$ is second-countable. Does it follow that the space $X$ is Dugundji? Is $X$ dyadic?

10.3.4. Is it true that every compact $G_\delta$-set in a Hausdorff (Tychonoff) paratopological group is a Dugundji space? Is dyadic?

10.3.5. Suppose that $G$ is an $\omega$-balanced topological group, and $H$ a closed subgroup of $G$ such that the quotient space $G/H$ is a Lindelöf $p$-space of countable tightness. Is $G/H$ second-countable?

## 10.4. Continuous actions of compact groups and some classes of spaces

We have already used the techniques developed in the preceding sections to prove a few important results such as Theorems 10.3.5 and 10.3.8. These techniques are also quite effective in answering basic questions on the behaviour of certain general properties of $G$-spaces, in particular, of cardinal invariants. We present a sample of results of this kind below.

PROPOSITION 10.4.1. *Suppose that a compact topological group $G$ acts continuously on a Hausdorff space $X$. Then the orbital space $X/G$ is Hausdorff, and if, in addition, $X$ is normal, then $X/G$ is normal as well.*

PROOF. Any image of a Hausdorff space under a perfect mapping is a Hausdorff space, and any image of a normal space under a continuous closed mapping is a normal space, by Theorems 3.7.20 and 1.5.20 of [**165**], respectively. Hence, it suffices to apply Theorem 10.2.13 saying that every orbital projection is perfect. □

THEOREM 10.4.2. *Suppose that a compact topological group $G$ acts continuously on a Hausdorff space $X$, and that the orbital space $X/G$ has one of the following properties:*

a) *$X/G$ is compact;*
b) *$X/G$ is countably compact;*
c) *$X/G$ is pseudocompact;*
d) *$X/G$ is Lindelöf.*

*Then $X$ has the same property.*

PROOF. This again follows from Theorem 10.2.13, since the preimage of a compact (countably compact, pseudocompact, Lindelöf) space under a perfect mapping is compact (countably compact, pseudocompact, Lindelöf), according to Theorems 3.7.2, 3.10.9, Problem 3.10.G, and Theorem 3.7.26 of [**165**], respectively. □

Similar statements are valid for a variety of topological properties that behave nicely under perfect mappings. We just formulate some of them, omitting the routine proofs.

THEOREM 10.4.3. *Suppose that a compact topological group $G$ acts continuously on a Hausdorff space $X$. Then the orbital space $X/G$ is locally compact (Čech-complete) if and only if $X$ is locally compact (respectively, Čech-complete).*

THEOREM 10.4.4. *If a compact topological group $G$ acts continuously on a paracompact Hausdorff space $X$, then the orbital space $X/G$ is also paracompact.*

The case when the orbital space $X/G$ is metrizable deserves special attention.

THEOREM 10.4.5. *If a compact topological group $G$ acts continuously on a metrizable space $X$, then the orbital space $X/G$ is also metrizable.*

PROOF. Indeed, the image of a metrizable space under a perfect mapping is metrizable [**165**, Theorem 4.4.15]. □

THEOREM 10.4.6. *Suppose that a compact topological group $G$ acts continuously on a Tychonoff space $X$ in such a way that the orbital space $X/G$ is metrizable. Then $X$ is a paracompact $p$-space.*

PROOF. Indeed, the preimage of a metrizable space under a perfect mapping is a paracompact $p$-space, provided this preimage is a Tychonoff space [**60**, Chapter V, no. 228]. Therefore, it suffices to apply Theorem 10.2.13. □

THEOREM 10.4.7. *Suppose that $G$ is a compact topological group acting continuously on a zero-dimensional Hausdorff space $X$. Then the orbital space $X/G$ is also zero-dimensional.*

PROOF. Again, Theorem 10.2.13 implies the conclusion, since continuous mappings that are both open and closed, obviously preserve zero-dimensionality [**165**, 6.2.H (a)]. □

Under some mild restrictions on a $G$-space $X$, we can prove some more delicate and less expected connections between the properties of $X$, where the action takes place, and the properties of the orbital space $X/G$. We start with an obvious auxiliary fact:

LEMMA 10.4.8. *The orbit of every point under a continuous action of a compact topological group is a dyadic compactum.*

THEOREM 10.4.9. *Assume that $\mathfrak{c} = 2^{\aleph_0} < 2^{\aleph_1}$, and let $G$ be a compact topological group acting continuously on a Hausdorff space $X$ such that $|X| \leq \mathfrak{c}$. Then each orbit under this action is metrizable.*

PROOF. If the cardinality of a dyadic compactum $C$ does not exceed $2^{\omega}$, then, under the assumption $2^{\omega} < 2^{\omega_1}$, $C$ is metrizable. Indeed, $C$ must contain a dense subset of points of countable character — otherwise, by the Čech–Pospíšil theorem (see [**165**, 3.12.11 (a)]), the cardinality of $C$ would be greater than or equal to $2^{\omega_1} > 2^{\omega}$, a contradiction. However, every dyadic compactum with a dense subset of points of countable character is metrizable, by [**165**, 3.12.12 (g)]. It remains to refer to Lemma 10.4.8. □

One cannot drop the assumption that $\mathfrak{c} < 2^{\aleph_1}$ in Theorem 10.4.9. Indeed, the compact group $D^{\omega_1}$ acts continuously and transitively on itself, while the orbit $D^{\omega_1}$ is not metrizable.

THEOREM 10.4.10. *Suppose that $G$ is a compact topological group acting continuously on a Hausdorff space $X$ of countable tightness. Suppose also that the orbital space $X/G$ is first-countable. Then $X$ is first-countable as well.*

PROOF. Every orbit $Gx$ under the action is a dyadic compactum, according to Lemma 10.4.8. Since every dyadic compactum of countable tightness is metrizable [**165**, 3.12.12 (h)], it follows that the orbit $Gx$ is first-countable. On the other hand, the set $Gx$ has a countable base of open neighbourhoods in $X$, since the space $X/G$ is first-countable, and $Gx$ is a fiber under the orbital projection which is perfect. Hence, by the transitivity of the character (see [**165**, 3.1.E]), $X$ is first-countable. □

Notice that in the above argument we have established the next lemma:

LEMMA 10.4.11. *The orbit $Gx$ of every point $x \in X$ under a continuous action of a compact topological group $G$ on a Hausdorff space $X$ of countable tightness is a metrizable compactum.*

THEOREM 10.4.12. *Suppose that $G$ is a compact topological group acting continuously on a Hausdorff space $X$ such that the tightness of $X$ does not exceed $2^{\omega}$. Suppose also that the orbital space $X/G$ is separable. Then $X$ is also separable.*

PROOF. Fix a countable dense subset $A$ in $X/G$, and let $\pi\colon X \to X/G$ be the orbital projection. Since $B_a = \pi^{-1}(a)$ is a dyadic compactum, by Lemma 10.4.8, it follows from [**165**, 3.12.12 (h)] that the weight of $B_a$ is not greater than $t(B_a) \leq \mathfrak{c} = 2^{\omega}$. Therefore, by [**165**, 3.12.12 (b)], $B_a$ is a continuous image of $D^{\mathfrak{c}}$. Hence, the density of $B_a$ does not exceed the density of $D^{\mathfrak{c}}$, that is, $B_a$ is separable, according to [**165**, Corollary 2.3.16]. For each $a \in A$, choose a countable dense subset $B_a$ of $\pi^{-1}(a)$. Since $\pi$ is open, the countable set $B = \bigcup\{B_a : a \in A\}$ is dense in $X$. □

THEOREM 10.4.13. *Suppose that $G$ is a compact topological group acting continuously on a Hausdorff space $X$ of countable tightness. Suppose also that the orbital space $X/G$ is separable. Then there exists a closed subgroup $H$ of $G$ such that $H$ is a $G_\delta$-set in $G$, and $hx = x$, for all $h \in H$ and $x \in X$.*

PROOF. It follows from Theorem 10.4.12 that the space $X$ is separable. Fix a countable dense subset $A$ of $X$. For each $a \in A$, consider the mapping $\theta^a\colon G \to Ga$ given by $\theta^a(g) = ga$. According to Lemma 10.4.11, the orbit $Ga$ is a metrizable compactum. Clearly, the stabilizer $G_a$ of $a$ is the preimage of $a$ under the mapping $\theta^a$. Since $\theta^a$ is latter is continuous, and the space $Ga$ is first-countable, $G_a$ is a closed $G_\delta$-set in $G$. We also know that $G_a$ is a subgroup of $G$.

Put $H = \bigcap\{G_a : a \in A\}$. Obviously, $H$ is a closed subgroup of $G$ and a $G_\delta$-set in $G$. Take any $h \in H$. From the definition of $H$ it follows immediately that $h(a) = a$, for each $a \in A$. Since $A$ is dense in $X$, and the action of $G$ on $X$ is continuous, it follows that $h(x) = x$, for every $x \in X$. □

Theorem 10.4.13 allows us to reduce the study of continuous actions of compact Abelian groups on separable spaces of countable tightness (in particular, on separable metrizable spaces) to the case when the acting group is metrizable. Indeed, assuming the commutativity of $G$ in Theorem 10.4.13, the quotient group $G/H$ is defined. It is metrizable, compact, and acts in an obvious way on $X$ so that the orbital space $X/(G/H)$ is naturally homeomorphic to the orbital space $X/G$.

There are many other topological properties, besides countable tightness, that can influence drastically the size and topological properties of orbits under continuous actions of topological groups. At least one statement, Theorem 10.4.17, is worth mentioning.

Recall that a space $X$ is said to be *$\tau$-monolithic*, for a given infinite cardinal $\tau$, if for each subset $A$ of $X$ with $|A| \leq \tau$, there is a network $S$ in $\overline{A}$ such that $|S| \leq \tau$, that is, the network weight of the closure of $A$ in $X$ does not exceed $\tau$. The space $X$ is called *monolithic* if it is $\tau$-monolithic, for each $\tau \geq \omega$.

The proof of Theorem 10.4.17 requires three auxiliary results, each of which is interesting in itself.

LEMMA 10.4.14. *For every infinite compact space $X$, the weight of the space $C(X)$ of continuous real-valued functions on $X$, endowed with the sup-norm topology, is equal to the weight of $X$.*

PROOF. Let $\kappa$ be the weight of $C(X)$. Clearly, $\kappa \geq \omega$, and $C(X)$ contains a dense subset $S$ satisfying $|S| \leq \kappa$. It is easy to see that the set $S$ separates points and closed sets in $X$, so that the diagonal product $\varphi$ of the functions in $S$ is a homeomorphic embedding of $X$ into $\mathbb{R}^S$. Therefore, $w(X) \leq |S| \cdot \omega \leq \kappa$.

Conversely, denote by $\tau$ the weight of $X$, and take a base $\mathcal{B}$ for $X$ with $|\mathcal{B}| = \tau$. For every pair $(U, V)$ such that $U, V \in \mathcal{B}$ and $\overline{U} \subset V$, choose $f = f_{U,V} \in C(X)$ such that $f(U) \subset \{1\}$ and $f(X \setminus V) \subset \{0\}$. Then the family $\mathcal{F}_0$ of the functions $f_{U,V}$ separates points of $X$ and, clearly, $|\mathcal{F}_0| \leq \tau$. For every rational number $r$, let $\mathbf{r}$ be the function on $X$ constantly equal to $r$. Put $\mathcal{F}_1 = \{\mathbf{r} : r \in \mathbb{Q}\}$ and $\mathcal{F} = \mathcal{F}_0 \cup \mathcal{F}_1$. Then $|\mathcal{F}| \leq \tau$. Denote by $S$ the minimal subset of $C(X)$ which contains $\mathcal{F}$ and is closed with respect to the usual pointwise operations of sum, subtraction, and multiplication of functions. By the Stone–Weierstrass theorem (see [**165**, Theorem 3.2.21]), $S$ is dense in the metric space $C(X)$. It is also clear that $|S| \leq \tau$. Therefore, the density of $C(X)$ is not greater than $\tau$. By [**165**, Theorem 4.1.15], the density and weight of $C(X)$ coincide, whence it follows that $w(C(X)) \leq \tau$.

Combining the above results, we conclude that $w(C(X)) = w(X)$. □

LEMMA 10.4.15. *Every non-metrizable compact space $X$ admits a continuous mapping onto a compact space $Y$ of weight $\aleph_1$.*

PROOF. According to Lemma 10.4.14, the space $C(X)$ has uncountable weight. Since, by [**165**, Theorem 4.1.15], the cellularity and weight of every metrizable space coincide, $C(X)$ contains a family $\{U_\alpha : \alpha < \omega_1\}$ of pairwise disjoint non-empty open sets. For every $\alpha < \omega_1$, choose an element $f_\alpha \in U_\alpha$, and let $\varphi$ be the diagonal product of the functions $f_\alpha$'s. Clearly, $\varphi$ is a continuous mapping of $X$ to $\mathbb{R}^{\omega_1}$, and $Y = \varphi(X)$ is a compact subspace of $\mathbb{R}^{\omega_1}$. In particular, $w(Y) \leq \aleph_1$. We claim that the cellularity of the metric space $C(Y)$ is not less than $\aleph_1$.

Indeed, consider the dual mapping $\varphi^*\colon C(Y) \to C(X)$ defined by $\varphi^*(f) = f \circ \varphi$, for each $f \in C(Y)$. It is clear that $\varphi^*$ is an isometric embedding of $C(Y)$ into $C(X)$. In particular, $\varphi^*(C(Y))$ is a homeomorphic copy of $C(Y)$. It follows from the definitions of $\varphi$ and $\varphi^*$ that $f_\alpha \in \varphi^*(C(Y)) \cap U_\alpha \neq \varnothing$, for each $\alpha < \omega_1$. Therefore, the cellularity of the spaces $\varphi^*(C(Y))$ and $C(Y)$ is uncountable.

We now apply Lemma 10.4.14 to deduce that $w(Y) \geq \aleph_1$. This, together with the inequality $w(Y) \leq \aleph_1$ established earlier, implies the conclusion of the lemma. □

PROPOSITION 10.4.16. *An $\omega$-monolithic dyadic compactum is metrizable.*

PROOF. Suppose by the way of contradiction that $X$ is a non-metrizable $\omega$-monolithic dyadic compactum. Since $X$ is dyadic, we can find a continuous mapping $g$ of the Cantor cube $D^\kappa$ onto $X$, for some cardinal $\kappa$. It is clear that $\kappa$ is not less than the weight of $X$. By Lemma 10.4.15, there exists a continuous mapping $g$ of $X$ onto a compact spaces $Y$ of weight $\aleph_1$. Since continuous mappings do not increase the weight in the class of compact spaces, the space $Y$ is also $\aleph_0$-monolithic. The composition $g \circ f$ is a continuous mapping of $D^\kappa$ onto $Y$, and Corollary 1.7.4 applies to find a subset $A \subset \kappa$ with $|A| \leq \aleph_1$ and a continuous mapping $h\colon D^A \to Y$ such that $g \circ f = h \circ \pi_A$, where $\pi_A\colon D^\kappa \to D^A$ is the projection.

$$\begin{array}{ccc} D^\kappa & \xrightarrow{\ f\ } & X \\ {\scriptstyle \pi_A}\downarrow & & \downarrow{\scriptstyle g} \\ D^A & \xrightarrow{\ h\ } & Y \end{array}$$

Since $|A| = \aleph_1 \leq 2^\omega$, it follows from the Hewitt–Marczewski–Pondiczery theorem (see [**165**, Theorem 2.3.15]) that the Cantor cube $D^A$ is separable. Hence $Y$ is also separable, as a continuous image of $D^A$. Finally, we apply the $\aleph_0$-monolithicity of $Y$ to conclude that $w(Y) \leq d(Y) \leq \omega$, a contradiction. We have thus proved that $X$ has countable weight. □

THEOREM 10.4.17. *Suppose that $G$ is a compact topological group acting continuously on an $\aleph_0$-monolithic Hausdorff space $X$. Then:*

a) *every orbit $Gx$ is a metrizable compactum;*
b) *the orbital space $X/G$ is also $\aleph_0$-monolithic;*
c) $d(X) = d(X/G)$;
d) *if, in addition, $X$ is monolithic, then $nw(X) = nw(X/G)$.*

PROOF. Item a) follows from Lemma 10.4.8, since every $\aleph_0$-monolithic dyadic compactum is metrizable, by Proposition 10.4.16.

To prove b), fix an arbitrary countable subset $A$ of $X/G$. For each $a \in A$, fix $b_a \in X$ such that $\pi(b_a) = a$, where $\pi$ is the orbital projection of $X$ onto $X/G$. Put $B = \{b_a : a \in A\}$ and $F = \overline{B}$. Then $nw(F) \leq |B| = |A| \leq \omega$, since $X$ is $\aleph_0$-monolithic. Since $\pi$ is closed and $\pi(B) = A$, we have that $\pi(F) = \overline{A}$. However, continuous mappings do not increase the network weight. Therefore, $nw(\overline{A}) \leq nw(F) \leq \omega$, that is, $X/G$ is $\aleph_0$-monolithic.

To prove c), fix a dense subset $A$ of $X$ such that $|A|$ is the density of $X$. We may assume that $A$ is infinite. According to a), all fibers under $\pi$ are separable and metrizable, so we can fix a countable dense subset $B_a$ in $\pi^{-1}(a)$, for each $a \in A$. Put $B = \bigcup\{B_a : a \in A\}$. Since the mapping $\pi$ is open and continuous, and $A$ is dense in $X/G$, the set $B$ is dense in $X$. Clearly, $|B| = |A|$. Hence, $d(X) \leq |B| = |A| = d(X/G)$. On the other hand, $d(X/G) \leq d(X)$, by the continuity of $\pi$. Hence, $d(X) = d(X/G)$.

For d), suppose that $X$ is monolithic. The same argument as in b) implies that $X/G$ is also monolithic. Therefore, $nw(X) = d(X)$ and $nw(X/G) = d(X/G)$. It follows that $nw(X) = nw(X/G)$. □

Clearly, if no restrictions are imposed on a space $X$, where $G$ acts, we can say nothing about the density or network weight of $X$, even if we know that the space $X/G$ of orbits is separable (just take the action of $G$ on itself by left translations). However, it turns out that we can always say something about the cellularity of $X$. Indeed, we have the following:

THEOREM 10.4.18. *Suppose that $G$ is a compact topological group acting continuously on a Hausdorff space $X$. Then the Souslin number of $X$ never exceeds the density of $X/G$. In particular, if $X/G$ is separable, then the Souslin number of $X$ is countable, and even more in this case, $\aleph_1$ is a precalibre of $X$, that is, every uncountable family of open sets in $X$ has a centered uncountable subfamily.*

PROOF. Suppose that $X/G$ is separable (in the general case, the argument is similar). Fix a countable dense subset $A$ in $X$ and, for each $a \in A$, consider the fiber $F_a = \pi^{-1}(a)$ of $a$ under the orbital projection $\pi\colon X \to X/G$. By Lemma 10.4.8, each $F_a$ is a dyadic compactum. The set $M = \bigcup\{F_a : a \in A\}$ is dense in $X$, since the mapping $\pi$ is open and continuous. Clearly, $M$ is $\sigma$-compact.

Let $\eta$ be an uncountable family of open sets in $X$. Clearly, there exists $a \in A$ such that the subfamily $\eta_a = \{U \in \eta : U \cap F_a \neq \varnothing\}$ is uncountable. Since $F_a$ is a dyadic compactum, it follows from Corollary 5.4.3 that some uncountable subfamily $\xi$ of $\eta_a$ is centered. □

An interesting general question is: Given a Hausdorff topological space, which compact groups admit a non-trivial continuous action on $X$? For example, it is clear that if $X$ is totally disconnected, then no connected group can act non-trivially on $X$. A related more precise question is: For which spaces $X$, every orbit under an action of a compact group on the space $X$ is finite? The following theorem gives a partial answer to this question.

THEOREM 10.4.19. *If $X$ is an extremally disconnected Hausdorff space, and $G$ is a compact topological group acting continuously on $X$, then every orbit $Gx$ is finite.*

PROOF. Indeed, by Lemma 10.4.8, $Gx$ is a dyadic compactum. If $Gx$ is infinite, then $Gx$ contains a non-trivial convergent sequence, by [**165**, 3.12.12 (i)]. However, there are no such sequences in any extremally disconnected Hausdorff space [**165**, 6.2.G (a)]. □

The cases when the compact group acting on a space $X$ is metrizable or just finite, deserve special consideration. We first make below an observation on the case of metrizable $G$ which is almost obvious after the above discussion. Much stronger results are in fact possible, but they are far from easily available. Then we will consider in detail the case of actions of compact groups with finite orbits.

THEOREM 10.4.20. *Suppose that a compact topological group $G$ is acting continuously on a Hausdorff space $X$ in such a way that each orbit $Gx$ is metrizable, and the orbital space $X/G$ is second-countable. Then $X$ is first-countable and separable.*

PROOF. The argument repeats some fragments of the proofs of Theorems 10.4.10 and 10.4.12. □

In the problems section below, much more advanced results are given compared to the last theorem (see Problems 10.4.B and 10.4.D).

Now we are going to establish a more delicate result on actions with finite orbits, Theorem 10.4.23. The key to it is the next purely topological statement on continuous mappings of spaces.

PROPOSITION 10.4.21. *Let $f$ be an open and closed continuous finite-to-one mapping of a Hausdorff space $X$ onto a space $Y$. Then $nw(X) = nw(Y)$ and $w(X) = w(Y)$.*

PROOF. Since $f$ is continuous and open, we obviously have that $nw(Y) \leq nw(X)$ and $w(Y) \leq w(X)$. Put $\tau = nw(Y)$, and fix a network $\mathcal{S}$ in $Y$ such that $|\mathcal{S}| = \tau$. For $n \in \omega$, put $Y_n = \{y \in Y : |f^{-1}(y)| = n\}$ and $X_n = f^{-1}(Y_n)$, and let $f_n$ be the restriction of $f$ to $X_n$, $f_n \colon X_n \to Y_n$. Then $X_0 = Y_0 = \varnothing$, and each $f_n$ is an open and closed continuous mapping, since $X_n$ is a full preimage of $Y_n$ under the mapping $f$, which is open, closed and continuous. Besides, each fiber $f_n^{-1}(y)$ under $f_n$ consists of exactly $n$ points, say, $f_n^{-1}(y) = \{x_1, \dots, x_n\}$. Since $X$ is Hausdorff, the subspace $X_n$ is also Hausdorff, and we can choose pairwise disjoint open neighbourhoods $Ox_1, \dots, Ox_n$ of points $x_1, \dots, x_n$ in the space $X_n$. Since $f_n$ is open and exactly $n$-to-one, the restriction of $f_n$ to each $Ox_i$ is a homeomorphism of $Ox_i$ onto $f(Ox_i)$. Therefore, each $f_n$ is a local homeomorphism, and each point $x$ of $X_n$ has an open neighbourhood in $X_n$ such that $nw(Ox) \leq \tau$.

Observe now that the Lindelöf degree $l(Y_n)$ of $Y_n$ does not exceed $nw(Y_n)$, by Proposition 5.3.3. Since the mapping $f_n$ is perfect, it follows from (an obvious generalization of) [**165**, Theorem 3.8.8] that $l(X_n) \leq \tau$. Using the fact that, locally, the network weight of $X_n$ does not exceed $\tau$, we can cover $X_n$ by a family $\eta_n$ of open in $X_n$ sets of network weight not greater than $\tau$. This clearly implies that $nw(X_n) \leq \tau$ (recall that, by the definition, cardinal invariants take only infinite values).

Since $f$ is finite-to-one, we have $X = \bigcup_{n=0}^{\infty} X_n$. Hence, $nw(X) \leq \tau \cdot \omega = nw(Y)$, whence the equality $nw(X) = nw(Y)$ follows.

To prove that $w(X) = w(Y)$, put $\tau = w(Y)$. Then, by the first part of the argument, $nw(X) = nw(Y) \leq w(Y) = \tau$. According to [**165**, Lemma 3.1.18], it follows that there exists a one-to-one continuous mapping $g$ of $X$ onto a Hausdorff space $M$ such that $w(M) \leq nw(X) \leq \tau$. Then the diagonal product $\phi$ of the mappings $f$ and $g$ is a homeomorphism of $X$ onto a subspace $Z$ of the product space $Y \times M$. Indeed, $\phi$ is perfect, since $f$ is perfect [**165**, Theorem 3.7.9], and $\phi$ is one-to-one, since $g$ is one-to-one. Clearly, $w(X) = w(Z) \leq w(Y \times M) = \tau$. Hence, $w(X) \leq \tau = w(Y)$ and, therefore, $w(X) = w(Y)$. □

THEOREM 10.4.22. *Suppose that a compact topological group $G$ is acting continuously on a Hausdorff space $X$ in such a way that each orbit $Gx$ is finite. Then $w(X) = w(X/G)$ and $nw(X) = nw(X/G)$.*

PROOF. By Theorem 10.2.13, it is enough to refer to Proposition 10.4.21. □

Here is an application of the above result:

THEOREM 10.4.23. *Suppose that a compact metrizable zero-dimensional topological group $G$ continuously acts on a Hausdorff space $X$ such that the orbital space $X/G$ is second-countable. Then $X$ is also second-countable.*

PROOF. By Theorem 3.1.11, we can choose a decreasing sequence $\xi = \{H_n : n \in \omega\}$ of open invariant subgroups with $H_0 = G$ such that $\xi$ is a local base at the neutral element of $G$. For every $n \in \omega$, put $X_n = X/H_n$ and let $p_n \colon X \to X_n$ be the orbital projection. Then $X$ is the limit space of the inverse spectrum $\mathcal{S} = \{X_n,\ p^n_m : m \leq n,\ m, n \in \omega\}$, where $p^n_m \colon X_n \to X_m$ are defined as the natural continuous mappings satisfying $p_m = p^n_m \circ p_n$, whenever $m \leq n$. Clearly, the quotient group $K_n = H_{n+1}/H_n$ is finite and, by Theorem 10.2.13, the orbital projections $p^n_m$ and $p_n$ are open and perfect. It follows that $X_n$ is naturally homeomorphic to the orbit space $X_{n+1}/K_n$, for each $n \in \omega$.

Since the group $K_n$ is finite, and the fibers of the mapping $p^{n+1}_n$ have cardinalities at most $|K_n|$, Theorem 10.4.22 applied to $p^{n+1}_n$ implies (by induction on $n$) that the space $X_{n+1}$ is second-countable. The diagonal product of perfect mappings $p_n$, where $n \in \omega$, is a homeomorphic embedding of $X$ into the product space $\prod_{n\in\omega} X_n$, since $X$ is the limit space of the spectrum $\mathcal{S}$. Hence, $X$ is second-countable as well. □

One of basic questions on a continuous action of a topological group $G$ on a topological space $X$ is whether there exists a continuous selection $s$ for the orbital projection $\pi \colon X \to X/G$. Recall that $s$ is a *continuous selection* for the mapping $\pi$ if $s$ is a continuous mapping of $X/G$ to $X$ such that $\pi(s(y)) = y$, for each $y \in X/G$. Unfortunately, such selections exist very seldom. This can be easily demonstrated with the help of the following statement:

PROPOSITION 10.4.24. *If $s$ is a continuous selection for the orbital projection $\pi$ corresponding to a continuous action of a topological group $G$ on a Hausdorff space $X$,* *then $s$ is a homeomorphism of the orbital space $X/G$ onto a closed subspace $s(X/G)$ of the space $X$.*

PROOF. Denote by $p$ the restriction of the orbital projection $\pi\colon X \to X/G$ to the subspace $Y = s(X/G)$ of $X$. Then clearly $p$ is continuous and we have that $p \circ s = id_{X/G}$ and $s \circ p = id_Y$. Therefore, $s$ is a homeomorphism between $X/G$ and $Y$. It is also clear that $r = \pi \circ s$ is a continuous retraction of $X$ onto its subspace $r(X) = s(X/G)$. Since $X$ is Hausdorff, $s(X/G)$ is closed in $X$. □

EXAMPLE 10.4.25. Consider the group $\mathbb{R}$ of reals, with the usual topology. The subgroup $G = \mathbb{Z}$ of integers acts continuously on the space $X = \mathbb{R}$; clearly, the space of orbits for this action is the circle $\mathbb{T}$, that is, $X/G = \mathbb{T}$. Since $\mathbb{T}$ is not homeomorphic to any subspace of $\mathbb{R}$, it follows that there is no continuous selection for the orbital projection in this case. However, one can easily define some partial selections locally. □

## Exercises

10.4.a. Give a detailed proof of Theorem 10.4.20.

10.4.b. Let $G = \{1, -1\}$ be the two-element multiplicative group. We define an action of $G$ on the Euclidean plane $X = \mathbb{R}^2$ as follows: $1x = x$ and $(-1)x = -x$, for each $x \in \mathbb{R}^2$ (that is, 1 acts as the identity mapping, and $-1$ acts as the symmetry with respect to the origin).
   a) Describe the space of orbits $X/G$.
   b) What are the topological properties of $X/G$? In particular, is $X/G$ metrizable? Is $X/G$ locally compact?
   c) Prove that for any open neighbourhood $U$ of the point $\{\bar{0}\} = \pi(0,0)$ of the space $X/G$, there is no continuous selection on $U$ for the orbital projection $\pi\colon X \to X/G$ restricted to $\pi^{-1}(U)$.

10.4.c. Suppose that the group $G = \{1, -1\}$ is acting continuously on a Hausdorff space $X$. Then either there exists a fixed point under the action of $G$, that is, $a \in X$ such that $Ga = \{a\}$, or the orbital projection $\pi\colon X \to X/G$ is a local homeomorphism, that is, for each $x \in X$, there is an open neighbourhood $U$ of $X$ such that $\pi$ restricted to $U$ is a homeomorphism of $U$ onto the open subspace $\pi(U)$ of $X/G$.

10.4.d. Suppose that $G$ is a finite group acting continuously on a Hausdorff space $X$ and that $|Gx| = |G|$, for each $x \in X$, that is, each orbit under the action of $G$ consists of exactly as many elements as $G$ itself. Prove that the orbital projection $\pi\colon X \to X/G$ is a local homeomorphism.

10.4.e. Suppose that a compact topological group $G$ acts continuously on a Hausdorff space $X$. Prove that for each $x \in X$, and for any open set $U$ in $X$ containing the orbit $Gx$, there exists an open neighbourhood $V$ of the orbit $Gx$ such that $V$ is $G$-invariant, that is, $GV = V$.

## Problems

10.4.A. Let $G$ be a metrizable compact topological group acting continuously on a Tychonoff space $X$ in such a way that the space $X/G$ of orbits is second-countable. Prove that $X$ is metrizable.

10.4.B. (V. V. Filippov [**170**]) Suppose that $G$ is a compact topological group acting continuously on a Tychonoff space $X$. Suppose further that each orbit $Gx$ is metrizable and that the space $X/G$ of orbits is second-countable. Prove that $X$ is metrizable.

10.4.C. Suppose that $G$ is a compact group acting continuously on a Hausdorff space $X$. Suppose further that the space $X/G$ of orbits is metrizable. Is $X$ Tychonoff?

10.4.D. Let $G$ be a compact topological group acting continuously on a compact Hausdorff space $X$ in such a way that the space $X/G$ of orbits is second-countable. Prove that $X$ is Dugundji.

10.4.E. Let $p$ be the natural projection of the two arrows space onto the closed unit interval. Prove that $p$ cannot be represented as the orbital projection of the two arrows space under a continuous action of a compact group.

10.4.F. Give an example of an open continuous mapping $p$ of a non-metrizable compact Hausdorff space $X$ onto a metrizable compact space $Y$. Observe that $p$ cannot be represented as the orbital projection of any space under a continuous action of a compact group.

10.4.G. Let $X = C_p(X)$ be the space of continuous real-valued functions on a compact Hausdorff space $X$, in the topology of pointwise convergence. Prove that if a compact group $G$ act continuously on $X$ in such a way that the orbital space $X/G$ is separable or first-countable, then $X/G$ is cosmic.

10.4.H. (V. V. Filippov [**170**]) Suppose that $G$ is a compact topological group acting continuously on a Tychonoff space $X$. Suppose further that each orbit $Gx$ is metrizable and that the space $X/G$ of orbits is metrizable. Prove that $X$ is metrizable.

10.4.I. (V. V. Filippov [**170**]) Suppose that $G$ is a compact group acting continuously on a Tychonoff space $X$. Suppose further that each orbit $Gx$ is metrizable. Prove that the weight of the orbital space $X/G$ is equal to the weight of the space $X$.

10.4.J. Suppose that $G$ is a compact group acting continuously on a space $X$ metrizable by a complete metric. Prove that the space $X/G$ of orbits is also metrizable by a complete metric.

10.4.K. Suppose that $G$ is a compact topological group acting continuously on a realcompact space $X$. Prove that the space $X/G$ of orbits is also realcompact.
*Hint.* See [**165**, 3.11.G].

10.4.L. Recall that $B(\tau) = D(\tau)^\omega$ is the Baire space of weight $\tau \geq \omega$, where $D(\tau)$ is a discrete space of cardinality $\tau$. The space $B(\tau)$ is zero-dimensional and metrizable by a complete metric (see Theorem 4.3.12 and Example 7.3.14 of [**165**]). Suppose that $G$ is a compact group acting continuously on the Baire space $X = B(\tau)$, for some $\tau \geq \omega$. Prove that the space $X/G$ of orbits is homeomorphic to $B(\tau)$.
*Hint.* See [**60**, Ch. 6, no. 149].

10.4.M. Recall that a *linearly ordered space* is a topological space $X$ such that the topology of $X$ is generated by some linear order on the set $X$. Suppose that $G$ is a compact topological group acting continuously on a linearly ordered space $X$. Prove that:
a) $w(X/G) = w(X)$;
b) if the space $X/G$ is metrizable, then $X$ is also metrizable.
*Hint.* Apply the fact that every linearly ordered dyadic compactum is metrizable [**165**, 3.12.12 (f)].

10.4.N. The following assertion is consistent with $ZFC$: If $\phi\colon G \times X \to X$ is a continuous and transitive action of a countably compact group $G$ on a compact space $X$, then $X$ is homeomorphic to the quotient space $G/H$, where $H$ is a closed subgroup of $G$.
*Hint.* Fix $x \in X$ and for each $g \in G$, put $f(g) = \phi((g, x))$. Then $f$ is a continuous mapping of $G$ onto $X$, since the action $\phi$ is continuous and transitive. Then $X$ is a dyadic compactum. Let $H = \{g \in G : g(x) = x\}$. Then $H$ is a closed subgroup of $G$, and $f$ is the composition of the canonical quotient mapping $\pi$ of $G$ onto the quotient space $G/H$ and of a one-to-one continuous mapping $i$ of $G/H$ onto $X$. The space $G/H$ is countably compact and Tychonoff. One can show now that, consistently, $i$ is a homeomorphism (see [**34**]).

10.4.O. If a countably compact topological group $G$ of countable tightness acts continuously and transitively on a compact space $X$, then $X$ is metrizable.
*Hint.* Clearly, there exists a one-to-one continuous mapping of the quotient space $G/H$, where $H$ is a closed subgroup of $G$, onto $X$. The tightness of $G/H$ is countable and $G/H$ is countably compact. It remains to refer to Problem 9.8.N.

## Open Problems

10.4.1. Let $G$ be a compact metrizable topological group acting continuously on a Tychonoff space $X$ in such a way that the space $X/G$ of orbits is submetrizable. Is $X$ submetrizable?

10.4.2. Let $G$ be a compact metrizable group acting continuously on a Tychonoff space $X$ in such a way that the space of orbits satisfies $iw(X/G) \leq \omega$. Is it true that $iw(X) \leq \omega$?

10.4.3. Let $G$ be a compact topological group acting continuously on a compact Hausdorff space $X$ in such a way that the space $X/G$ of orbits is a Dugundji compactum. Is $X$ Dugundji?

## 10.5. Historical comments to Chapter 10

Continuous actions of topological groups on topological spaces provide a natural general background for topological dynamics. At the heart of it is the fundamental idea of a topological group of transformations; and topological dynamics, in its general setting, is closely related to the theory of topological semigroups. Concrete prototypes of general concepts and problems of topological dynamics came from differential equations. V. V. Niemytzki and K. S. Sibirski were among those who contributed to this discipline at the early stage. The first monograph on general topological dynamics was [**200**]. For some important modern aspects of the theory see the works of D. V. Anosov, K. Kuperberg, and J. Kennedy.

We should warn the reader that we provide only a very sketchy introduction to this vast subject to which quite a few extensive monographs (see below) have been devoted. In particular, see [**530**].

The roots of the notion of 0-soft mapping can be found in the seminal work of R. Haydon [**221**]. The concept itself was conceived and introduced by E. V. Schepin in [**420**] (for a more advanced treatment of it, see [**422**]). In [**516**], V. V. Uspenskij showed how the methods created by Haydon and Schepin can be applied to prove that every compact topological group is a Dugundji space. He also proved a generalization of this theorem obtained by M. M. Choban: Every compact $G_\delta$-set in a topological group is a Dugundji compactum (see [**101**], [**516**], [**517**]). The important class of Dugundji compacta was introduced by A. Pełczyński in [**368**], where a series of influential questions was raised. R. Haydon proved in [**221**] that every Dugundji compactum is dyadic. It should be mentioned that the technique of inverse spectra introduced to topology by P. S. Alexandroff, was greatly developed in the fundamental papers of Haydon and Schepin mentioned above.

In connection with Proposition 10.1.1 and Corollary 10.1.2, see [**368**]. Theorem 10.1.3 is due to R. Haydon [**221**]. For Propositions 10.1.4 and 10.1.5 see [**420**] and [**422**], where a discussion of elementary results on inverse spectra, such as Propositions 10.1.6, 10.1.9, 10.1.12, and 10.1.13 can be also found. Theorem 10.1.8 and Example 10.1.10 are old results (see [**8**]); for further references see [**8**]. Theorem 10.1.16 belongs to Haydon [**221**]. In connection with Theorem 10.1.15, Propositions 10.1.17, 10.1.20, Corollaries 10.1.18, 10.1.21, 10.1.23, and for further references, see [**516, 517**].

The general references for continuous actions of topological groups on topological spaces are the books of J. de Vries [**530**], G. E. Bredon [**86**], and R. Ellis [**161**]. An older source is the book of W. Gottshalk and G. A. Hedlund [**200**]. A recent nice introduction to topological dynamics is given in [**271**]. See also [**2**]. In particular, for further references in connection with Propositions 10.2.1, 10.2.2, 10.2.7, 10.2.8, 10.2.12, Corollary 10.2.3, and Theorem 10.2.13 see [**530**], [**86**].

Theorems 10.3.5 and 10.3.8 are due to Uspenskij (see [**517, 516, 515**]). The proof of these results presented in the book, in particular, Proposition 10.3.1, Lemma 10.3.2, Example 10.3.3, closely follow Uspenskij's argument. In connection with Corollary 10.3.9 see [**101**]. Probably, Theorem 10.3.10 and Corollary 10.3.11 appear in print for the first time.

# Bibliography


[1] Ajtai, M., Havas, I., and Komlós, J. (1983). Every group admits a bad topology, *Studies in Pure Math.*, To the Memory of P. Turan, pp. 21–34.

[2] Akin, E. (1993). *The General Topology of Dynamical Systems*, Graduate Stud. in Math. **1** (Amer. Math. Soc., Providence, RI).

[3] Alas, O. T. (1971). Topological groups and uniform continuity, *Portugal. Math.* **50**, 3, pp. 137–143.

[4] Alas, O. T. and Sanchis, M. (2007). Countably compact paratopological groups, *Semigroup Forum* **74**, 3, pp. 423–438.

[5] Alas, O. T., Tkachenko, M. G., Tkachuk, V. V., and Wilson, R. G. (1996). Connectifying some spaces, *Topol. Appl.* **71**, pp. 203–215.

[6] Alexandroff, A. D. (1942). On the extension of a Hausdorff space to an $H$-closed space, *Dokl. Acad. Sci. URSS* **37**, pp. 118–121. Russian original in: *Dokl. Akad. Nauk SSSR* **37**, pp. 138–141.

[7] Alexandroff, P. S. (1936). To the theory of topological spaces, *Dokl. Akad. Nauk SSSR* **2**, pp. 51–54 (*in Russian*).

[8] Alexandroff, P. S. and Pasynkov, B. A. (1973). *Introduction to Dimension Theory* (Nauka, Moscow; *in Russian*).

[9] Alfsen, E. M. and Holm, P. (1962). A note on compact representations and almost periodicity in topological groups, *Math. Scand.* **10**, pp. 127–136.

[10] Antonyan, S. A. and Sanchis, M. (2002). Extension of locally group actions, *Annali di Matematica Pura ed Applicata* **181**, pp. 239–246.

[11] Anzai, H. and Kakutani, S. (1943). Bohr compactifications of a locally compact Abelian group I, *Proc. Imperial Acad. Tokyo* **19**, pp. 476–480.

[12] Arens, R. (1946). Topologies for homeomorphism groups, *Amer. J. Math.* **68**, pp. 593–610.

[13] Arens, R. (1951). The adjoint of a bilinear operation, *Proc. Amer. Math. Soc.* **2**, pp. 839–848.

[14] Arhangel'skii, A. V. (1963). A class of spaces which contains all metric spaces and all locally compact spaces, *Soviet Math. Dokl.* **4**, pp. 1051–1055. Russian original in: *Dokl. Akad. Nauk SSSR* **151**, pp. 751–754.

[15] Arhangel'skii, A. V. (1965). Bicompact sets and the topology of spaces, *Trans. Mosc. Math. Soc.* **13**, pp. 1–62. Russian original in: *Trudy Mosk. Mat. Obsch.* **13**, pp. 3–55.

[16] Arhangelskii, A. V. (1965). A class of spaces which contains all metric and all locally compact spaces, *Amer. Math. Soc. Transl.* **92** (1970) pp. 1–39. Russian original in: *Mat. Sb. (N.S.)* **67 (109)** (1965), pp. 55–88.

[17] Arhangel'skii, A. V. (1967). Groupes topologiques extremalement discontinus, *C. R. Acad. Sci. Paris* **265**, pp. 822–825.

[18] Arhangel'skii, A. V. (1968). Mappings connected with topological groups, *Soviet Math. Dokl.* **9**, pp. 1011–1015. Russian original in: *Dokl. AN SSSR* **181**, pp. 1303–1307.

[19] Arhangel'skii, A. V. (1969). *Topological spaces and continuous mappings. Notes on topological groups*, Moscow State Univ., Moscow (*in Russian*).

[20] Arhangel'skii, A. V. (1978). Structure and classification of topological spaces and cardinal invariants, *Russian Math. Surveys* **33**, pp. 33–96. Russian original in: *Uspekhi Mat. Nauk* **33**, 6, pp. 29–84.

[21] Arhangel'skii, A. V. (1979). Cardinal invariants of topological groups. Embeddings and condensations, *Soviet Math. Dokl.* **20**, pp. 783–787. Russian original in: *Dokl. Akad. Nauk SSSR* **247**, pp. 779–782.

[22] Arhangelskii, A. V. (1980). Relations among the invariants of topological groups and their subspaces, *Russian Math. Surveys* **35**, pp. 1–23. Russian original in: *Uspekhi Mat. Nauk* **35**, pp. 3–22.

[23] Arhangel'skii, A. V. (1981). Every topological group is a quotient group of a zero-dimensinal topological group, *Soviet Math. Dokl.* **23**, 3, pp. 615–618. Russian original in: *Dokl. AN SSSR* **258**, pp. 1037–1040.

[24] Arhangel'skii, A. V. (1981). Classes of topological groups, *Russian Math. Surveys* **36**, 3, pp. 151–174. Russian original in: *Uspekhi Mat. Nauk* **36**, pp. 127–146.

[25] Arhangel'skii, A. V. (1982). Factorization theorems and function spaces: Stability and monolicithity, *Soviet Math. Dokl.* **26**, pp. 177–181. Russian original in: *Dokl. Akad. Nauk SSSR* **265**, 5, pp. 1039–1043.

[26] Arhangel'skii, A. V. (1983). Functional tightness, $Q$-spaces, and $\tau$-embeddings, *Comment. Math. Univ. Carolin.* **24**, 1, pp. 105–120.

[27] Arhangel'skii, A. V. (1984). Continuous mappings, factorization theorems and function spaces, *Proc. Moscow Math. Soc.* **47**, pp. 1–22. Russian original in: *Trudy Mosk. Mat. Obsch.* **47**, pp. 3–21.

[28] Arhangel'skii, A. V. (1987). Topological homogeneity. Topological groups and their continuous images, *Russian Math. Surveys* **42**, 2, pp. 83–131. Russian original in: *Uspekhy Mat. Nauk* **42**, 2, pp. 69–105.

[29] Arhangel'skii, A. V. (1987). Algebraic objects generated by topological structure, *J. Soviet Math.* **45**, 1 (1989), pp. 956–990. Russian original in: *Itogi Nauki i Tekhiki, Ser. Algebra, Topologiya, Geometriya* **25** (1987), pp. 141–198.

[30] Arhangel'skii, A. V. (1988). *Cantor's Set Theory*, (Moscow State Univ., Moscow; *in Russian*).

[31] Arhangel'skii, A. V. (1990). On biradial topological spaces and groups, *Topol. Appl.* **36**, pp. 173–180.

[32] Arhangel'skii, A. V. (1992). *Topological Function Spaces* (Kluwer Academic Publ.).

[33] Arhangel'skii, A. V. (1994). Bisequential spaces, tightness of products, and metrizability conditions in topological groups, *Trans. Mosc. Math. Soc.* **55**, pp. 207–219.

[34] Arhangel'skii, A. V. (1994). On countably compact topologies on compact groups and on dyadic compacta, *Topol. Appl.* **57**, pp. 163–181.

[35] Arhangel'skii, A. V. (1999). On a theorem of W. W. Comfort and K. A. Ross, *Comment. Math. Univ. Carolin.* **40**, 1, pp. 133–151.

[36] Arhangel'skii, A. V. (1999). On $P$-spaces, extremal disconnectedness, and some results of J. R. Isbell and V. I. Malykhin, *Questions and Answers in Gen. Topol.* **17**, pp. 257–267.

[37] Arhangel'skii, A. V. (2000). Moscow spaces, Pestov–Tkachenko Problem, and $C$-embeddings, *Comment. Math. Univ. Carolin.* **41**, 3, pp. 585–595.

[38] Arhangel'skii, A. V. (2000). On topological and algebraic structure of extremally disconnected semitopological groups, *Comment. Math. Univ. Carolin.* **41**, 4, pp. 803–810.

[39] Arhangel'skii, A. V. (2000). Moscow spaces and topological groups, *Topol. Proc.* **25**, pp. 383–416.

[40] Arhangel'skii, A. V. (2001). Topological groups and $C$-embeddings, *Topol. Appl.* **115**, 3, pp. 265–289.

[41] Arhangel'skii, A. V. (2002). On power homogeneous spaces, *Topol. Appl.* **122**, 1, pp. 15–33.

[42] Arhangel'skii, A. V. (2002). Topological invariants in algebraic environment, In: *Recent Progress in General Topology II*, M. Hušek and J. van Mill, eds., pp. 1–57 (North-Holland, Amsterdam).

[43] Arhangel'skii, A. V. (2002). Diagonal structures on topological spaces, *Topol. Appl.* **125**, pp. 419–446.

[44] Arhangel'skii, A. V. (2002). When the Dieudonné completion of a topological group is a $p$-space? *Topol. Proc.* **25**, 2, pp. 1–15.

[45] Arhangel'skii, A. V. (2002). From classical topological invariants to relative topological properties, *Sci. Math. Japon.* **55**, 1, pp. 153–201.

[46] Arhangel'skii, A. V. (2003). Perfect mappings in topological groups, cross-complementary subsets, and quotients, *Comment. Math. Univ. Carolin.* **44**, 4, pp. 701–709.

[47] Arhangelskii, A. V. (2005). Quotients with respect to locally compact subgroups, *Houston J. Math.* **31**, 1, pp. 215–226.

[48] Arhangel'skii, A. V. (2005). Remainders in compactifications and generalized metrizability properties, *Topol. Appl.* **150**, pp. 79–90.

[49] Arhangel'skii, A. V., Private communication, November 2007.

[50] Arhangel'skii, A. V. and Bella, A. (2007). The diagonal of a first countable paratopological group, submetrizability, and related results, *Applied General Topology* **8**, 2, pp. 207–212.

[51] Arhangel'skii, A. V. and Burke, D. K. (2006). Spaces with a regular $G_\delta$-diagonal, *Topol. Appl.* **153**, 11, pp. 1917–1929.

[52] Arhangel'skii, A. V. and Collins, P. J. (1995). On submaximal spaces, *Topol. Appl.* **64**, pp. 219–241.

[53] Arhangel'skii, A. V. and Genedi, K. M. M. (1993). Properties of placement type: Relative strong pseudocompactness, *Proc. Steklov Inst. Math.* **3**, pp. 25–27.

[54] Arhangelskii, A. V. and Hušek, M. (2001). Extensions of topological and semitopological groups and the product operation, *Comment. Math. Univ. Carolin.* **42**, 1, pp. 173–186.

[55] Arhangel'skii, A. V. and Just, W. (1995). Dense, sequentially continuous maps on dyadic compacta, *Topol. Appl.* **64**, 1, pp. 95–99.

[56] Arhangel'skii, A. V., Just, W., and Plebanek, G. (1996). Sequential continuity on dyadic compacta and topological groups, *Comment. Math. Univ. Carolin.* **37**, 4, pp. 775–790.

[57] Arhangel'skii, A. V., Just, W., Reznichenko, E. A., and Szeptycki, P. (2000). Sharp bases and weakly uniform bases versus point-countable bases, *Topol. Appl.* **100**, pp. 39–46.

[58] Arhangel'skii, A. V., Okunev, O. G., and Pestov, V. G. (1989). Free topological groups over metrizable spaces, *Topol. Appl.* **33**, pp. 63–76.

[59] Arhangel'skii, A. V. and Ponomarev, V. I. (1968). On dyadic bicompacta, *Soviet Math. Dokl.* **9**, pp. 1220–1224. Russian original in: *Dokl. Akad. Nauk SSSR* **182**, pp. 993–996.

[60] Arhangel'skii, A. V. and Ponomarev, V. I. (1984). *Fundamentals of General Topology: Problems and Exercises* (Reidel, translated from Russian).

[61] Arhangel'skii, A. V. and Ranchin, D. V. (1982). Dense subspaces of topological products, and properties related to Lindelöfness, *Moscow Univ. Math. Bull.* **37**, 6, pp. 21–29. Russian original: *Vestnik Moskov. Univ. Ser. 1 Mat. Mekh.* **6**, pp. 21–28.

[62] Arhangel'skii, A. V. and Reznichenko, E. A. (2005). Paratopological and semitopological groups versus topological groups, *Topol. Appl.* **151**, pp. 107–119.

[63] Arhangel'skii, A. V. and Uspenskij, V. V. (2006). Topological groups: local versus global, *Appl. General Topology* **7**, 1, pp. 67-72.

[64] Arnautov, V. I. and Zenenyuk, E. G. (1991). On the problem of the completeness of maximal topological groups, *Ukrain. Math. J.* **43**, 1, pp. 15–20. Russian original in: *Ukrain. Mat. Zh.* **43**, 1, pp. 21–27.

[65] Baker, J. and Butcher, R. (1976). The Stone–Čech compactification of a topological semigroup, *Math. Proc. Cambridge Phil. Soc.* **80**, pp. 103–107.

[66] Banach, S. (1948). *Kurs Funkzional'nogo Analiza* (Kiiv; *in Ukrainian*).

[67] Banakh, T. (2000). Locally minimal topological groups and their embeddings into products of $o$-bounded groups, *Comment. Math. Univ. Carolin.* **41**, 4, pp. 811–815.

[68] Banakh, T. and Ravsky, O. (2003). On subgroups of saturated or totally bounded paratopological groups, *Algebra Discrete Math.* 4, pp. 1–20.

[69] Becker, H. (1998). Polish group actions: dichotomies and generalized elementary embeddings, *J. Amer. Math. Soc.* **11**, pp. 397–449.

[70] Bell, M. G. (1994). Special points in compact spaces, *Proc. Amer. Math. Soc.* **122**, pp. 619–624.

[71] Bel'nov, V. K. (1976). On zero-dimensional topological groups, *Dokl. AN SSSR* **226**, 4, pp. 749–753 (*in Russian*).

[72] Bel'nov, V. K. (1979). On dimension of free topological groups, *Proc. IVth Tiraspol Symposium on General Topology*, pp. 14–15 (*in Russian*).

[73] Berglund, J., Junghenn, H., and Milnes, P. (1978). *Compact right topological semigroups and generalizations of almost periodicity*, Lecture Notes in Math. **663** (Springer-Verlag, Berlin).

[74] Bessaga, C. and Pełczyński, A. (1975). *Selected Topics in Infinite-dimensional Topology* (PWN, Warszawa).

[75] Birkhoff, G. (1936). A note on topological groups, *Comput. Math.* **3**, pp. 427–430.

[76] Blair, R. L. (1976). Spaces in which special sets are $z$-embedded, *Canad. J. Math.* **28**, pp. 673–690.

[77] Blair, R. L. and Hager, A. W. (1975). Notes on the Hewitt realcompactification of a product, *Gen. Topol. Appl.* **5**, pp. 1–8.

[78] Bokalo, B. M. and Guran, I. I. (1996). Sequentially compact Hausdorff cancellative semigroup is a topological group, *Mat. Studii* **6**, pp. 39–40.

[79] Borges, C. R. (1985). Quotient topologies on free topological groups, *Math. Japonica* **30**, pp. 835–837.

[80] Bourbaki, N. (1942). *Éléments de Mathématique*, Premiere Partie, Livre 3, 3-m ed., Actualités Sci. et Ind. no. 916 (Hermann, Paris; *in French*).

[81] Bourbaki, N. (1966). *Éléments de Mathématique*, Fasc. XV. Livre 5: Espaces vectoriels topologiques, Actualités Sci. et Ind. no. 1189 (Hermann, Paris; *in French*).

[82] Bouziad, A. (1982). Another note on the continuity of the inverse, *Arch. Math.* **39**, pp. 241–245.

[83] Bouziad, A. (1993). The Ellis theorem and continuity in groups, *Topol. Appl.* **50**, pp. 73–80.

[84] Bouziad, A. (1996). Continuity of separately continuous group actions on $p$-spaces, *Topol. Appl.* **71**, pp. 119–124.

[85] Bouziad, A. (1998). Every Čech-analytic Baire semitopological group is a topological group, *Proc. Amer. Math. Soc.* **24**, 3, pp. 953–959.

[86] Bredon, G. E. (1972). *Introduction to compact transformation groups* (Academic Press, New York–London).

[87] Brouwer, L. E. J. (1910). On the structure of perfect sets of points, *Proc. Acad. Amsterdam* **12**, pp. 785–794.

[88] Brown, L. G. (1972). Topologically complete groups, *Proc. Amer. Math. Soc.* **35**, pp. 593–600.

[89] Bruguera, M. and Tkachenko, M. G. (2004). Extensions of topological groups do not respect countable compactness, *Questions and Answers in Gen. Topol.* **22**, 1, pp. 33–37.

[90] Bruguera, M. and Tkachenko, M. G. (2006). The three space problem in topological groups, *Topol. Appl.* **153**, 13, pp. 2278–2302.

[91] Bruguera, M. and Tkachenko, M. G. (2007). Bounded sets in topological groups and embeddings, *Topol. Appl.* **154**, 7, pp. 1298–1306.

[92] Buchwalter, H. (1972). Produit topologique, produit tensoriel et $c$-repletion, *Bull. Soc. Math. France Suppl. Mém.* **31**, pp. 51–71.

[93] Burke, D. K. (1984). Covering properties, In: *Handbook of Set-Theoretic Topology*, K. Kunen and J. Vaughan, eds., pp. 347–422 (North-Holland, Amsterdam).

[94] Buzyakova, R. Z. (1997). An example of a product of two normal groups that cannot be condensed onto a normal space, *Moscow Univ. Math. Bull.* **52**, 3, p. 42. Russian original in: *Vestnik Moskov. Univ. Ser. I Mat. Mekh.* 3, p. 59.

[95] Buzyakova, R. Z. (2004). Hereditary $D$-property of function spaces over compacta, *Proc. Amer. Math. Soc.* **132**, 11, pp. 3433–3439.

[96] Cartan, É. (1930). La Théorie des Groupes Finis et Continus et l'Analysis Situs, *Mémorial des Sci. Math.* **42** (Gauthier-Villars&Cie, Paris).

[97] Chen, Y. Q. (1999). Note on two questions of Arhangel'skii, *Questions and Answers in Gen. Topol.* **17**, pp. 91–94.

[98] Černavskii, A. V. (1969). Local contractibility of the homeomorphism group of a manifold, *Math. USSR Sb.* **8**, pp. 287–333. Russian original in: *Mat. Sbornik* **79**, pp. 307–356.

[99] Chevalley, C. (1946). *Theory of Lie Groups I*, Princeton Univ. Press (Princeton, New York).

[100] Choban, M. M. (1970). On completions of topological groups, *Vestnik Moskov. Univ. Ser. Mat. Mekh.* 1, pp. 33–38 (*in Russian*).

[101] Choban, M. M. (1977). Topological structure of subsets of topological groups and their quotients, In: *Topological Structures and Algebraic Systems*, pp. 117–163 (Shtiintsa, Kishinev; *in Russian*).

[102] Choban, M. M. (1985). On the theory of topological algebraic systems, *Trudy Moskov. Mat. Obshch.* **48**, pp. 106–149 (*in Russian*).

[103] Choban, M.M. and Kiriyak, L. L. (1992). Application of uniform structures in the investigation of free topological algebras, *Sib. Math. J.* **33**, 5, pp. 891–904. Russian original in: *Sib. Mat. Zh.* **33**, 5, pp. 159–175.

[104] Civin, P. and Yood, B. (1961). The second conjugate space of a Banach algebra as an algebra, *Pacific J. Math.* **11**, pp. 847–870.

[105] Cleary, J. and Morris, S. A. (1989). Locally dyadic topological groups, *Bull. Austral. Math. Soc.* **40**, pp. 417–419.

[106] Clemens, J. D., Gao, S., and Kechris, A. S. (2001). Polish metric spaces: their classification and isometry groups, *Bull. Symb. Logic* **7**, pp. 361–375.

[107] Comfort, W. W. (1968). On the Hewitt realcompactification of the product space, *Trans. Amer. Math. Soc.* **131**, pp. 107–118.

[108] Comfort, W. W. (1975). Compactness-like properties for generalized weak topological sums, *Pacific J. Math.* **60**, pp. 31–37.

[109] Comfort, W. W. (1984). *Topological groups*, In: *Handbook of Set-Theoretic Topology*, K. Kunen and J. E. Vaughan, eds., Ch. 24, pp. 1143–1263 (North-Holland, Amsterdam).

[110] Comfort, W. W. (1990). *Problems on topological groups and other homogeneous spaces*, In: *Open Problems in Topology*, J. van Mill and G. M. Reed, eds., pp. 313–348 (North-Holland, Amsterdam).

[111] Comfort, W. W. and García-Ferreira, S. (2000). Dense subsets of maximally almost periodic groups, *Proc. Amer. Math. Soc.* **129**, 2, pp. 593–599.

[112] Comfort, W. W., Hernández, S., and Trigos-Arrieta, F. J. (1996). Relating a locally compact Abelian group to its Bohr compactification, *Advances in Math.* **120**, pp. 322–344.

[113] Comfort, W. W., Hofmann, K.-H., and Remus, D. (1992). Topological groups, In: *Recent Progress in General Topology*, M. Hušek and J. van Mill, eds. (North-Holland, Amsterdam).

[114] Comfort, W. W. and Mill, J. van (1989). Concerning connected, pseudocompact Abelian groups, *Topol. Appl.* **33**, pp. 21–45.

[115] Comfort, W. W. and Mill, J. van (1994). Groups with only resolvable group topologies, *Proc. Amer. Math. Soc.* **120**, pp. 687–696.

[116] Comfort, W. W. and Mill, J. van (2007). Extremal pseudocompact Abelian groups are compact metrizable, Preprint.

[117] Comfort, W. W. and Negrepontis, S. (1966). Extending continuous functions on $X \times Y$ to subsets of $\beta X \times \beta Y$, *Fund. Math.* **59**, pp. 1–12.

[118] Comfort, W. W. and Remus, D. (1994). Compact groups of Ulam-measurable cardinality: partial converse to theorems of Arhangel'skii and Varopoulos, *Math. Japonicae* **39**, pp. 203–210.

[119] Comfort, W. W. and Robertson, L. C. (1985). Cardinality constraints for pseudocompact and for totally dense subgroups of compact groups, *Pacific J. Math.* **119**, pp. 265–285.

[120] Comfort, W. W. and Robertson, L. C. (1988). Extremal phenomena in certain classes of totally bounded groups, *Dissert. Math.* **272**, pp. 1–48.

[121] Comfort, W. W. and Ross, K. A. (1964). Topologies induced by groups of characters, *Fund. Math.* **55**, pp. 283–291.

[122] Comfort, W. W. and Ross, K. A. (1966). Pseudocompactness and uniform continuity in topological groups, *Pacific J. Math.* **16**, 3, pp. 483–496.

[123] Comfort, W. W. and Saks, V. (1973). Countably compact groups and finest totally bounded topologies, *Pacific J. Math.* **49**, pp. 33–44.

[124] Comfort, W. W. and Soundararajan, T. (1982). Psudocompact group topologies and totally dense subgroups, *Pacific J. Math.* **100**, pp. 61–84.

[125] Comfort, W. W. and Trigos-Arrieta, F. J. (1991). Remarks on a theorem of Glicksberg, In: *General Topology and Appl.*, S. J. Andima, R. Copperman, P. R. Misra, J. Z. Reichman, and A. R. Todd, eds., pp. 25–33 (Marcel Dekker, Inc., New York–Basel–Hong Kong).

[126] Comfort, W. W. and Trigos-Arrieta, F. J. (1995). Locally pseudocompact topological groups, *Topol. Appl.* **62**, 3, pp. 263–280.

[127] Cotlar, M. and Ricabarra, R. (1954). On the existence of characters in topological groups, *Amer. J. Math.* **76**, pp. 375–388.

[128] Dantzig, D. van, (1932). Zur topologischen Algebra 1. Komplettierungstheorie, *Math. Ann.* **107**, pp. 587–626.

[129] Day, M. (1957). Amenable semigroups, *Illinois J. Math.* **1**, pp. 509–544.

[130] de Leo, L. and Tkachenko, M. G. (2008). Maximal $\omega$-narrow topological group topologies, *Houston J. Math.*, to appear.

[131] Dieudonné, J. (1944). Sur la complétion des groupes topologiques, *C. R. Acad. Sci. Paris* **218**, pp. 774–776.

[132] Dieudonné, J. On topological groups of homeomorphisms, *Amer. J. Math.* **70** (1948), 659–680.

[133] Dierolf, S. and Schwanengel, U. (1979). Examples of locally compact non-compact minimal topological groups, *Pacific J. Math.* **82**, pp. 349–355.

[134] Dikranjan, D. (1993). Dimension and connectedness in pseudocompact groups, *C. R. Acad. Sci. Paris* **316**, Série I, pp. 309–314.

[135] Dikranjan, D. (1994). Zero-dimensionality of some pseudocompact groups, *Proc. Amer. Math. Soc.* **120**, pp. 1299–1308.

[136] Dikranjan, D. (1998). Compactness and connectedness in topological groups, *Topol. Appl.* **84**, pp. 227–252.

[137] Dikranjan, D. (1999). Countably compact groups satisfying the open mapping theorem, *Topol. Appl.* **98**, pp. 81–129.

[138] Dikranjan, D. and de Leo, L. (2008). Straightening theorem for bounded Abelian groups, *Topol. Appl.*, to appear.

[139] Dikranjan, D. N., Prodanov, I., and Stoyanov, L. (1989). *Topological groups: characters, dualities and minimal group topologies*, Monographs and Textbooks in Pure and Apllied Mathematics **130** (Marcel Dekker Inc., New York–Basel).

[140] Dikranjan, D. and Shakhmatov, D. (1998). Algebraic structure of the pseudocompact groups, *Memoirs Amer. Math. Soc.* **133/633**, April, pp. viii+83.

[141] Dikranjan, D. and Shakhmatov, D. (2005). Forcing hereditarily separable compact-like group topologies on Abelian groups, *Topol. Appl.* **151**, pp. 2–54.

[142] Dikranjan, D. and Tkachenko, M. G. (2001). Weakly complete free topological groups, *Topol. Appl.* **112**, 3, pp. 259–287.

[143] Dikranjan, D. and Tkachenko, M. G. (2002). Varieties generated by countably compact Abelian groups, *Proc. Amer. Math. Soc.* **130**, pp. 2487–2496.

[144] Dikranjan, D. and Tkachenko, M. G. (2003). Algebraic structure of small countably compact groups, *Forum Math.* **15**, 6, pp. 811–837.

[145] Dikranjan, D., Tkachenko, M. G., and Tkachuk, V. V. (2000). Topological groups with thin generating sets, *J. Pure Appl. Algebra* **145**, pp. 123–148.

[146] Dikranjan, D. and Watson, S. (2001). A solution to van Douwen's problem on Bohr topologies, *J. Pure Appl. Algebra* **163**, 2, pp. 147–158.

[147] Douwen, E. K. van (1979). Homogeneity of $\beta G$ if $G$ is a topological group, *Colloq. Math.* **41**, pp. 193–199.

[148] Douwen, E. K. van (1980). The weight of a pseudocompact (homogeneous) space whose cardinality has countable cofinality, *Proc. Amer. Math. Soc.* **80**, pp. 678–682.

[149] Douwen, E. K. van (1980). The product of two countably compact topological groups, *Trans. Amer. Math. Soc.* **262**, pp. 417–427.

[150] Douwen, E. K. van (1990). The maximal totally bounded group topology on $G$ and the biggest minimal $G$-space for Abelian groups $G$, *Topol. Appl.* **34**, pp. 69–91.

[151] Douwen, E. K. van (1991). The Čech–Stone compactification of a discrete grupoid, *Topol. Appl.* **39**, pp. 43–60.

[152] Douwen, E. K. van (1993). Applications of maximal topologies, *Topol. Appl.* **51**, pp. 125–139.

[153] Dranishnikov, A. N. (1988). Free actions of zero-dimesional compact groups, *Math. USSR Izv.* **32**, 1 (1989), pp. 217–232. Russian original in: *Izv. Akad. Nauk SSSR Ser. Mat.* **52**, 1 (1988), pp. 212–228.

[154] Eda, K., Kamo, S., and Ohta, H. (1993). Abelian groups of continuous functions and their duals, *Topol. Appl.* **53**, pp. 131–151.

[155] Eda, K. and Ohta, H. (1985). On Abelian groups of integer-valued continuous functions, their $\mathbb{Z}$-duals and $\mathbb{Z}$-reflexivity, In: *Abelian Group Theory*, R. Göbel and E. Walker, eds., pp. 281–257 (Gordon and Breach, London).

[156] Eda, K., Ohta, H., and Yamada, K. (1995). Prime subspaces in free topological groups, *Topol. Appl.* **62**, pp. 163–171.

[157] Edwards, R. E. (1965). *Functional Analysis* (Holt, Rinehart, and Winston, New York).

[158] Efimov, B. A. (1965). Dyadic bicompacta, *Trudy Moskov. Mat. Obshch.* **14**, pp. 211–247 (*in Russian*).

[159] Ellis, R. (1957). A note on the continuity of the inverse, *Proc. Amer. Math. Soc.* **8**, pp. 372–373.

[160] Ellis, R. (1957). Locally compact transformation groups, *Duke Math. J.* **24**, pp. 119–125.

[161] Ellis, R. (1969). *Lectures on Topological Dynamics* (Benjamin, New York).

[162] Ellis, R. and Gottshalk, W. (1960). Homomorphisms of compact transformation groups, *Trans. Amer. Math. Soc.* **94**, pp. 258–271.

[163] Ellis, R. and Keynes, H. B. (1972). Bohr compactifications and a result of Fölner, *Israel J. Math.* **12**, pp. 314–330.

[164] Engelking, R. (1965). Cartesian products and dyadic spaces, *Fund. Math.* **57**, pp. 287–304.

[165] Engelking, R. (1977). *General Topology* (PWN, Polish Scientific Publ., Warszawa).

[166] Engelking, R. (1978). *Dimension Theory* (PWN, Polish Scientific Publ., Warszawa).

[167] Fay, T. H., Ordman, E. T., and Thomas, B. V. S. (1979). The free topological group over rationals, *Topol. Appl.* **10**, pp. 33–47.

[168] Fernández, M. (2008). On some classes of paratopological groups, Preprint.

[169] Filippov, V. V. (1967). On perfect images of paracompact $p$-spaces, *Soviet Math. Dokl.* **176**, pp. 533–536.

[170] Filippov, V. V. (1979). On weight-type characteristics of spaces with a continuous action of a compact group, *Mat. Zametki* **25**, 6, pp. 939–947.

[171] Fletcher, P. (1971). Homeomorphism groups with the topology of quasi-uniform convergence, *Archiv der Mathematik* **22**, pp. 88–93.

[172] Flor, P. (1968). Zur Bohr-Konvergenz der Folgen, *Math. Scand.* **23**, pp. 169–170.

[173] Ford, L. R. (1954). Homeomorphism groups and coset spaces, *Trans. Amer. Math. Soc.* **77**, pp. 490–497.

[174] Franklin, S. P. and Thomas, B. V. S. (1977). A survey of $k_\omega$-spaces, *Topol. Proc.* **2**, pp. 111–124.

[175] Fréchet, M. (1910). Les dimensions d'un ensemble abstrait, *Math. Ann.* **68**, pp. 145–168.

[176] Fremlin, D. H. (1984). *Consequences of Martin's Axiom* (Cambridge University Press, Cambridge).

[177] Freudenthal, H. (1936). Einige Sätze über topologische Gruppen, *Ann. Math.* **37**, pp. 46–56.

[178] Frobenius, G. (1896). Über Gruppencharactere, *Sitzungsber Preuss. Akad. Wiss.*, phys.-math. K1., pp. 985-1021.
[179] Frolík, Z. (1960). The topological product of two pseudocompact spaces, *Czech. Math. J.* **10**, pp. 339–349.
[180] Frolík, Z. (1967). Homogeneity problems for extremally disconnected spaces, *Comment. Math. Univ. Carolin.* **8**, pp. 757–763.
[181] Fuchs, L. (1970). *Infinite Abelian groups*, Vol. I (Academic Press, New York).
[182] Galindo, J. and Hernández, S. (1998). On a theorem of van Douwen, *Extracta Math.* **13**, 1, pp. 115–123.
[183] García-Ferreira, S. and García-Maynez, A. (1994). On weakly pseudocompact spaces, *Houston J. Math.* **20**, 1, pp. 145–159.
[184] García-Ferreira, S., Sakai, M., and Sanchis, M. (2004). Free topological groups over $\omega_\mu$-metrizable spaces, *Houston J. Math.* **30**, 4, pp. 1039–1053.
[185] García-Ferreira, S., Sanchis, M., and Watson, S. (2000). Some remarks on the product of two $C_\alpha$-compact subsets, *Czech. Math. J.* **50**, pp. 249–264.
[186] García-Ferreira, S., Romaguera, S., and Sanchis, M. (1999). Bounded subsets and Groenthendieck's theorem for bispaces, *Houston J. Math.* **25**, 2, pp. 267–283.
[187] Gartside, P., Reznichenko, E. A., and Sipacheva, O. V. (1997). Maltsev and retral spaces, *Topol. Appl.* **80**, pp. 115–129.
[188] Gel'fand, I. M. (1939). On normed rings (O normirovannyh kol'zah), *Dokl. Akad. Nauk SSSR* **23**, pp. 430–432 (*in Russian*).
[189] Gel'fand, I. M. (1941). Normed rings (Normirovannye kol'za), *Matem. Sb.* **9**, pp. 3–24 (*in Russian*).
[190] Gel'fand, I. M. and Raĭkov, D. A. (1943). Irreducible unitary representations of locally bicompact groups, *Rec. Math. (Mat. Sbornik) N.S.* **13 (55)**, pp. 301–316 (*in Russian*).
[191] Gillman, L. and Jerison, M. (1976). *Rings of Continuous Functions* (Springer-Verlag, Berlin).
[192] Ginsburg, J. and Saks, V. (1975). Some applications of ultrafilters in topology, *Pacific J. Math.* **57**, pp. 403–418.
[193] Glasner, S. (1980). Divisibility properies and the Stone–Čech compactification, *Canadian J. Math.* **32**, pp. 993–1007.
[194] Glasner, S. (1998). On minimal actions of Polish groups, *Topol. Appl.* **85**, pp. 119–125.
[195] Gleason, A. M. (1950). Spaces with a compact Lie group of transformations, *Proc. Amer. Math. Soc.* **1**, pp. 35–43.
[196] Gleason, A. M. (1952). Groups without small subgroups, *Ann. Math.* **56**, 2, pp. 193–212.
[197] Glicksberg, I. (1959). Stone–Čech compactifications of products, *Trans. Amer. Math. Soc.* **90**, pp. 369–382.
[198] Glicksberg, I. (1962). Uniform boundedness for groups, *Canad. J. Math.* **14**, pp. 269–276.
[199] González, F. and Sanchis, M. (2006). Dieudonné completion and $b_f$-group actions, *Topol. Appl.* **153**, 17, 3320–3326.
[200] Gottshalk, W. H. and Hedlund, G. A. (1955). *Topological Dynamics*, Amer. Math. Soc. Publ. **36** (Amer. Math. Soc., Providence, RI).
[201] Graev, M. I. (1948). Free topological groups, In: *Topology and Topological Algebra*, Translations Series 1, **8** (1962), pp. 305–364 (American Mathematical Society). Russian original in: *Izv. Akad. Nauk SSSR Ser. Mat.* **12** (1948), pp. 279–323.
[202] Graev, M. I. (1950). Theory of topological groups I, *Uspekhi Mat. Nauk* **5**, pp. 3–56 (*in Russian*).
[203] Greenleaf, F. P. (1969). *Invariant Means on Topological Groups*, Van Nostrand Math. Studies **16** (van Nostrand, New York).
[204] Groot, J. de (1959). Groups represented by homeomorphism groups I, *Math. Ann.* **138**, pp. 80–102.
[205] Gruenhage, G. and Tanaka, Y. (1982). Products of $k$-spaces and spaces of countable tightness, *Proc. Amer. Math. Soc.* **273**, pp. 299–308.
[206] Gul'ko, A. S. (1996). Rectifiable spaces, *Topol. Appl.* **68**, 2, pp. 107–112.
[207] Gul'ko, S. P. (1979). On the structure of spaces of continuous functions and their hereditary paracompactness, *Russian Math. Surveys* **34**, 6, pp. 36–44.
[208] Guran, I. I. (1981). On topological groups close to being Lindelöf, *Soviet Math. Dokl.* **23**, pp. 173–175.
[209] Guran, I. I. (1981). Topological groups and properties of their subspaces, PhD thesis, 68 pp. (Moscow State University, Moscow; *in Russian*).
[210] Guran, I. I. (1998). Some open problems in the theory of topological groups and semigroups, *Mat. Studii* **10**, 2, pp. 223–224.

[211] Haar, A. (1933). Der massbegriff in der Theorie der kontinuerlichen Gruppen, *Ann. Math.* **34**, pp. 147–169.

[212] Hahn, H. (1927). Über lineare Gleichungsysteme in Lineare Räumen, *J. reine und angew. Math.* **157**, pp. 214–229.

[213] Hajnal, A. and Juhász, I. (1968). On hereditarily $\alpha$-Lindelöf and hereditarily $\alpha$-separable spaces, *Ann. Univ. Sci. Budapest Eötvös Sect. Math.* **11**, pp. 115–124.

[214] Hajnal, A. and Juhász, I. (1974). On hereditary $\alpha$-Lindelöf and $\alpha$-separable spaces II, *Fund. Math.* **81**, 2, pp. 147–158.

[215] Hajnal, A. and Juhász, I. (1976). A separable normal topological group need not be Lindelöf, *Gen. Topol. Appl.* **6**, pp. 199–205.

[216] Hardy, J. P., Morris, S. A., and Thompson, H. B. (1976). Application of the Stone–Čech compactification to free topological groups, *Proc. Amer. Math. Soc.* **55**, pp. 160–164.

[217] Hart, K. P., Junilla, H., and Mill, J. van (1985). A Dowker group, *Comment. Math. Univ. Carolin.* **26**, pp. 799–810.

[218] Hart, K. P. and Mill, J. van (1991). A countably compact group $H$ such that $H \times H$ is not countably compact, *Trans. Amer. Math. Soc.* **323**, pp. 811–821.

[219] Hart, K. P. and Mill, J. van (1991). Discrete sets and the maximal totally bounded group topology, *J. Pure Appl. Algebra* **70**, pp. 73–80.

[220] Hartman, S. and Mycielski, J. (1958). On embeddings of topological groups into connected topological groups, *Colloq. Math.* **5**, pp. 167–169.

[221] Haydon, R. (1974). On a problem of Pełczyński. Milutin spaces, Dugundji spaces, and $AE(0-dim)$, *Studia Math.* **52**, pp. 23–31.

[222] Heath, R. W. (1992). Some non-metric, first countable, cancellative topological semigroups that are generalized metric spaces, *Topol. Appl.* **44**, pp. 167–173.

[223] Henriksen, M. and Isbell, J. R. (1958). Some properties of compactifications, *Duke Math. J.* **25**, pp. 83–106.

[224] Hensel, K. (1913). *Zahlentheorie* (Berlin u. Leipzig, G. J. Göschensche Verlagshandlung).

[225] Hernández, C. (2000). Topological groups close to being $\sigma$-compact, *Topol. Appl.* **102**, pp. 101–111.

[226] Hernández, C. (2001). Condensations of Tychonoff universal topological algebras, *Comment. Math. Univ. Carolin.* **42**, 3, pp. 529–533.

[227] Hernández, C. and Tkachenko, M. G. (1998). Subgroups of $\mathbb{R}$-factorizable groups, *Comment. Math. Univ. Carolin.* **39**, pp. 371–378.

[228] Hernández, C. and Tkachenko, M. G. (2002). A note on $\omega$-modification and completeness concepts, *Bol. Soc. Mat. Mexicana* (3) **8**, pp. 93–96.

[229] Hernández, C. and Tkachenko, M. G. (2006). Three examples of pseudocompact quasitopological groups, *Topol. Appl.* **153**, 18, pp. 3615–3620.

[230] Hernández, S. and Sanchis, M. (1993). $G_\delta$-open functionally bounded subsets in topological groups, *Topol. Appl.* **53**, pp. 289–299.

[231] Hernández, S., Sanchis, M., and Tkachenko, M. G. (2000). Bounded sets in spaces and topological groups, *Topol. Appl.* **101**, pp. 21–43.

[232] Hernández, S. and Uspenskij, V. V. (2000). Pontryagin duality for spaces of continuous functions, *J. Math. Analysis Appl.* **242**, pp. 135–144.

[233] Hesse, G. (1979). *Zur Topologisierbarkeit von Gruppen*, Dissertation (Univ. Hannover, Hannover).

[234] Hewitt, E. (1948). Rings of real-valued continuous functions I, *Trans. Amer. Math. Soc.* **64**, pp. 45–99.

[235] Hewitt, E. (1962). Some applications of compactness in Harmonic Analysis, *General Topology and its Relations to Modern Analysis and Algebra*, pp. 204–210 (Proc. Prague Topol. Symp., September, 1961).

[236] Hewitt, E. and Ross, K. A. (1963). *Abstract Harmonic Analysis*, Vol. I (Springer-Verlag, Berlin–Heidelberg–New York).

[237] Hewitt, E. and Ross, K. A. (1970). *Abstract Harmonic Analysis*, Vol. II (Springer-Verlag, Berlin–Heidelberg–New York).

[238] Heyer, H. (1970). *Dualität lokalkompakter Gruppen*, Lecture Notes in Math. **150** (Springer-Verlag, Berlin-Heidelberg-New York).

[239] Hindman, N. (1979). Ultrafilters and combinatorial number theory, *Lecture Notes in Math.* **751**, pp. 49–184.

[240] Hindman, N., Protasov, I., and Strauss, D. (1998). Topologies on $S$ determined by idempotents in $\beta S$, *Topol. Proc.* **23**, pp. 155–196.

[241] Hindman, N. and Strauss, D. (1998). *Algebra in the Stone–Čech Compactification – Theory and Applications* (Walter de Gruyter and Co., Berlin).

[242] Hodel, R. E. (1984). Cardinal functions I, In: *Handbook of Set-Theoretic Topology*, K. Kunen and J. E. Vaughan, eds., pp. 1–62 (North-Holland, Amsterdam).

[243] Hofmann, K.-H. and Morris, S. A. (1998). *The Structure of Compact Groups: A Primer for Students — A Handbook for Experts* (Walter de Gruyter Publ., Berlin).

[244] Holm, P. (1964). On the Bohr compactification, *Math. Ann.* **156**, pp. 34–46.

[245] Hooper, R. C. (1968). Topological groups and integer-valued norms, *J. Functional Analysis* **2**, 3, pp. 243–257.

[246] Hulanicki, A. (1958). On cardinal numbers related with locally compact groups, *Bull. Acad. Polon. Sci. Ser. Sci. Math.* **6**, pp. 67–70.

[247] Hunt, D. C. and Morris, S. A. (1974). Free subgroups of free topological groups, In: *Proc. Second Internat. Conf. on the Theory of Groups* 1973, M. F. Newman, editor, Lecture Notes in Mathematics **372**, pp. 377–387 (Springer-Verlag, Berlin).

[248] Hurewicz, W. (1927). Über das Verhältnis separabler Räume zu kompakten Räumen, *Proc. Akad. Amsterdam* **30**, pp. 425–430.

[249] Husain, T. (1966). *Introduction to topological groups* (W. B. Saunders Comp., Philadelphia–London).

[250] Hušek, M. (1970). The Hewitt realcompactification of a product, *Comment. Math. Univ. Carolin.* **11**, pp. 393–395.

[251] Hušek, M. (1972). Realcompactness of function spaces and $v(P \times Q)$, *Gen. Topol. Appl.* **2**, pp. 165–179.

[252] Hušek, M. (1996). Sequentially continuous homomorphisms on products of topological groups, *Topol. Appl.* **70**, pp. 155–165.

[253] Hušek, M. and Vries, J. de (1987). Preservation of products by functors close to reflectors, *Topol. Appl.* **27**, pp. 171–189.

[254] Isbell, J. R. (1955). Zero-dimensional spaces, *Tohoku Math. J., Ser. 2* **7**, pp. 1–8.

[255] Isbell, J. R. (1964). *Uniform spaces* (Amer. Math. Soc. Colloq. Publ., Providence).

[256] Isiwata, T. (1967). Mappings and spaces, *Pacific J. Math.* **20**, pp. 455–480.

[257] Itzkowitz, G. (1974). Continuous measures, Baire category, and uniform continuity in topological groups, *Pacific J. Math.* **54**, pp. 115–125.

[258] Itzkowitz, G. and Tkachuk, V. V. (2003). Any pseudo-$pG$-compact group $G$ is fine, *New Zealand J. Math.* **32**, pp. 127–138.

[259] Ivanovskij, L. N. (1958). On a hypothesis of P. S. Alexandroff, *Dokl. Akad. Nauk SSSR* **123**, pp. 785–786.

[260] Jech, T. (1978). *Set Theory* (Academic Press, New York).

[261] Joiner, C. (1976). Free topological groups and dimension, *Trans. Amer. Math. Soc.* **220**, pp. 401–418.

[262] Juhász, I. (1971). *Cardinal Functions in Topology* (Math. Centre Tracts 34, Amsterdam).

[263] Just, W. and Weese, M. (1997). *Discovering Modern Set Theory II*, Graduate Studies in Mathematics, 18 (American Math. Soc.).

[264] Kakutani, S. (1936). Über die Metrization der Topologischen Gruppen, *Proc. Imp. Acad. Tokyo* **12**, pp. 82–84.

[265] Kakutani, S. (1944). Free topological groups and direct product of topological groups, *Proc. Imp. Acad. Tokyo* **40**, pp. 595–598.

[266] Kampen, E. van (1935). Locally bicompact Abelian groups and their character groups, *Ann. Math.* **36**, pp. 448–463.

[267] Kampen, E. van (1936). Note on a theorem of Pontryagin, *Amer. J. Math.* **58**, 1, pp. 177–180.

[268] Kargapolov, M. I. and Merzlyakov, Yu. I. (1977). *Fundamentals of the Group Theory* (Nauka, Moscow; *in Russian*).

[269] Kenderov, P., Kortezov, I. S., and Moors, W. B. (2001). Topological games and topological groups, *Topol. Appl.* **109**, pp. 157–165.

[270] Katetov, M. G. (1967). A theorem on mappings, *Comment. Math. Univ. Carolin.* **8**, pp. 431–433.

[271] Katok, A. and Hasselblatt, B. (1995). *Introduction to the Modern Theory of Dynamical Systems*, Encyclopedia Math. Appl. **54** (Cambridge Univ. Press, Cambridge).

[272] Katz, G. I. (1953). An isomorphic mapping of topological groups into a direct product of groups satisfying the first axiom of countability, *Uspekhi Mat. Nauk* **8**, 6, pp. 107–113 (*in Russian*).

[273] Katz, E., Morris, S. A., and Nickolas, P. (1982). A free subgroup of the free Abelian topological group on the unit interval, *Bull. London Math. Soc.* **14**, pp. 399–402.

[274] Katz, E., Morris, S. A., and Nickolas, P. (1983). Characterization of bases of subgroups of free topological groups, *J. London Math. Soc.* **27**, pp. 421–426.

[275] Keller, O. H. (1931). Die Homeomorphie der kompakten konvexen Mengen in Hilbertschen Raum, *Math. Ann.* **105**, pp. 748–758.

[276] Kelley, J. L. (1955). *General Topology* (van Nostrand, New York).

[277] Kertész, A. and Szele, T. (1953). On the existence of non-discrete topologies in infinite Abelian groups, *Publ. Math. Debrecen* **3**, pp. 187–189.

[278] Kister, J. M. (1962). Uniform continuity and compactness in topologcal groups, *Proc. Amer. Math. Soc.* **13**, pp. 37–40.

[279] Klee, V. L. (1953). Invariant metrics in groups (solution of a problem of Banach), *Proc. Amer. Math. Soc.* **3**, pp. 484–487.

[280] Kombarov, A. P. (1972). The product of normal spaces. Uniformities on $\Sigma$-products, *Dokl. AN SSSR* **205**, 5, pp. 1033–1035.

[281] Kopperman, R. D., Mislove, M. W., Morris, S. A., Nickolas, P., Pestov, V. G., and Svetlichny, S. (1996). Limit laws for wide varieties of topological groups, *Houston J. Math.* **22**, 2, pp. 307–328.

[282] Korovin, A. V. (1992). Continuous actions of pseudocompact groups and axioms of topological group, *Comment. Math. Univ. Carolin.* **33**, pp. 335–343.

[283] Krein, M. G. and Milman, D. P. (1940). On extreme points of regularly convex sets, *Studia Math.* **9**, pp. 123–138.

[284] Kulesza, J. (1993). Spaces which do not embed in topological groups of the same dimension, *Topol. Appl.* **50**, pp. 139–145.

[285] Kunen, K. (1980). *Set Theory* (North Holland, Amsterdam).

[286] Kunen, K. (1998). Bohr topologies and partition theorems for vector spaces, *Topol. Appl.* **90**, pp. 97–107.

[287] Kuz'minov, V. I. (1959). On a hypothesis of P. S. Alexandroff in the theory of topological groups, *Dokl. Akad. Nauk SSSR* **125**, pp. 727–729.

[288] Lankaster, P. (1969). *Theory of matrices* (Academic Press, New York–London).

[289] Lawson, J. (1972). Joint continuity in semitopological semigroups, *Illinois Math. J.* **18**, pp. 275–285.

[290] Leiderman, A., Morris, S. A., and Pestov, V. G. (1997). The free Abelian topological group and the free locally convex space on the unit interval, *Proc. London Math. Soc.* (2) **56**, 3, pp. 529–538.

[291] Leja, F. (1927). Sur la notion du groupe abstrait topologique, *Fund. Math.* **9**, pp. 37–44.

[292] Leptin, H. (1955). Abelsche Gruppen mit kompakten Charaktergruppen und Dualitätstheorie gewisser linear topologischer abelscher Gruppen, *Abh. Math. Sem. Univ. Hamburg* **19**, pp. 244–263.

[293] Levi, E. (1905). Sulla struttura dei gruppi finiti e continui, *Atti Accad. Torino* **40**, pp. 3–17 (*in Italian*).

[294] Lie, S. and Engel, F. (1888–1893). *Theorie der Transformationsgruppen* (Leipzig, Teubner 1, 2, 3).

[295] López, M. A. and Tkachenko, M. G. Topological group versions of realcompactness and Dieudonné completeness, In preparation.

[296] Lukács, G. (2008). *Compact-like topological groups*, Research and Exposition series, 30 (Helderman Verlag).

[297] Mack, J., Morris, S. A., and Ordman, E. T. (1973). Free topological groups and the projective dimension of locally compact Abelian group, *Proc. Amer. Math. Soc.* **40**, pp. 303–308.

[298] Malykhin, V. I. (1975). Extremally disconnected and similar groups, *Soviet Math. Dokl.* **16**, pp. 21–25. Russian original in: *Dokl. Akad. Nauk SSSR* **220**, pp. 27–30.

[299] Malykhin, V. I. (1979). On extremally disconnected topological groups, *Uspekhi Mat. Nauk* **34**, 6, pp. 59–66 (*in Russian*).

[300] Malykhin, V. I. (1987). Nonpreservation of properties of topological groups on taking their square, *Sib. Math. J.* **28**, pp. 639–645.

[301] Malykhin, V. I. and Shapiro, L. B. (1985). Pseudocompact groups without converging sequences, *Math. Notes* **37**, pp. 59–62.

[302] Marczewski, E. (1941). Remarque sur les produits des espaces topologiques, *Dokl. Akad. Nauk SSSR* **31**, pp. 525–527.

[303] Mardešić, S. (1960). On covering dimension and inverse limits of compact spaces, *Illinois J. Math.* **4**, pp. 278–291.

[304] Markov, A. A. (1935). Über endlich-dimensionale vektor räume, *Ann. of Math.* **36**, pp. 464–506.

[305] Markov, A. A. (1941). On free topological groups, *Dokl. Akad. Nauk SSSR* **31**, pp. 299–301 (*in Russian*).

[306] Markov, A. A. (1944). On the existence of periodic connected topological groups, *Izv. Akad. Nauk SSSR* **8**, pp. 225–232.

[307] Markov, A. A. (1944). On unconditionally closed sets, In: *Topology and Topological Algebra*, Translations Series 1, **8**, pp. 273–304. American Math. Society, 1962. Russian original in: *C. R. Dokl. Akad. Nauk SSSR (N.S.)* **44** (1944), pp. 180–181.

[308] Markov, A. A. (1945). On free topological groups, In: *Topology and Topological Algebra,* Translation Series 1, **8** (1962), pp. 195–272. American Math. Society. Russian original in: *Izv. Akad. Nauk SSSR* **9** (1945), pp. 3–64.

[309] Marxen, D. (1976). Neighborhoods of the identity of the free abelian topological groups, *Math. Slovaca* **26**, pp. 247–256.

[310] Martín-Peinador, E. (1995). A reflexive admissible topological group must be locally compact, *Proc. Amer. Math. Soc.* **123**, pp. 3563–3566.

[311] Matveev, M. V. (1985). Pseudocompactness, *Uspekhi Mat. Nauk* **40**, 2(242), pp. 189–190 (*in Russian*).

[312] Mazur, S. (1938). Sur les anneaux lineares, *C. R. Acad. Sci. Paris* **207**, pp. 1025–1027.

[313] Mazur, S. (1952). On continuous mappings of Cartesian products, *Fund. Math.* **39**, pp. 229–238.

[314] Megrelishvili, M. G. (2001). Every semitopological semigroup compactification of the group $H_+[0,1]$ is trivial, *Semigroup Forum* **63**, 3, pp. 357–370.

[315] Megrelishvili, M. G. and Scarr, T. (1998). Constructing Tychonoff $G$-spaces which are not $G$-Tychonoff, *Topol. Appl.* **86**, pp. 69–81.

[316] Michael, E. (1956). Continuous selections I, *Ann. Math.* **63**, pp. 361–382.

[317] Michael, E. (1956). Continuous selections II, *Ann. Math.* **64**, pp. 562–580.

[318] Michael, E. (1957) Continuous selections III, *Ann. Math.* **65**, 2, pp. 375–390.

[319] Michael, E. (1972). A quintuple quotient quest, *Gen. Topol. Appl.* **2**, pp. 91–138.

[320] Michael, E. (1973). On $k$-spaces, $k_{\mathbb{R}}$-spaces and $k(X)$, *Pacific J. Math.* **47**, pp. 487–498.

[321] Mill, J. van (1982). Representing countable groups by homeomorphism groups in Hilbert space, *Math. Ann.* **259**, pp. 143–148.

[322] Mill, J. van (1983). A topological group having no homeomorphisms other than translations, *Trans. Amer. Math. Soc.* **280**, 2, pp. 491–498.

[323] Mill, J. van (1985). Closed images of topological groups, *Colloq. Math. Soc. János Bolyai* **41**, pp. 419–426.

[324] Montgomery, D. (1936). Continuity in topological groups, *Bull. Amer. Math. Soc.* **42**, pp. 879–882.

[325] Montgomery, D. and Zippin, L. (1955). *Topological Transformation Groups*, Interscience Tracts in Pure and Applied Mathematics, Tract 1 (Interscience Publ., New York).

[326] Moran, W. (1971). On almost periodic compactifications of locally compact groups, *J. London Math. Soc.* **3** (2), pp. 507–512.

[327] Morris, S. A. (1977). *Pontryagin duality and the structure of locally compact Abelian groups*, London Math. Soc. Lecture Notes Series **29** (Cambridge Univ. Press).

[328] Morris, S. A. and Nickolas, P. (1982). Locally invariant topologies on free groups, *Pacific J. Math.* **103**, 2, pp. 523–537.

[329] Morris, S. A. and Thompson, H. B. (1973). Invariant metrics on free topological groups, *Bull. Austral. Math. Soc.* **9**, pp. 83–88.

[330] Morris, S. A. and Thompson, H. B. (1974). Free topological groups with no small subgroups, *Proc. Amer. Math. Soc.* **46**, pp. 431–437.

[331] Morris, S. A. and Thompson, H. B. (1986). Metrizability of subgroups of free topological groups, *Bull. Austral. Math. Soc.* **33**, pp. 103–112.

[332] Motorov, D. B. (1984). Bicompacta of countable character are (under $CH$) continuous images of homogeneous bicompacta, In: *Cardinal invariants and mappings of topological spaces*, pp. 48–50 (Udmurt State Univ., Izhevsk; *in Russian*).

[333] Mukherjea, A. and Tserpes, N. A. (1972). A note on countably compact semigroups, *J. Austral. Math. Soc.* **13**, pp. 180–184.

[334] Nachbin, L. (1965). *The Haar Integral*, D. van Nostrand (Princeton, New Jork).

[335] Nagami, K., (1969). $\Sigma$-spaces, *Fund. Math.* **61**, pp. 169–192.

[336] Nagami, K. (1972). Perfect classes of spaces, *Proc. Japan Acad.* **48**, pp. 21–24.

[337] Naimark, M. A. (1959). *Normed Rings*, Groningen. Translated from Russian, by P. Noordholff.

[338] Nakayama, T. (1943). Note on free topological groups, *Proc. Imp. Acad. Sci.* **19**, pp. 471–475.

[339] Namioka, I. (1974). Separate continuity and joint continuity, *Pacific J. Math.* **51**, 2, pp. 515–531.

[340] Nedev, S. J. (1971). $o$-metrizable spaces, *Trudy Mosk. Mat. Obsch.* **24**, pp. 201–236 (*in Russian*).

[341] Nedev, S. J. and Choban, M. M. (1972). On the theory of $o$-metrizable spaces I, *Vestnik Moskov. Univ. Ser. I Mat. Meh.* **27**, 1, pp. 8–15 (*in Russian*).

[342] Nedev, S. J. and Choban, M. M. (1972). On the theory of $o$-metrizable spaces II, *Vestnik Moskov. Univ. Ser. I Mat. Meh.* **27**, 2, pp. 10–17 (*in Russian*).

[343] Nedev, S. J. and Choban, M. M. (1972). On the theory of $o$-metrizable spaces III, *Vestnik Moskov. Univ. Ser. I Mat. Meh.* **27**, 3, pp. 10–15 (*in Russian*).

[344] Neumann, J. von (1934). Zum Haarschen Mass in Topologishen Gruppen, *Compositio Math.* **1**, pp. 106–114.

[345] Nickolas, P. (1976). *Free topological groups and free products of topological groups*, PhD thesis (University of New South Wales, Australia).

[346] Nickolas, P. (1976). Subgroups of the free topological group on [0, 1], *J. London Math. Soc.* **12**, pp. 199–205.

[347] Nickolas, P. and Tkachenko, M. G. (2003). Local compactness in free topological groups, *Bull. Austral. Math. Soc.* **68**, pp. 243–265.

[348] Nickolas, P. and Tkachenko, M. G. (2005). The character of free topological groups I, *Applied Gen. Topol.* **6**, 1, pp. 15–41.

[349] Nickolas, P. and Tkachenko, M. G. (2005). The character of free topological groups II, *Applied Gen. Topol.* **6**, 1, pp. 43–56.

[350] Noble, N. (1969). Products with closed projections, *Trans. Amer. Math. Soc.* **140**, pp. 381–391.

[351] Noble, N. (1970). The continuity of functions on Cartesian products, *Trans. Amer. Math. Soc.* **149**, pp. 187–198.

[352] Numakura, K. (1952). On bicompact semigroups, *Math. J. Okayama Univ.* **1**, pp. 99–108.

[353] Nummela, E. C. (1980). The completion of a topological group, *Bull. Austral. Math. Soc.* **21**, pp. 407–417.

[354] Nummela, E. C. (1982). Uniform free topological groups and Samuel compactifications, *Topol. Appl.* **13**, pp. 77–83.

[355] Nyikos, P. J. (1981). Metrizability and the Fréchet–Urysohn property in topological groups, *Proc. Amer. Math. Soc.* **83**, pp. 793–801.

[356] Nyikos, P. J. (1992). Subsets of $\omega^\omega$ and the Fréchet–Urysohn and $\alpha_i$-properties, *Topol. Appl.* **48**, pp. 91–116.

[357] Nyikos, P. J. and Reichel, H. C. (1975). Topologically orderable groups, *Topol. Appl.* **5**, pp. 195–204.

[358] Okunev, O. G. (1990). A method for constructing examples of $M$-equivalent spaces, *Topol. Appl.* **36**, 2, pp. 157–172.

[359] Okunev, O. G. (2005). Fréchet property in compact spaces is not preserved by $M$-equivalence, *Comment. Math. Univ. Carolin.* **46**, 4, pp. 747–749.

[360] Okuyama, A. (1967). Some generalizations of metric spaces, their metrization theorems and product spaces, *Sci. Rep. Tokyo Kyoiku Daigaku* **9**, pp. 236–254.

[361] Ol'shanskiĭ, A. I. (1980). A remark on a countable non-topologizable group, *Vestnik Mosk. Gos. Univ. Ser. I Mat. Mekh.* no. 3, p. 103 (*in Russian*).

[362] Papazyan, T. (1991). Extremal topologies on a semigroup, *Topol. Appl.* **39**, 3, pp. 229–243.

[363] Pasynkov, B. A. (1960). The coincidence of various definitions of dimensionality for locally bicompact groups, *Soviet Math. Dokl.* **1**, pp. 720–722. Russian original in: *Dokl. Akad. Nauk SSSR* **132**, pp. 1035–1037.

[364] Pasynkov, B. A. (1965). Almost-metrizable topological groups, *Soviet Math. Dokl.* **7** (1966), pp. 404–408. Russian original in: *Dokl. Akad. Nauk SSSR* **161** (1965), pp. 281–284.

[365] Pasynkov, B. A. (1967). Open mappings, *Soviet Math. Dokl.* **8**, pp. 853–856. Russian original in: *Dokl. Akad. Nauk SSSR* **175**, pp. 292–295.

[366] Pasynkov, B. A. (1976). Spaces with a bicompact transformation group, *Soviet Math. Dokl.* **17**, 6, pp. 1522–1526. Russian original in: *Dokl. Akad. Nauk SSSR* **231**, 1, pp. 39–42.

[367] Pasynkov, B. A. and Vylov, V. (1981). On free groups of topological spaces, *Dokl. Bulgar. Acad. Sci.* **34**, pp. 1049–1052.

[368] Pełczyński, A. (1968). Linear extensions, linear averagings and their applications to linear topological classification of spaces of continuous functions, *Diss. Math. (Rozprawy Mat.)* **58**, pp. 1–90.

[369] Pestov, V. G. (1982). Some properties of free topological groups, *Moscow Univ. Math. Bull.* **37**, pp. 46–49. Russian original in: *Vestnik Mosk. Univ. Ser. I Mat. Makh.* no. 1, pp. 35–37.

[370] Pestov, V. G. (1982). Embeddings and condensations of topological groups, *Math. Notes* **31**, 3-4, pp. 228–230. Russian original in: *Mat. Zametki* **31**, 3, pp. 443–446.

[371] Pestov, V. G. (1982). The coincidence of the dimensions *dim* of $l$-equivalent topological spaces, *Soviet Math. Dokl.* **26**, pp. 380–382. Russian original in: *Dokl. Akad. Nauk SSSR* **266**, 3, pp. 553–556.

[372] Pestov, V. G. (1984). Some topological properties preserved by the relation of $M$-equivalence, *Russian Math. Surveys* **39**, pp. 223–224. Russian original in: *Uspekhy Mat. Nauk* **39**, pp. 203–204.

[373] Pestov, V. G. (1985). Neighbourhoods of unity in free topological groups, *Mosc. Univ. Math. Bull.* **40**, pp. 8–12. Russian original in: *Vestnik Mosk. Univ. Ser. I Mat. Mekh.* no. 3, pp. 8–10.

[374] Pestov, V. G. (1985). The class of almost metrizable topological groups is not closed under extensions, *Mosc. Univ. Math. Bull.* **40**, 4, pp. 74–76. Russian original in: *Vestnik Mosk. Univ. Ser. I Mat. Mekh.* no. 4, pp. 72–73.

[375] Pestov, V. G. (1986). On compactly generated topological groups, *Math. Notes* **40**, pp. 880–882. Russian original in: *Mat. Zametki* **40**, 5, pp. 671–676.

[376] Pestov, V. G. (1986). On a theorem of M. M. Choban on the extension of pseudometrics to free universal algebras, In: *Continuous functions on topological spaces*, pp. 142–146 (Latv. Gos. Univ., Riga; *in Russian*).

[377] Pestov, V. G. (1995). Free Abelian topological groups and the Pontryagin–van Kampen duality, *Bull. Austral. Math. Soc.* **52**, pp. 297–311.

[378] Pestov, V. G. (1998). On free actions, minimal flows, and a problem by Ellis, *Trans. Amer. Math. Soc.* **350**, 10, pp. 4149–4165.

[379] Pestov, V. G. (1999–2000). Topological groups: where to from here?, *Topol. Proc.* **24**, pp. 421–502.

[380] Pestov, V. G. and Shakhmatov, D. B. (1986). Continuous homomorphic images of groups with a countable base do not exhaust all groups with a countable network, *Mosc. Univ. Math. Bull.* **41**, pp. 92–95. Russian original in: *Vestnik Mosk. Univ. Ser. I Mat. Mekh.* no. 3, pp. 98–101.

[381] Pestov, V. G. and Tkachenko, M. G. (1985). *Problem 3.28*, In: *Unsolved Problems of Topological Algebra*, p. 18 (Academy of Science, Moldova, Kishinev, Shtiinca; *in Russian*).

[382] Pestov, V. G. and Yamada, K. (1999). Free topological groups on metrizable spaces and inductive limits, *Topol. Appl.* **98**, pp. 291–301.

[383] Peter, F. and Weyl, H. (1927). Die Vollständigkeit der primitiven Darstellungen einer geschlossenen kontinuerlichen Gruppe, *Math. Ann.* **97**, pp. 737–755.

[384] Pfister, H. (1985). Continuty of the inverse, *Proc. Amer. Math. Soc.* **2**, pp. 312–314.

[385] Pontryagin, L. S. (1934). The theory of topological commutative groups, *Ann. Math.* **35**, pp. 361–388.

[386] Pontryagin, L. S. (1934). Sur les groupes abéliens continus, *C. R. Acad. Sci. Paris* **198**, pp. 328–330.

[387] Pontryagin, L. S. (1973). *Continuous Groups*, Third edition (Nauka, Moscow; *in Russian*).

[388] Prodanov, I. (1981). Elementary example of a group without characters, In: *Mathematics and Mathematical Education*, pp. 79–81. Proc. Ninth Spring Conf. (April, 1980) of the Bulgarian Math. Soc. at Slunchev brjag (Bulgarian Acad. Sci., Sophia, Bulgaria).

[389] Protasov, I. V. (1991). Functionally balanced groups, *Math. Notes* **49**, pp. 614–616. Russian original in: *Mat. Zametki* **49**, pp. 87–91.

[390] Protasov, I. V. (1993). Ultrafilters and topologies on groups, *Sib. Math. J.* **34**, pp. 163–180.

[391] Protasov, I. V. (1994). Discrete subsets of topological groups, *Math. Notes* **55**, 1–2, pp. 101–102. Russian original in: *Mat. Zametki* **55**, pp. 150–151.

[392] Protasov, I. V. (1997). *Combinatorics of Numbers*, Mathematical Studies. Monograph Series **2** (VNTL Publishers).

[393] Protasov, I. V. (1998). Maximal topologies on groups, *Siberian Math. J.* **39**, 6, pp. 1184–1194. Russian original in: *Sibirsk. Mat. Zh.* **39**, pp. 1368–1381.

[394] Protasov, I. V. and Zelenyuk, E. (1999). *Topologies on groups determined by sequences*, Mathematical Studies. Monograph Series **4** (VNTL Publishers).

[395] Raghavan, T. G. and Reilly, I. L. (1978). On the continuity of group operations, *Indian J. Pure Appl. Math.* **9**, 8, pp. 747–752.

[396] Raĭkov, D. A. (1946). On the completion of topological groups, *Izv. Akad. Nauk SSSR* **10**, pp. 513–528 (*in Russian*).

[397] Raimi, R. (1966). Homeomorphisms and invariant measures for $\beta\mathbb{N} \setminus \mathbb{N}$, *Duke Math. J.* **33**, pp. 1–12.

[398] Ravsky, O. V. (2001). Paratopological groups I, *Mat. Studii* **16**, 1, pp. 37–48.

[399] Ravsky, O. V. (2002). Paratopological groups II, *Mat. Studii* **17**, 1, pp. 93–101.

[400] Ravsky, O. V. (2002). *Topological and algebraic properties of paratopological groups*, Ph.D. thesis (Lviv University, Lviv; *in Ukrainian*).

[401] Ravsky, O. V. (2003). An example of a Hausdorff countably compact paratopological group which is not a topological group, IVth Internat. Algebraic Conf. in Ukraine, August 4–9, 2003, Lviv.

[402] Reid, G. A. (1967). On sequential convergence in groups, *Math. Z.* **102**, pp. 227–235.

[403] Reznichenko, E. A. (1989). A pseudocompact space in which all subsets of not full cardinality are closed and discrete, *Moscow Univ. Math. Bull.* **44**, 6, pp. 70–71. Russian original in: *Vestnik Moscov. Univ. Ser. I Mat. Mekh.* no. 6, pp. 69–70.

[404] Reznichenko, E. A. (1994). Extensions of functions defined on products of pseudocompact spaces and continuity of the inverse in pseudocompact groups, *Topol. Appl.* **59**, pp. 233–244.

[405] Reznichenko, E. A. and Uspenskij, V. V. (1998). Pseudocompact Mal'tsev spaces, *Topol. Appl.* **86**, pp. 83–104.

[406] Rice, M. (1977). A note on uniform paracompactness, *Proc. Amer. Math. Soc.* **62**, pp. 359–362.

[407] Rickert, N. W. (1967). Locally compact topologies for groups, *Trans. Amer. Math. Soc.* **126**, pp. 225–235.

[408] Robbie, D. and Svetlichny, S. (1996). An answer to A. D. Wallace's question about countably compact cancellative semigroups, *Proc. Amer. Math. Soc.* **124**, pp. 325–330.

[409] Robinson, D. J. F. (1982). *A Course in the Theory of Groups* (Springer-Verlag, Berlin).

[410] Roelcke, W. and Dierolf, S. (1981). *Uniform structures on topological groups and their quotients* (McGraw-Hill International Book Co., New York).

[411] Romaguera, S. and Sanchis, M. (2000). Locally compact topological groups and cofinal completeness, *J. London Math. Soc.* **62**, 2, pp. 451–460.

[412] Romaguera, S. and Sanchis, M. (2007). Continuity of the inverse in pseudocompact paratopological groups, *Algebra Colloq.* **14**, 1, pp. 167–175.

[413] Rudin, M. E. (1975). *Lectures on Set-Theoretic Topology*, Conf. Ser. in Math. **23** (Amer. Math. Soc., Providence).

[414] Rudin, M. E. (1977). Martin's Axiom, In: *Handbook of Mathematical Logic*, J. Barwise, ed., pp. 491–501 (North-Holland, Amsterdam).

[415] Ruppert, W. (1984). *Compact Semitopological Semigroups: An Intrinsic Theory*, Lecture Notes in Mathematics **1079** (Springer-Verlag, Berlin).

[416] Sanchis, M. (1998). Continuous functions on locally pseudocompact groups, *Topol. Appl.* **86**, pp. 5–23.

[417] Sanchis, M. and Tamariz-Mascarúa, A. (2002). A note on $p$-bounded and $p$-quasi-bounded subsets, *Houston J. Math.* **28**, 3, pp. 511–527.

[418] Sanchis, M. and Tkachenko, M. G. (2007). Totally Lindelöf and totally $\omega$-narrow paratopological groups, *Topol. Appl.* **155**, 4, pp. 322–334. Doi:10.1016/j.topol.2007.05.017.

[419] Schepin, E. V. (1976). Real-valued functions and canonical sets in Tychonoff products and topological groups, *Russian Math. Surveys* **31**, pp. 19–30. Russian original in: *Uspekhy Mat. Nauk* **31**, pp. 18–27.

[420] Schepin, E. V. (1976). Topology of limit spaces of uncountable inverse spectra, *Russian Math. Surveys* **31**, pp. 155–191.

[421] Schepin, E. V. (1979). On $\kappa$-metrizable spaces, *Izv. Akad. Nauk SSSR Ser. Mat.* **43**, 2, pp. 442–478.

[422] Schepin, E. V. (1981). Functors and uncountable powers of compacta, *Russian Math. Surveys* **36**, 3, pp. 1–71. Russian original in: *Uspekhi Mat. Nauk* **36**, 3, pp. 3–62.

[423] Schreier, O. (1925). Abstrakte kontinuerliche Gruppen, *Abh. Math. Sem. Univ. Hamburg* **4**, pp. 15–32.

[424] Semadeni, Z. (1971). *Banach spaces of continuous functions*, Vol. I, Monografie Matematyczne, 55 (PWN—Polish Scientific Publishers, Warsaw).

[425] Serre, J. P. (1949). Compacité locale des espaces fibrés, *C. R. Acad. Sci. Paris* **229**, pp. 1295–1297.

[426] Shakhmatov, D. B. (1983). Cardinal invariants of topological fields, *Soviet Math. Dokl.* **28**, pp. 290–294. Russian original in: *Dokl. AN SSSR* **271**, pp. 1332–1336.

[427] Shakhmatov, D. B. (1984). Condensations of universal topological algebras preserving continuity of operations and decreasing weights, *Moscow Univ. Math. Bull.* **39**, 2, pp. 57–60. Russian original in: *Vestnik Moskov. Univ. Ser. I Mat. Mekh.* no. 2, pp. 42–45.

[428] Shakhmatov, D. B. (1984). No upper bound for cardinalities of Tychonoff c.c.c. spaces with a $G_\delta$-diagonal exists (an answer to J. Ginsburg and R. G. Woods' question), *Comment. Math. Univ. Carolin.* **25**, 4, pp. 731–745.

[429] Shakhmatov, D. B. (1986). Pregauges (Precalibres) of $\sigma$-compact topological groups, *Math. Notes* **39**, pp. 465–470. Russian original in: *Mat. Zametki* **39**, 6, pp. 859–868.

[430] Shakhmatov, D. B. (1990). Imbeddings into topological groups preserving dimensions, *Topol. Appl.* **36**, pp. 181–204.

[431] Shakhmatov, D. B. (1990). A survey of current researches and open problems in the dimension theory of topological groups, *Q & A in Gen. Topol.* **8**, pp. 101–128.

[432] Shakhmatov, D. B. (1990). Dugundji spaces and topological groups, *Comment. Math. Univ. Carolin.* **31**, pp. 129–143.

[433] Shakhmatov, D. B. (1991). On zero-dimensionality of subgroups of locally compact groups, *Comment. Math. Univ. Carolin.* **32**, pp. 581–582.

[434] Shakhmatov, D. B. (1994). A direct proof that every infinite compact group $G$ contains $\{0, 1\}^{w(G)}$, *Ann. New York Acad. Sci.* **728**, pp. 276–283.

[435] Shakhmatov, D. B. (1999). A comparative survey of selected results and open problems concerning topological groups, fields, and vector spaces, *Topol. Appl.* **91**, 1, pp. 51–63.

[436] Shakhmatov, D. B. (2002). Convergence in the presence of algebraic structure, In: *Recent Progress in General Topology, II*, M. Hušek and J. van Mill, eds., pp. 463–483 (North-Holland, Amsterdam).

[437] Shanin, N. A. (1946). On mutual intersections of open sets in products of topological spaces, *Dokl. Akad. Nauk SSSR* **53**, pp. 503–506.

[438] Shanin, N. A. (1948). On products of topological spaces, *Trudy Matem. Inst. im. Steklova Akad. Nauk SSSR* **24**, pp. 1–112 (*in Russian*).

[439] Shapiro, L. B. (1993). On the homogeneity of dyadic compact Hausdorff spaces, *Math. Notes* **54**, 3-4, pp. 1058–1072. Russian original in: *Mat. Zametki* **54**, 4, pp. 117–139.

[440] Shapirovskiĭ, B. E. (1975). $\pi$-character and $\pi$-weight in bicompacta, *Soviet Math. Dokl.* **16**, pp. 999–1003. Russian original in: *Dokl. Akad. Nauk SSSR* **223**, 4, pp. 799–802.

[441] Shelah, S. (1978). On $P$-points, $\beta\omega$ and other results in general topology, *Notices Amer. Math. Soc.* **25**, A-365, 87T-G49.

[442] Shelah, S. (1980). On a problem of Kurosh, Jónsson groups, and applications, In: *Word Problems II*, S. I. Adian, W. W. Boone, and G. Higman, eds., pp. 373–394 (North–Holland, Amsterdam).

[443] Shibakov, A. (1996). Sequential group topology on rationals with intermediate sequential order, *Proc. Amer. Math. Soc.* **124**, pp. 2599–2607.

[444] Shibakov, A. (1999). Countable Fréchet topological groups under $CH$, *Topol. Appl.* **91**, pp. 119–139.

[445] Shibakov, A. (1998). Sequential topological groups of any sequential order under $CH$, *Fund. Math.* **155**, pp. 79–89.

[446] Shirota, T., (1952). A class of topological spaces, *Osaka Math. J.* **4**, pp. 23–40.

[447] Shkarin, S. A. (1999). On universal Abelian topological groups, *Sb. Math.* **190**, pp. 1059–1076. Russian original in: *Mat. Sb.* **190**, pp. 127–144.

[448] Simon, P. (1980). A compact Fréchet space whose square is not Fréchet, *Comment. Math. Univ. Carolin.* **21**, 4, pp. 749–753.

[449] Sipacheva, O. V. (1991). Compacta with a continuous Mal'tsev operation and retracts of topological groups, *Moscow Univ. Math. Bull.* **46**, 1, pp. 22–24. Russian original in: *Vestnik Moskov. Univ. Ser. I Mat. Mekh.* no. 1, pp. 33–36.

[450] Sipacheva, O. V. (1993). On the stratifiability of free Abelian topological groups, *Topol. Proc.* **18**, pp. 271–311.

[451] Sipacheva, O. V. (1996). On free topological groups with the inductive limit topologies, *Ann. New York Acad. Sci.* **788**, pp. 188–196.

[452] Sipacheva, O. V. (2000). Free topological groups of spaces and their subspaces, *Topol. Appl.* **101**, pp. 181–212.

[453] Sipacheva, O. V. and Tkachenko, M. G. (1990). Thin and bounded subsets of free topological groups, *Topol. Appl.* **36**, pp. 143–156.

[454] Sipacheva, O. V. and Uspenskij, V. V. (1987). Free topological groups with no small subgroups, and Graev metrics, *Moscow Univ. Math. Bull.* **42**, pp. 24–29. Russian original in: *Vestnik Moskov. Univ. Ser. I Mat. Mekh.* no. 4, pp. 21–24.

[455] Sirota, S. (1969). A product of topological groups and extremal disconnectedness, *Matem. Sb. (NS)* **79**, pp. 179–192 (*in Russian*).

[456] Sklyarenko, E. G. (1963). On topological structure of locally compact groups and their quotient groups, *Mat. Sbornik* **60**, pp. 57–82 (*in Russian*).

[457] Solecki, S. (2000). Actions of non-compact and non-locally compact Polish groups, *J. Symb. Logic* **65**, pp. 1881–1894.

[458] Solovay, R. M. (1971). Real-valued measurable cardinals, *Proc. Symp. Pure Math.* **13**, 1, pp. 397–428.

[459] Spanier, E. H. (1966). *Algebraic Topology*, New York.

[460] Stephenson, R. M. (1971). Minimal topological groups, *Math. Ann.* **192**, pp. 193–195.

[461] Stone, A. H. (1948). Paracompactness and product spaces, *Bull. Amer. Math. Soc.* **54**, pp. 977–982.

[462] Talagrand, M. (1979). Espaces de Banach faiblement $K$-analtiques, *Ann. Math.* **110**, 3, pp. 407–438.

[463] Teleman, S. (1957). Sur la représentation linéaire des groupes topologiques, *Ann. Sci. Ecole Norm. Sup.* **74**, pp. 319–339.

[464] Thomas, B. V. S. (1974). Free topological groups, *Gen. Topol. Appl.* **4**, pp. 51–72.

[465] Tkachenko, M. G. (1981). Some results on inverse spectra I, *Comment. Math. Univ. Carolin.* **22**, 3, pp. 621–633.

[466] Tkachenko, M. G. (1981). Some results on inverse spectra II, *Comment. Math. Univ. Carolin.* **22**, 4, pp. 819–841.

[467] Tkachenko, M. G. (1981). Examples of connected left-separated spaces and topological groups, *Acta Math. Acad. Sci. Hungar* **38**, 3, pp. 257–261.

[468] Tkachenko, M. G. (1983). The notion of $o$-tightness and $C$-embedded subsets of products, *Topol. Appl.* **15**, 1, pp. 93–98.

[469] Tkachenko, M. G. (1983). On the Souslin property in free topological groups over compact Hausdorff spaces, *Math. Notes* **34**, 3-4, pp. 790–793. Russian original in: *Mat. Zametki* **34**, pp. 601–607.

[470] Tkachenko, M. G. (1983). On the completeness of free Abelian topological groups, *Soviet Math. Dokl.* **27**, pp. 341–345. Russian original in: *Dokl. Akad. Nauk SSSR* **269**, pp. 299–303.

[471] Tkachenko M. G. (1984). On topologies of free groups, *Czech. Math. J.* **34**, pp. 541–551.

[472] Tkachenko, M. G. (1985). Some properties of free topological groups, *Math. Notes* **37**, pp. 62–66. Russian original in: *Mat. Zametki* **37**, pp. 110–118.

[473] Tkachenko, M. G. (1987). Strong collectionwise normality and countable compactness in free topological groups, *Sib. Math. J.* **28**, pp. 824–831. Russian original in: *Sib. Mat. Zh.* **28**, pp. 167–177.

[474] Tkachenko, M. G. (1987). On boundedness and pseudocompactness in topological groups, *Math. Notes* **41**, pp. 229–231. Russian original in: *Mat. Zametki* **41**, pp. 400–405.

[475] Tkachenko, M. G. (1988). Compactness type properties in topological groups, *Czech. Math. J.* **38**, pp. 324–341.

[476] Tkachenko, M. G. (1989). Generalization of a theorem of Comfort and Ross I, *Ukr. Math. J.* **41**, pp. 334–338. Russian original in: *Ukrain. Mat. Zh.* **41**, 3, pp. 377–382.

[477] Tkachenko, M. G. (1990). On dimension of locally pseudocompact groups and their quotients, *Comment. Math. Univ. Carolin.* **31**, pp. 159–166.

[478] Tkachenko, M. G. (1990). On countably compact and pseudocompact topologies on free Abelian groups, *Soviet Math. (Izv. VUZ)* **34**, 5, pp. 79–86. Russian original in: *Izv. VUZ. Ser. Mat.* **5**, pp. 68–75.

[479] Tkachenko, M. G. (1991). Subgroups, quotient groups and products of $\mathbb{R}$-factorizable groups, *Topol. Proc.* **16**, pp. 201–231.

[480] Tkachenko, M. G. (1991). Zero-dimensional groups and factorization of homomorphisms with respect to weight and dimension, *Sib. Math. J.* **32**, pp. 477–484. Russian original in: *Sib. Mat. Zh.* **32**, pp. 151–159.

[481] Tkachenko, M. G. (1991). Factorization theorems for topological groups and their applications, *Topol. Appl.* **38**, pp. 21–37.

[482] Tkachenko, M. G. (1994). On the linearity of the Graev extension of pseudometrics, In: *Algebraic and logical constructions*, pp. 84–90 (Tver. Gos. Univ., Tver; *in Russian*).

[483] Tkachenko, M. G. (1994). Free topological groups and inductive limits, *Topol. Appl.* **60**, pp. 1–12.

[484] Tkachenko, M. G. (1998). Introduction to topological groups, *Topol. Appl.* **86**, 3, pp. 179–231.

[485] Tkachenko, M. G. (2001). Complete $\aleph_0$-bounded groups need not be $\mathbb{R}$-factorizable, *Comment. Math. Univ. Carolin.* **42**, 3, pp. 551–559.

[486] Tkachenko, M. G. (2002). Provisional solution to a Comfort–van Mill problem, *Topol. Appl.* **125**, 1, pp. 171–182.

[487] Tkachenko, M. G. (2004). $\mathbb{R}$-factorizable groups and subgroups of Lindelöf $P$-groups, *Topol. Appl.* **136**, pp. 135–167.

[488] Tkachenko, M. G. (2006). Homomorphic images of $\mathbb{R}$-factorizable groups, *Comment. Math. Univ. Carolin.* **47**, 3, pp. 525–537.

[489] Tkachenko, M. G. (2008). Many Abelian groups admit a "good" pseudocompact Hausdorff topological group topology, Preprint.

[490] Tkachenko, M. G. (2008). Some remarks on the class of $PT$-groups, Preprint.

[491] Tkachenko, M. G. and Torres, Y. (2008). Subgroups of $\sigma$-products of second-countable groups, Preprint.

[492] Tkachenko, M. G. and Yaschenko, I. V. (2002). Independent group topologies on Abelian groups, *Topol. Appl.* **122**, 1-2, pp. 425–451.

[493] Tkachuk, V. V. (1983). On a method of constructing examples of $M$-equivalent spaces, *Russian Math. Surveys* **38**, pp. 135–136. Russian original in: *Uspekhi Mat. Nauk* **38**, pp. 127–128.

[494] Todorčević, S. (1989). *Partition problems in topology*, Contemporary Math. **84** (Amer. Math. Soc., Providence, RI).

[495] Todorčević, S. (1993). Some applications of $S$ and $L$ combinatorics, *Ann. New York Acad. Sci.* **705**, pp. 130–167.

[496] Tomita, A. H. (1996). The Wallace Problem: a counterexample from $MA_{countable}$ and $p$-compactness, *Canad. Math. Bull.* **39**, 4, pp. 486–498.

[497] Tomita, A. H. (1998). The existence of initially $\omega_1$-compact group topologies on free Abelian groups is independent of $ZFC$, *Comment. Math. Univ. Carolin.* **39**, pp. 401–413.

[498] Tomita, A. H. (2003). Two countably compact topological groups: One of size $\aleph_\omega$ and the other of weight $\aleph_\omega$ without non-trivial convergent sequences, *Proc. Amer. Math. Soc.* **131**, 8, pp. 2617–2622.

[499] Tomita, A. H. and Watson, S. (2004). Ultraproducts, $p$-limits and antichains on the Comfort group order, *Topol. Appl.* **143**, pp. 147–157.

[500] Torres, Y. (2001). Union of chains of subgroups of a topological group, *Applied Gen. Topol.* **2**, 2, pp. 227–235.

[501] Trigos-Arrieta, F. J. (1991). Continuity, boundedness, connectedness and the Lindelöff property for topological groups, *J. Pure Appl. Algebra* **70**, pp. 199–210.

[502] Trigos-Arrieta, F. J. (1991). Pseudocompactness on groups, *General Topology and Its Applications*, pp. 369–378; Fifth Northeast Conference (Marcel Dekker, New York).

[503] Trigos-Arrieta, F. J. (1994). Every uncountable Abelian group admits a non-normal group topology, *Proc. Amer. Math. Soc.* **122**, pp. 907–909.

[504] Tychonoff, A. N. (1930). Über die Topologische Erweiterung von Räumen, *Math. Ann.* **102**, pp. 544–561.

[505] Tychonoff, A. N. (1935). Ein Fixpunktsatz, *Math. Ann.* **111**, pp. 767–776.

[506] Ulam, S. M. (1930). Zur Masstheorie in der allgemeinen Mengenlehre, *Fund. Math.* **16**, pp. 140–150.

[507] Ursul, I. M. (1989). Embeddings of locally precompact groups in locally pseudocompact ones, *Izv. Akad. Nauk Moldav. SSR Ser. Fiz.-Tekh. Mat. Nauk* **3**, pp. 54–56 (*in Russian*).

[508] Uspenskij, V. V. (1982). A topological group generated by a Lindelöf $\Sigma$-space has the Souslin property, *Soviet Math. Dokl.* **26**, pp. 166–169. Russian original in: *Uspekhi Mat. Nauk* **265**, 4, pp. 823–826.

[509] Uspenskij, V. V. (1983). For any $X$, the product $X \times Y$ is homogeneous for some $Y$, *Proc. Amer. Math. Soc.* **87**, 1, pp. 187–188.

[510] Uspenskij, V. V. (1983). A characterization of realcompactness in terms of the topology of pointwise convergence, *Comment. Math. Univ. Carolin.* **24**, 1, pp. 121–126.

[511] Uspenskij, V. V. (1984). Extensions of topological groups with a countable network, *Moscow Univ. Math. Bull.* **39**, 5, pp. 84–85. Russian original in: *Vestnik Mosk. Univ.* no. 3, p. 69.

[512] Uspenskij, V. V. (1985). On continuous images of Lindelöf topological groups, *Soviet Math. Dokl.* **32**, pp. 802–806. Russian original in: *Dokl. Akad. Nauk SSSR* **285**, 4, pp. 824–827.

[513] Uspenskij, V. V. (1985). Subgroups of free topological groups, *Soviet Math. Dokl.* **32**, 3, pp. 847–849. Russian original in: *Dokl. Akad. Nauk SSSR* **285**, 5, pp. 1070–1072.

[514] Uspenskij, V. V. (1986). A universal topological group with a countable base, *Functional Analysis and its Appl.* **20**, pp. 160–161. Russian original in: *Funktz. Analiz y ego Prilozheniya* **20**, pp. 76–87.

[515] Uspenskij, V. V. (1987). Compact quotient spaces of topological groups and Haydon spectra, *Math. Notes* **42**, pp. 827–831. Russian original in: *Mat. Zametki* **42**, 4, pp. 594–602.

[516] Uspenskij, V. V. (1988). Why compact groups are dyadic? *General Topology and its relations to modern analysis and algebra, VI* (Prague, 1986), pp. 601–610; *Res. Exp. Math.* **16** (Heldermann Ferlag, Berlin).

[517] Uspenskij, V. V. (1989). Topological groups and Dugundji compact spaces, *Math. USSR-Sb.* **67**, 2 (1990), pp. 555–580. Russian original in: *Matem. Sb.* **180**, 8 (1989), pp. 1092–1118.

[518] Uspenskij, V. V. (1989). The Mal'tsev operation on countably compact spaces, *Comment. Math. Univ. Carolin.* **30**, pp. 395–402.

[519] Uspenskij, V. V. (1990). On the group of isometries of the Urysohn universal metric space, *Comment. Math. Univ. Carolin.* **31**, pp. 181–182.

[520] Uspenskij, V. V. (1990). Free topological groups of metrizable spaces, *Math. USSR Izv.* **37** (1991), pp. 657–680. Russian original in: *Izv. Akad. Nauk SSSR Ser. Mat.* **54** (1990), pp. 1295–1319.

[521] Uspenskij, V. V. (1995). On the Souslin number of subgroups of products of countable groups, *Acta Univ. Carolin. Math. Phys.* **36**, pp. 85–87.

[522] Varopoulos, N. T. 1964. A theorem on the continuity of homomorphisms of locally compact groups, *Proc. Cambridge Phil. Soc.* **60**, 3, pp. 449–463.

[523] Veech, W. A. (1977). Topological dynamics, *Bull. Amer. Math. Soc.* **83**, pp. 775–830.

[524] Velichko, N. V. (1998). The Lindelöf property is $l$-invariant, *Topol. Appl.* **89**, pp. 277–283.

[525] Vilenkin, N. Ya. (1948). Theory of weakly separable topological groups, *Mat. Sb.* **22**, pp. 135–177 (*in Russian*).

[526] Vilenkin, N. Ya. (1958). Dyadicity of the group space of bicompact commutative groups, *Uspehi Mat. Nauk* **13**, 6 (84), pp. 79–80 (*in Russian*).

[527] Venkataraman, M., Rajagopalan, M., and Soundararajan, T. (1972). Orderable topological spaces, *Gen. Topol. Appl.* **2**, pp. 1–10.

[528] Vries, J. de (1975). Pseudocompactness and the Stone–Čech compactification for topological groups, *Nieuw Archiev voor Wiskunde* **23**, pp. 35–48.

[529] Vries, J. de (1978). On the existence of $G$-compactifications, *Bull. Acad. Polon. Sci. Ser. Math.* **26**, pp. 275–280.

[530] Vries, J. de (1993). *Elements of Topological Dynamics* (Kluwer Academic Publ., Dordrecht–Boston–London).

[531] Wallace, A. D. (1955). The structure of topological semigroups, *Bull. Amer. Math. Soc.* **61**, pp. 95–112.

[532] Weil, A. (1937). *Sur les Espaces a Structure Uniforme et sur la Topologie Generale*, Publ. Math. Univ. Strasbourg (Hermann, Paris).

[533] Weil, A. (1950). *L'integration dans les groupes topologiques et ses applications*, Actualités Sci. Ind. **869**, Paris.

[534] Wilcox, H. J. (1966). Pseudocompact groups, *Pacific J. Math.* **19**, pp. 365–379.

[535] Wilcox, H. J. (1971). Dense subgroups of compact groups, *Proc. Amer. Math. Soc.* **28**, pp. 578–580.

[536] Williams, S. W. (1984). Box products, In: *Handbook of Set-Theoretic Topology*, K. Kunen and J. Vaughan, eds., Ch. 4, pp. 169–200 (North-Holland, Amsterdam).

[537] Yajima, Y. (1984). The normality of $\Sigma$-products and the perfect $\kappa$-normality of Cartesian products, *J. Math. Soc. Japan* **36**, 4, pp. 689–699.

[538] Yamada, K. (1993). Characterizations of a metrizable space $X$ such that every $A_n(X)$ is a $k$-space, *Topol. Appl.* **49**, pp. 75–94.

[539] Yamada, K. (1996). Free Abelian topological groups and $k$-spaces, *Glas. Mat.* **31**, 1, pp. 83–91.

[540] Yamada, K. (1997). Tightness of free Abelian topological groups and of finite products of sequential fans, *Topol. Proc.* **22**, pp. 363–381.

[541] Yamada, K. (1998). Metrizable subspaces of free topological groups on metrizable spaces, *Topol. Proc.* **23**, pp. 379–409.

[542] Yamada, K. (2002). Fréchet-Urysohn spaces in free topological groups, *Proc. Amer. Math. Soc.* **130**, 8, pp. 2461–2469.

[543] Yamanoshita, T. (1954). On the dimension of homogeneous spaces, *J. Math. Soc. Japan* **6**, pp. 151–159.

[544] Yur'eva, A. A. (1993). Countably compact sequential topological semigroup is a topological group, *Mat. Studii* **2**, pp. 23–24 (*in Russian*).

[545] Zelenyuk, E. G. (1997). Finite groups in $\beta\mathbb{N}$ are trivial, *Semigroup Forum* **55**, pp. 131–132.

[546] Zelenyuk, E. G. (2003). On group operations on homogeneous spaces, *Proc. Amer. Math. Soc.* **132**, 4, pp. 1219–1222.

[547] Zelenyuk, E. G. and Protasov, I. V. (1990). Potentially compact abelian groups, *Math. USSR-Sb.* **69**, 1, (1991), pp. 299–305. Russian original in: *Mat. Sb.* **181**, 2 (1990), pp. 279–285.

[548] Zelenyuk, Y. (2000). On topologies on groups with continuous shifts and inversion, *Visnyk Kyiv Univ. Ser. Fiz. Mat.* **2**, pp. 252–256 (*in Ukrainian*).

[549] Zel'manov, E. I. (1992). On periodic compact groups, *Israel J. Math.* **77**, pp. 83–95.

# List of symbols

# Author Index

# Subject Index

## C

**T**

**U**